Contents in Brief

Preface	xxiii
Introduction	1

PART 1

Biology of Cells — 21

SECTION 1 The Unity of Life — 21

1. Atoms and Molecules — 23
2. Water — 40
3. Organic Molecules — 55
4. Cells: An Introduction — 84
5. How Cells Are Organized — 102
6. How Things Get into and out of Cells — 127
7. How Cells Divide — 144

SECTION 2 Energetics — 159

8. The Flow of Energy — 161
9. How Cells Make ATP: Glycolysis and Respiration — 186
10. Photosynthesis, Light, and Life — 206

SECTION 3 Genetics — 233

11. From an Abbey Garden: The Beginning of Genetics — 235
12. Meiosis and Sexual Reproduction — 249
13. Genes and Gene Interactions — 263
14. The Chemical Basis of Heredity: The Double Helix — 281
15. The Genetic Code and Its Translation — 301
16. The Molecular Genetics of Prokaryotes and Viruses — 319
17. Recombinant DNA: The Tools of the Trade — 340
18. The Molecular Genetics of Eukaryotes — 355
19. Human Genetics: Past, Present, and Future — 382

PART 2

Biology of Organisms — 405

SECTION 4 The Diversity of Life — 405

20. The Classification of Organisms — 407
21. The Prokaryotes and the Viruses — 428
22. The Protists — 452
23. The Fungi — 479
24. The Plants — 493
25. The Animal Kingdom I: Introducing the Invertebrates — 518
26. The Animal Kingdom II: The Protostome Coelomates — 544
27. The Animal Kingdom III: The Arthropods — 565
28. The Animal Kingdom IV: The Deuterostomes — 587

SECTION 5 Biology of Plants — 611

29. The Flowering Plants: An Introduction — 613
30. The Plant Body and Its Development — 625
31. Transport Processes in Plants — 650
32. Plant Responses and the Regulation of Growth — 671

SECTION 6 Biology of Animals — 699

33. The Vertebrate Animal: An Introduction — 701
34. Energy and Metabolism I: Digestion — 715
35. Energy and Metabolism II: Respiration — 732
36. Energy and Metabolism III: Circulation — 749
37. Homeostasis I: Excretion and Water Balance — 765
38. Homeostasis II: Temperature Regulation — 777
39. Homeostasis III: The Immune Response — 791
40. Integration and Control I: The Endocrine System — 821
41. Integration and Control II: The Nervous System — 841
42. Integration and Control III: Sensory Perception and Motor Response — 858
43. Integration and Control IV: The Vertebrate Brain — 881
44. The Continuity of Life I: Reproduction — 900
45. The Continuity of Life II: Development — 919

PART 3

Biology of Populations — 959

SECTION 7 Evolution — 959

46. Evolution: Theory and Evidence — 961
47. The Genetic Basis of Evolution — 974
48. Natural Selection — 991
49. On the Origin of Species — 1010
50. The Evolution of the Hominids — 1031
51. Animal Behavior and Its Evolution — 1052

SECTION 8 Ecology — 1085

52. Population Dynamics: The Numbers of Organisms — 1087
53. Interactions in Communities — 1106
54. Ecosystems — 1131
55. The Biosphere — 1154

APPENDIX A Metric Table — 1182
APPENDIX B Temperature Conversion Scale — 1183
APPENDIX C Classification of Organisms — 1184
GLOSSARY — G-1
ILLUSTRATION ACKNOWLEDGMENTS — A-1
INDEX — I-1

Giraffe with oxpeckers

Biology FIFTH EDITION

PART 2   Biology of Organisms

HELENA CURTIS

N. SUE BARNES

WORTH PUBLISHERS, INC.

This book is dedicated to those whose creative and painstaking studies have contributed to our understanding of biology.

Red-eyed tree frogs (Agalychnis callidryas) *clinging to a branch. Native to the tropical forests of Central America, they are adapted to life in the treetops. The bulges at the ends of their fingers and toes are suction pads, which allow the frogs to cling to vertical tree trunks and even to the undersurfaces of leaves. Their elongated fingers and toes are able to grasp small branches. Adaptations, such as those of the tree frog to its arboreal life, have led to the enormous diversity among living things. (© Zig Leszczynski/Animals Animals)*

BIOLOGY, FIFTH EDITION, PART 2

COPYRIGHT © 1968, 1975, 1979, 1983, 1989 BY WORTH PUBLISHERS, INC.

ALL RIGHTS RESERVED

LIBRARY OF CONGRESS CATALOG CARD NUMBER: 88-51041

ISBN: 0-87901-394-X (CLOTHBOUND)

ISBN: 0-87901-435-0 (PART 1, PAPERBACK)

ISBN: 0-87901-436-9 (PART 2, PAPERBACK)

ISBN: 0-87901-437-7 (PART 3, PAPERBACK)

PRINTED IN THE UNITED STATES OF AMERICA

PRINTING: 1 2 3 4 5 6 7   YEAR: 9 0 1 2 3 4

EDITOR: SALLY ANDERSON

PRODUCTION: SARAH SEGAL

ART DIRECTOR: GEORGE TOULOUMES

LAYOUT DESIGN: PATRICIA LAWSON, DAVID LOPEZ

DESIGN: MALCOLM GREAR DESIGNERS

ILLUSTRATOR: SHIRLEY BATY

PICTURE EDITORS: DAVID HINCHMAN, ANNE FELDMAN, ELAINE BERNSTEIN

TYPOGRAPHY: NEW ENGLAND TYPOGRAPHIC SERVICE, INC.

COLOR SEPARATION: CREATIVE GRAPHIC SERVICES CORPORATION

PRINTING AND BINDING: VON HOFFMANN PRESS, INC.

COVER PHOTOGRAPH: TREE FROGS (© ZIG LESZCZYNSKI/ANIMALS ANIMALS)

FRONTISPIECE: TROPICAL FOREST, COSTA RICA (© GARY BRAASCH)

ILLUSTRATION ACKNOWLEDGMENTS BEGIN ON PAGE A-0, CONSTITUTING AN EXTENSION OF THE COPYRIGHT PAGE.

WORTH PUBLISHERS, INC.

33 IRVING PLACE

NEW YORK, NEW YORK 10003

BIOLOGY

Contents

Ladybug on flower of shepherd's needle

Preface   xiii

PART 2
Biology of Organisms   405

SECTION 4 The Diversity of Life   405

CHAPTER 20
The Classification of Organisms   407

- The Need for Classification   407
- What Is a Species?   407
  - The Naming of Species   408
- Hierarchical Classification   410
- Evolutionary Systematics   412
  - The Monophyletic Ideal   412
  - Homology and Phylogeny   413
- Taxonomic Methods   414
  - Alternative Methodologies   415
  - ESSAY: *How to Construct a Cladogram*   417
- Molecular Taxonomy   418
  - Amino Acid Sequences   418
  - Nucleotide Sequences   420
  - DNA-DNA Hybridization   420
  - ESSAY: *The Riddle of the Giant Panda*   422
- A Question of Kingdoms   423
- Summary   426

CHAPTER 21
The Prokaryotes and the Viruses   428

- The Classification of Prokaryotes   429
- The Prokaryotic Cell   431
  - The Cell Membrane   431
  - The Cell Wall   432
  - Flagella and Pili   433
  - ESSAY: *Navigation by the Poles*   435
- Diversity of Form   436
- Reproduction and Resting Forms   437
- Prokaryotic Nutrition   438
  - Heterotrophs   438
  - Chemosynthetic Autotrophs   439
  - Photosynthetic Autotrophs   440
  - ESSAY: *Two Unusual Photosynthetic Prokaryotes*   442
- Viruses: Detached Bits of Genetic Information   443
  - Viroids and Prions: The Ultimate in Simplicity   446
- Microorganisms and Human Ecology   447
  - Symbiosis   447
  - How Microbes Cause Disease   448
  - Prevention and Control of Infectious Disease   448
- Summary   450

CHAPTER 22
The Protists   452

- The Evolution of the Protists   452
  - Classification of Protists   454
- Photosynthetic Autotrophs   455
  - Characteristics of the Photosynthetic Protists   456
  - Division Euglenophyta: Euglenoids   458
  - Division Chrysophyta: Diatoms and Golden-Brown Algae   459
  - Division Dinoflagellata: "Spinning" Flagellates   460
  - Division Chlorophyta: Green Algae   461
  - Division Phaeophyta: Brown Algae   465
  - Division Rhodophyta: Red Algae   465
- Multinucleate and Multicellular Heterotrophs   466
  - The Slime Molds   466
  - The Water Molds   467
- Unicellular Heterotrophs   468
  - Phylum Mastigophora   468
  - Phylum Sarcodina   470
  - Phylum Ciliophora   470
  - ESSAY: *The Evolution of Mitosis*   472
  - Phylum Opalinida   473
  - Phylum Sporozoa   473
- Patterns of Behavior in Protists   474
  - Avoidance in *Paramecium*   475
- Summary   476

CHAPTER 23

The Fungi — 479

- **Characteristics of the Fungi** — 480
  - Reproduction in the Fungi — 481
- **Classification of the Fungi** — 481
- **Division Zygomycota** — 482
  - ESSAY: *Ready, Aim, Fire!* — 484
- **Division Ascomycota** — 484
- **Division Basidiomycota** — 486
- **Division Deuteromycota** — 488
- **Symbiotic Relationships of Fungi** — 488
  - The Lichens — 488
  - ESSAY: *Predaceous Fungi* — 489
  - Mycorrhizae — 490
- **Summary** — 491

CHAPTER 24

The Plants — 493

- **The Ancestral Alga** — 493
- **The Transition to Land** — 495
  - Subsequent Diversification — 496
- **Classification of the Plants** — 497
- **Division Bryophyta: Liverworts, Hornworts, and Mosses** — 498
  - Bryophyte Reproduction — 499
- **The Vascular Plants: An Introduction** — 501
  - Evolutionary Developments in the Vascular Plants — 501
- **The Seedless Vascular Plants** — 502
  - Division Pterophyta: The Ferns — 502
- **The Seed Plants** — 503
  - ESSAY: *Coal Age Plants* — 504
  - Gymnosperms — 505
  - Angiosperms: The Flowering Plants — 509
  - ESSAY: *The Ice Ages* — 510
- **The Role of Plants** — 515
- **Summary** — 516

CHAPTER 25

The Animal Kingdom I: Introducing the Invertebrates — 518

- **The Diversity of Animals** — 518
- **The Origin and Classification of Animals** — 519
- **Phylum Porifera: Sponges** — 522
  - Reproduction in Sponges — 523
- **Phylum Mesozoa: Mesozoans** — 524
- **Radially Symmetrical Animals** — 524
  - Phylum Cnidaria — 525
  - ESSAY: *The Coral Reef* — 530
  - Phylum Ctenophora — 531
- **Bilaterally Symmetrical Animals: An Introduction** — 531
- **Phylum Platyhelminthes: Flatworms** — 533
  - Class Turbellaria — 533
  - Classes Trematoda and Cestoda — 536
  - ESSAY: *The Politics of Schistosomiasis* — 537
- **Other Acoelomates** — 538
  - Phylum Gnathostomulida — 538
  - Phylum Rhynchocoela — 538
- **Pseudocoelomates** — 539
  - Phylum Nematoda — 539
  - Other Pseudocoelomate Phyla — 540
- **Summary** — 542

CHAPTER 26

The Animal Kingdom II: The Protostome Coelomates — 544

- **Phylum Mollusca: Mollusks** — 545
  - Characteristics of the Mollusks — 546
  - Minor Classes of Mollusks — 548
  - Class Bivalvia — 549
  - Class Gastropoda — 550
  - Class Cephalopoda — 551
  - ESSAY: *Behavior in the Octopus* — 552
  - Evolutionary Affinities of the Mollusks — 554
- **Phylum Annelida: Segmented Worms** — 554
  - Class Oligochaeta: The Earthworms — 555
  - Class Polychaeta — 558
  - Class Hirudinea — 558
- **Minor Protostome Phyla** — 559
- **The Lophophorates** — 562
- **Summary** — 563

CHAPTER 27

The Animal Kingdom III: The Arthropods — 565

- **Characteristics of the Arthropods** — 565
  - The Exoskeleton — 566
  - Internal Features — 567
  - The Arthropod Nervous System — 568
- **Subdivisions of the Phylum** — 568
  - The Chelicerates — 568
  - The Aquatic Mandibulates: Class Crustacea — 571
  - The Terrestrial Mandibulates: Myriapods — 572
  - The Terrestrial Mandibulates: Class Insecta — 573
- **Reasons for Arthropod Success** — 578
  - Arthropod Senses and Behavior — 579
  - ESSAY: *Firefly Light: A Warning, an Advertisement, a Snare* — 580
- **Summary** — 585

CHAPTER 28

The Animal Kingdom IV: The Deuterostomes — 587

- **Phylum Echinodermata: The "Spiny-Skinned" Animals** — 587
  - Class Stelleroidea: Starfish and Brittle Stars — 588
  - Other Echinoderms — 589

Phylum Chaetognatha: Arrow Worms	590	Modifications in the Pattern of Shoot Growth	642
Phylum Hemichordata: Acorn Worms	591	Vegetative Reproduction	642
Phylum Chordata: The Cephalochordates and Urochordates	591	**Secondary Growth**	**644**
Phylum Chordata: The Vertebrates	592	ESSAY: *The Record in the Rings*	646
Classes Agnatha, Chondrichthyes, and Osteichthyes: Fishes	593	**Summary**	**647**
Class Amphibia	595		
Class Reptilia	596		
Class Aves: Birds	598		
Class Mammalia	600		
Summary	**604**		
Suggestions for Further Reading	**605**		

SECTION 5 Biology of Plants — 611

CHAPTER 29

The Flowering Plants: An Introduction — 613

Sexual Reproduction: The Flower	**613**
The Pollen Grain	615
Fertilization	615
The Embryo	**619**
The Seed and the Fruit	**619**
Types of Fruits	620
Adaptations to Seasonal Change	**621**
Dormancy and the Life Cycle	621
ESSAY: *The Staff of Life*	622
Seed Dormancy	623
Summary	**624**

CHAPTER 30

The Plant Body and Its Development — 625

The Cells and Tissues of the Plant Body	**625**
Leaves	**627**
Leaf Structure	627
Leaf Adaptations and Modifications	629
Characteristics of Plant Growth	**630**
Roots	**631**
Root Structure	632
Primary Growth of the Root	634
Patterns of Root Growth	635
Stems	**636**
Stem Structure	636
Primary Growth of the Shoot System	**640**

CHAPTER 31

Transport Processes in Plants — 650

The Movement of Water and Minerals	**650**
Transpiration	650
The Uptake of Water	650
The Cohesion-Tension Theory	651
Factors Influencing Transpiration	652
The Uptake of Minerals	655
Mineral Requirements of Plants	656
The Movement of Sugars: Translocation	**657**
ESSAY: *Halophytes: A Future Resource?*	658
Evidence for the Phloem	660
ESSAY: *Radioactive Isotopes in Plant Research*	661
The Pressure-Flow Hypothesis	662
Factors Influencing Plant Nutrition	**662**
Soil Composition	663
The Role of Symbioses	664
ESSAY: *Carnivorous Plants*	666
Summary	**669**

CHAPTER 32

Plant Responses and the Regulation of Growth — 671

Phototropism and the Discovery of Plant Hormones	**671**
Hormones and the Regulation of Plant Growth	**673**
Auxins	673
Cytokinins	676
ESSAY: *Plants in Test Tubes*	678
Ethylene	679
Abscisic Acid	680
Gibberellins	680
Oligosaccharins	682
Gravitropism	**682**
Photoperiodism	**684**
Photoperiodism and Flowering	684
Photoperiodism and Phytochrome	686
Other Phytochrome Responses	687
ESSAY: *Is There a Flowering Hormone?*	688
Circadian Rhythms	**689**
Biological Clocks	689
Touch Responses	**691**
Twining and Coiling	691
Rapid Movements in the Sensitive Plant	692
Rapid Movements in Carnivorous Plants	692
Generalized Effects of Touch on Plant Growth	693
Chemical Communication among Plants	**693**
Summary	**694**
Suggestions for Further Reading	**696**

SECTION 6 Biology of Animals 699

CHAPTER 33

The Vertebrate Animal: An Introduction 701
- Characteristics of *Homo sapiens* 701
- Cells and Tissues 702
  - Epithelial Tissues 702
  - Connective Tissues 704
  - Muscle Tissue 706
  - ESSAY: *The Injury-Prone Knee* 707
  - Nerve Tissue 708
- Levels of Organization 709
- Functions of the Organism 710
  - Energy and Metabolism 710
  - Homeostasis 711
  - Integration and Control 711
  - Continuity of Life 713
- Summary 713

CHAPTER 34

Energy and Metabolism I: Digestion 715
- Digestive Tract in Vertebrates 715
  - The Oral Cavity: Initial Processing 716
  - The Pharynx and Esophagus: Swallowing 718
  - ESSAY: *The Heimlich Maneuver* 719
  - The Stomach: Storage and Liquefaction 720
  - The Small Intestine: Digestion and Absorption 720
  - ESSAY: *Aids to Digestion* 724
  - The Large Intestine: Further Absorption and Elimination 725
- Major Accessory Glands 725
  - The Pancreas 725
  - The Liver 725
- Regulation of Blood Glucose 726
- Some Nutritional Requirements 726
  - ESSAY: *Mother's Milk—It's the Real Thing* 727
  - The Price of Affluence 728
- Summary 730

CHAPTER 35

Energy and Metabolism II: Respiration 732
- Diffusion and Air Pressure 733
- Evolution of Respiratory Systems 734
  - Evolution of Gills 734
  - Evolution of Lungs 736
  - Respiration in Large Animals: Some Principles 737
- The Human Respiratory System 738
  - ESSAY: *Cancer of the Lung* 739
- Mechanics of Respiration 741
- Transport and Exchange of Gases 741
  - Hemoglobin and Its Function 741
  - ESSAY: *Diving Mammals* 744
  - Myoglobin and Its Function 745
- Control of Respiration 745
  - ESSAY: *High on Mt. Everest* 746
- Summary 747

CHAPTER 36

Energy and Metabolism III: Circulation 749
- The Blood 750
  - Plasma 750
  - Red Blood Cells 750
  - White Blood Cells 750
  - Platelets 751
  - Blood Clotting 751
- The Cardiovascular System 752
- The Blood Vessels 753
  - The Capillaries and Diffusion 753
- The Heart 754
  - Evolution of the Heart 754
  - The Human Heart 755
  - Regulation of the Heartbeat 756
- The Vascular Circuitry 758
- Blood Pressure 759
  - ESSAY: *Diseases of the Heart and Blood Vessels* 760
  - Cardiovascular Regulating Center 762
- The Lymphatic System 762
- Summary 763

CHAPTER 37

Homeostasis I: Excretion and Water Balance 765
- Regulation of the Chemical Environment 765
  - Substances Regulated by the Kidneys 766
- Water Balance 767
  - An Evolutionary Perspective 767
  - Sources of Water Gain and Loss in Terrestrial Animals 767
  - Water Compartments 768
- The Kidney 769
  - Function of the Kidney 770
  - Control of Kidney Function: The Role of Hormones 774
- Summary 775

CHAPTER 38

Homeostasis II: Temperature Regulation 777
- Principles of Heat Balance 778
  - Body Size and the Transfer of Heat 779
- "Cold-Blooded" vs. "Warm-Blooded" 780
- Poikilotherms 780
- Homeotherms 781
  - ESSAY: *Avian Mechanical Engineers* 783
  - The Thermostat 784
  - Cutting Energy Costs 786

Adaptations to Extreme Temperature	**787**	Stimulation of Smooth Muscle	835
Adaptations to Extreme Cold	787	Other Prostaglandin Effects	835
Adaptations to Extreme Heat	788	**Mechanisms of Action of Hormones**	**836**
Summary	**789**	Intracellular Receptors	836
		Membrane Receptors	837
		Summary	**839**

CHAPTER 39

Homeostasis III: The Immune Response — 791

- **Nonspecific Defenses** — **791**
  - Anatomic Barriers — 791
  - The Inflammatory Response — 791
  - Interferons — 793
- **The Immune System** — **793**
- **B Lymphocytes and the Formation of Antibodies** — **795**
  - The B Lymphocyte: A Life History — 795
  - ESSAY: *Death Certificate for Smallpox* — 796
  - The Action of Antibodies — 797
  - The Structure of Antibodies — 798
  - The Clonal Selection Theory of Antibody Formation — 799
- **T Lymphocytes and Cell-Mediated Immunity** — **801**
  - The T Lymphocyte: A Life History — 801
  - ESSAY: *Monoclonal Antibodies* — 802
  - ESSAY: *Children of the* Desaparecidos: *An Application of MHC-Antigen Testing* — 806
  - The Functions of the T Lymphocytes — 806
- **Cancer and the Immune Response** — **808**
- **Tissue Transplants** — **810**
  - Organ Transplants — 810
  - Blood Transfusions — 810
- **Disorders of the Immune System** — **812**
  - Autoimmune Diseases — 812
  - Allergies — 812
  - Acquired Immune Deficiency Syndrome (AIDS) — 813
- **Summary** — **818**

CHAPTER 40

Integration and Control I: The Endocrine System — 821

- **Glands and Their Products: An Overview** — **823**
- **The Pituitary Gland** — **825**
  - The Anterior Lobe — 825
  - The Intermediate and Posterior Lobes — 826
- **The Hypothalamus** — **826**
  - The Pituitary-Hypothalamic Axis — 826
  - Other Hypothalamic Hormones — 828
- **The Thyroid Gland** — **828**
- **The Parathyroid Glands** — **829**
- **Adrenal Cortex** — **829**
  - ESSAY: *The Regulation of Bone Density* — 830
  - Glucocorticoids — 830
  - Mineralocorticoids — 832
- **Adrenal Medulla** — **832**
- **The Pancreas** — **832**
- **The Pineal Gland** — **833**
- **Prostaglandins** — **833**
  - ESSAY: *Circadian Rhythms* — 834

CHAPTER 41

Integration and Control II: The Nervous System — 841

- **Evolution of Nervous Systems** — **841**
- **Organization of the Vertebrate Nervous System** — **842**
  - The Central Nervous System — 843
  - The Peripheral Nervous System — 844
- **The Nerve Impulse** — **847**
  - The Ionic Basis of the Action Potential — 848
  - Propagation of the Impulse — 850
- **The Synapse** — **851**
  - Neurotransmitters — 853
  - ESSAY: *Internal Opiates: The Endorphins* — 854
  - The Integration of Information — 855
- **Summary** — **856**

CHAPTER 42

Integration and Control III: Sensory Perception and Motor Response — 858

- **Sensory Receptors and the Initiation of Nerve Impulses** — **858**
  - Types of Sensory Receptors — 859
  - Chemoreception: Taste and Smell — 860
  - ESSAY: *Chemical Communication in Mammals* — 863
  - Mechanoreception: Balance and Hearing — 863
  - Photoreception: Vision — 866
  - ESSAY: *What the Frog's Eye Tells the Frog's Brain* — 870
- **The Response to Sensory Information: Muscle Contraction** — **873**
  - The Structure of Skeletal Muscle — 873
  - The Contractile Machinery — 874
  - The Neuromuscular Junction — 876
  - ESSAY: *Twitch Now, Pay Later* — 878
  - The Motor Unit — 878
- **Summary** — **879**

CHAPTER 43

Integration and Control IV: The Vertebrate Brain — 881

- **The Structural Organization of the Brain: An Evolutionary Perspective** — **881**
  - Hindbrain and Midbrain — 882
  - Forebrain — 883
- **Brain Circuits** — **884**
  - The Reticular Activating System — 885
  - The Limbic System — 885
- **The Cerebral Cortex** — **885**

Motor and Sensory Cortices	885
Left Brain/Right Brain	888
ESSAY: *Electrical Activity of the Brain*	890
Intrinsic Processing Areas	892
Learning and Memory	**893**
Anatomical Pathways of Memory	894
Synaptic Modification	895
ESSAY: *Alzheimer's Disease*	896
Summary	**898**

CHAPTER 44

The Continuity of Life I: Reproduction — 900

The Male Reproductive System	**901**
Spermatogenesis	901
Pathway of the Sperm	903
The Role of Hormones	905
The Female Reproductive System	**907**
ESSAY: *Sex and the Brain*	908
Oogenesis	910
Pathway of the Oocyte	912
Hormonal Regulation in Females	913
Contraceptive Techniques	**915**
Summary	**917**

CHAPTER 45

The Continuity of Life II: Development — 919

Development of the Sea Urchin	**920**
Fertilization and Activation of the Egg	920
From Zygote to Pluteus	921
ESSAY: *The Cytoplasmic Determination of Germ Cells*	924
The Influence of the Cytoplasm	925
Development of the Amphibian	**926**
Cleavage and Blastula Formation	927
Gastrulation and Neural Tube Formation	927
The Role of Tissue Interactions	930
Development of the Chick	**931**
Extraembryonic Membranes of the Chick	932
Organogenesis: The Formation of Organ Systems	935
Morphogenesis: The Shaping of Body Form	938
Development of the Human Embryo	**940**
ESSAY: *Genetic Control of Development: The Homeobox*	942
Extraembryonic Membranes	944
The Placenta	944
The First Trimester	945
The Second Trimester	948
The Final Trimester	948
Birth	950
Epilogue	**951**
Summary	**951**
Suggestions for Further Reading	**953**
APPENDIX A Metric Table	1182
APPENDIX B Temperature Conversion Scale	1183
APPENDIX C Classification of Organisms	1184
GLOSSARY	G–1
ILLUSTRATION ACKNOWLEDGMENTS	A–1
INDEX	I–1

Preface

In the twenty years since the first edition of *Biology* appeared, the science of biology has been characterized by ever accelerating change, including not only a flood of new information but also new ideas and unifying concepts. Some areas of biology have undergone metamorphoses before our very eyes, while others have attained a new maturity. This has presented us—now at our word processors—and you—in the classroom and laboratory—with new challenges and opportunities as we work together to provide students with a solid foundation in the principles of biology while simultaneously sharing with them the excitement of the contemporary science.

With this, the Fifth Edition, *Biology* becomes both one of the oldest and one of the newest introductory biology textbooks. One of our principal goals in preparing this edition has been to maintain a balance between the old and the new. This has required not only a willingness to discard material, but also considerable care that we not eliminate or slight material that, although not new, is essential if students are to be adequately prepared to understand current and future developments in biology. Simultaneously, we have, of course, wanted to be as up to the minute as possible, without becoming merely trendy. At a time when important discoveries are published almost continuously, there is a temptation to become so engrossed in the new that we lose sight of the fact that the majority of today's students are, like their predecessors, coming to the formal study of contemporary biology for the first time. The clear explication of the basic principles of biology, with pertinent and readily understood examples, has become increasingly important with each passing year. Thus, specific topics for detailed treatment have been selected on the basis of their centrality to modern biology, their utility in illuminating basic principles, their importance as part of the requisite store of knowledge of an educated adult as we approach a new century, and their inherent interest and appeal to students. Throughout, we have tried to provide the underlying framework and arouse student curiosity so that a foundation is laid for those areas—very diverse—in which you may wish to give more extensive coverage in the classroom or laboratory than is possible in any introductory textbook, regardless of its length.

The central, essential foundation of biology is, of course, evolution, the major organizing theme of this text as of all modern biology texts. As in previous editions, the stage is set in the Introduction, which focuses on the development of the Darwinian theory. New to the Introduction is a section previewing the other major unifying principles of modern biology that are also recurring themes throughout the text; also new is a brief overview of the diversity of life. Both are designed to provide students with a broad framework before they begin their study of the details on which modern biology is built. In this edition, we have also strengthened the introductory discussion of the nature of science, and, throughout the text, we have included more information about how biologists know what they know and how scientists in general go about their business.

After the Introduction, this edition, like previous editions, follows the levels-of-organization approach. Part 1 deals with life at the subcellular and cellular levels, Part 2 with organisms, and Part 3 with populations, ending with a survey of the distribution of life on earth. Each part is divided into two or three sections. A significant amount of restructuring has occurred in the sequence of chapters within certain sections and within the chapters themselves.

One of the most striking aspects of the enormous burst of new discoveries in molecular and cell biology since the Fourth Edition is the power of these discoveries to explain processes that previously could only be described—and in the most general terms. The immune response, olfaction and color vision, events at the synapse, the summing of information by individual neurons, and differentiation and morphogenesis in animal development are just a few of the many phenomena whose secrets are being revealed by studies at the molecular and cellular level. These revelations depend, in large part, on what is now a flood tide of reports identifying specific membrane proteins, their amino acid sequences, their three-dimensional structures, and, in many cases, the nucleotide sequences of the genes coding for the proteins and the location of these genes within the genome. Because these discoveries, although fascinating in and of themselves, are of such value in explaining organismal phenomena, we have generally chosen to defer their discussion to later sections of the text, where their significance will be most readily grasped by students. Molecular and cell biology have, in many ways, come of age, and it seems to us that the essential task in these early sections of the book has become the clear communication of the underlying principles on which so much is now being built—rather than a catalog of the latest new discoveries, which will soon be superseded by even more exciting ones.

The extraordinary pace of discovery in genetics, principally as a result of recombinant DNA technology, requires, with each new edition, a major rethinking of Section 3. Responses to our surveys indicate that, however tempting it might be to begin the section with molecular genetics and to reduce the coverage of classical genetics, doing so could make this most exciting area of modern biology less accessible to students. Thus, as in previous editions, we begin our consideration of genetics with Mendel and take an essentially historical approach to the development of the powerful science we know today. Within that overall framework, however, there has been the addition of a significant amount of new material, coupled with a number of internal reorganizations that we believe provide greater clarity and a smoother conceptual development in our coverage of molecular genetics.

Part 2, Biology of Organisms, has also undergone many changes, particularly in the early chapters of Section 4 and in Sections 5 and 6. In the previous edition, Section 4, The Diversity of Life, was significantly expanded. The enthusiasm with which the revised section was received—plus our own continuing awe at the incredible variety of living organisms—led to our decision to retain the expanded section intact. We have, however, made major revisions in the first chapter of the section, dealing with the classification of organisms, and minor revisions throughout the section.

The organization of Section 5, the Biology of Plants, has long been problematic. It has been difficult to find a sequence that would flow logically, coordinate well with laboratory programs, and—most important—captivate students with the beauty and biological accomplishments of plants without overwhelming them with the vocabulary necessary for an accurate description of the living plant. In this edition we have chosen to begin the section with the familiar—the flower—and with the dynamic process of plant reproduction, a sequence that flows directly from the discussion of plant evolution and diversity in Section 4. In the following chapter, the anatomy of the plant body is considered in conjunction with another dynamic process, the development of the embryo into the mature sporophyte. The two chapters on plant hormones and plant responses in the Fourth Edition have now been merged into one integrated chapter; new understandings of the physiological processes of plants have made such a separation increasingly artificial.

In Section 6, the Biology of Animals, we have retained the overall organizational scheme and problem-solving approach of the Fourth Edition, while significantly revising many chapters. As noted previously, animal physiology is one of the principal areas in which enormous and rapid progress is being made as a result of new discoveries at the molecular and cellular level, and we have tried to capture and share with students as much of the current excitement as possible. Although we have continued to use the human animal—inherently fascinating to most students—as our representative organism in these chapters, we have strengthened the comparative thread and made explicit much comparative material that was previously implicit.

Part 3, the Biology of Populations, covers what G. E. Hutchinson aptly described as "the ecological theater and the evolutionary play." Modern evolutionary theory and ecology are so intertwined that any separation of the two is arbitrary. We believe, however, that the student's understanding of modern ecology is deepened and enriched if it is preceded by a knowledge of the mechanisms of evolution.

In Section 7, Evolution, the five chapters of the previous edition have been reworked into four. As in the Fourth Edition, the section begins with a chapter that reviews the key points of Darwin's theory, examines the types of evidence that support evolution, and considers the changes that have occurred in evolutionary theory since Darwin's original formulation. This is followed by extensively revised chapters on the genetic basis of evolution, natural selection, and the origin of species. Then follow two chapters, also heavily revised, on the evolution of the hominids and on animal behavior and its evolution. Many of you have told us that you prefer to cover human evolution while the discussion of evolutionary mechanisms is still fresh in students' minds, and we have accordingly shifted that chapter to this section from the end of the book. Behavior is a topic for which little, if any, time is available in many courses, but it holds great interest for students, professors, and these authors alike. We have tried to provide students with a solid introduction to the contemporary study of behavior and then to focus on topics that we think are most likely to be of immediate interest and appeal to them.

Section 8, Ecology, has also been extensively revised, as we attempt to track the continual shifts, rethinkings, and controversies that characterize this most vibrant science. As in the Fourth Edition, the section moves from population dynamics, through the interactions of populations in communities and ecosystems, to the overall organization and distribution of life on earth. The text ends with a consideration of the tropical forests—the most complex and most seriously threatened of all ecological systems.

Each section ends with suggestions for further reading. Scientifically speaking, the selections are arbitrary. They were chosen not as documentation for statements in the book or as fuller presentations of difficult subjects, but rather because of their accessibility to students. Our hope is that at least some students will continue reading on their own, preferably reports not yet published about discoveries just now being dreamed of.

A number of new supplements accompany this edition of *Biology*. Of particular interest is *More Biology in the Laboratory*, by Doris R. Helms of Clemson University, an expanded version of *Biology in the Laboratory*, which accompanies the Fourth Edition of *Invitation to Biology*. A detailed Preparator's Guide accompanies the lab manual. Other supplements include *BioBytes*, a series of computer simulations by Robert Kosinski of Clemson University, a Study Guide and a Test

Bank by David J. Fox of the University of Tennessee, a new computerized test-generation system, a new and greatly expanded Instructor's Resource Manual by Debora Mann of Clemson University, and an extensive set of acetate transparencies, most of them in color.

As with previous editions, we have been deeply dependent on the advice of consultants and reviewers. In addition to her work on the new laboratory manual, Dori Helms played a major role in the revisions of the genetics section, the plant section, and the development chapter in animal physiology. She has generously shared with us her extensive knowledge, her wealth of experience in the classroom and laboratory, and her enthusiasm—all of which have been marvelous resources that we have greatly appreciated.

We are also deeply indebted to Rita Calvo of Cornell University, who reviewed a series of revisions of the genetics section; to Jacques Chiller of the Lilly Research Laboratories, who has been an invaluable source on contemporary immunology; to Mark W. Dubin of the University of Colorado, who guided us through our revision of the integration and control chapters of animal physiology; and to Manuel C. Molles, Jr., of the University of New Mexico, and Andrew Blaustein of Oregon State University, both of whom made major contributions to our revision of the evolution and ecology sections.

In addition, we have been greatly assisted by advice and counsel from the following reviewers:

BRUCE ALBERTS, University of California Medical School, San Francisco
WILLIAM E. BARSTOW, University of Georgia
CHARLES J. BIGGERS, Memphis State University
WILLIAM L. BISCHOFF, University of Toledo
ROBERT BLYSTONE, Trinity University
LEON BROWDER, University of Calgary
RALPH BUCHSBAUM, Pacific Grove, California
JAMES J. CHAMPOUX, University of Washington
JAMES COLLINS, Arizona State University
JOHN O. CORLISS, University of Maryland
MICHAEL CRAWLEY, Imperial College at Silwood Park, Ascot, England
CHARLES CURRY, University of Calgary
FRED DELCOMYN, University of Illinois, Urbana-Champaign
RUTH DOELL, San Francisco State University
RICHARD DUHRKOPF, Baylor University
DAVID DUVALL, University of Wyoming
JUDI ELLZEY, University of Texas, El Paso
ROBERT C. EVANS, Rutgers University, Camden
RAY F. EVERT, University of Wisconsin
KATHLEEN FISHER, University of California, Davis
ROBERT P. GEORGE, University of Wyoming
URSULA GOODENOUGH, Washington University, St. Louis
PATRICIA GOWATY, Clemson University
LINDA HANSFORD, Baltimore, Maryland
JEAN B. HARRISON, University of California, Los Angeles
STEVEN HEIDEMANN, Michigan State University
MERRILL HILLE, University of Washington
GERALD KARP, San Francisco, California
JOHN KIRSCH, University of Wisconsin
ROBERT M. KITCHIN, University of Wyoming
KAREL LIEM, Harvard University
JANE LUBCHENCO, Oregon State University
R. WILLIAM MARKS, Villanova University

LARRY R. McEDWARD, University of Washington
SUE ANN MILLER, Hamilton College
RANDY MOORE, Wright State University
BETTE NICOTRI, University of Washington
JAMES PLATT, University of Denver
FRANK E. PRICE, Hamilton College
EDWARD RUPPERT, Clemson University
TOM K. SCOTT, University of North Carolina, Chapel Hill
LARRY SELLERS, Louisiana Tech University
DAVID G. SHAPPIRIO, University of Michigan
JOHN SMARRELLI, Loyola University, Chicago
GILBERT D. STARKS, Central Michigan University
IAN TATTERSALL, American Museum of Natural History
ROBERT VAN BUSKIRK, State University of New York, Binghamton
ERIC WEINBERG, University of Pennsylvania
JOHN WEST, University of California, Berkeley
ARTHUR WINFREE, University of Arizona

As always, the preparation of a new edition is a staggering and complex task, and its successful completion has depended on the efforts of many highly talented individuals. In particular, we wish to thank Shirley Baty, who, in addition to preparing many new illustrations for this edition, has reworked virtually all of the Fourth Edition art as we converted the book to full color throughout; David Hinchman, Anne Feldman, and Elaine Bernstein, who have located an enormous number of marvelous new photographs and micrographs; John Timpane, who prepared the comprehensive index; George Touloumes and the members of his staff who are responsible for the design and layout of each page of the book; Sarah Segal, who has managed the production process and somehow kept us all on course; and Sally Anderson, our extraordinary editor, and her capable assistant, Lindsey Bowman. Sally's editorial expertise, her thorough knowledge of biology in general and of this text in particular, and her long experience in working with us both have played an incalculable role in the successful completion of this revision. And, a special thank you to Bob Worth, whose vision and constant support have made it all possible.

Finally, we want to thank all of the professors and students who have written to us, some with criticisms, some with suggestions, some with questions, and some simply because they enjoyed the book. These letters serve to remind us of how privileged we are to be writing for young people. We continue to appreciate their curiosity, their energy, their imaginativeness, and their dislike of the pompous and pedantic. We hope we serve them well.

New York

December, 1988

Helena Curtis

N. Sue Barnes

An added note: As you may have noticed, with this edition of *Biology,* N. Sue Barnes is listed as coauthor. This is a recognition long overdue. Sue has been a member of the team for eleven years now. Over this period of time, she has assumed increasing responsibility for the revisions of both *Biology* and *Invitation to Biology* (on which she has been listed as coauthor for the last two editions). The Fifth Edition of *Biology* would have been impossible without her. In addition, I wish to express my personal gratitude for her integrity, patience, fortitude, and good spirits—and for the fact that she always comes through.

H.C.

PART 2

Biology of Organisms

SECTION 4

The Diversity of Life

Suspended in the canopy of a tropical rain forest, a biologist encounters a vast variety of new species of plants and animals. According to a recent estimate, there may be as many as 30 million species of insects alone in the tropical rain forest, the great majority of them still unknown and unclassified. At present, the tropical rain forests are being destroyed so rapidly that, in 25 years, most of the inhabitants of this rich environment will have been lost forever.

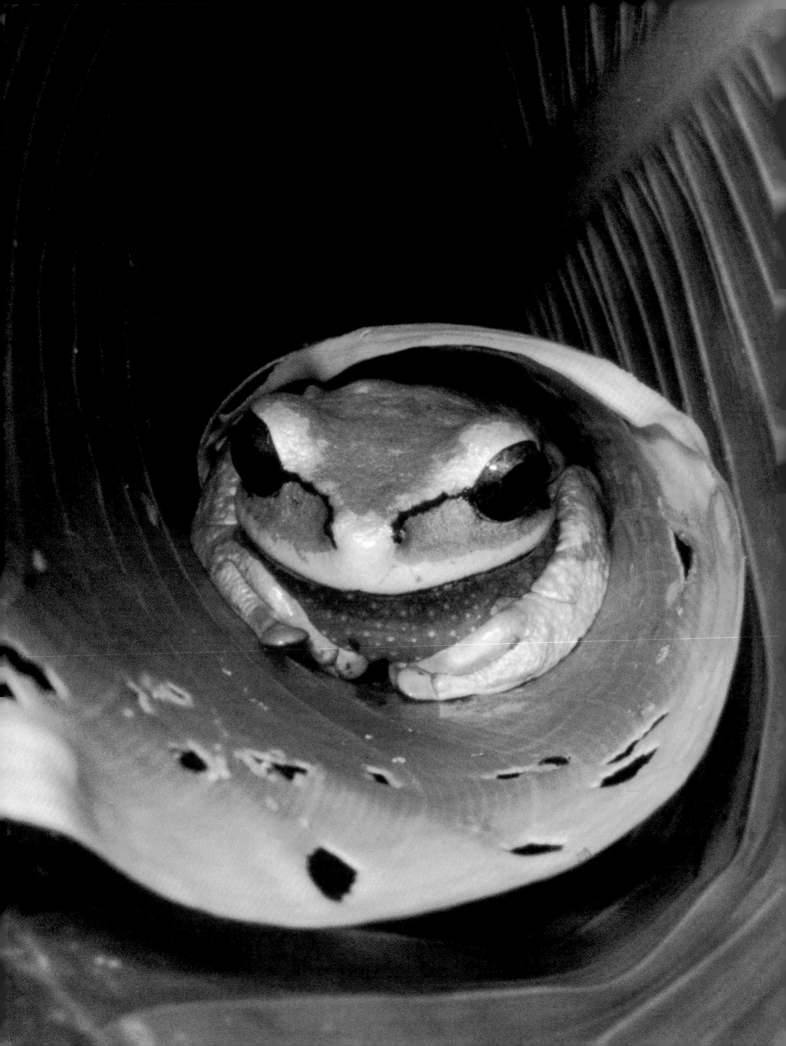

CHAPTER 20

The Classification of Organisms

Part 1 of this book dealt largely with cellular and subcellular aspects of living things. In Part 2, we shall be concerned with a higher level of organization, the whole organism. The chief focus will be on the plants, particularly the flowering plants, and on the more complex animals, particularly the vertebrates. These are the subjects of Sections 5 and 6, respectively. By way of introduction to these organisms, in the next few chapters we shall examine the tremendous diversity of life on our planet. Our discussion begins, in Chapter 21, with the simplest of single-celled organisms, the bacteria (which, as we have seen, are astonishingly gifted and complex), and ends, in Chapter 28, with the most familiar multicellular organisms, the mammals. At the same time, we shall trace the major developments in the course of evolution that have given rise to this diversity.

THE NEED FOR CLASSIFICATION

Most people have a limited awareness of the natural world and are concerned chiefly with the organisms that influence their own lives. For example, gauchos, the cowboys of Argentina, who are famous for their horsemanship, have some 200 names for different colors of horses but generally divide plants into four groups: *pasto*, or fodder; *paja*, bedding; *cardo*, wood; and *yuyos*, everything else.

Most of us are like the gauchos. Once beyond the range of common plants and animals, and perhaps a few uncommon ones that are of special interest to us, we usually run out of names and categories. Biologists, however, face the task of systematically identifying, studying, and exchanging information about the vast diversity of organisms—more than 5 million different species—with which we relative newcomers share this planet. In order to do this, they must have a system for naming all these organisms and for grouping them together in orderly and logical ways. The problems of developing such a system are immensely complicated and begin with the basic unit of biological classification, the species.

WHAT IS A SPECIES?

Species in Latin simply means "kind," and so species, in the simplest sense, are different kinds of organisms. A more rigorous definition of species was set forth in 1940 by Ernst Mayr of Harvard University, who said that species are "groups of actually or potentially interbreeding natural populations which are reproductively isolated from other such groups." The phrase "actually or potentially" allows for the fact that although members of the human population of Greenland are not likely to interbreed with those of Patagonia, they are still members of the human species; similarly, transporting a group of insects to some remote island does not automatically make them members of another species. The words "groups" and "populations" are important in this definition also. The possibility that single

20-1 *An inhabitant of the canopy. This masked puddle frog, like many other animals that dwell in the tropical treetops, seldom descends to the forest floor. Active at night, puddle frogs sleep during the day, preferably wedged into a tight place, such as an unfurling leaf, where they are well camouflaged. The great diversity of life on earth is a result of a long evolutionary history in which different kinds—species—of organisms have become adapted to different environments and different ways of life.*

407

individuals of different species may have occasional offspring—such as by the crossing of lions and tigers in a zoo—is unimportant in terms of the group. Mayr's definition conforms to common sense: if members of one species freely exchanged genes with members of another species, they could no longer retain those unique characteristics that identify them as different kinds of organisms.

This definition works well for animal species and is generally accepted by zoologists. Many plants, however, can reproduce asexually and also can form fertile hybrids with other species. Bacteria, with their variety of forms of genetic exchange, do not fit this definition neatly, nor do the many unicellular eukaryotes that reproduce by cell division, forming clones of identical cells. Thus, although botanists and microbiologists use the term "species," they are more likely to consider it a category of convenience, existing rather in the human mind than in the natural world.

For most practical purposes, a species is a category into which is placed an individual organism that conforms to certain fairly rigid criteria concerning its structure and other characteristics. From an evolutionary perspective, however, a species is a group or population of organisms, reproductively united but very probably changing as it moves through space and time. Splinter groups, reproductively isolated from the population as a whole, can undergo sufficient change that they become new species. This process is known as **speciation**. Occurring repeatedly in the course of more than 3.5 billion years, it has given rise to the diversity of organisms that have lived in the past and that live today.

The Naming of Species

A group of closely related species, presumably derived by speciation from a common ancestor, constitute a **genus** (plural, genera). According to the binomial system of nomenclature devised by the Swedish naturalist Linnaeus in the eighteenth century and still in use today, the scientific name of an organism consists of two parts—the name of the genus plus a specific epithet (an adjective or modifier). The genus name is always written first, as in *Drosophila melanogaster*, and it may be used alone when one is referring to members of the entire group of species making up that genus, such as *Drosophila* or *Paramecium*.

20-2 *These plants are all members of the genus* Viola: (**a**) *the common blue violet,* Viola papilionacea, (**b**) *the wild pansy,* Viola tricolor, *and* (**c**) *the long-spurred violet,* Viola rostrata. *Although there is an overall similarity among all three species, there are clear-cut differences in leaf shape, flower color and size, and other characteristics.*

(a)

(b)

(c)

A specific epithet is meaningless when written alone, however, because many different species in different genera may have the same specific epithet. For example, *Drosophila melanogaster* is the fruit fly that has played such an important role in genetics; *Thamnophis melanogaster*, however, is a semiaquatic garter snake. Thus, by itself, the specific epithet *melanogaster* ("black stomach") would not identify either organism. For this reason, the specific epithet is always preceded by the genus name, or, in a context where no ambiguity is possible, the genus name may be abbreviated to its initial letter. Thus *Drosophila melanogaster* may be designated *D. melanogaster*.

Whoever describes a genus or species first has the privilege of naming it. It may not be named after oneself, but often it is named after a friend or colleague. *Escherichia*, for example, is named after Theodor Escherich, a German physician (*coli* simply means intestinal); and *Rhea darwinii*, an ostrichlike bird found in Patagonia, is named after Charles Darwin.

Names may be descriptive. The first discovered early fossil of a horse—which does not superficially resemble a modern horse at all—was named *Hyracotherium*, the "hyrax-like beast." (A hyrax looks like a big guinea pig and is now thought to be a distant relative of the elephant.) When O. C. Marsh of Yale University began, in the 1870s, to study fossil horses, he recognized that the little dog-sized *Hyracotherium* was an early equine and gave it the charming name Eohippus ("dawn horse"). However, *Hyracotherium* remains its official designation because it was published first.

Some names are heartfelt. Thus, members of various mosquito genera have been given the specific epithets *punctor, tormentor, vexans, horrida, perfidiosus, abominator,* and *excrucians*. Others are frivolous. An English entomologist coined a whole series of generic names based on the pseudo-Greek ending *chisme*, pronounced "kiss me." Thus, there are squash bugs, stink bugs, and seed bugs known variously as *Polychisme, Peggichisme, Dolichisme*, and the promiscuous *Ochisme*. A new species of wasp in the genus *Lalapa* was recently given the specific epithet *lusa*; there is a (presumably treacherous) beetle named *Ytu brutus*, and a horsefly called *Tabanus balazaphyre*. So, although species names may appear formidable and be unpronounceable, they are not necessarily as pompous (or even as informative) as they may seem.

These binomials are a necessary tool for clear and unambiguous communication among biologists. Many species lack common names, and, even when common names do exist for a kind of organism, more than one name may be given to the same species, such as groundhog and woodchuck, gnu and wildebeest, pill bug and sow bug and wood louse. Or names may vary from place to place. As shown in Figure 20-3, a robin in North America is distinctly different from the

20-3 Though they are both called robins, (a) the North American robin, Turdus migratorius *("thrush, of the migratory habit"), is a distinctly different bird from (b) the English robin,* Erithacus rubecula.

(a)   (b)

20-4 *The eighteenth-century Swedish professor, physician, and naturalist Carolus Linnaeus (1707–1778), who developed the binomial system for naming species of organisms and established the major categories that are used in the hierarchical system of biological classification. (Linnaeus was born Carl von Linné but latinized his name in the scholarly fashion of the time.) When he was 25, Linnaeus spent five months exploring Lapland for the Swedish Academy of Sciences; he is shown here wearing his Lapland collector's outfit, which you might want to compare with those of modern-day collectors, as shown in the photograph on page 15.*

English bird of the same name. A yam in the southern United States is a totally different vegetable from a yam several hundred kilometers away in the West Indies. When different languages are involved, the problems of communication would be virtually insurmountable without a system of nomenclature universally recognized and agreed upon by biologists.

HIERARCHICAL CLASSIFICATION

One of the fundamental goals of observers of the natural world, from well before the time of Aristotle, has been to perceive order in the diversity of life. One way of achieving order is through **taxonomy,** which is the classification of organisms (or, for that matter, of any other item, such as books in the library or groceries on the shelf).

The taxonomy of organisms is a hierarchical system—that is, it consists of groups within groups, with each group ranked at a particular level. In such a system, a particular group is called a **taxon** (plural, taxa), and the level at which it is ranked is called a **category.** For example, in political geography, nation, state or province, and city are categories, whereas Canada, Ontario, and Toronto are taxa within those categories. Similarly, genus and species are categories, and *Homo* and *Homo sapiens* are taxa.

In the time of Linnaeus, three categories were in common use: the species, the genus, and a category of much higher level, the **kingdom.** Naturalists recognized three kingdoms: plant, animal, and mineral. Kingdom is still the highest category used in biological classification. Between the level of genus and the level of kingdom, however, Linnaeus and subsequent taxonomists have added a number of other categories. Thus, genera are grouped into **families,** families into **orders,** orders into **classes,** and classes into **phyla** or **divisions.** (The categories of division

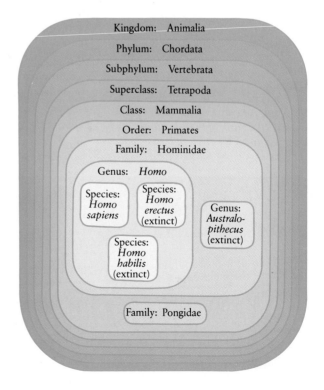

20-5 *The hierarchical nature of biological classifications, consisting of groups within groups, can be represented visually in what is known as a Venn diagram. This Venn diagram shows the classification of the genus* Homo.

and phylum are equivalent. The term "division" is generally used in the classification of prokaryotes, algae, fungi, and plants, whereas "phylum" is used in the classification of protozoa and animals.) These categories may be further subdivided or aggregated into a number of less frequently employed categories such as subphylum or superfamily. By convention, generic and specific names are written in italics, while the names of families, orders, classes, and other taxa whose categories rank above the genus level are not, although they are capitalized.

This system of classification makes it possible to generalize. For example, Table 20-1 shows the classification of two different organisms. Notice how the classification of an animal as a mammal, or a plant as Anthophyta, provides an index to a vast amount of information. Notice also that, in progressing downward from kingdom to species, there is an increase in detail, proceeding from the general to the particular. In short, hierarchical classification is a highly useful means of storing and retrieving information.

TABLE 20-1 Biological Classifications

Red Maple (*Acer rubrum*)

CATEGORY	TAXON	CHARACTERISTICS
Kingdom	Plantae	Multicellular organisms primarily adapted for life on land; usually have rigid cell walls and chlorophylls a and b contained in chloroplasts
Division	Anthophyta	Vascular plants (plants with conducting tissues) with seeds and flowers; ovules enclosed in ovary; seeds enclosed in fruit; the flowering plants
Class	Dicotyledones	Embryo with two seed leaves (cotyledons)
Order	Sapindales	Soapberry order; usually woody plants
Family	Aceraceae	Maple family; characterized by watery, sugary sap; opposite leaves; winged fruit; chiefly trees of temperate regions
Genus	*Acer*	Maples and box elder
Species	*Acer rubrum*	Red maple

Human Being (*Homo sapiens*)

CATEGORY	TAXON	CHARACTERISTICS
Kingdom	Animalia	Multicellular organisms requiring complex organic substances for food; food usually ingested
Phylum	Chordata	Animals with notochord, dorsal hollow nerve cord, gill pouches in pharynx at some stage of life cycle
Subphylum	Vertebrata	Spinal cord enclosed in a vertebral column, body basically segmented, skull enclosing brain
Superclass	Tetrapoda	Land vertebrates, four-limbed
Class	Mammalia	Young nourished by milk glands, skin with hair, body cavity divided by a muscular diaphragm, red blood cells without nuclei, three ear bones (ossicles), high body temperature
Order	Primates	Tree dwellers or their descendants, usually with fingers and flat nails, sense of smell reduced
Family	Hominidae	Flat face; eyes forward; color vision; upright, bipedal locomotion
Genus	*Homo*	Large brain, speech, long childhood
Species	*Homo sapiens*	Prominent chin, high forehead, sparse body hair

The fundamental category in the hierarchical classification of organisms is the species, which, despite the difficulties in its definition, may be considered a biological reality. The other categories, however, exist only in the human mind. To take a familiar group as an example, some taxonomists, "lumpers," would combine all the cats except one into the single genus *Felis*, excluding only the cheetah *(Acinonyx)*, because of its nonretractable claws. Others, "splitters," would reserve the designation *Felis* for the smaller cats, such as the cougar, the ocelot, and the domestic cat, and divide the others into the larger cats *(Panthera)*, including the lion, tiger, and leopard, and the bobtail cats *(Lynx)*. More extreme splitters favor separate genera for the clouded leopard *(Neofelis)* and the snow leopard *(Uncia)*. No one disagrees about the characteristics of the animals themselves but only about the weighing of similarities and differences. One taxonomist's family can easily be another's order.

EVOLUTIONARY SYSTEMATICS

For Linnaeus and his immediate successors, the goal of taxonomy was the revelation of the grand, unchanging design of creation. After 1859, however, differences and similarities among organisms came to be seen as products of their evolutionary history, or **phylogeny.** Thus genera came to be regarded as, ideally, groups of recently diverged (and therefore closely related) species, families as less recently divergent genera, and so on. Biologists now wanted their taxonomies to be not only convenient and useful but also an accurate reflection of the historical relationships among organisms. Such taxonomies are, in effect, hypotheses about evolutionary history. Like other hypotheses, they can be tested (through detailed study of the fossil record and of the structure and other characteristics of living organisms) and revised as necessary. This study of the historical relationships among organisms is known as evolutionary **systematics.**

The Monophyletic Ideal

In a classification scheme that accurately reflects evolutionary history, every taxon is, ideally, **monophyletic.** This means that the members of a taxon, at whatever

(a)

20-6 *A sampling of members of the class Mammalia, order Carnivora, family Felidae. According to some taxonomic schemes, all of these felines would be considered members of the genus* Felis. *Alternatively, the genus* Felis *is reserved for small cats, such as* (a) *the domestic cat,* (b) *the mountain lion, or cougar, and* (c) *the ocelot. The larger cats, including* (d) *the leopard, are classified as* Panthera, *and the bobtail cats, such as* (e) *the lynx, as* Lynx.

(b)

(c)

categorical level, should all be descendants of the nearest common ancestral species. In other words, taxa should be real historical units. Thus, a genus should consist of species descended from the most recent common ancestor—and only of species descended from that ancestor. Similarly, a family should consist of genera descended from a more distant common ancestor—and only of genera descended from that ancestor.

Although this ideal sounds relatively straightforward, it is often difficult to attain. In many cases, biologists do not know enough about the evolutionary history of the organisms to establish taxa that are, with a reasonable degree of certainty, monophyletic. In other cases, the two different functions of biological classification come into conflict, and convenience and utility may be deemed more important than an accurate reflection of phylogeny. Thus, as we shall see, some generally accepted taxa contain organisms descended from more than one ancestral line. Such taxa are said to be **polyphyletic.**

Homology and Phylogeny

The grouping of organisms into taxa from the categorical levels of genus through phylum or division is based on similarities in structure and other phenotypic characteristics. From Aristotle on, however, biologists have recognized that superficial similarities are not useful criteria for taxonomic decisions. To take a simple example, birds and insects should not be grouped together simply because both have wings. A wingless insect (such as an ant) is still an insect, and a flightless bird (such as the kiwi) is still a bird on the basis of overall structure. Linnaeus classified whales with mammals and not with fish, despite their external similarities.

A key question in evolutionary systematics is the origin of a similarity or difference. Does a similarity reflect inheritance from a common ancestor, or does it reflect adaptation to similar environments by organisms that do not share a common ancestor? A related question arises with differences between organisms: Does a difference reflect separate phylogenetic histories, or does it reflect instead the adaptations of closely related organisms to very different environments? A classic example is the vertebrate forelimb. The wing of a bird, the flipper of a whale, the foreleg of a horse, and the human arm have quite different functions

(d)

(e)

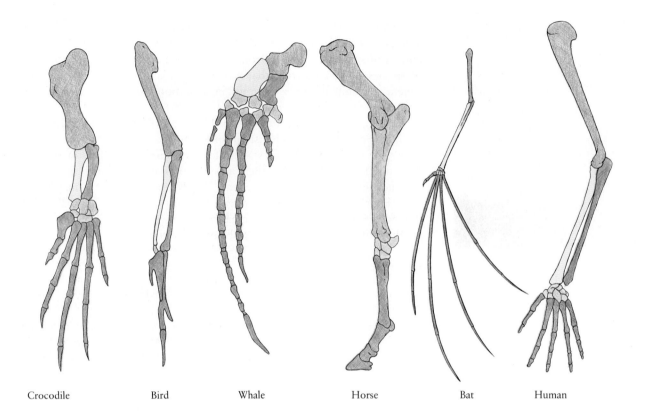

Crocodile Bird Whale Horse Bat Human

20-7 *The bones in these forelimbs are color-coded to indicate fundamental similarities of structure and organization. The crocodile is placed first because it is the closest to the ancestral type—the form from which all the others arose. (Note also the similarity between the forelimb of the crocodile and that of the human.) Structures that have a common origin but not necessarily a common function are known as homologues. Analogous structures, by contrast, are superficially similar but have an entirely different evolutionary background—for example, the spine of a cactus (a modified leaf) and the thorn of a rose (a modified branch).*

and appearances. Detailed study of the underlying bones reveals, however, the same basic structure (Figure 20-7). Such structures, which have a common origin but not necessarily a common function, are said to be **homologous.** These are the features upon which evolutionary classification systems are ideally constructed.

By contrast, other structures, which may have a similar function and superficial appearance, have an entirely different evolutionary background. Such structures are said to be **analogous.** Thus the wings of a bird and the wings of an insect are analogous, not homologous.

Decisions as to homology and analogy are seldom so simple. In general, the features most likely to be homologous—and thus useful in determining phylogenetic relationships—are those that are complex and detailed, consisting of a number of separate parts. This is true whether the similar feature is anatomical, as in the bones of the vertebrate forelimb, or is a biochemical pathway or a behavioral pattern. The more separate parts involved in a feature shared by several species, the less likely it is that the feature evolved independently in each.

TAXONOMIC METHODS

The traditional means of determining the classification of a newly discovered organism requires several different steps, involving different kinds of appraisals. First, the organism is tentatively assigned to a particular taxon on the basis of its

20-8 *Occasionally, newly discovered organisms cannot be classified in existing taxa and require the creation of new taxa. This tiny animal, Nanaloricus mysticus, dredged from the ocean bottom off the coast of France in 1982, is a case in point. With a plate-covered body, numerous spines projecting from its head, and a retractable tube for a mouth, it is unlike any previously known animal. It is the first member of a newly established phylum, Loricifera ("girdle-wearer"). Closely related species, also assigned to the new phylum, have now been found in the Coral Sea, in Greenland, and off the coast of Florida.*

overall outward similarities to other members of that taxon. Then these similarities are tested for homologies. Fossils are taken into account when possible. For instance, hares and rabbits (collectively known as lagomorphs) were long believed to be rodents, but the earliest fossil remains of lagomorphs and rodents show that the two groups had quite different origins. Conversely, bears, once considered a very distinct group of carnivores, are now known, on the basis of paleontology, to have diverged relatively recently from dogs. Various stages in the life cycle and the patterns of embryonic development are also compared; as you will see in the chapters that follow, some major decisions concerning the phylogenetic relationships of animal groups are based on similarities in early development.

Taxonomies constructed by the traditional methods reflect the consideration and weighing of a large number of factors. Some of these factors provide evidence of the genealogy, or branching patterns, that have characterized the evolutionary history of the organisms, while others reflect the degree to which the organisms have diverged since they began to travel separate evolutionary paths. Thus, traditional taxonomies contain information about both the sequence in which branchings occurred and the extent of the subsequent biological changes. Such taxonomies can be summarized in the form of phylogenetic trees, as shown in Figure 20-9.

Given the large number of factors that are taken into account in constructing an evolutionary taxonomy and the fact that different biologists have different views of the importance of various factors, it is not surprising that radically different classifications have sometimes been proposed for the same organisms.

Alternative Methodologies

Two alternative methodologies—numerical phenetics and cladistics—have been proposed as replacements for the traditional methods of evolutionary systematics. Both pheneticists and cladists have sought to develop a truly objective taxonomic method that would eliminate the subjectivity that seems unavoidable with the traditional methods. Both groups have also noted that it is not really possible for a

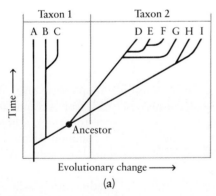

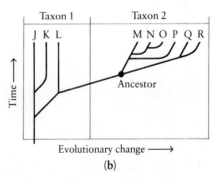

20-9 *The evolutionary history of a group of related organisms can be represented by a phylogenetic tree. The vertical locations of the branching points indicate when particular taxa diverged from one another; the horizontal distances indicate how much the taxa have diverged, taking into account a number of different characteristics.*

The two diagrams shown here represent the evolutionary histories of two different groups of taxa, labeled A through I in (a) and J through R in (b). Each group has been classified using traditional methods. In (a), the ancestor of taxa D through I is included in taxon 1 because of its close resemblance to taxa B and C.

In (b), the ancestor of taxa M through R is placed in taxon 2, because of its close resemblance to taxon M. In each case, taxa 1 and 2 would themselves be members of a taxon at a higher categorical level, which would probably include other taxa as well.

single classification scheme to indicate both overall similarity (the concern of the earliest taxonomists) and genealogy (the additional concern of taxonomists since the time of Darwin). They point out that some long-separated lineages have evolved in parallel and so continue to resemble one another more closely than organisms that have diverged rapidly from a recent common ancestor. Not only are the traditional methods suspect, but the goals are unattainable, according to this joint analysis. The remedies proposed by the two groups are, however, exactly opposite.

Numerical Phenetics

Numerical phenetics is based only on a species' observable characteristics. To begin, the characteristics of the species being studied are divided into unit characters, that is, characters of two or more states that cannot logically be subdivided further. These unit characters are assigned numbers and coded as plus or minus or 0 (data not available). As many different characters as possible—at least 100—are taken into consideration. The data are then processed by computer, which scores the taxa according to the number of unit characters they share.

Each character is given equal weight by this system with no subjective evaluation or prior knowledge taken into consideration. For instance, the possession of five fingers in a phenetic analysis would signify that lizards are more similar to humans than to snakes. The difference between homology and analogy is disregarded. Characters known to be subject to environmental pressures—such as the shape of a leaf—are weighted equally with more constant characters—for example, the structure of a flower. Pheneticists maintain that such problems are resolved if enough characters are taken into consideration. Thus, for instance, despite the fact that one has five fingers and one does not, the relationship between lizard and snake would emerge when the other characters are taken into account.

Cladistics

In contrast to numerical phenetics, which bases classification exclusively on degree of overall similarity, cladistics ignores overall similarity and is based exclusively on phylogeny. Cladists maintain that the branching of one lineage from another in the course of evolution is the one event that can be determined objectively. Such points are marked by the appearance of evolutionary novelties—that is, characteristics that were not present in the ancestral, or **primitive**, condition (see essay).

The goal of the cladists is the construction of **holophyletic** taxa. Whereas a monophyletic taxon includes only organisms descended from a common ancestor (but not necessarily all such organisms nor the ancestor itself), a holophyletic taxon must include all of the descendants from the common ancestor, plus the ancestor. None of the taxa in Figure 20-9, determined by traditional methods, would satisfy these requirements.

The substitution of cladistics for traditional methods of classification can produce revolutionary changes. Figure 20-10 shows a traditional phylogenetic tree of the principal land-dwelling vertebrates. Four of the modern groups (B through E), which include organisms such as crocodiles, turtles, snakes, and lizards, are placed in class Reptilia according to conventional schemes. Two of the groups, the birds and the mammals, are placed in separate classes because of their obvious biological differences. According to the cladist scheme, however, taxa must conform strictly to branching patterns, forming "nested sets" within the hierarchy of the more inclusive taxa. Thus, crocodiles (B) end up with the birds (A), rather than with the turtles, snakes, and lizards.

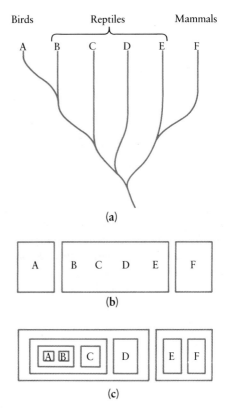

20-10 (a) *The phylogeny of the living representatives of the dominant terrestrial vertebrates. Branches that have become extinct, such as the dinosaurs, are not shown.* (b) *According to traditional classification, taxa B through E are grouped together (class Reptilia). Taxa A and F, the birds and the mammals, are placed in class Aves and class Mammalia because of their extensive biological differences from the other groups.* (c) *According to cladistic methodology, classification must be based solely on genealogy (branching patterns), with the more recent branches assigned lower categorical ranks in the hierarchical system. The biological changes that have occurred since the groups branched from one another are not taken into account.*

How to Construct a Cladogram

A cladogram is a hypothesis of branching sequences. It looks, at first glance, like a phylogenetic tree, but it is not one. It contains no ancestors, only branching points as determined by the appearance of evolutionary novelties. For example, consider the following organisms: lizard, mouse, trout, cow, shark. They have some shared characteristics—a nerve cord on the dorsal (top) side of the body, a chambered heart, four appendages, and jaws—that set them off, as a group, from other groups, such as clams or grasshoppers, for instance. Four of them have bony skeletons and one, the shark, does not. Three of them are characterized by an amniote egg (that is, an egg that contains its own water supply). Two of them have mammary glands. Thus, in terms of evolutionary novelties, the branching sequence shown in (a) is constructed.

The hypothesis can now be tested. For instance, cow and mouse not only both have mammary glands, but both also have hair, another evolutionary novelty. If the lizard and mouse had hair, but the cow lacked hair, the hypothesis would have to be reexamined. Or, if the mouse and trout had three ear bones, but the lizard did not, the cladogram would have to be reconstructed. Thus, every cladogram fulfills the scientific ideal of presenting a testable hypothesis.

From the cladogram, a hierarchical classification can be constructed:

Group: shark, trout, lizard, mouse, cow

Subgroup 1a: shark
Subgroup 1b: trout, lizard, mouse, cow

Subgroup 2a: trout
Subgroup 2b: lizard, mouse, cow

Subgroup 3a: lizard
Subgroup 3b: mouse, cow

Subgroup 4a: mouse
Subgroup 4b: cow

Cladists view such a hierarchical listing as representing "nested sets," that is, groups within groups, as shown in (b). Examination of this figure reveals that, following the logical steps outlined, we have been led to put sharks in one high-ranking taxon and trout in another, along with mammals. This sort of result has led traditionalists to note that cladists cannot tell a fish from a cow. In this we see the essence of the disagreement.

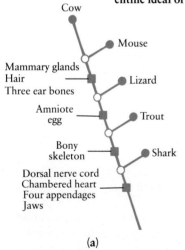

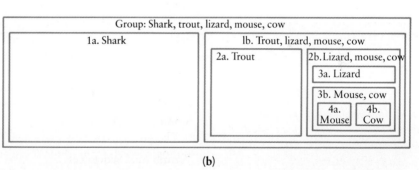

(a) *A cladogram for five types of vertebrates. Each branch point is defined by one or more characteristics interpreted as evolutionary novelties.* (b) *From the cladogram, the cladistic hierarchy, represented as "nested sets," can be constructed.*

Traditional systematists point out that such a classification ignores an important aspect of the evolutionary history of these organisms: the striking biological changes that occurred as the early birds diversified into a new life zone—the air—and as the early mammals diversified into a terrestrial environment made available by the demise of the dinosaurs. Given the obvious differences between birds and crocodiles and the similarity, for instance, of crocodiles and lizards (neither of which have changed significantly from the ancestral condition), a system that groups birds with crocodiles, instead of crocodiles with lizards, makes no biological sense. Cladists, however, are concerned with the lack of consistency of the traditional system; class Reptilia, for the cladist, simply should not exist. It is analogous to a political taxon made of the United States minus all states admitted since 1912. Such a taxon, which consists of a holophyletic group from which strikingly divergent members, and perhaps the common ancestor as well, have been removed, is said to be **paraphyletic.** Even in political taxonomy, however, paraphyletic taxa—such as the contiguous 48 states—make sense and are useful.

It has become apparent in recent years that no one approach to biological classification is adequate for all purposes. In their search for consistency and objectivity, both pheneticists and cladists must exclude a portion of the available knowledge about organisms. Moreover, neither phenetics nor cladistics completely escapes subjectivity; decisions about the characters to be tabulated or the identification of evolutionary novelties must still be made by fallible human beings on the basis of incomplete knowledge. These two methodologies, however, have provided important new perspectives in taxonomy and triggered many new questions about organisms and their evolutionary relationships. The answers to some of these questions are beginning to emerge as taxonomists apply the tools of molecular biology to their studies.

MOLECULAR TAXONOMY

Taxonomy by any methodology has been based largely on anatomy, and this will probably continue to be true in the future since a large body of data on comparative anatomy has been accumulated. As we have seen, however, it is often difficult to determine the appropriate weights to assign to different kinds of similarities. Moreover, such differences are of little help in the study of structurally dissimilar organisms—fish and fungi, for instance. New biochemical techniques are, however, becoming increasingly important in evolutionary systematics. They offer two distinct advantages: the results are objectively quantifiable, and very diverse organisms can be compared. Biochemical studies can reveal, for example, similarities and differences in enzymes, reaction pathways, hormones, and important structural molecules. With the development of techniques for sequencing the amino acids in proteins and the nucleotides in DNA and RNA molecules, it has become possible to compare organisms at the most basic level of all—the gene.

Amino Acid Sequences

One of the first proteins to be analyzed in taxonomic studies was cytochrome c, one of the carriers of the electron transport chain (page 196). Cytochrome c molecules from a great variety of organisms were sequenced, making it possible to determine the number of amino acids by which the molecules of the various organisms differ. Presumably, the greater the number of amino acid differences between any two organisms, the more distant their evolutionary relationship; conversely, the smaller the number of differences, the closer their relationship. Figure 20–11 illustrates a phylogeny based on cytochrome c data. The results conform well, but not perfectly, with phylogenies constructed by more traditional methods.

As data on protein variations accumulated, it became clear that although protein structure is a useful parameter of evolutionary relationships, there are difficulties in interpreting the results. Some biologists maintain that differences in

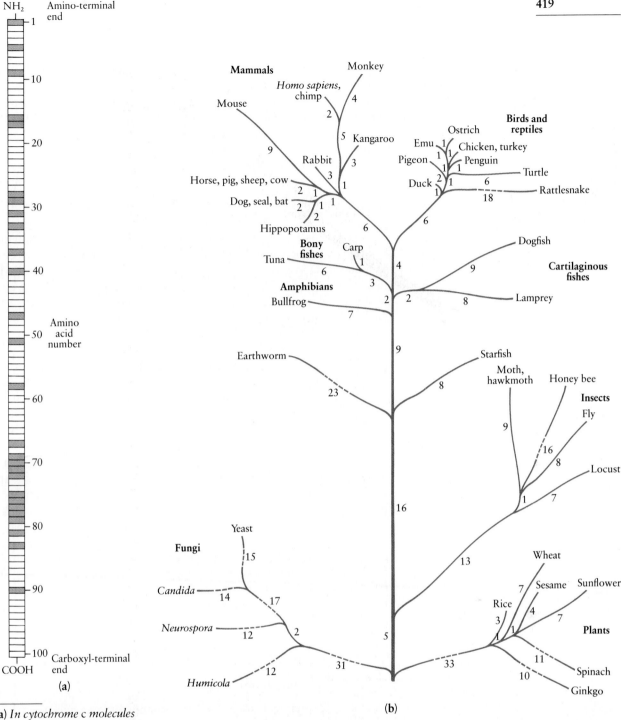

20–11 (a) *In cytochrome c molecules from the more than 60 species that have been studied, 27 of the amino acids are identical (color).* (b) *The main branches of a phylogenetic tree based on comparisons of the amino acid sequences of cytochrome c molecules. The numbers indicate the number of amino acids by which each cytochrome c differs from the cytochrome c at the nearest branch point. Dashes indicate that a line has been shortened and is therefore not to scale. Although based on comparisons of a single type of protein molecule, this tree is in fairly good agreement with phylogenies constructed by more conventional means.*

protein structure represent functional differences among the molecules, just as differences in the structure of the beaks of birds represent adaptations to different food sources. Other biologists believe that amino acid changes occur at random —as the result of random mutations—and that they do not represent the result of a selection process but merely mark off the passage of time, like grains of sand trickling through an hour glass or the decay of radioisotopes. From this viewpoint, the amino acid differences in the homologous proteins of different groups of organisms do not represent functional differences; instead, they are **molecular clocks** and can be used to determine the time at which various groups diverged.

In support of the "random-tick" hypothesis, biological clock makers point out that two frog species that diverged millions of years ago, but remained similar enough in appearance to be included in a single genus, differ from one another in amino acid substitutions as much as a bat does from a whale. *Homo sapiens* and the chimpanzee, on the other hand, two species that differ anatomically and in a number of other characteristics but that diverged recently, according to paleontological evidence, have identical amino acid sequences in cytochrome *c* and some other proteins as well.

Nucleotide Sequences

Now that it is possible to sequence nucleic acids (see page 348), the use of homologous proteins to estimate evolutionary relationships has been largely abandoned. One reason is that nucleic acid sequencing is technically far easier than protein sequencing, dealing, as it does, with only four different nucleotides as compared to 20 amino acids. Also, it is more sensitive, since changes in nucleotides may not be reflected in changes in amino acids because of the many synonyms in the genetic code.

As the sequences of nucleic acids from a variety of species have been determined, the information has been entered into computer data banks, making possible detailed comparisons. Such comparisons have demonstrated the value of nucleic acid sequences in taxonomic studies. For example, analyses of the rRNA and tRNA molecules of prokaryotes have made it possible, for the first time, to begin determining the evolutionary relationships among these organisms, which are extremely difficult to distinguish on the basis of structural features alone.

A number of studies have shown, however, that there are a variety of different molecular clocks, ticking at different rates. As you might expect, nucleotide changes that result in amino acid substitutions affecting the function of critical proteins are weeded out by natural selection; thus, clocks based on such proteins are a bit slow. Conversely, nucleotide changes in segments of DNA that are never translated—such as introns (page 361)—appear to be relatively free of functional constraints, leading to a faster ticking of the clock. In addition, for reasons that are only poorly understood, different portions of the DNA of an organism appear to be subject to different rates of mutation. Theoretically, these difficulties would be eliminated (or at least minimized) if many different homologous DNA segments were sequenced. Limitations of time and resources, however, make such an approach impractical and have led to the search for simpler ways of comparing large portions of the genome.

DNA-DNA Hybridization

One of the earliest techniques used in the study of large nucleic acid molecules was hybridization. As we saw earlier (page 345), when a solution of DNA is heated, the hydrogen bonds break and the double-stranded molecules separate into single strands. Upon cooling, the single strands reassociate with each other—or with homologous strands of DNA or RNA from another source. This property of nucleic acid molecules has been invaluable in locating particular genes of interest for further study. More recently, Charles G. Sibley and Jon E. Ahlquist of Yale University have devised an elegant taxonomic adaptation of the technique.

In their studies, Sibley and Ahlquist break organismal DNA into fragments of approximately 500 nucleotides each and remove the repeated DNA segments (page 363) that characterize the eukaryotic genome. In groups of two, solutions of the resulting single-copy DNA are mixed, heated, and cooled, allowing hybridization of homologous sequences to occur (Figure 20–12). The DNA from one source is radioactively labeled, and the other is unlabeled; the amount of unlabeled DNA is approximately 1,000 times the amount of labeled DNA. Because of

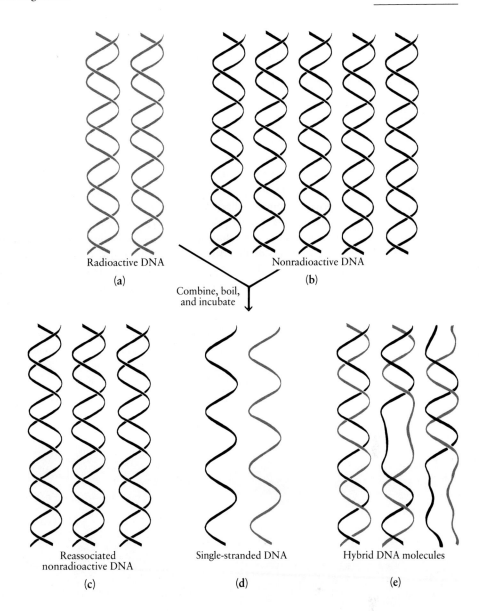

20-12 *In DNA-DNA hybridization experiments, radioactively labeled DNA (a) is mixed with unlabeled DNA (b) from the same species or a different species. The amount of unlabeled DNA is approximately 1,000 times the amount of labeled DNA. The solution is boiled to dissociate the double-stranded molecules and then incubated to allow the single strands to reassociate. Most of the unlabeled strands reassociate with other unlabeled strands (c), and some strands fail to reassociate at all (d). However, because of the vast excess of unlabeled DNA, about 99 percent of the radioactive strands reassociate with unlabeled strands, forming hybrid molecules (e). The more similar the nucleotide sequences of the strands forming a hybrid molecule, the more tightly they are held together—and the higher the temperature required to dissociate them again. As you can see, one of the hybrid molecules is a perfect match, while the other two have differing degrees of mismatch.*

this vast excess of unlabeled DNA, most of the reassociated double-stranded molecules that contain radioactivity are hybrids. As the solution is then gradually reheated, the molecules that dissociate into single strands are removed and tested for radioactivity. The temperature at which 50 percent of the hybrid molecules dissociate is a measure of how tightly the strands are bonded, which is, in turn, a measure of the similarity of the DNA sequences. The higher the temperature, the more similar the DNAs.

The 50-percent-dissociation temperature is determined individually for the DNA of each species to be studied (using radioactive and nonradioactive samples from that species). This provides a base line for each species, against which the 50-percent-dissociation temperatures of the hybrids its DNA forms with the DNA of other species can be compared. A 1°C lowering from the single-species 50-percent-dissociation temperature corresponds to a 1 percent difference in the nucleotide sequences of the two species being compared; this, in turn, corresponds to an evolutionary separation of approximately 4.5 million years. By correlating hybridization results with known dates of specific evolutionary and geologic events, as determined from the fossil record, it is possible to establish, in real time, the branching points in evolutionary lineages.

The Riddle of the Giant Panda

Since 1869, when the giant panda of China was discovered, its true identity has been a riddle. It was initially classified as a member of the bear family, but biologists almost immediately began to wonder if it might instead be most closely related to another unusual mammal indigenous to China, the lesser panda. The lesser panda was clearly a member of the raccoon family, even though there were no other living members of that family in the Old World—unless, perhaps, the giant panda was also a raccoon, albeit a very odd one. The two pandas share many unusual anatomical and behavioral characteristics, and although the giant panda does resemble the bears in certain features, it differs significantly in others. Over the years, biologists debated the question, opting in roughly equal numbers for "bear" or "raccoon." The answer seemed to be at hand in 1964 with the publication of a detailed anatomical study of the giant panda. This study demonstrated, to the satisfaction of most biologists, that the giant panda is a bear and that the features in which it resembles the lesser panda are adaptations to feeding on its exclusive energy source, bamboo.

This conclusion has now been confirmed by the application of four different techniques: DNA-DNA hybridization, comparison of the charge and size of homologous protein molecules (as determined by electrophoretic studies), comparison of the antibody-binding properties of homologous proteins, and detailed study of the banding patterns of the chromosomes. Each procedure provided the same answer: the giant panda is a bear, as children the world over have known all along.

(a)

(b)

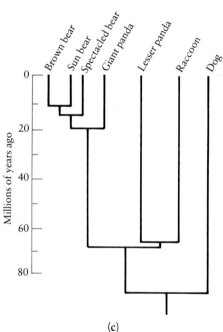

(c)

Although both native to China and sharing many unusual anatomical and behavioral adaptations, the lesser panda (a) and the giant panda (b) are not each other's closest relatives. The results of DNA-DNA hybridization studies (c), as well as studies using other techniques, have demonstrated that the lesser panda is most closely related to the raccoons, whereas the giant panda is most closely related to the bears. The hybridization results also confirm the fossil evidence indicating that the lineage that gave rise to the bears is a branch of the lineage that gave rise to the dogs (see page 417).

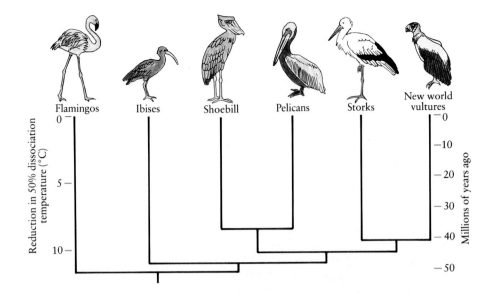

20-13 *The family tree of six closely related groups of living birds, as determined by DNA-DNA hybridization studies. The DNA used in these studies was extracted from red blood cells, which, in birds, contain nuclei. This particular set of experiments not only answered a long-standing question about the proper classification of the flamingos but also provided surprising new information about the history of the New World vultures, such as the California condor. On the basis of structural and behavioral similarities, these birds have traditionally been classified with the Old World vultures; it now appears that those similarities are the result of adaptation to similar life styles. As shown here, the New World vultures are most closely related to the storks. The Old World vultures, by contrast, belong to a completely different evolutionary lineage and are more closely related to the hawks and the eagles.*

Using this method, Sibley and Ahlquist have compared—two by two—more than 25,000 DNA samples from about 1,600 species of birds. In the process, they have solved a number of puzzles in bird taxonomy. For example, flamingos have long been classified with the storks by some authorities, with the geese and ducks by others; hybridization studies have now demonstrated unequivocally that flamingos are more closely related to the storks (Figure 20-13). To cite another example, the starlings were long classified with the crows. However, data from DNA-DNA hybridization studies have revealed that these two lineages separated some 60 million years ago and that starlings are much more closely related to mockingbirds, catbirds, and thrushes, from which they diverged a mere 25 million years ago.

Recent evidence indicates that the DNA of some widely separated groups (such as, for example, insects and mammals) evolves at quite different rates. Similarly, mitochondrial DNA appears to evolve much more rapidly than nuclear DNA. Thus even the molecular clock of DNA-DNA hybridization, which is based on the entire single-copy genome, is not the answer to all taxonomic problems. For groups of organisms that are not too distantly related, however, it is the molecular technique of choice. In conjunction with other methods of analysis, it provides taxonomists with the opportunity not only to answer long-standing questions about the relationships of organisms (see essay) but also to ask many new questions. The answers to these will undoubtedly upset some traditional classifications and generate additional questions, providing new insights into the biology of both familiar and unfamiliar organisms. Laboratory taxonomists are engaged in an adventure equal in excitement—if not in physical danger—to that of the tropical biologists who are almost daily discovering organisms previously unknown to science.

A QUESTION OF KINGDOMS

One problem of biological classification currently unresolved—and probably unresolvable, even with the battery of new methods available—is the placement of taxa in the category of kingdom. In Linnaeus's time, as we mentioned earlier, three kingdoms were recognized—animals, plants, and minerals—and until very recently it was common to classify every living thing as either an animal or a plant. Kingdom Animalia included those organisms that moved and ate things, and whose bodies grew to a certain size and then stopped growing. Kingdom Plantae comprised all living things that did not move or eat and that grew indefinitely. Thus the fungi, algae, and bacteria were grouped with the plants, and the protozoa—the one-celled organisms that ate and moved—were classified with the animals.

In the twentieth century, new data began to emerge. This was partly a result of improvements in the light microscope and, subsequently, the development of the electron microscope, and partly because of the application of biochemical techniques to studies of differences and similarities among organisms. As a result, the number of groups recognized as constituting different kingdoms has increased. The new techniques revealed, for example, the fundamental differences between prokaryotic and eukaryotic cells—differences sufficiently great to warrant placing the prokaryotes in a separate kingdom, Monera.

Other studies have provided new information about the evolutionary history of the major types of organisms. As we shall see in Chapter 22, there is strong evidence that different lineages of eukaryotes arose independently from different prokaryotic ancestors. Moreover, different lineages of unicellular eukaryotes appear to have given rise to the multicellular plants, fungi, and animals, and, at least in the case of photosynthetic organisms, multicellularity has arisen several times. This history makes it impossible, on the basis of current knowledge, to establish monophyletic kingdoms and still have the kingdoms reflect similarities and differences among major groups of living organisms.

Most contemporary proposals concerning kingdoms are based not on evolutionary history, but rather on the cellular organization and the mode of nutrition of the organisms. The proposal we shall follow recommends five kingdoms: Monera, Protista, Fungi, Plantae, and Animalia. In Appendix C, the major groups of organisms are classified in these five kingdoms. The members of kingdom Monera, the prokaryotes, are identified on the basis of their unique cellular organization and biochemistry. Members of the kingdom Protista are eukaryotes, both autotrophs and heterotrophs, and most are unicellular. A few groups of relatively simple multicellular organisms are also included in this kingdom, because they are more closely related to unicellular forms than to superficially similar fungi, plants, or animals. All the other multicellular eukaryotes are divided into three kingdoms, based primarily on their mode of nutrition: fungi absorb organic molecules from the surrounding medium, plants manufacture them by photosynthesis, and animals ingest them in the form of other organisms. These three groups of organisms have distinct ecological roles—plants are generally producers, animals are consumers, and fungi are decomposers. Table 20–2 summarizes some of the essential similarities and differences among these five kingdoms of organisms. We shall discuss each group in turn in the chapters that follow.

No system of kingdoms is completely satisfactory. For instance, as we shall see in Chapter 24, there is a clear evolutionary sequence leading from certain one-celled photosynthetic eukaryotes to the flowering plants. So if we group all of the one-celled eukaryotes together in kingdom Protista, as we do in this text, we break up what would appear to be a direct evolutionary line to the plants. However, at the one-celled level of organization, the capacity for photosynthesis is not always a useful criterion for determining the degree of relationship among organisms. Two species of single-celled, motile organisms may be virtually identical in most respects, except that one has chloroplasts and the other does not. In some cases, the one that has chloroplasts can lose them from time to time and still continue to survive and reproduce indefinitely. To separate these two closely related forms at the level of kingdom seriously misrepresents biological reality.

The number of kingdoms into which organisms are classified and the decisions made in assigning problematic taxa to particular kingdoms do not, fortunately, affect the genus and species designations of those organisms nor most of the other hierarchical categories in which they are now classified. Most important, the unique characteristics and evolutionary history of each group of organisms are totally unaffected by any difficulties we may have in ordering them to suit our purposes.

20–14 *Representatives of the five kingdoms.* (**a**) *Monera. Cells of the bacterium* Pseudomonas aeruginosa, *which thrives in standing water, such as that in wooden hot tubs. This bacterium produces a characteristic rash on the skin of its human cohabitants. It is also the cause of "swimmer's ear."* (**b**) *Protista.* Vorticella, *like most protists, is a single cell. It attaches itself to a substrate by a long stalk. A contractile fiber, visible in this micrograph, runs through the stalk.* (**c**) *Fungi. Inky cap mushroom. Fungi are characterized by a multicellular underground network and also spores and sporangia, which, in mushrooms, are borne in these familiar structures.* (**d**) *Plantae. California poppies and lupines. Flowers, attractive to pollinators, are among the principal reasons for the evolutionary success of the plants of division Anthophyta.* (**e**) *Animalia. A luna moth. A nervous system with a variety of sense organs is a chief characteristic of the animal kingdom.*

TABLE 20-2 **Characteristics of the Five Kingdoms**

	MONERA	PROTISTA	FUNGI	PLANTAE	ANIMALIA
Cell type	Prokaryotic	Eukaryotic	Eukaryotic	Eukaryotic	Eukaryotic
Nuclear envelope	Absent	Present	Present	Present	Present
Mitochondria	Absent	Present	Present	Present	Present
Chloroplasts	Absent (photosynthetic membranes in some types)	Present (some forms)	Absent	Present	Absent
Cell wall	Noncellulose (polysaccharide and peptidoglycan)	Present in some forms, various types	Chitin and other noncellulose polysaccharides	Cellulose and other polysaccharides	Absent
Means of genetic recombination	Conjugation, transduction, transformation, or none	Fertilization (syngamy) and meiosis, conjugation, or none	Fertilization and meiosis, dikaryosis (page 481), or none	Fertilization and meiosis	Fertilization and meiosis
Mode of nutrition	Autotrophic (chemosynthetic or photosynthetic) or heterotrophic	Photosynthetic or heterotrophic, or combination of these	Heterotrophic, by absorption	Photosynthetic	Heterotrophic, by ingestion
Motility	Bacterial flagella, gliding, or nonmotile	9 + 2 cilia and flagella, amoeboid, contractile fibrils	Nonmotile	9 + 2 cilia and flagella in gametes of some forms, none in most forms	9 + 2 cilia and flagella, contractile fibrils
Multi-cellularity	Absent	Absent in most forms	Present	Present	Present
Nervous system	Absent	Primitive mechanisms for conducting stimuli in some forms	Absent	Absent	Present, often complex

SUMMARY

More than 5 million different kinds of organisms are known to inhabit the earth, and many times that number await discovery. Scientists have sought order in this vast diversity of living things by classifying them, that is, by grouping them in meaningful ways.

The basic unit of classification is the species, most easily defined as an interbreeding group of organisms that does not breed with other groups of organisms. Species are named by a binomial system that includes the name of the genus (written first) and the specific epithet, a modifier that identifies the particular species within the genus.

CHAPTER 20 The Classification of Organisms

Taxonomy, the science of classifying groups (taxa) of organisms in formal groups, is hierarchical. Genera are groups of similar species, presumably species that have recently diverged. Genera are grouped into families, families into orders, orders into classes, classes into phyla or divisions, and divisions or phyla into kingdoms. The grouping of divisions or phyla into kingdoms is based on cellular organization and mode of nutrition. The classification system followed in this text consists of five kingdoms: Monera, Protista, Fungi, Plantae, and Animalia.

In classifying organisms in the categories of genus through phylum or division, evolutionary systematists seek to group the organisms in ways that reflect their phylogeny (evolutionary history). In a phylogenetic system, every taxon should be, ideally, monophyletic; that is, every taxon should consist only of organisms descended from a common ancestor. A major principle of such classification is that the similarities taken into account should be homologous—that is, the result of common ancestry rather than of adaptation to similar environments (analogous). Among the types of data used by evolutionary systematists are structural and biochemical characteristics, fossil evidence, stages in the life cycle, and patterns of embryonic development.

Two alternative methodologies used in classification are numerical phenetics and cladistics. Numerical phenetics relies solely on the scoring of equally weighted, objectively observable similarities and differences among groups of organisms, without regard to homology and analogy. Cladistics, by contrast, is based entirely on branching sequences (genealogy), determined by evolutionary novelties, and ignores overall similarities. The goal of cladistics is the creation of holophyletic taxa, in which the common ancestor and all of its descendants are included.

New techniques in molecular taxonomy are providing objective, numerical comparisons of organisms at the most basic level of all, the gene. A large body of evidence indicates that protein and nucleic acid molecules are molecular clocks, with changes in composition reflecting the time that has elapsed since different groups of organisms diverged from one another. Amino acid sequencing, nucleotide sequencing, and DNA-DNA hybridization are all making valuable contributions to more accurate classification schemes and, most important, to our understanding of organisms and their evolutionary history.

QUESTIONS

1. Distinguish among the following: taxonomy/evolutionary systematics; category/taxon; genus/species; division/phylum; homology/analogy; numerical phenetics/cladistics; monophyletic/holophyletic/paraphyletic/polyphyletic.

2. Identify which of the following are categories and which are taxa: undergraduates; the faculty of the University of Tennessee; the Washington Redskins; major league baseball teams; the U.S. Marine Corps; Mozart's symphonies.

3. The use of the terms "division" and "phylum" for the same category in biological classification is an accident of history, resulting from the human tendency to place different subjects of study in separate compartments. What compartmentalization produced these two terms?

4. It is generally thought that nucleic acid molecules are more reliable molecular clocks than protein molecules. Give at least three reasons why this should be the case.

5. Based on your knowledge of DNA structure, the genetic code, and protein structure, what sorts of random mutations would you expect to persist in a lineage of organisms, generation after generation, unaffected by natural selection? What sorts of mutations would you expect to be harmful to the organisms and thus suppressed by natural selection?

6. What are the major identifying characteristics of each of the five kingdoms?

CHAPTER **21**

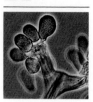

The Prokaryotes and the Viruses

In the next three chapters, we are going to describe three diverse kingdoms of organisms: Monera (the prokaryotes), Protista (unicellular and some simple multicellular eukaryotes), and Fungi. Many of the organisms in these three kingdoms are major agents of disease not only in *Homo sapiens,* but also in the plants and animals on which we depend. Thus, although they are dissimilar in many important ways, and some multicellular protists and fungi are quite large, these organisms have conventionally been studied together in the discipline known as microbiology. Because of their disease-causing properties, viruses, which are, by most definitions, not living organisms at all and thus are not classified in any of the five kingdoms, have also traditionally been studied as a part of microbiology. We shall therefore examine their diversity in the latter part of this chapter.

The prokaryotes are, in evolutionary terms, the oldest group of organisms on earth. And, despite their relative simplicity, contemporary prokaryotes are the most abundant organisms in the world. Although there are sometimes difficulties in defining prokaryotic species unambiguously, about 2,700 distinct species are currently recognized. Prokaryotes are the smallest cellular organisms; a single gram (about 1/28 of an ounce) of fertile soil can contain as many as 2.5 billion individuals.

The success of the prokaryotes, biologically speaking, is undoubtedly due to their great metabolic diversity and their rapid rate of cell division. Growing under

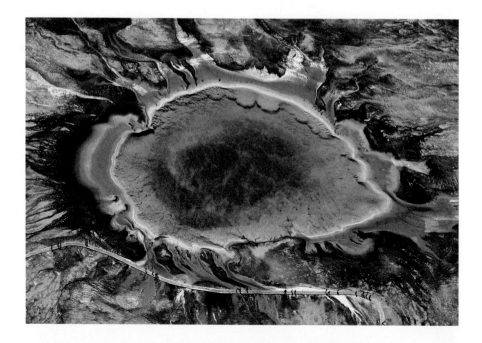

21–1 *Thermophilic ("heat-loving") bacteria thrive at 92°C, a temperature close to the boiling point of water. Shown here is a boiling hot spring, Grand Prismatic Spring, in Yellowstone National Park in Wyoming. Carotenoid pigments of the thickly growing thermophilic bacteria and cyanobacteria color the run-off channels a brownish orange.*

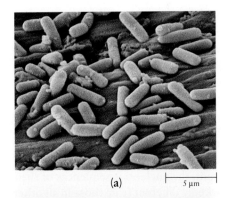

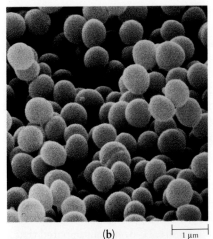

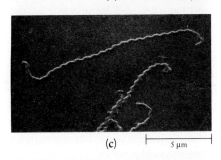

21–2 *The cells of many familiar genera of bacteria have one of three readily distinguished shapes:* (a) *straight rods,* (b) *spheres, or* (c) *long, spiral rods. The rod-shaped bacteria (bacilli) include those microorganisms that cause lockjaw (Clostridium tetani),* diphtheria *(Corynebacterium diphtheriae),* and tuberculosis *(Mycobacterium tuberculosis), as well as the familiar* E. coli. *Among the spherical cocci, which may form pairs, clusters, or chains, are* Streptococcus pneumoniae, *a cause of bacterial pneumonia;* Streptococcus lactis, *which is used in the commercial production of cheese; and* Nitrosococcus, *soil bacteria that oxidize ammonia to nitrates. The helically coiled spirilla are less common. Cell shape is a relatively constant feature in many species of bacteria.*

optimum conditions, a population of *Escherichia coli*, probably the best-known prokaryote, can double in size every 20 minutes. Prokaryotes can survive in many environments that support no other form of life. They have been found in the icy wastes of Antarctica, the dark depths of the ocean, and even in the near-boiling waters of natural hot springs. Some prokaryotes are among the very few modern organisms that can survive without free oxygen, obtaining their energy by anaerobic processes (see page 190). Oxygen is lethal to some types (obligate anaerobes), whereas others can exist with or without oxygen (facultative anaerobes).

When conditions are unfavorable, some types of prokaryotes can form thick-walled spores. These spores are inactive, resistant forms that enable the cells to survive for long periods of time without water or nutrients or in conditions of extreme heat or cold. They may stay dormant for years, and some remain viable even when boiled in water for several hours.

From an ecological point of view, prokaryotes are most important as decomposers, breaking down organic material to forms in which it can be used by plants. They also play a major role in the process known as nitrogen fixation, by which nitrogen gas (N_2) is reduced to ammonia (NH_3) or ammonium ion (NH_4^+). Although nitrogen is abundant in the atmosphere, eukaryotes are not able to use atmospheric nitrogen, and so the crucial first step in the incorporation of nitrogen into organic compounds depends largely on certain species of prokaryotes; some of these species are free-living, while others are found only in close association with plants. Some prokaryotes are photosynthetic, and a few species are both photosynthetic and nitrogen-fixing.

THE CLASSIFICATION OF PROKARYOTES

Until quite recently, the possibility of classifying the prokaryotes on an evolutionary basis seemed remote. Most of the characteristics used to determine phylogenetic relationships among eukaryotes—for example, intricate anatomical structures composed of interconnected parts and complex patterns of reproduction, development, and growth—simply do not exist in prokaryotes. Biologists studying the prokaryotes were forced to rely upon differences in the size and shape of individual cells, the general appearance of the colonies formed by some prokaryotes, type of movement, mode of nutrition, the presence or absence of spores, the presence or absence of a cell wall, and the characteristics of the cell wall, as revealed by its ability to retain certain chemical dyes. Many of these characteristics do not reflect phylogenetic relationships. The prokaryotes have been in existence—and evolving in response to diverse selection pressures—for a very long time. Certain features, such as cell shape and colony form, have probably evolved over and over again. In contrast, other features, such as a cell wall, the capacity for photosynthesis, or the ability to form spores, have been lost independently in a number of lineages. As a result of these difficulties, the classification schemes for the prokaryotes that have long been in use are not hierarchical; the only category above the level of genus is division, with the number of divisions varying in different schemes.

In recent years, studies of cell ultrastructure and biochemistry (particularly the details of metabolic pathways) have enabled biologists to begin unraveling the evolutionary relationships of the prokaryotes. Crucial breakthroughs have come with the development of the molecular techniques described in the previous chapter: amino acid sequencing, nucleotide sequencing, and DNA-DNA hybridization. In general, as with eukaryotes, the smaller the number of differences in the nucleotide sequences of two organisms, the more recent their common ancestor. An additional technique, determination of the percentage of guanine and cytosine (G + C) and of adenine and thymine (A + T) in the DNA has also proved valuable in identifying the degree of relationship among prokaryotic species.

These techniques are rapidly leading to a revolution in our understanding of prokaryotic phylogeny. One of the most striking discoveries thus far has been that a few genera of prokaryotes, not previously classified together, differ very dramatically from all other organisms. These genera include the methanogens (bacteria that synthesize methane from carbon dioxide and hydrogen gas), the salt-loving halobacteria (see page 221), and the thermoacidophiles (bacteria that thrive in very acidic environments at high temperatures). They differ from all other prokaryotes in several significant ways. First, their cell walls have a unique composition. Second, they do not use the Calvin cycle for carbon reduction. Third, they require several unique coenzymes that serve the function of NAD^+ and FAD as electron acceptors. Fourth, and perhaps most important, the nucleotide sequences of their transfer RNAs and ribosomal RNAs are markedly different from those in all other organisms. A particularly interesting feature of the methanogens is that they would have been excellently suited to the conditions prevailing on earth during the earliest stages of biological evolution. It appears that the methanogens, extreme halophiles, and thermoacidophiles may be the few surviving representatives of a lineage that diverged very early from the common ancestor that gave rise to the other prokaryotes. Recently it has been proposed that they be placed in a new kingdom, Archaebacteria; all of the other prokaryotes, collectively known as the Eubacteria, would remain in kingdom Monera. (A corollary of this proposal recommends the establishment of two superkingdoms: Prokaryota, to include kingdoms Archaebacteria and Monera, and Eukaryota, to include kingdoms Protista, Fungi, Plantae, and Animalia.)

At the present time, molecular taxonomic studies of the prokaryotes are proceeding rapidly in a number of laboratories, and the probable evolutionary relationships of the different groups are being worked out. The outlines of the emerging prokaryotic phylogeny are shown in Figure 21–3. It is anticipated that a detailed, evolutionary classification of the prokaryotes will be possible in the relatively near future, but it is unlikely that the older classification schemes will become obsolete. Although those schemes do not necessarily reflect phylogeny, they do group prokaryotes on the basis of similarities that are of immense practical importance in medical and veterinary diagnosis and treatment, agriculture, and industrial microbiology. Two different schemes, each with different uses, each illuminating different aspects of the incredible diversity of prokaryotes, are likely to exist side by side. For our purposes, however, the best introduction to this diversity is not through the details of a classification scheme but through an examination of the characteristics found among the prokaryotes.

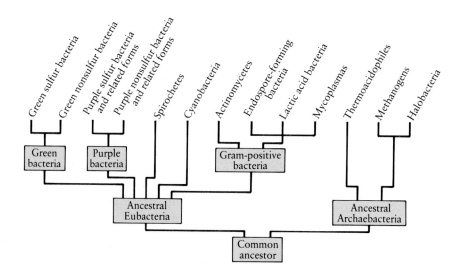

21–3 An outline of the phylogenetic relationships of the prokaryotes, as revealed by molecular taxonomy. Eubacteria that are photosynthetic include the cyanobacteria, the green bacteria, and many of the purple bacteria. Closely related to the purple bacteria are many familiar nonphotosynthetic forms, including E. coli, the pseudomonads (Figure 20–14a), and the nitrogen-fixing bacteria of genus Rhizobium, which live in close association with the roots of plants. The lactic acid bacteria are of considerable economic importance in the production of such fermented foods as cheese, yogurt, pickles, and sauerkraut. The other lineages include numerous animal pathogens, as well as species that play vital roles in the cycling of minerals through ecosystems.

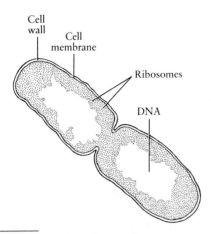

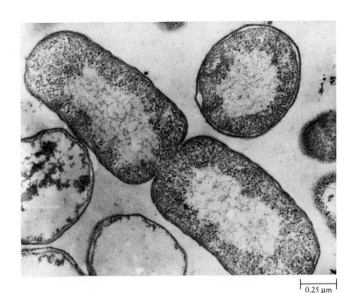

21–4 *Cells of* Escherichia coli, *a modern prokaryote that is a common inhabitant of the human digestive tract. The DNA is localized in the nucleoid—the less dense (lighter-appearing) region in the center of each cell. The small, dense bodies in the cytoplasm are ribosomes. The two cells in the center have just finished dividing and have not yet separated completely.*

THE PROKARYOTIC CELL

The essential features of a prokaryotic cell were outlined in Chapter 4 and are reviewed in Figure 21–4. The cell shown here is the familiar *Escherichia coli*, the most thoroughly studied of the prokaryotes.

The most prominent feature within the prokaryotic cell is the nucleoid, the region in which the chromosome is localized. All prokaryotic chromosomes analyzed so far have proved to consist of one single, continuous ("circular") molecule of DNA associated with a small amount of RNA and non-histone proteins. A prokaryotic cell may also contain one or more plasmids (page 326). As we saw in Section 3, studies of the prokaryotic chromosome have contributed greatly to our understanding of genetic mechanisms.

The cytoplasm of most prokaryotes is relatively unstructured, although it often has a fine granular appearance due to its many ribosomes. These are somewhat smaller than eukaryotic ribosomes but have the same general shape. Generally the cytoplasm is not divided or compartmentalized by membranes and does not contain any membrane-bound organelles. The principal exception occurs in the cyanobacteria, which contain an extensive membrane system bearing chlorophyll and other photosynthetic pigments.

The Cell Membrane

The membrane enclosing the cytoplasm of a prokaryotic cell is formed from a lipid bilayer; it is similar in chemical composition to that of a eukaryotic cell (see page 104). However, except in the mycoplasmas (the smallest free-living cells known), the membranes of prokaryotes lack cholesterol or other steroids. In the aerobic prokaryotes, the cell membrane incorporates the electron transport chain found in the mitochondrial membrane of eukaryotic cells. In the photosynthetic green and purple bacteria (but not in the cyanobacteria), the sites of photosynthesis are found in the cell membrane. In the photosynthetic purple bacteria and in aerobes with large energy requirements, the membrane is often extensively convoluted, with folds extending into the interior. These, of course, greatly increase the working surface of the membrane. Also, as we noted on page 145, the membrane appears to contain specific attachment sites for the DNA molecules; these sites are believed to play a role in ensuring the separation of the replicated chromosomes at cell division.

The Cell Wall

Almost all prokaryotes are surrounded by a cell wall, which gives the different types their characteristic shapes. Many prokaryotes have rigid walls, some have flexible walls, and only the mycoplasmas have no cell walls at all. Because most bacterial cells are hypertonic in relation to their environment, they would burst without their walls. (The mycoplasmas live as intracellular parasites in an isotonic environment.)

Chemical Structure of the Cell Wall

The cell walls of prokaryotes are complex and contain many kinds of molecules not present in eukaryotes. Except for the Archaebacteria (the methanogens and their close relatives), the walls of prokaryotes contain complex polymers known as peptidoglycans, which are primarily responsible for the mechanical strength of the wall.

Prokaryotic cell walls occur in two different configurations, which are readily distinguished by their capacity to combine firmly with such dyes as gentian violet. Those that combine with the dyes are known as gram-positive, whereas those that do not combine with them are known as gram-negative, after Hans Christian Gram, the Danish microbiologist who discovered the distinction. In gram-positive cells (Figure 21–6a), the wall consists of a homogeneous layer of peptidoglycans and polysaccharides that ranges from 10 to 80 nanometers in thickness. In gram-negative cells (Figure 21–6b), by contrast, the wall consists of two layers: an inner peptidoglycan layer, only 2 to 3 nanometers thick, and an outer layer of lipoproteins and lipopolysaccharides. These molecules are arranged in the form of a bilayer, about 7 to 8 nanometers in thickness, similar in structure to the cell membrane. Thus, the boundary of a gram-negative cell is actually a sandwich within a sandwich: an inner cell membrane, a thin peptidoglycan layer, and an outer membrane.

Gram staining is widely used as a basis for classifying bacteria, since it reflects a fundamental difference in the architecture of the cell wall. As shown in Figure 21–3, all gram-positive bacteria are thought to constitute a distinct phylogenetic lineage. The architecture of the cell wall, in turn, affects various other characteristics of the bacteria, such as their patterns of susceptibility to antibiotics (see Figure 8–22, page 178). Gram-positive bacteria, such as *Staphylococcus,* are much more susceptible to some types of antibiotics than are gram-negative bacteria, such as *E. coli.* Their walls are also more readily digested by lysozyme, an enzyme found in the nasal secretions, saliva, and other body fluids of many animals.

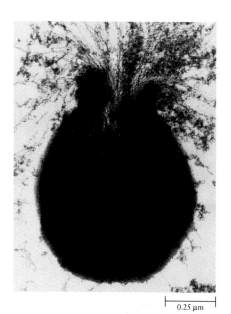

21–5 *This unusual micrograph, taken by Victor Lorian at the Bronx-Lebanon Hospital Center in New York, shows a bacterial cell exploding. This bacterium is a member of the species* Staphylococcus aureus, *the cause of many human infections. Because water tends to move into the cell by osmosis, the contents of bacterial cells are under pressure. When this cell was treated with an antibiotic that damaged the cell wall, it exploded.*

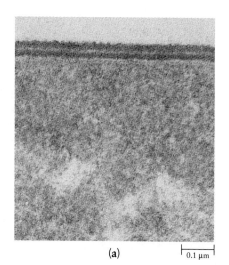 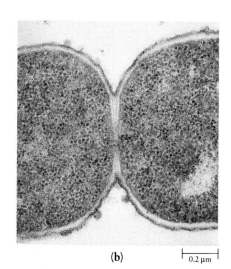

21–6 *Electron micrographs of sections through the cell walls of* (a) Bacillus polymyxa, *an endospore-forming gram-positive bacterium, and* (b) E. coli, *a gram-negative bacterium. The wall of a gram-positive bacterium consists of a homogeneous layer of peptidoglycans and polysaccharides, seen here as the lower dark band. The upper dark band represents a layer of surface proteins. In a gram-negative bacterium, a layer of peptidoglycan (visible here as a light band) is sandwiched between the cell membrane and an outer membrane, similar in composition to the cell membrane.*

21-7 *Most forms of meningitis, an infection of the membranes covering the brain and spinal cord, are caused by the gram-negative bacterium* Neisseria meningitidis. *In this electron micrograph of* N. meningitidis, *you can distinguish the cell membrane (visible here as a bilayer), the peptidoglycan layer, the outer membrane, and a fuzzy-appearing polysaccharide capsule. Different strains of* N. meningitidis *are characterized by polysaccharide capsules of different composition. Scientists at the Walter Reed Army Institute of Research and Rockefeller University have used purified capsule polysaccharides to develop effective vaccines against two of the three major strains of this bacterium.*

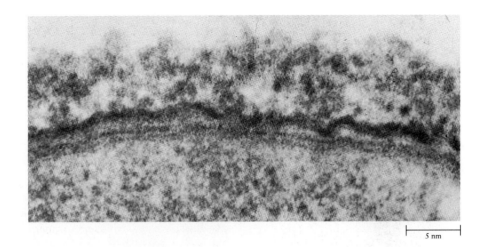

In some bacteria, a gluey polysaccharide capsule, which is secreted by the bacterium, is present outside the cell wall (Figure 21–7). The function of the capsule is not entirely clear, but its presence is associated with pathogenic activity in certain organisms. For example, as shown in Figure 14–2, the encapsulated form of *Streptococcus pneumoniae* is virulent, whereas the nonencapsulated form is generally nonvirulent. It appears that the capsule may interfere with phagocytosis by host white blood cells.

Flagella and Pili

Some types of bacteria have long, slender extensions, known as flagella and pili. Each bacterial flagellum is made up of monomers of a small globular protein, flagellin, assembled into chains that are wound in a triple helix (three chains) with a hollow central core. Bacterial flagella grow from the tip; the individual flagellin molecules pass down a channel in the center of the helix and are assembled onto the ends of the chains. The flagella of different species differ slightly in diameter (12 to 18 nanometers), probably due to slight differences in the composition of their flagellin. In some species, the flagella are distributed over the entire surface of the cell (Figure 21–8); in others, they occur in tufts at one end, or pole, of the cell.

21-8 *A bacterial cell* (Proteus mirabilis) *with numerous flagella—176, to be exact.*

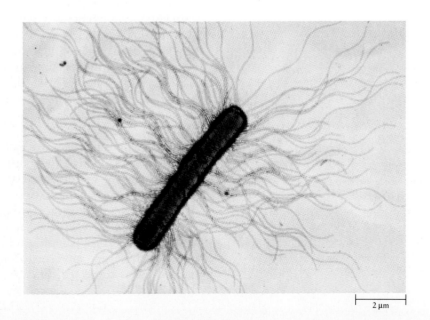

21–9 (a) A flagellum from a gram-negative bacterium, showing the basal end. (b) Diagram of a flagellum from E. coli. The basal body, which serves to anchor the flagellum, consists of two pairs of rings surrounding a rod. The M ring is integrated in the cell membrane, the S ring in the peptidoglycan layer, and the P and L rings in the outer membrane. The filament is made up of several protein chains that form a helix with a hollow core.

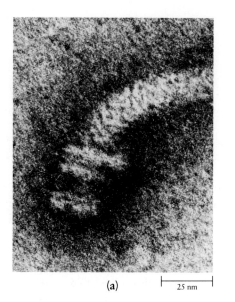
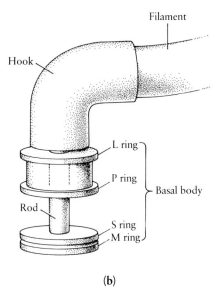

(a) 25 nm (b)

The bacterial flagellum is not enclosed within the cell membrane, as eukaryotic flagella are, but protrudes from the cell as a naked, corkscrew-shaped protein filament. It is anchored into the cell membrane and wall by a complicated assembly (Figure 21–9). The filament (the helix of flagellin monomers with its hollow core) terminates in a hook made up of a different protein. The hook is inserted into a basal body that consists of a rod and, in gram-negative bacteria, two pairs of rings, one pair of which is embedded in the inner membrane and one pair in the outer membrane. In gram-positive bacteria, only the inner pair of rings is present, embedded in the cell membrane.

Bacterial flagella beat with a rotary movement, but they are so fine and the beat so fast that the motion of the flagellum itself cannot be seen. (The flagella of *Spirillum serpens,* for instance, have been clocked at 2,400 rpm.) Ingenious methods have been devised by which the cells can be tethered in place by the flagella. As a result, the cells rotate instead of the flagella, and the movement can be observed. The basal end of the flagellum, with its complex structure, apparently acts as a motor, running on chemiosmotic power.

Flagellated bacteria typically move rapidly in gently curved lines, or "runs," each lasting approximately a second. In species in which the flagella originate at sites all over the cell surface, a series of runs is interrupted frequently by periods of tumbling, lasting about a tenth of a second. After each tumble, the bacterium starts off in a new direction on its next run. The frequency of tumbling controls the length of the runs and hence the overall direction of movement. Whether such a cell moves in a run or tumbles is determined by the direction in which its flagella rotate. When the flagella rotate in a counterclockwise direction, they work together, producing sustained swimming in one direction. But when the flagella reverse direction and rotate in a clockwise manner, they work independently, and the cell tumbles. In species with tufts of polar flagella, clockwise rotation does not produce tumbling but rather a 180° reversal in the direction of the cell's movement. As we saw in Chapter 6 (see page 131), concentration gradients of attractive and noxious chemicals in the surrounding medium are the critical stimuli affecting the frequency with which the flagella reverse their direction of rotation. Receptors in the cell membrane measure changes in the concentration of specific molecules from one moment to the next; in some way, as yet unknown, this information is translated into counterclockwise or clockwise rotation of the flagella.

Navigation by the Poles

For heterotrophic prokaryotes, as for heterotrophic eukaryotes, adequate supplies of digestible organic molecules are essential. As we have seen, flagellated prokaryotes actively swim toward food sources, typically guided by concentration gradients of attractive chemicals in the surrounding medium. In 1975, however, the navigation of a bacterial species from a Massachusetts swamp was shown to be guided not by concentration gradients but rather by the magnetic field of the earth. All of the Massachusetts bacteria observed were found to swim toward magnetic north.

Detailed studies revealed that each bacterium accumulates about 20 opaque, roughly cubic granules of magnetite (Fe_3O_4), the iron ore from which magnets are made. The granules are arranged in a linear fashion, forming a single magnet with a north-seeking pole and a south-seeking pole. The interaction of the earth's magnetic field and the bacterium's internal magnet orients it toward the north; the rotation of its flagella, which are in a cluster at the end of the bacterium opposite the north-seeking pole of its magnet, then propel it in that direction. Because the earth's magnetic field has a vertical component as well as a horizontal component, this interaction also orients the bacterium downward—toward sediments rich in decaying matter.

After the discovery of these bacteria, scientists predicted that if similar magnet-containing bacteria existed in the Southern Hemisphere, their polarity would be reversed. In 1980, such bacteria were found in New Zealand and Tasmania, and, as predicted, their internal magnets are oriented so that the south-seeking poles are opposite the end at which the flagella are attached. Similar bacteria have also been isolated from sediments near the earth's magnetic equator. These bacteria are equally divided between north-seeking and south-seeking forms. There is virtually no vertical component to the earth's magnetism at the equator, so neither form has an advantage over the other. Their internal magnets, however, do prevent them from swimming upward, away from rich sources of food.

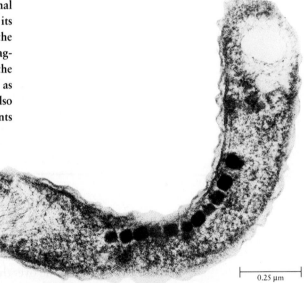

A chain of tiny magnets is clearly visible in this swamp-dwelling bacterium. The capacity to manufacture a magnetic chain is inherited. When such a cell divides, the magnetic particles are divided equally between the two daughter cells. Each daughter cell then manufactures additional particles. Although not visible in this micrograph, a membrane surrounds the magnetic chain and holds the particles in place.

Pili (singular, pilus) are assembled from protein monomers in much the same way that the filaments of flagella are. (You will not be surprised to learn that the protein is called pilin.) They are rigid, cylindrical rods that extend out from the cell, sometimes to a considerable distance. Pili range from 4 to 35 nanometers in diameter, and they are often present in large numbers (hundreds on a single cell). They serve to attach bacteria to a food source, to the surface of a liquid (where oxygen is present), or, in the case of conjugating bacteria, to one another (see Figure 16–9, page 324).

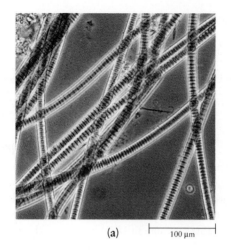
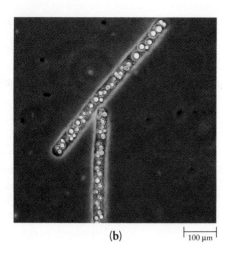

21-10 (a) Oscillatoria, *a filamentous cyanobacterium. Although many cyanobacteria are nonmotile, the filamentous forms typically glide on a slime secreted by the cells. As you may have guessed from the green color, all cyanobacteria are photosynthetic.*

(b) Beggiatoa, *a genus of gliding bacteria, consists of relatively large cells that form filaments and move in a fashion similar to that of many cyanobacteria. The members of this genus are all chemosynthetic autotrophs. The conspicuous granules in the cells are sulfur, produced by the oxidation of hydrogen sulfide—a process utilized by the bacteria for the production of energy. Other types of gliding bacteria are heterotrophic.*

DIVERSITY OF FORM

The oldest method of identifying microorganisms is by their physical appearance. The shape of individual prokaryotes is a result, as we noted previously, of their cell wall. Bacteria exhibit considerable diversity of form, but many of the most familiar species fall into one of three form-groups (see Figure 21-2). Straight, rod-shaped forms like *E. coli* are known as **bacilli;** spherical ones are called **cocci;** and long, spiral rods are called **spirilla.** A fourth form-group, the **vibrios,** consists of short, curved rods that are thought to be incomplete spirals.

Different types of bacteria have characteristic patterns of growth, producing filaments, clusters, or colonies that also have a distinctive shape. For example, cocci may stick together in pairs after division (diplococci), they may occur in clusters (staphylococci), or they may form chains (streptococci). One bacterium that causes pneumonia is a streptococcus, while the staphylococci are responsible for many serious infections characterized by boils or abscesses. The rod-shaped bacilli usually separate after cell division. When they do remain together, they spread out end to end in filaments, since they always divide in the same plane (transversely). In some genera, these filaments are funguslike in appearance and the combining form *myco-* (from the Greek word for "fungus") is part of the generic name. *Mycobacterium tuberculosis,* for example, the cause of tuberculosis, is a bacillus that forms a filamentous, funguslike growth.

Many of the cyanobacteria (Figure 21-10a) as well as members of a nonphotosynthetic group, the gliding bacteria (Figure 21-10b), also form filamentous structures consisting of numerous individual cells. These cells typically secrete a slime or mucus that attaches to a solid surface and provides a pathway along which they can glide from one place to another. Other types of bacteria are characterized by distinctive sheaths or coatings outside their cell walls (Figure 21-11).

Spirochetes are among the easiest microorganisms to identify (Figure 21-12). They are very long (5 to 500 micrometers) and slender (about 0.5 micrometer in diameter) and have an unusual structure, known as an axial filament, made up of

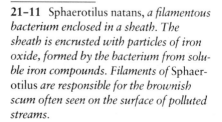

21-11 Sphaerotilus natans, *a filamentous bacterium enclosed in a sheath. The sheath is encrusted with particles of iron oxide, formed by the bacterium from soluble iron compounds. Filaments of* Sphaerotilus *are responsible for the brownish scum often seen on the surface of polluted streams.*

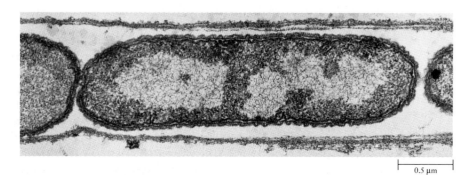

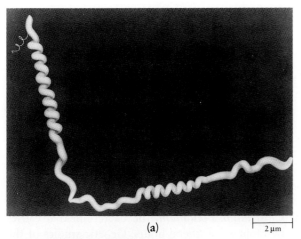
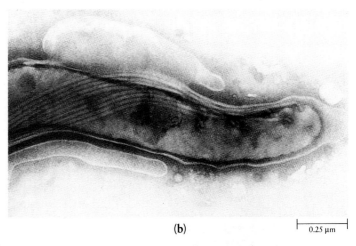
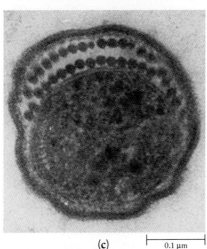

21-12 (a) *The spirochetes range in size up to 500 micrometers long, which is an enormous size for prokaryotes.* Treponema pallidum, *two of which are shown here, is the causative agent of syphilis.* (b) *The end of a spirochetal cell that has been stained to show the insertion points of two fibrils of the axial filament.* (c) *Cross section of a spirochete showing the fibrils of the axial filament between the cell membrane and the cell wall.*

two sets of fibrils attached at each end of the cell. The fibrils are identical to flagella in structure and so are recognized as modified flagella, or endoflagella. They are wrapped around the cell between the cell membrane and the delicate wall, with the fibrils of each set overlapping at the middle of the cell. Rotation of the axial filament is thought to produce the corkscrew movement characteristic of the spirochetes. All members of this group are thought to belong to the same phylogenetic lineage.

Rickettsiae, which are the smallest cells known, constitute still another group of prokaryotes distinguishable by their form (Figure 21-13).

REPRODUCTION AND RESTING FORMS

Most prokaryotes reproduce by simple cell division (page 145), also called **binary fission.** In some forms, reproduction is by budding or by the fragmentation of filaments of cells. As they multiply, these prokaryotes, barring mutations, produce clones of genetically identical cells. Mutations do occur, however; it has been estimated that in a culture of *E. coli* that has divided 30 times, about 1.5 percent of the cells are mutants. Mutations, combined with the rapid generation time of prokaryotes, are responsible for their extraordinary adaptability. Further adaptability is provided by the genetic recombinations that take place as a result of conjugation, transformation, transduction, and exchanges of plasmids. It is not known how common such genetic recombinations are in nature or whether they occur in all types of prokaryotes.

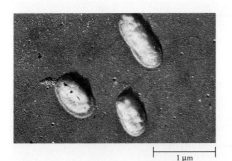

21-13 *The tiny cells of* Rickettsia prowazekii, *shown here, are the cause of typhus. They are transmitted by human body lice, and under crowded conditions, large numbers of people can be infected in a very short time. More human lives have been taken by rickettsial diseases than by any other infection except malaria. At the siege of Granada in 1489, 17,000 Spanish soldiers were killed by typhus, but only 3,000 in combat. In the Thirty Years War, the Napoleonic campaigns, and the Serbian Campaign during World War I, typhus was also the decisive factor.*

Many prokaryotes have the capacity to form spores, which are dormant, resting cells. This process has been studied most extensively in the bacilli. It characteristically occurs when a population of cells, growing very rapidly, begins to use up its food supply. Each cell, at the beginning of sporulation (spore formation), contains two duplicate chromosomes. A cell membrane grows around one of the chromosomes, separating it from the rest of the cell, which then engulfs the newly formed cell. Thus the spore-to-be is now surrounded by two membranes, its own and that of the larger cell (these events are shown in Figure 21-14). A spore coat, consisting of two layers, then forms around this smaller cell. The inner layer (the cortex) contains a peptidoglycan that is completely different from that present in the bacterial cell wall; the outer layer consists of proteins that are composed largely of hydrophobic amino acids. The mature spore is released from

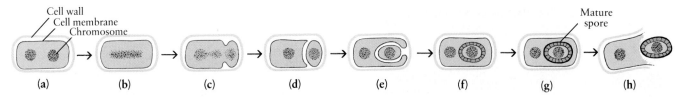

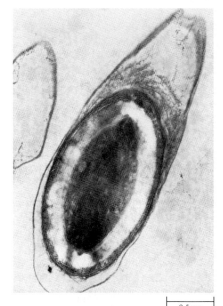

21–14 *Endospore formation in* Bacillus cereus. **(a, b)** *The two chromosomes within the vegetative cell have condensed into a rod-shaped form.* **(c)** *The transverse wall begins to form,* **(d)** *cutting off the spore material from the vegetative cell.* **(e, f)** *The vegetative cell grows around the spore, and the spore coat is formed.* **(g, h)** *The spore matures and is released from the cell. The micrograph shows an endospore of* Clostridium.

the cell in this protected state, and it remains dormant until appropriate events trigger its germination. Germination takes place rapidly, with the uptake of water, the dissolution of the spore coat, and the formation of a new cell wall.

Spore formation greatly increases the capacity of prokaryotic cells to survive. For example, the spores of *Clostridium botulinum*, the bacterium that causes botulism, are not destroyed by boiling for several hours. Genetic studies of the spore-making *Bacillus subtilis* indicate that some 50 genes, clustered in about five segments of the chromosome, are involved in spore formation.

The myxobacteria, a type of gliding bacteria, form fruiting bodies, which are brightly colored collections of spores and slime large enough to be seen by the unaided eye (Figure 21–15). Although the cells of the myxobacteria retain their independent function, their association represents a rudimentary form of multicellularity, described by one whimsical biologist as "multicellularity by committee."

PROKARYOTIC NUTRITION

Heterotrophs

Although the members of kingdom Monera exhibit tremendous metabolic diversity, most prokaryotes are heterotrophs. Of these, the vast majority are **saprobes** (from the Greek *sapros*, "rotten" or "putrid"), feeding on dead organic matter.

21–15 *A fruiting body of* Chondromyces crocatus, *a myxobacterium. Each fruiting body, which may contain as many as 1 million cells, consists of a central stalk that branches to form clusters of single-cell spores.*

Bacteria and other microorganisms are responsible for the decay and recycling of organic material in the soil. Typically, different groups of bacteria play different, specific roles—such as the digestion of cellulose, starches, or other polysaccharides, or the hydrolysis of specific peptide bonds, or the breakdown of amino acids. Because of their high degree of nutritional specialization, bacteria are able to live in large numbers in the same small area with little competition and, indeed, with mutual assistance, as the activities of one group make food molecules available to another group. These combined activities release the nutrients and make them available to plants and, through plants, to animals. Thus, the bacteria are an essential part of ecological systems.

Some heterotrophic bacteria live in close association with other organisms. Some of these are parasites that break down organic material in the bodies of living organisms (Figure 21–16). The disease-causing (pathogenic) bacteria belong to this group, as do a number of nonpathogenic forms. Some of these bacteria have little effect on their hosts, and some are actually beneficial. Cows and other ruminants can utilize cellulose only because their stomachs contain bacteria and protists that have cellulose-digesting enzymes. Our own intestines contain a number of types of generally harmless bacteria (including *E. coli*). Some supply vitamin K, which is necessary for blood clotting. Others prevent us from developing serious infections. When the normal bacterial inhabitants of the human intestinal tract are destroyed—as can happen, for example, following prolonged antibiotic therapy—our tissues are much more vulnerable to disease-causing microorganisms.

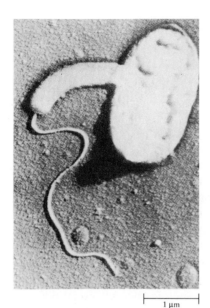

21–16 *Pathogenic bacteria are themselves vulnerable to other pathogenic bacteria. This electron micrograph captured a cell of* Bdellovibrio bacteriovirus *(left) attacking a cell of* Erwinia amylovora *(upper right), the bacterium that causes fire blight of pears and apples.* Bdellovibrio bacteriovirus, *which is abundant in soil and sewage, moves by means of a single posterior flagellum. Like other members of the genus* Bdellovibrio *(from* bdello, *the Greek word for "leech"), it makes a hole in the host cell wall and multiplies between the wall and the cell membrane, digesting the host cell as it multiplies.*

Chemosynthetic Autotrophs

Chemosynthetic autotrophs obtain their energy from the oxidation of inorganic compounds. Only prokaryotes are able to use inorganic compounds as an energy source.

An unusual group of chemosynthetic prokaryotes are the methanogens (page 90). Although their existence was long known, they were not studied extensively until quite recently because they are obligate anaerobes—that is, they are poisoned by oxygen—and hence are difficult to isolate and impossible to grow under ordinary laboratory conditions. The methanogens are the final participants in decomposition processes involving organic matter in anaerobic environments, including marshes, lake sediments, and the digestive tracts of animals. They convert CO_2 and H_2 formed by the fermentation processes of other anaerobes to methane (CH_4).

Certain chemosynthetic bacteria are essential components of the nitrogen cycle, the process by which nitrogen compounds are cycled and recycled through ecosystems. One group oxidizes ammonia or ammonium (derived from the breakdown of organic materials, the activities of nitrogen-fixing prokaryotes, or, to a minor extent, from lightning or volcanic activities). The products of this reaction are nitrite (NO_2^-) and energy. Another group oxidizes nitrites, producing nitrate (NO_3^-) and energy. Nitrate is the form in which nitrogen moves from the soil into the roots of plants.

Sulfur is also required by plants for amino acid synthesis. Like nitrogen, it is converted to the form in which it is taken up by plant roots by the activities of chemosynthetic bacteria that oxidize elemental sulfur to sulfate:

$$2S + 2H_2O + 3O_2 \longrightarrow 2H_2SO_4$$

Other sulfur bacteria, such as *Thiothrix* and *Beggiatoa*, obtain energy by oxidizing hydrogen sulfide.

These important biogeochemical cycles and the roles of microorganisms in them will be discussed further in Chapter 54.

(a)

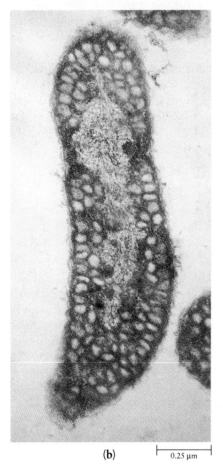

(b) |—— 0.25 μm ——|

21-17 *Purple bacteria.* (a) *Mats of the purple sulfur bacterium* Thiopedia roseopersicinia *growing in a spring in Wisconsin. The bacteria grow below the surface of the water. Gas vesicles within the cells cause them to rise to the surface when disturbed.* (b) *A purple nonsulfur bacterium,* Rhodospirillum rubrum. *The structures that resemble vacuoles are intrusions of the cell membrane, containing the photosynthetic pigments. This cell, which has a very high content of bacteriochlorophyll, is from a culture that was grown in dim light; in cells grown in bright light, the membrane intrusions are less extensive.*

Photosynthetic Autotrophs

Green and Purple Bacteria

Among the eubacteria are five photosynthetic types, which are thought to represent three distinct phylogenetic lineages: the cyanobacteria, the green bacteria, and the purple bacteria (see Figure 21-3). The green bacteria and the purple bacteria include both sulfur and nonsulfur forms. A number of nonphotosynthetic eubacteria, including the familiar *Escherichia,* are believed to belong to the same lineage as the purple bacteria.

The distinctive colors of the green and purple bacteria result, of course, from the pigments they contain. The reaction centers of the photosynthetic green bacteria contain bacteriochlorophyll *a*, which differs only slightly from the chlorophyll *a* of eukaryotes (page 211). The antennae pigments are bacteriochlorophylls *c*, *d*, or *e*, which differ more significantly. In the two groups of photosynthetic purple bacteria, the reaction centers contain bacteriochlorophylls *a* or *b*, and the same molecules also function as antennae pigments. Bacteriochlorophyll *b*, which differs chemically in several details from chlorophyll *a*, is a pale blue-gray. The colors of the purple bacteria are due to the presence of several different yellow and red carotenoids, which function as additional accessory pigments. In the purple nonsulfur bacteria, the colors may actually range from purple to red or brown.

In the photosynthetic sulfur bacteria, as we noted in Chapter 10, the sulfur compounds are the electron donors, playing the same role in bacterial photosynthesis that water does in photosynthesis in eukaryotes.

$$CO_2 + 2H_2S \xrightarrow{\text{Light}} (CH_2O) + H_2O + 2S$$

In the photosynthetic nonsulfur bacteria, other compounds, including alcohols, fatty acids, and a variety of other organic substances, serve as electron donors for the photosynthetic reactions.

Photosynthesis by green and purple bacteria is carried out anaerobically, and it never results in the production of molecular oxygen. However, in all species except the extreme halophiles (which, as we have seen, are members of a completely different phylogenetic lineage), carbon is fixed by means of the Calvin cycle.

Cyanobacteria

Unlike other photosynthetic prokaryotes, but like all photosynthetic eukaryotes, the cyanobacteria contain chlorophyll *a* and lyse water during photosynthesis, producing molecular oxygen. For this reason, they were classified with the eukaryotic algae until biochemical studies and electron microscopy revealed their prokaryotic nature. Although the name has no taxonomic status, they are still often referred to as the "blue-green algae." Cyanobacteria thrive in freshwater environments, and they are a principal component of the bluish-green scum seen on many ponds in late summer.

Cyanobacteria have several kinds of accessory pigments, including xanthophyll, which is a yellow carotenoid, and several other carotenoids. The cells also contain one or two pigments known as phycobilins: phycocyanin, a blue pigment, which is always present, and phycoerythrin, a red one, which is often present. Chlorophyll and the accessory pigments are not enclosed in chloroplasts, as they are in eukaryotic cells, but are part of a membrane system distributed in the peripheral portion of the cell (Figure 21-18).

Cells of the cyanobacteria, like those of many other prokaryotes, have an outer polysaccharide sheath, or coating. In some species, particularly those that spread up onto the land, this outer sheath is deeply pigmented; in most species, however,

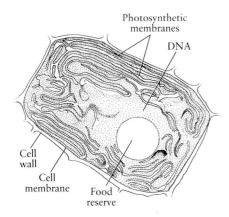
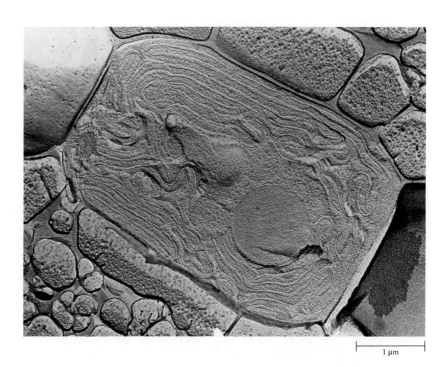

21–18 *Electron micrograph of the cyanobacterium* Anabaena cylindrica. *Cyanobacteria have no chloroplasts and no membrane-bound nucleus, such as are found in eukaryotic photosynthetic cells. Photosynthesis takes place in chlorophyll-containing membranes within the cell, and the chromosome is a single molecule of DNA. The three-dimensional quality of this electron micrograph is due to freeze-fracturing (page 135).*

it is not pigmented, and the colors of the cells result from the carotenoids and phycobilins that they contain. Different species of cyanobacteria are golden yellow, brown, red, emerald green, blue, violet, or blue-black. The Red Sea was so named because of the dense concentrations, or "blooms," of red-pigmented cyanobacteria that float on its surface.

Some species of cyanobacteria are capable of nitrogen fixation. The photosynthetic, nitrogen-fixing cyanobacteria are among the most self-sufficient of organisms. They have the simplest nutritional requirements of any living thing, needing only nitrogen and carbon dioxide, which are always present in the atmosphere, a few minerals, and water. On a worldwide scale, the ecological importance of the nitrogen-fixing cyanobacteria appears to be less than that of the nitrogen-fixing bacteria, at least for agriculture. However, in Southeast Asia, such cyanobacteria make the productivity of rice paddies about 10 times greater than that of other types of farmland.

21–19 *Women planting rice in a paddy in Perak, Malaysia. In Southeast Asia, rice can often be grown on the same land continuously without the addition of fertilizers because of nitrogen-fixing cyanobacteria. The cyanobacteria are found not in the water but rather in the tissues of a small water fern* (Azolla) *that grows in the rice paddies.*

Two Unusual Photosynthetic Prokaryotes

Virtually all of the contemporary photosynthetic eubacteria can be readily identified as members of one of three lineages: cyanobacteria, green bacteria, or purple bacteria, all of which are gram-negative. Moreover, many nonphotosynthetic gram-negative bacteria, including *E. coli*, are thought to be descended from purple bacteria that lost the capacity for photosynthesis. The recent serendipitous discoveries of two photosynthetic prokaryotes with unique combinations of characteristics suggest the existence of other lineages.

The first of these unusual prokaryotes, discovered in 1975, exhibits some important similarities to the cyanobacteria. Its principal photosynthetic pigment is chlorophyll *a*, it contains the same types of carotenoids, and it lyses water during photosynthesis, releasing molecular oxygen. However, unlike the cyanobacteria, it contains no phycobilins. The principal accessory pigment is chlorophyll *b*, found elsewhere only in the chloroplasts of green algae and plants. As in the cyanobacteria, the photosynthetic membranes are not invaginations of the cell membrane but are instead separate structures. They consist, however, of two membranous sacs pressed together—a structure intermediate between the individual membranes of the cyanobacteria (Figure 4–9, page 92) and the stacks of thylakoids in the chloroplasts of the green algae and the plants (Figure 10–10, page 213). This constellation of characteristics has led to the name *Prochloron*. As we shall see in the next chapter, there is reason to believe that *Prochloron* is a living representative of an ancient lineage of prokaryotes that gave rise to the chloroplasts of the green algae and, ultimately, of the plants.

The second unusual prokaryote was discovered in 1981. An undergraduate at the University of Indiana made a mistake in a laboratory exercise designed to isolate photosynthetic bacteria from samples of pond water and soil, and none of the desired organisms multiplied in the experimental flasks. However, as the laboratory instructors were discarding the flasks at the end of the exercise, they noticed a green film on the bottom of one of the "botched" flasks. Happily for biological knowledge (and, undoubtedly for the student as well), they saved the contents and set about determining the source of the green film. It turned out to be a photosynthetic bacterium quite unlike any previously known. This organism, which has been named *Heliobacterium chlorum*, is a photosynthetic, nitrogen-fixing bacterium that is an obligate anaerobe. It contains a distinct type of bacteriochlorophyll that, unlike all other types of bacteriochlorophyll and chlorophyll, contains no oxygen atoms. This suggests that *Heliobacterium chlorum* is a representative of a lineage that evolved very early, well before the presence of even small amounts of free oxygen in the earth's atmosphere. Although *Heliobacterium chlorum* is gram-negative, analyses of its rRNA molecules indicate that it is most closely related to the group of gram-positive bacteria that includes *Clostridium*. This raises the possibility that the contemporary gram-positive bacteria (all of which are nonphotosynthetic) are the descendants of photosynthetic ancestors. And, where had the student obtained the soil sample that contained *Heliobacterium chlorum*? Right in front of the biology building.

Because of their nutritional independence, the cyanobacteria are able to colonize bare areas of rock and soil. A dramatic example of such colonization was seen on the island of Krakatoa in Indonesia, which was denuded of all visible plant life by a cataclysmic volcanic explosion in 1883. Filamentous cyanobacteria were the first living things to appear on the pumice and volcanic ash; within a few years they had formed a dark-green gelatinous growth. The layer of cyanobacteria eventually became thick enough to provide a substrate for the growth of plants. It is very probable that the cyanobacteria were similarly the first colonizers of land in the course of biological evolution.

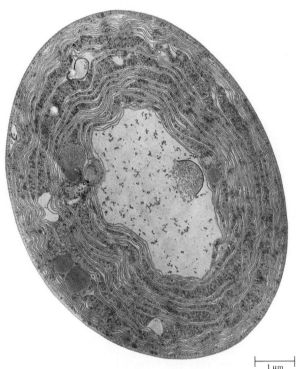

A single cell of the genus Prochloron. *Analyses of* Prochloron *DNA and rRNA molecules indicate that its closest prokaryotic relatives are the cyanobacteria. The natural habitat of* Prochloron *cells is the body surface of sea squirts, soft-bodied animals that live along seashores and that belong to the lineage from which vertebrate animals arose.*

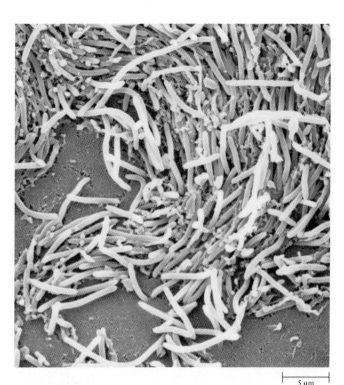

Scanning electron micrograph of Heliobacterium chlorum *cells. These rod-shaped cells form motile masses similar to those of the gliding bacteria. The cell membrane of* Heliobacterium *is not amplified and intruded into the cytoplasm, which suggests a photosynthetic process more primitive than that of the green and purple bacteria.*

VIRUSES: DETACHED BITS OF GENETIC INFORMATION

Viruses do not fit easily into any of the kingdoms of living organisms. Because of their small size and infectious capacities, however, they have usually been studied with the prokaryotes. As we saw in Section 3, viruses consist of a nucleic acid core—either DNA or RNA—surrounded by a protein coat, or capsid. They reproduce only within living cells, commandeering the enzymes and other metabolic machinery of their hosts. Without this machinery, they are as inert as any other macromolecule, lifeless by most criteria. As our knowledge of viruses has

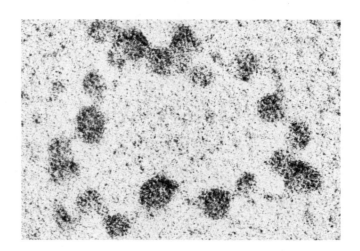

21-20 *One of the more striking evidences of the cellular origin of viruses is provided by the chromosome of simian virus 40 (SV40), shown here. As in the eukaryotic chromosome, its double-stranded DNA is complexed with histone proteins to form nucleosomes (see Figure 18-4, page 357). The histones are those of the host cell; SV40 not only uses host-cell machinery for its own synthetic activities but also directly incorporates products of the host cell's synthetic activities.*

grown, and particularly as the nucleotide sequences of increasing numbers of viruses have been determined—and shown to be very similar to sequences in the chromosomes of cellular organisms—biologists have come to regard viruses as cellular fragments that have set up a partially independent existence (Figure 21-20). Once this partial independence is gained, however, viruses appear to pursue their own evolutionary course, distinct, in many important aspects, from that of the cells from which they were originally derived.

In size, viruses range from about 17 nanometers (a hemoglobin molecule is 6.4 nanometers in diameter) to about 300 nanometers, larger than small bacteria. The larger ones are at the limits of resolution of the light microscope. Viruses can be characterized and classified on the basis of their usual host cells, their nucleic acid content (DNA or RNA, single-stranded or double-stranded), and their specific shapes, which are determined by the structure of the protein capsid (Figure 21-21). The proteins of the capsid may take the form of a helix, as in tobacco mosaic virus (see Figure 15-6, page 305) and influenza virus, or the form of triangular plates arranged in a polyhedron, as in adenovirus and the T-even bacteriophages. The capsid may be surrounded by additional layers or have other complex protein structures attached to it.

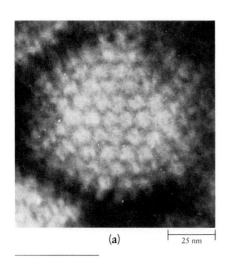
(a) 25 nm

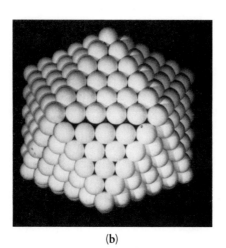
(b)

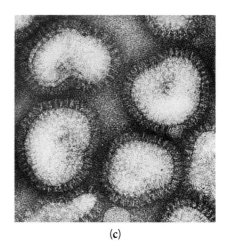
(c)

21-21 *Representative viral structures. (a) Adenovirus, one of the many viruses that cause colds in humans. This virus is an icosahedron. Each of its 20 sides is an equilateral triangle composed of identical protein subunits. Many viruses, and also Buckminster Fuller's geodesic domes, are constructed on this principle. There are 252 subunits in all. Within the icosahedron is a core of double-stranded DNA. (b) A model of the adenovirus, made up of 252 tennis balls.*

(c, d) Electron micrograph and diagram of influenza virus. The virus is composed of a core of RNA surrounded by a helical protein capsid and a lipoprotein membrane through which protrude stubby protein spikes. The influenza virus, for reasons

The proteins of the capsid determine the specificity of a virus; a cell can be infected by a virus only if viral protein can fit into one of the specific receptor sites in the cell membrane of that type of cell. Thus bacteriophages attack bacterial cells; tobacco mosaic virus infects leaf cells of the tobacco plant; adenoviruses and rhinoviruses, causes of the common cold, invade cells in the mucous membranes of the respiratory tract; and polio virus infects cells of the upper respiratory tract, the intestinal lining, and sometimes the nervous system. Apparently all types of cells—both prokaryotic and eukaryotic—are susceptible to infection by specific viruses capable of interacting with their membrane receptors.

In some virus infections, the protein coat is left outside the cell while the nucleic acid enters (see Figure 14-7, page 286); in others, the intact virus enters the cell, but once inside, the protein is destroyed by enzymes, freeing the viral nucleic acid. In the DNA viruses, the DNA of the virus replicates and is also transcribed into messenger RNA. The mRNA codes for viral enzymes, viral coat protein, and probably, at least in some cases, repressors and other regulatory chemicals. For its synthetic activities, the virus uses the equipment of the host cell, including ribosomes, transfer RNA molecules, amino acids, and nucleotides. Many viruses use host enzymes as well as those coded by their own nucleic acids, and some break up host DNA and recycle the nucleotides as viral DNA. In most RNA viruses, the viral RNA both replicates and serves directly as messenger RNA. In other RNA viruses, however, the viral RNA is transcribed into DNA, from which mRNA is later transcribed. This phenomenon of **reverse transcription** is characteristic of the cancer-causing viruses and the virus responsible for AIDS (acquired immune deficiency syndrome, to be discussed further in Chapter 39).

Virus particles are assembled within the host cell. In viruses with helical capsids, the protein subunits of the capsid come together around the newly synthesized nucleic acid. In other types of viruses, the capsid is formed separately, and then the nucleic acid is inserted into it, apparently drawn in by a protein functioning as an internal spool. Some viruses, such as influenza virus, code for additional proteins that are inserted into the membrane of the host cell; the newly formed viruses bud off from portions of the host cell membrane containing the viral proteins, and, in so doing, become wrapped in fragments of it (Figure 21-22).

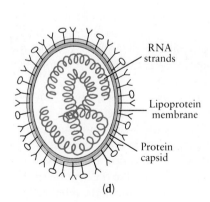
(d)

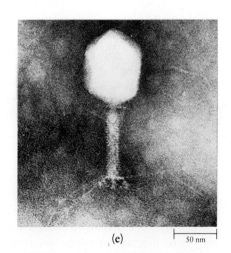
(e)

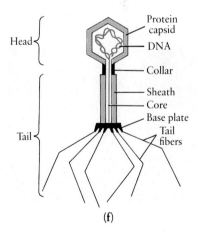
(f)

that are not understood, mutates frequently. The changes in its nucleic acid alter the proteins of the outer envelope and, hence, previously formed antibodies no longer "recognize" it. New strains of influenza viruses are likely to arise more rapidly than new vaccines can be produced to combat them.

(e, f) Electron micrograph and diagram of a T-even bacteriophage, showing its many different structural components. The DNA of the virus codes for all of the necessary proteins. The capsid head, the major structures of the tail, and the tail fibers are assembled separately. After the DNA has been inserted into the capsid head, the pre-formed tail assembly is attached to it. Addition of the tail fibers completes the viral particle.

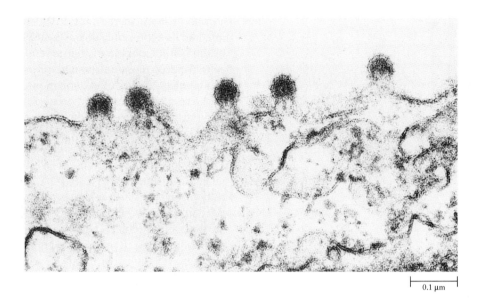

21–22 *In this electron micrograph, particles of Semliki Forest virus are budding from the cell membrane of an animal cell, acquiring a lipoprotein envelope in the process. Semliki Forest virus is named for the rain forest in Uganda where it was first isolated from the tissues of mosquitoes. It infects a variety of animals and is related to the virus that causes yellow fever.*

When assembly of the viral particles is completed, they are released from the host cell, often lysing its membrane in the process. Each new viral particle is capable of setting up a new infection cycle in an uninfected cell.

Viroids and Prions: The Ultimate in Simplicity

For many years, the agents causing certain serious diseases of plants and animals remained mysterious. These diseases exhibit the characteristics of viral infection and yet no viruses could be identified. Recent studies have now demonstrated that even simpler infectious agents are at work: **viroids,** which are naked RNA molecules found in plants, and **prions,** which are small proteinaceous particles found in animals. Both viroids and prions replicate in susceptible cells.

The first viroid to be characterized, known as PSTV, is the causative agent of potato spindle-tuber disease; in this disease, the plant produces elongated, gnarled potatoes (tubers) that sometimes have deep crevices in the surface. PSTV can also infect tomato plants—close relatives of potato plants—producing stunted growth and twisted leaves. PSTV is a single-stranded RNA molecule containing 359 nucleotides; the molecule can be either linear or a closed circle. The linear form folds back on itself in a hairpin-shaped structure, whereas the circular form is flattened. In each case, complementary base pairs are joined by hydrogen bonds, resulting in a double-stranded RNA structure similar to that of DNA. Under the electron microscope, both forms of PSTV appear as rods about 50 nanometers long (Figure 21–23).

The way in which viroids replicate is an enigma, as is the means by which they cause disease. They are found almost exclusively in the nuclei of infected cells, and experiments have shown that the RNA of viroids does not function as messenger RNA. Unlike the DNA or RNA of viruses, it is not translated into enzymes that participate in its own replication. The location of viroids in the nucleus and their inability to act as messenger RNAs has led to the hypothesis that they cause their symptoms by interfering with gene regulation in the infected host cells. This hypothesis is supported by the fact that certain plant proteins, found in healthy cells, are present in significantly larger quantities in infected cells. New evidence suggests that the interference may be at the stage of mRNA editing, particularly the removal of introns and the splicing together of exons (see page 369). All viroids studied thus far share important similarities with a group of introns that includes "self-splicing" members—that is, introns that can remove

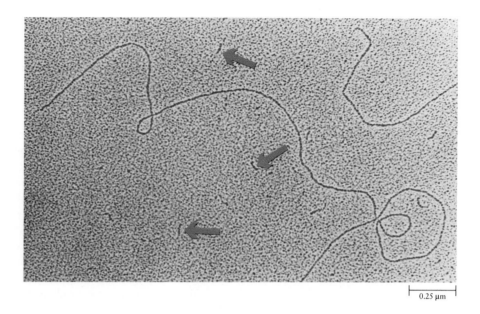

21-23 *The short, rodlike structures indicated by arrows in this electron micrograph are potato spindle-tuber viroids. The much larger, looping structure is a portion of the DNA molecule from a bacteriophage, included to show the tremendous difference in size between a viroid and a virus. Viroids, unlike viruses, have no protein coat. Other viroids have now been identified that cause diseases in citrus trees, cucumbers, and chrysanthemums.*

themselves from an RNA molecule, splicing together the exons, without the assistance of any enzymes. Viroids have certain nucleotide sequences that are identical to sequences found in these introns; among the identical segments are nucleotide sequences that play a major role in determining the configuration of the excised intron. Are viroids actually "escaped" introns that have established a partially independent existence, analogous to that of viruses? The present evidence suggests that this may be the case; it also suggests that excised introns may have previously unsuspected functions in the cell.

The discovery of viroids in plant cells led to the suspicion that they were also the cause of scrapie, an infectious neurological disease in sheep that is similar in its symptoms to several devastating human diseases, including Alzheimer's disease. The infectious agent of scrapie, however, has now been identified as a prion—a protein particle not associated with any detectable nucleic acid (Figure 21-24). Amino acid sequencing of the prion has made it possible to develop DNA probes to search for its gene. Although no gene with a corresponding sequence has been identified in any purified preparation of the infectious agent itself, such a gene has been located in the chromosomes of both infected and healthy animals. Moreover, mRNA transcribed from the gene has been identified in the cells of both types of animals. The normal function of this gene and its product, the effect of prions on either the gene or its product, and the way in which the prions are replicated remain totally mysterious—but perhaps not for long, given the intense interest in the human diseases similar to scrapie.

21-24 *A collection of prions in a section of the brain of a hamster infected with scrapie. This section is from a region of the brain near the hippocampus, a structure associated with memory (see Chapter 43). Prions are known to be the infectious agent of Creutzfeldt-Jakob disease, a degenerative neurological disease of humans, but whether they play a role in Alzheimer's disease remains an open question.*

MICROORGANISMS AND HUMAN ECOLOGY

Symbiosis

Symbiosis ("living together") is a close and long-term association between organisms of different species. Although there is some disagreement as to precisely what constitutes a symbiotic relationship, and the details of the relationship between two closely associated species are often difficult to determine, symbiotic relationships are generally considered to be of three kinds. If the relationship is beneficial to both species, it is called **mutualism**. If one species benefits from the association while the other is neither harmed nor benefited, it is called **commensalism**. If one species benefits and the other is harmed, the relationship is known as **parasitism**.

An enormous variety of microorganisms live symbiotically with human beings, but the lines of demarcation between the different categories of symbiosis are not clear-cut. A bacterium such as *E. coli,* which lives in the lower intestine, may be commensal, depending on its host for food and shelter and neither helping nor harming. If it produces a needed vitamin or digestive enzyme, the relationship is mutualistic. If it gets into the bloodstream and causes septicemia (blood poisoning), it is a parasite and a pathogen.

Evolutionary success is measured in terms of surviving progeny. A symbiotic microorganism that destroys its host before the reproduction and dispersal of its progeny to new hosts is less likely to be successful by the evolutionary criterion than one that enjoys a long and comfortable relationship with its protector. Thus disease is generally the result of a sudden change in the microorganism, in the host, or in their relationship. For instance, many people harbor small numbers of *Mycobacterium tuberculosis* without any symptoms of disease; however, factors such as malnutrition, fatigue, or other diseases may weaken host defenses so that the signs of tuberculosis appear. Similarly, the herpes simplex viruses that cause fever blisters or genital lesions may remain latent for months or years at a time, with an outbreak occurring only in response to some change in the condition of the host.

How Microbes Cause Disease

The pathogenic effects of microbes are produced in a variety of ways. Viruses, as we have seen, enter particular types of cells and often destroy them. Bacteria may produce cell destruction also. Frequently, however, the effects we recognize as disease are caused not by the direct action of the pathogens but by toxins, or poisons, produced by them. For instance, diphtheria is caused by a bacillus, *Corynebacterium diphtheriae.* The organisms are inhaled and establish infection in the upper respiratory tract, where they produce a powerful toxin that is transported through the bloodstream to body cells. This toxin, which is made only when the bacterium is harboring a particular prophage (page 331), inhibits protein synthesis.

Some diseases are the result of the body's reaction to the pathogen. In pneumonia caused by *Streptococcus pneumoniae,* the infection causes a tremendous outpouring of fluid and cells into the air sacs of the lungs, thus interfering with breathing. The symptoms caused by fungus infections of the skin similarly result from inflammatory responses.

A single disease agent can cause a variety of diseases. Skin infections of *Streptococcus pyogenes* cause the disease known as impetigo. Throat infections by the same bacterium are the familiar strep throat. Throat infections with strains of the bacterium that produce toxins (again as a result of a bacteriophage) are known as scarlet fever. Among persons with untreated strep throat or scarlet fever, about 0.5 percent develop rheumatic fever, which is characterized by inflammatory changes in the joints, heart, and other tissues, apparently as a result of reactions involving the body's own immune system. Conversely, many agents may cause the same symptoms; the "common cold" can result from infection with any one of a large number of viruses.

Prevention and Control of Infectious Disease

Although microorganisms were first seen and depicted with remarkable accuracy by Antony van Leeuwenhoek in the late seventeenth century, they were not generally recognized as a cause of disease until about 100 years ago. This recognition opened the way to control measures, among the most important of which was the introduction of sterile procedures in hospitals. Even more important than improvement in medical practices was the institution of public health

21-25 *Until the 1950s, poliomyelitis was one of the most feared of all childhood diseases. It was brought under control by immunization, initially with a killed-virus vaccine (Salk vaccine) that was subsequently replaced by a live-virus vaccine (Sabin vaccine), the one now in wide general use. The possible aftereffects of polio are depicted in this Egyptian hieroglyph, dated at approximately 1400 B.C.*

measures—for example, the eradication of fleas, lice, mosquitoes, and other agents that carry disease; disposal of sewage and other wastes; protection of public water supplies; pasteurization of milk; and quarantine. Not long ago, for instance, infant mortality during the first years of life was often as high as 50 percent in some localities owing to infant diarrhea caused by contaminated milk and water; in some regions in the developing countries, it still remains quite high.

Many infectious diseases, both bacterial and viral, can be prevented by immunization (to be discussed in Chapter 39). Bacteria, in particular, are also susceptible to antimicrobial drugs, such as sulfa and penicillin, for which the battlefields of World War II were the proving grounds. Penicillin, which is synthesized by the fungus *Penicillium*, was the first known antibiotic—by definition, a chemical that is produced by a living organism and is capable of inhibiting the growth of microorganisms. Many antibiotics are produced by bacteria, especially the actinomycetes; some are formed by fungi. Many, including penicillin, can now be synthesized in the laboratory. Antibiotics and other chemotherapeutic agents are effective because they interfere with some essential process of the pathogen without affecting the cells of the host.

Antimicrobial drugs made possible not only the treatment of battlefield wounds and of common infectious diseases but also the widespread and often life-saving use of major surgery as a treatment for cancer and other diseases. The tremendous decrease in deaths from infectious disease over the last decades is a chief cause of the present population explosion. Ironically, with the advent of strains of bacteria resistant to these drugs (page 329) and endemic in hospitals, the latter are once more becoming reservoirs of serious bacterial diseases.

With a few exceptions, viruses have been impervious to attack by chemotherapeutic agents; drugs that successfully disrupt viral replication generally have devastating effects on cellular processes. However, recent advances in determining the three-dimensional structure of macromolecules are providing the first realistic hopes for the development of antiviral agents. Membrane receptors and viral capsids (Figure 21-26) are beginning to yield the structural secrets that explain their interactions, raising the possibility of devising molecules that will block either the relevant receptors or the portions of the capsid that fit into them.

New approaches to the control of both bacterial and viral diseases are also likely to emerge from the wealth of information being generated by recombinant DNA studies. Almost as a by-product, these studies are shedding new light on disease processes, for example, by identifying bacterial and viral genes that affect virulence. If previous patterns of biological history hold, it is reasonable to expect that as molecular biologists learn more about the organisms themselves—and about the genes dictating their properties—medical scientists will apply that knowledge in practical ways.

(a)

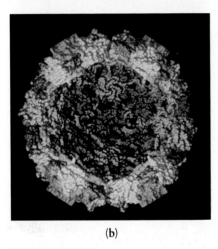

(b)

21-26 *Computer-generated images of the protein capsid of type I polio virus, as seen* (a) *from the outer surface and* (b) *in a cut-away view. The capsid is formed from four protein subunits, each of which is repeated 60 times. Three of the subunits are visible on the exterior surface, and one of them occurs only on the inner surface of the hollow capsid. The structure of the capsid of a human rhinovirus was also determined recently, and it is likely that the details of other capsid structures will follow in the near future.*

SUMMARY

The prokaryotes are both the most ancient group of organisms and the most abundant. The application of molecular techniques is enabling biologists to begin determining the phylogenetic relationships in this diverse kingdom. Two distinct lineages of prokaryotes, the Archaebacteria and the Eubacteria, have been identified. Among the major lineages of Eubacteria are green bacteria, purple bacteria and related forms, spirochetes, cyanobacteria, and gram-positive bacteria.

The prokaryotes are the only organisms with cells in which the DNA is not associated with histones and in which there is no membrane-bound nucleus or membrane-bound organelles. Their cell membrane is formed from a lipid bilayer that does not contain cholesterol and that often incorporates enzyme systems such as—in aerobic forms—the electron transport system. The varied forms of prokaryotes are imparted by their cell walls, which may be rigid or flexible. In gram-positive bacteria, the cell wall consists of a relatively thick layer of polysaccharides and complex polymers known as peptidoglycans; in gram-negative bacteria, a thin layer of peptidoglycans is enclosed by an outer membrane, similar in composition to the cell membrane. A polysaccharide capsule, surrounding the cell wall, is also present in many strains of bacteria.

Many species have long, slender extensions, flagella and pili, extending from the surface of the cell. Flagella are locomotor structures, whereas pili serve to attach the cells to surfaces or to each other. Bacterial flagella, which have a complex structure quite different from that of eukaryotic flagella, propel the cell through the medium by a rotary movement. In spirochetes, the flagella are modified to form an axial filament between the cell wall and the cell membrane; rotation of this filament imparts a corkscrew motion to the cells. Nonflagellated forms are nonmotile or move by gliding on the surface of a secreted mucus or slime.

Reproduction of prokaryotes is usually by binary fission, but some forms reproduce by budding or fragmentation. Genetic variability is introduced by mutation and by genetic exchanges and recombinations. Many types of prokaryotes form tough, resistant spores.

Prokaryotes comprise both heterotrophs and autotrophs; the autotrophs include both chemosynthetic and photosynthetic forms. Chemosynthetic autotrophs are found only among prokaryotes. The photosynthetic green and purple bacteria use a variety of substances, including sulfur compounds, as electron donors and do not produce oxygen gas. The cyanobacteria have a type of photosynthesis essentially like that in plants. Photosynthetic prokaryotes lack chloroplasts, and their photosynthetic pigments are embedded in the cell membrane or in internal membranes.

Viruses consist of nucleic acid (DNA or RNA) enclosed in a protein capsid.

They are now thought to be cellular fragments that have set up a partially independent existence. Viral reproduction can occur only within a host cell; the nucleic acid of the virus replicates and directs the formation of new protein capsids, utilizing the host cell's enzymes and other metabolic equipment. Infectious agents even simpler than viruses have recently been isolated and identified: viroids (small molecules of RNA with which no protein is associated) and prions (proteinaceous particles with which no nucleic acid is associated). Viroids are the causative agent of certain plant diseases, and prions transmit scrapie, a disease of sheep. The mechanisms by which viroids and prions exert their pathogenic effects remain unknown.

Microorganisms and human beings live in a multitude of symbiotic relationships, which may take the form of (1) commensalism, in which one partner benefits and the other is neither harmed nor benefited, (2) mutualism, in which both benefit, and (3) parasitism, in which one partner benefits and the other is harmed. A symbiotic relationship may change from one type to another as a result of changes in the host or, sometimes, in the microorganism. Pathogenic microorganisms may cause disease by direct tissue destruction, by producing toxins, and by provoking host defenses.

QUESTIONS

1. Distinguish among the following: Archaebacteria/Eubacteria; gram-positive/gram-negative; eukaryotic flagella/bacterial flagella/pili; axial filament/fibril; pathogen/toxin; virus/viroid/prion.

2. Label the drawings below.

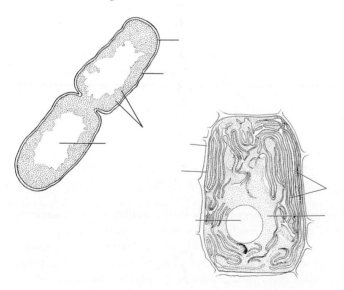

3. Identify the types of bacteria shown in the drawings below.

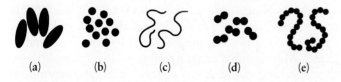

4. Many microorganisms produce antibiotics. What do you think their function might be for the organisms that produce them?

5. Five types of Eubacteria are photosynthetic: the green sulfur bacteria, the green nonsulfur bacteria, the purple sulfur bacteria, the purple nonsulfur bacteria, and the cyanobacteria. Describe the similarities and differences among these groups of organisms and in their photosynthetic processes.

6. In addition to the photosynthetic Eubacteria, a few genera of salt-loving prokaryotes are photosynthetic. How do these cells differ from the other photosynthetic prokaryotes? (In thinking about this question, you may wish to review the essay on page 221 as well as material in this chapter.)

7. Before the structure of any virus was known, Crick and Watson predicted that the protein coats of viruses would prove to be made up of large numbers of identical subunits. Can you explain the basis of their prediction?

8. Some biologists consider viruses to be living organisms. By what criteria might viruses be considered alive?

9. Which of the three types of symbiotic association would, in your opinion, apply to the following relationships: dog and flea; human being and dog; maggot and wound; *E. coli* and human being; diphtherial microorganism and human being?

10. Prokaryotes are considered more primitive than eukaryotes. Does this mean they are all identical to forms of life that existed before eukaryotes arose? Explain your answer.

CHAPTER 22

The Protists

22-1 Stentor, *a ciliated protist. Extended, it looks like a trumpet (this genus is named after "bronze-voiced" Stentor of the* Iliad, *"who could cry out in as great a voice as 50 other men"). Stentor is crowned with a wreath of membranelles. These beat in rhythm, creating a powerful vortex that draws edible particles up to and into the funnel-like oral groove leading to the buccal cavity ("mouth"). Elastic protein threads, myonemes ("little muscles"), run the length of the cell. When contracted,* Stentor *is an almost perfect sphere.*

The protists are all eukaryotic. The majority of protist species are unicellular, and those that are not have relatively simple multicellular or multinucleate structures. Apart from these two characteristics, they are an extremely diverse group of organisms. The kingdom Protista includes heterotrophs, among them a number of parasitic forms; photosynthetic autotrophs; and, as you will see, a few versatile organisms that are both heterotrophic and photosynthetic.

Protists are far from simple. Indeed, among one group, the ciliates, are found probably the most complex of all cells, with an astonishing variety of highly specialized structures (Figure 22–1). That single cells should be so complicated is actually not surprising. Among the protists, each cell is a self-sufficient organism, as capable of meeting all of the requirements of living as any multicellular plant or animal.

THE EVOLUTION OF THE PROTISTS

The microfossil record indicates that the first eukaryotes evolved at least 1.5 billion years ago. Eukaryotes, as we saw in Section 1, are distinguished from prokaryotes by their larger size, the separation of nucleus from cytoplasm by a nuclear envelope, the association of the DNA with histone proteins and its organization into a number of distinct chromosomes, and complex organelles, among which are chloroplasts and mitochondria.

The step from the prokaryotes to the first eukaryotes (the protists) was one of the major evolutionary transitions, second in importance only to the origin of life itself. The question of how it came about is a matter of current and lively discussion. One interesting hypothesis, gaining increasing acceptance, is that larger, more complex cells evolved when certain prokaryotes took up residence inside other cells.

As we noted previously, oxygen began slowly to accumulate in the atmosphere about 2.5 billion years ago, as a result of the photosynthetic activity of the cyanobacteria. Those prokaryotes that were able to use oxygen in ATP production gained a strong advantage, and so such forms began to prosper and increase. Some of these cells evolved into modern forms of aerobic bacteria. Others, according to this hypothesis, became symbionts within larger cells and evolved into mitochondria.

Several lines of evidence support the idea that mitochondria are descended from specialized bacteria. Mitochondria contain their own DNA, and this DNA is present in a single, continuous ("circular") molecule, like the DNA of bacteria. Many of the enzymes contained in the cell membranes of bacteria are found in mitochondrial membranes. Mitochondria contain ribosomes that resemble those of bacteria both in their small size and in details of their chemical composition, including some of the nucleotide sequences in the constituent rRNA molecules.

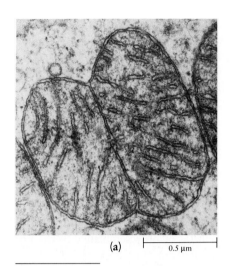 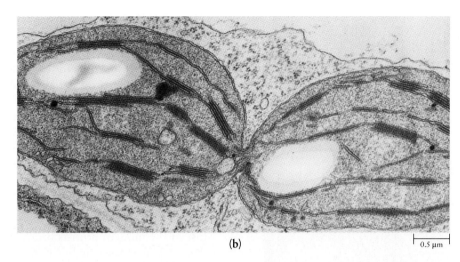

22–2 (a) *Mitochondria* and (b) *chloroplasts may have originated as symbiotic prokaryotes. This hypothesis is supported by the facts that both contain their own DNA and both, as shown here, replicate by binary fission.*

Further, mitochondria appear to be produced only by other mitochondria, which divide within their host cell (Figure 22-2). However, to make the situation more complex, there is not nearly enough DNA in mitochondria to code for all the mitochondrial proteins. Host-cell DNA is also required for the synthesis of mitochondrial enzymes and structural proteins.

It is thought that mitochondria are most likely derived from a lineage of purple nonsulfur bacteria in which the capacity for photosynthesis had been lost. Little is known, however, about the original cells in which these bacteria first set up housekeeping—or, indeed, if they actually existed. Contemporary prokaryotes provide few clues about the origins of two key features of all eukaryotic cells—a nuclear envelope and chromosomes containing both DNA and histone proteins. On the basis of nucleotide sequences in transfer and ribosomal RNAs, it has been suggested that very early in the history of life a common ancestral lineage gave rise not only to the two lineages (Archaebacteria and Eubacteria) shown in Figure 21-3 (page 430) but also to a third lineage. Members of this third lineage, which have been dubbed the "urkaryotes," are hypothesized to have been the host cells in the evolution of the eukaryotes. If such cells existed, they probably had no means of using oxygen for cellular respiration. Thus they were dependent entirely on glycolysis and fermentation, anaerobic processes that are, as we have seen (page 190), relatively inefficient. Cells with oxygen-utilizing respiratory assistants would have been more efficient than those lacking them and so would have reproduced at a faster rate.

In an analogous fashion, photosynthetic prokaryotes ingested by larger, nonphotosynthetic cells are believed to be the forerunners of chloroplasts. Such symbioses are thought to have occurred independently in a number of lineages, giving rise to the various groups of modern photosynthetic eukaryotes.

The endosymbiotic hypothesis accounts for the presence in eukaryotic cells of complex organelles not found in the far simpler prokaryotes. It gains support from the fact that many modern organisms contain intracellular symbiotic bacteria, cyanobacteria, or photosynthetic protists, indicating that such associations are not difficult to establish and maintain. Different species of cyanobacteria, for example, are found living within other cyanobacteria—and within protists, fungi, plants, and animals. These symbiotic cyanobacteria typically lack cell walls and are, functionally, chloroplasts. Moreover, they divide at the same time as the host cell, by a process similar to chloroplast division. A number of species of photosynthetic protists also form symbiotic associations with invertebrate animals (Figure 22-3). In such symbioses, whether ancient or modern, the smaller cells gain nutrients and protection, and the larger cells are given a new energy source.

22-3 *Plakobranchus, a marine mollusk. The tissues of this animal contain chloroplasts, which it obtains by eating certain green algae.*

(a) *Ordinarily, structures called parapodia are folded over the animal's back, hiding the chloroplast-containing tissues.* **(b)** *When the parapodia are spread apart, however, the deep green tissues become visible. The chloroplasts carry on photosynthesis so efficiently that within a 24-hour cycle of light and darkness some individuals produce more oxygen than they consume.*

The greater complexity of the eukaryotic cell brought with it a number of advantages that ultimately made possible the evolution of multicellular organisms. The eukaryotic cell is capable of carrying vastly more genetic information than the prokaryotic cell—enough, for example, to specify an oak tree or a human being. Because of the compartmentalization of functions by membranes, eukaryotic cells are more efficient metabolically and can be larger. Although a larger cell requires more energy, it can obtain it more easily, either because of an increased surface area for photosynthesis or absorption of food molecules or because of a greater ability to catch and subdue prey organisms. Moreover, larger cells are generally better able to make necessary adjustments to potentially life-threatening environmental changes, for example, in temperature or available water.

Classification of Protists

Most of the modern protists probably bear little or no resemblance to the first eukaryotes. As their complexity reminds us, they are not beginnings. They, like the modern prokaryotes, are the end products of millions of years of evolution, and in that time, they have undergone tremendous diversification. Those that have not changed much—amoebas might be an example—have survived over the millennia because they are exquisitely adapted to the relatively unchanging environment in which they live. Biologists generally agree (1) that the modern protists represent a number of quite different evolutionary lineages, and (2) that all other eukaryotic organisms—fungi, plants, and animals—originated from primitive protists (Figure 22–4).

The classification of the protists, like that of the prokaryotes, is currently undergoing a major upheaval, as the electron microscope and modern biochemical and molecular techniques make available a vast amount of new information about these organisms. Such questions as the number of divisions or phyla constituting the kingdom, the relationships among the various divisions and phyla, the proper placement of particular species, and the criteria to be used in making these decisions are all topics of discussion and, in some cases, of heated

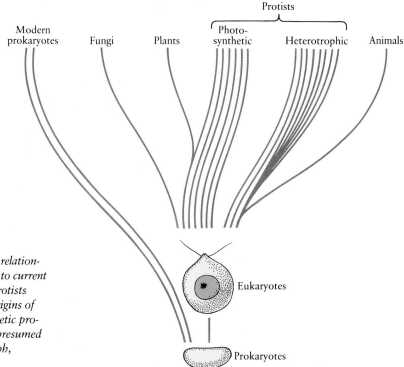

22-4 *A simplified scheme of the possible evolutionary relationships among the major groups of organisms. According to current hypotheses, the fungi and different groups of modern protists evolved separately from single-celled eukaryotes. The origins of the plants have been traced to one group of photosynthetic protists, the Chlorophyta, or green algae. The animals are presumed to have arisen from a single-celled eukaryotic heterotroph, although direct evidence is lacking.*

TABLE 22-1	**The Kingdom Protista**
Photosynthetic Autotrophs	
Division Euglenophyta	*Euglena* and related algae, all unicellular, mostly freshwater
Division Chrysophyta	Diatoms and related algae, all unicellular, marine and freshwater
Division Dinoflagellata	Dinoflagellates, all unicellular, mostly marine
Division Chlorophyta	Green algae, including unicellular, colonial, and multicellular forms; freshwater and marine
Division Phaeophyta	Brown algae, including the kelps; all multicellular; almost all marine
Division Rhodophyta	Red algae, all multicellular, mostly marine
Heterotrophs	
Division Myxomycota	Plasmodial slime molds
Division Acrasiomycota	Cellular slime molds
Division Chytridiomycota	Water molds (chytrids)
Division Oomycota	Water molds (oomycetes)
Phylum Mastigophora	Flagellates, parasitic or free-living
Phylum Sarcodina	Amoeba-like organisms, free-living or parasitic
Phylum Ciliophora	Ciliates, free-living or parasitic
Phylum Opalinida	Opalinids, intestinal parasites of lower vertebrates
Phylum Sporozoa	Sporozoans, parasitic

controversy. At present, no comprehensive classification of kingdom Protista exists that is widely accepted by biologists concerned with these organisms. For our purposes, we shall consider 15 principal divisions and phyla, which are grouped informally in Table 22-1 on the basis of their mode of nutrition.

In the past, it was common to regard the photosynthetic protists (the algae) as "lower plants" and the unicellular heterotrophs (the protozoa) as "lower animals." Similarly, the slime molds and water molds (unusual heterotrophic organisms that resemble the fungi in some features but differ from them in others) were considered "lower fungi." However, it is now increasingly apparent that, with one notable exception, the contemporary protists represent lineages that are not closely related to the lineages that gave rise to the members of the three multicellular kingdoms. (The notable exception is to be found among the green algae of division Chlorophyta, from which the plants are thought to have arisen.) Moreover, as we shall see, certain divisions of algae appear to be more closely related to certain phyla of protozoa than to other divisions of algae, and vice versa. Thus "algae" and "protozoa" have been abandoned as formal terms in modern classification; they remain, however, convenient informal terms and will be used as such in this text.

PHOTOSYNTHETIC AUTOTROPHS

By the time of the earliest known eukaryotes—about 1.5 billion years ago—a number of distinct lines of simple photosynthetic eukaryotes had already evolved. The approximately 30,000 described species now in existence can be grouped into six divisions. Three of these divisions (Euglenophyta, Chrysophyta, and Dinoflagellata) consist almost entirely of unicellular organisms. The other three divisions (Chlorophyta, Rhodophyta, and Phaeophyta) include groups that are multicellular.

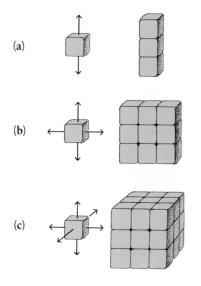

22-5 *Different planes of cell division can produce different patterns of growth, all represented among the algae.* **(a)** *Cell divisions in one plane produce a filament.* **(b)** *Divisions in two planes produce a one-celled layer.* **(c)** *Divisions in three planes produce a three-dimensional solid body. The arrows indicate the direction of growth.*

22-6 *Rocks along the coast of Cornwall, in Britain, showing the zonation of algal species from lower levels to higher levels. Zonation is based on a variety of factors, including not only light requirements and resistance to desiccation but also predation and competition.*

Nearly all members of these divisions are photosynthetic. In general, they have a relatively simple structure, which may be a single cell, a filament of cells, a plate of cells, or a solid body that may begin to approach the complexity of a plant body (Figure 22–5).

The unicellular algae are usually found floating near the surface of the oceans and inland waters, where light is abundant. Each cell is a totally self-sufficient individual, dependent only on light from the sun and carbon dioxide and minerals from the water that surrounds it. These cells may live in colonies, and sometimes the cells of a single colony may be attached to one another as the gliding bacteria, for instance, are held together by sticky secretions; occasionally there is even a division of labor among the cells. Together with small invertebrates and immature forms of larger animals they form the **plankton** (from *planktos*, the Greek word for "wanderer"). The photosynthetic members of the plankton community—sometimes called the **phytoplankton**—carry out most of the photosynthesis that takes place in the oceans. They are the ultimate energy source for most of the oceans' other inhabitants and, as a by-product of their photosynthetic activities, contribute enormous amounts of oxygen to the atmosphere.

Most of the multicellular algae are adapted to living in shallow waters and along the shores. Here the waters are usually rich in nutrients, washed down from the land or swept up in currents from the deeper waters. Living conditions are otherwise difficult, however. Along a rocky shore, for instance, multicellular algae—the seaweeds—are subject to great fluctuations of humidity, temperature, salinity, and light; to the pounding of the surf; and to the abrasive action of sand particles churned up by the waves. Life close to the shore is also crowded and competitive. Under these pressures, different groups of algae have become specialists at exploiting particular areas along tidal shores. The result is the zonation of life forms visible in many coastal areas (Figure 22–6).

Characteristics of the Photosynthetic Protists

The six divisions of algae vary widely in their biochemical characteristics, especially pigmentation, nature of stored food reserves, and cell-wall components, and in the number and position of their flagella, when present (Table 22–2). Typically, the cell walls of algae have a cellulose matrix, often with massive amounts of other polysaccharides that give certain algae a mucilaginous consistency.

The names of some divisions are derived from the colors of the predominant accessory pigments, which mask the bright green of the chlorophylls. A wide variety of carotenoids is found in the chloroplasts of algae. The xanthophylls are yellowish-brown carotenoids; the xanthophyll fucoxanthin gives the brown algae their characteristic color and name. It is also found in the golden-brown algae and diatoms (division Chrysophyta). The red algae (Rhodophyta) owe their colors to several kinds of phycobilins, accessory pigments that are also characteristic of the cyanobacteria. In the green algae the color of the chlorophylls is usually not masked by accessory pigments.

The wide variety of pigments found in the chloroplasts of the various divisions of algae suggests that (1) different types of oxygen-producing prokaryotes were in existence prior to the development of eukaryotic cells, and (2) the various divisions of algae may have evolved as a result of the establishment of symbiotic relationships with these different photosynthetic prokaryotes, which then evolved into modern chloroplasts. The chloroplasts of the red algae (Rhodophyta) are clearly derived from primitive cyanobacteria. Similarly, the chloroplasts of the Chlorophyta and Euglenophyta are thought to be derived from a prokaryotic lineage represented by the newly discovered *Prochloron* (page 442). It has also been suggested that the chloroplasts of the Chrysophyta, Dinoflagellata, and Phaeophyta may be derived from a lineage represented by *Heliobacterium chlorum* (page 442), although many authorities regard this as highly unlikely.

TABLE 22-2 **Comparative Summary of Characteristics of the Photosynthetic Protists**

DIVISION	NUMBER OF SPECIES	PHOTOSYNTHETIC PIGMENTS	FOOD RESERVE	FLAGELLA	CELL WALL COMPONENT	REMARKS
Euglenophyta (euglenoids)	1,000	Chlorophylls *a* and *b*, carotenoids	Paramylon and fats	1, 2, or 3 per cell, apical (at end of cell)	No cell wall; have a proteinaceous pellicle	Mostly freshwater; sexual reproduction unknown
Chrysophyta (diatoms, golden-brown algae, and yellow-green algae)	13,000	Chlorophylls *a* and *c*, carotenoids, including fucoxanthin	Laminarin and oils	None, 1, or 2, apical, equal or unequal	Cellulose, frequently impregnated with silica; cell wall lacking in some	Marine and freshwater
Dinoflagellata (dinoflagellates)	2,000	Chlorophylls *a* and *c*, carotenoids	Starch and oils	None or 2, lateral	Cellulose	Mostly marine; sexual reproduction rare
Chlorophyta (green algae)	9,000	Chlorophylls *a* and *b*, carotenoids	Starch	Usually 2 per cell, identical	Polysaccharides, cellulose in some	Freshwater and marine
Phaeophyta (brown algae)	1,500	Chlorophylls *a* and *c*, carotenoids, including fucoxanthin	Laminarin and oils	2, lateral, in reproductive cells only	Cellulose and algin	Almost all marine, flourish in cold ocean waters
Rhodophyta (red algae)	4,000	Chlorophyll *a*, carotenoids, phycobilins	Floridean starch	None	Cellulose, pectin compounds, impregnated with calcium carbonate in some	Mostly marine; complex sexual cycle; many species tropical

A diversity of storage products is found among the algae, with most having distinctive carbohydrate food reserves and many containing lipids as well. The green algae and dinoflagellates store their carbohydrates as starch, as do the plants. In the brown algae, golden-brown algae, and diatoms, another glucose polymer, laminarin (Figure 22-7), takes the place of starch. The carbohydrate reserves of the red algae are biochemically similar to starch.

When algal cells divide, the cell membrane generally pinches inward from the margin of the cell (furrowing), just as in heterotrophic protists, fungi, and animals (see page 153). Cell plates, like those of plants (also on page 153), are formed during cell division only in one brown alga and a few genera of filamentous green algae. Unlike plant cells, most algal cells—except those of the red algae—have centrioles.

22-7 *Laminarin, the principal storage product in the protists of divisions Chrysophyta and Phaeophyta. Like starch, it is made up of glucose residues, but unlike starch, there are only about 15 to 30 glucose units per molecule, and their linkage is different (see page 65).*

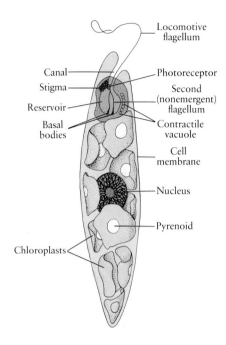

22-8 *Euglena is one of the most versatile of all one-celled organisms. Containing numerous chloroplasts, it is photosynthetic, but it can also absorb organic nutrients from the surrounding medium and can live without light. The pyrenoid is believed to be a food storage body. Euglena is pulled through the water by the whiplike motion of the single locomotive flagellum.*

Division Euglenophyta: Euglenoids

Euglenophyta are a small group of unicellular organisms (about 1,000 species), most of which are found in fresh water. They are named for the genus *Euglena*, the most common member of the group (Figure 22-8). About a third of the approximately 40 genera of euglenoids contain chloroplasts. These chloroplasts, which contain chlorophylls *a* and *b*, are very similar to those of the green algae but are enclosed by a triple—rather than a double—membrane. This suggests that euglenoids originally acquired chloroplasts by ingesting green algal cells; over long periods of time, most elements of the symbiotic green algae were lost, ultimately leaving only the chloroplast and the cell membrane. An alternative hypothesis is that euglenoid chloroplasts were derived independently from symbiotic photosynthetic prokaryotes resembling *Prochloron*. The nonphotosynthetic euglenoids are frequently able to absorb dissolved organic substances, and certain forms can ingest living prey, including other euglenoids. Euglenoids store their food as paramylon, an unusual polysaccharide that is found almost exclusively in this group.

Euglenoid cells are complex. *Euglena* is characteristically an elongated cell with a nucleus, two flagella (one of which is short and inactive), and numerous small emerald-green chloroplasts that give the cell its bright color. The flagella are attached at the base of a flask-shaped opening, the reservoir, at the anterior end of the cell. Emptying into the reservoir is the contractile vacuole, which collects excess water from all parts of the cell. Contractile vacuoles, which are found in flagellated cells of other algal groups as well, are also common features among the ciliates and other protozoa (see page 133). Euglenoid cells lack a cell wall but have a flexible series of protein strips, which make up the pellicle, inside the cell membrane (Figure 22-9). Unlike the stiff wall of other algal cells, the flexible pellicle permits *Euglena* to change its shape, providing an alternative form of locomotion—wriggling—for mud-dwelling forms.

If one leaves a culture of *Euglena* near a sunny window, a clearly visible green cloud will form in the water, and this will follow the light as it moves. If the light is too bright, however, the cells will swim away from it. *Euglena* is probably able to orient with respect to the light because of a pair of special structures: the stigma, or eyespot, which is a patch of pigment on the reservoir, and a photoreceptor on the locomotive flagellum. When the cell is in certain orientations with respect to light, the photoreceptor is shaded by the stigma. The amount of light striking the

22-9 *(a) A living cell of* Euglena, *and* (b) *a cell broken open, showing the flexible protein strips that make up the pellicle. The large organelle at the upper right is a chloroplast. The spongy network below the chloroplast is endoplasmic reticulum. At the bottom left is a broken mitochondrion, identifiable by its characteristic cristae.*

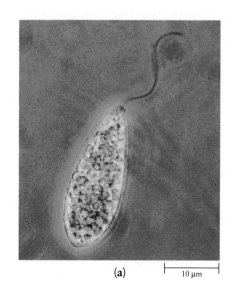

(a) 10 μm

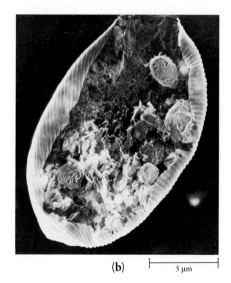

(b) 5 μm

photoreceptor influences the action of the flagellum and therefore directs the movement of the cell.

Euglenoids reproduce asexually, dividing longitudinally to form two new cells that are mirror images of one another. If some strains of *Euglena* are kept at an appropriate temperature and in a rich medium, the cells may divide faster than the chloroplasts, producing nonphotosynthetic cells, which can survive indefinitely in a suitable medium containing a carbon source. It is tempting to speculate that some modern heterotrophs, including the nonphotosynthetic euglenoids, arose from autotrophs in an analogous manner.

Division Chrysophyta: Diatoms and Golden-Brown Algae

This division includes two major classes, the diatoms (almost 10,000 species) and the golden-brown algae (about 3,000 species), and one minor class, the yellow-green algae (about 600 species). The chrysophytes have several identifying characteristics: (1) their photosynthetic pigments are chlorophyll *a* and chlorophyll *c*, which closely resembles the chlorophyll *b* of green algae and plants; (2) all species except the yellow-green algae contain a yellow-brown carotenoid, fucoxanthin, which functions as an accessory pigment and gives them their characteristic color; (3) their cell walls, which contain cellulose, are often impregnated with silicon compounds and thus are very rigid; (4) they characteristically store food in the form of oils rather than starch. Because of these oil reserves, fresh water containing large numbers of chrysophytes may have an unpleasant oily taste, as may fish caught in such waters. It has, in fact, been suggested that chrysophytes might be a potential source of oil for fuel.

Diatoms are a major component of the plankton and are an important source of food for small marine animals. Recently, the golden-brown algae have been found to be of extraordinary importance in the nanoplankton—components of the plankton so small that they pass through the fine mesh of an ordinary plankton net. It is now thought that the golden-brown algae may be the major food-producing organisms of the oceans.

Diatoms are enclosed in a fine double shell, the two halves of which fit together, one on top of the other, like a carved pillbox (Figure 22-10). Electron microscopy has shown that the fine tracings in diatom shells actually represent minute, intricately shaped pores or passageways connecting the interior of the cell to the exterior environment. The piled-up silica shells of diatoms, which have

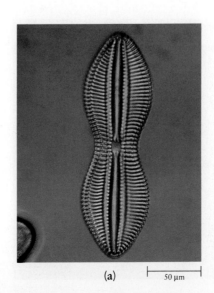

22-10 *Diatoms (division Chrysophyta). (a) Side view showing the characteristic intricately marked shell. (b) Pillbox type seen from above and from the side. Notice that one cell is dividing. Each new cell will get half of the "pillbox" and then construct the other half. When the parent cell divides, one half is always smaller than the other. When this smaller cell divides, one of its daughter cells is, in turn, still smaller. When the individual cells are reduced in size to about 30 percent of the maximum diameter characteristic of the species, a cycle of sexual reproduction may be triggered. The offspring, which are full size, then undergo a new series of cell divisions.*

(a) 50 μm (b) 100 μm

460   SECTION 4   The Diversity of Life

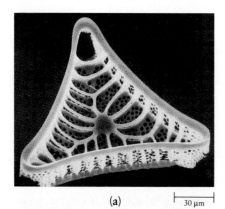

(a) 30 μm

(b) 1 μm

22-11 *Scanning electron micrographs of* (a) *the silicon-containing shell of a diatom and* (b) *the shell of a golden-brown alga, which is made up of scales containing calcium carbonate. Each diatom species has its characteristic pattern of perforations in the walls. The delicate markings of these shells, by which the species are identified, were traditionally used by microscopists to test the resolving power of their lenses.*

22-12 *Dinoflagellates.* (a) Ceratium tripos, *an armored dinoflagellate.* (b) Noctiluca scintillans, *a bioluminescent marine dinoflagellate. Yellow-brown diatoms that have been ingested can be seen inside the cell.*

collected over millions of years, form the fine, crumbly substance known as "diatomaceous earth," used as an abrasive in silver polish and toothpaste and for filtering and insulating materials.

Some other members of this division lack cell walls and are amoeboid. Except for the presence of chloroplasts, the amoeboid chrysophytes are indistinguishable from some species of phylum Sarcodina (page 470), and the two groups may be closely related.

Reproduction in the chrysophytes usually takes place by cell division, an asexual process, but diatoms sometimes reproduce sexually. Gametes are produced by meiosis and then undergo fusion, or **syngamy**. (Syngamy in plants and animals is usually referred to as fertilization.) The resulting zygote expands to the full size characteristic of the species and undergoes cell division; the daughter cells then produce new silica shells. Diatoms are diploid, except for the gametes.

Division Dinoflagellata: "Spinning" Flagellates

The Dinoflagellata, of which about 2,000 different species are known, are also largely composed of single-celled algae, most of them marine forms. Like the chrysophytes, the dinoflagellates are important components of the phytoplankton. Other members of the division are heterotrophs clearly related to the photosynthetic forms.

Many of the dinoflagellates are bizarre in appearance, with a stiff cellulose wall (theca) that often looks like a strange helmet or an ancient coat of armor. There are also "naked" forms, including some parasitic species. Dinoflagellates usually have two flagella that beat within grooves; one encircling the body like a belt and a second lying perpendicular to the first. The beating of the flagella in their respective grooves causes the cells to spin like tops as they move through the water. Many dinoflagellates are bioluminescent.

Dinoflagellates are often red in color, and the infamous red tides, in which thousands of fish die, are caused by great blooms of red dinoflagellates. The poison in these red tides has been traced to several species of dinoflagellates, including the armored *Gessnerium catenellum*. It is such an extraordinarily powerful nerve toxin that 1 gram of it would be enough to kill 5 million mice in 15 minutes. Blooms of *Gessnerium catenellum* appear regularly on the Pacific Coast and the Gulf of Mexico and have been reported off the New England coast. Mussels can also ingest the algae responsible for the red tides, concentrating the poison; these mollusks then become dangerous for consumption by vertebrates, including humans.

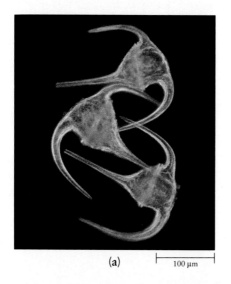

(a) 100 μm

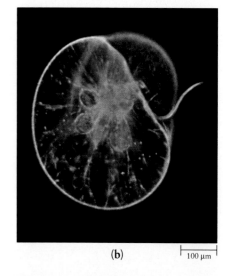

(b) 100 μm

CHAPTER 22 The Protists

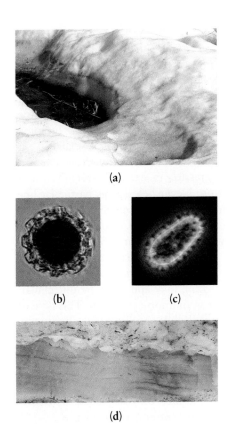

22–13 *Snow algae.* (a) *In many parts of the world, the presence of large numbers of snow algae produces "red snow" in high-altitude snow banks during the summer. This photograph was taken near Cedar Breaks National Monument in Utah.* (b) *Dormant zygote of the snow alga* Chlamydomonas nivalis. *The red color results from pigments that serve to protect the chlorophyll within the zygote following fertilization (see Figure 12–5, page 251).* (c) *Resting zygote of the alga* Chloromonas brevaspina, *found in green snow.* (d) *Green snow occurs just below the surface, usually near tree canopies. It is widespread, occurring as far south as Arizona and as far north as Alaska.*

Division Chlorophyta: Green Algae

The chlorophytes are the most diverse of all the algae, comprising about 9,000 species. They are thought to represent four distinct evolutionary lineages, one with only a few living species and three with the vast majority of contemporary species. The three principal classes of Chlorophyta, corresponding to these three lineages, are Chlorophyceae, Charophyceae, and Ulvophyceae. Although most green algae are aquatic, others occur in a wide variety of habitats, including the melting surface of snow (Figure 22–13), as green patches on tree trunks, and as symbionts in a variety of organisms. Of the aquatic species, most of the Ulvophyceae are marine, but the great majority of the Chlorophyceae and Charophyceae are found in fresh water. Many green algae are unicellular and microscopic in size. For example, *Chlamydomonas* (Figure 5–3, page 104), a member of class Chlorophyceae, is visible only as a fleck of green under the light microscope. Some of the Ulvophyceae are large; *Codium magnum* in the Gulf of Mexico, for example, sometimes attains a breadth of 25 centimeters and a length of more than 8 meters.

The plants are believed to have originated from the same lineage of green algae as the contemporary Charophyceae. Both the plants and the green algae—but no other group of algae—contain chlorophylls *a* and *b* and beta-carotene as their photosynthetic pigments and store their foods as starch. In addition, most of the Charophyceae also have cellulose-containing cell walls, and cell division in a few members of this class is characterized by the formation of a cell plate.

The Increase of Complexity

The green algae are of particular interest to students of evolution not only because of their relationship to the plants but also because of the wide range of complexity they exhibit. Whereas the three divisions previously discussed contain only unicellular forms and the members of the two divisions that follow are almost entirely multicellular, division Chlorophyta contains both unicellular and multicellular organisms, as well as a number of intermediate forms.

One form intermediate between the unicellular and the multicellular involves the association of individual cells in colonies. Colonies differ from true multicellular organisms in that, in colonies, the individual cells preserve a high degree of independent function. The cells are often connected by cytoplasmic strands, which integrate the colony sufficiently that it may be regarded as a single organism. The classic example of increasing complexity among colonial organisms is called the volvocine line, after one of its more spectacular members, *Volvox*. It is based on the cohesion of *Chlamydomonas*-like cells in motile colonies. The simplest member of the volvocine line is *Gonium* (Figure 22–14a). The *Gonium* colony consists of 4 to 32 separate cells (depending on the species) arranged in a shield-shaped disk. The flagella of each cell beat separately, pulling the entire colony forward. Each cell in *Gonium* divides to produce an entire new colony.

A closely related colonial organism is *Pandorina* (Figure 22–14b), which consists of 16 or 32 cells in a tightly packed ovoid or ellipsoid shape. The colony is polar; that is, one end is different from the other: the eyespots are larger in the cells at one pole of the colony. Each cell has two flagella, and because all the flagella point outward, *Pandorina* rolls through the water like a ball. When the cells attain their maximum size, the colony sinks to the bottom, and each of the cells divides to form a daughter colony. The parent colony then breaks open like Pandora's box (which suggested its name), releasing new daughter colonies.

Volvox (Figure 22–14c), the culmination of the volvocine line, is a hollow sphere that is made up, according to species, of a single layer of 500 to 60,000 tiny biflagellate cells. The flagella of each cell beat in such a way as to turn the entire colony around its axis, spinning it majestically through the water. Like *Pandorina*, *Volvox* is polar; it orients its anterior end toward the light but moves away from light that is too strong. Most of the cells are vegetative and do not take part in reproduction. Only some of the cells of the lower hemisphere can form daughter colonies; in other words, there is specialization of function. These cells, which appear identical to others in the young colony, become larger, greener, and structurally distinct as the colony matures. Then they enlarge and divide, forming new spheres of hundreds or thousands of cells. Daughter colonies remain inside the mother colony until the mother colony eventually breaks apart. In some species, the reproductive cells produce gametes and reproduction is sexual.

In these four extant genera of class Chlorophyceae, *Chlamydomonas*, *Gonium*, *Pandorina*, and *Volvox*, there is a steady progression in size and complexity and also a trend toward specialization of function. Nevertheless, this particular line represents an evolutionary "dead end," in that it has not given rise to a more complex group of organisms.

Another type of intermediate form, characteristic of some members of class Ulvophyceae, results from repeated nuclear divisions that are not accompanied by a corresponding division of the cytoplasm and the formation of cell walls. An entire organism, such as *Valonia* (Figure 22–15a), may appear to be unicellular, but, in fact, it consists of many nuclei within a common cytoplasm. This type of organization is neither truly unicellular nor multicellular but is known as **coenocytic** (pronounced "see-no-sit-ik"; from the Greek *koinos*, "shared in common"). Some coenocytic green algae, such as *Cladophora*, are filamentous, while others, such as *Valonia* and *Codium magnum*, form more massive structures. A number of common seaweeds, especially tropical ones, are coenocytic.

True multicellularity is exemplified by a number of green algae in classes Charophyceae and Ulvophyceae, including *Spirogyra* and *Ulva* (Figure 22–15b and c). In these organisms, nuclear division is followed by cytokinesis and the formation of cell walls, but the daughter cells do not separate from one another. As we saw in Figure 22–5, such cell division can lead to the formation of filaments, sheets, or a three-dimensional body.

Life Cycles in the Green Algae

A variety of different life cycles are observed among the green algae. In Figure 12–5 (page 251), we examined the life cycle of the unicellular green alga *Chlamydomonas*. The haploid cells of this alga usually reproduce asexually, with each cell undergoing two successive mitotic divisions, resulting in four new haploid cells. However, when essential nutrients, particularly nitrates, are in short supply, *Chlamydomonas* reproduces sexually. Haploid cells of different mating types function as gametes—they fuse, forming a diploid zygote around which a protective coat forms. This cell, known as a **zygospore**,* remains dormant until conditions are once more favorable for growth. It then undergoes meiosis, producing four haploid cells. The correlation between nutrient deficiency and sexual reproduction in *Chlamydomonas* and in many other organisms suggests that the rate of genetic recombination is subject to environmental influences.

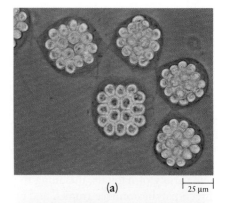

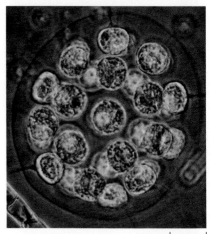

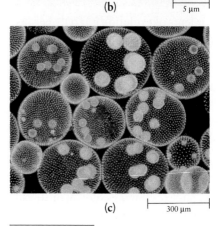

22–14 *The volvocine line.* **(a)** *Gonium. Each colony is composed of a shield-shaped mass of* Chlamydomonas-*like cells, held together in a gelatinous matrix.* **(b)** *Pandorina, in which the individual cells form an egg-shaped colony.* **(c)** *Volvox, the largest and most complex member of this evolutionary line. Each colony is made up of hundreds or thousands (depending on the species) of individual bright green biflagellate cells, attached to one another by fine threads of cytoplasm. Each colony forms a hollow sphere that spins through the water as a result of coordinated beating of the flagella. Daughter colonies form inside the mother colony. Many such mother-daughter combinations are visible in this micrograph.*

---

* Note that the word "spore" has more than one meaning in biology. Among the prokaryotes, spores are produced by a transformation of the original cell contents. Among the protists and fungi, spores may contain haploid nuclei produced by meiosis or diploid nuclei produced by syngamy ("zygospores"); such diploid nuclei typically undergo meiosis, producing haploid nuclei, before they are released from the protective coat. Among the plants, spores are always haploid cells produced by meiosis. All spores, however, have some essential features in common: they are tiny, light, and able to withstand desiccation and extremes of temperature, and, under favorable circumstances, they are able to germinate without combining with another cell. Thus, in four of the five kingdoms, spores serve as resting and dispersal forms, providing the organism with a means of waiting out hostile environmental conditions and of colonizing new territories.

 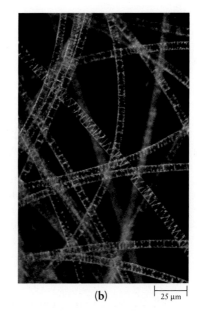 

22–15 *Division Chlorophyta includes a variety of multinucleate and multicellular organisms in addition to colonial forms.* (a) Valonia, *a coenocytic green alga. About the size of a hen's egg,* Valonia *contains many nuclei but has no partitions separating them. It is common in tropical waters.* (b) Spirogyra *is a freshwater alga in which the cells all elongate and then are divided by transverse cross walls so that they are strung together in long, fine filaments. The chloroplasts form spirals that look like strips of green tape within each cell.* (c) Ulva, *or sea lettuce, is a marine alga in which the cells divide both longitudinally and laterally, with a single division in the third plane. This produces a broad thallus (vegetative body) two cells thick.*

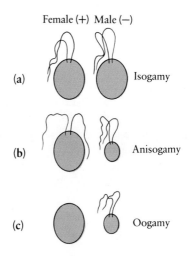

22–16 *Evolution of gametes.* (a) *Isogamy. The gametes are similar in size and shape.* (b) *Anisogamy. One gamete, conventionally termed male, is smaller than the other.* (c) *Oogamy. One gamete, usually the larger, is not motile. This is where sexual differentiation, with its mixed blessings and many complications, all began.*

In most species of *Chlamydomonas*, the cells of the two different mating types, conventionally designated as plus (+) and minus (−), are identical in size and structure; this condition is known as **isogamy** (Figure 22–16). In other species of *Chlamydomonas*, one gamete is larger than the other, but both are motile, a condition known as **anisogamy**. In still other species, one gamete, usually the larger, is not motile, a condition known as **oogamy**. The larger, nonmotile gametes are specialized for storing nutrients for the zygote, whereas the smaller gametes are specialized for seeking out and finding the first type of gamete. An individual that produces gametes that are nonmotile and (usually, though not always) larger is known as female. The entire range of differences between gametes that occurs among the algae (and among other types of organisms, as well) is exhibited in the various species of the single genus *Chlamydomonas*.

A more complex life cycle, characterized by alternation of generations, is found in some multicellular green algae, as well as in all plants. As we saw in Chapter 12, in this type of life cycle, a diploid, spore-producing generation alternates with a haploid, gamete-producing generation. The gamete-producing form is known as the **gametophyte;** the individual cells of which it is composed are haploid *(n)* and so are the gametes it produces. The gametes fuse to form a diploid (2*n*) zygote. The zygote develops into the spore-producing form, the **sporophyte,** in which all the cells are diploid. Spores, which are produced in specialized structures of the sporophyte by meiosis, are always haploid. Spores differ from gametes in that when a spore germinates it forms a new organism, while a gamete must first unite with another gamete before further development occurs. The gametophyte and the sporophyte are always genetically different, since one is composed of haploid cells and the other of diploid cells. Although all gametes from a given gametophyte are genetically identical (since they are produced by mitosis from haploid cells), genetic recombination occurs both in the fusion of

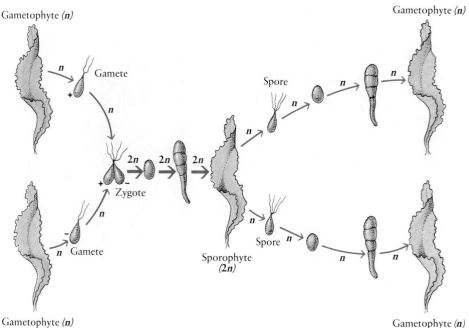

22–17 *In the sea lettuce,* Ulva, *we can see the reproductive pattern known as alternation of generations, in which one generation produces spores, the other gametes. The haploid (n) gametophyte produces haploid isogametes, and the gametes fuse to form a diploid (2n) zygote. A sporophyte, a multicellular body in which all the cells are diploid, develops from the zygote. The sporophyte produces haploid spores by meiosis. The haploid spores develop into haploid gametophytes, and the cycle begins again. The micrograph shows isogametes of* Ulva *before cytoplasmic fusion.*

gametes of different types to form the zygote and in the meiotic divisions that produce spores. Moreover, the spores, with their new genetic combinations, may be dispersed quite widely from the parent sporophyte, enabling the organism to spread into new environments.

In some green algae, such as the sea lettuce, *Ulva* (Figure 22–17), the two generations look alike and are said to be **isomorphic** (from the Greek word *morphe*, meaning "shape" or "form"). In other species, the sporophyte and the gametophyte do not resemble one another, and the generations are said to be **heteromorphic**. In fact, in some cases, the two generations of the same alga were once considered to represent two entirely different genera until the organism was studied in the laboratory (Figure 22–18).

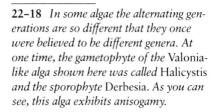

22–18 *In some algae the alternating generations are so different that they once were believed to be different genera. At one time, the gametophyte of the* Valonia*-like alga shown here was called* Halicystis *and the sporophyte* Derbesia. *As you can see, this alga exhibits anisogamy.*

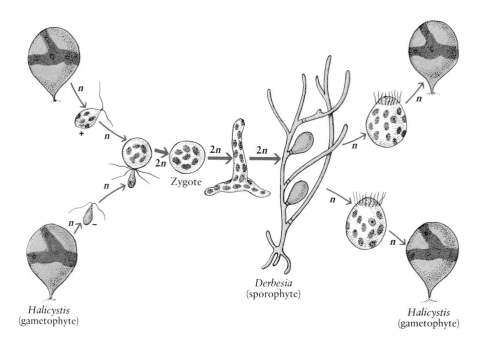

Division Phaeophyta: Brown Algae

The 1,500 species of brown algae are the principal seaweeds of temperate and polar regions. An almost exclusively marine group, they dominate the rocky shores throughout the cooler regions of the world, and some, like the kelps, often form extensive beds offshore. The brown algae contain chlorophylls *a* and *c* and fucoxanthin, as do the Chrysophyta. They store their food as an unusual polysaccharide (laminarin), or sometimes as oil, but never as starch, as do green algae and plants. Their cell walls contain cellulose.

The brown algae are often very large, and many have a variety of specialized tissues. Some of the giant kelps are nearly 60 meters long; many of these are annuals, reaching their full size in a single season. The body of a protist (or a plant) that is multicellular, but relatively unspecialized, is known as a **thallus.** In kelps, the thallus is well differentiated into **holdfast** ("root"), **stipe** ("stalk"), and **blade** ("leaf"). The words are put into quotation marks because, although the corresponding parts of the alga superficially resemble the structures of plants, they are not really comparable in their internal organization. However, most kelps have strands of elongated conducting cells in the center of the stipe that are similar to the cells that conduct sugars in the vascular plants. Carbohydrates produced in the blades, which are exposed to sunlight, are thus transported to the stipe and holdfast, which may be far below the surface of the water.

There are no modern unicellular forms in this group, except for the gametes. The life cycles of most brown algae involve alternation of generations. In some species the two generations are isomorphic, whereas in others they are heteromorphic. In the larger kelps, the gametophyte is much smaller than the conspicuous, familiar sporophyte. In the rockweed *Fucus* and its relatives, the life cycle is superficially similar to that in higher animals (see Figure 12–4c, page 250). Meiosis occurs in specialized cells of the diploid organism, producing haploid cells that immediately undergo mitosis, giving rise to gametes.

Division Rhodophyta: Red Algae

Most of the seaweeds of the world are red algae, of which there are some 4,000 species. They are most commonly found in warm marine waters; fewer than 2 percent of the species are freshwater forms. Red algae usually grow attached to rocks or other algae. Their red color indicates that they absorb blue light, which is the color with the greatest penetration in water. As a result, red algae can grow at greater depths than other algae; some have been found attached 175 meters below the ocean surface in the clear water of the tropics. Although some grow to lengths of several meters, red algae never attain the size of the largest of the brown algae.

The red algae contain chlorophyll *a* and carotenoids, and also certain phycobilins, which give them their distinctive colors. The cell walls of most forms include an inner layer of cellulose and an outer layer of mucilaginous carbohydrates, from which agar, used in culturing bacteria in the laboratory, is derived. In addition, some red algae have the capacity to deposit calcium carbonate in their cell walls. Such algae are called coralline algae, and they play an important role in building coral reefs.

The basic life cycle of the red algae involves alternation of generations. In most species, the gametophyte and sporophyte are isomorphic, but an increasing number of heteromorphic life cycles are being discovered. The gametes are unusual in that neither type is motile; the male gamete is carried to the fixed female reproductive cell by the movement of the water. The red algae are one of the few groups of organisms that have no flagellated cells; they also lack centrioles.

22-19 (a) *A brown alga*, Laminaria, *showing holdfasts, stipes, and portions of the blades.* (b) *Unlike the brown algae, most red algae are made up of filaments. The branched filaments of this red alga are hooked, enabling it to cling to other seaweeds.*

MULTINUCLEATE AND MULTICELLULAR HETEROTROPHS

The Slime Molds

The slime molds are a group of curious organisms, usually classified with the protists because of their similarity to the amoebas. Two main groups are known, the plasmodial slime molds (division Myxomycota), with about 550 species, and the cellular slime molds (division Acrasiomycota), with 65 species.

Most slime molds live in cool, shady, moist places in the woods—on decaying logs, dead leaves, or other damp organic matter. One of the common plasmodial species *(Physarum cinereum)*, however, is sometimes found creeping across city lawns. Plasmodial slime molds come in a variety of colors and can be spectacularly beautiful. The function of the pigments is not known with certainty, but they are probably photoreceptors because only pigmented species require light for spore production.

During their nonreproductive stages, the plasmodial slime molds are thin, streaming masses of protoplasm, which creep along in amoeboid fashion. As one of these plasmodia travels, it engulfs bacteria, yeast, fungal spores, and small particles of decayed plant and animal matter, which it digests. It may grow to weigh as much as 50 grams or more, and, since slime molds are spread thinly, this mass can cover an area more than a meter in diameter. The plasmodium is coenocytic; as it grows, the nuclei divide repeatedly and, in the early stages, synchronously.

Plasmodial growth continues as long as an adequate food supply and moisture are available. When either of these is in short supply, the plasmodium separates into many mounds of protoplasm, each of which develops into a mature **sporangium** (a structure in which spores develop) borne at the tip of a stalk. Meiosis takes place and cell walls form around the individual haploid nuclei, producing the spores.

The spores germinate under favorable conditions, and each spore, depending on the species, produces one to four haploid, flagellated cells. Some of these cells fuse to form a zygote, from which a new plasmodium develops.

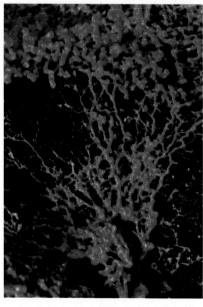

22–20 (a) *The plasmodium—a streaming mass of protoplasm—of a slime mold. Such a plasmodium, with its multiple nuclei, can pass through a piece of silk or filter paper and come out the other side apparently unchanged.* (b) *Sporangia of a plasmodial slime mold on a rotting log.*

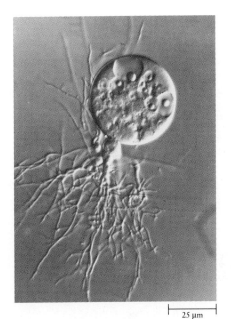

22–21 Chytridium confervae, *a common chytrid, as photographed through a differential-interference microscope. Note the slender, nutrient-absorbing rhizoids extending downward from the growing sporangium. The sporangium of this coenocytic organism contains numerous nuclei, vacuoles, and organelles, such as mitochondria. The rhizoids contain vacuoles and organelles but no nuclei.*

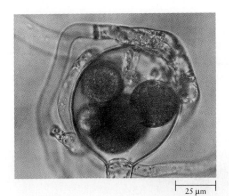

22–22 *Mating in the oomycete* Achyla ambisexualis. *The large spherical structure is the female gametangium. The dark bodies within it contain eggs. Encircling the female gametangium is the male gametangium. Fertilization tubes extending from the male into the female gametangium are barely visible. Sperm nuclei pass through these tubes to the egg nuclei. Development of the male gametangia and their attraction to the female are controlled by the production of a steroid hormone remarkably similar in structure to the human sex hormones.*

The life cycle of an acrasiomycete, or cellular slime mold, is shown on page 141. These slime molds also begin as amoeba-like organisms but differ from the plasmodial slime molds in that the amoebas, on swarming together, do not lose their cell membranes but retain their identity as individual cells.

The Water Molds

The water molds are coenocytic organisms, many of which resemble fungi in their basic structure. Until quite recently, they were usually classified with the fungi, but their biochemical characteristics and the presence of flagellated reproductive cells (which do not occur in any of the fungi) indicate that they represent distinct lineages. The two principal divisions are Chytridiomycota, with about 900 species, and Oomycota, with about 800 species.

Division Chytridiomycota

Most chytrids consist of a small vegetative body (a thallus), differentiated into a coenocytic sporangium and rootlike anchoring structures, called **rhizoids** (Figure 22–21). The rhizoids, which contain no nuclei, absorb dissolved nutrients from the substrate. When the chytrid is mature, the sporangium cleaves into flagellated spores, each of which contains a single nucleus. Following free-swimming and encysted stages, each spore can germinate to form a new thallus.

Other chytrids are much more complex in their structure and reproduction. Sexually reproducing forms have flagellated gametes, which may be similar or different in size and motility. In the genus *Allomyces*, the larger, less active gametes produce a hormone, appropriately named sirenin, that attracts the smaller, more active gametes. Like some algae and all plants, *Allomyces* and one other closely related genus have a life cycle characterized by alternation of generations.

Chytrids are found in both fresh and salt water and in moist soil. They live as parasites on algae, plants, and fungi, or as saprobes, feeding on dead algae, pollen grains, and other plant debris.

Division Oomycota

Structurally, most oomycetes resemble fungi, consisting of coenocytic filaments known as **hyphae**. As in the fungi, portions of the hyphae form specialized structures, the **gametangia**, in which gametes are produced. However, the cell walls of oomycetes, like those of many algae but unlike those of fungi, contain cellulose.

The oomycetes derive their name from *oion*, the Greek word for "egg." All oomycetes display oogamy—the differentiation of gametes into forms that are clearly male (sperm) and female (egg). However, the gametes of oomycetes, unlike those of chytrids, have no flagella and are nonmotile. In sexual reproduction, the sperm nucleus and the egg, each of which is borne in its own type of gametangium, fuse to produce a zygote (Figure 22–22). Oomycetes can also reproduce by forming asexual spores, each of which bears two flagella. Many oomycetes are aquatic (and thus they also are known as water molds), but even the terrestrial forms can produce flagellated spores that require free water.

Most oomycetes are saprobes, living on dead organic matter. Some forms are parasitic and pathogenic, however, and as mycologist C. J. Alexopoulos has said, "At least two of them have had a hand—or should we say a hypha!—in shaping the economic history of an important portion of mankind." The first of these is *Phytophthora infestans* (*phytophthora* literally means "plant destroyer"), the cause of the "late blight" of potatoes, which produced the great potato famines in Ireland (Figure 22–23). The second economically important member of this group

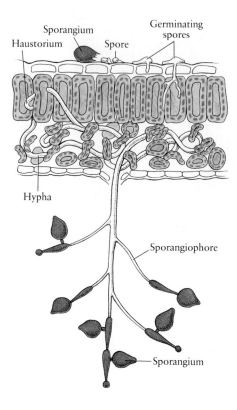

22–23 Phytophthora infestans, *cause of potato blight. Infection begins when an airborne sporangium alights on a leaf, releasing spores that move about in the film of water on the leaf's surface. These spores germinate, producing haustoria (specialized hyphae) that penetrate the epidermis and attack the photosynthetic cells in the interior of the leaf. Eventually aerial hyphae—sporangiophores—bear sporangia from which a new generation of asexual spores is released.*

is *Plasmopara viticola,* the cause of downy mildew of grapes. This mildew threatened the entire French wine industry during the latter part of the nineteenth century.

UNICELLULAR HETEROTROPHS

In addition to the multinucleate and multicellular heterotrophs we have just considered, kingdom Protista includes a vast number of unicellular heterotrophs (Table 22–3). These organisms, known informally as the protozoa, are classified in five principal groups. Three phyla, which contain both free-living and parasitic members, are distinguished on the basis of mode of locomotion: (1) by flagellar movement (phylum Mastigophora, the mastigophorans), (2) by pseudopodia (phylum Sarcodina, the sarcodines), and (3) by ciliary movements (phylum Ciliophora, the ciliates). Two phyla are entirely parasitic: (1) Opalinida, the opalinids, which have ciliary movement, and (2) Sporozoa, the sporozoans, in which the motility of the cells is much reduced.

Protozoans usually reproduce asexually, by binary fission. Many also have sexual cycles, involving meiosis and the fusion of gametes, which results in a diploid ($2n$) zygote. The zygote often takes the form of a thick-walled, resistant zygospore, especially during periods of drought or cold. Some protozoans, notably the ciliates, undergo conjugation, in which nuclei are exchanged between cells.

Phylum Mastigophora

The Mastigophora are regarded as the most primitive of the heterotrophic protists. Some authorities believe they are derived from photosynthetic flagellated cells, such as *Euglena,* that lost their chloroplasts; others think they may be the descendants of protists that never acquired photosynthetic symbionts. Almost all of the smaller cells have one or two flagella, and larger cells frequently have many. These flagella, like those of other protists, have the characteristic 9 + 2 structure (page 122). The mastigophorans multiply asexually by binary fission (mitosis and cytokinesis), and, in some forms, sexually, by syngamy. They generally have no

TABLE 22-3 Comparative Summary of Characteristics of the Heterotrophic Protists

DIVISION OR PHYLUM	NUMBER OF SPECIES	LOCOMOTOR STRUCTURES	MODE OF REPRODUCTION	REMARKS
Myxomycota (plasmodial slime molds)	550	Pseudopodia; flagella on reproductive cells	Asexual (spores) and sexual (fusion of germinated spores)	Coenocytic, except in reproductive stages
Acrasiomycota (cellular slime molds)	65	Pseudopodia	Asexual (spores); also sexual in some species	Unicellular, with multicellular aggregates forming prior to reproduction
Chytridiomycota (chytrids)	900	Flagella on spores and gametes	Asexual (spores); also sexual in some forms	Coenocytic; mostly aquatic; cell walls contain chitin
Oomycota (oomycetes)	800	Flagella on spores	Asexual (spores) and sexual; all exhibit oogamy	Coenocytic; many aquatic; cell walls contain cellulose; cause blights and mildews of plants
Mastigophora (flagellates)	1,500	Flagella; some also form pseudopodia	Asexual (binary fission); sexual (meiosis and syngamy) in some species	Unicellular; mostly parasitic; some free-living
Sarcodina (amoebas and related forms)	11,500	Pseudopodia; a few develop flagella at some stages of life cycle	Asexual (binary fission) or sexual	Unicellular; mostly free-living; many have outer shells, or tests; about 33,000 fossil species known
Ciliophora (ciliates)	8,000	Cilia	Asexual; genetic exchanges through conjugation	Unicellular; mostly free-living; have micronuclei and macronuclei
Opalinida (opalinids)	400	Cilia or flagella; flagellated gametes	Asexual or sexual	Unicellular; all intestinal parasites of lower vertebrates
Sporozoa (sporozoans)	5,000	None	Complex life cycle, involving both asexual and sexual reproduction	Unicellular; all parasitic; cause devastating diseases, including malaria, in humans and other animals

outer wall, and some are able to form pseudopodia, which are, as you will recall, temporary extensions of the cell body that are used in locomotion and in engulfing food particles.

Most mastigophorans are parasitic, but some are free-living. The former include *Trypanosoma gambiense* and *Trypanosoma rhodesiense*, flagellates that cause African sleeping sickness, and members of the genus *Trichonympha*, complex and beautiful flagellates that live as symbionts in the digestive tracts of termites, where they digest the wood ingested by the termite. A number of other mastigophorans cause debilitating diseases in humans and their poultry and livestock; "hiker's diarrhea" is caused by members of the genus *Giardia*, which is also endemic in some day-care centers.

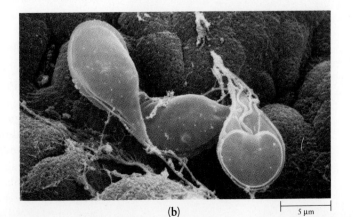

22-24 (a) Trichonympha. *This flagellate, which breaks down cellulose, is responsible for the well-known proficiency of its termite hosts to digest wood.* (b) Three cells of Giardia lamblia, *one of the many protists that cause diarrhea in humans. G. lamblia cells are characterized by four pairs of flagella and an adhesive disk with which they adhere to the lining of the small intestine. Using his primitive microscope, van Leeuwenhoek first identified* Giardia *about 300 years ago and hypothesized its pathogenic role; confirmation of his hypothesis has come only in recent years, with the development of methods to study this protist in the laboratory and in other animals. Encysted* Giardia *are transmitted from host to host in the feces.*

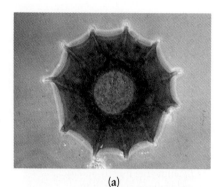

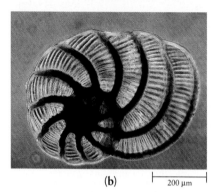

22–25 *Outer shells, or tests, are characteristic of certain groups of sarcodines.* (a) *The brilliantly colored test of* Arcella dentata *consists of a proteinaceous material secreted by the organism.* (b) *The shell of a foraminiferan. In addition to as many as 7,000 living species of foraminiferans, there are about 30,000 extinct species, known only from their fossilized shells.*

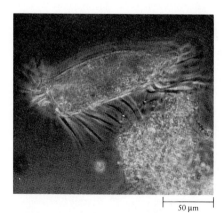

22–26 *The cilia of some protists are clumped to form cirri. In* Euplotes patella, *shown here, the cirri move individually, propelling the cell in a jerky motion.*

Phylum Sarcodina

The sarcodines include the amoebas and related forms. They have no coat or wall outside their cell membrane and generally move and feed by the formation of pseudopodia (see Figure 1–15a, page 37). They take their name from the word "sarcode," coined in the early nineteenth century to describe the "simple, glutinous, and homogeneous jelly" of which, at one time, simple cells were thought to be composed. Despite their uncomplicated appearance, they are complex cells and, as we shall see (page 474), are even capable of complex behavior patterns—for example, when sensing and pursuing prey organisms.

Sarcodines are thought to have originated from mastigophorans; some sarcodines may develop flagella during particular stages of their life cycle or under particular environmental conditions. They are found in both fresh and salt water. Some are parasites, such as those that cause amoebic dysentery in humans.

Reproduction may be asexual or sexual. Asexual reproduction takes place by cell division accompanied by mitosis in which the nuclear envelope usually does not break down. In sexual reproduction, the cells, which are diploid, undergo meiosis, forming gametes, which then fuse to form zygotes.

Many of the sarcodines have outer shells, or tests, sometimes brightly colored. Some, like *Arcella*, secrete a proteinaceous material that hardens on exposure. Others, like *Difflugia*, exude a sticky organic substance in which they deposit silicon-containing particles. These particles, which were previously ingested by the organism, are divided between the two daughter cells at cell division; each daughter cell then arranges them into an almost exact replica of the parental shell. Such shells, which are produced primarily by amoebas that live in sand or soil, on the mosses of bogs and forest floors, or in crevices in tree trunks, are thought to have evolved as a protection against abrasion or dehydration of the organism within. Other sarcodines, the heliozoans ("sun animals"), resemble pincushions, with fine pseudopodia stiffened with microtubules (page 78) radiating from their bodies; they are found in both salt and fresh water. *Actinosphaerium*, the protist shown in Figure 5–2 (page 103), may reach as much as a millimeter in diameter and may sometimes be seen as a tiny white speck floating on the surface of a pond.

Another group of sarcodines, the Foraminifera (numerically the most common of all unicellular heterotrophs), have snail-like shells and live in the sea. Their shells are made of calcium carbonate, extracted from the sea water. The white cliffs of Dover and similar chalky deposits throughout the world are the result of the long accumulation of these shells. The shells of the Foraminifera have been accumulating on the ocean bottom for millions of years, and in many areas, as a result of geologic changes, thick deposits of their skeletons (the "foraminiferan ooze") can be found on the surface of the land or under later rock formations. Since the skeletons have evolved over this long period of time, it is possible to date a particular stratum by the type of Foraminifera that it contains, a fact that has proved of immense practical value in locating oil-bearing strata.

Phylum Ciliophora

The ciliophores, or ciliates, are the most highly specialized and complicated of the protozoans. About 8,000 species of ciliates are known, both freshwater and saltwater forms; most are free-living. They are believed to have been derived from primitive mastigophorans, which, traveling in the opposite evolutionary direction from the sarcodines, developed elaborate ciliary systems (see Figure 5–30, page 122). In some species, the cilia adhere to each other in rows, forming brushlike structures called membranelles (see Figure 22–1) or clumps of cilia called cirri, which can be used for walking or jumping, as well as in feeding. Cilia, membranelles, and cirri move in a coordinated fashion, although the way in which they are coordinated is not entirely understood. Some ciliates also have myonemes, contractile threads. All have a complex "skin," the cortex, which is bound on the outer side by the cell membrane. In a few groups, the cortex contains small barbs known as trichocysts, which are discharged when the cell is stimulated in certain ways.

The ciliates have another unusual feature: they have two kinds of nuclei, micronuclei and macronuclei. One or more of each kind is present in all cells. They also have a complex process for exchange of genetic information, in which cells conjugate and the micronuclei undergo meiosis. Cells then exchange haploid micronuclei that fuse, so that each cell has a new diploid micronucleus, which then divides. The old macronucleus dissolves, and a new macronucleus develops from one of the daughter micronuclei. The macronucleus in certain ciliates contains 50 to 100 times as much DNA as the micronucleus and so is believed to represent multiple copies of it. This view is supported by the fact that, in many ciliates, a cell can survive indefinitely without a micronucleus if even a small portion of a macronucleus is left, although the cell cannot conjugate. However, it cannot live without a macronucleus, even if it has a micronucleus. The macronucleus does not divide mitotically but is apportioned approximately equally between dividing cells as they constrict and separate.

The complexity of the conjugation process among ciliates reminds us again of the expenditure of energy and other resources involved in effecting exchanges of genetic information among individuals of the same species, from bacteria to *Homo sapiens*. The high biological cost of these activities is an indication of the great survival value, in evolutionary terms, of the genetic variability produced by such exchanges.

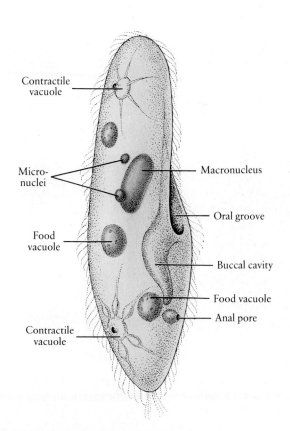

22-27 *Drawing of* Paramecium, *a ciliate. The body of this protist is completely covered by 9 + 2 cilia, although only a relative few are shown here. Like other ciliates,* Paramecium *feeds largely on bacteria, smaller microorganisms, and other particulate matter. The beating of specialized cilia drives particles into the buccal cavity, where they are formed into food vacuoles that enter the cytoplasm. The food is digested in the vacuoles, and the undigested matter, still in vacuoles, is emptied out through the anal pore. The contractile vacuoles serve to eliminate excess water from the cell.*

The Evolution of Mitosis

Among modern prokaryotes and protists, it is possible to observe varying patterns in the division of the genetic material. These patterns are believed to reflect evolutionary history. Among prokaryotes, you will recall, the replicated chromosomes (mostly DNA) attach to the cell membrane and so are separated as the newly formed cells divide. Among plant and animal cells, the nuclear envelope breaks down at mitosis, and the spindle fibers (composed of microtubules) are apparently involved in the mechanism of separation. Detailed analysis of some of the protists has revealed a number of intermediate stages between these two. In dinoflagellates, for instance, the chromosomes—which have no histones attached to the DNA and are always condensed—are anchored permanently to the nuclear envelope. The envelope remains intact during mitosis, and the chromosomes, as in bacteria, are separated as the enveloping membrane elongates. At the time of cell division, cytoplasmic channels containing bundles of microtubules form between the two sets of attached chromosomes; these microtubules are all oriented in the same direction and are thought to regulate the separation process.

The macronucleus of ciliated protozoa appears to be apportioned in just this same way, but in the micronucleus of ciliates, both kinetochore and polar spindle fibers are seen, as they are in slime molds. The spindle fibers appear to push the poles apart as the fibers grow, elongating the nucleus in preparation for its division. The nuclear envelope, however, does not break down.

As the history is reconstructed, in the earliest eukaryotes the nuclear envelope remained intact, and the chromosomes separated as it elongated. Next, microtubules, which first played an extranuclear role, entered the nuclear envelope, and, in an increasingly more organized way, aided in its elongation. Finally, the nuclear envelope, no longer useful in mitosis, began to fragment at prophase, and spindle fibers took over as a more efficient means for separating pairs of chromosomes in a diploid organism.

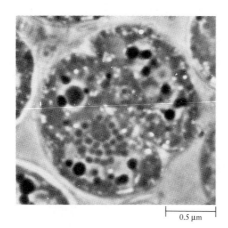

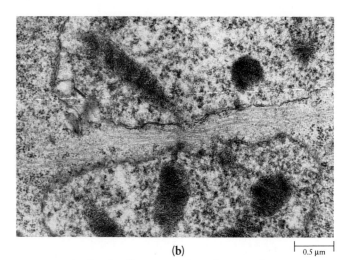

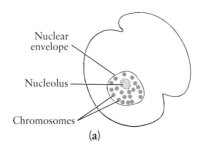

Mitosis in the dinoflagellate Cryptothecodinium cohnii. **(a)** *Photomicrograph and diagram of interphase chromosomes.* **(b)** *Microtubules in the cytoplasmic channel separating the two nuclei. The dark areas are chromosomes.*

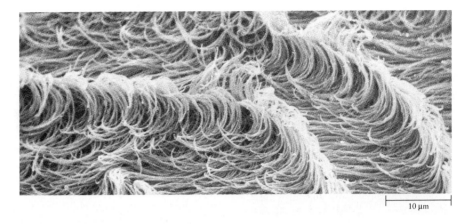

22–28 *The surface of* Opalina, *one of the parasitic opalinids, as seen through the scanning electron microscope. Its cilia beat continuously, about 40 to 60 times per second, in a coordinated beat that produces synchronous waves of motion.*

Phylum Opalinida

The opalinids have been found most often in the digestive tracts of frogs and toads, and, occasionally, in fishes and reptiles. They are structurally simple, with a uniform covering of cilia or flagella (Figure 22–28). At least two nuclei are present; in some species, the number of nuclei increases as the cell gets larger. The nuclei are not, however, differentiated into macronuclei and micronuclei. The opalinids may be distant relatives of the ciliates, or they may represent a distinct evolutionary lineage. Like certain groups of mastigophorans, opalinids produce gametes that fuse to form a zygote; some authorities consider this an indication of a close relationship with those particular mastigophoran groups.

Phylum Sporozoa

All sporozoans are parasites. They are characterized by the lack of cilia or flagella and by complex life cycles. The best-known sporozoans are members of the genus *Plasmodium*, which cause malaria in many species of birds and mammals. A typical *Plasmodium* life cycle is shown in Figure 22–29. Malaria has generally been

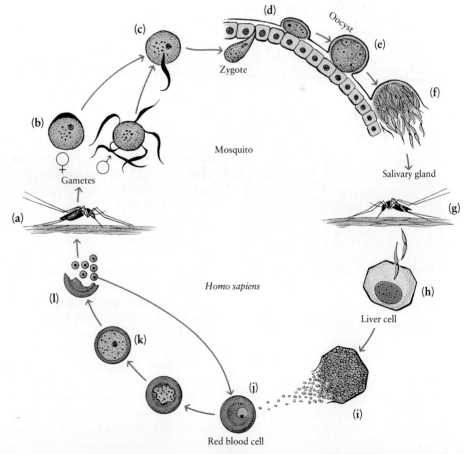

22–29 *Life cycle of* Plasmodium vivax, *one of the sporozoans that cause malaria in humans. The cycle begins* (a) *when a female* Anopheles *mosquito "bites" a person with malaria, and, along with the blood, sucks up gametes* (b) *of the sporozoan. In the mosquito's digestive tract, the gametes unite* (c) *and form a zygote* (d)*. From the zygotes, multinucleate structures called oocysts develop* (e)*, which, within a few days, divide into thousands of very small, spindle-shaped cells, sporozoites* (f)*, which then migrate to the mosquito's salivary glands. When the mosquito "bites" another victim* (g)*, she infects the person with the sporozoites. These first enter liver cells* (h)*, where they undergo multiple division* (i)*. The products of these divisions (merozoites) enter the red blood cells* (j)*, where again they divide repeatedly* (k)*. They break out of the blood cells* (l) *at regular intervals of about 48 hours, producing the recurring episodes of fever characteristic of the disease. After a period of asexual reproduction, some of these merozoites become gametes, and, if they are ingested by a mosquito at this stage, the cycle begins anew.*

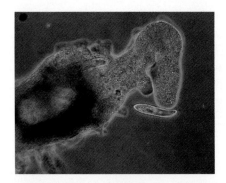
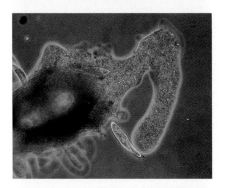
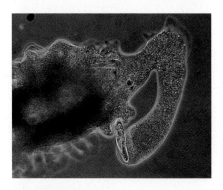
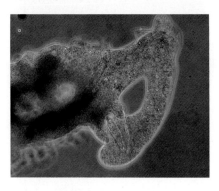

22-30 *The giant amoeba* (Chaos chaos) *capturing its prey. The initial stimulus produced by the prey—a* Paramecium*—induces pseudopodia from the amoeba and causes the amoeba to move toward the* Paramecium. *A large pseudopodium surrounds the* Paramecium, *drawing it toward the amoeba and engulfing it. A vacuole forms around the prey.*

controlled by drugs that act on the parasite at various stages of its life cycle and by insecticides that kill the mosquitoes that transmit it. Recently, however, both the parasite and the mosquitoes have begun to develop resistance to the chemicals used to attack them, and malaria remains a major cause of human death and disability. It is estimated that some 200 to 400 million people, principally in tropical regions of the world, are infected with the *Plasmodium* parasite. In Africa, about 10 percent of its victims die as a direct consequence of the infection and the remaining 90 percent experience repeated episodes of severe illness; mortality rates in young children often approach 50 percent.

With the advent of techniques for culturing *Plasmodium* in the laboratory, it has become possible to begin detailed studies of this parasite. Attention has been focused on the sporozoites, the merozoites, and the gametes—stages in which the life cycle is potentially most vulnerable to interruption—and particularly on the surface proteins of these cells. With recombinant DNA techniques, genes coding for the surface proteins are being identified, as are the regulatory mechanisms governing their expression. The hope is to use recombinant DNA techniques to prepare synthetic versions of key proteins that will function as effective vaccines, stimulating the immune system to attack the parasite. The development of such vaccines, however, is an extraordinarily complex task because, in the course of its evolution, the parasite has developed ways of altering these proteins, thereby evading attacks by the immune system. Despite the inherent difficulties and the relatively small amount of money devoted to research on malaria, considerable progress has been made in the last few years and there is hope that an effective vaccine—probably incorporating a variety of surface proteins—will eventually be available.

PATTERNS OF BEHAVIOR IN PROTISTS

As we have seen in previous chapters, even at the prokaryotic level of organization, organisms are capable of behavioral responses to environmental stimuli, including light, chemicals, and magnetic fields. Slightly more complex patterns of behavior are seen among the protists, at the lowest level of eukaryotic organization. Photosynthetic protists, such as *Euglena,* are especially sensitive to light intensity, moving into areas with optimal levels of light but out of areas where the light is too bright. Nonphotosynthetic protists, such as amoebas, may also exhibit **phototaxis,** or movement in response to light. If a bright pinpoint of light is focused on the advancing pseudopodium of an amoeba, the pseudopodium withdraws. If the entire body of the amoeba is exposed to bright light, the cell contracts suddenly and will even extrude any half-digested food. However, if it cannot escape from the light, the amoeba, after a few moments' hesitation, will resume its normal activities. This type of response, by which the stimulus comes to be ignored and a previous behavior pattern restored, is known as **habituation.** It is an important component of the behavior of multicellular eukaryotes, especially animals. (If our own cells were not capable of habituation, we would be continuously responding to and distracted by, for example, the touch of our own clothing or background noises.)

Amoebas are also **chemotactic,** that is, responsive to chemical stimuli. When something edible, an algal cell or a fellow protozoan, is in the vicinity of an amoeba, the amoeba can sense it at some distance, at least the length of its own body away. It then sends out a pseudopodium that is shaped quite specifically for its intended victim. A fine pincerlike projection will be formed for a small, quiet morsel; a much stronger and more massive pseudopodium will reach for a large ciliate or vigorously moving object (Figure 22–30). If the intended victim moves away, the amoeba will remain in pursuit so long as it is close enough to the prey to receive stimuli from it.

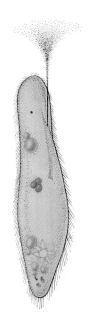

22–31 Paramecium *samples a drop of India ink. Currents created by the strongly beating cilia draw particles of the ink toward the protist's buccal cavity.*

Avoidance in *Paramecium*

Other patterns of protist behavior have been studied extensively in ciliates. *Paramecium,* for example, shows behavior that appears more complex than that of amoebas but is just as simple in principle. These protists are responsive to a variety of stimuli, including very subtle changes in temperature and chemistry that we can detect only by finely calibrated instruments. But though the detection methods are highly sensitive, the responses are fixed (stereotyped).

As a *Paramecium* swims, the strongly beating cilia around the oral groove create currents that constantly bring a sample of water to the buccal cavity, which seems to be the testing area. In this way, the *Paramecium* continuously explores the environment that lies ahead, shifting and sampling (Figure 22-31).

In general, *Paramecium* responds to changes in its environment by avoidance reactions. If it receives a negative stimulus, it turns away. If it turns, it always turns in the same direction, to its left, which is the aboral side—the side away from the oral groove. This occurs because of the relaxation of the beat of the body cilia. It does not matter which side the stimulus comes from. If the microscopist takes a blunt needle and jabs a *Paramecium* on the aboral side, it will still turn toward that side; and once it has turned, it will continue in this new direction indefinitely.

As shown in Figure 22-32, if the negative stimulus is powerful, such as a poisonous chemical, the *Paramecium* will stop short, reverse its ciliary beat and back up, turn toward the aboral side about 30°, and then start forward again, testing. If necessary, it will repeat this performance. Under a strong negative stimulus, it will turn a full 360° and will continue to turn until an avenue of escape is found. It can find its way around solid objects in the same way.

The end result of this, of course, is about the same as if the organisms were "attracted" to a favorable situation. For example, *Paramecium* is extremely temperature-sensitive. If you place the cells in a culture dish that can be made warmer at one end than another, you will find, by trying different combinations of temperatures, that they will tend to congregate in water that is about 27°C (80°F). If you place them on a slide with water that is below 27°C and then warm one little portion of it gently (you can do this by placing a hot needle on the cover slip), the cells will all gather in this warm spot. They reach it by chance as they move about through the water. But once they get into it, they literally cannot get out again, since as soon as they reach the edge of the warm water, they receive a sample of cooler water, which will turn them right back again. Similarly, on a slide that contains water that is too hot, they will trap themselves in an area that has been cooled slightly, even by only a few degrees.

22–32 *Avoidance behavior in* Paramecium. *The colored area at the top of the figure represents a drop of a toxic substance, and the arrows indicate the direction of movement. The* Paramecium (a) *moves up to the substance,* (b) *tests it,* (c) *backs up,* (d) *turns 30°, and* (e) *moves forward in the new direction. Many protists are capable of this sort of simple behavior.*

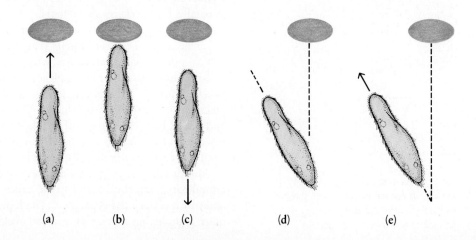

They have also been shown to congregate in the same way in a spot that is slightly acidic, by avoiding the neutral or alkaline areas. Bacteria, which are a primary food of *Paramecium,* create a slightly acidic environment by their metabolism, and indeed, *Paramecium* itself creates a slightly acidic environment by giving off carbon dioxide; so their avoidance of neutral or alkaline areas enhances the tendency of these protists to group together and also to gather where there is food. This tendency is strengthened further by the fact that they are more likely to cling to some other object—such as the leaf or stalk of a plant, a pile of debris, or even a shred of filter paper—if they are in a slightly acidic medium.

As described here, the behavior of *Paramecium,* helpless to escape from a drop of warm water or forced to cling by the presence of carbon dioxide, seems very unlifelike. Yet, as one watches the cell negotiating the changes in its environment, its actions seem purposeful. Both of these impressions are true. The individual *Paramecium* has no choice as to its behavior, but the behavior of these protists, in general, is a result of millions of years of choice, the choice by natural selection.

SUMMARY

The kingdom Protista comprises an enormous variety of eukaryotic organisms, mostly unicellular with some relatively simple multicellular forms. A major factor in the evolution of the eukaryotes may have been the establishment of symbiotic relationships with prokaryotic cells that, internalized, eventually became specialized as mitochondria and chloroplasts. It is thought that the protists represent a number of quite distinct phylogenetic lineages and, moreover, that all other eukaryotic organisms are derived from primitive protists. We have considered 15 divisions and phyla of protists, informally grouped into photosynthetic autotrophs (algae), multinucleate and multicellular heterotrophs (slime molds and water molds), and unicellular heterotrophs (protozoa).

The photosynthetic protists are classified into six divisions on the basis of their photosynthetic pigments, reserve food supply, and cell wall composition. Euglenophyta represent a small group of unicellular algae, mostly found in fresh water. They contain chlorophylls *a* and *b* and store carbohydrates in an unusual starch-like substance, paramylon. The cells lack a wall but have a flexible series of protein strips, which make up the pellicle, inside the cell membrane. The cells are highly differentiated, containing chloroplasts, a contractile vacuole, an eyespot, and flagella. No sexual cycle is known. This division also contains nonphotosynthetic forms.

The Chrysophyta—diatoms and golden-brown algae—are important components of freshwater and marine phytoplankton. They are unicellular. Diatoms are characterized by fine, double silicon-containing shells. They usually reproduce asexually, but syngamy also occurs.

The Dinoflagellata are unicellular biflagellates, many of which are marine. They are characterized by two flagella that beat in different planes, causing the organism to spin; dinoflagellates often have stiff, bizarrely shaped cellulose walls.

The Chlorophyta, or green algae, are thought to represent four distinct evolutionary lineages. The plants appear to have originated from the same lineage as the green algae of class Charophyceae. All green algae and plants have chlorophylls *a* and *b* and beta-carotene as their photosynthetic pigments and store their food reserves in the form of starch. Members of class Charophyceae, like plants, have cellulose-containing cell walls, and a few also form a cell plate during cell division.

Among the green algae are a variety of forms, representing different degrees of complexity: unicellular organisms, colonies, coenocytic (multinucleate) forms, and true multicellular organisms. The reproductive cycles of green algae are often quite complex. In species with sexual cycles, the gametes of different mating types may be similar in size and structure (isogamy), different in size but both motile (anisogamy), or different in size, with one, usually the larger, not motile (oogamy). Some multicellular green algae have a life cycle known as alternation of generations, in which a haploid phase alternates with a diploid phase. The haploid (n) generation, known as the gametophyte, produces haploid gametes. The gametes fuse to form the zygote, which develops into a diploid $(2n)$ sporophyte. The sporophyte produces spores by meiotic division. A spore is a single cell that, unlike a gamete, can develop into an adult organism without combining with another cell. In organisms with alternation of generations, the spore, which is haploid, germinates to produce the haploid gametophyte.

The Phaeophyta (brown algae) and Rhodophyta (red algae) are the principal seaweeds. The brown algae, which include the kelps, are found more commonly in cooler water; the red algae, in the tropics. In many brown algae the thallus is differentiated into holdfast, stipe, and blade, analogous to the root, stem, and leaf of plants. Most kelps have tissues specialized for the conduction of sugar from the blades to nonphotosynthetic parts of the thallus.

The slime molds are heterotrophic, amoeba-like organisms that reproduce by the formation of spores. There are two principal divisions: Myxomycota (plasmodial slime molds), which are coenocytic during the nonreproductive stages, and Acrasiomycota (cellular slime molds), in which the aggregating amoeboid cells retain their individual identity. The water molds (divisions Chytridiomycota and Oomycota) are coenocytic heterotrophs that superficially resemble fungi. They reproduce both asexually and sexually. In chytrids, both the spores and the gametes are flagellated, whereas in oomycetes, only the spores are flagellated; all oomycetes exhibit oogamy.

Unicellular heterotrophic protists (the protozoa) are thought to have evolved from nonphotosynthetic, flagellated ancestors. Among the protozoa are some of the largest known cells and also the most complex. Three phyla—the Mastigophora (flagellates), the Sarcodina (amoebas and their relatives), and the Ciliophora (ciliates)—contain both free-living and parasitic species, and their members may be identified on the basis of their locomotor structures. Two phyla—Opalinida and Sporozoa—contain only parasitic forms. The opalinids are ciliated, but the sporozoans have no locomotor organelles. In the course of their complex life cycles, however, sporozoans are transported quite efficiently from host to host.

Protists exhibit a variety of simple behavioral responses that prefigure the complex behaviors characteristic of multicellular eukaryotes. Among the protist responses are phototaxis (in both photosynthetic and nonphotosynthetic organisms), chemotaxis, avoidance, and habituation.

QUESTIONS

1. Describe the similarities and differences among the Euglenophyta, the Chrysophyta, and the Dinoflagellata.

2. To which other divisions or phyla of protists may the Euglenophyta, the Chrysophyta, and the Dinoflagellata be related? Describe the evidence for the relationships you cite.

3. In some classification schemes, the Chlorophyta, the Phaeophyta, and the Rhodophyta are placed in the plant kingdom. What similarities between the plants and these three divisions of algae might justify such a placement? What differences between the plants and these divisions justify their placement in kingdom Protista? Which of these similarities and differences are most likely to be homologous and which analogous?

4. Describe three different pathways to multicellularity, as exemplified by organisms discussed in this chapter.

5. In the previous chapter, the fruiting bodies of the myxobacteria were described as "multicellularity by committee." To which protists could such a description also apply? Explain your answer.

6. Distinguish among the following: colonial organism/multicellular organism; syngamy/fertilization; isogamy/anisogamy/oogamy; isomorphic/heteromorphic; sporophyte/gametophyte; spore/gamete/zygospore.

7. Depending on the stage in the life cycle of an organism at which meiosis occurs, it may result in one of three types of haploid cells. All three types are found among the various protists.

Identify the three types of haploid meiotic products, and for each give an example of a protist in which it is formed.

8. Explain alternation of generations, using as your example the sea lettuce, *Ulva*.

9. Name the five phyla of unicellular heterotrophic protists, and give the distinguishing characteristics of each.

10. Label the drawing below.

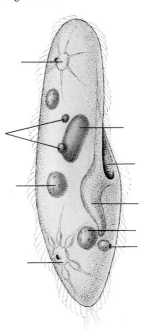

11. Consider the life cycle of *Plasmodium*. At what stages in the cycle do its numbers increase? Why might a parasite that requires several hosts find it advantageous to evolve a life cycle in which its numbers increase at several stages? Why might it be advantageous to a parasite to have a second host, such as the mosquito?

12. Early work on malaria vaccines has suggested that the *Plasmodium* life cycle may be most vulnerable to disruption at the stage of gamete fusion, which occurs in the *Anopheles* mosquito. A synthetic vaccine, based on surface proteins of the gametes, triggers the production of circulating antibodies in the blood of experimental animals with malaria. When a mosquito feeds on these animals, it ingests both *Plasmodium* gametes and the antibodies. In the mosquito, the antibodies cause the male gametes to clump together, preventing fusion of gametes to form zygotes. However, medical scientists involved in malaria research feel that a vaccine that had *only* this effect would be unethical, and they are also seeking vaccine components that will disrupt the life cycle at the sporozoite and merozoite stages as well. Why?

13. Given that light is essential for photosynthesis, it is easy to understand the adaptive significance of behavioral responses to light by photosynthetic protists. The significance of such responses for nonphotosynthetic organisms is not so obvious. What might be the advantages to an amoeba of a capacity to respond to light?

14. Arrange in order of evolutionary development: shell or skeleton, phagocytosis, active transport, multicellularity, symbiosis. (Like the rest of us, you will only be guessing, but be prepared to defend your guess.)

C H A P T E R 23

The Fungi

As we saw in the last chapter, the eukaryotic cell, by virtue of its size and complexity, has a number of properties that made possible the great diversification of the protists in both structure and mode of life. These properties include the capacity to carry a great deal of genetic information and to transmit it reliably from generation to generation; the compartmentalization and specialization of different parts of the cell for different functions, leading to greater efficiency; the ability to acquire more food; and greater adaptability to life-threatening environmental change.

There are, however, limits to the size that a single cell can attain and still function efficiently. One critical factor is the surface-to-volume ratio (page 102); the larger the cell, the greater the quantity of materials that must be moved in and out, but this movement can occur only through the cell surface. Another critical factor is the capacity of the nucleus to regulate a large amount of cytoplasm and the diverse functions of a complex cell. One solution to these problems is multicellularity: the repetition of individual units—cells—each with an efficient surface-to-volume ratio and each with its own nucleus. Another solution is the flattening or spreading out of the cells and the presence of multiple nuclei in a common cytoplasm, exemplified most strikingly in the fungi (Figure 23–1). Unlike true multicellularity, this solution does not open the way for great diversification of form, but it is nonetheless an evolutionary success. Fungi, like bacteria, are literally everywhere; thus far about 100,000 distinct species have been identified, and it is estimated that another 200,000 species await discovery.

23–1 A fungus growing on a fallen tree trunk. The bulk of its body consists of masses of thin filaments through which nutrients are absorbed. In many fungi, the cells are incompletely separated by perforated cell walls; both cytoplasm and nuclei flow through the filaments.

23-2 *The cell walls of fungi contain the polysaccharide chitin (a) rather than cellulose (b), the polysaccharide found in the cell walls of plants. Chitin resembles cellulose in that it is tough, inflexible, and insoluble in water. As you can see, the two molecules are also structurally very similar; in chitin, the hydroxyl (OH) group on the carbon atom in position 2 of each glucose subunit is replaced by a nitrogen-containing group.*

CHARACTERISTICS OF THE FUNGI

The fungi are so unlike any other organisms that, although they were long classified with the plants, biologists now assign them to a separate kingdom. Although some fungi, including the yeasts, are unicellular, most species are composed of masses of coenocytic or multicellular filaments. A fungal filament is called a hypha, and all the hyphae of a single organism are collectively called a **mycelium**. The walls of the hyphae are composed primarily of chitin (Figure 23-2), a polysaccharide that is never found in plants. (It is, however, the principal component of the exoskeleton—the hard outer covering—of insects and other arthropods.) The visible structures of most fungi represent only a small portion of the organism; these structures, such as mushrooms, are tightly packed hyphae, specialized for the production of spores.

A mycelium normally arises by the germination and outgrowth of a single spore, with growth taking place only at the tips of hyphae. All fungi are nonmotile throughout their life cycle, although spores may be carried great distances by the wind. Growth of the mycelium substitutes for motility, bringing the organism into contact with new food sources and different mating strains. This growth can be quite rapid—some fungi can grow a mass of new hyphae in 24 hours that, if placed end-to-end, would total more than a kilometer in length.

All fungi are heterotrophs, either saprobes or parasites. Because of their filamentous form, each fungal cell is no more than a few micrometers from the soil, water, or other substance in which the fungus lives, and is separated from it by only a thin cell wall. Because their cell walls are rigid, fungi are unable to engulf small microorganisms or other particles. They obtain food by absorbing dissolved inorganic or organic materials. Typically a fungus will secrete digestive enzymes onto a food source and then absorb the smaller molecules that are released. The mycelium may appear as a mass on the surface of the food source or may be hidden beneath the surface. Parasitic fungi often have specialized hyphae, called **haustoria** (singular, haustorium), that absorb nourishment directly from the cells of the host organism. (As we shall see on page 691, some parasitic plants have analogous structures, also known as haustoria.)

23-3 *Gill fungi on the trunk of a dead tree in southern Ontario. The only portions of these fungi visible are the spore-producing structures, composed of tightly packed hyphae; the bulk of the mycelium is below the surface of the dead trunk. This fungus,* Flammulina velutipe, *is commonly known as velvet foot or as winter mushroom.*

The fungi, together with the bacteria, are the principal decomposers of organic matter. It is estimated that the top 20 centimeters of fertile soil contain, on the average, nearly 5 metric tons of fungi and bacteria per hectare (2.47 acres). As we shall see in Section 8, the activities of these organisms are as vital to the continued function of the earth's ecosystems as are those of the food producers. Some fungi are, from the human standpoint, destructive, attacking our foodstuffs, our domestic plants and animals, our shelter, our clothing, and even our persons. Others, however, are essential for the production of bread, cheese, and wine. Moreover, fungi are the source of a number of antibiotics and other life-saving medications.

23-4 *A sporangium, an asexual reproductive structure of a fungus, from the black bread mold* Rhizopus. *The contents of the sporangium are segregated from the rest of the mycelium by a cell membrane and a cell wall.*

Reproduction in the Fungi

Most fungi reproduce both asexually and sexually. Asexual reproduction takes place either by the fragmentation of the hyphae (with each fragment becoming a new individual) or by the production of spores. In some of the fungi, spores are produced in sporangia, which are borne on specialized hyphae called **sporangiophores**. Fungal spores are often, but not necessarily, resting forms, surrounded by a tough, resistant wall. Like the spores of other organisms, these spores are able to survive during periods of drought or extreme temperatures. The airborne spores of some fungi are very small and therefore can remain suspended in the air for long periods and be widely dispersed. Often the sporangia are raised above the mycelium by the sporangiophores; thus the spores are easily caught up and transported by air currents. The bright colors and powdery textures associated with many types of molds are the colors and textures of the spores and sporangia.

Sexual reproduction in many fungi involves the specialization of portions of the hyphae to form gametangia (page 467). The contents of a gametangium, like those of a sporangium, are separated from the hypha from which it is formed by a cell membrane and a complete cell wall, known as a **septum**. Sexual reproduction can occur in a variety of ways: (1) by fusion of gametes that have been released from the gametangia, (2) by fusion of gametangia, or (3) by fusion of unspecialized hyphae.

Sometimes the fusion of fungal hyphae is not followed immediately by the fusion of nuclei. Thus strains of fungi may exist with two or more genetically distinct kinds of nuclei operating simultaneously. When such a combination contains two nuclei of complementary mating types, it is known as a **dikaryon**. Dikaryons are found uniquely among the fungi.

CLASSIFICATION OF THE FUNGI

The members of the kingdom Fungi are generally classified into three principal divisions: Zygomycota, the zygomycetes; Ascomycota, the ascomycetes; and Basidiomycota, the basidiomycetes. The criteria used in distinguishing these three divisions involve both features of basic structure and patterns of reproduction, particularly sexual reproduction (Table 23–1). An additional taxon, Deuteromycota, or the Fungi Imperfecti, includes fungi in which sexual reproduction is unknown, either because it has been lost in the course of evolution or because it has not been observed. Also included in this taxon for convenience are certain

TABLE 23-1 **The Kingdom Fungi**

DIVISION	NUMBER OF SPECIES	EXAMPLES	DISTINCTIVE CHARACTERISTICS	DISEASES	ECONOMIC USES
Zygomycota	About 600	Black bread mold	Formation of zygospores (tough, resistant spores resulting from a fusion of gametangia)	Few	None
Ascomycota	30,000	*Neurospora*, yeasts, morels, truffles	Formation of fine asexual spores (conidia); sexual spores in asci; hyphae divided by perforated septa; dikaryons	Powdery mildews of fruits, chestnut blight, Dutch elm disease, ergot	Food (morels, truffles); wine-, beer-, and bread-making (yeasts)
Basidiomycota	25,000	Toadstools, mushrooms, rusts, smuts	Sexual spores in basidia; hyphae divided by perforated septa; dikaryons	Rusts, smuts	Food (mushrooms)
Deuteromycota (Fungi Imperfecti)	25,000	*Penicillium*	Fungi with no known sexual cycles	Ringworm, thrush	Cheeses, antibiotics

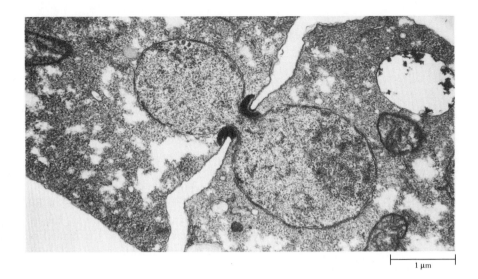

23-5 *The perforated cross wall of an ascomycete showing a nucleus squeezing through the perforation. The fungus is* Neurospora crassa, *the red bread mold.*

other closely related fungi whose sexual stages are known. Taxonomists refer to groups like this as "waste baskets" because species are included here only because they do not fit in other taxa. Because of similarities between the patterns of asexual reproduction in many of the Fungi Imperfecti and the ascomycetes, this taxon is sometimes considered a class of division Ascomycota; most authorities, however, rank it at the level of division.

The ancestors of the fungi were probably unicellular eukaryotic organisms that apparently have no living counterparts. These organisms are thought to have given rise to three distinct lineages, one leading to the modern chytrids (page 467), a second leading to the oomycetes (page 467), and a third leading to the zygomycetes. All three of these groups are characterized by a coenocytic organization, and with the exception of the reproductive structures, there is no compartmentalization of the hyphae. However, the chytrids and the oomycetes differ so significantly from the fungi in a number of basic features that most authorities believe that they should be classified in kingdom Protista, a recommendation that we have followed in this text.

In both the ascomycetes and the basidiomycetes, the hyphae are septate—divided by transverse cell walls—but the walls are perforated, and the cytoplasm and even the nuclei (Figure 23-5) are able to flow through the septa. It is generally thought that the ascomycetes and the basidiomycetes are derived from a common ancestor, and that this ancestor and the zygomycetes evolved from an earlier common ancestor.

DIVISION ZYGOMYCOTA

The zygomycetes are terrestrial fungi, most of which are saprobes, living in the soil and feeding on dead plant or animal matter. Some are parasites of plants, insects, or small soil animals. Their sexual reproduction is characterized by the formation of zygospores, which, as we saw in the last chapter, are thick-walled, resistant spores that develop from a zygote.

One of the most common members of this division is *Rhizopus stolonifer*, the black bread mold. Infection begins when a spore germinates on the surface of bread, fruit, or some other organic matter and forms hyphae. Some of the hyphae extend rhizoids, which anchor the fungus to the substrate, secrete digestive enzymes, and absorb dissolved organic materials. Other specialized hyphae, the sporangiophores, push up into the air, and sporangia form at their tips. As the sporangia mature, they become black, giving the mold its characteristic color.

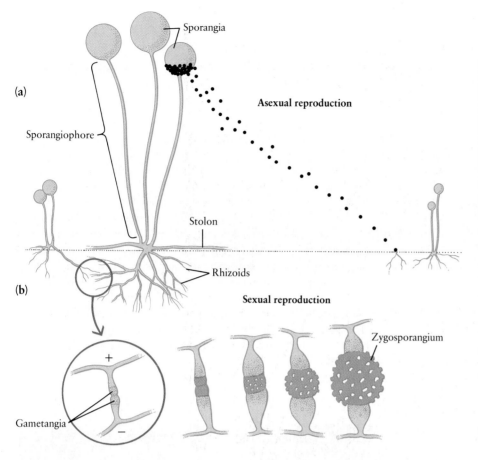
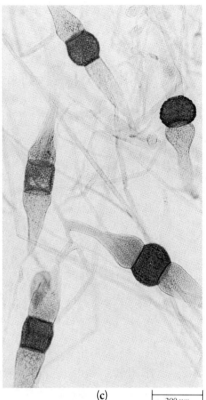

23-6 *Asexual and sexual reproduction in a black bread mold. The mycelium consists of branched hyphae, including rhizoids, which anchor the organism and absorb nutrients; stolons, which run above the surface of the bread; and sporangiophores, which elevate the sporangia.* **(a)** *At maturity, the fragile wall of the sporangium disintegrates, releasing the asexual spores, which are carried away by air currents. Under suitable conditions of warmth and moisture, the spores germinate, giving rise to new masses of hyphae.* **(b)** *Sexual reproduction occurs when two hyphae from different mating strains come together, forming gametangia. After fusion, the gametangia develop into a thick-walled, resistant structure, the zygosporangium, which contains a number of zygotes. After a period of dormancy, the zygotes undergo meiosis, and the zygosporangium germinates, producing a new sporangium from which haploid spores are released.*

(c) *Zygosporangia of the black bread mold* Rhizopus stolonifer.

They eventually break open, releasing numerous airborne spores, each of which can germinate to produce a new mycelium.

Sexual reproduction in *Rhizopus* occurs when the specialized hyphae of two different mating strains meet and fuse, attracted toward one another by hormones that diffuse in the form of gases. The two strains are designated + and −, since there are no structural differences between them on which to base male and female designations. Septa, or cross walls, form behind the tips of the touching hyphae; the two tip cells thus formed are gametangia, one containing numerous + nuclei, the other containing numerous − nuclei. The two gametangia fuse, and then many pairs of + and − nuclei fuse, producing diploid nuclei; any unfused haploid nuclei degenerate. The resulting multinucleate cell, containing a number of zygotes, forms a hard, warty wall and becomes a dormant zygosporangium. During this dormant stage, the zygosporangium can survive periods of extreme heat or cold or desiccation. At the end of dormancy, just prior to germination, the diploid nuclei undergo meiosis. Half of the haploid nuclei resulting from meiosis are + nuclei, and half are − nuclei. Following meiosis, the zygosporangium germinates, producing a sporangium from which both + and − airborne spores are released.

Ready, Aim, Fire!

Over the millennia, the fungi have evolved a variety of methods that ensure wide dispersal of their spores. One of the most ingenious is found in *Pilobolus,* a zygomycete that grows on dung. The sporangiophores of this fungus, which attain a height of 5 to 10 millimeters, are positively phototropic—that is, they grow toward light. An expanded region of the sporangiophore located just below the sporangium (known appropriately as the subsporangial swelling) functions as a lens, focusing the rays of the sun on a photoreceptive area at its base. The region of the sporangiophore *away* from the focused light grows more rapidly than other regions, causing the sporangiophore to curve toward the light.

A vacuole in the subsporangial swelling contains a high concentration of solutes, with the result that water moves in by osmosis. Eventually the pressure becomes so great that the swelling splits, firing the intact sporangium in the direction of the light. The initial velocity approaches 50 kilometers per hour, and the sporangia often travel farther than 2 meters—an enormous distance, when you consider that mature sporangia are only 80 micrometers in diameter. Each sporangium adheres where it lands, and if it should land on a blade of grass—a reasonable possibility, since it was fired toward the light—it may be eaten by a grazing animal. It then passes through the digestive tract of the animal unharmed and is deposited in the dung, where the cycle begins anew.

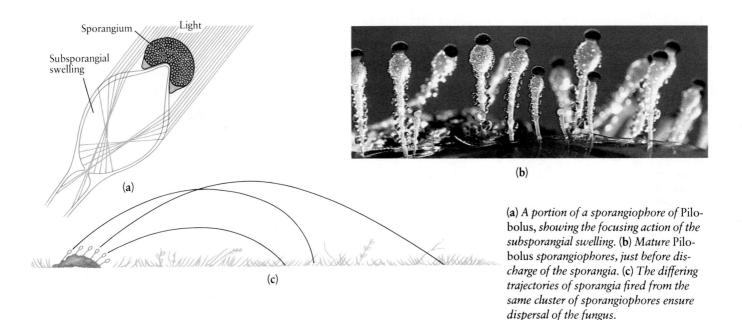

(a) *A portion of a sporangiophore of* Pilobolus, *showing the focusing action of the subsporangial swelling.* (b) *Mature* Pilobolus *sporangiophores, just before discharge of the sporangia.* (c) *The differing trajectories of sporangia fired from the same cluster of sporangiophores ensure dispersal of the fungus.*

DIVISION ASCOMYCOTA

The ascomycetes are the largest division of fungi, with some 30,000 species, plus an additional 25,000 species that are found only in lichens (page 488). Among the ascomycetes are the yeasts and powdery mildews, many of the common black and blue-green molds, and the morels and truffles prized by gourmets. Members of this group of fungi are the cause of many plant diseases, such as chestnut blight and Dutch elm disease, and are the source of many of the antibiotics. The red bread mold *Neurospora,* which played a major role in the history of modern genetics, is an ascomycete.

23–7 *Two ascomycetes.* **(a)** *A common morel,* Morchella esculenta. *These (and the truffles) are among the most prized of the edible fungi. The structure recognized as the morel is the ascocarp, in which asci and ascospores are produced.* **(b)** *Scarlet cup,* Sarcoscypha coccinea, *a harbinger of spring in hardwood forests throughout the United States. It is usually found arising from a fallen branch.*

(a)   (b)

In ascomycetes, as we noted earlier, the hyphae are divided by cross walls, or septa. Each compartment generally contains a separate nucleus, but the septa have pores in them through which the cytoplasm and the nuclei can move (see Figure 23–5). The life cycle of an ascomycete (Figure 23–8) typically includes both asexual and sexual reproduction. Asexual spores are commonly formed either singly or in chains at the tip of a specialized hypha. They are characteristically very fine and so are often called **conidia,** from the Greek word for "dust."

Sexual reproduction in the ascomycetes always involves the formation of an **ascus** ("little sac"), a structure that characterizes this division (Figure 23–9a).

23–8 *Life cycle of an ascomycete. An ascospore (upper left) germinates to produce a haploid monokaryotic mycelium, which reproduces through the formation of asexual spores (conidia). When monokaryotic mycelia of different mating strains form gametangia, the stage is set for sexual reproduction. A bridge forms between female (color) and male (black) gametangia, allowing the haploid male nuclei to enter the female gametangium. The hyphae that proliferate from this gametangium are dikaryotic—that is, each cell contains a pair of haploid nuclei, one of each parental type. These dikaryotic hyphae, with interspersed monokaryotic hyphae, give rise to the ascocarp (bottom). In the ascocarp, the dikaryotic hyphae grow and differentiate to form the asci, within which the haploid nuclei fuse. The resulting diploid nucleus undergoes meiosis, producing four new haploid nuclei. These nuclei then divide mitotically, and the mature ascus thus contains eight haploid ascospores. With the release and germination of the ascospores, the cycle begins once more.*

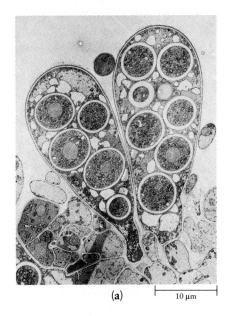

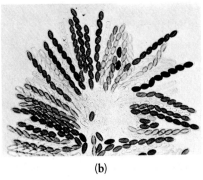

23-9 (a) *Electron micrograph of two asci in which ascospores are maturing. The closed ascus within which the sexually produced spores develop is the "trademark" of the ascomycete.* (b) *Asci of the red bread mold* Neurospora. *Each ascus contains eight haploid ascospores, lined up in the order in which they were produced by meiosis and the subsequent mitotic division.*

Depending on the species, ascus formation is preceded by the fusion of gametes, gametangia, or unspecialized hyphae of different mating strains. The nuclei form pairs—dikaryons—that divide synchronously as the hypha grows. Eventually some of the nuclei fuse; this is the only truly diploid stage in the life cycle. The diploid nuclei immediately undergo meiosis, producing four haploid nuclei, which then usually divide mitotically, producing eight haploid nuclei (Figure 23-9b). Each of these nuclei becomes surrounded by a tough wall; a mature ascus contains eight of these spores (ascospores). In most ascomycetes, the asci are formed in complex structures called ascocarps. At maturity the asci become turgid and finally burst, releasing their ascospores explosively into the air.

Single-celled ascomycetes are known as yeasts. Many yeasts are adapted to environments with high sugar content, such as the nectar of flowers or the surface of fruits and, as we noted in Chapter 9, they are responsible for the fermentation of fruit juice to wine. Yeasts are characteristically small, oval cells that reproduce asexually by budding. Sexual reproduction in yeasts occurs when two cells (or two ascospores) unite and form a zygote. The zygote may produce diploid buds or may undergo meiosis to produce four haploid nuclei. There may be a subsequent mitotic division. Within the zygote wall, which is now an ascus, walls are laid down around the haploid nuclei, forming ascospores. The ascospores are liberated when the ascus wall breaks down.

Many ascomycetes are plant parasites. Ergot, for instance, one of the most famous fungus-produced diseases, is caused by *Claviceps purpurea,* a parasite of rye. Although ergot seldom causes serious damage to a crop of rye, it is dangerous because a small amount mixed with rye grains is enough to cause severe illness among domestic animals that eat the grain or among people who eat bread made with the flour. Ergotism is often accompanied by gangrene, nervous spasms, psychotic delusions, and convulsions. It occurred frequently during the Middle Ages, when it was known as St. Anthony's fire. In one epidemic in the year 994, more than 40,000 people died. Ergot, which causes muscles to contract and blood vessels to constrict, has various medical uses. It is also the initial source for the psychedelic drug lysergic acid diethylamide (LSD).

DIVISION BASIDIOMYCOTA

The most familiar basidiomycetes are mushrooms. The mushroom, or basidiocarp, which is the spore-producing body, is composed of masses of tightly packed hyphae. The mycelium from which the mushrooms are produced forms a diffuse mat, which may grow as large as 35 meters in diameter. Mushrooms usually form at the outer edges, where the mycelium grows most actively, since this is the area in which there is the most nutritive material. As a consequence, the mushrooms appear in rings, and as the mycelium grows, the rings become larger and larger in diameter. Such circles of mushrooms, which might appear in a meadow overnight, were known in European folk legends as "fairy rings." They can make such a rapid appearance because most of the new protoplasm is produced underground, in the mycelium. The protoplasm then streams into the new hyphae of the fruiting body as it forms above ground.

The basidiomycetes, like the ascomycetes, have hyphae subdivided by perforated septa. Sexual reproduction is initiated by the fusion of haploid hyphae to form a dikaryotic mycelium (Figure 23-10). The dikaryotic mycelium may persist for years, forming an elaborate structure. Eventually, some of the nuclei fuse to form diploid nuclei that immediately undergo meiosis. Fusion and meiosis always take place in a specialized hypha called a **basidium** (from the Greek word for "club"). The spores (basidiospores) are formed externally on the basidium. Many of the larger basidiomycetes seem to have lost the capacity to produce asexual spores.

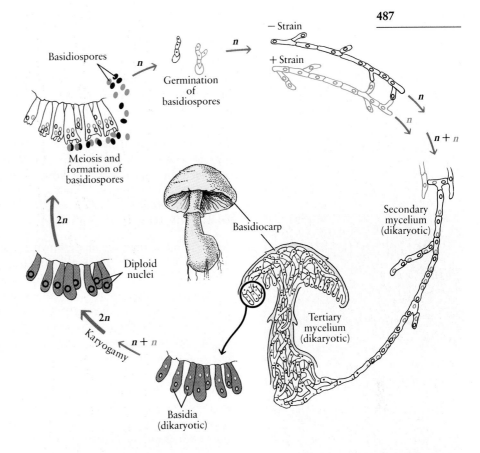

23–10 *Life cycle of a basidiomycete. Basidiospores (upper left) germinate to produce primary monokaryotic mycelia. Secondary dikaryotic mycelia are formed by the fusion of hyphae from different mating types. The secondary mycelia grow and differentiate to form the reproductive structures (basidia). In mushrooms, the basidia form within the gills. After the basidium enlarges, the two nuclei, one from each mating strain, fuse. Meiosis follows almost immediately, resulting in the formation of four nuclei, from each of which a basidiospore develops. After the basidiospores are released, the basidiocarp disintegrates.*

23–11 *Three basidiomycetes.* **(a)** *Corn smut, a common fungal disease of corn. The black, dusty-looking masses are spores.* **(b)** *A shelf fungus, which grows on decaying wood.* **(c)** *Amanita bisporigera. Members of the genus* Amanita *include the most beautiful and also the most poisonous of the mushrooms. The "skirt" near the top of the stalk is one of the identifying characteristics of this genus. One mushroom has been picked to show the gills, on which the sexual spores are formed. The toxin in* Amanita bisporigera *consists of two linked cyclopeptides, each containing eight amino acids. This protein binds to an RNA polymerase in liver cells and causes acute liver damage, resulting in death.*

The best-known mushrooms belong to the group known as the gill fungi. The spores of these fungi are found in the furrows, or "gills," under the cap. If you cut off the cap of a mature mushroom and place it on a piece of white paper, it will release fine spores that trace out a negative copy of the gill structure. The spores, which come in a wide range of colors, are a useful means of identifying various mushrooms. Varieties of the gill fungus *Agaricus campestris*, the common field mushroom, are among the most familiar of the mushrooms commercially cultivated in North America. Most of the known poisonous mushrooms are also gill fungi. Mushrooms of the genus *Amanita* are the most highly poisonous of all mushrooms; even one bite of the white *Amanita bisporigera*, a "destroying angel," can be fatal. Some species of toxic mushrooms, such as *Psilocybe mexicana* (the source of psilocybin), are eaten for their hallucinogenic effects.

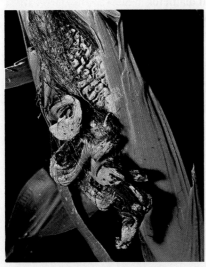
(a)

(b)

(c)

Other types of basidiomycetes include puffballs (a few of which are a meter in diameter), earthstars, stinkhorns, jelly fungi, and the parasitic rusts and smuts, some of which cause severe losses among cereal crops. Another basidiomycete, white rot fungus, a voracious destroyer of wood, is currently being investigated for its potential in destroying toxic wastes. This fungus contains a complex enzyme that is able to degrade not only the tough polymers found in wood but also such substances as DDT, dioxin, and a wide range of organic pollutants.

DIVISION DEUTEROMYCOTA

As we noted earlier, the deuteromycetes, or Fungi Imperfecti, are generally fungi in which sexual reproduction is unknown. About 25,000 species of deuteromycetes have been described, among them parasites that cause diseases of plants and animals. The most common human diseases caused by this group are infections of the skin and mucous membranes, such as ringworm (including "athlete's foot") and thrush (to which infants are particularly susceptible). A few deuteromycetes are of economic importance due to the part they play in the production of certain cheeses (Roquefort and Camembert, for example) and of antibiotics, including penicillin. Cyclosporin, a compound that suppresses the immune reactions involved in the rejection of organ transplants (to be discussed in Chapter 39), is synthesized by a soil-dwelling deuteromycete. Consisting of 13 amino acids, one of which appears to be synthesized only by this fungus, cyclosporin has made possible a dramatic increase in successful heart transplants.

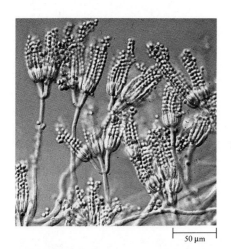

23-12 Penicillium, *an "imperfect fungus," showing conidiophores ("conidia-bearers"), which have formed at the tips of hyphae. Conidia are dust-fine asexual spores, characteristic of the ascomycetes.*

SYMBIOTIC RELATIONSHIPS OF FUNGI

Although most fungi are saprobes, living on dead organic matter, a large number of fungi are parasitic on plants and animals, causing a variety of diseases. Fungi are also involved in other types of symbioses. Two of these—lichens and mycorrhizae—have been and are of extraordinary importance in enabling photosynthetic organisms to become established in previously barren terrestrial environments.

The Lichens

A lichen is a combination of a specific fungus and a green alga or a cyanobacterium. The product of such a combination is very different from either the photosynthetic organism or the fungus growing alone, as are the physiological conditions under which the lichen can survive. The lichens are widespread in nature. They occur from arid desert regions to the Arctic and grow on bare soil, tree trunks, sun-baked rocks, fence posts, and windswept alpine peaks all over the world; they have even been found growing in air spaces within Antarctic rocks, a few millimeters below the frigid rock surface. Lichens are often the first colonists of bare rocky areas; their activities begin the process of soil formation, gradually creating an environment in which mosses, ferns, and other plants can gain a foothold.

Lichens do not need an organic food source, as do their component fungi, and unlike many free-living algae and cyanobacteria, they can remain alive even when desiccated. They require only light, air, and a few minerals. They apparently absorb some minerals from their substrate (this is suggested by the fact that particular species are characteristically found on specific kinds of rocks or soil or tree trunks), but minerals reach the lichen primarily through the air and in rainfall. Because lichens rapidly absorb substances from rainwater, they are particularly susceptible to airborne toxic compounds. Thus, the presence or absence of lichens is a sensitive index of air pollution.

Predaceous Fungi

Among the most highly specialized of the fungi are the predaceous fungi that have developed a number of mechanisms for capturing small animals that they use as food. Some secrete a sticky substance on the surface of their hyphae in which passing protists, rotifers, small insects, or other animals become glued. More than 50 species of Fungi Imperfecti capture small roundworms (nematodes) that abound in the soil. In the presence of a population of roundworms (or even of water in which the worms have been growing), the hyphae of the fungi produce loops that swell rapidly, closing the opening like a noose when a nematode rubs against its inner surface. The stimulation of the cell walls is thought to increase the amount of osmotically active material in the cells, causing water to enter and expand them rapidly.

(a) The predaceous imperfect fungus *Arthrobotrys dactyloides* has trapped a nematode. The traps consist of rings, each comprising three cells, which swell rapidly to about three times their original size and garrote the nematode. Once the worm has been trapped, fungal hyphae grow into its body and digest it. When triggered, the ring cells can expand completely in less than a tenth of a second. This species was appropriately called the "nefarious noose fungus" by the late W. H. Weston of Harvard University, who made vast contributions to our present knowledge of the fungi. (b) Another nematode-trapping fungus, *Dactylella drechsleri*. This species traps the worms with small adhesive knobs and was dubbed the "lethal lollipop fungus" by Weston.

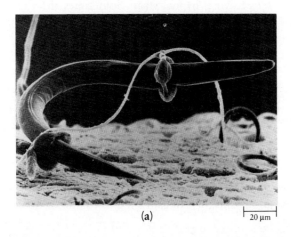
(a) 20 μm

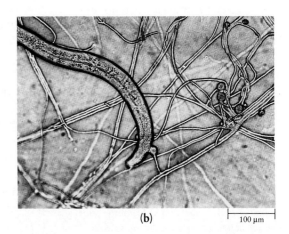
(b) 100 μm

Many of the algae and cyanobacteria found in lichens are also commonly found as free-living species. Lichen fungi are thought to have free-living hyphal stages, but these fungi can generally be detected and identified only after they have encountered a suitable photosynthetic organism and formed a lichen. For these reasons, lichens are generally named and classified according to the species of the fungal component. There are about 25,000 species of lichens, in almost all of which the fungus is an ascomycete. Photosynthetic organisms from some 26 different genera are found in symbiotic association with these fungi. The most frequent are the green algae *Trebouxia* and *Trentepohlia* and the cyanobacterium *Nostoc;* members of one of these three genera are found in about 90 percent of all lichens.

23-13 (a) *Crustose ("encrusting") lichen growing on bare rock in central California.* (b) *Foliose ("leafy") lichen growing on a tree in northern Ontario.* (c) *British soldier lichen* (Cladonia cristatella), *a fruticose ("shrubby") lichen. Each soldier (so called because of the scarlet color) is 1 to 2 centimeters tall.*

Lichens reproduce most commonly by the breaking off of fragments containing both fungal hyphae and photosynthetic cells. New individuals can also be formed by the capture of an appropriate alga or cyanobacterium by a lichen fungus in its free-living, hyphal stage. Sometimes the captured photosynthetic cells are destroyed by the fungus, in which case the fungus also dies. If the photosynthetic cells survive, a lichen is produced.

Mycorrhizae

Mycorrhizae ("fungus-roots") are symbiotic associations between fungi and the roots of vascular plants. The importance of mycorrhizae was first recognized in connection with efforts to grow orchids in greenhouses. Orchids have microscopic seeds that germinate to form a tiny pad of tissue called a protocorm. Cultivators of orchids found that the plants seldom developed beyond the protocorm stage unless they were infected with a particular kind of fungus. Subsequently it was found that if seedlings of many forest trees are grown in nutrient solutions and then transplanted to prairie or other grassland soils, they

23-14 *Scanning electron micrographs showing the establishment of the British soldier lichen* (Cladonia cristatella) *in sterile laboratory culture.* (a) *An early interaction between fungal and algal components of the lichen.* (b) *Penetration of an algal cell by a fungal haustorium (arrow).*

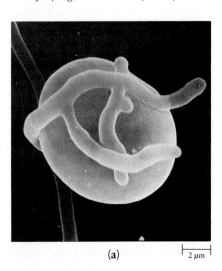 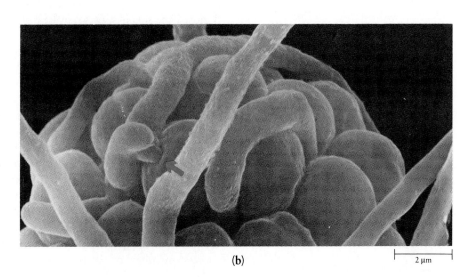

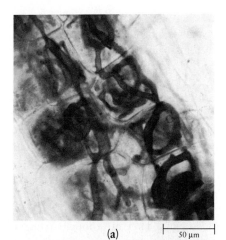

23-15 (a) *Surface view of an endomycorrhizal rootlet of fescue, a perennial grass, showing coiled hyphae in some of the cells. Mycorrhizal associations are especially important for grasses growing on nutrient-poor soils or at high elevations.* (b) *Ectomycorrhizal rootlets from a western hemlock. Hormones secreted by the fungus cause the root to branch in a special pattern. This growth pattern and the swollen hyphal sheath impart a characteristic appearance to the ectomycorrhizae. The narrow strands extending from the mycorrhizae are bundles of hyphae that function as extensions of the root system.*

fail to grow. Eventually they may die from malnutrition, even if analysis shows that there are abundant nutrients in the soil. If a small amount (0.1 percent by volume) of forest soil containing fungi is added to the soil around the roots of the seedlings, however, they will grow promptly and normally. Mycorrhizae are now thought to occur in more than 90 percent of all families of plants.

In some mycorrhizal associations, known as endomycorrhizae, the fungal hyphae penetrate the cells of the root, forming coils, swellings, or branches (Figure 23–15a). The hyphae also extend out into the surrounding soil. Endomycorrhizae occur in about 80 percent of all vascular plants, and the fungal component is usually a zygomycete. In other associations, known as ectomycorrhizae, the hyphae form a sheath around the root but do not actually penetrate its cells (Figure 23–15b). Ectomycorrhizae are characteristic of certain groups of trees and shrubs, including pines, beeches, and willows; the fungus is usually a basidiomycete, but some associations involve ascomycetes, including truffles.

The exact relationship between roots and fungi is not known. Apparently the roots secrete sugars, amino acids, and possibly some other organic substances that are used by the fungi. Although the evidence is just now accumulating, it appears that the fungi convert minerals in the soil and decaying material into an available form and transport them into the root. It has been shown experimentally that mycorrhizae transfer phosphorus from the soil into roots, and there is evidence that water uptake is facilitated by the fungi. One of the most intriguing recent observations is that, under certain circumstances, mycorrhizae appear to function as a bridge through which phosphorus, carbohydrates, and probably other substances pass from one host plant to another.

A study of the fossils of early vascular plants has revealed that mycorrhizae occurred as frequently in them as they do in modern vascular plants. This has led to the interesting suggestion that the evolution of mycorrhizal associations may have been the critical step allowing plants to make the transition to the bare and relatively sterile soils of the then-unoccupied land.

SUMMARY

Fungi are typically composed of masses of filaments called hyphae; the principal component of the hyphal walls is the polysaccharide chitin. Fungi are heterotrophs, deriving their nutrition by absorption of organic compounds digested extracellularly by secreted enzymes.

Fungi form both asexual and sexual spores, although not all fungi form both kinds. The asexual spores may be formed in sporangia. The sexual cycle is initiated by the fusion of hyphae of different mating strains. In some fungi, the nuclei in the fused hyphae combine immediately, and a zygote is formed. In others, the two genetically distinct nuclei usually remain separate, forming pairs—known as dikaryons—that divide synchronously, sometimes over prolonged periods. Once the nuclei fuse, meiosis always follows immediately.

The kingdom Fungi includes three principal divisions—Zygomycota, Ascomycota, and Basidiomycota—as well as another major taxon, Deuteromycota, or Fungi Imperfecti, which is usually ranked at the level of division. The zygomycetes, which are thought to share a common ancestor with the lineage leading to the ascomycetes and basidiomycetes, are characterized by thick-walled zygospores. The hyphae of both ascomycetes and basidiomycetes are compartmentalized by perforated septa. In the ascomycetes, the sexual spores develop within an ascus (sac), and in the basidiomycetes, the sexual spores develop on a basidium (club). Most of the Deuteromycota have no known sexual cycle.

Fungi have an important ecological role as decomposers of organic material. They are also parasitic on many types of organisms, particularly plants, in which they often cause serious disease. Fungi participate in two additional types of symbioses that are of ecological significance—lichens and mycorrhizae.

Lichens are combinations of fungi and green algae or cyanobacteria that are structurally and physiologically different from either organism as it exists separately. They are able to survive under adverse environmental conditions where neither partner could exist independently. The lichen represents a symbiotic relationship in which a fungus encloses photosynthetic cells and is dependent upon them for nourishment.

Mycorrhizae ("fungus-roots") are associations between soil-dwelling fungi and plant roots. There are two principal types: endomycorrhizae and ectomycorrhizae. Mycorrhizal associations facilitate the uptake of minerals by the roots of the plant and provide organic molecules for the fungus. They are thought to have played a key role in enabling plants to make the transition to land.

QUESTIONS

1. Distinguish between the following: hypha/mycelium; chitin/cellulose; sporangia/gametangia; conidia/ascospores; ascus/basidium; monokaryotic/dikaryotic; endomycorrhizae/ectomycorrhizae.

2. In a multicellular organism, all of the individual cells must be supplied with food and water. What evolutionary solution to this problem is exemplified by the fungi? How have the fungi, which have no motile cells at any stage of the life cycle, solved the problem of obtaining new food supplies when they have exhausted a particular source?

3. Give the distinguishing characteristics of the Zygomycota, Ascomycota, and Basidiomycota.

4. Some authorities believe that the Chytridiomycota and Oomycota should be placed in kingdom Fungi, while others believe they are better placed in kingdom Protista. What characteristics do the members of these two divisions share with the fungi? In what ways do the chytrids and oomycetes differ from the fungi?

5. As you can see from the last two chapters, our method of classification into kingdoms does not produce entirely satisfactory results. What are some of the difficulties unresolved (or even created) by this system? Can you suggest alternative solutions?

6. Most of the fungi that have no sexual reproduction cannot currently be classified with their sexually-reproducing relatives. Is this likely to change in the future? Why or why not?

7. Coenocytic organisms, such as the zygomycetes, show little differentiation. When differentiation does occur, as in gametangium formation, it is preceded by construction of a septum. In your opinion, why?

8. Most fungi can reproduce both asexually and sexually. What are the advantages and disadvantages of each type of reproduction?

9. What type of symbiotic relationship would you say exists between the fungus and the alga or cyanobacterium of a lichen? Between the fungus and the plant roots of a mycorrhizal association?

10. How are fungi thought to have aided photosynthetic organisms in the transition to land?

CHAPTER 24

The Plants

24-1 *Although plants are primarily adapted for life on land, some, such as the water lily,* Nymphaea odorata, *have returned to an aquatic existence. Like whales and dolphins,* Nymphaea *retains the traces of its ancestors' terrestrial sojourn. These include a water-resistant cuticle, stomata (openings through which gas exchange occurs), and a highly developed internal transport system.*

Plants are, quite simply, multicellular photosynthetic organisms primarily adapted for terrestrial life. Their characteristics are best understood in terms of the transition from water to land, an event that occurred some 500 million years ago. The land offered a wealth of advantages to photosynthetic organisms. On land, light is abundant from daybreak to dusk and is not blocked by turbulent water. Carbon dioxide, needed for photosynthesis, is plentiful in the atmosphere and circulates more freely in air than in water. And, most important, the land was then unoccupied by competing forms of life.

Terrestrial life, however, presented photosynthetic organisms with a new and major difficulty, that of obtaining and retaining adequate amounts of water. The solutions to this problem that gradually evolved in the plants depended on multicellularity, which made possible specialization on a broad scale. As we saw in Chapter 18, the selective activation or inactivation of particular genes during development results in cells or groups of cells specialized for particular functions. Specialization brings with it the potential for a vast increase in the size of the organism, which, in turn, creates additional problems to be solved: the organism must synthesize increased amounts of food to supply its numerous cells with energy; larger quantities of materials must be moved into and out of the organism and transported to the individual cells; a larger body requires more physical support; the activities of all of the cells must be integrated for the smooth functioning of the entire organism; and a greater length of time is required for the full development of the organism, often requiring protection and nourishment of the immature stages. The story of plant evolution represents a series of natural experiments at solving these problems; the experiments have been enormously successful, producing a diversity of multicellular photosynthetic organisms that, collectively, have been able to occupy most of the earth's land surface for at least 300 million years.

THE ANCESTRAL ALGA

As we noted in Chapter 22, all plants appear to have arisen from the green algae (division Chlorophyta). Like the plants, the green algae contain chlorophylls *a* and *b* and beta-carotene as their photosynthetic pigments, and they accumulate their food reserves in the form of starch. In plants and green algae—but in no other organisms—the starch is stored in plastids (page 120), rather than in the cytoplasm. Beyond these similarities, however, the green algae exhibit a great diversity of characteristics, some of which are shared with the plants and some of which are not. The constellation of characteristics that could have given rise to the plants is found among contemporary green algae only in some members of class Charophyceae, most strikingly in the genus *Coleochaete* (Figure 24-2).

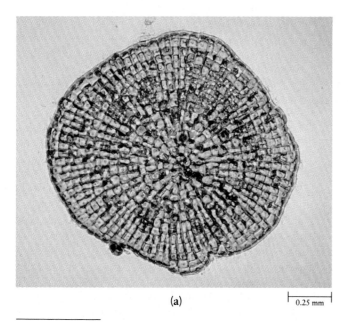

(a)   ⊢0.25 mm⊣   (b)

24-2 (a) Coleochaete, *a green alga of class Charophyceae that grows on the surface of submerged freshwater plants in shallow water. This multicellular alga takes the form of a disk, generally one cell thick.* (b) *Fossils of* Parka decipiens, *an extinct green alga, dated at about 380 million years ago.* Parka decipiens *closely resembles* Coleochaete *in its shape, tissue structure, and chemistry. It has many features suggesting that it may be closely related to the alga that gave rise to the plants.*

Although *Coleochaete* itself does not seem to have been the alga from which the plants evolved, it is thought to be closely related to it. Several sets of data lead to this conclusion. The cells of *Coleochaete*, like those of plants, have cellulose in their walls and contain peroxisomes (page 119), in which the key enzyme involved in photorespiration (page 225) is synthesized. Additional evidence is found in its pattern of cytokinesis. In almost all other organisms, including most green algae, division of the cytoplasm takes place by constriction and pinching off of the cell membrane. In plants and in *Coleochaete*, the cytoplasm is divided by the formation of a cell plate at the equator of the spindle. Similarities are also found in the pattern of microtubules underlying the flagella in *Coleochaete* and in those plant cells that have flagella.

One important characteristic shared by all plants, but absent in *Coleochaete*, is a well-defined alternation of generations. As we saw in Chapter 22, this type of life cycle is found not only in many multicellular green algae but also in the red algae and the brown algae, which biochemical evidence indicates originated from totally different symbiotic events than the green algae. This suggests that alternation of generations has arisen independently on several occasions, and *Coleochaete* provides a significant clue as to how it may have occurred. *Coleochaete*, like *Chlamydomonas* and a number of other green algae, is haploid for most of its life cycle. It is oogamous, producing clearly differentiated egg and sperm that fuse to form a zygote that subsequently undergoes meiosis, producing haploid cells from which new individuals develop. In *Coleochaete*, however, fusion of the gametes occurs not in the open waters but rather on the surface of the parent organism, and neighboring cells grow around the zygote, enclosing and protecting it (Figure 24–3). Prior to meiosis, additional rounds of DNA replication occur in the zygote,

24-3 *The large, dark cells in this micrograph of* Coleochaete *are the diploid zygotes. As you can see, they are protected by a layer of the smaller, haploid cells of the parent organism. The hair cells that extend outward from the disk are ensheathed at its base and are the source of the organism's name;* Coleochaete *means "sheathed hair." The hairs are thought to discourage aquatic animals from feeding on the alga.*

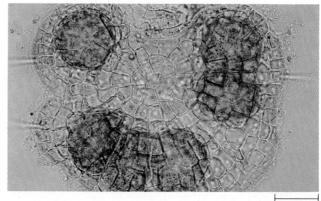

24–4 *The aboveground surfaces of plants are characteristically covered with a waxy cuticle that retards water loss. Epidermal pores permit gas exchange. The pore shown in this scanning electron micrograph is from* Marchantia, *a liverwort. Unlike the stomata of vascular plants (Figure 10–17, page 222), it does not open and close in response to environmental cues.*

with the result that the number of haploid cells ultimately released is not a mere 4, as in *Chlamydomonas*, but anywhere from 8 to 32. Only slight modifications of this life cycle—division of the diploid cells of the zygote by mitosis and cytokinesis *prior to* meiosis, followed by meiotic division of some, but not all, of the diploid cells—would be required to produce alternation of generations.

THE TRANSITION TO LAND

By the time the immediate ancestor of the plants moved from shallow waters onto the land, it had apparently evolved a well-defined alternation of heteromorphic generations. After the transition to land, new adaptations began to evolve in the life cycle and in other features as well. These adaptations were critical to the ultimate success of the plants on land and must have occurred early in their evolutionary history—most modern plants, even though very diverse, share them. One such characteristic, clearly associated with the transition to land, is the protective cuticle that covers the aboveground surfaces of plants and retards the loss of water from the plant body. The cuticle is formed of a waxy substance called cutin, secreted by the epidermal cells. Associated with the cuticle—in fact, made necessary by it—are pores through which the gas exchanges necessary for photosynthesis can take place (Figure 24–4).

Another adaptation was the development of multicellular reproductive organs (gametangia and sporangia) that were surrounded by a protective layer of sterile (nonreproductive) cells. The gametangia are called **archegonia** if they give rise to egg cells and **antheridia** if they give rise to sperm (Figure 24–5). A correlated adaptation was the retention of the fertilized egg (the zygote) within the female gametangium (the archegonium) and its development there into an embryo. Thus, during its critical early stages of development, the embryo, or young sporophyte, is protected by the tissues of the female gametophyte.

24–5 *Multicellular gametangia of the liverwort* Marchantia, *a member of division Bryophyta.* (a) *Female gametangia, or archegonia, in various stages of development. The archegonia are flask-shaped, with a single egg developing in the base of the flask.* (b) *Antheridium developing in the male gametophyte. The spermatogenous tissue will give rise to sperm cells, which, when mature, will swim to the egg through the neck canal of the archegonium.*

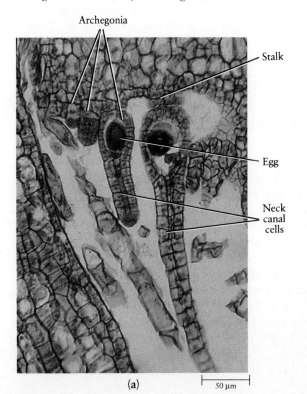

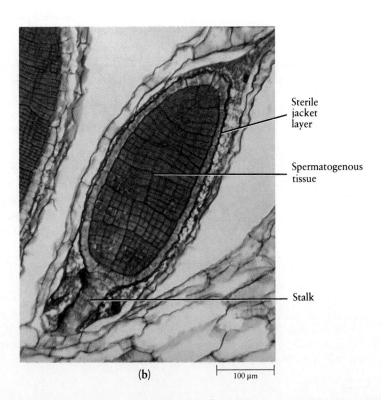

TABLE 24-1 **Major Physical and Biological Events in Geologic Time**

MILLIONS OF YEARS AGO	PERIOD	EPOCH	LIFE FORMS	CLIMATES AND MAJOR PHYSICAL EVENTS
Cenozoic Era				
0–2	Quaternary	Recent Pleistocene	Planetary spread of *Homo sapiens*; extinction of many large mammals and birds. Deserts on large scale.	Fluctuating cold to mild. Numerous glacial advances and retreats; uplift of Sierra Nevada.
2–5	Tertiary	Pliocene	Large carnivores. First known appearance of hominids (humanlike primates).	Cooler. Continued uplift and mountain building, with widespread glaciation in Northern Hemisphere. Uplift of Panama joins North and South America.
5–25		Miocene	Whales, apes, grazing mammals. Spread of grasslands as forests contract.	Moderate. Extensive glaciation begins again in Southern Hemisphere. Moderate uplift of Rockies.
25–38		Oligocene	Large, browsing mammals; monkey-like primates appear. Origin of many modern families of flowering plants.	Rise of Alps and Himalayas. Lands generally low. Volcanoes in Rockies. South America separates from Antarctica.
38–55		Eocene	Primitive horses, tiny camels, modern and giant types of birds. Formation of grasslands.	Mild to very tropical. Many lakes in western North America. Australia separates from Antarctica; India collides with Asia.
55–65		Paleocene	First known primitive primates and carnivores.	Mild to cool. Wide, shallow continental seas largely disappear.
Mesozoic Era				
65–144	Cretaceous		Extinction of dinosaurs at end of period. Marsupials, insectivores, and flowering plants become abundant.	Tropical to subtropical. Elevation of Rockies at end of period. Africa and South America separate.
144–213	Jurassic		Dinosaurs' zenith. Flying reptiles, small mammals. Birds appear. Gymnosperms, especially cycads, and ferns.	Mild. Continents low, with large areas covered by seas. Mountains rise from Alaska to Mexico.
213–248	Triassic		First dinosaurs. Primitive mammals appear. Forests of gymnosperms and ferns.	Continents mountainous and joined in one mass. Large areas arid. Eruptions in eastern North America. Appalachians uplifted and broken into basins.

Subsequent Diversification

Not long after the transition to land, the plants diverged into at least two separate lineages. One gave rise to the bryophytes, a group that includes the modern liverworts, hornworts, and mosses, and the other to the vascular plants, a group that includes all of the larger land plants. A principal difference between the bryophytes and the vascular plants is that the sporophytes of the latter, as their name implies, have a well-developed vascular system that transports water, minerals, sugars, and other nutrients throughout the plant body. The bryophytes first appear in the fossil record during the Devonian period, about 370 million years ago (see Table 24-1, which gives the major events in the evolution of plants and animals). These ancient fossils are quite similar to bryophytes living today. The oldest fossils of vascular plants are from the early part of the Silurian period, about 430 million years ago. The vascular plants, as we shall see, subsequently underwent a great diversification.

TABLE 24-1 **Major Physical and Biological Events in Geologic Time** *(Continued)*

MILLIONS OF YEARS AGO	PERIOD	EPOCH	LIFE FORMS	CLIMATES AND MAJOR PHYSICAL EVENTS
Paleozoic Era				
248–286	Permian		Reptiles diversify. Origin of conifers, cycads, and ginkgos; possible origin of flowering plants; earlier forest types wane.	Extensive glaciation in Southern Hemisphere. Seas drain from land; worldwide aridity. Appalachians formed by end of Paleozoic.
286–360	Carboniferous Pennsylvanian Mississippian		Age of amphibians. First reptiles. Variety of insects. Sharks abundant. Great swamps, forests of ferns, gymnosperms, and horsetails.	Warm; conditions like those in temperate or subtropical zones—little seasonal variation, water plentiful. Lands low, covered by shallow seas or great coal swamps. Mountain building in eastern U.S., Texas, Colorado.
360–408	Devonian		Age of fishes. Amphibians appear. Mollusks abundant. Lunged fishes. Extinction of primitive vascular plants. Origin of modern groups of vascular plants.	Europe mountainous with arid basins. Mountains and volcanoes in eastern U.S. and Canada. Rest of North America low and flat. Sea covers most of land.
408–438	Silurian		Rise of fishes and reef-building corals. Shell-forming sea animals abundant. Invasion of land by arthropods. Earliest vascular plants. Modern groups of algae and fungi.	Mild. Continents generally flat. Again flooded. Mountain building in Europe.
438–505	Ordovician		First primitive fishes. Invertebrates dominant. First fungi. Possible invasion of land by plants.	Mild. Shallow seas, continents low; sea covers U.S. Limestone deposits.
505–590	Cambrian		Shelled marine invertebrates. Explosive diversification of eukaryotic organisms.	Mild. Extensive seas, spilling over continents.
Precambrian Era				
590–4,500			Origin of life. Prokaryotes. Eukaryotic cells and multicellularity by close of era. Earliest known fossils, including soft-bodied marine invertebrates.	Dry and cold to warm and moist. Planet cools. Formation of earth's crust. Extensive mountain building. Shallow seas. Accumulation of free oxygen.

CLASSIFICATION OF THE PLANTS

According to the classification scheme we follow, the modern plants are placed in 10 separate divisions (Table 24-2). The liverworts, hornworts, and mosses of division Bryophyta are quite different from one another, and there is some question as to whether they represent three distinct lineages from the ancestral plant or subsequent branchings of one lineage from that ancestor. By contrast, each of the nine divisions of vascular plants is, by all available evidence, monophyletic—that is, all of its members are descended from a common ancestor.

The vascular plants are frequently grouped, for convenience, in ways that may or may not reflect evolutionary relationships. For instance, these plants, as a group, are often referred to as tracheophytes. They can be grouped into those without seeds (divisions Psilophyta, Lycophyta, Sphenophyta, and Pterophyta) and those with seeds. The seed plants also form two informal groups, the gymnosperms and the angiosperms. The gymnosperms are those with "naked," unprotected seeds (divisions Coniferophyta, Cycadophyta, Ginkgophyta, and Gnetophyta); the angiosperms, with enclosed, protected seeds, are, formally speaking, the Anthophyta, the flowering plants.

TABLE 24-2 **A Classification of Living Plants**

TAXON	COMMON NAME	NUMBER OF SPECIES
Division Bryophyta	Bryophytes	16,000
Class Hepaticae	Liverworts	6,000
Class Anthocerotae	Hornworts	100
Class Musci	Mosses	9,500
Division Psilophyta	Whisk ferns	Several
Division Lycophyta	Club mosses	1,000
Division Sphenophyta	Horsetails	15
Division Pterophyta	Ferns	12,000
Division Coniferophyta	Conifers	550
Division Cycadophyta	Cycads	100
Division Ginkgophyta	Ginkgos	1
Division Gnetophyta	Gnetophytes	70
Division Anthophyta	Flowering plants (angiosperms)	235,000
Class Monocotyledones	Monocots	65,000
Class Dicotyledones	Dicots	170,000

DIVISION BRYOPHYTA: LIVERWORTS, HORNWORTS, AND MOSSES

Lacking water-gathering roots and the kind of specialized tissues that transport water up the body of a vascular plant, the bryophytes must absorb moisture through aboveground structures. As a consequence, they grow most successfully in moist, shady places and in bogs. Some of them, like sphagnum (peat moss), are able to absorb and hold large amounts of water; in effect, they maintain a watery existence even on land. The majority of the bryophytes are tropical, but some species occur in temperate regions, and a few even reach the Arctic and Antarctic.

Most bryophytes are comparatively simple in their structure and are relatively small, usually less than 20 centimeters in length. A single moss plant may sprawl over a considerable area, but most liverworts are so small that they are noticeable only to a keen observer. In the damp environments frequented by the bryophytes, individual cells can absorb water and nutrients directly from the air or by diffusion from nearby cells. Like the lichens, they are sensitive indicators of air pollution.

Although bryophytes do not have true roots, they are generally attached to the substrate by means of rhizoids, which are elongate single cells or filaments of cells. Many bryophytes also have small leaflike structures in which photosynthesis takes place. These structures lack the specialized tissues of the "true" leaves of the vascular plants and are only one or a few cell layers thick. For these reasons, the leaflike structures of the bryophytes and the leaves of the vascular plants are believed to have evolved separately. As in other plants, the body of a bryophyte is specialized for support and food storage.

24–6 *Representative bryophytes, the only plants in which the gametophyte, which is haploid (n), is the dominant, nutritionally independent generation.* (a) *A young gametophyte of the liverwort* Marchantia, *growing on a rock.* (b) Anthoceros, *a hornwort. The "horns" are the diploid (2n) sporophytes, which are attached to a dish-shaped gametophyte.* (c) *A haircap moss with spore capsules. The lower green structures are the gametophytes; the nonphotosynthetic stalks and capsules are the sporophytes. Depending on the species, the sporophytes of mosses growing in temperate regions take 6 to 18 months to reach maturity.*

Bryophyte Reproduction

Bryophytes, like all plants, have a life cycle with alternation of generations. In contrast to the vascular plants, however, the bryophytes are characterized by a haploid gametophyte that is usually larger than the diploid sporophyte.

The life cycle of a moss begins when a haploid spore germinates to form a network of horizontal filaments known as protonemata (singular, protonema) (Figure 24–7). Individual gametophytes grow up like branches from this network. The multicellular antheridia and archegonia are borne on the gametophyte. When sufficient moisture is present, the sperm, which are biflagellate, are released from the antheridium and swim to the archegonium, to which they are attracted chemically. Without free water in which the sperm can swim, the life cycle cannot be completed.

24–7 *Protonema of a moss, with a budlike structure from which the gametophyte will develop. Protonemata are characteristic of mosses and liverworts. They often resemble filamentous green algae. For the subsequent stages of the life cycle, see Figure 24–8.*

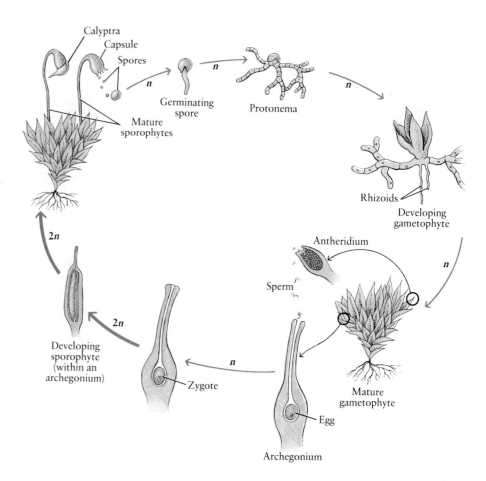

24–8 *The moss life cycle begins with the release of spores from a capsule, which opens when a small lid bursts (upper left). The spore germinates to form a branched, filamentous protonema, from which a leafy gametophyte develops. Sperm cells, which are expelled from the mature antheridium, are attracted into the archegonium, where one fuses with the egg cell to produce the zygote. The zygote divides mitotically to form the sporophyte and, at the same time, the base of the archegonium divides to form the protective calyptra. The mature sporophyte consists of a capsule, which may be raised on a stalk—also part of the sporophyte—and a foot. Meiosis occurs within the capsule, resulting in the formation of haploid spores.*

In this moss, the gametophytes bear both antheridia and archegonia. In other species, a single gametophyte may bear either antheridia or archegonia, but not both.

Fusion of sperm and egg takes place within the archegonium. Inside the archegonium, the zygote develops into a sporophyte, which remains attached to the gametophyte and is nutritionally dependent upon it. Typically the sporophyte consists of a foot, a stalk, and a single, large sporangium (or capsule), from which the spores are discharged.

Asexual reproduction, often by fragmentation, is also common. Many mosses and liverworts also produce minute bodies, known as gemmae, that can give rise to new plants (Figure 24–9).

24–9 (a) *The bowl-shaped gemma cups visible on this gametophyte of the liverwort* Marchantia *contain minute bodies, the gemmae, which are splashed out by the rain and grow in the vicinity of the parent plant. Asexual reproduction by fragmentation or by gemmae is common among the liverworts.*

In Marchantia, *the elevated antheridia and archegonia are formed on different plants, the male (b) and the female (c) gametophytes. The zygote, formed in the archegonium, develops into the sporophyte, which remains attached to the female gametophyte. Gemma cups develop on both types of gametophyte.*

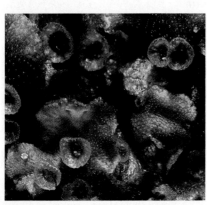

(a)

(b)

(c)

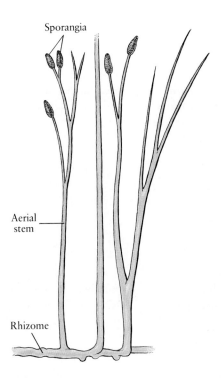

24-10 Rhynia major, *one of the earliest known vascular plants. It lacked leaves and roots. Its aerial stems, which were photosynthetic, were attached to an underground stem, or rhizome. The aerial stems were covered with a cuticle and contained stomata. The dark structures at the tips of the stems are sporangia, which apparently released their spores by splitting lengthwise.*

THE VASCULAR PLANTS: AN INTRODUCTION

Evolutionary Developments in the Vascular Plants

Rhynia major, now extinct, is an example of the earliest known vascular plants, dating back some 400 million years. As you can see in Figure 24–10, it is much more primitive in appearance than a bryophyte. However, it differs from the bryophytes in one important respect: within its stem is a central cylinder of vascular tissue, specialized for conducting water and dissolved substances up the plant body and products of photosynthesis down.

Beginning with a very simple vascular plant, such as *Rhynia,* it is possible to trace some major evolutionary trends, as well as several key innovations. One early innovation was the root, a structure specialized for anchorage and the absorption of water. Another was the leaf, a structure specialized for photosynthesis. Two distinct types of leaves evolved, the microphyll and the megaphyll. The microphyll contains only a single strand of vascular tissue, whereas the megaphyll typically contains a complex system of veins. These two types of leaves seem to have originated in different ways (Figure 24–11). Of the modern plants, only the Psilophyta, the Lycophyta, and the Sphenophyta have microphylls.

One of the most striking trends in plant evolution has been the development of increasingly efficient conducting systems between the two portions of the plant body. The conducting system in modern vascular plants consists of two distinct tissues: the **xylem,** which transports water and ions from the roots to the leaves, and the **phloem** (pronounced "flow-em"), which carries dissolved sucrose and other products of photosynthesis from the leaves to the nonphotosynthetic cells of the plant. The conducting elements of the xylem are **tracheids** and **vessel members,** and the conducting elements of the phloem are **sieve cells** or **sieve-tube members.** In stems, the longitudinal strands of xylem and phloem are side by side, either in vascular bundles or arranged in two concentric layers (cylinders), in which one tissue (typically the phloem) occurs outside the other. With the development of roots, leaves, and efficient conducting systems, the plants effectively solved the most basic problems confronting multicellular photosynthetic organisms on land—acquiring adequate supplies of water and food and delivering them to all of the cells making up the organism.

Another pronounced trend is the reduction in size of the gametophyte. In all vascular plants, the gametophyte is smaller than the sporophyte. However, in the more primitive vascular plants, the gametophyte is separate and nutritionally independent of the sporophyte. In the more recently evolved groups—the gymnosperms and the angiosperms—the gametophyte has been reduced to microscopic size and dependent status.

Also related to the reproductive cycle is a trend toward heterospory. The earliest vascular plants produced only one kind of spore (homospory) in one kind of sporangium. Upon germination, such spores typically produce gametophytes on which both antheridia and archegonia form. In plants that are heterosporous, two different kinds of gametophytes develop: one bearing archegonia and the other, antheridia.

24-11 *According to one widely accepted theory,* (a) *microphylls evolved as lateral outgrowths of the stem. Representatives of the three divisions of modern plants with microphylls are shown in Figure 24–12.*

(b) *Megaphylls evolved by fusion of branch systems and thus have a complex vascular network. The great majority of vascular plants have megaphylls.*

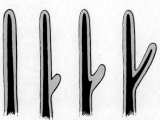

Evolution of microphylls
(a)

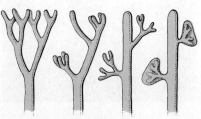

Evolution of megaphylls
(b)

24-12 *Representatives of three divisions of seedless vascular plants.* **(a)** *The whisk fern,* Psilotum, *one of the two living genera of division Psilophyta. The bulbous structures are the sporangia, which occur in fused groups of three.* Psilotum *is unique among living vascular plants in that it lacks roots and leaves. If you look closely, however, you can see small, scalelike outgrowths below the sporangia.*

(b) *The club mosses of the genus* Lycopodium *are the most familiar members of division Lycophyta. In this genus, the sporangia are borne on specialized leaves, sporophylls, which are aggregated into a cone at the apex (top) of the branches, as shown in this running ground pine,* Lycopodium complanatum. *The airborne, waxy spores give rise to small, independent subterranean gametophytes. The sperm, which are biflagellate, swim to the archegonium, where the young sporophyte, or embryo, develops.*

(c) *The horsetails, division Sphenophyta, of which there is only one living genus (*Equisetum*), are easily recognized by their jointed, finely ribbed stems, which contain silica. At each node, there is a circle of small, scalelike leaves. Spore-bearing structures are clustered into a cone at the apex of the stem. The gametophytes are independent, and the sperm are coiled, with numerous flagella.*

As the gametophytes became reduced, archegonia and antheridia decreased in size until, in the angiosperms, they disappeared altogether, made obsolete by the seed—perhaps the innovation most important in the enormous success of the vascular plants on land. The seed is a complex structure in which the young sporophyte, or embryo, is contained within a protective outer covering, the seed coat. The seed coat, which is derived from tissues of the parent sporophyte, protects the embryo while it remains dormant, sometimes for years, until conditions are favorable for its germination. The earliest known seeds were fossilized in late Devonian deposits some 360 million years ago.

THE SEEDLESS VASCULAR PLANTS

Four divisions of seedless vascular plants have living representatives: Psilophyta (whisk ferns), Lycophyta (club mosses), Sphenophyta (horsetails), and Pterophyta (ferns), the largest group.

Division Pterophyta: The Ferns

Ferns are vascular plants that can usually be distinguished from most other plants by their large feathery leaves, which, in most species, unroll from base to tip as they develop. According to the fossil record, ferns first appeared almost 400 million years ago, and they are still relatively abundant. Most of the 12,000 living species are found in the tropics, but many also occur in temperate and even arid regions. Because they have flagellated sperm and need free water for fertilization, those species growing in arid regions exploit the seasonal occurrence of water for sexual reproduction.

The stems of ferns are usually not as complex as those of gymnosperms and angiosperms, and they are often reduced to a creeping underground stem (rhizome). Although ferns do not exhibit secondary growth—the type of growth that results in increase of girth and formation of bark and woody tissue—some grow

(a)

(b)

(c)

24–13 *Ferns. (a) The immature sporophyte of many common ferns develops as a "fiddle head," which uncoils and spreads as it elongates. (b) A cinnamon fern. The sporangia are borne on separate stalks, visible in the center of the photograph.*

24–14 *Fern spores develop on the sporophyte in sporangia, which are usually found in clusters (sori) on the underside of a leaf (sporophyll), as shown here. Fern spores give rise to tiny gametophytes that, although photosynthetic, are barely visible to the naked eye.*

very tall. For instance, *Cyathea australis*, a tree fern found on Norfolk Island in the South Pacific, sometimes reaches 28 meters in height.

The leaves (fronds) of ferns are often finely divided into leaflets (pinnae). With a high surface-to-volume ratio, these widely spread, divided leaves are very efficient light collectors, well adapted to growing on the forest floor in diffuse light. The sporangia commonly are on the undersurface of the leaves or, sometimes, on specialized leaves. Sporangium-bearing leaves are called **sporophylls.** (As we shall see, the reproductive structures of flowers are also sporophylls.) The sporophylls may resemble the other green leaves of the plant or may be nonphotosynthetic stalks (modified leaves). The sporangia of ferns commonly occur in small clusters known as sori (singular, sorus).

In the ferns, as in all the vascular plants, the dominant generation is the sporophyte. The gametophyte of the homosporous ferns begins development as a small algalike filament of cells, each filled with chloroplasts, and then develops into a flat structure, often only one layer of cells in thickness. Although this gametophyte is small, it is nutritionally independent, as is the sporophyte. All but a few genera of ferns are homosporous, and a single gametophyte usually produces both antheridia and archegonia. The sperm are coiled and multiflagellate. The life cycle of a fern is shown in Figure 12–6 on page 251.

THE SEED PLANTS

The humid Carboniferous period, which ended some 286 million years ago, was the age when most of the earth's coal deposits were formed from lush vegetation that sank so swiftly into the warm, marshy soil that there was no chance for much of it to decompose. Seed plants were in existence by the close of this period; according to the fossil record, some of the fernlike plants and even some of the club mosses had seedlike structures.

In the Permian period (248 to 286 million years ago), there were worldwide changes of climate, with the advent of widespread glaciers and drought. Terrestrial plants and animals were under strong selection pressures for specialized structures that would enable them to survive during periods when no water was available.

Coal Age Plants

About 100 billion metric tons of carbon dioxide are fixed in photosynthesis every year. Almost the same amount is released into the atmosphere by living organisms in the course of respiration—less only 1 part in 10,000. This very slight imbalance is caused by the burying of organisms in sediment or mud under conditions in which oxygen is excluded and decay is only partial. This accumulation of partially decayed material is known as peat. The peat may eventually become covered with sedimentary rock and so be placed under pressure. Depending on time, temperature, and other factors, peat may become compressed into coal, petroleum, or natural gas—the so-called fossil fuels.

During certain periods in the earth's history, the rate of fossil fuel formation was greater than at other times. One such period, aptly called the Carboniferous, extended from 360 to 286 million years ago. What are now the temperate regions of Europe and North America were then tropical to subtropical, supporting year-round growth. The lands were low, covered by shallow seas or swamps. The dominant plants of the Carboniferous period were lycopod trees and giant horsetails; these disappeared during the Permian period, a time of worldwide drought and extensive glaciation. Other plants with surviving descendants include several families of ferns and one group of gymnosperms, the conifers. The flowering plants had not yet put in an appearance.

A reconstruction of a Carboniferous swamp forest, dominated by the lycopod tree Lepidodendron. *The young, unbranched trees that resemble bottle brushes matured into the many-branched trees that formed the canopy of the forest. Giant horsetails of the genus* Calamites *are at the left and in the center foreground. Seed plants also thrived in this environment; the plant at the far right with forked branches is the seed fern* Medullosa. *As you can see, these plants had shallow root systems, as do modern swamp plants, and so were easily toppled by the winds.*

24–15 *Representative gymnosperms.* **(a)** *Male and female plants of* Zamia pumila, *the only cycad native to the United States. It is common in the sandy woods of Florida. The stems are mostly or entirely underground and, along with the roots, were used by the Seminole Indians as food. The two large gray cones in the foreground are female cones; the smaller brown cones are male cones.*

(b) *Leaves and fleshy seeds of* Ginkgo biloba, *the only surviving species of the Ginkgophyta, a lineage that extends back to the late Paleozoic era.* Ginkgo *is especially resistant to air pollution and is commonly cultivated in urban parks and along city streets. The fleshy coating of the seeds has a putrid smell, similar to that of rancid butter. However, the inner "kernel" of the seed, which has a fishy taste, is a much-prized delicacy in the Orient.*

(c) *A branch of a conifer, the Ponderosa pine (also called western yellow pine), bearing a female cone. When the cone was mature, it opened and released its winged seeds, two of which have become caught between the scales. Like other pines, the Ponderosa has flexible, needlelike leaves, which are held together in a bundle (fascicle). Ponderosa pine, one of the principal forest trees of the Rockies from Canada to Mexico, is a staple of the Northwest lumber industry.*

(d) *A large seed-producing plant of* Welwitschia mirabilis, *a gnetophyte, growing in the Namib Desert of southern Africa.* Welwitschia *produces only two adult leaves, which continue to grow for the life of the plant. As growth continues, the leaves break off at the tips and split lengthwise; hence, older plants appear to have numerous leaves.*

Those organisms in which water-conserving, protective structures had previously evolved were at a great advantage. For example, the amphibians gave way to the reptiles with their scaly skins and shelled eggs, which may have been better suited to the harsh climate. Similarly, by the close of this period, which ended the Paleozoic era, plants with seeds gained a major evolutionary advantage and came to be the dominant plants of the land.

The two major modern groups of plants are both seed plants: the **gymnosperms,** which have naked seeds, and the **angiosperms** (from the Greek word *angio,* meaning "vessel"—literally, a seed borne in a vessel), which have protected seeds.

Gymnosperms

It was during the Permian period that the gymnosperms diversified. Four groups of gymnosperms have living representatives—three small divisions (Cycadophyta, Ginkgophyta, and Gnetophyta) and one large and familiar division (Coniferophyta). The conifers ("cone-bearers") include the pines, firs, spruces, hemlocks, cypresses, junipers, and the giant coastal redwoods of California and Oregon.

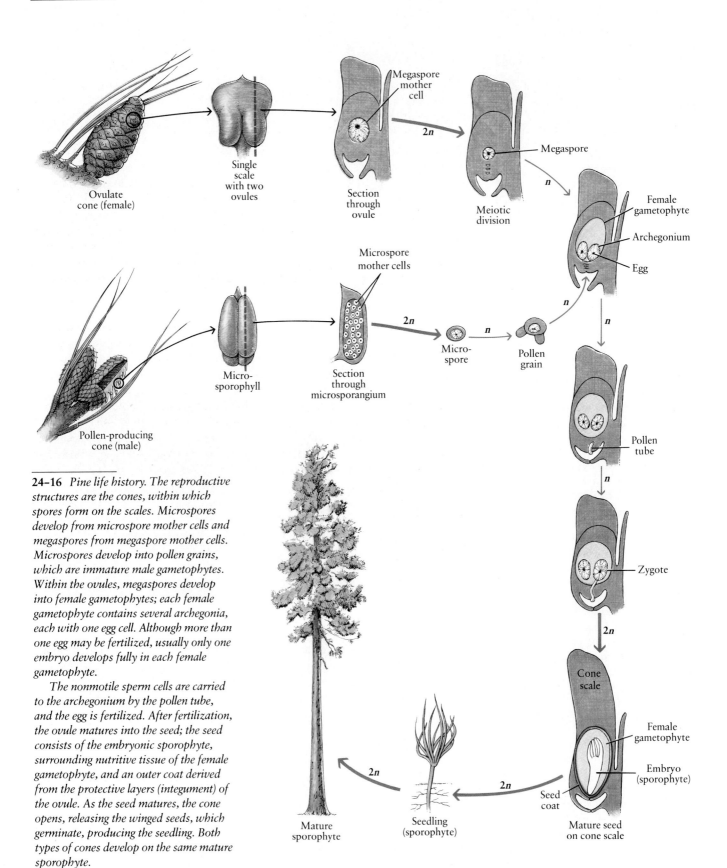

24–16 *Pine life history. The reproductive structures are the cones, within which spores form on the scales. Microspores develop from microspore mother cells and megaspores from megaspore mother cells. Microspores develop into pollen grains, which are immature male gametophytes. Within the ovules, megaspores develop into female gametophytes; each female gametophyte contains several archegonia, each with one egg cell. Although more than one egg may be fertilized, usually only one embryo develops fully in each female gametophyte.*

The nonmotile sperm cells are carried to the archegonium by the pollen tube, and the egg is fertilized. After fertilization, the ovule matures into the seed; the seed consists of the embryonic sporophyte, surrounding nutritive tissue of the female gametophyte, and an outer coat derived from the protective layers (integument) of the ovule. As the seed matures, the cone opens, releasing the winged seeds, which germinate, producing the seedling. Both types of cones develop on the same mature sporophyte.

Formation of the Seed

The seed is a protective structure in which the embryonic plant can be dispersed and lie dormant until conditions become favorable for its survival. Thus, in its functions, it parallels the spores of bacteria or the resistant zygotes of the freshwater algae. In structure it is far more elaborate, however. A seed includes the embryo (the young, dormant sporophyte), a store of nutritive tissue, and an outer protective coat.

To understand the structure of the seed, it is necessary to return for a moment to the alternation of generations found in all plants and in some of the green algae (see pages 463–464). In the green algae, the two generations, sporophyte and gametophyte, are independent and are generally about the same size. In the seedless vascular plants, including the ferns, the gametophytes, although still independent, are smaller than the sporophytes. In the seed plants, the gametophytic generation is reduced still further and is totally dependent on the sporophyte.

All gymnosperms are heterosporous, producing two different types of spores in two different types of sporangia. Spores that give rise to male gametophytes are known as **microspores,** and they are formed in structures known as microsporangia. Spores from which female gametophytes develop are **megaspores,** and they are formed in megasporangia. A megasporangium contains a single megaspore mother cell, which gives rise by meiosis to a megaspore, and it is surrounded by one or two layers of tissue, the integument. The entire structure—the megasporangium, its protective integument, and its contents—is known as the **ovule.**

With this information in mind, let us look at a specific example, the formation of a pine seed (Figure 24–16). A pine tree—the mature sporophyte—has two types of cones, which produce the two types of spores. The small, male cones superficially resemble those of the club mosses, but the large, female cones have ovule-bearing scales that are much thicker and tougher than the sporophylls of the male cones.

In the male cones, specialized microspore mother cells inside the microsporangia undergo meiosis to produce haploid microspores. Each microspore differentiates into a microscopic, windborne pollen grain, an immature male gametophyte. The wind is an unreliable messenger, disseminating the pollen grains at random, and wind-pollinated plants characteristically produce pollen in great quantities (Figure 24–17).

24-17 (a) *Male cones of Scotch pine* (Pinus sylvestris) *shedding pollen. The pollen grains are immature male gametophytes, which complete their maturation when they reach the ovules, embedded in the female cones. There they produce pollen tubes that carry the nonmotile sperm cells to the egg cells.*

(b) *Female cone of Scotch pine. The female gametophytes develop in ovules on the base of a scale of the cone, and the eggs are fertilized there. Each scale contains two ovules. When the seeds are mature, they drop from the cone.*

(a) (b)

Within the ovules of the female cones, the megaspores are formed by meiosis. Of the four cells produced within the ovule by each meiotic sequence, three disintegrate and the remaining one—the megaspore—develops into a tiny female gametophyte. This haploid gametophyte grows within the ovule and develops two or more archegonia, each of which contains a single egg cell. The development from the megaspore into the gametophyte with its egg cells may take many months—slightly more than a year, for example, in some common pines.

As the ovule ripens, it secretes a sticky liquid. When the female cone becomes dusted with pollen, some of the pollen sifts down between the cone scales and comes into contact with the liquid. Pollen grains, caught in the sticky liquid, are drawn to the ovule as the liquid dries. Here, some three months later, the pollen grain develops into a mature male gametophyte. This gametophyte produces two nonmotile cells, the male gametes, or sperm. These are carried toward the egg within the pollen tube, which is produced by the male gametophyte and slowly grows through the tissues of the ovule—a process that takes almost a year.

Because the drought-resistant pollen is blown to the female cones by the wind, and the sperm are carried to the egg by the pollen tube, the pines and other conifers are not dependent on free water for fertilization. Thus they are able to reproduce sexually when (and where) ferns and bryophytes cannot.

Following fertilization, the zygote begins to divide and forms the embryo, or young sporophyte. As the ovule matures, its integument hardens into a seed coat, enclosing both the embryo and the female gametophyte (the latter provides food for the embryo when the seed germinates). After the cone matures, it opens, releasing its seeds; this typically occurs in the fall of the second year after the initial appearance of the cones. In most conifers the seeds are winged and are dispersed by the wind.

Figure 24-18 shows a longitudinal section of a pine seed, which has been described as "three generations under one roof." The seed coat and the wing on which the seed is carried arise from the hardened integument of the ovule, derived from the mother sporophyte. The seed coat surrounds the tissue of the female gametophyte; swollen and packed with stored food reserves, the gametophyte tissue grows and displaces the tissue of the original megasporangium. Innermost in the seed is the embryo with its several **cotyledons,** or seed leaves, which will appear as the first leaves of the shoot of the new sporophyte when the seed germinates. The lower part of the embryo will develop into the first root.

Under favorable conditions, the seed—the mature ovule and its contents—will germinate and give rise to a seedling. If conditions remain favorable, the seedling will ultimately become a mature pine tree, producing spores that give rise to gametophytes that, in turn, produce gametes.

The Conifer Leaf

Another feature commonly associated with conifers, although not with all gymnosperms, is a needlelike leaf. Figure 24-19 shows a cross section of a pine leaf, which may be 10 or more centimeters long but only 1 to 3 millimeters in diameter. In the center are the veins, which carry water in one set of conducting cells (the tracheids) and sugars in another (the sieve cells). Outside the region containing the veins are the cells in which photosynthesis takes place. The ducts on the flat sides of the needles carry resin, a substance that is released when the plant is wounded and apparently serves to close the break. The outer layer of cells, the epidermis, is very hard but contains stomata through which gas exchange takes place.

The needlelike leaf, despite its slender form, is a megaphyll (page 501). It is well adapted to long periods of low humidity, as in regions with seasonal rainfall or long, cold winters, and to moisture-losing, sandy soils. One or more of those are characteristic of many regions in which modern conifers are abundant.

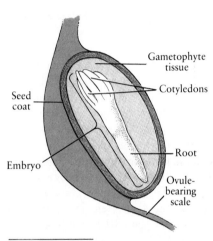

24-18 *Pine seed. The outer layers (integument) of the ovule have hardened into a seed coat, enclosing the female gametophyte and the embryo, which now consists of an embryonic root and a number of embryonic leaves, the cotyledons.*

When the seed germinates, the root will emerge from the seed coat and penetrate the soil. When the root absorbs water, the tightly packed cotyledons will elongate and swell with the moisture, rising above ground on the lengthening stem and forcing off the seed coat. During this period, the cotyledons absorb nutrients that are stored in the gametophyte tissue and are essential for the growth of the embryo into a seedling.

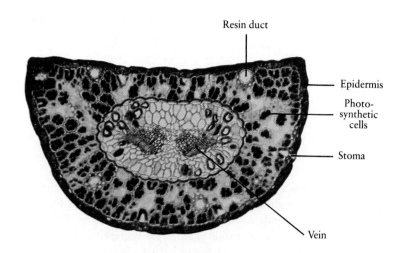

24-19 *Cross section of a pine leaf stained to show the tissues and cell types. The hard outer covering and compact shape protect the leaf from water loss, an essential factor in the survival of these trees in areas in which there is little rainfall, or in which the water is locked in the ground as ice during many months of the year.*

Angiosperms: The Flowering Plants

It is believed that the angiosperms—plants with enclosed, protected seeds—evolved from a now-extinct group of gymnosperms. They appear in the fossil record in abundance during the Cretaceous period, about 120 million years ago, as the dinosaurs were declining. Of the numerous angiosperm genera that appeared at that time, many seem to have been very similar to our modern genera.

Daniel Axelrod, a paleobotanist at the University of California, believes that angiosperms probably arose long before the Cretaceous, during the Permian period. They are likely to have originated on the less fertile hills and uplands of tropical areas, the richer lowlands being crowded with club mosses, ferns, and gymnosperms. Once established, they spread into the lowlands, where they became the dominant plant forms and were deposited as fossils. During this mid-Cretaceous period, the climate of the earth was warmer and more uniform than it is at present, and by the end of the Cretaceous period, much of the land was covered with a rich forest of angiosperms, reaching almost as far north as the Arctic Circle.

Angiosperms, like gymnosperms, have megaphylls, stomata, and a cuticle impervious to water. The modern forms, however, have a more highly evolved vascular system than is found in the gymnosperms (Figure 24–20). They also have two new, interrelated structures that distinguish them from all other plants: the flower and the fruit. Both are devices by which animals are induced, rewarded, tricked, and even seduced into carrying out the plants' reproductive strategies.

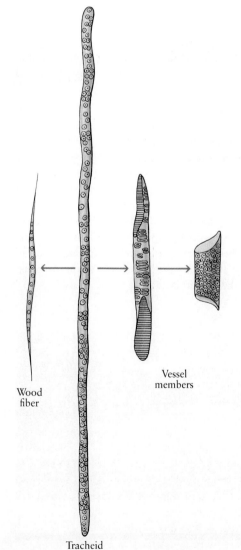

24-20 *Evolutionary relationships among some cells of the xylem of vascular plants. Tracheids, which are the only water-conducting cells of the conifers, are believed to resemble more primitive cells. Tracheids are elongated cells with thin areas (pits) in their lateral walls through which water moves from one tracheid to another up the trunk from the roots. Tracheids also provide mechanical support. Wood fibers, specialized for support, and vessel members, specialized for conducting water, are presumed to have evolved from the primitive water-conducting and supporting tracheids of early vascular plants. In the most highly evolved vessels, the end walls of the individual cells (vessel members) disintegrate during development, and the members are stacked on top of one another, leaving a continuous tube.*

The Ice Ages

During most of our planet's history, its climate appears to have been warmer than it is at the present time. However, these long periods of milder temperatures have been interrupted periodically by Ice Ages, so called because they are characterized by glaciations, or persistent accumulations of ice and snow. Such glaciations occur whenever the summers are not hot enough and long enough to melt ice that has accumulated during the winter. In many parts of the world, an alteration of only a few degrees in temperature is enough to begin or end a glaciation.

An early Ice Age appears to have occurred at the beginning of the Paleozoic era, some 590 million years ago. Another, marked by extensive glaciations in the Southern Hemisphere, closed the Paleozoic, some 248 million years ago. The conifers evolved during this period and possibly the angiosperms, as older forest types disappeared. A more recent, less severe cold, dry period occurred at the end of the Mesozoic, about 65 million years ago, and may be associated with the events causing the extinction of the dinosaurs (to be discussed in Chapter 49).

The most recent Ice Age began during the Pleistocene epoch, about 1.5 million years ago. The Pleistocene has been marked by four extensive glaciations that have covered large areas of North America, Great Britain, and northern Europe. Between the glaciations there have been intervals, called interglacials, during which the climate has become warmer. In each of these four Pleistocene glaciations, sheets of ice, thicker than 3 kilometers in some regions, spread out locally from the poles, scraped their way over much of the continents—reaching as far south as southern Illinois in North America and covering Scandinavia, most of Great Britain, northern Germany, and northern Russia—and then receded again. We are living at the end of the fourth glaciation, which completed its retreat only some 8,000 years ago.

The fossil record shows that during these periods of violent climatic change the populations of these regions were under extraordinary evolutionary pressures. Plant and animal populations moved, changed, or became extinct. In the interglacial periods, during which the average temperatures were at times warmer than those of today, the tropical forests and their inhabitants spread up through today's temperate zones. During the periods of glaciations, only animals of the northern tundra could survive in these same locations. Rhinoceroses, great herds of horses, large bears, and lions roamed Europe in the interglacial periods. In North America, as the fossil record shows, there were camels and horses, saber-toothed cats, and great ground sloths, one species as large as an elephant. In the colder periods, reindeer ranged as far south as southern France, while during the warmer periods, the hippopotamus reached England.

The reason for these large changes in temperature is one of the most controversial issues in modern science. They have been variously ascribed to changes in the earth's orbit, variations in the earth's angle of inclination toward the sun, migration of the magnetic poles, fluctuations in the solar energy, higher elevations of the continental masses, continental drift, and combinations of these and other causes. Also under debate is the question of whether this period of glaciations is over—perhaps for another 200 million years—or whether we are merely enjoying a brief interglacial before the ice sheets begin to creep toward the equator once more.

Glaciers are accumulations of snow and ice that flow across a land surface as a result of their own weight. This Alaskan glacier is flowing into the sea. As the glacier moves forward, its edges melt; the rocks carried within it have become concentrated in the dark band at the right.

About 235,000 different species of angiosperms are known. They dominate the tropical and temperate regions of the world, occupying well over 90 percent of the earth's vegetative surface. The angiosperms include not only the plants with conspicuous flowers but also the great hardwood trees, all the fruits, vegetables, nuts, and herbs, and grains and grasses that are the staples of the human diet and the basis of agricultural economy all over the world. These tremendously diverse plants are classified in two large groups: class Monocotyledones (the monocots), with about 65,000 species, and class Dicotyledones (the dicots), with about 170,000 species. Among the monocots are such familiar plants as the grasses, lilies, irises, orchids, cattails, and palms. The dicots include many of the herbs, almost all the shrubs and trees (other than conifers), plus many other plants. The major differences between these two classes are summarized in Table 24-3 and will be discussed further in Section 5.

TABLE 24-3 **Principal Differences between Monocots and Dicots**

CHARACTERISTIC	MONOCOTS	DICOTS
Flower parts	Usually in threes	Usually in fours or fives
Pollen grains	Have one furrow or pore	Have three furrows or pores
Cotyledons ("seed leaves")	One	Two
Leaf venation	Major veins usually parallel	Major veins usually netlike
Vascular bundles in young stem	Scattered	In a ring
Secondary (woody) growth	Absent	Usually present

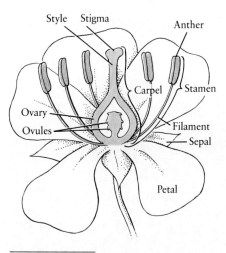

24-21 Structure of a flower. The reproductive structures are shown in color. Some flowers have only the male structures, some only the female; such flowers are said to be imperfect. A flower that possesses both stamens and carpel, such as this flower, is known as a perfect flower.

The petals and sepals are, like the stamens and carpel, modified leaves.

The Flower

Figure 24-21 is a diagram of a simple flower. The central structure is the **carpel,** the female reproductive structure, which is thought to be a modified sporophyll (sporangium-bearing leaf). (A single carpel or a group of fused carpels is also known as a pistil, because of its resemblance to an apothecary's pestle.) The swollen base of the carpel is the **ovary,** within which is the ovule, or ovules, in which the female gametophyte develops from a megaspore. The tip of the carpel is specialized as the **stigma,** a sticky surface to which pollen grains adhere. The stigma and ovary are connected by a slender column of tissue, the **style.**

The pollen grains (immature male gametophytes) develop in the **stamen,** which, like the carpel, is a sporophyll. It consists of the **anther,** which contains microsporangia in which the pollen grains develop, and a supporting **filament.**

As shown in Figure 24-22, pollen grains produced in the anthers are carried to the stigma of (usually) another flower, where they germinate, developing pollen tubes that grow through the style toward the ovule. The nonmotile sperm cells are carried by the pollen tubes to the female gametophyte, which typically consists of only seven cells. The extraordinary events of angiosperm fertilization, which give rise not only to an embryonic sporophyte but also to a special nutritive tissue, will be discussed in detail in Chapter 29 (page 615).

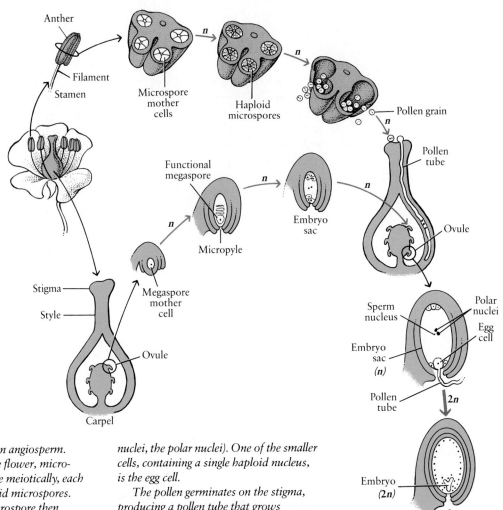

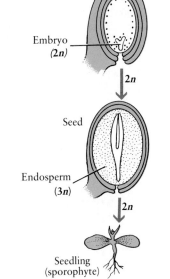

24–22 *Life history of an angiosperm. Within the anther of the flower, microspore mother cells divide meiotically, each giving rise to four haploid microspores. The nucleus in each microspore then divides mitotically, and the microspore develops into a two-celled pollen grain, which is an immature male gametophyte. One of the cells subsequently divides again, usually after germination, resulting in three haploid cells per pollen grain: two sperm cells and the tube (pollen tube) cell.*

Within the ovule, a megaspore mother cell divides meiotically to produce four haploid megaspores. Three of the megaspores disintegrate; the fourth divides mitotically, developing into an embryo sac—the female gametophyte—consisting of seven cells with a total of eight haploid nuclei (the large central cell contains two nuclei, the polar nuclei). One of the smaller cells, containing a single haploid nucleus, is the egg cell.

The pollen germinates on the stigma, producing a pollen tube that grows through the style into the ovary. The growing pollen tube enters the ovule through a small opening known as the micropyle. The two sperm cells pass through the tube into the embryo sac; one sperm nucleus fertilizes the egg cell, the other merges with the polar nuclei, forming a triploid (3n) cell that develops into a nutritive tissue, the endosperm. The embryo undergoes its first stages of development while still within the ovary of the flower, and the ovary itself matures to become a fruit. The seed, released from the mother sporophyte in a dormant form, eventually germinates, forming a seedling.

Evolution of the Flower

Plants are, generally speaking, immobile. Thus, uniting the sperm of one individual with the egg of another individual of the same species presents a problem, to which the flower—that apt symbol of spring and romance—is a solution. The early gymnosperms from which the angiosperms evolved were probably wind-pollinated, as are modern gymnosperms. And, as in the modern gymnosperms, the ovule probably exuded droplets of sticky sap in which pollen grains were caught and drawn to the female gametophyte. Insects, principally beetles and flies, that fed on plants must have come across the protein-rich pollen grains and the sticky, sugary droplets. As they began to depend on these new-found food supplies, they inadvertently carried pollen from plant to plant.

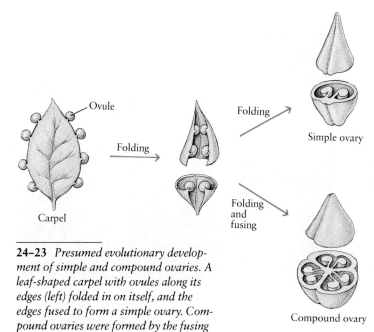

24–23 *Presumed evolutionary development of simple and compound ovaries. A leaf-shaped carpel with ovules along its edges (left) folded in on itself, and the edges fused to form a simple ovary. Compound ovaries were formed by the fusing of separate infolded carpels. A single flower may contain one or more carpels, which may be separate, as in blackberry flowers, or fused, as in the flowers of tomato plants and apple trees.*

24–24 *Flower of a round-leaved hepatica (Hepatica americana). Notice the open, bowl shape and the numerous and separate floral parts. The insect is a pollen-eating beetle.*

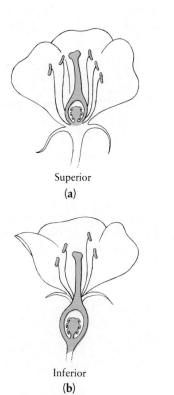

24–25 *(a) Superior and (b) inferior ovaries. In flowers with superior ovaries, the floral parts are attached below the ovaries. In flowers with inferior ovaries, the floral parts are attached above, and the ovaries are protected.*

Insect pollination must have been more efficient than wind pollination for some plant species because, clearly, selection began to favor those plants that had insect pollinators. The more attractive the plants were to the insects, the more frequently they would be visited and the more seeds they would produce. Any chance variations that made the visits more frequent or that made pollination more efficient offered immediate advantages; more seeds would be produced, and thus more offspring would be likely to survive. Nectaries (nectar-secreting structures) evolved, which lured the pollinators. Plants developed white or brightly colored flowers that called attention to the nectar and other food supplies. The carpel, originally a leaf-shaped structure, became folded on itself, enclosing and protecting the ovule from hungry pollinators (Figure 24–23). By the beginning of the Cenozoic era, some 65 million years ago, the first bees, wasps, butterflies, and moths had appeared. These are insects for which flowers are often the only source of nutrition for the adult forms. From this time on, flowers and certain insect groups have had a profound influence on one another's history, each shaping the other as they evolved together.

The primitive flower is believed to have resembled that of the modern hepatica, shown in Figure 24–24, which has numerous floral parts, each separate from the other. By comparing this type of flower with some of the more specialized ones shown on page 616, it is possible to see four main trends in flower evolution:

1. Reduction in number of floral parts. Most specialized flowers have few stamens and few carpels.

2. Fusion of floral parts. Carpels and petals, in particular, have become fused, sometimes elaborately so.

3. Elevation of free floral parts above the ovary. In the primitive flower, the floral parts arise at the base of the ovary (Figure 24–25a). Such ovaries are said to be superior. In the more advanced flowers, the free portions of the floral parts are above the ovary (Figure 24–25b); such ovaries are said to be inferior. This is an important adaptation by which the ovules are believed to be protected from foraging insects.

4. Changes in symmetry. The radial symmetry of the primitive flower has given way, in more advanced forms, to bilaterally symmetrical forms.

24–26 *Pollinators.* **(a)** *A honey bee foraging in a flower of Gaillardia pulchella. Seen frequently along roadsides in the American Southwest, this flower is commonly known as firewheel because of its resemblance to a Fourth of July pinwheel. It is also known as Indian blanket.*

(b) *A longhorn beetle pollinating a lily. The pollen-covered head of the beetle is brushing against a stigma of the flower. Notice how precisely the pollen grains were deposited on the beetle's head by the flowers it visited previously.*

(c) *A gossamer-winged butterfly sipping nectar in a daisy. Notice the long, sucking tongue of the butterfly. The head of a daisy and other similar flowers consists of numerous separate florets.*

(d) *Female rufous hummingbird probing for nectar in a columbine. Her head is collecting pollen, which she will carry to another flower. Flowers pollinated by birds are scentless, bright red or orange, and have copious nectar that makes the visit worthwhile.*

(e) *By thrusting its face into the center of an organ-pipe cactus flower, a* Leptonycteris *bat is able to lap up nectar with its long, bristly tongue. Pollen grains clinging to its face and neck are transferred to the next flower visited by the bat. Bat-pollinated flowers have dingy colors and a musty scent (similar to that produced by bats to attract one another), and they open at night. Other mammalian pollinators include a few rodents, marsupials, and primates.*

A flower that attracts only a few kinds of animal visitors and attracts them regularly has an advantage over flowers visited by more promiscuous pollinators: its pollen is less likely to be wasted on a plant of another species. In turn, it is an advantage for an animal to have a "private" food supply that is relatively inaccessible to competing species. Many of the distinctive features of modern flowers are special adaptations that encourage regular visits (constancy) by particular pollinators. The varied shapes, colors, and odors allow sensory recognition by pollinators. The diverse, sometimes bizarre, structures such as deep nectaries and complex landing platforms that are found, for example, in orchids, snapdragons, and irises, represent ways of excluding indiscriminate pollinators and of increasing the precision of pollen deposits on the animal messengers.

Evolution of the Fruit

A fruit is the mature, ripened ovary of an angiosperm and contains the seeds. A great variety of fruits, adapted for many different dispersal mechanisms, have evolved in the course of angiosperm history (Figure 24-27). In most cases, the chief requirement is that the seed be transported some distance from the parent plant, where it is more likely to find open ground and sunlight. Many fruits, like flowers, evolved as a payment to an animal visitor for transportation services. A familiar example is provided by the edible fleshy fruits that become sweet and brightly colored as they ripen, attracting the attention of birds and mammals, including ourselves. The seeds within the fruits pass through the digestive tract hours later and are often deposited some distance away. In some species the seed coat's exposure to digestive juices is a prerequisite for germination. The seeds themselves may be bitter or toxic, as in apples, discouraging animals from grinding up and digesting them.

Biochemical Evolution

Another factor in the dominance of the flowering plants has been the evolution of characteristic bad-tasting or toxic compounds. The mustard family (Brassicaceae), for example, is characterized by the pungent taste and odor associated with cabbage, horseradish, and mustard. Milkweeds (family Asclepiadaceae) contain cardiac glycosides, substances that act as heart poisons in vertebrates, potential predators of this group of plants. Bitter-tasting quinine is derived from tropical trees and shrubs of the genus *Cinchona*. Nicotine and caffeine are plant products. Mescaline comes from the peyote cactus; tetrahydrocannabinol from *Cannabis sativa*; opium from a poppy; cocaine from the coca leaf. All of these substances were, at one time, thought of as by-products of plant metabolism and termed "secondary plant substances"; they are now recognized as products of angiosperm evolution that provide powerful defenses against animal predators.

Angiosperms represent the most successful of all plants in terms of numbers of individuals, numbers of species, and their effects on the existence of other organisms. In Section 5, we shall consider the structure and physiology of this dominant group of plants, paying particular attention to the processes of reproduction, growth, and development, the transport of materials by the vascular system, and the integration of the activities of the diverse types of cells forming the plant body.

(a)

(b)

24-27 *Angiosperms are characterized by fruits. As we shall see in Chapter 29, a fruit is a mature ovary, enclosing the seed or seeds. It often also includes accessory parts of the flower. The fruit aids in dispersal of the seed. Some fruits are borne on the wind, some are carried from one place to another by animals, some float on water, and some are even forcibly ejected by the parent plant.*

(a) In milkweed, the fruit bursts open when it is ripe, releasing seeds with tufts of silky hair that aid in their dispersal.

(b) The tough seeds in these blackberries will pass unharmed through the digestive tract of the harvest mouse. If they are deposited in a suitable environment, they will germinate, giving rise to new blackberry plants.

THE ROLE OF PLANTS

The only forms of life on land that do not depend on plants for their existence are a few kinds of autotrophic prokaryotes and protists. For all other terrestrial organisms, the chloroplast of the plant cell is the "needle's eye" through which the sun's energy is channeled into the biosphere. Even those animals that eat only other animals—the carnivores—could not exist if their prey, or their prey's prey, had not been nourished by plants.

24–28 *Poison ivy* (Toxicodendron radicans) *produces an organic alcohol that causes an irritating rash on the skin of many people. The ability to produce this compound is thought to have evolved under the selection pressure exerted by herbivorous animals. Fortunately, the plant is easily identified by its characteristic compound leaves with their three leaflets. Leaves growing in the sun are generally shiny, while those growing in the shade are dull.*

Moreover, plants are the channels by which many of the simple inorganic substances vital to life enter the biosphere. Carbon in the form of carbon dioxide is taken from the atmosphere and incorporated into organic compounds during photosynthesis. Elements such as nitrogen and sulfur are taken from the soil in the form of simple inorganic compounds and incorporated into proteins, vitamins, and other essential organic compounds within green plant cells. Animals cannot make these organic compounds from inorganic materials and so are entirely dependent on plants for these compounds as well as for their energy supply.

It was only after the plants had successfully invaded the land that members of a number of different animal groups could, in turn, carry out their own invasions of this vast new environment. As we shall see in subsequent chapters of this section, both invertebrates and vertebrates underwent great diversification as they took advantage of the variety of habitats and food supplies made available by the plants.

SUMMARY

Plants are multicellular photosynthetic organisms adapted for life on land. Among their adaptations are a waxy cuticle, pores through which gases are exchanged, protective layers of cells surrounding the reproductive cells, and retention of the young sporophyte within the female gametophyte during its embryonic development.

The ancestor of the plants is believed to have been a multicellular green alga of class Charophyceae, similar to the modern genus *Coleochaete*. It was oogamous (that is, its gametes were differentiated into egg and sperm), and it evolved a life cycle characterized by the alternation of heteromorphic generations, two features present in all plants. From this common ancestor, two principal lineages diverged: the bryophytes and the vascular plants.

The mosses and other members of division Bryophyta are relatively small plants usually found in moist locations. Most lack specialized vascular tissues and all lack true leaves, although the plant body is differentiated into photosynthetic, food-storing, and anchoring tissues. In sexual reproduction, the female gamete (egg) develops in the archegonium; the male gametes (sperm) develop in the antheridium. The sperm, which are flagellated, swim to the archegonium and fertilize the egg cell. The resulting zygote develops into an embryo within the archegonium. In the bryophytes—and in no other members of the plant kingdom—the gametophyte (n) is dominant and the sporophyte $(2n)$ is smaller, attached, and often nutritionally dependent.

Although the bryophytes seem to have changed little in the course of their history, the vascular plants have undergone a great diversification. Major developments in their evolution include better conducting systems, a progressive reduction in the size of the gametophyte, and the "invention" of the seed. The nine divisions of vascular plants can be informally grouped into the seedless vascular plants (divisions Psilophyta, Lycophyta, Sphenophyta, and Pterophyta) and the seed plants. The seed plants can be grouped into the gymnosperms, or naked-seed plants (divisions Coniferophyta, Cycadophyta, Ginkgophyta, and Gnetophyta), and the angiosperms, or flowering plants (division Anthophyta).

Among living seedless vascular plants, the ferns (division Pterophyta) are the most numerous. They are characterized by large, often finely dissected leaves called fronds. These leaves are, like the leaves of the seed plants, megaphylls. The sporophyte is the dominant generation, but in most ferns the gametophytes are independent. The sperm are flagellated, and free water is needed for fertilization. Sporangia are typically formed on the underside of specialized leaves (sporophylls) of the sporophyte.

Gymnosperms and angiosperms are seed plants. The most numerous modern gymnosperms are the conifers ("cone-bearers"). On the scales of the smaller, male cones, microspores differentiate into male gametophytes, which are released in the form of windblown pollen. In the ovules that form on the scales of the larger, female cones, the female gametophytes develop from megaspores. Within the female gametophyte, archegonia form. The male gametophyte germinates and produces a pollen tube through which the nonmotile sperm enter the archegonium and fertilize the egg cell. The seed, or mature ovule, consists of the seed coat, the young embryo, and the female gametophyte, which serves as nutritive tissue. The seed, which is shed from the female cone, can remain dormant for long periods of time and hence is adapted to withstanding cold and drought.

The angiosperms, of which there are about 235,000 species, are characterized by the flower and the fruit. Flowers attract pollinators, and fruits enhance the dispersal of seeds. The reproductive structures of the flower are the stamens, composed of filament and anther, and the carpels, composed of ovary, style, and stigma. The stamens and the carpels are highly specialized sporophylls. The ovule or ovules are enclosed in the ovary, which is the base of the carpel or group of fused carpels. Pollen grains, the immature male gametophytes, are formed in the anthers and germinate on the sticky surface of the stigma, the tip of the carpel.

The various shapes and colors of flowers evolved under selection pressures for more efficient pollinating mechanisms. Major trends in flower evolution include reduction and fusion of floral parts, a change in the position of the ovary relative to the other flower parts to a more protected (inferior) position, and a shift from radial to bilateral symmetry.

In addition to the flower and the fruit, a third factor in the success of the angiosperms has been the evolution of bad-tasting or toxic chemicals that discourage the predations of foraging animals. Angiosperms are the dominant plants of the modern landscape, providing a diversity of habitats and foods for terrestrial animals.

QUESTIONS

1. Distinguish among the following: bryophytes/vascular plants; moss/club moss; spore/sporangium/sporophyte/sporophyll; homosporous/heterosporous; gametophyte/antheridium/archegonium; xylem/phloem; microphyll/megaphyll; gymnosperm/angiosperm; microspore/megaspore; carpel/stamen.

2. Consider the green alga *Coleochaete*. What advantages does the alga gain from the production of "extra" haploid cells, made possible by additional replication of the DNA of the zygote prior to meiosis? What might be the advantages—to *Coleochaete*—of retaining the zygote on the body of the parent alga?

3. Bryophytes, among the plants, and amphibia, among the animals, often live in habitats intermediate between fresh water and dry land, rather than between salt water and land. Propose a physiological argument (referring back to Chapter 6) to explain why invasions of the land were more likely by organisms previously adapted for life in fresh water.

4. Where did the liverworts get their name? (The answer is not in the text, but a little research should tell you.) What does this name tell you about the human perspective of the rest of the biosphere?

5. Vascular plants are characterized by the presence of lignin, which stiffens and supports the plant body as it grows upward toward the light. What is the primary source of support for the photosynthetic portions of the multicellular seaweeds?

6. Tall trees today are not appreciably taller than plants of the Devonian forests. What factors select for tallness in trees? What factors, by contrast, select against tallness? What kind of evolutionary innovation might be required to alter the optimal balance among these various factors?

7. In some areas on earth, large gymnosperms are either dominant or manage to coexist with angiosperms. List some advantages for a big plant's being a gymnosperm.

8. Describe the sporophylls of a club moss, a horsetail, a fern, a pine, and an angiosperm.

9. Sketch a pine seed and label it. Indicate the origin of each of its components, and whether the component is haploid or diploid. How many generations are represented in the tissues of the seed?

10. Sketch a flower and label the structures. What is the function of each floral part?

11. A major trend in the evolution of angiosperms is a reduction in the number of floral parts. Why should reduction, rather than greater elaboration, be so common?

12. How have plants solved the problem of obtaining adequate food and water? Of supplying food and water to the individual cells? Of protecting and nourishing the immature stages?

CHAPTER 25

The Animal Kingdom I: Introducing the Invertebrates

25-1 *Animals are characterized by their motility, usually a result of the contraction of assemblies of protein fibers within specialized (muscle) cells. Aequorea victoria, a jellyfish, is shown here swimming actively, its bell contracted by muscles around its margin.*

Animals are many-celled heterotrophs, and their principal mode of nutrition is ingestion. They depend directly or indirectly for their nourishment on photosynthetic autotrophs—algae or plants. Typically they digest their food in an internal cavity and store food reserves as glycogen or fat. Their cells, unlike those of most other eukaryotes, do not have walls. Generally, animals move by means of contractile cells (muscle cells) containing actin, myosin, and associated proteins (see page 122). The most complex animals—the octopuses, the insects, and the vertebrates—have many kinds of specialized tissues, including elaborate sensory and neuromotor mechanisms not found in any other kingdom.

For most of us, animal means mammal, and mammals are, in fact, the chief focus of attention in Section 6. However, the mammals, or even the vertebrates as a whole, represent only a small fraction of the animal kingdom. More than 1.5 million different species of animals have been described, of which more than 95 percent are invertebrates—that is, animals without backbones. Indeed, the enormous variety displayed by the invertebrates is partly why they are so endlessly fascinating to study. There are also two practical reasons for the study of invertebrates. First, by virtue of their variety and their sheer numbers, they are of great ecological importance. Second, the invertebrates demonstrate a wide spectrum of ingenious solutions to the biological problems confronting multicellular heterotrophic organisms. In this way, they illuminate the fundamental nature of these problems and so lead us to a deeper understanding of the physiology of the vertebrate animals, including ourselves.

THE DIVERSITY OF ANIMALS

The unifying characteristic among all the animals is their mode of nutrition. Unlike plants, which are passive recipients of energy from the sun, animals must seek out food sources or, alternatively, devise strategies for ensuring that the food comes to them. Thus motility—of the entire organism, or its parts, or both—is a requirement for animal survival. Motility in general and the hunting of prey in particular require efficient systems of integration and control of the multitude of cells that make up the organism. In short, muscle and nerve, distinguishing features of the animal kingdom, have their origin in heterotrophy.

Paradoxically, these unifying characteristics of the animal kingdom are also the key to its diversity. Given the properties of living tissue and the characteristics of our planet—particularly the force of gravity and the physical properties of water and air—there are only a few basic ways that locomotion, the capture of food, self-defense, and coordination can be accomplished. In the course of evolution, however, as animals have adapted to new or changing environments, those few basic ways have been "reinvented," refined, and elaborated upon, resulting in the great diversity of structural and functional detail that we see today.

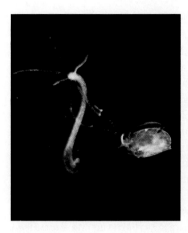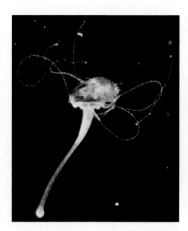

25–2 Some animals actively seek their food, whereas others arrange to have it delivered (see Figure 25–3). Here a Hydra encounters a small crustacean (Daphnia), grasps it with its tentacles, engulfs it, and digests it. These activities are coordinated by the animal's simple but effective nervous system.

25–3 The adult barnacle is nonmotile, depending on its six pairs of bristly appendages to deliver its food supply. Louis Agassiz, America's leading nineteenth century naturalist, characterized a barnacle as "nothing more than a little shrimp-like animal, standing on its head in a limestone house and kicking food into its mouth."

A similar diversification has occurred in reproductive patterns, providing mechanisms not only for genetic recombination but also for dispersal into and exploitation of habitats with abundant food supplies. Although sexual reproduction is the usual pattern in animals, many different types of animals are also capable of rapid asexual reproduction when conditions are suitable. Most animals are diploid, with gametes the only haploid stage in the life cycle (see Figure 12–4c, page 250). In the simplest and most primitive form of sexual reproduction, sperm and egg cells produced by different individuals are shed into the water, where they unite. Although external fertilization is retained in many animals that are otherwise highly evolved (for example, most frogs and toads), a variety of different methods of internal fertilization have also evolved. Additional versatility, particularly in feeding and dispersal, is provided by the immature forms, or **larvae,** that are characteristic of the life cycles of many different types of animals. As adults, most animals are fixed in size and shape (in contrast to plants, in which growth often continues for the lifetime of the organism); as we shall see, however, larvae often differ dramatically in size and shape from their adult counterparts.

In Chapter 21, we commented upon the tremendous diversity of the prokaryotes, as exemplified by the wide range of environments they inhabit and the many ways in which they satisfy their energy requirements. Among the more than 1.5 million species of animals that have been described, we see a similar pattern of adaptation to many different ways of life. Thus, for instance, on and around a single coral head, only a meter or two in diameter, one finds a dazzling array of different forms—reef fish, crustaceans, sponges, jellyfish, starfish, sea urchins, anemones, and the coral animals themselves. Similarly, to take a terrestrial example, a single spadeful of soil turns up earthworms, pillbugs, spiders, nematodes, and various other tiny animals. The branch of a single tree may harbor a dozen different kinds of insects, both adults and larvae, differing from one another in their nutritional specializations.

THE ORIGIN AND CLASSIFICATION OF ANIMALS

Animals, like plants, presumably had their origins among the protists. In the case of animals, however, we have fewer clues about which protists most closely resemble the ancestral ones, and multicellularity may well have arisen more than once. By the Cambrian period (590 to 505 million years ago), two major—and distinct—diversifications of animal life had already occurred, only one of which was to survive. The earliest evidences of animal life are provided by the fossilized animals found in greatest abundance in the Ediacaran Hills of Australia and a rock formation known as the Burgess Shale in the Canadian Rockies. The Ediacaran animals (Figure 25-4), which arose some 670 million years ago and flourished for about 100 million years, were quite unlike later animals and apparently were not their ancestors. As we shall see, the fundamental structure of virtually all living animals is a tube (or, in some cases, a cavity) within a tube, modified and elaborated in a variety of ways that maintain a high ratio of surface to volume. Ediacaran animals, by contrast, lacked tubular internal structures and were flat, leaflike, or, occasionally, quilted like an air mattress. Although this evolutionary solution to the problem of maintaining an adequate surface-to-volume ratio was ultimately unsuccessful, it produced a diversity of exquisitely beautiful organisms.

The Burgess Shale, which is dated at about 530 million years ago, includes members of at least 10 extinct phyla as well as fossil representatives of virtually all contemporary phyla. Some of these animals (Figure 25-5) are very similar indeed to animals living today. Although the fossils of the Burgess Shale and other Cambrian formations around the world attest to the ancient origins of the modern phyla of animals, they shed little light on the chronological order of their appearance. Most of the evidence for charts such as Figure 25-6 comes from studies of living animals.

Modern animals are classified in about 30 phyla, each of which is thought to be monophyletic. A number of criteria are used in classifying an animal and in attempting to elucidate the evolutionary relationships among the different phyla. Among the factors considered are the number of tissue layers into which the cells are organized, the basic plan of the body and the arrangement of its parts, the presence or absence of body cavities and the manner in which they form, and the pattern of development from fertilized egg to adult animal. Table 25-1 provides a short outline of animal classification and indicates the key features that are used in grouping the various phyla. In our examination of the diversity of animals, we shall devote most of our attention to the phyla indicated in boldface type, which exemplify the most significant developments and adaptations in animal evolution.

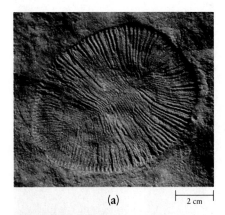

25-4 *Among the unusual organisms of the Ediacaran deposits are* (a) *Dickinsonia, a segmented animal that lacked a clearly defined head and appendages, and* (b) *Spriggina, a segmented animal with a prominent "head shield." Although these organisms superficially resemble some modern invertebrates, they appear to have lacked mouths, and their flattened bodies contained no tubular structures.*

TABLE 25-1 **An Outline of Animal Classification***

I. Subkingdom Parazoa: phylum **Porifera**

II. Subkingdom Mesozoa: phylum Mesozoa

III. Subkingdom Eumetazoa

    A. Radially symmetrical animals: phyla **Cnidaria** and Ctenophora

    B. Bilaterally symmetrical animals

        1. Acoelomates (animals that lack a body cavity): phyla **Platyhelminthes**, Gnathostomulida, and **Rhynchocoela**

        2. Pseudocoelomates (animals with the type of body cavity known as a pseudocoelom): phyla **Nematoda**, Nematomorpha, Acanthocephala, Kinorhyncha, Gastrotricha, Loricifera, Rotifera, and Entoprocta

        3. Coelomates (animals with the type of body cavity known as a coelom)

            a. Protostomes (animals in which the mouth appears at or near the first opening that forms in the developing embryo): phyla **Mollusca, Annelida,** Sipuncula, Echiura, Priapulida, Pogonophora, Pentastomida, Tardigrada, Onychophora, and **Arthropoda**

            b. Lophophorates (protostomes but with some deuterostome characteristics): phyla Brachiopoda, Phoronida, and Bryozoa

            c. Deuterostomes (animals in which the anus appears at or near the first opening that forms in the developing embryo): phyla **Echinodermata**, Chaetognatha, Hemichordata, and **Chordata**

---

* A summary of the distinctive characteristics of each phylum can be found in Appendix C.

CHAPTER 25 The Animal Kingdom I: Introducing the Invertebrates

25-5 *Some animals of the Burgess Shale.* (a) *Amiskwia, a gelatinous worm with prominent fins, and* (b) *Dinomischus, a stalked animal, represent phyla known only from fossil deposits of the Cambrian period. Members of living phyla include* (c) *Hyolithes, a mollusk distinguished by a cone-shaped shell with a protective cap;* (d) *Aysheaia, an onychophoran that bears a striking resemblance to the modern Peripatus (see Figure 26–22, page 561); and* (e) *Pikaia, an early chordate, readily identified by the stiffening rod, or notochord, that runs the length of its body and by the segmental pattern of its muscles (see Figure 28–10, page 591, for its modern counterpart, Branchiostoma). This specimen is 4 centimeters long.*

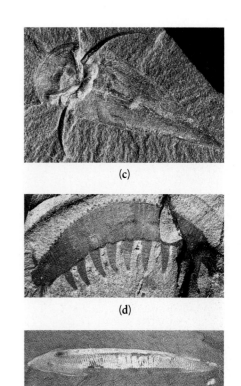

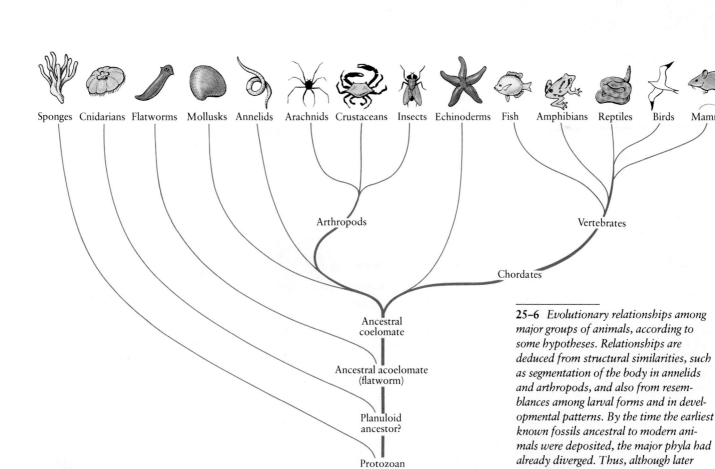

25-6 *Evolutionary relationships among major groups of animals, according to some hypotheses. Relationships are deduced from structural similarities, such as segmentation of the body in annelids and arthropods, and also from resemblances among larval forms and in developmental patterns. By the time the earliest known fossils ancestral to modern animals were deposited, the major phyla had already diverged. Thus, although later fossils illuminate evolutionary history within phyla (for example, within the arthropods or the chordates), we lack clear documentary evidence of the relationships among the different phyla.*

In the remainder of this chapter, we shall look at the so-called lower invertebrates, characterized by relatively simple body plans. In the three subsequent chapters, we shall consider the more complex animals, all of which possess the type of body cavity known as a coelom.

PHYLUM PORIFERA: SPONGES

Sponges may have had a different origin from other members of the animal kingdom and seem to have traveled a solitary evolutionary route. For this reason, they are often placed in a subkingdom of their own, the Parazoa ("beside the animals"). In fact, until the nineteenth century, the sponges were classified as plant-animals ("zoophytes"), since during their adult life they are all **sessile** (attached to a substrate). Sponges are common on ocean floors throughout much of the world. Most live along the coasts in shallow water, but some, such as the fragile glass sponges, are found at great depths, where currents are relatively slow. A few types are found in fresh water.

A sponge is essentially a water-filtering system, made up of one or more chambers through which water is driven by the action of numerous flagellated cells. Sponges are made up of a relatively few cell types, the most characteristic of which are the **choanocytes,** or collar cells, the flagellated cells that line the interior cavity of the sponge (Figure 25–8). The protozoan choanoflagellates have similar cells, and it is possible that sponges arose from colonial forms of such organisms.

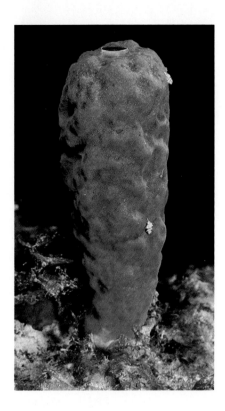

25-7 *A purple tube sponge. Pigments within the cells of this sponge are responsible for its brilliant color, which may also be enhanced by the refraction, or bending, of light as it passes through the water.*

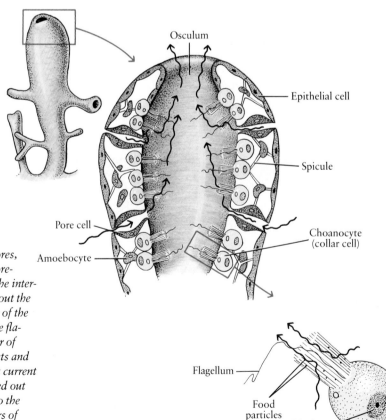

25-8 *The body of a simple sponge is dotted with tiny pores, from which the phylum derives its name (Porifera, or "pore-bearers"). Water containing food particles is drawn into the internal cavity of the sponge through these pores and is forced out the osculum. The water is moved by the sucking effect of flow of the local currents across the osculum and by the beating of the flagella protruding from the collars of choanocytes. The collar of each choanocyte is made up of about 20 retractile filaments and surrounds a single flagellum, the lashing of which directs a current of water through the filaments. Minute particles are filtered out and cling to one or more filaments and are then drawn into the cell. A sponge 10 centimeters high filters more than 20 liters of water a day.*

A sponge represents a level of organization somewhere between a colony of cells and a true multicellular organism. The cells are not organized into tissues or organs, yet there is a form of recognition among the cells that holds them together and organizes them. If the body of a living sponge is squeezed through a fine sieve or cheesecloth, it is separated into individual cells and small clumps of cells. Within an hour, the isolated sponge cells begin to reaggregate, and as these aggregations get larger, canals, flagellated chambers, and other features of the body organization of the sponge begin to appear. This phenomenon has been used as a model for the analysis of cell adhesion, recognition, and differentiation, all of which are basic features of embryonic development in higher organisms.

The outer surface of a sponge is covered with epithelial cells, some of which contract in response to touch or to irritating chemicals, and in so doing, close up pores and channels. Each cell acts as an individual, however; there is little coordination among them. Between the epithelial cells and the choanocytes is a middle, jellylike layer, and in this layer are amoebocytes, amoeba-like cells that carry out various functions. Amoebocytes play several roles in reproduction, secrete skeletal materials, and, most important, carry food particles from the choanocytes to the epithelial and other nonfeeding cells. Because all the digestive processes of sponges are carried out within single cells, even a giant sponge—and some stand taller than a man—can consume nothing larger than microscopic particles.

25-9 *A section of the skeleton of* Euplectella aspergillum, *a species of glass sponge picturesquely known as Venus's flower basket. These fragile sponges, with their delicate silica-containing skeletons, are usually found at great depths. According to the fossil record, glass sponges were present in the Ordovician period, which began 505 million years ago.*

The sponge shown in Figure 25-8 is a small and simple one. In larger sponges, the body plan, although essentially the same, looks far more complex. These sponges, which need correspondingly more food, have highly folded body walls that greatly increase the filtering and feeding surfaces. We have already encountered this evolutionary stratagem for increasing biological work surfaces at the cellular level—as in the inner membrane of the mitochondrion—and we shall be encountering it repeatedly in different animal structures.

The approximately 5,000 species of sponges are grouped into four classes, according to their skeletal structure, which serves for protection, stiffening, and support. In class Calcarea, the skeleton consists of individual spicules of calcium carbonate. Members of class Hexactinellida, the glass sponges, have spicules of silica fused in a continuous and often very beautiful latticework (Figure 25-9). The largest class, Demospongiae, has unfused silica spicules, or a tough, keratin-like protein called spongin, or a combination of the two. The cleaned and dried proteinaceous skeletons of this group are the "natural" sponges available commercially. Members of the fourth and smallest class, Sclerospongiae, have skeletons that contain all three kinds of material—calcium carbonate, silica, and spongin.

Reproduction in Sponges

The reproduction of sponges exhibits many of the features characteristic of sessile or slow-moving animals. Asexual reproduction is quite common, either by fragments that break off from the parent animal, or by gemmules, aggregations of amoebocytes within a hard, protective outer layer. Production of such resistant forms occurs most frequently in freshwater organisms. In the ocean, conditions are relatively unchanging, but the freshwater environment is much more variable. Invertebrates that live in freshwater are more likely to have protected embryonic forms than even closely related marine species.

Sexual reproduction in sponges is highly specialized. As we noted earlier, the simplest and most primitive form of fertilization is external, with sperm and egg cells shed into the water. In most sponges, however, fertilization is internal.

25-10 *A breadcrumb sponge* (Halichondria panacea) *showing some of the many openings (oscula) through which water leaves the animal. This sponge grows as a flat, encrusting mass on the undersurfaces of rocks where it is protected from direct sunlight at low tide. The oscula usually protrude from the rest of the animal. This arrangement allows the natural water flow in the habitat to draw water through the sponge.*

Gametes appear to arise from enlarged amoebocytes, but there are reports that choanocytes can also form gametes. The sperm cells, which resemble those of other animals, are carried by the water currents out of the osculum of one sponge and into the interior cavity of another sponge. There they are captured by choanocytes and transferred to amoebocytes, which then transfer them to ripe eggs—a method of fertilization unique to the sponges. Most sponges even provide a certain amount of maternal care, retaining the young during the early stages of development. The embryonic sponge develops into a flagellated, free-swimming larva that, after a short life in the plankton, locates an appropriate site, settles, and develops into an adult sponge.

Most kinds of sponges are **hermaphrodites** (from Hermes and Aphrodite); that is, the same individual has both male and female reproductive structures and produces both sperm and egg cells. This is a great advantage for animals with little or no motility. In animals with separate sexes, a given individual can mate only with members of the opposite sex, but for a hermaphrodite any partner—or the gametes of any partner—will suffice.

PHYLUM MESOZOA: MESOZOANS

Like the sponges, the mesozoans are so different from all other animals that they are placed in a subkingdom of their own. About 50 species of mesozoans are known; they are extremely simple wormlike animals that live as parasites inside a variety of marine invertebrates. The 20 to 30 cells that make up the body of a mesozoan are organized in two layers—a mass of reproductive cells surrounded by a single layer of ciliated cells (Figure 25-11). There are no organs or body cavities of any kind. As their name ("middle animals") suggests, the mesozoans may be primitive forms, transitional between protists and multicellular animals. Some authorities believe, however, that they are flatworms (phylum Platyhelminthes) that have become simplified as an adaptation to a parasitic way of life.

RADIALLY SYMMETRICAL ANIMALS

Two phyla, Cnidaria (jellyfishes, sea anemones, and corals) and Ctenophora (comb jellies and sea walnuts), consist of gelatinous animals in which the adult form is generally radially symmetrical. In radial symmetry, the body parts are arranged around a central axis, like spokes around the hub of a wheel (Figure 25-12).

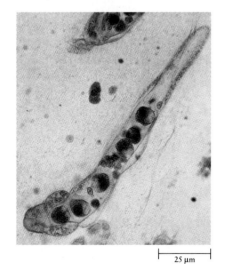

25-11 Dicyema typoides *is a mesozoan found in the kidneys of octopuses living in the Gulf of Mexico. Mesozoans typically reproduce asexually during one phase of the life cycle and sexually during another phase. This individual is a sexually reproducing adult, forming both sperm and egg cells.*

The animals in these two phyla are also characterized by a **gastrovascular cavity,** which has only one opening. Within this cavity, enzymes are released that break down food, partially digesting it extracellularly, as our own food is digested within the stomach and intestinal tract. The food particles are then taken up by the cells lining the cavity; they complete the digestive process and pass the products on to the other cells of the animal. Water circulating through the gastrovascular cavity supplies dissolved oxygen to the lining cells and carries carbon dioxide, other waste products, and the inedible remains of food particles out through the single opening.

Phylum Cnidaria

The cnidarians are a large and often strikingly beautiful group of aquatic organisms. The cells of cnidarians, unlike those of sponges and mesozoans, are organized into distinct tissues, and their activities are coordinated by a nervous system.

As you can see in Figure 25–13a, the basic body plan is simple: the animal is essentially a hollow container, which may be either vase-shaped, the **polyp,** or bowl-shaped, the **medusa.** The polyp is usually sessile; the medusa, motile. Both consist of two layers of tissue: **epidermis** and **gastrodermis** (from the Greek *epi,* "on" or "over," and *gaster,* "stomach," plus *derma,* "skin"). Between the two layers is a gelatinous filling, the **mesoglea** ("middle jelly"), which is made of a collagenlike material. In the polyp form, the mesoglea is sometimes very thin, but in the medusa, it often accounts for the major portion of the body substance. The two tissues of cnidarians are derived from two embryonic tissues, the **ectoderm** and the **endoderm** (from the Greek *ektos,* "outer," and *endon,* "inner"). Because they have two embryonic tissue layers, cnidarians are said to be diploblastic.

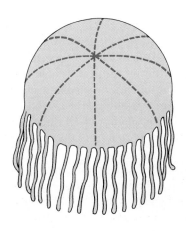

25-12 *In organisms with radial symmetry, any plane through the animal that passes through the central axis divides the body into halves that are mirror images of one another.*

25-13 (a) *Among cnidarians, there are two basic body forms: the vase-shaped polyp (left) and the bowl-shaped medusa (right). The gastrovascular cavity has a single opening. The cnidarian body has two tissue layers, epidermis and gastrodermis, with gelatinous mesoglea between them.*

(b) *Cnidocytes, specialized cells located in the tentacles and body wall, are a distinguishing feature of cnidarians. The interior of the cnidocyte is filled by a nematocyst, which consists of a capsule containing a coiled tube, as shown on the left. A trigger on the cnidocyte, responding to chemical or mechanical stimuli, causes the tube to shoot out, as shown on the right. The capsule is forced open and the tube turns inside out, exploding to the outside. The cnidocyte cannot be "reloaded"; it is absorbed and a new cell grows to take its place.*

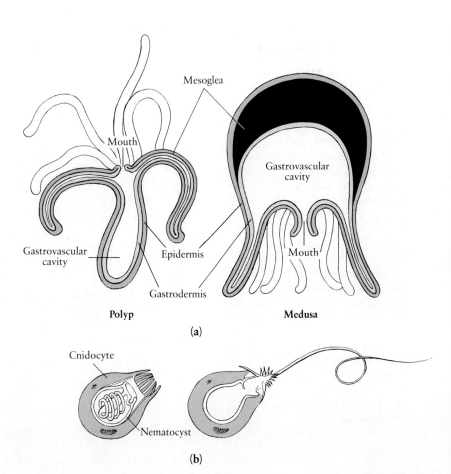

A distinctive feature of these animals, the **cnidocyte** (Figure 25–13b), gives the phylum its name. Cnidarians are carnivores. They capture their prey by means of tentacles that form a circle around the mouth. These tentacles are armed with cnidocytes, special cells that contain **nematocysts** (thread capsules). Nematocysts are discharged in response to a chemical stimulus or touch. The tubular nematocyst threads, which are often poisonous and may be sticky or barbed, can lasso prey, harpoon it, or paralyze it—or some useful combination of all three. The toxin apparently produces paralysis by attacking the lipoproteins of the nerve cell membranes of the prey.

Nematocysts occur only in this phylum, with some interesting exceptions. Certain other invertebrates, including nudibranchs (a kind of mollusk) and flatworms, can eat cnidarians without triggering the nematocysts. The nematocysts are then moved to the surface of the predator and can be fired in their new host's defense.

The cnidarian life cycle is characterized by an immature larval form, known as the **planula,** which is a small, free-swimming ciliated organism. Following this larval stage, some cnidarians go through both a polyp and a medusa stage in their life cycles (Figure 25–14). In such species, polyps reproduce asexually and medusas sexually. This sort of life cycle, in which the sexually reproductive form is distinctly different from the asexual form, superficially resembles alternation of generations in plants. There is, however, no alternation between haploid and diploid forms as there is in plants; the only haploid forms are the gametes. The cnidarian life cycle allows for rapid asexual reproduction (by the polyp), dispersal and genetic recombination (by the medusa), and habitat selection (by the planula larva).

The approximately 9,000 species of cnidarians are grouped in three major classes: Hydrozoa, in which the polyp is usually the dominant form; Scyphozoa, predominantly medusoid, exemplified by the large jellyfishes; and Anthozoa, which includes the sea anemones and the reef-building corals, and has only the polyp.

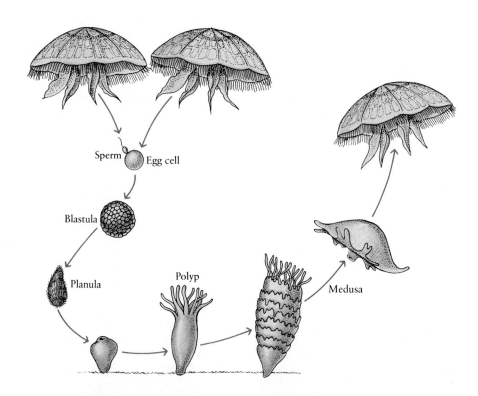

25–14 *The life cycle of the cnidarian* Aurelia. *Sperm and egg cells are released from adult medusas into the surrounding water. Fertilization takes place, and the resulting zygote develops first into a hollow sphere of cells, the blastula. It then elongates and becomes a ciliated larva called a planula. After dispersal, the planula settles to the bottom, attaches by one end to some object, and develops a mouth and tentacles at the other end, thus transforming into the polyp stage. The body of the polyp grows and, as it grows, begins to form medusas, stacked upside down like saucers. In a phase of rapid asexual reproduction, these bud off, one by one, and grow into full-sized jellyfish.*

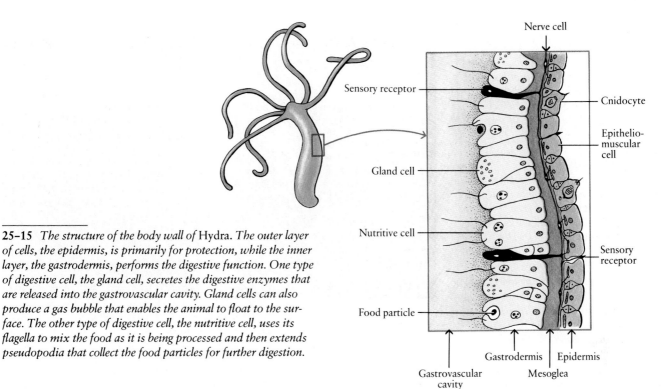

25-15 *The structure of the body wall of* Hydra. *The outer layer of cells, the epidermis, is primarily for protection, while the inner layer, the gastrodermis, performs the digestive function. One type of digestive cell, the gland cell, secretes the digestive enzymes that are released into the gastrovascular cavity. Gland cells can also produce a gas bubble that enables the animal to float to the surface. The other type of digestive cell, the nutritive cell, uses its flagella to mix the food as it is being processed and then extends pseudopodia that collect the food particles for further digestion.*

Class Hydrozoa

Among the most thoroughly studied of the cnidarians are species of *Hydra* and related genera, which are small, common freshwater forms, convenient to keep in the laboratory. Figure 25–15 shows a small section of the body wall of *Hydra*. The epidermis is composed largely of epitheliomuscular cells, which perform a covering, protective function and also serve as muscle cells. Each cell has contractile fibers, myonemes, at its base and so can contract individually, like the contractile epithelial cells of the sponge. The gastrodermis is mostly made up of cells concerned with digestion; these cells also contain contractile fibers. In *Hydra*, as in other polyps, the contractile fibrils of the epidermis run lengthwise in the animal and the fibrils of the gastrodermis run circularly, so the body wall can stretch or bulge, depending on which group contracts.

In addition to cnidocytes and epitheliomuscular cells, which are independent effectors—cells that both receive and respond to stimuli—*Hydra* contains two other types of nerve cells: sensory receptor cells and cells connected into a network, the nerve net. Sensory receptor cells are more sensitive than other epithelial cells to chemical and mechanical stimuli, and when stimulated they transmit their impulses to an adjacent cell or cells, which then respond. Note that this system is one step more complicated than the individual epitheliomuscular cell or cnidocyte, which acts as both receptor and effector. The nerve net, a loose connection of nerve cells lying at the base of the epidermal layer, is the simplest example of a nervous system that links an entire organism into a functional whole. It coordinates the muscular contractions of *Hydra*, making possible a wide variety of activities (Figures 25–2 and 25–16). However, there is no center of operations for the nervous system. This type of conducting system occurs in *Hydra* and certain other cnidarians.

Hydra, a relatively simple hydrozoan that lives as a solitary polyp and has no medusa stage, is thought to be descended from more complex ancestors. Most hydrozoans are colonial marine animals and have both hydroid (polyp) and medusoid forms at different times in their life cycles. Members of the genus *Obelia*, for example, spend most of their lives as colonial polyps (Figure 25–17a).

25-16 Hydra *has a nervous system that integrates the body into a functional whole, making possible a range of fairly complex activities. For instance,* Hydra *may float, glide on its base, or, as shown here, it may travel by a somersaulting motion.*

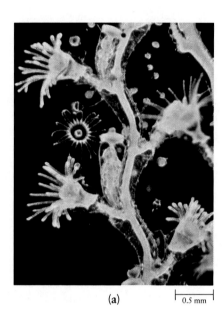 

(a) 0.5 mm (b)

25-17 *Colonial hydrozoans.* **(a)** *Obelia is made up of two types of polyp, a* Hydra-*like form—shown here with tentacles extended—which is the feeding polyp, and a reproductive form, which lacks tentacles. You can see two of these reproductive forms in the axils of the branches. At the left is a newly released free-swimming medusa.*

(b) *Cnidarians of the order Siphonophora are large, floating colonies made up of both polyps and medusas. The polyps are feeding forms and the medusas are reproductive forms; in some species, nonreproductive medusas are swimming bells. The colony produces a gas-filled float, or pontoon. In the Portuguese man-of-war, shown here with a captured fish, the large float serves also as a sail. The blue strands are composed of reproductive and feeding individuals; the purple strands, which may grow as long as 15 meters, are made up of stinging, food-gathering polyps armed with nematocysts. A large colony can inflict serious injury on a human being, leading, in some cases, to death by drowning or as a consequence of a severe allergic reaction.*

The colony arises from a single polyp, which multiplies by budding. The new polyps do not separate but remain interconnected so that their gastrovascular cavities form a continuous channel, through which food particles and oxygen-bearing water are circulated. Within the colony are two types of polyps: feeding polyps with tentacles and cnidocytes, and reproductive polyps from which tiny medusas bud off. These medusas produce eggs or sperm that are released into the water and fuse to form zygotes. Many colonial hydrozoans, with their division of labor between feeding and reproductive forms, are very like a single organism (Figure 25-17b).

Class Scyphozoa

A second major class of cnidarians is Scyphozoa, or "cup animals," in which the medusa form is dominant. Although the fairly simple polyp form reproduces itself asexually, in addition to budding off medusas, elaborate colonies are not formed. Scyphozoan medusas, known more commonly as jellyfishes, range in size from less than 2 centimeters up to 4 meters in diameter; the largest jellyfishes trail tentacles as long as 10 meters. In the adult animal, the mesoglea is so firm that a large, freshly beached jellyfish can easily support the weight of a man. The mesoglea of some jellyfishes contains wandering, amoeba-like cells, which serve to transport food from the nutritive cells of the gastrodermis.

In scyphozoans, the epitheliomuscular cells, which are more specialized than those of *Hydra*, underlie the epidermis, contracting rhythmically to propel the medusa through the water (see Figure 25-1). The contractions are coordinated by concentrations of nerve cells in the margin of the bell. These nerve cells connect with fibers innervating (providing the nerve supply for) not only the musculature but also the tentacles and the sense organs.

The bell margin is liberally supplied with sensory receptor cells sensitive to mechanical and chemical stimuli. In addition, jellyfishes have two types of multicellular sense organs: statocysts and light-sensitive ocelli. **Statocysts** are specialized receptor organs that provide information by which an animal can orient itself with respect to gravity (Figure 25-19a). The statocyst, which seems to have been one of the first types of sensory organ to have evolved, has persisted apparently unchanged to the present day. This simple structure, providing essential information for motile organisms, is found in many animal phyla and is

25-18 *The moon jellyfish* (Aurelia aurita) *has four frilly mouth lobes that gather minute organisms into the gastrovascular cavity. It has short tentacles and usually grows to only about 30 centimeters in diameter.*

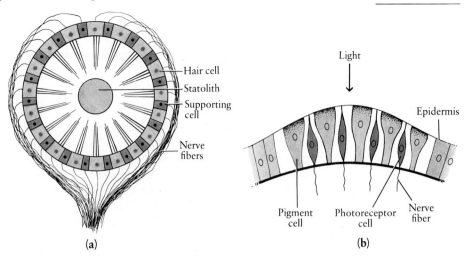

25–19 (a) *The statocyst is a specialized receptor organ that orients the jellyfish with respect to gravity. When the bell tilts, gravity pulls the statolith, a grain of hardened calcium salts, down against the hair cells. This stimulates the nerve fibers and signals the animal to right itself.* (b) *Ocellus (eyespot), the simplest type of photoreceptor organ. Ocelli of this sort are common among the cnidarians.*

thought to have arisen independently on a number of occasions. (As we shall see in Chapter 32, similar structures may also play a role in orienting the growth of plant roots with respect to gravity.) **Ocelli**, which may have evolved even earlier, are groups of pigment cells and photoreceptor cells (Figure 25–19b). They are typically located at the bases of the tentacles.

Class Anthozoa

Anthozoans ("flower animals")—the sea anemones and corals—are cnidarians that, like *Hydra*, have no medusa stage. These polyps reproduce both asexually, by budding, division, or fragmentation, and sexually, by the production of gametes. The zygote develops into a planula larva that may travel some distance from the parents before attaching itself to a suitable substrate and developing into an adult animal.

The gastrovascular cavity of anthozoans, unlike that of hydrozoan polyps, is divided by vertical partitions. In most corals, which are colonies of anthozoans, the epidermal cells secrete protective outer walls, usually of calcium carbonate (limestone), into which each delicate polyp can retreat. The limestone-forming polyps are the most ecologically important of the cnidarians. A coral reef is composed primarily of the accumulated limestone skeletons of the corals, covered by a thin crust occupied by the living colonial animals. A reef is both the structural and nutritional basis of the complex coral reef community (see essay).

25–20 *Ghost anemones (Diadumene leucolena), photographed in the waters off the coast of Rhode Island. The flowerlike appearance is deceptive; sea anemones are carnivorous animals belonging to a class of cnidarians that, like* Hydra, *have dropped the medusa stage. In common with other cnidarians, their tentacles are equipped with stinging nematocysts. The tentacles move food into the gastrovascular cavity, which is divided longitudinally by partitions.*

The Coral Reef

The coral reef is the most diverse of all marine communities. The reef structure itself is formed by colonial anthozoans. Each polyp in the colony secretes its own calcium-containing skeleton, which then becomes part of the reef. The photosynthetic activity of the reef is carried out almost entirely by symbiotic dinoflagellates and green algae living within the tissues of the corals; in fact, as much as half of the living substance of a coral reef may consist of green algae. Carbon, oxygen, and dissolved minerals flow over the reef as a result of the movement of waves and the ocean currents. The reef furnishes both food and shelter for other sea animals, including numerous species of reef fishes and a tremendous variety of invertebrates, such as sponges, sea urchins, marine worms, and crustaceans. The coral polyps and algae that form the reef can grow only in warm, well-lighted surface water, where the temperature seldom falls below 21°C.

The largest coral-created land masses in the world are the Marshall Islands in the Pacific and the 2,000-kilometer-long Great Barrier Reef, off the northeast shores of Australia. Other reefs are found throughout tropical waters and as far north as Bermuda, which is warmed by the Gulf Stream.

(a)

(b)

(a) *Living gorgonian coral, photographed in the Caribbean. Notice the individual polyps, extended from their limestone skeletons.*

(b) *A portion of a coral reef in the New Hebrides, islands in the Pacific, east of Australia. Staghorn coral is in the foreground.*

25–21 *Mnemiopsis leidyi, a ctenophore that is common along the Atlantic and Gulf coasts of the United States. In adults of* Mnemiopsis *and related genera, the two long tentacles characteristic of other ctenophores are greatly reduced in size. Muscular mouth lobes and short tentacles around the mouth work together in the capture and movement of prey organisms into the gastrovascular cavity.*

Phylum Ctenophora

About 90 species, commonly known as comb jellies and sea walnuts, make up phylum Ctenophora ("comb-bearers"). These small animals are readily identified by comblike plates of fused cilia that are arranged in eight longitudinal bands on the body surface (Figure 25–21). The coordinated beating of the cilia is their primary means of locomotion, although a few species also use muscular movements of the body wall for more rapid swimming. Ctenophores are abundant in the plankton of the open ocean; comb jellies and sea walnuts are bioluminescent, giving off bright flashes of light that are particularly visible at night.

Ctenophores, like cnidarians, are characterized by a gastrovascular cavity and two tissue layers, epidermis and gastrodermis, with a gelatinous mesoglea between. Amoeboid cells and muscle fibers are scattered through the mesoglea, which makes up most of the body substance. In many species, two long tentacles, containing specialized cells that secrete a sticky substance, are used to capture food.

Reproduction is sexual, and all individuals are hermaphrodites. The fertilized eggs develop into free-swimming larvae that gradually change into the adult form.

BILATERALLY SYMMETRICAL ANIMALS: AN INTRODUCTION

All of the remaining animals, including ourselves, are bilaterally symmetrical. The apparent exceptions, such as the echinoderms, which are radially symmetrical as adults, pass through a bilateral stage during their development. In bilaterally symmetrical animals, the body is organized along a longitudinal axis, with the right half an approximate mirror image of the left half (Figure 25–22). Bilateral symmetry makes possible more efficient locomotion than does radial symmetry, which is typically found in animals that move slowly or are sedentary.

A bilaterally symmetrical animal also has a top and a bottom or, in more precise terms, a **dorsal** and a **ventral** surface. These terms are applicable even when the organism is turned upside down, or, as with humans, stands upright, in which case dorsal means back and ventral means front. Most bilateral organisms also have distinct "head" and "tail" ends—**anterior** and **posterior**. Having one end that goes first is characteristic of actively moving animals. In such animals, many of the sensory cells are collected into the anterior end, enabling the animal to assess an area before entering it. With the aggregation of sensory cells, there came a concomitant gathering of nerve cells that is the forerunner of the brain. Structures useful in capturing and consuming prey are also generally located in the anterior region of the animal, whereas digestive, excretory, and reproductive structures

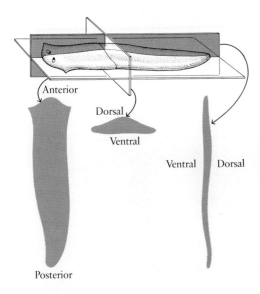

25–22 *In a bilaterally symmetrical organism, such as the planarian shown here, the right and left halves of the body are mirror images of one another. The upper and lower (or back and front) surfaces are known as dorsal and ventral. With some exceptions, the end that goes first is termed anterior and the one that brings up the rear, posterior.*

tend to be located toward the posterior. The concentration of sensory and nerve cells and of structures associated with feeding at the anterior end of an animal is known as **cephalization.**

The bilateral animals are all triploblastic; that is, they have three embryonic tissue layers. The third layer is the **mesoderm,** a layer of cells located between the ectoderm and the endoderm. These three layers can be detected very early in the development of bilateral animals and give rise to the various specialized tissues of the adult animal. Thus they are known as the "germ layers." Generally speaking, covering and lining tissues, as well as nerve tissues, are derived from ectoderm; digestive structures from endoderm; and muscles and most other parts of the body from mesoderm. This general pattern makes functional sense. When the sensory and nerve cells evolved, they did so as specializations of the outer layer, the ectoderm, where sensation was most important. Similarly, digestive structures evolved in the inner layer, the endoderm, which surrounds the food-containing cavity. As animals became more complex and additional structures evolved for locomotion, internal transport, excretion, and reproduction, they did so from the "new," middle layer, the mesoderm.

The triploblastic animals can be grouped in three categories, according to the presence or absence of a body cavity—a **coelom** (pronounced "see-loam")—in addition to the digestive cavity. In the simplest arrangement (Figure 25–23b), tissues derived from the three germ layers are packed together and there is no body cavity other than the digestive cavity. Animals of this type are known as **acoelomates.** A more complex arrangement is found in the **pseudocoelomates.** These animals have an additional cavity that develops between the endoderm and the mesoderm (Figure 25–23c); this cavity is known as a **pseudocoelom** because of its location and because it lacks the epithelial lining characteristic of a coelom. The **coelomates** (mollusks, annelids, and all of the more complex animals) have a true coelom, which is a fluid-filled cavity that develops *within* the mesoderm (Figure 25–23d). Within the coelom, the digestive tract ("gut") and other internal organs—lined with epithelium—are suspended by double layers of mesoderm known as mesenteries. The advantages of these increasingly complex arrangements will become apparent as we examine the animals themselves.

25–23 *Basic arrangements of animal tissue layers, as shown in cross section.*
(a) *A body that consists of only two tissue layers is characteristic of cnidarians and ctenophores.*
(b) *Acoelomates, such as flatworms and ribbon worms, have three-layered bodies, with the layers closely packed on one another.*
(c) *Pseudocoelomates, such as nematodes, have three-layered bodies in which a pseudocoelom develops between the endoderm and mesoderm.*
(d) *Coelomates—mollusks, annelids, and most other animals, including vertebrates—have bodies that are three-layered with a cavity, the coelom, that develops within the middle layer (mesoderm). The mesodermal mesenteries suspend the gut within the body wall.*

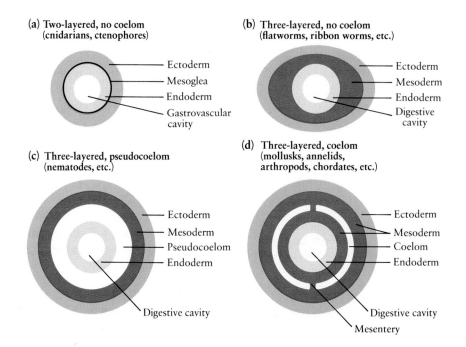

PHYLUM PLATYHELMINTHES: FLATWORMS

The flatworms are the simplest animals with bilateral symmetry and three distinct germ layers. Moreover, not only are their tissues specialized for various functions, but also two or more types of tissue cells may combine to form organs. Thus, while sponges are made up of aggregations of cells and cnidarians are largely limited to the tissue level of organization, flatworms can be said to exemplify the organ level of complexity.

Like the cnidarians, but unlike most other bilaterally symmetrical animals, the flatworms have a digestive cavity with only one opening. Since the animal cannot feed, digest, and eliminate undigested residues simultaneously, food cannot be processed continuously.

Flatworms are acoelomates with solid bodies, and they have no circulatory system for the transport of oxygen and food molecules. Thus all cells must be within diffusion distance of sources of oxygen and of food. Flatworms have solved this problem in two ways. First, the body is flattened, which keeps the cells close to the external oxygen supply, and second, the digestive cavity is branched, carrying food particles to all regions of the body.

The flatworms are believed by some zoologists to have evolved from the cnidarians (or, perhaps, vice versa), not by way of either adult form, however, but from a ciliated planula larva that became sexually mature without going through polyp or medusa stages. Others are persuaded that the flatworms had independent origins among the ciliates. Still another group maintains that they are degenerate annelids. The new molecular techniques for determining phylogenetic relationships (page 418) may ultimately resolve this question.

About 13,000 species of flatworms have been described, and they are placed in three classes. Class Turbellaria contains mostly free-living forms, whereas the members of classes Trematoda (flukes) and Cestoda (tapeworms) are parasitic.

Class Turbellaria

The free-living flatworms form a large and varied group, and we shall single out just one type for examination, the freshwater planarian (Figure 25–24). The ectoderm of a planarian gives rise to cuboidal epithelial cells, many of which are ciliated, particularly those on the ventral surface. Ventral epithelial cells secrete mucus, which provides traction for the planarian as it moves by means of its cilia along its own slime trail. Planarians are among the largest animals that can use cilia for locomotion. The cilia in larger species are usually employed for moving water or other substances along the surface of the animal, as in the human respiratory tract, rather than for propelling the animal.

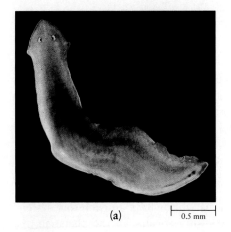

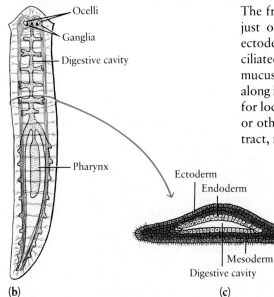

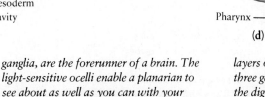

25–24 (a) *Example of a flatworm: a freshwater planarian.* (b) *The nervous system is indicated in color. Note that some of the fibers are aggregated into two cords, one on each side of the body. Clusters of nerve cells in the head, known as ganglia, are the forerunner of a brain. The light-sensitive ocelli enable a planarian to see about as well as you can with your eyes closed.* (c) *Planarians, like other flatworms, but unlike cnidarians, have three layers of body tissues, derived from the three germ layers. The only body cavity is the digestive cavity; thus flatworms are acoelomates.* (d) *A carnivore, the planarian feeds by means of its extensible pharynx.*

25–25 *A free-living marine flatworm,* Pseudoceros zebra, *on the surface of a red sponge. The many species belonging to phylum Platyhelminthes, though widely varied in color and shape, are nearly all small and flat-bodied. This flatworm, a member of the order Polycladida, has a highly branched digestive cavity, extending into all parts of its body.*

Like other turbellarians, the planarian is carnivorous. It eats either dead meat or other slow-moving animals, including smaller planarians. It feeds by means of a muscular organ, the pharynx, which is free at one end. The free end can be stretched out through the mouth opening. Muscular contractions in the tubular pharynx cause strong sucking movements, which tear the meat into microscopic bits and draw them into the digestive cavity, where they are phagocytized by the cells of the lining. Because this digestive cavity has three main branches, planarians are placed in the order Tricladida.

Unlike the sponges or cnidarians, most flatworms have an excretory system, which is especially well developed in freshwater species. In the planarian, the system is a network of fine tubules that runs the length of the animal's body (Figure 25–26). Side branches of the tubules contain flame cells, each of which has a hollow center in which a tuft of cilia beats, moving water along the tubules to the exit pores between the epithelial cells. The flame-cell system appears to function largely to regulate water balance; most of the metabolic waste products probably leave the flattened body by diffusion.

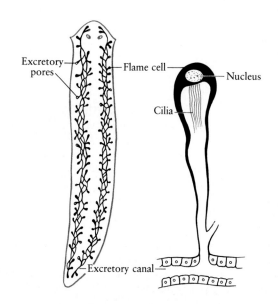

25–26 *Flatworms have a tubular excretory system. The system usually consists of two or more branching tubules running the length of the body. In the planarian and its relatives, the tubules open to the body surface through a number of tiny pores. At the ends of the side branches are small bulblike structures known as flame cells. Within each of these cells, a tuft of cilia in constant motion resembles the flickering of a flame. Water and some waste materials from the tissue fluids are moved by the cilia through tubules to the excretory pores, where the collected liquid leaves the body.*

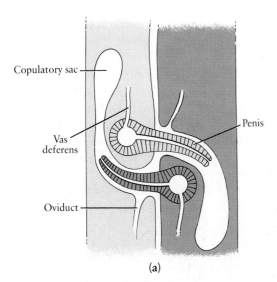
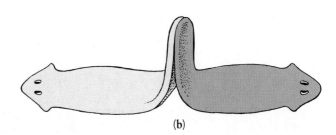

25–27 (a) *Planarians have both male and female reproductive structures, and mating involves a mutual exchange of sperm. The erect penis of each partner is inserted into the copulatory sac of the other. The vas deferens is a duct leading from the testes, where sperm are produced by meiosis, to the penis.* (b) *Two planarians mating.*

Planarians can also reproduce asexually, by fragmentation or by fission. Fragments of the tail, for example, may regenerate new heads.

In most respects, the free-living flatworms are relatively simple, and the rest of the bilaterally symmetrical animals may have evolved from something quite like them. Their reproductive systems, however, are complex, and fertilization is internal. Planarians, for example, are hermaphrodites, like most other flatworms, and when they mate, each partner deposits sperm in the copulatory sac of the other (Figure 25–27). These sperm then travel along special tubes, the **oviducts,** to fertilize the eggs as they become ripe.

The Planarian Nervous System

The evolution of bilateral symmetry brought with it marked changes in the organization of the nervous system as well as of other systems. Even among primitive flatworms, the neurons (nerve cells) are not dispersed in a loose network, as in *Hydra,* but are instead condensed into longitudinal cords. In the planarians, this condensation is carried further, and there are two main conducting channels, one on each side of the flat, ribbonlike body. These channels carry impulses to and from the aggregation of nerve cells in the anterior end of the body (see Figure 25–24b). Such aggregations of nerve cell bodies are known as **ganglia** (singular, ganglion).

The ocelli of the planarian are usually inverted pigment cups (Figure 25–28). They have no lenses, and they cannot form an image. However, they can distinguish light from dark and can identify the direction from which the light is coming. Planarians are photonegative; if you shine a light from the side on a dish of planarians, they will move steadily away from the source of light.

Among the epithelial cells are receptor cells sensitive to certain chemicals and to touch. The head region in particular is rich in chemoreceptors. If you place a small piece of fresh liver in the culture water so that its juices diffuse through the medium, the planarians will raise their heads off the bottom and, if they have not eaten recently, will lope directly and rapidly (on a planarian scale) toward the meat, to which they then attach themselves to feed. The animal locates the food source by repeatedly turning toward the side on which it receives the stimulus more strongly until the stimulus is equal on both sides of its head. This can be demonstrated experimentally; if the chemoreceptor cells are removed from one side of the head, the animal will turn constantly toward the intact side.

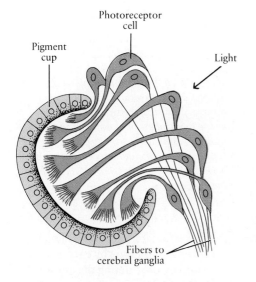

25–28 *Ocellus of the planarian. The light stimulus is received by the ends of the photoreceptor cells adjacent to the pigment cup. Thus the light travels first through the fibers carrying the signals to the cerebral ganglia. In this inside-out organization, the planarian ocellus resembles the vertebrate eye (see Figure 42–12).*

Classes Trematoda and Cestoda

Phylum Platyhelminthes includes also the trematodes, or flukes (class Trematoda), and the tapeworms (class Cestoda), parasitic forms that can cause serious and sometimes fatal diseases among vertebrates. Members of both of these parasitic classes have a tough outer layer of cells that is resistant to the body fluids of their hosts, particularly digestive fluids; most also have suckers or hooks on their anterior ends by which they fasten to their victims. Trematodes feed through a mouth, but the tapeworms—which have no mouths, digestive cavities, or digestive enzymes—merely hang on and absorb predigested food molecules through their skin (Figure 25-29).

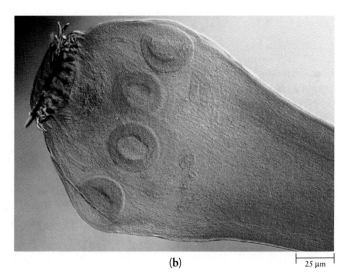

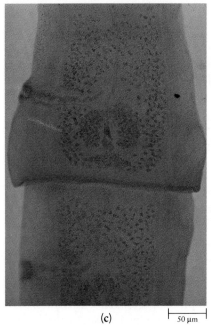

25-29 (a) Dipylidium caninum, *a common tapeworm of dogs and cats. This specimen was removed from a cat. Tapeworms are intestinal parasites that lack any digestive system of their own. They cling by their heads, which, as shown in* (b) Taenia pisiformis, *another common tapeworm of dogs, are equipped with hooks and suckers. They absorb food molecules—digested by their hosts—through their body walls.* (c) *Their bodies, posterior to the heads, are divided into segments known as proglottids. In these mature proglottids from* Taenia pisiformis, *the dark areas are the reproductive structures; the opening is the genital pore. Each proglottid is a sexually complete hermaphroditic unit in which sperm and eggs are produced. Proglottids break off when the eggs are mature, and new ones are formed. Typically, the eggs develop into larvae in members of one species that eat food contaminated by the eggs, and the larvae develop into adult worms in members of another species that eat the flesh of the first species.*

Tapeworms are found in the intestines of many vertebrates, including humans, and may grow as long as 5 or 6 meters. They cause illness not only by encroaching on the food supply but also by producing wastes and by obstructing the intestinal tract. The most common human tapeworm, the beef tapeworm, infects people who eat the undercooked flesh of cattle that have grazed on land contaminated by human feces containing tapeworm segments.

All parasites, including parasitic flatworms, are believed to have originated as free-living forms and to have lost certain tissues and organs (such as the digestive cavity) as a secondary effect of their parasitic existence, while developing adaptations of advantage to the parasitic way of life. Such adaptations often include a complex life cycle involving two or more hosts (see essay).

The Politics of Schistosomiasis

Many of the parasitic platyhelminths have a complex life cycle involving two different hosts. In the case of three species of the trematode genus *Schistosoma,* the hosts are freshwater snails and humans. The life cycle begins when small, swimming larvae are released from a snail; a single infected snail can shed some 100,000 larvae in its six-month lifetime. The larvae, if successful, attach to human skin and penetrate it. They feed, mature, and mate in the bloodstream. The female lays her eggs, which have sharp spines on the surface, in the capillaries of the bladder wall or the intestine (depending on the species). The symptoms of schistosomiasis (also sometimes called bilharzia, or snail fever) are caused by the eggs; they lodge in the liver and spleen, blocking blood vessels, and their sharp spines tear the surrounding tissues, causing hemorrhages.

Eggs leave the human host by passing through the blood vessel wall into the bladder or intestinal tract, where they are flushed out with the urine or feces. If the eggs are deposited in fresh water, they hatch immediately into ciliated larvae that seek out the particular species of snail in which they multiply asexually.

Schistosomiasis now affects some 200 to 300 million people in 71 countries in Asia, Africa, and South America. Ironically, western-style progress has contributed to its spread. Snails thrive in the still waters of irrigation canals and artificial lakes. The disease spread rapidly through Upper Egypt following the construction of the Aswan High Dam and the digging of permanent irrigation canals. Before the creation of the huge man-made Lake Volta in Ghana, the schoolchildren of the villages along the riverbank had an incidence of schistosomiasis of 1 percent. Now the prevalence in some lakeside villages is 100 percent.

Medical progress has not followed technology into the tropics. When the colonial period ended, the major research programs into tropical diseases ended too. Western medical research is geared overwhelmingly to the conquest of diseases affecting mainly rich, urban, older men and women. For instance, more than $1.4 billion was spent on cancer research by the U.S. government alone in 1987, as compared to $8 million worldwide on schistosomiasis and $24 million on malaria. Cancer affects some 10 million people, schistosomiasis 200 million, malaria 200 to 400 million. The challenge of schistosomiasis and the other tropical parasitic diseases lies not so much in the disease process itself as in providing incentives (and therefore funding) to cure—or, better yet, prevent—the diseases of the politically powerless rural poor.

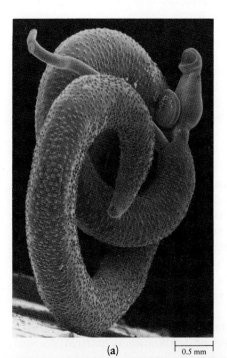

(b)

(a) *A scanning electron micrograph of male and female worms of* Schistosoma mansoni, *one of the three species that cause schistosomiasis in humans. The slender female is enveloped by the much larger body of the male. The two worms remain intertwined except when the female moves away to lay her eggs—about 300 per day. Adult worms live from 3 to 30 years, during which egg production continues unabated.* (b) *In the snail-infested waters near Aswan, Egypt, bathing children are rapidly infected by Schistosoma larvae. During the years that pass before the symptoms of disease become apparent, large quantities of eggs are shed by the children, continually infecting new generations of snails.*

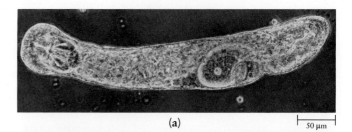

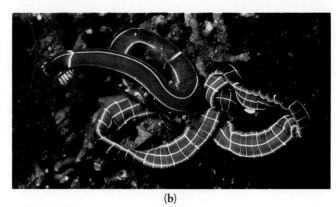

25–30 (a) *A living gnathostomulid,* Problognathia minima, *as seen through the phase-contrast microscope. This species lives in the intertidal sand flats along the coast of Bermuda. It moves by slowly gliding between the sand grains, moving its head rhythmically from side to side.*

(b) *Two ribbon worms of the species* Tubulanus superbus, *photographed in Scotland. Ribbon worms range in length from less than 2 centimeters to 30 meters and are virtually all the colors of the spectrum. All members of the phylum, however, have thin bodies (seldom more than 0.5 centimeter thick), a mouth-to-anus digestive tract, a circulatory system, and a long, muscular tube than can be thrust out to grasp prey.*

OTHER ACOELOMATES

Phylum Gnathostomulida

This small phylum contains about 80 species of tiny marine worms. Their distinguishing characteristic is a unique pair of hard jaws, from which the phylum derives its name (from the Greek *gnathos,* "jaw," plus *stoma,* "mouth"). Gnathostomulids are abundant along coastal shorelines, where they live in the spaces between particles of sand and silt, using their hard jaws to scrape bacteria and fungi from the particles.

These tiny worms have no coelom or pseudocoelom, and the digestive cavity has only one opening. Some zoologists believe that they are closely related to the free-living flatworms, whereas others think they are degenerate forms, derived from pseudocoelomates.

Phylum Rhynchocoela

Phylum Rhynchocoela ("snout" plus "hollow") consists of about 650 species of acoelomate worms, commonly called ribbon worms or nemertines. They are characterized by a long, retractile, slime-covered hollow tube (proboscis). The proboscis, sometimes armed with a barb, seizes prey and draws it to the mouth, where it is engulfed. Some inject a paralyzing poison into their prey.

The ribbon worms are of special interest to biologists attempting to reconstruct the evolution of the invertebrates. They appear to be closely related to the flatworms, but they exhibit two significant new features. Ribbon worms have a one-way digestive tract beginning with a mouth and ending with an anus. This is a far more efficient arrangement than the digestive cavity of the cnidarians and flatworms, with its single opening. In the two-opening tract, food moves assembly-line fashion, always in the same direction, with the consequent possibilities (1) that eating can be continuous and (2) that various segments of the tract can become specialized for different stages of digestion. These worms also have a circulatory system, typically consisting of one dorsal and two lateral blood vessels that carry the colorless blood.

Reproduction in the ribbon worms, however, is simpler than in most flatworms. Ordinarily the sexes are separate, and fertilization is external. Asexual reproduction by fragmentation of the body and regeneration of whole worms from the parts is also fairly common.

PSEUDOCOELOMATES

One major phylum (Nematoda) and seven minor phyla are characterized by a pseudocoelom, a body cavity that develops between the endoderm and the mesoderm (see Figure 25-23c). The pseudocoelom, which is essentially a sealed, fluid-filled tube, increases the effectiveness of the animal's muscular contractions. In addition to working against the water or a substrate, muscles must also have something to work against in an animal's body—otherwise the body just bends in the direction of contraction and a floppy, uncoordinated motion results. Because it resists bending, the pseudocoelom functions as a **hydrostatic skeleton** within the animal's body, causing the body to return to its original shape after the muscles have contracted. It makes possible a significant advance over the simple, rather flaccid movements of the acoelomate worms and most cnidarians.

All of the pseudocoelomates have a one-way digestive tract, but they lack a circulatory system. However, the movement of fluids within the pseudocoelom, enhanced by muscular contractions of the body wall, compensates for this.

Phylum Nematoda

About 12,000 species of nematodes (roundworms) have been described and named, but some authorities think that there may be as many as 400,000 to 500,000. Most are free-living, microscopic forms. It has been estimated that a spadeful of good garden soil usually contains about a million nematodes. Some are parasites; most species of plants and animals are parasitized by at least one species of nematode.

Nematodes (Figure 25-31) are cylindrical, unsegmented worms and are covered by a thick, continuous cuticle, which is molted periodically as they grow. An interesting, and unique, feature of nematode construction is the absence of circular muscles. The contraction of the longitudinal muscles acting against both the tough, elastic cuticle and the internal hydrostatic skeleton gives the worm its characteristic whipping motion in water. The mouth of a nematode has a muscular pharynx and often is equipped with piercing stylets. Reproduction is sexual, and the sexes are usually separate.

Humans are hosts to about 50 species of parasitic nematodes, which are a major cause of death and disability throughout the world. In North America, the most common parasitic nematodes are pinworm *(Enterobius)*, whipworm *(Trichuris)*, hookworm *(Ancylostoma)*, intestinal roundworm *(Ascaris)*, and *Trichinella*. The latter causes trichinosis, which is contracted by eating uncooked or undercooked pork, a single gram of which may contain 3,000 cysts (resting forms)

25-31 *A free-living marine nematode.*

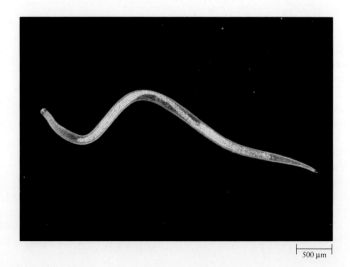

of *Trichinella*. Ingestion of only a few hundred of these cysts can be fatal. On a worldwide basis, it is estimated that 650 million people are infected by *Ascaris*, 450 million by *Ancylostoma*, 250 million by *Filaria*, the cause of filariasis, and 50 million by *Onchocerca*, the cause of "river blindness" in fertile regions of West Africa and in the coffee-growing highlands of Mexico and Guatemala. Research is revealing that many of these worms have unique enzymes and metabolic pathways directly related to such features of the parasitic life style as migration within the body of the host, attachment to host tissues, and the massive production of eggs.

Other Pseudocoelomate Phyla

The members of seven other phyla, quite diverse, have body plans based upon the pseudocoelom. Phylum Nematomorpha consists of about 230 species of horsehair worms that resemble nematodes. The adults, which do not feed, are free-living and reproduce in water (Figure 25-32a); the juveniles are parasites of arthropods. The 500 species of spiny-headed worms, phylum Acanthocephala, are parasitic throughout their life cycle. The larval forms develop in the tissues of arthropods, and the adults live and reproduce in the intestines of vertebrates. The adult worms, which superficially resemble tapeworms, lack a digestive tract and have a proboscis bearing hooks that attach to the intestinal wall of the host (Figure 25-32b and c).

The members of four pseudocoelomate phyla are often found, along with the acoelomate gnathostomulids, in the spaces between sand and silt particles along shorelines. These include the three species of the new phylum Loricifera (see Figure 20-8, page 415), discovered clinging tightly to sand grains in ocean waters at a depth of 25 to 30 meters. The larvae of these animals are free-swimming, but the adults are sedentary. By contrast, the adults of the other three phyla (Figure 25-33) are motile. Phylum Kinorhyncha contains about 100 species of tiny, burrowing marine worms that feed on diatoms in muddy ocean shores. Short-bodied, these worms are covered with spines and have a spiny, retractile proboscis. The sexes are separate. Most of the approximately 400 species in phylum Gastrotricha, however, are hermaphrodites. They live in sandy shores of both fresh and salt water, where they feed on protists and dead organic matter. Many of the 1,500 to 2,000 species of phylum Rotifera are found in the same environment in fresh water, as well as around the vegetation in ponds and along lake shores. Other members of the phylum are marine, and some are terrestrial. Rotifers are sometimes called "wheel animalcules" because the beating of a crown of cilia around the mouth causes them to spin through the water like tiny wheels.

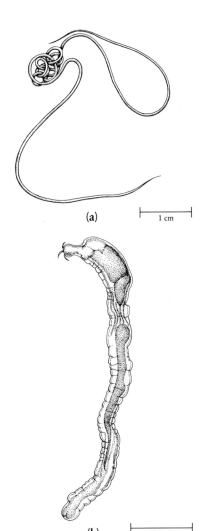

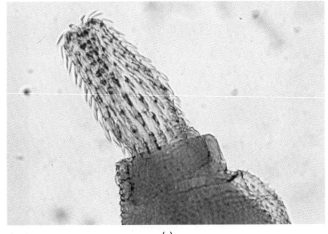

25-32 *Two phyla of pseudocoelomates consist of worms that are parasitic during at least one stage of the life cycle.* **(a)** *A horsehair worm, phylum Nematomorpha. The specimen from which this drawing was prepared was found in a garden hose on Long Island, New York. It was probably carried into the hose by a cricket or a grasshopper that it parasitized during its juvenile stage. Because of the intricate knots into which they tie themselves, nematomorphs are also known as gordian worms, after Gordius, King of Phrygia, who tied the gordian knot severed by Alexander the Great.*

(b) *An adult male of phylum Acanthocephala. Known commonly as spiny-headed worms, acanthocephalans are characterized by a hook-bearing proboscis* **(c)**.

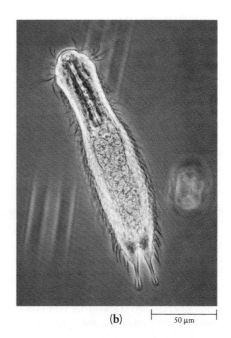
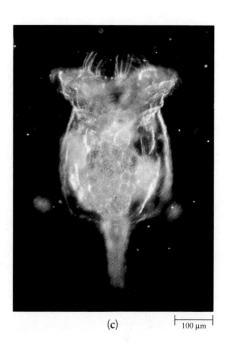

25–33 *Among the residents of the sand and silt of shorelines are members of three pseudocoelomate phyla—Kinorhyncha, Gastrotricha, and Rotifera.* **(a)** *Centroderes spinosus, a kinorhynch. Unable to swim, a kinorhynch burrows by forcing fluid into its head; when the head is anchored in the mud by its spines, the animal can pull the rest of its body forward.* **(b)** *Chaetonotus, a common gastrotrich. Gastrotrichs can both swim and crawl, clinging to surfaces by means of adhesive tubes that project from the sides of their bodies.* **(c)** *A rotifer of the genus* Brachionus, *commonly found in freshwater plankton. The cilia that propel the rotifer through the water are visible at the anterior end of the organism.*

Rotifers have a muscular pharynx with hard jaws; they feed on protists, bits of vegetation, and other animals even smaller than themselves. The sexes are separate in rotifers, but in some groups the females produce eggs that can develop without fertilization, a phenomenon known as **parthenogenesis.**

Phylum Entoprocta contains about 75 species of stalked, usually sessile animals that superficially resemble both hydrozoans and a group of coelomate animals known as bryozoans, or ectoprocts (see page 562). Their internal structure, however, identifies them as pseudocoelomates. They have a U-shaped digestive tract, and both the mouth and anus are located within a circle of tentacles (Figure 25–34). (In the bryozoans, by contrast, the anus is located outside a circle of tentacles, and the tentacles are structurally different from those of the entoprocts.) Reproductive patterns vary among the entoprocts. Some species have separate sexes, some are simultaneous hermaphrodites, and in others a single reproductive organ produces sperm at one stage in the life cycle and eggs at a later stage. As we shall see in the next chapter, this phenomenon, known as sequential hermaphroditism, is not limited to the entoprocts.

25–34 *A marine entoproct,* Barentsia benederi. *Most entoprocts are colonial, but members of a few species live as solitary individuals.*

SUMMARY

Animals are multicellular heterotrophs that depend directly or indirectly on plants or algae as their source of food energy. Almost all digest their food in an internal cavity. Most are motile. Reproduction is usually sexual. More than 95 percent of the animal species are invertebrates, animals without backbones.

Modern animals are classified into about 30 phyla. Criteria used in classification include the number of embryonic tissue layers, the basic body plan and the arrangement of body parts, the presence or absence of body cavities, and the pattern of development from fertilized egg to adult.

Both the sponges (phylum Porifera) and the extremely simple, parasitic mesozoans (phylum Mesozoa) are so different from all other animals that they are placed in separate subkingdoms. Sponges are composed of a number of different cell types, including choanocytes, which are the feeding cells; epithelial cells, some of which are contractile; and amoebocytes, which perform a variety of functions in reproduction, the secretion of skeletal materials, and the transport of food particles within the animal. In the sponge, there is little coordination among the various cells. Sponges reproduce both asexually and sexually, with a highly specialized form of internal fertilization. Most sponges, like many other sessile or slow-moving animals, are hermaphrodites.

The animals in phylum Cnidaria and phylum Ctenophora share three major characteristics: (1) radial symmetry; (2) a gastrovascular cavity, in which food is partially digested extracellularly and oxygen-bearing water is circulated; and (3) a two-layered body plan, in which the two tissue layers, the epidermis and the gastrodermis, are separated by a noncellular jellylike substance, the mesoglea. The jellyfishes, sea anemones, and corals of phylum Cnidaria are distinguished from all other animals by their special stinging cells, the cnidocytes; the sea walnuts and comb jellies of phylum Ctenophora are identified by comblike plates of fused cilia, arranged in eight rows on the body surface. Adult cnidarians may take the form of either polyps or medusas; in many species, the life cycle includes an asexually reproducing polyp, a sexually reproducing medusa, and a ciliated planula larva. Cnidarians have a simple nervous system that coordinates the movements of the animal, a variety of sensory cells, and two types of specialized sensory organs, statocysts and ocelli.

The animals in all of the remaining phyla have primary bilateral symmetry, with distinct "head" and "tail" ends and a concomitant clustering of nerve cells in the anterior region (cephalization). They also have three distinct embryonic tissue layers—ectoderm, mesoderm, and endoderm. The simplest of the bilateral animals are the acoelomates (phyla Platyhelminthes, Gnathostomulida, and Rhynchocoela), which have no body cavity other than the digestive cavity.

The flatworms, phylum Platyhelminthes, are characterized by a flattened body, a branched digestive system with only one opening, and an excretory system, involving flame cells, that serves largely to maintain water balance. Ocelli are present, as are sensory cells responsive to touch and to various chemicals; the nerve cells are organized into longitudinal cords. Flatworms may be free-living (class Turbellaria) or parasitic (classes Trematoda and Cestoda).

Other acoelomates include the tiny marine worms of phylum Gnathostomulida and the ribbon worms of phylum Rhynchocoela. The ribbon worms, which are characterized by a retractile, prey-seizing proboscis, are the most primitive animals with a one-way (mouth-to-anus) digestive tract and a circulatory system.

Eight phyla of animals have body plans based on the pseudocoelom, a fluid-filled cavity that develops between the endoderm and the mesoderm. The pseudocoelom functions as a firm hydrostatic skeleton, enabling these animals to move more efficiently than the cnidarians or the acoelomate worms. The pseudocoelomates all have a one-way digestive tract but lack a circulatory system.

The roundworms (phylum Nematoda) are the largest and ecologically most important group of pseudocoelomates. Most are free-living, but many are parasitic, causing a variety of serious diseases in plants and animals, including humans. Unlike other types of worms, roundworms have only longitudinal muscles and move in a characteristic whipping manner.

Other pseudocoelomates include the horsehair worms (phylum Nematomorpha) and the spiny-headed worms (phylum Acanthocephala), which are both parasitic; the tiny marine and freshwater animals of phyla Loricifera, Kinorhyncha, Gastrotricha, and Rotifera, most of which are bottom-dwellers; and the Entoprocta, which superficially resemble the hydrozoans of phylum Cnidaria. Among the various pseudocoelomates, the entire range of sexual reproductive patterns in animals is seen: separate sexes, simultaneous hermaphrodites, sequential hermaphrodites, external fertilization, internal fertilization, and parthenogenesis.

QUESTIONS

1. Distinguish among the following: cnidocyte/nematocyst; polyp/medusa; diploblastic/triploblastic; endoderm/mesoderm/ectoderm; radial symmetry/bilateral symmetry; acoelomate/pseudocoelomate/coelomate.

2. In which phylum or phyla do you find each of the following and what is its function: choanocyte; gastrovascular cavity; cnidocyte; statocyst; planula larva; flame cell?

3. Describe the similarities and differences among the three classes of phylum Cnidaria.

4. How are the ocelli of jellyfish and planarians similar? How are they different?

5. What advantages does radial symmetry provide to sessile organisms? What advantages does bilateral symmetry provide to motile organisms?

6. A pseudocoelom (or, as we shall see in the next chapter, a coelom) provides hydrostatic support for an animal. Why is a gastrovascular cavity less likely to fill this same function?

7. How do the parasitic flatworms of phylum Platyhelminthes differ from the free-living flatworms? What general features are characteristic of adaptation to a parasitic way of life?

8. Virtually all of the animals in phyla Gnathostomulida, Loricifera, Kinorhyncha, and Gastrotricha and many animals in phylum Rotifera (as well as some animals we have yet to meet) live in the spaces between sand and silt grains along shorelines. What specializations enable such a large number of different animals to occupy the same microenvironment?

9. On the basis of your knowledge of sexual reproduction, what do you think might be the advantages to an organism of functioning as a male in the earlier stages of its life and then as a female in the later stages, or vice versa?

10. What might be the advantages of parthenogenetic reproduction?

11. What might be the advantages and disadvantages of internal fertilization as compared to external fertilization?

CHAPTER 26

The Animal Kingdom II: The Protostome Coelomates

The coelom, a fluid-filled cavity that develops within the mesoderm, characterizes all the remaining phyla in the animal kingdom. Although it may seem less dramatic than other evolutionary innovations, the coelom is extremely important. Within this cavity, organ systems—suspended by the mesenteries—can bend, twist, and fold back on themselves, increasing their functional surface areas and filling, emptying, and sliding past one another, surrounded by lubricating coelomic fluid. Consider the human lung, constantly expanding and contracting in the chest cavity, or the 6 or 7 meters of coiled human intestine; neither of these could have evolved before the coelom. The coelom, like the pseudocoelom, also constitutes a hydrostatic skeleton, stiffening the body in somewhat the same way water pressure stiffens and distends a fire hose.

The various phyla of coelomate animals fall into two broad groups that roughly correspond to major branches of the phylogenetic tree. These groups are based on characteristic features of embryonic development. When a fertilized egg—the zygote—begins to divide, the early cell divisions usually follow one of two patterns. In mollusks, annelids, arthropods, and a number of other coelomate phyla (as well as in acoelomates and pseudocoelomates), the early cleavages are spiral, occurring in a plane oblique to the long axis of the egg. In echinoderms, chordates, and a few other coelomate phyla, the cleavage pattern is radial, parallel to and at right angles to the axis of the egg (Figure 26-1). With both types of cleavage, the embryo gradually develops to a stage known as the blastula, which is a hollow ball of cells. Next, an opening, the blastopore, appears. Among the animals with spiral cleavage, the mouth (stoma) develops at or near the blastopore, and this group is called the **protostomes**—"first the mouth." In the animals with radial cleavage, the anus forms at or near the blastopore and the mouth forms secondarily elsewhere; thus these animals are known as **deuterostomes**—"second the mouth." These differences are so fundamental that they are believed to have originated very early in animal evolution, before the branchings that gave rise to the modern coelomate phyla.

Another characteristic difference between the two groups of animals is the manner in which the coelom forms (Figure 26-2). In the protostomes, the coelom usually forms by a splitting of the mesoderm, and the process is said to be

26-1 *Egg cleavages, showing spiral and radial patterns at the third cleavage. Spiral cleavage is characteristic of protostomes (acoelomates, pseudocoelomates, mollusks, annelids, and arthropods); radial cleavage is characteristic of deuterostomes (echinoderms and chordates).*

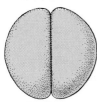
First cleavage

Second cleavage

Third cleavage—spiral

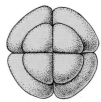
Third cleavage—radial

CHAPTER 26 The Animal Kingdom II: The Protostome Coelomates

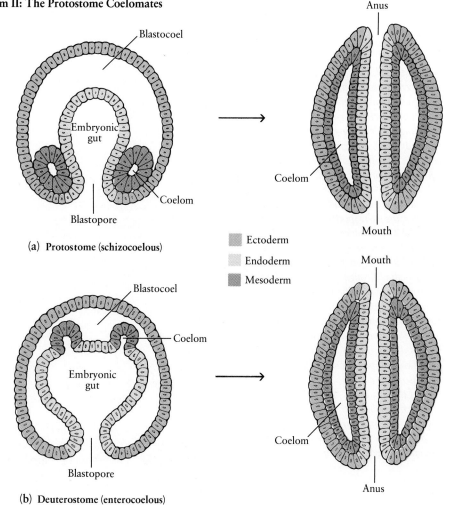

26-2 *Key features of embryonic development in coelomate animals.* **(a)** *In protostomes, the mesoderm takes shape between the endoderm and the ectoderm, in the region around the blastopore. The coelom arises from a splitting of the solid mesoderm. As the cells continue to multiply, the blastocoel—the original embryonic cavity of the blastula—is obliterated. The blastopore of protostomes becomes the mouth.*

(b) *In deuterostomes, the mesoderm originates from outpocketings of the embryonic gut. These outpocketings create cavities within the mesoderm that become the coelom. The blastopore becomes the anus, and the mouth develops elsewhere.*

schizocoelous (from the Greek *schizo,* "split"). In the deuterostomes, however, the coelom is usually formed by an outpouching of the cavity of the embryonic gut, a process said to be **enterocoelous** (from the Greek *enteron,* "gut"). Enterocoelous formation of the coelom is thought to have arisen from the schizocoelous process several times in the course of animal evolution.

In the next two chapters, we shall focus on the protostome coelomates—the mollusks, annelids, and a variety of smaller groups in this chapter, and the arthropods in the next. In the final chapter of this section, our attention will be devoted to the deuterostomes.

PHYLUM MOLLUSCA: MOLLUSKS

The mollusks constitute one of the largest phyla of animals, both in numbers of living species (at least 47,000, and perhaps many more) and in numbers of individuals. Their name is derived from the Latin word *mollus,* meaning "soft," and refers to their soft bodies, which are generally protected by a hard, calcium-containing shell. In some forms, however, the shell has been lost in the course of evolution, as in slugs and octopuses, or greatly reduced in size and internalized, as in squids.

Mollusks exhibit a tremendous diversity of form and behavior. The three major classes range from largely sedentary or sessile filter-feeding animals, such as clams and oysters (class Bivalvia), through aquatic and terrestrial snails and slugs (class Gastropoda), to the predatory cuttlefish, squids, and octopuses (class Cephalopoda), which are not only the most active of the mollusks but also the most intelligent of all invertebrates.

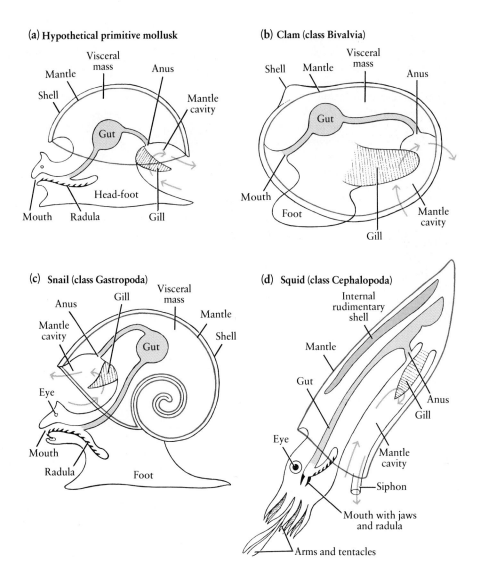

26–3 *Mollusks are characterized by soft bodies composed of a head-foot, a visceral mass, and a mantle, which can secrete a shell. They exchange gases with the surrounding water through gills, except for the land snails, in which the mantle cavity has been modified for air breathing. A hypothetical primitive mollusk is shown in (a).*

The three major modern classes are the bivalves, the gastropods, and the cephalopods. (b) The bivalves, such as the clam shown here, are generally sedentary and feed by filtering water currents, created by beating cilia, through large gills. (c) In the gastropods, exemplified by the snail, the visceral mass has become coiled and rotated through 180° so that mouth, anus, and gills all face forward and the head can be withdrawn into the mantle cavity. (d) In the cephalopods, such as the squid, the head is modified into a circle of arms, and part of the head-foot forms a tubelike siphon through which water can be forcibly expelled, providing for locomotion by jet propulsion. The arrows indicate the direction of water movement.

Characteristics of the Mollusks

Structurally, mollusks are quite distinct from all other animals (Figure 26–3). As you can see, the hypothetical primitive mollusk displayed a clear bilateral symmetry. Among modern mollusks, only the polyplacophorans (chitons) and monoplacophorans, both relatively small classes, bear any obvious resemblance to the primitive model. Modern mollusks, however, all have the same fundamental body plan. There are three distinct body zones: a **head-foot,** which contains both the sensory and motor organs; a **visceral mass,** which contains the well-developed organs of digestion, excretion, and reproduction; and a **mantle,** a specialized tissue formed from folds of the dorsal body wall, that hangs over and enfolds the visceral mass and that secretes the shell. The **mantle cavity,** a space between the mantle and the visceral mass, houses the gills; the digestive, excretory, and reproductive systems discharge into it. Water sweeps into the mantle cavity (usually propelled by cilia on the gills), passing across the surface of the gills and aerating them. Water leaving the mantle cavity carries excreta and, in season, gametes, both of which are discharged downstream from the gills.

A characteristic organ of the mollusk, found only in this phylum, and in all classes except the bivalves, is the **radula** (Figure 26–4), a movable, tooth-bearing strap of chitinous material suggestive of a tongue. The radular apparatus, which can be projected out of the mouth and then drawn back in with a licking

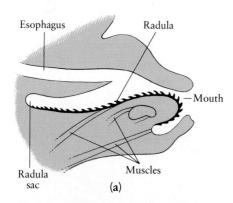

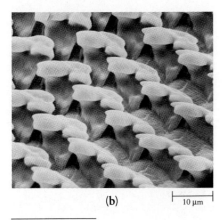

26–4 (a) *A vertical section through the head of a land snail to show the radula. Teeth on the radula rasp and tear food materials and then convey them to the esophagus, a narrow tube leading to the stomach.* (b) *Scanning electron micrograph of radular teeth from a minute African land snail that feeds on bits of dead leaves. The large teeth (upper left) cut and tear the leaf; the smaller teeth (lower right) pull the pieces into the snail's mouth.*

movement, serves both to scrape off algae and other food materials and also to convey them backward to the digestive tract. In some species, it is also used in combat.

Supply Systems

In the protists and in the smaller and simpler animals, oxygen and food molecules are supplied to cells—and waste products removed from them—largely by diffusion, aided by the movement of external fluids. Internal transport may also be assisted, as we have seen, by wandering, amoeboid cells, as in the cnidarians, or by the movement of fluids in a body cavity, as in the pseudocoelomates. For larger, thicker animals, a more effective method of providing each cell with a direct and rapid line of supply is a circulatory system that propels extracellular fluid—blood—around the body in a systematic fashion.

The molluscan circulatory system consists of a muscular pumping organ, the heart, and vessels that carry the blood to and from the heart. The heart usually has three chambers: two of them (atria) receive blood from the gills, and the third (the ventricle) pumps it to the other body tissues. Except for the cephalopods, mollusks have what is known as an open circulation; that is, the blood does not circulate entirely within vessels—as it does in the annelids, for example—but is collected from the gills, pumped through the heart, and released directly into spaces in the tissues from which it returns to the gills and then to the heart. Such a blood-filled space is known as a **hemocoel** ("blood cavity"). In the mollusks, the hemocoel has largely replaced the coelom, which is reduced to a small area around the heart and to the cavities of the organs of reproduction and excretion. Cephalopods, whose vigorous activities require that the cells be supplied with large quantities of oxygen and food molecules, have a closed circulatory system of continuous vessels and accessory hearts that propel blood into the gills.

Oxygen enters the body of a mollusk through the moist surface of the mantle and the gills. A **gill** is an external structure with an increased amount of surface area, through which gases can diffuse, and a rich supply of blood for the transport of the gases to and from the rest of the body. Oxygen travels inward by diffusion, down the concentration gradient. The gradient exists because the surface film, exposed to the dissolved oxygen in the passing water, contains more oxygen than does the blood within the gills, which was depleted of oxygen as it passed through the body tissues. Carbon dioxide, produced by cellular respiration, moves out to the surface film and then into the surrounding water by the same mechanism. In fact, all gas exchange in animals, whether water-dwelling or land-dwelling, takes place by diffusion across moist membranes.

The digestive tract of all mollusks is extensively ciliated and has many different working areas. Food is taken up by cells lining the digestive glands arising from the stomach and the anterior intestine, and then is passed into the blood. Undigested materials are compressed into mucus-coated fecal pellets, which are discharged through the anus into the mantle cavity and are carried away from the animal in the water currents. This packaging of digestive wastes in solid form prevents fouling of the water passing over the gills.

Nitrogenous wastes produced by the metabolic activities of the cells are removed by one or two tubular structures known as **nephridia**. One opening of each nephridium is in the coelom surrounding the heart, and the other opening discharges into the mantle cavity. Coelomic fluid is forced, under pressure, into the opening near the heart; as the fluid passes through the tubule, water, sugar, salts, and other needed materials are returned to the tissues through its walls, while waste products are secreted into it for excretion. Thus the excretory system is concerned not only with the problem of water balance, as are the contractile vacuoles of *Paramecium* and the flame cells of planarians, but also with the regulation of the chemical composition of body fluids.

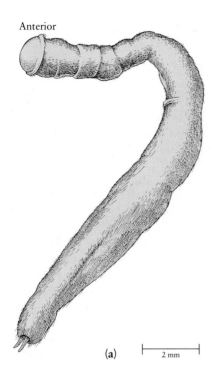
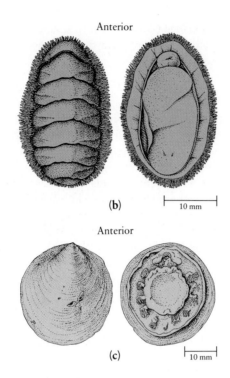
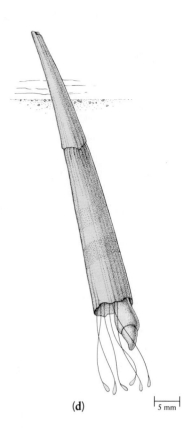

26–5 *Representatives of the four minor classes of mollusks.* **(a)** *A solenogaster, class Aplacophora. Its wormlike body is covered with fine bristles, or setae. The pronglike structures at the posterior of the animal are the gills. This specimen, a member of the species* Chaetoderma nitidulum, *was found in St. Margaret's Bay, Nova Scotia.*

(b) *Dorsal (left) and ventral (right) views of a chiton, class Polyplacophora. The dorsal shell of eight plates is fringed by a girdle of hard spicules. As seen from the ventral surface, the bulk of the chiton's body is its foot, which has been pulled aside at the left to reveal the gills within the mantle cavity. At the anterior of the animal is the mouth, through which it extends the radula when grazing.*

(c) *Dorsal and ventral views of* Neopilina galatheae, *the first known living representative of class Monoplacophora. The large central structure of the ventral surface is the foot, which is surrounded by five pairs of gills.*

(d) *A tusk shell, class Scaphopoda. This mollusk burrows in sand or mud with the large foot that extends from the wide end of the shell. The specialized tentacles bring food particles, such as diatoms, to the mouth hidden within the shell.*

Minor Classes of Mollusks

Four of the seven classes of mollusks with living representatives are fairly small. About 250 species of wormlike marine animals, known as solenogasters (Figure 26–5a), constitute the class Aplacophora ("no plates"). Although they have no shell and the foot is greatly reduced, they are unequivocally identified as mollusks by the presence of a radula.

The approximately 600 species of chitons (Figure 26–5b) are placed in class Polyplacophora. As we noted previously, they bear some resemblance to the hypothetical primitive mollusk (Figure 26–3a). Common as grazers on surf-swept ocean shores, chitons have a somewhat flattened body, covered by a dorsal shell formed from a series of eight plates. On either side of the body, a series of gills are suspended between the mantle and the foot and are continuously aerated by water flowing through the mantle cavity from the anterior to the posterior of the animal.

Class Monoplacophora contains just eight living species. The first species, *Neopilina galatheae* (Figure 26–5c), was discovered in the 1950s in a deep ocean trench off the coast of Costa Rica. The only previously known representatives of the class were fossils from the Cambrian period, which ended 505 million years ago. *Neopilina,* which is little more than 2.5 centimeters long, also has an organization suggestive of the hypothetical primitive mollusk. It has a large, single dorsal shell, but is unusual in having five pairs of gills, six pairs of nephridia, and eight pairs of retractor muscles, with which it can pull the shell down securely over its soft body.

About 350 species of tusk or tooth shells (Figure 26–5d) make up class Scaphopoda. These familiar residents of the seashore have long, tubular shells, open at both ends. They live a sedentary life, with the wide end of the shell, containing the head (which bears structures used in feeding) and the foot, buried in the sand or mud. Water, carrying dissolved gases, enters and leaves the mantle cavity through the exposed, narrow end of the shell.

Class Bivalvia

The approximately 7,500 living species of bivalves include such common animals as clams, oysters, scallops, and mussels. They derive their name from the two parts, or valves, into which the shell is divided. The left and right valves are connected dorsally by a hinge with a flexible ligament. One or two large **adductor muscles,** familiar as the delectable portion of the scallop, are used to close the shell tightly in times of danger. These muscles are so strong in scallops that the animals can move swiftly through the water by clapping their shells together, thereby eluding predators such as starfish.

In this class of mollusks, the body has become flattened between the valves, and a distinct "head" has generally disappeared (see Figure 26-3b). The bivalves are sometimes called Pelecypoda—"hatchet foot"—because the muscular foot is often highly developed in this group. A clam, using its "hatchet foot," can dig itself into sand or mud with remarkable speed. However, many bivalves are sessile, and some of them secrete strands of protein by which they anchor themselves to rocks. The adhesive polymer synthesized by mussels, which consists of a repeated sequence of 10 amino acids, is as powerful as epoxy—and, as you might expect, it is resistant to damage by salt water. Work is currently underway, using recombinant DNA techniques, to produce sufficient quantities of this adhesive for use in dentistry and medicine, for example, in reattaching broken teeth and bones.

Abundant in both salt and fresh water, most bivalves are filter-feeding herbivores; they live largely on microscopic algae. Their gills, which are large and elaborate, collect food particles. Water is circulated through the sievelike gills by the beating of gill cilia. Small organisms and particles of food are trapped in mucus on the gill surface and swept toward the mouth by the cilia; the gills also sort particles by size, rejecting sand and other larger particles.

Throughout the molluscan phylum, there is a wide range of development of the nervous system. The bivalves have three pairs of ganglia of approximately equal size—cerebral, visceral, and pedal (supplying the foot)—and two long pairs of nerve cords interconnecting them. They have statocysts, usually located near the pedal ganglia, and sensory cells for discrimination of touch, chemical changes, and light. The scallop has quite complex eyes; a single individual may have a hundred or more eyes located among the tentacles on the fringe of the mantle (Figure 26-6). The lens of this eye cannot focus images, however, so it does not appear to serve for more than the detection of light and dark, including passing shadows cast by other moving organisms.

26-6 *A bivalve, the calico scallop* (Aequipecten gibbos). *Its bright blue eyes are visible among its tentacles. This scallop lives in the waters off the coast of Florida.*

In the earliest mollusks, the sexes were separate and fertilization was external. This primitive condition is retained in most bivalves, but internal fertilization has evolved in a number of different bivalve lineages. Specialized "brood pouches," in which the young are protected during their early development, are found in hermaphroditic species as well as in the females of some species with separate sexes.

Class Gastropoda

The gastropods, which include the snails, whelks, periwinkles, abalones, and slugs, are the largest group of mollusks (at least 37,500 living species). They have either a single shell or, as a secondary evolutionary development, no shell. Gastropods are common in both salt and fresh water and on land. The group is unusual in that some members are able to digest cellulose and other structural carbohydrates. In addition to herbivores, class Gastropoda includes omnivores, a wide variety of specialized carnivores, scavengers, and even some parasites.

Gastropods have lost the bilateral symmetry characteristic of other mollusks and have become asymmetrical through a curious anatomical rearrangement called **torsion**. Torsion, which is a separate phenomenon from the coiling of the shell, is a twisting through 180° of the rest of the body relative to the head-foot. It occurs at a specific stage of larval development and, in many cases, requires only a few minutes. Torsion is initiated by the contraction of a large muscle that runs from the right side of the shell to the left side of the head-foot; it is generally completed by asymmetrical growth in which one side of the larval gastropod's body grows more rapidly than the other side. As a result of torsion, the shell, mantle cavity, and visceral mass are moved so that parts that once were located at the rear of the body now lie over the head (see Figure 26–3c). Other events in gastropod development usually result in a spiral coiling of the visceral mass and the shell. In response to the displacement and consequent crowding of the internal organs, the gill and nephridium of the right side have been completely lost in many species. In cases where the shell has been lost in the course of subsequent evolution and snails have evolved into slugs, the mantle cavity is generally moved back toward its original position by a process of detorsion—but the missing gill and other organs have not been regained.

Land-dwelling snails do not have gills, but the area in their mantle cavities once occupied by gills is rich in blood vessels, and the snail's blood is oxygenated there.

26–7 *Land-dwelling gastropods.* **(a)** *A terrestrial snail. The shell, which is secreted by the mantle and grows as the soft body grows, covers and protects the visceral mass. The head contains sensory organs, including two eyes at the tips of the longer tentacles. Note the thick trail of slime secreted by the snail as it glides along.* **(b)** *A banana slug,* Ariolimax columbianus, *photographed at Point Reyes, California.*

(a)   (b)

Thus the mantle cavity has become, in effect, a lung. Moreover, as with all lungs, the opening is reduced to retard evaporation. Some snails that were probably once land dwellers have returned to the water, but they have not regained gills. Instead, they bob up to the surface at intervals to entrap a fresh bubble of air in their mantle cavities.

Gastropods, which lead a more mobile, active existence than bivalves, have a ganglionated nervous system with as many as six pairs of ganglia connected by nerve cords (Figure 26–8). There is a concentration of nerve cells at the anterior end of the animal, where the tentacles, which have chemoreceptors and touch receptors, are located. In some of the animals, the eyes are quite highly developed in structure; they appear, however, to function largely in the detection of changes in light intensity, like the eyes of the scallop.

In some gastropods, the primitive form of reproduction—separate sexes with external fertilization—is retained. In most gastropods, however, fertilization is internal, and hermaphroditism has evolved repeatedly in this group. The simultaneous hermaphroditism found in many snails and slugs probably resulted from the difficulties of such slow-moving animals in finding mates. In some species, the animals are sequential hermaphrodites: they are males when they are younger, then females when they are older and larger.

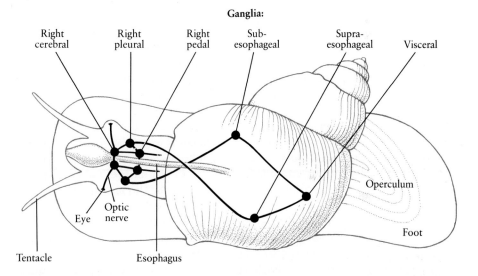

26–8 *The ganglionated nervous system of a gastropod. The cerebral ganglia supply the tentacles and eyes; the pleural ganglia, the mantle; the pedal ganglia, the foot muscles; and the subesophageal, supraesophageal, and visceral ganglia supply the visceral mass. Counterclockwise torsion of the visceral mass during development results in the figure-eight pattern seen here, in which one major nerve passes over and another under the esophagus, a tubular structure leading from the mouth to the stomach.*

The operculum, a hardened plate attached to the foot, functions as a trapdoor when the snail withdraws into its shell.

Class Cephalopoda

The cephalopods (the "head-foots") are, in many respects, the most evolutionarily advanced animals to be found among the invertebrates. The 600 living species in this strictly marine class rival the vertebrates in complexity and, in some cases, in intelligence. Active predators, they compete quite successfully with fish.

Although obviously mollusks, the cephalopods have become greatly modified (see Figure 26–3d). The large head has conspicuous eyes and a central mouth surrounded by arms, some 90 in the chambered *Nautilus*, 10 in the squid, and 8 in the octopus. The octopus body seldom reaches more than 30 centimeters in diameter (except on the late late show), but giant squids sometimes attain sea-monster proportions. One caught in the Atlantic in 1861 was about 6 meters long, not counting the tentacles.

Nautilus, as the only modern shelled cephalopod, offers an indication of some of the steps by which this class disposed of the shell entirely. The animal occupies only the outermost portion of its elaborate and beautiful shell, the rest of which serves as a flotation chamber. In the squid and its relative, the cuttlefish, the shell has become an internal stiffening support, and in the octopus, it is lacking entirely.

Behavior in the Octopus

Behaviorally, the octopus is the best studied of the cephalopods. The octopus is a sea dweller that creeps about actively on its arms or swims rapidly through the water by strong rhythmic muscular contractions that expel water from the mantle cavity. The octopus, like other cephalopods, is carnivorous, living on smaller sea animals, usually crabs. It bites the crab, or other prey, with its parrotlike beak, injecting a toxin from its salivary glands. It then bundles up the paralyzed animal in its web and carries it home to eat. When it is not actively in pursuit of food, the octopus, lacking any protective shell, lives in small caves behind rocks or in reefs or wreckage.

Curious and able to use its arms with great dexterity, the octopus makes an extremely apt experimental subject. In a seawater tank, it will gather together bricks, shells, and any movable debris into a crude sort of house, and there it sits and watches, often bobbing its head up and down. Although the eye is remarkably similar to our own and equally acute, apparently the octopus does not have stereoscopic vision. Head bobbing seems to be the way it estimates distance, fixing on an object from two points, just as surveyors triangulate a distant landmark.

When confronted by an object larger than itself, an octopus literally pales. It flattens and turns almost white except for a dark area around the eyes and a dark trim along the edges of its web. This makes the octopus seem larger than it is and probably serves to deter would-be predators. By contrast, the sight of a crab can so excite an octopus that its arms weave about and its skin color darkens, breaking out in patches of bright blue, pink, or purple, depending on the species. These color changes are brought about by the contraction of muscles that draw out small sacs of pigment, the chromatophores, to form flat plates of color. When the muscles relax, the chromatophores return to their original size and the patches of color disappear.

By using a system of rewards (crabs, for example) and punishments (mild electric shock), the investigator can readily teach the octopus to seize certain objects and not to seize others. Such experiments approximate what an octopus must learn in nature—that some objects are edible, some are not, and some may bite or sting. A small crab, for example, is a meal, but if it is carrying a sea anemone on its shell, it had better be left alone. Because of the comparative ease with which such tests can be performed, owing to the octopus's natural curiosity and appetite, the animal has been the subject of a great many experiments on vision, touch, and learning.

The manipulatory powers of the octopus are great, and the arms are very sensitive to texture and are rich in chemoreceptors. However, the animal seems to have difficulty processing and coordinating sensory data. Martin Wells, of Churchill College, Cambridge, who has carried out many studies on octopus behavior, points out that the problems of handling data from eight extensible arms, each of which has several hundred suckers and can move separately and in any direction, are probably insurmountable. As we shall see in the next chapter, the severe limitations of movement imposed by an articulated skeleton prove, in contrast, to be a great advantage for many types of specialized activities.

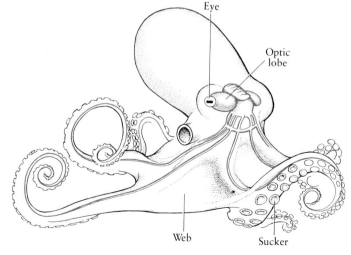

A simplified view of the brain and nervous system of the octopus. The large lobes behind the eyes are the optic lobes, concerned with collecting and analyzing visual data. The octopus often finds its prey by reaching its arms into a crevice into which it cannot see. The suckers on the arms of an octopus are softer and more flexible than our fingertips and so can make fine distinctions among textures. They also contain chemoreceptors that can detect sugars, salts, and other chemicals in dilutions well below the range of discrimination of the human tongue.

Freedom from the external shell has given the mantle more flexibility. The most obvious effect of this is the jet propulsion by which cephalopods dart through the water. Usually, water taken into the mantle cavity bathes the gills and is then expelled slowly through a tube-shaped structure, the siphon; but when the cephalopod is hunting or being hunted, it can contract the mantle cavity forcibly and suddenly, thereby squirting out a sudden jet of water. Contraction of the mantle-cavity muscles usually shoots the animal backward, head last, but the squid and the octopus can turn the siphon in almost any direction they choose. Cephalopods also have sacs from which they can release a dark fluid that forms a cloud, concealing their retreat and confusing their enemies. These colored fluids were at one time a chief source of commercial inks. *Sepia* is the name of the genus of cuttlefish from which a brown ink used to be obtained.

26-9 *A cephalopod. Jet-propelled, this multicolored cuttlefish is moving to the right. You can see its siphon extending from the mantle, below its eye. On either side, an undulating fin runs the length of the broad, flattened body.*

26-10 *Eledone cirrhosa, the lesser octopus, from the British seacoast. Note its siphon, its well-developed eyes, and the suckers on the undersurface of its arms. The suckers, which contain a variety of sensory receptors, are also used to immobilize prey organisms, such as crabs, so the octopus can bite them and inject its paralyzing toxin.*

The cephalopods have well-developed brains, composed of many groups of ganglia, in keeping with their highly developed sensory systems and their lively, predatory behavior. These large brains are covered with cartilaginous cases. The rapid responses of the cephalopods are made possible by a bundle of giant nerve fibers that control the muscles of the mantle. Many of the studies on conduction of nerve impulses are made with the giant axons of squids, which are large enough to permit the insertion of electrodes.

In cephalopods, the sexes are always separate, and fertilization is internal. Courtship and mating behavior are complex, and males often fight for access to females. Fertilization occurs when the male uses one of his arms to transfer packets of sperm, called spermatophores, from his mantle cavity to the mantle cavity of the female. The female produces an intricate, gelatinous egg mass in which the developing embryos are protected until they hatch as miniature adults. In *Octopus* and some other cephalopod genera, the mother guards the egg masses, cleaning and aerating them.

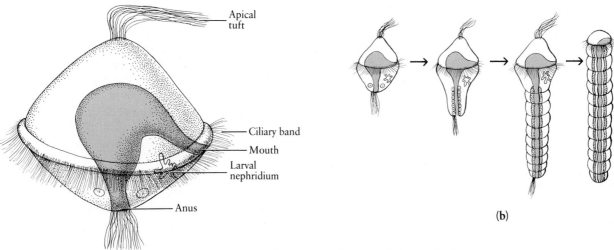

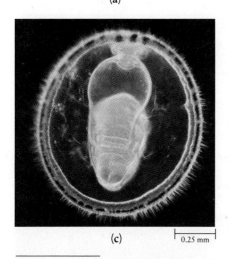

26-11 (a) *Diagram of a trochophore larva. Although their adult forms are very different, certain annelids and mollusks have larvae of this type.* (b) *Development of the annelid trochophore into a segmented worm. The process begins with the elongation of the lower part of the trochophore. The elongated region then becomes constricted into segments, which soon develop bristles. The apical tuft disappears, and the upper part of the trochophore becomes the head. The worm's growth will continue throughout its lifetime by the addition of new segments just in front of the rear segment.*

(c) *The trochophore larva of a marine annelid; this larva will develop into a polychaete worm.*

Evolutionary Affinities of the Mollusks

Although the mollusks and the annelids (segmented worms) are, as we are about to see, quite different in their basic body plans, they have an important similarity that suggests an evolutionary link. This is the trochophore larva (Figure 26-11). Most marine mollusks (except the cephalopods) and marine annelids pass through this very distinct larval form during their development. With the exception of *Neopilina* and *Nautilus*, however, adult mollusks do not exhibit signs of the segmentation so characteristic of the annelids, and no traces of segmentation are seen in the larval development of any known mollusk. Most, but not all, authorities think that the segmental patterns seen in the gills, nephridia, and some other structures of *Neopilina*, *Nautilus*, and some fossil bivalves were late evolutionary developments, unrelated to the development of segmentation in the annelids. If so, the lineages giving rise to the mollusks and to the annelids diverged from an unsegmented coelomate ancestor that was characterized by the trochophore larva.

PHYLUM ANNELIDA: SEGMENTED WORMS

This phylum includes almost 9,000 species of marine, freshwater, and terrestrial worms, including the familiar earthworms. The term annelid is from the Latin for "ringed" and refers to the most distinctive feature of this group: the division of the body into segments, or **metameres**. The metameres are visible as rings on the outside and are separated by partitions (septa) on the inside. This segmented pattern is found in a modified form in arthropods, too, such as millipedes, crustaceans, and insects, which are thought to have evolved from the same ancestors that gave rise to modern annelids.

The annelids have a segmented coelom, a tubular gut, and a closed circulatory system that transports oxygen (diffused through the skin or through fleshy extensions of the skin) and food molecules (from the gut) to all parts of the body. The excretory system consists of paired nephridia, which typically occur in each segment of the body except the head. Annelids have a centralized nervous system and a number of special sensory cells, including touch cells, taste receptors, light-sensitive cells, and cells concerned with the detection of moisture. Some also have well-developed eyes and sensory antennae.

The three classes of annelids are Oligochaeta (terrestrial worms, with some freshwater and marine relatives), Polychaeta (mostly marine worms), and Hirudinea (leeches). It is generally agreed that the hirudineans are derived from oligochaetes, but the question of whether the polychaetes or the oligochaetes came first remains unresolved. However, we shall begin with class Oligochaeta, since its most familiar members—the earthworms—are such clear exemplars of annelid structure.

CHAPTER 26 The Animal Kingdom II: The Protostome Coelomates

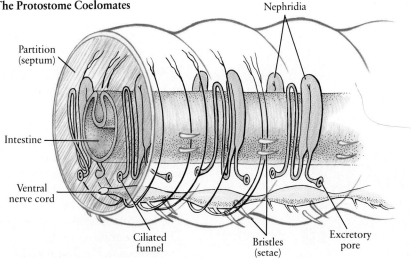

26-12 *Three segments of the earthworm, an annelid. On each segment are four pairs of bristles, which are extended and retracted by special muscles. These are used by the worm to anchor one part of its body while it moves another part forward. Two excretory tubes, or nephridia, are in each segment (except the first three and the last). Each nephridium really occupies two segments, since it opens externally by a pore in one segment and internally by a ciliated funnel in the segment immediately in front of it. The intestine, nephridia, and other internal organs are suspended in the large fluid-filled coelom, which also serves as a hydrostatic skeleton.*

Class Oligochaeta: The Earthworms

Figure 26-12 shows a portion of the body of an earthworm, a representative oligochaete. Note how the body is compartmentalized into regular segments. Most of these segments, particularly the central and posterior ones, are identical, each exactly like the one before and the one after. Each identical segment contains two nephridia, three pairs of nerves branching off from the central nerve cord running along the ventral surface, a portion of the digestive tract, and a left and right coelomic cavity; each segment also bears four pairs of bristles, or **setae.** The chief exceptions to this rule of segmented structure are found in the most forward segments. In these, specialized areas of the nervous, digestive, circulatory, and reproductive systems are found.

The tubelike body is wrapped in two sets of segmental muscles, one set running longitudinally and the other encircling the segments. When the earthworm moves, it anchors some of its segments by its setae, and the circular muscles of the segments anterior to the anchored segments contract, thus extending the body forward. Then its forward setae take hold, and the longitudinal muscles contract while the posterior anchor is released, drawing the posterior segments forward. The ultimate in a hydrostatic skeleton is provided by the coelom of the earthworm. Not only is the coelom partitioned by the septa between segments, but it is also divided into left and right compartments within each segment. This arrangement, in combination with the two sets of segmental muscles, allows exquisite control over movements of small parts of the body.

Digestion in Earthworms

The digestive tract of the earthworm (Figure 26-13) is a long, straight tube. The mouth leads into a strong, muscular **pharynx,** which acts like a suction pump, helping the mouth to draw in decaying leaves and other organic matter, as well as dirt, from which organic materials are extracted. The earthworm makes burrows in the earth by passing such material through its digestive tract and depositing it outside in the form of castings, a ceaseless activity that serves to break up, enrich, and aerate the soil. Darwin, in his study of earthworms *(The Formation of Vegetable Mould through the Action of Worms),* estimated 15 tons as the weight of castings thrown up annually on an acre of land—perhaps 20 ounces per worm per year, he calculated.

The narrow section of digestive tract posterior to the pharynx, the **esophagus,** leads to the **crop,** where food is stored. In the **gizzard,** which has thick, muscular walls lined with protective cuticle, the food is ground up with the help of the ever-present soil particles. The rest of the digestive tract is made up of a long intestine, which has a large fold along its upper surface that increases its surface area. The intestinal epithelium consists of enzyme-secreting cells and ciliated absorptive cells.

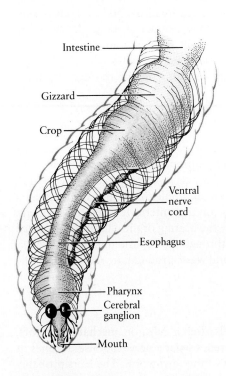

26-13 *The digestive tract of an earthworm. The mouth leads into a muscular pharynx, which sucks in decaying vegetation and other materials. These are stored in the crop and ground up in the gizzard with the help of soil particles. The rest of the tract is a long intestine in which food is digested and absorbed.*

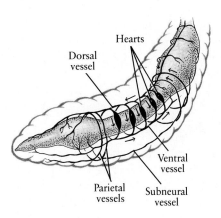

26–14 *The circulatory system of the earthworm is made up of longitudinal vessels running the entire length of the animal, one dorsal and several ventral. Smaller vessels (the parietal vessels) in each segment collect the blood from the tissues and from the subneural vessel and feed it into the muscular dorsal vessel, through which it is pumped forward. In the anterior segments are five pairs of hearts—muscular pumping areas in the blood vessels—whose irregular contractions force the blood downward to the ventral vessel, from which it returns to the posterior segments. The arrows indicate the direction of blood flow.*

Circulation in Earthworms

The circulatory system of the earthworm (Figure 26–14) is composed of longitudinal vessels running the entire length of the worm, one dorsal and several ventral. The largest ventral vessel underlies the intestinal tract, supplying blood to it, to the smaller ventral vessels that surround the nerve cord, and, by means of many small branches, to all the tissues of the body. Numerous small capillaries in each segment carry blood from the tissues to the dorsal vessel. Also in each segment are larger parietal ("along the wall") vessels transporting blood from the subneural vessel to the dorsal vessel, which also collects nutrients from the intestinal tract. The muscular dorsal vessel propels forward the fluids collected in this way from all over the animal's body.

Connecting the dorsal and ventral vessels, and so completing the circuit, are five pairs of hearts, muscular pumping areas in the blood vessels. Their irregular contractions force the blood down to the ventral vessels and also forward to the vessels that supply the more anterior segments. Both the hearts and the dorsal vessel have valves that prevent backflow. Note that annelids have a closed circulatory system in which the blood flows entirely through vessels. Evolution of a closed circulatory system made possible a degree of control, not previously feasible, over the composition of the circulating body fluid.

Respiration in Earthworms

Earthworms, unlike mollusks, have no special respiratory organs; respiration takes place by simple diffusion through the body surface. The gases of the atmosphere dissolve in the liquid film on the earthworm's body, which is kept moist by secreted mucus and by coelomic fluid released through dorsal pores. Oxygen diffuses inward to the network of capillaries just underlying the body surface and is consumed by body cells as the blood circulates. Carbon dioxide, which is picked up by the blood, diffuses out into the air through the body surface.

Excretion in Earthworms

In annelids, as in mollusks, the nephridia remove nitrogenous wastes from the coelomic fluid and regulate its chemical composition. The excretory system of the earthworm consists of one pair of nephridia for each segment (see Figure 26–12). Each nephridium consists of a long, convoluted tubule that begins with a ciliated funnel opening into the coelomic cavity of the anteriorly adjacent segment. Coelomic fluid is carried into the funnel by the beating of the cilia and is excreted through an outer pore. During its passage through the nephridium, needed water, other molecules, and ions are reabsorbed and waste products are secreted into the fluid.

The Nervous System of Earthworms

Earthworms have a variety of sensory cells. Touch cells, or **mechanoreceptors,** contain tactile hairs, which, when stimulated, trigger a nerve impulse. Patches of these hair cells are found on each segment of the earthworm. The hairs probably also respond to vibrations in the ground, to which the earthworm is very sensitive. The earthworm does not have ocelli or eyes—as one might expect, since it rarely emerges from underground—but it does have light-sensitive cells. Such cells are more abundant in its anterior and posterior segments, the parts of its body most likely to be outside of the burrow. These cells are not responsive to light in the red portion of the spectrum, a fact exploited by anglers who search for worms in the dark using red-lensed flashlights.

Among the earthworm's most sensitive cells are those that detect moisture. The cells are located on its first few segments. If an earthworm emerging from its burrow encounters a dry spot, it swings from side to side until it finds dampness; failing that, it retreats. However, when the anterior segments are anesthetized, the earthworm will crawl over dry ground. The animal also appears to have taste cells. In the laboratory, worms can be shown to select, for example, celery in preference to cabbage leaves and cabbage leaves in preference to carrots.

Each segment of the worm is supplied by nerves that receive impulses from sensory cells and by nerves that cause muscles to contract. The cell bodies for these nerves are grouped together in clusters (ganglia). The movements of each segment are directed by a pair of ganglia and are triggered by movement in the adjacent anterior segment; thus a headless earthworm can move in a coordinated manner. However, an earthworm without its cerebral ganglia moves ceaselessly; in other words, the cerebral ganglia inhibit and modulate activity.

There are also, as in planarians, conducting channels made up of nerve fibers bound together in bundles, like cables, which run lengthwise through the body. These nerve fibers are gathered together in a fused, double nerve cord that runs along the ventral surface of the body and then divides to encircle the pharynx at the anterior end of the animal (see Figure 26-13). The nerve cords contain fast-conducting fibers that make it possible for the earthworm to contract its entire body very quickly, withdrawing into its burrow when disturbed.

Reproduction in Earthworms

Earthworms are hermaphrodites, and some can reproduce parthenogenetically. In most species, however, two earthworms, held together by mucous secretions from the clitellum (a special collection of glandular cells), exchange sperm and separate (Figure 26-15). Two or three days later, the clitellum forms a second mucous sheath surrounded by an outer, tougher protective layer of chitin. This sheath is pushed forward along the animal by muscular movements of its body. As it passes over the female gonopores, it picks up a collection of mature eggs, and then, continuing forward, it picks up the sperm deposited in the sperm receptacles, or spermathecas. Once the mucous band is slipped over the head of the worm, its sides pinch together, enclosing the now fertilized eggs in a small capsule from which the infant worms hatch.

26-15 *Earthworms mating. The worms' heads are facing in opposite directions, and their ventral surfaces are in contact. The clitellum, a thickened band that surrounds the body of each, secretes mucus. The mucus holds the worms together during copulation, which may take as long as two hours. Sperm cells are released through pores in specialized segments of one worm into the sperm receptacles of its partner. After the partners separate, the clitellum secretes a mucous band, or cocoon, into which first the eggs and then the sperm are released. The eggs are fertilized within this cocoon.*

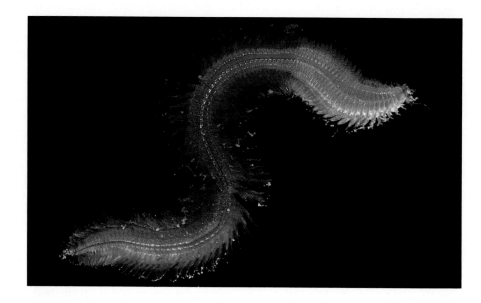

26–16 *A deep-sea polychaete worm. Unlike the more familiar earthworm, this annelid has a well-differentiated head with sensory appendages and lateral parapodia ("side feet") with many setae.*

Class Polychaeta

The polychaetes, which are almost all marine, differ from the earthworms and other oligochaetes in a number of ways (Figure 26–16). The most striking difference is that they typically have a variety of appendages, including tentacles, antennae, and specialized mouthparts. Each segment contains two fleshy extensions, **parapodia**, which function in locomotion and also, because they contain many blood vessels, are important in gas exchange. In polychaetes, as in other segmented animals, there is a tendency for the division of labor between segments to lead to **tagmosis**—the formation of groups of segments into body regions with functional differences. This tendency is more pronounced in the polychaetes than in the oligochaetes and often results in distinct head, trunk, and tail regions.

Polychaetes have diverse life styles. Some are motile predators, using their strong jaws to feed on small animals. Others are more sedentary, feeding on materials suspended in the water or deposited in bottom sediments. Many polychaetes live in elaborately fashioned tubes constructed in the mud or sand of the ocean bottom. Usually the sexes are separate, fertilization is external, and there is a free-swimming trochophore larva.

26–17 *Leeches are primarily bloodsuckers with digestive tracts specially adapted for storage of blood. They range in size from 1 to 30 centimeters and are found mostly in inland waters or in damp places on land, where they parasitize fish, turtles, and other vertebrates. A few species are predatory, feeding on worms and insects that they swallow whole.*

Shown here is a medicinal leech, genus Hirudo, of the type commonly used for bloodletting. Bloodletting, one of the most common forms of medical treatment as late as the 1800s, was used for patients suffering from whooping cough, gout, drunkenness, rheumatism, sore throat, and asthma, among other ailments. Patients were commonly "bled to faintness," losing as much as 1.5 liters of blood.

Class Hirudinea

Hirudineans are the leeches, which have flattened, often tapered, bodies with a sucker at each end (Figure 26–17). In most species, the setae have been lost, and the animals either creep along with a loping movement or swim with undulating motions of the body. Like the earthworm and other oligochaetes, leeches are hermaphrodites.

Bloodsucking leeches attach themselves to their hosts by their posterior sucker, and then, using their anterior sucker, either slit the host's skin with their sharp jaws or digest an opening through the skin by means of enzymes. Finally, they secrete chemicals into the host's blood that prevent the formation of clots. One of these substances, appropriately called hirudin, is the most powerful natural anticoagulant known. The gene coding for hirudin, which consists of 65 amino acids, has recently been cloned, and it is expected that synthetic hirudin will soon be available for treating heart attack victims and patients with severe atherosclerosis. Another substance in leech saliva has been demonstrated to inhibit the spread of malignant cells from lung cancers. In keeping with their rather unpleasant image, however, leeches secrete no anesthetic in their saliva.

MINOR PROTOSTOME PHYLA

In addition to the three major phyla of protostome coelomates (Mollusca, Annelida, and Arthropoda), there are seven minor phyla with living representatives. Four of these phyla contain bottom-dwelling marine worms that show varying degrees of similarity to the annelids.

The peanut worms, phylum Sipuncula, range from 1 to more than 60 centimeters in length and are characterized by a long proboscis that can be retracted into the stout, bulblike body (Figure 26-18a). Although the approximately 300 species of this phylum have neither segmentation nor setae, their trochophore larvae are very similar to those of the polychaete annelids.

Phylum Echiura contains about 100 species that are sometimes called spoon worms. Most have a long proboscis that, unlike the proboscis of the peanut worms, cannot be retracted into the stout body; it can contract, however, forming a structure that resembles the bowl of a spoon (Figure 26-18b). As in the peanut worms, the sexes are separate, fertilization is usually external, and there is a trochophore larva. The adult worms have setae but show no traces of segmentation.

Fifteen species of burrowing worms, ranging from 0.5 millimeter to 20 centimeters in length, make up phylum Priapulida. They are characterized by a retractile proboscis bearing spines that are used to capture soft-bodied prey (Figure 26-18c). Priapulids are not segmented internally, but the external surface of the body may be marked by many superficial rings; setae are found only on the males of one genus. These worms most closely resemble the pseudocoelomate kinorhynchs (see page 541) but may have a true coelom. Very little is known about either their embryonic development or their phylogenetic relationships.

(a)

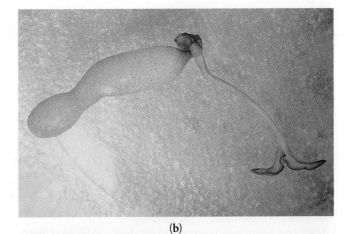

(b)

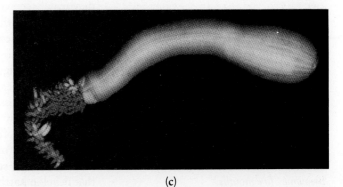

(c)

26-18 *Representatives of the smaller phyla of protostome worms.* **(a)** Dendrostomum pyroides, *a sipunculan. The retractile proboscis represents about one-third of the worm's total length. Food particles, trapped by the mucus-covered tentacles, are moved to the mouth by cilia. When disturbed, these worms contract into the shape of a peanut.* **(b)** *An echiuran,* Bonellia viridis, *from the bottom sediments off the Mediterranean coast of France. The algal cells on which this species feeds are caught in the cilia of the long, forked proboscis and are carried to the mouth.* **(c)** Priapulis caudatus, *the first priapulid described, with its anterior proboscis everted. The bushy tail at the posterior of the animal consists of tubular structures that are thought to have a respiratory function. In some species the tail is muscular and bears hooks; it presumably anchors the worm when it is burrowing.*

26-19 *The most stunning members of phylum Pogonophora are the giant tube worms that live near fissures in the earth's crust, deep in the Pacific Ocean. These worms, first discovered in 1977, sometimes reach 1.5 meters in length. Their bodies harbor large quantities of symbiotic chemosynthetic bacteria that provide them with organic molecules.*

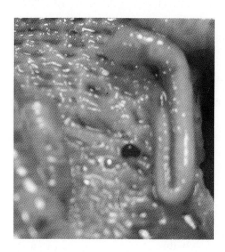

26-20 *Porocephalus crotali, a pentastomid. Adults of this genus live as parasites in the lungs of snakes, attaining lengths of 10 to 13 centimeters. Their life cycle includes two larval stages, each parasitic in a different host, which may be a snake or a mammal. Although not visible in this photograph, the mouth of a pentastomid is surrounded by four projections. At one time, these projections were erroneously thought to be additional mouths; hence the name "five mouths."*

The 100 species of beard worms, phylum Pogonophora, have a segmented posterior end with setae but have no mouth or digestive tract. These very slender worms live in long tubes buried in deep-sea sediments (Figure 26-19). The anterior region of the body bears a crown of tentacles that either provide surfaces for the uptake of nutrients or, in some species, contain symbiotic chemosynthetic bacteria that supply the worms with nutrients.

The remaining three phyla—Pentastomida, Tardigrada, and Onychophora—have varying combinations of annelid and arthropod characteristics. These animals are distinguished from all others by unjointed appendages bearing claws; arthropod appendages, by contrast, are jointed. Like the arthropods, however, they have an external cuticle that is molted periodically.

The 70 species of wormlike pentastomids (Figure 26-20) are all parasites of vertebrate respiratory systems. Pentastomids are so highly specialized for their parasitic existence—lacking, for example, circulatory, respiratory, and excretory organs—that it has been difficult to determine the primitive form from which they may have evolved. Although their larvae resemble young tardigrades, recent studies indicate that they are most closely related to the arthropods. Pentastomids have separate sexes and internal fertilization and, like many other parasites, produce tremendous numbers of eggs.

Phylum Tardigrada contains about 350 species of tiny segmented animals often called "water bears" (Figure 26-21). Common in fresh water and in the film of moisture on mosses, they lumber along on their four pairs of stubby legs. Their protective cuticle is thin, but they are able to survive drying out; when conditions are unfavorable, they enter a state of suspended animation and remain dormant—in some cases, for years—until moisture is once more available. The sexes are separate, but in some species males are unknown and the females produce eggs that develop parthenogenetically.

26–21 *A "water bear," phylum Tardigrada. These remarkable animals are able to survive extremes of temperature as well as extreme desiccation. They are found not only in temperate climates, moving in their slow, deliberate way over the surfaces of mosses and lichens, but also in the Arctic, in the tropics, and even in hot springs. Their mouths are equipped with piercing stylets, through which they suck the juices of plant cells or of such small animals as rotifers and nematodes.*

50 μm

The caterpillarlike animals of phylum Onychophora have a particularly striking combination of annelid and arthropod characteristics. The annelid characteristics shared by the 70 species in this phylum include relatively soft bodies, segmentally arranged nephridia, muscular body walls, and ciliated reproductive tracts. On the other hand, their jaws (derived from appendages), protective cuticle, relatively large brains, and circulatory and respiratory systems resemble those of arthropods. Their antennae and eyes are similar to those of both the polychaete annelids and the arthropods. Although it is reasonable to consider onychophorans primitive animals, suggestive of a stage in the early evolution of the arthropods, their reproduction is very advanced. Although some onychophorans lay eggs, most give birth to live young (Figure 26–22). Moreover, in some species the embryos develop within a uterus and are nourished through a placenta-like structure, analogous to that of the mammals. Modern onychophorans are all terrestrial, living inconspicuously in moist habitats, mainly in the Southern Hemisphere. Fossil evidence, however, indicates that the earliest species were marine (see Figure 25–5d, page 521). As onychophorans made the transition to land, they, like the plants (and, as we shall soon see, the arthropods and the vertebrates), successfully solved the problem of protecting and nourishing the developing young.

26–22 *The birth of an onychophoran. With a strong contraction of its body, this young* Peripatus *has just freed itself from its mother's genital orifice. Although a newborn* Peripatus *is able to walk immediately on its short, unjointed legs and to eject the "glue" with which these carnivorous animals ensnare prey, it usually remains with its mother for a few days.*

26-23 *The extensive lophophore of a phoronid worm,* Phoronopsis harmeri. *The lophophore is in the shape of a horseshoe, with each of the ends forming a spiral coil. Cilia on the tentacles of the lophophore direct water currents through the groove between the two coils. Food particles are trapped in mucus on the tentacles and carried by the cilia to the mouth, which is at the base of the groove. Although phoronids are not attached to the chitinous or leathery tubes in which they live, they never leave them. When disturbed, the animal can completely withdraw into the safety of the tube.*

26-24 *The brachiopod* Terebratulina septen, *with its shell closed and open. When the shell is open, the only structure visible is the large, spiral lophophore, which constitutes about two-thirds of the animal's body. Water is circulated through the shell by the moving tentacles of the lophophore. Its cilia sweep small organisms in the water toward the mouth, which is under the lophophore, toward the left. The brachiopod is anchored to the substrate by a stalk, or pedicel.*

SECTION 4 The Diversity of Life

THE LOPHOPHORATES

The animals in three additional phyla, known collectively as the lophophorates, are, strictly speaking, protostomes (that is, the mouth develops at or near the blastopore). Nevertheless, they possess other characteristics that lead many authorities to regard them as primitive deuterostomes. Lophophorates do not display the segmentation characteristic of annelids and arthropods, but the coelom of the adult forms of some species and of the embryos of most species is partitioned into several major compartments. A similar sort of partitioning is characteristic of many deuterostomes. Other deuterostome characteristics shared by the lophophorates include radial cleavage in the early stages of embryonic development and, in some species, enterocoelous formation of the coelom (see page 545).

The lophophorates, which are all aquatic, include the lamp shells (phylum Brachiopoda), the phoronid worms (phylum Phoronida), and the bryozoans, or "moss animals" (phylum Bryozoa, or Ectoprocta). Although the three phyla have diverged considerably, all of these animals have a characteristic food-gathering organ known as the **lophophore** (Figure 26-23). Located at the anterior end of the animal, the lophophore consists of a crown of hollow, ciliated tentacles used to capture small organisms and bits of organic debris suspended in the surrounding water. The cavity within the tentacles is an extension of the coelom, and gases are exchanged between the coelomic fluid and the external environment through the thin walls of the tentacles.

The most familiar of the lophophorates are the lamp shells, phylum Brachiopoda. Until well into the nineteenth century, these animals, which resemble a Greek or Roman oil-burning lamp (Figure 26-24), were classified with the bivalve mollusks. The lophophore, however, clearly distinguishes the brachiopods from the bivalves. Moreover, the two valves of the brachiopod shell are dorsal and ventral, rather than left and right as in the bivalves. There are about 250 species of brachiopods living today, but the fossil record reveals an additional 30,000 extinct species.

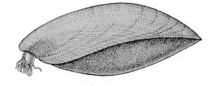 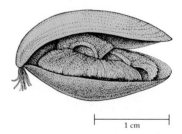

Simpler in structure are the 18 species of phylum Phoronida (see Figure 26-23). These worms, which range from 4 to 25 centimeters in length, live in tubes in or on soft ocean bottoms in shallow waters. The lophophore projects above the top of the tube but can be withdrawn into the tube when the animal is disturbed. Although phoronids are often found clustered together, each animal is independent of the others. Most species are hermaphroditic, and one species is known to reproduce asexually.

In contrast to the phoronids, the tiny bryozoans, or ectoprocts, are colonial animals, often with a considerable division of labor among the members of the colony. Their colonies can be found on virtually any type of firm surface in salt water and, less frequently, in fresh water. Bryozoans secrete a hard, protective covering around themselves, from which only the individual lophophores extend (Figure 26-25). Their name derives from the superficial resemblance of their colonies to patches of moss. As we noted earlier (page 541), the pseudocoelomate

26-25 *Colonies of the bryozoan* Hippodiplosia insculpta, *photographed in Monterey Bay, California. Note the extended lophophores with which the colony members catch passing algal cells. Because of the protective covering that bryozoans secrete around themselves, their colonies are also known as "sea mat."*

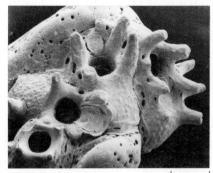

26-26 *The skeletal remains of a microscopic colony of the bryozoan* Trematooecia psammophila *on the edge of a sand grain. The stalklike extensions are spines. When the bryozoan was alive, the lophophores extended through the large openings visible in this scanning electron micrograph as dark holes. The thumbprint-like depression adjacent to the second hole is a developing brood chamber. Within each brood chamber an embryo develops into a motile larva.*

Entoprocta have a similar appearance, and before the differences in internal structure were fully understood, the entoprocts were also included in phylum Bryozoa. About 4,000 living species of bryozoans are known, and most are hermaphrodites; the freshwater forms, however, reproduce asexually as well as sexually.

The bryozoan life cycle includes a motile larva that, in many species, is such a poor swimmer that biologists have long wondered how new colonies can become established at relatively great distances from the parent colony. Recent studies have demonstrated that, in at least some species, the secret is a series of short-lived sexually reproducing generations. The larvae attach themselves to individual sand grains, rapidly mature, and reproduce again, without investing either the time or the resources to form an elaborate colony—for which there is, in any case, insufficient space (Figure 26–26). By jumping from sand grain to sand grain, one generation at a time, descendants of the original colony ultimately reach a distant site suitable for the establishment of a new colony. Among the many mechanisms of genetic recombination and dispersal in the animal kingdom, this may be one of the most ingenious.

SUMMARY

One of the most significant innovations in the course of animal evolution was the coelom, a fluid-filled cavity that develops within the mesoderm. The coelom not only functions as a hydrostatic skeleton but also provides space within which the internal organs can be suspended by the mesenteries.

The coelomate animals are divided into two broad groups on the basis of their embryonic development. In the protostomes, the cleavage of the fertilized egg is usually spiral, the mouth develops at or near the blastopore, and the coelom is formed by a splitting of the solid mesoderm. In the deuterostomes, the cleavage pattern is radial, the anus develops at or near the blastopore, and the coelom is formed by outpocketings of the primitive gut.

The soft-bodied animals of phylum Mollusca are classified into four relatively small classes and three major classes: Bivalvia (oysters and clams), Gastropoda (snails), and Cephalopoda (octopuses and their relatives). The basic body plan of all mollusks is the same—a head-foot, a visceral mass, and a shell-secreting mantle—but it has been modified in the course of adaptation to different ways of life. Shells are often present but may be reduced or absent. Mollusks are also characterized by a toothed tongue (the radula), an open circulatory system, and a greatly reduced coelom. In most mollusks, respiration is carried out by means of gills, thin-walled structures that are an extension of the epidermis and are located in the mantle cavity. Excretion is accomplished by special organs, the nephridia, tubular structures that collect fluids from the coelom and exchange salts and other substances with body tissues as the fluid passes along the tubules for excretion. Nervous systems and behavior vary among the species, reaching a zenith of complexity in the brainy octopus.

On the basis of similarities in their trochophore larvae, the mollusks and the segmented worms of phylum Annelida are thought to have diverged from a common coelomate ancestor. Annelids are characterized by bodies that are conspicuously segmented, both internally and externally; well-developed coeloms; tubular digestive tracts; paired nephridia in each segment; and closed circulatory systems, often with contractile vessels. The phylum includes terrestrial worms, such as the earthworms (class Oligochaeta), marine worms (class Polychaeta), and leeches (class Hirudinea). Unlike the oligochaetes, the polychaetes have a variety of appendages and exhibit tagmosis—the formation of groups of segments into body regions with functional differences. Annelids have relatively complex nervous systems, consisting of a ventral nerve cord and pairs of ganglia (clusters of nerve cells), one pair to each segment, which receive impulses from a variety of sensory receptors and trigger motor activities in each segment.

Of the seven minor phyla of protostome coelomates, four consist of bottom-dwelling marine worms that show varying similarities to the annelids. These groups include the peanut worms (phylum Sipuncula), the spoon worms (phylum Echiura), the priapulids (phylum Priapulida), and the beard worms (phylum Pogonophora). Three other phyla—Pentastomida, Tardigrada, and Onychophora —all characterized by unjointed, clawed appendages and an external cuticle, have combinations of annelid and arthropod features. The onychophorans are of particular interest, for they suggest a stage in the evolution of arthropods from a segmented, coelomate ancestor common to both the annelids and the arthropods.

Lophophorates, although protostomes, resemble deuterostomes in the radial cleavage of the fertilized egg and, in some species, the enterocoelous formation of the coelom. The lamp shells (phylum Brachiopoda), the marine worms of phylum Phoronida, and the colonial Bryozoa are all lophophorates. They are characterized by the lophophore, a crown of hollow tentacles bearing cilia that sweep food particles into the mouth.

QUESTIONS

1. Distinguish between the following terms: protostome/deuterostome; schizocoelous/enterocoelous; coelom/hemocoel; planula/trochophore.

2. What modifications of the basic molluscan body plan have occurred in bivalves, gastropods, and cephalopods?

3. Many mollusks have lost or may be in the evolutionary process of losing their shells. What are the advantages to an organism of having a shell? Of losing one?

4. Smallness may also be an advantage to an organism. What are some of the advantages of smallness?

5. Why does any heart truly worthy of the name have at least two chambers?

6. Label the drawing at the right.

7. Nematodes have only longitudinal muscles in their body walls but have very high internal fluid pressures; earthworms, with both longitudinal and circular muscles, have low fluid pressures. Can you suggest a mechanism by which internal fluid at high pressure can circumvent the need for certain muscles?

8. Describe the structure and function of a lophophore. How does it differ from the tentacles of a cnidarian, such as *Hydra*?

9. A larval or medusoid stage occurs in the life histories of many marine invertebrates. Yet in group after group of freshwater and terrestrial invertebrates, each presumably separately evolved from marine ancestors, the free-living larva or medusa has been lost. What might be the selective factor underlying the loss?

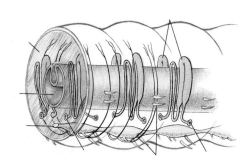

CHAPTER 27

The Animal Kingdom III: The Arthropods

Phylum Arthropoda, the "joint-footed" animals, is by far the largest of the animal phyla. More than 1 million species of insects and other arthropods have been classified to date, and estimates of the total number are as high as 50 million. As to the number of individuals, it has been calculated that of insects alone, as many as 10^{18}—a billion billion—are alive at any one time. Arthropods are abundant in virtually all habitats. It has been estimated that over every square kilometer in the temperate zone there are, at certain seasons, some 20 million individual arthropods, layered in the atmosphere like plankton.

CHARACTERISTICS OF THE ARTHROPODS

Arthropods are characterized by their jointed appendages. In the more highly evolved members of the phylum, the number of appendages is reduced, but they are more specialized and efficient. Particularly among the insects and crustaceans, these specialized appendages include not only walking legs but also a kit of marvelously adapted tools—jaws, gills, tongs, egg depositors, sucking tubes, claws, antennae, paddles, and pincers.

All arthropods are segmented, a characteristic that strongly suggests a common ancestry with the annelids (Figure 27–2). In the course of arthropod evolution, however, the body has become shorter, and it has fewer segments, which have become fixed in number and more specialized. In many arthropods, tagmosis

27–1 A harlequin beetle in flight. Arthropods are characterized by a variety of specialized appendages. Notice the three pairs of jointed legs, the jointed antennae, the intricately colored wings, and the segmented body.

27-2 *Arthropods, like the annelids, are segmented. In the more primitive arthropods, such as this millipede, the segmented pattern remains clearly visible in the adult animal. However, unlike annelids, adult arthropods have rigid, jointed exoskeletons and appendages. In millipedes, which are members of class Diplopoda, most of the body segments are fused into "diplosegments," each of which has two pairs of legs.*

(page 558) has been carried much farther than in the polychaete annelids, with the fusion of segments to form distinct body regions—a head, a thorax (sometimes fused with the head to form a cephalothorax), and an abdomen. But the basic segmented pattern is often still clearly evident in the immature stages (witness the caterpillar) and can be discerned in the adult by examination of the appendages, the musculature, and the nervous system.

At some point well after the lineage leading to the arthropods diverged from that leading to the annelids, further major branchings occurred (see Figure 25-6, page 521). These branchings gave rise to three principal types of arthropods: chelicerates, aquatic mandibulates, and terrestrial mandibulates. The conspicuous differences in the appendages of these three types can be clearly distinguished by even an inexperienced eye. In both the aquatic mandibulates (class Crustacea) and the terrestrial mandibulates (class Insecta and four smaller classes), the most anterior appendages are one or two pairs of **antennae,** and the next are **mandibles** (jaws). Differences in the development and structure of the mandibles suggest that they evolved independently in the two groups and do not reflect a common mandibulate ancestor. The chelicerates, which include class Merostomata (horseshoe crabs), class Pycnogonida (sea spiders), and class Arachnida (spiders, mites, scorpions, and their relatives), have no antennae and no mandibles. Their first pair of appendages consist of **chelicerae** (singular, chelicera), which take the form of pincers or fangs. Chelicerates may also have **book lungs** or **book gills;** these respiratory structures, which are not present in mandibulates, derive their name from their resemblance to the leaves of a partially opened book.

Despite the huge number of arthropods and the richness of their diversity, there are a number of features shared by all members of this phylum.

The Exoskeleton

The most striking characteristic of all arthropods is their articulated (jointed) exoskeleton. This exoskeleton, or cuticle, is secreted by the underlying epidermis and is attached to it; it is made up of an outer, often waxy layer, composed of lipoprotein, a hardened middle layer, and an inner flexible one, both composed principally of chitin and proteins. The exoskeleton not only covers the surface of the animal but also extends inward at both ends of the digestive tract; in insects, it lines the tracheae (breathing tubes) as well. The cuticle serves as protection against predators, and it is often waterproof, keeping exterior water out and interior water in. It is used for food grinders in the foregut, for wings, and for tactile hairs. Cuticle even forms the lens of the arthropod eye.

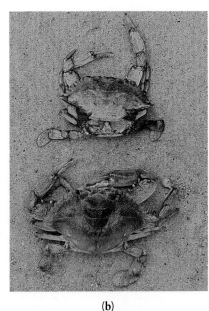

27-3 (a) *A many-jointed exoskeleton, as in this South American katydid, is characteristic of the arthropods. The slits in each foreleg are the insect's ears.*

(b) *Molting. An exoskeleton, once formed, does not grow and so must be periodically discarded. After shedding the old cuticle, an arthropod expands rapidly by taking in air or water, stretching the new exoskeleton before it hardens. The old cuticle is at the top, and below is the newly emerged, already expanded "soft-shelled" blue crab. Its new exoskeleton has begun to harden.*

The exoskeleton may form a veritable coat of armor, as it does in beetles and in some of the crustaceans (in which it is often infiltrated with calcium salts), but at the joints it is flexible and thin, permitting free movement. Muscles are attached to the various portions of the exoskeleton, just as they are attached to the various bones of the endoskeleton in vertebrates. When the muscles contract, the exoskeleton moves at its joints; such movements can be exquisitely precise because the force is brought to bear on very small areas, for example, on the different sections of an insect's leg.

The exoskeleton has certain disadvantages. It does not grow (as the bony vertebrate endoskeleton does), and so it must be discarded and re-formed many times as the animal grows and develops. At molting time, the animal secretes an enzyme that dissolves the inner layer of the exoskeleton, and a new skeleton, not yet hardened, is formed beneath the old one. Molting is dangerous; the newly molted animal is particularly vulnerable to predators and, in the case of terrestrial forms, subject to water loss. Many arthropods go into hiding until their new cuticle has hardened. Molting is also costly in terms of metabolic expenditures, although a number of insects and some freshwater crustaceans limit their losses by thriftily reabsorbing the dissolved inner layer and eating the outer layers of the old exoskeleton.

The fact that the exoskeleton can be waterproofed made possible the evolution of terrestrial forms among the arthropods, a process that was well underway by the late Ordovician. Fossil traces of animal burrows in relatively dry soil, thought to have been made by arthropods similar to millipedes, were recently discovered in deposits near Potters Mills, Pennsylvania; dated at 440 to 445 million years ago, these fossils are the earliest evidence of animals adapted to a semiarid or arid environment. Of all the invertebrate phyla, only Arthropoda contains a multitude of species well adapted to withstand the drying action of the air.

Internal Features

Arthropods, like annelids, have a tubular mouth-to-anus gut. They also have a coelom, but the arthropod coelom—like that of the mollusks—is markedly reduced, consisting only of the cavities of the reproductive and excretory organs.

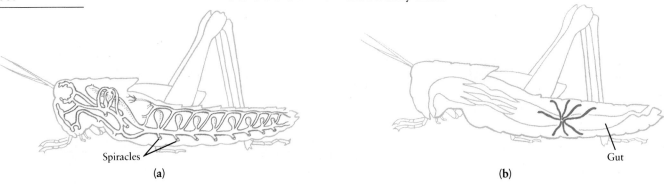

(As you will recall, the coelom serves as a hydrostatic skeleton in annelids, but the arthropods, with their exoskeleton, require less internal stiffening.) Like most mollusks, arthropods also have an open circulatory system in which blood flows through free spaces in the tissues—the hemocoel—as well as through vessels. Blood returns from the hemocoel to the tubular heart through special valved openings.

The insects and some other terrestrial forms have an unusual means of respiration, consisting of a system of cuticle-lined air ducts, known as **tracheae,** that pipe air directly into the various parts of the body (Figure 27–4a). Gases must pass through the tracheae largely by diffusion, thus placing a limit on the size of insects. Some terrestrial arthropods, such as spiders, have book lungs, either in addition to or instead of tracheae. Excretion in terrestrial forms is by means of **Malpighian tubules** (Figure 27–4b), which are attached to and empty into the midgut or hindgut. These tubules absorb wastes from the body cavities. Respiration by tracheae, book lungs, or book gills and excretion by Malpighian tubules are found almost exclusively in the arthropods (although not all arthropods have these features).

27–4 (a) *Respiration by means of a system of internal tubes (called tracheae) is found almost exclusively among terrestrial arthropods. Tracheae are usually branched, like those shown here, and open to the outside by spiracles that may be closed to conserve water. The tubes are lined with spiral rings and cuticle, which keeps them open. Because it delivers oxygen directly to the cells, a tracheal system is one of the most efficient respiratory systems in the animal kingdom. As the size of the animal increases, however, its efficiency decreases.*

(b) *Malpighian tubules represent another exclusively arthropod characteristic, although, like tracheae, they are not found in all classes. These tubules collect water and nitrogenous wastes from the hemocoel and empty them into the gut. The wastes (in the form of uric acid or guanine) are excreted with the feces.*

The Arthropod Nervous System

The bulk of the arthropod nervous system consists of a double chain of segmental ganglia running along the ventral surface (Figure 27–5). The double chains part to encircle the esophagus, ending in three fused pairs of dorsal ganglia. These fused dorsal ganglia, which perform a number of specialized functions, constitute a brain. However, many arthropod activities are controlled at the segmental level, as in annelids. For example, even after the brain has been removed, members of a number of species can move, eat, and carry on other functions normally. In fact, in the arthropods generally, the brain appears to act not so much as a stimulator of the action of the animal but as an inhibitor, as in the earthworms. The grasshopper, for example, can walk, jump, or fly with its brain removed; indeed, the brainless grasshopper responds to the slightest stimulus by jumping or flying.

SUBDIVISIONS OF THE PHYLUM

The Chelicerates

As we noted previously, chelicerates have neither antennae nor mandibles. The first pair of appendages, the chelicerae, which bear pincers or are sharp and fanglike, are used for biting prey. The second pair are the **pedipalps,** which may bear pincers, be modified as walking legs, or serve as sensory organs. Posterior to the pedipalps are a series of jointed walking legs. The body segments have become fused into two tagmata—an anterior cephalothorax and a posterior abdomen, which is unsegmented in most chelicerates but conspicuously segmented in scorpions. Except for some mites, all chelicerates are carnivorous. The sexes are almost always separate.

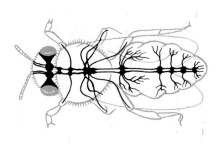

27–5 *The arthropod nervous system, as exemplified in the bee. The brain consists of three fused pairs of dorsal ganglia at the anterior end of a double chain of ganglia. The ganglia are interconnected by two bundles of nerve fibers running along the ventral surface. Because of the arthropod's segmental type of nervous system, many functions are controlled at a local level, and a number of species can carry on some of their normal activities after the brain has been removed.*

CHAPTER 27 The Animal Kingdom III: The Arthropods

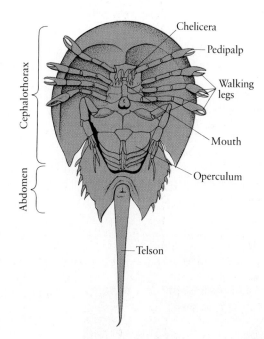

27-6 *Because the body of the horseshoe crab is largely covered with a heavy shield, or carapace, its segmented body plan is only evident from the underside. The operculum is a flat, movable plate that covers and protects the book gills. Horseshoe crabs are often called living fossils because remarkably similar organisms date back to the Silurian period (408 to 438 million years ago). There are now only four living species.*

Two classes of chelicerates are relatively small. Class Merostomata consists of only four species of horseshoe crabs (Figure 27-6). In viewing a horseshoe crab from below, one can see the chelicerae, pedipalps, and four pairs of walking legs. Posterior to the walking legs are a series of flaplike book gills, derived from modified limbs. *Limulus*, a bottom dweller that feeds upon such small animals as annelids and clams, is common in the shallow waters along the East Coast of the United States.

The approximately 500 species of sea spiders (class Pycnogonida) have slender bodies and four (or, rarely, five) pairs of legs, which are often very long (Figure 27-7). They feed on soft-bodied invertebrates—particularly cnidarians—by sucking juices from their bodies. Although sea spiders are common in coastal waters, most species are quite small and inconspicuous.

All of the remaining chelicerates belong to class Arachnida, with about 57,000 species.

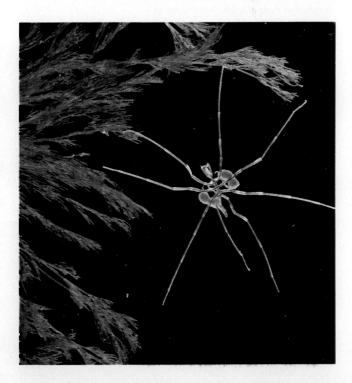

27-7 *A male sea spider,* Nymphon gracile, *carrying two large masses of developing eggs. Among pycnogonids, the eggs are always carried by the males on a subsidiary pair of specialized legs. Sea spiders have neither respiratory nor excretory systems; gases and waste products are thought to diffuse through the large surface area provided by the long, thin legs.*

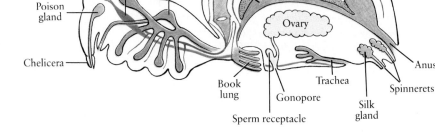

27–8 *A spider, an arachnid. Ducts from the poison glands open at or near the tips of the chelicerae. The flow of poison is voluntarily controlled by the spider. Only a few spiders are dangerous to human beings; perhaps the most dangerous are members of the species shown here, the black widow.*

Class Arachnida

The arachnids, which include spiders, ticks, mites, scorpions, and daddy longlegs, are almost all terrestrial, except for a few species that have returned to the water. Arachnids have four pairs of walking legs, and their chelicerae and pedipalps are often highly specialized. In spiders, ducts from a pair of poison glands lead through the chelicerae, which are sharp and pointed and are used for biting and paralyzing prey (Figure 27–8). Scorpions use their pedipalps to handle and to tear food. Male spiders also use the pedipalps to transfer semen to the female.

Spiders, like most arachnids, live on a completely liquid diet. All are predatory. The prey is bitten and often paralyzed by the chelicerae, and then enzymes from the midgut are poured out over the torn tissues to produce a partially digested broth. The liquefied tissues of the prey are pumped into the stomach and then to the intestine, where digestion is completed and the juices absorbed. Arachnids respire by means of tracheae or book lungs, or both. Book lungs are a series of leaflike plates within a chitin-lined chamber into which air is drawn and expelled by muscular action.

On the posterior portion of the spider's abdominal surface is a cluster of spinnerets, modified appendages from which a fluid protein exudes that polymerizes into silk as it is exposed to air. Silk is used not only for the variety of

(a)

27–9 *Some arachnids.* (a) *A giant desert hairy scorpion,* Hadrurus hirsutus, *photographed in Arizona. The southwestern United States and Mexico are also home to the highly toxic genus* Centruroides. *Scorpions sting to immobilize their prey and in self-defense, releasing a venom that attacks the nervous system of the victim.* (b) *An adult red velvet mite of the genus* Trombidium. (c) *A female hunting spider, photographed in a rain forest in Costa Rica as she guarded her cocoon of fertilized eggs. Note how the segments of the body have become fused into two regions —cephalothorax and abdomen.* (d) *An orb-weaving spider (*Neoscona oaxacensis*) has caught a grasshopper in its web. The prey, immobilized by venom, is being wrapped in silk. If you look closely, you can see the silk being spun out from the spider's abdomen. Webs, which may be as much as a meter in diameter, are reconstructed daily.* (e) *A daddy longlegs. These familiar arachnids are also known as harvestmen; most species mature in late summer, and they are seen in greatest numbers at harvest time.*

(b)

(c)

(d)

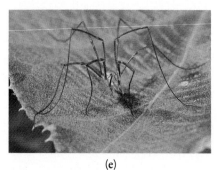

(e)

prey-snaring webs made by the different species but for a number of other purposes as well—such as a drop line, on which the spider can make a defensive dive, draglines for marking a course, gossamer threads for ballooning, hinges for trap doors, an egg case, lining for a burrow, the shroud of a victim, or a wrapping for an edible offering presented to the female of certain species by the courting male. Most spiders can spin several kinds and thicknesses of silk. Webs are species-specific, and web-building is a genetically programmed behavior.

The Aquatic Mandibulates: Class Crustacea

The 25,000 species of crustaceans include crabs, crayfish, lobsters, shrimp, prawns, barnacles, *Daphnia* (water fleas), and a number of smaller forms. Some crustaceans, such as the familiar pillbugs, or sowbugs, are adapted to life in moist land environments. Crustaceans differ from the terrestrial mandibulates, such as insects, in that they have legs or leglike appendages on the abdomen as well as the thorax and have two pairs of antennae as compared to the insects' one pair.

Among the crustaceans, the sexes are usually separate, but there are exceptions. Barnacles, which are sessile as adults, are simultaneous hermaphrodites. Some species of shrimp are sequential hermaphrodites; when the animal is small, it is male, but when it reaches a size that is effective for carrying eggs, it becomes female. Most marine crustaceans have larval stages that swim about before developing into the adult animals.

The Lobster

Figure 27–10 shows the structure of a lobster, a representative crustacean. A crayfish, which is a freshwater form, differs anatomically from the lobster only in minor respects.

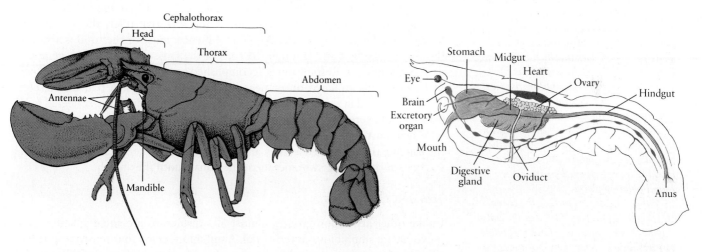

27–10 *A representative crustacean, the American lobster,* Homarus americanus. *The lobster has 19 pairs of appendages, including antennae, mouthparts, and legs specialized for feeding, walking, and swimming.*

The lobster has 19 segments, the first 13 of which are united on the dorsal side in a combined head and thorax, or cephalothorax. A heavy shield, or carapace, arises from the head and covers the thorax. Carapaces are common among crustaceans. The abdomen consists of six distinct segments. (To lobster eaters, the abdomen is the "tail," but to biologists, the designation "tail" is usually reserved for areas of the body posterior to the anus.)

The various appendages have special functions. The antennae, of which there are two pairs, are sensory. The mandibles, or jaws, are used for crushing food; like all arthropod jaws, they move laterally, opening and closing from side to side like a pair of ice tongs rather than up and down like the jaws of vertebrates. Crustacean mandibles are thought to have arisen from a pair of limbs, the bases of which took

(a) (b)

(c)

27-11 *Some crustaceans.* (a) *A cleaner shrimp of the Indian and Pacific Oceans, photographed near the Philippine Islands.* (b) *A female copepod, an inhabitant of the plankton. The two large structures at the posterior are egg sacs in which zygotes are developing; they will be released as free-swimming larvae. Copepods are the most numerous animals not only in the plankton but also in the world; the individuals of a single genus (*Calanus*) are thought to outnumber all other animals put together.* (c) *Pillbugs (also known as sowbugs or wood lice) are terrestrial crustaceans found in damp places.*

Other crustaceans include the barnacles (Figure 25-3, page 519) and the crabs (Figure 27-3b, page 567). Crabs resemble lobsters, but the relative proportions of the body have changed; the cephalothorax is broader and larger than the abdomen, which is tucked under the rest of the body.

on a chewing function. The thorax of the lobster bears five pairs of walking legs, the first pair of which are modified as claws. The claws are unequal in size in the full-grown lobster; the larger claw is used for defense and for crushing food, and the smaller claw, which has the sharper teeth, seizes and tears the prey. The next two pairs of walking legs have small pincers, which can also seize prey. The last pair of walking legs also serve to clean the appendages of the abdomen. These flattened appendages, or swimmerets, are used, like flippers, for swimming. The claws and abdomen are filled almost entirely with the large striated muscles many of us enjoy eating. These muscles are extremely powerful. A lobster can snap its abdomen ventrally with enough force to shoot backward through the water, and a large lobster can shatter the shell of a clam or oyster with its crushing claw. Any one of its appendages can be regenerated over a series of molts if lost, and a lobster will often drop off (autotomize) a claw or leg that is held by a predator, in order to escape. Severely damaged legs are also autotomized, reducing blood loss.

The anterior and posterior regions of the digestive tract, the foregut and hindgut, are lined with cuticle. As a consequence, most of the food is absorbed through the midgut and through the cells of the large digestive gland, an appendage of the midgut. Respiration is accomplished by the flow of water over 20 pairs of feathery gills attached on or near the bases of the legs. A pair of excretory organs is located in the head; wastes extracted from the blood are collected into a bladder and excreted from a pore at the base of each of the second antennae.

When lobsters mate, the sperm are deposited in or near the female gonopore; the eggs are then fertilized as they are laid. The fertilized eggs cling by means of a sticky secretion to the swimmerets of the female until they hatch.

Terrestrial Crustaceans

Unlike other arthropod groups, almost all crustaceans are aquatic; some crabs, however, are amphibious or terrestrial. Amphibious crabs continue to respire with gills, carrying water in their thoracic cavities with which to keep the gills wet. The true land crabs have lost some of the gill structures but have an area of highly vascularized epithelial tissue through which gases are exchanged. The land snails, you will recall, solved the respiratory problems involved in the transition from water to land in an analogous way.

The Terrestrial Mandibulates: Myriapods

Terrestrial mandibulates are identified by their single pair of antennae and by their mandibles, which, as we noted earlier, differ from those of the crustaceans. They respire through tracheae, and excretion is by means of Malpighian tubules. In addition to the insects, there are four smaller classes of relatively unspecialized terrestrial mandibulates. There has been little tagmosis in these arthropods, and

27–12 *A centipede of the genus* Scolopendra, *photographed in southern Africa. In centipedes, the appendages of the first segment are modified as poison claws. Prey is killed with the claws and then chewed with the mandibles.*

their bodies consist of a head region followed by an elongated trunk with many distinct segments, all more or less alike. With a few exceptions, all segments have paired appendages, and the members of these four classes are known collectively as myriapods ("many-footed").

The most familiar myriapods are the centipedes (class Chilopoda) and the millipedes (class Diplopoda). The approximately 3,000 species of centipedes prefer damp places—under logs or rocks or in basements. They are all carnivorous, feeding on cockroaches and other insects, as well as on soft-bodied annelids. A centipede (Figure 27–12) has one pair of appendages on each body segment, unlike a millipede (see Figure 27–2, page 566), which appears to have two pairs of appendages per segment. Each body ring of the millipede, however, actually represents two segments fused into a double unit. About 7,500 species of millipedes have been described, and they too prefer damp environments. Unlike centipedes, they are herbivorous, usually feeding on bits of decaying vegetation.

Less familiar are the approximately 300 species of class Pauropoda and the 130 species of class Symphyla. Unlike other arthropods, they are soft-bodied, living in moist soil, leaf litter, and other decaying matter. Although both pauropods and symphylans are abundant, they are quite small and, with the exception of one species of symphylan that is a common pest in greenhouses, they usually go about their business unnoticed by human eyes.

The Terrestrial Mandibulates: Class Insecta

The insects constitute the largest class, by far, of the arthropods. In fact, more than 70 percent of all animal species on earth are insects.

Insects are the only invertebrates capable of flight. When they began to fly—more than 240 million years ago—they were able to move into and exploit a life zone almost totally unoccupied by any other form of animal life. In terms of both numbers of species and numbers of individuals, they are the dominant terrestrial organisms of this planet.

There are about 30 orders of insects, of which the four largest are Diptera, Lepidoptera, Hymenoptera, and Coleoptera. The Diptera ("two-winged") include the familiar flies, gnats, and mosquitoes. The Lepidoptera ("scale-winged") are the moths and the butterflies. Hymenoptera ("membrane-winged") include ants, wasps, and bees, many species of which live in complex societies (to be described in Chapter 51). The Coleoptera ("shield-winged") are the beetles, most of which have a pair of hard protective forewings, which pivot forward out of the way during flight (see Figure 27–1, page 565), and a pair of membranous hindwings used for flying. Of the approximately 1 million classified species of insects, at least 300,000 are beetles.

27–13 *The late J. B. S. Haldane, who was noted for his crusty disposition, as well as for his scientific achievements, was once asked what his study of biology had revealed to him about the mind of God. "Madame," he replied, "only that He had an inordinate fondness for beetles."*

(a) Everybody's favorite beetle, a ladybird ("ladybug"), feeding on aphids. This is the seven-spot ladybird, Coccinella septempunctata. *(b) A stag beetle, so called because of the antlerlike mandibles characteristic of the pugnacious males. This species,* Durcus curvidens, *is common in the rain forests of Southeast Asia. (c) A scarab beetle,* Stephanorhina guttata, *photographed in the Congo.*

Insect Characteristics

Figure 27–14 shows a grasshopper. Here you can see many of the characteristic features of insects: three body regions—the head, the thorax, and the abdomen; three pairs of legs; one pair of antennae; and a set of complex mouthparts. In the less specialized insects, such as the grasshopper, the mouthparts are used for handling and masticating food (Figure 27–15), but in the more specialized groups, the mouthparts are often modified into sucking, piercing, slicing, or sponging organs. Some are exquisitely adapted to draw in nectar from the deep, tubular nectaries of specialized flowers.

Most adult insects have two pairs of wings made up of light, strong sheets of chitin; the veins in the wings are chitinous tubules that serve primarily for strengthening. In some primitive orders, wings never evolved, and in other orders, such as fleas and lice, wings were lost secondarily, returning the insect to the condition of its wingless ancestors. Other species may have short, nonfunctional wings in one or both sexes.

Digestive, Excretory, and Respiratory Systems

The foregut and hindgut of the insect digestive tract are lined with cuticle. Salivary gland fluids are carried with food into the crop, where digestion begins. The stomach, or midgut, which lies mainly in the abdomen, is the chief organ of absorption. Insects have digestive enzymes as specialized as their mouthparts; the structure of the enzymes depends on whether the insect dines on blood, seeds, other insects, eggs, flour, cereal, glue, wood, paper, or your woolen clothes.

27–14 *In the grasshopper, an insect, the head consists of six fused segments that have appendages specialized for tasting and biting. Each of the three segments of the thorax carries a pair of legs (three pairs in all), and two of them carry wings (in the grasshopper, the forewings are hardened as protective covers). The spiracles in the abdomen open into a network of chitin-lined tubules through which air circulates to various tissues of the body. This sort of tubular breathing system is found in the terrestrial mandibulates and some arachnids. Excretion takes place by Malpighian tubules that empty into the hindgut.*

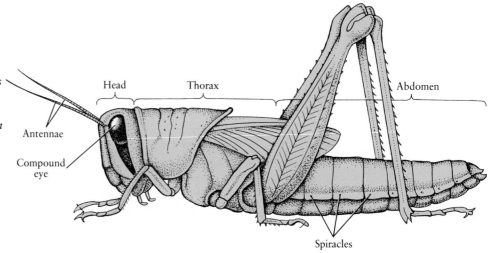

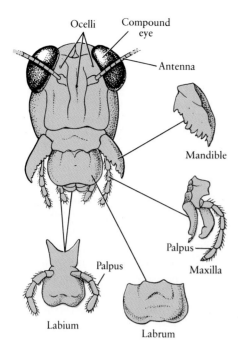

27–15 *Mouthparts of a grasshopper. The mandibles are crushing jaws. The labium and the labrum are the lower and upper lip. The maxillae move food into the mouth, and the palpi assist in tasting.*

Excretion is carried out through Malpighian tubules. In the grasshopper and many other insects, the nitrogenous wastes are eliminated in the form of nearly dry crystals of uric acid, an adaptation that promotes water conservation.

The respiratory system consists of a network of cuticle-lined tubules through which air circulates to the various tissues of the body, supplying each cell directly. Muscular movements of the animal's body improve the internal circulation of air. The amount of incoming air and also the degree of water loss is regulated by the opening and closing of the spiracles.

Insect Life Histories

The life histories of insects are fundamentally different from those of marine invertebrates, particularly forms that are sessile or slow-moving as adults. Although many marine invertebrate larvae feed, their most important biological function seems to be the invasion and selection of new habitats at some distance from the parental habitat. Such larvae may migrate great distances, resulting in the dispersal of the species over a wide area. The immature forms of most insects, by contrast, have limited mobility and usually pass through the various stages of their development very close to the location where the adult female originally laid the eggs—sometimes in the exact spot. Insect young are voracious feeders, acquiring the resources needed for their own growth and development and, in some species, storing up reserves for an adult stage in which they will not feed. The capacity to

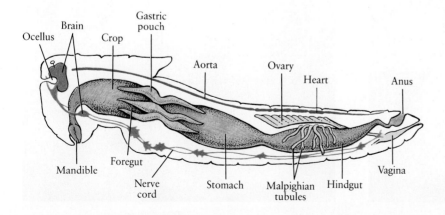

27–16 *Some immature insect forms.* (a) *Scarab beetle (white grub) larva in soil.* (b) *The larva (caterpillar) of the cecropia moth,* Hyalophora cecropia. (c) *Mosquito larvae and a pupa (on right). Mosquito larvae are aquatic, hanging onto the undersurface of the water with respiratory tubes.* (d) *Tent caterpillars on their protective web.*

(a)

(b)

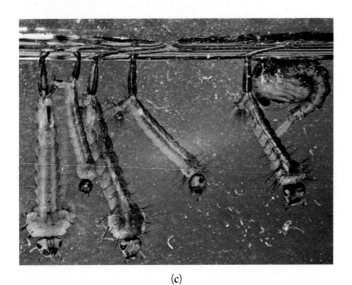

(c)

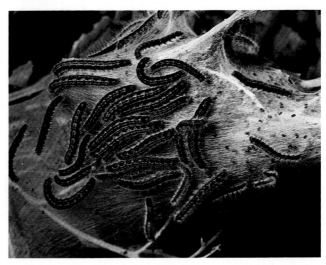

(d)

fly has, of course, given most adult insects an extraordinary mobility, and they have little difficulty finding mates or locating new habitats in which to lay their eggs—thereby ensuring dispersal of the species. One corollary of their mobility is that the sexes are separate in all insects. In many insects, some degree of parental care has evolved, with one or both parents protecting or feeding the young, or both. Among the termites, ants, wasps, and bees, such care has led ultimately to complex societies with a substantial division of labor.

Young growing insects change not only in size but often in form, a phenomenon known as **metamorphosis.** The extent of change varies. In some species, the young, although sexually immature, look like small adults; they grow larger by a series of molts until they reach full size. In others, like the grasshopper, the newly hatched young are wingless but may gain wingpads in later immature stages; otherwise they are similar to the adult. These immature, nonreproductive forms are known as **nymphs.** In almost 90 percent of insect species, however, a complete metamorphosis occurs, and the adults are drastically different from their immature forms. These immature feeding forms are all correctly referred to as larvae, although they are also commonly known as caterpillars, grubs, or maggots, depending on the species. Following the larval period, the insect undergoing complete metamorphosis enters a pupal stage, in which extensive remodeling of the organism occurs. The adult (sexually mature) insect emerges from the pupa. Both eggs and pupae (which are nonfeeding) can endure lengthy cold or dry seasons, a critical adaptation for terrestrial organisms.

An insect that undergoes complete metamorphosis exists in four different forms in the course of its life history. The first form is the egg and the embryo. The second form is the larva, the animal that hatches from the egg; larvae eat and grow. In many larvae, such as those of flies, growth takes place not by an increase in the number of cells, as in most animals, but by an increase in the size of the cells, in somewhat the same way that growth takes place in certain plant tissues. During the course of its growth, the larva molts a characteristic number of times—twice in the fruit fly, for example. The stages between molts are known as **instars.** Then, when the larva is full-grown, it molts to form the pupa. In many, but not all species, the pupa is enclosed within a cocoon or some other protective covering. During the outwardly lifeless pupal stage, many of the larval cells break down, and entirely new groups of cells, set aside in the embryo, begin to proliferate, using the degenerating larval tissue as a culture medium. These groups of cells develop into the complicated structures of the adult.

27–17 *Life history of* Heliconius ismenius, *a butterfly of Central America. The life history begins* (a) *with the mating of male and female butterflies. Fertilization is internal, and the fertilized eggs* (b) *undergo the early stages of their development within a rubbery egg shell. The newly hatched caterpillar* (c), *the larval form, eats, grows, and continues its development;* (d) *and* (e) *show the second and third instars—that is, the caterpillar after its first and second molts.* (f) *When ready to pupate, the caterpillar attaches itself upside down to a leaf with a patch of silk. Inside the pupa* (g), *metamorphosis into the butterfly occurs through a recycling of the tissues of the caterpillar. The emergent adult butterfly* (h) *is the sexually reproductive form, ready to start the cycle once more.*

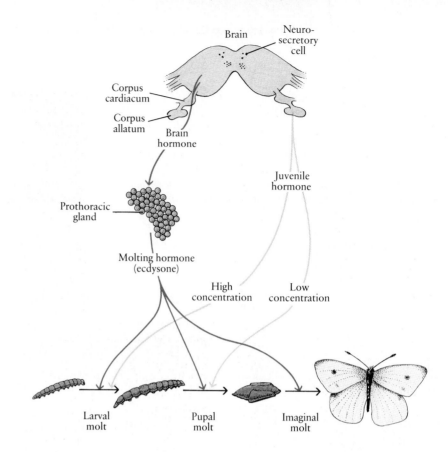

27-18 *Molting is under hormonal control. In insects, a hormone (brain hormone) produced by neurosecretory cells in the brain and released from the corpora cardiaca stimulates the prothoracic gland, which, in turn, produces molting hormone (ecdysone). Although all molts require ecdysone, whether or not metamorphosis occurs depends on a third hormone, juvenile hormone, produced by the corpora allata. Continued presence of juvenile hormone at high concentrations ensures that larval molts occur during the first portion of the life history. In later larval life, production of juvenile hormone declines, permitting adult structures to develop.*

Aristotle and his contemporaries thought of the adult insect as the imago, *the perfect form or ideal image that the immature form was "seeking to express." Although this concept has long since been abandoned by biologists, the molt giving rise to the adult is still known as the imaginal molt.*

Molting and metamorphosis are under the control of **hormones,** which are organic molecules secreted by one tissue of an organism that regulate the functions of another tissue or organ of the same organism. Like most hormone-controlled processes, molting and metamorphosis are the end result of an interplay of several substances: brain hormone, molting hormone (ecdysone), and juvenile hormone (Figure 27–18). At intervals during larval growth, brain hormone, produced by neurosecretory cells in the brain, is released into the blood. It stimulates the release, in turn, of molting hormone from a gland in the thorax. The molting hormone stimulates not only molting but also the formation of a pupa and the development of adult structures. The latter are held in check, however, by a third hormone, the juvenile hormone. Only when the production of juvenile hormone declines, in later larval life, can metamorphosis to the adult form take place.

REASONS FOR ARTHROPOD SUCCESS

Among all the invertebrates, why are the arthropods, in general, and the insects, in particular, so spectacularly successful? One important reason is undoubtedly the nature of the exoskeleton, which waterproofs, provides protection, and makes possible the evolution of the many finely articulated appendages characteristic of this phylum.

Another reason, which applies especially to insects, is their small size and the high specificity of diet and other requirements of each species. As a consequence, many different species can live in a single small environment—in a few cubic centimeters of soil, on a small plant, or on or within the egg or body of a single animal—without competing with one another. The varied and highly specialized mouthparts are a reflection of this specificity of diet.

The diversity of insects and the specificity of their requirements may be, in part, an evolutionary response to the great diversity of microenvironments provided by the vascular plants. Not only are there 235,000 species of angiosperms alone, but also the structure of each individual plant is so complex that it provides a variety of resources that can be exploited by insects with different requirements and adaptations. This is a much richer environment than open waters, the seashore, or the soil. And, virtually every cubic centimeter of it is available to insects because of the extraordinary mobility that results from their small size and their capacity for flight. The resources provided by vascular plants were, in effect, an evolutionary laboratory in which great numbers of variations could be tested, leading to the diversity of adaptations we see today. This process, of course, continues.

Another factor in the success of insects is the complete metamorphosis that occurs in the vast majority of insect species. In these insects, the adaptations for feeding and growth that are found in the larvae are separated from the adaptations for dispersal and reproduction found in the adults. The adaptations for each function can be more finely honed because compromises between conflicting needs are minimal. Another consequence of complete metamorphosis is that the dietary and other requirements of the larvae and the adults of the same species are so different that they do not compete with each other. In effect, they occupy different environments.

A final reason for success is undoubtedly the arthropod nervous system, with its fine control over the various appendages and the many highly sensitive sensory organs found in great diversity throughout the phylum. Sensory perception among arthropods, especially insects, is so important that we shall devote the remainder of this chapter to that topic and to several examples of arthropod behavior.

Arthropod Senses and Behavior

Vision: The Compound Eye

The most conspicuous sensory organ of the arthropods is the compound eye, which is an evolutionary development characteristic of this one phylum. The basic structural unit of this eye is the **ommatidium** (Figure 27–20), repeated over and over. A dragonfly, for example, has some 30,000 ommatidia. Each ommatidium is covered by a cornea, usually with a round or hexagonal surface; these are visible under low-power magnification as individual facets of the eye. Underlying the cornea is a group of eight retinular cells surrounded by pigment cells. The light-sensitive portion of the ommatidium is the **rhabdom,** which is the central core of the ommatidium. Nerve fibers carry the stimulus from each ommatidium to the brain. The pigment cells prevent light from traveling from one ommatidium to another. An ommatidium is much larger than a vertebrate photoreceptor, and so there are far fewer in an equivalent space. Hence, the image has less resolution, like a newspaper picture under high magnification.

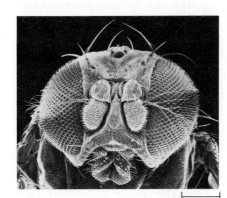

27–19 *The head of* Drosophila, *as shown by the scanning electron microscope. Note the large compound eyes on either side of the head. Although insect eyes cannot change focus, they can define objects only a millimeter from the lens, a useful adaptation for an insect.*

27–20 *Structure of the compound eye. The eye is composed of a large number of structural and functional units called ommatidia. Each ommatidium has its own cornea, which forms one of the facets of the compound eye, and its own light-focusing lens. The light-sensitive part is the rhabdom, which is surrounded by the retinular cells, which transmit the stimulus. The ommatidium is surrounded by pigment cells that prevent light from traveling from one ommatidium to another.*

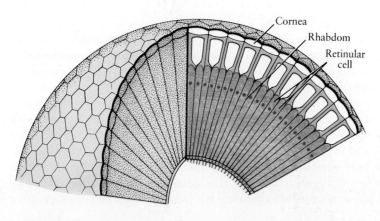
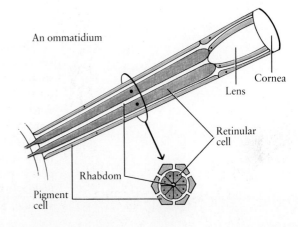

Firefly Light: A Warning, an Advertisement, a Snare

Fireflies are beetles of the family Lampyridae. As is characteristic of most other insects, they undergo a complete metamorphosis. The eggs are laid in moist soil, hatching into larvae in about three weeks. They spend up to two years as larvae; the larvae are carnivorous, feeding on annelids, insects, and mollusks, which they subdue with poison from their mandibles.

Firefly larvae are luminescent, producing a glow that waxes and wanes over a period of seconds. (Even the embryos glow a little.) The survival value of larval luminescence is not clear. The best hypothesis is that the larvae taste bad and that the glow is a warning signal to predators. Compounds related to the skin poisons of toads have been found in some species. Experiments reported by Albert Carlson of the State University of New York at Stony Brook have shown that mice that bite into larvae reject them violently, flinging the carcass away and scrubbing their mouths with their forefeet. It has also been shown that mice can be readily trained to avoid glowing larvae.

After molting through a number of instars, the larvae pupate over a period of two to three weeks. During this time the adult luminescent organ—a lantern—develops on the ventral surface of the terminal abdominal segments. This luminescent organ is under neural control and is used to solve the all-important (for the firefly) problem of finding a mate and reproducing. The adult lives only one to four weeks, so time is of the essence. Each species of firefly has its own signal pattern composed of light flashes lasting only a fraction of a second. Males emit their signals in flight; the females are stationary, near or on the ground. Females flash in response to the signals from males of their own species, aiming their lantern toward the flashing male. When the male sees the

(a) *Two fireflies of the species* Photuris hebes *mating as they hang beneath a goldenrod leaf.* (b) *A female of the species* Photuris versicolor *devours a male of the species* Photinus tanytoxus.

answering flash, he alights and walks toward the female, still flashing. In some species, this courtship flashing is different from the previous flashing pattern of seeking. Copulation lasts minutes to hours, after which the female lays her eggs. The male immediately resumes his signaling and searching activities, looking for another mate. On any given summer evening, there are many more males than females seeking mates—perhaps a ratio of 50 to 1—and competition among males is very intense.

Unlike the larvae, most adult flies are not carnivorous, and some species do not feed at all. Females of the genus *Photuris,* however, are exceptions. They are predaceous carnivores, and their prey is the male firefly of other species. *Photuris versicolor* is one of the most extensively studied species. Within three days after mating, the female undergoes a behavioral

Although the compound eye is deficient in acuity, offering less detail than the vertebrate eye, it is better for detecting motion because each ommatidium is stimulated separately and so has a separate visual field. Also, each ommatidium responds to stimuli more rapidly than does a vertebrate photoreceptor. Ability to detect motion can be measured accurately in the laboratory by testing a phenomenon known as flicker fusion. In this test, a light is flicked on and off with increasing rapidity until the observer sees the flicker as a continuous beam. The beam is perceived as continuous because stimulation of any photoreceptor cell persists for a brief period even after the stimulus disappears. So, in effect, the flicker-fusion test is a measurement of how quickly the cell recovers from one stimulus and becomes sensitive to another. It is possible to test flicker-fusion rates

(a)

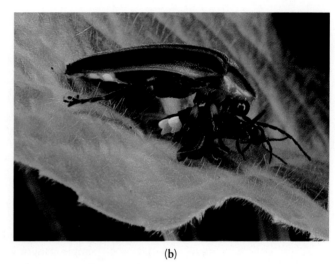
(b)

change. Instead of responding only to the flashing pattern of a male of her own species, she becomes an active hunter, flying to an area where the prey species is active. When a male firefly flashes nearby, she responds—not with her own signal, but rather with a signal characteristic of the female of the male's own species. A *Photuris versicolor* female can respond appropriately to the male signals of at least five different firefly species. The male's dilemma is acute. If he hesitates, he loses his chance to mate, given the 50 to 1 competition. If he rushes in, he risks being devoured by the aggressive female.

Males of the species *Photinus macdermotti* have responded to this dilemma by finding a means of stalling for time while holding off the competition. These tiny males are preyed upon by the females of at least three different *Photuris* species, so their incentives are particularly strong. Their weapons are also false flashes. As they approach a female, they emit flashes that mimic the female predator of the other species, thereby deterring the rival males of their own species. Also, by injecting their flashes into the flash patterns of rival males, they can disrupt their signals and get the females to respond to them rather than to their rivals.

James Lloyd, of the University of Florida, who has been a leader in these studies for almost 20 years, reminds us that on the same summer evenings that the fireflies are engaged in these rituals of courtship, mating, deceit, treachery, and death, thousands of other species of insects are also occupied in similar activities. Are fireflies more complex than other insects, he asks, or are comparable but unilluminated dramas taking place throughout the insect world?

in animals by training experiments in which the animal learns to associate a flickering light with a reward (usually food) and a steady beam with no reward, or vice versa. Such tests have demonstrated that the compound eye greatly exceeds the camera eye of vertebrates in this respect. A bee would see in clear outline a moving figure that we would see as blurred, and if the bee went to the movies, the film seen by us as a continuous picture would jerk along from frame to frame for the bee. The ability to perceive motion is extremely important for an insect since it must be able to make out objects when it is flying at high speed (which, as far as the visual apparatus is concerned, presents the same problems as following a moving object).

27–21 *The hairs on the legs of this red and green tiger beetle* (Cicindela scutellaris) *are touch receptors, or sensilla. At the base of the hairs are sensory cells. When a hair is touched or bent, nerve impulses are initiated.*

In addition to, or instead of, compound eyes, many of the arthropods possess simple eyes, or ocelli, which seem generally to serve only for light detection. Most insects have two or three ocelli, and spiders, which do not have compound eyes, may have as many as eight ocelli, depending on the species.

Touch Receptors

The body surfaces of terrestrial arthropods are often covered with sensory-receptor units known as **sensilla** (singular, sensillum), or "little sense organs." Most of the sensilla take the form of fine spines, or setae, composed of hollow shafts of cuticle. At the bases of these shafts are sensory cells. In their simplest form, the sensilla are touch receptors. In these, when the spine is touched or bent, the sensory cell responds and initiates nerve impulses. Such receptors are found, in particular, on the antennae and the legs. In addition to being stimulated by direct contact, they can also be stimulated by vibrations and air currents. A spider monitors what is going on in its web by sensing vibrations transmitted through the threads when the web is touched. Soldier termites of certain species strike the ground or the walls of their nest with their heads when threatened or disturbed; the vibrations they produce warn their colony mates. A fly perceives the air currents from the movement of a hand or fly swatter and so escapes; a fly in a glass jar is much less likely to be disturbed by such movements.

Proprioceptors

Proprioceptors are sensory receptors that provide information about the position of various parts of the body and the stresses and strains on them. A type common in the arthropods is the campaniform sensillum (Figure 27–22a). Campaniform sensilla are located in thin, stretchable areas of the cuticle. When the cells are twisted or stretched, a nerve fiber signals the central nervous system.

Touch receptors can also serve as proprioceptors. The praying mantis, for example, is capable of making a lightning-swift strike at a moving object. When it sights a potential victim, the insect moves its entire head to bring it into binocular range, since the eyes themselves do not move (Figure 27–22b). Movement of the head results in the stimulation of proprioceptive hairs on the head and thorax of the insect. On the basis of the impulses received from these hairs, the position of the prey and the movement of its own legs are automatically coordinated by the brain and nervous system of the mantis. If these hairs are removed, the mantis can strike a moving object only if the object is directly in front of it.

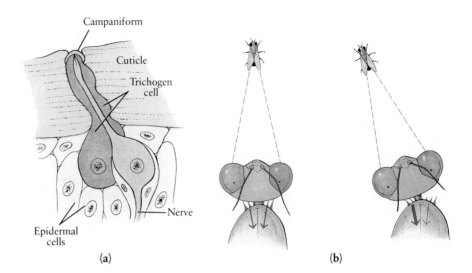

27-22 (a) *Campaniform sensillum, a type of proprioceptor common in arthropods.*

(b) *Since its eyes do not move, the praying mantis must move its entire head to bring its victim into binocular range. The movement of its head sends impulses through proprioceptive hairs on its head and thorax to its legs. The movement of its legs is thus automatically coordinated with the position of its prey, giving the mantis the ability to strike at a victim swiftly and at the proper range.*

Communication by Sound

Arthropods, particularly insects, have developed complex forms of communication. A number of species, such as the locusts, grasshoppers, and crickets, call to one another with sounds made by rubbing their legs or wings together or against their bodies (Figure 27–23). Five distinct types of calls are known: (1) calling by males and (2) calling by females, both of which are long-range sounds; (3) courtship sounds by males and (4) aggressive sounds by males, both of which are short-range; and (5) alarm sounds, which may be given by either males or females. Recognition of and response to the sound may be based on its pattern, rhythm, or frequency (pitch). Some, but not all, insects are unable to distinguish frequencies (the differences between high and low notes) and so are essentially "tone deaf." The effectiveness of calling songs is often increased by group singing, such as the famous chorus of male seventeen-year cicadas, which can attract females from distances far greater than an individual "voice" would reach. Insects produce songs and respond to appropriate songs without ever having heard a song before.

27-23 *The katydid produces its characteristic loud, shrill sounds by rubbing the scraper at the base of its right wing against the file at the base of its left wing.*

Sound waves are set in motion by the vibration of some object, for example, a beating wing, a flexible area of an arthropod's skeleton, a human larynx, or a violin string. The vibrations not only cause the molecules of the surrounding medium (air or water) to travel away from the object in small bursts but also produce a series of changes in air (or water) pressure. The diverse sound receptors of arthropods may be sensitive either to the impact of the moving molecules or to the pressure changes, enabling the animal to detect the calls produced by other members of its species or the sounds of approaching predators.

The simplest of the arthropod sound receptors are sensilla with tactile hairs that vibrate when they are struck by moving molecules in the air; the vibrations of these hairs, in turn, trigger nerve impulses in the animal. The antennae of the male mosquito, for example, contain thousands of such hairs, which are sensitive to the sounds produced by the vibrating wings of the female mosquito in flight. The response of the male to these sounds serves to bring the sexes together. When the male mosquito first emerges from its pupal shell, it is sexually immature and also deaf, with its antennal hairs lying flat along the shafts. When the male matures sexually, some 24 hours later, the hairs almost simultaneously become erect and free to vibrate when a female approaches.

Other insects have developed special structures that respond to the pressure changes in sound waves; these structures, which may be located on the legs (see Figure 27-3a), the thorax, or the abdomen, are known as tympanic organs. In these organs, a fine membrane, the tympanum (or eardrum), is stretched across one or more closed air sacs (Figure 27-24). The tympanic membrane vibrates in response to the pressure differences in sound waves of certain frequencies, and this vibration is transmitted to underlying receptor cells. In many insects, the tympanic air sacs abut portions of the tracheal system and, in some species, are adjacent to the air sacs of the tympanic organ on the opposite side of the body. As a result, the pressure changes of the sound waves are received by the tympanic membrane not only on its outer surface but also on its inner surface, after transmission from the outside through the tracheae or the opposite tympanic organ. The differences in the pressures on the two sides of the membrane enable the insect to detect the direction from which the sound is coming. Many moths, in particular, have a remarkable ability to identify precisely the direction of approach of predatory bats, which locate their prey by emitting sounds that reflect off the prey, back to the bat.

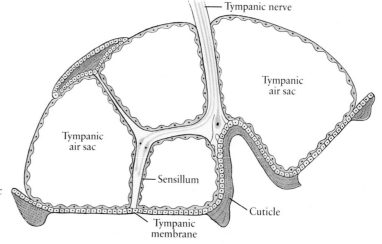

27-24 *The most elaborate of the insect sound receptors is the tympanic organ. The tympanic air sacs are covered by a membranous drum, and the sensory cells are so arranged in the organ that they are stimulated by movements of the drum or the air-sac walls. Tympanic organs respond to pressure changes in the air or in water. This is a cross section of the tympanic organ of a noctuid moth.*

27-25 *The two large appendages extending from either side of the head of this cricket are palpi, which contain special chemoreceptors (taste organs).*

27-26 *The lavishly plumed antennae of the male cecropia moth are receptors for the alluring pheromones released by the females.*

Communication by Pheromones

The use of chemicals for communication is common among organisms, and the substances employed range from the sex attractants of the little algal cell *Chlamydomonas* to Chanel No. 5. Many insects communicate by chemicals known as **pheromones.** These chemical messengers, usually produced in special glands, are discharged into the environment, where they act on other members of the same species.

Among the best studied of the pheromones are the mating substances of moths. Female gypsy moths, by the emission of minute amounts of a pheromone commonly known as disparlure, can attract male moths that are several kilometers downwind. One female produces about one-millionth (10^{-6}) of a gram of disparlure, enough to attract more than a billion males if it were distributed with maximum efficiency. The male moth characteristically flies upwind, and the pheromone, of course, disperses downwind. When a male moth detects the pheromone of a female of the species, he will fly toward the source. Since the male can detect as little as a few hundred molecules per milliliter of the attractant, disparlure is still potent even when it has become widely diffused. If the male loses the scent, he flies about at random until he either picks it up again or abandons the search. It is not until he is quite close to the female that he can fly "up the gradient" and use the intensity of the odor as a locating device. Figure 27–26 shows the antennae of a male cecropia moth by which the pheromone emitted by the female is detected.

Programmed Behavior

Complex patterns of unlearned, genetically transmitted behavior—such as web-building in spiders—are another arthropod characteristic. This relatively rigid programming of behavior may be a necessary correlate of the shortness of the life span of the smaller arthropods. It provides an interesting contrast to the more flexible behavior patterns of the mammals, with their comparatively long life spans, long periods of learning, and more complex brains.

SUMMARY

Arthropoda is the largest animal phylum in both number of species and of individuals. Arthropods are segmented animals with jointed chitinous exoskeletons and a variety of highly specialized appendages and sensory organs. In most groups, the segments are combined, forming a head, a thorax (sometimes fused with the head as a cephalothorax), and an abdomen. The arthropods are also characterized by an open circulatory system and a nervous system consisting of a series of ganglia, a pair per segment, interconnected by a double ventral nerve cord. Tracheae (cuticle-lined breathing tubes), book gills, book lungs, and Malpighian tubules (excretory ducts leading into the hindgut) are found almost exclusively among arthropods.

There are three major groups of arthropods: the chelicerates, characterized by chelicerae (fangs or pincers) and pedipalps; the aquatic mandibulates, with two pairs of antennae and a pair of mandibles (jaws); and the terrestrial mandibulates, with one pair of antennae and a pair of mandibles that differ from those of aquatic mandibulates. The chelicerates include the horseshoe crabs (class Merostomata), the sea spiders (class Pycnogonida), and the spiders, scorpions, mites, and ticks (class Arachnida). The aquatic mandibulates (some of which actually live in moist environments on land) all belong to class Crustacea and include such familiar animals as lobsters, crabs, shrimp, and barnacles. Terrestrial mandibulates include four relatively small classes (Chilopoda, Diplopoda, Pauropoda, and Symphyla) and the largest class in the animal kingdom, Insecta, with about 1 million species. The insects are the only invertebrates capable of flight.

In the life histories of most insects, dispersal and habitat selection are carried out by the highly mobile adult forms, whereas the immature forms have limited mobility and feed voraciously in a fairly restricted area. In the course of their development, most insects pass through a complete metamorphosis that is controlled by the interaction of at least three hormones. The stages in the life history are egg, larva, pupa, and adult. In a minority of species, including the grasshopper, the hatchling looks much like a miniature adult; in these, the immature form is known as a nymph.

Among the factors contributing to the extraordinary success of the arthropods are their exoskeleton, their generally small size, and their great specialization in both diet and habitat. Additional factors in the success of the insects are the capacity for flight and complete metamorphosis, which allows for greater refinement of adaptations for feeding and for reproduction and dispersal, as well as reducing competition between adults and immature forms.

The finely tuned arthropod nervous system, with its diverse sensory organs, has also been important in arthropod success. Among the most important sensory receptors are the compound eye, touch receptors, proprioceptors, and tympanic organs. Arthropods communicate with members of the same species by sound and also by pheromones—chemicals released by one individual that affect the behavior or physiology of another. Arthropod behavior is notable not only for its complexity and diversity but also for the extent to which it is programmed in the nervous system, that is, not learned.

QUESTIONS

1. Distinguish among the following terms: chelicerae/mandibles; nephridia/Malpighian tubules; nymph/larva/pupa; hormone/pheromone; brain hormone/molting hormone/juvenile hormone; ommatidium/sensillum.

2. Describe the following arthropod structures and explain their functions: book gills, tracheae, spiracles, spinnerets, compound eye, tympanic organ.

3. How would you distinguish an insect from an arachnid? From a crustacean?

4. Note that both the mollusks and the arthropods have a greatly reduced coelom. What two features—one obvious, the other not—are also shared by most members of these two groups? What is the correlation between these structural similarities?

5. Describe gas exchange in a clam, a terrestrial snail, an earthworm, a lobster, an arachnid, and an insect. How does gas exchange in these animals differ from that in a cnidarian? How is it similar?

6. Compare the nervous systems of a clam, an octopus, an annelid, and an arthropod. How do they differ from those of a *Hydra* and a planarian?

7. The arthropods, which include some of the most active animals, have open circulatory systems, often considered inefficient. Annelids, which may share a common ancestor with the arthropods, have closed circulatory systems. Presumably, half of the annelid system must have been lost. How could the acquisition of a relatively rigid exoskeleton make superfluous the vessels returning blood to the heart?

CHAPTER 28

The Animal Kingdom IV: The Deuterostomes

On the basis of characteristic features of embryonic development, the coelomate animals fall into two broad groups. In the protostomes—the subject of the two preceding chapters—the cleavage pattern in the early cell divisions of the embryo is spiral, the mouth develops at or near the blastopore, and the coelom results from a splitting of the mesoderm (schizocoelous formation). By contrast, in the deuterostomes, the early cleavage pattern is radial, the anus develops at or near the blastopore, the mouth forms secondarily elsewhere, and the coelom is formed by outpocketings of the embryonic gut (enterocoelous formation).

These features of embryonic development are shared by four strikingly different phyla of animals: Echinodermata (starfish, sea urchins, and related forms), Chaetognatha (arrow worms), Hemichordata (acorn worms and their relatives), and Chordata. The largest subphylum of Chordata includes all the vertebrate animals—fishes, amphibians, reptiles, birds, and mammals, among them *Homo sapiens*.

PHYLUM ECHINODERMATA: THE "SPINY-SKINNED" ANIMALS

The echinoderms include the sea lilies and feather stars (class Crinoidea), starfish and brittle stars (class Stelleroidea), sea urchins and sand dollars (class Echinoidea), and sea cucumbers (class Holothuroidea). The majority of the echinoderm species are known only through fossils, but the 6,000 living species are abundant throughout the oceans of the world. Particularly in deep waters, echinoderms often make up the bulk of living tissue.

Most adult echinoderms are radially symmetrical, like most cnidarians, but the symmetry is imperfect with some traces of bilaterality. The larvae, however, are bilaterally symmetrical. During their development, different parts of the larval body grow at different rates, preparing the way for the change in body symmetry. At the time of metamorphosis, the larvae, which have drifted in the ocean currents for weeks, feeding all the while, temporarily attach to a solid surface; they then rapidly transform into adults with a fivefold radial symmetry. The echinoderms are believed to have evolved from an ancestral, bilateral, motile form that settled down to a sessile life and then became radially symmetrical. The feather stars (Figure 28–1) represent this second hypothetical stage. In the third evolutionary stage, some of the animals, as represented by the starfish and sea urchins, became motile again. Following this line of reasoning, one might expect an eventual return to bilateral symmetry in this group, and, in fact, this is seen to some extent in the soft, elongated bodies of sea cucumbers.

The characteristic features of this phylum are clearly visible in the most familiar of the echinoderms, the starfish.

28–1 *According to available evidence, echinoderms evolved from bilaterally symmetrical, motile animals into radially symmetrical, sessile forms, such as feather stars and sea lilies. This is the feather star* Nemaster rubiginosa, *photographed off South Caicos Island in the British West Indies.*

28-2 *Echinoderms are characterized by a calcium-containing skeleton of rods, plates, or spicules embedded just below the skin, a water vascular system of canals, and tube feet. The five-part body plan, clearly visible in the candy-cane sea star shown here, is characteristic of most living echinoderms.*

28-3 *The water vascular system of the starfish supports its locomotion. Five radial canals, one for each arm, connect the ring canal with many pairs of tube feet, which are hollow, thin-walled cylinders ending in suckers. At the other end of each tube foot is a rounded muscular sac, the ampulla. When the ampulla contracts, the water in it, prevented by a valve from flowing back into the radial canal, is forced under pressure into the tube foot. This stiffens the tube, making it rigid enough to walk on, and extends the foot until it attaches to the substrate by its sucker. The muscles of the foot then contract, forcing the water back into the sac and creating the suction that holds the foot to the surface.*

Class Stelleroidea: Starfish and Brittle Stars

The body of a starfish consists of a central disk from which radiate a number of arms. Most starfish have five arms, which was the ancestral number, but some have more. Like all echinoderms, a starfish has an interior skeleton that typically bears projecting spines, the characteristic from which the phylum derives its name. The skeleton is made up of tiny, separate calcium-containing plates held together by the skin tissues and by muscles.

The central disk of a starfish contains a mouth on the lower surface, above which is the stomach. A starfish has no head or brain, but rings of nerves around the mouth provide coordination of the arms, any one of which may lead the animal in its sluggish, creeping movements along the sea bottom. Each arm contains a pair of digestive glands and also a nerve cord, with an eyespot at the end. These eyespots are the only sensory organs, strictly speaking, of the starfish, but the epidermis contains thousands of neurosensory cells (as many as 70,000 per square millimeter) concerned with touch, photoreception, and chemoreception. The sexes are separate in most starfish, and each arm also has its own pair of sperm-producing or egg-producing organs, which open directly to the exterior through small pores. Although reproduction is usually sexual, with external fertilization, some starfish multiply asexually, regenerating whole animals after division of the central disk. A few species can reproduce asexually by regeneration of shed arms.

The coelom forms a complicated system of cavities and tubes and helps provide for circulation. Respiration is accomplished by many small, fingerlike projections, the skin gills, which are protected by spines. Waste removal is carried out by amoeboid cells that circulate in the coelomic fluid, picking up the wastes and then escaping through the thin walls of the skin gills, where they are ejected.

The **water vascular system** (Figure 28–3) is a unique feature of this phylum. This system, which is a modified coelomic cavity, creates hydrostatic support for the **tube feet,** unusual locomotor structures found only in the echinoderms. Each arm of a starfish contains two or more rows of fluid-filled tube feet, which are interconnected by radial canals and a central ring canal. At one end of each tube foot is a sucker, with which the foot attaches to the substrate, and at the other end, a rounded muscular sac, the ampulla. When the ampulla contracts, water is forced under pressure through a valve into the soft, hollow tube, extending the foot and making it rigid enough to walk on.

When the tube feet are planted on a hard surface, such as a rock or a clam shell, contractions of the muscles at the base of each tube foot collectively exert enough force to pull the starfish forward or to pull open a bivalve mollusk, a feat that will be appreciated by anyone who has ever tried to open an oyster or a clam. When attacking bivalves, which are its staple diet, the starfish everts its stomach through its mouth opening and then squeezes the stomach tissue through the opening that it has made between the bivalve shells. The stomach tissues can insinuate themselves through a slit as narrow as 0.1 millimeter to digest the soft tissue of the prey.

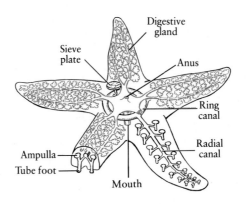

28-4 *Brittle stars,* Ophiothrix fragilis. *The echinoderm class Stelleroidea includes both the brittle stars (subclass Ophiuroidea) and the starfish (subclass Asteroidea).*

28-5 *Among the most familiar echinoderms are sea urchins and sand dollars.* **(a)** *The edible sea urchin,* Echinus esculentus, *resting on the surface of a brown alga of the genus* Laminaria. *Sea urchins are common inhabitants of tidal pools.* **(b)** *A sand dollar,* Echinarachnius parma, *photographed on a beach on the Isles of Shoals, Maine. Sand dollars are often seen on sandy shores.*

Brittle stars (Figure 28-4), which are also known as serpent stars, look like starfish with particularly skinny arms. The arms bend from side to side in a snakelike motion, enabling the animal to crawl about on the ocean bottom. Their tube feet, which lack suckers, are used in gathering and handling food, rather than for locomotion. As you might suspect from their name, the brittle stars have arms that break easily. These echinoderms, like many crustaceans (see page 572), can autotomize body parts seized by a predator, thereby making good their escape.

Other Echinoderms

The members of the other classes of echinoderms show a number of variations on the body plan of the starfish. The sea lilies and feather stars (Figure 28-1), which are thought to be the most ancient of the echinoderms, spend most of their lives either attached or clinging to the substrate. As we have seen with so many sessile animals, the food-gathering structures (in this case, the arms, tube feet, and mouth) are directed upward, enabling the animal to gather food items from the surrounding water.

In the sea urchins and sand dollars (Figure 28-5) and in the tiny members of a newly discovered group of echinoderms (Figure 28-6), arms are not present. The five-part body plan of sea urchins and sand dollars is clearly visible in five paired rows of tube feet that extend through the strong skeleton. As in the starfish, the tube feet are used primarily for locomotion. These animals are formidably armed

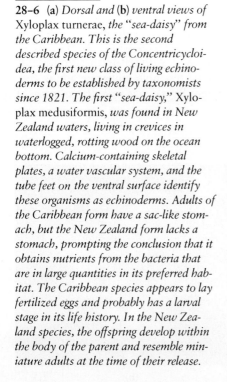

28-6 **(a)** *Dorsal and* **(b)** *ventral views of* Xyloplax turnerae, *the "sea-daisy" from the Caribbean. This is the second described species of the Concentricycloidea, the first new class of living echinoderms to be established by taxonomists since 1821. The first "sea-daisy,"* Xyloplax medusiformis, *was found in New Zealand waters, living in crevices in waterlogged, rotting wood on the ocean bottom. Calcium-containing skeletal plates, a water vascular system, and the tube feet on the ventral surface identify these organisms as echinoderms. Adults of the Caribbean form have a sac-like stomach, but the New Zealand form lacks a stomach, prompting the conclusion that it obtains nutrients from the bacteria that are in large quantities in its preferred habitat. The Caribbean species appears to lay fertilized eggs and probably has a larval stage in its life history. In the New Zealand species, the offspring develop within the body of the parent and resemble miniature adults at the time of their release.*

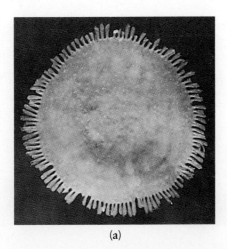 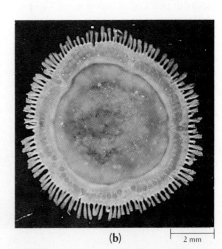

28–7 *A sea cucumber,* Parastichopus californicus, *photographed at Point Loma, California. Depending on the species, sea cucumbers range in length from about 5 millimeters to more than 1 meter. Some tropical species attain a length of 3 meters.*

with movable spines and, in some species, with poison glands hidden among the spines. Typically they are also equipped with complex and powerful jaws that are used to graze on algae and other organic materials adhering to the substrate.

Sea cucumbers (Figure 28–7), with their relatively soft bodies and partial return to bilateral symmetry, bear little superficial resemblance to the other echinoderms, but they too have five rows of tube feet on the body surface. The calcium-containing skeleton is greatly reduced, usually consisting of ossicles in the skin. The body wall is often tough and leathery. Modified tube feet around the mouth look like tentacles and are used in gathering food. Some sea cucumbers feed on plankton in the surrounding water, while others feed primarily on organic matter in the bottom deposits. Some simply ingest the bottom sediment, extracting whatever nutrients it may contain as it passes through the digestive tract.

PHYLUM CHAETOGNATHA: ARROW WORMS

Although there are only about 60 species in phylum Chaetognatha ("bristle jaws"), these arrow-shaped animals are among the most abundant predators in marine plankton. They range from 1 to 10 centimeters in length and, with a flick of the tail, can shoot forward to capture other small planktonic animals (Figure 28–8). Arrow worms are unsegmented, but their bodies have three distinct regions—head, trunk, and tail. Their embryonic development identifies them as deuterostomes, but they do not seem to be closely related to any other deuterostome phylum. All arrow worms are hermaphroditic, with self-fertilization occurring in some species. The newly hatched young resemble miniature adults and, without passing through a dramatically different larval stage, soon begin a life of active predation.

28–8 **(a)** *The predatory arrow worms are an important component of marine plankton, feeding on copepods and, occasionally, small fish. They usually feed near the surface at night, descending to deeper waters during the day.*

(b) *An arrow worm is marvelously equipped for its predaceous activities. The head bears two large eyes on the dorsal surface and numerous sharp spines, used to spear prey. Within the chamber leading to the mouth are strong teeth. This particular arrow worm, however, was captured by a medusa of the cnidarian* Obelia *(page 528). When photographed, it was dead, and most of its body was in the gastrovascular cavity of its captor.*

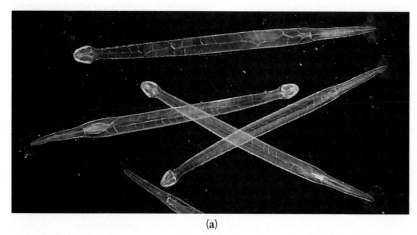

(a)

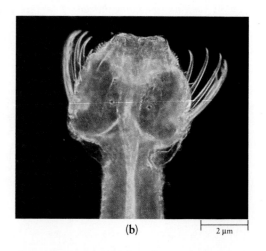

(b) 2 μm

28-9 *An acorn worm,* Glossobalanus sarniensis, *burrowing in shell gravel. Its body consists of three regions—a proboscis, a short collar, and a long trunk, only a portion of which is visible here. Notice the gill slits in the anterior portion of the trunk. Such slits are one of the identifying characteristics of the chordates.*

PHYLUM HEMICHORDATA: ACORN WORMS

Most of the 80 species in phylum Hemichordata are acorn worms (Figure 28-9). As you can see, the body is divided into three regions—a proboscis, with which the animal burrows in ocean sediments in shallow waters, a short collar, and a long trunk. A few species of sessile hemichordates, known as pterobranchs, look more like bryozoans than acorn worms, but they have the same three body regions and share other hemichordate characteristics.

The hemichordates are of particular interest to evolutionary biologists because they have features characteristic of both the echinoderms and the chordates. Some of the hemichordates have ciliated larvae that are almost identical to the larvae of starfish. Moreover, the coelomic cavities of the tentacles of the pterobranchs provide a hydrostatic support similar to the water vascular system of echinoderms, although there are no tube feet. The hemichordate nervous system is, in some respects, similar to that of the echinoderms, but it includes both ventral and dorsal nerve cords that are joined by a ring at the posterior limit of the collar. Elsewhere in the animal kingdom, a principal nerve cord on the dorsal side of the body is found only among the chordates. The dorsal nerve cord of chordates is hollow, unlike the ventral nerve cords in other animals, which are solid; in some hemichordates, the anterior portion of the dorsal cord is also hollow. The strongest evidence of a close relationship between the hemichordates and the chordates is provided by the pharynx, a structure in the anterior portion of the trunk, which is perforated by holes known as pharyngeal gill slits. As we are about to see, such a pharynx is one of the identifying characteristics of the chordates.

PHYLUM CHORDATA: THE CEPHALOCHORDATES AND UROCHORDATES

The phylum Chordata includes some 43,000 species, grouped in three subphyla: the Cephalochordata, or lancelets; the Urochordata, or tunicates, of which the most familiar are the sea squirts; and the Vertebrata, or vertebrates. Thus the term "invertebrate" refers to all animals except the members of one subphylum of the Chordata.

Subphylum Cephalochordata includes only 28 species. The best-known cephalochordate is *Branchiostoma* (Figure 28-10), a small, blade-shaped, semitransparent animal found in shallow marine waters all over the warmer parts of the world. Although it can swim very efficiently, it spends most of its time buried in the sandy bottom, with only its mouth protruding above the surface. This animal exemplifies all four of the salient features of the chordates. The first is the **notochord,** a rod that extends the length of the body and serves as a firm but flexible axis. The notochord is a structural support. Because of it, *Branchiostoma* can swim with strong undulatory motions that move it through the water with a speed unattainable by the flatworms or aquatic annelids.

28-10 (a) *Branchiostoma, a lancelet, exemplifies the four distinctive chordate characteristics: (1) a notochord, the dorsal rod that extends the length of the body; (2) a dorsal, tubular nerve cord; (3) pharyngeal gill slits; and (4) a tail.* Branchiostoma *retains these four characteristics throughout its life; many chordates, however, have a notochord, pharyngeal gill slits or pouches, and a tail only during their immature stages.*

(b) *Lancelets, with their anterior ends protruding from the substrate. Much of the body tissue consists of segmental blocks of muscle. The square, yellowish structures along the side of the body are the reproductive organs. The sexes are separate in lancelets, and sperm or egg cells are released into the atrium. The gametes pass out through the atriopore, with fertilization occurring externally.*

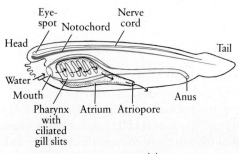
(a)

(b)

The second chordate characteristic is the **dorsal, hollow nerve cord,** a tube that runs beneath the dorsal surface of the animal, above the notochord. (The principal nerve cords in other phyla, as we have noted, are solid and are almost always near the ventral surface.)

The third characteristic is a **pharynx with gill slits.** The pharyngeal apparatus becomes highly developed in fishes, in which it serves a respiratory function, and traces of the gill pouches remain even in the human embryo. In *Branchiostoma,* the pharynx serves primarily for collecting food. The cilia on the sides of the gill slits pull in a steady current of water, which passes through the slits into a chamber known as the atrium and then exits through the atriopore. Food particles are collected in the sievelike pharynx, mixed with mucus, and channeled along ciliated grooves to the intestine.

The fourth characteristic is a **tail,** posterior to the anus, consisting of blocks of muscle around an axial skeleton. Most of the body tissue of *Branchiostoma* is made up of blocks of muscles, the myotomes.

Although *Branchiostoma* usefully exemplifies the chordates, many biologists believe it is more likely to be a degenerate form of fish rather than a truly primitive member of the phylum. Other possible candidates for the ancestral form are found among the tunicates of the subphylum Urochordata. Although adult tunicates do not have all of the typical chordate features, the larvae, which resemble *Branchiostoma,* are clearly chordates possessing the four identifying characteristics (Figure 28–11). About 1,300 urochordate species are known, and they are found in the plankton and on ocean bottoms throughout the world in both shallow and deep waters. They are commonly known as tunicates because the body is covered by a firm, protective tunic that contains cellulose.

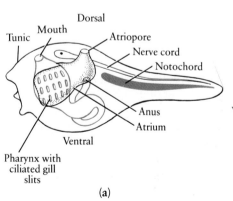
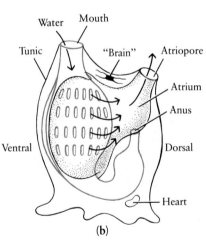
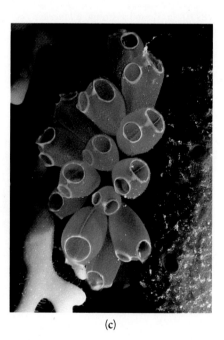

28–11 *Two stages in the life of a tunicate:* **(a)** *the larva, and* **(b)** *the adult form. In the larva, the tunic covers the mouth and atriopore, preventing the flow of water through the pharynx and its ciliated gill slits. Thus, even in species in which the larval pharynx is well developed, as shown here, the larva is unable to feed. After a brief free-swimming existence, it settles to the bottom and attaches at the anterior end. Metamorphosis then begins. The larval tail, with the notochord and dorsal nerve cord, disappears, and the animal's entire body is turned 180°. The mouth is carried backward to open at the end opposite that of attachment, and all the other internal organs are also rotated back. It has been hypothesized that the ancestral vertebrates arose from tunicate larvae that became sexually mature—and thus capable of reproducing—without undergoing metamorphosis.*

(c) *Living adult tunicates, or sea squirts, photographed in the U.S. Virgin Islands.*

PHYLUM CHORDATA: THE VERTEBRATES

The vertebrates constitute the largest (about 41,700 species) and most familiar subphylum of the chordates. Vertebrates typically have a vertebral column, or backbone, as their structural axis. This is a flexible, usually bony support that develops around the notochord, supplanting it entirely in most species. Dorsal projections of the vertebrae encircle the nerve cord along the length of the spine.

The brain is similarly enclosed and protected by a cranium, usually consisting of bony skull plates. Between the vertebrae are cartilaginous disks, which give the vertebral column its flexibility. Associated with the vertebrae are segmental muscles by which sections of the vertebral column can be moved separately. This segmental pattern persists in the embryonic forms of higher vertebrates but is largely lost in the course of development.

One of the great advantages of a bony endoskeleton is that it is composed of living tissue that can grow with the animal. In the developing vertebrate embryo, the skeleton is largely cartilaginous; in most vertebrates, bone gradually replaces cartilage in the course of maturation (Figure 28-12). The growing portions of the bones characteristically remain cartilaginous until the animal reaches its full size.

There are seven living classes of vertebrates: the fishes (comprising three classes), the amphibians, the reptiles, the birds, and the mammals. Their evolution is clearly documented in the fossil record.

Classes Agnatha, Chondrichthyes, and Osteichthyes: Fishes

The first fishes were jawless and had a strong notochord running the length of their bodies. Today these jawless fishes (class Agnatha), once a large and diverse group, are represented only by the hagfish and the lampreys. They have a notochord throughout their lives, like *Branchiostoma*. Although their ancestors had bony skeletons, modern agnaths have a cartilaginous skeleton. Lacking true bones, they are very flexible; a hagfish can actually tie itself in a knot. Many cyclostomes ("round mouths"), as the agnaths are also called, are highly predatory, attaching to other fish by their suckerlike mouths (Figure 28-13), and rasping through the skin into the viscera of their hosts. The juvenile lamprey, which resembles *Branchiostoma*, however, feeds by sucking up mud containing microorganisms and organic debris—as, most probably, did the primitive Agnatha.

The sharks (including the dogfish) and skates, the Chondrichthyes, the second major class of fishes, also have a cartilaginous skeleton. Their ancestors, like those of the agnaths, were bony animals. Their skin is covered with small, pointed teeth (denticles), which resemble vertebrate teeth structurally and give the skin the texture and abrasive quality of coarse sandpaper.

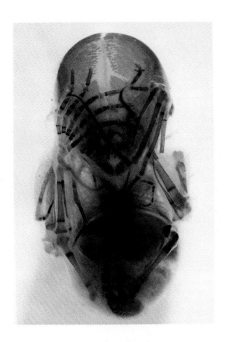

28-12 *The fetus of a long-legged bat. The bones have been stained red and the cartilage blue, so that you can see the extent to which the skeleton is still cartilaginous. Notice the legs, for example. Only the red areas are bone; these will gradually grow and replace the cartilage as the animal matures.*

28-13 *A sea lamprey* (Petromyzon marinus), *which normally attaches to other fish and feeds on their blood, has attached itself to a rock. The majority of lampreys are detritus feeders, eating decayed organic material. Respiration occurs through the prominent gill pouches, seen here as seven indentations.*

28-14 *Rays, like skates and sharks, are cartilaginous fish that have existed in their present form for about 350 million years. Their flattened body is an adaptation to bottom living. Shown here is a spotted eagle ray, which feeds on clams and oysters that it locates by rooting in the sand with its snout.*

28-15 *The evolution of the jaw.* (a) *In the earliest filter-feeding vertebrates, water currents were created by the beating of cilia around the mouth. As the water moved through the pharyngeal apparatus, food items were filtered out.* (b) *In some groups, the cilia were lost, and modified filters began to function in both feeding and respiration; they were not only pumps, moving food-bearing water through strainers, but also gills, across the surface of which gases were exchanged. At the same time, scales migrated to the region of the mouth. Subsequently,* (c) *the bones of the first gill arch evolved into the upper and lower jaws, and* (d) *the second gill arch moved forward, bracing the jaws at the back of the skull. The scales around the mouth evolved into teeth, and the muscles attached to the jaws became highly developed.*

The result was fish, such as (e) *the deep-sea viperfish (Chauliodus), able to feed on a wide variety of large prey organisms.*

(e)

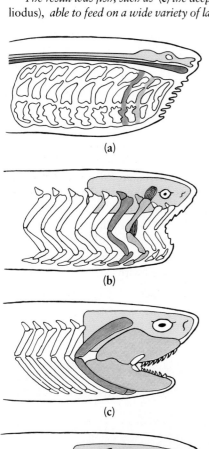

The third major class of fishes includes those with bony skeletons, the Osteichthyes. This group includes the trout, bass, salmon, perch, and many others—almost all of the familiar freshwater and saltwater fishes.

One of the major events in the evolution of fishes—and of the vertebrate groups descended from them—was the transformation of the anterior gill arches of filter-feeding fish into jaws (Figure 28–15). The development of powerful jaws, often armed with formidable teeth, greatly increased the range of other organisms on which the fish could feed. With more efficient feeding on larger and more concentrated sources of energy came the possibility of significant increases in size.

According to present evidence, fishes evolved in fresh water. The chondrichthyans moved to the sea early in their evolution, while the bony fishes went through most of their evolution in fresh water and spread to the seas at a much later period. Some still make this difficult physiological transition in each lifetime. Salmon, for example, hatch in fresh water, spend most of their lives in salt water, and return to fresh water to spawn. At breeding time, eels travel from the fresh waters of Europe and North America to the Sargasso Sea (an area of the south Atlantic), from which distant point the young begin the long, difficult journey, often lasting many years, back to the rivers and lakes.

The Transition to Land

Another characteristic of the bony fishes is that the early forms seem to have had lungs or lunglike structures, as well as gills. These lungs, however, were not efficient enough to serve as more than accessory structures to the gills. They were a special adaptation to fresh water, which, unlike ocean water, may become stagnant (depleted of oxygen) because of decay of organic matter or of algal bloom. Lunged fishes apparently evolved independently several times, and they were the most common fishes in the later Devonian period, a time of recurring drought. In most of them, the lung evolved into an air bladder, or swim bladder; many modern osteichthyans have gas-filled swim bladders that serve as flotation chambers or organs of sound production. A fish raises or lowers itself in the water by adding gases to or removing them from the air bladder via the bloodstream. Still other primitive fishes evolved into the modern lungfish (Figure 28–17). These fish can live in water that does not have sufficient oxygen to support other fish life. Lungfish surface and gulp air into their lungs in much the same way that certain aquatic but air-breathing snails bob to the surface to fill their mantle cavities.

28–16 *The heavily armored placoderms were the ancestors of two major classes of present-day fishes—the Chondrichthyes (cartilaginous fishes) and the Osteichthyes (bony fishes).*

28–17 *A modern lungfish,* Protopterus annectens. *When the dry seasons come, members of this African genus wriggle downward into the mud, which eventually hardens around them. Mucus glands under the skin secrete a watertight film around the body, preventing evaporation. Only the mouth is left exposed. During this period, the fish takes a breath only about once every two hours.*

28–18 *Many toads are clearly fishlike in their larval (tadpole) stages. As adults, they require water to reproduce, and their moist skins are an important accessory respiratory organ. Toads, like all adult amphibians, are carnivores. They catch insects with a flick of their long tongues, which are attached at the front of their mouths and which have a sticky, flypaper-like surface. This common toad has just captured a beetle larva (a grub).*

In yet others, skeletal supports evolved that served to prop up the thorax of the fish. These fish could gulp air even when their bodies were not supported by water. It is thought that these osteichthyans could waddle, dragging their bellies on the ground, along the muddy bottom of a drying stream bed to seek deeper water or perhaps even make their way from one water source to another one nearby. Thus the transition to land may have begun as an attempt to remain in the water.

Class Amphibia

Amphibians descended from air-breathing lunged fish. Modern amphibians include frogs and toads (which typically lack tails as adults) and salamanders (which have tails throughout their lives). They can readily be distinguished from the reptiles by their thin, usually scaleless skins, which serve as respiratory organs. Adult frogs also have lungs, into which they gulp air, but some salamanders respire entirely through their skins and the mucous membranes of their throats. Because water evaporates rapidly through their skins, amphibians can die of desiccation in a dry environment. Those found in deserts spend the drier times of the day far below the surface of the sand.

Most frogs in cold climates have two life stages, one in water and the other on land (hence their name, from *amphi* and *bios*, meaning "both lives"). The eggs are laid in water and are fertilized externally. They hatch into gilled larvae (tadpoles). The tadpoles later develop into adults that lose their gills and develop lungs. The adults may live out of the water, at least in the summer. However, there are many variations on this theme. Some of the American salamanders fertilize their eggs internally; the males deposit sperm packets, either in water or on moist land, and these packets are picked up by the females. Many modern amphibians skip the free-living larval stage. The eggs, which may be laid on land, in a hollow log or cupped leaf, or may even be carried by the parent, hatch into miniature versions of the adult. Some salamanders, such as the mud puppy and the axolotl, never complete their metamorphosis, remaining essentially aquatic larvalike forms (see Figure 35–4, page 735), even as sexually mature adults. In some species, these larvalike forms can be induced to metamorphose into "adult" forms by administration of hormones, indicating that the genetic capacity for this later developmental stage has not been lost.

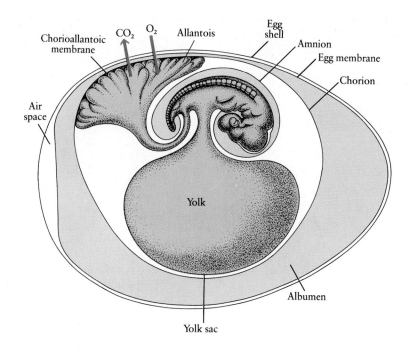

28-19 *Amniote egg. The membranes, which are produced as outgrowths from the embryo as it develops, surround and protect the embryo and the yolk (its food supply). The egg shell and egg membrane, which are waterproof but permeable to gases, are added as the early embryo passes down the maternal reproductive tract.*

Class Reptilia

As you will recall, the vascular plants were freed from the water by the evolution of the seed. Analogously, the vertebrates became truly terrestrial with the evolution in the reptiles of the **amniote egg** (Figure 28–19), an egg that retains its own water supply and so can survive on land. The reptilian egg, which is much like the familiar hen's egg in basic design, contains a large yolk, the primary food supply for the developing embryo, and abundant albumen (egg white), which supplies additional nutrients and water. A membrane, the amnion, surrounds the developing embryo with a liquid-filled space that substitutes for the ancestral pond. A gill-like stage is passed in a shelled egg or in the maternal oviduct or uterus. In mammals also, although their eggs typically develop internally, the embryos are enclosed in water within the amnion and pass through a stage with gill pouches before birth.

Reptiles are characteristically four-legged, although the legs are absent in snakes and some lizards. They have a dry skin, usually covered by protective scales, that makes possible their terrestrial existence. Modern reptiles, of which there are about 6,000 species, include lizards, snakes, turtles, and crocodiles.

28-20 (a) *A painted turtle,* Chrysemys picth. *The dorsal carapace of turtles and tortoises is partly fused to the vertebral column and the ribs, with a mosaic of horny plates on the surface. Unlike other reptiles, most tortoises do not molt but add new epidermal scales on the undersurface. This results in the addition of a growth ring each year.*

(b) *Alligators and crocodiles, the largest modern reptiles, lay their eggs on land, and their skins are reinforced with horny epidermal scales. Crocodiles, which are essentially tropical animals, have more slender snouts than alligators. Alligators have jaws that are broader and more rounded anteriorly; they are also reported to be less aggressive. The animal shown here is the American alligator,* Alligator mississipiensis.

(a)

(b)

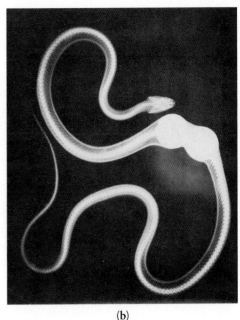

28–21 (a) *Eastern hognose snakes*, Heterodon playrhinos, *hatching from their leathery egg shells. Even as they hatch, these young snakes are sunning themselves, a behavior characteristic of ectotherms—animals that take in heat from the environment. Eastern hognose snakes, like all other snakes, are carnivorous, devouring their prey whole. Such behavior is made possible by the structure of the jaws; the upper jaw is loosely articulated with the skull, and the lower jaw consists of two bones joined only by muscle and skin, which can be easily stretched. Swallowing prey whole can, however, lead to untoward consequences, as befell one rattlesnake. As shown in its x-ray* (b), *taken at the hospital of the University of Florida, it swallowed two golf balls, which it apparently misidentified as eggs. Following surgery, the snake was returned to the wild, well away from the nearest golf course.*

Evolution of the Reptiles

By late in the Carboniferous period (see Table 24–1, page 496), the first reptiles had begun to evolve from closely similar amphibian ancestors. During the succeeding Permian period, there was an explosive increase in the number of reptilian species. (During this same period, conifers began to replace the ferns and other "amphibious" plants, suggesting that a drier climate may have been a primary selective force in both of these events.) Throughout the Permian and much of the Triassic, the dominant land vertebrates were mammal-like reptiles, an abundant and diverse group, members of which were later to give rise to the mammals. Around the beginning of the Triassic period, several of the more specialized reptile groups arose, including turtles, lizards, and the thecodonts, ancestors of the Archosauria, or ruling reptiles, perhaps the most spectacular of all the land's inhabitants so far.

The origin and rise of the Archosauria were associated with improvements in reptilian locomotion. Vertebrates first came to the land with all four limbs splayed far out to the side; turtles have retained this sprawling gait. Among the early Archosauria, there was a progressive tendency toward bipedalism, with a concomitant freeing of the front legs for other purposes—flight, for example. There were three major groups of archosaurs: the pterosaurs, or flying reptiles; the crocodilians, which, reverting to (or perhaps retaining) four-legged posture, became our modern crocodiles and alligators; and the dinosaurs, a varied and splendid group of reptiles.

Dinosaurs were long assumed to have been **ectothermic**—that is, to have maintained their body temperatures within broad limits by taking in heat from the environment or by giving it off to the environment. Some biologists contend, however, that at least some groups of dinosaurs were **endothermic**—that is, their body temperatures were maintained by heat generated internally, as are the body temperatures of birds and mammals. Only a few modern reptiles, such as leatherback turtles, show any degree of endothermy.

Fossil remains of the largest dinosaur known thus far, dubbed *Seismosaurus* ("earthshaker"), were discovered near Albuquerque, New Mexico, in 1985. It is estimated that this animal was 30 to 40 meters in length and weighed between 70 and 95 metric tons, far larger than any land animal that has succeeded it.

Throughout the long Mesozoic era, the dinosaurs dominated the life of the land, rulers of the earth for 150 million years. Then, about 65 million years ago, they vanished, leaving only a single line of descendants, the birds. The cause of their extinction has been the subject of speculation since the first dinosaur fossils were discovered in the nineteenth century, but it is only recently that new data have emerged, making possible the formulation of testable hypotheses. In Chapter 49, we shall consider these data and current hypotheses about the phenomenon of extinction and its role in the process of evolution. As we shall see, extinction is a common fate; it is now estimated that at least 99.9 percent of all the species that have ever lived have become extinct.

Class Aves: Birds

Birds are essentially reptiles specialized for flight (Figure 28–22). Their bodies contain air sacs, and their bones are hollow. The frigate bird, a large seagoing bird with a wingspread of more than 2 meters, has a skeleton that weighs only 110 grams (about 4 ounces). The most massive bone in the bird skeleton is the keel, or breastbone, to which are attached the huge muscles that operate the wings. Flying birds have jettisoned all extra weight; the female's reproductive system has been trimmed down to a single ovary, and even this becomes large enough to be functional only in the mating season.

Birds have feathers, which is their outstanding, unique physical characteristic. They are endothermic, generating heat by internal metabolic processes and maintaining a high and constant body temperature. In modern birds, feathers make flight possible and also serve as insulation. (Only animals that are endothermic require insulation; insulation would be a disadvantage for animals that warm their bodies by exposure to the environment.) Birds also have scales on their legs, a reminder of their reptilian ancestry. Many birds hatch at a very immature stage, and virtually all birds require a long period of parental care.

Evolution of Flight

How did flight evolve? Biologists agree that evolution occurs by a series of changes, each of which, to be conserved by natural selection, must be of survival value. Being able to fly not very well is a dubious advantage. The most widely accepted hypothesis for the origin of flight is that the ancestors of the birds were tree-dwelling reptiles and that flight evolved from gliding, as a way to extend or brake jumps from branch to branch.

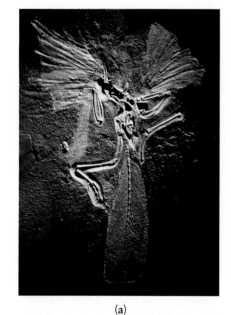

(a)

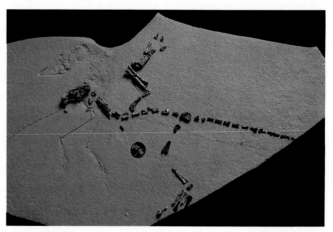

(b)

28–22 (a) *One of the five known specimens of* Archaeopteryx, *the most extensively studied fossil bird, which dates from the late Jurassic period, about 150 million years ago. It still had many reptilian characteristics. The teeth and the long, jointed tail are not found in modern birds. The clearly evident feathers may have been related as much to endothermy as to flight.*

(b) *Although clear evidence of feathers is lacking, these fossil bones, discovered in 1984 in western Texas, are thought to be those of a bird that lived 75 million years before* Archaeopteryx. *Tentatively named* Protoavis, *it has hollow bones, a well-developed "wishbone," and a breastbone with a keel, all characteristic of modern birds. Its skull is also similar to that of a bird, but its hind legs, bony tail, and pelvis resemble those of ground-dwelling dinosaurs.*

28–23 *A red-shouldered hawk capturing a mouse.*

However, John Ostrom of Yale University, having studied the anatomy of the known specimens of *Archaeopteryx* and many related forms as well, supports a second hypothesis: that the ancestors of birds were ground-dwelling reptiles. *Archaeopteryx*, according to the fossil evidence, is a close relative of small, bipedal, carnivorous dinosaurs known as theropods, one of the groups of dinosaurs now thought to have been endothermic. The only major distinctions between the fossil theropods and *Archaeopteryx* are that *Archaeopteryx* has feathers and fused collar bones (a "wishbone"), like modern birds. But why should feathers evolve in a ground-dwelling animal? According to Ostrom, feathers were originally an adaptation not for flight but for insulation.

Evidence that the original function of feathers was not flight is provided by anatomical studies showing that the wing feathers apparently were not attached to the bones of the "hand" in *Archaeopteryx*, as they are in modern birds, but were merely embedded in the skin. Also the breastbone and its keel are lacking, indicating that the wing muscles were not well developed. *Archaeopteryx* was clearly not a good flyer, if indeed it could fly at all. In Ostrom's reconstruction, the early stages of flight began with a feathered, endothermic, carnivorous dinosaur running after its prey, flapping its long feathered arms, and leaping. The fact that the feathered forelimbs of *Archaeopteryx* end in claws, as did the elongated arms of the theropods, lends support to this image. Natural selection may have favored the evolution of the long "wing" feathers because they increased the speed of the running predator or because they served as cagelike traps—natural nets—for capturing prey; some modern predatory birds use their wings in this way (Figure 28–23). Thus flight is seen, in this hypothesis, as the culmination of a long, successful predatory leap.

Ostrom's reconstruction, first set forth in 1974, focused new attention on the evolution of flight, which remains the subject of active inquiry. In addition to continuing study of the available fossils, including the newly discovered *Protoavis*, computer models are now being used to analyze the aerodynamic properties of wings of different sizes, shapes, and patterns of movement. Although the evolutionary question may not be soon resolved, this ferment should provide new understanding of the marvelous feats of aerial skill exhibited by contemporary birds.

Whether flight evolved from the "trees down" or from the "ground up," it made available to the birds a new and vast life zone, filled with a previously inaccessible food supply—airborne insects. An enormous diversification followed, producing not only the 9,000 species of living birds but also some 14,000 species that have become extinct.

Class Mammalia

Mammals, like the birds, descended from the reptiles. Although their ancestors predated the dinosaurs, their great diversification did not occur until after the demise of the dinosaurs. When the dinosaurs' long reign ended, previously occupied terrestrial habitats became available to the mammals. Characteristics distinguishing these animals from other vertebrates are that mammals (1) have hair, (2) provide milk for their young from specialized glands (mammary glands), and (3) like birds, but unlike other vertebrates, maintain a high body temperature by generating heat metabolically.

Nearly all mammalian species bear live young, as do some fish and reptiles, which retain the eggs in their bodies until they hatch. However, some very primitive mammals, the **monotremes,** such as the duckbilled platypus, lay eggs with shells but nurse their young after hatching. By contrast, the **marsupials,** which include the opossums and the kangaroos, bear live young. They differ from the largest group of mammals, however, in that the infants are born at a tiny and extremely immature stage and are often kept in a special protective pouch in which they suckle and continue their development (Figure 28–24). Most of the familiar mammals are **placentals,** so called because they utilize their efficient nutritive connection, the placenta, between the uterus and the embryo for a relatively long period of time. As a result, the young develop to a much more advanced stage before birth. Thus the young are afforded protection during their most vulnerable period.

The earliest placentals were small, secretive, and probably nocturnal, thus avoiding the carnivorous dinosaurs that were active during the daylight hours. They undoubtedly lived mostly on insects, grubs, worms, and eggs. Shrews, which are believed to closely resemble these primitive mammals, have retained their elusive habits.

28-24 Marsupial infants are born at an immature stage and continue their development attached to a nipple in a special protective pouch of the mother. (a) This tiny kangaroo accidentally became dislodged from its mother's pouch. As you can see, it is still attached to the nipple. After the picture was taken, the baby was restored to the pouch, with no apparent ill effects from its premature introduction to the outside world. (b) Opossum infants spend about two weeks in the womb and about three months in their mother's pouch. A newborn opossum is much smaller than a honey bee.

(a)

(b)

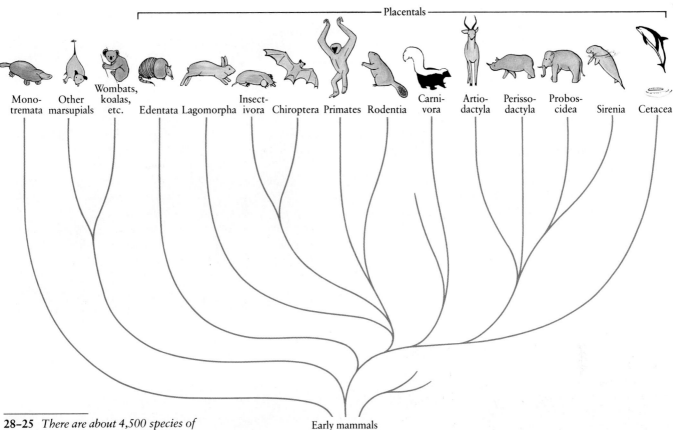

28–25 *There are about 4,500 species of mammals, divided into three subclasses: the monotremes, the marsupials, and the placentals. Twelve of the sixteen orders of placentals are shown here.*

Mammals have fewer, but larger, skull bones than the fishes and reptiles, an example of the fact that "simpler" and "more primitive" may have quite opposite meanings. In the mammals, as in some other vertebrates, a bony platform or partition has developed that separates nasal and food passages far back in the throat, preventing food from entering the lungs. The lower mammalian jaw, unlike that of reptiles, consists of a single bone. Moreover, mammals, unlike snakes or lizards, cannot move the upper jaw in relation to the brain case. As a result, mammals lack the ability of snakes to swallow food items larger than themselves (see Figure 28–21b); mammals must either feed on organisms smaller than themselves or tear the food into pieces small enough to be swallowed.

The major evolutionary lines of mammals are summarized in Figure 28–25. The primates, the order to which we belong, are placental mammals that retain all four kinds of teeth (canines, incisors, premolars, and molars) and have opposable first digits (thumbs and usually big toes), two mammary glands in the chest (rather than in the abdomen), frontally directed eyes, and a relatively large brain with a convoluted cerebral cortex. We are distinguished from the other primates by our upright posture, long legs and short arms, high forehead and small jaw, and sparse body hair.

In many respects, humans are among the least specialized of the mammals. Unlike the carnivores, which are meat eaters, and the several orders of herbivores, we are omnivores, eating a wide variety of fruits, vegetables, and other animals. Our hands closely resemble those of a primitive reptile (see Figure 20–7, page 414), in contrast to the highly specialized forelimbs developed by, for example, whales, bats, and horses. We cannot see as well as an owl monkey. Our sense of smell is much less keen than a dog's, and our sense of taste far less sensitive than that of an enormous variety of other animals. Many animals can run faster, swim more powerfully, and climb trees with more agility (though few can do all three).

28–26 An assortment of mammals. (**a**) A female three-toed sloth, photographed with her baby in a rain forest in Panama. Young sloths, like all young mammals, are dependent on mother's milk for nourishment. Unlike most members of order Edentata, the three-toed sloth has teeth—five molars on each side of the upper jaw and four on each side of the lower jaw. (**b**) Lagomorphs, such as the snowshoe hare shown here in its summer coat, have, in addition to molars and premolars, two pairs of upper incisors. By contrast, rodents, such the beaver (**c**), have only one pair of upper incisors. In both lagomorphs and rodents, the teeth grow continuously.

(**d**) Carnivores, such as these two young male lions chasing a herd of zebras and springbok, are adapted to hunt and kill for food. The zebras are perissodactyls (odd-toed ungulates), and the springbok are artiodactyls (even-toed ungulates).

(a)

(b)

(c)

(d)

(e) *Hippopotamuses, which are also artiodactyls, graze by night and spend most of the daylight hours resting in the water.*

(e)

(f) *Elephants, the Proboscidea, are the largest land mammals living today; some reach a weight of 7.5 metric tons.*

(f)

(g) *Among the mammals that have returned to the water are the whales and dolphins of order Cetacea. This is the common dolphin,* Delphinus delphis.

(g)

Humans have, however, one area of extreme specialization: the brain. Because of our brain, we are unique among all the other animals in our capacity to reason, to speak, to plan, to learn, and so, to some extent, to control our own future and that of the other organisms with which we share this planet.

SUMMARY

The four deuterostome phyla are Echinodermata, Chaetognatha, Hemichordata, and Chordata. Echinoderms include the sea lilies and feather stars, starfish and brittle stars, sea urchins and sand dollars, and sea cucumbers. Although echinoderms have bilaterally symmetrical larvae, the adult forms of most species are radially symmetrical, with a five-part body plan. Echinoderms have an internal, calcium-containing skeleton that typically bears spines. Their most unusual characteristic is the water vascular system, a modified coelomic cavity that provides hydrostatic support for the tube feet and creates suction for the clinging and pulling activities associated with locomotion and feeding.

The arrow worms, phylum Chaetognatha, have three distinct body regions and are among the most active predators in the marine plankton. Hemichordates, of which the majority are acorn worms, have a mixture of echinoderm and chordate characteristics. Their larvae, the coelomic cavities in some members of the phylum, and portions of the nervous system resemble those of echinoderms. In the hemichordates, however, the nerve cord is dorsal (rather than ventral), a feature found elsewhere only among the chordates. Hemichordates also have a pharynx with gill slits, another of the identifying features of the chordates.

The phylum Chordata comprises three subphyla: the Cephalochordata (lancelets), the Urochordata (tunicates), and the Vertebrata. The primary characteristics of the chordates are the notochord, a flexible longitudinal rod running just ventral to the nerve cord and serving as the structural axis of the body (present only in embryonic life in most vertebrates); the nerve cord, which is a hollow tube located dorsally; a pharynx with gill slits; and a tail. *Branchiostoma* best illustrates the basic chordate body plan. It is hypothesized that the tunicate larva, which also shows these characteristics, resembles the primitive chordate from which the vertebrates evolved.

The vertebrates, the largest subphylum of the chordates, are characterized by a vertebral column, a flexible and usually bony support that develops around and supplants the notochord and encloses the nerve cord, and a cranium enclosing the brain. The vertebrates include the fishes (three living classes), the amphibians, the reptiles, the birds, and the mammals.

Some primitive bony fishes, forerunners of the amphibians, were aided in the transition to land by the development of lungs or lunglike structures and by strong skeletal supports that could prop up the body of the fish out of water. Most modern amphibians remain incompletely adapted to life on land and must spend part of the life cycle in water. Vertebrates became truly terrestrial with the development, in the reptiles, of the amniote egg. The diversification of the reptiles that followed their conquest of the land ultimately gave rise not only to a great variety of reptiles (most of which have been extinct for about 65 million years) but also to their descendants, the birds and the mammals.

QUESTIONS

1. Distinguish among the following: exoskeleton/endoskeleton; tube foot/ampulla; ectotherm/endotherm; monotreme/marsupial/placental.

2. Describe the water vascular system of a starfish. What are its similarities to and differences from the hydrostatic skeleton of an earthworm?

3. Describe the identifying characteristics of the phylum Chordata. What is the functional significance of each?

4. How do the three living classes of fishes differ from one another?

5. Consider the graceful swimming of a squid, a lancelet, or a fish. What is the role of the semirigid beam running the length of the animal, whether the "pen" of a squid, the notochord of a *Branchiostoma,* or the vertebral column of a fish?

6. What anatomical features make possible the flight of birds?

7. Among fish and reptiles, some species are oviparous (that is, they lay eggs from which the young hatch) and some are viviparous (giving birth to live young). Name some of the advantages of each alternative. Why are all birds oviparous?

SUGGESTIONS FOR FURTHER READING

Classification

Books

AYALA, F. J., and J. A. KIGER, JR.: *Modern Genetics,* 2d ed., The Benjamin/Cummings Publishing Company, Menlo Park, Calif., 1984.

Contains a good chapter on phylogeny and the comparison of proteins and nucleotides.

BARNES, R. S. K.: *A Synoptic Classification of Living Organisms,* Sinauer Associates, Inc., Sunderland, Mass., 1984.*

A concise presentation of the diversity of life on earth, organized according to a five-kingdom system. A comparison of the taxonomic decisions made in this book, which incorporates the judgment of many British systematists, with those made by Margulis and Schwartz (see below) reveals many of the current disagreements about the classification of particular groups of organisms.

DOBZHANSKY, THEODOSIUS, et al.: *Evolution,* W. H. Freeman and Company, San Francisco, 1977.

Although now becoming somewhat dated, this text remains a good general introduction to evolution. It contains valuable chapters on taxonomy and phylogeny.

KESSEL, R. G., and C. Y. SHIH: *Scanning Electron Microscopy in Biology: A Students' Atlas on Biological Organization,* Springer-Verlag, New York, 1976.

An atlas filled with marvelous scanning electron micrographs of specialized structures of prokaryotes, protists, fungi, plants, and animals, as well as of whole organisms.

MARGULIS, LYNN, and KARLENE V. SCHWARTZ: *Five Kingdoms: An Illustrated Guide to the Phyla of Life on Earth,* 2d ed., W. H. Freeman and Company, New York, 1987.

A concise presentation of the diversity of life on earth, organized according to a five-kingdom system. Although many biologists disagree with some of the taxonomic decisions made by the authors, this book contains a wealth of fascinating information about organisms and their life styles, coupled with outstanding micrographs and diagrams.

MAYR, ERNST: *The Growth of Biological Thought: Diversity, Evolution, and Inheritance,* Harvard University Press, Cambridge, Mass., 1982.*

This is the first of two projected volumes on the history of biology and its major ideas, written by one of the leading figures in the study of evolution. Following an outstanding introductory analysis of the philosophy and methodology of the biological sciences, the first major section of the book provides a thorough treatment of the history and current status of taxonomy.

PERRY, DONALD: *Life Above the Jungle Floor,* Simon and Schuster, New York, 1986.*

Many, and perhaps most, of the earth's undiscovered organisms reside high in the trees of tropical forests. In a lively book, Perry, who devised the climbing and suspension apparatus that is making possible the exploration of this life zone, describes his adventures in the treetops. For a photograph of the author at work, see page 404.

Articles

BRITTEN, R. J.: "Rates of DNA Sequence Evolution Differ Between Taxonomic Groups," *Science,* vol. 231, pages 1393-1398, 1986.

DIAMOND, JARED: "How Many Unknown Species Are Yet to Be Discovered?" *Nature,* vol. 315, pages 538-539, 1985.

LEWIN, DONALD A.: "The Nature of Plant Species," *Science,* vol. 204, pages 381-384, 1979.

LEWIN, ROGER: "Molecules vs. Morphology: Of Mice and Men," *Science,* vol. 229, pages 743-745, 1985.

LOWENSTEIN, J. M.: "Molecular Approaches to the Identification of Species," *American Scientist,* vol. 73, pages 541-547, 1985.

MAY, ROBERT M.: "How Many Species Are There?" *Nature,* vol. 324, pages 514-515, 1986.

MAYR, ERNST: "Biological Classification: Toward a Synthesis of Opposing Methodologies," *Science,* vol. 214, pages 510-516, 1981.

MAYR, ERNST: "Uncertainty in Science: Is the Giant Panda a Bear or a Raccoon?" *Nature,* vol. 323, pages 769-771, 1986.

O'BRIEN, STEPHEN J.: "The Ancestry of the Giant Panda," *Scientific American,* November 1987, pages 102-107.

SIBLEY, CHARLES G., and JON E. AHLQUIST: "Reconstructing Bird Phylogeny by Comparing DNA's," *Scientific American,* February 1986, pages 82-92.

WILSON, EDWARD O.: "Time to Revive Systematics," *Science,* vol. 230, page 1227, 1985.

---

* Available in paperback.

Prokaryotes and Viruses

Books

AUSTRIAN, ROBERT: *Life with the Pneumococcus: Notes from the Bedside, Laboratory, and Library,* University of Pennsylvania Press, Philadelphia, 1985.

> The history of the efforts to understand, control, and prevent pneumococcal pneumonia is lucidly recounted in this collection of lectures, reviews, and research reports. Although the unexpected results of Griffith's experiments (see page 282) and the introduction of antibiotics diverted attention from the quest for an effective vaccine against pneumococcal pneumonia, work has gone forward, with Austrian one of the major participants. This book provides many insights into the challenges and rewards of a career devoted to one specific biological problem.

BURNET, MACFARLANE, and DAVID O. WHITE: *Natural History of Infectious Disease,* 4th ed., Cambridge University Press, New York, 1972.*

> A general introduction to the ecology of infectious diseases and their influence on human activities.

STANIER, R. Y., J. L. INGRAHAM, M. L. WHEELIS, and P. R. PAINTER: *The Microbial World,* 5th ed., Prentice-Hall, Inc., Englewood Cliffs, N.J., 1986.

> An introduction to the biology of microorganisms, with special emphasis on the properties of bacteria. It is widely considered one of the most authoritative accounts.

ZINSSER, HANS: *Rats, Lice, and History,* The Atlantic Monthly Press/Little, Brown and Company, Boston, 1935.*

> A classic popular account of the influence of infectious diseases and their animal vectors on the course of human history, reissued in 1984.

Articles

BLAKEMORE, R. P., and R. B. FRANKEL: "Magnetic Navigation in Bacteria," *Scientific American,* December 1981, pages 58–65.

BROCK, T. D.: "Life at High Temperatures," *Science,* vol. 230, pages 132–138, 1985.

BUTLER, P. J. G., and A. KLUG: "The Assembly of a Virus," *Scientific American,* November 1978, pages 62–69.

COSTERTON, J. W., G. G. GEESEY, and K. J. CHENG: "How Bacteria Stick," *Scientific American,* January 1978, pages 86–95.

DIENER, T. O.: "The Viroid—A Subviral Pathogen," *American Scientist,* vol. 71, pages 481–489, 1983.

DWORKIN, M., and D. KAISER: "Cell Interactions in Myxobacterial Growth and Development," *Science,* vol. 230, pages 18–24, 1985.

FOX, G. E., et al.: "The Phylogeny of Prokaryotes," *Science,* vol. 209, pages 457–463, 1980.

GALLO, ROBERT C.: "The First Human Retrovirus," *Scientific American,* December 1986, pages 88–98.

GALLO, ROBERT C.: "The AIDS Virus," *Scientific American,* January 1987, pages 46–56.

HIRSCH, MARTIN S., and JOAN C. KAPLAN: "Antiviral Therapy," *Scientific American,* April 1987, pages 76–85.

HOGLE, J. M., M. CHOW, and D. J. FILMAN: "The Structure of Poliovirus," *Scientific American,* March 1987, pages 42–49.

LERNER, R. A.: "Synthetic Vaccines," *Scientific American,* February 1983, pages 66–74.

PENNY, DAVID: "What Was the First Living Cell?" *Nature,* vol. 331, pages 111–112, 1988.

PRUSINER, S. B.: "Prions," *Scientific American,* October 1984, pages 50–59.

SHAPIRO, JAMES A.: "Bacteria as Multicellular Organisms," *Scientific American,* June 1988, pages 82–89.

SIMONS, K., H. GAROFF, and A. HELENIUS: "How an Animal Virus Gets into and out of Its Host Cell," *Scientific American,* February 1982, pages 58–66.

WOESE, CARL R.: "Archaebacteria," *Scientific American,* June 1981, pages 98–122.

Protists

Books

BOLD, HAROLD C., and MICHAEL J. WYNNE: *Introduction to the Algae: Structure and Reproduction,* 2d ed., Prentice-Hall, Inc., Englewood Cliffs, N.J., 1985.

> A detailed reference work on the algae that contains a wealth of information on all groups; taxonomically oriented.

BONNER, JOHN T.: *The Cellular Slime Molds,* 2d ed., Princeton University Press, Princeton, N.J., 1968.

> A record of experimental work with a small but fascinating group of organisms.

CURTIS, HELENA: *The Marvelous Animals,* Natural History Press, Garden City, N.Y., 1968.

> An informal introduction to one-celled eukaryotes.

JURAND, A., and G. C. SELMAN: *The Anatomy of* Paramecium aurelia, The Macmillan Company, New York, 1964.

> An exploration, mainly by electron microscopy, of the astonishing complexity of a single-celled organism.

LEE, J. J., S. H. HUTNER, and E. C. BOVEE (eds.): *An Illustrated Guide to the Protozoa,* Society of Protozoologists, Lawrence, Kansas, 1985.

> The most recent comprehensive reference on the protozoan protists. This well-illustrated guide to an enormous diversity of fascinating organisms also includes a glossary of terms used by students of the protists.

MARGULIS, LYNN: *Symbiosis in Cell Evolution: Life and Its Environment on the Early Earth,* W. H. Freeman and Company, New York, 1981.*

> A fascinating discourse proposing the origin of eukaryotic cells by serial symbiotic events.

PICKETT-HEAPS, JEREMY D.: *Green Algae: Structure, Reproduction and Evolution in Selected Genera,* Sinauer Associates, Inc., Sunderland, Mass., 1975.

---

* Available in paperback.

A beautifully illustrated book providing much insight into the variety of form and function in the cells of the green algae.

ROUND, F. E.: *The Ecology of Algae*, Cambridge University Press, New York, 1981.

A comprehensive account of the ecology of both freshwater and marine algae.

Articles

CORLISS, JOHN O.: "The Kingdom Protista and Its 45 Phyla," *BioSystems*, vol. 17, pages 87–126, 1984.

COX, F. E. G.: "Malaria Vaccines: The Shape of Things to Come," *Nature*, vol. 333, page 702, 1988.

DONELSON, J. E., and M. J. TURNER: "How the Trypanosome Changes Its Coat," *Scientific American*, February 1985, pages 44–51.

FRIEDMAN, M. J., and W. TRAGER: "The Biochemistry of Resistance to Malaria," *Scientific American*, March 1981, pages 154–164.

GODSON, G. N.: "Molecular Approaches to Malaria Vaccines," *Scientific American*, May 1985, pages 52–59.

SCHOPF, J. W.: "The Evolution of the Earliest Cells," *Scientific American*, September 1978, pages 110–138.

SCHWARTZ, R. M., and M. O. DAYHOFF: "Origins of Prokaryotes, Eukaryotes, Mitochondria, and Chloroplasts," *Science*, vol. 199, pages 395–403, 1978.

VIDAL, G.: "The Oldest Eukaryotic Cells," *Scientific American*, February 1984, pages 48–57.

YATES, G. T.: "How Microorganisms Move through Water," *American Scientist*, vol. 74, pages 358–365, 1986.

Fungi

Books

AHMADJIAN, V., and S. PARACER: *Symbiosis: An Introduction to Biological Associations*, University Press of New England, Hanover, N.H., 1986.

In this small but fascinating book, Ahmadjian describes the nature of the relationship between the fungal and algal components of a lichen.

ALEXOPOULOS, C. J., and C. W. MIMS: *Introductory Mycology*, 3d ed., John Wiley & Sons, Inc., New York, 1979.

A thorough introduction to the fungi, the slime molds, and the water molds.

LARGE, E. C.: *The Advance of the Fungi*, Dover Publications, Inc., New York, 1962.*

A fascinating popular account of the closely interwoven histories of fungi and humans, first published in 1940.

SMITH, A. H.: *The Mushroom Hunter's Field Guide*, The University of Michigan Press, Ann Arbor, Mich., 1980.

A clear, concise, well-illustrated guide to edible mushrooms, enlivened with good advice and pertinent anecdotes.

Articles

AHMADJIAN, VERNON: "The Nature of Lichens," *Natural History*, March 1982, pages 30–37.

FRIEDMANN, E. I.: "Endolithic Microorganisms in the Antarctic Cold Desert," *Science*, vol. 215, pages 1045–1053, 1982.

LITTEN, W.: "The Most Poisonous Mushrooms," *Scientific American*, March 1975, pages 90–101.

MATOSSIAN, MARY K.: "Ergot and the Salem Witchcraft Affair," *American Scientist*, vol. 70, pages 355–357, 1982.

RUEHLE, J. L., and D. H. MARX: "Fiber, Food, Fuel, and Fungal Symbionts," *Science*, vol. 206, pages 419–422, 1979.

STROBEL, G. A., and G. N. LANIER: "Dutch Elm Disease," *Scientific American*, August 1981, pages 56–66.

Plants

Books

BARTH, FRIEDRICH G.: *Insects and Flowers: The Biology of a Partnership*, Princeton University Press, Princeton, N.J., 1985.

A well-written and beautifully illustrated introduction to the interrelationships of flowers and insects. Incorporating many recent discoveries, the text considers not only the diverse structures of flowers but also the sensory, navigational, and communication abilities of pollinating insects.

CONARD, HENRY S., and PAUL L. REDFEARN, JR.: *How to Know the Mosses and Liverworts*, 2d ed., William C. Brown Publishing Co., Dubuque, Iowa, 1979.*

A good beginner's guide for identifying many of the more common bryophytes. It includes a profusely illustrated key and an excellent glossary.

MILNE, DAVID, et al. (eds.): *The Evolution of Complex and Higher Organisms*, NASA Special Publication 478, U.S. Government Printing Office, 1985.

A concise summary of the history of multicellular life on earth, including an up-to-date review of the history of fossil plants.

RAVEN, PETER H., RAY F. EVERT, and SUSAN E. EICHHORN: *Biology of Plants*, 4th ed., Worth Publishers, Inc., New York, 1986.

This general botany text contains an excellent presentation of the evolution of plants and related organisms, as well as a wealth of information on prokaryotes, protists, and fungi.

Articles

CREPET, W. L.: "Ancient Flowers for the Faithful," *Natural History*, April 1984, pages 38–45.

DILCHER, D., and P. R. CRANE: "In Pursuit of the First Flower," *Natural History*, March 1984, pages 56–61.

GENSEL, PATRICIA G., and HENRY N. ANDREWS: "The Evolution of Early Land Plants," *American Scientist*, vol. 75, pages 478–489, 1987.

GRAHAM, LINDA E.: "The Origin of the Life Cycle of Land Plants," *American Scientist*, vol. 73, pages 178–186, 1985.

---

* Available in paperback.

MULCAHY, DAVID L.: "Rise of the Angiosperms," *Natural History*, September 1981, pages 30–35.

NIKLAS, KARL J.: "Aerodynamics of Wind Pollination," *Scientific American*, July 1987, pages 90–95.

NIKLAS, KARL J.: "Computer-simulated Plant Evolution," *Scientific American*, March 1986, pages 78–86.

NIKLAS, KARL J.: "Wind Pollination—A Study in Controlled Chaos," *American Scientist*, vol. 73, pages 462–470, 1985.

NORSTOG, KNUT: "Cycads and the Origin of Insect Pollination," *American Scientist*, vol. 75, pages 270–279, 1987.

ROSENTHAL, GERALD A.: "The Chemical Defenses of Higher Plants," *Scientific American*, January 1986, pages 94–99.

VALENTINE, JAMES W.: "The Evolution of Multicellular Plants and Animals," *Scientific American*, September 1979, pages 140–158.

WIENS, D.: "Secrets of a Cryptic Flower," *Natural History*, May 1985, pages 70–77.

Animals

Books

BARNES, ROBERT D.: *Invertebrate Zoology*, 5th ed., Saunders College/Holt, Rinehart and Winston, Philadelphia, 1987.

  One of the best general introductions to protozoans and invertebrates.

BUCHSBAUM, RALPH, MILDRED BUCHSBAUM, JOHN PEARSE, and VICKI PEARSE: *Animals without Backbones,* 3d ed., University of Chicago Press, Chicago, 1987.*

  A delightful introduction to the invertebrates, for the general reader, with a multitude of photographs.

DESMOND, ADRIAN J.: *The Hot-Blooded Dinosaurs: A Revolution in Paleontology,* The Dial Press, Inc., New York, 1976.*

  Desmond, a historian of science, describes the development of evolutionary theories, particularly as they were influenced by the discovery of dinosaur fossils, and presents a lively review of the evidence that is leading an increasing number of paleontologists to the conclusion that dinosaurs were endothermic. (If you are not interested in the history of paleontology, you may want to begin in the middle.) The text is well written and the illustrations are wonderful.

EVANS, HOWARD E.: *Life on a Little-Known Planet,* University of Chicago Press, Chicago, 1984.*

  Professor Evans is the author of many popular articles and books on insects. This book, originally published in 1966, profits from his wide knowledge, clarity, and humor.

FEDUCCIA, ALAN: *The Age of Birds,* Harvard University Press, Cambridge, Mass., 1980.*

  A history of the birds, tracing their evolution from reptilian ancestors through the diversifications that gave rise to the major groups of modern birds. Well-illustrated with many photographs and drawings.

FENTON, M. BROCK: *Just Bats,* University of Toronto Press, Toronto, 1983.*

  A short, well-illustrated account of the fascinating lives of the bats, the only mammals capable of flight.

HANSON, EARL D.: *The Origin and Early Evolution of Animals,* Wesleyan University Press, Middletown, Conn., 1977.

  Recommended to the serious student who is interested in how the experts make decisions concerning evolutionary relationships.

KLOTS, ALEXANDER B., and ELSIE B. KLOTS: *Living Insects of the World,* Doubleday & Company, Inc., Garden City, N.Y., 1975.

  A spectacular gallery of insect photos. The text is informal but informative, written by experts for the general public.

MCMAHON, T. A., and J. T. BONNER: *On Size and Life, Scientific American* Library, W. H. Freeman and Company, New York, 1985.*

  Written by an engineer and a biologist, this beautifully illustrated book explores the relationship of the size of an organism to its shape, its speed, its physiological functions, its evolution, and its ecology.

PEARSE, VICKI, JOHN PEARSE, MILDRED BUCHSBAUM, and RALPH BUCHSBAUM: *Living Invertebrates,* Blackwell Scientific Publications, Palo Alto, Calif., 1987.

  An introduction to the invertebrates, for the more advanced student, by the authors of Animals without Backbones *(see above). Although this text is more detailed and contains a great deal more information than* Animals without Backbones, *it retains the same delightful writing style. A wonderful book for both serious study and browsing.*

ROMER, ALFRED: *The Procession of Life,* World Publishing Co., Cleveland, 1968.*

  A history of evolution, written by an expert, but as readable as a novel.

WELLS, M. J.: *Brain and Behavior in Cephalopods,* Stanford University Press, Stanford, Calif., 1962.

  Experimental analyses of behavior in the octopus and squid.

WHITTINGTON, H. B.: *The Burgess Shale,* Yale University Press, New Haven, Conn., 1985.

  The Burgess Shale, a rock formation in the Canadian Rockies, contains one of the world's largest fossil assemblages of soft-bodied invertebrates from the Cambrian period. This beautifully illustrated book summarizes two decades of study of these fossilized animals, many of which were very different from the animals that succeeded them.

---

* Available in paperback.

WILFORD, JOHN NOBLE: *The Riddle of the Dinosaur,* Alfred A. Knopf, Inc., New York, 1985.*

A well-written history of dinosaur discovery and research, with balanced discussions of the many controversies that surround these fascinating animals.

Articles

AUSTAD, STEVEN N.: "The Adaptable Opossum," *Scientific American,* February 1988, pages 98–104.

BUFFETAUT, ERIC: "The Evolution of the Crocodilians," *Scientific American,* October 1979, pages 130–144.

CAMERON, J. N.: "Molting in the Blue Crab," *Scientific American,* May 1985, pages 102–109.

CRACRAFT, JOEL: "Early Evolution of Birds," *Nature,* vol. 331, pages 389–390, 1988.

DOMNING, D. P.: "Sea Cow Family Reunion," *Natural History,* April 1987, pages 64–71.

FIELD, KATHARINE G., et al.: "Molecular Phylogeny of the Animal Kingdom," *Science,* vol. 239, pages 748–753, 1988.

GHISELIN, MICHAEL T.: "A Movable Feaster," *Natural History,* September 1985, pages 54–61.

GOSLINE, J. M., and M. E. DeMONT: "Jet-propelled Swimming in Squids," *Scientific American,* January 1985, pages 96–103.

GRIFFITHS, MERVYN: "The Platypus," *Scientific American,* May 1988, pages 84–91.

HADLEY, N. F.: "The Arthropod Cuticle," *Scientific American,* July 1986, pages 104–112.

HORN, MICHAEL H., and ROBIN N. GIBSON: "Intertidal Fishes," *Scientific American,* January 1988, pages 64–70.

HORNER, J. R.: "The Nesting Behavior of Dinosaurs," *Scientific American,* April 1984, pages 130–137.

HORRIDGE, G. A.: "The Compound Eye of Insects," *Scientific American,* July 1977, pages 108–122.

---

* Available in paperback.

KIRSCH, JOHN A. W.: "The Six-Percent Solution: Second Thoughts on the Adaptedness of the Marsupialia," *American Scientist,* vol. 65, pages 276–288, 1977.

KOEHL, M. A. R.: "The Interaction of Moving Water and Sessile Organisms," *Scientific American,* December 1982, pages 124–134.

LaBARBERA, MICHAEL, and STEVEN VOGEL: "The Design of Fluid Transport Systems in Organisms," *American Scientist,* vol. 70, pages 54–60, 1982.

LENHOFF, HOWARD M., and SYLVIA G. LENHOFF: "Trembley's Polyps," *Scientific American,* April 1988, pages 108–113.

LENT, CHARLES M., and MICHAEL H. DICKINSON: "The Neurobiology of Feeding in Leeches," *Scientific American,* June 1988, pages 98–103.

LEWIN, ROGER: "On the Origin of Insect Wings," *Science,* vol. 230, pages 428–429, 1985.

LLOYD, JAMES E.: "Mimicry in the Sexual Signals of Fireflies," *Scientific American,* July 1981, pages 138–145.

MACURDA, D. B., JR., and D. L. MEYER: "Sea Lilies and Feather Stars," *American Scientist,* vol. 71, pages 354–365, 1983.

McMENAMIN, MARK A. S.: "The Emergence of Animals," *Scientific American,* April 1987, pages 94–102.

MICHELSEN, AXEL: "Insect Ears as Mechanical Systems," *American Scientist,* vol. 67, pages 696–706, 1979.

MORRIS, SIMON CONWAY: "The Search for the Precambrian-Cambrian Boundary," *American Scientist,* vol. 75, pages 156–167, 1987.

OSTROM, JOHN H.: "Bird Flight: How Did It Begin?" *American Scientist,* vol. 67, pages 46–56, 1979.

PANCHEN, ALEC L.: "In Search of the Earliest Tetrapods," *Nature,* vol. 333, page 704, 1988.

RICHARDSON, J. R.: "Brachiopods," *Scientific American,* September 1986, pages 100–106.

SEBENS, K. P.: "The Anemone Below," *Natural History,* November 1986, pages 48–53.

WEBB, P. W.: "Form and Function in Fish Swimming," *Scientific American,* July 1984, pages 72–82.

SECTION 5

Biology of Plants

Branch of an apple tree, showing one of the major characteristics of the angiosperms, the most successful group of plants. Flowers lure the insects and other animals that carry pollen from plant to plant.

CHAPTER 29

The Flowering Plants: An Introduction

For most of earth's history, the land was bare. A billion years ago, algae clung to the shores at low tide and may even have begun to cover a few moist surfaces farther inland. But, had anyone been there to observe it, the earth's surface would generally have appeared almost as barren and forbidding as the bleak Martian landscape. According to the fossil record, plants began to invade the land a mere half billion years ago. Not until then did the earth's surface truly come to life. As a film of green spread from the edges of the waters, other forms of life—heterotrophs—were able to follow. The shapes of these new forms and the ways in which they lived were determined by the plant life that preceded them. Plants supplied not only their food—their chemical energy—but also their nesting, hiding, stalking, and breeding places.

And so it is today. In all terrestrial communities except those created by human activities, the character of the plants still determines the character of the animals and other forms of life that inhabit a particular area. Even we members of the human species, who have seemingly freed ourselves from the life of the land and even, on occasion, from the surface of the earth, are still dependent on the photosynthetic events that take place in the green leaves of plants.

In Chapter 24, we traced the evolution of the plants; if you have not already read that chapter, we recommend that you do so now. In this section, we shall focus on the group of plants that evolved most recently, the angiosperms (division Anthophyta). The angiosperms are, by far, the most abundant plants on earth today, with about 235,000 living species. As we noted in Chapter 24, the group is divided into two large classes, Monocotyledones (the monocots), with about 65,000 species, and Dicotyledones (the dicots), with about 170,000 species. (See Table 24–3 on page 511 for the principal differences between these two classes.)

Angiosperms are characterized by specialized reproductive structures—flowers—in which sexual reproduction occurs, in which the seeds are formed, and from which the fruits develop. For these plants, a new cycle of life begins when a pollen grain—often brushed from the body of a foraging insect—comes into contact with the stigma of a flower of the same species. As we saw in Chapter 24, flowers are exquisitely adapted to achieve this act of pollination. Unlike the reproductive organs of animals, which are permanent structures that develop in the embryo, flowers are transitory. After fertilization, some parts of the flower become the fruit, protecting and enclosing the seed or seeds; other parts die and are discarded.

29-1 *The second major characteristic of the angiosperms is the fruit. Seeds are dispersed from the fruit, sometimes with the assistance of other organisms, such as this obliging porcupine.*

SEXUAL REPRODUCTION: THE FLOWER

Most flowers consist of four sets of floral parts—sepals, petals, stamens, and carpels (Figure 29–2). Each floral part is thought to be, evolutionarily speaking, a modified leaf (see page 511). The floral parts may be arranged spirally on a more or

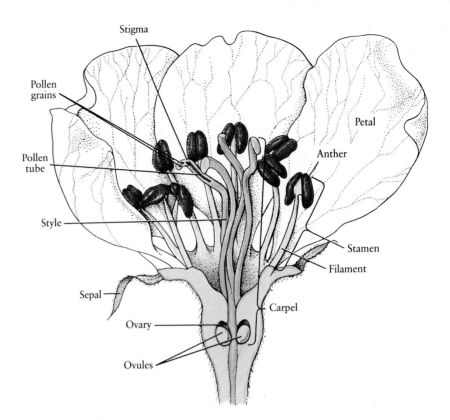

29-2 *The flower of a pear tree. This flower is a complete flower, which means that it contains all four floral parts—sepals, petals, stamens, and carpels. It is also a perfect flower, which means that it contains both male structures (stamens) and female structures (carpels). Each stamen consists of a pollen-bearing anther and its filament. Each carpel consists of a stigma, a style, and an ovary. In the pear flower, each set of floral parts is located at one level, in a whorl.*

Pollen grains have been deposited on the sticky surface of a stigma of this flower, and the pollen tube of one is growing down the style to an ovule.

less elongated stalk, or similar parts—such as the petals—may be located at one level in a whorl. In monocots, each of the floral parts is usually present in multiples of three; in dicots, by contrast, the floral parts typically occur in multiples of four or five.

The outermost parts of the flower are the **sepals,** which are usually green and leaflike. The sepals, collectively known as the **calyx,** enclose and protect the developing flower bud. Next are the **petals,** which together are called the **corolla.** Petals may also be leaf-shaped, but they are often brightly colored. They advertise the presence of the flower among the green leaves, attracting insects or other animals that visit flowers for their nectar or for other edible substances. As these animals forage for food, they are likely to carry pollen from flower to flower (see Figure 24–26, page 514).

Within the corolla are the stamens. Each stamen consists of a single elongated stalk, called the filament, and at the end of the filament, the anther. The pollen grains, formed within the anther, are the immature male gametophytes. (For a review of alternation of generations in angiosperms, see Figure 24–22 on page 512). When ripe, the pollen grains are released, often in large numbers, through slits or pores in the anther.

The centermost parts of the flower are the carpels, which contain the female gametophytes. A single flower may have one carpel or several carpels, which may be separate or fused together. Typically a single carpel or fused carpels consist of a stigma, which is a sticky surface to which pollen grains adhere; a stalk, the style, through which the pollen tube grows; and a swollen base, the ovary. Within the ovary are one or more ovules, each of which encloses a female gametophyte, or **embryo sac,** containing a single egg cell. After the egg is fertilized, the ovule develops into a seed and the ovary into a fruit.

A flower that contains both stamens and carpels is known as a perfect flower. In some species, however, the flowers are imperfect—that is, they are either male (staminate) or female (carpellate). Male and female flowers may be present on the same plant, as in corn, squash, oaks, and birches; such plants are said to be monoecious ("one house"). Species in which the male and female flowers are on separate plants, such as the tree of heaven *(Ailanthus),* American mistletoe, and holly, are known as dioecious ("two houses"). As gardeners know, in order for a female holly plant to produce berries, a male holly—which never produces berries—must be planted nearby.

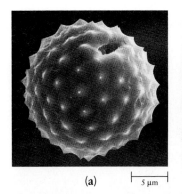 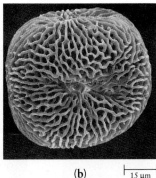 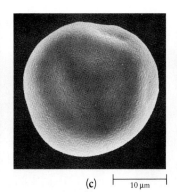 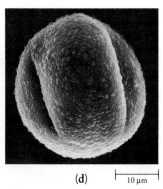

29–3 *Pollen grains. The walls of the pollen grain protect the male gametophyte on its journey from the anther to a stigma. These outer surfaces, which are remarkably tough and resistant, are often elaborately sculptured. As you can see, the pollen grains of different species are distinctly different: (a) common ragweed (such spiny pollen grains trigger the misery of hay fever in susceptible individuals); (b) Geranium, a dicot; (c) timothy grass, a monocot (smooth pollen grains are found in most wind-pollinated plants); and (d) oak, a dicot.*

Upon germination, the pollen tube emerges through a pore or slit in the outer coat, clearly visible here in the ragweed and oak pollen grains. The pollen grains of monocots typically have one pore or slit, whereas those of dicots have three.

The Pollen Grain

By the time the pollen grain is released from the anther, it usually consists of three haploid cells—two sperm cells contained within a larger cell, known as a tube cell. The tube cell, in turn, is enclosed by the thick outer wall of the pollen grain. The pollen grain contains its own nutrients and has so tough an outer coating that intact grains thousands of years old have been found in peat bogs.

As you will recall, in many multicellular algae and in the bryophytes and seedless vascular plants, there is a distinct alternation of generations in which the sporophyte produces spores that produce gametophytes that produce gametes, with the gametophyte and sporophyte having separate existences. In the course of plant evolution, the gametophyte has been steadily reduced in size. In the angiosperms, all that remains of the male gametophyte is the tough, tiny pollen grain and the pollen tube that grows from it. The sperm cells are the gametes.

Fertilization

Once on the stigma, the pollen grain germinates, and, under the influence of the tube nucleus, a pollen tube grows through the style into an ovule (Figure 29–4).

 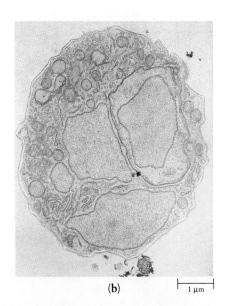 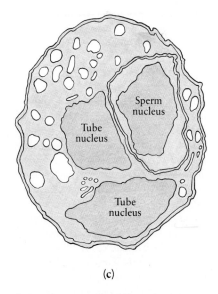

29–4 *(a) Pollen tube growth in* Geranium maculatum, *a species in which the stigma is divided into distinct lobes. Germination of pollen grains is thought to depend on species-specific recognition, perhaps involving the interaction of chemical substances on the sticky surface of the stigma with the pollen grain. It has also been suggested that sugary substances on the stigma are used by the pollen grain to provide energy for the rapid growth of the pollen tube. In* Geranium, *the pollen tube from a grain 0.1 millimeter in diameter can grow about 1 centimeter in length within 20 minutes.*

(b, c) Electron micrograph and diagram of a cross section of a pollen tube. Two lobes of the single tube nucleus are visible. This nucleus directs the formation of the pollen tube and eventually disintegrates. Numerous mitochondria are visible in the tube cell, as are several plastids. At the right is a sperm cell, which is actually a cell within a cell.

29–5 *Flowers—forms and variations.* **(a)** *The sweet bay magnolia,* Magnolia virginiana. *The carpels form a cone-shaped receptacle from which the curved styles emerge. The cream-colored stamens, some of which have dropped off, surround the carpels. Magnolias are very primitive flowers with numerous separate floral parts arranged in a spiral pattern.* **(b)** *The lady's slipper orchid,* Paphiopedilum invincible. *Orchids are highly specialized flowers, with the style and filaments fused into a single structure that bears both a stigma and, a short distance away, an anther. The lip of the flower is a modified petal that can serve as a landing platform for insects. Orchidaceae, with at least 17,000 species, is the second largest family of flowering plants.* **(c)** *The sunflower,* Helianthus annuus *("annual flower of the sun"). In the sunflower and other composites, numerous individual flowers make up the head, which acts as a large, single flower in attracting insects. The central portion of the head consists of separate florets, each comprising a pair of fused carpels forming a single ovary and fused anthers enclosed in a small corolla of fused petals. This central portion is surrounded by ray flowers (with yellow petals), which are often sterile. Composites (family Asteraceae), with some 22,000 species, are the largest family of flowering plants.* **(d)** *In* Hibiscus, *a column of stamens is fused around the style.* **(e)** *Stamens and stigma of a purple crocus.* **(f)** *Corn (*Zea mays*), a monoecious species. Separate male and female flowers are borne on the same plant. The tassels, at the top of the stem, are the male (pollen-producing) flowers. Each thread of "silk," seen emerging from the ear of corn, is the combined stigma and style of a female flower.* **(g)** *Water hemlock, a member of the carrot family, provides an example of inflorescences, or flower clusters.*

This may be a long distance; in corn, for instance, the pollen tube may grow to a length of 40 centimeters. The number of grains reaching the stigma is often greater than the number of ovules available for fertilization, creating intense competition among pollen tubes, the race going to the swift.

Each ovule contains a female gametophyte, which has also become reduced in size in the course of evolution. In many species, the female gametophyte consists of seven cells, with a total of eight haploid nuclei (Figure 29-6). One of the seven cells is the egg, containing a single haploid nucleus. On either side of the egg is a small cell known as a synergid. At the opposite end of the gametophyte are three small cells, the antipodal cells, whose function, if any, is unknown. The large central cell contains two haploid nuclei, called the polar nuclei because they move to the center from each end, or pole, of the gametophyte.

The two synergids, one on each side of the egg cell, attract the growing pollen tube. When the pollen tube reaches one of the synergids, it fuses with it, releasing the tube nucleus and the two sperm nuclei into the synergid. From the synergid, one sperm nucleus enters the egg cell and unites with the egg nucleus. The fertilized egg, or zygote, develops into the embryo—the young diploid sporophyte. The second sperm nucleus is released into the large central cell, where it unites with the two polar nuclei in a process of triple fusion. From the resulting $3n$ (triploid) cell, a specialized tissue called the **endosperm** develops. The endosperm surrounds and nourishes the developing embryo. These extraordinary phenomena of fertilization and triple fusion—together called "double fertilization"—take place, in all the natural world, only in the flowering plants.

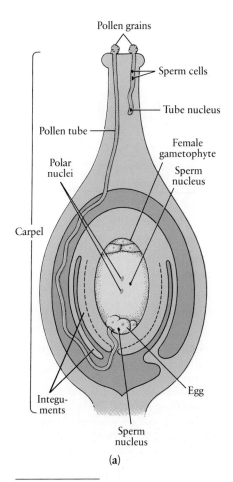

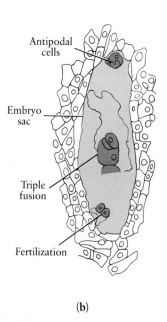

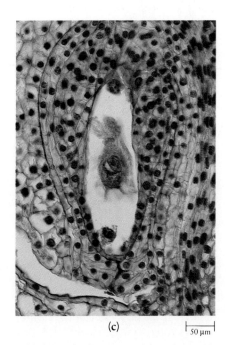

29-6 (a) *Fertilization in angiosperms. The pollen tube of the male gametophyte, or pollen grain, grows through the style and enters an ovule, which contains a seven-celled female gametophyte (the embryo sac). One of the sperm nuclei unites with the egg, forming the zygote. The other sperm nucleus fuses with the two polar nuclei that are present in a single large cell (which in the drawing fills most of the ovule). This triple fusion produces a triploid (3n) cell, from which the endosperm will develop. The carpel shown here contains a single ovule.*

(b, c) Diagram and micrograph of the embryo sac of a lily (Lilium) shortly after "double fertilization" has occurred.

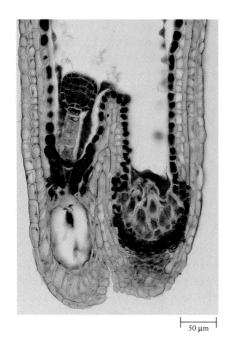
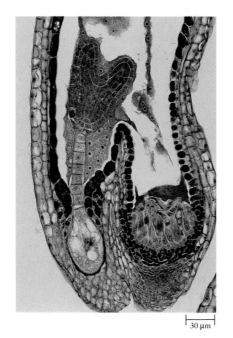
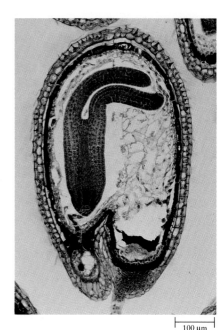

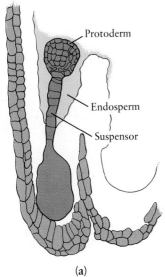

(a)

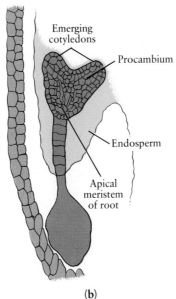

(b)

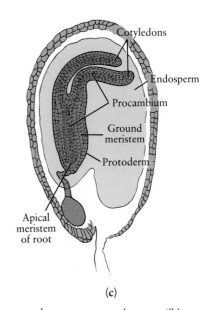

(c)

29–7 *Some stages in the development of the embryo of shepherd's purse* (Capsella bursa-pastoris), *a dicot.* **(a)** *The first embryonic tissue to differentiate is the protoderm, from which the outer covering of the young plant will develop. The large cell at the bottom of the embryo is the basal cell of the suspensor.* **(b)** *The cotyledons ("seed leaves") are beginning to emerge. The procambium, a second embryonic tissue, will later give rise to the vascular tissues of the plant.* **(c)** *The cotyledons have developed further. Additional differentiation has produced a third embryonic tissue, the ground meristem, from which the bulk of the tissue of the young leaves, stems, and roots will be derived. The three embryonic tissues, known as the primary meristems, are continuous between the cotyledons and the axis of the embryo.* **(d)** *The mature embryo within its protective seed coat. The apical meristems of the root and the shoot are clearly differentiated.*

CHAPTER 29 The Flowering Plants: An Introduction

THE EMBRYO

Following "double fertilization," the 3n cell divides mitotically to produce endosperm. The zygote also divides mitotically to form the embryo. As the embryo grows, its cells begin to **differentiate**—that is, they become different from one another. In its earliest stages, the embryo consists of a globular mass of cells on a stalk known as the suspensor. The cells of the suspensor, also formed from divisions of the fertilized egg, are actively involved in the delivery of nutrients to the embryo.

As development of the embryo proceeds, changes in its internal structure result in the formation of three distinct embryonic tissues. At the same time, or slightly later, emergence of the one cotyledon ("seed leaf") in monocots, or the two in dicots, occurs. Gradually the embryo takes on its characteristic form, a process known as **morphogenesis**. The developmental stages of a dicot embryo are shown in Figure 29-7.

In the early stages of embryonic growth, cell division takes place throughout the body of the young plant. As the embryo grows older, however, the addition of new cells becomes gradually restricted to certain parts of the plant body: the **apical meristems** (from the Greek *merizein*, "to divide"), located near the tips of the root and the shoot. During the rest of the life of the plant, primary growth—which chiefly involves the elongation of the plant body—originates in the apical meristems of roots and shoots.

THE SEED AND THE FRUIT

The seed consists of the embryo, which develops from the fertilized egg; the stored food, which consists of or derives from the endosperm; and the seed coat, which develops from the outermost layer or layers (integuments) of the ovule (Figure 29-8). At the same time, the fruit develops from the wall of the ovary (the base of the carpel). As the ovary ripens into fruit and the seeds form, the petals, stamens, and other parts of the flower may fall away.

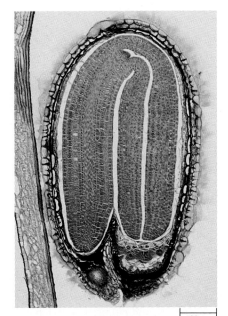

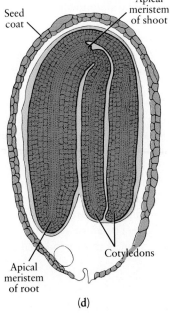
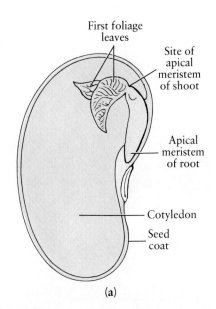
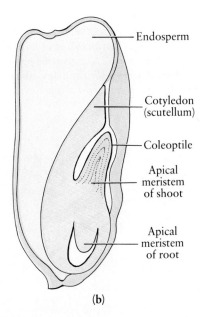

29-8 *Seeds.* **(a)** *In dicots such as the common bean, the endosperm is digested as the embryo grows, and the food reserve is transferred to the fleshy cotyledons.* **(b)** *In corn and other monocots, the single cotyledon, known as the scutellum in corn and other grains, absorbs food reserves from the endosperm. The coleoptile is a sheath that encloses the apical meristem of the shoot; it is the first structure to appear above ground after the seed germinates.*

29-9 *A simple fruit, an aggregate fruit, and, on the right, a multiple fruit.*

Types of Fruits

As you know from your own observations, fruits take many different forms. These forms are adaptations for a variety of dispersal mechanisms, as we noted in Chapter 24. Fruits are generally classified as simple, aggregate, or multiple (Figure 29–9), depending on the arrangement of the carpels in the parent flower. Simple fruits develop from one carpel or the fused carpels of a single flower, whereas aggregate fruits, such as magnolia, raspberry, and strawberry, develop from several separate carpels of a single flower. Multiple fruits consist of the carpels of more than one flower. A pineapple, for example, is a multiple fruit formed from an inflorescence, or flower cluster. The ovaries of the individual flowers fuse as they mature.

Simple fruits are by far the most diverse. When ripe, they may be soft and fleshy or dry. The three principal types of fleshy fruit are the berry, the drupe, and the pome. In berries, such as tomatoes, dates, and grapes, there are one to several carpels, each of which may have one or many ovules. The inner layer of the fruit wall is usually fleshy.

In drupes, there are also one to several carpels, but usually only a single seed develops. The inner wall of the fruit is stony and usually adheres tightly to the seed. Some familiar drupes are peaches, cherries, olives, and plums. The peach is a typical drupe; the skin, the succulent, edible portion of the fruit, and the stone are three distinct layers of the wall of the mature ovary. The almond-shaped structure within the stone is the seed.

Pomes are highly specialized fleshy fruits, characteristic of the subfamily of roses that produces rose hips. A pome is derived from an inferior ovary (page 513) in which the fleshy portion comes largely from the floral tube (Figure 29–10). Apples and pears are pomes.

Dry fruits are classified into two groups, dehiscent and indehiscent. Mature dehiscent fruits break open while still attached to the parent plant, releasing the seeds; the seeds of indehiscent fruits, by contrast, are still within the fruit when it is shed from the parent plant (Figure 29–11). Among the most familiar dehiscent fruits are those of legumes (such as the pea family), in which the ovary splits down two sides; the pod is the mature ovary wall and the peas are the seeds. In other dehiscent fruits, the fruit is derived from a single carpel in which the ovary wall splits down one side; the fruits of columbines and milkweeds are examples.

Indehiscent fruits occur in many plant families. The most common is the achene, a small, single-seeded fruit; some achenes, such as those produced by the

29-10 *Development and structure of the pear, a pome.* (a) *Flower of the pear. The ovary is the basal portion of the carpel.* (b) *Older flower, after petals have fallen.* (c) *Longitudinal section and* (d) *cross section of the mature fruit. The core of the pear is the ripened ovary wall. The fleshy, edible part develops from the floral tube.*

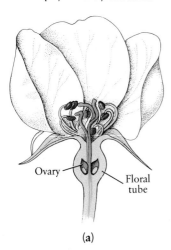
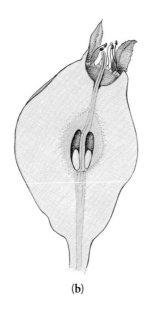
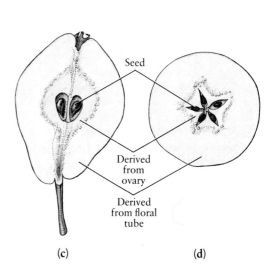

(a) (b) (c) (d)

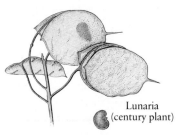

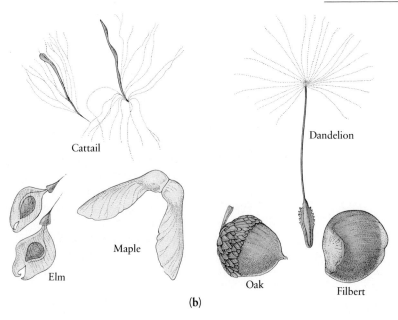

29–11 *Examples of* (a) *dehiscent and* (b) *indehiscent fruits. The distance that the seeds of dehiscent fruits are dispersed from the parent plant often depends on the force with which the fruit splits open. Many indehiscent fruits are carried a great distance from the parent plant, either by the wind or by hungry animals, before the seeds are released.*

elm and the ash, are winged. The most familiar kind of indehiscent fruit is the nut, which resembles the achene but has a stony coat and is derived from a compound ovary (an ovary formed from fused carpels). Examples of nuts are acorns and hazelnuts. Note that the word "nut" is used very indiscriminately in common speech: peanuts are the seeds of legumes; pine nuts are conifer seeds; almonds and coconuts are drupes.

ADAPTATIONS TO SEASONAL CHANGE

As we noted in Chapter 24, the angiosperms are thought to have evolved in upland regions of the tropics during a relatively mild period in the earth's history. As the climate grew colder, some angiosperms became extinct, while others were able to survive only near the equator. Some, presumably because of adaptations to drought (perhaps reflecting their highland origins), were able to survive in the cold, when water is locked up in ice. Chief among such adaptations is the capacity to remain dormant.

Dormancy and the Life Cycle

Depending on their characteristic patterns of active growth, dormancy, and death, modern plants are classified as annuals, biennials, and perennials. Among **annual** plants, the entire life cycle from seed to vegetative plant to flower to seed again takes place within a single growing season. Annuals include many familiar weeds, wild flowers, garden flowers, and vegetables. All vegetative organs (roots, stems, and leaves) die, and only the seeds, which are characteristically highly resistant to cold, desiccation, and other environmental hazards, bridge the gap between one generation and the next. Annual plants are usually soft-stemmed (nonwoody), or **herbaceous**. (An herb, botanically speaking, is a nonwoody seed plant, a definition that differs from the more common, culinary designation.)

In **biennial** plants, the period from seed germination to seed formation spans two growing seasons. The first season of growth often results in a short stem, a rosette of leaves near the soil surface, and a root. The root is often modified for food storage; sugar beets and carrots are examples of such storage roots. During the second growing season, the stored food reserves are mobilized for flowering, fruiting, and seed formation, after which the plant dies. Species that are characteristically biennials may, in some locations, complete their cycle in a single season or may require three or more years under adverse conditions.

The Staff of Life

Grains are the small, one-seeded fruits of grasses. Because they are relatively dry, they can be stored for long periods of time. The collecting and storing of grains from wild grasses is believed to have been an important impetus to the agricultural revolution of some 11,000 years ago (see Chapter 54). Today we are heavily dependent on cultivated wheat, rice, corn, rye, and other grains. In many countries, they constitute the principal component of the human diet. Wheat is about 9 to 14 percent protein. Its protein value is diminished, however, by its deficiency in certain essential amino acids, notably lysine (see page 73).

The grain of wheat, sometimes known as the kernel, is made up of the embryo, the endosperm, and, fused together, the mature ovary wall and the remains of the seed coat. More than 80 percent of the bulk of the wheat kernel and 70 to 75 percent of its protein are in the endosperm. White flour is made from the endosperm. Wheat germ, which is the embryo, forms about 3 percent of the kernel. It is usually removed as wheat is processed because it contains oil, which makes the flour more likely to spoil. Bran is the mature ovary wall, the remains of the seed coat, and the aleurone layer (the outer part of the endosperm); it constitutes about 14 percent of the kernel and is comprised mostly of cellulose. The bran is also removed when wheat is milled to make white flour. Bran somewhat decreases the caloric value of the wheat kernel, because we are unable to digest cellulose; bran therefore tends to speed the passage of food through our intestinal tracts, resulting in decreased absorption. Until fairly recently, wheat germ and bran, which contain most of the vitamins, were sometimes used for human consumption but more often were fed to livestock. With new evidence of the importance of adequate amounts of fiber in the human diet, many more cereals and breads are made from whole grains in which the bran is retained.

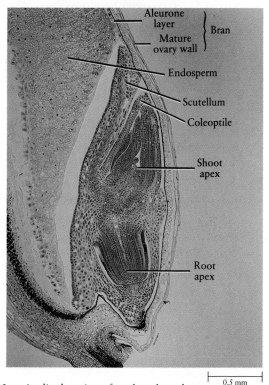

Longitudinal section of a wheat kernel.

Perennials are plants in which the vegetative structures persist year after year. The herbaceous perennials remain dormant as modified underground structures during unfavorable seasons, while the woody perennials, which include vines, shrubs, and trees, survive above ground. Woody perennials flower only when they become adult plants; a horse chestnut, for instance, may not flower until it is 25 years old.

The great advantage to being woody and perennial is that height can be added each growing season. Thus leaves can overtop and shade those of neighboring plants, flowers can be more conspicuously displayed, and seeds can be dispersed over a wider area. Woody perennials in favorable climates, such as the tropical rain forest, may live year after year with little change during the annual cycle, just as did the ancestral angiosperms. Perennials that live in areas where part of the year is unfavorable to growth have a variety of adaptations. Some, such as cacti and most gymnosperms, undergo little apparent change, although their rates of metabolism,

(a) (b) (c)

29–12 *Annual, biennial, and perennial plants.* (a) *An annual,* Linanthus dianthiflorus, *photographed in Baja California in the Sonoran desert. Many desert plants are annuals, racing through the entire life cycle from seed to flower to seed during a brief period following the seasonal rains.*

(b) *A biennial, the black-eyed Susan,* Rudbeckia hirta, *photographed on a July day in a Michigan field.*

(c) *An ancient perennial, the dawn redwood,* Metasequoia. *Twenty-five million years ago, the temperate zone of the Northern Hemisphere was covered with great forests of* Metasequoia. *Long thought to be extinct, living trees were discovered in western China in 1944. This tree in Philadelphia grew from seeds planted in 1947. Like the ginkgo, the bald cypress, and the larch, but unlike other gymnosperms, the dawn redwood is deciduous, shedding all its leaves in the fall.*

and therefore of growth, change with the seasons. Among dicots, many of the common vines, shrubs, and trees drop their leaves annually; such plants are said to be **deciduous**. They are typically found in regions where there is a marked seasonal variation in available water. The broad leaf of a deciduous dicot presents a much more efficient light-collecting surface than the needlelike leaf of a conifer (see page 508). The annual replacement of leaves is expensive, however, in terms of energy and other resources and can only take place in areas where the soil is fertile enough to supply the necessary nutrients and where the growing season is long enough to show an energy profit.

Seed Dormancy

The seeds of many wild plants require a period of dormancy before they will germinate. For example, the seeds of almost all plants growing in areas with marked seasonal temperature variations require a period of cold prior to germination. This physiological requirement ensures that the seed will "wait" at least until the next favorable growth period. The seeds of a few species can remain dormant and yet viable—with the embryo in a state of suspended animation—for hundreds of years. The record for dormancy, as far as is known, has been set by some seeds of East Indian lotus from a deposit at Pulantien, Manchuria. As determined by radioisotope dating, their age at the time of germination, in 1982, was 466 years.

The seed coat often plays a major role in maintaining dormancy. In some species, it acts primarily as a mechanical barrier, preventing the entry of water and gases, without which growth is not possible. In these plants, growth is initiated when the seed coat is worn away—for example, by being abraded by sand or soil, burned away in a forest fire, or partially digested as it passes through the digestive tract of a bird or other animal. In other species, dormancy is maintained chiefly by chemical inhibitors in the seed coat. These inhibitors undergo chemical changes in response to various environmental factors, such as light or prolonged cold or a sudden rise in temperature, which neutralize their effects, or they may be eroded or washed away by rainfall. Eventually, dormancy is broken and the stage is set for germination.

The dormancy requirement of seeds apparently evolved only recently, geologically speaking, among groups of plants subjected to the environmental stress of increasing winter cold characteristic of the most recent Ice Age. By this time—only 1.5 to 2 million years ago—the angiosperms were already a highly diversified group, and different populations responded to these pressures in different ways.

This explains why even closely related types of plants may have different mechanisms for maintaining and breaking dormancy.

A dormant seed is usually very dry—only about 5 to 20 percent of its total weight is water. Germination cannot begin until the seed has imbibed the water required for the metabolic activities of the growing embryo. Once germination is underway, the seed coat ruptures and the young sporophyte emerges. As we shall see in the next chapter, a series of growth processes immediately begin that give rise to the plant body and continue throughout its life.

SUMMARY

Flowers are the structures of sexual reproduction in the angiosperms. The pollen grains, which contain the male gametophytes, are produced by the anthers. Anther plus filament is known as the stamen. The carpel typically consists of the stigma (an area on which the pollen grains germinate), a style, and, at its base, the ovary. The ovary contains one or more ovules, and within each ovule is a female gametophyte that contains an egg cell.

A new cycle of life begins when a pollen grain germinates on the stigma of a flower of the same species, sending a pollen tube through the style and into an ovule. One of the sperm nuclei of the pollen grain fertilizes the egg cell in the female gametophyte; the other sperm nucleus unites with the two polar nuclei of the female gametophyte to form a triploid ($3n$) cell. Division of this triploid cell produces a special nutritive tissue, the endosperm. The phenomena of fertilization and triple fusion, called "double fertilization," occur only in the flowering plants. Following "double fertilization," the ovule develops into a seed and the ovary into a fruit.

The angiosperm seed, or mature ovule, consists of the embryo, the seed coat, and stored food. The petals, stamens, and other floral parts of the parent plant may fall away as the ovary ripens into fruit and the seeds form.

Dormancy—of the seed, of vegetative parts of the plant body, or of both—enables angiosperms to bridge periods of drought or cold unsuitable for plant growth. Angiosperms are classified as annuals, biennials, or perennials, depending on whether the plant body dies at the end of one growing season (annual) or after two seasons (biennial) or whether vegetative portions of the plant body persist from year to year (perennial). In regions where there is marked variation in the availability of water, perennials are frequently deciduous, dropping their leaves at the end of each growing season.

QUESTIONS

1. Distinguish among the following: ovary/ovule/egg cell/seed; male gametophyte/female gametophyte/embryo sac; pollination/fertilization; protoderm/procambium/ground meristem; annual/biennial/perennial.

2. In many plants, pollen production in the anthers occurs prior to or after the full development of the carpel of the same flower. What are the consequences of this shift in time frames?

3. J. B. S. Haldane, a mathematician who made major contributions to biology and whose pithy comment about beetles we quoted earlier (see page 574), once remarked: "A higher plant is at the mercy of its pollen grain." Explain.

4. Which of the following is a berry: blackberry, strawberry, mulberry, grape, pumpkin?

5. Seeds of the jack pine (a gymnosperm) are released from the female cones only after exposure to intense heat, as in a forest fire. Following release, the tough seed coats rupture, and the seeds germinate. What might be the advantage for the jack pine of this delay in seed release and germination?

6. Sketch a dicot embryo at the time of seed release. Identify each part in terms of the future development of the plant body.

7. What are the major adaptations that enable plants to survive the periodic drought (winter) of temperate regions?

CHAPTER 30

The Plant Body and Its Development

As we discussed in Chapter 24, plants are multicellular photosynthetic organisms primarily adapted for life on land. The ancestor of the plants is thought to have been a multicellular green alga similar to *Coleochaete* (page 494). As in modern plants, its principal photosynthetic pigments were chlorophylls *a* and *b* and beta-carotene, all of which were contained in chloroplasts. Each of its cells had a membrane-bound nucleus and mitochondria and other organelles, as well as an external cell wall containing cellulose. Its energy source was sunlight, and it obtained oxygen, carbon dioxide, and minerals from the water in which it lived.

The photosynthetic cells of plants have the same few and relatively simple requirements: light, water, oxygen, carbon dioxide, and certain minerals. From these simple materials they, like their ancestors, make the sugars, fatty acids, amino acids, nucleotides, and other organic substances on which all plant and animal life depends. In algae and in small, simple plants (such as bryophytes) living in a moist environment, each of the required materials is immediately available to every cell. Most plants, however, live in a very different environment, and their photosynthetic cells require a complex life-support system in order to function. The plant body is, in effect, that life-support system.

Figure 30-1 diagrams the external structure of an economically important angiosperm, the potato plant. This plant, like other vascular plants, is characterized by a root system that anchors the plant in the ground and collects water and minerals from the soil; a stem or trunk that raises the photosynthetic parts of the plant toward the sun; and structures highly specialized for light capture and photosynthesis, the leaves. Its cells, unlike those of a multicellular alga, are not autonomous and can survive only through cooperation and a division of labor. This division of labor is made possible by processes of differentiation that begin during embryonic development and continue throughout the life of the plant.

THE CELLS AND TISSUES OF THE PLANT BODY

As we saw in the last chapter, the cells of the angiosperm embryo differentiate early in its development into three distinct tissues: the protoderm, the procambium, and the ground meristem (see Figure 29-7, page 618). These embryonic tissues, known as the primary meristems, give rise to three **tissue systems** that are continuous throughout the plant body. The protoderm, the first tissue to differentiate, is the origin of the **dermal tissue** system, which provides an outer protective covering for the entire plant body. The procambium, the next tissue to differentiate, gives rise to the **vascular tissue** system, composed of xylem and phloem. As we saw in Chapter 24, xylem transports water and dissolved minerals, while phloem transports dissolved sugars and other organic compounds from the photosynthetic (autotrophic) cells of leaves and green stems to the nonphoto-

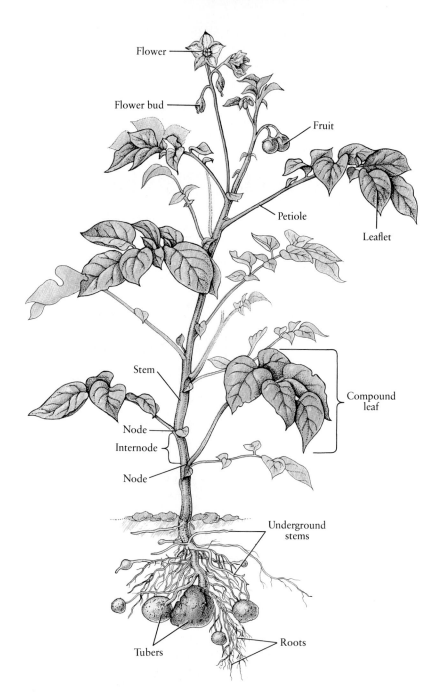

30–1 *The body plan of a flowering plant, the potato* (Solanum tuberosum). *The aboveground shoot system consists of the stem, the leaves, whose primary function is photosynthesis, and the flowers, the reproductive structures. After fertilization, the flower petals fall away and the ovaries mature to form the fruits, as we saw in the last chapter. Leaves, which may be simple or, as in the potato, compound, appear at regions on the stem known as nodes. The portions of the stem between successive nodes are called internodes.*

In many plants, the only belowground structures are the roots, which supply water and minerals to the stem, leaves, flowers, and fruits. In the potato, however, the most conspicuous underground structures are the tubers, which are enlarged stems adapted for food storage.

synthetic (heterotrophic) cells of the plant. The vascular tissues are embedded in the **ground tissue** system, which is derived from the embryonic ground meristem. As we shall see in the course of this chapter, the principal differences in the structure of leaves, stems, and roots lie in the relative distribution of the vascular and ground tissue systems.

The most frequently encountered cells in the plant body are of a type known as **parenchyma**. These cells, which occur in all three tissue systems and predominate in the ground tissues, are typically many-sided, with thin, flexible walls. In addition to a nucleus, mitochondria, and other organelles, parenchyma cells generally contain plastids, which, depending on the location of the cell, may be chloroplasts, leucoplasts, or chromoplasts (see page 120). The plant cell shown in Figure 5–16 (page 113) is a photosynthetic parenchyma cell, containing chloroplasts. In addition to photosynthesis, parenchyma cells perform a variety of essential functions in the plant, including respiration and storage of food and water. Each of the tissue systems also contains additional cell types, specialized for the particular functions of the tissue.

CHAPTER 30 The Plant Body and Its Development

Photosynthesis is, of course, the most fundamental activity of a plant, on which all else depends. We shall therefore begin our examination of the plant body with the leaves, the primary photosynthetic organs, which are essentially the same in a seedling as in a mature plant. Following our consideration of leaves, we shall shift our attention back to the germinating seed, and examine the structure and development of roots and stems, the other two organs of the plant body, without which the leaves could not survive.

LEAVES

Leaf Structure

The structure of a leaf (Figure 30–2) is a compromise between three conflicting evolutionary pressures: to expose a maximum photosynthetic surface to sunlight, to conserve water, and, at the same time, to provide for the exchange of gases necessary for photosynthesis.

The photosynthetic cells of leaves are parenchyma cells of two types: **palisade parenchyma**, which are densely packed, columnar cells located just below the upper surface of the leaf, and **spongy parenchyma**, which are irregularly shaped cells in the interior of the leaf, often with large spaces between them. These spaces are filled with gases, including water vapor, oxygen, and carbon dioxide. Most of the photosynthesis occurs in the palisade cells, which are specialized for intercepting light.

Palisade and spongy parenchyma make up the ground tissue of the leaf, known as the **mesophyll**, or "middle leaf." The mesophyll is enclosed in an almost airtight wrapping of epidermal cells, which secrete a waxy substance called **cutin.** The cutin forms a coating, the **cuticle,** over the outer surface of the epidermis. The epidermal cells and the cuticle are transparent, permitting light to penetrate to the photosynthetic cells.

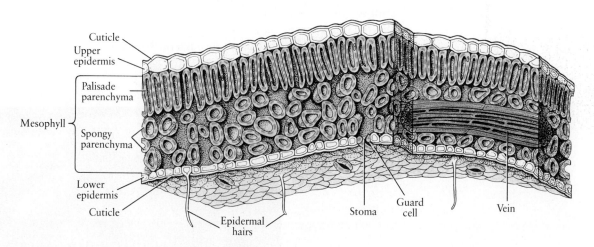

30–2 *The structure of a leaf. Photosynthesis takes place in the palisade cells and, to a lesser extent, in the spongy parenchyma. The chloroplasts are indicated in bright green. Note that the cytoplasm, which contains the chloroplasts, is concentrated near the cell surface, and the centers of the cells are filled with large vacuoles. The chloroplasts move within the cytoplasm, orienting themselves to the sun. The veins carry water and solutes to and from these mesophyll cells. The interior of the leaf is enclosed by epidermal cells covered with a waxy layer, the cuticle. Openings in the epidermis, the stomata, permit the exchange of gases. The guard cells surrounding the stomata also have chloroplasts, which may (or may not) play a role in opening and closing the stomata (see pages 653–655).*

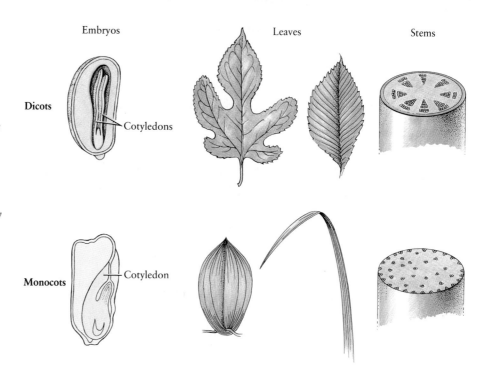

30–3 *As we have noted previously, the two classes of angiosperms are the dicots and the monocots. The names refer to the fact that the dicot embryo has two cotyledons ("seed leaves"), and the monocot embryo has one. Other characteristic differences are visible in the plant body. The principal veins of dicot leaves are usually netted; those of monocot leaves are usually parallel. In dicot stems, bundles of vascular tissue are arranged around a central core of ground tissue; in monocot stems, the vascular bundles are scattered throughout the ground tissue. As we saw in the last chapter, there are also characteristic differences in the number of floral parts and in the structure of the pollen grains.*

Substances move into and out of leaves through two quite different structures: vascular bundles and stomata. Water and dissolved minerals are transported into leaves—and the products of photosynthesis are transported out—by way of the vascular bundles. The vascular bundles, which are known in leaves as **veins,** pass through the petioles (leaf stalks) and are continuous with the vascular tissues of the stem and root. Veins form distinctive patterns in leaf blades, which are conspicuously different in monocots and dicots. In monocots, the major veins are usually all parallel; in dicots, the venation (vein pattern) is netted (Figure 30–3).

Gases—oxygen and carbon dioxide—move into and out of leaves by diffusion through stomata (plural of *stoma,* the Greek word for "mouth"). A stoma consists of a small opening, or pore; it is surrounded by two specialized cells in the leaf epidermis, called guard cells, that open and close the pore (Figure 30–4). The exchange of oxygen and carbon dioxide is, as we saw in Chapter 10, necessary for photosynthesis. However, as these gases are exchanged between the atmosphere and the leaf interior, water also escapes from the leaf. About 90 percent of the water loss from the plant body is through the stomata; the remaining 10 percent is through the epidermal cells.

30–4 *An open stoma. The stomata lead into air spaces within the leaf that surround the thin-walled spongy parenchyma cells. The air in these spaces, which make up 15 to 40 percent of the total volume of the leaf, is saturated with water vapor that has evaporated from the photosynthetic cells.*

10 μm

Stomata are typically most abundant on the undersurface of leaves. They may be very numerous. For example, on the lower surface of tobacco leaves there are about 19,000 stomata per square centimeter; on the upper surface there are about 5,000 stomata per square centimeter.

Leaf Adaptations and Modifications

Leaves come in a variety of shapes and sizes, ranging from broad fronds to tiny scales. Some of these differences can be correlated with the environments in which the plants live. Large leaves with broad surfaces are often found in plants that grow under the canopy in a tropical rain forest, where water is plentiful but competition for light is intense. Leaves of such plants sometimes have "drip tips," which facilitate the runoff of rainwater. Leaves with small surfaces are usually associated with dry climates. In conifers, for example, the photosynthetic surfaces are greatly reduced and there is an extra-thick layer of epidermis and cuticle (see page 508). In such plants, photosynthesis is reduced but so is water loss. The trees are thus able to survive long periods of drought, including the drought of winter, when water is frozen in snow or ice and is unavailable to plants. Similarly, angiosperms in dry habitats often have small, leathery leaves. This reduction in leaf surface reaches an extreme in those desert cacti in which the leaves are modified as spines—hard, dry, nonphotosynthetic structures. (The terms "spine" and "thorn" are often used interchangeably; however, thorns are technically modified branches.) In these plants, photosynthesis takes place in the fleshy stems, which are also water-storage organs.

In many plants, leaves are succulent; that is, they are adapted for water storage. Among the most interesting examples of this adaptation are the "window" plants native to the deserts of South Africa. Their leaves grow almost entirely underground, with only the transparent "window" tip of the leaf protruding above the soil surface. The transparent water-storage tissue of the leaf provides a conduit for light to reach the underground photosynthetic cells.

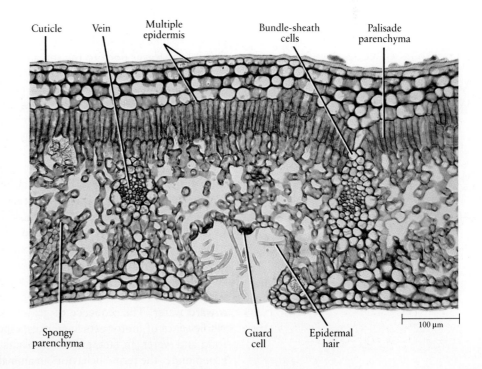

30-5 *Cross section of the leaf of an oleander, an angiosperm adapted to a dry climate. At the top, a very thick cuticle covers the multiple epidermis, so called because it consists of four layers of cells. The stoma is contained within a stomatal crypt, which is lined with epidermal hairs. The veins are seen in cross section, which is why they look different from the vein in Figure 30–2.*

(a)   (b)   (c)

30-6 *Modified leaves.* (a) *Spines on a giant prickly-pear cactus, photographed in the Galapagos Islands.* (b) *Succulent leaves, adapted for water storage (Sedum).* (c) *Tendrils of a pea plant. In the pea plant, which has compound leaves, only individual leaflets are modified as tendrils; other leaflets of a given compound leaf are flattened, providing a broad surface for photosynthesis.*

Leaves may also be specialized for other functions, such as food storage or support. For example, a bulb, such as the onion, is a large bud consisting of a short stem with many leaves modified for storing food. The "head" of a cabbage also consists of a compressed stem bearing numerous thick, overlapping leaves. In some plants, the petioles become thick and fleshy: celery and rhubarb are two familiar examples. The tendrils of some climbing plants—the garden pea, for instance—are modified leaves or leaflets.

CHARACTERISTICS OF PLANT GROWTH

As we discussed in the last chapter, the embryos of many angiosperms pass through a dormant stage prior to germination of the seed. With germination, growth resumes, the seed coat ruptures, and the young sporophyte emerges. The first foliage leaves open to the sun and begin photosynthesizing, while internally the growth processes that give rise to the plant body continue.

The primary growth of the plant involves differentiation of the three tissue systems, elongation of roots and stems, and the formation of lateral roots and of branches. As we noted earlier, after development of the embryo is complete, subsequent primary growth originates in the apical meristems of the root and shoot.

The existence of such meristematic areas, which add to the body of the plant throughout its life, is one of the principal differences between plants and animals. Higher animals stop growing when they reach maturity, although the cells of certain "turnover" tissues, such as the skin or the lining of the intestine, continue to divide. Plants, however, continue to grow during their entire life span. Growth in plants is the counterpart, to some extent, of motility in animals. Plants "move" by extending their roots and shoots, both of which involve changes in size and form. As a result of these changes, a plant modifies its relationship with the environment, for example, by curving toward the light and extending its roots toward water. The sequence of growth stages in plants thus corresponds to a whole series of motor acts in animals, especially those associated with obtaining food and water. In fact, growth in plants serves many of the functions that we group under the term "behavior" in animals.

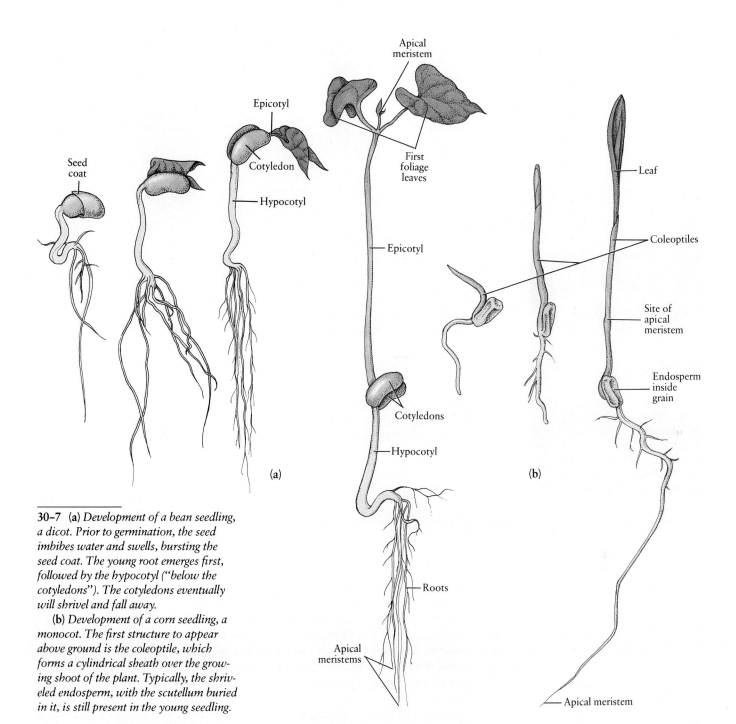

30–7 (a) *Development of a bean seedling, a dicot. Prior to germination, the seed imbibes water and swells, bursting the seed coat. The young root emerges first, followed by the hypocotyl ("below the cotyledons"). The cotyledons eventually will shrivel and fall away.*

(b) *Development of a corn seedling, a monocot. The first structure to appear above ground is the coleoptile, which forms a cylindrical sheath over the growing shoot of the plant. Typically, the shriveled endosperm, with the scutellum buried in it, is still present in the young seedling.*

ROOTS

Roots are specialized structures that anchor the plant and take up water and essential minerals. The embryonic root is the first structure to break through the seed coat, and, in an older plant, the root system may make up more than half of the plant body. The lateral spread of tree roots is usually greater than the spread of the crown of the tree. In a study made on a four-month-old rye plant, the total surface area of the root system was calculated to be 639 square meters, 130 times the surface area of the leaves and stem. Root growth is affected by soil conditions and availability of water. The deepest known roots were those of a mesquite (a desert shrub) found growing in a new open-pit mine near Tucson, Arizona, in 1960; they had penetrated to a depth of 53.3 meters.

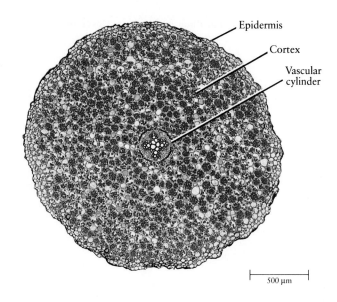

30-8 *Root of a buttercup, a dicot, in cross section. The plastids in the parenchyma cells of the cortex contain starch grains, stained purple in this preparation. The vascular cylinder is shown in more detail in Figure 30-11a.*

30-9 *A radish seedling. Note the discarded seed coat, the cotyledons, and the primary root with its numerous root hairs. Most of the uptake of water and minerals occurs through the root hairs, which form just behind the growing tip of the root.*

Root Structure

The internal structure of the angiosperm root is comparatively simple. In dicots and most monocots, the three tissue systems (dermal, ground, and vascular) are arranged in three concentric layers: the **epidermis**, the **cortex**, and the **vascular cylinder** (Figure 30-8).

The Epidermis

The epidermis, which covers the entire surface of the young root, absorbs water and minerals from the soil and protects the internal tissues. The cuticle is either absent or very thin compared with that found on the surface of a leaf.

The epidermal cells of the root are characterized by fine, tubular outgrowths, known as **root hairs** (Figure 30-9). Root hairs are slender extensions of the epidermal cells; in fact, the nucleus of the epidermal cell is often found within the root hair. In the rye plant previously mentioned, the roots were estimated to have some 14 billion root hairs. Placed end to end, they would have extended more than 10,000 kilometers. Most of the water and minerals that enter the root are absorbed by these delicate outgrowths of the epidermis. In the mature plants of many species, however, mycorrhizal associations (page 490) substitute for root hairs.

The Cortex

As you can see in Figure 30-8, the cortex occupies by far the greatest volume of the young root. The cells of the cortex are parenchyma cells, as in the ground tissue of the leaf. However, root parenchyma usually lacks functional chloroplasts; instead, the plastids are specialized for food storage and contain starch and other organic substances. (In some species, the roots are highly specialized for this function; beets and carrots are examples of roots with an abundance of storage parenchyma.) There are many air spaces in the cortex, and oxygen from the soil enters these spaces through the epidermal cells and is used by the cortical cells in respiration.

Unlike the rest of the cortex, the cells of the innermost layer, the **endodermis**, are compact and have no spaces between them. Each endodermal cell is encircled by a continuous band of wax, known as the **Casparian strip** (Figure 30-10). The Casparian strip, which is located within the cell wall and adheres tightly to the cell membrane, is not permeable to water. Therefore, water and dissolved substances, which can move freely around the other cortical cells and through their cell walls, must pass through the cell membranes of endodermal cells. As you will recall (page 130), water, oxygen, and carbon dioxide pass easily through cell membranes, but many ions and other substances do not. Thus, the membranes of the endodermal cells regulate the passage of such substances into the vascular tissues of the root, thereby determining what is transported to the rest of the plant body.

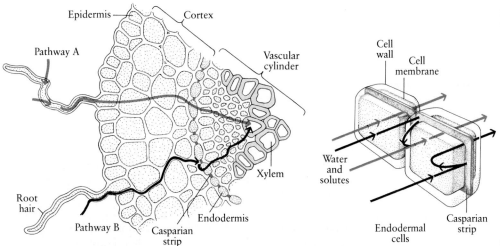

30–10 *Diagrammatic cross section of a root, showing the two pathways of uptake of water and dissolved substances. Most of the solutes and some of the water entering the root follow pathway A, indicated in color; the solutes move by active transport and diffusion and the water by osmosis through the cell membranes and plasmodesmata (page 140) of a series of living cells. Most of the water and some of the solutes entering the root follow pathway B, indicated in black, flowing through the cell walls and along their surfaces. Notice, however, the location of the Casparian strip and how it blocks pathway B all around the vascular cylinder of the root. In order to pass the Casparian strip, both water and solutes must be transported through the cell membranes of the endodermal cells, as in pathway A. After the water and solutes have crossed the endodermis, most of the solutes continue along pathway A to the conducting cells of the xylem, and most of the water returns to pathway B for the remaining distance to the conducting cells.*

The Vascular Cylinder

The vascular cylinder of the root consists of the vascular tissues (xylem and phloem) surrounded by one or more layers of cells, the **pericycle,** from which branch roots arise. In most species, the vascular tissues of the root are grouped in a solid cylinder, similar to that shown in Figure 30–8. Figure 30–11a shows the details of the vascular cylinder of a buttercup root (a dicot). In some monocots, however, the vascular tissues form a cylinder around a **pith,** a central core of ground tissue (Figure 30–11b).

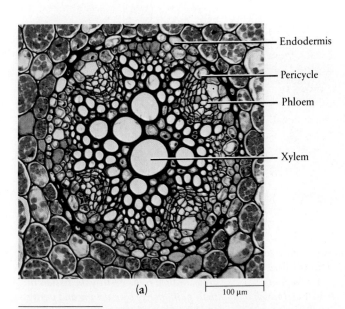

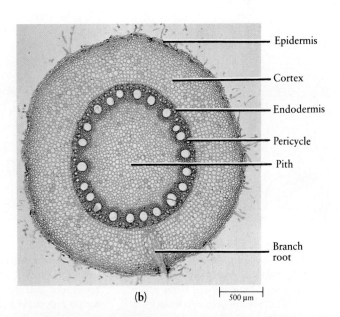

30–11 *(a) Details of the vascular cylinder of the buttercup root (a dicot) shown in Figure 30–8. The endodermis, which is outside the pericycle, is considered part of the cortex. The endodermis contains the Casparian strips.*

(b) Cross section of the root of a corn plant (a monocot), showing the vascular cylinder enclosing the pith. Part of a branch root, which emerges from the pericycle, can be seen in the lower portion of the micrograph.

SECTION 5 Biology of Plants

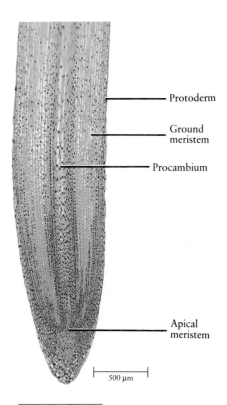

30–12 *Longitudinal section of the root tip of an onion, a monocot. The primary meristematic tissues—protoderm, ground meristem, and procambium—can be distinguished close to the apical meristem.*

Primary Growth of the Root

The first part of the embryo to break through the seed coat, in nearly all seed plants, is the embryonic root, or **radicle**. At its tip is the root cap, which protects the apical meristem as the root is pushed through the soil. Cells of the root cap wear away and are constantly replaced by new cells from the meristem.

Certain cells in the meristem retain the capacity to produce new cells and thus perpetuate the meristem. All the other cells in the root—which are the progeny of these relatively few meristematic cells—eventually differentiate, some becoming cells of the root cap and others forming the tissue systems of the root. The maximum rate of cell division occurs at a point well above the tip of the meristem. Then, just above the point where cell division is greatly reduced, the cells gradually elongate, growing to 10 or more times their original length, often within the span of a few hours. This elongation is the principal cause of primary growth in roots, although, of course, growth is ultimately dependent on the production of new cells that become part of the zone of elongation.

As the cells elongate, they begin to differentiate. The first cells to differentiate in roots are the conducting cells of the phloem, followed by the conducting cells of the xylem. In the region of the root where the xylem first forms, the endodermis also differentiates. To the inside of the endodermis, the pericycle forms. At approximately the same level in the elongating root, the epidermal cells differentiate and begin to extend root hairs into crevices between the soil grains.

This sequence of growth, as shown in Figure 30–14, occurs in the first root of a seedling and is repeated over and over again in all the growing root tips of a plant—even those of a tree 50 meters tall.

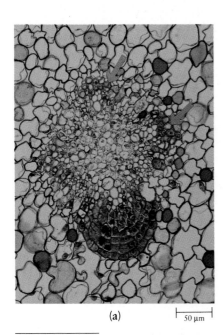
(a)

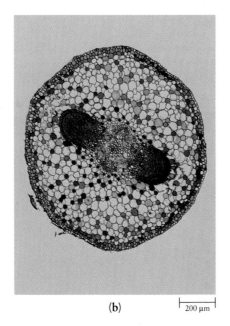
(b)

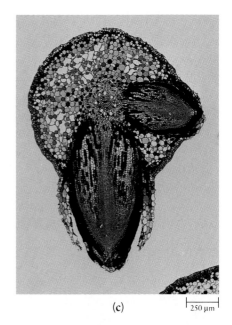
(c)

30–13 *Sections from three different willow (Salix) roots, showing stages in the development of branch roots. In the upper portion of (a), the origin of two branch roots in the pericycle is marked by arrows; in the lower portion of the micrograph, a slightly older branch root is beginning to grow through the cortex. This is a very young root, and the vascular cylinder is not yet fully developed. In (b), two branch roots have traveled almost halfway across the cortex. In (c), one branch root has broken through the epidermis to the outside, and another is about to do so. As they grow, branch roots destroy the tissues in their path, partly by crushing and partly by digestion with enzymes.*

CHAPTER 30 The Plant Body and Its Development

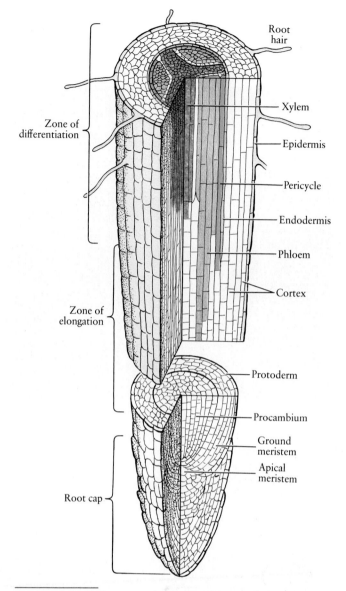

30-14 *The growth regions of a dicot root. New cells are produced by the division of cells within the apical meristem. The cells above the meristem undergo a characteristic series of changes as the distance increases between them and the root tip. First, there is a maximum rate of cell division, followed by cell elongation with little further division. As the cells elongate, they differentiate into the three primary meristems that give rise to the three tissue systems of the root. The protoderm becomes the epidermis, the ground meristem becomes the cortex, and the procambium becomes the primary xylem and primary phloem. Some of the cells produced by the apical meristem differentiate to form the protective root cap.*

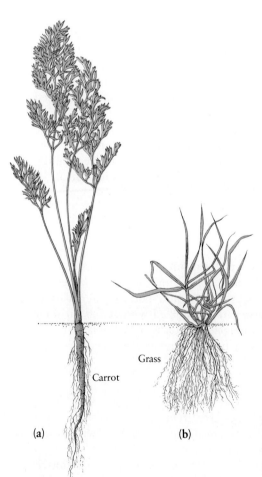

30-15 *Types of root systems.* (a) *Taproot system of a carrot (a dicot).* (b) *Fibrous root system of rye grass (a monocot).*

Patterns of Root Growth

In many dicots, the primary root develops into a large **taproot,** which, in turn, gives rise to lateral, or branch, roots (Figure 30–15a). In monocots, the primary root is usually short-lived and the final root system develops from the base of the stem; roots of this kind are called **adventitious roots.** ("Adventitious" describes any structure growing from other than its "usual" place.) These adventitious roots and their branches form a fibrous root system (Figure 30–15b).

635

30–16 (a) *Prop roots of corn. These are adventitious roots, arising from the stem.* (b) *Air roots (pneumatophores) of a white mangrove. The root tips grow up out of the mud in which these trees grow and take up oxygen needed by the roots for respiration.*

Aerial roots are adventitious roots produced from aboveground structures. Some aerial roots, such as those of English ivy, cling to vertical surfaces and thus provide support for the climbing stem. In other plants, such as corn, aerial roots serve as prop roots (Figure 30–16a). Trees that grow in swamps, such as the red mangrove and the bald cypress, often have prop roots.

In swampy areas the soil is usually low in oxygen. Some trees that grow in these habitats develop roots that grow out of the water and serve not only to anchor the plant but also to supply the root cells with the oxygen needed for respiration. White mangroves, for example, have roots whose tips grow upward out of the mud and serve this aerating function (Figure 30–16b).

STEMS

Stems display leaves to the light and are the pathway through which substances are transported between the roots and the leaves. They may also be adapted for storing food or water. As we saw in Figure 30–1, white potatoes are enlarged, underground stem tips, known as tubers, that are filled with starch. In some species of plants that grow in arid environments, water is stored in large parenchyma cells in the stems; the stem of a cactus, for example, may be 98 percent water by weight.

Stem Structure

The outer surface (dermal tissue) of young green stems, like that of leaves and roots, is made of epidermal cells. Like leaves, green stems are covered with a waxy cuticle, contain stomata, and are photosynthetic.

The bulk of the tissue in a young stem is ground tissue. As in leaves and roots, it is composed mostly of parenchyma cells. The turgor (page 134) of these cells provides the chief support for young green stems. The ground tissue of stems also may contain specialized supporting tissues known as **collenchyma** and **sclerenchyma**. Unlike the thin-walled parenchyma cells, collenchyma cells (Figure 30–

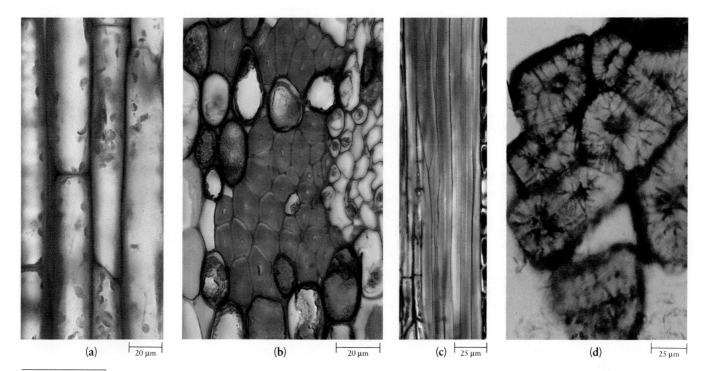

30–17 *Some types of cells found in the ground tissue of stems.* **(a)** *Collenchyma cells, viewed in longitudinal section. Their irregularly thickened cellulose walls are rich in pectin and contain much water. The walls are plastic, and so the cell can continue to grow.* **(b)** *Cross section and* **(c)** *longitudinal section of phloem fibers from the stem of a linden tree* (Tilia americana). *Only a portion of the length of the fibers can be seen in* **(c)**. *These sclerenchyma cells have thickened, often lignified cell walls that give them strength and rigidity. Many fibers, but not all, are dead at maturity.* **(d)** *Sclereids, another type of sclerenchyma cell, have very thick lignified walls. Sclereids are often found in seeds and fruits, as well as in stems. These sclereids, called stone cells, are from a pear; they give the fruit its characteristic gritty texture.*

17a) have primary walls (see page 107) that are thickened at the corners or in some other uneven fashion. Their name derives from the Greek word *colla*, meaning "glue," which refers to their characteristic thick, glistening walls. Collenchyma cells are often located just inside the epidermis, forming either a continuous cylinder or distinct vertical strips of tissue. They provide support for the growing regions of young stems and branches.

Sclerenchyma cells are of two types: fibers and sclereids. Fibers, which are extremely elongated, somewhat elastic cells (Figure 30–17b and c), typically occur in strands or bundles arranged in a definite pattern characteristic of the particular species. They are often associated with the vascular tissues. Plant fibers such as flax, hemp, jute, sisal, and raffia have long been used in human artifacts, including baskets, rope, and cloth. Sclereids (Figure 30–17d), which are variable in form, are also common in stems. Layers of sclereids are found in seeds, nuts, and fruit stones as well, where they form the hard outer coverings.

The name sclerenchyma is derived from the Greek *skleros*, meaning "hard." This property of sclerenchyma cells is a result of the impregnation of their cell walls with lignin, a complex macromolecule that toughens and hardens cellulose. Sclerenchyma cells differ from collenchyma in three other respects: (1) they have secondary walls (page 107) in addition to primary walls; (2) the cells are often dead at maturity, with only their cell walls remaining; and (3) they usually occur in regions of the plant body that have completed primary growth.

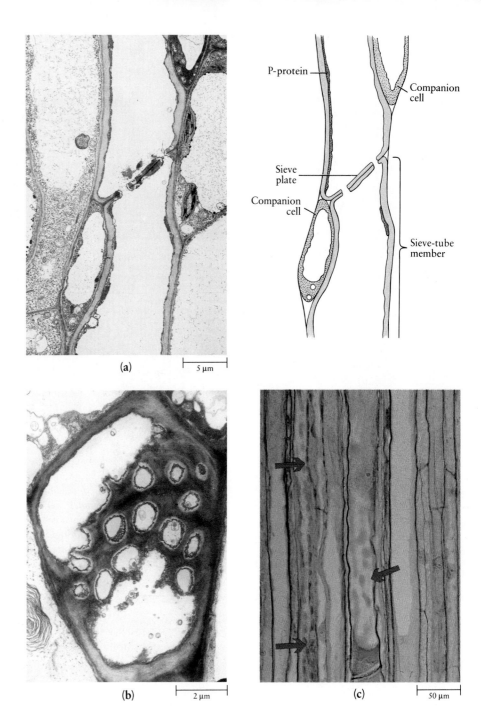

30–18 *In angiosperms, the conducting elements of the phloem are sieve tubes, made up of individual cells, the sieve-tube members. These cells, which lack nuclei at maturity, are usually found in close association with companion cells, which do have nuclei. Sieve-tube members are joined to other sieve-tube members at their ends by sieve plates. (a) Longitudinal view of a sieve tube in the stem of the squash (Cucurbita maxima). P-protein lines the inner surface of the cell walls of the sieve-tube members. (b) Face view of a sieve plate between two mature sieve-tube members. In this electron micrograph, the pores of the sieve plate are open. In most cut sections of phloem tissue, however, the pores are plugged either with callose, a polysaccharide deposited by the sieve-tube members in response to injury, or with P-protein. Both callose and P-protein prevent leakage from the sieve tube. Callose is also deposited as part of the normal aging process. (c) Longitudinal section of phloem, showing mature and immature sieve-tube members. The arrows point to P-protein bodies in immature cells.*

Vascular Tissues

The vascular tissues, phloem and xylem, consist of specialized conducting cells, supporting fibers, and parenchyma cells, which store food and water. The conducting cells of the phloem, as we noted previously, transport the products of photosynthesis, chiefly in the form of sucrose, from the leaves to the nonphotosynthetic cells of the plant. In gymnosperms, these conducting cells are sieve cells; in angiosperms, they are sieve-tube members (Figure 30–18a). A **sieve tube** is a vertical column of sieve-tube members joined by their end walls. These end walls, called **sieve plates,** have pores leading from one sieve-tube member to the next (Figure 30–18b).

The sieve-tube members, which are alive at maturity, are filled largely with a watery fluid called sieve-tube sap. In dicots and some monocots, they also contain a proteinaceous substance called slime, or P-protein (the "P" stands for phloem). The function of P-protein is unknown, although some botanists believe that, along with the polysaccharide callose, it seals sieve-plate pores in response to injury. In mature sieve-tube members, P-protein lies along the inner surface of the cell wall and is continuous from one cell to the next through the sieve plates.

As a sieve-tube member matures, its nucleus and many of its organelles disintegrate. Sieve-tube members, however, are characteristically associated with specialized parenchyma cells called **companion cells,** which contain all of the components commonly found in living plant cells, including a nucleus. Companion cells may be responsible for the secretion of substances into and out of the sieve-tube members and are thought also to provide nuclear functions for the sieve tubes and fulfill their energy requirements. A sieve-tube member and its companion cell arise from the same mother cell.

Specialized cells in the xylem conduct water and minerals from the roots to other parts of the plant body. It is customary to think of xylem as transporting water up and phloem as transporting sugars down, but if you think of the various shapes of plants you can see that water must also often be transported laterally, as along a tendril, or even down, as to the branches of a weeping willow. Conversely, sugars must often go upward, as into a flower or fruit.

In angiosperms, the conducting cells of the xylem are tracheids and vessel members. Both of these cell types have thick secondary walls impregnated with lignin, and both are dead at functional maturity. Tracheids are long, thin cells that overlap one another on their tapered ends (Figure 30–19a). These overlapping surfaces contain thin areas, known as pits, where no secondary wall has been deposited. Water passes from one tracheid to the next through the pits. Vessel members, which are much shorter and wider, also differ from tracheids in that their end walls contain perforations or are entirely absent (Figure 30–19b and c). Thus, the vessel members form a continuous **vessel,** which is a more effective conduit than a series of tracheids. Seedless vascular plants and most gymnosperms have only tracheids; most angiosperms have both tracheids and vessel members.

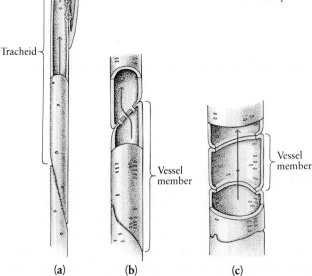
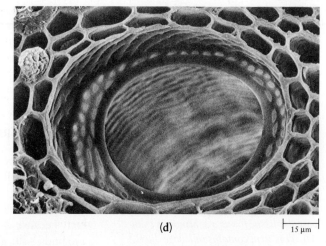

30–19 *Tracheids and vessel members are the conducting cells of the xylem in angiosperms.* (a) *Tracheids are a more primitive and less efficient type of conducting cell. Water moving from one tracheid to another passes through pits. Pits are not perforations but simply areas in which there is no secondary cell wall. Water moving from one tracheid to another passes through two primary cell walls and the middle lamella (see page 106).*

Vessel members differ from tracheids in that the primary walls and middle lamellae of vessel members are perforated at the ends where they are joined with other vessel members. (b) *There may be numerous perforations in adjoining walls of vessel members, or* (c) *the adjoining walls may dissolve completely as the cells mature, forming a single opening. Vessel members are also characteristically shorter and wider than tracheids, and their adjoining walls are less oblique. Vessel members are connected with other vessel members and also with other cells by pits in the side walls.*

(d) *A view into a vessel in the xylem of a prop root in a corn plant.*

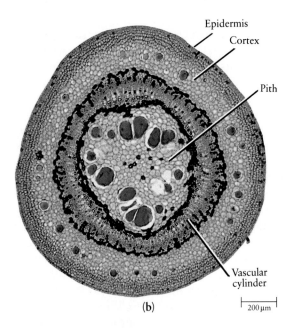
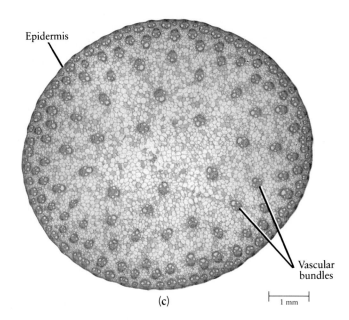
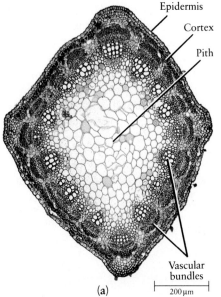

30–20 *Cross sections of two dicot stems and one monocot stem.* **(a)** *In alfalfa, a dicot, the vascular cylinder is made up of separate vascular bundles.* **(b)** *In this young stem of the linden, also a dicot, the vascular tissue forms a continuous cylinder. This stem contains mucilage ducts, which stain red.* **(c)** *In corn, a monocot, numerous vascular bundles are scattered throughout the ground tissue.*

Stem Patterns

In green stems, the xylem and the phloem are usually arranged in longitudinal parallel strands, the vascular bundles, which are embedded in the ground tissue. In young dicot stems, the vascular bundles form a ring, the vascular cylinder, around a central pith (Figure 30–20a and b). The cylinder of ground tissue outside the vascular bundles is the cortex. Within each bundle, the xylem is characteristically on the inside, adjacent to the pith, and the phloem is on the outside, adjacent to the cortex. In monocots, the vascular bundles are usually scattered throughout the ground tissue (Figure 30–20c).

The vascular tissues of the stem are continuous with those of the root, and yet, as we have seen, their arrangement in the stem is quite different from that in the root. The change from the patterns observed in the root to those found in the stem is a gradual one. The region of the plant axis between root and stem in which the change occurs is known as the transition region.

PRIMARY GROWTH OF THE SHOOT SYSTEM

The shoot system includes the stem and all of the structures that develop from it—typically, all of the aboveground parts of the plant. The pattern of growth of the developing shoot tip is similar to the pattern we saw earlier in the root: first, cell division takes place; next, cell elongation; and finally, differentiation. However, because of the regular occurrence of nodes and their appendages—leaves and buds—the growth zones are not as distinct in the shoot as they are in the root. Also, no covering analogous to the root cap is produced over the shoot tip.

As in the root, the outermost layer of cells develops into the epidermis. In the shoot, these cells are covered by a relatively conspicuous cuticle. Underlying cells differentiate to form the ground tissues and the primary vascular tissues—the primary xylem and the primary phloem. The pattern of development is more complicated, however, than in the root tip since the apical meristem of the shoot is the source of tissues that give rise to new leaves, branches, and flowers.

Figure 30–21 shows the shoot tip of the familiar house plant *Coleus*. In the center is the apical meristem, which is very small, with the beginnings—primordia

—of two leaves. Flanking the apical meristem are the two previously formed leaves, which are completing their growth and development. The leaves originate by cell division in localized areas along the side of the apical meristem. As growth progresses, the vascular tissue of the stem differentiates upward into the leaf primordia, becoming part of the general vascular system that connects the plant from root to leaf tip. Leaves are formed in an orderly sequence at the shoot tip. In some species, they arise simultaneously in pairs opposite one another, as in *Coleus*. In other species, they form spirally or in circles (whorls) at the nodes. As the internodes elongate, the young leaves become separated and thus spaced out along the stem of the plant.

As the shoot tip elongates, small masses of meristematic tissue are left just above the points at which the leaves are attached to the stem (the leaf axils). These new meristematic regions, the axillary buds, remain dormant until after growth of the adjacent leaf and internode is complete, or, in perennials, until the next growing season. In many species, including *Coleus*, development of the axillary buds is suppressed by the influence of the terminal bud. (This phenomenon, known as apical dominance, will be discussed further on page 675.) In some species, specific buds are destined to become lateral branches or specialized shoots, such as tubers or flowers. In other species, the fate of the buds is determined by environmental conditions, particularly day length, as we shall see in Chapter 32. When a flower forms—from either an axillary bud or a terminal bud—the meristem gives rise to the floral parts and ceases to exist as a source of new cells.

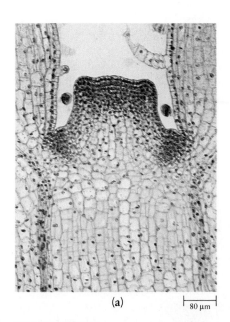
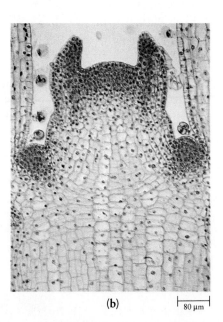
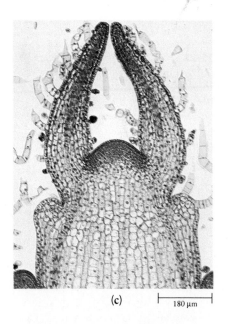

30-21 *Stages in leaf development at the shoot apex of* Coleus, *as seen in longitudinal section.* Coleus *leaves develop in pairs, opposite one another, with each successive pair at right angles to the preceding pair. In these micrographs, the nuclei are stained purple; thus, the meristematic regions, densely packed with small, rapidly dividing cells, appear purple.* **(a)** *Two small bulges, or leaf buttresses, appear on opposite sides of the stump-shaped apical meristem. In addition, buds are developing in the axils of the two previously formed leaves.* **(b)** *Two erect, peg-like leaf primordia have developed from the leaf buttresses. Strands of vascular tissue are extending upward into the leaf primordia.* **(c)** *As the leaf primordia elongate, the vascular tissues continue their upward differentiation. The epidermal hairs, visible on the outer surface of the developing leaves, originate from certain cells of the protoderm very early, long before the protoderm matures to become the epidermis.*

Modifications in the Pattern of Shoot Growth

In most plants, primary growth of the shoot proceeds from its tip, the apical meristem, and from the axillary buds. Grass plants, however, also have meristems at the bases of the leaves. The grass leaf consists of a blade (the broad part) and a sheath that fits around the plant stem (Figure 30–22). When the blade is cut off, cells in the sheath, in response to some unknown signal, are activated, producing more blade.

The shoots of some climbing plants coil themselves around the structures on which they are growing, a process that we shall examine in Chapter 32. Others produce modified branches in the form of tendrils. (As we saw previously, tendrils may also be modified leaves or leaflets.) The tendrils of grape, English ivy, and Virginia creeper are all modified branches.

Vegetative Reproduction

Among the specialized shoots that originate from the axillary buds of many species are runners and rhizomes. Runners, such as those found in most varieties of strawberry, are long, slender stems that grow along the surface of the soil. Rhizomes are also horizontal stems, growing either along or below the surface of the soil. Both runners and rhizomes develop adventitious roots and are the source of new plants, genetically identical to the parent plant. Strawberries (Figure 30–23) are a familiar example of plants that propagate vegetatively by runners, as are spider plants, commonly grown as hanging plants. Plants that reproduce by rhizomes include potatoes, many flowering garden perennials, such as irises and lilies-of-the-valley, the sod-forming grasses of lawns and pastures, and bamboos and reed grasses (Figure 30–24). In grasses, for example, rhizomes form buds that produce upright stems bearing leaves and flowers.

The asexual production of genetically identical clones from runners and rhizomes is a very efficient way for a plant to spread quickly and invade new territory. The young plants that develop in this way have a continuous source of nourishment from the parent plant and a lower mortality than seedlings. The production of clones can also provide a mechanism for reproducing in environments that are only periodically hospitable to the germination of seeds. Creosote bushes, common in the North American deserts, produce clones that form a circle or an ellipse around a bare area where the parent plant first took root (Figure

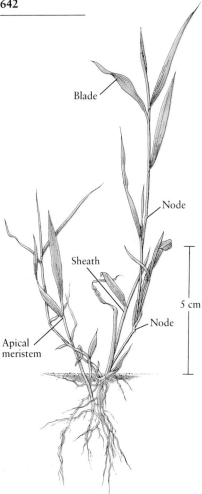

30–22 *The structure of a grass plant. The capacity of a grass leaf to grow from the sheath at its base is a useful adaptation to herbivore grazing. It has also resulted, over the eons, in the piteous spectacle of thousands of hominids vibrating behind their lawnmowers from spring to fall.*

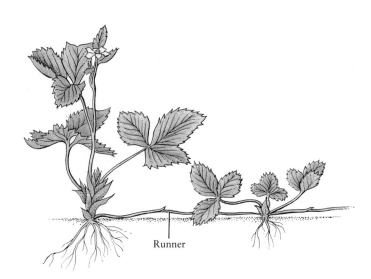

30–23 *Strawberry plants reproduce asexually by means of runners, thin horizontal stems that grow along the surface of the soil. Roots and leaves develop at every second node along the modified stems. These plants also form flowers and reproduce sexually.*

30–24 *Common reed grass* (Phragmites communis) *propagates asexually by means of rhizomes. Each stand of reed grass is, in effect, a single individual. The tassels at the tops of the stalks are flowers; the seeds produced through sexual reproduction give rise to new stands of reed grass, genetically different from the parent plants.*

30–25). At ages approaching 12,000 years, some creosote clones are the oldest living plants known; these extraordinary plants are thought to have started from seeds that germinated near the end of the last glacial expansion.

The capacity of plants to reproduce vegetatively has been exploited in the development of cultivated varieties of plants for food or ornamental use. Because such plants are genetically identical to the parent stock, vegetative reproduction is a way of preserving uniformity. Many plants are reproduced by stem cuttings, which involves placing young stems in soil or some other planting medium and protecting them from drying out until adventitious roots appear. Rooting can often be facilitated by treating the cuttings with hormones (see page 674). Many stem cuttings will also form roots when placed in water. Another artificial form of plant propagation is grafting, in which a stem cutting is inserted in a slit in the stem or trunk of a rooted woody plant. The wound is sealed with tape, and if the stem cutting and the rooted plant have not been damaged too much by the surgery, the grafted shoot will "take" and begin growing. Most fruit trees and roses are propagated in this way.

Many economically important plants are sterile and can only be propagated vegetatively; these include pineapples, bananas, seedless grapes, navel oranges, sugar cane, and numerous ornamental plants.

30–25 *Creosote bushes of the King Clone in the Mojave desert in California. The bushes form an elliptical ring that is about 23 meters by 8 meters. Standing just within the ring is Andrew Sanders of the University of California, Riverside.*

When a creosote seed germinates, it gives rise to a small stem, known as the crown. As the crown grows thicker, it produces many new branches at its base. Over the first 40 to 90 years of its life, the crown splits into lobes, each of which forms a new crown and produces more branches. Eventually the original crown dies, leaving a central area of bare sand surrounded by the genetically identical bushes of the clone. Although a single crown has a life expectancy of a mere 80 to 100 years, the clone derived from the original seedling may live far longer. The parent crown of the King Clone has been located and dated using radioactive isotope methods. It is roughly 11,700 years old, indicating that the King Clone is, by far, the oldest living plant known. Recently, the Nature Conservancy purchased 17 acres of land surrounding the King Clone, ensuring its protection.

SECONDARY GROWTH

As you know from your own observations, many plants not only grow taller with age (or longer, as they expand into new territory by vegetative reproduction) but also become thicker. The process by which woody dicots increase the thickness of their trunks, stems, branches, and roots is known as secondary growth (Figure 30–26). The so-called "secondary tissues" are not derived from the apical meristems; instead, they are produced by **lateral meristems** known as the vascular cambium and the cork cambium.

The **vascular cambium** is a thin, cylindrical sheath of tissue between the xylem and the phloem. In stems, its cells are derived from the embryonic procambium and from parenchyma cells in the pith rays; in roots, the cells are derived from the procambium and the pericycle. In plants with secondary growth, the cambial cells divide continually during the growing season, adding secondary xylem toward the inside of the cambium and secondary phloem toward the outside. Some meristematic cells remain as a cylinder of undifferentiated cambium, in which cell division will resume at the start of the next growing season.

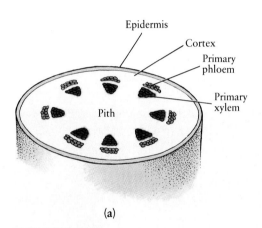

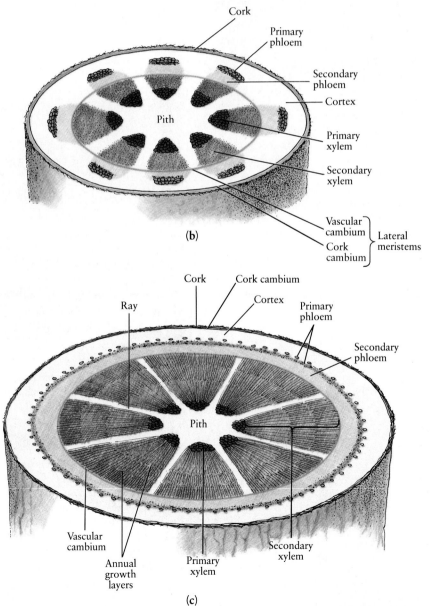

30–26 (a) *Stem of a dicot before the onset of secondary growth.*

(b) *Beginnings of secondary growth. Secondary xylem and secondary phloem are produced by the vascular cambium, a meristematic tissue formed late in primary growth. As the trunk increases in diameter, the epidermis is stretched and torn. This is accompanied by the formation of the cork cambium, from which cork is formed, replacing the epidermis.*

(c) *Cross section of a three-year-old stem, showing annual growth rings. Rays are rows of living cells that transport nutrients and water laterally (across the trunk). On the perimeter of the outermost growth layer of secondary xylem is the vascular cambium, encircled by a band of secondary phloem. The primary phloem and the cortex will eventually be sloughed off. In an older stem, the thin cylinder of active secondary phloem is immediately adjacent to the vascular cambium. The tissues outside the vascular cambium, including the phloem, constitute the bark.*

CHAPTER 30 The Plant Body and Its Development 645

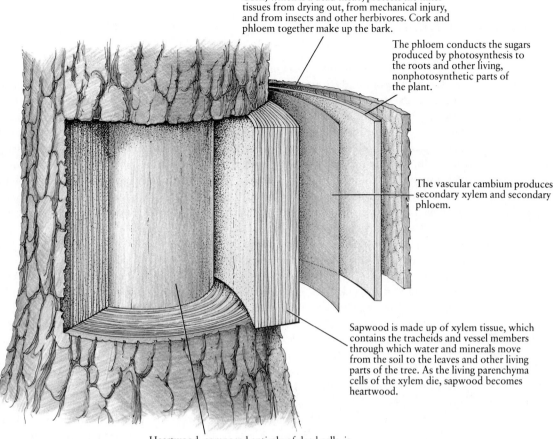

30-27 *A tree trunk showing the relationships of the successive concentric layers. The heartwood is composed entirely of dead cells. The cortex and epidermis, which are outside the phloem in a green stem, are sloughed off during secondary growth.*

Cork, which is a dead tissue, protects the inner tissues from drying out, from mechanical injury, and from insects and other herbivores. Cork and phloem together make up the bark.

The phloem conducts the sugars produced by photosynthesis to the roots and other living, nonphotosynthetic parts of the plant.

The vascular cambium produces secondary xylem and secondary phloem.

Sapwood is made up of xylem tissue, which contains the tracheids and vessel members through which water and minerals move from the soil to the leaves and other living parts of the tree. As the living parenchyma cells of the xylem die, sapwood becomes heartwood.

Heartwood, composed entirely of dead cells, is the central supporting column of the mature tree.

As secondary growth increases the girth of stems and roots, the epidermis becomes stretched and torn. In conjunction with this tearing process, a new type of cambium, the **cork cambium,** forms from the cortex. The cork cambium produces cork (phellem), which replaces the epidermis as the protective covering of woody stems and roots. Unlike the epidermis, cork is a dead tissue at maturity. The cork cambium often forms anew each year, moving further inward until, finally, there is no cortex left. The tissues outside the vascular cambium—a thin layer of phloem, the cork cambium, and the cork—constitute the bark.

As the plant grows older, the parenchyma cells of the xylem in the center of the stem and root die, and their neighboring vessels become clogged and cease to function. This nonconducting xylem, called heartwood, forms the center of the trunk and major roots of a tree, providing the support and anchorage required as the height of the tree continues to increase through primary growth. The living parenchyma cells and open vessels just inside the vascular cambium constitute the sapwood, through which water and minerals flow from the root tips to the leaves.

Season after season the new xylem forms visible growth layers, or rings. Each growing season leaves its trace, so that the age of a tree can be estimated by counting the number of growth rings in a section near its base. Since the growth rate of a tree depends on climatic conditions, the width of the annual growth layers can be used to estimate fluctuations in temperature and rainfall that occurred many years ago (see essay).

The Record in the Rings

As we have seen, each growing season in the life of a tree leaves its record in the ring of secondary xylem formed in the trunk. The growth rings are visible because of differences in the density of wood produced early in the growing season and that produced late in the season. The early wood has large cells with thin walls, whereas the late wood has smaller cells with proportionately thicker walls. Within a given growth layer, the change from early wood to late wood may be quite gradual, but a distinct change is visible where the small, thick-walled cells of the late wood of one growing season abut the larger, thin-walled cells of the early wood of the next growing season.

The width of the individual growth layers may vary greatly from year to year, depending on such environmental factors as light, temperature, rainfall, available soil water, and the length of the growing season. Under favorable conditions, the growth rings are wide; under unfavorable conditions, they are narrow. In semiarid regions, where there is very little rain, trees are sensitive rain gauges. An excellent example is the bristlecone pine *(Pinus longaeva)* of the western Great Basin. Each growth ring is different, and a study of the rings tells a story that dates back thousands of years.

The oldest known living specimen of bristlecone pine is 4,900 years old. Dendrochronologists—scientists who conduct historical research using the growth rings of trees—have been able to match samples of wood from living and dead trees. In this way, they have built a library containing a continuous series of rings dating back more than 8,200 years. This record of average ring width has provided a valuable guide to past conditions of precipitation and temperature. The widths of the growth rings of bristlecone pines at higher elevations (the upper tree line) have been found to be closely related to temperature changes. The rings of these trees have revealed that the summers in the White Mountains of California were relatively warm from 3500 B.C. to 1300 B.C., and the tree line was about 150 meters higher than it is at present. Summers were cool from 1300 B.C. to 200 B.C.

Information provided by the growth rings of both gymnosperms and angiosperms is being used not only to reconstruct past climatic conditions but also to help predict future conditions. With more accurate knowledge of past climates than is provided by human records—even those of the past few centuries—it is possible to determine cyclic patterns of temperature change and of rainfall and drought. Such information is of considerable importance, for example, in planning the wise management and allocation of our finite resources of fresh water.

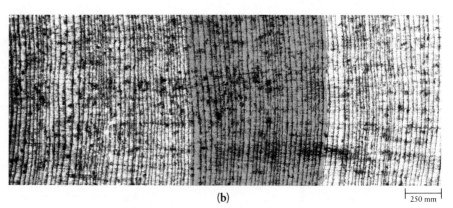

(a) *A bristlecone pine. These pines, which grow near the timberline, are the oldest living trees (but not, as we saw in Figure 30–25, the oldest living plants).* (b) *A portion of a cross section of wood from a bristlecone pine, showing the variation in width of the annual rings. This section begins approximately 6,260 years ago; the band of rings in color represents the years from 4240 B.C. to 4210 B.C.*

By the continuous formation of secondary xylem and (to a lesser extent) phloem, woody dicots increase their diameter as primary growth increases their height. Moreover, newly formed tracheids, vessel members, and sieve-tube members provide fresh conduits—undamaged by the activities of the numerous parasites and herbivores that attack plants—for the transport of water and nutrients from one part of the plant body to another. In the next chapter, we shall examine the mechanisms by which materials move through these cells.

SUMMARY

Plants are multicellular photosynthetic organisms adapted to life on land. The plant body has specialized photosynthetic areas (leaves), conducting and supporting structures (stems), and organs that anchor the plant in the soil and absorb water and minerals from it (roots).

When a seed germinates, growth of the root and shoot proceeds from the apical meristems of the embryo. Certain cells within the meristem retain the capacity to divide. Others elongate and then differentiate, forming, according to their position, the various specialized cells of the plant (see Table 30-1). These

TABLE 30-1 **Summary of Main Cell Types in Angiosperms**

CELL TYPE	ORIGIN	LOCATION	CHARACTERISTICS	FUNCTION
Meristematic (in apical meristem)	Embryonic cells	Apices of shoots and roots	Many-sided, small, thin-walled cells; vacuoles usually small	Origin of primary meristematic tissues and of root cap cells
Meristematic (in vascular cambium)	Procambium and parenchyma of pith rays in stems; procambium and pericycle in roots	Lateral, between secondary phloem and xylem tissues	Elongate, often spindle-shaped	Produces secondary xylem and phloem
Epidermal	Protoderm	Surface of entire primary plant body	Flattened, variable in shape, overlaid by cuticle; some specialized as guard cells	Protective covering; prevents desiccation yet allows gas exchange
Parenchyma	Protoderm, ground meristem, procambium, vascular cambium, cork cambium, wound tissues	Everywhere, usually dominant in pith, cortex, mesophyll	Many-sided, usually thin-walled; abundant air spaces between cells	Photosynthesis, respiration, storage, wound healing, among others
Collenchyma	Ground meristem of leaf and stem	Peripheral in cortex of stem and in leaves	Elongate, with irregularly thickened primary walls	Support for young stems and leaves
Sclereid	Protoderm, ground meristem, procambium, vascular cambium, and cork cambium	In pith and cortex of stems; in leaves and flesh of fruits; seed coats	Irregular; massive secondary wall; alive or dead at maturity	Produces hard texture, mechanical support
Fiber	Procambium or vascular cambium; ground meristem	Primary and secondary xylem and phloem; cortex	Very long, narrow cell, with secondary cell wall; usually dead at maturity	Support
Tracheid	Procambium or vascular cambium	Primary or secondary xylem	Elongate, with pits in walls; dead at maturity	Conduction of water and solutes
Vessel member	Procambium or vascular cambium	Interconnected series (= vessels) in primary or secondary xylem	Elongate, with pits in walls and end walls perforated; dead at maturity	Conduction of water and solutes
Sieve-tube member	Procambium or vascular cambium	Primary or secondary phloem, usually with companion cells; form interconnected series (= sieve tubes)	Elongate, with specialized sieve plates; nucleus lacking at maturity	Conduction of organic solutes
Cork (phellem)	Cork cambium	Surface of stems and roots with secondary growth	Flattened cells, compactly arranged; dead at maturity, the cells often air-filled	Restricts gas exchange and water loss

cells, in varying combinations, form the three tissue systems—dermal, ground, and vascular—that are continuous throughout the plant body. Parenchyma cells, which are thin-walled and many-sided, are the most common cells in plants.

The ground tissue of the leaf is composed primarily of photosynthetic parenchyma cells. Palisade cells, in which most of the photosynthesis takes place, are elongated cells with large central vacuoles. Spongy parenchyma cells, also photosynthetic, are surrounded by large air spaces. The palisade and spongy parenchyma cells make up the mesophyll, or "middle leaf."

The upper and lower surfaces of the leaf consist of one or more layers of transparent epidermal cells covered with a waxy layer, the cuticle. Specialized pores, the stomata, open and close, regulating the exchange of gases and the release of water vapor. Veins, the vascular bundles of the leaf, conduct water and minerals to the mesophyll cells of the leaf (through the xylem) and transport sugars away from them (through the phloem). The vascular tissues of leaves are continuous with those of the stem and roots.

The embryonic root is the first structure to break out of the germinating seed. Cells produced by its apical meristem form a root cap, which protects the root tip as it is pushed through the soil. Young roots have an outer layer of epidermis and, at most, a very thin cuticle. Extensions of the epidermal cells form root hairs, which greatly increase the absorptive surface of the root. Inside the epidermis is the ground tissue of the root, the cortex, composed mostly of parenchyma cells, often modified for storage. The innermost layer of the cortex is the endodermis, a single layer of specialized cells whose walls contain a waterproof zone, the Casparian strip. Just inside the endodermis is another layer of cells, the pericycle, from which branch roots arise. Within the pericycle are the xylem and phloem.

Green stems, like leaves, have an outer layer of epidermal cells covered with a cuticle. The bulk of the young stem is ground tissue, which may be divided into an outer cylinder (the cortex) and an inner core (the pith). The ground tissue is composed largely of parenchyma cells but also may contain collenchyma cells and sclerenchyma cells (fibers and sclereids).

The vascular tissues consist of phloem and xylem. In angiosperms, the conducting cells of the phloem are the sieve-tube members, living cells with perforated end walls that form continuous sieve tubes. Associated closely with each sieve-tube member is a companion cell. The conducting tissue of the xylem is made up of a series of tracheids or vessel members. Tracheids and vessel members characteristically have thick secondary walls and are dead at functional maturity. Phloem and xylem also contain parenchyma cells and fibers.

The height (or length) of the aboveground parts of a plant is increased through primary growth of the shoot system. Leaf primordia arise from the apical meristem of the shoot. As the nodes are separated by elongation of the internodes, small masses of meristem (buds) form in the axils of the leaves. These axillary buds may remain dormant or may give rise to branches or specialized shoots. Among the specialized shoots that may form are tubers, runners, and rhizomes, through which many species reproduce asexually, producing populations of genetically identical individuals.

Secondary growth is the process by which woody plants increase their girth. Such growth arises primarily from the vascular cambium, a sheath of meristematic tissue completely surrounding the xylem and completely surrounded by phloem. The cambial cells divide during the growing season, adding new xylem cells (secondary xylem) on their inner surfaces and new phloem cells (secondary phloem) on their outer surfaces. As the trunk increases in girth, the epidermis is eventually ruptured and replaced by cork.

CHAPTER 30 The Plant Body and Its Development

QUESTIONS

1. Distinguish among the following: dermal tissue/ground tissue/vascular tissue; epidermis/cuticle; parenchyma/collenchyma/sclerenchyma; palisade parenchyma/spongy parenchyma; sieve-tube member/tracheid/vessel member; apical meristem/lateral meristem; primary growth/secondary growth.

2. Describe the distinguishing characteristics of monocots and dicots.

3. Sketch the structure of a leaf, labeling the principal cells and tissues. Describe the function of each of the labeled parts.

4. Where might you expect to find a leaf with stomata only on the bottom? Only on the top?

5. Reexamine the oleander leaf in Figure 30–5, relating its various structural features to the ecology of the oleander.

6. Some leaves are specialized for functions other than photosynthesis. List three other functions for which leaves may be specialized, and give an example of each specialization.

7. We have seen that the tissue of the root cortex contains many air spaces. Also, we noted that some plants growing in swampy areas have aerial roots. You may have discovered that overwatering can kill house plants. What do these data indicate about roots?

8. Why are there no root hairs or lateral roots at the extreme tip of a root?

9. Sketch cross sections of (a) a root, (b) a dicot stem, and (c) a monocot stem. Label the principal cells and tissues in each sketch, and describe the function of each of the labeled parts.

10. What two cell layers are present in roots but absent in stems? What functions do the cells of these layers perform in the roots, and why are they necessary for the survival of the plant?

11. It has often been said that the principal differences among root, stem, and leaf are quantitative in nature rather than qualitative. Explain.

12. Suppose you carve your initials 1 meter above the ground on the trunk of a mature tree that is growing vertically at an average of 15 centimeters per year. How high will your initials be at the end of 2 years? At the end of 20 years?

13. As a result of secondary growth, most of our common trees increase in girth as they increase in height. What limitations would a lack of secondary growth impose on the form and mechanical support of a tree?

14. Sketch a tree trunk with secondary growth. Compare your sketch with Figure 30–26c.

15. Porcupines often eat all of the bark off young trees, at the height they can reach. When they do this, the tree dies. Why?

16. Consider a mature tree. Which parts of the tree are composed of living cells? Why might it be advantageous for living cells to make up such a small fraction of the mass of the tree? Similarly, why might it be advantageous for a deciduous tree to invest as little material as possible in making leaves?

CHAPTER **31**

Transport Processes in Plants

As we noted in Chapter 24, the plants that have come to dominate the modern landscape are those that have evolved the best solutions (so far) to two critical problems: conveying the sperm to the egg in the absence of free water, and transporting needed materials to the various cells and tissues of the plant body. In this chapter, we shall discuss first the movement of water and minerals in the xylem; second, the transport of dissolved sugars and other organic substances in the phloem; and third, a related subject, factors affecting the availability of nutrients to the plant. You may find it helpful to review the properties of water (especially pages 42 to 44) and the ways in which it moves (pages 128 to 134) before proceeding. The transport processes of plants depend on the extraordinary properties of this very common liquid.

THE MOVEMENT OF WATER AND MINERALS

Transpiration

A plant needs far more water than an animal of comparable weight. In an animal, most of its water remains in its body and recirculates; by contrast, more than 90 percent of the water entering a plant's roots is released into the air as water vapor. This loss of water vapor from the plant body is known as **transpiration.** It is a necessary consequence of the opening of the stomata that must occur for a plant to obtain the carbon dioxide required for photosynthesis (Figure 31–1), but the price is high. For every gram of carbon dioxide incorporated into organic matter, a plant utilizing the C_3 pathway of photosynthesis loses 400 to 500 grams of water, and a plant utilizing the C_4 pathway (pages 223 to 228) loses 250 to 300 grams of water.

Transpiration creates an enormous "thirst" in a plant. Thus, for example, a single corn plant requires 160 to 200 liters of water to grow from seed to harvest, and 1 hectare (2.47 acres) of corn requires almost 5 million liters of water a season. British ecologist John L. Harper describes the terrestrial plant as a "wick connecting the water reservoir of the soil with the atmosphere."

The Uptake of Water

As we saw in the last chapter, water enters most plants almost entirely through the roots, principally through the root hairs. During periods of rapid transpiration, water may be removed from around the roots so quickly that the soil water in the vicinity of the roots becomes depleted. Water then moves slowly by diffusion and capillary action through the soil toward the depleted region near the roots. Roots

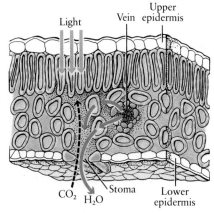

31–1 *As carbon dioxide, essential for photosynthesis, enters the leaf through the stomata, water vapor is lost. Although this water loss poses serious problems for plants, it provides both the motive force for the uptake of water by the roots and a mechanism for cooling the leaf. The temperature of a leaf may be as much as 10 to 15°C lower than that of the surrounding air, simply because evaporating water carries heat away (see page 44).*

31-2 *Guttation droplets on the edge of wild strawberry leaves. Guttation, the loss of liquid water, is a result of root pressure. The water escapes through specialized pores located near the ends of the principal veins of the leaves. Guttation, which is restricted to relatively small plants, usually occurs at night when the air is moist and transpiration is greatly reduced.*

also obtain additional water by growing beyond the depleted region; the main roots of corn plants, for example, grow an average of 52 to 63 millimeters a day.

Root cells, like other living parts of the plant, contain a higher concentration of solutes (both organic and inorganic) than does soil water. As a consequence, water from the soil enters the roots by osmosis. The resulting pressure, known as root pressure, is sufficient to move water a short distance up the stem. Guttation, the loss of liquid water through the leaves (Figure 31-2), is a visible consequence of root pressure. But how can water reach 20 meters high to the top of an oak tree, travel three stories up the stem of a vine, or move 125 meters up in a tall redwood?

One important clue is the observation that during times when the most rapid transpiration is taking place—which is when the flow of water up the stem must be the greatest—xylem pressures are negative (less than atmospheric pressure). This negative pressure can be demonstrated. If you peel a piece of bark from a transpiring tree and make a cut in the xylem, no liquid runs out. In fact, if you place a drop of water on the cut, the drop will be drawn in.

What is the pulling force? It is not simple suction, as the negative pressure might indicate. Suction simply removes air from a system so that the water (or other liquid) is pushed up by atmospheric pressure. But atmospheric pressure is only enough to raise water (against no resistance) approximately 10 meters at sea level, and many trees are much taller than 10 meters.

The Cohesion-Tension Theory

According to the now generally accepted theory, the explanation for the movement of water is to be found not only in the properties of plants but also in the properties of water, to which plants have become exquisitely adapted. As we pointed out in Chapter 2, in every water molecule, two hydrogen atoms are covalently bonded to a single oxygen atom. Each hydrogen atom is also held to the oxygen atom of a neighboring water molecule by a hydrogen bond. The cohesion resulting from this secondary attraction is so great that the tensile strength in a thin column of water can be as much as 140 kilograms per square centimeter (2,000 pounds per square inch). This means that a negative pressure of more than 140 kilograms per square centimeter is required to pull the column of water apart.

In a leaf, water evaporates, molecule by molecule, from the walls of the parenchyma cells into the air spaces of the leaf. As the water potential of a leaf cell decreases, water from the vessels and tracheids moves, molecule by molecule, into the cell. But each water molecule in a tracheid or vessel is linked to other water molecules in the tracheid or vessel. They, in turn, are linked to others, forming a long, narrow, continuous stream of water reaching down all the way to a root hair and, in fact, into the surrounding soil solution. As a molecule of water moves through the stem and into the leaf, it tugs the next molecule along behind it. In plants with large vessels, this process moves water at a rate of 30 to 40 meters per hour; in plants with smaller vessels, the rate is about 5 to 10 meters per hour.

Because the diameter of the vessels is relatively small and because water molecules adhere to the cell walls of the vessel members (even as they are cohering to one another), gas bubbles, which could rupture the water column, do not usually form. The pulling action, molecule by molecule, causes the negative pressure observed in the xylem. The technical term for a negative pressure is tension, and this theory of water movement is known as the **cohesion-tension theory**.

The principle of cohesion-tension is illustrated in Figure 31-4. As indicated in the diagram, the power for this process comes not from the plant, which plays only a passive role in transpiration, but from the energy of the sun.

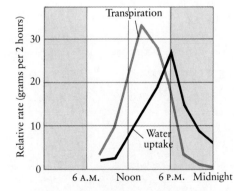

31-3 *Measurements in ash trees show that a rise in water uptake follows a rise in transpiration. These data suggest that the loss of water generates forces for its uptake.*

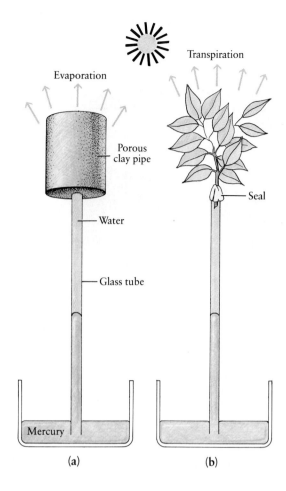

31–4 (a) *A simple model that illustrates the cohesion-tension theory. A piece of porous clay pipe, closed at both ends, is filled with water and attached to the end of a long, narrow glass tube also filled with water. The water-filled tube is placed with its lower end below the surface of a volume of mercury contained in a beaker. As water molecules evaporate from the pores in the pipe, they are replaced by water "pulled up" through the narrow glass tube in a continuous column. As the water evaporates, the mercury rises in the tube to replace it. (b) Transpiration from plant leaves results in sufficient water loss to create a similar negative pressure.*

Factors Influencing Transpiration

Transpiration is, as we have seen, costly, especially when water supply is limited. Several factors affect the rate of water loss. One is temperature; the rate of evaporation doubles for every 10°C increase in temperature. Humidity is also important. Water is lost much more slowly into air already laden with water vapor. Air currents also affect transpiration. Wind blows away the water vapor from leaf surfaces, which makes the concentration gradient of water steeper and so hastens the evaporation of water molecules from the leaf. Leaves of plants that grow in exposed, windy areas are often hairy; these hairs are believed to protect the leaf surface from the wind and so retard transpiration (Figure 31–5). By far the most important factor affecting transpiration, however, is the regulatory effect exercised by the opening and closing of the stomata.

31–5 *Leaf hairs create a layer of still air over the epidermis, restricting gas exchange and water loss from the stomata (visible in the lower portion of the micrograph).*

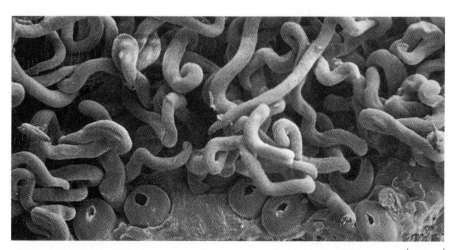

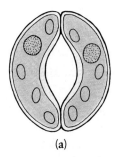 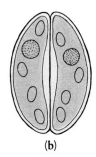 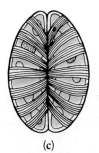 

(a)   (b)   (c)   (d)

31–6 *Mechanism of stomatal movement. A stoma is bordered by two guard cells that (a) open the stoma when they are turgid and (b) close it when they lose turgor. In many species, the inner walls of the guard cells are slightly thickened. For many years it was thought that as turgor increased, the thinner parts of the cell wall were stretched more than the thicker parts, causing the cells to bow out and the stoma to open. It is now known that the key to stomatal opening lies instead in cellulose microfibrils that are arranged in hoops around the circumference of the guard cells (c). The arrangement is similar to that of the belts in radial tires. When water enters the guard cells, the microfibrils prevent radial expansion, and the only direction in which the cells can expand is lengthwise. Because the two cells are attached to each other at the ends, this lengthwise expansion forces them to bow out and the stoma to open (d).*

The Mechanism of Stomatal Movements

It has long been known that the osmotic movement of water is involved in the opening and closing of the stomata. As shown in Figure 31–6, each stoma has two surrounding guard cells. Stomatal movements are caused by changes in the turgor of these cells. When the guard cells are turgid, they bow out, opening the stoma; when they lose water, they relax, and the stoma closes.

Turgor, as we saw on page 134, is maintained or lost due to the osmotic movement of water into or out of cells. As you will recall, water moves across the cell membrane from a solution of low solute concentration into a solution of high solute concentration. The active accumulation of solutes in the guard cells causes water to move into them by osmosis; conversely, a decrease in the solute concentration of guard cells results in the osmotic movement of water out of the cells.

Techniques that make it possible to measure ionic concentrations within individual guard cells have revealed that the critical solute affecting the osmotic movement of water into and out of these cells is the potassium ion (K^+). With an increase in the K^+ concentration, the stomata open, and with a decrease, they close. Current evidence indicates that potassium ions are actively transported between the guard cells and the reservoir provided by the surrounding epidermal cells (Figure 31–7). Water follows by osmosis, causing the observed changes in the turgor and shape of the guard cells.

In many, but not all, species, chloride ions, Cl^-, accompany the K^+ ions across the membrane, thus maintaining electrical neutrality. In other, perhaps all, species, H^+ ions are transported in the opposite direction, producing a decreased hydrogen ion concentration within the guard cells of open stomata.

The active transport of K^+ ions between the guard cells and the surrounding epidermal cells is, of course, an energy-requiring process. The energy source is not yet known, but, on the basis of present evidence, one of two possibilities seems likely. As you will recall, guard cells contain chloroplasts, and it may be that K^+ transport is powered by ATP produced by photophosphorylation reactions (page 218) occurring in these chloroplasts. Another possibility is that the transport of H^+ ions, in a direction opposite to that of the K^+ ions, establishes an electrochemical gradient down which the K^+ ions move. If this should be the case, the transport of K^+ ions into and out of the guard cells—and thus, the opening and closing of the stomata—would be yet another example of a vital process powered by a chemiosmotic mechanism (page 197).

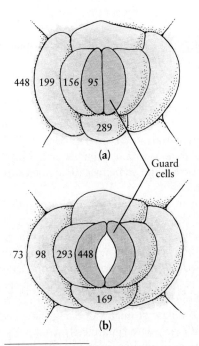

31–7 *Quantitative changes in potassium ion (K^+) concentrations in the guard cells and neighboring epidermal cells of (a) a closed stoma and (b) an open stoma of the dayflower (*Commelina communis*) leaf. The measurements were made with K^+-sensitive microelectrodes that were inserted into the individual cells.*

31-8 *Corn plants, photographed in a Wisconsin field in June of 1988, during a severe drought that affected many of the prime agricultural areas of North America. The leaves of corn plants have numerous stomata on both the upper and lower surfaces, with slightly more on the lower surface. The cuticle on the lower surface is significantly heavier than on the upper surface. When corn plants are under severe water stress, and closure of the stomata is inadequate to protect the plants against further water loss, the leaves roll up, as shown here. Leaf rolling reduces transpiration, protecting virtually all of the stomata on the upper surface, as well as many on the lower surface. All that remains exposed to the environment is a portion of the lower surface, with its heavy cuticle.*

Factors Influencing Stomatal Movements

A number of environmental factors affect—either directly or indirectly—the movement of potassium ions into and out of the guard cells and thus influence the opening and closing of the stomata. The principal one is the availability of water. When the water available to a leaf drops below a certain critical point (which varies from species to species), the stomata close, thereby limiting evaporation of the remaining water. This is not a direct effect; stomatal closure occurs before the leaf loses turgor and wilts.

A plant's capacity to anticipate water stress depends on the action of the hormone abscisic acid, about which we shall have more to say in the next chapter. Several lines of evidence link abscisic acid with stomatal movements. First, during the initial stage of water stress, levels of the hormone increase markedly in many species of plants. Second, application of abscisic acid to a leaf causes stomatal closure within a few minutes. Third, a strain of tomato plant, known as the "wilty mutant," produces very little abscisic acid and fails to close its stomata as the water supply decreases; application of abscisic acid to a "wilty mutant" tomato plant enables it to regulate its stomata normally. Abscisic acid acts by binding to specific receptors in the guard cell membrane; the receptor-hormone complex then triggers a change in the cell membrane, making it more permeable to potassium ions.

Stomatal movement also occurs quite independently of the loss or gain of water by the plant. Among the other factors that affect stomatal movement are the carbon dioxide concentration, temperature, and light. In most species, an increase in carbon dioxide concentration in the intercellular spaces of a leaf causes the stomata to close. The magnitude of this response varies greatly from species to species and with the degree of water stress that a particular plant has undergone. In corn, the stomata may respond to changes in the internal carbon dioxide concentration within seconds. The site for sensing the level of carbon dioxide is inside the guard cells.

Within normal ranges, temperature changes have little effect on the stomata, but temperatures higher than 35°C (95°F) can cause stomatal closure in some species. This temperature-induced closure can be prevented, however, by maintaining the plant in air that contains no carbon dioxide. This suggests that temperature changes affect stomatal movements primarily by altering the carbon dioxide concentration within the leaf. An increase in temperature results in an increase in respiration and thus in the concentration of intercellular carbon dioxide. Many plants in hot climates close their stomata regularly at midday, apparently because of both increased water stress and the effect of temperature on the concentration of carbon dioxide in the leaves.

In most species, the stomata close regularly in the evening, when photosynthesis is no longer possible, and open again in the morning. These movements in response to light occur even when there are no changes in the water available to the plant. Opening in the light and closing in the dark may be a result of changes in the carbon dioxide concentration—photosynthesis uses carbon dioxide and thus reduces the internal concentration, whereas respiration causes an increase in the carbon dioxide concentration. Light may, however, have another, more direct effect. Blue light has long been known to stimulate stomatal opening independently of the carbon dioxide concentration. When isolated guard cells from onion are illuminated with blue light in the presence of potassium ions, they swell. The blue-absorbing pigment (which is thought to be located in either the cell membrane or the vacuole membrane) promotes the uptake of K^+ ions by the guard cells. Recent experiments have demonstrated that this uptake of K^+ ions is a secondary effect, and that the primary effect of the interaction of blue light with the pigment is to stimulate the pumping of H^+ ions *out of* the guard cells. Although these experiments provide support for the hypothesis that an electrochemical gradient of H⁺ ions powers K⁺ transport, they do not rule out the possible involvement of ATP. The blue-absorbing pigment, which would be yellow in color, is thought to be a flavin; as you may recall (page 195), flavin-containing nucleotides are among the electron carriers involved in cellular respiration in the mitochondria.

Crassulacean Acid Metabolism

Although the stomata of most plants are open during the day and closed at night, the stomata of some species close in the daytime and open at night. Not only is the temperature lower at night but also the humidity is usually higher; both factors reduce the rate of transpiration. The species that open their stomata only at night include a variety of plants adapted to hot, dry climates, for example, cacti, pineapples, and members of the stonecrop family (Crassulaceae). These plants take in carbon dioxide at night, converting it to malic and isocitric acids. During the day, when the stomata are closed, the carbon dioxide is released from these organic acids and used immediately in photosynthesis. This process is known as Crassulacean acid metabolism, and plants that employ it are known as CAM plants. It is analogous to the C_4 pathway in photosynthesis described on page 224, although it apparently evolved independently. CAM plants achieve a considerable savings in water; in contrast to C_3 and even C_4 plants (see page 650), they lose only 50 to 100 grams of water for each gram of carbon dioxide incorporated into organic matter.

31-9 Among the most familiar CAM plants are the stonecrops of genus Sedum. *These stonecrops, flourishing on otherwise barren cliffs high above the surf, were photographed in Redwood National Park, California.*

The Uptake of Minerals

In addition to water and the energy-rich sugars and other compounds produced by photosynthesis, plant cells require a number of different chemical elements, which are found in the earth in the form of minerals. A **mineral** is a naturally occurring inorganic substance, usually solid, with a definite chemical composition. Some minerals, such as diamond, sulfur, and copper, consist of single elements. Others, such as quartz (SiO_2) and calcite ($CaCO_3$), are compounds. Of the 92 different elements found in rocks and soil, oxygen and silicon are by far the most abundant, followed by aluminum and iron.

Mineral ions are taken up in water solution by the roots and travel through the xylem in the transpiration stream. As you saw in Figure 30-10 (page 633), the cells of the endodermis play a major role in determining which substances enter the xylem. The ionic composition of plant cells is far different from the ionic composition of the medium in which the plant grows. For example, in one study, cells of pea roots were found to have a concentration of potassium ions (K^+) 75 times greater than that of the nutrient solution. Similarly, in another study, the vacuoles of rutabaga cells were shown to contain 10,000 times more K^+ than the external solution.

Thus it seems evident that mineral ions are brought into plant cells by active transport. Support for this hypothesis comes from observations indicating that the uptake of ions is an energy-requiring process. For instance, if roots are deprived of oxygen or poisoned so that respiration is curtailed, ion uptake is drastically decreased. Also, if a plant is deprived of light, it stops absorbing ions and eventually releases them back into the soil.

One way that roots move ions such as K^+ against a concentration gradient is by the action of transport proteins in the cell membrane (see page 136). Such proteins combine with the ions and then, utilizing the energy of ATP, carry them to the other side of the membrane and release them. Because the membrane is only slowly permeable to ions such as K^+, the continued action of the "pump proteins" results in differences in the concentration of the transported ions on one side of the membrane as compared to the other.

31–10 *Some plants have special mineral requirements.* (a) *Plants of the mustard family use sulfur in the synthesis of the mustard oils that give the plants their characteristic sharp taste. These mustard plants were growing along the side of a highway near Lima, Ohio.* (b) *Horsetails incorporate silicon into their cell walls, making them indigestible to most herbivores but useful, at least in colonial America, for scouring pots and pans. These bushy vegetative shoots, as well as several stalk-like fertile shoots, of the horsetail* Equisetum telmateia *were photographed in California.*

As a result of these differences in ion concentrations, there is often a difference not only of osmotic potential but also of electric potential between the inside and the outside of the membrane. As we have seen, such differences may play a role in the transport of K^+ ions into and out of the guard cells. They may also have other functions. For example, in the nerve cells of animals, differences in electric potential on either side of the membrane provide the basis for the nerve impulse. Stimulation of the nerve cell membrane results in an abrupt change in its permeability to ions, allowing ions to move rapidly down the gradient of potential energy. This rapid movement of ions across the membrane constitutes the nerve impulse. Similar events are apparently involved in certain plant responses, as we shall see in the next chapter.

Mineral Requirements of Plants

Plants require minerals for many different functions. One of the most critical, of course, is regulation of water balance. Because osmotic potential is determined by the number of solute particles in a solution, rather than by their specific chemical identity, several kinds of ions may serve interchangeably in this role. Thus, the requirement is described as nonspecific. On the other hand, a specific element may be an essential component of a critical biological molecule, which will not function properly in its absence. In such cases, the requirement is highly specific. For example, magnesium is an essential component of the chlorophyll molecule (see Figure 10–6, page 211). Some minerals—phosphorus and calcium, for example—are required constituents of cell membranes, and some control membrane permeability. Others are indispensable components of a variety of enzyme systems that catalyze chemical reactions in the cell. Still others provide a proper ionic environment in which such reactions can occur. Because mineral elements are involved in many fundamental processes, the effects of mineral deficiencies are typically very wide-ranging, affecting a number of structures and functions in the plant body.

Table 31–1 lists the mineral elements required by plants, the form in which they are usually absorbed, and some of the uses plants make of them. Depending on the concentration of a mineral element typically found in plants, it is classified as either a macronutrient or a micronutrient. A macronutrient may constitute as much as 0.5 to 3 or 4 percent of the dry weight of a plant; by contrast, only a few parts per million of a micronutrient may be present. For many elements, the amount present is determined by burning the plant completely—which permits the carbon, hydrogen, oxygen, nitrogen, and sulfur to escape as gases—and then analyzing the remaining ash. Proportions of each element vary in different species and in the same species grown under different conditions. Also, the ash often contains elements, such as silicon, which are present in the soil and are taken up by plants but which are not generally thought to be required for growth.

Mineral requirements can also be identified by studying the capacity of plants to grow in distilled water to which small amounts of various minerals are added. This sounds easier than it is. Sometimes it has been found that a substance—molybdenum, for example—is needed in such small amounts that it is almost impossible to set up experimental conditions that exclude it. Thus, it is difficult to prove that its absence is lethal.

You might anticipate that organisms make use of what is most readily available; indeed, they seem to have done this when life originated from elements in the gases of the primitive atmosphere. But Table 31–1 reveals some findings that you might not expect. Sodium, for instance, which is one of the most abundant of the elements, is apparently not required at all by most plants. The fact that plants evolved with no functions requiring sodium is even more striking when you consider that sodium is vital to animals. (Sodium is the principal osmoregulator in

TABLE 31-1 **A Summary of Mineral Elements Required by Plants**

ELEMENT	PRINCIPAL FORM IN WHICH ABSORBED	APPROXIMATE CONCENTRATION IN HEALTHY WHOLE PLANTS (AS % OF DRY WEIGHT)	SOME FUNCTIONS
Macronutrients			
Nitrogen	NO_3^- (or NH_4^+)	1–4%	Component of amino acids, proteins, nucleotides, nucleic acids, chlorophyll, and coenzymes
Potassium	K^+	0.5–6%	Involved in osmosis and ionic balance and in opening and closing of stomata; activator of many enzymes
Calcium	Ca^{2+}	0.2–3.5%	Component of cell walls; enzyme cofactor; involved in cell membrane permeability and transport of ions and hormones
Phosphorus	$H_2PO_4^-$ or HPO_4^{2-}	0.1–0.8%	Component of energy-carrying phosphate compounds (ATP and ADP), phospholipids, nucleic acids, and several essential coenzymes
Magnesium	Mg^{2+}	0.1–0.8%	Part of the chlorophyll molecule; activator of many enzymes
Sulfur	SO_4^{2-}	0.05–1%	Component of some amino acids, proteins, and coenzyme A
Micronutrients			
Iron	Fe^{2+}, Fe^{3+}	25–300 parts per million (ppm)*	Required for chloroplast development; component of cytochromes and nitrogenase
Chlorine	Cl^-	100–10,000 ppm	Involved in osmosis and ionic balance; essential in photosynthesis in the reactions in which oxygen is produced
Copper	Cu^{2+}	4–30 ppm	Activator or component of some enzymes
Manganese	Mn^{2+}	15–800 ppm	Activator of some enzymes; required for integrity of chloroplast membrane and for oxygen release in photosynthesis
Zinc	Zn^{2+}	15–100 ppm	Activator or component of many enzymes
Boron	BO^{3-} (borate) or $B_4O_7^{2-}$ (tetraborate)	5–75 ppm	Influences Ca^{2+} utilization, nucleic acid synthesis, and membrane integrity
Molybdenum	MoO_4^{2-}	0.1–5.0 ppm	Required for nitrogen metabolism
Elements Essential to Some Plants or Organisms			
Cobalt	Co^{2+}	Trace	Required by nitrogen-fixing microorganisms
Sodium	Na^+	Trace	Involved in osmosis and ionic balance; for many plants, probably not essential; required by some desert and salt-marsh species and may be required by all plants that utilize C_4 pathway of photosynthesis

* Parts per million (ppm) equal units of an element by weight per million units of oven-dried plant material; 1% equals 10,000 ppm.

animals and, as we shall see in Chapter 41, is necessary for transmission of the nerve impulse.) In the seas, where both plant and animal life seem to have originated, sodium is the most abundant mineral element and is far more available than potassium, which it closely resembles in its essential properties. Potassium, however, is the principal osmoregulator in plants. Similarly, silicon and aluminum are usually present in large amounts in soils, but only a few plants (specifically, grasses) require silicon and apparently none requires aluminum. Conversely, most plants need molybdenum, which is relatively rare.

THE MOVEMENT OF SUGARS: TRANSLOCATION

In addition to water and minerals, the cells of a plant also need energy. As we have seen, the photosynthetic cells of a plant, which are typically most abundant in the leaves, not only capture the energy of sunlight for their own use but also provide the organic molecules that are the energy source for all of the other cells of the plant. The process by which the products of photosynthesis are transported to other tissues is known as **translocation**.

Halophytes: A Future Resource?

Unlike most animals, most plants do not require sodium and, moreover, cannot survive in brackish waters or saline soils. In such environments, the solution surrounding the roots often has a higher solute concentration than the cells of the plant, causing water to move out of the roots by osmosis. Even if the plant is able to absorb water, it faces additional problems from the high level of sodium ions. If the plant takes up water and excludes sodium ions, the solution surrounding the roots becomes even saltier, increasing the likelihood of water loss through the roots. The salt may even become so concentrated that it forms a crust around the roots, effectively blocking the supply of water to the roots. Another problem is that sodium ions may enter the plant in preference to potassium ions, depriving the plant of an essential nutrient as well as inhibiting some enzyme systems.

Some plants, however—known as halophytes—can grow in saline environments such as deserts, salt marshes, and coastal areas. All of these plants have evolved mechanisms for dealing with high sodium concentrations, and for some of them sodium appears to be a required nutrient. The adaptations of halophytes vary. In many halophytes, the sodium-potassium pump (page 136) seems to play a major role in maintaining a low sodium concentration within the cells while simultaneously ensuring that a sufficient supply of potassium ions enters the plant. In some species, the pump operates primarily in the root cells, pumping sodium back to the environment and potassium into the root. The presence of calcium ions (Ca^{2+}) in the soil solution is thought to be essential for the effective functioning of this mechanism.

Other halophytes take in sodium through the roots but then either secrete it or isolate it from the living cytoplasm of the plant body. In *Salicornia* (pickleweed), a sodium-potassium pump (or a variant of it) operates in the membranes of the vacuoles of leaf cells. Sodium ions enter the cells but are immediately pumped into vacuoles and isolated from the cytoplasm. In such plants, the solute concentration of the vacuoles is higher than that of the environment, establishing the necessary osmotic potential for the movement of water into the roots. In other genera, the salt is pumped into the intercellular spaces of the leaves and then secreted from the plant. In *Distichlis*

(a) Atriplex *(saltbush) is one of several halophytes being evaluated as potential crop plants.*

(a)

palmeri (Palmer's grass), the salt exudes through specialized cells (not the stomata) onto the surface of the leaf. In *Atriplex* (saltbush), it is concentrated by special salt glands and pumped into bladders. The bladders expand as the salt accumulates, finally bursting. Rain or the passing tide washes the salt away.

Halophytes are of current interest not only because of the light they may shed on the osmoregulatory mechanisms of plants, but also because of their potential as crop plants. In a world with an ever-increasing need for food, vast areas are unsuitable for agricultural purposes because of the salinity of the soil. For example, there are over 30,000 kilometers of desert coastline and about 400 billion hectares of desert with potential water supplies that are, however, too salty for crop plants. Moreover, each year about 200,000 hectares of irrigated crop land become so salty that further agriculture is impossible. When arid land is heavily irrigated, as in large areas of the western United States, salts from the irrigation waters accumulate in the soil. This occurs because in both evaporation from the soil and transpiration from plants, essentially pure water is given off, leaving all solutes behind. Over many years, the salt concentration of the soil increases, eventually reaching levels that cannot be tolerated by most plants. It has been suggested that the ancient civilizations of the Near East ultimately fell because their heavily irrigated land became so salty that food could no longer be grown on it.

One way of extending the life of irrigated crop lands and of bringing presently barren areas into agricultural use would be to breed salt-tolerance into conventional crop plants. Thus far, however, such efforts have met with little success. Scientists at the University of Arizona's Environmental Research Laboratory are taking what appears to be a more promising approach. They have gathered halophytes from all over the world and are engaged in an extensive research program to determine optimal growing conditions, potential yields, and the nutritional value and palatability of the seeds and vegetative parts of the various species. Their results suggest that a number of halophyte species have great potential for use in livestock feed and, quite possibly, for human consumption as well.

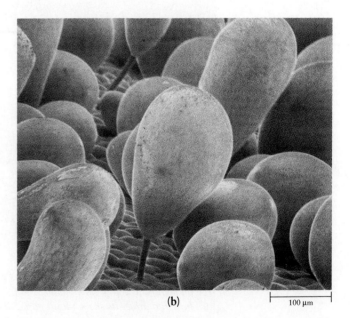

(b) *The surface of a leaf of* Atriplex. *Salt is pumped from the leaf tissues through narrow stalk cells into the large, expandable bladder cells.*

Evidence for the Phloem

The fact that water is transported in the tracheids and vessels of the xylem was recognized by botanists 300 years ago. The role of the phloem in the movement of sugar was not generally agreed upon, however, until well into the twentieth century. Early evidence suggesting that the phloem is involved in sugar transport came from observations of trees that had a complete ring of bark removed from them. As we saw in the last chapter, bark contains the phloem but not the xylem. When a photosynthesizing tree is "girdled" in this manner, the tissue above the ring becomes swollen, indicating that fluid moving downward in the phloem from the photosynthesizing leaves has accumulated there. Convincing evidence for the role of the phloem was obtained when radioactive tracers became available. If plants carry out photosynthesis in air containing carbon dioxide made with carbon 14 ($^{14}CO_2$), the sugars produced will carry a radioactive tag. Studies of the passage of radioactivity through the plant have shown conclusively that sugars are transported in the sieve tubes (see essay).

Aphids, which are very small sap-sucking insects, have provided valuable information on the movement of substances through the sieve tubes (Figure 31-11). Data gathered with the assistance of aphids indicate that sieve-tube sap contains (by weight) 10 to 25 percent solutes, more than 90 percent of which are sugars, mostly sucrose. Low concentrations of amino acids and other nitrogen-containing substances are also present. Tracer stains indicate that the rate of movement of the solutes along the sieve tube is remarkably fast: in one set of experiments, for example, it was estimated that the sap was moving at a rate of about 100 centimeters per hour, far faster than could be accounted for by diffusion alone. At this rate, each individual sieve-tube member was emptying and completely refilling every two seconds.

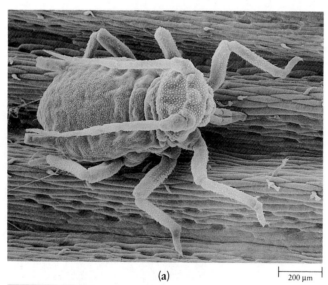
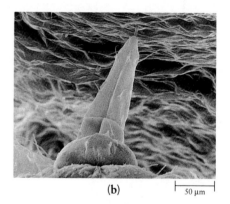
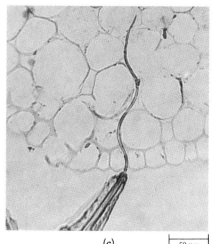

31-11 *Assistance by aphids.* (a) *Aphids are small insects that feed on plant sap; you probably have seen them on rose bushes.* (b) *The aphid drives its sharp mouthparts, or stylets, like a hypodermic needle through the epidermis.* (c) *As this micrograph reveals, the stylets traverse the cortical or mesophyll cells and then tap the contents of a single sieve-tube member. If the aphid is anesthetized, it is possible to sever its body from the stylets, leaving the latter undisturbed in the cell. The sieve-tube sap often continues to exude through the stylets for many hours, and pure samples can be collected for analysis without damaging the sieve tube or interfering with its function.*

Radioactive Isotopes in Plant Research

Radioactive isotopes can be used in a number of ways to study the synthesis, transport, and use of materials within plants. Initially, the radioactive isotope must be incorporated into the plant. Radioactive carbon, for example, will be taken up by a plant if its leaves are exposed to carbon dioxide containing carbon 14. Or, radioactive phosphorus will be taken up if the roots are exposed to a solution containing phosphorus-32 ions.

The length of time the plant is exposed to the radioactive material is determined by the information the investigators hope to obtain. For example, in studies to determine the time required for carbon dioxide to be incorporated into the various products of photosynthesis, a sequence of exposure times would be used. In studies focusing on the location of a particular product formed from the radioactive substance, the length of exposure would depend on the time required for the chemical reactions under study.

After exposure to the radioactive substance, the plant is quick-frozen and freeze-dried. In whole-plant autoradiography, the plant is flattened and then pressed against a sheet of x-ray film. Radiation given off by the isotope exposes the film adjacent to the portions of the plant in which it is located. By comparing the flattened plant with the developed film, investigators can determine the location of the radioactive substance within the plant.

In tissue autoradiography (histoautoradiography), the freeze-dried plant tissues are embedded in paraffin, resin, or a similar material. Next, they are sliced into very thin sections, which are mounted on microscope slides. The tissue sections are then placed in contact with a photographic emulsion or film. As in whole-plant autoradiography, the radiation from the isotope exposes the film in contact with the portions of the tissue section containing radioactive material. After an appropriate interval of time, the film is developed. Comparison, under the microscope, of the developed film and the underlying tissue section reveals the exact location of the radioactive substance in the plant tissues.

In the study illustrated here, three leaflets of a bean plant were enclosed in a flask and were exposed to $^{14}CO_2$ and light for 35 minutes (a). During that time, the radioactive carbon dioxide was incorporated into sugars, which were then being transported to other parts of the plant. A cross section (b) and a longitudinal section (c) from the stem were placed in contact with autoradiographic film for 32 days. When the film was developed and compared with the underlying tissue sections, it was apparent that the radioactivity (visible as dark specks on the film) was confined almost entirely to the sieve tubes.

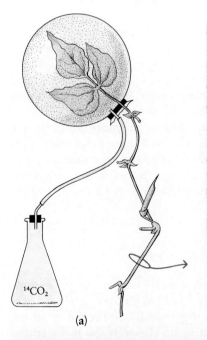
(a)

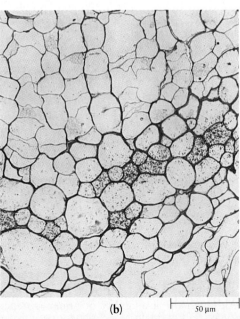
(b) 50 μm

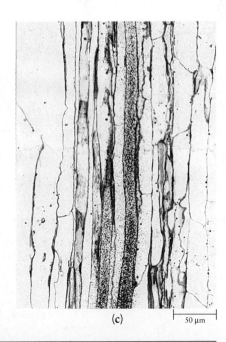
(c) 50 μm

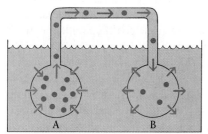

- Sugar molecule
- → Movement of water
- → Movement of sugar solution

31-12 *Model of the pressure-flow hypothesis. Bulbs A and B, which are interconnected and are permeable to water, are placed in a bath of distilled water. Bulb A contains a higher concentration of sucrose than bulb B. Water enters bulb A from the medium, increasing the hydrostatic pressure and pushing the solution to bulb B. If B were connected to a third bulb, C, with a still lower concentration of sucrose, as sieve-tube members are connected in a series, hydrostatic pressure building up in B would push the solution on to C, and so on.*

The Pressure-Flow Hypothesis

The movement of sugars and other organic solutes in translocation follows what is known as a **source-to-sink pattern**. The principal sources of the solutes are the photosynthesizing leaves, but storage tissues may also serve as important sources. All plant parts unable to meet their own nutritional needs may act as sinks, that is, as importers of organic solutes. Thus, storage tissues act as sinks when they are importing solutes and as sources when they are exporting solutes.

The most widely accepted explanation for the source-to-sink movement in translocation is the **pressure-flow hypothesis.** According to this hypothesis, the solutes move in solutions that, in turn, move because of differences in water potential caused by concentration gradients of sugar. The principle underlying this hypothesis can be illustrated by a simple physical model consisting of porous bulbs permeable only to water and connected by glass tubes (Figure 31-12). The first bulb contains a solution in which there is a dissolved material, such as sucrose, and the second, to make the example as simple as possible, contains only water. When these interconnected bulbs are placed in distilled water, water will enter the first by osmosis. The entry of water will increase the hydrostatic pressure within this bulb and cause the water and the solutes in it to move along the tube to the second bulb, where the pressure again builds up. If the second bulb is connected with a third bulb containing water or a sucrose concentration lower than that now in the second bulb, the solution will flow from the second to the third by the same process, and so on indefinitely down the sucrose gradient. This hypothesis is supported by the fact that distinct gradients in the concentration of sucrose have been demonstrated along the phloem tissues. Moreover, it can account for the known rates of movement in the phloem.

As shown in Figure 31-13, sugars from the photosynthetic cells of the leaf are moved into sieve tubes against a concentration gradient. In sugar beets, it has been shown that sucrose molecules are moved from the mesophyll cells of the leaf to the phloem of the veins where they are "loaded" into the sieve tubes by active transport. This loading process appears to involve a cotransport of sucrose molecules and hydrogen ions by way of a specific transport protein on the sieve-tube membrane. The incoming sugar decreases the water potential in the sieve tube and causes water to move into the sieve tube from the xylem by osmosis. At a sink—for example, a storage root—sugar molecules leave the sieve tube. Water molecules then follow the sugar molecules out, again by osmosis. Thus the water flows in at one end of the sieve tube and out at the other. Between these two points, the water and its solutes, including sugar, move passively by bulk flow. The speed of transport depends on the differences in concentration between source and sink. At the sink, sugars may be either utilized or stored, but most of the water returns to the xylem and is recirculated in the transpiration stream.

Companion cells, because of their dense appearance and many mitochondria, are believed to be very active metabolically, and it is hypothesized that one of their functions is to meet the energy requirements of the sieve-tube members with which they are associated. Note, however, that the actual flow of the sugar solution is a passive process, requiring no metabolic activity on the part of the phloem cells.

FACTORS INFLUENCING PLANT NUTRITION

As we saw earlier, the mineral elements needed by plants are taken up by the roots in solution and are transported through the plant body in the transpiration stream. Although the availability of minerals depends principally on the nature of the surrounding soil, the activities of symbiotic fungi and bacteria also play a crucial role.

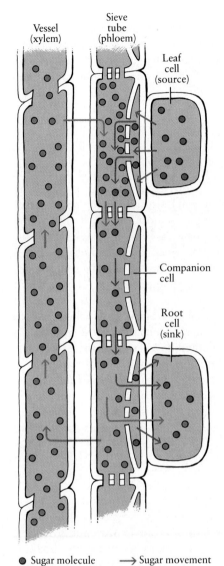

- Sugar molecule → Sugar movement
- Water molecule → Water movement

31-13 *The pressure-flow mechanism as it is thought to occur in the plant body. Sugar molecules enter a companion cell at the source by active transport and then move into the sieve tube through the many cytoplasmic connections in the common cell wall of the sieve-tube member and its companion cell. As a consequence of the increased concentration of sugar, the water potential is decreased, and water enters the sieve tube by osmosis. Sugar molecules leave the sieve tube at the sink, and the sugar concentration in the sieve tube falls; as a result, water moves out of the sieve tube by osmosis. Because of the active secretion of sugar molecules into the sieve tube at the source and their removal from the sieve tube at the sink, a flow of sugar solution takes place along the tube between source and sink.*

Soil Composition

Soil, the uppermost layer of the earth's crust, is composed of rock fragments associated with organic material, both living and in various stages of decomposition. It typically has three layers: the A horizon, the B horizon, and the C horizon. The A horizon, or topsoil, is the zone of maximum accumulation of organic matter (humus). The B horizon, or subsoil, consists of inorganic particles in combination with mineral nutrients that have leached (washed) down from the A horizon. The C horizon is made up of loose rock that extends down to the bedrock beneath it. As shown in Figure 31-14, the depth and composition of these three layers—and consequently, the fertility of the soil—vary considerably in different environments.

The mineral content of soil depends in part on the parent rock from which the soil was formed. These differences in mineral content can be extremely localized, with sharp lines of demarcation. Geologists sometimes use types of vegetation or changes in color or growth patterns of plants as indicators of mineral deposits.

In most soils, however, the mineral content is more dependent on biological factors. In an undisturbed environment, most of the mineral nutrients stay within the system—the soil itself and the plants, microorganisms, and small soil animals that it supports and contains. If, however, the vegetation is repeatedly removed, as when crops are harvested, grasslands are overgrazed, or the top, humus-rich layer of the A horizon is eroded, the soil rapidly becomes depleted of nutrients. It can then be used for agricultural purposes only if it is heavily fertilized.

Another factor influencing the mineral content of soil is the size of the soil particles. The smaller fragments of rock are classified as sand, silt, or clay (Table

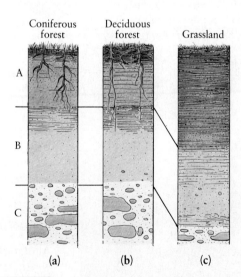

31-14 *Diagrams of soil layers of three major soil types. (a) The litter of the northern coniferous forest is acidic and slow to decay, and the soil has little accumulation of humus, is very acidic, and is leached of minerals. (b) In the cool, temperate deciduous forest, decay is somewhat more rapid, leaching less extensive, and the soil more fertile. Such soils have been widely used for agriculture, but they need to be prepared by adding lime (to reduce acidity) and fertilizer. (c) In the grasslands, almost all of the plant material above the ground dies each year, as do many of the roots, and thus large amounts of organic matter are constantly returned to the soil. In addition, the finely divided roots penetrate the soil extensively. The result is highly fertile soil, often black in color, with a topsoil sometimes more than a meter in depth.*

TABLE 31-2 Soil Classification	
	DIAMETER OF PARTICLES (MICROMETERS)
Coarse sand	200–2,000 (0.2–2 millimeters)
Sand	20–200
Silt	2–20
Clay	Less than 2

31-2). Water and minerals drain rapidly through soil composed of large particles (sandy soil). Soil composed of small particles (clay) holds the water against gravity. Moreover, the small clay particles are negatively charged and therefore bind positively charged ions, such as calcium (Ca^{2+}), potassium (K^+), and magnesium (Mg^{2+}). However, a pure clay soil is poorly suited for plant growth because it is usually too tightly packed to let in enough oxygen for the respiration of plant roots, soil animals, and most soil microorganisms. Clay soils that contain enough particles of sand and silt to keep the soil from packing are known as loams, and these are generally the best soils for plant growth.

The pH of the soil also affects its capacity to retain minerals. In acidic soil, hydrogen ions replace other positively charged ions clinging to clay particles, and these nutrient ions leach out of the soil. Soil pH also affects the solubility of certain nutrient elements. Calcium, for example, is more soluble (and therefore more available to plant roots) as pH increases, whereas iron becomes less available to plants as pH increases. Crops such as alfalfa, sweet clover, and other legumes have high calcium requirements and therefore grow best in alkaline soil. Rhododendrons and azaleas, on the other hand, have high requirements for iron, which is abundant only when the soil is acidic.

Soils and plant life interact. Plants secrete hydrogen ions, which help to degrade rock surfaces and release positively charged ions from those surfaces. As they decay, plant parts constantly add to the humus, thereby changing not only the content of the soil but also its texture and its capacity to hold minerals and water. In turn, the plants are dependent upon the mineral content of the soil and its holding capacity. As these improve, plants increase in both number and size and also often change in kind, thereby producing further changes in the soil. Thus, under natural conditions, the soil—and the availability of nutrients and water to plant roots—are constantly changing.

The Role of Symbioses

Mycorrhizae

Two types of symbiotic relationships play important roles in plant nutrition. One of these is mycorrhizae ("fungus-roots"), the associations between fungi and roots that we described in Chapter 23. In these associations, the fungi extract nutrients from the soil and make them available to plants, thus enabling plants to prosper on nutrient-poor soils. Recent studies indicate that the fungi may also screen out chemicals, making it possible for plants to live in soils that would otherwise be toxic.

Rhizobia and Nitrogen Fixation

All but one of the elements listed in Table 31-1 are derived principally from the weathering of rocks. The exception is nitrogen, the nutrient for which plants have the greatest requirement. Although nitrogen constitutes about 78 percent of the air, most plants cannot use gaseous nitrogen, in which each molecule consists of two atoms of nitrogen held together by a triple covalent bond (page 33), an exceptionally strong bond. Plants are dependent upon nitrogen-containing ions—ammonium (NH_4^+) and nitrate (NO_3^-)—from the soil.

In plant cells, nitrate ions are reduced to ammonium ions, and the ammonium ions are then combined with carbon-containing compounds to form amino acids, nucleotides, chlorophyll, and other nitrogen-containing compounds. These nitrogen-containing compounds are then returned to the soil with the death of the plants (or of the animals that have eaten the plants) and are reprocessed by soil organisms, taken up by plant roots in the form of nitrate dissolved in soil water,

31-15 *Indian pipe* (Monotropa uniflora). *It was once believed that this non-photosynthetic angiosperm, with its strange white waxy flowers, lived on decaying matter in the soil. It is now known, however, that it is dependent on mycorrhizal fungi that transfer nutrients from other plants to the parasitic Indian pipe.*

31–16 *Effects of mycorrhizae on tree nutrition. Nine-month-old seedlings of white pine were grown for two months in a sterile nutrient solution and then transplanted to prairie soil. The seedlings on the left were transplanted directly. The seedlings on the right were placed in forest soil containing fungi for two weeks before being transplanted to the prairie.*

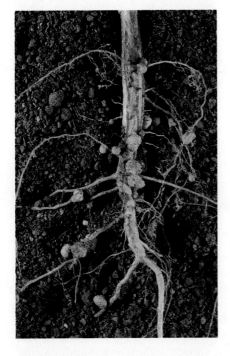

31–17 *Nitrogen-fixing nodules on the roots of a soybean plant, a legume. These nodules are the result of a symbiotic relationship between a soil bacterium (Rhizobium) and root cells.*

and reconverted to organic compounds. In the course of this cycle (to be discussed in more detail in Chapter 54), a certain amount of nitrogen is always "lost," in the sense that it becomes unavailable to plants.

The main source of nitrogen loss is the removal of plants from the soil. Soils under cultivation often show a steady decline of nitrogen content. Nitrogen may also be lost when topsoil is carried off by soil erosion or when ground cover is destroyed by fire. Nitrogen-containing ions are also leached away by water percolating down through the soil. In addition, numerous types of bacteria are present in the soil that, when oxygen is not available, break down nitrates, releasing nitrogen into the air and using the oxygen for cellular respiration.

If the nitrogen lost from the soil were not steadily replaced, virtually all life on this planet would finally flicker out. The "lost" nitrogen is returned to the soil by **nitrogen fixation,** the process by which atmospheric nitrogen is incorporated into organic nitrogen-containing compounds. On a worldwide basis, most nitrogen fixation is carried out by a few kinds of prokaryotes, including both free-living and symbiotic forms of cyanobacteria and heterotrophic bacteria. Of the various classes of nitrogen-fixing organisms, the symbiotic bacteria are by far the most important in terms of total amounts of nitrogen fixed. The most common of the nitrogen-fixing symbiotic bacteria is *Rhizobium*, which invades the roots of leguminous plants, such as clover, peas, beans, vetches, and alfalfa (Figure 31–17).

The beneficial effects to the soil of growing leguminous plants are so obvious that they have been recognized for hundreds of years. Where leguminous plants are grown, some of the "extra" nitrogen is usually released into the soil; it then becomes available to other plants. In modern agriculture, it is common practice to rotate a nonleguminous crop, such as corn, with a leguminous one, such as alfalfa. The leguminous plants are then either harvested, leaving behind the nitrogen-rich roots, or, better still, plowed back into the field. A crop of alfalfa that is plowed back into the soil may add as much as 350 kilograms of nitrogen to the soil per hectare, frequently enough to grow a crop of a nonleguminous plant without any additional fertilization.

Carnivorous Plants

More than 350 species of plants are meat-eaters; their diets include insects, other invertebrates, and even some vertebrates, such as small birds and frogs. Unlike carnivorous animals, they do not utilize their prey for energy but rather as a source of mineral elements, particularly nitrogen, phosphorus, and calcium. As Darwin noted in his book on this subject, published in 1875, these plants are usually found in swamps, bogs, and peat marshes where acids leach the soil of nutrients.

(a) The sundew is a tiny plant, often only 2 to 5 centimeters across, with club-shaped tentacles on the upper surface of its leaves. These tentacles secrete a clear, sticky liquid that attracts insects. When an insect is caught by one tentacle, the other tentacles bend toward it until the insect drowns, its air passages filled with a mucilage-like fluid. The tentacles also secrete digestive enzymes.

(b) Pitcher plants attract insects into their flower-like tubular leaves by means of nectar. Following the

(a)

(b)

The Symbiotic Relationship

The symbiosis between a species of *Rhizobium* and a legume is quite specific; for example, the species of bacteria that invade and induce nodule formation in clover roots will not induce nodules on the roots of soybeans. The recognition process on which this specificity depends is thought to involve the interaction of plant proteins known as lectins on the root surface with polysaccharides of the bacterial cell wall. Rhizobia enter the root-hair tips of legumes while the plants are still seedlings (Figure 31–18a). The bacteria usually induce these epidermal cells to produce internal cellulose tubes, or infection threads, through which the bacteria move to the cortical cells of the root (Figure 31–18b). Within the cortex, the infection threads branch, becoming populated with multiplying bacteria. Soon there is a proliferation of the surrounding cortical cells, presumably the result of

nectar over the rim, the insect finds itself on a carpet of fine, transparent hairs. When the nectar trail ends, and the insect turns to exit, it encounters the points of these fine hairs, pointing downward and blocking its way. If it moves inward, it encounters a slick waxed surface from which it slides or drops into a foul-smelling broth of rainwater, digestive enzymes, and bacteria at the pitcher's base.

(c, d) The Venus flytrap has leaves that close around any insect moving on the surface of the leaf. The closing of the leaf is triggered by touching one or two of the three trigger hairs in the middle of each leaf lobe. Venus flytraps are found in nature only on the coastal plain of North and South Carolina, usually at the edges of wet depressions and pools. It has long been believed that the Venus flytrap lures its victims by exuding nectar, but recent studies indicate that insects visit the leaves by chance. Despite its name, a Venus flytrap's usual diet in the wild consists of crawling invertebrates, such as ants.

(c)

(d)

hormones released during the growth of the rhizobia. The infection threads typically pass close to the nuclei of the cortical cells and seem to cause them to degenerate. Near the nuclei, vesicles form in the threads and then break open, releasing the bacteria into the cytoplasm of the host cells (Figure 31–18c).

Soon after their release, the bacteria begin to grow, increasing in size some tenfold and synthesizing an enzyme complex called **nitrogenase**. Nitrogenase, which consists of two different polypeptides, one containing iron and the other molybdenum, catalyzes the following reaction:

$$\underset{\text{Nitrogen}}{N_2} + \underset{\substack{\text{Hydrogen} \\ \text{ions}}}{6H^+} + \underset{\text{Electrons}}{6e^-} \xrightarrow{\text{Nitrogenase}} \underset{\text{Ammonia}}{2NH_3}$$

 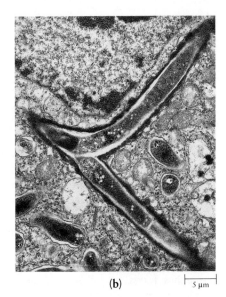 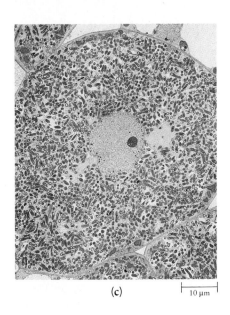

31–18 (a) *Scanning electron micrograph of an emerging root hair of a soybean seedling, with several rhizobia (arrows) attached.* (b) *Two branches of an infection thread in a soybean nodule cell. As you can see, the thread is passing near the cell nucleus, at the top of the micrograph.* (c) *Cross section of a nodule, showing a cell containing numerous bacteria. The plant supplies the bacteria with an energy source; the bacteria supply the plant with fixed nitrogen.*

The energy required to break the strong triple bond of the nitrogen molecule is supplied by ATP; some 15 to 20 molecules of ATP are hydrolyzed to ADP for each molecule of gaseous nitrogen incorporated into ammonia. The ammonia produced by this reaction is combined with carbon compounds synthesized by the photosynthetic cells of the plant, producing amino acids.

The legumes are by far the largest group of plants that enter into nitrogen-fixing partnerships with symbiotic bacteria. There are, however, numerous nitrogen-fixing symbioses that involve plants other than legumes. Sweet fern (an angiosperm), for example, forms nodules that are induced by and contain nitrogen-fixing actinomycetes (moldlike bacteria), rather than *Rhizobium*. Like rhizobia, the actinomycetes enter the host plant by way of a root-hair infection. The resulting symbiosis allows the plants to carry out a pioneering role in the revegetation of barren land. Sweet fern, for instance, is commonly planted along highways in Massachusetts for this purpose. There is also evidence that the roots of some grasses, such as sorghum, are involved in less intimate symbiotic associations with nitrogen-fixing bacteria. The bacteria involved, the most prevalent of which is *Azospirillum*, usually reside in the transition zone between the root and the soil. Some of the nitrogen fixed by the bacteria is absorbed by the roots, and carbohydrates released from the roots provide energy for the bacteria.

31–19 *Sweet fern,* Comptonia peregrina, *growing on West Rattlesnake Mountain in New Hampshire. Sweet fern, which obtains its needed nitrogen from symbiotic nitrogen-fixing actinomycetes, is able to colonize nutrient-poor soils on which most other plants cannot grow.*

Recombinant DNA and Nitrogen Fixation

Genetic mapping of a free-living nitrogen-fixing bacterium has shown that the 17 genes known to be involved in nitrogen fixation are clustered on one portion of the chromosome. In experiments at the University of Sussex, biologists have succeeded in transferring this gene cluster to a plasmid and then introducing the plasmid into *Escherichia coli* cells. The *E. coli* cells were then able to synthesize nitrogenase and to fix nitrogen.

In other experiments at Cornell University, the gene cluster for nitrogenase has been transferred to yeast cells (eukaryotes) and has remained intact during repeated cell divisions. These experiments raise the hope, of course, that nitrogen-fixing genes can be transferred to the cells of plants such as corn, which could then, in effect, make their own nitrogen-containing compounds. Thus far, however, the yeast cells have not demonstrated the capacity to fix nitrogen, perhaps because of differences in start/stop signals or in the cytoplasmic makeup of prokaryotic and eukaryotic cells. An alternative way of conferring nitrogen-fixing capability on plants would be to transfer to nonlegumes the genes involved in the leguminous plant's contribution to the symbiotic association. This might be accomplished either by plasmids or by somatic cell hybridization. However, because of the number of genes involved and the complexity of the factors controlling nitrogen fixation, many botanists believe that such an achievement is, at best, many years away.

About 250 million metric tons of nitrogen are added to the soil each year, of which some 200 million tons are biological in origin. The other 50 million tons are largely in the form of chemical fertilizers, produced by commercial nitrogen-fixation processes in which the required energy is supplied by fossil fuels. More than one-third of the total amount of energy needed to produce a crop of corn is used in the manufacture, transportation, and application of chemical fertilizers. As the cost of the fossil fuels required for all of these processes fluctuates, so does the cost of producing the crop. From this perspective, the possible use of recombinant DNA techniques to increase biological nitrogen fixation takes on great practical significance. For, just as all animals, including ourselves, are ultimately dependent on photosynthesis for carbon-containing molecules, they—and we—depend on nitrogen fixation for the nitrogen-containing molecules without which living cells cannot function.

SUMMARY

Transpiration is the loss of water vapor by plants. As a consequence of transpiration, plants require large amounts of water. Water enters the plant from the soil through the roots and travels through the plant body in the conducting cells of the xylem (vessel members and tracheids). According to the cohesion-tension theory, water moves through the tracheids and vessels under negative pressure (tension). Because the molecules of water cling together (cohesion), a continuous column of water molecules is pulled from the soil solution and into the root, molecule by molecule, by the evaporation of water above.

Diffusion of gases, including water vapor, into and out of the leaf is regulated by the stomata. The stomata are opened and closed by the guard cells due to changes in turgor. Turgor is increased or decreased by the osmotic movement of water, which follows the movement of potassium ions into or out of the guard cells. The active transport of potassium ions is regulated by a variety of factors, including water stress, abscisic acid, carbon dioxide concentration, temperature, and light.

Minerals, which are naturally occurring inorganic substances, are brought into the plant from the soil and are carried with the transpiration stream in the xylem. They fulfill a variety of functions in plants, some of which are nonspecific, such as the effects on osmotic potential. Other functions are specific, such as the presence of magnesium in the chlorophyll molecule. A number of minerals are essential components of enzyme systems.

The movement of organic compounds from the photosynthetic parts of the plant is known as translocation. It takes place in the phloem and follows a source-to-sink pattern. According to the pressure-flow hypothesis, sugars are loaded into the sieve tubes in the leaf by active transport and are removed from them in other parts of the plant body where they are needed for growth and energy. Water moves into and out of the sieve tubes by osmosis, following the sugar molecules. These processes create a difference in water potential along the sieve tube, which causes water and the sugars dissolved in it to move by bulk flow along the sieve tube.

Characteristics of the soil affect the availability of minerals to plants. These characteristics include the rock from which the soil was formed, the size of the soil particles, the amount of humus present, and the soil pH.

Two types of symbiotic relationships are important in plant nutrition: mycorrhizae (discussed in Chapter 23) and relationships involving nitrogen-fixing bacteria. The associations between nitrogen-fixing bacteria, such as rhizobia, and the roots of certain plants, particularly legumes, result in the incorporation of gaseous nitrogen from the atmosphere into organic nitrogen-containing compounds.

QUESTIONS

1. Distinguish among the following: transpiration/translocation; cohesion-tension theory/pressure-flow hypothesis; source/sink; A horizon/B horizon/C horizon; rhizobia/mycorrhizae.

2. What properties of water discussed in Section 1 are important to the movement of water and solutes through plants?

3. Transpiration has often been described as a "necessary evil" to the plant. Why is it necessary? How is it evil?

4. Gardeners advise removing many leaves of a plant after transplanting. How does this help the plant to survive?

5. Consider a tree transpiring most rapidly at midday and an investigator with a sensitive instrument for measuring changes in the diameter of the trunk. If water is pulled up from the top (cohesion-tension theory), what changes in diameter should be observed from night to day?

6. In Figure 31–4, why doesn't air enter the tube through the top of the enclosed porous pipe or the leaf, even though water vapor can easily escape? How can the porous pipe (and, by analogy, the leaf) be permeable to air or water but effectively impermeable at an air-water interface?

7. Identify the cells and tissues through which a molecule of water travels from the time it enters the root until it is used in photosynthesis.

8. When K^+ ions move out of the guard cells, they move into adjacent epidermal cells. How would the influx of K^+ ions affect the epidermal cells? What effect would this have on the stomata?

9. Using the techniques described in this chapter for the analysis of sieve-tube sap, how would you measure the rate of movement? (*Hint:* Aphids are available.)

10. A particular sugar molecule was produced by photosynthesis in a leaf of a perennial plant late one August. Throughout the following winter, it was stored in a root of the plant. The next spring, it was oxidized in the process of respiration, providing energy for the growth of a new shoot tip. Identify the cells and tissues through which this sugar molecule traveled between its synthesis and its ultimate use.

11. As you will recall from Chapter 10, the immediate product of the Calvin cycle is glyceraldehyde phosphate, a three-carbon sugar (page 222). What chemical reactions did the sugar molecule in Question 10 probably undergo between its synthesis and its use?

12. (a) Experts in flower arranging advise recutting the stems of flowers while holding them under water. Explain why. (b) Some florists advise adding ordinary table sugar to the water in which cut flowers are placed. When this is done, certain types of flowers will remain fresh for several weeks. What is the explanation?

13. As we have seen, nitrogen fixation is an energy-requiring process. What is the source of the energy used by *Rhizobium*? If new symbiotic associations are developed in which the nitrogen-fixing bacteria produce greater yields of nitrogen-containing compounds, what price is the host plant likely to pay? Why?

C H A P T E R 32

Plant Responses and the Regulation of Growth

32-1 *This young bean seedling has just broken through the soil. Its growth into a mature plant will depend on light, water, temperature, and minerals from the soil and also on the interaction of many internal factors, among which are plant hormones.*

The capacity to respond to stimuli—both internal and external—is one of the essential properties of all living organisms (page 27). However, perhaps because of an inevitable bias based on our own experience, we tend to think of the perception of and active response to stimuli as phenomena limited to motile organisms, which rapidly and visibly alter their relationship with their environment when stimulated in various ways. Although plants cannot literally move from one place to another in response to a stimulus, they can significantly alter their relationship with the environment through their patterns of growth. As we noted in Chapter 30, growth in plants serves many of the same functions as motility in animals and is, to some extent, a counterpart of behavior. Many plant responses are slower than those of animals, but they are no less effective in solving the problems with which the environment confronts the organism.

We saw earlier that as a plant grows it does far more than simply increase its mass and volume. In a process that begins in the embryo and continues throughout the lifetime of a plant, cells produced in the meristematic tissues differentiate and elongate to form the specialized tissues and organs of the plant body. The rate at which this process occurs—in the plant as a whole and in different parts of the plant body—is affected by such environmental factors as temperature, water, sunlight, gravity, and physical contact with other objects, including other organisms. Growth rates, in turn, affect both the size of the whole plant and the relative sizes of its parts, thus determining its shape, or form. Moreover, many of a plant's activities—especially its cycles of active growth, reproduction, and dormancy—are finely tuned to the pattern of the changing seasons. Not only are plants able to sense and react to a range of factors in the immediate environment, but also, and perhaps more important, they are able to anticipate environmental changes and prepare for them.

Many of the details of how plants sense their environment and then respond with altered patterns of growth and development are, despite many years of research, poorly understood. Application of the techniques of molecular biology to plant physiology is beginning to yield new insights into many of the mechanisms underlying specific plant responses, but a complete and coherent picture of the functioning plant remains elusive. It is clear, however, that regulation of the physiological activities of a plant depends on the interplay of a number of internal and external factors. Chief among the internal factors are the plant hormones.

PHOTOTROPISM AND THE DISCOVERY OF PLANT HORMONES

A **tropism** is a growth response that involves the curvature of a plant part toward or away from an external stimulus that determines the direction of movement. If the plant part curves toward the stimulus, the tropism is said to be positive; if it curves away, the tropism is negative. One of the most obvious and useful

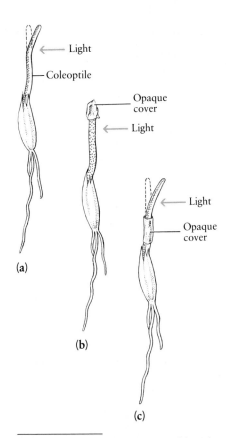

32-2 *The Darwins' experiment.* (a) *Light striking a growing coleoptile (such as the tip of this oat seedling) causes it to curve toward the light.* (b) *Placing an opaque cover over the tip of the seedling inhibits this curvature, but* (c) *an opaque collar placed below the tip does not. These experiments indicate that something produced in the tip of the seedling and transmitted down the stem causes the curvature.*

responses of plants is their positive **phototropism,** that is, their curvature toward light. It was with the study of phototropism that our knowledge of plant hormones and their effects on growth began, and it remains an appropriate starting point in the consideration of plant responses.

Charles Darwin and his son Francis performed some of the first experiments on phototropism. Working with grass seedlings, they noted that the curvature occurs below the tip, in a lower part of the coleoptile (the hollow sheath that surrounds the shoot tip in the embryos and seedlings of grasses). Then they showed that if they covered the tip of the coleoptile with a cylinder of metal foil or a hollow tube of glass blackened with India ink and exposed the plant to a light coming from the side, the characteristic curvature of the seedling did not occur. If, however, light was permitted to penetrate the cylinder, curving occurred normally. Curving also occurred normally when the lightproof cylinder was placed below the tip (Figure 32-2). "We must therefore conclude," they wrote, "that when seedlings are freely exposed to a lateral light some influence is transmitted from the upper to the lower part, causing the material to bend."*

In 1926, the Dutch plant physiologist Frits W. Went succeeded in separating this "influence" from the plants that produced it. Went cut off the coleoptile tips from a number of oat seedlings. He placed the tips on a slice of agar (a gelatinlike substance), with their cut surfaces touching the agar, and left them there for about an hour. He then cut the agar into small blocks and placed a block off-center on each stump of the decapitated plants, which were kept in the dark during the entire experiment. Within one hour, he observed a distinct curvature *away* from the side on which the agar block was placed (Figure 32-3).

Agar blocks that had not been previously in contact with a coleoptile tip produced either *no* curvature or only a slight curvature *toward* the side on which the block had been placed. Agar blocks that had been exposed to a section of coleoptile lower on the shoot produced no physiological effect.

Went interpreted these experiments as showing that the coleoptile tip exerted its effects by means of a chemical stimulus rather than a physical stimulus, such as

---

* Writing in 1881, the Darwins used the term "bend" to describe the phototropic response of a plant. Contemporary botanists, however, make a distinction between "bending," which does not involve growth, and "curvature," which does. As we shall see, the phototropic response is caused by differential cell elongation, and thus is, strictly speaking, a "curvature."

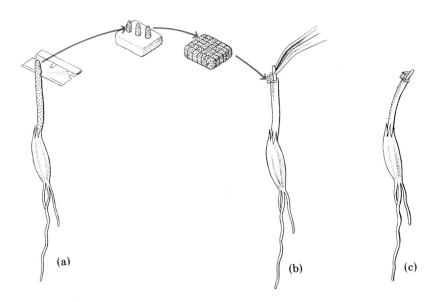

32-3 *Went's experiment.* (a) *He cut the coleoptile tips from oat seedlings and placed them on a slice of agar. After about an hour, he cut the agar in small blocks and* (b) *placed each block, off-center, on a decapitated seedling (the leaf is pulled up).* (c) *The seedlings curved away from the side on which the block was placed.*

an electrical impulse. This chemical stimulus came to be known as **auxin**, a term coined by Went from the Greek word *auxein*, "to increase." Auxin was one of the first plant hormones to be discovered.

The phototropism observed by the Darwins, it is now known, results from the fact that, under the influence of light, auxin migrates from the light side to the dark side of the shoot tip. The cells on the dark side, which contain more auxin, elongate more rapidly than those on the light side, causing the plant to curve toward the light—a response with high survival value for young plants. As with the influence of light on stomatal opening, only blue light—that is, light less than 500 nanometers in wavelength—is effective. In the phototropic response, as in the stomatal response, the blue-absorbing pigment is thought to be a flavin (see page 655); when stimulated, it apparently alters the permeability of the cell membrane, facilitating the movement of auxin to the shaded side of the stem.

HORMONES AND THE REGULATION OF PLANT GROWTH

A hormone, by definition, is a chemical substance that is produced in one tissue and transported to another, where it exerts one or more highly specific effects. Hormones integrate the growth, development, and metabolic activities of the various tissues of the plant. Typically they are active in very small quantities. In the shoot of a pineapple plant, for example, only 6 micrograms of auxin are found per kilogram of plant material. One enterprising plant physiologist calculated that the weight of the hormone in relation to that of the shoot is comparable to the weight of a needle in 20 metric tons of hay.

The term "hormone" comes from the Greek word *hormaein*, meaning "to excite." It is now clear, however, that many hormones also have inhibitory effects. So, rather than thinking of hormones as stimulators, it is perhaps more useful to consider them as chemical regulators. But this term also needs qualification. As we shall see, the response to a particular regulatory "message" depends not only on its content (that is, its chemical structure) but also on the identity of its recipient (that is, the specific tissue) and when and how it is received. Moreover, the response to any particular hormone is influenced by a variety of other factors in the internal environment of the plant, chief among which are likely to be other hormones.

Five principal types of plant hormones are known: auxins, cytokinins, ethylene, abscisic acid, and gibberellins. Recent evidence suggests that short carbohydrate chains, known as oligosaccharins, also function as plant hormones. Although we shall, for clarity, discuss these groups of hormones one at a time, keep in mind that we are always, in fact, dealing with interactions among them.

Auxins

The substance isolated by Went and given the name auxin is indoleacetic acid, or IAA. Several different substances with activity similar to that of IAA have now been isolated from plant tissues, and others have been synthesized in the laboratory. These substances are known collectively as auxins (Figure 32-4).

32-4 *Natural and synthetic auxins. IAA (indoleacetic acid), isolated from plant tissues, is the most common natural auxin. Naphthalenacetic acid, a synthetic auxin, is commonly used to induce the formation of adventitious roots in cuttings and to reduce fruit drop in orchards. 2,4-D, also a synthetic auxin, is used as an herbicide.*

IAA

Naphthalenacetic acid

2,4-D

32-5 *The stalk of the African violet leaf on the left was placed in a solution containing the synthetic auxin naphthalenacetic acid for 10 days before the picture was taken. The stalk of the leaf on the right was placed in pure water. Note the growth of adventitious roots on the stalk of the hormone-treated leaf.*

IAA is synthesized in the plant by enzymatic conversion of the amino acid tryptophan (see Figure 3–18, page 73). It is produced principally by the apical meristems of shoots and is transported to other parts of the plant, moving in one direction only, from shoot to root. (If a cut portion of a stem is turned upside down, IAA moves from bottom to top.) IAA causes cells in the growing region of the shoot to elongate; if the shoot apex is removed, growth stops. If the hormone is applied to the cut surface, growth resumes. If an auxin is applied to the intact plant, no growth effect on the stem occurs until relatively high concentrations are reached; then growth is inhibited. In short, it appears as if the apex produces the optimal concentration of auxin to which the shoot can respond positively.

Roots, which also receive their auxin from the shoot apex, are more sensitive to the hormone. Auxin in very small amounts is required for root growth; however, even a slight increase in auxin concentration inhibits root growth. As we shall see later in this chapter, such increases in auxin concentration along the lower side of horizontally oriented roots are thought to be involved in the downward growth of roots in response to gravity.

Synthetic auxins, unlike IAA, are not readily broken down by natural plant enzymes or by the enzymes of soil bacteria. Their long-lasting effects make them better suited for commercial purposes than IAA. The rooting preparations commonly used by gardeners to stimulate the growth of branch roots and adventitious roots in stem cuttings contain a synthetic auxin. Another synthetic auxin, 2,4-D, is used as an herbicide. For reasons not known, 2,4-D and related compounds are toxic to dicots at concentrations not harmful to monocots. Because of this selective effect and their low cost, synthetic auxins are commonly used on lawns to control broad-leaved weeds.

Mechanism of Action of Auxin

The stimulatory effects of auxin on the growth of shoots and roots are the result of a relatively rapid elongation of cells. Under the influence of auxin, the plasticity of the cell wall increases and the cell expands in response to the turgor exerted by the movement of water into the cell vacuole (see page 134). Current evidence indicates that IAA exerts its influence by activating a proton pump in the plant cell membrane that transports hydrogen ions (H^+) from the cell into the cellulose-containing cell wall. As a result, the cell wall becomes acidified and this, in turn, activates a pH-dependent enzyme in the cell wall that breaks the cross-links between the cellulose molecules. This allows the molecules to slide past one another as turgor acts against the cell wall. Subsequently, the cross-links re-form, and the wall becomes rigid once more. This response, known as "acid growth," is rapid, often beginning within 3 to 5 minutes, reaching a maximum in 30 minutes, and ending after 1 to 3 hours. As we shall see later in this chapter, a similar acid growth response is involved in the closing of Venus flytrap leaves.

The induction of acid growth, however, is only part of the auxin story. Auxin also stimulates long-term growth, triggering the expression of at least 10 specific genes. Long-term growth induced by auxin involves increased transcription of messenger RNA and ribosomal RNA, the assembly of many new ribosomes, and the biosynthesis of proteins different from those previously present in the cell. Although these proteins have not been identified with certainty, it is reasonable to believe, as a working hypothesis, that some of them are enzymes involved in the synthesis of the new cell-wall materials necessary for continued growth.

An increasing body of evidence indicates that calcium ion (Ca^{2+}) is intimately involved in the effects of auxin on growth. Inhibitors of auxin transport abolish not only auxin effects but also Ca^{2+} transport. Moreover, plants deficient in Ca^{2+} are often unresponsive to auxin. It has been hypothesized that auxin induces an increase in the amount of Ca^{2+} in treated cells, which, in turn, activates a regulatory substance known as calmodulin. Calmodulin, which has been found in a wide variety of eukaryotic organisms, affects membrane permeability, the transport of H^+ ions, and the activity of a number of enzymes.

Apical Dominance and Other Auxin Effects

In most dicot species, the growth of axillary buds is inhibited by auxin that moves down the stem from the shoot apex. This phenomenon is known as **apical dominance.** If you cut off the growing tip (apical meristem) of the stem of the house plant *Coleus,* for example, the axillary buds begin to grow vigorously, producing a plant with a bushier, more compact body and with more flowers. These derepressed buds can be repressed again by applying auxin to the cut surface of the shoot tip. Similarly, if you treat the "eyes" (actually axillary buds) of a potato with auxin, they will be inhibited from sprouting and the potato can therefore be stored longer.

Auxin was long thought to be directly responsible for apical dominance. However, recent experiments suggest that its effects are indirect. It now appears that auxin stimulates the production of ethylene in cells surrounding the axillary buds, and that this hormone, rather than auxin, inhibits bud growth. As we shall see shortly, cytokinins are also involved in apical dominance.

In woody plants, auxin also plays a role in the seasonal initiation of activity in the vascular cambium (page 644). When the meristematic region of the shoot begins to grow in the spring, auxin moving down from the shoot tip stimulates cambial cells to divide, forming secondary phloem and secondary xylem. Here, as elsewhere, its effects are modulated by other growth-regulating substances in the plant body.

Although auxin is produced principally in the apical meristems of shoots, it is also produced by young leaves, flowers, developing embryos, and fruits. Auxin promotes the maturation of the ovary wall and the development of fleshy fruits (Figure 32–6). Auxin treatment of the female floral parts of some species is used to produce fruits without fertilization, such as seedless tomatoes, cucumbers, and eggplants.

32–6 *Auxin, apparently produced by developing seeds, promotes the growth of fruit.* (a) *Normal strawberry,* (b) *strawberry from which all seeds have been removed, and* (c) *strawberry in which three horizontal rows of seeds were left. If a paste containing auxin is applied to* (b), *the strawberry grows normally.*

Cytokinins

The isolation of auxin spurred the search for other growth-promoting factors in plants, particularly for a hormone that would stimulate cell division. Such a factor was first detected in coconut "milk," which is a liquid endosperm. It or related compounds have now been found in all plants examined, especially in actively dividing tissues, such as meristems, germinating seeds, fruits, and roots. They are called **cytokinins** (from "cytokinesis"). The most active naturally occurring cytokinin is zeatin, which was originally isolated from corn *(Zea mays);* the synthetic cytokinin most commonly used in research is known as kinetin. Cytokinins resemble the purine adenine (Figure 32–7).

Studies of the way in which cytokinins promote cell division have shown that the hormone is required for some process that takes place after DNA replication is complete but before mitosis begins. Cytokinins increase the rate of protein synthesis, and it is thought that some of the resulting proteins may be necessary for cell division. The effect of cytokinins is not on the transcription of messenger RNA but rather on its translation into protein. The relationship, if any, between the mechanism of action of the cytokinins and their resemblance to adenine is not known.

32–7 *Cytokinins.* **(a)** *Zeatin and isopentenyl adenine have been isolated from plant material.* **(b)** *Kinetin and 6-benzylamino purine (BAP) are commonly used synthetic cytokinins. Note the resemblances between the purine adenine and a portion of the molecule of each of these cytokinins. The significance of the resemblance between adenine and the hormones of this group is not known. It may merely be another example of biological economy: the use of a single major biosynthetic pathway to produce a number of functionally different products.*

Responses to Cytokinin and Auxin Combinations

Studies of responses to combinations of auxin and cytokinins are helping physiologists glimpse how plant hormones work together to produce the total growth pattern of the plant. Apparently, an undifferentiated plant cell—such as a meristematic cell—has two courses open to it: either it can enlarge, divide, enlarge, and divide again, or it can elongate without cell division. The cell that divides repeatedly remains essentially undifferentiated, whereas the elongating cell tends to differentiate and become specialized. In studies of tobacco stem cells, the addition of IAA to the culture medium on which the cells were growing produced rapid cell expansion, so that giant cells were formed. Adding kinetin alone had little or no effect. However, adding IAA plus kinetin resulted in rapid cell division, so that large numbers of relatively small cells were formed (Figure 32–8).

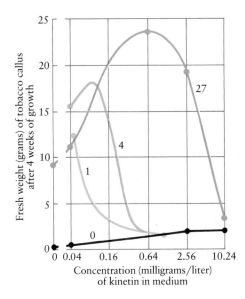

32–8 *The response of tobacco cells in tissue culture to combinations of auxin (IAA) and cytokinin (kinetin). The concentrations of IAA (in milligrams per liter) are indicated on the curves. Kinetin alone has little effect on the growth of undifferentiated tobacco tissue (a callus). IAA alone causes the culture to grow to a weight of about 10 grams, regardless of the concentration used. When both hormones are present, growth is greatly increased. Notice, however, that when optimum concentrations are exceeded, the growth rate declines.*

By slightly altering the relative concentrations of IAA and kinetin, investigators have been able to affect the development of undifferentiated cells growing in tissue culture. At roughly equal concentrations of the two hormones, the cells remain undifferentiated, forming a mass of tissue known as a **callus.** When a higher concentration of IAA is present, undifferentiated tissue gives rise to organized roots. With a higher concentration of kinetin, buds appear (Figure 32–9). Further, careful balancing of the two hormones can produce both roots and buds and, thus, an incipient plant.

However, lest you think this is simple, we shall describe another tissue culture study, in which tuber tissue of the Jerusalem artichoke was used. In this study, it was shown that a third factor, the calcium ion (Ca^{2+}), can modify the action of the auxin-cytokinin combination. Auxin plus low concentrations of kinetin favored cell enlargement, but as Ca^{2+} was added to the culture, there was a steady shift in the growth pattern from cell enlargement to cell division. High concentrations of Ca^{2+} apparently prevent the cell wall from relaxing and expanding, and at such concentrations the cell switches course and divides. Thus, not only do hormones modify the effects of hormones, but these combined effects may, in turn, be modified by nonhormonal factors, such as calcium ion and, undoubtedly, many others.

Other Cytokinin Effects

Cytokinins have also been shown to reverse the inhibitory effect of auxin in apical dominance. Local application of kinetin to repressed axillary buds releases them from inhibition. In intact plants, cytokinins are synthesized in the roots and travel upward, primarily through the xylem, reaching the lower buds first and in highest concentration. As the dominating apex of the plant grows and moves away from these lower buds, the influence of the cytokinins overcomes that of auxin and the buds begin to grow.

Another, apparently unrelated function of cytokinins is preventing the senescence (aging) of leaves. In many plants, the lower leaves turn yellow and drop off as the upper, new leaves develop; if kinetin is applied to the lower leaves, they remain green. Similarly, excised leaves remain green when maintained in a nutrient solution containing kinetin.

It is hypothesized that senescence in leaves, and probably in other plant parts as well, results from the progressive "turning off" of segments of DNA, with a consequent loss of messenger RNA production and protein synthesis. In experiments with excised leaves that contain radioactive amino acids and have been "spotted" with kinetin, the amino acids migrate to the kinetin-treated areas.

32–9 *Two buds forming on undifferentiated tissue (a callus) from a geranium following treatment with both an auxin and a cytokinin. Callus from some types of plants will continue to grow as undifferentiated tissue, or roots, or buds, depending on the relative proportions of auxin and cytokinins.*

Plants in Test Tubes

As long ago as the 1930s, scientists developed techniques for growing plant cells in test tubes. Tiny fragments of meristem are implanted under sterile conditions in a medium containing minerals and various combinations of organic compounds. Under these conditions, the meristematic cells proliferate to form clumps of undifferentiated cells. Subsequently, it was discovered that, by adjusting the hormone balance in the medium, it was possible to make these cells differentiate and grow into mature plants. Using these techniques, hundreds or even thousands of subcultures of meristematic tissue can be produced in a relatively short time and in a small space. Then, by altering the hormone balance, each of these can be turned into a small but perfect plant. This technique was initially employed in the culture of orchids and other hard-to-grow plants and is now being extended to many other plants.

Within the last 20 years, it has become possible, using similar techniques, to grow isolated protoplasts —plant cells without cell walls—in the laboratory. These protoplasts can be grown as isolated cells, like cultures of bacteria. Alternatively, if they are grown in a suitable medium, they will regenerate cell walls, multiply, and differentiate into whole plants. The young plants produced in this way are not only genetically uniform but also free of infectious disease.

Cultured cells are being used in a variety of ways. For instance, they are used in rapid screening tests to determine resistance to infectious diseases or to detect nutritional requirements. In this way, a scientist not only can work much more rapidly but can also do assays with millions of cells growing in a very small space, as compared to the far fewer number of plants that can be grown in a field or a greenhouse. Moreover, protoplasts from two different species of plant can be fused to create a hybrid. Protoplast fusion holds particular promise for combining the desirable characteristics of species that are sexually incompatible and therefore cannot be crossbred by conventional techniques.

The first plant created by a fusion of protoplasts was a hybrid of tobacco that had already been produced by conventional crossbreeding. (Producing a plant whose characteristics were known was a necessary preliminary test for the method.) Subsequently, protoplast fusion was used to create hybrids of species within the genera that include petunia *(Petunia)*, carrot *(Daucus)*, and potato *(Solanum)*. More recently, hybrids involving species belonging to different genera have been obtained by protoplast fusion. Of particular interest is the fusion of the potato *(Solanum tuberosum)* and the tomato *(Lycopersicon esculentum)*, both of which are members of the nightshade family (Solanaceae). Tomato plants are resistant to the oomycete that causes potato blight (see page 468), a disease that remains a serious threat to potato crops. It is hoped that, through protoplast fusion, it will be possible to develop potato plants that incorporate the tomato plants' resistance genes. Similar work, funded by the Campbell Soup Company, is under way with protoplast fusion involving tomato plants and strains of tobacco that are resistant to diseases to which tomato plants are particularly vulnerable.

The regeneration of whole plants from cultured cells, protoplast fusion procedures, and, most recently, the application of recombinant DNA techniques to plants (see page 352) together raise hopes for the production of entirely new "superplants." Such plants might be not only resistant to diseases that destroy a large proportion of the world's food supply each year, but also, for instance, capable of simultaneously carrying out C_4 photosynthesis and nitrogen fixation.

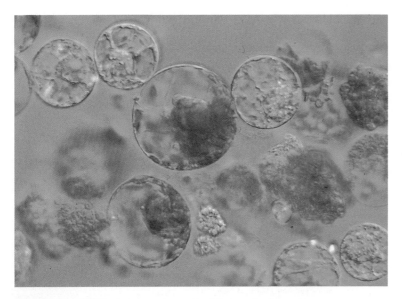

Protoplasts of wild tobacco (green) and commercial tobacco (clear). As you can see, some of the protoplasts have fused, giving rise to hybrid cells. With suitable treatment, the hybrid cells will regenerate cell walls, multiply, and differentiate into whole plants.

These experiments have led to the proposal that cytokinins prevent the DNA from being turned off and so promote continued enzyme synthesis and the production of such compounds as chlorophyll.

Ethylene

Ethylene is an unusual hormone in that it is a gas, a simple hydrocarbon, $H_2C = CH_2$. Its effects have been known for a long time. In the early 1900s, many fruit growers made a practice of improving the color and flavor of citrus fruits by "curing" them in a room with a kerosene stove. (Long before this, the Chinese used to ripen fruits in rooms where incense was being burned.) It was long believed that it was the heat that ripened the fruits. Ambitious fruit growers, who went to the expense of installing more modern heating equipment, found to their sorrow that this was not the case. As experiments later showed, the incomplete combustion products of kerosene were actually responsible for ripening the fruits. The most active gas was identified as ethylene. As little as 1 part per million of ethylene in the air will speed the ripening process. Subsequently, it was found that ethylene is produced by plants, fungi, and bacteria, as well as by kerosene stoves. The ethylene-synthesizing system is apparently located on the cell membrane, from which the hormone is released. Ethylene can be detected just before and also during fruit ripening and is responsible for a number of changes in color, texture, and chemical composition that take place as fruits mature. It also is involved in the senescence of floral parts that follows fertilization and precedes fruit development.

Auxin at certain concentrations causes a burst of ethylene production in some plants, and it is now believed that some of the effects on fruits and flowers generally attributed to auxin are related to the release of ethylene. As we noted earlier, ethylene is thought to be an effector of apical dominance. Auxin induces ethylene production in or near the axillary buds, whereas the cytokinins may inhibit it.

Ethylene and Leaf Abscission

Plants drop their leaves at regular intervals, either as a result of normal aging of the leaf or, in the case of deciduous shrubs and trees, in response to environmental cues. This process of **abscission** ("cutting off") is preceded by changes in the **abscission zone,** at the base of the petiole. In woody dicots, the abscission zone consists of two cell layers: (1) a structurally weak layer in which the actual abscission occurs, and (2) a protective layer that forms a leaf scar on the stem (Figure 32–10).

Once leaf senescence begins, ethylene produced in the abscission layer is the principal regulator of leaf drop. It acts by promoting the synthesis and release of cellulase, an enzyme that breaks down plant cell walls (and which is also involved in the ripening of fruits). Auxin inhibits abscission if applied to the leaf before senescence begins; however, once the abscission layer is formed, auxin promotes abscission by stimulating ethylene production.

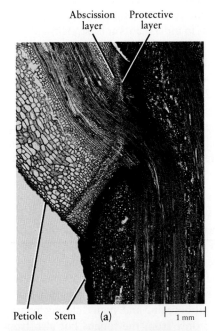

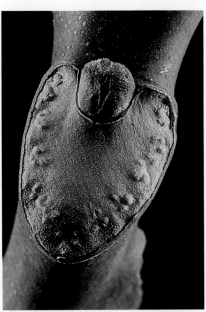

32–10 (a) *Abscission zone in a maple leaf, as seen in a longitudinal section. The abscission layer, which forms across the base of the petiole, consists of structurally weak cells. Under the influence of ethylene, enzymes are produced that cause the walls of these cells to dissolve.* (b) *After the leaf drops off, the protective layer forms a covering, the leaf scar, on the stem. This leaf scar is on a branch of* Ailanthus, *the tree of heaven.*

32-11 *Abscisic acid, an inhibitor that blocks the action of the growth-promoting hormones. In some species, it produces dormancy in buds and leaves. As we saw in the last chapter, it also plays a major role in stomatal closing.*

Abscisic acid

32-12 *Winter buds of an American basswood. These buds, or young shoots, form in one growing season and elongate in the next. The buds contain small amounts of water and high concentrations of proteins and lipids, which prevent the formation of ice crystals that could kill the cells. Bud scales, which are small, tough modified leaves, protect the delicate tissues of the bud from mechanical injury and desiccation. As the first shoots emerge in the spring, the bud scales are forced off. Note the leaf scar at the base of each bud.*

Abscisic Acid

Soon after the discovery of the growth-promoting hormones, plant physiologists began to speculate that growth-inhibiting hormones would be found, since it is clearly advantageous to the plant not to grow at certain times and in certain seasons. Not long afterwards, an inhibitory hormone, which was called dormin, was isolated from dormant buds. Subsequently, the same hormone was discovered in leaves, where it was thought to promote abscission. Thus it was called **abscisic acid,** or ABA (Figure 32–11). The latter name proved dominant, which is somewhat unfortunate since it is now known that, in most plants at least, ABA has little to do with abscission.

Abscisic acid may, however, induce dormancy. For example, application of ABA to vegetative buds changes them to winter buds, converting the outermost leaf primordia to bud scales (Figure 32–12). ABA is also present in the seeds of many species, where it is a major factor in maintaining seed dormancy. Moreover, as we saw in the last chapter, ABA brings about the closing of stomata under conditions of impending water shortage. Thus, ABA has come to be known as the stress hormone in recognition of its role as a protector of the plant against unfavorable environmental conditions.

Gibberellins

The **gibberellins** were discovered slightly earlier than auxin by a Japanese scientist who was studying a disease of rice plants called "foolish seedling disease." The diseased plants grew very rapidly but were spindly and tended to fall over under the weight of the developing grains. The cause of the symptoms, it was found, was a chemical produced by a fungus, *Gibberella fujikuroi*, which infected the seedlings. The substance, which was named gibberellin, and many closely related substances were subsequently isolated not only from the fungus but also from bacteria and many species of plants. More than 70 different gibberellins are now known (Figure 32–13); of these, gibberellic acid (GA_3) has been studied most thoroughly. Gibberellin and auxin (to a lesser extent) control elongation in mature trees and shrubs. The mechanism by which gibberellin affects elongation remains unknown; however, it does not involve the transport of H^+ ions or acid growth and thus would appear to be different from that of auxin.

Gibberellic acid (GA_3) GA_7 GA_4

32-13 *Three of the more than 70 gibberellins that have been isolated from natural sources. Gibberellic acid (GA_3) is the most abundant in fungi and the most biologically active in many tests. The minor structural differences that distinguish the other two gibberellins are indicated by arrows.*

In many species of plants, gibberellins characteristically produce hyperelongation of the stem, such as that seen in the "foolish seedlings." Particularly striking effects are seen in some plants that are genetic dwarfs. In some of these, application of gibberellin makes them indistinguishable from normal plants, suggesting that these dwarfs lack a gene coding for an enzyme needed for gibberellin production.

Gibberellins can also produce bolting, a phenomenon observed in many plants, especially biennials, that first grow as rosettes. Just before flowering, the flower stem elongates rapidly ("bolts"). Bolting and flowering, which normally occur only after an environmental cue (such as a cold season), occur in some plants following treatment with gibberellin (Figure 32–14). Abscisic acid inhibits the effects of gibberellin in promoting bolting; this inhibition is, in turn, reversed by cytokinins.

Gibberellins can also induce cellular differentiation. In woody plants, gibberellins stimulate the vascular cambium to produce secondary phloem. Both phloem and xylem develop in the presence of gibberellins and auxins. In intact plants, interactions between the two types of hormones are thought to determine the relative rates of production of secondary phloem and secondary xylem.

Gibberellins and Seed Germination

The highest concentrations of gibberellins have been found in immature seeds, although they are present in varying amounts in all parts of plants.

In grains (the one-seeded fruits of grasses), there is a specialized layer of cells, the aleurone layer, just inside the seed coat (see page 622). These cells are rich in protein. During the early stages of germination, following water uptake, the embryo produces gibberellin, which diffuses to the aleurone layer. In response to gibberellin, the aleurone cells produce and release enzymes that hydrolyze the starch, lipids, and proteins in the endosperm, converting them to sugars, fatty acids, and amino acids that the embryo and, later, the emerging seedling can use (Figure 32–15). In this way, the embryo itself calls forth the substances needed for its metabolism and growth at just the times they are required.

32–14 Bolting and flowering in cabbage plants following gibberellin treatment. The plants on the left were not treated.

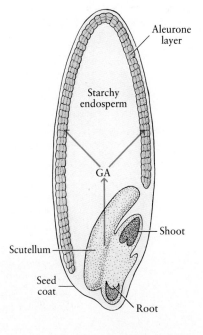

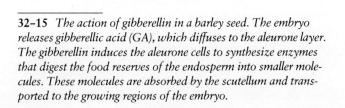

32–15 The action of gibberellin in a barley seed. The embryo releases gibberellic acid (GA), which diffuses to the aleurone layer. The gibberellin induces the aleurone cells to synthesize enzymes that digest the food reserves of the endosperm into smaller molecules. These molecules are absorbed by the scutellum and transported to the growing regions of the embryo.

The production of the hydrolytic enzymes is the result of the synthesis of specific new messenger RNA molecules from the DNA template. For example, transcription of the gene coding for alpha-amylase, which digests starch in the endosperm, increases in response to gibberellin. Thus it is hypothesized that gibberellins, in this instance at least, are acting as derepressors, turning on certain genes.

Oligosaccharins

Like the gibberellins, the most recently discovered plant regulatory substances came to the attention of botanists in the course of studies of diseased plants. Plants infected by bacteria, fungi, and some protists defend themselves by producing antibiotics at the site of infection. These antibiotics are known as phytoalexins (from the Greek *alexein*, "to ward off"), having been originally discovered in plants infected by oomycetes of the genus *Phytophthora*, the cause of potato blight (see page 468).

Plant physiologists studying this defensive response hypothesized that some signal released by the pathogen was triggering the expression of the plant genes coding for the enzymes involved in antibiotic synthesis. Studies to test this hypothesis and to identify the signal were begun in the early 1970s by Peter Albersheim and his colleagues at the University of Colorado. They began with a fungus that attacks soybeans and found that the signal recognized by the plant was an oligosaccharide—a short chain of sugar molecules—released from the fungal cell wall. Related studies of antibiotic synthesis in response to bacterial infections also indicated that oligosaccharides were the signaling molecules, but, surprisingly, these oligosaccharides were released not from the bacterial cell wall but from the plant cell wall. Later investigations of plant responses to other pathogens and to wounding revealed a number of different, specific oligosaccharide molecules released from plant cell walls in response to different stimuli. In all cases, these molecules, dubbed **oligosaccharins,** were released from the noncellulose matrix of the primary cell wall and stimulated the transcription of messenger RNA.

These discoveries raised the possibility that oligosaccharins might have a more general role as plant regulatory substances. Could it be that the other plant hormones, such as auxin and gibberellin, work indirectly, by activating enzymes that release specific oligosaccharins from the cell wall, and that these molecules are the real regulators of the physiological processes of the plant? Studies published in 1985 by Kiem Tran Thanh Van and her colleagues in France indicate that oligosaccharins can indeed alter the growth and development of plants in cultures in which hormone levels and pH are carefully controlled (Figure 32–17). Her studies have generated considerable excitement, and the pace of research on oligosaccharins is quickening. There is hope that this new class of plant substances will provide the key to understanding what, as you have seen, is a bewildering diversity of responses to the five classic categories of plant hormones.

GRAVITROPISM

As we noted when we began our consideration of plant hormones, phototropism —curvature toward the light—has high survival value for a young plant. Another response with high survival value is the capacity of a young plant to respond to gravity, righting itself so that the shoot grows up and the root grows down. This response is known as **gravitropism;** like phototropism, gravitropism is thought to involve auxin.

When young monocot shoots are oriented horizontally, auxin migrates to the lower side of the coleoptile and calcium ion (Ca^{2+}) accumulates along the upper

32–16 *An experiment investigating the action of gibberellin in barley seeds. Forty-eight hours before the picture was taken, each of these seeds was cut in half and the embryo removed. The seed at the bottom was treated with plain water, the seed in the center was treated with a solution of 1 part per billion of gibberellin, and the seed at the top was treated with 100 parts per billion of gibberellin. As you can see, digestion of the starchy storage tissue has begun to take place in the seeds treated with gibberellin.*

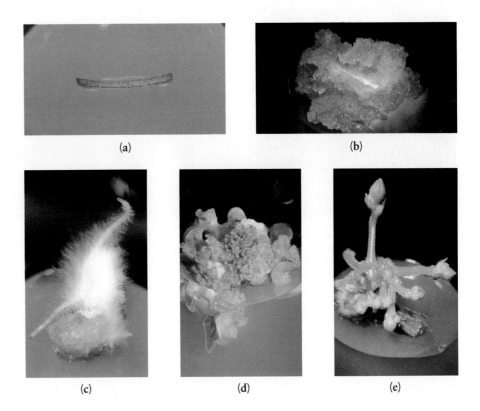

32–17 *The effects of oligosaccharins on plant development have been studied using thin cell layers of stem from tobacco plants (a). The thin cell layers, which include the epidermis, the layer of cells immediately below the epidermis, and one to three layers of parenchyma cells, are maintained in a nutrient medium containing glucose and salts. The ratio of auxin to cytokinin, the pH, and the particular oligosaccharins added to the culture medium determine how the thin cell layer develops. Under the influence of different oligosaccharins, it may form (b) an undifferentiated callus, (c) roots, (d) vegetative shoots and leaves, or (e) floral shoots.*

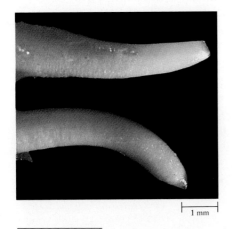

32–18 *A demonstration of the importance of the root cap in gravitropism. The root cap of the corn root at the top of the micrograph has been removed; the root cap of the root at the bottom is intact. When the two roots were oriented horizontally, only the root with an intact cap responded with downward curvature. Application of a small amount of auxin to one side of a root from which the cap has been removed will cause it to curve toward the side on which the auxin is applied.*

side. The increased concentration of Ca^{2+} inhibits growth along the upper side, while auxin stimulates elongation of cells on the lower side. The result is an upward curvature of the shoot. When the shoot becomes vertical, the differences in auxin and Ca^{2+} concentrations disappear, and growth continues in an upright direction. Whether a similar mechanism is involved in the gravitropism of dicot shoots remains unknown. Thus far, experiments designed to determine concentrations of auxin and Ca^{2+} in different parts of horizontally oriented dicot shoots have yielded conflicting results.

As you will recall, roots are exquisitely sensitive to even slight increases in auxin concentration. Their downward growth is thought to result from the inhibitory effects of increased auxin concentrations along the lower side. As cells along the upper side of the root elongate more rapidly than those along the lower side, the root curves downward. The transport of auxin to the lower side of roots is dependent upon Ca^{2+}, but nothing is known of the mechanism by which it occurs.

Another major question concerns the way in which the influence of gravity is detected. How does a seedling "know" it is on its side? Hormones and ions are soluble, and so gravity itself should have no effect on their distribution. The answer to this question has long been thought to lie in specialized cells of the shoot and the root cap (Figure 32–18). The inner, or core, cells of the root cap seem to be analogous to the statocysts found in many animals (see page 528). Like statocysts, these cells contain statoliths—particles that move in response to gravity. In jellyfish, the statoliths are grains of hardened calcium salts; in the core cells of the root cap, they are starch-containing plastids, known as amyloplasts. When a root is growing vertically, the amyloplasts collect near the lower walls of the core cells (Figure 32–19a). If the root is placed in a horizontal position, however, the amyloplasts slide downward and come to rest near what were previously vertically oriented walls (Figure 32–19b). Within minutes, the root begins to curve down, gradually returning the amyloplasts to their original

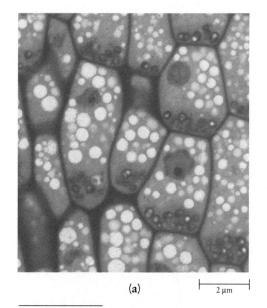
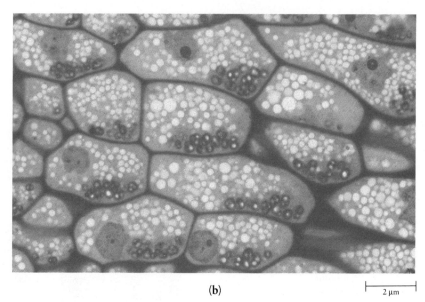

32–19 *Core cells of the root cap* (a) *in their normal vertical orientation and* (b) *after the root has been turned horizontally. The amyloplasts are the dark globular bodies containing white starch grains. Note how their position within the cells changes when the orientation of the root is changed.*

position. Roots from which the amyloplasts have been removed are unable to respond to gravity, suggesting that their movement is indeed critical in the detection of gravity. Recently, however, a mutant plant has been discovered in which the roots lack amyloplasts and yet exhibit nearly normal responses to gravity. Thus, the mechanism of gravity detection, like so many other mechanisms in plants, remains in question.

How the detection of gravity—by whatever means—is translated into chemical gradients of auxin and calcium ion is similarly unknown. Many of the investigators of root gravitropism think that both the detection of gravity and the translation of that information into chemical gradients probably involve alterations in both electrical properties and membrane permeability.

PHOTOPERIODISM

In many regions of the biosphere, the most important environmental changes affecting plants (and, indeed, land organisms in general) are those that result from the changing seasons. Plants are able to accommodate to these changes because of their capacity to anticipate the yearly calendar of events: the first frost, the spring rains, long dry spells, long growing periods, and even the time that nearby plants of the same species will flower. For many plants, all of these determinations are made in the same way: by measuring the relative periods of light and darkness. This phenomenon is known as **photoperiodism.**

Photoperiodism and Flowering

The effects of photoperiodism on flowering are particularly striking (Figure 32–20). Plants are of three general types: day-neutral, short-day, and long-day. Day-neutral plants flower without regard to day length. Short-day plants flower in early spring or fall; they must have a light period *shorter* than a critical length—for instance, the cocklebur flowers when exposed to 16 hours or *less* of light. Other short-day plants are poinsettias, strawberries, primroses, ragweed, and some chrysanthemums. Long-day plants, which flower chiefly in the summer, will flower only if the light periods are *longer* than a critical length. Spinach, potatoes, clover, henbane, and lettuce are examples of long-day plants. With both short-day and long-day plants, the photoperiodic initiation of flowering can take place only if the plant has passed from its juvenile stage into a phase designated as "ripeness to flower." In woody perennials, it may take decades for a plant to reach this stage.

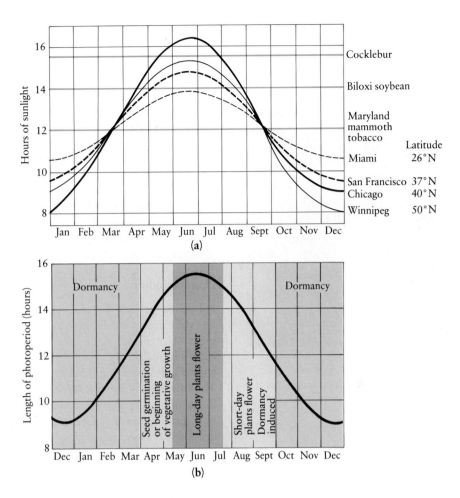

32–20 (a) *The relative length of day and night determines when many plants flower. The four curves depict the annual change in day length in four North American cities at four different latitudes. The green lines indicate the effective photoperiod of three different short-day plants. The cocklebur, for instance, requires 16 hours or less of light. In Chicago, it can flower as soon as it matures, but in Winnipeg the buds do not appear until early in August, so late that frost usually kills the plants before the seeds are mature.*

(b) *Relationship between day length and the developmental cycle of plants in the temperate zone.*

The discovery of photoperiodism explained some puzzling data about the distribution of common plants. Why, for example, is there no ragweed in northern Maine? The answer, investigators found, is that ragweed starts producing flowers when the day length is 14.5 hours. The long summer days do not shorten to 14.5 hours in northern Maine until August, and then there is not enough time for ragweed seed to mature before the frost. For somewhat similar reasons, spinach cannot produce seeds in the tropics. Spinach needs 14 hours of light a day for a period of at least two weeks in order to flower, and days are never this long in the tropics.

Note that ragweed and spinach will both bloom if exposed to 14 hours of daylight, yet one is designated as short-day and one as long-day. The important factor is not the absolute length of the photoperiod but rather whether it is longer or shorter than a particular critical interval for that variety. Detection of the interval can be very precise; in some varieties 5 or 10 minutes' difference in exposure can determine whether or not a plant will flower.

The first field studies on photoperiodism in plants were performed more than 50 years ago by scientists at the U.S. Department of Agriculture. These scientists have since become known as the Beltsville group, for the small town in Maryland where they carried out their studies.

Measuring the Dark

Subsequently, other investigators, Karl C. Hamner and James Bonner, began a laboratory study of photoperiodism. They used the cocklebur as the experimental organism. As we mentioned previously, the cocklebur is a short-day plant, requiring 16 hours or less of light per 24-hour cycle to flower. It is particularly useful for experimental purposes because a single exposure under laboratory conditions to a short-day cycle will induce flowering two weeks later, even if the plant is immediately returned to long-day conditions.

32–21 *The cocklebur, a short-day plant that can stand a lot of abuse (see essay on page 688), has been important in experimental studies of photoperiodism. Each "burr" is an inflorescence with two flowers.*

In the course of these studies, in which they tested a variety of experimental conditions, the investigators made a crucial and totally unexpected discovery. If the period of darkness is interrupted by as little as a 1-minute exposure to the light of a 25-watt bulb, flowering does not occur. Interruption of the light period by darkness has no effect whatsoever on flowering. Subsequent experiments with other short-day plants showed that they, too, required periods not of uninterrupted light but of uninterrupted darkness (Figure 32–22).

What about long-day plants? They also measure darkness. A long-day plant that will flower if it is kept in a laboratory in which there is light for 16 hours and dark for 8 hours will also flower on 8 hours of light and 16 hours of dark if the dark is interrupted by even a brief exposure to light.

Photoperiodism and Phytochrome

Following up on the clues from the Hamner and Bonner experiments, the Beltsville group was able to detect and eventually to isolate the light-detecting pigment involved in photoperiodism. This pigment, which they called **phytochrome,** exists in two different forms. One form, known as P_r, absorbs red light with a wavelength of 660 nanometers; phytochrome is synthesized in the P_r form. The other form, P_{fr}, absorbs far-red light with a wavelength of 730 nanometers. P_{fr}, which is the biologically active form of the pigment (that is, it can trigger a biological response), promotes flowering in long-day plants and inhibits flowering in short-day plants.

When P_r absorbs red light, it is converted to P_{fr} (Figure 32–23). This conversion takes place in sunlight or in incandescent light; in both of these types of light, red wavelengths predominate over far-red. The persistence of the effect of even a very brief red-light exposure is due to the fact that P_{fr}, the active form of phytochrome, is relatively stable in the dark. When P_{fr} absorbs far-red light, it is converted back to P_r. In the dark, P_{fr} is converted slowly to P_r, is degraded and replaced by newly synthesized P_r, or both.

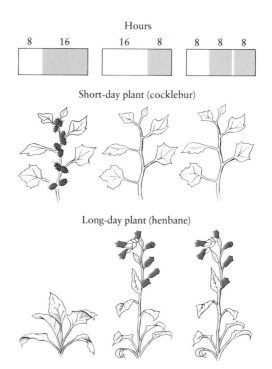

32–22 *Experiments on photoperiodism showed that plants measure the length of darkness rather than that of light. Short-day plants flower only when the dark period exceeds some critical value. Thus, the cocklebur, for instance, will flower on 8 hours of light and 16 hours of darkness. If the 16-hour period of darkness is interrupted even very briefly, as shown on the right, the plant will not flower.*

The long-day plant, on the other hand, which will not flower on 16 hours of darkness, will flower if the dark period is interrupted. Long-day plants flower only when the dark period is less than some critical value.

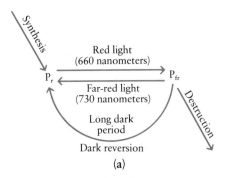
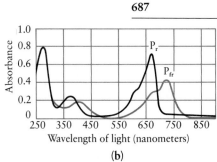

32–23 (a) *Phytochrome is synthesized (from amino acids) in the P_r form. P_r changes to P_{fr} when exposed to red light, which is present in sunlight. P_{fr} is the active form that induces the biological response. P_{fr} reverts to P_r when exposed to far-red light. In darkness, P_{fr} slowly reverts to P_r or is degraded.* (b) *Absorption spectra of the two forms of phytochrome. The similarities between the action spectra of the biological responses and the absorption spectra of the pigment provided important evidence that phytochrome is the pigment responsible for the responses. The reversible changes in absorbance provided the necessary clues both for detecting the pigment in plant extracts by spectrophotometry (page 210) and for isolating it.*

Other Phytochrome Responses

The $P_r \rightleftharpoons P_{fr}$ conversion, it is now known, acts as an off-on switch for a number of plant responses to light. Many kinds of small seeds, such as lettuce, germinate only when they are in loose soil, near the surface. This ensures that the seedling will reach the light before it runs out of stored food. Red light, a sign that sunlight is present, stimulates seed germination by converting phytochrome to the active form (P_{fr}). As with flower induction, exposure of the seeds to far-red light prevents the effect. The exposures can be alternated repeatedly and the seeds respond only to the final stimulus of the series. The inhibition of seeds by far-red light is interpreted as an adaptation that prevents them from germinating in the soil under a leaf canopy. As light filters through green leaves, the red is absorbed, leaving far-red predominant. Light-requiring plants have a better chance of survival if the seeds remain dormant, for years if necessary, until the required light is available.

Phytochrome is also involved in the early development of seedlings. When a seedling develops in the dark, as it normally does underground, the stem elongates rapidly, pushing the shoot up through the soil layers. A seedling grown in the dark will be elongated and spindly with small leaves (Figure 32–24). It will also be almost colorless, because the chloroplasts do not synthesize chlorophyll until they are exposed to light. Such a seedling is said to be etiolated. When the seedling tip reaches the light, normal growth begins. Phytochrome is involved in the switching from etiolated to normal growth. If a dark-grown bean seedling is exposed to only 1 minute of red (660 nanometers) light, it will respond with normal growth. If, however, the exposure to red light is followed by a 1-minute exposure to far-red light, thus negating the original exposure, etiolated growth continues.

The way in which phytochrome acts is not known. One recent suggestion is that it alters the permeability of the cell membrane, permitting particular substances to enter the cell, or, perhaps, inhibiting their entry, and that these substances, which probably include hormones, regulate the cell's activities. An increase in the concentrations of all the growth-promoting hormones can be detected almost immediately after phytochrome activation.

32–24 *Dark-grown seedlings, such as the bean plants on the left, are thin and pale with longer internodes and smaller leaves than the normal seedlings on the right. This group of characteristics, known as etiolation, has survival value because it increases the seedling's chances of reaching light before its stored energy supplies are used up.*

Is There a Flowering Hormone?

One of the most persistent—and still unanswered—questions in plant physiology concerns the existence of a flowering hormone. Since the 1930s, numerous experiments have suggested that such a hormone is formed in the leaves in response to an appropriate cycle of light and dark and then travels to the apical meristem of the plant and initiates flowering.

For example, (a) when certain plants, such as the cocklebur shown in the illustration below, are exposed to an appropriate light cycle, those with leaves flower and those without leaves do not. When even one-eighth of a leaf remains on a plant, flowering occurs, and the illumination of a single leaf—not necessarily the whole plant—suffices. These experiments indicate that chemical signals originating in the leaves cause the plant to flower. (b) This conclusion is supported by experiments on branched plants. Exposure of one branch to the light induces flowering on the other branch as well, even when only a portion of a leaf is present on the lighted branch. (c) When two plants are grafted together, exposure of one of the plants to the light cycle induces flowering in both the lighted plant and the grafted one.

Despite the evidence of these and other experiments, no specific flowering hormone has ever been isolated and identified. However, both auxin and gibberellin have been shown to induce flowering in some plants under some circumstances. These results suggest that flowering does not involve a single, unique hormone but rather a combination of hormones, probably acting in conjunction with other, as yet unknown, chemical factors. Recent tissue culture studies indicate that among these factors may be oligosaccharins (see Figure 32–17).

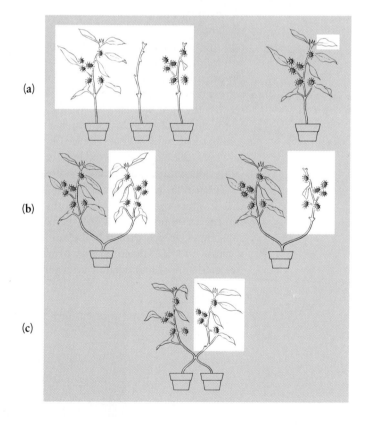

32–25 *Leaves of the oxalis plant, day (a) and night (b). Darwin believed that this folding of the leaves conserved heat energy. A more recent but also unproved hypothesis is that the "sleep movements" prevent the leaves from absorbing moonlight on bright nights, thus protecting photoperiodic phenomena.*

CIRCADIAN RHYTHMS

How can a spinach plant distinguish a 14-hour day from a 13.5-hour day? Seeking an answer to this question leads us to another group of readily observable phenomena. Some species of plants, for instance, have flowers that open in the morning and close at dusk. Others spread their leaves in the sunlight and fold them toward the stem at night (Figure 32–25). As long ago as 1729, the French scientist Jean-Jacques de Mairan noticed that these diurnal (daily) movements continue even when the plants are kept in constant dim light. More recent studies have shown that less evident activities, such as photosynthesis, auxin production, and the rate of cell division, also have daily rhythms. The rhythms continue even when all environmental conditions are kept constant. These regular day-night cycles are called **circadian rhythms,** from the Latin words *circa,* meaning "about," and *dies,* "day." They have been found throughout the four kingdoms of eukaryotic organisms.

Biological Clocks

Are these rhythms internal—that is, caused by factors within the organism itself—or is the organism keeping itself in tune with some external factor? For a number of years, biologists debated whether there might not be some environmental force, such as cosmic rays, the magnetic field of the earth, or the earth's rotation, that was setting the rhythms. Attempts to identify such an external factor have led to numerous expeditions under an extraordinary variety of conditions. Organisms have been taken down into salt mines, shipped to the South Pole, flown halfway around the world in airplanes, and, more recently, orbited in satellites—all without any detectable effect on their circadian rhythms.

Virtually all biologists now agree that circadian rhythms are endogenous—that is, they originate within the organism itself. The mechanism by which they are controlled is known as a **biological clock.** Strong evidence in support of an internal biological clock is that circadian rhythms are not exact. Different species and different individuals of the same species often have slightly different, but consistent, rhythms, often as much as an hour or two longer or shorter than 24 hours.

Biological clocks play a role in many aspects of plant and animal physiology, synchronizing internal and external events. For example, some flowers secrete nectar or perfume at certain specific times of the day or night. As a result, pollinators—which have their own biological clocks—have become programmed to visit these flowers at these times, thereby ensuring maximum rewards for both

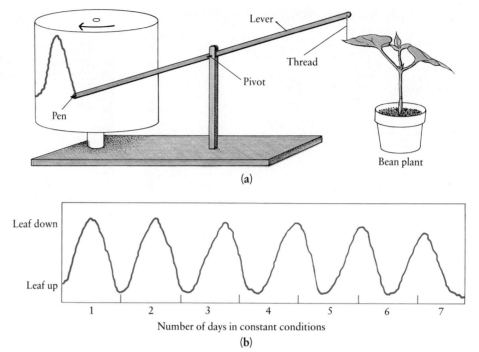

32–26 *"Sleep movement" rhythms in the bean plant. Many legumes, such as the bean, orient their leaves perpendicular to the rays of the sun during the day and fold them up at night. These movements can easily be transcribed on a rotating chart by a delicately balanced pen-and-lever system attached to the leaf by a fine thread (a). The rhythm will persist for several days in continuous dim illumination. A representative recording is seen in (b).*

the pollinators and the flowers. However, for most organisms, the "use" of the clock for such purposes is thought to be a secondary development. The primary function of the clock appears to be in enabling organisms to recognize the changing seasons of the year by "comparing" external rhythms of the environment, such as changes in day length, to their own internal rhythms.

Resetting the Clock

Although circadian rhythms originate within the organisms themselves, they can be modified by external conditions—which is, of course, important in keeping organisms in tune with their environment. For instance, a plant whose natural daily rhythm shows a peak every 26 hours when grown under continuous dim light will reset its rhythm, as if the hands of its clock had been turned, when exposed to an unusual interval of light and darkness. For example, exposed to 14-hour intervals of light alternating with 10-hour intervals of darkness, the plant resets its clock repeatedly, eventually becoming synchronized with that cycle. It can also adjust to 11 hours of light and 11 hours of darkness (or 22 hours). Such adjustment to an externally imposed rhythm is known as **entrainment**. If the new rhythm is too far removed from the original one, however, the organism will "escape" the entrained rhythm and revert to its natural one. A plant that has been kept on an artificial or forced rhythm, even for a long period of time, immediately reverts to its normal internal rhythm when returned to continuous dim light.

The Nature of the Clockwork

The chemical nature of the biological clock—or indeed whether there is just one kind of clock or many—is still not known. When the on-off switch of phytochrome was first discovered, it was hypothesized that the reversion of P_{fr} to P_r might be the time-measuring system of the plant, functioning by the same principle as an hourglass. Experiments have disproved this hypothesis, however; for instance, in many plants, P_{fr} disappears within three or four hours of darkness.

32-27 *Tendrils of a wild grape plant. Twining is caused by varying growth rates on different sides of the tendril.*

However, in some plants, the circadian rhythm of sleep movements can be rephased by conversion of P_r to P_{fr}. The best evidence to date indicates that the timing mechanism involves rhythmic changes in the cell membrane, either in the protein components, in the phospholipids, or both. Thus, phytochrome conversion might reset the clock by producing changes in membrane structure or permeability.

TOUCH RESPONSES

Twining and Coiling

Many plants respond to touch. One of the most common examples is seen in tendrils—the modified stems or leaves with which many plants support themselves and climb (Figure 32–27). Tendrils often move in a spiral as they grow; this process, called circumnutation, increases the tendrils' chances of finding a support. When the apex of a tendril touches any object, it responds to the touch by forming a tight coil. Cells touching the support shrink slightly, and those on the outer side elongate. In the garden pea, stroking the tendril with a glass rod for 2 minutes can induce a coiling response that lasts more than 48 hours. Coiling requires light, apparently for ATP production, since ATP will substitute for light in excised tendrils. Specialized epidermal cells of the tendrils are the sensors for touch, but the mechanism by which these cells induce coiling is unknown. Current evidence indicates that auxin and ethylene are probably involved; these hormones cause excised tendrils to coil even in the absence of touch.

Studies by M. J. Jaffe of Wake Forest University in North Carolina have shown that tendrils can store the "memory" of tactile stimulation. For example, if pea tendrils are kept in the dark for three days and then stroked, they will not coil, perhaps because of the requirement for ATP. If, however, they are illuminated within two hours of the stroking, they will show the coiling response.

A complex repertoire of coiling behavior is seen in the parasitic plant called dodder *(Cuscuta)*. A mature dodder plant looks like a tangle of cooked spaghetti (Figure 32–28). The plants, which have no leaves and no chlorophyll, range in color from almost white to yellow or orange and have a tiny waxy flower. When a dodder seed germinates, the seedling anchors itself with a small temporary root and begins to circumnutate. It can sense a host plant up to 8 centimeters away, presumably by chemical stimulus, and, in response, moves toward it. When it touches its new host (or any other object), it coils itself rapidly around it, pulling

32-28 (a) *Dodder* (Cuscuta), *a parasitic plant that wraps itself around and around its host. The vegetative parts of the plant are usually bright orange or yellow; the flowers are tiny, waxy, and absolutely colorless. One of the plants frequently parasitized by dodder is the halophyte* Salicornia *(see page 658), visible here as green stalks.* (b) *A salt marsh near Mill Valley, California, in which the host* Salicornia *plants are almost totally obscured by a sea of parasitic dodder.*

(a)

(b)

up its roots behind it. It then develops haustoria (see page 480) that it sinks into the vascular tissues of the host plant, tapping its food and water supply. Once its feeding network is established, the tip begins to circumnutate again. Thus alternately winding and looping, a single dodder plant can crochet itself into a mat that covers as much as a kilometer.

Rapid Movements in the Sensitive Plant

A more rapid response to touch occurs in the sensitive plant, *Mimosa pudica*. A few seconds after a leaf is touched, the petiole droops and the leaflets fold (Figure 32–29). This response is a result of a sudden change in turgor in specialized "motor cells" of the jointlike thickenings called pulvini (singular, pulvinus) at the base of the leaflets and leaves. These same cells are involved with sleep movements, which also occur in *Mimosa*. Depending on the variety and the stimulus, a single leaf or all the leaves on the plant may be affected. A series of reactions appears to be involved in spreading the stimulus. The sensory stimulus is apparently rapidly translated into an electrical signal, similar to a nerve impulse in animals, that passes along the petiole and can be detected by microelectrodes placed in the plant. The electrical signal, in turn, triggers a chemical signal that makes the cell membrane of the motor cells more permeable to potassium (K^+) and chloride (Cl^-) ions. Movement of these ions out of the motor cells causes water to leave the cells by osmosis. As a result of this water loss, the motor cells collapse, thereby causing movement of the leaf or leaflet. Tannins are also secreted by the motor cells into extracellular spaces and may protect the plant against predation.

Rapid Movements in Carnivorous Plants

Rapid responses to touch also occur in the capture of prey by the carnivorous Venus flytrap and the sundew (see essay, pages 666 to 667). The leaves of the Venus flytrap have two hinged lobes, each of which is equipped with three sensitive hairs. When an insect alights on a leaf, it brushes against the hairs, setting off an electrical impulse that triggers the closing of the leaf. The toothed edges mesh like a bear trap, snapping shut in less than half a second. Once the insect is trapped, the leaf halves gradually squeeze closed, and the captive animal is pressed against digestive glands located on the inner surface of the trap. The movement of the insect stimulates the trap to close more tightly and glands on the upper leaf surface to secrete digestive enzymes.

32–29 *Sensitive plant* (Mimosa pudica). *(a) Normal position of leaves and leaflets. (b) Response to touch. It has been hypothesized that these reactions may prevent wilting (when they occur in response to strong winds), startle insects, or dismay larger herbivores. Collapse of the leaflets is caused by rapid turgor changes in cells of the petioles. These changes are accompanied by the release from the cells of substances known as tannins. Tannins have an astringent taste and are repellent to herbivores. In the evolution of the touch response in* Mimosa, *the release of tannins may have been more important than the rapid movements.*

(a)   (b)

It was long thought that the response of the Venus flytrap, like that of the sensitive plant, involved turgor changes. Recent research, however, has revealed that the response is instead another example of acid growth (page 675). The electrical impulse set off when an insect brushes against the trigger hairs activates an enzyme that, using ATP for energy, pumps H^+ ions into the walls of epidermal cells along the outer surface of the trap's hinge; almost 30 percent of the ATP in these cells is consumed in the 1 to 3 seconds required to close the leaf completely. The rapid, irreversible cell expansion that results from acidification of the cell walls closes the trap. During this process, the cells on the inner surface of the hinge continue to grow at their usual, slow rate; some 10 hours later their growth catches up with the earlier growth of the cells on the outer surface, and the trap opens. As a result of both growth processes, the leaf is slightly larger than it was before it closed.

Studies by investigators at Washington University in St. Louis have shown that an electrical impulse is also involved in the trapping of insects by the sundew. As you saw on page 666, the club-shaped leaves of the plant are covered with tiny tentacles. A sticky droplet surrounding the tip of each tentacle attracts insects. When an insect is caught on the tip of a tentacle, the surrounding tentacles bend in, carrying the prey to the center of the leaf, where it is digested. Microelectrodes placed in the tentacles (Figure 32–30) have revealed that the touch of an insect's feet triggers an electrical impulse that moves down the tentacle.

The electrical impulses that have been observed in plants are the same, in principle, as the nerve impulses of animals (see page 847). Many botanists expect that electrical signals will be found to coordinate a variety of activities involving cells in different parts of the plant body.

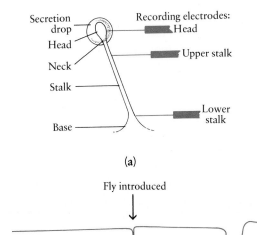

32–30 (a) Diagram of a sundew tentacle with electrodes in place. (b) A record of electrical impulses produced by positioning a fruit fly so that its feet stroked the head of a tentacle. The recording ended when the contact was broken as the tentacle bent.

Generalized Effects of Touch on Plant Growth

In addition to the specialized responses of some plants, touch and other mechanical stimuli may also have widespread effects on patterns of growth. Although botanists have long known that plants grown in a greenhouse tend to be taller and more spindly than plants of the same species grown outside, it was not until the 1970s that systematic studies revealed that regular rubbing or bending of stems inhibits their elongation and results in shorter, stockier plants. Plants in a natural environment are, of course, subjected to similar stimuli, in the form of wind, raindrops, and the movements of passing animals. This response also involves both electrical signals and a change in cell membrane permeability. The change in membrane permeability is believed to affect the relative proportions of hormones available to the plant cells.

CHEMICAL COMMUNICATION AMONG PLANTS

As we noted in Chapter 24 (page 515), many angiosperms produce toxic or bad-tasting compounds that function as powerful defenses against herbivorous animals. In some species, the production of such compounds is initiated or increased in response to damage inflicted on the plant by herbivores, for example, chewing insects. The resulting higher concentrations of these substances deter further predation and thereby protect the plant from severe damage. Studies by scientists at the University of Washington and at Dartmouth College have shown that the plants do more than protect themselves: they apparently also "warn" neighboring plants of the same species to mobilize their defenses prior to attack.

The effect was first observed in Sitka willows; three days after high levels of defensive chemicals were detected in the leaves of trees being directly attacked by tent caterpillars, high levels of the same chemicals were found in the leaves of trees that had not yet been touched by the caterpillars, some as far as 60 meters from the damaged trees. Presumably, the damaged leaves release an airborne substance that, upon reaching the leaves of another plant, triggers their synthesis of defensive chemicals. Similar effects have been observed in carefully controlled laboratory studies in which mechanical damage was inflicted on the leaves of poplar and maple seedlings. Studies currently in progress are designed to isolate and identify the airborne substance or substances involved in this communication.

As we saw in Chapter 27, chemical communication among individuals of the same species is a familiar phenomenon in animals, particularly insects. This recent research is the first hint that there may be many more interactions among individual plants than are apparent to even the well-trained eye.

SUMMARY

Plants respond to stimuli in both their internal and external environments. Such responses enable the plant to develop normally and to remain in touch with changing external conditions.

Hormones are important factors in plant responses. A hormone is a chemical that is produced in particular tissues of an organism and carried to other tissues of the organism, where it exerts one or more specific influences. Characteristically, it is active in extremely small amounts. Table 32-1 summarizes the five major groups of hormones that have been isolated from plants: auxins, cytokinins, ethylene, abscisic acid, and gibberellins. Other growth-regulating substances, such as the recently discovered oligosaccharins, may also be present.

Auxin is produced principally in rapidly dividing tissues, such as apical meristems. It causes lengthening of the shoot, chiefly by promoting cell elongation. In conjunction with cytokinin and ethylene, it is involved in apical dominance, in which the growth of axillary buds is inhibited, thus restricting growth principally to the apex of the plant. At low concentrations, auxin promotes the initiation of

TABLE 32-1 **Plant Hormones and Their Effects**

HORMONE	PHYSIOLOGICAL EFFECTS AND ROLES
Auxin	Stimulates cell elongation; involved in phototropism, gravitropism, apical dominance, and vascular differentiation; inhibits abscission prior to formation of abscission layer; stimulates ethylene synthesis; stimulates fruit development; induces adventitious roots on cuttings
Cytokinin	Stimulates cell division, reverses apical dominance, involved in shoot growth and fruit development, delays leaf senescence
Ethylene	Stimulates fruit ripening, leaf and flower senescence, and abscission; may be effector of apical dominance
Abscisic acid	Stimulates stomatal closure; may be necessary for abscission and dormancy in certain species
Gibberellin	Stimulates shoot elongation, stimulates bolting and flowering in biennials, regulates production of hydrolytic enzymes in grains

branch roots and adventitious roots; at higher concentrations, it inhibits the growth of the main root system. In developing fruits, auxin produced by seeds stimulates growth of the ovary wall; diminished production of auxin is correlated with abscission of fruits and leaves. The capacity of auxin to produce such varied effects appears to result from different responses of the various target tissues and the presence of other factors, including other hormones.

The cytokinins promote cell division. It is possible, by altering the concentrations of auxin and cytokinins, to alter patterns of growth in undifferentiated plant tissue in tissue culture.

Ethylene is a gas produced by fruits during the ripening process, which it promotes. It plays a major role in leaf abscission and is thought to be an effector of apical dominance. Abscisic acid, a growth-inhibiting hormone, affects stomatal closure and may be involved in the induction of dormancy in vegetative buds and the maintenance of dormancy in seeds.

Gibberellins stimulate shoot elongation, induce bolting and flowering in many plants, and are also involved in embryo and seedling growth. In grasses, they stimulate the production of hydrolyzing enzymes that act on the stored starch, lipids, and proteins of the endosperm, converting them to sugars, fatty acids, and amino acids, which nourish the seedling.

Oligosaccharins, short carbohydrate chains released from plant cell walls in response to various stimuli, may also be important growth regulators. Recent studies have shown that specific oligosaccharins can determine the course of development in tissue culture.

Plants respond to a number of environmental stimuli. Two responses with high survival value for young plants are phototropism, the curvature of a plant toward the light, and gravitropism, the capacity of the shoot to grow up and the root to grow down. Phototropism is mediated by auxin and gravitropism by auxin and calcium ion (Ca^{2+}).

Photoperiodism is the response of organisms to changing periods of light and darkness in the 24-hour day. Such a response controls the onset of flowering in many plants. Some plants, known as long-day plants, will flower only when the periods of light exceed a critical length. Other plants, short-day plants, flower only when the periods of light are less than some critical period. Day-neutral plants flower regardless of photoperiod. Experiments have shown that the dark period rather than the light period is the critical factor.

Phytochrome, a pigment commonly present in small amounts in the tissues of plants, is the receptor molecule for detecting transitions between light and darkness. The pigment exists in two forms, P_r and P_{fr}. P_r absorbs red light, whereas P_{fr} absorbs far-red light. P_{fr} is the biologically active form of the pigment; among its many known effects, it promotes flowering in long-day plants, inhibits flowering in short-day plants, promotes germination in lettuce seeds, and promotes normal growth in seedlings. Its mechanism of action appears to involve changes in permeability of the cell membrane.

Circadian rhythms are regular cycles of growth and activity occurring approximately on a 24-hour basis. Many of these rhythms are independent of the organism's environment and are controlled by some endogenous regulator—a biological clock. The principal function of the biological clock is apparently to provide the timing mechanism necessary for photoperiodic phenomena. Its chemical nature is not known.

Some species of plants exhibit specific, rapid movements in response to touch. Examples include the winding of tendrils, the collapse of the leaves of the sensitive plant *(Mimosa),* the triggering of the carnivorous Venus flytrap, and the bending of the tentacles of the sundew. In addition, all vascular plants appear to respond to touch and other mechanical stimuli with altered growth patterns, resulting in shorter, stockier plants. Plant responses to touch involve various combinations of electrical impulses, changes in turgor, and chemical changes that result in differential rates of growth.

Some species of plants respond to mechanical damage, such as may be caused by herbivorous animals, by synthesizing chemicals that deter further predation. They apparently also release airborne substances that "communicate" with other individuals of the same species, triggering in them the synthesis of defensive chemicals before any actual damage occurs.

QUESTIONS

1. Describe Went's experiments and the conclusions that can be drawn from them.

2. Describe the principal roles played by each of the following hormones: auxin, cytokinin, ethylene, abscisic acid, and gibberellin.

3. Describe the phenomenon of apical dominance and the role of each of the three hormones involved.

4. The distinctions we make among the plant hormones refer to general classes of their effects rather than to the precise chemical names of the substances actually produced by particular plants. Why might it be relatively easy to test chemicals for hormonal activity but be particularly difficult to determine just which chemical a plant actually produces and uses?

5. One bad apple can spoil the whole barrel. Explain.

6. Label the drawing at the right. In terms of the labeled structures, describe the action of gibberellin in initiating growth of the embryo.

7. Distinguish between the following: phototropism/photoperiodism; circadian rhythm/biological clock.

8. Why is a biological clock necessary for photoperiodism?

9. Plants must, in some manner, synchronize their activities with the seasons. Of the cues that they might use, day (or night) length seems to have been selected. What might be the advantage of using photoperiod rather than, say, temperature as a season detector?

10. Photoperiodic systems are not nearly as sensitive to low levels of illumination as are many visual systems. Why might extreme sensitivity be a disadvantage in a photoperiodic system? What might be the effect of the widespread use of bright lights for street illumination?

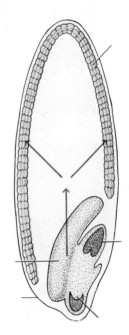

11. Suppose you were given a chrysanthemum plant, in bloom, one autumn, and you decided to keep it indoors as a house plant. What precautions would you need to take the following autumn to ensure that it would bloom again?

12. Carefully examine the graph in Figure 32-26. What is happening to the rhythm of the plant's "sleep movements" toward the end of the week? Speculate as to why this might be happening. It has also been observed that as the plant is maintained over time under the experimental conditions, it becomes rather sickly. What factors do you suppose might be involved in this change?

SUGGESTIONS FOR FURTHER READING

Books

BRADY, JOHN: *Biological Clocks, Studies in Biology, No. 104*, University Park Press, Baltimore, Maryland, 1979.*

An interesting and well-written introduction to the subject of biological clocks and their experimental study.

CUTLER, DAVID F.: *Applied Plant Anatomy*, Longman, Inc., New York, 1978.*

An interestingly written textbook on the fundamentals of plant anatomy showing some of the ways in which a knowledge of plant anatomy can be applied to solve many important everyday problems.

---

* Available in paperback.

GALSTON, ARTHUR W., PETER J. DAVIES, and RUTH L. SATTER: *The Life of the Green Plant*, 3d ed., Prentice-Hall, Inc., Englewood Cliffs, N.J., 1980.*

  A comprehensive and up-to-date description of the functioning of the green plant, especially strong in plant physiology. Written for students without an advanced background in biology or chemistry.

GREULACH, VICTOR: *Plant Structure and Function*, 2d ed., The Macmillan Company, New York, 1983.

  A short, lucid introduction to botany, with emphasis on physiology.

HEYWOOD, VERNON H. (ED.): *Flowering Plants of the World*, Prentice-Hall, Inc., Englewood Cliffs, N.J., 1985.

  The best available guide for students to the families of flowering plants.

KENDRICH, R. E., and B. FRANKLAND: *Phytochrome and Plant Growth*, Studies in Biology, No. 68, 2d ed., Edward Arnold Publishers, London, 1983.*

  A concise, well-illustrated introduction to the molecular properties and physiology of phytochromes.

LEDBETTER, M. C., and KEITH PORTER: *Introduction to the Fine Structure of Plant Cells*, Springer-Verlag, New York, 1970.

  An excellent atlas of electron micrographs of plant cells, with a detailed explanation of each.

RAVEN, PETER H., RAY F. EVERT, and SUSAN E. EICHHORN: *Biology of Plants*, 4th ed., Worth Publishers, Inc., New York, 1986.

  An up-to-date and handsomely illustrated general botany text, especially strong in evolution and ecology.

SALISBURY, FRANK B., and CLEON W. ROSS: *Plant Physiology*, 3d ed., Wadsworth Publishing Co., Inc., Belmont, Calif., 1985.

  A good, modern plant physiology text for more advanced students.

SCHNELL, DONALD E.: *Carnivorous Plants of the United States and Canada*, John F. Blair, Winston-Salem, N.C., 1976.

  Descriptions of 45 species of carnivorous plants with many color photographs. This book is not only interesting reading but also useful as a field guide, and it contains a chapter on growing techniques.

TROUGHTON, JOHN H., and F. B. SAMPSON: *Plants: A Scanning Electron Microscope Survey*, John Wiley & Sons, Inc., New York, 1973.*

  A scanning electron microscope study of some anatomical features of plants and the relationship of these features to physiological processes.

ZIMMERMANN, MARTIN H.: *Xylem Structure and the Ascent of Sap*, Springer-Verlag, New York, 1983.

  A delightfully written "idea" book on xylem structure and function by one who contributed a great deal to our understanding of functional xylem anatomy and sap movement.

ZIMMERMANN, MARTIN H., and CLAUDE L. BROWN: *Trees: Structure and Function*, Springer-Verlag, New York, 1975.

  An up-to-date discussion of how trees work, with emphasis on structure as it relates to function.

Articles

ALBERSHEIM, PETER, and ALAN G. DARVILL: "Oligosaccharins," *Scientific American*, September 1985, pages 58–64.

BRILL, WINSTON J.: "Nitrogen Fixation: Basic to Applied," *American Scientist*, vol. 67, pages 458–466, 1979.

BURNHAM, CHARLES R.: "The Restoration of the American Chestnut," *American Scientist*, vol. 76, pages 478–487, 1988.

CHILTON, MARY-DELL: "A Vector for Introducing New Genes into Plants," *Scientific American*, June 1983, pages 50–59.

EVANS, MICHAEL L., RANDY MOORE, and KARL-HEINZ HASENSTEIN: "How Roots Respond to Gravity," *Scientific American*, December 1986, pages 112–119.

EVERT, RAY F.: "Sieve Tube Structure in Relation to Function," *BioScience*, vol. 32, pages 789–795, 1982.

FOLKERTS, GEORGE W.: "The Gulf Coast Pitcher Plant Bogs," *American Scientist*, vol. 70, pages 260–267, 1982.

GOODMAN, R. M., H. HAUPTLI, A. CROSSWAY, and V. C. KNAUF: "Gene Transfer in Crop Improvement," *Science*, vol. 236, pages 48–54, 1987.

HITCH, CHARLES J.: "Dendrochronology and Serendipity," *American Scientist*, vol. 70, pages 300–305, 1982.

MARX, JEAN L.: "How Rhizobia and Legumes Get It Together," *Science*, vol. 230, pages 157–158, 1985.

MULCAHY, DAVID L., and GABRIELLA B. MULCAHY: "The Effects of Pollen Competition," *American Scientist*, vol. 75, pages 44–50, 1987.

OLSEN, RALPH A., RALPH B. CLARK, and JESSE H. BENNETT: "The Enhancement of Soil Fertility by Plant Roots," *American Scientist*, vol. 69, pages 378–384, 1981.

PETTITT, JOHN, SOPHIE DUCKER, and BRUCE KNOX: "Submarine Pollination," *Scientific American*, March 1981, pages 134–143.

RICK, CHARLES M.: "The Tomato," *Scientific American*, August 1978, pages 76–87.

RUEHLE, JOHN L., and DONALD H. MARX: "Fiber, Food, Fuel, and Fungal Symbionts," *Science*, vol. 206, pages 419–422, 1979.

SHIGO, ALEX L.: "Compartmentalization of Decay in Trees," *Scientific American*, April 1985, pages 96–103.

TOMLINSON, P. B.: "Tree Architecture," *American Scientist*, vol. 71, pages 141–149, 1983.

TORREY, JOHN G.: "The Development of Plant Biotechnology," *American Scientist*, vol. 73, pages 354–363, 1985.

WALKER, DAN B.: "Plants in the Hostile Atmosphere," *Natural History*, June 1978, pages 74–81.

---

* Available in paperback.

SECTION 6
Biology of Animals

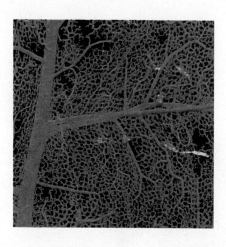

Blood carries the oxygen required for the energy-yielding reactions of the body—the reactions needed for maintenance, growth, reproduction, and the winning of gold medals. During exertion, heart output may increase tenfold, pumping up to 40 liters—about 10 gallons—of blood from the heart's left ventricle every minute, thus meeting the increased demand for oxygen.

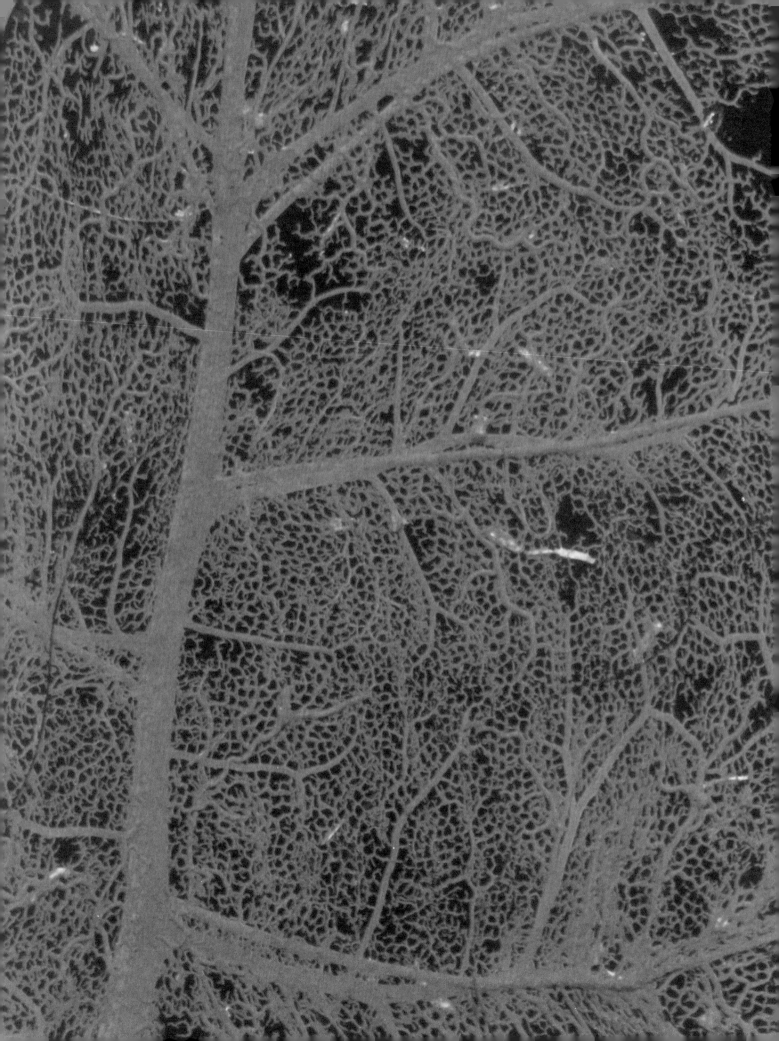

C H A P T E R 33

The Vertebrate Animal: An Introduction

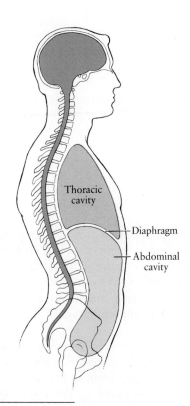

33–1 *An 80,000-kilometer network of capillaries delivers the blood to the muscles and other tissues of the body. Oxygen and nutrients move by diffusion across the thin, permeable capillary walls to enter the body cells, no one of which is more than 0.01 centimeter from one of these minute blood vessels. Depleted of oxygen, the blood returns to the heart again, propelled by muscle contraction.*

33–2 *Humans, like other vertebrates, are characterized by a dorsal central nervous system (spinal cord and brain) enclosed in vertebrae and the skull. As in other mammals, a muscular diaphragm divides the coelom into the thoracic cavity and the abdominal cavity.*

The animal kingdom, as we saw in Chapters 25 through 28, is made up of a vast array of organisms, ranging from microscopic aquatic forms to extremely complicated and highly organized multicellular creatures that are, by far, the most complex of all living systems. In this section of the book, we shall consider in more detail the principles of vertebrate anatomy and physiology, using *Homo sapiens* as our representative organism. Because we are part of a continuum of nature, it is logical to study the human animal (now one of the best understood) in order to gain a deeper understanding of animal life in general. This one species, however, will not command our full attention, for it is equally logical to study other animals in order to understand the human animal. Thus we shall rely on numerous examples from other organisms to gain insight into important physiological principles and adaptations.

Even those of us who have most marveled at the exquisite architecture of an orchid flower, or who are as content as Leeuwenhoek to watch the intricate and varied movements of a *Paramecium* and its neighbors, approach the subject of the biology of our own systems with a quickened interest. Indeed for some, student and scientist alike, the principal and perhaps the only reason for studying "lower forms" is the extent to which such studies bear directly on human welfare. But as *Escherichia coli*, T2 and T4 bacteriophages, and *Drosophila* remind us, there is no way to study only the human species, any more than it is possible to study only a fruit fly.

CHARACTERISTICS OF *HOMO SAPIENS*

The human being is a vertebrate and as such has a bony, articulated (jointed) endoskeleton that supports the body and grows as it grows. The dorsal nerve cord (the spinal cord) is surrounded by bony segments, the vertebrae, and the brain is enclosed in a protective casing, the skull.

As in other vertebrates, and most invertebrates as well, the human body contains a coelom—a cavity that forms within the mesoderm of the developing embryo (see page 532). In humans and other mammals, the coelom is divided into compartments, of which the two largest are the thoracic cavity and the abdominal cavity (Figure 33–2). These are separated by a dome-shaped muscle, the diaphragm. The thoracic cavity contains the heart, lungs, and esophagus (the upper portion of the digestive tract). The abdominal cavity contains a large number of organs, including the stomach, intestines, and liver.

Human beings are, of course, mammals. One of the most important characteristics of mammals is that they are warm-blooded. More precisely, they are endotherms (page 597), generating heat internally to maintain a high and relatively constant body temperature. As a consequence, mammals (and birds, which are also endotherms) are able to achieve and sustain levels of physical activity and mental alertness generally far greater than those of ectothermic animals, which can regulate their body temperature only within broad limits by absorbing heat from or releasing it to the environment. A concomitant of endothermy is a high metabolic rate, which requires relatively large and constant supplies of food (fuel) molecules and oxygen.

Mammals have other important characteristics. They have hair or fur rather than scales or feathers. In keeping with their high levels of activity and mental alertness, they have highly developed systems for receiving, processing, and reacting to information from the environment. All mammals (except the monotremes) give birth to live young, as distinct from laying eggs, which all birds and most fish, amphibians, and reptiles do. Mammals nurse their offspring, a process that involves a relatively long period of parental care and makes possible a long learning period. Contrast this, for example, with most insects and nearly all species of fish, amphibians, and reptiles, in which the young are independent from the moment they hatch from the egg. There is a tendency among the large mammals, in particular, toward fewer offspring per litter and prolongation of parental care. Humans, for instance, rarely have more than two surviving young per birth, only two mammary glands with which to nurse them, and an extraordinarily long period of infancy and childhood, with dependency on parents often lasting well past physical maturity. Finally, for better or worse, *Homo sapiens* is by far the most intelligent of all mammals.

CELLS AND TISSUES

The body of a vertebrate, like that of all complex multicellular organisms, is made up of a variety of different, specialized cells. Although these cells greatly resemble one-celled organisms in their requirements, they differ from one-celled organisms in that they develop and function as part of an organized whole. Cells are organized into tissues, which are groups of cells carrying out a unified function. Different kinds of tissues, united structurally and coordinated in their activities, form organs, such as the stomach.

Experts can distinguish about 200 different cell types in the human body, which are customarily classified into only four tissue types: (1) epithelial, (2) connective, (3) muscle, and (4) nerve.

Epithelial Tissues

Epithelial tissues consist of continuous sheets of cells that provide a protective covering over the whole body and contain sensory nerve endings. They also provide a protective wrapping for individual internal organs and form the interior lining membranes of organs, cavities, and passageways. As a moment's reflection will reveal, everything that goes into and out of the body and its various organs must pass through epithelium, which thus plays an important regulatory role in the movement of molecules and ions.

Epithelial tissues are classified according to the shape of the individual cells as squamous, cuboidal, or columnar (Figure 33–3). They may consist of only a single layer of cells (simple epithelium), as found in the inner lining of the circulatory system, or several layers (stratified epithelium), as found in the outer layer (epidermis) of the skin (Figure 33–4). One surface of the epithelial sheet is always attached to an underlying layer, called the basement membrane. This layer is composed of glycoproteins in which a fibrous protein derived from collagen is embedded; both the glycoproteins and the fibrous protein are produced by the epithelial cells themselves.

The epithelium of the body cavities and passageways frequently contains modified epithelial cells that secrete mucus, which lubricates the surfaces (see

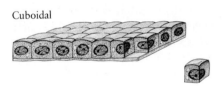
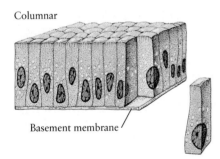

33–3 *The three types of epithelial cells that cover the inner and outer surfaces of the body. Squamous cells, which usually perform a protective function, make up the outer layers of the skin and the lining of the mouth and other mucous membranes. There are usually several layers of these flat cells piled on top of one another. Cuboidal and columnar cells, which line many internal passageways, are often involved in active transport and other energy-requiring processes. They also perform much of the chemical work of the body.*

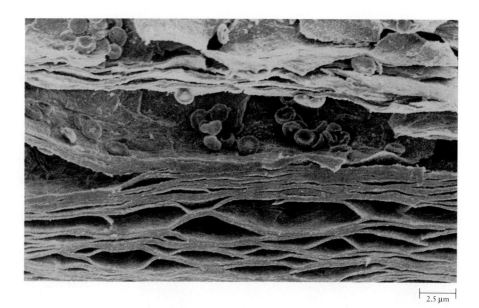

33–4 *The outer surface, or epidermis, of the skin is composed of stratified squamous epithelium. In the lower portion of this scanning electron micrograph, the layers of epithelium are clearly visible. On the skin surface, some of the older cells are being sloughed off. This is a continuous process in epithelial tissue that is subjected to wear and tear. Replacement cells are produced by cell division in the underlying layers of the epidermis. A number of red blood cells can be seen just below the top layer of epithelial cells.*

Figure 4–13, page 94). Other epithelial cells, specialized for the synthesis and secretion of specific substances for export, are often clustered together to form **glands**. Among the substances produced by glands are perspiration, saliva, milk, hormones, and digestive enzymes. Glandular epithelium is composed of cuboidal or columnar cells.

Desmosomes and Tight Junctions

In epithelial tissues that function as coverings and linings, the physical integrity of the tissue as a whole—no tears, no leaks—is often of paramount importance. Two general types of cell-cell junctions play an essential role in maintaining this integrity: **desmosomes** and **tight junctions** (Figure 33–5). Desmosomes have often been compared to spot welds between cells. They consist of plaques of dense fibrous material between cells, with clusters of filaments from the cytoplasm of the neighboring cells looping in and out of them. Desmosomes attach cells to one another and give tissues mechanical strength. They are found in especially large numbers in tissues subjected to mechanical stress, such as the skin.

Tight junctions, which appear to involve the fusion of adjacent cell membranes, form a continuous seal around each cell in a layer of tissue, preventing leakage between cells. For example, intestinal epithelial cells are surrounded by tight junctions that keep the intestinal contents from seeping between the cells. Similar arrangements are found in the epithelial cells that form the linings of other internal organs, such as the bladder and the kidney. Tight junctions occur only in vertebrates, but slightly different structures, known as septate junctions, perform the same function in invertebrates.

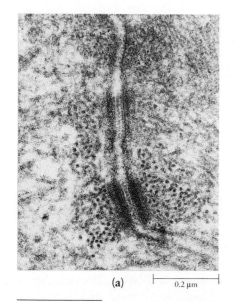

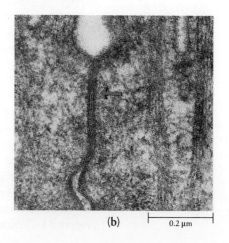

33–5 *Junctions between epithelial cells maintain the structural integrity of the tissue.* **(a)** *Desmosomes cement adjacent cells together. Here three desmosomes connect the cell membranes of two adjacent epithelial cells of the large intestine of a rat. The upper structure is a belt desmosome, and the lower two structures are spot desmosomes. Filaments cut in cross section as the sample was prepared for the electron microscope appear as black dots. As you can see, the filaments within belt desmosomes have a smaller diameter than those within spot desmosomes; the belt desmosome filaments are thought to contain actin (page 121).* **(b)** *A tight junction (arrow) seals cells together and prevents materials from leaking between them. These cells are from the epithelium lining the small intestine of a rat.*

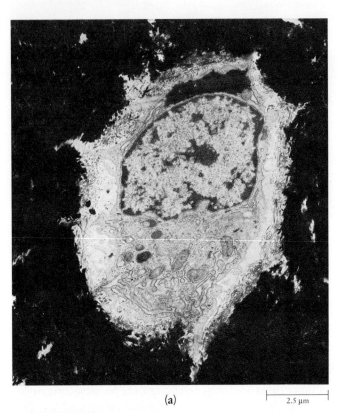
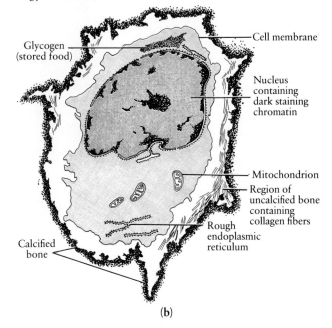

33–6 (a) *Electron micrograph and* (b) *diagram of a bone cell (osteocyte). Young bone cells (osteoblasts) produce the extracellular matrix, an organic material consisting of collagen fibers and ground substance. This matrix gradually hardens as calcium compounds in the ground substance crystallize. Cytoplasmic processes extend from the cells through channels in the hardened bone, as shown in* (c).

Connective Tissues

Connective tissues bind together, support, and protect the other three kinds of tissues. Unlike epithelial tissue cells, the cells of connective tissues are widely separated from one another by large amounts of extracellular material, the matrix, which anchors and supports the tissue. The matrix, synthesized by the cells of the tissue, consists of a ground substance, which is more or less fluid and amorphous (formless), and, in many connective tissues, fibers. There are several different types of fibers, which vary according to the particular tissue: (1) connecting and supporting fibers, such as collagen fibrils (Figure 3–23, page 77), which are a major component of skin, tendons, ligaments, cartilage, and bone; (2) elastic fibers, which are found, for example, in the walls of large blood vessels; and (3) reticular fibers, which form networks inside solid organs, such as the liver. While epithelial tissues are classified according to cell shape and arrangement, connective tissues are grouped by the characteristics of their extracellular matrix.

The principal connective tissues, by volume, in the human body are bone, blood, and lymph. In blood and lymph, the extracellular matrix is a watery fluid, called plasma, that contains numerous ions and molecules. A variety of specialized cells (to be discussed in Chapters 36 and 39) circulate through the body in this fluid matrix. The extracellular matrix of bone, by contrast, is impregnated with hard crystals of calcium compounds. Like other connective tissues, however, bone is living matter, consisting of cells, fibers, and ground substance (Figure 33–6). Bone tissue, despite its strength, is amazingly light; the human skeleton (Figure 33–7) makes up only about 18 percent of our weight.

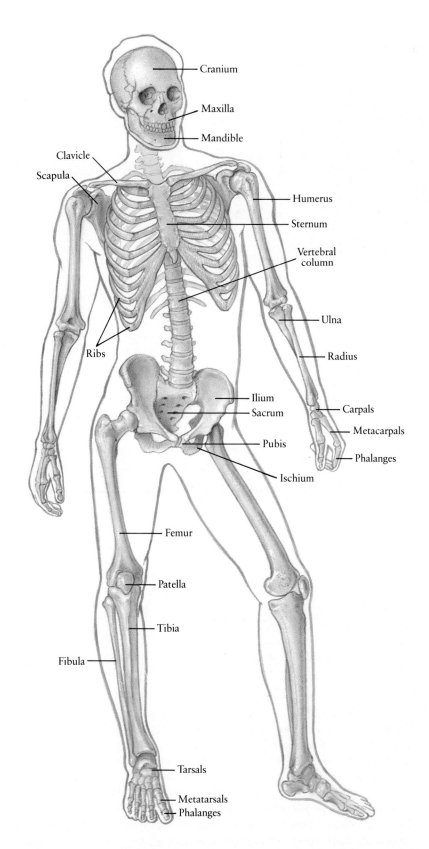
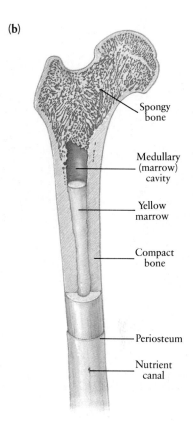

33–7 (a) *The skeleton of a human adult contains 206 bones. Twenty-nine are in the skull, including 14 face bones and six small bones (ossicles) of the ears. There are 27 bones in each hand and 26 in each foot. Bones are living organs, made up not only of connective tissue but of the other tissue types as well, including nerve, muscle, and the epithelial lining of the blood vessels that traverse the bones.*

(b) *The ends of long bones, such as this femur, consist of spongy bone, in which there are large spaces, surrounded by compact bone. The shaft, which is hollow, is composed of compact bone. A central cavity containing bone marrow extends through the center of the shaft. The marrow of long bones is yellow because of stored fat. The periosteum is a fibrous sheath, which contains the blood vessels that supply oxygen and nutrients to the bone tissues. Blood vessels emerge from bone through openings known as nutrient canals.*

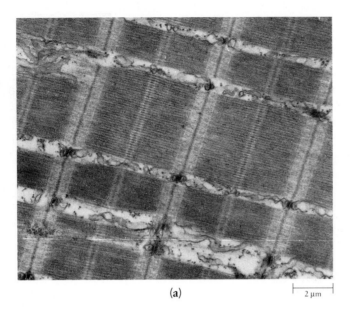
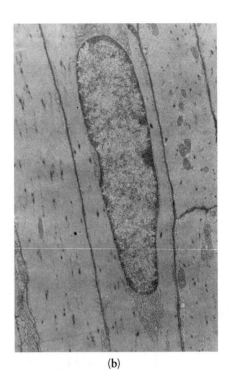

33-8 *Muscle cells are characterized by very thin fibrils of contractile proteins that run lengthwise through the cells and make up the bulk of the cytoplasm. These fibrils are arranged in a regular pattern in striated muscle, but they are irregular in smooth muscle and form no apparent pattern.*

(a) Electron micrograph of skeletal muscle, showing the striated pattern. A muscle fiber is a single large cell, containing many nuclei. The individual fibrils within the cell are separated from each other by a specialized endoplasmic reticulum, visible here as relatively wide, approximately horizontal, transparent bands.

(b) Electron micrograph of smooth muscle from the spermatic duct of a mouse. Smooth muscle is made up of long, spindle-shaped cells; portions of several cells are visible in this micrograph. Unlike the cells of skeletal muscle, but like most cardiac muscle cells, each smooth muscle cell contains only a single nucleus. One cell's nucleus, distinguishable by its granular texture, can be seen in the middle of the micrograph.

Muscle Tissue

Muscle cells are specialized for contraction. Every function of muscle—from running, jumping, smiling, and breathing to propelling the blood through the body and ejecting the fetus from the uterus—is carried out by the contraction of muscle cells in concert.

There are two general types of muscle tissue, as shown in Figure 33–8: striated muscle, which has a striped appearance under the microscope, and smooth muscle (no stripes). Smooth muscle surrounds the walls of internal organs, such as the digestive organs, uterus, bladder, and blood vessels. The muscles that move the skeleton are striated muscle and are sometimes called voluntary muscles since we can move them at will. A special type of striated muscle, cardiac muscle, makes up the wall of the heart. Smooth muscle and cardiac muscle are not, except in rare individuals, under direct voluntary control and are thus categorized as involuntary.

Contraction of muscle cells depends on the interaction of two proteins: actin and myosin (page 121). In skeletal and cardiac muscle, these proteins are arranged in regular, repeating assemblies, resulting in the characteristic striations.

Striated Muscle

Some 40 percent of a man's body, by weight, consists of the striated muscle that moves the skeleton; women characteristically have less, around 20 percent. A skeletal muscle is typically attached to two or more bones, either directly or, more often, indirectly, by means of the tough strands of connective tissue known as tendons. Some tendons, such as those that connect the finger bones with their muscles in the forearm, are very long. When the muscle contracts, the bones move around a joint, which is held together by ligaments and generally contains a lubricating fluid. Most of the skeletal muscles of the body work in antagonistic groups, one flexing, or bending, the joint, and the other extending, or straightening, it (Figure 33–9). Also, two antagonistic groups may contract together to stabilize a joint. Such muscle action makes it possible for us (and others) to stand upright.

The Injury-Prone Knee

A joint is a structure, usually movable, at which two bones are opposed. The knee is the hinge joint between two strong, inflexible bones, the femur and the tibia. It is covered by the "kneecap," or patella (removed in this drawing, but visible in Figure 33-7a). The patella acts as a stop so that the hinge can open to 180° but no farther.

As in most joints, the bones of the knee are bound together by ligaments. The collateral ligaments bind the joint externally; they are slack when the knee is flexed and taut when the knee is extended. The cruciate ligaments cross within the joint, stabilizing it. The menisci are C-shaped wedges of cartilage, resting on the upper surface of the tibia and cushioning the joint.

Knees are particularly susceptible to injury, as Billie Jean King, Joe Namath, Wilt Chamberlain, and others will sadly attest. The reason for the knee's vulnerability becomes clear when you contemplate the analogous problem of fastening together two match sticks end-to-end with several rubber bands, to produce a hinge that is both strong and flexible and can not only swing back and forth but also, to some extent, twist, bend, and rotate. Common knee injuries are torn ligaments, especially the collateral ligaments (as a result of the knee's being hit from the side), and

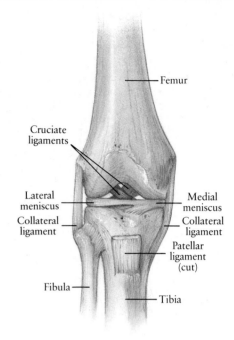

The right knee, viewed from the front, with the patella removed.

crushed menisci. When the shock-absorbing wedges of menisci are lost, bone chips begin to accumulate in the joint, making it less flexible and much more painful.

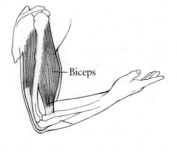

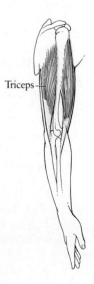

33–9 *Muscles attached to bone move the vertebrate endoskeleton. They often work in antagonistic pairs, with one relaxing as the other contracts. Muscles cannot lengthen spontaneously; they lengthen only when the joint moves in the opposite direction, due to contraction of the antagonistic muscles. For example, when you move your hand toward your shoulder, as shown here, the biceps contracts and the triceps relaxes. When you move your hand down again, the triceps contracts, while the biceps relaxes. The muscles that move the skeleton, such as those diagrammed here, are known as skeletal muscles. They are striated, as shown in Figure 33–8a.*

A skeletal muscle, such as the biceps, consists of bundles of muscle fibers—often hundreds of thousands of fibers—held together by connective tissue. Each fiber is a single cell, cylindrical or spindle-shaped (tapering at both ends), with many nuclei, formed by the fusion of a large number of small, mononucleate cells during embryonic development. These fibers are often very large cells—50 to 100 micrometers in diameter and many centimeters long. Their internal structure and the mechanism by which they contract will be described in Chapter 42.

Cardiac muscle resembles skeletal muscle in its assemblies of actin and myosin and, thus, in its striated appearance. However, its cells, which are shorter than those of skeletal muscle, are usually mononucleate and have branched, rather than tapered, ends. As we shall see in Chapter 36, certain cardiac muscle cells have the capacity to contract spontaneously some 70 times per minute, thus initiating the heartbeat hour after hour, day after day, throughout an entire lifetime.

Smooth Muscle

Although smooth muscle cells also contain actin and myosin, the molecules are not arranged in regular assemblies and do not form a striated pattern. Smooth muscle cells are spindle-shaped and mononucleate. Functionally, smooth muscle contracts much less rapidly than skeletal striated muscle, and its contractions are more prolonged. As we noted previously, it is generally not under voluntary control.

In most hollow organs, such as the intestines, smooth muscle fibers are organized into sheets arranged in two layers—an outer, longitudinal layer and an inner, circular layer—which can contract alternately. As you will recall, such a juxtaposition of circular and longitudinal muscles makes possible the locomotion of the earthworm (page 555). In the internal organs of a vertebrate, the changes in diameter or shape that result from alternating contraction of the two sets of muscles are utilized to propel fluids or semisolid materials (such as a food mass) through the organ. In blood vessels, bundles of smooth muscle fibers encircle the walls, constricting the vessels as the fibers contract. Depending on the location of the vessel and the strength of the contraction, these muscles either provide added propulsion to the moving blood or regulate its pressure and rate of flow.

Nerve Tissue

The fourth major tissue type is nerve tissue. The essential functional units of nerve tissue are the **neurons,** which transmit nerve impulses. Nerve tissue also contains cells of another type, known as glial cells, that physically support and insulate the neurons. These cells are called neuroglia when they are in the central nervous system (brain and spinal cord) and Schwann cells when they are outside the central nervous system. Glial cells are believed to supply the neurons with nutrients and other molecules and may play an important part in maintaining the ionic composition of nerve tissue, which is, as we shall see in Chapter 41, central to its function.

As shown in Figure 33–10, neurons come in a diversity of sizes and shapes. Typically, a neuron consists of a **cell body,** which contains the nucleus and much of the metabolic machinery of the cell; **dendrites,** usually numerous, short, threadlike cytoplasmic extensions—processes—that, with the cell body, receive stimuli from other cells; and an **axon,** a long process that is capable of rapidly conducting an electrochemical signal—the nerve impulse—over great distances. Axons are also known as nerve fibers.

Neurons are specialized to receive signals—from the external environment, the internal environment, or other neurons—to integrate the signals received, and to transmit the integrated information to other neurons, muscles, or glands. Func-

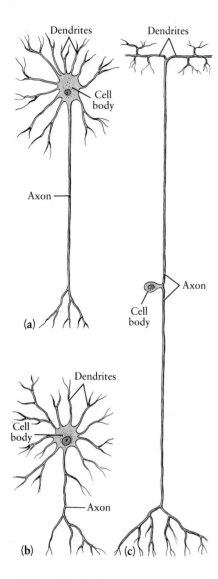

33–10 *Three of the many different forms characteristic of vertebrate neurons.* (a) *Motor neurons and relay neurons are characterized by a cell body with numerous dendrites and a long axon that travels without interruption to its terminal, where it branches.* (b) *Interneurons, which are found within localized regions of the central nervous system, typically have a complex system of dendrites and, as shown here, a short axon with branches —or no axon at all.* (c) *In sensory neurons, which transmit impulses from sensory receptors at the ends of the dendrite branches, the cell body is often off to one side of the long axon. All of these neurons form connections, known as synapses, with other neurons.*

33–11 *A representative reflex arc, showing the functions of the major types of neurons. Sensory receptor cells stimulate a sensory neuron, which relays a signal to an interneuron within a localized region of the central nervous system (often the spinal cord). From the interneuron, the signal is transmitted to a motor neuron, which, in turn, stimulates an effector, shown here as a muscle cell. Not shown in this diagram are relay neurons, which simultaneously transmit to other regions of the central nervous system (typically in the brain) the information received from the sensory neuron. These basic components of the reflex arc are found in all vertebrates, from the simplest to the most complex. Reflex arcs play an essential role in the regulation of many internal processes, as well as making possible almost instantaneous responses to numerous environmental stimuli.*

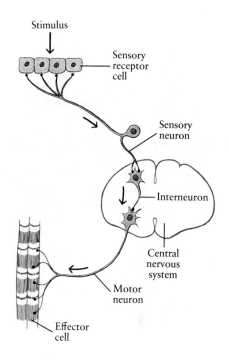

tionally, there are four classes of neurons: **sensory neurons,** which receive sensory information and relay it to the central nervous system; **interneurons,** which transmit signals within localized regions of the central nervous system; **relay neurons,** which relay signals between different regions of the central nervous system; and **motor neurons,** which transmit signals from the central nervous system to effectors, such as muscles or glands. These four types of neurons are linked in a variety of circuits, ranging from simple reflex arcs (Figure 33–11) to the extremely complex interconnections that characterize the human brain.

Neurons may reach astonishing lengths. For example, the axon of a single motor neuron may extend from the spinal cord down the whole length of the leg to the toe. Or the axon of a sensory neuron, the cell body of which is located just outside the spinal cord, may extend from the toe to the cell body and then up the entire length of the spinal cord to the lower part of the brain, where it terminates. In a giraffe, such a cell might be close to 5 meters long, and in an adult human, close to 2 meters.

Nerves are bundles of many axons from many neurons—usually hundreds and sometimes thousands. Each axon is capable of transmitting a separate message, like the wires in a telephone cable.

LEVELS OF ORGANIZATION

In animals, as in all other organisms, the basic structural unit is the living cell. As we have just seen, the vertebrate body comprises a variety of cells, organized into four types of tissues—groups of cells with a similar function. At the next level of organization, different kinds of tissues, united structurally and coordinated in their activities, form organs. The stomach, for example, is an organ made up of layers of glandular epithelium (in the stomach lining), connective tissue, nerves, and smooth muscle. The structure of the largest organ of the human body, the skin, is shown in Figure 33–12 on the next page.

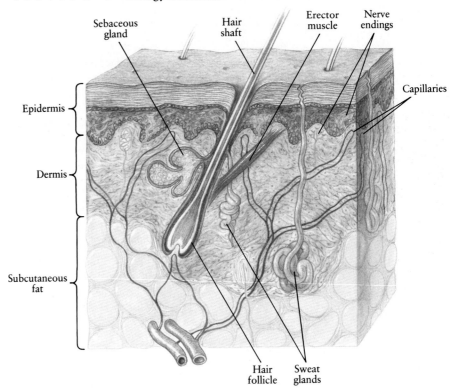

33–12 *A section of human skin, the body's largest organ, showing the epidermis and the dermis, which together make up the skin, and the underlying subcutaneous fat.*

The epidermis is made up mostly of epithelial cells and consists of two layers: an inner layer of living cells, and an outer layer of dead cells filled with keratin (page 77). The epidermis is a turnover system: cells produced in the inner layer migrate toward the surface and die. At the base of the epidermis are pigment cells that produce the granules responsible for skin color.

The dermis, consisting mostly of connective tissue, contains sensory nerve endings, small blood vessels known as capillaries, erector muscles, which raise the hair when contracted, and sweat and sebaceous glands, which are composed of modified epithelial cells. The latter produce a fatty substance that lubricates the skin surface. Fatty tissue, which makes up the insulating layer below the dermis, is also a form of connective tissue, as is the blood found in the capillaries.

Organs that work together in an integrated fashion to perform a particular function make up **organ systems,** the next level of organization. The digestive system, for example, is composed of the stomach and a number of other organs, each of which carries out specific activities that contribute toward the overall process. The organ systems together constitute the complete animal, a living organism that interacts with its external environment, including not only the physical environment but also other individuals, both of its own species and of other species. As we examine the structure and physiology of the vertebrate animal in the subsequent chapters of this section, we shall do so principally in terms of organ systems and the functions they perform.

FUNCTIONS OF THE ORGANISM

In Chapter 30, we examined the structure of the plant body in terms of the challenges imposed by the transition from life in the water to life on the land. Similarly, the structures of the animal body "make sense" only when seen as adaptations that solve particular problems presented by the relationship between the organism and its environment.

Before we further employ this metaphor of problem and solution, however, we should stop for a moment to see what we really mean by biological problem solving. An organism confronts its "problems" with a set of genetic instructions. If they work, the organism lives and passes on these instructions to its offspring. If its instructions are better than those carried by its neighbor, its offspring will probably be more numerous. It is in this way that "problems" get "solved."

Energy and Metabolism

A major problem for any living system—indeed, *the* major problem—is that posed by the second law of thermodynamics: to maintain the high level of organization characteristic of such systems in the face of the universal trend toward disorder

33–13 *Members of the shrew family, such as this pygmy shrew, have a regular three-hour cycle of activity and rest. Throughout the day and night, winter and summer, they alternate three hours of rest with three hours of feeding on insects and other small invertebrates. Pygmy shrews are common in Europe and central Asia. Their bodies are about 5 centimeters long, and their tails are almost 4 centimeters long.*

(page 164). To do this, organisms need sources of energy and raw materials to maintain and operate their energy-extracting machinery. Animals, as we saw in Chapter 25, are multicellular heterotrophs that ingest their food, from which energy is ultimately released by the oxygen-requiring reactions of cellular respiration. In large animals, a critical problem is transforming the ingested food into molecules that can be used by individual cells and then providing those molecules, along with oxygen, to each of the multitude of cells that make up the organism's body. The complexities of the vertebrate digestive, respiratory, and circulatory systems, to be discussed in Chapters 34 through 36, represent particular evolutionary solutions to this problem.

Homeostasis

An equally critical but less obvious problem facing all animals is the maintenance of a relatively constant internal environment. This problem is also imposed by the laws of physics and chemistry. The chemical reactions of a living cell and the synthesis and maintenance of its constituent structures require a tightly controlled chemical environment, a fairly narrow range of temperature, and protection from foreign invaders, such as bacteria and viruses, that may feed on the substance of the cell, poison its enzymes with toxins, or subvert its genetic machinery for their own replication.

As we saw in Chapter 1, **homeostasis,** the maintenance of a relatively constant internal environment, is one of the identifying characteristics of living systems. Such constancy is, in many ways, more difficult for one-celled organisms and small multicellular organisms than it is for large animals, such as *Homo sapiens*. Small organisms are extremely vulnerable to changes in the temperature or chemical composition of the medium in which they live, and it is likely that the strongest evolutionary pressures toward larger size and multicellularity were related to homeostasis. Larger organisms, with their lower surface-to-volume ratios (page 103) and, often, protective outer coverings, have a greater resistance to external change and foreign invaders. The world in which your individual cells function and flourish is decidedly different from the world around you—but not so different from the warm primordial soup in which we all began.

Maintaining a relatively constant environment throughout the body of a large animal is a complex process. It involves not only continuous monitoring and regulation of many different factors but also ready defenses against an enormous diversity of microorganisms. As we shall see, virtually all of the organ systems participate in homeostasis. Particularly important roles are played by the urinary system (Chapter 37), the immune system (Chapter 39), and the two organ systems responsible for integration and control.

Integration and Control

The third problem faced by animals has two aspects. First, homeostasis requires coordinating the activities of the numerous cells that constitute the organism, so that tissues and organs respond to overall physiological needs, which change with fluctuations in the environment. Second, animals are typically quite active, moving about as they seek mates and search for and capture food, while simultaneously trying to avoid being captured by other animals. A life of active movement requires receiving and processing information from the external environment, followed by coordinated and appropriate contractions of the skeletal muscles.

There are two major control systems in animals, the endocrine system (the hormone-secreting glands and their products) and the nervous system. Speaking very generally, the endocrine system is responsible for changes that take place over

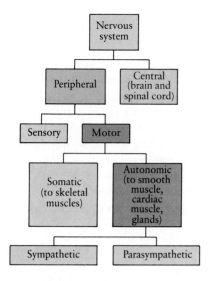

33-14 *The nervous system of vertebrates, including* Homo sapiens, *consists of a central nervous system (the brain and spinal cord) and a peripheral nervous system, a vast network of nerves connecting the central nervous system with all other parts of the body. Sensory neurons carry information to and motor neurons carry information from the central nervous system. The motor neurons are organized into the somatic and autonomic systems, and the autonomic system contains two divisions, the sympathetic and the parasympathetic.*

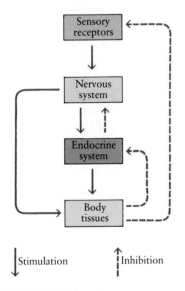

33-15 *A schematic representation of the feedback control pathways that regulate virtually all physiological processes. As we shall see in subsequent chapters, the details—the type of sensory stimuli perceived, the regions of the nervous system activated, the hormones released, and the target tissues of nervous and endocrine control—vary from process to process, but the underlying principle is always the same.*

a relatively long time period—minutes to months—whereas the nervous system is involved with more rapid responses—milliseconds to minutes. However, the more we learn about these systems, the more they are seen to be closely interrelated. For instance, the production of sex and other hormones was once believed to be under the control of the "master" pituitary gland; it is now known that the pituitary, though perhaps a master in some aspects of the endocrine world, is actually an executive secretary of the **hypothalamus,** a major brain center. And within little more than the last decade, it has been discovered that the embryonic development of certain major brain centers is profoundly influenced by hormones regulated by the pituitary.

Even anatomically, the endocrine and nervous systems are not distinct. One of the body's most important glands, the adrenal medulla, a major source of adrenaline (also known as epinephrine), is not, strictly speaking, a gland. That is, it is not modified epithelial tissue but is rather a large ganglion—a collection of nerve cell bodies—whose nerve endings secrete the hormone.

Within the nervous system, there are a number of structural and functional subdivisions (Figure 33-14). One subdivision, the **somatic** system, innervates skeletal muscle. Another, the **autonomic** ("involuntary") system, innervates smooth muscle, cardiac muscle, and glands. The autonomic system is further subdivided into the **sympathetic** and the **parasympathetic** divisions, which interact with one another in a finely tuned system of checks and balances. The sympathetic division is most active in times of stress or danger. Among the major effectors of the sympathetic division is the adrenal medulla, and the overall effects of general sympathetic stimulation are those we associate with a "rush of adrenaline." The parasympathetic division plays its major role in supporting everyday activities such as digestion and excretion. The functions of the endocrine and nervous systems will be described in greater detail in Chapters 40 through 43, but we introduce them now because every organ system to be discussed is under the influence of both the endocrine system and the sympathetic and parasympathetic divisions of the autonomic nervous system.

Feedback Control

The body's integration and control systems characteristically act through feedback loops, both negative and positive. The simplest example from everyday life of a negative feedback system is the thermostat that regulates your furnace. When the temperature in your house drops below the preset thermostat level, the thermostat turns the furnace on. When the temperature rises above the preset level, the thermostat turns the heat off. In a living organism, systems are seldom completely on or off, and homeostatic control is much more finely modulated. The principle, however, is the same: a deviation from a "preset" condition stimulates a response that reduces the deviation.

Feedback loops typically involve both the nervous and endocrine systems (Figure 33-15). One important homeostatic function of the body, for example, is keeping the blood volume constant. This is accomplished, in large part, by regulation of the rate at which water is removed from the bloodstream by the kidneys. A hormone known as antidiuretic hormone (ADH), which is produced by the hypothalamus and released from the pituitary gland, acts on the tubules of the kidney to decrease water excretion. The production of ADH is controlled by sensory receptors in the circulatory system, particularly in the heart, that measure blood pressure—an indirect measure of blood volume. When blood pressure goes up, firing of these receptors inhibits the release of ADH, and more water is excreted, reducing blood volume. As blood pressure goes down, the stimulus from the receptors decreases, ADH production increases, water is retained by the kidney, and blood pressure and blood volume increase.

Some feedback systems involve additional relay loops. The thyroid gland, for instance, produces the thyroid hormone, thyroxine, which, among other effects, increases the rate of cellular metabolism. The amount of thyroxine produced depends on the production of another hormone, thyroid-stimulating hormone (TSH) from the pituitary gland. A major factor regulating the production of TSH is the concentration of thyroid hormone in the circulating blood. Thus, although an additional step is involved, the principle is the same: the concentration of the hormone itself or the response to the hormone by a target tissue inhibits the synthesis or secretion of the hormone in question.

Many functions are controlled by a number of separate feedback control pathways. An example is temperature regulation, which we shall consider in Chapter 38. Core body temperature is controlled by a thermostat located in the hypothalamus. This thermostat collects information from a number of sensory receptors, integrates it, compares the results with its thermostat setting, and marshals a complex of appropriate responses.

Continuity of Life

The fourth challenge an organism faces—which may or may not be a problem—is to multiply. The biological imperative to reproduce, following the dictates of the genes, is enormous. As we shall see when we consider evolution and ecology later in this text, animals devote much of their energy and resources to meeting this challenge. Reproduction, as we saw in Section 4, may be carried out in a variety of ways, but in mammals it is always sexual and always involves formation of gametes, their union to form a zygote, or fertilized egg, and the development of the zygote into an adult. The last two chapters of this section will describe these processes, with particular emphasis on the formation of human gametes, their fusion in the fertilized egg, and the almost miraculous, though oft-repeated, phenomena by which this single cell develops into a human being.

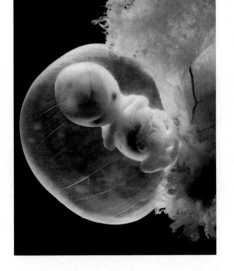

33-16 *Genus,* Homo; *species,* sapiens. *Embryo at 7 weeks.*

SUMMARY

Vertebrates, including *Homo sapiens,* are characterized by a bony, jointed endoskeleton that includes a skull and a vertebral column enclosing the central nervous system (the brain and spinal cord). The human body, like that of other mammals, contains a coelom that is divided by a muscle, the diaphragm, into two major compartments, the abdominal cavity and the thoracic cavity.

The cells of the vertebrate body are organized into tissues, groups of cells that carry out a unified function. Various types of tissues are grouped in different ways to form organs, and organs are grouped to form organ systems. The four principal tissue types of which the vertebrate body is made are epithelial tissue, connective tissue, muscle, and nerve. Epithelium serves as a covering or lining for the body and its cavities. Glands are composed of specialized epithelial cells; their secretions include mucus, perspiration, milk, saliva, hormones, and digestive enzymes. Connective tissues are characterized by their capacity to secrete substances, which often include collagen and other fibers, that make up the extracellular matrix. They serve to support, strengthen, and protect the other tissues of the body. Muscle cells are specialized for contraction, which is accomplished by assemblies of two proteins, actin and myosin. In striated muscle, which includes skeletal muscle and cardiac muscle, these assemblies form a striped pattern, visible under the microscope. In smooth muscle, no such pattern is apparent. Nerve cells, or neurons, are specialized for the reception, processing, and transmission of information. Neurons typically consist of a cell body, dendrites, and an axon. Signals, in the form of electrochemical impulses, can be conducted rapidly over great distances by the axon. Neurons are surrounded and supported by glial cells.

Four basic functions are essential to the on-going life of a multicellular animal. Food must be obtained and processed to yield molecules that can be used by individual cells, and these molecules, as well as oxygen, must be delivered to the cells. The internal environment must be closely regulated, a process known as homeostasis. The contractions of its skeletal muscles in locomotion and the physiological activities of its many tissues and organs must be coordinated in response to changes in both the internal and external environments. Integration and control involve both the endocrine and nervous systems and characteristically function through feedback loops. Finally, as dictated by its genes, the organism reproduces.

QUESTIONS

1. Distinguish among the following terms: animal/vertebrate/mammal; coelom/diaphragm; endoskeleton/exoskeleton; muscle cell/muscle fiber; striated muscle/smooth muscle; skeletal muscle/cardiac muscle; nerve cell/neuron/glial cell; neuroglia/Schwann cell; dendrite/axon; sensory neuron/interneuron/relay neuron/motor neuron; nerve fiber/nerve; somatic/autonomic; sympathetic/parasympathetic.

2. What is the functional significance of each of the four tissue types? Give an example of each type.

3. What are the three types of epithelial tissue? What is the basis for classification of epithelial cells?

4. In addition to protection, what is the other major function of epithelial tissue?

5. What is the major structural difference between connective tissue and epithelial tissue? What functions of connective tissue account for this structural difference?

6. What is meant by antagonistic paired muscles?

7. Give some examples of homeostasis. Why is it so important?

8. What are the major survival problems of an organism? Compare the solutions to these problems achieved by plants with those achieved by vertebrate animals.

9. Label the drawing below. Explain the process that is taking place, as signified by the four arrows.

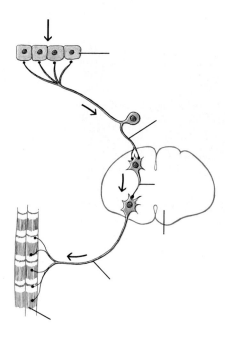

CHAPTER 34

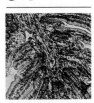

Energy and Metabolism I: Digestion

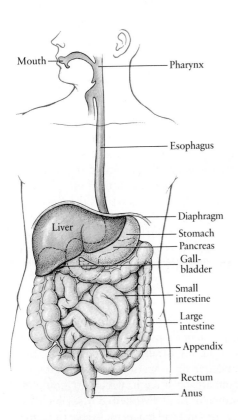

34-1 *The human digestive tract. Food passes from the mouth through the pharynx and esophagus to the stomach and small intestine, where most digestion takes place. Undigested materials pass through the large intestine, are stored briefly in the rectum, and are eliminated through the anus. Accessory organs of the digestive system are the salivary glands (shown in Figure 34-6), pancreas, liver, and gallbladder.*

Digestion is the breakdown of ingested food materials into molecules that can be delivered to and utilized by the individual cells of the animal body. These molecules serve a variety of functions. For example, they may be energy sources; they may provide essential chemical elements, such as calcium, nitrogen, or iron; or they may be molecules—such as certain amino acids, fatty acids, and vitamins —that cells need but cannot synthesize for themselves.

In Section 4, we traced the evolution of digestive systems from the feeding cells of sponges (page 522), through the gastrovascular cavity of jellyfish (page 525) and the digestive cavity of flatworms (page 533), to the major evolutionary breakthrough, exemplified by ribbon worms, the invention of the one-way digestive tract (page 538). As animals became larger and more complex, the tract became more convoluted, providing increased working surfaces, and different portions became specialized for different stages of the digestive process.

DIGESTIVE TRACT IN VERTEBRATES

In vertebrates, the digestive tract—known rather less elegantly but just as accurately as the gut—consists of a long convoluted tube, extending from mouth to anus (Figure 34-1). The digestive system also includes the salivary glands, the pancreas, the liver, and the gallbladder, accessory organs that provide the enzymes and other substances essential for digestion.

The inner surface of the gut is continuous with the outer surface of the body, and so, technically, the cavity of the gut is outside the body. Because the gut contents are thus sequestered, they can be subjected to the action of enzymes and bacteria and to conditions of pH that, although optimal for the breakdown of food, would rapidly destroy the living cells and tissues of the body proper. Nutrient molecules actually enter the body only when they pass through the epithelial lining of the digestive tract. Thus, the process of digestion involves two components: the breakdown of food molecules and their absorption into the body.

The digestive tract begins with the oral cavity and includes the mouth, pharynx, esophagus, stomach, small intestine, large intestine, and anus. Each of these areas is specialized for a particular phase in the overall process of digestion, but the fundamental structure of each is similar. As shown in Figure 34-2, the tube, from beginning to end, has four layers: (1) the innermost layer, the **mucosa,** which is made up of epithelial tissue, an underlying basement membrane, and connective tissue, with a thin outer coating of smooth muscle in some places; (2) the **submucosa,** which is made up of connective tissue and contains nerve fibers and blood and lymph vessels; (3) the **muscularis externa,** muscle tissue; and (4) the **serosa,** an outer coating of connective tissue.

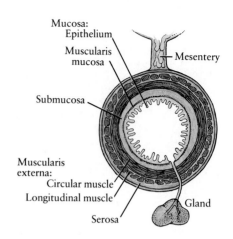

34-2 *The layers of the digestive tract include (1) the mucosa, (2) the submucosa, which contains nerves and blood and lymph vessels, (3) the muscularis externa, and (4) the serosa (also known as the visceral peritoneum), which covers the outer surfaces of the organs in the abdominal cavity. A mesentery is a fold of the peritoneum that connects the digestive tract with the posterior abdominal wall. Glands outside the digestive tract, principally the pancreas and liver, discharge digestive enzymes and bile into the tract through various ducts.*

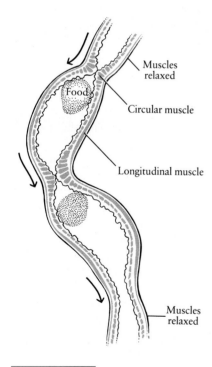

34-3 *Food is moved through the vertebrate digestive tract by several different patterns of contraction of the smooth muscle in its walls. One of the most important patterns, known as peristalsis, consists of progressive waves of contraction of the circular muscles.*

The epithelium of the digestive tract contains many mucus-secreting goblet cells and, in some parts of the gut, glands that secrete digestive enzymes. Along most of the digestive tract, the muscularis is made up of two layers of smooth muscle, an inner layer, in which the orientation is circular, and an outer layer, in which the cells are longitudinally arranged. Coordinated contractions of these muscles produce ringlike constrictions that mix the food, as well as the wavelike motions, known as **peristalsis,** that move food along the digestive tract (Figure 34-3). At several points the circular layer of muscle thickens into heavy bands, called **sphincters.** These sphincters, by relaxing or contracting, act as valves to control passage of food from one area of the digestive tract to another.

The Oral Cavity: Initial Processing

As we noted in Chapter 25, the great diversification of kingdom Animalia was determined, in large part, by the differing adaptations for obtaining food. In vertebrates, the seizing and ingesting of food is typically accomplished by the mouth, the structure in which the mechanical breakdown of food also begins.

Many vertebrates, including most mammals, have teeth, which are complex structures adapted for the tearing and grinding of food. (Modern birds, which are toothless, have gizzards containing particles of sand and gravel that serve the same tearing and grinding function.) The crown of a tooth—the visible part—is covered with enamel, the hardest substance in the body (mostly calcium phosphate); the root, within the gum, is covered with cement, a substance similar to bone. The bulk of the tooth is composed of dentine, another bonelike material, which forms slowly during the life of the tooth. The pulp cavity within the tooth contains the cells that produce dentine, as well as the nerve endings and blood vessels.

Figure 34-4 illustrates the pattern of dentition in five different mammalian orders. As you can see, the dentition of humans is relatively unspecialized. Children have 20 teeth, which are gradually lost and replaced by a second set of 32 teeth as the jaw grows larger. Of the 16 adult teeth in each jaw, four are incisors, flat chisel-like structures specialized for cutting; two are canines, used by carnivores for stabbing and tearing; four are premolars ("in front of the molars"), each of which has two cusps, or protuberances, and so are also called bicuspids; and six are molars, each of which has four or five cusps. The premolars and molars are used for grinding. Molars do not replace temporary teeth but are added as the jaw grows.

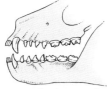
Generalized mammal

Rodent (Rodentia)

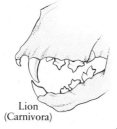
Lion (Carnivora)

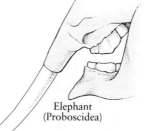
Elephant (Proboscidea)

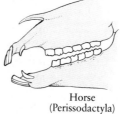
Horse (Perissodactyla)

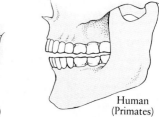
Human (Primates)

34-4 *The basic pattern of dentition in mammals includes, in each quadrant, two incisors, one canine ("eye tooth"), two to four premolars, and three molars. The various orders of mammals have modifications of this pattern that reflect their dietary habits. Rodents (whose name comes from the Latin* rodere, *"to gnaw") are characterized by their sharp, chisel-shaped incisors that grow throughout their lives. Most rodents have no canines. Large predatory carnivores, such as the lion, have canines adapted for stabbing and slicing, and large molars, which can easily crush the bones of large herbivores. In elephants, tusks are modified incisors; they are used for attack and defense and for rooting food from the ground or breaking branches. (The tusks of most other mammals, such as walruses, are modified canines.) The modern horse is a grazing animal; its incisors clip off grasses, which are then ground by the large, flat molars. Humans, with a relatively unspecialized diet, have a correspondingly unspecialized dentition. Similar, though less spectacular, differences in anatomy and physiology of other components of the digestive system also reflect adaptation to different diets.*

The tongue, also a vertebrate development, serves largely to move and manipulate food in mammals. However, some vertebrates, such as the jawless hagfish and lampreys (page 593), have tongues equipped with horny "teeth." The sticky tongues of frogs and toads, which are attached at the front rather than the back of the mouth, flip out to catch insects (see Figure 28-18, page 595). Mammalian tongues carry taste buds (Figure 34-5). In humans, the tongue has acquired a secondary function of formulating sounds for communication.

As the food is being chewed, it is moistened by saliva, a watery secretion produced by three pairs of large salivary glands (Figure 34-6) plus numerous minute glands, the buccal glands, which are located in the mucosal lining of the mouth. The saliva, which contains mucus, lubricates the food so that it can be swallowed easily. Saliva is slightly alkaline, owing to the presence of sodium

34-5 *(a) The outer pore of a human taste bud, as revealed by a scanning electron micrograph of the surface of the tongue. The sensory receptor cells of the taste bud are just visible within the pore. Other taste buds are found on the roof of the mouth and the pharynx. (b) Longitudinal section of a taste bud. As we shall see in Chapter 42, different receptor cells respond to the molecules responsible for different tastes. These cells perform a critical function in enabling an animal to distinguish what is good to eat from what is not.*

(a) 5 μm

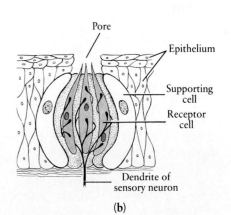
(b)

bicarbonate. In humans and other mammals that chew their food, saliva also contains a digestive enzyme, salivary amylase, that begins the breakdown of starches. Like all digestive enzymes, amylase works by hydrolysis; that is, the breaking of each bond involves addition of a molecule of water (see page 64). Carnivores, such as dogs, which characteristically tear and gulp their food, have no digestive enzymes in their saliva.

The secretion of saliva is controlled by the autonomic nervous system. It can be initiated by the presence of food in the mouth, which triggers reflexes originating in taste buds and in the walls of the mouth, and also by the mere smell or anticipation of food. (Think hard, for a moment, about eating a lemon.) Fear inhibits salivation; at times of great danger or stress, the mouth may become so dry that speech is difficult. Conversely, the presence of a noxious substance in the mouth or stomach stimulates the copious production of a watery saliva as a protective reaction. On the average, we produce 1 to 1.5 liters of saliva every 24 hours.

The Pharynx and Esophagus: Swallowing

From the mouth, food is propelled backward toward the esophagus, a muscular tube about 25 centimeters long in adult humans. Swallowing is the passing of food to the esophagus and through the esophagus on to the stomach (Figure 34–7). It begins as a voluntary action but, once under way in humans, it continues involuntarily. In humans, the upper part of the esophagus is striated muscle, but the lower part is smooth muscle. In dogs, cats, and other animals that gulp their food, the entire length of the esophagus is striated muscle. Both liquids and solids are propelled along the esophagus by peristalsis. This process is so efficient that we can swallow water while standing on our heads.

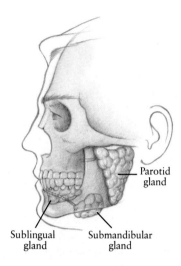

34–6 *The bulk of the saliva is produced by three pairs of salivary glands. Additional amounts are supplied by minute glands, the buccal glands, in the mucous membrane lining the mouth. The parotid glands are the sites of infection by mumps virus.*

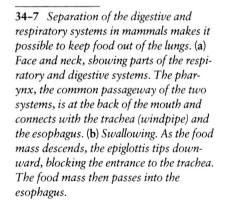

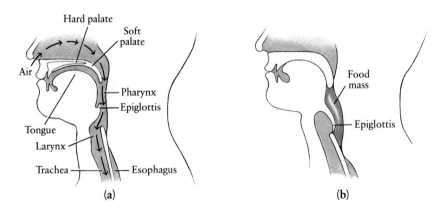

34–7 *Separation of the digestive and respiratory systems in mammals makes it possible to keep food out of the lungs. (a) Face and neck, showing parts of the respiratory and digestive systems. The pharynx, the common passageway of the two systems, is at the back of the mouth and connects with the trachea (windpipe) and the esophagus. (b) Swallowing. As the food mass descends, the epiglottis tips downward, blocking the entrance to the trachea. The food mass then passes into the esophagus.*

The esophagus passes through the diaphragm, which separates the thoracic and abdominal cavities, and opens into the stomach, which, with the remaining digestive organs, lies in the abdomen. The abdominal cavity is completely lined by the **peritoneum,** a thin layer of connective tissue covered by moist epithelium. (The portion of the peritoneum that covers the outer surface of the digestive tract is, as we noted earlier, known as the serosa.) The stomach, intestines, and other organs in the abdominal cavity are suspended by folds of peritoneum known as **mesenteries,** as shown in Figure 34–2. Mesenteries consist of double layers of tissue, with blood vessels, lymph vessels, and nerves lying between the two layers.

The Heimlich Maneuver

Accidental choking on food claims the lives of almost 3,000 persons a year in this country alone, more than accidents involving firearms or airplanes. It occurs when a food mass enters the trachea rather than the esophagus (Figure 34–7). If the food becomes lodged, the victim cannot speak or breathe and, if the airway is blocked for four or five minutes, the victim will die. (The fact that the victim cannot speak helps onlookers to distinguish such blockage from a heart attack. Although the symptoms are similar, heart attack sufferers can talk.)

The food can nearly always be dislodged by the Heimlich maneuver, a procedure so simple that it has been carried out successfully in at least two instances by eight-year-olds. There are three steps: (1) Stand behind the victim and wrap your arms around his or her waist. (2) Make a fist with one hand, grasp it with your other hand, and then place the fist against the victim's abdomen, slightly above the navel and below the rib cage. (3) Press your fist into the victim's abdomen with a quick upward thrust. The sudden elevation of the diaphragm compresses the lungs and forces air up the trachea, pushing the food out. Repeat several times if necessary.

If the victim is sitting, the procedure can be carried out in the same way. If the victim is lying on his back, face him, and kneel astride the hips. Put the heel of one hand on the abdomen above the navel and below the rib cage, put the other hand on top of the first hand, and press with a quick upward thrust. You can even do it on yourself: place your hands at your waist, make a fist, and press quickly upward.

The victim should be seen by a physician immediately after the emergency treatment because it is possible to break a rib or cause other internal injuries, especially if the movements are performed incorrectly. But, considering the alternative, it is well worth the risk.

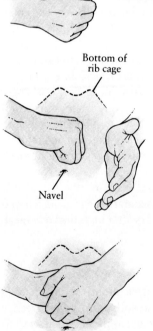
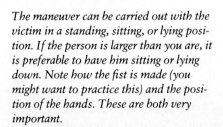
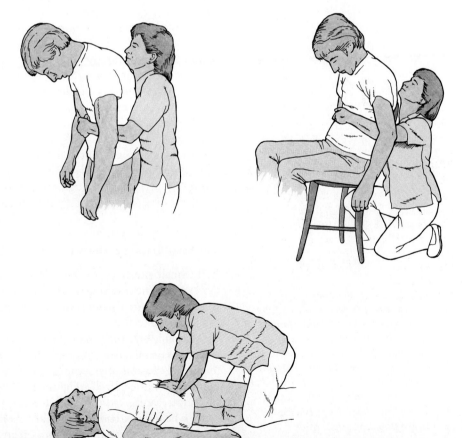

The maneuver can be carried out with the victim in a standing, sitting, or lying position. If the person is larger than you are, it is preferable to have him sitting or lying down. Note how the fist is made (you might want to practice this) and the position of the hands. These are both very important.

SECTION 6 Biology of Animals

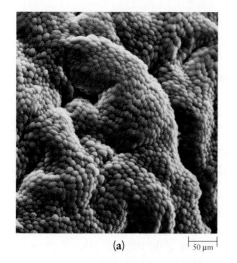

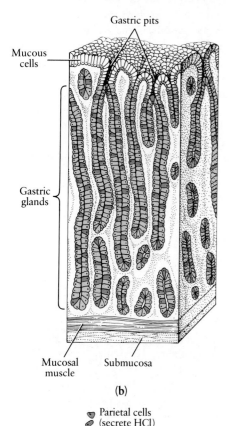

34-8 (a) *Surface of the stomach, as shown in a scanning electron micrograph. The numerous grooves and indentations are gastric pits.* (b) *A cross section of stomach mucosa. The parietal cells secrete hydrochloric acid, and the chief cells produce pepsinogen. Mucus, secreted by other cells of the epithelium, coats the surface of the stomach and lines the gastric pits, protecting the stomach surface from digestion.*

The Stomach: Storage and Liquefaction

The stomach is a collapsible, muscular bag, which, unless it is fully distended, lies in folds. Stomachs vary widely in capacity. A hyena's stomach can hold an amount equivalent to one-third of the animal's own body weight, for instance, whereas a mammal that eats small, regularly available, and highly nutritious food items—such as seeds or insects—needs only a small stomach. The human stomach, distended, holds 2 to 4 liters of food.

The mucosal layer of the stomach is very thick and contains numerous gastric pits (Figure 34–8). Mucus-secreting epithelial cells cover the surface of the stomach and line the gastric pits. Opening into the lower portions of the gastric pits are the gastric glands, whose walls contain the parietal cells, which produce hydrochloric acid (HCl), and the chief cells, which produce pepsinogen, the precursor of the digestive enzyme pepsin. These secretions, with the water in which they are dissolved, constitute gastric juice.

As a consequence of the HCl secretion, the pH of gastric juice is normally between 1.5 and 2.5, far more acidic than any other body fluid. The burning sensation you feel if you vomit is caused by the acidity of gastric juice acting on unprotected membranes. Normally, the mucus in the stomach forms a barrier between the epithelium and the gastric juices and so prevents the stomach from digesting itself. The HCl, which kills most bacteria and other living cells in the ingested food, loosens the tough, fibrous components of plant and animal tissues and erodes the cementing substances between cells. HCl also initiates the conversion of pepsinogen to its active form, pepsin, by splitting off a small portion of the molecule. Once pepsin is formed, it acts on other molecules of pepsinogen to form more pepsin. Pepsin, which breaks proteins down into peptides, is active only at the low pH of the normal stomach.

The stomach is influenced by both the nervous and endocrine systems. Anticipation of food and the presence of food in the mouth stimulate churning movements of the stomach and the production of gastric juice. Fear and anger decrease the stomach's motility. When protein-containing food reaches the stomach, its presence causes the release of a hormone, gastrin, from gastric cells into the bloodstream. This hormone acts on the epithelial cells of the stomach mucosa to increase their secretion of gastric juice and on the muscle cells of the stomach wall to increase their contractions.

In the stomach, food is converted into a semiliquid mass, which is gradually moved by peristalsis through the pyloric sphincter, separating the stomach and small intestine. The stomach is usually empty four hours after ingestion of a meal.

The Small Intestine: Digestion and Absorption

In the small intestine, the breakdown of food begun in the mouth and stomach is completed. The resulting nutrient molecules are then absorbed from the digestive tract into the circulatory system of the body, through which they are delivered to the individual cells.

Anatomically, the small intestine is characterized by circular folds in the submucosa; numerous microscopic fingerlike projections, **villi,** on the mucosa (Figure 34–9); and tiny cytoplasmic projections, **microvilli,** on the surface of the individual epithelial cells (Figure 34–10). All of these structural features increase the surface area of the small intestine. If the small intestine of an adult human were to be fully extended, it would be some 6 meters long; the total surface area of the human small intestine is approximately 300 square meters, about the size of a doubles tennis court. The **duodenum,** the upper 25 centimeters of the small intestine, is the most active in the digestive process; the rest is principally concerned with absorption of nutrients.

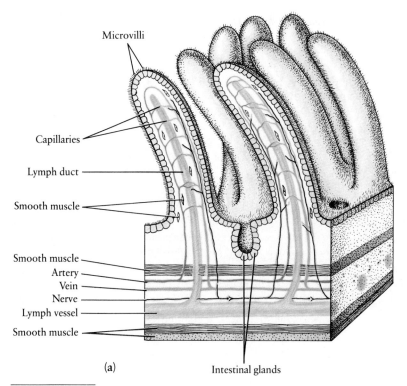
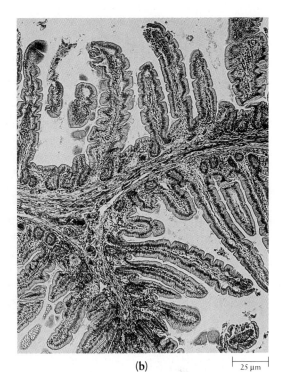

34–9 (a) *Diagram and* (b) *photomicrograph of villi of the small intestine, as seen in longitudinal section. Nutrient molecules are absorbed through the walls of the villi and, with the exception of fat molecules, enter the bloodstream by means of the capillaries. Fats—hydrolyzed to fatty acids and glycerol, resynthesized into new fats, and packaged into particles known as chylomicrons—are taken up by the lymphatic system. The villi can move independently of one another; their motion increases after a meal.*

In the micrograph, the columnar epithelial cells that are responsible for the absorption of nutrient molecules are stained pink, and their nuclei are stained dark red. Mucus-secreting goblet cells are stained blue, as is the layer of smooth muscle at the base of the villi.

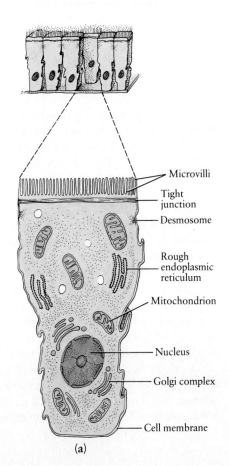
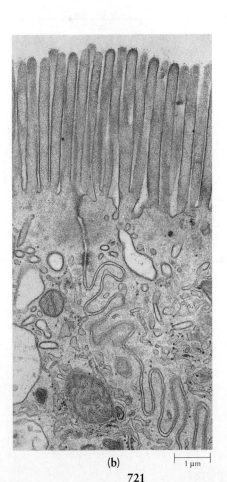

34–10 (a) *A columnar epithelial cell from the lining of the intestine. Notice, in particular, the tight junctions and desmosomes that bind these cells together in a continuous sheet.* (b) *Microvilli on the surface of two adjacent intestinal epithelial cells. These cellular extensions greatly increase the absorptive surface of the intestine. They also contain digestive enzymes, which are part of the cell membrane, and proteins involved in the transport of nutrient molecules across the cell membrane. Notice the tight junction at the surface where the two cells meet and, slightly lower, a desmosome. Although not visible in this micrograph, intestinal epithelial cells contain numerous mitochondria. Their abundance indicates that the cells have high energy requirements, probably due, in large part, to active transport processes.*

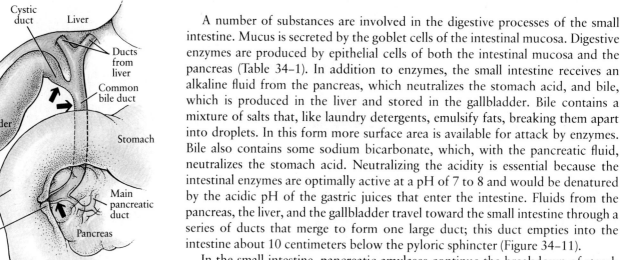

TABLE 34-1 Principal Digestive Enzymes

ENZYME	SOURCE	SUBSTRATE	SITE OF ACTION
Salivary amylase	Salivary glands	Starches	Mouth
Pepsin	Stomach mucosa	Proteins, pepsinogen	Stomach
Pancreatic amylase	Pancreas	Starches	Small intestine
Lipase	Pancreas	Fats	Small intestine
Trypsin	Pancreas	Polypeptides, chymotrypsinogen	Small intestine
Chymotrypsin	Pancreas	Polypeptides	Small intestine
Carboxypeptidase	Pancreas	Polypeptides	Small intestine
Deoxyribonuclease	Pancreas	DNA	Small intestine
Enterokinase	Small intestine	Trypsinogen	Small intestine
Aminopeptidase	Small intestine	Polypeptides	Small intestine
Dipeptidase	Small intestine	Dipeptides	Small intestine
Maltase	Small intestine	Maltose	Small intestine
Lactase*	Small intestine	Lactose	Small intestine
Sucrase	Small intestine	Sucrose	Small intestine
Phosphatases	Small intestine	Nucleotides	Small intestine

* Often absent in adults, especially those of recent African origin.

A number of substances are involved in the digestive processes of the small intestine. Mucus is secreted by the goblet cells of the intestinal mucosa. Digestive enzymes are produced by epithelial cells of both the intestinal mucosa and the pancreas (Table 34-1). In addition to enzymes, the small intestine receives an alkaline fluid from the pancreas, which neutralizes the stomach acid, and bile, which is produced in the liver and stored in the gallbladder. Bile contains a mixture of salts that, like laundry detergents, emulsify fats, breaking them apart into droplets. In this form more surface area is available for attack by enzymes. Bile also contains some sodium bicarbonate, which, with the pancreatic fluid, neutralizes the stomach acid. Neutralizing the acidity is essential because the intestinal enzymes are optimally active at a pH of 7 to 8 and would be denatured by the acidic pH of the gastric juices that enter the intestine. Fluids from the pancreas, the liver, and the gallbladder travel toward the small intestine through a series of ducts that merge to form one large duct; this duct empties into the intestine about 10 centimeters below the pyloric sphincter (Figure 34-11).

In the small intestine, pancreatic amylases continue the breakdown of starch begun in the mouth, producing disaccharides. Lipases hydrolyze fats into glycerol and fatty acids. Three types of enzymes break down proteins. The members of one group break apart the long protein chains; each enzyme in this group acts only on the bonds linking particular amino acids, so that several enzymes are required to break a single large protein into shorter peptide fragments. A second type of enzyme acts only on the end of a chain, some on the amino end and some on the carboxyl end, splitting off dipeptides. A third group of enzymes then comes into action, breaking the remaining dipeptides into single amino acids.

The digestive activities of the small intestine are coordinated and regulated by hormones (Table 34-2). In the presence of acidic gastric juice, the duodenum releases secretin, a hormone that stimulates the pancreas and liver to secrete alkaline fluids. Fats and amino acids in the food stimulate the production of

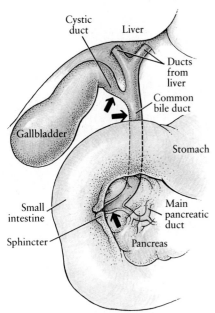

34-11 *Ducts from the liver, the gallbladder, and the pancreas merge just before emptying into the small intestine through a sphincter in its wall. The arrows indicate the locations in which gallstones typically lodge. Gallstones, which consist primarily of cholesterol, form when the delicate balance in the relative concentrations of bile ingredients is disturbed.*

another hormone, cholecystokinin, which triggers the release of enzymes from the pancreas and the emptying of the gallbladder. At least eight other substances are suspected of being gastrointestinal hormones. In addition to hormonal influences, the intestinal tract is also regulated by the autonomic nervous system; stimulation by parasympathetic nerve fibers increases intestinal contractions, whereas inhibition by sympathetic fibers decreases contractions. Thus, a complex interplay of stimuli and checks and balances serves to activate digestive enzymes, to adjust the chemical environment, and to regulate movements of the intestines.

TABLE 34-2 **Major Gastrointestinal Hormones**

HORMONE	SOURCE	MAJOR STIMULUS FOR PRODUCTION	MAJOR ACTIONS
Gastrin	Stomach	Protein-containing food in stomach, also parasympathetic nerves to stomach	Stimulates secretion of gastric juices and muscular contractions of stomach and intestine
Secretin	Duodenum	HCl in duodenum	Stimulates secretion of alkaline pancreatic fluids and bile
Cholecystokinin	Duodenum	Fats and amino acids in duodenum	Stimulates release of enzymes from pancreas and of bile from gallbladder

Absorption of Nutrients

The food molecules released by the digestive processes are absorbed through the epithelial cells of the intestinal mucosa. Specialized enzymes—lactase, sucrase, and maltase—embedded in the epithelial cell membranes cleave disaccharides to monosaccharides, which are then rapidly absorbed by active transport and facilitated diffusion (page 136). Amino acids and dipeptides are absorbed by active transport. These molecules all enter the bloodstream by way of the capillaries of the villi.

Small fatty acids also enter the blood vessels of the intestine directly, but large fatty acids, glycerol, and cholesterol travel an indirect route. First, these molecules enter mucosal cells by passive diffusion. Within the cells, fatty acids and glycerol are resynthesized into fats (three molecules of fatty acid combined with one molecule of glycerol, as shown on page 68), which are then packaged into protein-coated droplets, known as chylomicrons. Similarly, cholesterol is packaged into low-density lipoprotein (LDL) complexes (see page 71). Both the chylomicrons and the LDLs are secreted into the lymph vessels and ultimately enter the bloodstream through ducts that empty into veins in the chest. In the bloodstream, the chylomicrons are gradually broken apart. Fats are delivered in the form of fatty acids to cells such as muscle cells, where they are oxidized for energy, or to fat cells, where they are stored. (One gram of fat, as we noted on page 67, contains more than twice as many calories as 1 gram of protein or glycogen; hence it is a very useful storage form for motile organisms.) The LDL particles are picked up by cells in the liver, where the cholesterol is stored, secreted in the bile, or repackaged for delivery to other body cells. It is used by all cells in the synthesis of cell membranes and, by specialized cells, for the production of steroid hormones.

Aids to Digestion

Compared to the microscopic photosynthetic cells of the open waters, plant material is tough and hard to digest. By logic, one would expect terrestrial herbivores to have sets of enzymes capable of breaking down cellulose and other structural polysaccharides found in plants. Not so. Terrestrial animals have evolved another solution: symbiotic relationships with bacteria and protists that perform these activities on their behalf, receiving, in return, a free food supply, protection (except from the host itself), and good housing.

In the warm, wet, airless interior of a mammalian digestive tract, symbionts break down plant products anaerobically. The glucose they use themselves; the host's share is mainly fatty acids, which are absorbed through the host's intestine. In addition, the microorganisms synthesize many vitamins, especially vitamin K and those of the B group, which are utilized by the host.

The fermentation process, the anaerobic oxidation of organic molecules, takes place in different areas in the intestinal tracts of different species. In the horse family, it occurs in the colon, which is enlarged, and in the cecum, a blind sac at the junction of the small and large intestines. (The cecum persists in humans in a vestigial form as the trouble-making appendix.) Rabbits and other lagomorphs have an enlarged cecum in which bacterial fermentation takes place and also a curious adaptation, peculiar to this group. At night, in their burrows, rabbits produce, from the cecum, a special type of feces, different from those produced in the daytime, that consist almost entirely of bacteria. They eat these feces, thus digesting and absorbing additional nutrients, obtaining vitamins produced by the bacteria, and recycling their beneficial symbionts.

The most elaborate specialization, anatomically speaking, is found among the ruminants, such as cattle, antelopes, sheep, and deer. It is probably not a coincidence that the ruminants are among the most successful terrestrial herbivores, with their rise in evolutionary prominence taking place at the time the grasslands were undergoing rapid expansion. Ruminants hastily shred grass blades and other plant parts and store them, regurgitating them as a cud and completing the chewing process at leisure. The actual fermentation takes place in the rumen, a large esophageal pouch. The contents of the rumen are then passed to the stomach where absorption begins. Bacteria and ciliates in the rumen not only break down the cellulose and other polysaccharides but also synthesize proteins, using ammonia and urea as nitrogen sources. Some of these microorganisms are also digested by the host, thus providing it with protein.

Ruminant digestion is a major activity. In cattle, the rumen-stomach complex makes up about 15 percent of the weight of the animal. To keep the mixture moist, a cow secretes 60 liters of saliva a day. She also produces about 2 liters of gas a minute, most of which escapes in sweet, chlorophyll-scented belches.

Omnivorous humans have no such symbionts and so are unable to utilize cellulose, which is eliminated as roughage. However, like other mammals, we are dependent on the bacteria of our digestive tracts for the synthesis of vitamins, and, as in the rabbit, a major component of our feces is symbiotic bacteria.

The Large Intestine: Further Absorption and Elimination

The absorption of water, sodium, and other minerals, a process that occurs primarily in the small intestine, continues in the large intestine. In the course of digestion, large amounts of water—approximately 7 liters per day—enter the stomach and small intestine as secretions of the glands emptying into and lining the digestive tract, by osmosis from the body fluids, and directly in the ingested food and drink. When the absorption of this water and the minerals it contains is disrupted, as in diarrhea, severe dehydration can result. Mortality from infant diarrhea, still the chief cause of infant death in many countries, is principally a consequence of water loss.

The large intestine harbors a considerable population of symbiotic bacteria (including the familiar *E. coli*), which break down food substances that escaped digestion and absorption in the small intestine. Living on these food substances, largely materials we lack the enzymes to digest, the bacteria synthesize amino acids and vitamins, some of which are absorbed into the bloodstream. These bacteria are our chief source of vitamin K.

A blind pouch off the large intestine, the appendix, is an evolutionary remnant from herbivorous ancestors (see essay). As many individuals know from personal experience, it may become irritated, inflamed, and then infected. If it ruptures, as a consequence of inflammation and swelling, it spills its bacterial contents into the abdominal cavity. Serious, and even fatal, infection can result. Although the appendix plays no known role in human digestion, it is one of the sites of interaction of cells involved in the immune response.

The bulk of the fecal matter consists of water, bacteria (mostly dead cells), and cellulose fibers, along with other indigestible substances. It is lubricated by mucus, which is secreted by some of the epithelial cells lining the large intestine, stored briefly in the rectum, and then eliminated through the anus as feces.

MAJOR ACCESSORY GLANDS

The Pancreas

As we have seen, the pancreas is the source of a number of substances essential to the digestive process. It is a specialized secretory organ, developing both in the embryo and in the course of evolutionary history from the upper portion of the small intestine. The bulk of the tissue of the pancreas (Figure 34–12) resembles salivary gland tissue and, like the salivary glands, secretes an amylase that plays a major role in the breakdown of starch. In addition, it secretes a number of other digestive enzymes.

The pancreas is also an endocrine (hormone-producing) gland. Clusters of pancreatic cells, known as the islets of Langerhans (named after their discoverer, not some romantic Pacific atoll), secrete insulin, glucagon, and somatostatin, which are released into the bloodstream and participate in the regulation of blood glucose.

The Liver

The liver also plays vital roles in digestion. The largest internal organ of the body, the liver is a three-pound chemical factory with an extraordinary variety of processes and products. It stores and releases carbohydrates, playing a central role in the regulation of blood glucose. It processes amino acids, converting them to carbohydrates, channeling them to other tissues of the body, and synthesizing essential proteins such as enzymes and clotting factors from them. It manufactures the plasma proteins that make the blood hypertonic in relation to the

34–12 *Portion of a pancreatic cell involved in the production of digestive enzymes. Notice the extensive rough endoplasmic reticulum characteristic of cells that synthesize proteins for export. The cell nucleus can be seen at the bottom of the micrograph, and, just outside the nuclear envelope, there are several mitochondria. The dark, spherical objects are secretory granules, moving toward the periphery of the cell.*

Other types of pancreatic cells synthesize the hormones insulin, glucagon, and somatostatin, which play major roles in the regulation of blood glucose.

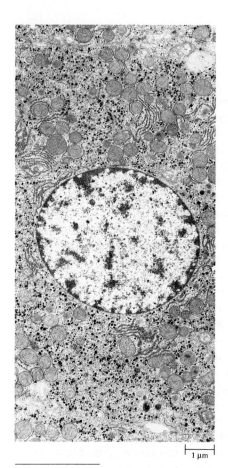

34-13 *Electron micrograph of a liver cell. The dark granules are glycogen. Despite the variety of activities carried out by the liver, the cells of the liver all resemble one another in appearance and in function, with no division of labor among cell groups, as there is in the pancreas. Liver cells, as you would expect, have many mitochondria and abundant rough endoplasmic reticulum.*

interstitial ("between-the-tissues") fluids and so prevents the osmotic movement of water from bloodstream to tissues. It is the major source of the plasma lipoproteins, including LDLs and HDLs, that transport cholesterol, fats, and other water-insoluble substances in the bloodstream and is of central importance in the regulation of blood cholesterol (see page 71). It stores fat-soluble vitamins, such as A, D, and E. It produces bile (which is then stored in the gallbladder). It breaks down the hemoglobin from damaged and dead red blood cells to bilirubin, a yellow pigment; bilirubin is released into the bile and excreted through the intestinal tract. The liver inactivates a number of hormones, thus playing an important role in hormone regulation. It also breaks down a variety of foreign substances, some of which—alcohol, for instance—may form metabolic products that damage liver cells and interfere with their functions (see page 202).

REGULATION OF BLOOD GLUCOSE

The major function of digestion is, of course, to provide organic molecules that can serve as energy sources and raw materials for each cell of the body. Although vertebrates rarely eat 24 hours a day, their blood glucose—the major cellular energy supply and the fundamental building-block molecule—remains extraordinarily constant. As we have noted previously, the liver plays a central role in this critical process. Glucose and other monosaccharides are absorbed into the blood from the intestinal tract and are passed directly to the liver by way of the hepatic portal vein. The liver converts some of these monosaccharides to glycogen and fat, storing enough glycogen to satisfy the body's needs for about four hours. The fat is stored in fat cells, which can also form fat from glucose. Similarly, the liver breaks down excess amino acids (which are not stored) and converts them to glucose. The nitrogen from the amino acids is excreted in the form of urea, and the glucose is stored as glycogen.

Whether the liver takes up or releases glucose and the amount it takes up or releases are determined primarily by the concentration of glucose in the blood. As we shall see in Chapter 40, the concentration of glucose is, in turn, regulated by a number of hormones and is influenced by the autonomic nervous system. Among the hormones involved in this process are insulin, glucagon, and somatostatin, all produced by the pancreas. Insulin stimulates the uptake of glucose by cells, thus decreasing blood glucose. Glucagon promotes the breakdown of glycogen, thus increasing blood glucose. Somatostatin, which was the first vertebrate protein synthesized using recombinant DNA technology (page 351), has a variety of inhibitory effects that collectively help to regulate the rate at which glucose and other nutrients are absorbed from the digestive tract.

SOME NUTRITIONAL REQUIREMENTS

Because of the liver's activities in converting various types of food molecules into glucose and because most tissues can use fatty acids as an alternative fuel, the energy requirements of the body can be met by carbohydrates, proteins, or fats—the three principal types of food molecules. Energy requirements are ordinarily met by a combination of the three. Carbohydrates and proteins supply about the same number of calories per gram of dry weight, and fats supply about twice as many as either of them. However, because carbohydrates and proteins are always combined with water in the form in which they are eaten, the actual bulk of ingested material is usually nine or ten times greater.

Mother's Milk—It's the Real Thing

Throughout this chapter, we have been concerned with the digestive functions and nutritional requirements of mammalian adults. The digestive system of the mammalian infant, however, requires additional development and growth before it can perform its full range of functions. In the meantime, the infant is totally dependent on the complex, highly nutritive fluid known as milk, which is produced by the modified sweat glands from which the Mammalia derive their name.

In the course of evolution, the various species of mammals have developed milks especially suited to the nourishment of their own young. Human milk, for instance, contains about the same amount of fat (3.7 percent) as cow's milk, but it has less protein and more carbohydrate (in the form of lactose, milk sugar). The milk of a harp seal, on the other hand, is twice as concentrated as either cow's milk or human milk (45 percent water as compared to 87 percent water), has no carbohydrate, is 43 percent fat, and so fulfills the higher energy requirements of an animal that maintains a high body temperature in water. A kangaroo can provide an even more specialized diet; one mother carrying two joeys of different ages in her pouch supplies different milks (through different teats) to each.

In addition, each species produces milk with particular vitamins and minerals and, perhaps most important, a number of antibodies, other immunologically active proteins, and various white blood cells that protect the infant of that species against infection. For example, human milk contains antibodies that protect against *Staphylococcus, E. coli,* and polio, in particular. This immunological protection is especially important in the first few months of life, before the infant's own immune system is functioning fully. Another important feature of mother's milk is that it comes in a container that keeps it fresh, temperature-controlled, and free from bacterial contamination.

The length of time mammalian infants depend on milk for nourishment also varies with the species. Rabbits nurse for only a few days, cats for six to ten weeks, hyenas and humans for a year, elephants for four or five. However, with the exception of Western humans, once infancy is over, not only does milk consumption stop but there is also a decline in the level of lactase, the enzyme that breaks down lactose. This enzyme, present in most human infants, is absent in adults of most animal species and is much diminished in most human beings after the age of four. Lactose ingestion by individuals lacking the enzyme may cause intestinal cramps and diarrhea, as the lactose is broken down by bacterial fermentation. Natural selection for lactose digestion in adult humans is thought to have begun some 10,000 years ago when certain groups began to domesticate and milk mammals. Individuals of northern European descent are much more likely to have this enzyme as adults than are those of African or Mediterranean descent. This distribution of the allele (or of the gene that turns it on or off) is apparently correlated with the fact that northerners were more likely to drink milk (including reindeer milk) as adults, whereas people from hot climates, if they used milk at all, were likely to use fermented milk products, such as cheese or yogurt, in which lactose has already been broken down.

Various commercially produced formulas exist that, mixed with water, offer a substitute for human milk. Where families have access to clean water, sterilized bottles, good pediatric care, and money for an adequate amount of formula, milk substitutes can be used successfully by women who cannot or choose not to nurse their babies. However, at a World Health Organization assembly in 1981, 118 countries voted to regulate the promotion of infant formula (the United States cast the only vote against regulation). These regulations were aimed particularly at the promotion of infant formula in the Third World, where access to clean water or to facilities for boiling water is often simply not available. In fact, bottles are often not even washed between feedings. Another reason for the regulations is that the formula may be overdiluted to save money. Adding this to the absence in formulas of antibodies and other protective substances, it can be seen that formula-fed infants run a much higher risk of infectious disease than breast-fed babies. Many die of "bottle-baby disease," severe diarrhea, and some starve due to watered-down formula. Ironically, women in the Third World often use milk substitutes not as a matter of necessity but because bottle feeding is associated with progress, social status, and the adoption of Western conventions.

SECTION 6 Biology of Animals

In addition to calories, the cells of the body need the 20 different kinds of amino acids required for assembling proteins. When any one of the amino acids necessary for the synthesis of a particular protein is unavailable, the protein cannot be made, and the other amino acids are converted to carbohydrates and oxidized or stored. Vertebrates are not able to synthesize all 20 amino acids. Humans (and albino rats) can synthesize 12, either from a simple carbon skeleton or from another amino acid. The other eight, which must be obtained in the diet, are known as essential amino acids (see essay, page 76). Plants are the ultimate source of the essential amino acids, but it is difficult (although by no means impossible) to obtain sufficient quantities of them by eating a completely vegetable diet, largely because plant proteins are relatively deficient in lysine and tryptophan.

Mammals also require but cannot synthesize certain polyunsaturated fats that provide fatty acids needed for the synthesis of fats and a group of hormonelike compounds known as prostaglandins (see page 833). These essential fatty acids can be obtained by eating plants or insects (or eating other animals that have eaten plants or insects).

Vitamins are an additional group of molecules required by living cells that cannot be synthesized by animal cells. Many of them function as coenzymes, and they are characteristically required only in small amounts. Table 34–3 indicates some of the vitamins required in the human diet and their functions. Severe vitamin deficiencies, such as may occur in regions where malnutrition is chronic, can have appalling consequences. Studies triggered by an extraordinary incidence of blindness among children who were victims of the famines in Ethiopia in the mid-1980s have revealed that, in addition to its effects on vision, severe vitamin A deficiency dramatically increases deaths due to measles, diarrhea, and other childhood diseases. At the present time, the World Health Organization is mounting a massive campaign to supply high doses of supplementary vitamin A to children in poverty-stricken regions in Africa, Asia, and Latin America. Although such supplements are vital in cases of severe malnutrition, there is no clear evidence that the ingestion of amounts of any particular vitamin in excess of the amounts available in a well-balanced diet has any beneficial effect on a normally healthy individual. Some, including the fat-soluble vitamins A, D, and K, which can accumulate in body tissues, are toxic in large doses. One of the most concentrated sources of vitamin A known is polar bear liver, a half pound of which contains about 2,600 times the recommended daily allowance. For centuries, Eskimos and Arctic explorers have known that consumption of polar bear liver causes illness and can be fatal—a knowledge shared by sled dogs and Arctic birds, which also refuse to eat it.

The body also has a dietary requirement for a number of inorganic substances, or minerals. These include calcium and phosphorus for bone formation, iodine for thyroid hormone, iron for hemoglobin and cytochromes, and sodium, chloride, and other ions essential for ionic balance. Most of these are present in the ordinary diet or in drinking water. Like the vitamins, however, they must be given in supplementary form when the dietary intake is inadequate or when the individual is not able to assimilate them normally.

The Price of Affluence

The major nutritional problem among North Americans (and also Europeans) is obesity. In the United States, 30 percent of middle-aged women and 15 percent of middle-aged men are obese—that is, they weigh more than 120 percent of their appropriate weight. Obesity is correlated with a significant increase in coronary artery disease, diabetes, and other disorders.

34–14 (a) *Vitamin D is a steroid-like substance (see page 70) that is produced in the skin by the action of ultraviolet rays from the sun on cholesterol (the absorbed energy opens the second ring of the molecule). Further chemical processing in the liver and kidneys produces the active form of the molecule shown here.*

According to one current hypothesis, the ancestors of modern Homo sapiens *originated in the tropics and were all dark-skinned. In those populations that moved northward, however, the screening effects of the darker pigment, which inhibits the production of vitamin D, caused selection for lighter skin color, whereas no such selection pressures occurred in the sun-drenched areas nearer the equator. In fact, in the tropics, selection would favor dark skin, which prevents carcinogenic ultraviolet rays from reaching the dermis. Nor did such selection occur among the Eskimos (b), who eat a diet rich in fish oils, a major source of vitamin D.*

TABLE 34-3 **Vitamins**

DESIGNATION: LETTER AND NAME	MAJOR SOURCES	FUNCTION	DEFICIENCY SYMPTOMS
A, carotene	Egg yolk, green or yellow vegetables, fruits, liver, butter	Formation of visual pigments, maintenance of normal epithelial structure	Night blindness; dry, flaky skin
B-complex vitamins:			
B_1, thiamine	Brain, liver, kidney, heart, pork, whole grains	Formation of coenzyme involved in Krebs cycle	Beri-beri, neuritis, heart failure
B_2, riboflavin	Milk, eggs, liver, whole grains	Part of electron carrier FAD	Photophobia, fissuring of skin
B_3, niacin (nicotinic acid)	Whole grains, liver and other meats, yeast	Part of electron carriers NAD, NADP, and of CoA	Pellagra, skin lesions, digestive disturbances
B_5, pantothenic acid	Present in most foods	Forms part of CoA	Neuromotor and cardiovascular disorders, gastrointestinal distress
B_6, pyridoxine	Whole grains, liver, kidney, fish, yeast	Coenzyme for amino acid metabolism and fatty acid metabolism	Dermatitis, nervous disorder
B_{12}, cyanocobalamin	Liver, kidney, brain, eggs, dairy products	Maturation of red blood cells, coenzyme in amino acid metabolism	Anemia, malformed red blood cells
Biotin	Egg yolk, synthesis by intestinal bacteria	Concerned with fatty acid synthesis, CO_2 fixation, and amino acid metabolism	Scaly dermatitis, muscle pains, weakness
Folic acid	Liver, leafy vegetables	Nucleic acid synthesis, formation of red blood cells	Failure of red blood cells to mature, anemia
C, ascorbic acid	Citrus fruits, tomatoes, green leafy vegetables, potatoes	Vital to collagen and ground (extracellular) substance	Scurvy, failure to form connective tissue fibers
D_3, calciferol	Fish oils, liver, fortified milk and other dairy products, action of sunlight on lipids in the skin	Increases Ca^{2+} absorption from gut, important in bone and tooth formation	Rickets (defective bone formation)
E, tocopherol	Green leafy vegetables, wheat germ, vegetable oils	Maintains resistance of red cells to hemolysis, cofactor in electron transport chain	Increased red blood cell fragility
K, naphthoquinone	Synthesis by intestinal bacteria, leafy vegetables	Enables synthesis of clotting factors by liver	Failure of blood coagulation

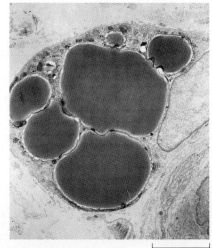

34-15 *A fat cell from the skin of a newborn rat. The nucleus is to the right. The large stored droplets of fat that nearly fill the cell have been fixed with osmium for electron microscopy. When excess calories are taken in in the diet, fat accumulates in these specialized cells, and when caloric intake is less than sufficient, fat is mobilized, broken down to glycerol and fatty acids, and released into the bloodstream.*

In addition to excess calories, our diets appear to contain a number of health hazards as compared to those of other cultures. We consume, on the average, about twenty times as much salt as our bodies need. Excess salt has been correlated with hypertension (high blood pressure). Another hazard is animal fat, such as that present in beef and pork. It has been known for some time that diets high in animal fat interfere with the regulation of blood cholesterol (page 71), implicated in atherosclerosis and heart attacks. More recently, attention has been focused on the role of dietary fibers in decreasing colon cancer. As a result of a diet high in fat and protein and low in carbohydrate, the feces of North Americans have only about half the bulk of their Far Eastern or African counterparts, and are eliminated from the intestinal tract much more slowly. Also, the rate of colon cancer is far higher among North Americans, suggesting that the retention of a fat-laden, compact fecal mass in the lower bowel may contribute to cancer development.

A final nutritional hazard, also related to affluence, is our willingness to experiment with our own bodies and to adopt extreme diets and follow nutritional fads—fasting, liquid protein, "macrobiotic," high protein, low protein, high fiber, megavitamin—without consideration of our requirements as biological organisms and the dictates of our long, omnivorous evolutionary history.

SUMMARY

Digestion is the process by which food is broken down into molecules that can be taken up by the cells lining the intestine, transferred to the bloodstream, and so distributed to the individual cells of the body. It occurs in successive stages, regulated by an interplay of hormones and nervous stimuli. In mammals, food is processed initially in the mouth, where the breakdown of starch begins in humans. It moves through the esophagus to the stomach, where gastric juices destroy bacteria and begin to break down proteins.

Most of the digestion occurs in the upper portion of the small intestine, the duodenum; here, digestive activity, which is performed by enzymes, is almost completely under hormonal regulation. The breakdown of starch by amylases continues, fats are hydrolyzed by lipases, and proteins are reduced to dipeptides or single amino acids. Monosaccharides, amino acids, and dipeptides are absorbed into the blood vessels of the villi; fats are absorbed into the lymph vessels and ultimately enter the bloodstream. Hormones secreted by duodenal cells stimulate the functions of the pancreas and the liver. The pancreas releases an alkaline fluid containing digestive enzymes; the liver produces bile, which is also alkaline and emulsifies fats.

Much of the water that enters the stomach and small intestine in the course of digestion is reabsorbed in the small intestine itself. Most of the remaining water is reabsorbed from the residue of the food mass as it passes through the large intestine. The large intestine contains symbiotic bacteria, which are the source of certain vitamins. Undigested residues are eliminated from the large intestine.

The chief energy source for cells in the mammalian body is glucose circulating in the blood. The organ principally responsible for maintaining a steady supply of glucose is the liver, which stores glucose (in the form of glycogen) when glucose levels in the blood are high and breaks down glycogen, releasing glucose, when the levels drop. These activities of the liver are regulated by a number of different hormones.

Requirements for good nutrition include molecules for fuel (which can be obtained from carbohydrates, fats, or proteins), essential amino acids, essential fatty acids, vitamins, and certain minerals.

QUESTIONS

1. Distinguish, in terms of function, between: gastrointestinal hormones/gastrointestinal enzymes; gastric juice/bile; amylases/lipases; digestion/absorption; villi/microvilli; vitamins/essential amino acids.

2. Diagram the human digestive system, including all tubes, chambers, valves, and accessory organs. Describe the function of each structure.

3. Trace the chemical processing of a hamburger on a bun, with lettuce and tomato, as it passes through your digestive tract.

4. Most of the protein-digesting enzymes are secreted in an inactive form and are themselves activated by special enzymes secreted into the digestive tract. Explain the adaptive value of this two-step process.

5. If you oxidize a pound of fat, whether butter or the fat in your own body cells, about 4,200 kilocalories are released. Suppose that you go on a weight-reducing diet, limiting yourself to 1,000 kilocalories a day. You spend most of your time sitting in a well-heated library (studying, of course) and so you expend only about 3,000 kilocalories a day. How many pounds will you lose each week?

6. Four principal symptoms of cirrhosis of the liver and other liver diseases are jaundice (a yellowing of the eyes and skin), uncontrolled bleeding, increased sensitivity to drugs, and swelling of parts of the body, such as the legs and abdominal cavity. Explain each of these symptoms in terms of liver function.

7. Humans are omnivores rather than strict herbivores or carnivores or specialists that use a single organism as a food source. Humans have an unusually large number of substances specifically required in their diets. How might these two phenomena be related?

C H A P T E R 35

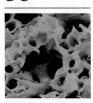

Energy and Metabolism II: Respiration

Digestion, as we saw in the last chapter, is the process by which foodstuffs ingested by an animal are broken down into smaller organic molecules and absorbed into the bloodstream, through which they can be transported to the diverse cells of which the organism is composed. The release of energy from these molecules—the sole source of energy for heterotrophic cells—depends upon their oxidation. This process usually (although not always) requires oxygen, and when it does, it is called respiration. Actually, respiration has two meanings in biology. At the cellular level, it refers to the oxygen-requiring chemical reactions, discussed in Chapter 9, that take place in the mitochondria and are the chief source of energy for eukaryotic cells. At the level of a whole organism, it designates the process of taking in oxygen from the environment and returning carbon dioxide to it. The latter process—which is, of course, essential for the former—is the subject of this chapter.

Oxygen consumption is directly related to energy expenditure (Figure 35–1), and, in fact, energy requirements are usually calculated by measuring oxygen intake or the release of carbon dioxide. The energy expenditure at rest is known as basal metabolism. Metabolic rates increase sharply with exercise. A person exercising consumes 15 to 20 times the amount of oxygen that he or she consumes sitting still, with the oxygen consumption increasing in proportion to the energy expenditure.

35–1 *Oxygen is required for the energy-yielding reactions that take place in the mitochondria. The oxygen consumption of running animals—such as these female impalas—increases linearly with their speed.*

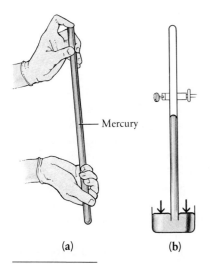

35–2 *Atmospheric pressure is usually measured by means of a mercury barometer.* (a) *To make a simple mercury barometer, put on a pair of protective gloves and then fill a long glass tube, open at one end, with mercury. Closing the tube with your finger,* (b) *invert it into a dish of mercury. Remove your finger and clamp the tube to a stand. The mercury level will drop until the pressure of its weight inside the tube is equal to the atmospheric pressure outside. At sea level, the height of the column will be about 760 millimeters (29.9 inches).*

TABLE 35-1	Composition of Dry Air
GAS	% OF VOLUME
Oxygen	21
Nitrogen	77
Argon	1
Carbon dioxide	0.03
Other gases*	0.97

* Includes hydrogen, neon, krypton, helium, ozone, xenon, and now, unfortunately, in some environments, radon.

DIFFUSION AND AIR PRESSURE

In every organism from an amoeba to an elephant, gas exchange—the exchange of oxygen and carbon dioxide between cells and the surrounding environment—takes place by diffusion. Diffusion, you will recall (page 129), is the net movement of particles from a region of higher concentration to a region of lower concentration as a result of their random motion. In describing gases, scientists speak of the pressure of a gas rather than its concentration. At sea level, the air around us exerts a pressure on our skin of 1 atmosphere (about 15 pounds per square inch). This pressure is enough to support a column of water about 10 meters high or a column of mercury 760 millimeters high (Figure 35–2). Atmospheric pressure is generally measured in terms of mercury simply because mercury is relatively heavy—so the column will not be inconveniently tall.

The total pressure of a mixture of gases, such as air, is the sum of the pressures of the separate gases in the mixture. The pressure of each gas is proportional to its concentration. Oxygen, for instance, makes up about 21 percent by volume of dry air (Table 35–1). Thus 21 percent of the total air pressure, or 160 millimeters of mercury (mm Hg), results from the pressure of oxygen in the air. This is known as the partial pressure of oxygen and is abbreviated P_{O_2}. In air that contains water vapor (also a gas), the volume of O_2 is proportionately less and so the P_{O_2} is less—around 155 mm Hg.

If a liquid containing no dissolved gases is exposed to air at atmospheric pressure, each of the gases in the air diffuses into the liquid until the partial pressure of each gas in the liquid is equal to the partial pressure of the gas in the air. Thus when one speaks of the P_{O_2} of a liquid, such as blood, one means the pressure of dry gas with which the dissolved O_2 in the liquid is in equilibrium. For example, blood with a P_{O_2} of 40 mm Hg would be in equilibrium with air in which the partial pressure of oxygen was 40 mm Hg. If, however, blood with a P_{O_2} of 40 mm Hg is exposed to the usual mixture of air, oxygen moves from the air into the blood until the blood P_{O_2} equals 155 mm Hg and equilibrium is reached. Conversely, if a liquid containing a dissolved gas is exposed to air in which the partial pressure of that gas is lower than in the liquid, the gas will leave the liquid until the partial pressures in the air and the liquid are equal. In short, gases move from a region of higher partial pressure to a region of lower partial pressure.

We are so accustomed to the pressure of the air around us that we are unaware of its presence or of its effects on us. However, if you visit a place, such as Mexico City, that is at a comparatively high altitude and therefore has a lower atmospheric pressure (and of course, a lower P_{O_2}), you will feel light-headed at first and will tire easily. As we shall see in an essay later in this chapter, a series of physiological adaptations are required for life at high altitude—and for successful mountain climbing.

The consequences of the opposite situation—higher gas pressures—can be seen in deep-sea divers. Early in the history of deep-sea diving, it was found that when divers come up from the bottom too quickly, they get the "bends," which are always painful and sometimes fatal. The bends develop as a result of breathing compressed air. Under high pressure, more nitrogen diffuses from the air in the lungs into solution in the blood and tissues. If the body is rapidly decompressed, the gas comes out of solution, and nitrogen bubbles form in the blood, like the carbon dioxide bubbles that appear in a bottle of soda when you decompress it by removing the top. The nitrogen bubbles lodge in the capillaries, stopping blood flow, or they invade nerves or other tissues. Diving mammals, such as whales and seals, have special adaptations that allow them to stay submerged at high pressures for relatively long periods of time. These adaptations—and their surprising human counterparts—will be the subject of another essay later in the chapter. First, however, we must consider the structure and function of respiratory systems.

EVOLUTION OF RESPIRATORY SYSTEMS

Oxygen enters and moves within cells by diffusion. Within the cell, as we have noted, it takes part in the oxidation of organic compounds that serve as cellular energy sources. In this process, carbon dioxide is produced, which then diffuses out of the cell down the concentration (partial pressure) gradient. This is true of all cells, whether an amoeba, a *Paramecium,* a liver cell, or a brain cell. But, substances can move efficiently by diffusion only for very short distances (less than 1 millimeter). These limits pose no problem for very small animals, in which each cell is quite close to the surface, or for animals in which much of the body mass is not metabolically active—like the mesoglea ("jelly") of jellyfishes. Many eggs and embryos also obtain oxygen in this simple way, particularly in the early stages of development.

Diffusion, however, cannot possibly meet the needs of large organisms in which cells in the animal's interior may be many centimeters from the air or water serving as the oxygen source. As organisms increased in size in the course of evolution, there also evolved circulatory and respiratory systems that transport large numbers of gas molecules by bulk flow. (Remember that whereas diffusion is the result of the random movement of individual particles, bulk flow is the overall movement of a gas or liquid in response to pressure or gravity.)

An early stage in the evolution of gas-transport systems is exemplified by the earthworm (page 556). As is the case with most other kinds of worms, an earthworm has a network of capillaries just one cell layer beneath the surface of its body. Oxygen and carbon dioxide diffuse directly through the moist body surface into and out of the blood as it travels through these capillaries. The blood picks up oxygen by diffusion as it travels near the surface of the animal and releases oxygen by diffusion as it travels past the oxygen-poor cells in the interior of the earthworm's body. Conversely, blood picks up carbon dioxide from the cells and releases it by diffusion as it travels near the surface of the animal. Thus the gases move into and out of the earthworm by diffusion but are transported within the animal by bulk flow.

This system is particularly suitable for worms because their tube shape exposes a proportionately large surface area. Some worms can adjust their surface area in relation to oxygen supply. If you have an aquarium at home, you may be familiar with tubifex worms, which are frequently sold as fish food. When these worms are placed in water low in oxygen, such as a poorly aerated aquarium, they stretch out to as much as 10 times their normal length, increasing the surface area through which oxygen diffusion occurs and decreasing the diffusion distance.

Insects and some other arthropods, as we have seen (page 568), have evolved a different strategy. Air is piped directly into the tissues by a network of chitin-lined tubules (Figure 35–3). In large insects, diffusion is assisted by body movements, which propel the air into and out of the spiracles by bulk flow. This system is fine for small organisms but is a major limitation on the size that can be obtained by insects and other tracheal-system breathers. The planet may eventually be taken over by cockroaches or ants, but it is safe to bet they will not be the giant forms of science fiction.

Evolution of Gills

Gills and lungs are other ways of increasing the respiratory surface. Gills are usually outgrowths (Figure 35–4), whereas lungs are ingrowths, or cavities. The respiratory surface of the gill, like that of the earthworm, is a layer of cells, one cell thick, exposed to the environment on one side and to circulatory vessels on the other. The layers of gill tissue may be spread out flat, stacked, or convoluted in various ways. The gill of a clam, for instance, is shaped like a steam-heat radiator (which is also designed to provide a high surface-to-volume ratio).

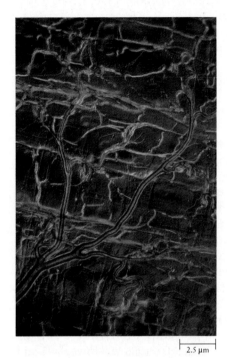

2.5 μm

35–3 *Insects and some other terrestrial arthropods breathe by means of tracheae. These are inward tubular extensions of the exoskeleton, which are thickened in places, forming spiral supports that hold the tubes open. The tracheae are lined with chitin, which is molted when a new exoskeleton is formed. Gas exchange takes place at the thin, moist terminal ends of the tubules. This micrograph shows a portion of the tracheal system that services cells in the wall of the digestive tract of a cockroach.*

The insect respiratory tract, like that of humans, is a fertile breeding ground for infectious organisms. For example, the tracheae of honey bees are often parasitized by mites, leaving the bees somewhat lethargic and greatly reducing their production of honey. Scientists at the U.S. Department of Agriculture have found that treatment with menthol, the soothing ingredient in most cough drops and vapor rubs, kills the mites, enabling the bees to breathe easier and get back to work.

CHAPTER 35 Energy and Metabolism II: Respiration

35–4 *Gills are essentially outpocketings of the epithelium that increase the surface area exposed to water. Often gills are covered by an exoskeleton, as in crustaceans, or a flaplike gill cover, as in fish. In the axolotl, the amphibian shown here, the external nature of the gills is clearly evident. The source of their bright red color is blood flowing through dense capillary networks just one cell layer beneath the gill surface.*

The vertebrate gill is believed to have originated primarily as a feeding device. Primitive vertebrates respired mostly through their skin. They filtered water into their mouths and out of what we now call their gill slits, extracting bits of organic matter from the water as it went through. *Branchiostoma* (page 591), which is believed to resemble closely the ancestral vertebrate, feeds in this way.

In the course of time, numerous selection pressures, chiefly involved with predation, came into operation. As one consequence, there was a trend toward an increasingly thick skin, even one armored or covered with scales. Such a skin is not, of course, useful for respiratory purposes. At the same time, related forces were operating to produce animals that were larger and swifter and so more efficient at capturing prey and escaping predators. Such animals had higher energy requirements—and consequently higher oxygen requirements. These problems were solved by the "capture" of the gill for a new purpose: respiratory exchange. The surface area and blood supply of the gill epithelium have slowly increased over the millennia. The modern gill is the result of this evolutionary process.

In most fishes, the water (in which oxygen is dissolved) is pumped in at the mouth by oscillations of the bony gill cover and flows out across the gills (Figure 35–5). In the gills of fish, the circulatory vessels are arranged so that the blood is pumped through them in a direction opposite to that of the oxygen-bearing water. This countercurrent arrangement (see page 130) results in a far more efficient transfer of oxygen to the blood than if the blood flowed in the same direction as the water. Also, the fish can regulate the rate of water flow, and sometimes assist it, by opening and closing its mouth. Fast swimmers, such as mackerel, obtain enough oxygen to meet their energy needs by keeping their mouths open as they swim. As a result, water moves rapidly over the gills. Such fish have become so dependent on this method of respiration that if they are kept in an aquarium or any other space where their motion is limited, they will suffocate.

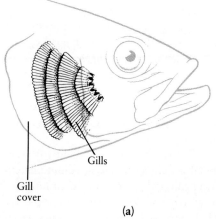

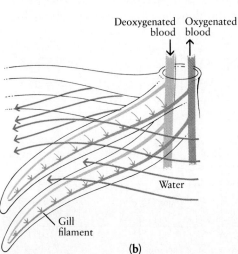

35–5 *In fish, oxygen enters the blood by diffusion from water flowing through the gills (a). The anatomical structure of the gills maximizes the rate of diffusion, which is proportional not only to the surface areas exposed but also to differences in concentration of the diffusing molecules. The greater the difference in concentration of a molecule, the more rapid its diffusion. (b) The circulatory vessels are arranged so that the blood is pumped through them in a direction opposite to that of the oxygen-bearing water. This countercurrent arrangement results in a far more complete transfer of oxygen to the blood than if the blood flowed in the same direction as the water.*

Evolution of Lungs

Lungs are internal cavities into which oxygen-containing air is taken. They have a disadvantage as compared with gills; it is more efficient from the point of view of diffusion to have a continuous flow across the respiratory surface. Air, however, is a far better source of oxygen than is water; 21 percent of the air of the modern atmosphere is oxygen, as compared to 0.5 percent by volume in water at 15°C. Not only must more water be processed to obtain a given amount of oxygen, but water also has a much higher viscosity than air. A fish spends up to 20 percent of its energy in the muscular work associated with respiration, whereas an air breather expends only 1 or 2 percent of its energy in respiration. Also, oxygen diffuses about 300,000 times more rapidly through air than through water, and so can be replenished much more quickly in air as it is used up by respiring organisms.

Lungs are not essential for air breathing. As we have seen, earthworms are air breathers. The overwhelming advantage of lungs, however, is that the respiratory surfaces can be kept moist without a large loss of water by evaporation. Although lungs are largely a vertebrate "invention," they are found in some invertebrates. Land-dwelling snails, for example, have independently evolved lungs that are remarkably similar to the lungs of some amphibians.

Some primitive fishes had lungs as well as gills, although, as we noted in Chapter 28, these lungs were not efficient enough to serve as more than accessory respiratory structures. They were probably a special adaptation to life in fresh water, which, unlike ocean water, may stagnate (become depleted of oxygen). A few species of lungfish still exist (Figure 28–17, page 595). By coming to the surface and gulping air into their lungs, they can live in water that does not have sufficient oxygen to support other fish life.

Amphibians and reptiles have relatively simple lungs, with small internal surface areas, although their lungs are far larger and more complex than those of the lungfish. The lungs of lungfish developed directly from the pharynx, the posterior portion of the mouth cavity, which leads to the digestive tract. In amphibians, reptiles, and other air-breathing vertebrates, we see the evolution of the windpipe, or **trachea,** guarded by a valve mechanism, the glottis, and of nostrils, which make it possible for the animal to breathe with its mouth closed. Amphibians still rely largely on their skin for gas exchange (Figure 35–6), but reptiles breathe almost entirely through their lungs.

An important feature of all vertebrate lungs is that the exchange of air with the atmosphere takes place by bulk flow as a result of changes in lung volume. Such lungs are known as ventilation lungs. Frogs gulp air and force it into their lungs in a swallowing motion; then they open the glottis and let the air out again. In reptiles, birds, and mammals, air is moved into and out of the lungs as a consequence of changes in the size of the thoracic cavity, brought about by muscular contractions and relaxations.

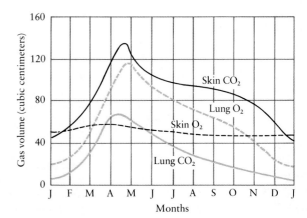

35–6 *Frogs have relatively small and simple lungs, and a major part of their respiration takes place through the skin. The chart shows the results of a study of a frog in which oxygen and carbon dioxide exchanges through the skin and lungs were measured simultaneously for one year. Can you explain why the respiratory activity peaks in April and May and is lowest in December and January?*

35–7 (a) *Although the lungs of birds are small, they are extraordinarily efficient. Each lung has several air sacs attached to it. With each cycle of expiration and inspiration, the air sacs empty and fill like balloons as they are compressed and expanded by movements of the body wall. No gas exchange takes place in the sacs. They appear rather to act as bellows, flushing fresh air through the lungs at every breath, always in the same direction. As a result, there is little residual "dead" air left in the lungs, as there is in mammals.* (b) *Scanning electron micrograph of lung tissue from a 14-day-old chicken. The tubes visible here are ventilated by air drawn in by the air sacs. Gas exchange takes place in the broad meshwork of air capillaries and blood capillaries, which make up the spongelike respiratory tissue seen here surrounding the tubes.*

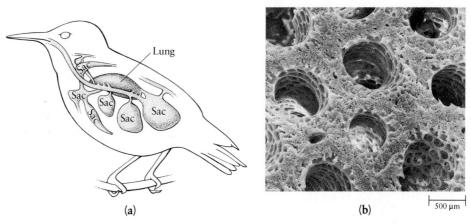

Respiration in Large Animals: Some Principles

Figure 35–8 summarizes the types of respiratory systems found among the animals. In large animals, both diffusion and bulk flow move oxygen molecules between the external environment and actively metabolizing tissues. This movement occurs in four stages:

1. Movement by bulk flow of the oxygen-containing external medium (air or water) to a thin, moist epithelium close to small blood vessels in lungs or gills.
2. Diffusion of the oxygen across this epithelium into the blood.
3. Movement by bulk flow of the oxygen with the circulating blood to the tissues where it will be used.
4. Diffusion of the oxygen from the blood into the interstitial fluids, from whence it diffuses into the individual cells.

Carbon dioxide, which is produced in the tissue cells, follows the reverse path as it is eliminated from the body.

35–8 *Respiratory systems.* (a) *Gas exchange across the entire surface of the body is found in a wide range of small organisms from protists to earthworms.* (b) *Gas exchange across the surface of a flattened body is seen, for example, in flatworms. Flattening increases the surface-to-volume ratio and also decreases the distance over which diffusion has to occur within the body.*

(c) *External gills increase the surface area but are unprotected and therefore easily damaged. External gills are found in polychaete worms and some amphibians. Gas exchange usually takes place across the rest of the body surface as well.* (d) *With internal gills, a ventilation mechanism draws water over the highly vascularized gill surfaces, as in fish.*

(e) *Gas exchange at the terminal ends of fine tracheal tubes that branch through the body and penetrate all the tissues is characteristic of insects and some other arthropods.* (f) *Lungs are highly vascularized sacs into which air is drawn by a ventilation mechanism. Lungs are found in all air-breathing vertebrates and some invertebrates, such as terrestrial snails.*

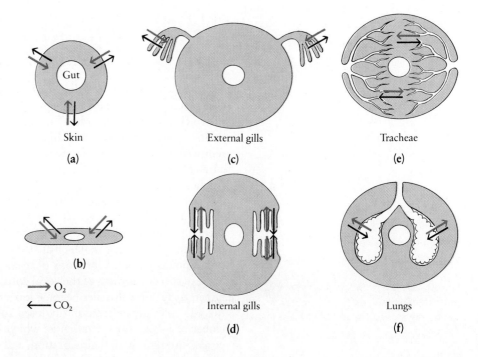

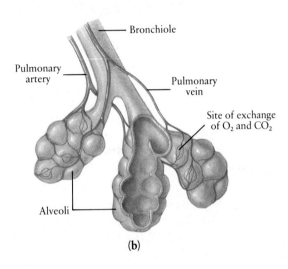
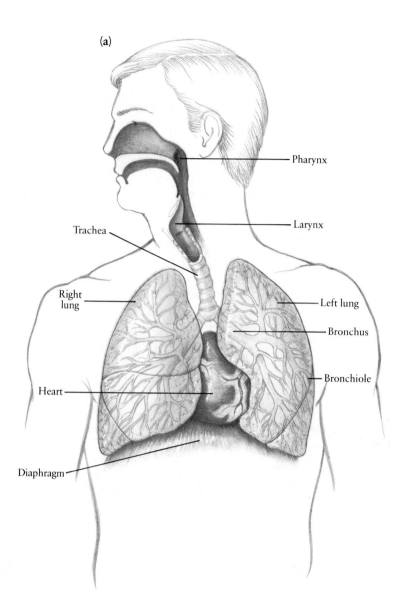

35-9 *The human respiratory system.* (a) *Air enters through the nose or mouth and passes into the pharynx, past the larynx, and down the trachea, bronchi, and bronchioles to the alveoli* (b) *in the lungs. The alveoli, of which there are some 300 million in a pair of lungs, are the sites of gas exchange. Oxygen and carbon dioxide diffuse into and out of the bloodstream through the capillaries in the walls of the alveoli.*

THE HUMAN RESPIRATORY SYSTEM

In *Homo sapiens,* inspiration (breathing in) and expiration (breathing out) usually take place through the nose. The nasal cavities are lined with hairs and cilia, both of which trap dust and other foreign particles. The epithelial cells that line the cavities secrete mucus, which humidifies the air and collects debris that can be removed by swallowing, sneezing, blowing the nose, or spitting. The cavities have a rich blood supply, which keeps their temperature high, warming the air before it reaches the lungs.

From the nasal passages, the air goes to the pharynx and from there to the larynx, located in the upper front part of the neck. An adult human larynx is shaped somewhat like a triangular box, with its point downward. Across it are stretched the vocal cords, which are two ligaments drawn taut across the lumen of the respiratory tract. Vibrations of these cords, by expired air, cause the sounds made in speech. Growth of the vocal cords is influenced by male sex hormones. At puberty, the cords in males become longer and thicker, sometimes so rapidly that the adolescent male temporarily loses control over them, occasionally emitting embarrassing squeaks. Laryngitis, which is simply an inflammation of the vocal cords, interferes with their vibration, and so you "lose your voice."

Cancer of the Lung

Lung cancer is the most rapidly increasing form of cancer in the United States. The death rate from lung cancer among U.S. men and women more than tripled during the period from 1950 to 1985. During that same period, the overall cancer death rate increased in the United States due entirely to an increase in the number of deaths from lung cancer. It is now the most common cause of death from cancer in both men and women, having recently surpassed breast cancer as the most common cause of death from cancer in women.

It is expected that well over 150,000 men and women will develop lung cancer in the United States this year, and more than 85 percent of them will die in less than five years. Most of these patients will be cigarette smokers.

The middle and lower lobes of a cancerous lung are shown in (a). The cancer is the solid grayish-white mass. Its rapid growth has replaced most of the normal lung tissue in the middle lobe. It has also probably begun to spread to other parts of the body. Notice the bronchial branch that leads into the cancer and is destroyed by it. The lung tissue remaining around the cancer is compressed, airless, and dark red. Away from the cancer, in the lower lobe, the lung is filled with air and is light pink.

The lung is normally protected by ciliated cells lining the trachea and bronchi. The cilia sweep out particles in the respired air that are caught in the mucus secreted by the goblet cells. These cilia are paralyzed by chemicals in cigarette smoke. Moreover, cancer cells arising in the bronchi generally do not have cilia. The scanning electron micrographs show (b) the normal ciliated surface of the bronchus and (c) the surface of a bronchus with cancer.

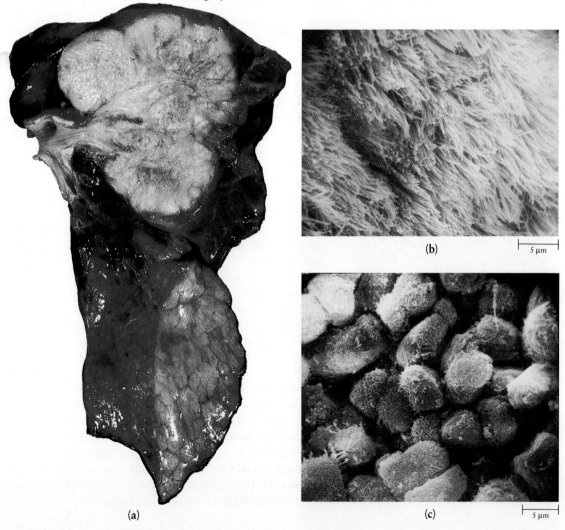

(a)

(b) 5 μm

(c) 5 μm

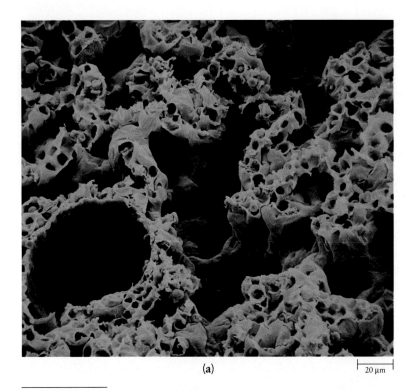 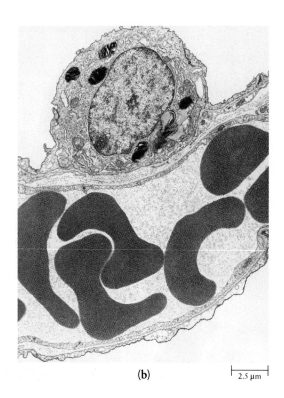

35-10 (a) *Scanning electron micrograph of lung tissue, showing numerous alveoli.* (b) *Longitudinal section of an alveolar capillary. The large, dark, irregularly shaped structures are red blood cells. An alveolar cell is visible at the top.*

35-11 *Gases are exchanged by diffusion as a consequence of the different partial pressures of oxygen and carbon dioxide in the alveolus and the alveolar capillary. The figures indicate millimeters of mercury.*

From the larynx, inspired air travels through the trachea, which is a long membranous tube, also lined with ciliated epithelial cells. The walls of the trachea are strengthened by rings of cartilage that prevent it from collapsing during inspiration or when food from the adjacent esophagus presses on it. The trachea leads into the **bronchi** (singular, bronchus), which subdivide into smaller and smaller passageways, the **bronchioles.** The bronchi and bronchioles are surrounded by thin layers of smooth muscle. Contraction and relaxation of this muscle alters resistance to air flow.

Cilia along the trachea, bronchi, and bronchioles beat continuously, pushing mucus and foreign particles embedded in mucus up toward the pharynx, from which it is generally swallowed. We are usually aware of this production of mucus only when it is increased above normal as a result of an irritation of the membranes.

The actual exchange of gases takes place in small air sacs, the **alveoli,** which are clustered in bunches like grapes around the ends of the smallest bronchioles. Each alveolus is about 0.1 or 0.2 millimeter in diameter, and each is surrounded by capillaries. The walls of the capillaries and of the alveoli each consist of a single layer of flattened epithelial cells separated from one another by a thin basement membrane; thus the barrier between the air in an alveolus and the blood in its capillaries is only about 0.5 micrometer. Gases are exchanged between the air and the blood by diffusion (Figure 35-11). A pair of human lungs has about 300 million alveoli, providing a respiratory surface of some 70 square meters—approximately 40 times the external surface area of the entire human body.

The lungs are surrounded by a thin membrane known as the pleura, which also lines the thoracic cavity. The pleura secretes a small amount of fluid that lubricates the surfaces so that they slide past one another as the lungs expand and contract. Pleurisy is an inflammation of these membranes that causes them to secrete excess fluid that collects in the thoracic cavity.

35-12 *A model illustrating the way air is taken into and expelled from the lungs, showing the action of the diaphragm.*

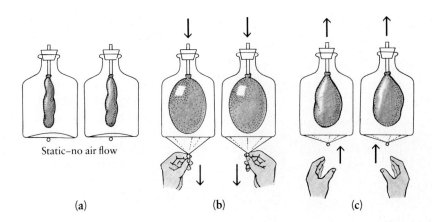

MECHANICS OF RESPIRATION

Air flows into or out of the lungs when the air pressure within the alveoli differs from the pressure of the external air (atmospheric pressure). When alveolar pressure is greater than atmospheric pressure, air flows out of the lungs, and expiration occurs. When alveolar pressure is less than atmospheric pressure, air flows into the lungs, and inspiration occurs.

The pressure in the lungs is varied by changes in the volume of the thoracic cavity, as illustrated by the model in Figure 35–12. These changes are brought about by the contraction and relaxation of the muscular diaphragm and of the intercostal ("between-the-ribs") muscles. We inhale by contracting the dome-shaped diaphragm, which flattens it and lengthens the thoracic cavity, and by contracting those intercostal muscles that pull the rib cage up and out. These movements enlarge the thoracic cavity; the pressure within it falls, and air moves into the lungs. Air is forced out of the lungs as the muscles relax, reducing the volume of the thoracic cavity. Usually, only about 10 percent of the air in the lung cavity is exchanged at every breath, but as much as 80 percent can be exchanged by deliberate deep breathing.

Whales and other large aquatic mammals suffocate on land because of the inability of their intercostal muscles to expand their massive chests to inhale when compressed under the weight of their bodies.

TRANSPORT AND EXCHANGE OF GASES

Hemoglobin and Its Function

Oxygen is relatively insoluble in blood plasma (the liquid part of blood)—only about 0.3 milliliter of oxygen will dissolve in 100 milliliters of plasma at normal atmospheric pressure. In insects, which do not depend on their blood to transport oxygen to the individual cells, this low solubility is of little consequence. In other animals, it would be a severe limitation were it not for the presence of special oxygen-carrying protein molecules, known as respiratory pigments, that raise the oxygen-transporting capacity of the blood as much as seventyfold. Such pigments are found in the blood of virtually all active animals except insects, including even the earthworm.

Hemoglobin is the respiratory pigment found among vertebrates and in a wide variety of invertebrate species representing many different phyla. A form of

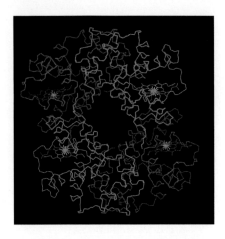

35–13 *A computer-generated model of the hemoglobin molecule. The turquoise lines represent the carbon backbones of the two alpha polypeptide chains, and the dark blue lines represent the carbon backbones of the two beta polypeptide chains. The porphyrin rings are shown in red, and the iron atoms as yellow starbursts.*

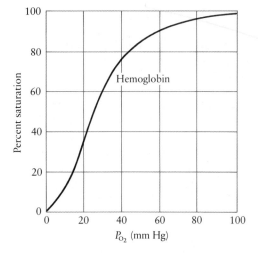

35–14 *Oxygen-hemoglobin association-dissociation curve. This curve represents figures for normal adult human hemoglobin at 38°C and at a normal pH. As the partial pressure of oxygen rises, hemoglobin picks up oxygen. When oxygen pressure reaches 100 mm Hg—the pressure usually present in the human lung—the hemoglobin becomes totally saturated with oxygen. As the P_{O_2} drops, the oxygen and the hemoglobin dissociate. Therefore, when oxygen-carrying blood reaches the capillaries, where pressure is only about 40 mm Hg or less, it gives up some of its oxygen to the tissues.*

oxygen carrier (hemocyanin), which contains copper rather than iron, is the most common respiratory pigment of mollusks and arthropods. (Unlike hemoglobin, which is red when combined with oxygen, hemocyanin is blue.) Other respiratory pigments are known, all of which are a combination of a metal-containing unit and a protein chain. In most invertebrates, respiratory pigments are simply dissolved in the blood plasma; in vertebrates and echinoderms, the pigments are carried in red blood cells. As we shall see in the next chapter, these cells are highly specialized for their transport function; a mature human red blood cell carries some 265 million molecules of hemoglobin.

Hemoglobin, as you will recall, is made up of four subunits, each of which comprises a heme unit and a polypeptide chain (Figure 35–13). The heme unit consists of a porphyrin ring with one atom of iron in the center. The iron in each heme unit can combine with one molecule of oxygen; thus each hemoglobin molecule can carry four molecules of oxygen. The oxygen molecules (O_2) are added one at a time:

$$Hb_4 + O_2 \rightleftharpoons Hb_4O_2$$
$$Hb_4O_2 + O_2 \rightleftharpoons Hb_4O_4$$
$$Hb_4O_4 + O_2 \rightleftharpoons Hb_4O_6$$
$$Hb_4O_6 + O_2 \rightleftharpoons Hb_4O_8$$

Combination of the first subunit (Hb) with O_2 increases the affinity of the second for O_2, and oxygenation of the second increases the affinity of the third, and so on. (As O_2 is taken up, the two beta polypeptide chains of hemoglobin move closer together, and this movement is apparently the reason the shift in affinity takes place.) As a consequence of the change in affinity of the hemoglobin molecule for oxygen, the curve relating the uptake of oxygen to P_{O_2} is not a straight line, but has a characteristic sigmoid shape (Figure 35–14). When hemoglobin is fully oxygenated, it enables our bloodstreams to carry about 65 times as much oxygen as could be transported by an equal volume of plasma alone.

Whether oxygen combines with hemoglobin or is released from it depends on the partial pressure of oxygen (P_{O_2}) in the surrounding blood plasma. Oxygen diffuses from the air into the alveolar capillaries. In these capillaries, where the P_{O_2} is high, most of the hemoglobin is combined with oxygen. In the tissues, however, where the P_{O_2} is lower, oxygen is released from the hemoglobin molecules into the plasma and diffuses into the tissues. This system compensates automatically for the oxygen requirements of the tissues. For example, in adult humans, the P_{O_2} as the blood leaves the lungs is about 100 mm Hg; at this pressure, the hemoglobin is saturated with oxygen. As the hemoglobin molecules travel through the tissue capillaries, the P_{O_2} drops, and as it drops, the oxygen bound to the hemoglobin molecules is given up. Little oxygen is yielded as the P_{O_2} drops from 100 mm Hg to 60 mm Hg. (This built-in safety factor ensures that oxygen is delivered to the tissues, even when the maximum blood P_{O_2} is lower than normal, as, for example, in individuals who live at high altitudes or who have heart and lung diseases.) However, as the P_{O_2} drops below 60 mm Hg, oxygen is given up much more readily (Figure 35–15).

The P_{O_2} of the blood in the tissue capillaries is normally about 40 mm Hg. As a consequence, even after the blood has passed through the tissue capillaries, its hemoglobin is still usually 70 percent saturated. The oxygen still carried by hemoglobin represents a reserve supply of this precious gas should the demand increase—as a result, for example, of exercise. In such a situation, the cells respire more rapidly, more oxygen is consumed, the P_{O_2} decreases, and more oxygen is released from the hemoglobin molecules.

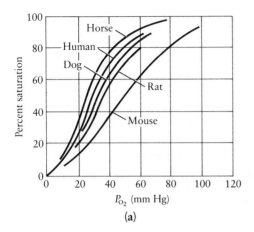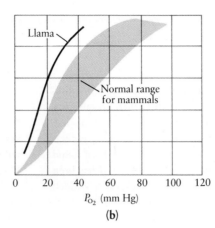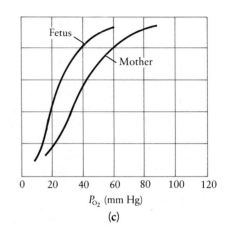

35–15 *These curves show how the amount of oxygen carried by the hemoglobin is related to oxygen pressure in a number of different mammals. A curve located to the right of another curve signifies that the oxygen is given up more readily at a given pressure.* **(a)** *Small animals have higher metabolic rates and so need more oxygen per gram of tissue than larger animals. Therefore, they have blood that gives up oxygen more readily.* **(b)** *The llama, which lives in the high Andes of South America, has a hemoglobin that enables its blood to take up oxygen more readily at the low atmospheric pressures.* **(c)** *The fetus must take up all its oxygen from the maternal blood. The hemoglobin of mammalian fetuses has a greater affinity for oxygen than does the hemoglobin of adult mammals, and so the oxygen tends to diffuse from the maternal blood to the fetal blood.*

Carbon dioxide is more soluble than oxygen in the blood, and a small amount of it is simply dissolved in the plasma. In addition, about 25 percent of it is bound to the amino groups of the hemoglobin molecules. However, most of the carbon dioxide (about 65 percent) is carried in the blood as bicarbonate ion (HCO_3^-). Bicarbonate ion is produced in a two-stage reaction. First, carbon dioxide combines with water to form carbonic acid. This reaction is catalyzed by the enzyme carbonic anhydrase found in red blood cells. Then carbonic acid, a weak acid, dissociates to yield bicarbonate and hydrogen ions:

$$CO_2 + H_2O \xrightleftharpoons[]{\text{Carbonic anhydrase}} \underset{\text{Carbonic acid}}{H_2CO_3} \rightleftharpoons \underset{\text{Bicarbonate ion}}{HCO_3^-} + \underset{\text{Hydrogen ion}}{H^+}$$

As you can see by the arrows, this reaction can go in either direction. The direction it actually takes depends on the partial pressure of carbon dioxide in the blood. In the tissues, where the partial pressure of carbon dioxide is high, bicarbonate and hydrogen ions are formed. In the lungs, where the partial pressure of carbon dioxide is low, carbonic acid dissociates to form carbon dioxide and water. Once it is released, the carbon dioxide diffuses from the plasma into the alveoli and flows out of the lung with the expired air.

The partial pressure of carbon dioxide in the blood also affects hemoglobin's affinity for oxygen. In the tissues, where more carbon dioxide is taken up by the blood, the blood becomes increasingly acidic. As the acidity increases, hemoglobin's affinity for oxygen decreases and so it gives up its oxygen more readily.

Table 35–2 provides a quantitative illustration of the gas exchanges occurring in the lung during the respiratory process.

TABLE 35–2 **Composition of Respiratory Gas at Standard Atmospheric Pressure**

	INSPIRED AIR		EXPIRED AIR		ALVEOLAR AIR	
GAS	% OF VOLUME	PARTIAL PRESSURE (mm Hg)	% OF VOLUME	PARTIAL PRESSURE (mm Hg)	% OF VOLUME	PARTIAL PRESSURE (mm Hg)
O_2	20.71	157	14.6	111	13.2	100
CO_2	0.04	0.3	4.0	30	5.3	40
H_2O	1.25	9.5	5.9	45	5.9	45
N_2	78.00	593	75.5	574	75.6	574

Diving Mammals

Most mammals have about the same oxygen requirements as humans. Structurally, their respiratory organs are very similar. Yet marine mammals can dive to great depths and stay submerged for long periods of time. A sperm whale, for example, has been found at a depth of more than 1,000 meters and has been known to stay submerged for 75 minutes. A Weddell seal, a much smaller animal, can dive to 600 meters and stay submerged for 70 minutes, or can swim submerged for between 2,000 and 4,000 meters—1 to 2 miles. Are their lungs different from ours? Is their blood different? Do they have special oxygen reserves, like a submarine?

Studies in a variety of diving mammals have shown than none has lungs significantly larger, in proportion, than ours. In fact, seals (and perhaps other diving mammals as well) exhale before diving or early in the dive. Blood volume is greater, however. In humans, blood is about 7 percent of the body weight, whereas in diving marine mammals, it is 10 to 15 percent. The blood vessels are proportionately enlarged, and they appear to serve as a reservoir of oxygenated blood. Moreover, the proportion of red blood cells is higher and myoglobin is more concentrated, giving the muscles a very dark color. Still, physiologists calculate, these adaptations would not provide enough oxygen for long dives.

A major survival factor in diving mammals, studies have shown, is a group of automatic reactions known collectively as the diving reflex. During a dive, the heart rate slows and blood supply is reduced to one-tenth or one-twentieth of normal to tissues that are tolerant of oxygen deprivation, such as the digestive organs, skin, and muscles. Muscles obtain energy anaerobically, by glycolysis, producing large amounts of lactic acid and building up an oxygen debt (see page 190). Most of the oxygen is shunted to the heart and brain, whose cells would begin to die after about four minutes without oxygen. Blood supply is reduced only slightly to the adrenal glands, which produce hormones that are involved in regulation of the metabolic rate, the ionic balance of body fluids, and the rate of heartbeat. In a pregnant female, the fetus is also given high priority for the available oxygen.

Studies by Martin Nemiroff of the University of Michigan Medical School indicate that the diving reflex is also present in human beings. In humans, its primary survival value is during birth, when the infant may be cut off from an oxygen supply during the later stages of labor. Operation of the reflex can be demonstrated in a very simple way: immersing the face in cold water causes the heart rate to slow down.

The observations by Nemiroff came in the course of investigating 60 near-drownings. It has been generally assumed that a person who is submerged for four minutes or more will die, or, at best, suffer irreversible brain damage. In the group of 60, 15 had been rescued after four minutes or more in the chilly waters of Michigan lakes. Of these 15, 11 survived without brain damage. One of these, a college student who crashed through the ice in an automobile accident, was under water for 38 minutes; he finished the semester with a 3.2 average. Another survivor resumed his career as a practicing physician. All the survivors had been submerged in cold (below 21°C) water, which slowed the metabolic demands of their cells and therefore reduced their need for oxygen. Some of those rescued required assisted breathing for as long as 13 hours.

Moral: Even if a drowning victim is cold, blue, with no pulse, no heartbeat, and the eyes fixed in a glassy stare—as was the case with the student—if he or she has been in cold water, efforts at resuscitation should be started immediately and continued indefinitely in a hospital. We are not Weddell seals, but owing to an ancient survival strategy, we have unexpected powers of underwater survival.

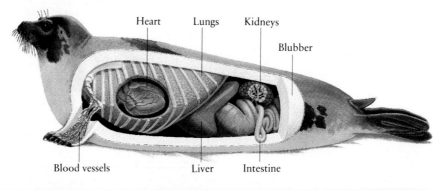

A seal exhales as it dives, reducing the likelihood of decompression sickness (the "bends"). It has 1.5 times as much blood as a land mammal of the same size and, when it dives, can reduce its heartbeat from 120 to 30 beats per minute.

35–16 *Tertiary structure of myoglobin, as deduced from x-ray diffraction analyses. Myoglobin closely resembles a single subunit of the four-subunit hemoglobin molecule. The heme group, to which the oxygen molecule binds, is shown in red.*

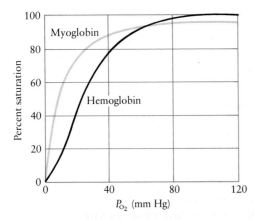

35–17 *Comparison of the oxygen association-dissociation curves of myoglobin and hemoglobin. Note that myoglobin remains almost 80 percent saturated with oxygen until the partial pressure of oxygen falls below 20 mm Hg. Therefore, myoglobin retains its oxygen in the resting cell and relinquishes it only when strenuous muscle activity uses up the available oxygen provided by hemoglobin.*

Myoglobin and Its Function

Myoglobin is a respiratory pigment found in skeletal muscle. Structurally, it resembles a single subunit of the hemoglobin molecule (Figure 35–16). Myoglobin's affinity for oxygen is greater than hemoglobin's and so it picks up oxygen from hemoglobin. It begins to release significant amounts of this oxygen only when the P_{O_2} in skeletal muscle falls below 20 mm Hg (Figure 35–17). Thus, when the muscle is at rest or engaged in only moderate activity, the myoglobin holds on to its oxygen. During strenuous exercise, however, when muscle cells are using oxygen rapidly and the partial pressure of oxygen in the muscle cells drops to zero, myoglobin gives up its oxygen. Thus myoglobin provides an additional reserve of oxygen for active muscles.

CONTROL OF RESPIRATION

The rate and depth of respiration are controlled by respiratory neurons in the brainstem. These neurons are responsible for normal breathing, which is rhythmic and automatic, like the beating of the heart. Unlike the beating of the heart, however, which few of us can control voluntarily, breathing may be brought under voluntary control within certain limits.

The respiratory neurons in the brain activate motor neurons in the spinal cord, causing the diaphragm and intercostal muscles to contract. This activity of the respiratory neurons is thought to occur spontaneously. Periodically, however, these neurons are inhibited, allowing expiration to occur. In addition to their own spontaneous activity, the respiratory neurons receive signals from receptors sensitive to carbon dioxide, oxygen, and hydrogen ions, as well as from receptors sensitive to the degree of stretch of the lungs and chest. Chemoreceptor cells located in the carotid arteries (Figure 35–18, page 747), which supply oxygen to the brain, signal the respiratory neurons when the concentration of oxygen in the blood decreases. The concentration of dissolved carbon dioxide and of hydrogen ion is simultaneously monitored by centers in the brain and also by chemoreceptors in the carotid arteries. Thus, information is provided by a number of different, independent sensors.

High on Mt. Everest

High altitudes are harsh, dangerous environments, and their most dangerous feature is not the cold, the precipitous slopes, the threat of avalanche, or the blinding wind and snow; it is oxygen deprivation. For this reason, the ability of humans and other animals to adjust to high altitudes has long been of particular interest to physiologists.

Until very recently, about 6,000 meters was thought to be the limit for human survival. In 1978, however, two European climbers reached the summit of Mt. Everest (8,848 meters) without supplemental oxygen, raising new questions about physiological adaptability. Three years later, the American Medical Research Expedition to Everest set off with the goal of collecting the first data on human physiological function above 6,000 meters. The team included six highly experienced Himalayan climbers; a group of six "climbing scientists," all physicians with much climbing experience and an interest in high-altitude physiology; and eight physiologists who worked at the base camp at 5,400 meters and at a laboratory at 6,300 meters. Members of the expedition were able to make measurements at above 8,000 meters on human subjects (themselves), including continuous monitoring of the electrical activity of the heart and sampling of the gases in the alveoli—even at the summit itself, which two of the monitored "climbing scientists" reached.

Survival at this extreme altitude, the scientists found, depends largely on hyperventilation—extremely deep breathing. In fact, there seems to be a correlation between the capacity to hyperventilate and the capacity to be a mountain climber. This extremely deep breathing results in an astonishing decrease in the partial pressure of carbon dioxide in the lungs and bloodstream to less than one-fifth normal levels, with a corresponding increase in the pH of the blood. Even with hyperventilation, the partial pressure of oxygen in arterial blood is less than one-third the partial pressure at sea level, and work capacity is greatly diminished. The European climbers, for example, reported that they moved only 2 meters per minute as they approached the summit.

There were also changes in metabolism and brain function. At 6,300 meters, there was a striking loss of body weight, with two of the expedition members each losing 15 kilograms (33 pounds). The concentration of thyroid hormone, which increases the rate of cellular respiration, increased with increasing altitude. The levels of noradrenaline, a compound that functions both as a hormone and as a transmitter in the nervous system, were also elevated. Verbal learning and short-term memory, as measured by standard tests, declined at high altitudes but were normal one year later. In a simple brain function test, all 16 men tested showed a decrease in finger-tapping speed. In 15 of the 16, this abnormality persisted after the expedition, and in 13 of them, it was still present a year later.

In short, despite the remarkable adaptations of which the body is capable, extremely high altitudes have profound and, in some instances, permanent physiological effects. The summit of Mt. Everest is very close to the limits of human survival.

The main laboratory of the American Medical Research Expedition to Everest, at an altitude of 6,300 meters. An altitude of about 5,300 meters appears to be the limit for long-term human habitation.

This system is extremely sensitive to even the smallest change in the chemical composition of the blood, particularly to the concentration of hydrogen ion, which reflects the carbon dioxide concentration (P_{CO_2}), as shown in the equation on page 743. If P_{CO_2}—and therefore the concentration of H^+ ions—increases only slightly, breathing immediately becomes deeper and faster, permitting more carbon dioxide to leave the blood until the concentration of H^+ ions has returned to normal. If you deliberately hyperventilate (breathe deeply and rapidly) for a few moments, you will feel faint and dizzy because of the blood's (and, therefore, the brain's) increased alkalinity.

You can, as we have noted, deliberately increase your breathing rate by contracting and relaxing your diaphragm and chest muscles, but breathing is normally involuntary. It is impossible to commit suicide by deliberately holding your breath; as soon as you lose consciousness and P_{CO_2} rises, the involuntary controls take over once more and breathing resumes. The receptors sensitive to P_{O_2} provide a kind of backup system to the P_{CO_2} and H^+ sensors. In some cases of drug poisoning—for example, by morphine or barbiturates—the brainstem cells sensitive to H^+ become depressed. This causes a decrease in the breathing rate, leading ultimately to a reduction in the P_{O_2} of the blood. The oxygen sensors are then stimulated, and they maintain breathing. (Massive overdoses of these drugs, however, depress the activity of the respiratory neurons themselves.) This complex system of sensors, monitoring different factors in different locations, underlines the critical importance of an uninterrupted supply of oxygen to the cells of an animal's body—particularly those of the brain.

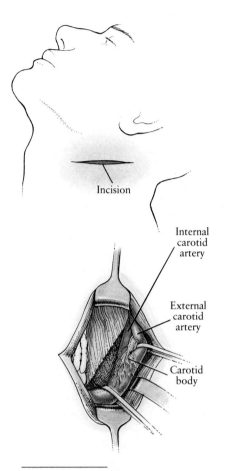

35-18 *Location of the carotid body, one of the receptors that monitor the concentration of dissolved oxygen (P_{O_2}) in the blood. Monitoring at this location is of critical importance, because the carotid arteries supply blood to the brain. To a lesser extent, the carotid body also monitors the concentrations of hydrogen ion (H^+) and carbon dioxide (P_{CO_2}). Hydrogen ion concentrations are also measured directly by neurons in the brain.*

SUMMARY

Heterotrophic cells obtain energy from the oxidation of carbon-containing compounds. This process releases carbon dioxide and, for maximum energy yields, requires oxygen. Respiration is the means by which an animal obtains oxygen for its cells and rids itself of carbon dioxide.

Oxygen is available in both water and air. It enters cells and body tissues by diffusion, moving from regions of higher partial pressure to regions of lower partial pressure. However, efficient movement of oxygen by diffusion requires a relatively large surface area exposed to the source of oxygen and a short distance over which the oxygen has to diffuse. Selection pressures for increasingly efficient means of gas exchange led to the evolution of gills and lungs. Both gills and lungs present enormously increased surface areas for the exchange of gases. They also have a rich blood supply for transporting these gases to and from other parts of the animal's body.

Respiration in large animals involves both diffusion and bulk flow. Bulk flow brings air or water to the lungs or gills and circulates oxygen and carbon dioxide in the bloodstream. Gases are exchanged by diffusion between the blood and the air in the lungs or the water around the gills and between the blood and the tissues.

In humans, air enters the lungs through the trachea, or windpipe, and goes from there into a network of increasingly smaller tubules, the bronchi and bronchioles, which terminate in small air sacs, the alveoli. Gas exchange actually takes place across the alveolar walls. Air moves into and out of the lungs as a result of changes in the pressure within the lungs, which, in turn, result from changes in the size of the thoracic cavity.

Respiratory pigments increase the oxygen-carrying capacity of the blood. In vertebrates, the respiratory pigment is hemoglobin, which is packed within red blood cells. Each hemoglobin molecule has four subunits, each of which can combine with one molecule of oxygen. The addition of each molecule of oxygen increases the affinity of the molecule for each subsequent molecule of oxygen.

Conversely, the loss of each molecule of oxygen facilitates the loss of the subsequent one. Carbon dioxide is transported in the blood plasma principally in the form of bicarbonate ion.

The rate and depth of respiration are controlled by respiratory neurons in the brainstem. These neurons, which activate motor neurons in the spinal cord that cause the diaphragm and intercostal muscles to contract, respond to signals caused by very slight changes in the hydrogen ion, carbon dioxide, and oxygen concentrations of the blood.

QUESTIONS

1. Distinguish among the following: gills/lungs; trachea/pharynx/larynx; bronchi/bronchioles/alveoli; hemoglobin/myoglobin/hemocyanin.

2. Why are calculations of partial pressure used in determining the movement of gases between liquids and the atmosphere rather than concentrations? When does a gas move from an area of lower concentration to one of higher concentration?

3. What are the advantages and disadvantages of obtaining oxygen from air rather than from water? You might be able to think of several of each besides those mentioned in the text.

4. Sketch and label a diagram of the human respiratory system. When you have finished, compare your drawing to Figure 35–9.

5. Explain the following statement: Frogs ventilate their lungs by positive pressure, whereas mammals, birds, and reptiles ventilate theirs by negative pressure.

6. One of the results of long-term smoking is the loss of bronchial cilia. What effects would you expect this to have on normal lung function?

7. Suppose a new cold remedy is guaranteed to suppress completely the secretion of mucus in the respiratory tract. Explain why you would or would not use this remedy during a cold.

8. Carbon monoxide (CO), which is extremely poisonous, has a greater affinity for hemoglobin than does oxygen. The resulting compound, which is a brighter red than normal hemoglobin, can no longer combine with oxygen. From these facts, suggest how you might recognize and give assistance to a victim of carbon monoxide poisoning.

CHAPTER 36

Energy and Metabolism III: Circulation

In all animals except those of very small size or with an extremely simple body plan, blood is the chemical "highway" interconnecting the multitude of cells that form the organism's body. It is the medium in which the nutrient molecules processed by digestion and the oxygen molecules taken in by respiration are delivered to the individual cells. It also carries away the waste materials, including carbon dioxide and urea, produced by cells in the course of their metabolic activities. As we have seen, carbon dioxide picked up by the blood leaves the body by diffusion across the respiratory surfaces. In vertebrates, urea and other wastes are processed in the kidney and excreted from the body, as we shall see in the next chapter. In addition, the blood transports other important substances, such as hormones, enzymes, and antibodies, and has among its basic constituents the cells that defend the body against foreign invaders.

In this chapter, we shall first look at the composition of the blood and then consider the structure and dynamics of the systems by which blood and another fluid, lymph, are transported in the vertebrate body.

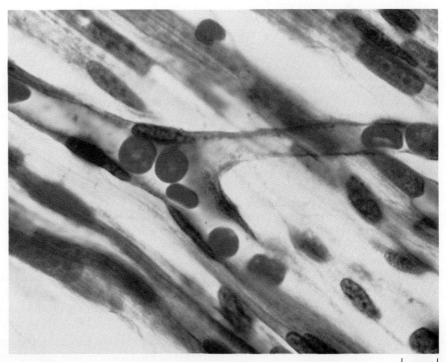

36-1 *The exchange of substances between the circulating blood and the body tissues takes place through the walls of the smallest blood vessels, the capillaries. The capillaries are so narrow that the red blood cells, with their oxygen-laden hemoglobin molecules, must move in single file.*

THE BLOOD

An individual weighing 75 kilograms (165 pounds) has about 6 liters of blood, which is approximately 8 percent of body weight. About 60 percent of the blood is a straw-colored liquid called **plasma,** which is 90 percent water. The other 40 percent of the blood that is not plasma is made up of the red blood cells, white blood cells, and platelets. Blood, as we noted in Chapter 33, is classified as a type of connective tissue.

Plasma

With the exception of the oxygen and carbon dioxide carried by hemoglobin, most of the molecules needed by individual cells, as well as waste products from the cells, are carried along in the heavy traffic of the bloodstream dissolved in the plasma. In addition, the plasma contains protein molecules, known as plasma proteins, that are not nutrients or waste products but function instead in the bloodstream itself. Because of the dissolved plasma proteins, the osmotic potential (page 133) of the blood is greater than that of the surrounding interstitial fluid. Thus these proteins have the effect of preventing excessive loss of fluid from the bloodstream to the tissues. They also serve to bind certain ions and small molecules, preventing them from leaving the bloodstream, and to transport fats, cholesterol, and otherwise insoluble molecules in the bloodstream.

Plasma proteins are of three major types: albumin, fibrinogen, and globulins. The chief function of albumin is to maintain the high osmotic potential of the plasma in relation to the interstitial fluid, whereas fibrinogen is responsible for blood clotting. Among the most important of the globulins are the immunoglobulins, or antibodies (to be discussed in Chapter 39).

Red Blood Cells

Red blood cells, or **erythrocytes** (Figure 36–2), are specialized for the transport of oxygen. As a mammalian red blood cell matures, it extrudes its nucleus and mitochondria, and its other cellular structures dissolve. Almost the entire volume of a mature red blood cell is filled with hemoglobin, about 265 million molecules per cell in humans.

There are about 5 million red blood cells per cubic millimeter of blood—some 25 trillion (25×10^{12}) in the adult human body. Because red blood cells, lacking a nucleus, cannot repair themselves, their life span is comparatively short, some 120 to 130 days. At this moment, in your body, red blood cells are dying at a rate of about 2 million per second; to replace them, new ones are being formed in the bone marrow at the same rate.

White Blood Cells

For every 1,000 red blood cells in the human bloodstream, there are 1 or 2 white blood cells, or **leukocytes,** for a total of about 6,000 to 9,000 per cubic millimeter of blood. These cells are nearly colorless, are larger than red blood cells, contain no hemoglobin, and have a nucleus.

The chief function of the white blood cells is the defense of the body against invaders such as viruses, bacteria, and other foreign particles. Unlike red blood cells, white blood cells are not confined within the blood vessels but can migrate out into the interstitial fluid. They appear spherical in the bloodstream, but in the tissues they become flattened and amoeba-like. Like amoebas, they move by means of pseudopodia, and many are phagocytic (Figure 36–3). As we shall see in Chapter 39, certain types of white blood cells play key roles in the immune response.

36–2 In vertebrates, oxygen is transported in red blood cells, shown here in a scanning electron micrograph. Red blood cells are about 7 or 8 micrometers in diameter, significantly larger than some of the smallest capillaries, which are only 5 micrometers in diameter. The passage of the red blood cells through the capillaries is made possible by their "donut without a hole" shape, which enables them not only to twist and turn but also to fold. This triumph of biological form also provides a large surface area through which oxygen can diffuse to and from the hemoglobin molecules contained within the cell.

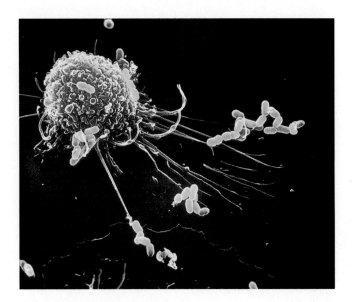

36-3 *A scanning electron micrograph of a human white blood cell entrapping bacterial cells. This type of cell defends the body against pathogens and other harmful particles by extending pseudopodia to the foreign objects, which are then engulfed within phagocytic vacuoles and destroyed with the help of enzymes from the cell's lysosomes (page 118).*

White blood cells are often destroyed in the course of fighting infection. Pus is composed largely of these dead cells. New white blood cells to take the place of those that are destroyed are formed constantly in the spleen, in bone marrow, and in certain other tissues.

Platelets

Platelets, so called because they look like little plates, are colorless, oval or irregularly shaped disks smaller than red blood cells (about 3 micrometers in diameter). Platelets are membrane-bound cytoplasmic fragments of unusually large cells, megakaryocytes, found in the bone marrow. They are, in effect, little bags of chemicals that play an essential role in initiating the clotting of blood and in plugging breaks in the blood vessels.

Blood Clotting

The clotting of blood is a complex phenomenon, requiring platelets and at least 15 factors normally present in the bloodstream or on cell membranes. The sequence of events begins when plasma encounters a rough surface or a protein molecule known as tissue factor. Tissue factor, which is thought to have many regulatory roles throughout the body, is found on the outer surface of many different cell types—but not on the cells of the inner lining of the blood vessels. When tissue factor reacts with a specific circulating plasma protein—as can occur when a blood vessel is broken—a cascade of chemical reactions is initiated. In this cascade, the product of each step of the reaction series acts as a catalyst for the next step, and the molecules involved are, like enzymes, used over and over. The result is that at each step in the series the number of molecules is amplified. Ultimately, a molecule called thromboplastin is activated. Thromboplastin acts to convert prothrombin, a plasma protein produced in the liver, to its active form, the enzyme thrombin:

$$\text{Prothrombin} \xrightarrow{\text{Thromboplastin}} \text{Thrombin}$$

Thrombin, in turn, converts fibrinogen, a soluble plasma protein, to fibrin:

$$\text{Fibrinogen} \xrightarrow{\text{Thrombin}} \text{Fibrin}$$

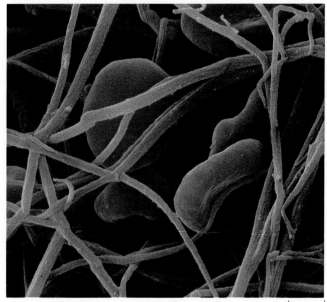

36-4 *An early stage in the formation of a blood clot. Note the fibers, composed of a protein known as fibrin, in which the red blood cells become enmeshed. Ultimately, the completed clot forms an impenetrable barrier, preventing both the loss of vital fluids and the entry of infectious organisms.*

The fibrin molecules clump together, forming an insoluble network that enmeshes red blood cells and platelets to form a clot (Figure 36-4). The clot contracts, pulling together the edges of the wound. In a typical clotting reaction, some 2 million thrombin molecules are produced from prothrombin, leading to the production of about 160 billion molecules of fibrin.

Clotting reactions also take place when blood is removed from the body and placed in a test tube, apparently triggered in some way by the contact with a foreign surface. The clear fluid that remains after the clot contracts is known as the blood serum. In a test tube, the clotting reactions always proceed inexorably to their conclusion unless anticoagulant chemicals are added. In the body, however, the reactions are precisely regulated by mechanisms that still remain mysterious. Were they not so regulated, we could, according to some calculations, clot from head to toe in 90 seconds after initiation of the reaction cascade.

Hemophilia is a group of genetically determined diseases that affect clotting (see page 392). In the most common type of hemophilia, one of the factors involved in the chain of reactions required to activate prothrombin, Factor VIII, occurs in a defective form. As a result, the blood does not clot easily. Hemophiliacs lacking Factor VIII are treated with Factor VIII extracted from normal blood. Such treatment, although lifesaving, also carries the risk of hepatitis, AIDS, and other blood-borne infectious diseases. Work is now under way, using recombinant DNA techniques, to develop a genetically engineered Factor VIII that will carry no risk of infection.

THE CARDIOVASCULAR SYSTEM

As we saw in Section 4, the systems through which blood is transported in animals vary in their structure and complexity. In the earthworm, for example, blood circulates in a closed system of continuous vessels; it is propelled by the contractions of "hearts" that are little more than expanded, muscular portions of particular blood vessels (see Figure 26-14, page 556). Mollusks, by contrast, have a much more complex heart, consisting of several chambers, but most of them have an open circulatory system: blood is pumped through vessels that open into spaces within the tissues, from which it returns to the gills and then to the heart.

CHAPTER 36 Energy and Metabolism III: Circulation

The vertebrate heart, like that of the mollusks, is a muscular organ consisting of several chambers, but the blood vessels form a closed system, similar in principle to that of the earthworm, but considerably more elaborate. The heart and vessels together are known as the **cardiovascular system** (from *cardio*, meaning "heart," and *vascular*, meaning "vessel").

THE BLOOD VESSELS

In the cardiovascular system, the heart pumps blood into the large arteries, from which it travels to branching, smaller arteries, then to the smallest arteries (the **arterioles**), and then into networks of very small vessels, the **capillaries**. From the capillaries, the blood passes into small veins, the **venules**, then into larger veins, and through them, back to the heart.

In humans, the diameter of the opening of the largest artery, the **aorta**, is about 2.5 centimeters, that of the smallest capillary only 5 micrometers, and that of the largest vein, the **vena cava**, about 3 centimeters. Arteries, veins, and capillaries differ not only in their size but also in the structure of their walls (Figure 36–5). Of the three types of vessels, the arteries have the thickest, strongest walls, made up of three layers. The inner layer, or **endothelium** (a type of epithelial tissue), forms the lining of the vessels; the middle layer contains smooth muscle and elastic tissues; the outer layer, also elastic, is made of collagen and other supporting tissues. Because of their elasticity, the arteries stretch when blood is pumped into them and then recoil slowly. The pulsation felt when the fingertips are placed over an artery close to the body surface—as in the wrist—represents the alternating expansion and recoil of an elastic arterial wall. The walls of the veins, like those of the arteries, are three-layered, but they are thinner, less elastic, and more pliable (Figure 36–6). An empty vein collapses, whereas an empty artery remains open.

The Capillaries and Diffusion

The heart, arteries, and veins are, in essence, the means for getting the blood to and from the capillaries, where the actual function of the circulatory system is carried out. The walls of the capillaries consist of only one layer of cells (Figure 36–7), the endothelium, and the lumen (passageway) of the smallest capillaries is, as we have noted, 5 micrometers in diameter, just wide enough for twisted and folded red blood cells to move in single file (see Figure 36–1). The total length of the capillaries in a human adult is more than 80,000 kilometers (50,000 miles).

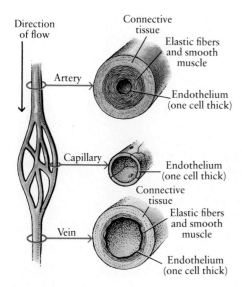

36–5 Structure of blood vessels. Arteries have thick, tough, elastic walls that can withstand the high pressure of the blood as it leaves the heart. Capillaries have walls only one cell thick. Exchange of gases, nutrients, and wastes between the blood and the cells of the body takes place through these thin capillary walls. Veins usually have larger lumens (passageways) and always have thin, readily extensible walls that minimize resistance to the flow of blood on its return to the heart.

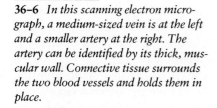

36–6 In this scanning electron micrograph, a medium-sized vein is at the left and a smaller artery at the right. The artery can be identified by its thick, muscular wall. Connective tissue surrounds the two blood vessels and holds them in place.

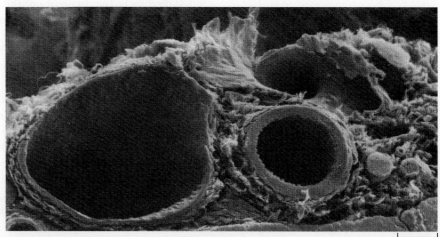

50 μm

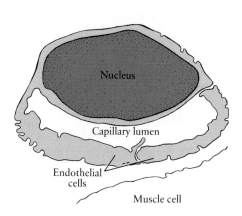
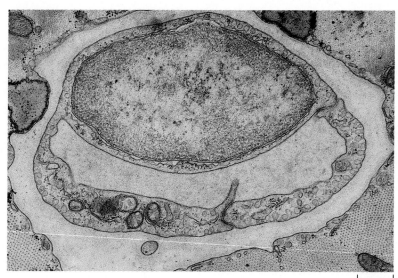

36-7 *Electron micrograph of a capillary from cardiac muscle. Portions of two endothelial cells can be seen, fitting together to form the capillary lumen. The size of the tiny gaps between the endothelial cells can be varied, and it is through these gaps that white blood cells migrate out into the tissues. The space between the capillary and the surrounding muscle contains interstitial fluid. If you look closely, you can see numerous pinocytic vesicles in the cytoplasm of the endothelial cells.*

Because of both of these facts, blood moves slowly through the capillary system. As it moves, gases (oxygen and carbon dioxide), hormones, and other materials are exchanged with the surrounding tissues by diffusion through junctions between the endothelial cells and through the cytoplasm of these cells. Organic molecules, such as glucose, are probably moved by transport systems of the endothelial cells. Pinocytic vesicles (page 138) have been observed in the endothelial cells; they presumably serve to ferry dissolved material into or out of the capillaries.

No cell in the human body is farther than 130 micrometers—a distance short enough for rapid diffusion—from a capillary. Even the cells in the walls of the large veins and arteries depend on capillaries for their blood supply, as does the heart itself.

THE HEART

Evolution of the Heart

In the course of vertebrate evolution, the heart has undergone some structural adaptations, as shown in Figure 36-8. Fish have a single heart divided into an **atrium,** the receiving area for the blood, and a **ventricle,** the pumping area from which the blood is expelled into the vessels. The ventricle of the fish heart pumps blood directly to the capillaries of the gills, where it picks up oxygen and releases carbon dioxide. From the gills, oxygenated blood is carried to the tissues. By this time, however, most of the propulsive force of the heartbeat has been dissipated by the resistance of the capillaries in the gills, so that the blood flow through the rest of the tissues (the systemic circulation) is relatively sluggish.

In amphibians, there are two atria; one receives oxygenated blood from the lungs, and the other receives deoxygenated blood from the systemic circulation. Both atria empty into a single ventricle. Although the ventricle is not divided, its internal structure ensures that the two kinds of blood remain relatively unmixed. The oxygenated blood is pumped into the systemic circulation under high pressure at the same time that the deoxygenated blood is pumped through the lungs. Branches of the vessels leading through the lungs also carry some of the deoxygenated blood to the moist skin, a major site of gas exchange in amphibians. Blood oxygenated in the skin enters the systemic circulation before returning to the heart.

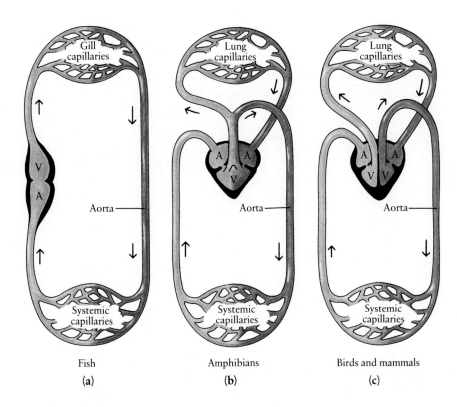

36-8 *Vertebrate circulatory systems. Oxygen-rich blood is shown as red, and oxygen-poor blood as blue.* (a) *In the fish, the heart has only one atrium (A) and one ventricle (V). Blood oxygenated in the gill capillaries goes straight to the systemic capillaries without first returning to the heart.* (b) *In amphibians, the single primitive atrium has been divided into two separate chambers. Oxygen-rich blood from the lungs enters one atrium, and oxygen-poor blood from the tissues enters the other. Little mixing of the blood occurs in the ventricle, despite its lack of a structural division. From the ventricle, oxygen-rich blood is pumped to the body tissues at the same time that oxygen-poor blood is pumped to the lungs. Some of the oxygen-poor blood is diverted from the lungs to the skin, a major respiratory organ in amphibians.* (c) *In birds and mammals, both the atrium and the ventricle are divided into two separate chambers, so that there are, in effect, two hearts—one for pumping oxygen-poor blood through the lungs and one for pumping oxygen-rich blood through the body tissues.*

In birds and mammals, the heart is separated longitudinally into two functionally distinct organs, the right heart and the left heart, each with an atrium and a ventricle. The right heart receives blood from the tissues and pumps it into the lungs, where it becomes oxygenated. From the lungs, the oxygenated blood returns to the left heart, from which it is pumped at high pressure into the body tissues. This efficient, high-pressure circulatory system, with its full separation of oxygenated and deoxygenated blood, makes possible the high metabolic rate of both birds and mammals, with their constant body temperature and their generally high level of activity.

The Human Heart

Figure 36-9 shows a diagram of the human heart. Its walls are made up predominantly of a specialized type of muscle—cardiac muscle. Blood returning from the body tissues enters the right atrium through two large veins, the superior and inferior venae cavae. Blood returning from the lungs enters the left atrium through the pulmonary veins. The atria, which are thin-walled compared to the ventricles, expand as they receive the blood. Both atria then contract simultaneously, assisting the flow of blood through open valves into the ventricles. Then the ventricles contract simultaneously; the valves between the atria and ventricles are closed by the pressure of the blood in the ventricles. The right ventricle propels deoxygenated blood into the lungs through the pulmonary arteries; the left ventricle propels oxygenated blood into the aorta, from which it travels to the other body tissues. Valves between the ventricles and the pulmonary artery and the aorta close after the ventricles contract, thus preventing backflow of blood. In a healthy adult at rest, this rhythmic process takes place about 70 times a minute; under strenuous exercise, the rate more than doubles.

If you listen to a heartbeat, you hear "lubb-dup, lubb-dup." The deeper, first sound ("lubb") is the closing of the valves between the atria and the ventricles; the second sound ("dup") is the closing of the valves leading from the ventricles to the

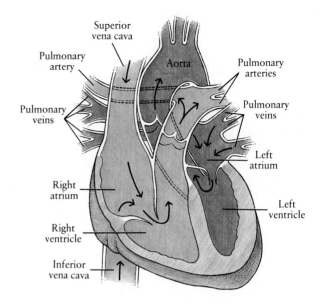

36–9 *The human heart. Blood returning from the systemic circulation through the superior and inferior venae cavae enters the right atrium and passes to the right ventricle, which propels it through the pulmonary arteries to the lungs, where it is oxygenated. Blood from the lungs enters the left atrium through the pulmonary veins, passes to the left ventricle, and then is pumped through the aorta to the body tissues.*

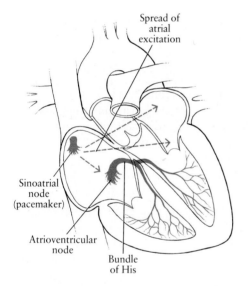

36–10 *The beat of the mammalian heart is controlled by a region of specialized muscle tissue in the right atrium, the sinoatrial node, which functions as the heart's pacemaker. Some of the nerves regulating the heart have their endings in this region. Excitation spreads from the pacemaker through the atrial muscle cells, causing both atria to contract almost simultaneously. When the wave of excitation reaches the atrioventricular node, its conducting fibers pass the stimulation to the bundle of His, which triggers almost simultaneous contraction of the ventricles. Because the fibers of the atrioventricular node conduct relatively slowly, the ventricles do not contract until after the atrial beat has been completed.*

arteries. If any one of the four valves is damaged, as from rheumatic fever, blood may leak back through the valve, producing the noise characterized as a "heart murmur" (a "ph-f-f-t" sound).

The total volume of blood pumped by the heart per minute is called the **cardiac output**. It is defined as:

$$\text{Cardiac output (liters per minute)} = \text{Heart rate (beats per minute)} \times \text{Stroke volume (liters per beat)}$$

Thus, if the heart beats 72 times per minute and ejects 0.07 liter of blood into the aorta with each beat, then

Cardiac output = 72 beats per minute × 0.07 liter of blood per beat
Cardiac output = 5 liters per minute

In addition to its function as a pump, the heart is also a hormone-secreting organ. In humans and other mammals, many of the cardiac muscle cells of the atria synthesize and release a peptide known as atrial natriuretic ("salt-excretion") factor or, more simply, as cardiac peptide. Receptors for cardiac peptide have been identified in the kidneys, blood vessels, adrenal glands, and brain. Although the full range of functions of this hormone—first sequenced in 1983—remain unknown, it appears to play a major role in the regulation of blood volume and blood pressure.

Regulation of the Heartbeat

Most muscle contracts only when stimulated by a motor nerve, but the stimulation of cardiac muscle cells originates in the muscle itself. A vertebrate heart will continue to beat even after it is removed from the body if it is kept in an oxygenated nutrient solution. In vertebrate embryos, the heart begins to beat very early in development, before the appearance of any nerve supply. In fact, isolated embryonic heart cells in a test tube will beat.

The contraction of cardiac muscle is initiated by a special area of the heart, the **sinoatrial node,** which is located in the right atrium (Figure 36–10). This region of tissue functions as the pacemaker. It is composed of specialized cardiac muscle cells that can spontaneously initiate their own impulse and contract. From the pacemaker the impulse spreads throughout the right and left atria. As it passes along the surface of the individual cardiac muscle cells, it activates their contractile machinery, and they contract. The impulse travels very quickly through gap junctions (page 142) that connect the cytoplasm of adjacent cardiac muscle cells.

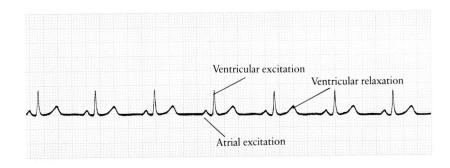

36-11 *An electrocardiogram showing seven normal heartbeats. Each beat is denoted by a series of waves that record the electrical activity of the heart during contraction. Analysis of the rate and character of these waves can reveal abnormalities in heart function. The first small hump represents the current generated by the passage of the impulse through the atria, the spike by its passage through the ventricles, and the final hump by the return of the ventricles to their resting state. Atrial relaxation occurs during ventricular excitation; therefore the record of its electrical activity is obscured.*

As a consequence, many cells in both atria are activated almost simultaneously.

About 100 milliseconds after the pacemaker fires, impulses traveling through both special conducting fibers and the atrial muscle itself stimulate a second area of nodal tissue, the **atrioventricular node**. From the atrioventricular node, impulses are carried by special muscle fibers, the **bundle of His** (named after its discoverer), to the walls of the right and left ventricles, which then contract almost simultaneously. The bundle of His is the only electrical bridge between the atria and the ventricles. Although its fibers conduct impulses very rapidly, the atrioventricular node consists of slow-conducting fibers. As a consequence, a delay is imposed between the atrial and ventricular contractions, ensuring that the atrial beat is completed before the beat of the ventricles begins.

When the impulses from the conducting system travel across the heart, electric current generated on the heart's surface is transmitted to the body fluids, and from there some of it reaches the body surface. Appropriately placed electrodes on the surface connected to a recording instrument can measure this current. The output, an electrocardiogram (Figure 36–11), is important in assessing the heart's capability to initiate and transmit the impulses.

Although the autonomic nervous system does not initiate the vertebrate heartbeat, it does modify its rate (Figure 36–12). Fibers from the parasympathetic division travel in the vagus nerve (a large nerve that runs through the neck) to the pacemaker. Parasympathetic stimulation has a slowing effect on the pacemaker and thus decreases the rate of heartbeat. Sympathetic nerves stimulate the pacemaker, increasing the rate of heartbeat. Adrenaline from the adrenal medulla (the interior portion of the adrenal gland) affects the heart in the same way that the sympathetic nerves do.

36-12 *Autonomic regulation of the rate of heartbeat. Sympathetic fibers stimulate the sinoatrial node, whereas parasympathetic fibers, which are contained in the vagus nerve, inhibit it.*

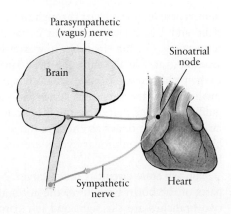
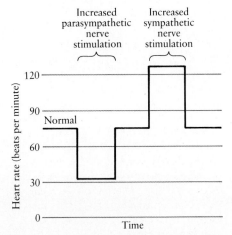

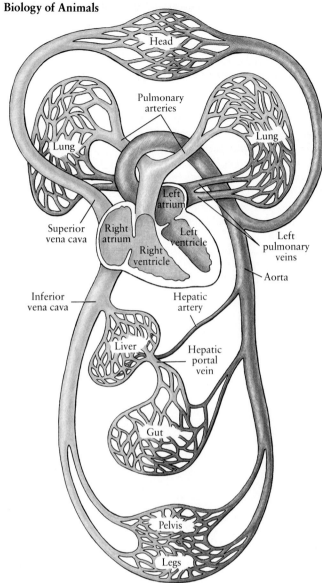

36–13 *Some of the principal circuits in the human cardiovascular system. Oxygenated blood is shown as red, and deoxygenated blood as blue. The portions of the lungs in which gas exchange occurs are served by the pulmonary circulation. Every other tissue of the body is served by the systemic circulation. Blood traveling through the capillaries supplies oxygen and nutrients to every cell of these tissues and carries off carbon dioxide and other wastes. At the venous ends of the capillary beds, blood passes into the venules (the smallest veins), then into larger veins, and finally back to the heart through either the superior (anterior) or the inferior (posterior) vena cava.*

THE VASCULAR CIRCUITRY

As we have seen, there are two principal circuits in the cardiovascular system of an air-breathing vertebrate: the pulmonary circuit and the systemic circuit. Figure 36–13 shows, in more detail, these circuits as they occur in a representative mammal, *Homo sapiens*. In the pulmonary circuit, deoxygenated blood leaves the right ventricle of the heart through the pulmonary artery. This artery divides into right and left branches, which carry the blood to the right and left lungs, respectively. Within the lungs, the arteries divide into smaller arteries, then into arterioles, and finally into the even smaller alveolar capillaries through which oxygen and carbon dioxide are exchanged. Blood flows from the capillaries into small venules and then into larger and larger veins, finally draining into the four pulmonary veins that carry the blood, now oxygenated, to the left atrium of the heart. The pulmonary arteries are the only arteries that carry fully deoxygenated blood, and the pulmonary veins are the only veins that carry fully oxygenated blood.

The systemic circuit is much larger. Many major arteries, supplying different parts of the body, branch off the aorta after it leaves the left ventricle. The first two branches are the right and left coronary arteries, which bring oxygenated blood to the heart muscle itself. Another major subdivision of the systemic circulation supplies the brain. If the circulation of freshly oxygenated blood to the brain is cut off for even five seconds, unconsciousness results; after four to six minutes, brain cells are damaged irreversibly.

Among the features of the systemic circulation are several **portal systems,** in which blood flows through two distinct capillary beds, connected by either veins or arteries, before entering the veins that return it to the heart. For example, in the hepatic portal system, venous blood collected from the capillaries of the digestive tract is shunted via the hepatic portal vein through the liver. There it goes through a second capillary network before it is emptied into the inferior vena cava. In this way, the products of digestion can be directly processed by the liver. The liver also receives freshly oxygenated blood directly from a major artery, the hepatic artery. As we shall see in subsequent chapters, other portal systems play important roles in the chemical processing of blood in the kidneys and in the functions of the pituitary gland.

BLOOD PRESSURE

The contractions of the ventricles of the heart propel the blood into the arteries with considerable force. Blood pressure is a measure of the force per unit area with which blood pushes against the walls of the blood vessels. It is conventionally described in terms of how high it can push a column of mercury. For medical purposes, it is usually measured at the artery of the upper arm. Normal blood pressure in a young adult is generally about 120 millimeters of mercury (120 mm Hg) when the ventricles are contracting (the systolic blood pressure) and 80 mm Hg when the ventricles relax (diastolic pressure); this is stated as a blood pressure of 120/80. The pressure is generated by the pumping action of the heart and changes with the rate at which it contracts. The strength of these contractions, the elasticity of the arterial walls, and the rate at which blood flows from the arteries also play important roles in determining the blood pressure.

The rate of blood flow is directly proportional to blood pressure; the greater the pressure, the greater the rate of flow. The regulation of blood flow depends on a very simple physical principle: Fluid flow through a tube is proportional to the fourth power of the radius of the tube (r^4). As you may recall, the radius of a tube is one-half of its diameter. The diameter of the arterioles, which directly supply the capillaries, can be altered by rings of smooth muscle in the vessel walls. As the smooth muscle contracts, the opening of the arteriole gets smaller (vasoconstriction), and blood flow through the arteriole (and the capillary bed it feeds) decreases (Figure 36–14). Conversely, when the smooth muscle relaxes, the arteriole opens wider (vasodilation), and blood flow into the capillaries increases. These smooth muscles are influenced by autonomic nerves (chiefly sympathetic nerves), the hormones adrenaline and noradrenaline (norepinephrine), cardiac peptide (page 756), and the levels of other chemicals that are produced locally in the tissues themselves.

Constriction and dilation of the arterioles in different parts of the body regulate the blood flow—and thus the supply of oxygen and nutrients—according to the varying requirements of the animal. For example, blood flow through skeletal muscle increases during exercise, flow to the stomach and intestines increases during digestion, and flow through the skin increases at high temperatures and decreases at low temperatures. Of particular importance is a constant flow of blood to the brain. One of the mechanisms that prevents serious damage to human brain cells as a result of an inadequate blood supply is fainting, which causes the person to fall; as a consequence, the force of gravity does not have to be overcome for blood to flow to the brain. This response is often thwarted by well-meaning bystanders anxious to get the affected individual "back on his feet."

36–14 *Diagram of a capillary bed. The muscular wall of the arteriole controls the blood flow through the capillaries. The arteriolar muscles, which are innervated by the sympathetic division of the autonomic nervous system, make it possible to regulate the blood supply to different regions of the body depending on the physiological state of the organism at different times.*

Diseases of the Heart and Blood Vessels

Cardiovascular diseases cause almost as many deaths in the United States as accidents and all other diseases combined. According to recent estimates, about 65 million people in this country (more than 25 percent of the total population) have some form of cardiovascular disease.

About 55 percent of the cardiovascular deaths are caused by heart attacks. A heart attack is the result of an insufficient supply of blood (ischemia) to an area of heart muscle; with their oxygen supply cut off, the cardiac muscle cells may die. A heart attack can be caused by a blood clot—a thrombus—that forms in the blood vessels of the heart itself or by a clot that forms elsewhere in the body and travels to the heart and lodges in a vessel there. (A wandering clot such as this is known as an embolus.) A heart attack can also result from blockage of a blood vessel due to atherosclerosis. Recovery from a heart attack depends on how much of the heart tissue is damaged, where the damage occurs, and whether or not other blood vessels in the heart can enlarge their capacity and form new branches that supply these tissues, which then may recover to some extent. One-fourth of all deaths in the United States are caused by ischemic heart disease.

Angina pectoris is a related condition in which the heart muscle receives an insufficient blood supply (but not so little that muscle dies)—often a result of a narrowing of the vessels. Its symptoms, like those of a heart attack, are pain in the center of the chest and, often, in the left arm and shoulder.

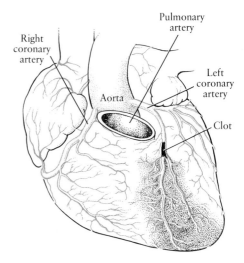

A heart attack. When a clot forms in a blood vessel, the cells in the area supplied by the vessel are deprived of oxygen and may die. The severity of the heart attack depends, in part, on the extent of damage to the heart muscle.

A stroke is caused by interference with the blood supply to the brain. This may be the result of a thrombus, an embolus, or the bursting of a blood vessel in the brain. Its effects depend on the extent of the damage and where it occurs in the brain.

(In fact, holding a fainting person upright can lead to severe shock and even death.) Psychological events also influence vasoconstriction and dilation and thus the distribution of blood flow. Familiar examples are blushing, turning pale with fear, angina pectoris (pain in the chest and left arm) precipitated by emotion, and erection of the penis or clitoris as a result of erotic stimulation.

As blood flows through the vascular circuitry, its pressure gradually drops as a consequence of both the damping caused by the recoil of the elastic arterial walls (see page 753) and the resistance of the arterioles and capillaries. Thus, blood pressure is not the same in the various parts of the cardiovascular system (Figure 36–15). As you would expect, it is highest in the aorta and other large systemic arteries, much lower in the veins, and lowest in the right atrium.

Atherosclerosis contributes to both heart attacks and strokes. In this disease, the linings of the arteries thicken due to the accumulation of abnormal smooth muscle cells, and their inner surfaces become roughened by deposits of cholesterol, fibrin, and cellular debris, as shown in the photos on page 71. Such arteries, becoming inelastic, no longer expand and contract, and blood moves with increasing difficulty through the narrowed vessels. Thrombi and emboli thus form more easily and are more likely to block the vessel.

The causes are not clear-cut. Atherosclerosis is associated with high blood pressure and with high levels of cholesterol in the blood. As we saw on page 71, a critical factor appears to be the relative proportions of the two principal forms in which cholesterol is carried in the bloodstream: high-density lipoproteins (HDLs), which appear to protect against atherosclerosis, and low-density lipoproteins (LDLs), which contribute to the disease process. Cigarette smoking, a lack of physical activity, a high dietary intake of cholesterol and saturated fats, and, in some cases, hereditary factors, are associated with increased levels of LDLs and decreased levels of HDLs.

Atherosclerosis is much less common in premenopausal women than it is among men of the same age, suggesting that female hormones may protect against the disease. Paradoxically, however, men with elevated levels of female hormones appear to be at an increased risk for atherosclerosis. Whatever the underlying cause may be, the difference in susceptibility to atherosclerosis is a major reason that, in the United States, women now live, on the average, almost 10 years longer than men.

In North America, the mortality rate from ischemic heart disease reached a peak in the mid-1960s and then, quite unexpectedly, began a decline that is still continuing; the total decrease in the mortality rate has, astonishingly, exceeded 30 percent. The increase, when it occurred, was attributed to increased cigarette smoking, increased stress, and an increasingly sedentary existence. The decreases, however, which are occurring at all age levels and in both sexes, do not correlate well with any known environmental factor. Perhaps if we could find out what it is that we are doing right, we could do it even better.

Hypertension—chronically increased arterial blood pressure—affects about 54 million people in the United States. It places additional strain on the arterial walls and increases the chances of emboli. Despite lack of knowledge about the causes, hypertension can be treated by drugs that act upon the autonomic nervous system to produce arteriolar dilation, and by measures that decrease blood volume, reducing the blood pressure to safer levels. In the United States, hypertension is about twice as common among blacks as among whites. Related to this disturbing statistic is the fact that the incidence of deaths among black men and women from cardiovascular disease is even higher than in the white population.

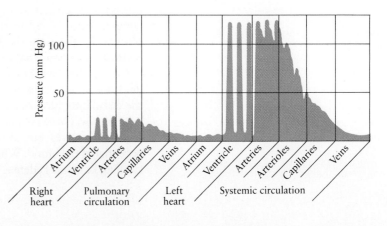

36-15 *In mammals, blood goes from the right heart to the lungs, from the lungs to the left heart, and, from the left heart, it enters the systemic circulation, moving from arteries to arterioles to capillaries to venules to veins. Blood pressure varies in the different areas of the cardiovascular system. The fluctuations in blood pressure produced by three heartbeats are shown in each section of the diagram. Note the fall in pressure as blood traverses the arterioles of the systemic circulation.*

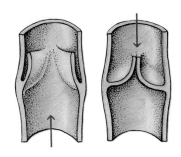

36-16 *Valves in the veins open to permit movement of blood toward the heart but close to prevent backflow.*

The veins, with their thin walls and relatively large diameters, offer little resistance to flow, making possible movement of the blood back to the heart despite its low pressure. Valves in the veins prevent backflow (Figure 36-16). The return of blood to the heart is enhanced by the contractions of skeletal muscles. For example, when you walk, your leg muscles bulge and squeeze the veins that lie between the contracting muscles, thus raising the pressure within the veins and increasing the flow. (If you have to stand still for long periods of time, try contracting your leg muscles periodically to move blood back toward the heart and prevent blood pooling.) Also, as the thoracic cavity expands on inspiration, the elastic walls of the veins in the chest dilate, venous pressure in the region of the heart decreases, and return of blood to the heart increases.

Cardiovascular Regulating Center

Activity of the nerves controlling the smooth muscle of the blood vessels is coordinated with the activity of nerves regulating heart rate and strength of heartbeat by the **cardiovascular regulating center.** This center is located in the medulla, a small part of the brain continuous with the spinal cord. It controls the sympathetic and parasympathetic nerves to the heart (see Figure 36-12) as well as the nerves to the smooth muscle in the arterioles. Thus, if there is a significant increase in blood flow due to dilation of blood vessels, the heart is simultaneously stimulated to beat faster, thus developing greater pressure to support the greater flow.

The cardiovascular regulating center integrates the reflexes that control blood pressure. It receives information about existing blood pressure from specialized stretch receptors in the carotid arteries (see Figure 35-18, page 747), the venae cavae, the aorta, and the heart. The effector organs of the reflex are, as we have indicated, the heart and blood vessels.

The blood-pressure reflex is an example of negative-feedback control. When pressure falls, the activity of the heart is increased and the blood vessels are constricted, raising the pressure again. Conversely, heart activity is decreased and the blood vessels dilated in response to high pressure.

THE LYMPHATIC SYSTEM

Earlier in this chapter, we discussed the role of plasma proteins in maintaining the high osmotic potential of the blood in relation to the interstitial fluid and thus in preventing excessive loss of fluid from the blood. However, as you may recall from Chapter 6, the osmotic potential of solutions separated by a selectively permeable membrane is only one of several factors influencing the movement of water across the membrane; another important factor is pressure. At the arteriolar end of the capillaries, the pressure of the circulating blood—the hydrostatic pressure—is greater than the osmotic potential, forcing some of the watery component of blood to leave the capillaries along with nutrient molecules and oxygen. By the time the blood reaches the venous end of the capillaries, the hydrostatic pressure has dropped, due to resistance in the capillary bed, to a point where it is less than the osmotic potential of the blood. Most—but not all—of the fluids then move back into the capillaries by osmosis. Fluids may also escape from the circulating blood into the tissues when the endothelium is damaged, as, for example, by a blow; damage to the endothelium allows the release not only of the fluids themselves but also of the plasma proteins that maintain the osmotic potential of the blood. When excessive fluid is lost, it tends to pool, producing the swelling known as edema.

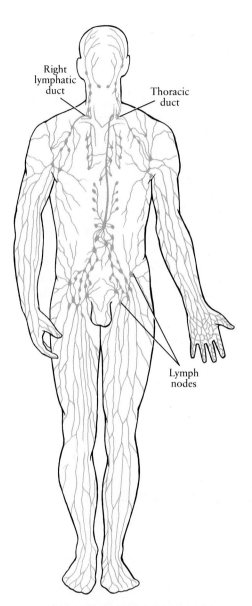

36-17 *The human lymphatic system consists of a network of lymph vessels and lymph nodes. Lymph reenters the bloodstream through the thoracic duct, which empties into the left subclavian vein, and through the right lymphatic duct, which empties into the right subclavian vein. These two veins empty into the superior vena cava.*

In higher vertebrates, the fluids lost from the blood to the tissues are collected by the **lymphatic system** (Figure 36-17), which routes them back to the bloodstream. The lymphatic system is like the venous system in that it consists of an interconnecting network of progressively larger vessels. The larger vessels are, in fact, similar to veins in their structure, and the small vessels are much like the capillaries through which the blood is carried. An important difference, however, is that the lymph capillaries begin blindly in the tissues, rather than forming part of a continuous circuit. Interstitial fluid seeps into the lymph capillaries, from which it travels to large ducts that empty into the two subclavian veins; these veins, located below the two clavicles (collar bones), empty into the superior vena cava. The fluid carried in the lymphatic system is known as **lymph**. As you may recall, lymph is also the medium in which fats absorbed from the digestive tract are transported to the bloodstream (see Figure 34-9, page 721).

Some nonmammalian vertebrates have lymph "hearts," which help to move the fluid. In mammals, lymph is moved by contractions of the body muscles, with valves preventing backflow, as in the venous system. Also, recent studies have shown that lymph vessels contract rhythmically; these contractions may be the principal factor propelling the lymph.

Lymph nodes, which are masses of spongy tissue, are distributed throughout the lymphatic system. They have two functions: they are the sites of proliferation of lymphocytes, specialized white blood cells that are the effectors of the immune response (to be discussed in Chapter 39), and they remove cellular debris and foreign particles from the lymph before it enters the blood. The removal of chemical wastes, however, requires processing of the blood itself; this function is performed by the kidneys, as we shall see in the next chapter.

SUMMARY

Oxygen, nutrients, and other essential molecules, as well as waste products, are carried in the blood. The blood is composed of plasma, red blood cells (erythrocytes), white blood cells (leukocytes), and platelets. Plasma, the fluid part of the blood, is chiefly water in which nutrients, waste products, ions, antibodies, hormones, enzymes, plasma proteins, and other substances are dissolved or suspended. Red blood cells contain oxygen-bearing hemoglobin, and white blood cells defend the body against foreign invaders. Platelets participate in the clotting of blood, which occurs as the result of a reaction cascade involving at least 15 factors.

In vertebrates the blood is pumped by muscular contractions of the heart into a closed circuit of arteries, arterioles, capillaries, venules, and veins. This network, which includes both pulmonary and systemic circuits, ultimately services every cell in the body. The essential function of the circulatory system is performed by the capillaries, through which substances are exchanged with the interstitial fluid surrounding the individual cells of the body.

Evolutionary changes in the structure of the vertebrate heart can be correlated with changes in metabolic rates and the level of activity of the animals. Fish have a two-chambered heart, whereas amphibians have a three-chambered heart. Birds and mammals have a double circulatory system, made possible by a four-chambered heart that functions as two separate pumping organs. One side pumps deoxygenated blood to the lungs, and the other pumps oxygenated blood to the body tissues under high pressure.

Synchronization of the heartbeat is controlled by the sinoatrial node (the pacemaker), located in the right atrium, and by the atrioventricular node, which delays the stimulation of ventricular contraction until the atrial contraction is completed. The rate of heartbeat is secondarily under neural and hormonal regulation. Parasympathetic stimulation slows the heartbeat; sympathetic stimulation and adrenaline accelerate it.

Blood pressure is a measure of the force per unit area with which blood pushes against the walls of the blood vessels. It is generated by the pumping action of the heart and drops precipitously as the blood traverses the arterioles. The magnitude of the pressure within the arteries is influenced by the rate and strength of the heartbeat, the elasticity of the arterial walls, and the rate at which blood flows from the arteries into the arterioles. The rate of flow is, in turn, influenced by the degree of dilation or constriction of the arterioles. Activity of the nerves regulating the rate and strength of the heartbeat is coordinated with the activity of the nerves controlling the smooth muscle of the arterioles by the cardiovascular regulating center, located in the medulla of the brain.

Because the hydrostatic pressure in the arteries is greater than the osmotic potential of the blood, some fluids are forced out of the capillaries at the arteriolar end of capillary beds. These fluids either reenter the capillaries at the venous end of the beds (where the hydrostatic pressure is much lower) or are returned to the blood by the lymphatic system. The lymph also picks up cellular debris and foreign particles, which are filtered out by the lymph nodes.

QUESTIONS

1. Distinguish between the following: blood/plasma; aorta/vena cava; atrium/ventricle; right heart/left heart; sinoatrial node/atrioventricular node; systolic/diastolic.

2. Blood serum is the portion of the plasma remaining after a clot is formed. Name some of the components of the plasma that would not be present or would be present in a lesser amount in serum.

3. Give two reasons why atherosclerotic arteries are much more susceptible to clot formation and blockage than are normal, healthy arteries.

4. Label the diagram below.

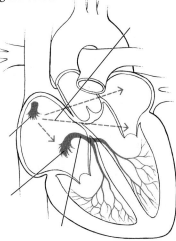

5. What is the advantage of the fibers of the atrioventricular node being slow-conducting?

6. The valves of the heart are not directly controlled by nerves. Yet, in most individuals, they open and shut at precisely the right points in the cardiac cycle for efficient heart operation. How is this precise timing possible? What does determine just when the valves will open and shut?

7. Trace the course of a single red blood cell from the right ventricle to the right atrium in a mammal. Trace the course of an oxygen molecule from the air to its arrival at a metabolizing cell.

8. How does the radius of a blood vessel affect the blood flow through it?

9. Explain the reasons for the changes in blood pressure shown in Figure 36–15.

10. When fair-skinned individuals are very frightened, they turn quite pale. What occurs to cause this change? Why is such an adaptation useful?

11. Individuals, particularly children, suffering from severe protein deprivation often have swollen, bloated bellies. What is the explanation for this phenomenon?

12. When an accident victim suffers blood loss, he or she is transfused with plasma rather than with whole blood. Why is plasma effective in meeting the immediate threat to life?

13. What is probably the immediate cause of death by crucifixion? (If you need a clue, see pages 759–760.)

C H A P T E R 37

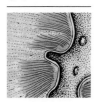

Homeostasis I: Excretion and Water Balance

As we noted in Chapter 33, one of the advantages of multicellularity is the increased capacity for homeostasis—the maintenance of a controlled internal environment in which the cells can live and function. In animals, a great variety of activities contribute to homeostasis. Among the specific examples we have seen in previous chapters are the regulation of blood sugar, the uptake and distribution of oxygen to the cells, and the elimination of carbon dioxide from the body. In this chapter and the two following, we shall examine three particularly noteworthy homeostatic functions: regulation of the chemical composition of body fluids, regulation of temperature, and defense of the body against foreign invaders.

REGULATION OF THE CHEMICAL ENVIRONMENT

Animals are about 70 percent water. About two-thirds of this water is within the cells, and one-third is in the extracellular fluid that surrounds, bathes, and nourishes the cells. Thus the extracellular fluid serves the same purpose for the cells of an animal's body that the Precambrian seas served for the earliest unicellular organisms. As animals became multicellular in the course of evolution, they began to produce their own extracellular fluid, similar in composition to the salty fluid of the sea. As they did so, they also evolved mechanisms for regulating its composition.

Although the blood plasma constitutes only about 7 percent of total body fluids, the regulation of its composition is a key factor in regulation of the chemical environment throughout the vertebrate body. As we saw in the last chapter, blood is the supply line for chemicals taken up by the individual cells, and it carries away the wastes released by these cells. Blood can function as an efficient supply and sanitation medium only because cellular wastes are constantly removed from it. Such removal is quite different in principle from the elimination of feces from the intestinal tract. In the latter case, the bulk of what is eliminated is material, such as cellulose, that was never actually in the body, since it never passed through the epithelium of the digestive tract. Excretion of substances carried within the bloodstream, by contrast, is a very selective process of monitoring, analysis, selection, and rejection.

In many invertebrates and all vertebrates, the composition of the blood—and thus the internal chemical environment—is to a large extent regulated by special excretory organs. These organs include the nephridia of mollusks and annelids (page 547), the Malpighian tubules of insects (page 568), and the kidneys of vertebrates. Although other organs, especially the liver, play important roles in regulating the chemical environment, it is possible to relate major advances in vertebrate evolution—particularly the transition to land—to increasing efficiency of kidney function.

37-1 The kangaroo rat, a common inhabitant of the American desert, may spend its entire life without drinking water, yet the composition of its body fluids remains essentially constant. As we shall see in this chapter, a number of physiological adaptations and mechanisms contribute to this remarkable feat.

$$H_2N-\underset{\underset{COOH}{|}}{\overset{\overset{CH_3}{|}}{C}}-H + 1/2\, O_2 \longrightarrow \underset{\underset{COOH}{|}}{\overset{\overset{CH_3}{|}}{C}}=O + NH_3$$

Alanine → Pyruvic acid

37-2 *The first step in the breakdown of amino acids is deamination—the removal of the amino group. The products of the reaction are ammonia and a carbon skeleton, which can be broken down to yield energy or converted to sugar or fat.*

Substances Regulated by the Kidneys

Regulation of the internal chemical environment of an animal involves solving three different—yet interwoven—problems: (1) excretion of metabolic wastes, (2) regulation of the concentrations of ions and other chemicals, and (3) maintenance of water balance.

The chief metabolic waste products that cells release into the bloodstream are carbon dioxide and nitrogenous compounds, mostly ammonia (NH_3), produced by the breakdown of amino acids (Figure 37–2). As we have seen, carbon dioxide diffuses out of the body across the respiratory surfaces, which may be skin, gills, or lungs. In simple aquatic animals, ammonia also leaves the body by diffusion into the surrounding water. Ammonia, however, is highly toxic, even in low concentrations. In more complex aquatic animals—and in all terrestrial animals—rapid diffusion of ammonia from the cells into an external water supply is not possible; thus it must be converted to nontoxic substances, which can be safely transported within the body to the excretory organs.

All birds, terrestrial reptiles, and insects convert their nitrogenous wastes into crystalline uric acid or uric acid salts, with the result that very little water is required for their excretion. In birds, the uric acid and uric acid salts are mixed with undigested wastes in the cloaca (the common exit chamber for the digestive, urinary, and reproductive tracts), and the combination is dropped as a semisolid paste, familiar to frequenters of public parks and admirers of outdoor statuary. This nitrogen-laden substance forms a rich natural fertilizer; guano, the excreta of seabirds, accumulates in such quantities on the small islands where great numbers of these birds gather that at one time it was harvested commercially.

In mammals, the ammonia resulting from the processing of nitrogenous wastes is quickly converted in the liver to urea (Figure 37–3), which diffuses into the bloodstream. This relatively nontoxic compound is then carried to the kidneys. Unlike uric acid, however, it must be dissolved in water for excretion.

Excretion is highly selective. For instance, although about half of the urea in the blood that enters a normal mammalian kidney is excreted, almost all the amino acids are retained. Glucose is not excreted unless it is present in high concentrations, as in diabetes mellitus. (The presence of glucose in the urine is, in fact, a basis for the diagnosis of this form of diabetes and the evaluation of its treatment.) Thus, although the kidneys have an excretory function, they are more accurately regarded as regulatory organs. Chemical regulation involves not only the retention of nutrient molecules such as glucose and amino acids but also the maintenance of closely controlled concentrations of ions. Such ions as Na^+, K^+, H^+, Mg^{2+}, Ca^{2+}, and HCO_3^- play vital roles in maintenance of protein structure, membrane permeability, and blood pH, in propagation of the nerve impulse, and in the contraction of muscles.

The concentration of a particular substance in the body depends not only on the absolute amount of the substance but also on the amount of water in which it is dissolved. Thus, regulation of the water content of body fluids is an important aspect of regulation of the chemical environment. The problem of water balance is such a universal one, biologically speaking, and is so important to the survival of

37-3 *Urea, the principal form in which nitrogen is excreted in most mammals. It is formed in the liver by the combination of two molecules of ammonia with one of carbon dioxide through a complex series of energy-requiring reactions. What would be the other product of this reaction?*

$$H_2N-\underset{\underset{O}{\|}}{C}-NH_2$$

Urea

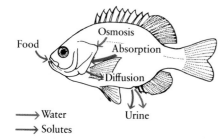

37–4 *Pathways by which water and solutes are gained and lost in freshwater fish. Because the body fluids are hypertonic to the surrounding environment, water enters the body of the fish by osmosis across the gill epithelium. Excess water is removed from the blood by the kidneys and excreted in the urine, which is much more dilute than the body fluids. Although the kidneys reabsorb the bulk of the essential solutes, some are nonetheless lost in the urine, and others leave the body by diffusion across the gills. These solutes are replaced principally by the action of specialized salt-absorbing cells in the gills, and, to a lesser extent, from the diet.*

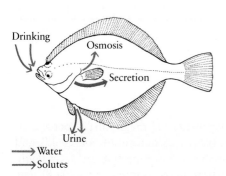

37–5 *Pathways by which water and solutes are gained and lost in bony saltwater fish. Because the body fluids are hypotonic to the surrounding environment, water leaves the body of the fish by osmosis across the gills. Water is also lost in the urine, where it is required to dissolve the urea removed from the blood by the kidneys. The fish maintains its internal fluid levels by drinking sea water, which entails the ingestion of solutes. Excess sodium and chloride ions are removed from the blood and excreted by specialized cells in the gills, while magnesium and sulfate ions are removed by the kidneys and excreted in the urine.*

the organism that we shall consider it in more detail before examining the structure and physiology of the kidney. Although the fundamental problem is always the same, the solution varies widely. A major factor is the availability of water to the organism.

WATER BALANCE

An Evolutionary Perspective

The earliest organisms probably had a salt and mineral composition much like that of the environment in which they lived. The early organisms and their surroundings were probably also isotonic; that is, each had the same total effective concentration of dissolved substances, so water did not tend to move either into or out of these organisms by osmosis. When organisms moved to fresh water (a hypotonic—less concentrated—environment), they had to develop systems for "bailing themselves out," since fresh water tended to move into their bodies; the contractile vacuole of *Paramecium* is an example of such a bailing device (see page 133).

If the earliest vertebrates—the fish—evolved in fresh water, as is generally believed, the first function of the kidneys, phylogenetically speaking, was probably to pump water out and to conserve salt and other desirable solutes, such as glucose. In freshwater fish today, the kidney works in just this way, primarily as a filter and reabsorber of solutes. The urine of these fish is hypotonic—that is, it has a concentration of solutes lower than that of body fluids. Some solutes are, however, inevitably lost both in the urine and by diffusion through the gills. This loss is counteracted by salt-absorbing cells in the gills, which actively transport salt back into the body (Figure 37–4).

When fish moved to the seas, they faced a different problem: the potential loss of water to their environment, principally by osmosis across the respiratory surfaces of the gills. The hagfish, a group of cartilaginous fish, have solved this problem by maintaining body fluids about as salty as the surrounding ocean waters. The body fluids of another group of cartilaginous fish, the sharks, are also isotonic with sea water, but this isotonicity is achieved in a different way. In the course of evolution, sharks developed an unusual tolerance for urea, so instead of constantly excreting it, as do most other fish, they retain a high concentration of it in their blood, which would otherwise be hypotonic to salt water. (This is a striking example of what has been called the "opportunism of evolution.")

The bony fish, which spread to the sea much later than the cartilaginous fish, have body fluids that are hypotonic to the marine environment, with a solute concentration only about one-third that of sea water. Thus they are constantly in danger of losing so much water to their environment that the solutes in their body fluids become so concentrated that the cells die. To compensate for their osmotic water loss, they drink sea water. This restores their water content but leads to a new problem—how to eliminate the excess salt ingested. This problem has been solved by the evolution of special gland cells in the gills that excrete excess salt. Magnesium and sulfate ions, also present in large quantities in sea water, are removed from the blood by the kidneys and excreted in the urine. Hence bony marine fish can drink freely from the water surrounding them and still remain hypotonic to it (Figure 37–5).

Sources of Water Gain and Loss in Terrestrial Animals

Since terrestrial animals do not always have ready access to either fresh or salt water, they must regulate water content in other ways, balancing gains and losses.

37-6 *Some marine animals, such as the turtle, have special glands in their heads that can excrete sodium chloride at a concentration about twice that of sea water. Since ancient times, turtle watchers have reported that these great armored reptiles come ashore, with tears in their eyes, to lay their eggs. It is only recently that biologists have learned that this is not caused by an excess of sentiment—as is the case with Lewis Carroll's mock turtle—but is, rather, a useful solution to the problem of excess salt from ingestion of sea water. Marine birds similarly excrete a salty fluid through their nostrils.*

They gain water by drinking fluids, by eating water-containing foods, and as an end product of certain metabolic reactions, such as the oxidative processes that take place in the mitochondria (see Figure 9-19, page 201). When 1 gram of glucose is oxidized, 0.6 gram of water is formed. When 1 gram of protein is oxidized, only about 0.3 gram of water is produced. Oxidation of 1 gram of fat, however, yields 1.1 grams of water because of the high hydrogen content of fat (the extra oxygen comes from the air).

Some animals can derive all their water from food and oxidation of nutrient molecules and therefore do not require fluids. The kangaroo rat (see Figure 37-1), for example, can live its entire existence without drinking water if it eats the right type of food. It is not surprising that it selects a diet of fatty seeds, which yield a large amount of water on oxidation. If it is fed high-protein seeds, such as soybeans—the oxidation of which produces a large amount of nitrogenous waste and a relatively small amount of water—the kangaroo rat will die of dehydration unless some other source of water is available. The kangaroo rat, like many other desert animals, is highly conservative in its water expenditures. It has no sweat glands, and, being nocturnal, it searches for food only when the surrounding air is relatively cool. Its feces have a very low water content, and its urine is highly concentrated (almost four times as concentrated as the most concentrated human urine). Its major water loss is through respiration, and even this loss is reduced by the animal's long nose in which some cooling of the expired air takes place, with condensation of water from it.

On the average, a human takes in about 2,300 milliliters of water a day in food and drink and gains an additional 200 milliliters a day by oxidation of nutrient molecules. Water is lost from the lungs in the form of moist exhaled air, is eliminated in the feces, is lost by evaporation from the skin, and is removed from the blood and excreted as urine. The latter is usually the major route of water loss. In a normal adult, the rate of water excretion in urine averages 1,500 milliliters a day. Although the actual amount of urine produced may vary from 500 to 2,300 milliliters per day, there will be a variation of less than 1 percent in the fluid content of the body. A minimum output of about 500 milliliters of water is necessary for health, since this much water is needed to remove potentially toxic waste products.

Water Compartments

The body has three principal water compartments: (1) the plasma (7 percent of body fluid), (2) the interstitial fluid and the lymph (28 percent of body fluid), and (3) the intracellular fluid, the fluid within the cells (65 percent of body fluid). Water is constantly moving from one compartment to another (Figure 37-8). The rates of exchange between compartments can be measured by administering a traceable substance, such as a harmless dye molecule, that passes readily through cell membranes, and then analyzing the concentration of the tracer material in each compartment.

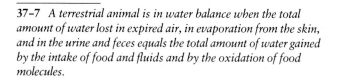

37-7 *A terrestrial animal is in water balance when the total amount of water lost in expired air, in evaporation from the skin, and in the urine and feces equals the total amount of water gained by the intake of food and fluids and by the oxidation of food molecules.*

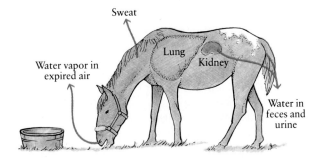

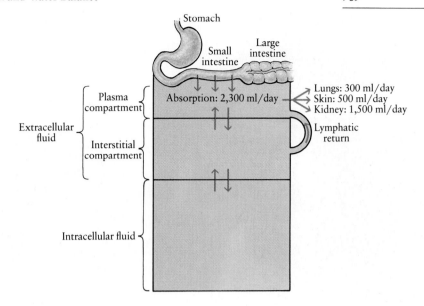

37-8 *The fluid compartments of the human body, showing the main routes of exchange among them. The interstitial fluid forms the environment in which the cells of the body live and multiply. The arrows indicate exchanges between various compartments. A relatively constant plasma volume is of extreme importance for maintenance of stable blood pressure and for normal cardiac function.*

Water absorbed from the digestive tract, the major source of water gain, passes largely into the intestinal capillaries and enters the plasma. This passage is mainly a result of osmosis. Because of the active transport of nutrient molecules and salts into the capillaries from the intestinal tract, the plasma becomes hypertonic in relation to the intestinal contents, and so water tends to follow the dissolved molecules into the plasma. Conversely, as blood moves through the systemic circulation, hydrostatic pressure forces watery fluid through the capillary walls into the interstitial fluid, as we saw on page 762. However, most of this fluid reenters the plasma osmotically or via the lymph vessels. Water remaining in the interstitial compartment comes in contact with the tissue cells. Since their membranes are permeable to water, water moves freely into the cells.

A number of factors affect the movement of water between compartments. Dehydration (water loss greater than water intake) increases the solute concentration of the extracellular fluid; water therefore moves out of the cells, including the cells of the mucous membranes of the oral cavity, producing the sensation of dryness that we associate with thirst. Physiological malfunctions, such as the retention of salts, which may occur as a consequence of kidney disease, or the loss of plasma proteins as a result of starvation, lead to the accumulation of interstitial fluids (edema).

Human sweat, unlike that of most other mammals, contains salt. The reason for salt excretion by the skin is not known, but it does tend to keep the solute concentration of the extracellular fluid relatively constant as water is lost through the skin. (It has also been suggested that the little crumbs of salt so produced clung to the fur of our primate ancestors and that these delicacies served as rewards for their companions who groomed them.) In cases of profuse sweating, if the water is replaced without the salt, water will move into body cells, diluting their contents. The effects of such dilution are particularly severe on the central nervous system, and water intoxication may produce disoriented behavior, convulsions, coma, and even death before the excess water can be excreted.

THE KIDNEY

In vertebrates, the complex functions involved in regulating the chemical composition of body fluids are performed chiefly by the kidney. The two human kidneys are dark red, bean-shaped organs about 10 centimeters long that lie at the back of the body, behind the stomach and liver.

The functional unit of the kidney is the **nephron** (Figure 37–9). It consists of a

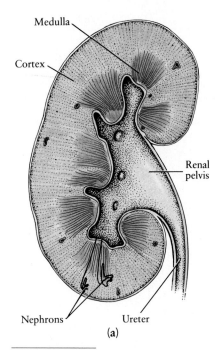

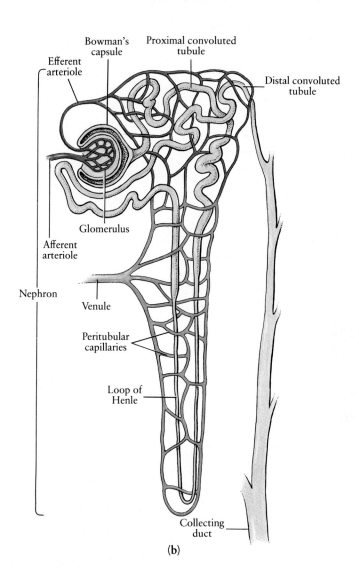

37–9 (a) *In longitudinal section, the human kidney is seen to be made up of two regions. The outer region, the cortex, contains the fluid-filtering mechanisms. The inner region, the medulla, is traversed by long loops of the renal tubules and by the collecting ducts carrying the urine. These ducts merge and empty into the funnel-shaped renal pelvis, which, in turn, empties into the ureter.*

(b) *The nephron is the functional unit of the kidney. Blood enters the nephron through the afferent arteriole leading into the glomerulus. Fluid is forced out by the pressure of the blood through the thin capillary walls of the glomerulus into Bowman's capsule. The capsule connects with the long renal tubule, which has three regions: the proximal convoluted tubule; the loop of Henle, which extends into the medulla; and the distal convoluted tubule. As the fluid travels through the tubule, almost all the water, ions, and other useful substances are reabsorbed into the bloodstream through the peritubular capillaries. Other substances are secreted from the capillaries into the tubules. Waste materials and some water pass along the entire length of the tubule into the collecting duct and are excreted from the body as urine.*

cluster of capillaries known as the **glomerulus** and a long narrow tube, the **renal tubule,** which originates as a bulb called **Bowman's capsule.** The renal tubule is made up of the **proximal** (near) and **distal** (far) **convoluted tubules,** which in humans and other mammals are connected by the **loop of Henle.** The nephron ends as the straight **collecting duct.** Each of the two human kidneys contains about a million nephrons with a total length of some 80 kilometers (50 miles) in an adult.

Urine is formed in the nephrons and passed from the collecting ducts into the renal pelvis, which is, in essence, a funnel. From this funnel, the urine trickles continuously through the ureter to the bladder, which stores the urine until it is passed out of the body through the urethra (Figure 37–10).

Function of the Kidney

Blood enters the kidney through the renal artery, which divides into progressively smaller arteries, leading ultimately to the arterioles, each of which feeds into a glomerulus. Unlike most other capillary beds, a glomerulus lies between two arterioles—the one leading in is the afferent arteriole, and the one leading out is the efferent arteriole. The efferent arteriole then divides again into capillaries, the peritubular ("around-the-tubes") capillaries, which surround the renal tubule and then merge to form a venule that empties into a small vein, leading ultimately to the renal vein. The efferent arteriole is thus a portal vessel (see page 759), providing a direct link between two distinct capillary beds.

Constriction of the afferent and efferent arterioles keeps the blood within the

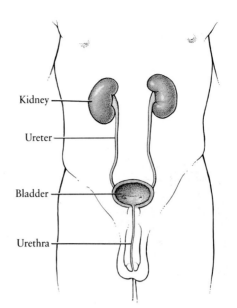

37-10 *Urinary system of the human male. Fluids are processed in the kidneys, which lie at the back of the body, behind the stomach and liver. Waste products and water—the urine—pass along a pair of tubes, the ureters, to the bladder, where they are stored. Urine leaves the body by way of the urethra. In the male mammal, a portion of the urethra also serves as the passageway for semen.*

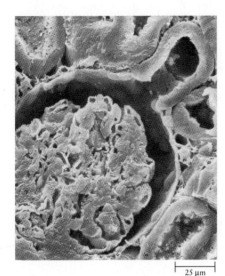

37-11 *The section of kidney tissue shown in this scanning electron micrograph was prepared by the freeze-fracture technique. A glomerulus occupies the center of the micrograph, enclosed at the top and sides by Bowman's capsule. The space within the capsule is continuous with the lumen of the proximal tubule, the beginning of which you can see at the upper right.*

glomerulus at a pressure about twice that in other capillaries. As a consequence, about one-fifth of the blood plasma that enters the kidney is forced through the walls of the glomerular capillaries and the wall of Bowman's capsule into the lumen of the renal tubule (Figure 37-11). This crucial first process in the formation of urine is called **filtration,** and the fluid entering the capsule is the **filtrate.** Except for the absence of large molecules, such as proteins, which cannot cross the capillary wall, the filtrate has the same chemical composition as the plasma.

The filtrate then begins its long passage through the renal tubule, the wall of which is made up of a single layer of epithelial cells specialized for active transport. In a second process in the formation of urine, **secretion,** molecules remaining in the plasma after filtration are selectively removed from the peritubular capillaries and actively secreted into the filtrate. Penicillin, for example, is removed from the circulation in this way.

A third major process in urine formation, **reabsorption,** occurs simultaneously. Most of the water and solutes that initially entered the tubule during filtration are transported back into the peritubular capillaries. For example, glucose, most amino acids, and most vitamins are returned to the bloodstream. Secretion and reabsorption of many solutes take place by active transport; as a consequence, the kidney has a high energy requirement, higher on a per-gram basis than even the heart.

Finally, the remaining fluid—now the urine—leaves the nephron and passes into the renal pelvis. This process is **excretion.**

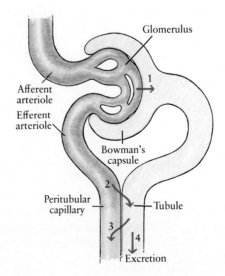

37-12 *The four basic components of renal function: (1) filtration through the glomerular capillaries into Bowman's capsule; (2) secretion from the peritubular capillary into the tubule; (3) reabsorption from the tubule to the peritubular capillary; (4) excretion. The pressure in the glomerular capillaries is regulated by constriction of the afferent and efferent arterioles.*

37-13 *The formation of hypertonic urine in the human nephron. The filtrate entering the proximal convoluted tubule is isotonic with the blood plasma. Although sodium ions are pumped from the tubule here, with chloride ions following passively, the filtrate remains isotonic because water also moves out by osmosis. As the filtrate descends the loop of Henle, it becomes increasingly concentrated as water moves by osmosis into the surrounding zone of high solute concentration. This zone is created by the action of the wall of the ascending branch of the loop of Henle, which pumps out sodium and chloride ions, and by the diffusion of urea out of the lower portion of the collecting duct. Because the wall of the ascending branch of the loop is impermeable to water, the filtrate becomes less and less concentrated as sodium chloride is pumped out. By the time it reaches the distal convoluted tubule, it is hypotonic in relation to the blood plasma, and it remains hypotonic throughout the distal tubule. The filtrate then passes down the collecting duct, once more traversing the zone of high solute concentration.*

From this point onward, the urine concentration depends on antidiuretic hormone (ADH). If ADH is absent, the wall of the collecting duct is not permeable to water, no additional water is removed, and a less concentrated urine is excreted. If ADH is present, the cells of the collecting duct are permeable to water, which moves by osmosis into the surrounding fluid, as shown in the diagram. In this case a concentrated (hypertonic) urine is passed down the duct to the renal pelvis, the ureter, the bladder, and finally out the urethra.

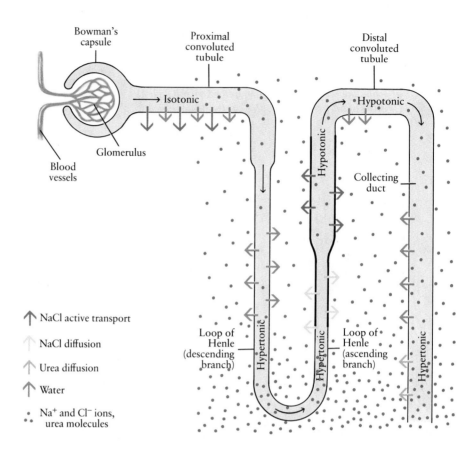

Water Conservation: The Loop of Henle

The control of urinary water loss is a major mechanism by which body water is regulated. Animals with free access to fresh water typically excrete a copious urine that is hypotonic in relation to their blood. The daily urinary output of a frog, for instance, totals 25 percent of its body weight. Most terrestrial animals cannot afford to be so profligate with water, however. In response to evolutionary pressures, birds and mammals have developed the ability to excrete hypertonic urines—that is, urines that are more concentrated than their body fluids. In mammals, this ability is associated with the hairpin-shaped section of the renal tubule known as the loop of Henle. By sampling fluids in and around different portions of the nephron, including different regions of the loop, and then analyzing the ions and molecules in each sample, physiologists have been able to determine how such a deceptively simple-looking structure makes this function possible.

Two structural features are the keys to the formation of a hypertonic urine. The first is the pathway formed by the basic structure of the nephron (Figure 37–13). From Bowman's capsule, the filtrate first enters the proximal convoluted tubule, descends the loop of Henle, ascends it, and then passes through the distal convoluted tubule into the collecting duct. The second feature is the differing permeability to water, salt, and urea of the wall in different parts of the nephron and the localized presence of membrane proteins for the active transport of salt. The wall of the proximal tubule is freely permeable to water and contains proteins that pump sodium ions (Na^+) out of the tubule by active transport, with chloride ions (Cl^-) following passively. The wall of the descending branch of the loop of Henle is also freely permeable to water but is relatively impermeable to salt. In the ascending branch of the loop the situation is quite different. The wall of the ascending branch is impermeable to water but permits the movement of salt: in the lower portion, diffusion of salt can occur, whereas in the upper portion, membrane proteins actively transport sodium and chloride ions out of the tubule. In the distal tubule, the wall is once more permeable to water and relatively permeable to salt. And, except under certain circumstances, so is the wall of the collecting duct. Moreover, the wall of the lower portion of the collecting duct is permeable to urea.

To understand the formation of a hypertonic urine, we must consider the consequences of the permeability of both the collecting duct and the ascending branch of the loop of Henle. First, because the lower portion of the collecting duct is permeable to urea, about half of the urea contained within the fluid passing through the duct diffuses into the surrounding interstitial fluid, from which it is ultimately reabsorbed into the peritubular capillaries. Second, in the upper portion of the ascending branch of the loop of Henle, sodium and chloride ions are actively pumped from the renal tubule into the interstitial fluid, through what is thought to be a cotransport process (see page 137). Water, however, cannot move through the wall of the ascending branch. As a result, sodium and chloride ions—as well as urea molecules—are present in high concentrations in the interstitial fluid bathing both branches of the loop of Henle and the collecting duct. Urea is most concentrated in the lower regions, and salt in the upper regions. With that in mind, let us now follow the filtrate as it moves through the nephron.

The fluid entering the proximal tubule from Bowman's capsule is isotonic with the blood plasma; that is, it has the same solute concentration as does the plasma. In the proximal tubule, sodium ions are pumped out, with chloride ions following passively. Water also moves out of the tubule by osmosis, following after these ions. Thus, as the fluid enters the descending branch of the loop of Henle, it is still isotonic with the blood plasma. Its volume is dramatically reduced, however; some 60 to 70 percent of the solutes and water contained within the original filtrate are removed from it during the passage through the proximal tubule. This water and the solutes it contains are taken up almost immediately by the peritubular capillaries and returned to the bloodstream.

The pathway of the loop then carries the remaining filtrate through a zone of high solute concentration—a result of the properties of both the ascending branch and the collecting duct. As we have noted, the walls of the descending branch are freely permeable to water, so large quantities of water move by osmosis from the tubule into the interstitial fluid (and from there into the peritubular capillaries), with only very small amounts of salt diffusing into the tubule. The fluid within the tubule is now hypertonic in relation to the plasma.

As the fluid travels through the ascending branch, some salt diffuses from the lower portion of the tubule and larger quantities are removed by the action of the transport proteins in the upper portion. As a consequence, the fluid within the tubule becomes hypotonic in relation to the plasma. In humans it remains hypotonic during its passage through the distal convoluted tubule. However, the fluid must again pass through the zone of high solute concentration as it flows down the collecting duct. Whether additional water is removed or not depends on the presence or absence of a hormone, antidiuretic hormone (ADH). If ADH is absent, the wall of the collecting duct is impermeable to water and a hypotonic urine is excreted. If, however, ADH is present, the wall is freely permeable to water. Water moves by osmosis through the wall, leaving within the duct a urine that is isotonic with the surrounding interstitial fluid but hypertonic in relation to the body fluids as a whole. In this way, mammals that need to conserve water are able to excrete a fluid, the urine, far more concentrated than the plasma from which it is derived.

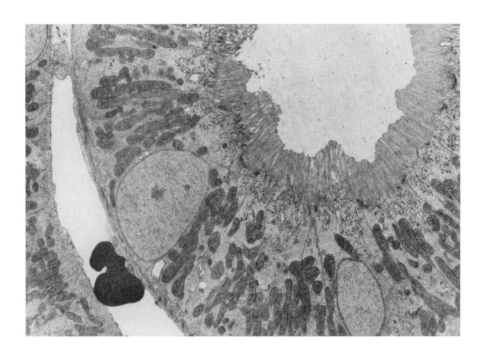

37-14 *The cells of the kidney require large amounts of energy to power their active transport processes. This electron micrograph shows cells of the proximal tubule, where the active secretion and reabsorption of many solutes take place. Notice the numerous mitochondria and their position in relation to the brush border on the surface of the cells bordering the lumen of the tubule (the light area at the upper right of the micrograph). This border is made up of many very fine, hair-like extensions that, like microvilli, greatly increase the absorptive capacity of the cell. It is assumed that the mitochondria provide a sufficient nearby energy source to power active transport mechanisms residing in the convoluted cell surface. A red blood cell can be seen within the peritubular capillary, in the lower left of the micrograph.*

Although the ion pumps in the ascending branch of the loop of Henle and the permeability of the lower portion of the collecting duct to urea make possible the concentration differences on which this system depends, a key role is played by the hairpin structure of the loop as a whole. The flow of fluid through the loop provides another example of a countercurrent system. As we noted earlier (see page 130), a countercurrent flow can help to maintain differences in solute concentration from one end to the other of a system. When it is coupled to an active transport system, as in the loop of Henle, it can actually multiply the differences and is known as a countercurrent multiplier. The longer the loop, the greater the concentration differences that can be established. Since the primary factor limiting the concentration of the urine is the solute concentration surrounding the collecting duct, it should come as no surprise to learn that those mammals that excrete the most hypertonic urine also have the longest loops of Henle.

Control of Kidney Function: The Role of Hormones

In mammals, several different hormones act on the nephron to affect the composition of the urine. One of these is ADH, which is formed in the hypothalamus, a major regulatory center in the brain, and is stored in and released from the pituitary gland. As we have seen, ADH acts on the membranes of the collecting ducts of the nephrons and increases their permeability to water, so more water moves, by diffusion, back into the blood from the nephron. The amount of ADH released depends on the osmotic concentration of the blood and also on the blood pressure. Osmotic receptors that monitor the solute content of the blood are located in the hypothalamus. Pressure receptors that detect changes in blood volume are found in the walls of the heart, in the aorta, and in the carotid arteries. Stimuli received by these receptors are transmitted to the hypothalamus. Factors that increase the concentration of solutes in the blood or decrease blood pressure, or both, increase the production of ADH and conservation of water. Such factors include dehydration and hemorrhage. Factors that decrease blood concentration, such as the ingestion of large amounts of water, or that increase blood pressure—adrenaline, for instance—signal the hypothalamus to decrease ADH production, and so more water is excreted. Cold stress inhibits ADH secretion and thus increases urinary flow. Alcohol also suppresses ADH secretion and increases urinary flow, a phenomenon familiar to imbibers of beer and other alcoholic beverages. Pain and emotional stress trigger ADH secretion and thus decrease urinary flow.

A second hormone, aldosterone, which is produced by the adrenal cortex, stimulates reabsorption of sodium ions from the distal tubule and collecting duct and secretion of potassium ions into them. When the adrenal glands are removed, or when they function poorly (as in Addison's disease), excessive amounts of sodium chloride and water are lost in the urine, and the tissues of the body become depleted of them. Generalized weakness results, and if a patient with Addison's disease is not given hormone replacement therapy, the fluid loss can eventually be fatal. Aldosterone production is controlled by an extremely complex feedback circuit involving potassium ion levels in the bloodstream and processes initiated in the kidneys themselves.

Other hormones are also involved in the regulation of kidney function, particularly in response to increases in blood volume or blood pressure. The most intriguing of these substances is cardiac peptide (page 756), which inhibits the reabsorption of sodium from the distal tubule and thus increases the excretion of both sodium and water. This hormone, released from the atria of the heart, apparently exerts its effects both directly on the nephron itself and indirectly by inhibiting the release of aldosterone from the adrenal cortex. Receptors for cardiac peptide have been identified in both locations. There is also evidence that it may interact with other factors in the feedback circuit controlling aldosterone production. The complexity of the mechanisms regulating excretion provides additional evidence of the critical importance to the organism of maintaining a constant internal chemical environment.

SUMMARY

Homeostasis—the maintenance of a constant internal environment—is the result of a variety of processes within the animal body. One of the most critical homeostatic functions is regulation of the chemical composition of body fluids. This function, which in vertebrates is carried out primarily by the kidneys, involves (1) excretion of toxic waste products, especially the nitrogenous compounds produced by the breakdown of amino acids, (2) control of the levels of ions and other solutes in body fluids, and (3) maintenance of water balance.

Animals living in saltwater, freshwater, and terrestrial environments face different problems in maintaining the composition of body fluids. Terrestrial animals generally need to conserve water. Reptiles and birds excrete nitrogenous wastes in the form of crystalline uric acid, whereas mammals excrete them as urea, which must be dissolved in water.

Maintaining water balance involves equalizing gains and losses. The principal source of water gain in most mammals is in the diet; water is also formed as a result of oxidation of nutrient molecules. Water is lost in the feces and the urine, by respiration, and from the skin. Although the amount of water taken in and given off may vary widely from animal to animal and also from time to time in the same animal, depending largely on environmental circumstances, the volume of water in the body remains very nearly constant. The principal water compartments of the body are the plasma, the interstitial fluids, including the lymph, and the intracellular fluids. The major factor determining exchange of water among compartments of the body is osmotic potential.

The functional unit of the kidney is the nephron. Each nephron consists of a long tubule attached to a closed bulb (Bowman's capsule), which encloses a twisted cluster of capillaries, the glomerulus. Blood entering the glomerulus is under sufficient pressure to force plasma (minus the larger proteins) through the capillary walls into Bowman's capsule. As the filtrate makes its long passage through the nephron, cells of the renal tubule selectively reabsorb molecules from the filtrate and secrete other molecules into it. Glucose, amino acids, most ions, and a large amount of water are returned to the blood through the peritubular capillaries. Excess water and waste products, including about half of the urea present in the original filtrate, are excreted from the body as urine.

An important means of water conservation in mammals is the capacity for excreting a urine that is hypertonic in relation to the blood. The loop of Henle is the portion of the mammalian nephron that makes this possible.

The function of the nephron is influenced by hormones, chiefly antidiuretic hormone (ADH), which is produced by the hypothalamus and released from the pituitary gland; aldosterone, a hormone of the adrenal cortex; and cardiac peptide, released from the atria of the heart. ADH increases the return of water to the blood and so decreases water excretion. Aldosterone increases the reabsorption of sodium ions and water and the secretion of potassium ions. Cardiac peptide, by contrast, inhibits the reabsorption of sodium ions and water. All of these hormones play a role in regulation of both blood pressure and blood volume.

QUESTIONS

1. Distinguish among the following: afferent arteriole/efferent arteriole; Bowman's capsule/glomerulus; extracellular fluid/intracellular fluid; ADH/aldosterone/cardiac peptide.

2. Explain the following terms in relation to kidney function: filtration, secretion, reabsorption, and excretion.

3. Sketch a nephron, indicating the pathways of the blood and the glomerular filtrate.

4. Diagram the paths of glucose and urea through the human kidney.

5. Reexamine the micrographs in Figures 37–11 and 37–14. Describe the function of each structure visible in them in terms of the overall function of the nephron.

6. Why does a high-protein diet require an increased intake of water? (You should be able to think of two different reasons.) Why does a person lose some weight after shifting to a low-salt diet, even without reducing the caloric intake? Given the fact that amino acids in excess of the body's requirements are broken down by the liver, not stored, what is the advantage of a high-protein diet? What might be a disadvantage?

7. Sodium and chloride ions are actively pumped from the renal tubule in both the proximal convoluted tubule and the upper portion of the ascending branch of the loop of Henle. How does the mechanism of active transport in the proximal convoluted tubule differ from the mechanism in the ascending branch of the loop of Henle?

8. The text asserts that the primary factor limiting the concentration of the urine is the solute concentration surrounding the collecting duct. Why is this so?

9. In Figure 37–13, the labels within the tubule indicate the concentration of solutes in the filtrate in relation to their concentration in the blood plasma. From the information given about the movement of ions and of water, determine, for each labeled location, the concentration of the filtrate in relation to the concentration of the surrounding interstitial fluid.

10. Viewed as a whole, the kidney selectively removes substances from the bloodstream. A closer look, however, reveals that it does so by filtering out all small molecules through the glomerulus and then reabsorbing those that are needed as the filtrate passes along the renal tubule. Thus, the system need not identify wastes as such; rather, it must identify useful substances. Why would such an arrangement have been beneficial to early mammals? Why is it especially beneficial to modern mammals, including ourselves?

11. Could a human being on a life raft survive by drinking sea water? By catching and eating bony saltwater fish? Explain your answers.

C H A P T E R 38

Homeostasis II: Temperature Regulation

Some 200 years ago, Dr. Charles Blagden, then secretary of the Royal Society of London, went into a room that had been heated to a temperature of 126°C (260°F), taking with him a few friends, a small dog in a basket, and a steak. The entire group remained there for 45 minutes. Dr. Blagden and his friends emerged unaffected. So did the dog (the basket had kept its feet from being burned by the floor). But the steak was cooked. As this episode illustrates, the capacity of a living animal to regulate the internal temperature of its body is a critical factor in its well-being.

Physiological processes depend on both the maintenance of cellular structures and a multitude of biochemical reactions, virtually all controlled by enzymes. As we saw in Chapter 8, the rate at which an enzymatic reaction occurs is determined by a number of factors, one of the most important of which is temperature. In general, the reaction rate approximately doubles with every 10°C rise in temperature. However, for every protein, including every enzyme, there is a relatively narrow temperature range in which it maintains the three-dimensional conformation that is optimal for its particular functions. Above the upper limit of this range, the protein begins to lose its conformation, in the process known as denaturation. Once denaturation occurs, enzymes—and other proteins whose

38-1 *Temperature regulation involves behavioral responses, as well as physiological and anatomical adaptations. In the cold, many small animals, such as these penguin chicks, huddle together for warmth. Huddling decreases the effective surface-to-volume ratio and thus the loss of heat from the body. A few of the chicks have lifted their heads from the warmth of the huddle to observe the photographer at work.*

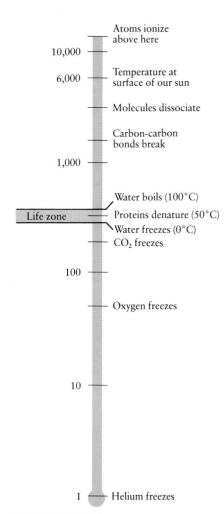

38-2 *Life processes can take place only within a very narrow range of temperature. The temperature scale shown here is the Kelvin, or absolute, scale. Absolute zero (0 K) is equivalent to −273.1°C, or −459°F, and is the temperature at which all molecular motion ceases.*

SECTION 6 Biology of Animals

function depends on a specific shape—are inactivated. Low temperatures can also bring physiological processes to a halt. The slightly salty water that is the principal constituent of living tissue freezes at −1 or −2°C. Molecules that are not immobilized in the ice crystals are left in such a highly concentrated state that their normal interactions are completely disrupted. As a consequence of these biochemical facts, most animals must either occupy environments in which the temperature ranges from just below freezing to between 45 and 50°C, or through their own physiological processes create an internal environment within that range. The few exceptions, such as the fishes and arthropods of polar regions, have special adaptations, such as antifreeze molecules in their body fluids or the capacity to enter a state of near-total dormancy, that enable them to survive extreme conditions.

PRINCIPLES OF HEAT BALANCE

Water balance, as we saw in the preceding chapter, requires that the water gained equal the water lost through the body surface and in the urine and feces. Similarly, maintaining a constant temperature requires that heat gains equal heat losses. For animals, as for other living organisms, there are two primary sources of heat: the radiant energy of the sun and exothermic chemical reactions within the cells. However, heat can be lost from the body by any one of several different mechanisms (Figure 38-3).

In accordance with the second law of thermodynamics, heat moves from a warmer body to a colder body (see Figure 8-4, page 164). If the two bodies are in physical contact, so that kinetic energy can be transferred directly from molecule to molecule, the movement of heat is called **conduction**. Some materials are better conductors of heat than others. When you step out of bed barefoot on a cold morning, you probably prefer to step on a wool rug than on the bare floor. Although both are at the same temperature, the rug feels warmer. If you touch something metallic, such as a brass doorknob, it will feel even colder than the floor. These apparent differences in temperature are actually differences in the speed at which the different materials conduct heat away from your body.

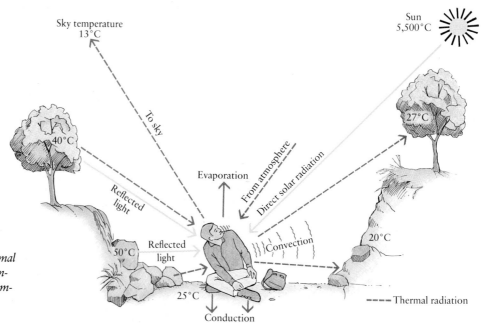

38-3 *Heat exchanges between a mammal and its environment. The core body temperature of the man is 37°C. The air temperature is 30°C and there is no wind movement.*

Water is an excellent conductor of heat, whereas air is a poor conductor. You are quite comfortable in air at 21°C (70°F) but may be uncomfortable in water at the same temperature. Conduction in fluids (air or water) is always influenced by **convection,** the movement of air or water in currents. Because both air and water become lighter as they get warmer, they move away from a heat source and are replaced by colder air or water, which again moves away as it warms. Fat, like air, is a poor conductor of heat, and both can serve as insulators. Animals that need to conserve heat are typically insulated with either fur or feathers, which trap air close to the body, or with fat or blubber.

Another route of heat loss is evaporation. As we saw in Chapter 2, every time a gram of water changes from a liquid to a gas, it takes more than 500 calories away with it. Many organisms, including ourselves, have exploited this property of water as a means for rapid adjustment of the heat balance.

Radiation, the transfer of energy by electromagnetic waves in the absence of direct contact, is a route not only of heat gain but also of heat loss. Depending on the wavelength of the radiation, energy may be transferred as light or heat (see Figure 10–3 on page 207). Light energy falling on an object is either absorbed as heat or reflected. As you undoubtedly know from your own experience, dark objects absorb heat more than do light-colored ones.

Body Size and the Transfer of Heat

Regardless of the mechanism, heat is transferred into or out of any object, animate or inanimate, across the body surface. Thus, the transfer of heat, like the diffusion of oxygen and carbon dioxide across the respiratory surface, is proportional to the surface area exposed. The smaller the object, as we saw in Chapter 5, the larger its surface-to-volume ratio. Ten pounds of ice, separated into individual ice cubes, will melt far faster than the equivalent volume in a solid block. For the same reason, it is much more difficult for a small animal to maintain a constant body temperature than it is for a large one. One of the ways in which many small animals deal with this problem in the face of extreme cold is togetherness, as illustrated in Figures 38–1 and 38–4.

38–4 *Within the wintering hive, bees maintain their temperature by clustering together in a dense ball; the lower the temperature, the denser the cluster. The clustered bees produce heat by constant muscular movements of their wings, legs, and abdomens. In very cold weather, the bees on the outside of the cluster keep moving toward the center, while those in the center move to the colder outside periphery. The entire cluster moves slowly about on the combs, eating the stored honey, which is their energy source.*

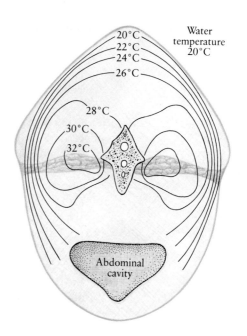

38–5 *The body of a 70-kilogram big-eye tuna, as shown in cross section. The internal temperature of the fish forms a gradient, ranging from a high of about 32°C deep within its body to 20°C (the temperature of the surrounding water) at its surface. Its body heat is produced by metabolic reactions and muscular activity, particularly in the hard-working, red, myoglobin-containing muscles that run the length of the body, parallel to the vertebral column.*

"COLD-BLOODED" VS. "WARM-BLOODED"

In common parlance, animals are often characterized as "cold-blooded" or "warm-blooded." This fits in with everyday experience: a snake is usually cold to the touch, and a bird feels warm. Actually, however, a "cold-blooded" animal may create an internal temperature for itself that is warmer than that of a "warm-blooded" animal. Another approach is to classify animals as ectotherms or endotherms. As we saw in Chapter 28, ectotherms are warmed from the outside in, and endotherms are warmed from the inside out. These categories correspond approximately but not completely with "cold-blooded" and "warm-blooded." Fish, for example, are considered "cold-blooded," yet some large fish, such as tuna, generate large amounts of metabolic heat and are, strictly speaking, endotherms (Figure 38–5).

When considering temperature regulation—rather than the actual temperature of an animal or the means by which its body is heated—the most useful classification of an animal is as a poikilotherm or a homeotherm. A **poikilotherm** (from the Greek word *poikilos*, meaning "changeable") has a variable temperature, and a **homeotherm,** a constant one. These terms, like ectotherm and endotherm, are generally used as synonymous with "cold-blooded" and "warm-blooded," though again the correspondence is not perfect. Fish that live deep in the sea, where the temperature of the water remains constant, have body temperatures far more constant than bats or hummingbirds, whose temperatures, as we shall see, fluctuate widely depending on their state of activity.

POIKILOTHERMS

Most aquatic animals are poikilotherms, maintaining a body temperature that is the same as the temperature of the surrounding water. Although the metabolic processes of such animals generate heat, it is usually quickly dissipated, even in large animals. In most fish, for example, heat is rapidly carried from the core of the body by the bloodstream and is lost by conduction into the water. A large proportion of this heat is lost from the gills. Exposure of a large, well-vascularized surface to the water is necessary in order to acquire oxygen, as you will recall from Chapter 35. This same process rapidly dissipates heat, so fish usually cannot maintain a body temperature significantly higher than that of the water. They also cannot maintain a temperature lower than that of the water, since they have no means of unloading heat.

CHAPTER 38 Homeostasis II: Temperature Regulation

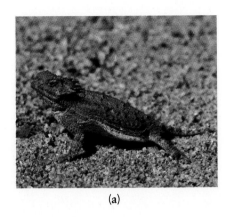

(a)

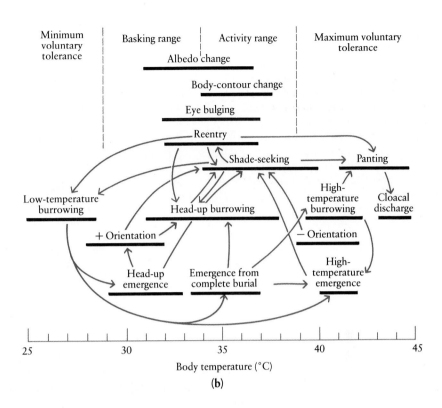

(b)

38-6 *In laboratory studies, the internal temperature of reptiles was shown to be almost the same as the temperature of the surrounding air. It was not until observations were made of these animals in their own environment that it was found that they have behavioral means of temperature regulation that give them a surprising degree of temperature independence. By absorbing solar energy, reptiles can raise their temperature well above that of the air around them. (a) Shown here is a horned lizard (often but not accurately known as a horned toad) that, having been overheated by the sun, has raised its body to allow cooling air currents to circulate across its underside. (b) This chart is based on field studies of the behavior of a horned lizard in response to temperature fluctuations. Changes in albedo (whiteness) are produced by pigment changes in the epithelial cells of the skin. Albedo changes affect the reflection versus absorption of light energy.*

In general, large bodies of water, for the reasons we discussed in Chapter 2, maintain a very stable temperature. At no place in the open ocean does the temperature vary more than 10°C in a year. In smaller bodies of water, where greater temperature changes occur, fish seek a level of optimal temperature, presumably the one to which their metabolic processes are adapted. However, because they can do almost nothing to make their own temperature different from that of the surrounding water, they can be quickly victimized by any rapid, drastic changes in water temperature.

Temperatures on land, in contrast to those in water, may vary annually in a given area by as much as 60 to 70°C. Thus temperature regulation, like water conservation, became a problem for animals when they moved from water to land. The first animals to become truly terrestrial, the reptiles (snakes, lizards, and tortoises), are, like fish, poikilotherms. Nonetheless, they are able to maintain remarkably stable body temperatures during their active hours by varying the amount of solar radiation they absorb (Figure 38–6). By careful selection of suitable sites, such as the slope of a hill facing the sun, and by orienting their bodies with a maximum surface exposed to sunlight, they can heat themselves rapidly. Such heating can occur as quickly as 1°C per minute, even on mornings when the air temperature, as on deserts or in the high mountains, is close to 0°C. By frequently changing position, these reptiles are able to keep their temperature within quite a narrow range as long as the sun is shining. When it is not, they seek the safety of their shelters. Although their body temperatures then drop and they become sluggish, they are not immobilized in an exposed position, where they are vulnerable to predators.

HOMEOTHERMS

Homeotherms are animals that maintain a constant body temperature despite fluctuations in their environment, and most maintain a body temperature well above that of their surroundings. Among modern animals, only birds and mam-

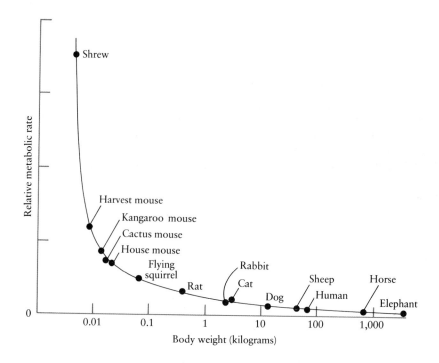

38–7 *Observed metabolic rates of mammals. Each division on the abscissa represents a tenfold increase in weight. The metabolic rate of very small mammals is much higher than that of larger mammals, owing principally to their greater surface-to-volume ratios.*

mals are true homeotherms. All homeotherms are endotherms, with the oxidation of glucose and other energy-yielding molecules within body cells as their primary source of heat.

Homeothermy brought with it a number of advantages, of which perhaps the most important was the capacity to function at peak efficiency even at low external temperatures. This capacity, for example, would have enabled the early mammals to be out and about at night, seeking food and mates, when the dominant animals of that era—the reptiles—were most likely inactive. It also made possible the invasion of less hospitable environments, particularly those with a relatively low temperature much of the year. In terms of energy requirements, however, the cost of homeothermy is high: the metabolic rate of a mammal is about 10 times that of a reptile of similar size at similar temperature. Moreover, the smaller the size, the higher the price (Figure 38–7). As a consequence of its high energy requirements, a mammal must absorb from its digestive tract about 10 times the quantity of nutrients as a reptile of comparable size. It does this by eating more and processing it faster, which is made possible by a more convoluted digestive tract with greater numbers of villi and microvilli.

Birds have even higher metabolic requirements. Although comparatively small, they maintain a higher body temperature—40 to 42°C—than most mammals. Also, unlike most small mammals, they spend much of their time exposed. Flight compounds their problems. To fly, birds must keep their weight down, and so they cannot store large amounts of fuel. Because of this and the high energy requirements of their way of life, birds need to eat constantly. A bird eating high-protein foods, such as seeds and insects, commonly consumes as much as 30 percent of its body weight per day. Bird migrations are dictated not so much by a need to seek warmer weather as by a need for longer days with more daylight hours for feeding themselves and their young.

Because it is heated from within, a homeotherm is warmer at the core of its body than at the periphery. Our temperature, for example, usually does not reach 37°C, or 98.6°F, until some distance below the skin surface. Heat is transported from the core to the periphery largely by the bloodstream. At the surface of the body, the heat is transferred to the air, as long as air temperature is less than body temperature. Temperature regulation involves increasing or decreasing heat production and increasing or decreasing heat loss at the body surface.

Avian Mechanical Engineers

Although we generally think of engineering as a profession invented and practiced only by representatives of *Homo sapiens,* birds have been practicing civil engineering in their design and construction of nests for far longer. Members of the family Megapodiidae, the mound-building birds of Australia, New Guinea, and other Pacific islands, are also accomplished mechanical engineers, harnessing the energy of external chemical reactions for their own purposes.

The developing young of homeotherms, both birds and mammals, must be maintained at a constant temperature, usually a temperature fairly close to that of the adult animal. In all mammals except the monotremes, this problem is solved automatically, as the young develop either within the body of the mother or, in the case of marsupials, in the mother's pouch. Birds generally keep their eggs warm by sitting on them, or in the case of penguins, tucking them on top of the feet, where they can be held close to the parent's body. In desert environments, however, there is a twofold problem: keeping the eggs cool during the day and yet warm at night, when the temperature drops dramatically. The mound-building mallee fowl of the Australian desert solve this problem by incubating their eggs in large compost heaps.

The parent birds, which are about the size of chickens, begin the process in the autumn by digging a pit, 3 meters wide by 1 meter deep, and raking plant litter into it. Next they build a huge mound of insulating sand around it, and then they wait for the spring rains. Following the rains, when the litter begins to decompose, the birds cover the fermenting heap with a layer of sand up to a meter in depth. Gradually, the fermentation reactions in the litter heat the mound from beneath. The male waits a few days and then begins checking its temperature with his beak, which contains a heat sensor. When the temperature is holding close to 33°C, he allows the female to begin laying her eggs. During the seven-week incubation period, he constantly checks and regulates the temperature in the mound by scraping away the sand around the eggs to expose them to cooling air currents or by piling up warm sand around them at night, maintaining a nearly constant 33°C, day and night.

The entire cycle from the beginning of mound building until the last egg hatches takes about a year. When the chicks hatch, completely feathered, their own internal temperature regulating mechanism is fully functional and they are able to feed and fend for themselves.

A pair of mallee fowl, surveying the result of their labors. All of the digging and moving of sand and litter is done with the large feet, from which this family of birds derive their name.

38-8 *Body temperature in mammals is regulated by a complex network of activities involving both the nervous and endocrine systems. The chief regulatory center is the hypothalamus, a brain center that, as we have seen, controls many physiological processes. In some animals (although not, apparently, in humans), the hormonal pathway—indicated by dashed lines—is of major importance. TRH is thyroid-releasing hormone, which is produced by the hypothalamus and which stimulates the production of TSH, thyroid-stimulating hormone, by the pituitary. TSH, in turn, stimulates the production of thyroxine, the thyroid hormone. Thyroxine increases cellular metabolism, apparently by acting directly on the mitochondria. Many behavioral responses, such as seeking sun or shelter, are also involved.*

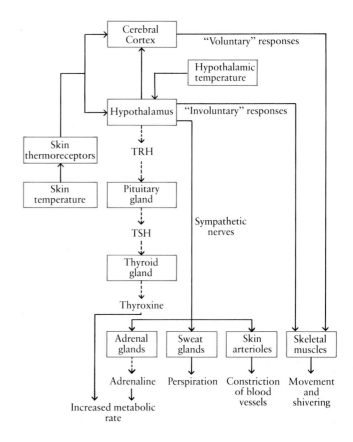

The Thermostat

The remarkable constancy of temperature characteristic of humans, and other homeotherms as well, is maintained by an automatic system—a thermostat—in the hypothalamus. This thermostat receives and integrates information from widely scattered temperature receptors, compares it to the set point of the thermostat, and, on the basis of this comparison, initiates appropriate responses. Unlike a furnace thermostat, which controls a simple on-off switch, the hypothalamic thermostat has a variety of complex responses at its command, as summarized in Figure 38-8.

Under ordinary conditions, the skin receptors for hot and cold are probably the most important sources of information about temperature change. However, the hypothalamus itself contains receptor cells that monitor the temperature of the blood flowing through it. As some interesting experiments have demonstrated, information received by these hypothalamic receptors can override that from other sources, ensuring that the core body temperature remains constant. For example, in a room in which the air is warmer than body temperature, if the blood circulating through a person's hypothalamus is cooled, he or she will stop perspiring, even though the skin temperature continues to rise.

The elevation of body temperature known as fever is due not to a malfunction of the hypothalamic thermostat but to a resetting. Thus, at the onset of fever, an individual typically feels cold and often has chills; although the body temperature is rising, it is still lower than the new thermostat setting. The substance primarily responsible for the resetting of the thermostat is a protein released by white blood cells in response to pathogens. The adaptive value of fever is not known with certainty, but current evidence indicates that moderate temperature increases stimulate the immune system, increasing not only the synthesis of antibodies but also the production of other substances and specialized cells that work together in the body's defenses against infection.

(a)

(b)

(c)

38-9 *The size of the extremities in a particular type of animal can often be correlated with the climate in which it lives.* (a) *The fennec fox of the North African desert has large ears, rich with blood vessels. As blood flows through the network of capillaries just below the skin surface, excess heat is dissipated from the body.* (b) *The red fox of the eastern United States has ears of intermediate size, and* (c) *the Arctic fox has relatively small ears. Similar correlations of characteristics such as size, weight, and color with environment can also sometimes be made among animals of a single species living over an extended geographic range.*

Regulating as Body Temperature Rises

As the body temperature of a mammal rises above its thermostat setting, the blood vessels near the skin surface are dilated, and the supply of blood to the skin increases. If the air is cooler than the body surface, heat can be transferred from the skin directly to the air. Animals that live in hot climates characteristically have larger exposed surface areas than animals that live in the cold (Figure 38-9).

Heat can also be lost from the surface by the evaporation of saliva or perspiration; the heat required for evaporation comes from blood traveling in vessels just below the skin surface. In humans and most other large mammals, when the external temperature rises above body temperature, perspiration begins. Horses and humans sweat from all over their body surfaces. Dogs and some other animals pant, so that air passes rapidly over their large, moist tongues, evaporating saliva. Cats lick themselves all over as the temperature rises, and evaporation of their saliva cools their body surface. For animals that dissipate heat through evaporation, temperature regulation at high temperatures necessarily involves water loss. This, in turn, stimulates thirst and water conservation by the kidneys. Dr. Blagden and his friends were probably very thirsty.

(a)

(b)

(c)

38-10 *Animals have various devices for utilizing the heat-absorbing properties of evaporating water.* (a) *Dogs unload heat by panting, which involves short, shallow breaths and the production of a copious saliva. When it is hot, a dog pants at a rate of about 300 to 400 times a minute, compared with a respiration rate of from 10 to 40 times a minute in cool surroundings. Panting, in contrast to sweating, does not result in the loss of salt or other important ions.* (b) *Among the animals that can sweat over their entire body surface are hippopotamuses. Their sweat, unlike that of humans and horses, is pink.* (c) *Elephants, lacking sweat glands, wet down their thick, dry skins with mud or water. They also unload heat by flapping their ears, which are highly vascularized.*

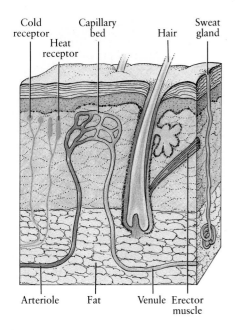

38–11 *Cross section of human skin showing structures involved in temperature regulation. In cold, the arterioles constrict, reducing the flow of blood through the capillaries, and the hairs, each of which has a small erector muscle under nervous control, stand upright. Beneath the skin is a layer of subcutaneous fat that serves as insulation, retaining the heat in the underlying body tissues. With rising temperatures, the arterioles dilate and the sweat glands secrete a salty liquid. Evaporation of this liquid cools the skin surface, dissipating heat (approximately 540 calories for every gram of H_2O).*

Regulating as Body Temperature Falls

When the temperature of the circulating blood begins to fall below the thermostat setting, blood vessels near the skin surface are constricted, limiting heat loss from the skin. Metabolic processes increase. Part of this increase is due to increased muscular activity, either voluntary (shifting from foot to foot) or involuntary (shivering). Part is due to direct stimulation of metabolism by the endocrine and nervous systems. Adrenaline stimulates the release and oxidation of glucose. Autonomic nerves to fat increase its metabolic breakdown. In some mammals exposed to prolonged cold, the thyroid gland increases its release of thyroxine, the thyroid hormone. Thyroxine appears to exert its effects directly on the mitochondria.

Most mammals have a layer of subcutaneous fat that serves as insulation. Homeotherms also characteristically have hair or feathers that, as temperature falls, are pulled upright by erector muscles in the skin, trapping air, which insulates the surface. All we get, as our evolutionary legacy, are goose pimples.

Cutting Energy Costs

Less energy is consumed by a sleeping animal than by an active one. Bears, for example, nap for most of the winter, living on fat reserves, permitting their temperature to drop several degrees, and keeping their energy requirements down.

Turning down the thermostat saves more fuel. Some small animals have different day and night settings. A hummingbird, for instance, has an extremely high rate of fuel consumption, even for a bird. Its body temperature drops every evening when it is resting, thereby decreasing its metabolic requirements and its fuel consumption. Another effect of this resetting is a reduction in the loss of water across the respiratory surfaces.

Hibernation (from *hiber,* the Latin word for "winter") is another means of adjusting energy expenditures to food supplies. Hibernating animals do not stop regulating their temperatures altogether—like the hummingbird, they turn down their thermostats. Hibernators are mostly small animals, including a few insectivores, hamsters, ground squirrels, some South American opossums, some bats, and a few birds, such as the poor-will of the southwestern United States. In these animals, the thermostat is set very low, often close to the temperature of the surrounding air (if the air is above 0°C). The heartbeat slows; the heart of an active ground squirrel, for instance, beats 200 to 400 times a minute, whereas that of a hibernating one beats 7 to 10 times a minute. The metabolism, as measured by oxygen consumption, is reduced 20 to 100 times. Apparently, even aging stops; hibernators have much longer life spans than similar nonhibernators. However, despite these profound physiological changes, hibernators do not cease to monitor the external environment. If the animals are exposed to carbon dioxide, for example, they breathe more rapidly, and those of certain species wake up. Similarly, hibernating animals wake up if the temperature in their burrows drops to a life-threatening low of 0°C. Hibernators of some species can also be awakened by sound or by touch.

Arousal from hibernation can be rapid. In one experiment, bats kept in a refrigerator for 144 days without food were capable of sustained flight after 15 minutes at room temperature. As indicated, however, by the arousals at 0°C, arousal is a process of self-warming rather than of collecting environmental heat. Breathing becomes more regular and then more rapid, increasing the amount of oxygen available for consumption; subsequently, the animal "burns" its stored food supplies as it returns to its normal temperature and the fast pace of a homeothermic existence.

(a) (b) (c)

38–12 *Some energy conservation measures. (a) This dormouse has stored food reserves in body fat in preparation for hibernating. Hibernation saves energy by turning the thermostat down very low, slowing the heartbeat, and reducing the metabolic rate. (b) Note the heat-conserving reduction in surface-to-volume ratio in this hibernating golden-mantled ground squirrel. (c) Bats have a number of adaptations that enable them to conserve heat despite their small body size. Some hibernate. In others, the temperature drops when they are at rest. Members of the species shown here also roll up their ears when they get cold.*

ADAPTATIONS TO EXTREME TEMPERATURE

We rely mainly on our technology, rather than our physiology, to allow us to live in extreme climates, but many animals are comfortable in climates that we consider inhospitable or, indeed, uninhabitable.

Adaptations to Extreme Cold

Animals adapt to extreme cold largely by increased amounts of insulation. Fur and feathers, both of which trap air, provide insulation for Arctic land animals. Fur and feathers are usually shed to some extent in the spring and regrow in the fall.

Aquatic homeotherms, such as whales, walruses, seals, and penguins, are insulated with fat (neither fur nor feathers serve as effective insulators when wet). These marine animals, which survive in extremely cold water, can tolerate a very great drop in their skin surface temperature; measurements of skin temperature have shown that it is only a degree or so above that of the surrounding water. By permitting the skin temperature to drop, these animals, which maintain an internal temperature as high as that of a person, expend very little heat outside of their fat layer and so keep warm—like a diver in a wet suit.

Some large Arctic mammals, such as fur seals and polar bears, are so well insulated that, when they are on land, their principal physiological problem is unloading excess heat. One solution to this problem is to keep the fur moist; another is to sleep, which in some species of fur seals reduces heat production by almost 25 percent.

Countercurrent Exchange in Heat Conservation

Numerous animals conserve heat by permitting temperatures in the extremities to drop. The extremities of such animals are actually adapted to a different internal temperature than the rest of the animal. For example, the fat in the foot of an Arctic fox has a different thermal behavior than the fat in the rest of its body, so that its footpads are soft and resilient even at temperatures of −50°C. Also, for many of these animals, particularly those that stand on the ice, this capacity is essential for another reason. If, for example, the feet of an Arctic seabird were as warm as its body, they would melt the ice, which might then freeze over them, trapping the animal until the spring thaws set it free.

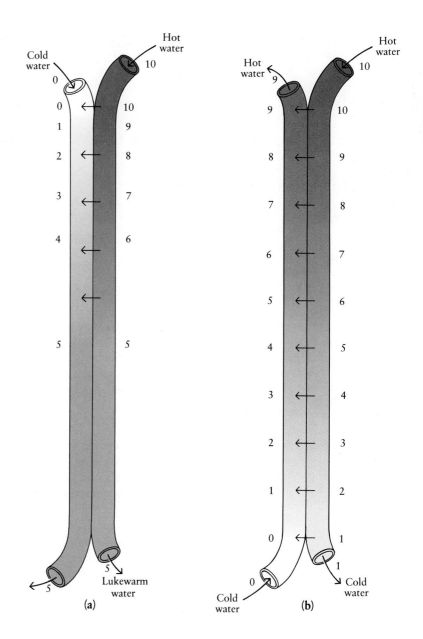

38–13 *The principle of countercurrent heat exchange as it occurs in the extremities of many animals is illustrated here by a hot-water pipe and a cold-water pipe placed side by side. In (a), the hot water and cold water flow in the same direction. Heat from the hot water warms the cold water until both temperatures equalize at 5 on this arbitrary scale. Thereafter, no further exchange takes place, and the outflowing water in both pipes is lukewarm. In (b), the flow is countercurrent, so heat transfer continues for the length of the pipes. The result is that the hot water transfers most of its heat as it travels through the pipe, and the cold water is warmed to almost the initial temperature of the hot water at its source.*

In many Arctic animals, the lower internal temperature of the extremities is made possible by a countercurrent arrangement that keeps heat in the body and away from the extremities (Figure 38–13). The arteries and veins leading to and from the legs (or fins or tail) are juxtaposed in such a way that the chilled blood returning through the veins picks up heat from the blood entering through the arteries. The veins and arteries are closely apposed to give maximum surface for heat transfer. Thus the body heat carried by the blood is not wasted by being dissipated to the cold air at the extremities. Instead it serves the useful function of warming the chilled blood that would otherwise put a thermal burden on the body.

Adaptations to Extreme Heat

We are efficient homeotherms at high external temperatures, as Dr. Blagden's experiment showed. Our chief limitation in this regard is that we must evaporate a

38-14 *By facing the sun, a camel exposes as small an area of body surface as possible to the sun's radiation. Its body is insulated by fat on top, which minimizes heat gain by radiation. The underpart of its body, which has much less insulation, radiates heat out to the ground. Note the loose-fitting garments worn by the camel driver. When such garments are worn, sweat evaporates from the skin surface. With tight-fitting garments, the sweat evaporates from the surface of the garment instead, and much of the cooling effect is lost. The loose robes, which trap air, also serve to keep desert dwellers warm during the cold nights. Other adaptations of the camel to desert life include long eyelashes, which protect its eyes from the stinging sand, and flattened nostrils, which retard water loss.*

great deal of water in order to unload body heat, and so our water consumption is high. The camel, the philosophical-looking "ship of the desert," has several advantages over human desert dwellers. For one thing, a camel excretes a much more concentrated urine. Also, a camel can lose more water proportionally than a human and still continue to function. If a person loses 10 percent of the body weight in water, he or she becomes delirious, deaf, and insensitive to pain. If the loss is as much as 12 percent, the individual is unable to swallow and so cannot recover without assistance. Laboratory rats and many other common animals can tolerate dehydration of up to 12 to 14 percent of body weight. Camels can tolerate the loss of more than 25 percent of their body weight in water, going without drinking for as long as one week in the summer months, three weeks in the winter.

Probably most important, the camel can tolerate a fluctuation in internal temperature of as much as 6°C. This tolerance means that it can let its temperature rise during the daytime (which the human thermostat would never permit) and drop during the night. The camel begins the next day at below its normal temperature—storing up coolness, in effect. It is estimated that the camel saves as much as 5 liters of water a day as a result of these internal temperature fluctuations.

Camels' humps were once thought to be water-storage tanks, but actually they are localized fat deposits. Physiologists have suggested that the camel carries its fat in a dorsal hump, instead of distributed all over the body, because the hump, acting as an insulator, impedes heat flow into the body core. A uniform fat distribution, which is decidedly useful in Arctic animals, would not be in inhabitants of hot climates.

Most small desert animals are, like the earliest mammals, nocturnal. The avoidance of direct heat is their principal means of temperature regulation. Small desert animals usually do not unload heat by sweating or panting; because of their relatively large surface areas, such mechanisms would be extravagant in terms of water loss. As these animals remind us, physiological problems, even those that may seem quite distinct at first glance, are always interrelated. A successful solution to any specific physiological problem invariably involves the interactions of different systems, balancing the various demands on and needs of the animal as a whole.

SUMMARY

Life can exist only within a very narrow temperature range, from about 0°C to about 50°C, with few exceptions. Animals must either seek out environments with suitable temperatures or create suitable internal environments. Heat balance requires that the net heat loss from an animal equal the heat gain. The two primary sources of heat gain are the radiant energy of the sun and cellular metabolism. Heat is lost through conduction, which, in fluids, is aided by convection, through evaporation, and through radiation.

Water-dwelling animals usually maintain a body temperature that is the same as the relatively constant temperature of the water. On land, environmental temperatures are much more variable. Animals—whether aquatic or terrestrial—whose internal temperature fluctuates with that of the external environment are known as poikilotherms, whereas those that maintain a constant internal temperature despite external temperature changes are known as homeotherms. Terrestrial poikilotherms are generally ectotherms, taking in heat energy from the environment; through behavioral means, they are able to adjust the amount of heat taken in and thus regulate their internal temperature during daylight hours. All true homeotherms—birds and mammals—are endotherms, generating heat internally from the oxidation of glucose and other fuel molecules and by friction from muscular activity.

In mammals, temperature is regulated by a thermostat in the hypothalamus. This thermostat receives and integrates information from temperature-sensitive receptors in the skin and in the hypothalamus itself and triggers appropriate responses. As body temperature rises, blood vessels in the skin dilate, increasing the blood flow to the surface of the body. Evaporation of water from body surfaces increases heat loss. As body temperature falls, blood vessels in the skin constrict, reducing heat loss. Energy production is increased by muscular activity and by nervous and hormonal stimulation of metabolism.

The energy costs of homeothermy are high. Sleeping is one means of reducing energy consumption; in some small homeotherms, additional energy is conserved by a lowering of the thermostat setting when the animal is resting. For example, hummingbirds have different thermostat settings for day and night. Other homeotherms make seasonal adjustments, entering the state known as hibernation during the winter when temperatures are low and food is scarce.

Adaptations to extreme cold principally involve insulation by fur or feathers and heavy layers of subcutaneous fat, coupled with countercurrent mechanisms that prevent the loss of heat from the extremities. Adaptations to extreme heat in large animals such as the camel include a relatively wide tolerance for fluctuations in temperature and total water content. In smaller animals, water conservation measures and behavioral adaptations are of primary importance.

QUESTIONS

1. Distinguish among the following terms: endotherm, ectotherm, poikilotherm, and homeotherm.

2. Compare the surface-to-volume ratio of an Eskimo igloo with that of a California ranch house. In what way is the igloo well suited to the environment in which it is found?

3. Describe what happens in the human body as its temperature rises. As it falls.

4. In terms of temperature regulation, what is the advantage of having temperature monitors that are located externally, as on the outside of a building or in the skin? Why is it also important to monitor the internal temperature? Which should be the principal source of information for your hypothalamic thermostat?

5. Compare hibernation in animals with dormancy in plants. In what ways are they alike? In what ways are they different?

6. Plants, as well as animals, are warmed above air temperature by sunlight, and plants cannot actively seek shade. Why might the leaves in the sunlight at the top of an oak tree be smaller and more extensively lobed (more fingerlike) than the leaves in the shady areas lower on the tree?

7. Diagram a heat-conserving countercurrent mechanism in the leg of an Arctic animal.

8. In the first few patients to receive an artificial heart, a cardiac output (page 756) of 6 liters per minute was found to be optimal and was thus considered the standard to which the device should be adjusted. A later patient, weighing 220 pounds, in whom the artificial heart was used as a temporary measure while awaiting a heart transplant, seemed also to do well at this level. However, his core body temperature was above normal and the temperature of his skin and extremities was below normal. What adjustment did his doctors make to restore both core and surface temperatures to normal? Why was it needed?

CHAPTER 39

Homeostasis III: The Immune Response

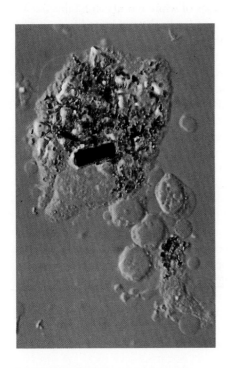

39-1 *The defense mechanisms of contemporary animals must deal with pollutants of human origin as well as with infectious microorganisms. This macrophage, a type of phagocytic white blood cell, was removed from a human lung. It had ingested a number of foreign particles, including a large rectangular fragment that is probably silicon, a principal component of computer chips. White blood cells are important in both nonspecific and specific responses to microorganisms and other possible pathogens.*

Every living thing, including yourself at this moment, is surrounded by potentially harmful microorganisms. Many of these microorganisms have the capacity not only to destroy individual cells but also to disrupt the numerous interrelated processes on which the continuing life of the organism depends; thus the defense against such microorganisms is an essential aspect of homeostasis.

Over the long course of evolution, organisms have developed a variety of defenses that function to exclude would-be invaders or to overcome them should they gain entry. In vertebrates, such defenses have evolved into an elaborate interacting network involving both nonspecific responses (in particular, inflammation) and the highly specific responses of the immune system, precisely tailored for each different invader. An understanding of these responses is of such practical importance for human health that most research has been specifically focused on mammals, with information derived from both laboratory animals, such as mice, and human patients. As a consequence, we now know a great deal about mammalian defensive mechanisms, but surprisingly little about those of other vertebrates.

NONSPECIFIC DEFENSES

Anatomic Barriers

The body's first line of defense against foreign invaders is its outer wrapping of skin and mucous membranes. The skin, with its tough layer of keratin, is an impregnable barrier as long as it is intact. When it is not, large numbers of microorganisms gain ready entry to the body. This is starkly illustrated in individuals who have suffered extensive burns; for such persons, the most immediate and greatest threats to life are the severe infections to which they are vulnerable.

The epithelium that forms mucous membranes is more fragile than the skin, but it is constantly flushed with fluids, such as mucus, saliva, and tears, that contain antimicrobial substances. The epithelium lining the respiratory tract is carpeted with cilia (page 122), which sweep away inhaled microorganisms, dirt, and debris trapped in the protective layer of mucus. The extremely acidic pH of the stomach contents creates an inhospitable environment for potential immigrants accompanying ingested food. In addition, the lower intestinal tract harbors resident populations of bacteria that defend their home territory against other microorganisms. Despite these defenses, the mucous membranes are the most common sites of entry of microorganisms or their toxins, typically through rips or tears in the epithelium.

The Inflammatory Response

If a microorganism penetrates the outer barrier, it encounters a second line of defense, consisting of a variety of agents carried by the circulating blood and

SECTION 6 Biology of Animals

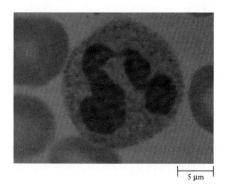

39-2 *A neutrophil, the most common type of granulocyte; note the numerous small granules in the cytoplasm. The nucleus of a neutrophil is many-lobed; as a result, it appears to have many different shapes in different microscopic sections. Thus neutrophils are said to be polymorphonuclear, and they are often called "polymorphs" or "polys" for short.*

TABLE 39-1 Types of White Blood Cells

	PERCENT OF TOTAL
Granulocytes	
Neutrophils	50–70
Eosinophils	1–4
Basophils	0.1
Lymphocytes	20–40
Monocytes	2.8

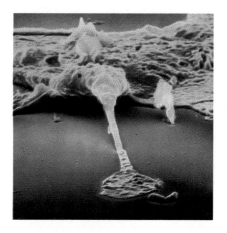

39-3 *Macrophages, as well as other types of white blood cells, move through the tissues by amoeboid motion, forming and retracting pseudopodia. They are also phagocytic, engulfing bacteria and other small particles. This patrolling macrophage has detected the chemical trail of an* Escherichia coli *cell, which it is about to capture with its precisely targeted pseudopod.*

lymph. Suppose, for example, you cut your skin. Cells in the area immediately release histamine and other chemicals that increase both blood flow into the area and the permeability of nearby capillaries. Circulating white blood cells, attracted by these chemicals, move through the capillary walls, crowding into the site of the injury. These cells engulf the foreign invaders, often literally eating themselves to death. Blood clots begin to form, walling off the injured area. The local temperature often rises, creating an environment unfavorable to the multiplication of microorganisms, while accelerating the motion of white blood cells. As a consequence of this series of events, known as the **inflammatory response,** the injured area becomes swollen, hot, red, and painful.

Both the inflammatory response and the more specific immune response depend on the interaction of a variety of types of white blood cells (Table 39-1). These cells, like the oxygen-carrying red blood cells, have a finite life span and must be continuously replenished. All of the different types of white blood cells, as well as the red blood cells, result from the differentiation and division of common, self-regenerating **stem cells** located in the marrow of long bones.

The principal cells involved in the inflammatory response are **granulocytes,** circulating white blood cells that are classified by their staining properties as neutrophils, eosinophils, or basophils. Neutrophils (Figure 39-2) are by far the most numerous, making up 50 to 70 percent of all white blood cells. When the first signs of inflammation appear, neutrophils, which are usually round and move freely through the bloodstream, begin to stick to the inner surface of the endothelium lining the blood vessels. They then develop amoeboid projections that enable them to push their way between the endothelial cells of the capillaries and move out into the infected tissues. Here they phagocytize microorganisms and other foreign particles, just as an amoeba engulfs its prey (page 474). The cytoplasmic granules from which granulocytes derive their name are actually lysosomes, and once the offending microparticle is within the neutrophil, lysosomes fuse with the phagocytic vacuole. Most phagocytized microbes are killed and digested within the vacuole by microbicidal proteins and lytic enzymes from the lysosomes. Some microbes—such as the encapsulated pneumococcus that played such a significant role in the earliest days of molecular genetics (page 282)—have evolved their own defenses against the defenses of their hosts and so remain virulent.

Basophils and eosinophils are also phagocytic. Basophils contain granules that rupture readily, releasing chemicals, such as histamine, that enhance the inflammatory response. Basophils are important components of allergic reactions; as we shall see later in this chapter, a key role in such reactions is played by mast cells, which are specialized, noncirculating basophils found in connective tissue. The role of eosinophils is not clear; they are found in increased numbers during infections involving internal parasites, such as worms.

Another type of circulating white blood cell that plays a major role in the inflammatory response is the **monocyte.** Monocytes, like neutrophils, are attracted to the site of an infection by chemicals released by both bacterial and host cells, generally arriving later than the neutrophils. Once on location, they are transformed into **macrophages,** becoming enlarged, amoeba-like, and phagocytic (Figure 39-3). Macrophages also lodge in the lymph nodes, spleen, liver, lungs, and connective tissues, where they entrap any microbes or foreign particles that may have penetrated the initial defenses (see Figure 39-1). They are also important, as we shall see, in activating the lymphocytes, other white blood cells that are the effectors of the immune response.

Inflammation can produce systemic as well as local effects. One common response to bacterial infection is as much as a fivefold increase in the proliferation and release of neutrophils. Another common response is fever (page 784), caused by a protein released by monocytes and macrophages in the course of the inflammatory response.

Interferons

Interferons differ from the other defense mechanisms of the body in two ways: (1) they are active only against viruses, and (2) they do not act directly on the invading viruses but rather stimulate the body's own cells to resist them.

The clue that led to the discovery of the interferons was the observation that an animal—human or mouse—infected with one virus is not usually susceptible to an infection with another virus. The source of this resistance was traced, with some difficulty, to a protein that was given the name of interferon. It is now known that at least three different classes of interferons exist, all small proteins that bind readily with other molecules and are active in very small amounts. As a result, the isolation of the first interferon took almost two decades.

When a cell is invaded by a virus, it releases interferon, which then interacts with receptor sites on the membranes of surrounding cells. Thus stimulated, these cells produce antiviral enzymes that block the translation of viral messenger RNA to protein. Only a very few molecules of interferon seem to be required to protect the surrounding cells from viral infection. Interferon molecules also interact with receptors on the surface of various types of white blood cells, stimulating both the inflammatory and immune responses.

Until recently, the only interferons available for research purposes were the minuscule amounts collected from mammalian cells exposed to virus in tissue culture. Now, as a result of recombinant DNA techniques, interferons are being produced in much larger quantities. These new supplies are being used to study clinical applications of interferons in the control of virus infections and to explore their effectiveness against certain forms of cancer. Among their many effects, interferons inhibit cell proliferation, which at one time led to hopes that they might be the long-sought "magic bullets" against this family of diseases. Thus far, however, only one extremely rare form of leukemia has been found to respond to an interferon.

THE IMMUNE SYSTEM

The immune response differs from the other defenses of the body in that it is highly specific, involving recognition of a particular invader and the tailoring of an attack against it. The response consists of two phases, a primary response to the initial attack of an invader and a rapid, secondary response to subsequent attacks by the same agent.

The specificity of the immune response derives from the actions and interactions of two remarkable groups of cells, known as **B lymphocytes** and **T lymphocytes** (or, more simply, as B cells and T cells). In mammals, the primary sites for the differentiation and proliferation of these cells are the bone marrow (B) and the thymus gland (T), a spongy, two-lobed organ that lies high in the chest.

The arena in which these cells operate is known as the immune system (Figure 39–4). As an organ system, it is more diffuse than the digestive system, for instance, or the excretory system, but it resembles these in being a functional, integrated unit. It includes not only the bone marrow and thymus gland, but also the lymph vessels, lymph nodes, spleen, and tonsils.

The lymph vessels, as you will recall, are the route for the return of interstitial fluid to the circulatory system. Strategically located within this system of vessels are the **lymph nodes**, which are masses of spongy tissue separated into compartments by connective tissue. Microorganisms, other foreign particles, and tissue

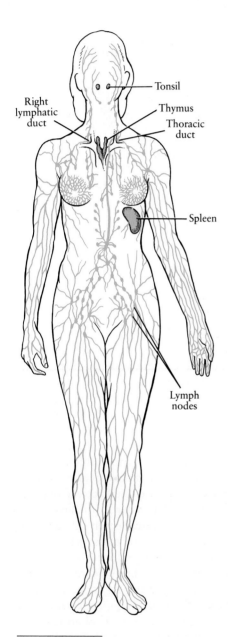

39–4 *The primary constituents of the human immune system are the bone marrow (not shown in this diagram) and the thymus, the sites of the initial proliferation of the B and T lymphocytes that are the effectors of the immune response. Other major components of the system are the lymph vessels, the numerous lymph nodes, the spleen, and the tonsils.*

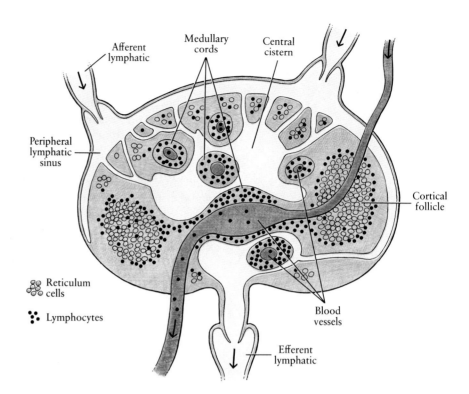

39-5 *Diagram of a lymph node. The cortical areas contain follicles, made up of reticulum cells, in which lymphocytes that have been exposed to foreign invaders are activated and undergo further proliferation. Thin, branched tubules, the medullary cords, traverse the node, adjacent to and within the central cistern. One cord is shown in longitudinal section, and the others are shown in cross section. Each cord consists of a small, central blood vessel and an outer network of cells, the mantle. The mantle, which is exposed to the lymphatic fluid on its outer surface, is typically distended with lymphocytes and other immunologically active cells that enter the bloodstream as the blood travels through the lymph nodes. Lymph nodes, which range in size from as small as a poppy seed to as large as a grape, typically increase in size during periods of exceptional immune system activity.*

debris entering the extracellular spaces of any tissue are caught up in the interstitial fluid, swept into the channels of the lymphatic system, and trapped in the lymph nodes. Single lymph nodes are distributed throughout the body, but most are found clustered in particular areas, such as the neck, armpits, and groin. Lymph nodes serve as filters, removing microbes, foreign particles, tissue debris, and dead cells from the circulation. Those near the respiratory system, for example, are often filled with particles of soot or tobacco smoke. Lymph nodes near a cancer site may contain malignant cells that have broken loose from the primary growth; in cancer operations, such lymph nodes are often excised not only to remove any malignant cells they may contain but also to determine if the disease has spread from the primary site. Lymph nodes also trap bacterial cells and other microorganisms that have managed to make their way past the first line of defense, the skin and mucous membranes. Lymph nodes are densely populated by lymphocytes and macrophages, and it is within these structures that essential interactions among the cells involved in the immune response occur.

The structure of a lymph node is shown in Figure 39–5. Lymph enters the node by the afferent lymphatic vessels, filters through the peripheral lymphatic sinuses into the central cistern, and trickles out through the efferent lymphatic vessel. The valves within the vessels prevent backflow. The peripheral cortical areas of the node contain follicles where lymphocytes proliferate after exposure to a foreign substance. The incoming lymph carries bacteria, viruses, and other potential pathogens into the node, exposes them to the lymphocytes in the follicles, and collects lymphocytes and other immunologically active cells, which are then circulated throughout the body.

The spleen and the tonsils are also rich in lymphocytes and particle-trapping cells. In the spleen, foreign materials enter by way of the blood rather than the lymph; hence this organ is most important in the case of blood-borne infections. The tonsils trap airborne particles. Patches of lymphoid tissue with large lymphoid

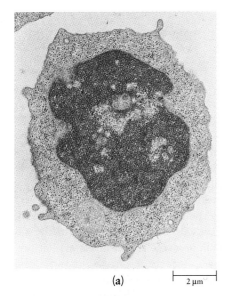

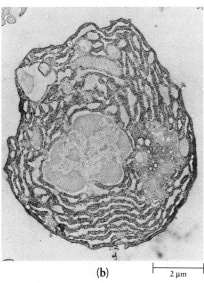

39-6 *As an activated B lymphocyte (a) differentiates into a plasma cell (b), it becomes highly specialized for the manufacture and secretion of antibodies. It grows larger, its nucleus becomes relatively smaller and less dense, and there is a large increase in endoplasmic reticulum and ribosomes.*

follicles—known as Peyer's patches—are also embedded in the wall of the intestine, lying between the inner lining of mucous membrane and the outer muscular coat and defending the body against the billions of microorganisms that inhabit the normal intestinal tract.

B LYMPHOCYTES AND THE FORMATION OF ANTIBODIES

B lymphocytes are the major protagonists in one type of immune response: the formation of **antibodies,** complex globular proteins that are also known as **immunoglobulins.** Antibodies make precise three-dimensional combinations with particular molecules or parts of molecules that the body recognizes as foreign, or "not-self." Any such molecular configuration that can trigger the synthesis of antibodies is known as an **antigen** (short for "antibody-generating substance"). Virtually all foreign proteins and most polysaccharides can act as antigens. The surface of a single cell, such as a bacterial cell, may have a number of different antigens, each of which can elicit production of a specific antibody.

The B Lymphocyte: A Life History

At any given time, about 2 trillion (2×10^{12}) B lymphocytes are on the alert in the human body. Many of these cells are on patrol, circulating with the bloodstream, squeezing out between the endothelial cells that form the walls of the capillaries, and migrating through the lymphatic system. Others are sessile, clustering in the lymph nodes, spleen, and other lymphoid tissues, where, as we have seen, they are exposed to circulating blood and lymph. Both the circulating and sessile B lymphocytes are small, round, nondividing, metabolically inactive cells. Embedded in the cell membrane of each B lymphocyte, and protruding from its surface, are antibodies with a specific three-dimensional structure. When a particular B lymphocyte meets its destiny in the form of an antigen with a three-dimensional structure complementary to the structure of the antibodies on its surface, the cell enlarges, its nucleolus swells, polysomes (page 314) form, and an increased synthesis of macromolecules begins. At the same time, microtubules form and the lymphocyte begins to divide. The proliferation of activated B lymphocytes often takes place in the follicles of the lymph nodes, which therefore enlarge during an infection.

The daughter cells resulting from B lymphocyte activation differentiate into two types, of which one is the **plasma cell** (Figure 39–6). Plasma cells, which rarely undergo further division, are, in essence, specialized antibody factories. A mature plasma cell can make 3,000 to 30,000 antibody molecules per second; these antibodies are released into the bloodstream and circulate throughout the body. However, it takes about five days to produce fully mature antibody-synthesizing cells working at this maximum capacity. Thus, if the microorganism is multiplying also, it may take about this long for the immune system to catch up. In the days before antibiotic therapy, about all the family doctor could do was to await the "break in the fever" that signaled this catching-up process and, often, forecast the patient's eventual recovery. Antibiotics, by suppressing the rate of multiplication of bacteria, enable antibody production to overtake the infection more rapidly.

The second type of cell produced by the antigen-stimulated B lymphocyte is the **memory cell.** Memory cells also produce antibodies, but they differ from plasma cells in their longevity: plasma cells live only a few days, whereas memory cells continue to circulate for long periods of time—up to a lifetime. Thus, the second time a particular pathogen gains entry to the body, large-scale production of antibodies to the invader begins immediately, often preventing any significant multiplication of the pathogen.

Death Certificate for Smallpox

In the early days of medicine, the human body was the laboratory and disease itself the instructor. Smallpox was one of the great teachers. Unlike other infections, which might go unreported or misdiagnosed, the pox left its unmistakable trace in history. It originated in the Far East and seems to have first been introduced into Europe by the returning Crusaders. Here it flourished and spread, until by the eighteenth century one in ten persons died of the pox, and 95 percent of those who survived their childhood had experienced it. About half had permanent scars and many were blinded. Young women studied their reflections in the mirror, waiting their almost inevitable turn. They learned to scratch their legs and feet at the first sign of the disease in the knowledge that the ugly lesions would then localize in these decorously concealed locations and so, perhaps, spare their faces and bosoms. No one could miss the fact that a person who had suffered an attack was thereby protected from a future one. Pox scars were required of domestic servants, particularly nursemaids, as a prime certificate of employability.

It came to be recognized that some outbreaks of pox were more severe than others (owing, we know now, to mutations of the virus). Since one had to have the disease some time, it was reasoned, it was advantageous to choose which pox and when. Intentional infection of children with material preserved from a mild attack was first practiced in the Far East. The Chinese did it in the form of powdered scabs, "heavenly flowers," used as snuff. The Arabs carried matter from pox pustules around in nutshells and injected it under the skin on the point of a needle. Smallpox inoculation, known as variolation, was introduced into England in 1717 by Lady Mary Wortley Montagu, wife of the British Ambassador to Turkey. In 1746, a Hospital for the Inoculation against Smallpox had been established for the poor of London, where they could be confined during the course of the deliberate infection. Although the induced disease was usually mild, it produced serious illness in some persons. Moreover, since the pox caused by variolation was as contagious as normally contracted smallpox, it appears to have been responsible for some epidemics.

Edward Jenner was an English country doctor who, despite the derision of his colleagues, listened to the tales of country folk. They told him that smallpox never infected milkmaids or other persons who had previously had cowpox, a mild disease of farm animals that sometimes was transmitted to humans. Jenner tried variolation on several persons who had had cowpox and was unable to induce the customary infection. Then, in 1796, he performed his classic experiment on the farm boy Jamie Phipps. First he inoculated the eight-year-old with fluid taken from a pustule of a milkmaid with cowpox. Subsequently he inoculated the boy with material from a smallpox lesion. Fortunately—for himself, for Jamie, and for us all—the child did not contract smallpox. (The success of the cowpox inoculation depended, of

---

This rapid response by the memory cells is the source of immunity to many infectious diseases—such as smallpox, measles, mumps, and polio—following an infection. It is also the basis for vaccination against a number of diseases (see essay). Vaccines may be prepared using a closely related pathogen, as in the smallpox vaccine; a killed pathogen, as in the Salk vaccine against polio; or a strain of the pathogen that has been weakened by growing in another host organism or in tissue culture, as in the widely used oral (Sabin) vaccine against polio.

The use of recombinant DNA technology to synthesize purified antigenic proteins found on the surfaces of pathogens has now become a powerful new tool in the preparation of vaccines. This technology was first used to prepare a vaccine against the virus causing hoof-and-mouth disease, a serious problem in the livestock industry. The synthetic proteins that make up the vaccine are identical to components of the protein coat of the virus. More recently, recombinant DNA technology has yielded a safe and effective vaccine against the hepatitis B virus, a major cause of liver disease throughout the world. It is also the principal tool in

course, on the antigenic similarity between the two naturally occurring pathogens.) Jenner called the process vaccination, from *vacca,* the Latin word for cow. By 1800, at least 100,000 persons had been vaccinated, and smallpox began to lose its hold on the Western world.

It was not until almost 100 years later that Louis Pasteur discovered serendipitously that a virus or bacterium grown in tissues other than those of its normal host would often lose its virulence while retaining its immunogenicity. (He retained the term vaccination, in recognition of the earlier work of Jenner.) Pasteur's discovery provided the basis for most of our modern vaccines, such as those against poliomyelitis, diphtheria, and measles.

Despite the availability of an effective vaccine for smallpox, by the 1950s there were still an estimated 2 million new cases of smallpox every year. These were largely confined to the Far East, mostly among the urban poor of India, whose crowded living conditions provided as fertile a soil as the European cities had two centuries earlier. In 1973, the World Health Organization (WHO) declared war on smallpox, which involved the mass production of vaccine and, more important, search-and-destroy missions charged with finding each new case of the disease and vaccinating every susceptible person who might be exposed to it; 150,000 persons were in the task force. On April 23, 1977, an international commission declared India free of smallpox and so it has remained.

A child with smallpox, showing the pustules characteristic of the disease. This photograph was taken in a relief camp in Bangladesh in 1973.

In that same year, WHO recommended that smallpox research laboratories be closed and their supply of virus destroyed. One such laboratory was at the University of Birmingham in England. On July 25, 1978, during the last six months of the laboratory's existence, a woman medical photographer working on the floor below contracted smallpox; the virus, as it turned out, had made its way out of the laboratory through a duct system and the photographer was unvaccinated. The disease was diagnosed by the head of the laboratory, a prominent virologist, who was responsible for the woefully inadequate safety precautions. He committed suicide by cutting his throat. The photographer died five days later. Those appear to have been the last two deaths from smallpox on this planet.

---

current efforts to develop vaccines against malaria (page 473), schistosomiasis (page 537), and other parasitic diseases. The use of recombinant DNA technology to prepare vaccines has the advantage of eliminating any possibility of disease-causing infection or of introducing contaminants that might produce adverse side reactions.

The Action of Antibodies

Antibodies commonly act against invaders in one of three ways: (1) they may coat the foreign particles and cause them to clump together (Figure 39-7) in such a way that they can be taken up by phagocytic cells; (2) they may combine with them in such a way that they interfere with some vital activity—for example, by covering the protein coat of a virus at the site where the virus attaches to the host cell membrane, thereby preventing attachment; or (3) they may themselves, in combination with other blood components, known collectively as **complement,** actually lyse and destroy foreign cells.

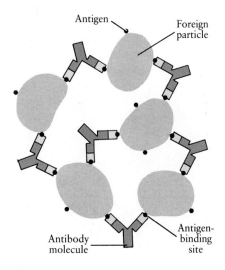

39–7 As we saw in Figure 18–18 (page 371), each antibody molecule has two antigen-binding sites. Thus one antibody can bind to antigens on two different cells or particles, causing them to stick together (agglutinate). Phagocytic white blood cells then consume these larger masses of foreign particles and antibodies.

Complement is a group of at least 11 different proteins found in blood. In particular combinations, these proteins function as lytic enzymes, acting at the point on the bacterial cell wall where antigen and antibody combine. By digesting holes in the foreign cells, complement causes them to burst. In addition to this lytic function, complement may itself coat the foreign cell and promote phagocytosis by other cells, and it mediates and enhances the inflammatory response.

The Structure of Antibodies

Determining the structure of antibodies was made possible by the peculiar properties of cancer cells. Virtually any type of cell in the body can turn cancerous. When it does, it produces a clone of rapidly multiplying malignant cells, all with similar properties. In the cancer known as multiple myeloma, the cancer cell is a plasma cell—in other words, a mature antibody-producing lymphocyte. As the cell multiplies, its descendants continue to turn out its one particular antibody, which, as the cells multiply, is produced in greater and greater quantities.

Using the antibodies produced by patients with multiple myeloma, Gerald Edelman and his colleagues at Rockefeller University were able to determine the amino acid sequence of a number of different antibody molecules. Each antibody, as we noted in Chapter 18 (page 371), is a complex protein consisting of four subunits, two identical light chains and two identical heavy chains (Figure 39–8). The light chains have about 214 amino acids each, and the heavy chains have about twice as many. Both the light chains and the heavy chains have constant (C) regions, in which the sequence of amino acids is identical from one molecule to the next. Each type of chain also has a variable (V) region, some 107 amino acids in

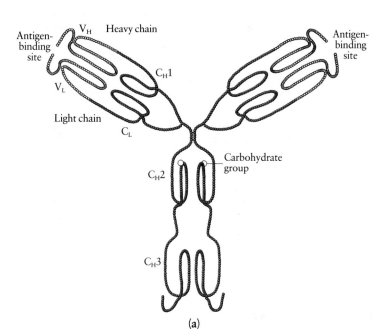

(a)

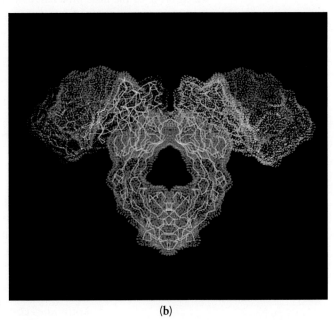

(b)

39–8 (a) A drawing of an antibody molecule, based on a bead model made by Gerald Edelman and his colleagues, that suggests how the four chains may be folded. Each bead represents an amino acid, of which there are more than 1,200. The variable regions are shown in yellow, and the constant regions in brown. The labeled portions of the molecule are independently folded segments, known as domains. The C_H2 domain is important in interactions with complement, whereas the C_H3 domain enables phagocytic cells to bind to the agglutinated mass of antibodies and foreign particles.

(b) A computer-generated model of an antibody molecule, showing the detailed structure of the surfaces of the heavy chains (blue dots) and light chains (green dots). The backbone of the constant regions is represented by the white and yellow lines, and the backbone of the variable regions by the red lines.

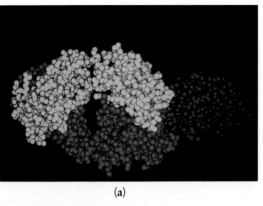
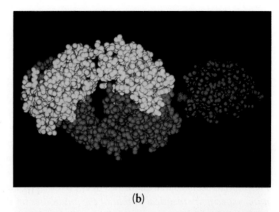
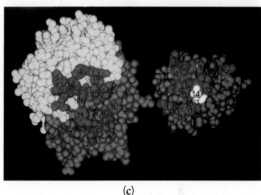

39-9 (a) *A space-filling model of the interaction of an antigen, the enzyme lysozyme, with the active site of an antibody specific for this protein. The heavy chain of the antibody active site is shown in blue, the light chain in yellow, and lysozyme in green. The red spheres represent a glutamine, located at position 121 in the lysozyme molecule, which plays a critical role in the precise binding of this particular antigen-antibody combination. In* (b), *the antigen-antibody complex has been pulled apart to reveal the surface contours of the two molecules. In* (c), *the separated molecules have been rotated to allow an end-on view of the combining sites. The glutamine at position 121 in lysozyme is shown in light purple, and the other amino acids that are in contact when the molecules are complexed are shown in red. The numbers on these amino acids are a computer code to their specific identity.*

length. Within this variable region, some of the sequences of amino acids are the same, but at approximately 40 positions the amino acids vary from one antibody to another. It is these variable amino acid sequences that, as the chains fold, come together to form the two active regions of the molecule, the sites that recognize and bind a specific antigen (Figure 39-9).

Biochemists have identified five distinct classes of immunoglobulins: IgG, IgA, IgD, IgM, and IgE. Structurally, these classes of antibodies are distinguished by the constant regions of their heavy chains; each class has a characteristic constant region, shared by all members of the class. There are functional differences as well. Antibodies of the IgG class, known as gamma globulin, are the principal type of circulating antibodies and are the type with which the great majority of the studies on antibody structure and function have been done. IgA antibodies are associated principally with mucosal immunity; they are found in such secretions as tears, saliva, milk, and the mucus on the interior lining of the digestive tract and the respiratory system. IgD molecules are found on the surface of B lymphocytes prior to activation and are the sites at which antigens initially bind to the cells. IgM molecules, which are also found on the surface of B lymphocytes, are the first to be secreted in the course of infection and last a relatively short time. IgE, according to present evidence, plays a major role in the expulsion of parasites, such as worms, from the intestinal tract. It also is involved in allergic reactions.

The Clonal Selection Theory of Antibody Formation

One of the most intriguing facts about the immune response is the great variety of antigens against which a single individual can produce antibodies. It is estimated that a mouse, for instance, can form antibodies against 10 million different antigens. Moreover, antibodies can be formed not only against the natural, common invaders that an individual organism might reasonably be expected to encounter in the course of its own life and those that its ancestors might have encountered, but also against synthetic antigens that are chemically unlike any substance found in nature.

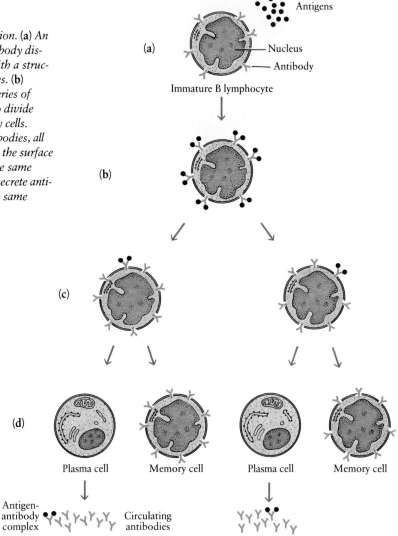

39–10 *The clonal selection model of antibody formation.* **(a)** *An immature B lymphocyte, with one specific type of antibody displayed on its surface, encounters antigen molecules with a structure complementary to the binding site of its antibodies.* **(b)** *Antigens bind to the antibodies, setting in motion a series of changes within the cell.* **(c)** *The B lymphocyte begins to divide and differentiate* **(d)**, *forming plasma cells and memory cells. Plasma cells secrete large quantities of circulating antibodies, all with a specificity identical to that of the antibodies on the surface of the original B lymphocyte. Memory cells bearing the same antibodies persist in the circulation indefinitely; they secrete antibodies only following a subsequent encounter with the same antigen.*

Our current understanding of antibody formation, first proposed in 1957 by the late Macfarlane Burnet of Australia, is known as the clonal selection theory. According to this model, each individual has a vast variety of different B lymphocytes, each genetically equipped with the capacity to synthesize only one type of antibody, which is displayed on its surface. Any given antigenic stimulus does not affect the great bulk of the B lymphocytes, but only those displaying an antibody that is able to bind that specific antigen. Thus, the antigen-antibody interaction "selects" particular lymphocytes. These cells then proliferate, producing clones of plasma cells and memory cells, all synthesizing the antibody displayed on the original B lymphocyte (Figure 39–10).

The clonal selection model predicted that (1) only a very small number of lymphocytes would respond to a given antigen, and (2) any one plasma cell would always form only one antibody. These predictions were initially confirmed by studies involving multiple myeloma cells. More dramatic confirmation has been provided by the development of monoclonal antibodies (see essay, page 802).

According to the clonal selection theory, antibodies are not "tailor-made" in response to an antigen. Rather, the antibodies displayed by the B lymphocytes are like the samples in the showroom of a huge "ready-to-wear" supplier; when a customer—an antigen—comes along that fits its antibodies, the manufacturer—the B lymphocyte—gears up its factory and begins mass production of that particular size and style.

The Genetics of Antibody Formation

In terms of genetics, the clonal selection theory would seem to predict that all the genes for making all the antibodies exist in each lymphocyte (indeed, if you follow the argument further, in every cell in the body), and that all but those coding for a single antibody are repressed in each lymphocyte's clone of plasma cells. Thus, for a number of years, the clonal selection model was in some difficulty because of simple mathematics. Each of us is apparently capable of making antibodies against 100 million different antigens. However, a human cell does not contain this many structural genes in its entire genome.

As we saw in Chapter 18, the seeming paradox was first resolved in work with mice, and further studies have indicated that the same principles apply in other mammals. In mice, the variable regions of both the light and heavy chains of antibodies are coded for by some 300 DNA sequences scattered through the genome, which are transposed and assembled into a variety of different arrangements as the B lymphocytes undergo their initial differentiation in the bone marrow. After transposition, the gene segments are ordered on the chromosome in the sequence of their appearance (Figure 18–19, page 372), separated by noncoding sequences (introns). The completed gene is then transcribed into RNA, and the introns are excised before the RNA is translated into the protein of the final antibody molecule. The number of possible combinations of these gene sequences has been calculated at 18 billion, enough to account for the enormous diversity of antibody molecules. Incredible as it may seem, this figure is actually low. Current evidence indicates that additional combinations are generated both by slight inaccuracies in the splicing and joining of the sequences that form the complete variable region genes and by somatic mutations that occur after the assembly of the genes.

39–11 *Four T lymphocytes attacking a much larger cancer cell. Note the numerous folds and microvilli that greatly increase the surface area of the T lymphocytes. According to present evidence, cell-to-cell interactions involving lymphocytes, macrophages, and other immunologically active cells depend upon recognition of the pattern of antigenic glycoproteins on the surface of the cell membrane.*

T LYMPHOCYTES AND CELL-MEDIATED IMMUNITY

Circulating antibodies were long believed to be the sole effectors of immunity. It is now known, however, that there is another category of highly specific immune response that is effected by cell-to-cell interactions involving the other class of lymphocytes, the T lymphocytes. This is sometimes known as the **cell-mediated response.**

Unlike the circulating antibodies produced by B lymphocytes, which are primarily active against viruses and bacteria and the toxins they may produce, T lymphocytes interact with other eukaryotic cells—specifically, the body's own cells. Functionally, three classes of T lymphocytes are known. The cells of two of these classes, the helper T cells and the suppressor T cells, are the principal regulators of the immune response, including the activities of the B lymphocytes. Members of the third class, the cytotoxic T cells, act against foreign eukaryotic cells and against cells of the body that are infected by viruses or other microorganisms. When a virus, for example, is multiplying within a cell, it is protected from the action of antibodies. However, its presence is reflected by the appearance of new antigens on the surface of the infected cell, which make it possible for cytotoxic T cells to find the infected cell and lyse it, exposing the viruses to antibody action.

The T Lymphocyte: A Life History

T lymphocytes, like red blood cells, granulocytes, monocytes, and B lymphocytes, are the offspring of self-regenerating stem cells in the marrow of long bones. Because they are indistinguishable from B lymphocytes under the microscope, their identification as a distinct population of cells with unique functions was long delayed. It came about as a result of research directed at uncovering the function

Monoclonal Antibodies

When a microorganism, such as a bacterial cell, invades the body fluids (blood and lymph) of a mammal, a number of different B lymphocytes are activated, giving rise to plasma cell clones, each producing a different antibody. As we have noted previously, the surface of a bacterium bears a variety of proteins and polysaccharides, each of which is potentially antigenic. In addition, a particular antigen may bind to the antibodies displayed by different B lymphocytes. This occurs because the structural variations between the binding sites of different antibodies are sometimes quite subtle. The strength of the binding varies according to the precision of the antigen-antibody fit, and this, in turn, influences the strength and effectiveness of the response by the B lymphocyte.

Until recently, this responsiveness of numerous B lymphocytes to a given antigenic challenge created considerable difficulties for scientists seeking to unravel the details of the immune response. With a great deal of work, it was possible to separate different clones of plasma cells and to obtain partially purified antibodies. The yields, however, were low, and the plasma cells would rapidly die out in tissue culture. Although these problems did not occur with myeloma cells, which would survive indefinitely in tissue culture, producing vast amounts of their specific antibody, there was no way to manipulate the cells to produce other antibodies of potentially greater interest.

The solution to these difficulties was provided by Cesar Milstein and Georges Köhler, working in the Laboratory of Molecular Biology in Cambridge, England. In 1975, they developed a technique for fusing mouse myeloma cells with activated B lymphocytes from mice that had been immunized with antigenic substances of the scientists' choosing. The myeloma cells used in the fusion experiments were mutants that lacked the capacity to synthesize their own antibodies; all of the antibodies produced by the hybrid cells were thus those of the activated B lymphocytes. Separation of the hybrid cells from unfused myeloma cells and B lymphocytes depended on two factors: (1) the mutant myeloma cells were missing certain enzymes necessary for growth and replication in the special tissue culture medium in which the fusion was carried out, and (2) the B lymphocytes, which on their own would quickly die out, contained the missing enzymes. Thus, only the hybrid cells survived, with the necessary enzymes supplied by the B lymphocyte partner and the capacity for longevity supplied by the myeloma partner. Screening procedures, utilizing the differences in the antibodies produced by different hybrid cells, made possible the separation of the hybrids from one another. Maintained in tissue culture, each cell subsequently gave rise to a clone of identical cells, all producing the same pure antibody. Hence the term, monoclonal antibodies.

The fusion technique and the screening procedures required to separate the hybrid cells have now become so refined that monoclonal antibodies can be prepared against virtually any antigen of interest. Moreover, the technique has been extended to human B lymphocytes and myeloma cells, making possible the preparation of human monoclonal antibodies. An enormous variety of monoclonal antibodies are available for research, medical diagnosis, and therapeutic trials against various diseases. Monoclonal antibodies have made possible more detailed study of antibody structure and function, including the models shown in Figures 39-8b and 39-9; purification of naturally occurring substances present in very small quantities, such as the interferons; and elucidation of the detailed structure and function of the T lymphocytes, the other major actors in the immune response. They are used in pregnancy tests and to diagnose perplexing infectious diseases and certain types of cancer. In addition, monoclonal antibodies hold particular promise in the treatment of cancer. Malignant cells bear unusual antigens on their surface, unlike those found on normal cells; there is hope that by coupling chemotherapeutic agents with monoclonal antibodies to those antigens, it will be possible to target toxic chemicals directly to the cancer cells, while sparing the normal cells of the body.

All things considered, the technique for producing monoclonal antibodies has been one of the most significant developments in the history of immunology—an accomplishment for which Milstein and Köhler were awarded the Nobel Prize in 1984.

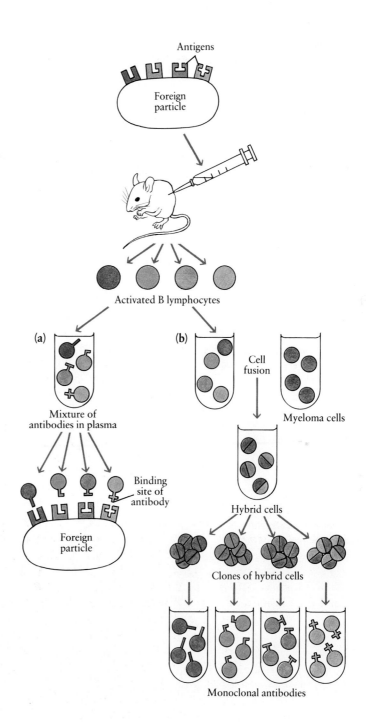

The preparation of monoclonal antibodies. When the immune system is challenged by foreign antigens, B lymphocytes bearing antibodies complementary to those antigens are activated and begin producing their specific antibodies. The result is (a) a mixture of circulating antibodies in the blood plasma. (b) Fusion of the antibody-producing lymphocytes (generally harvested from the spleen) with mutant myeloma cells produces hybrid cells, each of which synthesizes the antibody characteristic of its lymphocyte component. The hybrids can be separated from one another and maintained indefinitely in tissue culture. Each hybrid gives rise to a clone of cells, all producing the same specific antibody.

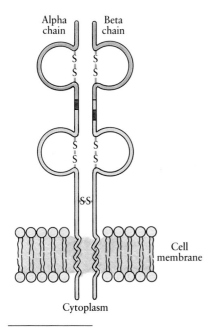

39-12 *A schematic representation of the structure of the T-cell receptor. Its polypeptide chains are, like those of an antibody molecule, coded by genes that are assembled from different components. Different portions of the chains are coded by variable genes (blue), diversity genes (yellow), joining genes (red), and constant genes (light green). The alpha chain of the resulting protein is characterized by acidic amino acids and the beta chain by basic amino acids (see Figure 3–18, page 73). The helical portions of the constant regions that anchor the receptor in the cell membrane are rich in hydrophobic amino acids. The binding site for antigens is a complex three-dimensional structure formed by the variable regions of the two chains.*

of the thymus gland. In humans, the thymus is located in the chest just behind the breastbone. The gland is large in infants and atrophies after puberty. In adults, surgical removal of the thymus has virtually no physiological effect. The first clue to its function came when a British investigator succeeded in removing the thymus gland in newborn mice—quite a feat when you consider that these infants are only 2 centimeters long and weigh about a gram. The animals survived, but they were sickly and runty due, further study showed, to their decreased resistance to infections. Following this lead, it was found that as early as the eleventh day of fetal life in the mouse (and the eighth week in the human) primitive blood cells creep into the embryonic thymus gland. These are the future T lymphocytes.

Within the thymus gland, the T-lymphocyte precursors go through a complex process of differentiation, selection, and maturation. Differentiation involves the synthesis of at least three different types of membrane glycoproteins, ultimately displayed on the surface of the mature T lymphocyte, that determine both its function and its antigenic specificity. The first type of membrane glycoprotein exists in one of two forms, known as T4 and T8,* and is correlated with function. Helper T cells bear the T4 molecule on their surface, whereas cytotoxic and suppressor T cells are characterized by the T8 molecule. The number of T4 cells formed is normally double or triple the number of T8 cells.

The second type of membrane glycoprotein is the receptor by which the T cell recognizes both the eukaryotic cells of the body itself and foreign antigens displayed on those cells. Although the T-cell receptor is not an antibody and is not secreted by the cell, there are important similarities. The receptor consists of two different protein chains, known as the alpha and beta chains (Figure 39–12). During the course of differentiation of each T lymphocyte, genes coding for these two chains are assembled by the selection and splicing together of variable, diversity, joining, and constant components—a process apparently identical to that occurring in differentiating B lymphocytes. The set of genes from which T-cell receptor genes are assembled is, however, distinct from the set from which antibody genes are assembled, and so are the resulting proteins.

Associated with the T-cell receptors on the surface of the differentiated cell are a set of five other proteins, known collectively as T3 (or as CD3). The structure and function of these molecules are still poorly understood, but they are thought to mediate the process of T-cell activation.

Differentiation of T lymphocytes is followed by a process of selection. As we have noted, the T-cell receptor recognizes foreign antigens displayed on the surface of the body's own cells. The process of gene rearrangement giving rise to the receptor is, however, random. As a consequence, some of the resulting T-cell receptors are incapable of recognizing "self"; others, by contrast, recognize it a little too well and thus have the potential to destroy healthy cells, wreaking havoc on the organism. Within the thymus, differentiated T lymphocytes with either of these characteristics are eliminated.

T lymphocytes that survive the selection process complete their maturation within the thymus and then are released to begin their duties throughout the body. Destiny for a T lymphocyte is like that for a B lymphocyte: the recognition and binding of antigens with a three-dimensional configuration that is complementary to the configuration of the T-cell receptors displayed on its surface. The result is also the same: cell division and differentiation, producing clones of active cells and memory cells. The functions of the active cells, however, are quite different from those of the plasma cells resulting from B lymphocyte activation. Before we consider these functions, we must pause to consider the mechanism by which T lymphocytes distinguish self from "not-self."

---

* In a more detailed system of nomenclature used to identify the membrane glycoproteins that are expressed during various stages of T-lymphocyte differentiation, these two types of molecules are known as CD4 and CD8; in this system, CD stands for "cluster of differentiation."

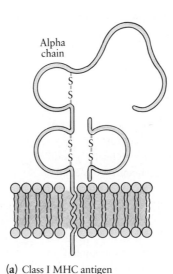
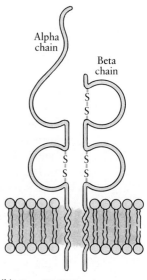

(a) Class I MHC antigen **(b)** Class II MHC antigen

39-13 Schematic representations of the structures of the antigens coded by genes of the major histocompatibility complex. (a) A Class I antigen consists of an alpha chain, coded by an MHC gene, combined with another protein, shown here in blue. This protein is coded by a gene that is not part of the major histocompatibility complex. (b) A Class II antigen consists of an alpha chain and a beta chain, both coded by MHC genes. Recent studies suggest that the three-dimensional structures formed by the outermost loops of these protein chains function as receptors to which foreign antigens bind. It is the combination of foreign antigen bound to MHC antigen that the T-cell receptor recognizes.

The Major Histocompatibility Complex

Studies of T lymphocyte functions and of tissue transplant rejections have revealed that recognition of the body's own cells depends on a group of glycoprotein antigens found on the surface of nucleated cells. The protein components of these antigens are coded by a group of genes known as the **major histocompatibility complex,** or MHC. *(Histo* is a Greek word signifying "tissue." The complex was first studied in the context of a search for compatible tissue grafts—that is, grafts that would not be rejected by the host.)

The major histocompatibility complex consists of at least 20 different genes, and within the human population, each of these genes has as many as 8 to 10 alleles. The total number of different combinations is astronomical, and it is predicted that no two persons—except identical twins—will ever be found to have the same major histocompatibility complex. Each is as individual as a fingerprint. Thus, these antigens provide for extremely accurate recognition of self. They also provide for an extremely accurate determination of family relationships (see essay, page 806).

Among the cells displaying the antigens coded by the specific MHC genes that an individual has inherited are the cells of the thymus. During the selection process that occurs within the thymus, developing T lymphocytes are exposed to the MHC antigens displayed by the cells of the thymus. Those T lymphocytes whose receptors bind optimally to the displayed MHC antigens—tight enough, but not too tight—are the ones selected to complete their development and maturation.

Current evidence indicates that there are two classes of MHC antigens, known simply as Class I and Class II (Figure 39-13). These molecules differ in both structure and function. Class I molecules are found on cells throughout the body and are necessary for recognition by cytotoxic T cells. Class II molecules are present only on cells of the immune system and identify such cells to each other. Recently, the detailed three-dimensional structure of a Class I molecule has been revealed in an elegant series of x-ray crystallography studies carried out at Harvard University. The most interesting feature of the structure is a groove in the exterior surface of the molecule (Figure 39-14), just the right size to hold a peptide consisting of 12 to 20 amino acids. Although the detailed three-dimensional structure of a Class II molecule has not yet been determined, it is expected to be similar. With this information in mind, let us turn our attention to the activities of the T lymphocytes.

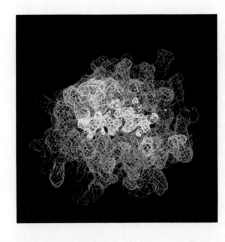

39-14 A computer-generated representation of the surface of a Class I MHC antigen, based on x-ray crystallography studies. Note the deep groove that is thought to be the site at which antigenic peptides are bound.

Children of the *Desaparecidos:* An Application of MHC-Antigen Testing

Because we are diploid organisms, we each inherit two copies of the loci coding for our Class I MHC antigens and two copies of the loci for our Class II MHC antigens. Although it is theoretically possible that a given individual could be homozygous at all of the loci coding for either class of antigen, it is extremely improbable. As we have noted, at least 20 different genes are involved in the complete code for the two MHC antigens, and each of these genes has numerous alleles within the human population. Consequently, virtually all of us are heterozygous for our MHC antigens. Both of the completed genes for each class of MHC antigen are expressed, and thus each of us has two different Class I MHC antigens and two different Class II antigens, one of each class inherited from each parent. This fact, coupled with the enormous diversity of MHC antigens within the population as a whole, makes MHC antigens an extraordinarily valuable tool in determining biological relationships.

One of the most poignant applications of MHC testing has occurred in Argentina. From 1975 to 1983, Argentina was governed by a military junta that, in the name of national security, was responsible for the disappearance of more than 9,000 citizens, most of whom were ultimately murdered. Among the *desaparecidos* ("disappeared ones") were at least 195 children, some of whom were abducted with their parents or were abducted independently. Most of the children, however, were born to women who were pregnant at the time of their abduction and who were kept alive until after they gave birth. In the years since the return of democratic government to the country, an Argentinian human rights organization, the Grandmothers of the Plaza de Mayo, has located dozens of children thought to be *desaparecidos.* In most cases, these children have been living in the families of former members of the military and police units that carried out the abductions. In a few cases, a child has been in the custody of a passerby to whom the mother had frantically handed her baby as she was being abducted. The tool used to determine the biological identity of the children—and thus to facilitate their return to the care of their grandparents—is MHC-antigen testing.

When a child is located who is thought to be a *desaparecido,* a blood sample is obtained (under court order) and the particular MHC-antigen combinations he or she carries is determined. This is known as the haplotype, and it consists of two sets of numbers, representing the two inherited sets of antigens. Haplotypes are similarly obtained from the persons thought to be the child's biological grandparents. The chart below shows the result of the 1984 haplotype determinations of the surviving parents of Claudio Ernesto Logares and Monica Grinspon, *desaparecidos* who were abducted in May of 1978, and of the child thought to be their daughter, Paula Eva, who was two years old at the time of the abduction. Although Monica Grinspon's father was deceased, it was possible to reconstruct his haplotype from that of her siblings. The policeman and his wife with whom the child was living refused to be tested, as was their legal right. Even without this information, however, the haplotypes indicated a probability of more than 99.9 percent that the child was indeed Paula Eva Logares.

The Functions of the T Lymphocytes

Functionally, the simplest of the T lymphocytes are the cytotoxic T cells. As we have noted previously, infection of a eukaryotic cell by a foreign microorganism, such as a virus, results in the appearance of new antigens on the surface of that cell. These antigens, which are short peptides derived from the antigens originally on the surface of the infectious particle, are thought to be bound to and displayed in the surface groove of the individual's Class I MHC antigens. When a cytotoxic

The case was clinched when she was returned, by court order, to the home of her grandparents. The eight-year-old immediately went to her former room and asked for a doll with which she had last played at the age of two.

Some 39 children, orphaned during the years of the junta, have already been restored to their grandparents. Argentinian scientists and the Grandmothers of the Plaza de Mayo are now establishing a data bank containing both the haplotypes of persons whose grandchildren—or pregnant daughters—disappeared during the years of the junta and the haplotypes of children who are thought to be survivors. It is hoped that this data bank will make it possible to reunite many other families.

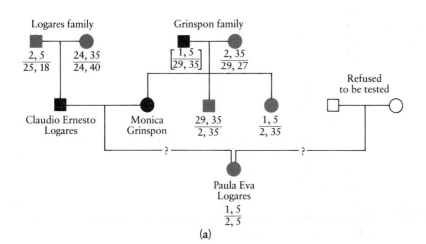

(a) *The haplotype determinations that indicated a 99.9 percent probability that the child being raised by an Argentinian policeman and his wife as their own daughter was actually Paula Eva Logares. The numbers above the lines represent one pair of Class I and Class II antigens, and the numbers below the line represent the other pair. Squares indicate males and circles females; individuals who were tested are shown in red, and those who were deceased in black.* (b) *A photograph of Liliana Pereyra, taken when she was 21. Abducted in 1977, when she was five months' pregnant, she was kept alive until February 1978, when she gave birth to a son. Forensic studies of her skeleton, found in 1985, indicate that she was killed by a shotgun blast to the head at close range. Her son has yet to be located.*

T cell encounters such a combination of Class I MHC antigen and foreign antigen to which its receptor can bind, it differentiates into active cells that attack and lyse the infected cells (Figure 39-15) and into memory cells that remain in the circulation indefinitely. In addition, the activated cytotoxic T cells release powerful chemicals, known as lymphokines, that attract macrophages and stimulate phagocytosis. Some of the cytotoxic T cells, known as "killer cells," secrete cytotoxins that destroy target cells directly; others secrete interferon. Thus, a whole battery of defenses is mobilized by activation of cytotoxic T cells.

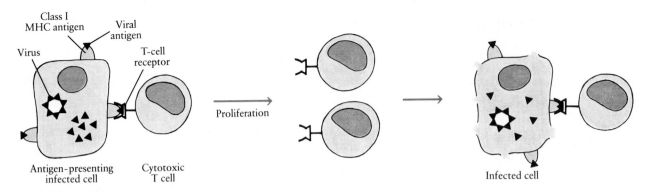

39-15 *When a virus invades a eukaryotic cell, its protein coat either remains on the surface of the cell or, as shown here, breaks apart within the cytoplasm. This is, of course, an essential step in the life of the virus, releasing its nucleic acid and enabling it to begin replication. As a consequence, however, telltale markers—viral antigens—appear on the surface of the infected cell and are displayed in conjunction with Class I MHC antigens. Cytotoxic T cells whose receptors are complementary to the specific antigenic combination that results bind to the cell and are thereby activated. This activation produces a clone of identical cytotoxic T cells, which then attack and destroy other infected cells.*

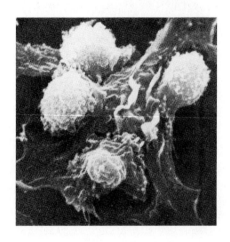

39-16 *This scanning electron micrograph shows four helper T cells attached to the surface of a macrophage (the large, flat cell). In the course of its duties in the inflammatory response, the macrophage has ingested cells of the bacterium* Listeria monocytogenes. *On its surface are antigens from the* Listeria *cells, displayed with Class II MHC antigens, a combination for which these particular helper T cells are specific. Thus macrophages constitute a key link between the inflammatory and immune responses. Macrophages also present foreign antigens to the membrane-bound antibodies of immature B lymphocytes, thereby participating in the first step in the production of circulating antibodies.*

The activities of the cytotoxic T cells, as well as those of the B lymphocytes, are regulated by the helper and suppressor T cells. Helper T cells are characterized by receptors that recognize and bind foreign antigens displayed in conjunction with the individual's Class II MHC antigens. Such combinations are found on the surface of both macrophages that have ingested foreign microorganisms (Figure 39-16) and activated B lymphocytes. A helper T cell encountering either type of cell bearing an antigenic combination to which it can bind becomes activated and begins producing proteins known as interleukins. These proteins act as hormones, stimulating the differentiation and proliferation of both B lymphocytes and cytotoxic T cells following activation. The actual binding of helper T cells to antigen-displaying B lymphocytes is also an essential step in the production of plasma cells and memory cells (Figure 39-17).

When an infection has been successfully eliminated, further activity of both B cells and T cells is suppressed. The mechanism of this suppression remains poorly understood. Moreover, although suppressor T cells are generally described as the "off" switch of the immune response, there are no reliable criteria by which to distinguish such cells from other T cells bearing the T8 glycoprotein. It is not yet known whether suppressor cells exist as a distinct class of T cells, or whether suppression is actually mediated by cytotoxic T cells and, perhaps, helper T cells.

Studies of T lymphocyte function have provided what is perhaps only the first glimpse of a vast cellular communication network, probably involving far more than the immune response. Various substances released by activated cells of the immune system appear to influence other cells of the body, particularly those of the endocrine and nervous systems. Conversely, the immune system seems to be influenced by the molecules responsible for cell-cell communication within the endocrine and nervous systems. The study of these complex interactions is one of the most intriguing areas of modern medical research, not only because of the basic biological knowledge that is being uncovered but also because of its potential bearing on many medical problems, a few of which we shall consider in the remainder of this chapter.

CANCER AND THE IMMUNE RESPONSE

Cancer cells resemble an individual's normal cells in many ways. Yet, within the body, they act like foreign organisms, invading and "choking off" or competing with normal tissues. Moreover, virtually all cancer cells have antigens on their cell surfaces that differ from the antigens of the normal cells of the individual and can be recognized as foreign. Does this mean that the body can mount an immune

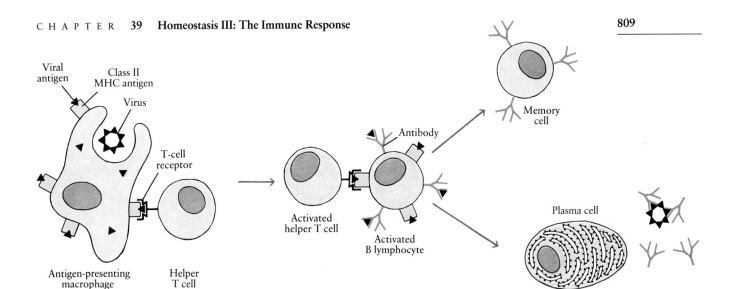

39-17 *A schematic representation of the interaction of macrophages, helper T cells, and B lymphocytes. Phagocytosis of a foreign microorganism or virus by a macrophage results in the display of foreign antigens on the surface of the macrophage. Helper T cells with a receptor that matches the combination of these specific antigens with Class II MHC antigens bind to the surface of the macrophage and are activated. Following activation, they bind to the same combination displayed on the surface of B lymphocytes that bear antibodies specific to the same foreign antigen and that have previously encountered the antigen. Binding of activated helper T cells to the activated B lymphocytes is an essential step in the differentiation and proliferation of memory cells and plasma cells and the subsequent production of large quantities of circulating antibodies. The agglutinated masses of foreign particles and antibodies that result from a successful attack on the pathogen are cleared from the system by phagocytizing macrophages.*

39-18 *The successful destruction of a cancer cell by cytotoxic T cells.* **(a)** *Recognition of the cancer cell as abnormal depends on the presence of unusual antigens displayed on its surface in conjunction with Class I MHC antigens.* **(b)** *As the cytotoxic T cells begin lysing the membrane of the cancer cell and consuming its contents, their shape changes.* **(c)** *When their work is completed, all that remains of the cancer cell is its cytoskeleton—a mass of lifeless protein fibers.*

response against its own cancers? A growing body of evidence suggests not only that cancer can induce an immune response (Figure 39–18) but also that it usually does so. In fact, according to this hypothesis, it usually does so successfully, overwhelming the cancer before it is ever detected; the cancers that are discovered represent occasional failures of the immune system. This conclusion suggests that bolstering the patient's immune response may provide a means for cancer prevention or control. It also suggests that the cell-mediated immune response may have had its evolutionary origins as a defense not only against parasitic invaders but also against the treacherous, malignant cells of the body itself.

(a)

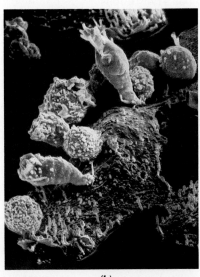

(b)

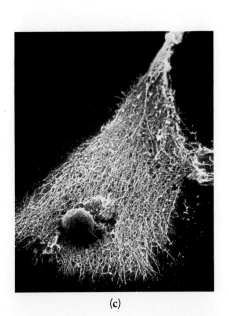

(c)

TISSUE TRANSPLANTS

Organ Transplants

People with extensive burns die from infection and because of the loss of body fluids from exposed areas. If skin is taken from one part of the patient's body and grafted to the burned area, the new tissue anneals to the exposed area, vascularization (the invasion of the transplanted tissue by blood vessels) takes place, and the tissue grows and spreads out. If a skin graft is taken from another individual (except the patient's identical twin), the initial stages of healing and vascularization take place, but then, on about the fifth to the seventh day, large numbers of white blood cells infiltrate the transplanted tissue, and by the tenth to twelfth day it has died. The infiltrating cells are mainly T lymphocytes and macrophages, and current evidence indicates that cytotoxic T cells and "killer cells" are responsible for transplant rejection.

The discovery and identification of the MHC antigens—a result of the search for compatible tissue grafts—is now making possible closer matches between donor and recipient in organ transplants. In the effort to further reduce rejection, transplant recipients are also generally given drugs to suppress the immune response. However, given that infection is a major complication among patients requiring skin grafts and is the leading cause of death among kidney transplant recipients, general suppression of the immune response is obviously not an ideal solution. Cyclosporin, a relatively new drug isolated from a soil fungus, acts selectively against the T cells involved in transplant rejection. Although the mechanism of its action is still poorly understood, cyclosporin has dramatically improved the success of organ transplants. Other cells of the immune system seem to be unaffected by this drug, with the result that the patient remains protected against infection.

Another approach to improving the success of organ transplants is to alter the transplanted tissue in some way to make it more acceptable to the recipient's natural defenses. The likelihood of rejection of a transplanted organ decreases greatly after several months have passed. Moreover, a second skin graft from the same donor is not more readily tolerated than the first, indicating that the ultimate acceptance of the transplant involves a change in the graft rather than in the immune response of the recipient. Identification of the exact nature of these changes may, in time, reduce the need for immunosuppressive drugs.

Blood Transfusions

The most frequently performed tissue transplants in modern medical practice are transfusions of blood. Blood transfusions are so routine that it is difficult to realize that in the past they often provoked severe—and sometimes fatal—immune responses.

At about the turn of the century, Karl Landsteiner set out to discover why blood transfusions between humans were sometimes safe and effective but, more often, led to serious complications. Mixing samples of blood taken from members of his laboratory staff, Landsteiner found that sometimes the red blood cells would agglutinate, and sometimes they would not. From these experiments, he came to realize that there were different categories of blood and that agglutination was caused by mixing blood of different categories. Soon after, the four major blood groups were determined: A, B, AB, and O.

Red blood cells, unlike nucleated cells, do not have MHC antigens on their surface. Instead, they display unique antigens, coded by an entirely different gene, which, in the human population, has three alleles (A, B, and O). The principal blood groups are defined by these antigens and by the presence of antibodies within the plasma (Table 39–2). The A and B alleles are codominant, whereas the

TABLE 39-2	Blood Groups				
GROUP (ANTIGENS PRESENT)	GENOTYPE (ALLELES PRESENT)	REACTION WITH ANTIBODIES		ANTIBODIES IN BLOOD PLASMA
		ANTIBODY A	ANTIBODY B	
O	O/O	No	No	Antibody A, antibody B
A	A/A, O/A	Yes	No	Antibody B
B	B/B, O/B	No	Yes	Antibody A
AB	A/B	Yes	Yes	None

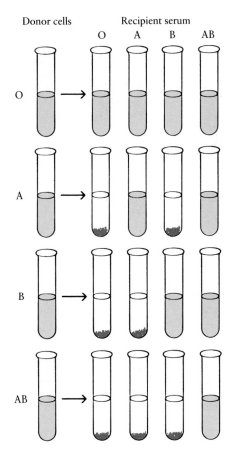

39-19 *Severe and sometimes fatal reactions can occur following transfusions of blood of a different type from the recipient's. These reactions are the result of agglutination of the red blood cells caused by antibodies present in the recipient's blood. Antibodies in the donor's blood are generally of little consequence because they are so diluted in the recipient's blood. Blood-group reactions to transfusions can be demonstrated equally well in test tubes, as shown here. The blood serum (plasma from which the proteins involved in clotting have been removed) in which agglutination occurs has natural antibodies against the donor blood. Persons with type O blood, whose red blood cells have neither A nor B antigens, used to be called universal donors; similarly, those with AB blood (neither A nor B antibodies in the plasma) used to be called universal recipients. Now other factors are checked as well.*

O allele is recessive. Thus, individuals with type A blood have either two A alleles or one A and one O, and their red blood cells bear the A antigen. Their plasma does not contain antibodies against the A antigen (self) but does carry antibodies against the B antigen ("not-self"). Individuals with type B blood have the B antigen and antibodies against the A antigen. Individuals with type AB blood have both antigens but neither A nor B antibodies; conversely, type O individuals have neither antigen but both A and B antibodies. For reasons that are still not clear, circulating antibodies against the "not-self" antigens are always present—even though an individual has never been exposed to those antigens.

For blood transfusions to be performed safely, the blood types must be matched. If a person receives a transfusion containing red blood cells bearing a foreign antigen, the antibodies in his or her plasma will react with those cells, causing them to agglutinate (Figure 39-19). The resulting clumps of blood cells and antibodies can clog capillaries, blocking the vital flow of blood through the body.

Because blood types are inherited and are also easy to determine, they have been used to settle questions of paternity in legal actions. As you can see by studying Table 39-2, however, they can only demonstrate that someone is not the father of a particular child, not that he is. Evidence of a positive nature indicating the probability of a particular relationship can now be provided by MHC-antigen testing (see page 806).

The Rh Factor

Since Landsteiner's initial discoveries, additional antigens have been identified on the surface of red blood cells. Modern blood typing takes these into account also, thus minimizing the possibility of immune responses that could complicate recovery from the illness or injury that necessitates the blood transfusion. Among the most important of these antigens is the Rh factor, named after the rhesus monkeys in which the research leading to its discovery was carried out.

A once common medical problem caused by the Rh factor is hemolytic anemia of the newborn. During the last month before birth, the human fetus usually acquires antibodies from its mother. Most of these antibodies are beneficial. An important exception, however, is found in the antibodies formed against the Rh factor. Like the other surface antigens of red blood cells, the Rh factor is genetically determined. If a woman who lacks the Rh factor (that is, an Rh-negative woman) has children fathered by a man homozygous for the Rh factor, all of the children will be Rh positive; if he is a heterozygote, about half of the children will be Rh positive.

During the birth of an Rh-negative mother's first Rh-positive child, fetal red blood cells bearing the Rh antigen are likely to enter her bloodstream. The

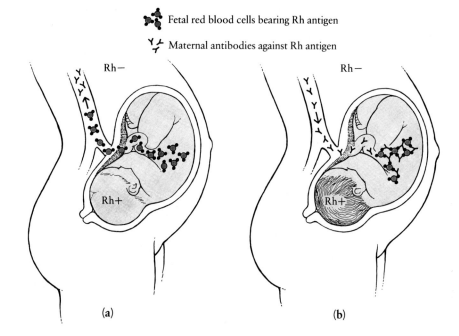

39-20 *The events leading to Rh disease. (a) Late in the first pregnancy—or during the birth of the baby—fetal blood cells spill across the barrier that separates the maternal and fetal circulations. Antigens on the red blood cells of the Rh-positive fetus stimulate the production of antibodies by the immune system of the Rh-negative mother. These antibodies remain in the mother's bloodstream indefinitely. (b) Late in the second pregnancy, the antibodies pass through the barrier from the maternal blood to the blood of the fetus. If the fetus is Rh positive, the antibodies react with the antigens on its red blood cells, destroying them.*

consequences are the same as would occur with a transfusion of Rh-positive blood: the mother's immune system produces antibodies against the foreign antigens, and these antibodies persist in her blood. In subsequent pregnancies, they may be transferred to the fetus. If that fetus is Rh positive, the antibodies will react with its red blood cells, destroying them (Figure 39–20). This reaction can be fatal either before or just after birth.

Now that its causes are recognized, Rh disease can be prevented by injecting the Rh-negative mother, within 72 hours of her delivery, with antibodies against the fetal Rh-positive red blood cells in her system. This destroys the cells and prevents them from triggering antibody production.

DISORDERS OF THE IMMUNE SYSTEM

Autoimmune Diseases

The immune response is a powerful bulwark against disease, but it sometimes goes awry. Usually the immune system can distinguish between self and "not-self." Substances that are present during embryonic life, when the immune system is developing, will not be antigenic in later life. This recognition occasionally breaks down, however, and the immune system attacks cells of the body. Certain disorders, among them myasthenia gravis, lupus erythematosus, and several types of anemia, have been identified as autoimmune diseases—that is, diseases in which an individual makes antibodies against his or her own cells. There is growing evidence that other disorders, such as multiple sclerosis, juvenile-onset diabetes, and some forms of rheumatoid arthritis, may have the same basis.

Allergies

Hay fever and other allergies are the result of immune responses to pollen, dust, or other substances, such as some foods, that are weak antigens to which most people do not react. When certain individuals are exposed to particular environmental antigens, the production of IgE antibodies by specific plasma cells is stimulated, as is the formation of memory cells. Upon reexposure to the same antigen, more IgE antibodies are formed. These antibodies circulate and attach themselves to the mast cells and other noncirculating basophils found in connec-

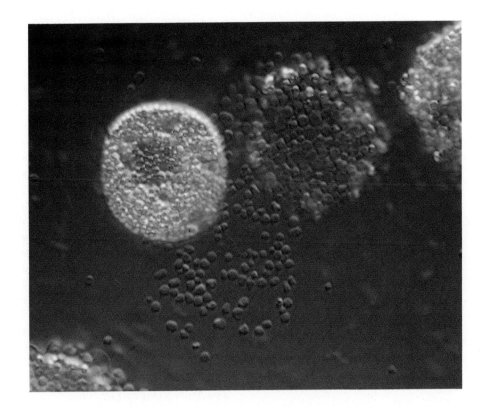

39-21 *Two mast cells, bearing receptors to which IgE antibodies to a specific antigen of birch pollen are bound. The cell on the left is "loaded" with granular inclusions, filled with histamine. The cell on the right, in response to the binding of the specific birch pollen antigen to the antibodies, is releasing its histamine-containing granules.*

tive tissue (page 792). Subsequent binding of the antigen to these attached antibodies triggers the release of histamine from the cells (Figure 39-21), which, in turn, induces an inflammatory response. This reaction typically occurs on an epithelial cell surface, producing increased mucus secretion, as in hay fever; hives or dermatitis; or cramps and diarrhea, as in the case of food allergies. Systemic reactions may result if the mast cells and other basophils release their chemicals into the circulation, causing dilation of the blood vessels, leading to a potentially dangerous fall in blood pressure, and constriction of the bronchioles (a syndrome known as anaphylactic shock). The evolutionary background of this decidedly maladaptive response—in which the response is far more dangerous than the antigen—remains a mystery.

Antihistamines, which counteract the histamine effect, suppress some of the symptoms of an allergic reaction. Most decongestants, on the other hand, promote the release of histamines, raising some questions about taking "allergy pills" that offer a combination of antihistamines and decongestants. Steroid hormones related to cortisone are used in more severe cases. These drugs act by suppressing the production of white blood cells and so of inflammation and the immune response in general.

Acquired Immune Deficiency Syndrome (AIDS)

Until fairly recently, epidemics of fatal infectious disease were a fact of life. For example, in 1918, at the close of World War I, a particularly deadly epidemic of influenza swept around the globe, leaving few families untouched. In the late 1940s and early 1950s, polio epidemics were a regular feature of summertime in the United States; most people who were alive at the time can remember more than one of their friends stricken by the disease. However, with the widespread introduction of antibiotics following World War II and, particularly, the development of polio vaccines in the mid-1950s, we entered an era in which it seemed that the conquest of infectious disease was virtually complete. That this is not so—and perhaps never can be so—has been powerfully demonstrated by the appearance of a virus that attacks the human immune system, leaving its victims susceptible to numerous types of opportunistic infections. The disease it causes is known as acquired immune deficiency syndrome (AIDS), and, thus far, it is invariably fatal.

AIDS was first identified in 1981, when epidemiologists noticed unusual clusters of a malignancy known as Kaposi's sarcoma, which affects the endothelial linings of the blood vessels. Previously, this rare cancer had been seen only in elderly men; its new victims, however, were young men. At about the same time—and also in young men—physicians began to observe an increasing incidence of fatal pneumonias and intestinal tract infections caused by ubiquitous but usually innocuous protists. In the past, such infections had been observed only in cancer patients and transplant recipients whose immune systems had been suppressed in the course of their treatment.

These facts suggested that the underlying cause of the new illnesses was a massive suppression of the immune system. Because most of the early victims were homosexual men, intravenous drug users, recipients of blood transfusions, or hemophiliacs who had received clotting factors prepared from donated blood, it appeared that the agent of this suppression was a microorganism transmitted sexually or through the exchange of blood. Within a remarkably short time—three years—the virus responsible was isolated and characterized. Since then, its principal effects on the immune system have been determined, its possible origins identified, and its modes of transmission traced.

The AIDS Virus and Its Effects

The AIDS virus is a retrovirus (page 342), known formally as human immunodeficiency virus (HIV). It is a particularly complex virus, as shown in Figure 39–22.

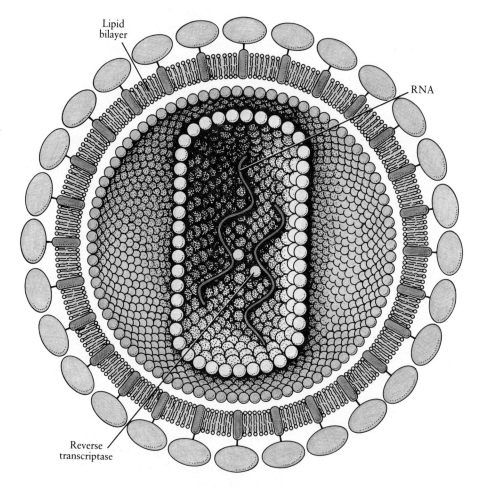

39-22 *The structure of the AIDS virus. The inner core consists of two molecules of RNA, accompanied by two or more molecules of the enzyme reverse transcriptase. Surrounding the core are envelopes formed of two distinct proteins. These are surrounded, in turn, by a lipid bilayer derived from the cell membrane of the host cell in which the virus previously replicated. Spanning this membrane are protein molecules to which are attached glycoprotein molecules that extend out from the surface.*

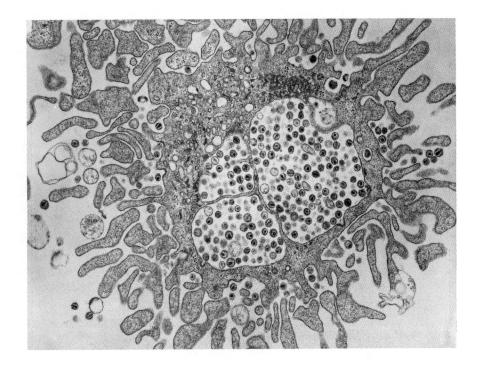

39-23 *An electron micrograph of a degenerating helper T cell from which newly replicated AIDS virus particles are being released. The T cell is the irregular structure with the granular appearance, and the virus particles are the small disks containing a dark core. Numerous virus particles are visible in the three large vacuoles in the center of the cell. AIDS virus particles bud from the host cell in a process essentially identical to exocytosis, taking with them portions of the host cell membrane. Because large numbers of particles bud off at the same time, holes are torn in the membrane, allowing the contents of the cell to leak out. As you would expect, the result is the death of the cell.*

The inner core consists of two molecules of RNA and several copies of the enzyme reverse transcriptase. Surrounding the core are two distinct protein envelopes, which are, in turn, surrounded by a lipid bilayer studded with glycoproteins. The protein portions of the surface molecules contain both constant regions, identical from one strain of the virus to another, and variable regions, the genes for which have an unusually high rate of mutation. The unique—and deadly—feature of the outer glycoproteins is that they are a perfect three-dimensional match to the T4 molecules that characterize helper T cells. When the AIDS virus meets a helper T cell, the T4 molecule functions as its receptor, enabling the virus to enter the cell by receptor-mediated endocytosis (page 138). The virus is also able to infect macrophages, which are now thought to be an important reservoir for the virus within the body; the interactions between macrophages and helper T cells, described previously, may play a major role in the spread of the virus to helper T cells.

Once within a helper T cell, the RNA is released from its multilayered capsule, and reverse transcriptase catalyzes the transcription of complementary DNA and its incorporation into a chromosome of the host cell. In the host cell chromosome, this complementary DNA may lie latent as a provirus (page 373) for some time. Sooner or later, however, replication of new viral particles begins—at a rate much higher than that of other known viruses. In a relatively short time, enormous numbers of new viruses burst from the infected helper T cell, which is often destroyed in the process (Figure 39-23). These viruses invade other helper T cells, and the process is repeated. The victim is eventually left with few functional helper T cells, and, as we have seen, these cells are critical for the proliferation and activities of both B lymphocytes and cytotoxic T cells. The body cannot mount an effective immune response against cells harboring the AIDS virus, against the virus itself, against other invading microorganisms, or against any malignant cells that may be present or may develop. However, in the initial phases of the infection, before the population of helper T cells is severely depleted, B lymphocytes respond to the foreign antigens of the virus, producing circulating antibodies. Although these antibodies are apparently ineffectual in controlling the infection, they do remain in the bloodstream. Their detection is the basis of screening tests for the presence of the virus.

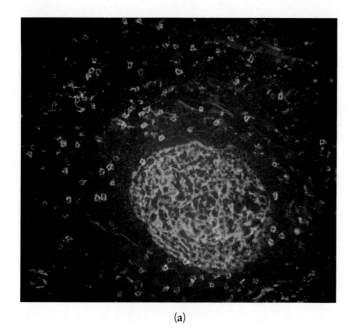
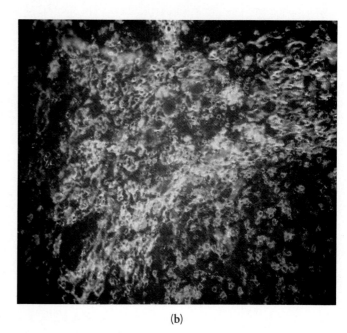

(a) (b)

39-24 *The effect of the AIDS virus on the cortical follicles of the lymph nodes.* (a) *An immunofluorescence micrograph of the germinal center of a normal cortical follicle; this is the region in which B lymphocytes proliferate following activation. The reticulum cells of the follicle, shown in green, form a regular network in which antigen-presenting cells are trapped and can be exposed to B lymphocytes. Cytotoxic and suppressor T cells, shown in orange, populate the region around the germinal center but are not found within the center.* (b) *The same region of a lymph node from a person with AIDS-related complex (ARC), a condition that often precedes full development of the disease. As you can see, the regular structure of the germinal center has been destroyed, and cytotoxic and suppressor T cells have invaded the entire region.*

The consequences of infection with the AIDS virus are many. One of the earliest signs of this infection, as of so many others, is a swelling of the lymph nodes, the principal sites of the interactions among macrophages, helper T cells, B lymphocytes, and other T lymphocytes that are so essential in the immune response. In persons infected by the AIDS virus, the structure of the cortical follicles of the lymph nodes (see Figure 39-5, page 794) is completely disrupted, as shown in Figure 39-24. Although the reasons for this change remain unknown, it appears to constitute an additional assault on the capacity to mount an immune response. As the immune system becomes steadily more crippled, the victim becomes increasingly vulnerable to other diseases. Among the most common are *Pneumocystis carinii* pneumonia, parasitic gastrointestinal infections accompanied by severe diarrhea, and Kaposi's sarcoma and other cancers. Weight loss is extreme. Ultimately, the person dies, most often as a result of the opportunistic diseases or of heart failure.

In many AIDS victims, the central nervous system is affected, usually in the latter stages of the disease. The result is atrophy of the brain, dementia, and, in some cases, symptoms that resemble those of multiple sclerosis. Current evidence indicates that the nervous system is attacked not only by microorganisms thriving in the absence of an immune response but also by the AIDS virus itself, most likely carried into the nervous system by macrophages.

The time span from diagnosis of full-blown AIDS to death varies from a few months to as long as five years. The time span from initial infection to the appearance of the first symptoms varies even more and may be associated with the absence or presence of other infections that trigger an immune response—an event that appears to activate latent AIDS proviruses. About 30 percent of infected individuals develop active disease within five years, and it appears probable that everyone infected with the virus will eventually become ill; cases have now been documented in which more than 10 years have elapsed between infection and illness. (Large numbers of homosexual men participated in the clinical trials of the vaccine for hepatitis B in the 1970s, and throughout the years of the trials, blood samples were taken at intervals and frozen for future study; testing these samples for antibodies to the AIDS virus has made it possible to determine when many AIDS victims first became infected.)

Transmission of the AIDS Virus

The AIDS virus is one of a family of primate retroviruses and appears to be most closely related to a relatively harmless virus that is endemic in the green monkeys of equatorial Africa. It has been hypothesized that this virus may have moved from the monkeys into the surrounding human population, and then undergone a series of mutations into its present deadly form. Tests of blood samples from individuals in equatorial Africa (saved for a variety of reasons) indicate that this may have occurred in the 1950s. According to this hypothesis, the virus gradually spread through the population, transmitted through sexual intercourse and, probably, through blood transfusions and other medical procedures in situations where sterilization was poor or nonexistent. Ultimately, it was acquired by visitors from Europe, North America, and the Caribbean and began its spread around the world.

Current evidence indicates that the virus can be transmitted through sexual intercourse, either vaginal or anal, through oral sex (perhaps including deep kissing), and through the exchange of blood. It is present at high levels in the semen and blood of infected individuals and can enter the body through any tear in the skin or mucous membranes—including those too small to be seen. Although the AIDS virus is one of the most virulent known, it is less readily transmitted than other viruses; without a surrounding environment of blood or semen or host cells, it quickly dies. There is no evidence that it can be transmitted through casual contact, hugs and light kisses, coughs or sneezes, on dishes used by an infected person, or on toilet seats. Thus far, no family member nursing an AIDS patient has become ill, and the few medical personnel who have tested positive for the virus were either accidentally pricked with a contaminated needle or exposed to large quantities of infected blood on portions of their body (for example, ungloved hands) where the skin was broken.

The initial spread of the virus in the West among homosexual men and intravenous drug users is thought to have occurred because these are relatively self-contained populations in which an individual might be repeatedly exposed. In equatorial Africa, AIDS is primarily a disease of heterosexuals, and its incidence among men and women is roughly equal. Continuing epidemiological studies suggest that members of the following groups in North America are presently at the greatest risk: (1) homosexual or bisexual men, (2) intravenous drug users, (3) sexual partners, whether male or female, of infected individuals, and (4) babies born to infected mothers. All available evidence indicates that once the virus has infected an individual, it remains for the rest of that person's life and can be transmitted in blood and semen—even if the person has no symptoms of illness.

The Prospects

As many as 2 million people in the United States alone are now thought to be infected by the AIDS virus. By November of 1988, 78,312 cases of AIDS had been documented in the United States, and 44,071 deaths had been attributed to the virus and its effects. It is estimated that by the end of 1991, an additional 270,000 cases will have been diagnosed, and another 178,760 deaths will have occurred. Although work is progressing on treatments and on the development of vaccines, any "quick fix" seems unlikely. Current treatments, which are directed at control of opportunistic infections and inhibition of reverse transcriptase (and thus of the replication of the virus), have serious side effects and are, at best, delaying actions.

The nucleotide sequences of the RNA of the AIDS virus have now been determined, and the codes for the various proteins of its coat have been identified. With this information, scientists are using recombinant DNA technology in the attempt to create synthetic vaccines against the virus. However, the fact that the

39-25 An African green monkey (Cercopithecus aethiops). These monkeys are host to a virus known as simian immunodeficiency virus (SIV). Although SIV shares important similarities with human immunodeficiency virus (HIV), the cause of AIDS, it seldom produces illness in its monkey hosts.

natural antibodies synthesized by the B lymphocytes are ineffectual suggests that development of a truly protective vaccine may be extraordinarily difficult. The task is also compounded by the high mutation rate of the genes coding for key parts of the protein coat.

Tests for the AIDS antibody, which came into use in 1985, have provided a means of identifying donated blood carrying the virus and now provide a high level of protection for the blood supply. As a consequence, the number of new cases of AIDS in transfusion recipients and hemophiliacs has dropped dramatically. The spread of the virus through the homosexual population has also slowed significantly, in part because the most susceptible individuals have already contracted the disease and, perhaps even more important, because of changes in sexual practices. Presently, the greatest increase in new cases is occurring among drug users sharing needles, their sexual partners (many of whom are prostitutes), and their children. Whether a similar increase will subsequently occur among the population at large remains to be seen.

AIDS, like smallpox in another era (page 796), is one of the great teachers on the subject of immunology. The price of its lessons—in the suffering of its victims, their families and their friends, in lives cut short in their prime, in the strains on medical and social services, and in economic terms—is extraordinarily high. Our response will say much about us as a people.

SUMMARY

Animals have evolved a number of responses that exclude or destroy microorganisms, other foreign invaders, and cells not typically self. These responses depend on a variety of types of white blood cells, all the progeny of self-regenerating stem cells in the bone marrow.

The inflammatory response, which is nonspecific, involves the release of histamine and other chemicals, causing capillary distension, a local increase in temperature, and the mobilization of phagocytic granulocytes and macrophages at the site of infection. Interferons provide another and very different type of nonspecific defense, directed against viruses. Interferons are small proteins produced by virus-infected cells that stimulate nearby cells to defend themselves against viral infection and that also stimulate cells involved in the immune response.

The immune response is highly specific and involves two types of cells: B lymphocytes and T lymphocytes. B lymphocytes are the major protagonists in the formation of antibodies, large protein molecules whose binding sites are complementary to foreign molecules called antigens. The combination of antigen and antibody immobilizes the invader, destroying it or rendering it susceptible to phagocytosis.

Five classes of antibodies (immunoglobulins) are known, of which the circulating IgG antibodies are the most intensively studied. Antibodies consist of four subunits: two identical light chains and two identical heavy chains. Each of the four chains has a constant (C) region—a region common to all antibodies of its class—and a variable (V) region, which differs from one antibody to another. The antigen-binding sites—of which there are two on each antibody molecule—are formed by foldings of the variable regions of the light and heavy chains.

The accepted model of antibody formation is the clonal selection theory. According to this theory, differentiation of B lymphocyte precursors, which occurs in the bone marrow, produces a vast variety of different B lymphocytes, each capable of synthesizing antibodies with one particular three-dimensional structure for the binding site. Upon encountering an antigen that binds to the antibodies displayed on its surface, the B lymphocyte matures and divides, resulting in a clone of plasma cells all synthesizing circulating antibodies against that particular antigen. Memory cells are also produced, which persist in the bloodstream following infection and produce antibodies immediately upon subsequent exposure to the same antigen. This memory-cell response is the cause of the rapid and enhanced immunity following vaccination or many viral infections. The capacity to produce a tremendous variety of B lymphocytes, each able to synthesize one specific antibody, is accounted for by the large number of gene sequences (at least 300) coding for the variable regions of antibodies, by the transposition of these gene sequences in the course of lymphocyte differentiation, and by subsequent somatic mutations.

T lymphocytes, which undergo differentiation and maturation in the thymus, are responsible for cell-mediated immunity. Although these cells neither display nor produce antibodies, they do bear several types of glycoprotein molecules on their surface. One type of molecule, which exists in two forms, T4 and T8, identifies cells as helper T cells (T4) or as cytotoxic or suppressor T cells (T8). Cytotoxic T cells work principally against the body's own cells that are harboring intracellular viruses or other parasites. Helper T cells promote immune responses involving B lymphocytes and cytotoxic T cells, whereas suppressor T cells moderate the activities of B lymphocytes and other T cells.

The ability of the T lymphocytes to perform their functions depends upon another type of surface molecule, known as the T-cell receptor. The T-cell receptor consists of two polypeptide chains, each with constant and variable regions coded by genes that, like those for antibodies, are rearranged in the course of differentiation. The result is an enormous diversity of T lymphocytes, each bearing T-cell receptors with a single antigenic specificity. T-cell receptors recognize and bind to complementary foreign antigens presented in conjunction with genetically determined antigens found on the surface of the body's own nucleated cells. These "self" antigens are coded by a group of genes known as the major histocompatibility complex (MHC). Two classes of MHC antigens are known. Class I molecules, which are found on cells throughout the body, are essential in the identification of diseased cells by cytotoxic T cells. Class II molecules are found on the surface of macrophages and B lymphocytes. They are essential in the presentation of foreign antigens to the helper T cells, which are, in turn, essential for the activation and proliferation of both B lymphocytes and cytotoxic T cells.

Skin and other organs transplanted between individuals other than identical twins evoke an immune response by cytotoxic T cells that can lead to rejection of the transplanted organ. The success of such transplants has been dramatically improved by the development of MHC-antigen testing, ensuring a closer match between donor and recipient, and the use of the selectively immunosuppressive drug cyclosporin. Similarly, blood transfusions can evoke an immune response by circulating antibodies to the A and B antigens found on the surface of red blood cells. Blood typing involves not only the ABO blood groups but also other surface antigens of red blood cells, such as the Rh factor.

Disorders associated with the immune system include allergies, autoimmune diseases caused by an individual's immune responses to his or her own tissues, and AIDS, a fatal infectious disease. The retrovirus responsible for AIDS (human immunodeficiency virus, or HIV) invades and destroys helper T cells, leaving the victim's immune system incapable of responding to other infections or to malignancies. In the final stages of the disease, it invades other cells and tissues of the body, including those of the nervous system. The AIDS virus is present in high levels in the blood and semen of infected individuals and is transmitted by sexual contact (heterosexual or homosexual, oral, vaginal, or anal) and through the exchange of blood or blood products.

QUESTIONS

1. Distinguish among the following: granulocytes/monocytes/lymphocytes; B lymphocyte/T lymphocyte; antigen/antibody/T-cell receptor; antibody light chain/heavy chain; T-cell receptor alpha chain/beta chain; T4 molecule/T8 molecule; MHC antigens/ABO blood groups/Rh factor; Class I MHC antigens/Class II MHC antigens.

2. Describe the life history of a B lymphocyte, beginning with a precursor cell in the bone marrow and ending with a clone of plasma cells and a clone of memory cells. Include its interactions with other types of white blood cells.

3. Describe the life history of a cytotoxic T cell, beginning with a precursor cell in the bone marrow and ending with a clone of active cells and a clone of memory cells. Include its interactions with other types of white blood cells.

4. Explain the functions of the macrophages in the inflammatory and immune responses. How do these cells provide the essential link between the inflammatory response and the immune response?

5. What prominent feature in the structure of a virus makes it more susceptible than a bacterial cell to control by vaccination?

6. An early hypothesis of antibody formation, the instructive theory, proposed that the antibody was molded into its special configuration by its encounter with the antigen. Since then, data on the structure of proteins have invalidated this model. What are these data?

7. Predict the results of the following experiment: A mouse of strain A receives a skin graft from another strain A mouse and one from a strain B mouse. Two weeks later, the same mouse is grafted with skin from strain B and strain C. What are the fates of these grafts?

8. If you are of blood type O and neither of your parents is, what are their possible genotypes? What is the probability that one of your siblings is also of type O? (You may find it helpful to review the use of Punnett squares, page 240).

9. Although it can never be proved that someone is the father of a particular child, it is possible to prove that someone could *not* be the father. Fill in the table below to see how this is so. In the famous Charlie Chaplin paternity case in the 1940s, the baby's blood was B, the mother's A, and Chaplin's O. If you had been the judge, how would you have decided the case?*

PHENOTYPES OF PARENTS		CHILDREN POSSIBLE	CHILDREN NOT POSSIBLE
A	A		
A	B		
A	AB		
A	O		
B	B		
B	AB		
B	O		
AB	AB		
AB	O		
O	O		

* As a matter of fact, Chaplin was judged to be the father. Blood-group data are not admitted as evidence by some states in cases of disputed parentage.

CHAPTER 40

Integration and Control I: The Endocrine System

As we have seen in the preceding chapters of this section, the maintenance of homeostasis in the body of an animal is a complex process, involving the regulation of diverse physiological activities. Internal and external conditions are constantly monitored, and the resulting information is transferred to centers in which it is integrated, leading to the initiation of appropriate responses. Throughout the living world, the integration and control necessary to maintain homeostasis are effected by chemical stimuli. Specific molecules interact with receptors that are either inside individual cells or embedded in their membranes, triggering changes in the cells that lead, directly or indirectly, to an appropriate response. In unicellular organisms, the response is direct and typically involves only the cell receiving the stimulus. In multicellular organisms, however, the response is more often indirect. Certain cells are specialized to receive and process information and to transmit it to other cells, thereby regulating the activities of the organism as a whole.

In small, simple animals and at many sites in larger, more complex animals, signalling molecules move from the cells in which they are produced to the cells on which they act by simple diffusion. Often, however, the target cells are a considerable distance away, and the signalling molecules are carried by bulk flow in the bloodstream, which can move them at a more rapid rate than would be possible by diffusion. This rate, although adequate for many of the processes involved in maintaining homeostasis, is nevertheless too slow for effective coordination of the numerous activities that characterize most animals. A much more rapid, direct channel of communication is provided by neurons, specialized cells that use an electrical signal—the nerve impulse—to conduct information over great distances.

For many years, the relatively slow, but long-acting regulation of the physiological activities of an animal by chemical means and the rapid conduction of information by electrical means were regarded as different phenomena, involving two distinct systems (Figure 40-1). Chemical regulation was considered the province of the endocrine system, the clusters of specialized secretory cells—glands—that synthesize and release chemical substances—hormones—that travel through the bloodstream. By contrast, the nervous system, the interconnected network of neurons and supporting tissues that attains its highest complexity in the vertebrates, was thought to be involved only in the conduction of nerve impulses. Recently, however, it has become apparent that there is enormous overlap between these two systems and they are more accurately viewed as different aspects of one overall system, the **neuroendocrine system.**

A variety of evidence supports this conception of one unified regulatory system. For example, although an individual neuron conducts information electrically, it transmits that information to other cells, including other neurons, through

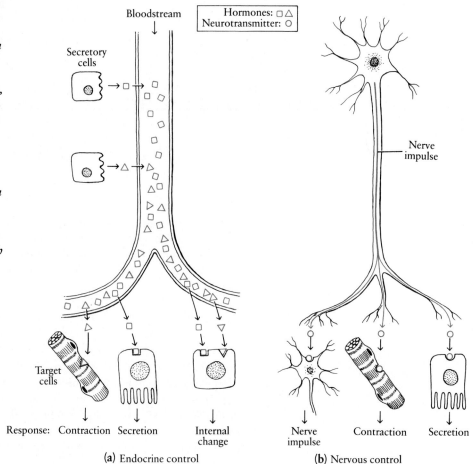

40-1 *The classical distinction between endocrine control and nervous control is based on the means by which information is carried over long distances. (a) In endocrine control, specific molecules, known as hormones, diffuse into the bloodstream, which carries them through the body to target tissues. This transport process may take minutes to hours, and the effects are typically long-lasting. (b) In nervous control, electrical signals—nerve impulses—are conducted along a neuron to its terminus where specific molecules, known as neurotransmitters, are released and diffuse to the target tissues. The entire process takes only a fraction of a second, and the effect is similarly short-lived. Both hormones and neurotransmitters interact with specific receptors on or in the target cells, leading to a response.*

the release of specific molecules known as **neurotransmitters.** A number of neurotransmitters are chemically identical to hormones. Moreover, some neurons, known as **neurosecretory cells,** release their signalling molecules into the bloodstream, which carries them—like hormones—to their target tissues. Although most endocrine glands are composed of columnar or cuboidal epithelial cells, some, such as the adrenal medulla, are clusters of neurosecretory cells. The molecules they release affect not only target tissues reached through the bloodstream but also other neurons that rapidly conduct the information to regulatory centers. In addition, certain substances that are, by all traditional definitions, hormones, such as insulin, have recently been found in the vertebrate brain.

Despite this evidence for a single, unified system, the older concepts of an endocrine system and a nervous system remain useful. By beginning our exploration of integration and control in the animal body with the endocrine system, as we shall do in this chapter, we can establish the basic principles of chemical communication among cells, which also apply to the nervous system. In Chapter 41, we shall focus our attention on the organization of the nervous system and the mechanisms by which information is transmitted through it. Chapter 42 will be devoted to the details of sensory perception, particularly of the external environment, and the subsequent motor response. In Chapter 43, all of these threads will come together again as we consider the culmination of vertebrate evolution, the brain.

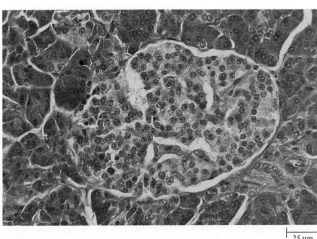

40-2 *A cross section of pancreatic tissue. The pancreas is both an endocrine and an exocrine gland. The group of small cells in the center of the micrograph are islet cells, which are the endocrine cells of the pancreas. Different types of islet cells secrete the hormones insulin, glucagon, and somatostatin. The surrounding exocrine cells produce digestive enzymes, which are carried through the pancreatic duct to the small intestine (see Figure 34-11, page 722).*

GLANDS AND THEIR PRODUCTS: AN OVERVIEW

Hormones are, by definition, organic molecules secreted in one part of an organism that diffuse or, in the case of vertebrates, are transported by the bloodstream to other parts of the organism, where they have specific effects on target organs or tissues. As we have seen in previous chapters, hormones are produced by a variety of different cell types: epithelial cells of the digestive tract (gastrin, secretin, cholecystokinin), cardiac muscle cells (cardiac peptide), white blood cells (histamine, lymphokines, interleukins), and even injured or infected cells (histamine, interferons). In these cell types and many others, the secretion of hormones is only one of several functions, although an essential one for the overall processes in which they—and the tissues of which they are a part—are involved. In glandular epithelial cells and neurosecretory cells, however, the secretion of hormones is the primary function, to which all other activities are subordinated.

Glandular epithelial cells and neurosecretory cells are often found in clusters, many of which are sufficiently large that they were readily identified by the early anatomists as distinct organs. With the discovery of their functions, which came much later, these organs were seen as constituting a distinct organ system. Still later, a distinction was drawn between exocrine glands and endocrine glands. **Exocrine glands** secrete their products into ducts; examples are digestive glands, the milk glands of the female mammal, and the sweat glands of the human skin. **Endocrine glands,** by contrast, secrete their products into the bloodstream (or, more precisely, into the extracellular fluids, from which they diffuse into the bloodstream); thus they are sometimes referred to as "ductless" glands.

The principal endocrine glands of the vertebrate body are shown in Figure 40-3. The hormones they secrete are of three general chemical types: steroids (see page 70), peptides or proteins, and amino acid derivatives. Hormones are characteristically active in very small amounts. It has been calculated, for example, that the concentration of adrenaline normally present in your bloodstream can be approximated by 8 milliliters (one teaspoonful) in a lake 2 meters deep and 100 meters in diameter. As befits such potent chemicals, playing key roles in the integration and control of the body's physiological functions, hormones are themselves under tight control. One aspect of this control is the regulation of their production. With very few exceptions, hormones are under negative feedback control, as described in Chapter 33. Another, equally important aspect is that they are rapidly degraded in the body; the steroids, peptides, and proteins are broken down by the liver, and the amines (amino acid derivatives) by enzymes in the blood.

Table 40-1 summarizes the principal endocrine glands of vertebrates. In this chapter, we shall discuss most of these glands, the hormones they secrete, and the regulation of their secretion. We shall, however, defer discussion of the glands and hormones involved in reproduction until Chapter 44.

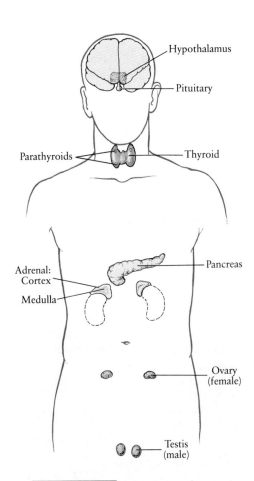

40-3 *Some of the hormone-producing (endocrine) organs. The pituitary releases hormones that, in turn, regulate the hormone secretions of the thyroid, the adrenal cortex (the outer layer of the adrenal gland), and the reproductive organs. The pituitary is itself under the regulatory control of a major brain center, the hypothalamus. The hypothalamus thus is the major link between the nervous system and the endocrine system.*

TABLE 40-1 Some of the Principal Endocrine Glands of Vertebrates and the Hormones They Produce

GLAND	HORMONE	PRINCIPAL ACTION	MECHANISM CONTROLLING SECRETION	CHEMICAL COMPOSITION
Pituitary, anterior lobe	Growth hormone (somatotropin)	Stimulates growth of bone, inhibits oxidation of glucose, promotes breakdown of fatty acids	Hypothalamic hormone(s)	Protein
	Prolactin	Stimulates milk production and secretion in "prepared" gland	Hypothalamic hormone(s)	Protein
	Thyroid-stimulating hormone (TSH) (thyrotropin)	Stimulates thyroid	Thyroxine in blood; hypothalamic hormone(s)	Glycoprotein
	Adrenocorticotropic hormone (ACTH)	Stimulates adrenal cortex	Cortisol in blood; hypothalamic hormone(s)	Polypeptide (39 amino acids)
	Follicle-stimulating hormone (FSH)*	Stimulates ovarian follicle, spermatogenesis	Estrogen in blood; hypothalamic hormone(s)	Glycoprotein
	Luteinizing hormone (LH)*	Stimulates ovulation and formation of corpus luteum in female, interstitial cells in male	Progesterone or testosterone in blood; hypothalamic hormone(s)	Glycoprotein
Hypothalamus (via posterior pituitary)	Oxytocin	Stimulates uterine contractions, milk ejection	Nervous system	Peptide (9 amino acids)
	Antidiuretic hormone (ADH, vasopressin)	Controls water excretion	Osmotic concentration of blood; blood volume; nervous system	Peptide (9 amino acids)
Thyroid	Thyroxine, other thyroxinelike hormones	Stimulate and maintain metabolic activities	TSH	Iodinated amino acids
	Calcitonin	Inhibits release of calcium from bone	Concentration of Ca^{2+} ions in blood	Polypeptide (32 amino acids)
Parathyroid	Parathyroid hormone (parathormone)	Stimulates release of calcium from bone; stimulates conversion of vitamin D to active form, which promotes calcium uptake from gastrointestinal tract; inhibits calcium excretion	Concentration of Ca^{2+} ions in blood	Polypeptide (34 amino acids)
Adrenal cortex	Cortisol, other glucocorticoids	Affect carbohydrate, protein, and lipid metabolism	ACTH	Steroids
	Aldosterone	Affects salt and water balance	Processes initiated in the kidney; K^+ ions in blood	Steroid
Adrenal medulla	Adrenaline and noradrenaline	Increase blood sugar, dilate or constrict specific blood vessels, increase rate and strength of heartbeat	Nervous system	Catecholamines (amino acid derivatives)
Pancreas	Insulin	Lowers blood sugar, increases storage of glycogen	Concentration of glucose and amino acids in blood; somatostatin	Polypeptide (51 amino acids)
	Glucagon	Stimulates breakdown of glycogen to glucose in the liver	Concentration of glucose and amino acids in blood; somatostatin	Polypeptide (29 amino acids)
Pineal	Melatonin	Involved in regulation of circadian rhythms	Light-dark cycles	Catecholamine
Ovary, follicle	Estrogens*	Develop and maintain sex characteristics in females, initiate buildup of uterine lining	FSH	Steroids
Ovary, corpus luteum	Progesterone and estrogens*	Promote continued growth of uterine lining	LH	Steroids
Testis	Testosterone*	Supports spermatogenesis, develops and maintains sex characteristics of males	LH	Steroid

* These hormones will be discussed in Chapter 44.

THE PITUITARY GLAND

The pituitary gland was once considered the master gland of the body, as it is the source of hormones stimulating the reproductive organs, the adrenal cortex, and the thyroid. However, it is now known that this "master" gland is itself regulated by a key brain center, the hypothalamus. Hormones from the hypothalamus stimulate or, in some cases, inhibit the production of pituitary hormones. The pituitary, about the size of a kidney bean, is located at the base of the brain in the geometric center of the skull. It consists of three lobes: the anterior, the intermediate, and the posterior.

The Anterior Lobe

The anterior lobe of the pituitary is the source of at least six different hormones, each produced by recognizably different cells. One of these is growth hormone, sometimes called somatotropin, which stimulates protein synthesis and promotes the growth of bone. As is the case with most of the hormones, growth hormone is best known by the effects caused by too much or too little. If there is a deficit in growth hormone production in childhood, a midget results, the so-called "pituitary dwarf." An excess of growth hormone during childhood results in a giant; most circus giants are the result of such an excess. Excessive growth hormone in the adult does not lead to giantism, since growth of the long bones has ceased; it leads instead to acromegaly, an increase in the size of the jaw and the hands and feet, adult tissues that are still sensitive to the effects of growth hormone. Growth hormone also affects glucose metabolism, inhibiting the uptake and oxidation of glucose by some types of cells. It also stimulates the breakdown of fatty acids, thus conserving glucose. This hormone is now being produced by recombinant DNA techniques, which will extend its use in medical treatment and facilitate the study of its role in glucose metabolism.

A second hormone produced by the anterior pituitary is prolactin, which stimulates the secretion of milk in mammals. Its production is controlled by an inhibitory hormone produced by the hypothalamus. As long as the infant continues to nurse, the nerve impulses produced by the suckling of the breast are transmitted to the hypothalamus, which decreases production of prolactin-inhibiting hormone. The pituitary then releases prolactin, which, in turn, acts upon the breast to maintain the production of milk. Once suckling ceases, the synthesis and release of prolactin decrease and so milk production stops. Thus supply is regulated by demand.

Four of the hormones secreted by the anterior pituitary are **tropic hormones**—hormones that act upon other endocrine glands to regulate their secretions. One of these tropic hormones is TSH, the thyroid-stimulating hormone, also known as thyrotropin. TSH stimulates cells in the thyroid gland to increase their production and release of thyroxine, the thyroid hormone. In a negative feedback loop involving both the pituitary and the hypothalamus, the increased concentration of thyroxine inhibits the further secretion of TSH by the pituitary. Adrenocorticotropic hormone (ACTH) has a similar regulatory relationship with the production of cortisol, one of the hormones produced by the adrenal cortex (the outer layer of the adrenal gland).

The other two tropic hormones secreted by the anterior pituitary are gonadotropins—hormones that act upon the **gonads,** or gamete-producing organs (the testes and the ovaries). These hormones, follicle-stimulating hormone (FSH) and luteinizing hormone (LH), will be discussed in Chapter 44.

(a)

(b)

(c)

40–4 *When a green anole* (Anolis carolinensis) *moves from a light background to a dark background, or vice versa, its pigmentation changes. The color change is the result of the movement of pigment-containing granules within the pigmented cells of the skin. When the granules are concentrated in the center of the cells, the skin color is light; when they are dispersed into extensions of the cells, the color is dark. Melanocyte-stimulating hormone plays an important role in triggering the dispersal of these granules. The change from light to dark* (a–c) *takes 5 to 10 minutes, while the change from dark to light takes 20 to 30 minutes.*

The Intermediate and Posterior Lobes

In many vertebrates, the intermediate lobe of the pituitary gland is the source of melanocyte-stimulating hormone. In reptiles and amphibians, this hormone stimulates color changes associated with camouflage (Figure 40–4) or with behavior patterns such as aggression or courtship. In humans, in which the secretion of melanocyte-stimulating hormone is greatly reduced, its functions are unknown.

The posterior lobe of the pituitary gland stores the hormones produced by the hypothalamus.

THE HYPOTHALAMUS

The Pituitary-Hypothalamic Axis

The pituitary gland lies beneath the hypothalamus and is directly under its influence. The pituitary is also under the influence, by way of the hypothalamus, of other parts of the brain. The hypothalamus is the source of at least nine hormones that act either to stimulate or inhibit the secretion of hormones by the anterior pituitary. These hormones are small peptides, one only three amino acids in length. They are unusual not only for their small size but also for the way in which they reach their target gland. Produced by neurosecretory cells of the hypothalamus, they travel only a few millimeters to the pituitary, apparently never entering the general circulation. However, they make this brief passage by way of a portal system (Figure 40–5).

The first of these hypothalamic hormones to be discovered was TRH, thyrotropin-releasing hormone. As its name implies, it stimulates the release of thyrotropin (TSH) from the pituitary. The second was gonadotropin-releasing hormone (GnRH), which controls the release of the gonadotropic hormones LH and FSH.

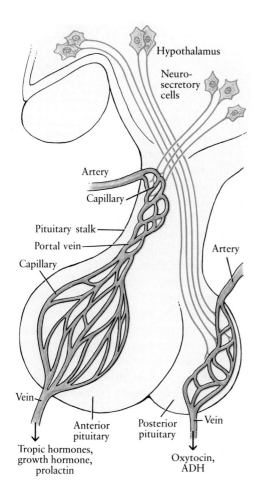

40–5 *Relationship between the hypothalamus and the pituitary. The hypothalamus communicates with the anterior lobe of the pituitary through a small portal system (see page 759). Neurosecretory cells of the hypothalamus secrete releasing or inhibiting hormones directly into capillaries that are linked by portal veins to a second capillary network in the anterior pituitary, where the hypothalamic hormones affect the production of pituitary hormones. Other hypothalamic neurosecretory cells produce oxytocin and ADH, which are transmitted to the posterior lobe of the pituitary through the nerve fibers. Following their release from the nerve endings in the posterior pituitary, these hormones diffuse into capillaries and thus enter the general circulation.*

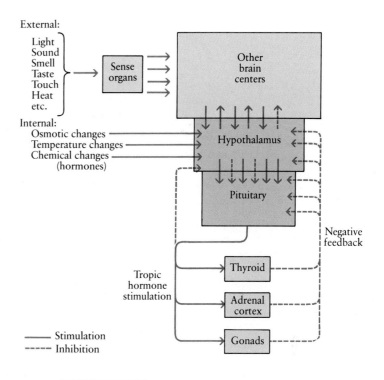

40–6 *The production of many hormones is regulated by complex negative feedback systems involving the pituitary and the hypothalamus. The hypothalamus controls the pituitary's secretion of tropic hormones, and these, in turn, stimulate the secretion of hormones from the thyroid, adrenal cortex, and gonads (the testes or the ovaries). As the concentration of the hormones produced by these target glands rises in the blood, the hypothalamus decreases its production of releasing hormones, the pituitary decreases its hormone production, and production of hormones by the target glands also slows. By way of the hypothalamus, which receives information from many other parts of the brain, hormone production is also regulated in response to other changes in the external and internal environments.*

These discoveries were made by two groups of scientists, one headed by Roger Guillemin at the Salk Institute (requiring the brains of 5 million sheep) and the other by Andrew Schally in New Orleans (using a comparable supply of pigs). The third hormone found was not a releaser but, surprisingly, an inhibitor. Because it inhibited release of the growth hormone somatotropin, it was given the name of somatostatin. As you may recall, somatostatin was the first mammalian hormone to be synthesized using recombinant DNA technology (see page 351). Several more releasing hormones and one other inhibitory hormone have now been isolated, and the search is not over.

Figure 40–6 summarizes the negative feedback systems linking the hypothalamus and pituitary with the thyroid, adrenal cortex, and gonads. These feedback control systems, generally involving both the pituitary and the hypothalamus, provide for both homeostasis and response to changing conditions. The pituitary feedback system usually provides for constancy. However, it can be overridden by the hypothalamic system, which takes into account not only the balance between output and input but also the changes elsewhere in the body and in the external environment.

Other Hypothalamic Hormones

The hypothalamus is also the source of two hormones stored in and released from the posterior pituitary: oxytocin and antidiuretic hormone (ADH). Oxytocin accelerates childbirth by increasing uterine contractions during labor; these contractions also cause the uterus to regain its normal size and shape after delivery. The release of oxytocin is under the control of the nervous system and may be triggered by increasing pressure within the uterine wall or by movements of the fetus. In experimental animals, labor can be induced by mechanical stimulation of the uterus or electrical stimulation of the hypothalamus. Oxytocin is also responsible for the "letting down" of milk that occurs when the infant begins to suckle. The hormone promotes contraction of the muscle fibrils around the milk-secreting cells of the mammary glands. Oxytocin is present in males as well as females, but its function in the male, if any, is unknown.

ADH, as we saw in Chapter 37, decreases the excretion of water by the kidneys. It achieves this effect by increasing the permeability of the membranes of cells in the collecting ducts of the nephrons so that more water passes through them and back into the blood from the urine. ADH is sometimes called vasopressin because it increases blood pressure in many vertebrates; in humans, however, it has this effect only in response to certain unusual circumstances, for example, the blood loss of a severe hemorrhage. Oxytocin has some ADH effect and ADH some oxytocin effect. This cross action is not surprising because each of these hormones consists of only nine amino acids, and differences in the two hormones involve only two amino acids among the nine (Figure 40–7).

40-7 *Primary structures of antidiuretic hormone (ADH) and oxytocin, hormones produced in the hypothalamus and released from the posterior lobe of the pituitary. Note that each consists of only nine amino acids and that they differ by only two. This similarity of structure gives rise to some cross action by these hormones.*

| cys | tyr | phe | glu | asn | cys | pro | arg | gly | ADH |

| cys | tyr | ile | glu | asn | cys | pro | leu | gly | Oxytocin |

THE THYROID GLAND

The thyroid, under the influence of thyroid-stimulating hormone (TSH) from the pituitary, produces thyroxine, which is an amino acid combined with four atoms of iodine (Figure 40–8). Thyroxine (or, most likely, its metabolic product, triiodothyronine) accelerates the rate of cellular respiration. In some animals, it also plays a major role in temperature regulation (see Figure 38–8, page 784).

40-8 *Thyroxine, the principal hormone produced by the thyroid gland. Note the four iodine atoms in its structure. Triiodothyronine differs from thyroxine by having one less iodine atom on the OH-bearing ring. Because iodine is needed for thyroxine, it is an essential component of the human diet. Where iodine is present in the soil, it is available in minute quantities in drinking water and in plants. In the United States, table salt is ordinarily iodized or must be specifically labeled as being uniodized.*

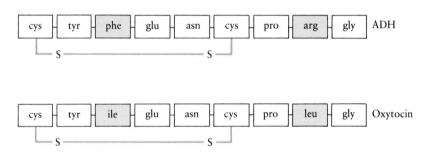

Thyroxine

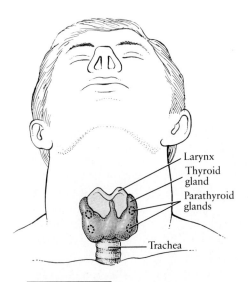

40–9 *The pea-sized parathyroid glands, the smallest of the known endocrine glands, are located behind or within the thyroid gland. They produce parathyroid hormone (parathormone), which increases concentrations of blood calcium. Calcitonin, a hormone produced by the thyroid gland, decreases blood calcium.*

40–10 *The chemical structures of representatives of the two major groups of hormones secreted by the adrenal cortex. Cortisol is a glucocorticoid, and aldosterone is a mineralocorticoid. The hormones of both groups are steroids, identified by their characteristic four-ring structure. As you can see, the differences in the molecular structures of cortisol and aldosterone are minor, and yet their physiological roles are profoundly different.*

Hyperthyroidism, the overproduction of thyroxine, results in nervousness, insomnia, and excitability; increased heart rate and blood pressure; heat intolerance and excessive sweating; and weight loss. Hypothyroidism (too little thyroxine) in infancy affects development, particularly of the brain cells; if not treated in time, it can lead to permanent mental deficiency and dwarfism. In adults, hypothyroidism is associated with dry skin, intolerance to cold, and lack of energy. Hypothyroidism may be caused by insufficient iodine, which is needed to make thyroxine, and, in these cases, it is often associated with goiter, an enlargement of the thyroid gland.

The thyroid gland also secretes the hormone calcitonin in response to rising calcium levels in the fluid surrounding the thyroid cells, which reflects the levels in the blood. Calcitonin's major action is to inhibit the release of calcium ion from bone.

THE PARATHYROID GLANDS

The pea-sized parathyroid glands, the smallest of the known endocrine glands, are located behind or within the thyroid gland (Figure 40–9). They produce parathyroid hormone (parathormone), which plays an essential role in mineral metabolism, specifically in the regulation of calcium and phosphate ions, which exist in a reciprocal relationship in the blood. Calcium is normally present in mammalian blood in concentrations of about 6 milligrams per 100 milliliters of whole blood. A rise or fall of more than 2 or 3 milligrams per 100 milliliters can lead to such severe disturbances in blood coagulation, muscle contraction, and nerve function that death may follow within hours.

Parathyroid hormone increases the concentration of calcium ion in blood in several different ways. It stimulates the conversion of vitamin D into its active form (see Figure 34–14, page 728); active vitamin D, in turn, increases the absorption of calcium ion from the intestine. Parathyroid hormone also reduces excretion of calcium ion from the kidneys. In addition, it stimulates the release into the bloodstream of calcium from bone, which contains 99 percent of the body's total calcium. Thus, parathyroid hormone and calcitonin work as a fine-tuning mechanism, regulating blood calcium, with parathyroid hormone apparently playing the principal role. The production of both hormones is regulated directly by the concentration of calcium ions in the blood.

Hyperparathyroidism, caused by tumors of the parathyroids, occasionally occurs in humans. When there is too much parathyroid hormone, the bones lose large amounts of calcium, becoming soft and fragile, and the vertebrae may shrink, producing a loss of height. Removal of the parathyroid glands without hormone replacement therapy results in violent muscular contractions and spasms, leading to death.

ADRENAL CORTEX

The adrenal cortex—the outer layer of the adrenal gland—is the source of a number of steroid hormones. The adrenal (ad-renal) glands, as their name implies, are on top of the kidneys. About 50 different steroids, all very similar in structure, have been isolated from the adrenal cortex of various mammals. Some of these corticosteroids, as they are known, undoubtedly represent steps in the synthesis of various hormones, though most of them have some hormonal activity. In humans, there are two major groups of adrenocortical steroids, the glucocorticoids and the mineralocorticoids (Figure 40–10).

The Regulation of Bone Density

Calcitonin and parathyroid hormone play the principal roles in the regulation of calcium levels in the blood. The major reservoir of calcium on which their activities depend is, as we have seen, bone. The regulation of bone density, however, is a considerably more complex process than the regulation of blood calcium and involves a number of hormones in addition to calcitonin and parathyroid hormone.

Despite its apparent solidity, bone, like skin and the intestinal lining, is a turnover tissue. As we saw in Figure 33-6 (page 704), it is formed by cells known as osteoblasts, which secrete the collagen and other fibrils in which calcium compounds become crystallized. However, other cells, known as osteoclasts, break down bone, dissolving the crystalline compounds and releasing calcium to the blood. The maintenance of bone mass depends on a balance between the activities of these two types of cells. One of the contributing factors to osteoporosis, the severe loss of calcium from bone that afflicts many older persons, particularly women, is greater activity by osteoclasts than by osteoblasts.

The symptoms of osteoporosis are similar to those of hyperparathyroidism: the bones lose large amounts of calcium, becoming soft and fragile, and the vertebrae shrink, producing a loss of height. The consequences frequently include fractured vertebrae and hips, both of which can lead to serious complications. Current evidence indicates that hormones secreted by

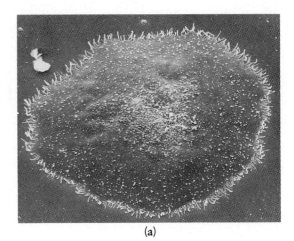

(a)

(a) *An osteoclast, growing in tissue culture. These large cells, which are multinucleate, are produced by the fusion of many smaller cells. The precursors of osteoclasts are among the progeny of the bone marrow stem cells and may be related to monocytes and macrophages. The thin processes extending from the cell surface pump acid onto the bone, and the acid, in*

stem cells in the bone marrow (page 792) play a major role in the activities of the osteoclasts. These substances, known as colony-stimulating factors, are the same molecules that stimulate the repeated divisions

Glucocorticoids

Cortisol is thought to be the most important glucocorticoid in humans. Cortisol and the other glucocorticoids promote the formation of glucose from protein and fat. At the same time, they decrease the utilization of glucose by most cells, with the notable exceptions of cells of the brain and the heart, thus favoring the activities of these vital organs at the expense of other body functions. Their release increases during periods of stress, such as facing new situations, engaging in athletic competition, and taking final exams. As we shall see in the next chapter, they work in concert with the sympathetic nervous system.

In addition to their effects on glucose metabolism, glucocorticoids suppress inflammatory and immune responses, an effect that may be a factor in the increased susceptibility to illness that often accompanies stress. Because of their immunosuppressive properties, cortisol and other glucocorticoids are sometimes used in the treatment of autoimmune diseases and severe allergic reactions.

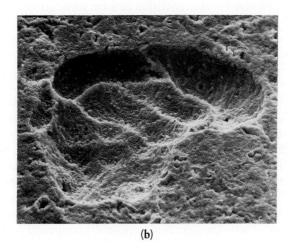
(b)

turn, leaches minerals from it. (b) *This crater, known as a resorption lacuna, was created on the surface of a bone by the activities of an osteoclast. Following the departure of the osteoclasts, osteoblasts migrate into resorption lacunae and begin their bone-creating activities to fill the lacunae.*

of the stem cells and regulate the differentiation of their progeny into red blood cells, granulocytes, monocytes, and lymphocytes. Stimulation of the osteoclasts also appears to involve molecules, particularly lymphokines and interleukins, produced in the course of the inflammatory and immune responses. There is evidence that prostaglandins (page 833) play a role as well. Damping the activity of the osteoclasts indirectly through the use of anti-inflammatory drugs is a component of medical therapy designed to halt the progress of osteoporosis.

Also contributing to the development of osteoporosis is a disruption of the pathways affecting absorption of calcium ion from the intestine, particularly those involved in the conversion of vitamin D into its active form. These pathways are influenced not only by parathyroid hormone but also by the female hormone estrogen, the production of which drops dramatically after menopause. Despite the claims of numerous television commercials, calcium supplements alone, in the absence of appropriate hormone replacement and anti-inflammatory agents, appear to be of little value in reversing or slowing the course of osteoporosis once it has begun. Accumulating evidence, however, indicates the importance of adequate calcium intake during childhood and adolescence, when bones are growing, and during the childbearing years. Drinking plenty of skim milk now, and even occasionally succumbing to that tempting chocolate milkshake, may help to provide the margin of safety —ample calcium reserves stored in the bones—that will enable you to be on the tennis court or golf course 50 years hence.

However, the serious side effects of the high doses often required limit their usefulness. Among these side effects are a reduced ability to combat infection; redistribution of body fat, resulting in "buffalo hump" and "moon face"; and mental disturbances, including both hyperactivity and depression.

The cortisol group of hormones are secreted in response to adrenocorticotropic hormone (ACTH). ACTH is secreted by the pituitary gland when it is stimulated by a releasing hormone from the hypothalamus. As with thyroxine, the secretion of the glucocorticoids is inhibited by negative feedback exerted on the pituitary and the hypothalamus. Recent studies indicate that positive feedback is also at work, involving an interleukin secreted by activated monocytes and macrophages. This molecule stimulates the secretion of ACTH, by acting either on the pituitary or the hypothalamus, or, perhaps, on both. Current evidence suggests that the immunosuppressive activities of corticosteroids may be part of the normal regulatory mechanism that turns off the inflammatory and immune responses when their work is done.

Mineralocorticoids

A second group of hormones secreted by the adrenal cortex comprises the mineralocorticoids, of which aldosterone is the primary example. These corticosteroids are involved in the regulation of ions, particularly sodium and potassium ions. The mineralocorticoids affect the transport of ions across the cell membranes of the nephrons and, as a consequence, have major effects both on ion concentrations in the blood and on water retention and loss. An increase in aldosterone secretion, as we noted in Chapter 37, results in greater reabsorption of sodium ions in the distal tubule and collecting duct of the nephron and increases the secretion of potassium ions into them. A deficiency in mineralocorticoids precipitates a critical loss of sodium ions from the body in the urine and, with it, a loss of water, leading, in turn, to a reduction in blood pressure.

In addition to glucocorticoids and mineralocorticoids, the adrenal cortex produces small amounts of male sex hormones in both males and females. An adrenal tumor may result in increased production of these hormones and, consequently, in women, the production of facial hair and other masculine characteristics. Bearded ladies in the circus were often the victims of such tumors. The adrenal cortex is also thought to secrete small amounts of female sex hormones.

ADRENAL MEDULLA

The adrenal medulla, the central portion of the adrenal gland, is a large cluster of neurosecretory cells whose nerve endings secrete adrenaline and noradrenaline (Figure 40–11) into the bloodstream. These hormones, known also as epinephrine and norepinephrine, increase the rate and strength of the heartbeat, raise blood pressure, stimulate respiration, and dilate the respiratory passages. They also increase the concentration of glucose in the bloodstream by promoting the activity of the enzyme that breaks down glycogen to glucose 6-phosphate. The adrenal medulla is stimulated by nerve fibers of the sympathetic division of the autonomic nervous system and so acts as an enforcer of sympathetic activity.

THE PANCREAS

The islet cells of the pancreas (Figure 40–2, page 823) are the source of insulin and glucagon, two hormones involved in the regulation of glucose metabolism. Insulin is secreted in response to a rise in blood sugar or amino acid concentration (as after a meal). It lowers the blood sugar by stimulating cellular uptake and utilization of glucose and by stimulating the conversion of glucose to glycogen.

When there is an insulin deficiency, as in persons with diabetes mellitus, the concentration of blood sugar rises so high that not all the glucose entering the kidney can be reabsorbed; the presence of glucose in the urine forms the basis of simple tests for diabetes. The loss of glucose is accompanied by loss of water. The resulting dehydration, which can lead to collapse of circulation, is one of the causes of death in an untreated diabetic.

Glucagon, produced by different islet cells of the pancreas, increases blood sugar. It stimulates the breakdown of glycogen to glucose in the liver and the breakdown of fats and proteins, which decreases glucose utilization.

Somatostatin, originally found in the hypothalamus, has now also been isolated from a third class of islet cells in the pancreas. It is released from the pancreas during digestion of a meal and exerts a variety of inhibitory effects on the digestive tract that, collectively, help to regulate the rate at which glucose and other nutrients are absorbed into the bloodstream. There is evidence that somatostatin also participates in controlling the synthesis of insulin and glucagon.

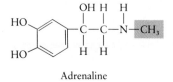

40–11 *The chemical structures of adrenaline and noradrenaline, hormones secreted by the adrenal medulla. These compounds are modified amino acids known as catecholamines, characterized by a six-sided ring bearing two hydroxyl groups. They are synthesized from the amino acid tyrosine (see page 73). As we shall see in the next chapter, noradrenaline is also an important neurotransmitter.*

Thus, as we have seen, at least seven different hormones are involved in regulating blood sugar: growth hormone, cortisol, adrenaline, noradrenaline, insulin, glucagon, and somatostatin. This tight control over blood glucose, illustrated in Figure 40–12, ensures that glucose is always available for brain cells. Unlike other cells in the body, which can derive energy from the breakdown of amino acids and fats, brain cells can utilize only glucose under most circumstances. Thus they are immediately affected by low blood sugar. After several days of fasting, however, the brain begins to use fatty acids as an energy source. An insulin overdose, which causes a rapid drop in blood sugar, can result in coma and death.

THE PINEAL GLAND

The pineal gland is a small lobe in the forebrain, lying near the center of the brain in humans. In lower vertebrates, it contains light-sensitive cells and so is sometimes called the third eye. The pineal gland secretes the hormone melatonin. In many species, including chickens, rats, and humans, the production of melatonin rises sharply at night and falls rapidly in the daytime. Exposure to light during the dark cycle interrupts production of melatonin, recalling the studies on photoperiodism in plants (page 686).

In larval amphibians, melatonin causes pigment-containing granules in cells of the skin to aggregate near the center of the cells, causing a blanching of the skin associated with darkness. In sparrows, injections of melatonin induce roosting and the lowering of body temperature, both of which are characteristic nighttime events in these birds. Melatonin also inhibits the development of the gonads in species as disparate as chickens and hamsters; its decreased production during long-day periods is believed to be associated with the seasonal enlargement of the gonads in preparation for mating. In humans, the pineal gland may be involved in sexual maturation; tumors of the pineal have been associated with precocious puberty.

Thus there are a few tempting clues suggesting that the pineal gland may function as a biological timekeeper, but the way in which light affects the gland and the way in which melatonin alters physiological responses—if it does—are yet to be discovered.

PROSTAGLANDINS

Among the most potent of all substances produced by and released from cells are a group of related chemicals first detected in semen. These substances were thought at the time to be produced by the prostate gland, a structure of the male reproductive system, and so they were called prostaglandins. Later research revealed, however, that most of the prostaglandins in semen are synthesized in other structures, the seminal vesicles. Since their initial discovery, a large number of prostaglandins have been identified, all related structurally but with a variety of different, and sometimes directly opposite, effects.

Although prostaglandins have hormonelike properties, they differ from other hormones in several significant ways: (1) Unlike any other hormones, they are fatty acids. Most are formed by the oxygenation of a 20-carbon polyunsaturated fatty acid known as arachidonic acid. (2) They are produced by cell membranes in most—if not all—organs of the body, as opposed to other hormones, which are produced by glandular epithelium or neurosecretory cells. (3) Their target tissues are generally either the same tissues in which they are produced or the tissues of another individual. (4) They produce marked effects at extremely low concentra-

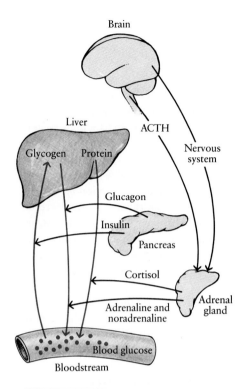

40–12 *Hormonal regulation of blood glucose. When blood sugar concentrations are low, the pancreas releases glucagon, which stimulates the breakdown of glycogen and the release of glucose from the liver. When blood sugar concentrations are high, the pancreas releases insulin, which removes glucose from the bloodstream by increasing its uptake by cells and promoting its conversion into glycogen, the storage form. Under conditions of stress, ACTH, produced by the pituitary, stimulates the adrenal cortex to produce cortisol and related hormones, which increase the breakdown of protein and its conversion to glucose in the liver. At the same time, the adrenal medulla releases adrenaline and noradrenaline, which also raise blood sugar.*

Growth hormone and somatostatin, not shown in this diagram, also affect blood glucose levels. Growth hormone inhibits the uptake and oxidation of glucose by many types of cells and stimulates the breakdown of fatty acids. Somatostatin influences the rate at which glucose is absorbed into the bloodstream from the digestive tract.

Circadian Rhythms

Virtually every living organism, from one-celled algae to bean plants (page 690) to *Homo sapiens*, exhibits circadian rhythms in many physiological functions. As we noted in Chapter 32, the chemical nature of the biological clock governing such rhythms remains unknown. Work with *Drosophila*, however, suggests that the products of one or more specific genes are involved. Mutations in a gene located on the X chromosome and known as *per* (for "period") have profound effects on the circadian rhythms of fruit flies. If the gene is deleted or rendered inoperative by a mutation, the flies become insomniacs, with no discernible sleep-wake cycle. Other mutations affect the length of the daily rhythms; for example, *per s* ("short periods") produces flies with daily cycles of 18 to 20 hours, and *per l* ("long periods") results in daily cycles of 28 to 30 hours. This same gene also affects the mating songs of male fruit flies, which beat their wings together in a rhythmic pattern that repeats at 60-second intervals. When the *per* gene is knocked out, the mating songs have no rhythms at all; *per s* results in songs at 40-second intervals, and *per l* in songs at 80-second intervals.

Investigators puzzled by how a single gene could affect both the daily 24-hour rhythm and the 60-second courtship song recently created mosaic flies in which cells in certain parts of the body contain the normal gene and cells in other parts of the body contain a mutant gene. They have found that the location in which *per* is expressed determines its effect; in cells of the head, it affects the daily rhythm, but in cells of the thorax, it affects the song rhythms. Moreover, the greater the quantities of the *per* gene product synthesized, the faster the clock runs. The gene has now been sequenced, and it turns out to contain repetitive sequences coding for alternating glycines and serines or threonines. Similar repetitive sequences have been identified in DNA from chickens, mice, and humans, but it is not yet known if these sequences are the portion of the gene controlling the rhythms.

Although the mechanism governing circadian rhythms remains mysterious, the effect of the rhythms in human physiology and behavior is not in doubt. For instance, we are more likely to be born between 3 and 4 A.M. and also to die in these same early morning hours. Body temperature fluctuates as much as 1°C (about 2°F) during the course of 24 hours, usually reaching a peak at about 4 P.M. and a low about 4 A.M. Alcohol tolerance is greatest, fortunately, at 5 P.M., while tolerance to pain is lowest at 6 P.M. in most persons. Respiration, heart rate, and urinary excretion of potassium, calcium, and sodium, all vary according to the time of day (heart rate by as much as 20 beats per minute). Secretion of various hormones follows circadian rhythms: serum levels of corticosteroids peak between 4 A.M. and 8 A.M. in people on a normal sleep-wake cycle; growth hormone levels rise about an hour after falling asleep; prolactin peaks around 3 A.M.; testosterone about 9 A.M.

Knowledge of circadian rhythms may have important practical consequences. A study by the Federal Aviation Agency showed that pilots flying from one time zone to another—from New York to Europe, for instance—exhibit "jet lag," a general decrease in mental alertness, an inability to concentrate, and an increase in decision time and physiological reaction time. Measurements of various functions show that the body may be "out of sync" for as much as a week after such a flight. This brings into question not only schedules for airline personnel but also present policies of speeding diplomats to foreign capitals at times of international crisis or of air transport of troops into combat. It may be significant that the accidents at the nuclear power plants at both Three Mile Island and Chernobyl occurred in the small hours of the morning local time; in the case of Three Mile Island, the workers on duty at the time of the accident had been alternating between the day shift and the night shift every week.

Medical researchers are finding that failure to take circadian fluctuations into account can lead to mistakes in diagnosis and treatment. For example, blood pressure may vary as much as 20 percent in the course of a day, so that a person might be found to be within the normal range at one time of day and diagnosed as hypertensive at another. Numbers of white blood cells may vary as much as 50 percent in a 24-hour period, rendering the same immunosuppressive therapy ineffectual at one time of day and life-threatening at another. Studies of cancer chemotherapy in mice indicate that proper timing of drug doses can mean a doubling of survival rate. Determination of a patient's chronobiology may someday become as routine in therapy as blood typing or taking a medical history.

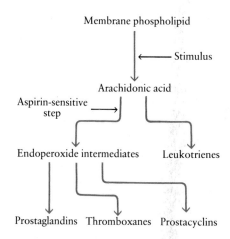

40-13 *A number of different substances, known collectively as prostaglandins, are synthesized from the phospholipids of cell membranes. Among the stimuli known to trigger the production of arachidonic acid, the precursor of all of these substances, are neurotransmitters, other hormones, various drugs and toxic agents, and allergens. Depending on the cell type and the stimulus, arachidonic acid may be converted to leukotrienes, a category of prostaglandins involved in inflammatory and immune responses, or to intermediates known as endoperoxides. These molecules, in turn, may be converted to the prostaglandins that act on smooth muscle (the group first discovered, and for which the entire class of molecules is named), or to thromboxanes or prostacyclins, both of which affect platelet aggregation and the dilation or constriction of blood vessels. Aspirin inhibits the conversion of arachidonic acid to endoperoxides.*

tions, much lower than those of most hormones. A concomitant of their extraordinary potency is the fact that they are released in very small amounts and are rapidly broken down by enzyme systems in the body. If it were not for their presence in unusually large amounts in semen, they might never have been discovered.

Stimulation of Smooth Muscle

Among the many effects of prostaglandins, one of the most striking is their capacity to induce contractions in smooth muscle. This is believed to play several important roles in reproduction. The walls of the uterus, which are composed of smooth muscle, normally contract in continuous waves. After sexual intercourse, prostaglandins from semen are found in the female reproductive tract, where they increase the rhythmic contractions of the uterine wall and oviducts. It is believed that this action assists both the sperm on its journey to the oviduct and the oocyte as it travels from the oviduct to the uterus. The semen of some infertile males has been found to be poor in prostaglandins, and the uterus of infertile females is often unresponsive to prostaglandins.

The contractions of the uterus also increase during a menstrual period and reach their greatest strength when a woman is in labor. Prostaglandins produced in the uterine lining are believed to play a key role in triggering the onset of both menstruation and labor.

Increased understanding of prostaglandins and their capacity to stimulate smooth muscle has shed light on a long-perplexing medical problem. Between 30 and 50 percent of all women of childbearing age experience painful uterine cramps during the first day or two of each menstrual period, a condition known as dysmenorrhea. In most of these women, no abnormalities of the reproductive organs can be detected. Recent research has revealed, however, that the menstrual fluid of such women contains concentrations of prostaglandins two to three times higher than the levels found in the menstrual fluid of women without dysmenorrhea. Present evidence indicates that the increased prostaglandin levels not only cause stronger, more rapid contractions of the uterine walls but also reduce the blood supply to the tissue. As a result, less oxygen is available to the actively contracting muscles, creating an oxygen debt (page 190) and its accompanying pain. It is also hypothesized that prostaglandins act directly on pain nerve endings, causing them to fire more rapidly.

Clinical studies in this country and in Europe have shown that several compounds that inhibit prostaglandin synthesis are highly effective in reducing or, in some cases, completely eliminating the symptoms of dysmenorrhea. This alleviation is accompanied by a marked reduction in the levels of prostaglandins found in the menstrual fluid. Other studies are investigating the use of these compounds in preventing labor in women who are in danger of giving birth prematurely.

Other Prostaglandin Effects

Prostaglandins that are closely related chemically may differ widely in their physiological effects. Although many of them stimulate contractions of smooth muscle, as previously described, others inhibit smooth muscle contraction. Furthermore, one may affect the smooth muscle of bronchioles, another the smooth muscle of blood vessels. One, produced by platelets, is a potent stimulus for platelet aggregation (see page 751) and constriction of the blood vessels; another, produced by the endothelial cells that line the blood vessels, is a potent inhibitor of platelet aggregation and a dilator of the vessels. The way in which prostaglandins exert these multitudinous effects is not known.

Among the prostaglandins are a group of substances known as leukotrienes, which are produced principally by the various white blood cells involved in the inflammatory and immune responses. The leukotrienes include the interleukins released by activated helper T cells (page 808), as well as a variety of molecules released by stimulated macrophages and mast cells. As this information might lead you to expect, increased levels of prostaglandins have been implicated in disorders of the immune system such as rheumatoid arthritis, asthma, and severe allergies. Discovery of the prostaglandins and their involvement in inflammation has also provided a solution to another long-standing medical puzzle. Aspirin is one of the oldest and most effective medications known, but how it acts was, for many years, a mystery. Now it has been discovered that aspirin exerts its effects, at least in part, by inhibiting the synthesis of prostaglandins, thus producing its well-known soothing effects on inflammation and fever.

MECHANISMS OF ACTION OF HORMONES

Prostaglandins, like many of the regulatory chemicals in small organisms, travel only short distances to the cells they affect; the communication is rather like that of one person talking to another in the same room. Neurotransmitters released from the nerve endings of neurons similarly travel only a short distance; as shown in Figure 40-1 (page 822), neurons make close, anatomical connections with the cells or organs they influence, like a conversation on the telephone. Most hormones, by contrast, broadcast their messages. Whether or not these messages are received and acted upon depends upon the receptivity of the target tissue as much as on the chemical characteristics of the hormone. Moreover, target tissues may be receptive under some circumstances and not under others. For example, prolactin causes the production of milk—but only by the mammary glands and only when it is acting in concert with the female hormones estrogen and progesterone, thyroxine, adrenal steroids, and growth hormone.

The key to this specificity of hormone action lies in protein receptor molecules that, like enzymes, transport proteins, and the surface receptors of lymphocytes, have quite precise configurations that allow them to bind with one particular molecule but not with another that differs only slightly in structure. The study of hormones and their receptors has revealed two quite different mechanisms of action. In one, the receptor molecules are intracellular, in the nucleus or cytoplasm; in the other, the receptor molecules are embedded in the cell membrane. The steroid hormones and thyroid hormone utilize the first mechanism; the catecholamine, peptide, and protein hormones act through the second mechanism.

Intracellular Receptors

Steroid hormones are relatively small, lipid-soluble molecules. Thus they pass easily through cell membranes and freely enter all cells of the body. However, in the cytoplasm of their target cells—and only in their target cells—these hormones encounter a specific protein receptor molecule with which they combine (Figure 40-14). The hormone-receptor complex moves to the nucleus, where it binds to a specific chromosomal protein, initiating mRNA transcription from the particular DNA sequence regulated by that protein. After appropriate processing, the mRNA moves into the cytoplasm and protein synthesis occurs. The newly synthesized proteins may be structural proteins, enzymes, or other hormones; the result is a functional change in the cell, in the substances released from it, or in the receptors displayed on its surface.

Thyroid hormone, although not soluble in lipids, also readily passes through cell membranes, apparently by facilitated diffusion through a transport protein. Its receptor is not in the cytoplasm but rather in the nucleus. As with the steroid hormones, the hormone-receptor complex binds to the chromosome, initiating mRNA transcription leading to protein synthesis.

40-14 *The mechanism of action of a steroid hormone. The lipid-soluble hormone passes through the cell membrane into the cytoplasm. In its target cell, the hormone encounters a specific receptor to which it binds. The hormone-receptor complex then passes into the nucleus, where it combines with a chromosomal protein, triggering the transcription of a segment of DNA into mRNA. After processing, the mRNA is translated into protein. Depending on the hormone and the particular target cell, the newly synthesized protein may be an enzyme, another hormone or other product that will be secreted from the cell, or a structural protein, for example, a receptor for a different hormone. The altered functions of the cell as a consequence of the newly synthesized protein constitute the cell's response to the hormone.*

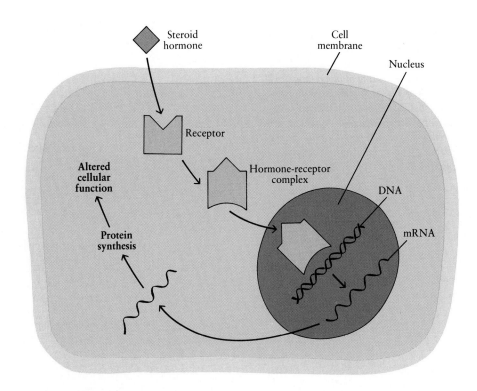

40-15 *Cyclic AMP (adenosine monophosphate) acts as a "second messenger" within a number of different types of vertebrate cells. Following stimulation by various hormones—the "first messengers"—cyclic AMP is formed from ATP. "Cyclic" refers to the fact that the atoms of the phosphate group form a ring. Cyclic AMP is also the chemical that attracts the amoebas of the cellular slime molds, causing them to aggregate into a slug-like body, which then behaves like a multicellular organism (see page 141).*

Membrane Receptors

The second group of hormones acts by combining with receptors in the membranes of target cells. The existence of these receptors has important medical implications. For example, it has long been known that juvenile diabetes—diabetes in young persons—is caused by a deficiency of the hormone insulin, which, as we have seen, promotes the uptake of glucose by cells. It was assumed by analogy that the diabetes found commonly in older persons and associated with obesity had the same cause. It has now been found, however, that adult diabetes often results from a decrease in the number of insulin receptor sites in target cell membranes rather than from a shortage of insulin. Such patients are treated most effectively by diet.

After a hormone combines with its membrane receptor, one of two events may follow, depending on the particular hormone. In some cases, the hormone-receptor complex is carried into the cytoplasm by receptor-mediated endocytosis (page 138). In other cases, the hormone never actually enters the cell; instead, its binding to the receptor sets in motion a "second messenger" that is responsible for the sequence of events inside the cell. The second messenger for many hormones is a chemical known as cyclic AMP (Figure 40-15).

An example of the second-messenger mechanism in action is the stimulation of the release of glucose from a liver cell by adrenaline (Figure 40-16). Adrenaline molecules bind to a receptor on the outer surface of the cell membrane. This event activates an enzyme, adenylate cyclase, that is bound on the inner surface of the membrane. Adenylate cyclase converts ATP to cyclic AMP (cAMP). The cyclic AMP binds to another enzyme, protein kinase, and activates it. This enzyme activates another enzyme, which, in turn, activates yet another. The final enzyme, phosphorylase *a*, then breaks down glycogen at a high rate, producing glucose 1-phosphate, which is further broken down to glucose and released from the cell. Since each enzyme greatly increases the rate of the particular reaction it catalyzes and can be used over and over again, the number of molecules involved in these

40–16 *Adrenaline triggers an amplification cascade in liver cells, similar in principle to the amplification cascade that occurs in the clotting of blood (page 751). Binding of a few molecules of adrenaline to its specific receptors on the outer surface of the cell membrane initiates a series of enzymatic reactions that result in the release of a very large amount of glucose into the blood.*

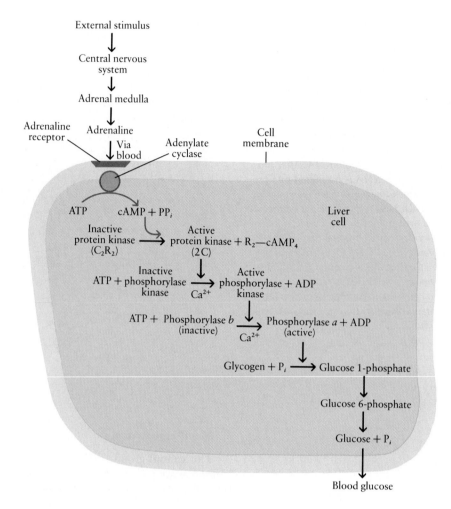

reactions is amplified at each step. Thus, the binding of a few molecules of adrenaline to cells in the liver leads to the activation of an estimated 25 million molecules of phosphorylase *a* and the consequent release of many grams of glucose into the blood.

Just about the same time that the role of cyclic AMP in mammalian cells was established—for which Earl W. Sutherland was awarded the Nobel Prize—biologists studying that peculiar group of organisms known as the cellular slime molds (see page 141) isolated a chemical of great importance in this biological system. As you will recall, the cells of the cellular slime mold begin as individual amoebas and then come together to form a single organism. The chemical that calls them together, which was named acrasin, was identified as cyclic AMP. You may also recall (page 325) that cyclic AMP plays a role in regulation of the *lac* operon in *Escherichia coli*.

More recently, it has been found that insulin is present in fruit flies, earthworms, protists, fungi, and even *E. coli*. Preliminary evidence indicates that other "mammalian" hormones, including ACTH, glucagon, and somatostatin, are present in unicellular organisms. Their function in these organisms is not known, but they may also, like acrasin, serve in cell-to-cell communication. These new discoveries of the universality of hormones provide other examples of the long thread of evolutionary history linking all organisms.

SUMMARY

Hormones are signalling molecules secreted in one part of an organism that diffuse or, in the case of vertebrates, are carried by the bloodstream to other tissues and organs, where they exert specific effects. Most hormones are secreted by glandular epithelial tissue or by neurosecretory cells. The principal endocrine glands of vertebrates include the pituitary, the hypothalamus, the thyroid, the parathyroids, the adrenal cortex and medulla, the pancreas (which is also an exocrine gland), the pineal, and the gonads (ovaries or testes).

The production of many hormones is regulated by negative feedback systems involving the anterior lobe of the pituitary gland and an area of the brain, the hypothalamus. Under the influence of hormones secreted by the hypothalamus, the pituitary produces tropic hormones that, in turn, stimulate the target glands to produce hormones. These hormones then act upon the pituitary or the hypothalamus (or both) to inhibit the production of the tropic hormones. Production of thyroid hormone and the steroid hormones of the adrenal cortex and gonads is regulated by the hypothalamus-pituitary system. The production of other hormones, such as calcitonin and parathyroid hormone, is regulated by the concentration in the bloodstream of other factors, such as ions.

Besides producing the tropic hormones, the anterior lobe of the pituitary also secretes somatotropin (growth hormone) and prolactin. In addition to producing at least nine peptide hormones (sometimes called releasing hormones) that act upon the anterior lobe of the pituitary, the hypothalamus produces the hormones ADH and oxytocin, which are stored in and released from the posterior lobe of the pituitary.

The islet cells of the pancreas are the source of three hormones involved in the regulation of blood glucose: insulin, which lowers blood sugar by stimulating cellular uptake of glucose; glucagon, which raises blood sugar by stimulating the breakdown of storage forms of glucose; and somatostatin. Blood sugar is also under the influence of adrenaline and noradrenaline, which are released from the adrenal medulla at times of stress; cortisol and other glucocorticoids, which are released from the adrenal cortex at times of stress; and somatotropin.

The pineal gland, located in the brain, is the source of melatonin and is believed to be involved with regulation of circadian and seasonal physiological changes. Its function in humans is not known.

Prostaglandins are a group of fatty acids that resemble other hormones in exerting effects on specific target tissues but that often act directly upon the tissues that produce them or, in some cases, on the tissues of another individual. Prostaglandins are formed in most, if not all, tissues of the body and affect such diverse functions as the contraction of smooth muscle, platelet aggregation, and the immune response.

Hormones act by at least two different mechanisms. Steroid hormones and thyroid hormone freely enter cells, where, after combining with an intracellular receptor, they exert a direct influence on the transcription of RNA. Catecholamine, peptide, and protein hormones, such as adrenaline, insulin, and glucagon, combine with receptor molecules on the surface of the target-cell membranes. The hormone-receptor combination may be carried into the cytoplasm by receptor-mediated endocytosis, or the combination may trigger the release of a "second messenger." The second messenger, in turn, sets off a series of events within the cell that is responsible for the end results of hormone activity. Cyclic AMP has been identified as the second messenger in many of these interactions.

QUESTIONS

1. Distinguish among the following: endocrine/exocrine; pituitary/hypothalamus; anterior pituitary/intermediate pituitary/posterior pituitary; thyroid gland/parathyroid glands; thyroxine/thyrotropin/triiodothyronine; thyroid-stimulating hormone/thyrotropin-releasing hormone; adrenal cortex/adrenal medulla; insulin/glucagon.

2. Diagram the feedback system regulating the production and release of thyroid hormone.

3. In terms of the feedback system diagrammed in Question 2, explain why a shortage of iodine produces goiter.

4. Describe how the concentration of calcium ion in the bloodstream is regulated by the thyroid and parathyroid glands.

5. Which hormones act to increase the level of blood glucose? To decrease it? How does each hormone exert its effects?

6. How does a steroid hormone exert its specific effects on a target cell? In what way is the action of thyroid hormone different?

7. How does a peptide or protein hormone exert its specific effects on a target cell?

8. What types of functions would you expect to be controlled by the endocrine system rather than by the nervous system? Does the reality, as described in this chapter, fulfill your expectations?

C H A P T E R **41**

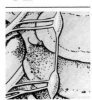

Integration and Control II: The Nervous System

As we saw in the last chapter, communication among cells—whether unicellular organisms or the component cells of a multicellular organism—depends upon chemical stimuli. Specific molecules are released from secretory cells and are transported to other cells, where they interact with protein receptors and trigger a response. The presence in prokaryotes, protists, and fungi of signalling molecules identical to those of vertebrates and the dual function of some vertebrate molecules as both hormones and neurotransmitters provide strong evidence that the endocrine and nervous systems had a common evolutionary origin in primitive cell-cell communication systems.

The specialization that distinguishes a nervous system from other communication systems is the neuron, a cell that converts appropriate stimuli into electrochemical signals that are rapidly conducted through the neuron itself, often over great distances. Neurons then transmit these signals to other neurons across junctions known as **synapses,** usually by the release of specific neurotransmitter molecules, and to effector cells, such as those of muscles and glands. The rapid communication provided by neurons and their arrangement in organized networks—nervous systems—are the keys to the integration and control underlying the active life style of animals.

EVOLUTION OF NERVOUS SYSTEMS

As we saw in Section 4, invertebrate nervous systems range from the very simple to the complex (Figure 41-1). In the cnidarian *Hydra,* the neurons form a diffuse network. They receive information from sensory receptor cells, which can be

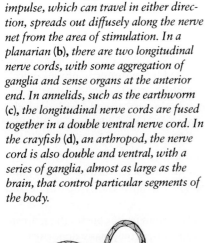

41-1 *In* Hydra (a), *a cnidarian, the nerve impulse, which can travel in either direction, spreads out diffusely along the nerve net from the area of stimulation. In a planarian* (b), *there are two longitudinal nerve cords, with some aggregation of ganglia and sense organs at the anterior end. In annelids, such as the earthworm* (c), *the longitudinal nerve cords are fused together in a double ventral nerve cord. In the crayfish* (d), *an arthropod, the nerve cord is also double and ventral, with a series of ganglia, almost as large as the brain, that control particular segments of the body.*

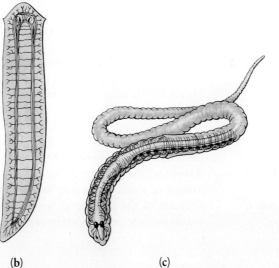

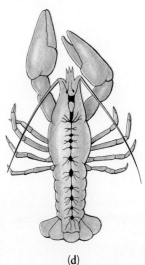

(a)　　　　　　　(b)　　　　　　　(c)　　　　　　　(d)

found among the epithelial cells on the inner (feeding) and outer surfaces of the animal. The neurons stimulate epitheliomuscular cells that cause movements in the body wall.

The nervous system of the planarian, a flatworm, is more highly organized than that of *Hydra,* providing a more efficient coordination that makes possible increased motility. Some of the nerve net is condensed into two cords, and there are two clusters of nerve cell bodies at the anterior end of the body. As you will recall (page 535), such clusters of nerve cell bodies are known as ganglia (singular, ganglion).

In the earthworm, the two cords have come together in a fused, double nerve cord that runs along the ventral surface of the body. Along the nerve cord are ganglia, one for each segment of the body. The nerve cord forks just below the pharynx, and the two forks meet again in the head, terminating in two large dorsal ganglia.

Arthropods, such as the crayfish, also have a double ventral nerve cord and, in addition, may have sizable clusters of nerve cell bodies in the head region. Collectively, these ganglia are large enough to be called a brain. The nervous system also contains many other ganglia interconnected by nerve fibers that run along the ventral surface. In animals with this type of nervous system, many quite complicated activities—for example, the complex movements of the finely articulated appendages—are coordinated by the nearest ganglion.

In vertebrates, the nervous system, which is dorsal rather than ventral, has been greatly elaborated. Its central processing centers—the spinal cord and the brain—are enclosed and protected by the bones of the vertebral column and the skull. The trend in vertebrate evolution has been toward increased centralization of control in the brain—cephalization—a trend that may be still continuing. The precise integration that accompanies such centralization makes possible such complex behaviors as the dive of a kingfisher for a bass (see page 42) and the fashioning of a tiny lure by a fly fisherman.

ORGANIZATION OF THE VERTEBRATE NERVOUS SYSTEM

As we noted in Chapter 33, the vertebrate nervous system has a number of subdivisions (Figure 41-2) that can be distinguished by anatomical, physiological, and functional criteria. The primary and most obvious is the subdivision of the system into the central nervous system (the brain and spinal cord) and the peripheral nervous system (the sensory and motor pathways that carry information to and from the central nervous system). The motor pathways are further divided into the somatic system, which stimulates skeletal muscle, and the autonomic system, which relays signals to smooth muscle, cardiac muscle, and glands. The autonomic system is, in turn, subdivided into the sympathetic and parasympathetic divisions.

The functional unit of the vertebrate nervous system is, of course, the neuron, a nerve cell characterized by a cell body, an axon, and, often, many dendrites (see Figure 33-10, page 708). Neurons are surrounded and insulated by glial cells; in the central nervous system, these are neuroglia, and in the peripheral nervous system, they are Schwann cells. In vertebrates, as in invertebrates, nerve cell bodies are often found in clusters. Such clusters outside the central nervous system are called ganglia; inside the central nervous system, they are generally called **nuclei**. Axons (nerve fibers) are also grouped together, forming bundles; these bundles are known as **tracts** when they are in the central nervous system, and **nerves** when they are in the peripheral nervous system. The individual axons in tracts and nerves are often enveloped in insulating myelin sheaths formed by specialized glial

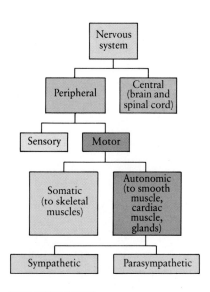

41-2 *A summary of the subdivisions of the vertebrate nervous system.*

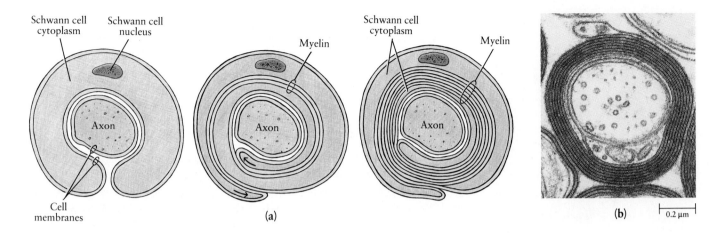

41–3 (a) *Formation of a myelin sheath by a Schwann cell, a type of glial cell found in the peripheral nervous system. As the Schwann cell grows, it wraps itself around and around the axon and gradually extrudes its cytoplasm from between the layers. The myelin sheath, which consists of layers of lipid-containing cell membranes, insulates the nerve fiber.* (b) *Electron micrograph of a cross section of a mature myelin sheath. Its dark appearance is the result of chemicals used to "fix" the specimen for electron microscopy (see page 98); without such treatment, the myelin sheath appears white.*

cells (Figure 41–3). These sheaths are rich in lipids, giving nerves and tracts a glistening white appearance.

The Central Nervous System

The central nervous system is made up of the brain and the spinal cord, which provides the critical link between the brain and the rest of the body. Your spinal cord, which is a slim cylinder about as big around as your little finger, can be seen in cross section to be divided into a central area of gray matter and an outer area of white matter (Figure 41–4). The gray matter is mostly interneurons (which, as you may recall from page 709, conduct signals within localized regions of the central nervous system), the cell bodies of motor neurons, and neuroglia. The white matter consists of fiber tracts running longitudinally through the spinal cord; the fibers comprising these tracts are primarily the axons of relay neurons, which conduct signals between regions of the central nervous system.

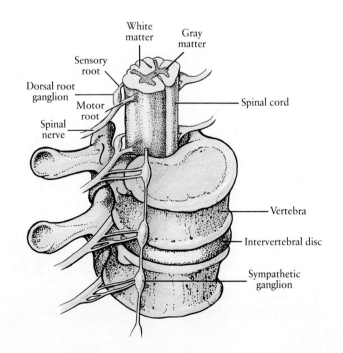

41–4 *A portion of the human spinal cord and vertebral column. Each spinal nerve divides into two fiber bundles, the sensory root and the motor root, at the vertebral column. The sensory root connects with the cord dorsally; the cell bodies of the sensory neurons are in the dorsal root ganglia. The motor root connects ventrally with the spinal cord; the cell bodies of the motor neurons are in the spinal cord itself. The sympathetic ganglia, which form a chain, are part of the autonomic nervous system.*

The butterfly-shaped gray matter within the spinal cord is composed mostly of interneurons, cell bodies of motor neurons, and glial cells. The surrounding white matter consists of ascending and descending fiber tracts.

SECTION 6 Biology of Animals

The spinal cord is continuous with the brainstem, which is the base of the brain. The brainstem contains fiber tracts conducting signals to and from the spinal cord and also the cell bodies of the neurons whose axons innervate the muscles and glands of the head. In addition, within the brainstem are centers for some of the important automatic regulatory functions, such as control of respiration and blood pressure, which are also influenced by other parts of the brain.

The Peripheral Nervous System

The peripheral nervous system is made up of neurons whose axons extend out of the central nervous system into the tissues and organs of the body. These include both motor (efferent) neurons, which carry signals out, and sensory (afferent) neurons, which carry signals in. The fibers of motor and sensory neurons are bundled together into nerves, which are classified as **cranial nerves,** those nerves that connect directly with the brain (such as the optic nerve), and **spinal nerves,** those that connect with the spinal cord. Pairs of spinal nerves enter and emerge from the cord through spaces between the vertebrae. The motor fibers of each pair innervate the muscles of a different area of the body, and the sensory fibers receive signals from sensory receptors in the same area. In humans there are 31 such pairs.

As you can see in Figure 41–4, the motor and sensory fibers of the spinal nerves separate from each other near the spinal cord. The cell bodies of the sensory neurons are in the dorsal root ganglia outside the spinal cord, and the sensory fibers feed into the dorsal side of the spinal cord. Here they may synapse with relay neurons, interneurons, or motor neurons, they may turn and ascend toward the brain, or they may do all of these. Fibers from the motor neurons emerge from the spinal cord on the ventral side. The cell bodies of motor neurons are located in the spinal cord, where they may receive signals from relay neurons, interneurons, and sensory neurons.

The four types of neurons are often interconnected in reflex arcs (Figure 41–6). A stimulus received by a sensory neuron is conducted to the central nervous system, where the neuron synapses with either a motor neuron (a monosynaptic reflex) or one or more interneurons (a polysynaptic reflex) that, in turn, synapse with a motor neuron that completes the arc, activating an effector that carries out the reflex action. Simultaneously, relay neurons conduct information concerning

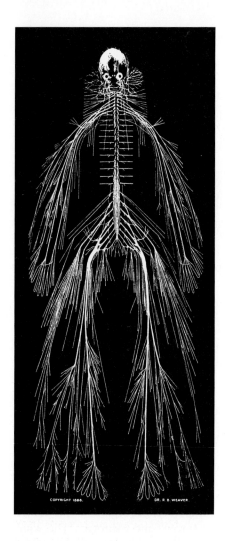

41–5 *The human nervous system, as dissected out in 1888 by Rufus B. Weaver, a Philadelphia physician. The subject, a maid employed by Dr. Weaver, had requested that her body be used to benefit science after her death. It took Dr. Weaver more than five months of daily work to remove the nerves.*

41–6 *A polysynaptic reflex arc. In this example, free nerve endings in the skin, when appropriately stimulated, transmit signals along the sensory neuron to an interneuron in the spinal cord. The interneuron transmits the signal to a motor neuron. As a result of the stimulation of the motor neuron, muscle fibers contract. Relay neurons, not shown here, are also stimulated by the sensory neuron and carry the sensory information to the brain.*

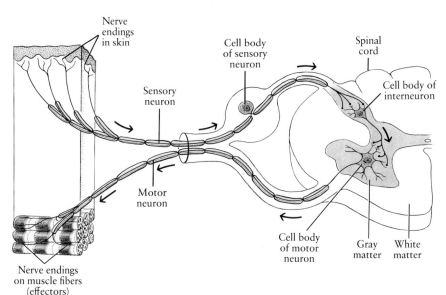

the event to other parts of the central nervous system. Thus, for instance, if your fingers touch a hot stove, your hand will automatically withdraw from the stove. Almost simultaneously, your brain will become aware of what has happened and you will take further action or make an appropriate comment.

Divisions of the Peripheral Nervous System: Somatic and Autonomic

As we noted earlier, there are two subdivisions of the motor pathways of the peripheral nervous system: the somatic and the autonomic. The autonomic ("involuntary") nervous system consists of the motor nerves that control cardiac muscle, glands, and smooth muscle (the type of muscle found in the walls of blood vessels and in the digestive, respiratory, excretory, and reproductive systems). The somatic ("voluntary") system controls the skeletal muscles—that is, the muscles that can be moved at will.

You will readily recognize that the distinction here between "voluntary" and "involuntary" is not clear-cut. Skeletal muscles—part of the somatic system—often move involuntarily, as in a reflex action. On the other hand, it has been reported that some individuals, such as practitioners of yoga or those who have had biofeedback training, can control their rate of heartbeat and the contractions of some smooth muscle, both of which are regulated by the autonomic system.

Anatomically, the motor neurons of the somatic system are distinct and separate from those of the autonomic nervous system, although axons of both types may be carried within the same nerve. The cell bodies of the motor neurons of the somatic system are located within the central nervous system, with long axons running without interruption all the way to the skeletal muscles. The pathways of the autonomic nervous system also include axons that originate in cell bodies inside the central nervous system, but these axons do not usually travel all the way to the target organs, or effectors. Instead, they synapse outside the central nervous system with motor neurons, which then innervate the effectors (see Figure 41–7 on the next page). These synapses occur within ganglia. Thus the neurons whose axons emerge from the central nervous system and terminate in the ganglia are known as **preganglionic,** while those whose axons emerge from the ganglia and terminate in the effectors are known as **postganglionic.** This two-neuron pathway constitutes a characteristic difference between the autonomic and somatic systems.

Another major difference between these systems is that, in vertebrates, the somatic system can only stimulate or not stimulate an effector; it cannot inhibit an effector. The autonomic system, by contrast, can stimulate or inhibit the activity of an effector. In the somatic system, sensory input often comes from neurons monitoring environmental change. The autonomic nervous system receives its sensory input from some of the same neurons as the somatic nervous system and also from sensory neurons monitoring changes in the interior of the body, such as the group signaling changes in blood pressure. These neurons are involved in reflexes similar to the one shown in Figure 41–6. One difference, however, is that, in reflex arcs involving the autonomic system, you are usually not conscious that the reflex action has taken place.

Divisions of the Autonomic Nervous System: Sympathetic and Parasympathetic

The autonomic nervous system itself has two divisions: the sympathetic division and the parasympathetic division. These two divisions are anatomically, physio-

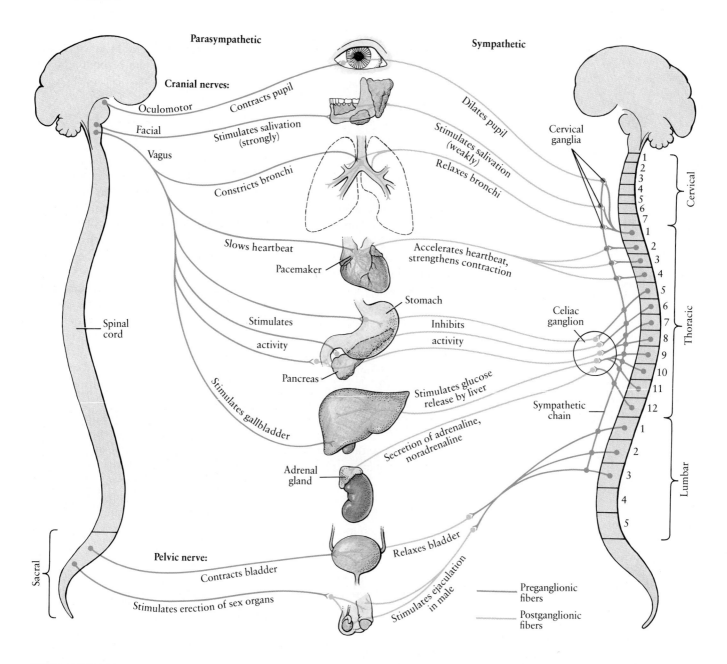

41–7 *The autonomic nervous system. It differs from the somatic system anatomically in that the axons emerging from the central nervous system do not travel without interruption to the effectors. Instead, they synapse outside the central nervous system with motor neurons, which then innervate the effectors. The fibers emerging from the central nervous system are known as preganglionic fibers, and those terminating in the effectors are known as postganglionic fibers.*

The autonomic nervous system consists of the sympathetic and the parasympathetic divisions. The preganglionic fibers of the parasympathetic division exit from the base of the brain and from the sacral region of the spinal cord and synapse with the postganglionic neurons at or near the target organs. The sympathetic division originates in the thoracic and lumbar regions. Preganglionic fibers of the sympathetic division synapse with postganglionic neurons in the chain of sympathetic ganglia or in other ganglia, such as the celiac ganglion, which is part of the solar plexus. The neurotransmitter at the effector is usually noradrenaline in the sympathetic division and acetylcholine in the parasympathetic division.

Most, but not all, internal organs are innervated by both divisions, which usually function in opposition to each other. In general, the sympathetic division stimulates functions involved in "fight-or-flight" reactions, and the parasympathetic division stimulates more tranquil functions, such as digestion.

$$H_3C-\overset{\overset{O}{\|}}{C}-O-CH_2-CH_2-\overset{\overset{CH_3}{|}}{\underset{\underset{CH_3}{|}}{N}}-CH_3$$

Acetylcholine

41-8 *Acetylcholine, the principal neurotransmitter in motor fibers of the peripheral nervous system. It is released by both the preganglionic and postganglionic nerve endings of the parasympathetic division of the autonomic nervous system and by the preganglionic nerve endings of the sympathetic division. Noradrenaline (Figure 40-11, page 832) is usually the neurotransmitter at the postganglionic nerve endings of the sympathetic division. As we saw in the last chapter, noradrenaline and a related molecule, adrenaline, are also secreted by the adrenal medulla, the central portion of the adrenal gland.*

41-9 *A cat prepared for "fight or flight," showing the effects of stimulation by the sympathetic nervous system. This drawing is from Darwin's* The Expression of the Emotions in Man and Animals, *published in 1872.*

logically, and functionally distinct. The major anatomical and physiological differences between them are:

1. Axons of the sympathetic division originate in the thoracic (chest) and lumbar (lower back) regions of the spinal cord. Axons of the parasympathetic division emerge through the cranial (brain) region and the sacral ("tail") region of the spinal cord.

2. As previously stated, in the autonomic nervous system, there is always a relay system of two neurons connecting the central nervous system and the effector organ. These neurons synapse at a ganglion. In the sympathetic division, the ganglia are usually close to the central nervous system. Thus, characteristically, the preganglionic axon is short, and the postganglionic axon is long. In the parasympathetic division, the opposite is true: the ganglia are close to or embedded in the target organ. Thus the preganglionic axon is long, and the postganglionic axon is short.

3. Although the preganglionic nerve endings of both divisions release acetylcholine (Figure 41-8) as their neurotransmitter, the postganglionic nerve endings of the two divisions utilize different neurotransmitters. The neurotransmitter released by most postganglionic sympathetic nerve endings is noradrenaline. All postganglionic parasympathetic endings release acetylcholine.

Functionally, the two divisions are generally antagonistic. As you can see in Figure 41-7, most of the internal organs are innervated by axons from both divisions. They work in close cooperation with each other and with hormones secreted by the endocrine glands for the ultimate homeostatic regulation of the body. The parasympathetic division is involved primarily in the restorative activities of the body; it is particularly active, for example, after a heavy meal or following orgasm. Parasympathetic stimulation slows down the heartbeat, increases the movements of the smooth muscle of the intestinal wall, and stimulates secretions of the salivary glands and the digestive glands of the stomach.

The sympathetic division, by contrast, prepares the body for action. The physical characteristics of fear, for example, result from the increased discharge of neurons of the sympathetic division. Some of these cause the blood vessels in the skin and intestinal tract to contract; this contraction increases the return of the blood to the heart, raising the blood pressure and allowing more blood to be sent to the muscles. The heart beats both faster and stronger, and the respiratory rate increases. The pupils dilate. The muscles attached to the hair follicles in the skin contract; this is probably a legacy from our furry forebears, which looked larger and more ferocious with their hair standing on end (Figure 41-9). The rhythmic movement of the intestines stops, and the sphincters, muscles at the end of the intestines and the opening of the bladder, relax. These reactions inhibit digestive operations, but the relaxing of the sphincters may also, in extreme cases, have the disconcerting consequence of allowing involuntary defecation or urination. Sympathetic stimulation causes the adrenal medulla to pour out adrenaline, which, with other hormones, causes the release of large quantities of glucose from the liver into the bloodstream, as we saw in Figure 40-16. This glucose is the extra energy source for the muscles. As a consequence of this constellation of responses, the body as a whole is prepared for "fight or flight"—or, at least, for action that would have been appropriate at some earlier stage of our cultural evolution.

THE NERVE IMPULSE

Some 200 years ago, Luigi Galvani observed that the passage of an electric current along the nerve of a frog's leg made the muscle twitch. Since that time, it has been known that nerve conduction is associated with electrical phenomena. As you will recall, there are two types of electric charge, positive and negative; like charges repel one another, and unlike charges attract. Thus, negatively charged particles tend to move toward a region of positive charge, and vice versa. (In a metal, such as copper, only negative charges move, but the principle is the same.) Materials that permit the movement of charged particles, such as a copper wire or a solution containing ions, are known as conductors. Materials that do not permit the movement of charged particles, such as fat and rubber, are insulators.

The difference in the amount of electric charge between a region of positive charge and a region of negative charge is called the **electric potential.** As we saw in Chapter 9 (page 198), an electric potential is a form of potential energy, like a boulder at the top of a hill or water behind a dam. This potential energy is converted to electrical energy when charged particles are allowed to move through a solution or along a wire between the two regions of differing charge. The difference in potential energy between the two regions is measured in volts or, if it is very small, in millivolts.

For a time, it was assumed that a nerve impulse was an electric current—that is, a flow of ions—traveling along an axon in the same way that electrons flow along a wire. However, this model did not hold up under critical scrutiny. First, it was shown that the axon is a poor conductor of electricity—meaning that ions flow through it for only a short distance before the current dies out. By contrast, the nerve impulse is undiminished from one end of the axon to the other. Second, although the impulse is fast by biological criteria, it is much slower than an electric current. Finally, unlike an electric current, the strength of the impulse is always the same. There is either no nerve impulse in response to the stimulation of a nerve fiber, or there is a maximum response—all or nothing.

Major advances in understanding the nature of the nerve impulse occurred when it became possible to monitor changes in electric potential in an individual neuron. The organism that first made this possible was the squid, which has motor neurons with large, long axons (Figure 41-10). Measurements within an axon are made with microelectrodes tiny enough to penetrate a living cell without seriously injuring it. The microelectrodes are connected with a very sensitive voltmeter called an oscilloscope, which measures voltage (in millivolts) in relation to time (in milliseconds).

When both electrodes are outside the neuron, no voltage difference is recorded (Figure 41-11a). When one electrode penetrates the axon, a voltage difference of about 70 millivolts can be detected between the outside and the inside of the axon. The interior of the membrane is negatively charged with respect to the exterior (Figure 41-11b). This is the **resting potential** of the membrane. When the axon is stimulated, the oscilloscope records a very brief reversal of polarity—that is, the interior becomes positively charged in relation to the exterior (Figure 41-11c). This reversal in polarity is called the **action potential.** The action potential traveling along the membrane is the **nerve impulse.** It travels more slowly than an electric current, because it is actually a series of miniature electric circuits, but it is undiminished in intensity during its conduction.

The action potentials recorded from any one neuron are almost always the same. Figure 41-12 shows, for example, the action potentials produced by a single sensory neuron in the skin of a cat in response to pressure. All the impulses are the same size (all-or-nothing response). The only variation—and it is a critical variation—is the frequency with which the impulses are produced. The message carried by nerve impulses is, in effect, a Morse code with dashes but no dots.

The Ionic Basis of the Action Potential

The action potential depends on the electric potential of the axon, which is, in turn, made possible by differences in the concentration of ions on either side of

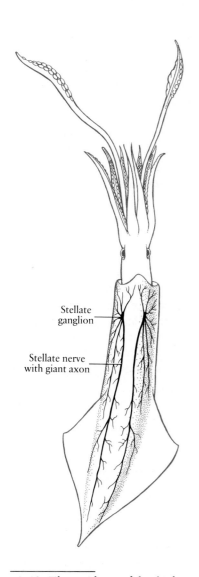

41-10 *The squid, one of the chief protagonists in research leading to an understanding of the nature of the nerve impulse. The stellate nerves contain the giant axons used in all the early studies of the nerve impulse. The giant axons innervate muscles in the wall of the mantle; powerful contractions of those muscles result in a rapid expulsion of water from the mantle cavity, producing the escape response described on page 553.*

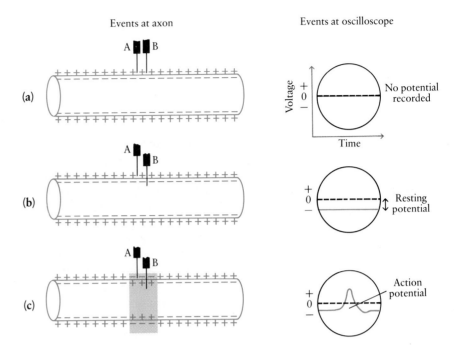

41-11 *The electric potential across the membrane of the axon is measured by microelectrodes connected to an oscilloscope. (a) When both electrodes are outside the membrane, no potential is recorded. (b) When one electrode penetrates the membrane, the oscilloscope shows that the interior is negative with respect to the exterior and that the difference between the two is about 70 millivolts. This is the resting potential. (c) When the axon is stimulated and a nerve impulse passes along it, the oscilloscope shows a brief reversal of polarity—that is, the interior becomes positive in relation to the exterior. This brief reversal in polarity is the action potential.*

the membrane. Such differences in concentration—and thus in electric potential—are characteristic of all cells, and, as we have seen in earlier chapters, they can be used to power a variety of cellular processes. Their use to carry information through the body of an animal is perhaps the most sophisticated application to have evolved thus far.

In axons, the critical concentration differences involve potassium ions (K^+) and sodium ions (Na^+). In the resting state, the concentration of K^+ ions in the cytoplasm of an axon is about 30 times higher than in the fluid outside; conversely, the concentration of Na^+ ions is about 10 times higher in the extracellular fluid than in the cytoplasm. The distribution of ions on either side of the membrane is governed by three factors: (1) the diffusion of particles down a concentration gradient (page 129), (2) the attraction of particles with opposite charges and the repulsion of particles with like charges, and (3) the properties of the membrane itself.

The lipid bilayer of the axon membrane is, like the lipid bilayer of other cell membranes, impermeable to ions and most polar molecules. The movement of such particles through the membrane depends on the presence of integral membrane proteins that provide channels through which the particles can move, either by facilitated diffusion or active transport. The membrane of the axon is rich in proteins that provide channels for the movement of specific ions, particularly Na^+

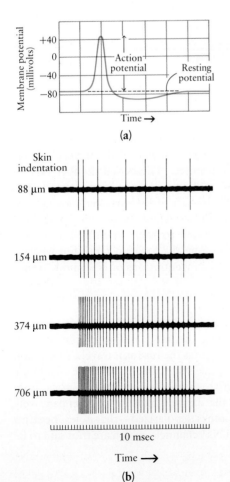

41-12 *(a) Nerve impulses can be monitored by electronic recording instruments. The impulses from any one neuron are all the same; that is, each impulse has the same duration and voltage change as any other.*

(b) Nerve impulses from a sensory neuron (touch receptor) in cat skin. The skin was touched and pressed in at varying depths, as indicated by the figures at the left. The more deeply the skin was pressed in, the more rapidly nerve impulses were produced. The vertical lines represent individual action potentials on a compressed time scale. As you can see, all the action potentials are the same size, but their frequency increases with the intensity of the stimulus.

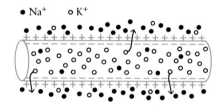

41-13 *Resting potential. The membrane of the axon is polarized, with a slight excess of negative charge inside the membrane caused by the outward diffusion of K^+ ions. The movement of ions across the axon membrane depends on the presence of ion channels formed by integral membrane proteins. In the resting state, the K^+ channels are open, but the Na^+ channels are closed.*

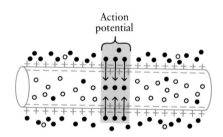

41-14 *The action potential. A portion of the membrane becomes momentarily permeable to Na^+ ions as Na^+ channels open. The Na^+ ions rush in, and the polarity of the membrane is reversed at that point.*

and K^+. An important feature of these proteins is that changes in their conformation result in either the opening or the closing of the ion channels; thus, the channels are said to be **gated**. Depending on the specific ion channel, the changes in conformation that open or close its gate may be regulated by the voltage across the membrane or by the action of neurotransmitters (or drugs that mimic neurotransmitters). Another significant feature of the axon membrane is the presence of the integral membrane protein known as the sodium-potassium pump (page 137), which pumps Na^+ ions out of the axon and K^+ ions in.

When the axon membrane is in its resting state, the Na^+ ion channels are mostly closed; as a consequence, the membrane is almost impermeable to Na^+ ions. The few Na^+ ions that diffuse in through open channels—moving down their concentration gradient—are promptly removed by the sodium-potassium pump. Many K^+ ion channels, however, are open, and the membrane is thus relatively permeable to K^+ ions. Because of the concentration gradient, K^+ ions tend to move out of the cell; if no other forces were at work, they would, of course, move down the concentration gradient until their distribution were equal on either side of the membrane. However, because of the impermeability of the lipid bilayer, negatively charged ions cannot follow the K^+ ions out of the cell. Thus, as K^+ ions leave, an excess negative charge builds up inside the cell. This excess of negative charge attracts the positive K^+ ions, impeding their further outward movement. As a result, an equilibrium is reached at which there is no net movement of K^+ ions across the membrane. At the point of equilibrium, there is a slight excess of negative charge inside the cell, and the membrane is said to be polarized. This point of equilibrium is the resting potential (Figure 41-13).

When the membrane is stimulated, it suddenly becomes permeable to Na^+ ions as Na^+ channels open at the site of stimulation. The Na^+ ions rush in, moving down their concentration gradient, attracted initially by the negative charge inside the axon. This influx of positively charged ions momentarily reverses the polarity of the membrane so that it becomes more positive on the inside than on the outside, producing the action potential (Figure 41-14). The change in Na^+ permeability lasts for only about half a millisecond; then the Na^+ channels close and the membrane regains its previous impermeability to Na^+ ions. During this time, more K^+ channels open, increasing the permeability to K^+ ions. The result is an outward flow of K^+ ions due to the concentration gradient and also to the positive charge inside the axon at the peak of the action potential. This outward flow of positive K^+ ions counteracts the previous inward flow of positive Na^+ ions, and the resting potential is quickly restored. The actual number of ions involved is very small. Only a very few Na^+ ions need enter to reverse the polarity of the membrane, and only the same small number of K^+ ions need move out of the cell to restore the resting potential. Subsequently, the sodium-potassium pump restores the Na^+ and K^+ concentrations to their original levels. As a consequence, action potentials can move along the axon in rapid fire without substantial changes occurring in the internal concentrations of Na^+ and K^+ ions.

Propagation of the Impulse

An important feature of the nerve impulse is that, once initiated, this transient reversal in polarity continues to move along the axon, renewing itself continuously, just as a flame traveling along a fuse ignites the fuse as it travels. The action potential is self-propagating because at its peak, when the inside of the membrane at the active region is comparatively positive, positively charged ions move from this region to the adjacent area inside the axon, which is still comparatively negative. As a result, the adjacent area becomes depolarized—that is, less negative (Figure 41-15). This depolarization opens Na^+ channels, which are thus said to be

41–15 *Propagation of the nerve impulse. In advance of the action potential, a small segment of the membrane becomes slightly depolarized, owing to the flow of positively charged ions along the inside of the membrane. When the membrane becomes depolarized in this way, Na^+ ion channels open and the permeability of the membrane to Na^+ increases. Na^+ ions move in through the open channels, creating an action potential in that region of the membrane and depolarizing the next adjacent segment of membrane. The nerve impulse is the action potential traveling along the membrane.*

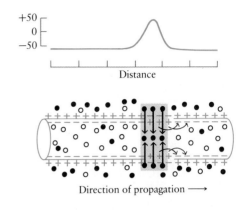

voltage-gated, and allows Na^+ ions to rush in. The resulting increase in the internal concentration of Na^+ ions depolarizes the next adjacent area of the membrane, causing its Na^+ ion channels to open, and allowing the process to be repeated. As a consequence of this renewal process, repeating itself along the length of the membrane, the axon—a very poor conductor of an ordinary electric current—is capable of conducting a nerve impulse over a considerable distance with absolutely undiminished strength. The nerve impulse moves in one direction only because the segment of the axon behind the action potential has a brief refractory period during which its Na^+ ion channels will not open; thus the action potential cannot go backwards.

The Role of the Myelin Sheath

As we saw in Figure 41–3 (page 843), long axons are generally enveloped in myelin sheaths formed by specialized glial cells. The myelin sheath, according to Nobel laureate John Eccles, is "a brilliant innovation." Because of it, the propagation of the nerve impulse is much more rapid in vertebrates than in invertebrates. The myelin sheath is not just an insulator. Its most important feature is that it is interrupted at regular intervals by openings, or nodes. Only at the nodes is it possible for Na^+ and K^+ ions to move into and out of the axon. Thus, in myelinated fibers—which include all the large nerve fibers of vertebrates—the impulse jumps from node to node (Figure 41–16), rather than moving continuously along the membrane like water along a wick. This saltatory (leaping) conduction greatly increases the velocity. Some large, myelinated nerve fibers, for example, conduct impulses as rapidly as 200 meters per second, compared with velocities of only a few millimeters per second in small, unmyelinated fibers. Also, because Na^+ and K^+ ions move across only a small portion of the axon's membrane, there is an enormous saving in energy expenditure by the sodium-potassium pump.

THE SYNAPSE

Signals travel from one neuron to another across the specialized junction known as the synapse, which may be electrical or chemical in nature (Figure 41–17). In the former, ions flow through gap junctions (page 142) connecting the cell membranes of closely juxtaposed neurons, and the nerve impulse moves directly from one neuron to the next. Such electrical synapses, which are common in lower invertebrates, have been identified at some sites in the mammalian brain. In chemical synapses, which make up the vast majority of connections between

41–16 *In an unmyelinated fiber (a), the action potential travels continuously along the axon, whereas in a myelinated fiber (b), the impulse jumps from node to node, greatly accelerating the conduction.*

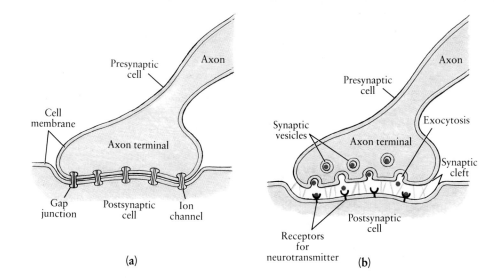

41–17 (a) *An electrical synapse. The arrival of an action potential at the axon terminal of the presynaptic cell is accompanied by changes in ion concentration. These changes are transmitted through gap junctions to the postsynaptic cell, where they depolarize the cell membrane and initiate a new action potential.* (b) *A chemical synapse. The arrival of an action potential at the axon terminal triggers the fusion of synaptic vesicles with the axon membrane, releasing neurotransmitter molecules into the synaptic cleft. These molecules diffuse to the postsynaptic cell, where they combine with specific receptors in the cell membrane. A network of protein fibers in the synaptic cleft anchors the presynaptic and postsynaptic membranes and, in some synapses, contains enzymes that rapidly degrade neurotransmitter molecules. Because of their shape, axon terminals are known as synaptic knobs or as "boutons" (French for "buttons").*

neurons in the mammalian nervous system, the two neurons never touch. As electron micrographs reveal (Figure 41–18), a space of about 20 nanometers, known as the **synaptic cleft,** separates the cell transmitting information (the presynaptic cell) from the cell receiving information (the postsynaptic cell). Information is transmitted across the synaptic cleft by means of the signalling molecules known as neurotransmitters. Unlike the nerve impulse along the axon —an all-or-nothing proposition—signals transmitted across chemical synapses are of varying strength and may have opposite effects. That is, some excite and some inhibit the postsynaptic cell.

Some neurotransmitters are synthesized in the cell body of the neuron and transported to the axon terminals, where they are packaged into synaptic vesicles and stored. Other neurotransmitters are both synthesized and packaged within the axon terminals. The release of neurotransmitter molecules is triggered by the arrival of an action potential at the axon terminal. The membrane in this region of the neuron is rich in membrane proteins that form channels for the transport of calcium ions (Ca^{2+}); these channels, like the Na^+ and K^+ channels, are regulated by the voltage across the axon membrane. Arrival of an action potential at the terminal alters the voltage, opening the channels and allowing Ca^{2+} ions to flow into the axon. This influx of Ca^{2+}, in turn, causes the synaptic vesicles to fuse with the cell membrane, emptying their contents into the synaptic cleft, in yet another example of exocytosis (page 138). The transmitter molecules diffuse from the presynaptic cell across the cleft and combine with receptor molecules on the membrane of the postsynaptic cell, setting in motion a series of events that, as we shall see shortly, may or may not trigger a nerve impulse in the postsynaptic cell.

41–18 (a) *Electron micrograph and* (b) *diagram of a cross section of a dendrite with which two axon terminals form synapses. Note the numerous synaptic vesicles, filled with neurotransmitter, in the axon terminals. Note also the fuzzy areas in the juxtaposed regions of the membranes of the presynaptic and postsynaptic cells. In presynaptic cells, these areas are specialized for exocytosis; in postsynaptic cells, they are rich in receptors for neurotransmitter molecules.*

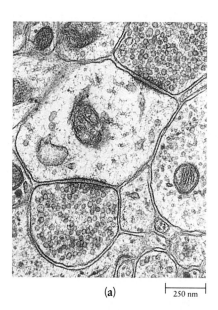
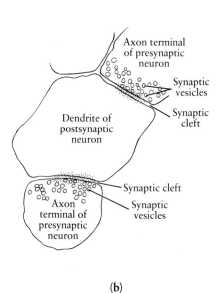

After their release, neurotransmitters are rapidly removed or destroyed, putting a halt to their effect; this is an essential feature in the control of the activities of the nervous system. The molecules may diffuse away or be broken down by specific enzymes, or they—or their breakdown products—may be taken up again by the axon terminal for recycling. At the same time, membrane from the synaptic vesicles, which fused with the cell membrane of the axon terminal, is apparently re-formed into vesicles by endocytosis, carried back into the cytoplasm, and recycled into new synaptic vesicles, filled with newly synthesized neurotransmitter. Membrane for the formation of new synaptic vesicles may also be supplied by smooth endoplasmic reticulum that extends from the cell body to the axon terminal.

The study of chemical synapses is one of the most active areas of contemporary neurobiological research, with a veritable tidal wave of new discoveries appearing in the scientific journals each week. Although the principal neurotransmitters have been known for many years, researchers are discovering an enormous number of chemicals that play a role in synaptic transmission, particularly in the central nervous system. Moreover, the specific receptors for different neurotransmitters are being identified, the details of their structure elucidated, and in some cases, their genes sequenced. Simultaneously, the biochemical events that occur when a neurotransmitter binds to its receptor are being revealed, as are the mechanisms by which a neuron integrates the information that it receives from the hundreds—or even thousands—of neurons that form synapses with it.

Neurotransmitters

A variety of chemical substances function as neurotransmitters. As we have seen, the principal neurotransmitters in the peripheral nervous system are acetylcholine (Figure 41-8) and noradrenaline. Acetylcholine is also found in the brain, although at relatively few synapses. Noradrenaline is a major neurotransmitter at some synapses in the hypothalamus and in other specific regions of the brain, where it is thought to play a role in arousal and attention. There is some evidence that severe depression may be related to an abnormally low level of noradrenaline at particular synapses; the two major types of antidepressants in clinical use apparently act by increasing the amount of noradrenaline at such synapses.

Many other neurotransmitters have been found in the central nervous system, including dopamine, serotonin (5-hydroxytryptamine), and gamma-aminobutyric acid (GABA), all of which, like noradrenaline, are amino acid derivatives (Figure 41-19). Dopamine is a transmitter for a relatively small group of neurons involved with muscular activity. Parkinson's disease, which is characterized by muscular tremors and weakness, is associated with a decrease in the number of dopamine-producing neurons and thus with the level of dopamine in certain areas of the brain. Serotonin is found in regions of the brain associated with arousal and attention; increasing levels of serotonin are associated with sleep. GABA is a major inhibitory transmitter in the central nervous system; the loss of GABA synapses is one of the features of Huntington's disease (page 395).

Almost all drugs that act in the brain to alter mood or behavior do so by enhancing or inhibiting the activity of neurotransmitter systems. Caffeine, nicotine, and the amphetamines, for example, stimulate brain activity by substituting for excitatory neurotransmitters at synapses. Chlorpromazine and related tranquilizers block dopamine receptors at many sites, whereas LSD inhibits brain serotonin.

Present evidence indicates two principal mechanisms by which neurotransmitters exert their effects on postsynaptic cells. In one mechanism, the binding of a neurotransmitter to its receptor triggers a change in the conformation of a membrane protein that functions as a channel for a specific ion. Depending on the

41-19 *The structures of some of the amino acid derivatives, known as biogenic amines, that function as neurotransmitters in the central nervous system. Biogenic amines are produced in neurons by slight chemical modifications of amino acids.*

Internal Opiates: The Endorphins

The word "opium" comes from the Greek *opion*, meaning "poppy juice." Since the time of the Greeks, poppy juice and its derivatives, such as morphine, have been used for the control of pain. They are the most potent painkillers known, and their physiological effects are greatly enhanced by the fact that they produce euphoria. They are also highly addictive, and although the pharmaceutical industry, urged on by the possibility of great profits, has made repeated attempts to develop an opium derivative that is nonaddictive, their efforts have been uniformly unsuccessful.

All substances with opiate action are related chemically and have similarities in their three-dimensional structures. Thus, it was long suspected that opiates act upon the brain by binding to specific membrane receptors. Using opium derivatives labeled with radioactive isotopes, investigators were able to show that the central nervous system does, indeed, have receptors for opiates. These receptors are located primarily in the spinal cord, brainstem, and brain regions in which drives and emotions are thought to be translated into complex actions, such as seeking food or a mate. When opiates bind to neurons bearing their receptors, they act as inhibitory neuromodulators, causing decreased production of nerve impulses by the neurons. Such receptors have been found not just in humans but in all other vertebrates tested.

Why would vertebrate brains have opiate receptors? Only one answer seemed logical: Because vertebrate brains themselves must produce opiates. This rather startling conclusion triggered a search for naturally occurring substances with opiate activity. Many such internal opiates have now been isolated, all of which act as neuromodulators. They have been given the name of endorphins, for endogenous morphine-like substances.

Two types of endorphins are recognized. One group, known as the enkephalins, is widespread in the central nervous system and is also abundant in the adrenal medulla. The enkephalins that have been identified are two pentapeptides (peptides containing five amino acids) that differ by only one amino acid (see illustration). According to recent evidence, the two enkephalins are produced in multiple copies on a single polypeptide chain.

The other endorphins are produced primarily by the pituitary gland and perhaps by other tissues as well. Recently it has been found that the most common of these, beta-endorphin, is synthesized as part of a long peptide chain that also contains ACTH, the hormone that is released by the anterior pituitary and

---

receptor, binding of the neurotransmitter may open the channel, allowing ions to flow between the cytoplasm of the neuron and the surrounding fluid, or it may close the channel, shutting off a previously existing flow of ions. The consequence is a change in the degree of polarization across the membrane of the postsynaptic cell. In the second mechanism, you will not be surprised to learn, binding of the neurotransmitter to its receptor activates an enzyme in the cell membrane and sets in motion a second messenger, generally cyclic AMP (see page 837) or a related compound, cyclic GMP (guanosine monophosphate). The events that follow activation of the second messenger are complex, but the ultimate effect is a change in the degree of polarization of the postsynaptic cell. This change, however, takes place at a slower pace than the changes triggered by the opening or closing of ion channels. Although it appears that some neurotransmitters utilize only one of these mechanisms, others, including acetylcholine, have two different types of receptors, activating the two different mechanisms. Such neurotransmitters produce both a rapid, short-term effect and a slower, long-term effect.

In addition to the principal neurotransmitters—acetylcholine and the amino acid derivatives—other molecules, mostly small peptides, also play a role in synaptic transmission. These molecules, which may be released from the same axon terminals as the principal neurotransmitters or from other cells, are known

stimulates the adrenal cortex. Although there is a great deal of overlap in the primary structure of the various endorphins, the functional relationships among them are not yet known.

The endorphins are of great interest to medical researchers because of the insight they may afford into the relief of two extremely serious (and related) medical problems, opiate addiction and pain. The endorphins are believed to function as natural analgesics (pain relievers). Individuals in stressful situations —soldiers in battle, athletes at crucial moments in a contest—have often reported being unaware of what later proved to be an extremely painful injury and so being able to continue to function in a life-threatening (or victory-threatening) situation. The recent discovery that macrophages are among the cell types bearing endorphin receptors suggests that these substances may also play a role in the stimulation of inflammatory and immune responses.

Morphine, heroin, and other exogenous opiates combine with the endorphin receptors, relieving stress, elevating mood, and soothing pain. However, it is hypothesized that these external opiates, acting by negative feedback, reduce the normal production of endorphins, resulting in ever-increasing dependence on the artificial source—or, in other words, in addiction.

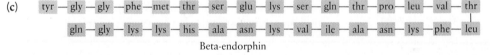

The amino acid sequences of three endorphins: (a) *methionine-enkephalin,* (b) *leucine-enkephalin,* and (c) *beta-endorphin. The enkephalins are found in the brain tissue and adrenal medulla of vertebrates, and beta-endorphin is isolated from the pituitary gland. The first four amino acids of each sequence are identical.*

as **neuromodulators**. Although neuromodulators may move directly across the synaptic cleft, they can also diffuse over a greater distance, affecting numerous cells within a local region of the central nervous system. Like neurotransmitters, they bind to specific membrane receptors and alter ion channels or set in motion second messengers; their effect is often to modulate the response of the cell to a principal neurotransmitter. Over 200 different substances that function as neuromodulators have been identified thus far. They include the endorphins (see essay), interferons and interleukins, hypothalamic releasing hormones, pituitary hormones, pancreatic hormones such as insulin, and even the digestive hormones gastrin and cholecystokinin (page 723).

The Integration of Information

As we noted previously, the dendrites and cell body of a single neuron may receive signals—in the form of neurotransmitter or neuromodulator molecules—from hundreds, or even thousands of synapses (see Figure 41–20 on the next page). The binding of each molecule to its receptor has some effect on the degree of polarization of the postsynaptic cell. If the effect is to make the interior of the cell less negative (depolarization), it is said to be excitatory. By contrast, if the effect is

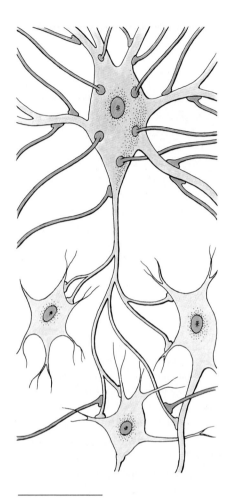

41–20 *A simplified representation of the many interconnections between neurons of the central nervous system. Note how numerous synapses, representing many different presynaptic neurons, converge on an individual neuron and how its axon, in turn, diverges to synapse on a number of other neurons. As you can see, most synapses are located on the dendrites and cell body of the postsynaptic neuron. Note, however, that there are also synapses on its axon, just before some of the axon terminals. These synapses are usually inhibitory or modulatory, influencing the response of the Ca^{2+} channels to the arrival of an action potential at the terminal and thus affecting the amount of neurotransmitter released.*

to maintain the membrane at or near the resting potential, or even to make the interior more negative (hyperpolarization), it is said to be inhibitory. The binding of acetylcholine, for example, is generally excitatory, causing a depolarization in the postsynaptic cell; GABA, on the other hand, is generally inhibitory.

The changes in polarity induced by neurotransmitters and neuromodulators spread from the synapses through the postsynaptic cell to a region known as the axon hillock. This is the region of the axon in which a nerve impulse can originate. If the collective effect is a sufficient depolarization to permit the influx of Na^+ ions that constitutes the beginning of an action potential, a nerve impulse is initiated in the axon of the postsynaptic cell and a new message is speeded on its way to the multitude of other neurons with which the axon synapses.

The processing of information that occurs within the cell body of each individual neuron plays a key role in the integration and control exercised jointly by the nervous and endocrine systems. It is affected not only by the specific neurotransmitters and neuromodulators received by the cell, but also by their quantity, the precise timing of their arrival, and the locations on the neuron of the various synapses and receptors. Each neuron is a tiny computer, summing an enormous quantity of information and issuing appropriate commands carried throughout the network with which we began this chapter.

SUMMARY

The nervous system, along with the endocrine system, integrates and controls the numerous functions that enable an animal to regulate its internal environment and to react to or deal with its external environment. The functional unit of the nervous system is the neuron, or nerve cell. A neuron consists of dendrites, which receive stimuli; a cell body, which contains the nucleus and metabolic machinery and also receives stimuli; and an axon, or nerve fiber, which relays stimuli to other cells.

The central nervous system consists of the brain and the spinal cord, which are encased, in vertebrates, in the skull and vertebral column. That part of the nervous system outside the central nervous system constitutes the peripheral nervous system. Cranial nerves enter and emerge from the brain in pairs, each pair consisting of motor fibers, sensory fibers, or both. Spinal nerves enter and emerge from the vertebral column, also in pairs. Each of these pairs innervates effectors and receives signals from sensory receptors of a different and distinct area of the body.

In vertebrates, the motor neurons of the peripheral nervous system are organized in two major divisions: (1) the somatic nervous system, which innervates the skeletal muscles, and (2) the autonomic nervous system, which controls cardiac muscle and the smooth muscles and glands involved in the digestive, circulatory, urinary, and reproductive functions. In the autonomic system, axons arising from neurons in the central nervous system synapse with motor neurons in ganglia outside the central nervous system. The postganglionic neurons stimulate or inhibit the effectors. The autonomic system has two divisions—sympathetic and parasympathetic—which are anatomically, physiologically, and functionally distinct (see Table 41–1).

Information received from the internal and external environments and instructions carried to effectors, such as muscles and glands, are transmitted in the nervous system as electrochemical signals. At rest, there is a difference in electric charge between the inside and outside of the axon membrane—the resting potential. With appropriate stimulation, an action potential, a transient reversal in membrane polarity, occurs. The action potential traveling along the axon membrane is the nerve impulse. Because all action potentials are the same size, the message carried by a particular axon can be varied only by a change in the

TABLE 41-1 **Comparison of Parasympathetic and Sympathetic Divisions**

	PARASYMPATHETIC DIVISION	SYMPATHETIC DIVISION
Preganglionic fibers	Long; emerge from cranial and sacral regions of spinal cord	Short; emerge from thoracic and lumbar regions of spinal cord
Location of synapse	In small local ganglia near or in innervated organs	In a series of ganglia close to the spinal cord or in ganglia halfway between spinal cord and innervated organs
Postganglionic fibers	Short	Long
Neurotransmitter	Preganglionic and postganglionic: acetylcholine	Preganglionic: acetylcholine; postganglionic usually noradrenaline
General effects	Promotes "vegetative" and restorative functions, such as digestion	Promotes "fight-or-flight" responses; inhibits "vegetative" functions

frequency or pattern of action potentials. In myelinated fibers, the nerve impulse leaps from node to node of the myelin sheath, thereby speeding conduction.

Neurons transmit signals to other neurons across a junction called a synapse. In most synapses, the signal crosses the synaptic cleft in the form of a chemical, a neurotransmitter, that binds to a specific receptor on the membrane of the postsynaptic cell. Other chemicals, known as neuromodulators, can diffuse through local areas of the central nervous system as well, binding to the membrane receptors of other neurons in the area. Binding of a neurotransmitter or neuromodulator to its receptor may open or close a membrane ion channel or set a second messenger in motion. The ultimate effect is a change in the membrane voltage of the postsynaptic cell. A single neuron may receive signals from many synapses, and, based on the summation of excitatory and inhibitory signals, an action potential will or will not be initiated in its axon. Thus individual neurons function as important relay and control centers in the integration of information by the nervous system.

QUESTIONS

1. Distinguish among the following: neuron/nerve/tract; ganglia/nuclei; central nervous system/peripheral nervous system; gray matter/white matter; afferent/efferent; preganglionic/postganglionic; resting potential/action potential/nerve impulse; presynaptic/postsynaptic; neurotransmitter/neuromodulator; adrenaline/noradrenaline/acetylcholine.

2. What are three significant differences between the somatic and autonomic nervous systems?

3. What are four significant differences between the sympathetic and parasympathetic divisions of the autonomic nervous system?

4. If a neuron is placed in a medium where ionic concentrations are the same as those of its own cytoplasm, what effect will this have on the resting potential?

5. Describe the way in which the nerve impulse propagates itself. Draw a diagram of this process.

6. Assume that three presynaptic neurons, A, B, and C, make adjacent synapses on the same postsynaptic neuron, D. No nerve impulse is initiated in the postsynaptic neuron as a result of single nerve impulses in A, B, or C, nor is one initiated if impulses arrive simultaneously in all three, in A and B, or in A and C. Only if impulses arrive together at the synapses of B and C will D fire an impulse. Explain these results in terms of excitatory and inhibitory neurotransmitters and their effects on the membrane voltage of the postsynaptic cell.

7. If you look closely at Figure 41-20, you will notice one neuron that has only very short dendrites and no axon at all. Axon terminals of other neurons form synapses on its cell body, but it forms no axonal synapses with other cells. Such neurons, which are located primarily in the brain, are incapable of initiating an action potential and are thus known as "nonspiking." Nevertheless, they affect the firing rate of other neurons. How do you think they might do this? Why might axons be superfluous for such neurons?

CHAPTER 42

Integration and Control III: Sensory Perception and Motor Response

An animal's sensory equipment is the means by which it knows the world around it. Through sensory perception, animals, including ourselves, are informed of predator and prey, friend and foe, whether something is good to eat or not, and changes in the weather and the seasons. Sensory perception identifies infant to mother, mother to infant, and mate to mate. For us, as perhaps for many other animals as well, it is also a source of pleasure and esthetic enjoyment and provides essential tools for learning. The capacity of an animal to act on the information provided by its sensory equipment—to move toward desirable objects, to move away from harmful or potentially harmful objects, and to perform the multitudinous activities of its daily life—depends on its skeletal muscles. In this chapter, we shall first examine the structures and processes involved in sensory perception. Then we shall consider the marvelous molecular machinery that not only enables animals to jump, run, swim, or fly but also permits us to smile, to frown, and to perform precise manipulative activities.

SENSORY RECEPTORS AND THE INITIATION OF NERVE IMPULSES

The information received, processed, and transmitted by the neurons and synapses of the vertebrate brain and spinal cord is carried into the central nervous system by sensory neurons. The triggering of nerve impulses in a sensory neuron

42–1 *The acquisition of food by a predatory animal depends not only on the reception and processing of detailed sensory information regarding the potential food item and the difficulties in obtaining it but also on exquisite muscular coordination. For this young chacma baboon, one false step or one slip of the hand could mean the difference between feasting on a stolen ostrich egg and disaster—particularly if it attracted the attention of the adult ostriches responsible for guarding the nest.*

depends on transduction, the conversion of one form of energy—the energy of a stimulus—into another form of energy—the energy of an action potential. Stimuli come in a variety of forms—pressure, heat or cold, chemical substances, vibrations, and light. Different types of sensory receptors are specialized to respond to different types of stimuli. In all cases, however, when a sensory receptor has been sufficiently stimulated, its membrane permeability or that of a nearby sensory neuron is altered, initiating the action potentials that start information on its way through the nervous system. The more intense the stimulus, the greater the frequency of the action potentials (see Figure 41-12, page 849).

The differences among the senses lie not in the form in which the signals are coded and conducted—the action potential—but rather in their pattern and its reception and interpretation in the central nervous system. As we shall see in the next chapter, information from different sensory receptors is transmitted to different regions of the vertebrate brain; the particular sensation experienced—a sunset, the song of a whip-poor-will, or a cooling breeze across the face—depends on the region of the brain that is stimulated.

Types of Sensory Receptors

Sensory receptors are many and varied. Most animals, including ourselves, have mechanoreceptors (touch, position, and hearing), chemoreceptors (taste and smell), photoreceptors (vision), temperature receptors, and receptors for the sensation recognized as pain. Some animals, but apparently not *Homo sapiens*, also have electroreceptors and magnetoreceptors. Like the other structures of the body of an animal, sensory receptors are the products of adaptation and evolution, and by these long processes, they have been tailored to the animal's specific requirements. Although we think we see what is visible and hear what is audible, visible and audible are not actually properties of objects; they are instead properties of our particular sensory equipment. What we see is very different from what an insect sees (Figure 42-2), and a bat or a fish, acoustically speaking, might as well be living on another planet. Moreover, it is likely that an animal's insensitivity to a great number of the potential stimuli that surround it is also as useful as its sensitivity to others.

Functionally, sensory receptors can be categorized as interoceptors, proprioceptors, and exteroceptors. **Interoceptors** include the mechanoreceptors and chemoreceptors that are sensitive to blood pressure and O_2, CO_2, and H^+ concentrations in the carotid arteries. The temperature sensors of the hypothalamus are also interoceptors. We are usually not conscious of signals from these receptors, although sometimes the signals result in perceptions of, for example, pain, hunger, thirst, nausea, or the sensations, produced by stretch receptors, of having a full bladder or bowel.

Proprioceptors (from the Latin *proprius*, meaning "self"), which are sometimes considered a subset of interoceptors, provide information about the orientation of the body in space and the position of arms, legs, and other body parts. Because of proprioceptors, a praying mantis can strike unerringly at its victim (see Figure 27-22, page 583), and you can tie your shoelace in the dark, or, with your eyes shut, touch your nose with your fingers. The semicircular canals of the ear are major proprioceptive organs in many vertebrates, performing a function similar to that of the statocysts of jellyfish (see Figure 25-19a, page 529).

The most familiar sensory receptors are **exteroceptors**, which provide information about the external environment. Some exteroceptors are small and relatively simple in structure. Look again at the human skin (Figure 42-3) as an example. The simplest receptors are the free nerve endings, which are receptors for pain, temperature, and perhaps other sensations as well. Slightly more complex are the combinations of free nerve endings with a hair and its follicle. Each of these little

42-2 *An evening primrose, as perceived (a) by the human eye, which responds to light in the visible portion of the electromagnetic spectrum, and (b) by the honeybee eye, which responds to light in the ultraviolet portion of the spectrum.*

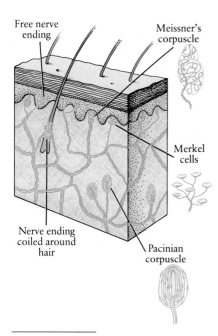

42–3 *Some sensory receptors present in human skin. The free nerve endings are largely pain and temperature receptors.*

The best understood of the skin receptors are the Pacinian corpuscles. The specialized nerve ending of a single myelinated fiber is encapsulated by the corpuscle, which is composed of many concentric layers of connective tissue. Pressure on these outer layers stimulates the firing of an action potential at this nerve ending.

Merkel cells and Meissner's corpuscles also respond to touch, as do the nerve endings surrounding the hair follicles. In regions where many hairs are present, the other touch receptors are sparse or absent.

organs is an exquisitely sensitive mechanoreceptor. When the hair is touched or bent, it causes changes in the nerve endings of a sensory neuron, setting off action potentials that are carried directly to the central nervous system. Three other types of mechanoreceptors are also shown, each a combination of one or more free nerve endings with an outer layer or layers of connective tissue. Meissner's corpuscles and Merkel cells are both involved with touch. They are found in particularly sensitive areas of the skin, such as the fingertips, palms, lips, and nipples, and are especially abundant where hairs are not present. They are responsible for the extraordinary cutaneous sensitivity of these parts of the human body and are associated with the ability, for example, to read Braille, do certain magic tricks, crack a safe, or enjoy a kiss. The Pacinian corpuscles, lying deeper within the tissues, respond to pressure and vibrations. The free nerve ending of the corpuscle is surrounded by layers of connective tissue and fluid. This layered structure is easily deformed, so it responds to even very slight pressure changes. However, a Pacinian corpuscle also adjusts quickly to pressure changes, in part because of its structure, and the nerve ending stops firing when the pressure is sustained. Of all the sensory receptors shown here, the simplest in structure—the free nerve endings associated with pain—are the least understood in terms of how they function. It seems likely that some of them are not mechanoreceptors, as are the others, but rather chemoreceptors, responding to very small amounts of some chemical released by cells when they are injured.

Chemoreception: Taste and Smell

If we were to lose our sense of taste or smell, our lives would lose many of their pleasures but we would not be greatly handicapped in our essential activities. Blindness or deafness is a far more serious threat. For many animals, however, the sense of smell is their window on the world. Many of the smaller mammals are nocturnal, and most of them live close to the ground in habitats in which their highly developed sense of smell provides the greatest amount of information about their environment. Presumably, it was the tree-dwelling habits of our immediate ancestors that made the sense of smell less useful for the primates.

Taste

Fish, particularly bottom feeders, have sensory cells for taste—the chemoreception of waterborne substances—scattered over the surface of their bodies. These cells play a major role in determining their behavior. In the carp, for example, the part of the brain that receives information from the taste cells is larger than all the other sensory centers combined. The catfish, also a bottom feeder, has taste cells on its body and, in addition, trails barbels, or "whiskers," along the bottom. These are richly supplied with gustatory nerve endings, and when they are stimulated, the fish turns and snaps at the source of the stimulus.

In terrestrial vertebrates, taste cells are located inside the mouth, where they act as sentinels, providing the information on which a final judgment can be made on what is and is not to be swallowed. The taste receptors and the supporting cells around them form the lemon-shaped clusters known as taste buds (see Figure 34–5, page 717). We are able to distinguish four primary tastes: sweet, sour, salty, and bitter. While each primary taste (or, more accurately, the molecules associated with the taste) appears to stimulate a different type of taste receptor, a single taste bud may respond to more than one category of substance.

Most animals seem to have the same general range of taste discrimination, although there is some difference in what animals "like" and "don't like." Birds, for example, will readily eat many seeds and insects with a bitter taste, whereas for most other vertebrates, bitterness serves as a warning signal. Cats are among the few animals that do not prefer substances with a sweet taste, and very few taste receptors that respond to sugar can be found in the cat's tongue.

42–4 *Reptiles and amphibians have, in addition to nostrils, a special olfactory organ, the vomeronasal organ, in the roof of the mouth. Food substances are apparently tested in this organ. When a snake lashes its forked tongue in the air, it is collecting samples for testing in its vomeronasal cavities. The snake shown here is a boa constrictor.*

Smell

In fish, taste and smell are operationally very similar since both involve the detection of substances dissolved in the surrounding water. However, taste receptors and smell receptors differ anatomically in fish as in higher vertebrates, and the centers of taste and smell within the brain are entirely distinct.

In terrestrial animals, smell can be defined as the chemoreception of airborne substances. To be detected, however, these substances must first be dissolved in the watery layer of mucus overlying a specialized tissue, the **olfactory epithelium.** In humans, this tissue, which is located high within the nasal passages, is comparatively small. The part within each nasal passage is only about as large as a postage stamp. Each of these areas contains some 50 million receptor cells (see Figure 42–5 on the following page), characterized by numerous cilia whose membranes contain specific receptor molecules into which odor molecules of a complementary shape fit. Current evidence indicates that the binding of an odor molecule to its receptor triggers a series of events involving, once more, cyclic AMP. In the best-studied examples of olfactory transduction, binding of an odor molecule to its receptor activates an enzyme that phosphorylates GDP (guanosine diphosphate) to GTP. GTP, in turn, activates the enzyme adenylate cyclase, which, as we have seen, catalyzes the conversion of ATP to cyclic AMP. Cyclic AMP then binds to and opens a membrane channel through which Na^+ ions flow into the cilium, and from the cilium into the body of the olfactory cell. Other receptors on the cilia are thought to bind inhibitory and modulatory molecules, triggering other sequences of events that have their own effects on the membrane potential across the axon hillock. This membrane potential ultimately determines whether action potentials are or are not initiated by the olfactory cell and the frequency with which they occur.

Even with our relatively insensitive olfactory equipment, we are able to discriminate some 10,000 different odors. One of the current models of odor discrimination was developed in the 1960s by John Amoore while he was still an undergraduate at Oxford University in England. According to the present version of Amoore's model, all scents are made up of combinations of "primary" odors: the seven proposed by Amoore are camphoric, musky, floral, pepperminty, etherlike, pungent, and putrid. There are also, according to this hypothesis, different types of olfactory cells corresponding to the primary odors, with each type bearing membrane receptors for molecules of a particular shape. All substances that have a pungent odor, for example, have the same general molecular shape and fit into the same type of receptor. Some molecules, depending on which way they are oriented, can fit into more than one receptor and so can evoke in the brain two different types of signals. From the signals received from the various different types of cells, the brain constructs a "picture" of an odor.

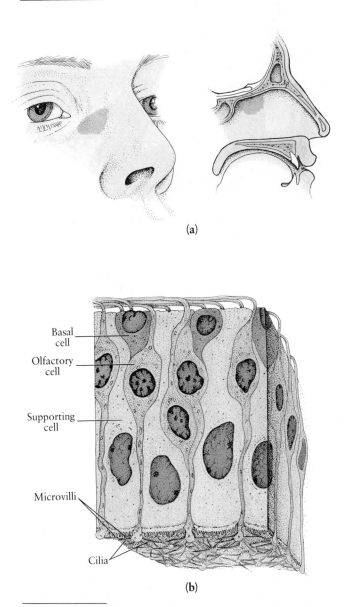
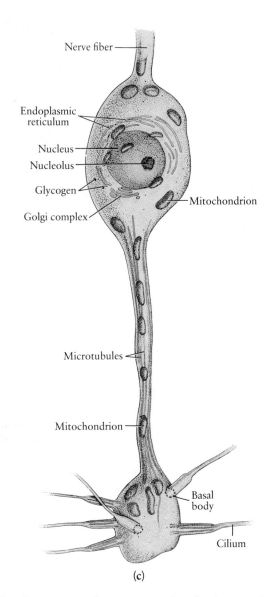

42–5 (a) A patch of specialized tissue, the olfactory epithelium, arching over the roof of each nasal cavity, is responsible for our sense of smell. Also, much of what we call flavor in food is actually a result of volatile substances reaching this tissue. (b) The olfactory epithelium is composed of three types of cells: supporting cells, basal cells, and olfactory cells, which are the sensory receptors. Supporting cells are tall and columnar, wider near the surface than they are deep within the tissue, and their outer surfaces are covered with microvilli, similar to the microvilli found on the surface of intestinal cells. Movements of these microvilli, in conjunction with proteins in the surrounding mucus, are thought to play a role in transporting odor molecules to the olfactory cells and, subsequently, in removing them. The basal cells, triangular in shape, are found along the innermost layer of the epithelium. Their function is unknown, but it is thought that they may give rise to new supporting cells when these are needed. (c) The olfactory cell is long and narrow. The cilia protruding from its exposed surface bear membrane receptors to which odor molecules, inhibitors, and modulators bind. This outermost part of the cell is connected with the cell body by a long stalk containing microtubules; changes in ion concentrations and second messengers set in motion by the binding of molecules to their receptors are transmitted to the cell body through this stalk. From the opposite end of the cell body, a nerve fiber extends through the surrounding tissue, conducting the computed sensory message to the brain.

Chemical Communication in Mammals

As we saw in Chapter 27 (page 585), the exchange of chemical messages—pheromones—among members of the same species is of great importance in the behavior of many insects. Chemoreception also plays a role in the behavior of most mammals. The males of many species—including domestic dogs and cats—scent-mark their territories with urine as a warning signal to other males. Males are attracted to females by special odors associated with estrus (being "in heat").

Among mice, males release substances in their urine that alter the reproductive cycles of the females. Juvenile females reach sexual maturity earlier when exposed to these substances, and the odor of urine from a strange male can cause a pregnant female to resorb her fetuses, leaving her free to mate with the newcomer.

In female rhesus monkeys, vaginal secretions of a volatile fatty acid have been shown to increase near the midpoint of the estrus cycle and to be attractive to male monkeys. Secretions of a similar compound also increase at the time of ovulation in the human female, but its attractiveness to human males has not been demonstrated. It has also been hypothesized that the chemicals (also fatty acids), which are the sources of "body odor" and are produced, beginning at puberty, by the apocrine sweat glands, originally played such a role. The pubic and axillary (underarm) hair that develops at about the same time would appear to have the function of retaining and amplifying these odors.

The most convincing evidence to date of pheromonal interactions in humans was assembled by Martha McClintock while still an undergraduate at Wellesley College. McClintock found a significant

A male dik-dik, a herbivore of the African plains, marking his territory by depositing scent on a grass stem. The glands from which the scent is secreted are located near the base of the eyes.

tendency toward synchronization of menstrual cycles over the school year among women who were roommates and close friends. The effect appeared to correlate directly with the time these women spent together and not with other external factors, such as food habits or photoperiodism. This finding was later extended by demonstrating that the odor from one woman's underarm secretions affected the timing of another woman's cycle, independent of any other contact.

Mechanoreception: Balance and Hearing

The mammalian ear is an organ containing two distinct mechanoreceptors, each of which transduces the energy of an enclosed, moving fluid into the energy of action potentials. One of these structures consists of three interlocking **semicircular canals** that provide information about the orientation of the head in space; this information is critical for the maintenance of physical balance. The other structure, the **cochlea** (snail), is a coiled chamber in which the sensory responses involved in hearing occur. Additional structures convert vibrations of the air (sound waves) into vibrations of the fluid contained within the cochlea. In both the semicircular canals and the cochlea, the primary sensory receptors are sensitive hair cells that vibrate in response to movements of the surrounding fluid.

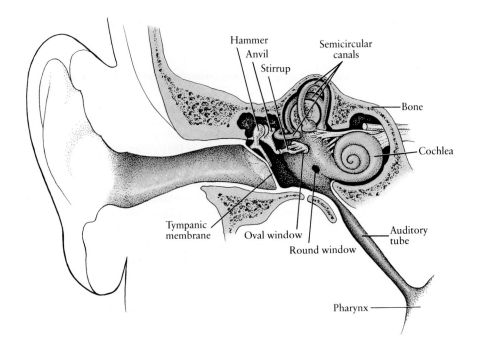

42–6 *The structure of the human ear. Sound waves entering the outer ear are funneled to the tympanic membrane, which they cause to vibrate. These vibrations are transmitted through three small bones of the middle ear—the hammer, anvil, and stirrup—to another membrane, the oval window membrane. Vibrations of this membrane, in turn, set off vibrations in the fluids in the cochlea, the structure of the inner ear concerned with hearing.*

The three semicircular canals are additional fluid-filled chambers within the bony labyrinth of the inner ear. Each one is in a plane perpendicular to the other two. Their function is to monitor the position of the head in space and to maintain equilibrium. Movements of the head set the fluid in these canals in motion, activating sensitive hair cells and triggering action potentials in sensory neurons with which they synapse.

The middle ear, as you can see, is connected with the upper pharynx by the auditory tube (also known as the Eustachian tube). This makes it possible to equalize the air pressure in the middle ear with atmospheric pressure; it also unfortunately makes the middle ear a fertile breeding ground for infectious microorganisms that enter the body through the nose or mouth.

Figure 42–6 diagrams the structures of the human ear. Sound travels through the ear canal to the tympanic membrane (the eardrum), in which it sets up a vibration. This vibration is transferred to a series of three very small and delicate bones in the middle ear, which are called, because of their shapes, hammer (malleus), anvil (incus), and stirrup (stapes). Vibrations in the tympanic membrane cause the stirrup to push gently and rapidly against the membrane covering the oval window, which leads to the fluid-filled cochlea, located in the inner ear.

The pressure must be amplified as it passes from the tympanic membrane to the cochlea because fluid is more difficult to move than air. This amplification is made possible by a difference in the sizes of the tympanic and oval window membranes. The oval window membrane is smaller than the tympanic membrane, resulting in more force per unit area—that is, more pressure—as the total force is transferred.

Figure 42–7 shows a diagram of the cochlea partially uncoiled. It consists essentially of three fluid-filled canals separated by membranes. The upper and lower canals are connected with one another at the far end of the spiral. At the near end of each of these two canals are movable membranes, the oval window membrane and the round window membrane. The stirrup vibrates against the oval window membrane at the near end of the upper canal, causing pressure waves in the fluid contained in these canals. The waves travel the length of the cochlea, around the far end, and back again to the round window membrane at the near end of the lower canal. As the oval window membrane moves in, the round window membrane moves out, keeping the pressure equalized.

The central canal contains the organ of Corti, which rests on the basilar membrane of the canal and contains the individual sensory cells, the hair cells. The movement of the fluid waves along the outer surface of the central canal causes vibrations in the basilar membrane, which, in turn, cause vibrations in the hair cells. Vibration of a hair cell opens ion channels in its membrane, causing a depolarization that, although it does not trigger an action potential, does alter the release of neurotransmitter molecules from the cell. Hair cells synapse with sensory neurons, which, when adequately stimulated at these synapses, initiate action potentials. The axons of these neurons form the auditory nerve, which links the ear to a region of the brain known as the cochlear nucleus. In the cochlear nucleus, neurons of the auditory nerve synapse with other neurons that synapse with still other neurons, ultimately carrying the information to the region of the brain associated with the conscious perception of sound.

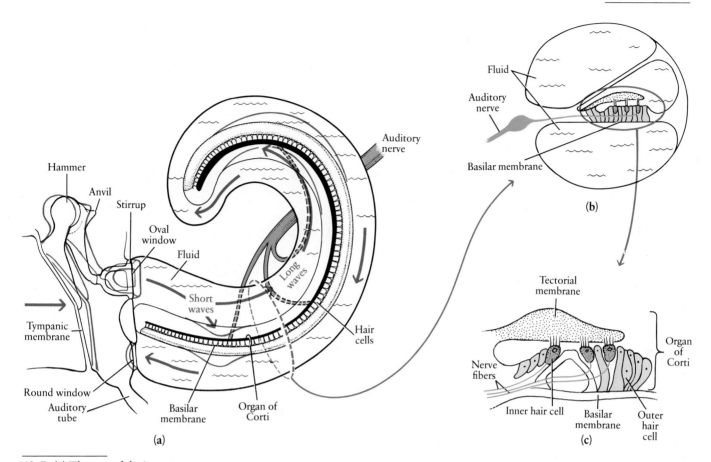

42–7 (a) *The part of the inner ear concerned with hearing is the cochlea, a coiled tube of 2.75 turns—shown here as if it were partially uncoiled. Vibrations transmitted from the tympanic membrane to the stirrup cause the stirrup to push against the oval window membrane, resulting in pressure waves in the fluid that fills the cochlear canals. Pressure waves in the fluid set up vibrations in the basilar membrane, stimulating the sensory cells in the organ of Corti, which rests on the basilar membrane. Sounds at different frequencies (or pitch) have their maximum effect on different areas of the membrane. The round window prevents the pressure from building up in the cochlea. A cross section of the cochlea is shown in (b), and a close-up of the organ of Corti in (c).*

The basilar membrane is narrower and less elastic at the end nearer the middle ear. Hence it does not vibrate uniformly along its length; instead different areas of the membrane respond to different frequencies of sound. Thus, different hair cells are stimulated by different frequencies. Human beings are capable of very fine discriminations of sound; such discriminations are made by the brain on the basis of the signals transmitted by the different sensory neurons with which the different hair cells synapse. In general, the human ear can detect sounds ranging from 16 to 20,000 cycles per second, although children can hear up to 25,000 cycles per second. (Middle C is 256 cycles per second.) From middle age onward, there is a progressive loss in ability to hear the higher frequencies.

Our inability to hear very low-pitched sounds (below 125 cycles per second at normal levels of intensity) is a useful adaptation. If we could hear at such low pitches, we would be barraged by sounds caused by internal movements within our own bodies, such as the flow of blood, as well as by sounds conducted through our bones. We do hear some sounds conducted directly through the skull—for example, the noise we make when we chew celery. We also hear our own voices mainly through the skull. This is why a recording of one's own voice sounds so startlingly unfamiliar.

Dogs, as all dog trainers know, can hear very high sounds (40,000 cycles per second) and mice squeak to one another at 80,000 cycles per second. An elephant named Lois, tested by investigators at the University of Kansas and the Ralph Mitchell Zoo in Independence, Kansas, could not hear sounds higher than 12,000 cycles per second but could hear sounds at 16 cycles per second at levels of intensity inaudible to human beings. (When Lois heard a sound, she pressed a lever with her trunk and was rewarded with half a cup of Kool-Aid.) Recent evidence indicates that elephants communicate with one another at pitches that would be the envy of any operatic basso—except that they are outside the range of human hearing.

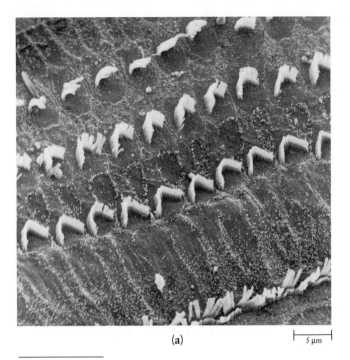
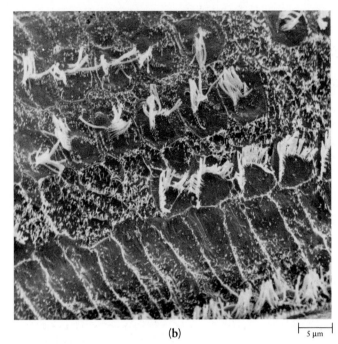

42–8 *Scanning electron micrographs of hair cells from the organ of Corti of a guinea pig before (a) and after (b) exposure to 24 hours of loud noise, comparable to that of a typical rock concert. On some of the hair cells, the orderly arrangement of cilia has been disrupted. Other hair cells have degenerated, losing their cilia.*

Photoreception: Vision

Eyes have evolved independently several times in the course of evolutionary history. Among the most highly developed of modern photoreceptor systems are the compound eye of arthropods (see page 579), the eye of the octopus (see page 552), and the vertebrate eye, of which the human eye, shown in Figure 42–9, is an example.

The vertebrate type of eye is often called a camera eye. In fact, it has a number of features in common with an ordinary camera equipped with several expensive accessories, such as a built-in cleaning and lubricating system, an exposure meter, and an automatic focus. Light from the object being viewed passes through the

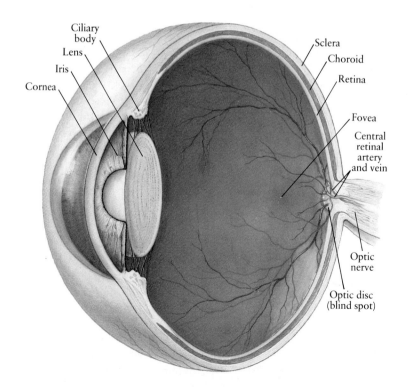

42–9 *The eye is a complex organ composed of three layers of tissue, which form a fluid-filled sphere. The outer layer, the sclera, is white fibrous connective tissue that serves a protective function. The anterior portion of the sclera, the cornea, is transparent. The middle layer, the choroid, contains blood vessels. Its anterior portion is modified into the ciliary body, the suspensory ligament, and the iris. The ciliary body is a circle of smooth muscle from which extend the suspensory ligaments that hold the lens in place. The colored part of the eye, the iris, is a circular structure attached to the ciliary body. The pupil is a hole in the center of the iris, which regulates its size. The innermost layer of the eye, the retina, contains the photoreceptor cells, the rods and cones. The fovea, near the center of the retina, is the region of greatest visual acuity. Only the front of the eye is exposed; the rest of the eyeball is recessed in and protected by the bony socket of the skull.*

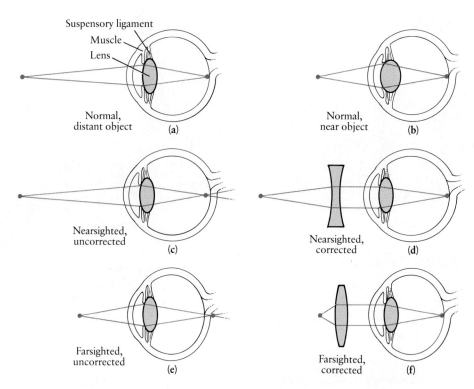

42-10 *Focusing the eye. Preliminary focusing is done by the cornea, and fine focusing by the lens, which is like a jelly-filled, transparent rubber balloon. The lens is held in place by suspensory ligaments attached to the ciliary body, which encircles the lens. The ciliary body contains the ciliary muscle, which is the muscle of accommodation. When it contracts, the lens is released, and, as it returns to its more nearly spherical shape, its curvature is increased. (a) The normal eye views distant objects with the lens stretched and flattened; (b) for viewing near objects, the lens is relaxed and more convex, bending the light rays more sharply. (c) Nearsightedness occurs when the eyeball is too long for the lens to focus a distant image on the retina; (d) it is corrected by a concave lens that bends the light rays out enough to focus the image. (e) Farsightedness is the result of an eyeball too short for the lens to focus a near object; (f) it is corrected by a convex lens that bends the light rays in before they reach the lens of the eye.*

transparent cornea and lens, which focus an inverted image of the object on the light-sensitive retina in the back of the eyeball. In mammals, fine focusing of the image on the retina is brought about by contracting or relaxing the ciliary muscles, thus changing the curvature, and hence the focal length, of the lens (Figure 42–10). Fish and amphibians, which lack ciliary muscles, focus their eyes the same way one focuses a camera with a lens of fixed focal length. Muscles within the eye change the position of the lens, drawing it back toward the retina in order to focus more distant objects.

Stereoscopic vision—that is, vision in three dimensions—depends on viewing the same visual field with both eyes simultaneously. When both eyes are trained on a distant object, both eyes see almost the same image. A nearby object, however, will present a slightly different image to each eye, a phenomenon known as parallax. The brain computes the disparity in the two images of a nearby object in such a way that we are able to judge its distance from us. When the eyes are trained on a very distant object, however, the disparity is too small; hence, we are not able to make such judgments for faraway objects. Instead, we estimate their distance on the basis of known sizes of the objects or such visual clues as the size and relative position of houses, people, or cars in the same field of vision. A young child viewing a distant object—an airplane, for example—may see it as little rather than as far away.

Tree-dwelling animals, such as the probable forerunners of *Homo sapiens*, usually have overlapping visual fields in the front that provide them with the stereoscopic vision essential for the distance judgments involved in moving from branch to branch. Predatory animals also tend to have stereoscopic vision, but animals that are more likely to be hunted than to hunt usually have an eye placed on each side of the head, giving a wide total visual field (Figure 42–11). Some birds with laterally placed eyes—the woodcock, the cuckoo, and certain species of crow—have binocular fields of vision both in front of them and behind them.

42–11 *Some vertebrate eyes.* **(a)** *The eyes of a mud skipper, a bottom-dwelling fish, are directed upward like periscopes. Mud skippers live in the shallow waters of mudflats and mangrove swamps in India and Asia. They can see objects in the air above the water, even when their bodies are totally submerged.* **(b)** *Eagles and other predatory birds often have eyes as large as ours (although the skull and brain are much smaller), two foveas in each retina, and strong powers of near and far accommodation. Shown here is a peregrine falcon.* **(c)** *Rabbits have eyes set on each side of their skulls, a placement that allows them to watch on both sides for predators while they are feeding.* **(d)** *The slow loris, a tree-dwelling primate of Southeast Asia, is nocturnal. Like many other nocturnal animals, it has large eyes. The photoreceptors of nocturnal animals are almost all rods, which are more light-sensitive than cones and are necessary for night vision.*

The Retina

The retina of the vertebrate eye contains the photoreceptor cells, which capture light energy and begin the process of transduction. These cells are of two types, named, because of their shapes, **rods** and **cones** (Figure 42–12). Rods are responsible for black-and-white vision; cones, for color vision.

Rods do not provide as great a degree of resolution as cones do, but they are more light-sensitive than cones. Dim light does not stimulate the cones, which is why the world becomes colorless to us at night. Nocturnal animals, which include most mammals, have retinas made up almost entirely of rods and thus have no color vision. Some diurnal animals, such as some reptiles and squirrels, have almost entirely cones. Higher primates, including humans, have both rods and cones.

As you can see in Figure 42–13, the retina of the vertebrate eye is anatomically inside out—that is, the photoreceptors of the eye are pointed toward the back of the eyeball. To reach them, light must pass through several layers of other neurons in the retina. Only about 10 percent of the light falling on the cornea reaches the retina. Of this light, that which is not captured by the photoreceptors is absorbed by pigmented epithelium that lines the back of the eyeball, just behind the photoreceptors. Some nocturnal vertebrates have a reflecting layer (tapetum) behind the photoreceptors that increases the likelihood of dim light stimulating the photoreceptors. This reflecting layer makes the animal's eyes seem to shine at night when light is directed into them.

The photoreceptor cells communicate, by way of intervening neurons, with the ganglion cells, whose axons form the optic nerve. When light is captured by the photoreceptor cells, a series of reactions occur that produce a change in their membrane polarity. This change influences their release of neurotransmitters at synapses they form with a group of cells known as bipolar cells. The release of neurotransmitters at these synapses causes, in turn, a change in the membrane polarity of the bipolar cells, and influences their release of neurotransmitters at synapses with the ganglion cells. The ultimate result of the stimulation by light is a change in the pattern in which action potentials fire in the axons of the ganglion cells.

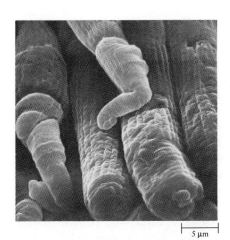

42–12 *Rods and cones as shown by the scanning electron microscope. A single photon of light (see page 208) is sufficient to cause a response in a photoreceptor; as few as five to eight photons in the blue-green frequency range can result in conscious perception.*

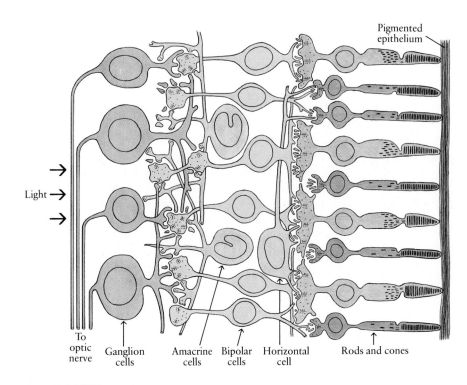

42–13 *The retina of the vertebrate eye. Light (shown here as entering from the left) must pass through several layers of cells to reach the photoreceptors (the rods and the cones) at the back of the eye. Signals from the photoreceptor cells are then transmitted through the bipolar cells to the ganglion cells, whose axons converge to become the optic nerve. Horizontal and amacrine cells, other neurons in the retina, also participate in the elaborate transmission paths. Some processing of information occurs in these pathways before nerve impulses leave the retina.*

Ganglion cell axons from all over the retina converge at the rear of the eyeball and bundle together like a cable to form the optic nerve, which connects the retina with the rest of the brain. The point at which the axons pass out of the retina is a blind spot, since photoreceptor cells are absent there. We generally are not aware of the existence of the blind spot since we usually see the same object with both eyes and the "missing piece" is always supplied by the other eye. You can demonstrate the existence of the blind spot with the help of Figure 42–14.

Information Processing in the Retina

There are about 125 million photoreceptors in the retina and about 1 million ganglion cell axons in the optic nerve, a reduction of 125 to 1. This reduction occurs principally with the rods. Many rods converge on each bipolar cell, and several bipolar cells, in turn, feed into each ganglion cell. Also, other neurons in the retina—horizontal and amacrine cells—participate in these connections and interactions. Therefore, a point-to-point representation is not transmitted from the rods—as with a television camera, for instance—but there is processing of the information before it even leaves the retina (see essay on the next page).

42–14 *With this diagram, you can prove to yourself that you have a blind spot in each eye. Hold the book about 30 centimeters (12 inches) from your face, cover your left eye, and gaze steadily at the X while slowly moving the book toward your face. Note that at a certain distance, the image of the dot becomes invisible. Then cover your right eye and gaze steadily at the dot while you move the book toward your face. What happens to the X?*

X   ●

What the Frog's Eye Tells the Frog's Brain

The frog's eye differs from the human eye in that it has no central fovea and the rods and cones are distributed uniformly over the surface of the retina. It resembles the human eye, however, in that these photoreceptor cells transmit their signals to a far fewer number of ganglion cells, whose axons make up the optic nerve leading from the eye to the brain.

By inserting microelectrodes into these axons while exposing the frog's eye to various kinds of stimuli, Jerome Lettvin and coworkers at the Massachusetts Institute of Technology found that different ganglion cells responded to different stimuli—light on, light off, or a big moving shadow, for instance.

Most interesting, one type of ganglion cell responded only to a small moving object; in other words, it was a bug detector. An object bigger than a bug would not stimulate these particular ganglion cells even if it was in motion, and a bug-sized object would not stimulate them if it was motionless. The existence of the bug detector corresponds nicely with certain well-known features of the animal's behavior. A frog will strike only—and virtually always—at a small moving object and will literally starve surrounded by dead insects. Thus, the information about an extremely important aspect of the frog's world is processed right in the retina itself.

A European tree frog (a) observing and (b) capturing a blue-bottle fly.

(a)

(b)

42–15 Eye-hand coordination is a characteristic of the higher primates, such as the young olive baboon shown here grooming itself.

The area of the retina in which the sharpest image is formed is known as the **fovea** (see Figure 42–9). In the fovea, the photoreceptor cells consist entirely of closely packed cones. These cones, instead of having the 125:1 relationship of the rods and ganglion cells, make one-to-one connections with the bipolar and ganglion cells. The one-to-one connections and the close packing of the cones provide greater resolution, giving a crisper picture.

Birds, which rely on vision above all other senses, may have two or three foveas. Also, the photoreceptors of birds tend to be more tightly packed. We have about 160,000 cones per square millimeter of retina, or some 5 to 10 million cones per eye. A hawk, with an eye of about the same size, has some 1 million cones per square millimeter and therefore a visual acuity about eight times that of a human being.

Among the mammals, only primates have a central fovea for sharp vision. In *Homo sapiens*, highly developed vision is undoubtedly closely correlated with the development of the use of the hands for fine manipulative movements. This reasonable hypothesis is supported by the fact that the human eye is so constructed that the sharpest images—images within the fovea—can be made of objects that are within the reach of the hand (Figure 42–15).

Visual Pigments and the Capture of Light

Both rods and cones contain light-sensitive compounds—visual pigments—embedded in a series of stacked membranes that are segregated in one portion of the cell (Figure 42–16). The pigments consist of a protein, called an opsin, and a carotenoid, called retinal, derived from vitamin A. (The relationships among retinal, vitamin A, and beta-carotene, the principal accessory photosynthetic pigment in plants, are summarized in Figure 10–7, page 212.) The opsin part of the molecule differs in different types of photoreceptors; the retinal is the same. As we noted previously, the eyes of arthropods, mollusks, and vertebrates all evolved separately, and yet each of the three types contains almost identical visual pigments. Since no animal can synthesize carotenoids, all vision depends on substances derived in the diet directly or indirectly from plants. Similar light-sensitive compounds occur in the cell membrane of the salt-loving halobacteria (see page 221), and they have also been identified in the eyespot of *Chlamydomonas*, suggesting a long evolutionary history in which the same, or closely related, light-sensitive molecules have been utilized for different functions.

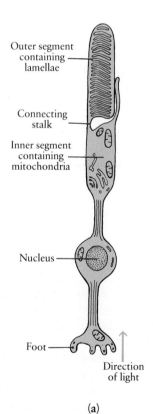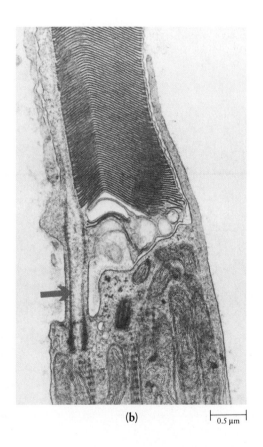

42–16 (a) *Diagram of a rod from the human retina. The molecules of light-sensitive pigment, rhodopsin, are located on the lamellae. The foot makes synaptic contact with bipolar and horizontal cells.* (b) *An electron micrograph of a rod, showing the region in which the outer and inner segments connect. The outer segment of the rod (the portion of the cell pointing toward the back of the eyeball) consists of a stack of membranes, piled up one on top of the other like poker chips. The light-sensitive pigment is built into these membranes. In the inner segment of the cell, new pigment molecules and other substances required by the light-receptor area are synthesized. These molecules are transported through a narrow connecting stalk (indicated by the arrow) to the outer, membrane-packed segment, where the actual work of this highly specialized cell is done. A cross section of the stalk reveals, surprisingly, that it has the internal structure of a cilium, lacking only the two central microtubules. Cones exhibit the same general structure as rods.*

In rods, the retinal-opsin combination is rhodopsin, sometimes called visual purple. When rhodopsin absorbs light, the retinal changes shape (Figure 42–17), triggering a change in shape in the opsin. This sets in motion a complex cascade of chemical reactions leading to a change in the membrane potential of the rod cell, and thus in its release of neurotransmitter to the bipolar cells. The reactions of the cascade involve at least two distinct enzymes and utilize cyclic GMP as a second messenger. Another participant is GTP, which, like ATP, is a ready source of

42–17 *When the retinal portion of the rhodopsin molecule is excited by light, rotation occurs at one bond (arrow), resulting in a significant change in its overall shape. The retinal is located in a cleft within the opsin portion of the molecule, and its change in shape causes a deformation in the protein. The resulting activated rhodopsin sets in motion the series of reactions that alter the membrane polarity of the rod cell and thus its release of neurotransmitter. One of the reactions in the series, catalyzed by an enzyme and powered by the hydrolysis of GTP to GDP, has the effect of restoring the opsin and retinal to their original conformations.*

cellular energy; energy supplied by GTP plays a key role in returning rhodopsin to its original shape, ready once more to respond to light and to initiate another series of reactions. The consequence of these reactions is not a depolarization of the rod cell membrane, but rather a hyperpolarization, resulting from the closure of membrane channels for Na^+. It is this hyperpolarization that alters the release of neurotransmitter from the cell.

Color vision in humans depends on the presence of three different types of cones, each containing one of three different visual pigments. Each of these pigments is made up of retinal and a slightly different opsin. As long ago as 1802, a three-receptor hypothesis for color vision was proposed; this hypothesis, supported by a number of psychological studies, was greatly bolstered by the identification of the three separate cone types and their pigments. Each pigment is most sensitive to the wavelength of one of the three fundamental colors—blue, green, or red. Different shades of color stimulate different combinations of these cones, and the signals, after being processed in the retina, are further processed by the brain and so become what we perceive as color. Most recently, the genes for each of the three opsins have been identified and sequenced, and the location of the genes for the pigments sensitive to red and green have been determined. As you would expect on the basis of your knowledge of the inheritance of color blindness, these two genes are located on the X chromosome, where multiple copies occur in tandem. Comparison of the sequences of all three genes strongly suggests that they, like the human globin genes, evolved from a single gene through processes of duplication and mutation (see page 367).

Visible light is only a small portion of the vast electromagnetic spectrum (see Figure 10–3, page 207). For the human eye, the visible spectrum ranges from violet light, which is made up of comparatively short light waves, to red light, the longest visible to us. Fortunately, we cannot see in the infrared (heat-wave) spectrum. Otherwise, we would see everything through an infrared glow emitted by our own bodies. Our photoreceptors are, however, sensitive to ultraviolet. Ordinarily, these waves are filtered out by yellow pigment in the lens, but persons who have had cataracts removed can read by ultraviolet light.

The perception of color often plays an important role in vertebrate behavior. Studies of the three-spined stickleback, a small freshwater fish, have shown that fighting behavior among the males is elicited by the red color that develops on their underbellies during the mating season. Niko Tinbergen, who kept a laboratory in England full of aquariums containing sticklebacks, found that the male sticklebacks would rush to the sides of their tanks and assume threatening postures every time a red mailtruck passed on the street outside. Similarly, it has been found that fighting behavior among English robins (see page 409) can be evoked by a small tuft of red feathers. Among birds, bright plumage, especially of the males, plays an important part in attracting the opposite sex and in courtship ceremonies.

THE RESPONSE TO SENSORY INFORMATION: MUSCLE CONTRACTION

The activities an animal undertakes as a result of the information received and processed by its sensory receptors and brain depend upon skeletal muscle, the effector of the somatic nervous system. Moreover, many of the adjustments in its internal environment depend upon cardiac muscle and smooth muscle, two of the effectors of the autonomic nervous system. Muscle is the principal tissue, by weight, in the vertebrate body. However, even if there were only a single muscle cell per organism, it would be worthy of our attention and admiration. Skeletal muscle is the one tissue of the biological world whose function on a macroscopic scale has been traced down to, and actually visualized in terms of, the very molecules that make it possible, a wonderful correspondence between structure and function.

The Structure of Skeletal Muscle

As we noted in Chapter 33, a skeletal muscle consists of bundles of muscle fibers—often hundreds of thousands of fibers—held together by connective tissue. Each fiber is a single, multinucleate cell 10 to 100 micrometers in diameter and, often, several centimeters long. Each muscle fiber is surrounded by an outer cell membrane called the **sarcolemma**. Like the membrane of the axon, the sarcolemma can propagate an action potential.

Embedded in the cytoplasm of each muscle fiber (cell) are some 1,000 to 2,000 smaller structural units, the **myofibrils** (from *myo*, the prefix for "muscle"). Tightly packed myofibrils run parallel for the length of the cell, crowding the nuclei to its periphery, where they are typically found just beneath the sarcolemma.

Each myofibril is encased in a sleevelike membrane structure, the **sarcoplasmic reticulum,** which is a specialized endoplasmic reticulum (Figure 42-18). The sacs of the sarcoplasmic reticulum contain calcium ions (Ca^{2+}), which, as we shall see, play an essential role in muscle contraction. Running perpendicular to the myofibrils is a system of transverse tubules, the T system. The membrane forming the T

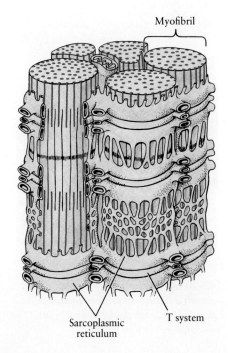

42-18 *A group of myofibrils, each of which is surrounded by the specialized endoplasmic reticulum of muscle cells, the sarcoplasmic reticulum. The sacs of the sarcoplasmic reticulum contain calcium ions, which, when released, trigger muscle contraction. Traversing the sarcoplasmic reticulum, perpendicular to the myofibrils, is a system of transverse tubules, the T system. The membrane forming the T system is continuous with the cell membrane; like the cell membrane, it can propagate an action potential.*

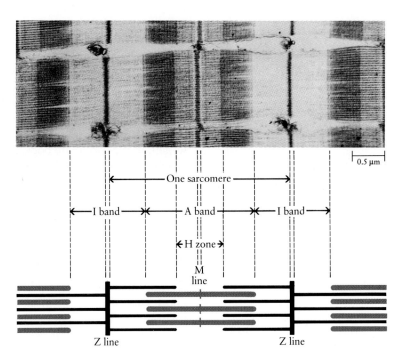

42-19 *Electron micrograph and diagram of a sarcomere, the contractile unit of muscle. Each sarcomere is composed of an array of thick and thin protein filaments arranged longitudinally. The Z line is where thin filaments from adjoining sarcomeres are anchored. The I band is a region that contains only thin filaments. The A band marks the extent of the thick filaments. The part of the A band where there are no thin filaments is called the H zone. The thick filaments are interconnected and held in place at the M line. Muscle contraction involves the sliding of the thin filaments between the thick ones.*

system is actually a complex invagination of the sarcolemma; it forms a series of channels (the interior of the transverse tubules) through which the solution of ions and molecules surrounding the muscle cell can flow. Despite the large diameter of the cell, no myofibril is separated from the extracellular fluid by more than the thickness of a lipid bilayer.

Myofibrils are composed of units called **sarcomeres,** each of which is about 2 or 3 micrometers in length. The repetition of these units gives the muscle its characteristic striated pattern. Figure 42-19 shows a sarcomere as seen in a longitudinal section of muscle. As the diagram shows, each sarcomere is composed of two types of filaments running parallel to one another. The thicker filaments in the central portion of the sarcomere are composed of the protein myosin; the thinner filaments are primarily actin, also a protein. The dense black line seen in the electron micrograph is the Z line, in which the thin filaments from adjacent sarcomeres interweave. The I band is the relatively clear, broad stripe that the Z line bisects. The large dense stripe in the center of the sarcomere, formed by thick filaments, is the A band; it is bisected by the central H zone, which contains thick filaments but no thin filaments. A transverse section of the region of the A band containing both types of filaments reveals that each thick filament is surrounded by six thin filaments (Figure 42-20).

The Contractile Machinery

The sarcomere is the functional unit of skeletal muscle, the mechanism by which contraction occurs. When the muscle is stimulated, the thin (actin) filaments of the sarcomere slide past the thick (myosin) filaments. Since the thin filaments are anchored into the Z line, this causes each sarcomere to shorten, and thus the myofibril as a whole to contract. As explained by the well documented sliding-filament model, cross bridges between the thick and thin filaments form, break, and re-form rapidly, as one filament "walks" along the other (Figure 42-21).

The actin strands of the thin filaments are composed of many globular actin molecules assembled in a long chain. As shown in Figure 42-21e, each thin filament consists primarily of two such actin chains wound around one another.

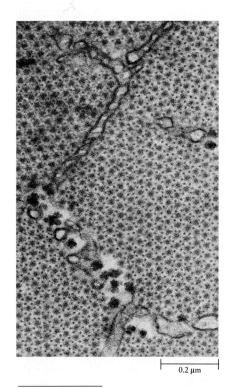

42-20 *A transverse section of myofibrils reveals a hexagonal array of filaments. As you can see, each thick filament is surrounded by six thin filaments.*

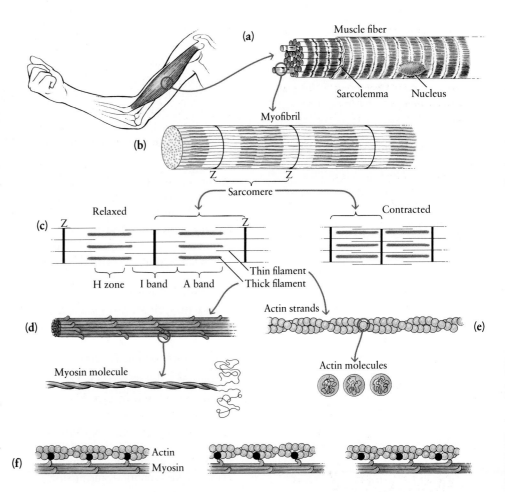

42–21 (a) *Skeletal muscle is composed of individual muscle cells, the muscle fibers. These are cylindrical cells, often many centimeters long, with numerous nuclei.* (b) *Each muscle fiber is made up of many cylindrical subunits, the myofibrils. These fibrils, which contain contractile proteins, run from one end of the cell to the other.* (c) *The myofibril is divided into segments, sarcomeres, by thin, dark partitions, the Z lines. The Z lines appear to run from myofibril to myofibril across the fiber. The sarcomeres of adjacent myofibrils are in line with each other, giving the muscle cell its striated appearance. Each sarcomere is made up of thick and thin filaments.* (d) *Chemical analysis shows that the thick filaments consist of bundles of a protein called myosin. Each individual myosin molecule is composed of two protein chains wound in a helix; the end of each chain is folded into a globular "head" structure.* (e) *Each thin filament consists primarily of two actin strands coiled about one another in a helical chain. Each strand is composed of globular actin molecules.* (f) *The globular myosin heads protruding from the thick filaments serve as hooks or levers, attaching to the actin molecules of the thin filaments and pulling them toward the center of the H zone, shortening the sarcomere and contracting the myofibril. When the myofibrils contract, the fiber shortens, and when enough fibers shorten, the entire muscle shortens, producing skeletal movements.*

The thick filaments are composed of bundles of myosin molecules. A myosin molecule consists of two long protein chains, each containing some 1,800 amino acids and each with a globular "head" at one end. In the molecule these two chains are wound around each other, with the globular "heads" free. These heads have two crucial functions: they are the binding sites at which force is exerted on the thin filaments during contraction, and they also act as enzymes that split ATP to ADP, thus providing the energy for muscle contraction.

When a muscle fiber is stimulated, the heads of the myosin molecules move away from the thick (myosin) filament toward the thin (actin) filament, to which they attach themselves. The heads move with a swiveling, oarlike motion, pulling the thick filament, pushing the thin one. Thus, a repeated cycle of attachment, breaking away, and reattachment moves the two filaments, ratchetlike, past one another. The thin filaments on opposite sides of the H zone move toward one another, so the Z lines bordering the sarcomere are pulled together.

The contraction of the sarcomeres is dependent on ATP in two ways: hydrolysis of ATP by the myosin molecule provides the energy for the cycle, and combination of a new ATP molecule with the myosin molecule releases the myosin head from the binding site on the actin molecule. The stiffened muscles of a corpse—rigor mortis—are due to the absence of ATP and the subsequent locking of all the actin-myosin cross bridges.

The contractile machinery of cardiac muscle, like that of skeletal muscle, consists of sarcomeres and is thought to function in the same fashion. Contraction of smooth muscle also depends on assemblies of actin and myosin molecules, but the exact mechanism remains unknown. Recent studies of smooth muscle using tiny resin beads attached to the outside of the cells revealed a corkscrew pattern of contraction rather than the linear contraction observed in skeletal muscle. This suggests that the actin and myosin molecules, or perhaps elements of the cytoskeleton with which they interact, are arranged within the smooth muscle cell in the form of a helix.

The Regulation of Contraction

The regulation of contraction in skeletal muscle depends upon two other groups of organic molecules, troponin and tropomyosin, plus calcium (Ca^{2+}) ions. As shown in Figure 42–22, tropomyosin molecules are long, thin double cables that lie along the actin molecules of the thin filament, blocking the cross-bridge binding sites on those molecules. The troponin molecules are complexes of globular proteins that are located at regular intervals on the tropomyosin chains. When Ca^{2+} combines with the troponin molecules, they undergo conformational changes that result in shifting of the tropomyosin chains and exposure of the cross-bridge binding sites. The availability of Ca^{2+}, and thus the initiation of contraction, depend on stimulation of the muscle by a signal received from a motor neuron.

The Neuromuscular Junction

A motor neuron (Figure 42–23a) typically has a single long axon that branches as it reaches the muscle (Figure 42–23b). At the end of each branch, the axon emerges from the myelin sheath and becomes embedded in a groove on the surface of a muscle fiber, forming the **neuromuscular junction** (Figure 42–23c). As is the case with most synapses between neurons, the signal travels across the neuromuscular junction by means of a chemical transmitter—in this case, acetylcholine. However, unlike synaptic transmission between neurons, this is a direct, one-to-one relationship involving only excitation. The acetylcholine combines with receptors on the sarcolemma, depolarizing the muscle cell membrane and initiating an action potential that sweeps along the sarcolemma, including the invaginations that form the T system. As the electrochemical impulse moves through the T system, it alters the sarcoplasmic reticulum, which then releases Ca^{2+} ions. These ions continue to be released only as long as the fiber is stimulated; once stimulation stops, the ions are pumped back into the sacs of the sarcoplasmic reticulum by active transport. Thus it is the Ca^{2+} ions that turn the contractile machinery on and off.

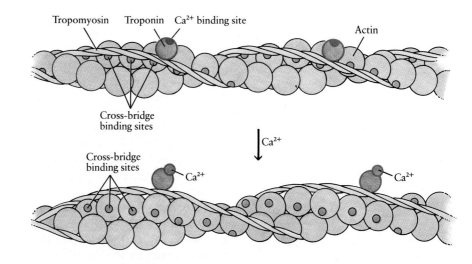

42–22 *Tropomyosin and troponin molecules, both proteins, have a regulatory role in muscle contraction. When calcium ions are not present, tropomyosin molecules, long, thin double cables, block the cross-bridge binding sites on the thin (actin) filaments. Troponin molecules, which are globular proteins, are situated at regular intervals on the long tropomyosin chain. When calcium ion binds to troponin, the tropomyosin molecule shifts position, exposing the binding sites and permitting the cross bridges to form.*

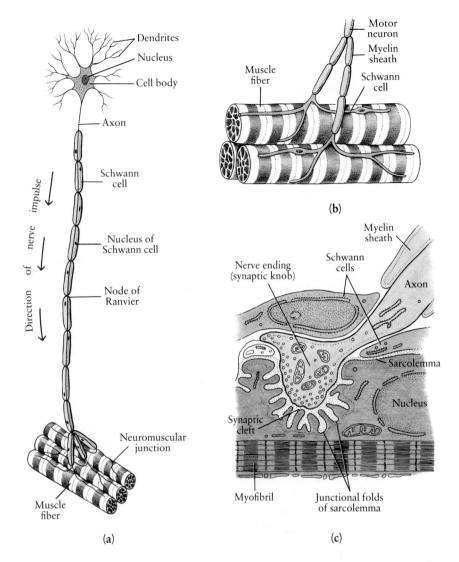

42–23 (a) *A motor neuron. The stimulus is received by the dendrites and cell body, which carry the signal to the axon hillock. The action potential initiated there travels along the axon, which is insulated by a myelin sheath composed of Schwann cell membranes.* (b) *The axon of each motor neuron divides into branches, each forming a neuromuscular junction with a different muscle fiber (cell). The motor neuron and the numerous muscle fibers that it innervates are known as a motor unit. Stimulation of a motor axon stimulates all of the fibers in that motor unit. Within a given muscle, fibers of different motor units are intermingled.* (c) *A neuromuscular junction. An action potential conducted along the axon of a motor neuron releases acetylcholine from synaptic vesicles into the synaptic cleft (see Figure 42–24). This neurotransmitter combines with receptor sites on the sarcolemma, altering the membrane permeability and initiating an action potential in the muscle fiber.*

Stimulation of smooth and cardiac muscle also triggers the movement of Ca^{2+} ions into the cytoplasm, not only from the sarcoplasmic reticulum but also from the extracellular fluid. However, as you will recall (page 845), neurons of the autonomic system can either stimulate or inhibit their targets. A smooth muscle cell is typically postsynaptic to neurons of both the sympathetic and parasympathetic divisions; whether it contracts or not depends on the summed effect of the neurotransmitters released at excitatory, inhibitory, and modulatory synapses.

42–24 (a) *A motor neuron terminal. The larger, darker bodies are mitochondria. Below them are the spherical synaptic vesicles. At the bottom of the micrograph is a portion of a muscle cell. The light area immediately below the neuron terminal is the synaptic cleft, and below that is a junctional fold in the sarcolemma. The muscle fiber underlying the sarcolemma is shown in a transverse section, as in Figure 42–20.* (b) *Synaptic vesicles discharging into the synaptic cleft.*

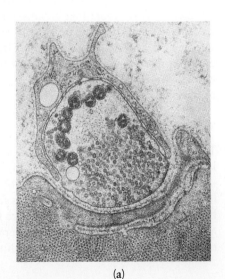

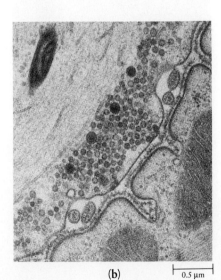

0.5 µm

Twitch Now, Pay Later

Vertebrate muscle fibers differ in their speed of contraction, which depends on the ATPase activity of their myosin heads, and in their energy metabolism. There are two general categories, red fibers and white fibers, most familiar to poultry fans as dark meat and light meat.

Red fibers are slow-twitch, pay-as-you-go fibers. They are relatively small, with large surface-to-volume ratios, a good blood supply, numerous mitochondria, and an abundance of myoglobin, which is mainly responsible for their color. As these characteristics would indicate, they have an abundant supply of oxygen, and oxidative phosphorylation provides most of their energy. In general, they use fatty acids as their fuel. (These fatty acids are stored in red fibers, which is why drumsticks tend to be greasy.) They are adapted for relatively continuous use that requires endurance. Thus in the domestic chicken, which walks some 14 hours a day, the leg muscles are mostly red fibers and the wing muscles—the "breast"—are white; whereas in a migratory bird—a wild duck, for instance—the "breast" is mostly red fibers.

White fibers, on the other hand, are fast-twitch, high-powered, twitch-now-pay-later muscles. They have fewer capillaries, fewer mitochondria, and less myoglobin. Their fuel supply is glucose or glycogen, which they break down by anaerobic glycolysis, resulting in lactic acid accumulation and oxygen debt. They are adapted for quick spurts of power. The strong, fast muscles of the legs of a rabbit or a frog are mostly white fibers.

The average human being has a fifty-fifty distribution of fast-twitch (white) and slow-twitch (red) fibers in the major skeletal muscles. However, recent studies have shown that the muscles of long-distance runners and swimmers average about 80 percent slow-twitch fibers, whereas sprinters tend to have about 75 percent fast-twitch fibers.

The proportion of slow-twitch fibers to fast-twitch fibers is genetically determined. (Identical twins have identical fiber patterns in their muscles, regardless of the type of muscular activity they engage in.) Training can cause fast-twitch fibers to behave more like slow-twitch fibers, but the reverse is not true. Some trainers recommend the use of fiber typing as part of a total body profile to determine what type of sport a would-be athlete should pursue. Russians and East Germans, who take their sports very seriously, are doing so already.

As we saw in Chapter 36, the contraction of cardiac muscle is initiated by the specialized cells of the pacemaker; signals transmitted to the pacemaker from the autonomic nervous system influence the rate at which the pacemaker fires but are not responsible for the actual initiation of contraction.

Many drugs act specifically on the neuromuscular junction. Curare, for instance, a plant extract that Indians of South America used to poison their arrow tips, blocks excitation of muscle by binding to acetylcholine receptors on the sarcolemma and produces paralysis. It is used medically as a muscle relaxant during surgery. The bacterial toxin of botulism—the most poisonous substance known—prevents nerve endings from liberating acetylcholine and so kills by paralysis of the muscles controlling breathing.

The Motor Unit

The axon of a single motor neuron and all the muscle fibers it innervates are known as a motor unit (Figure 42–25). The number of muscle fibers in a motor unit determines the fineness of control. In a muscle that moves the eyeball, for instance, a motor unit may contain as few as three muscle fibers, whereas in the biceps, each motor unit contains more than a thousand. Within a given muscle, fibers of different motor units are intermingled. A slight movement may involve the contraction of only a few motor units. The strength of contraction of a muscle

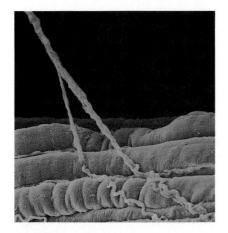

42–25 *A scanning electron micrograph showing the axon of a motor neuron and three skeletal muscle fibers.*

as a whole depends on the number of motor units that have been activated and the frequency with which they are stimulated.

Complex activities require the precisely timed contractions of different groups of fibers in antagonistic groups of muscles, often in different parts of the body. Moreover, many of an animal's most complex—and often most critical—activities are performed in relation to rapidly moving objects in its environment. Consider, for example, the dive of a hawk from high altitude in pursuit of a mouse scampering for cover. Less essential from a biological point of view, but equally marvelous as an example of the exquisite integration of sensory perception and motor response, is an event that happens numerous times each summer: a human being swings a piece of wood so that it connects with a small round ball 400 milliseconds after that ball was set in motion 18.4 meters away and drives it over the right-field fence.

SUMMARY

Sensory receptors are cells or groups of cells that transduce the energy of environmental stimuli into the energy of action potentials, and thus into information that can be transmitted and acted upon by the nervous system. Receptors may be classified according to the type of stimuli to which they respond as mechanoreceptors, chemoreceptors, photoreceptors, temperature receptors, and pain receptors. They can also be classified functionally as interoceptors, which monitor internal conditions; proprioceptors, which provide information about the orientation of the body and the position of its parts; and exteroceptors, which respond to stimuli from the external environment.

Chemoreception is the detection of specific chemical substances. In mammals, taste buds in the oral cavity are specialized for the discrimination of four primary tastes: sweet, sour, salty, and bitter. Detection of airborne molecules takes place in the olfactory epithelium. Chemoreception depends on the three-dimensional, hand-in-glove binding of molecules to specific membrane receptors, triggering changes in the sensory cell that lead to the initiation of action potentials.

The senses of balance and hearing result from mechanoreception, involving different structures in the ear. In humans, the outer ear functions as a funnel for capturing and routing sound waves to the tympanic membrane (eardrum). Vibrations in the tympanic membrane are transmitted via the bones of the middle ear (the hammer, anvil, and stirrup). The motion of the stirrup against the oval window membrane, which leads to the cochlea, causes movements in the fluid of the cochlea in the inner ear. The resulting movements of the basilar membrane stimulate the hair cells of the organ of Corti, which release neurotransmitters at synapses with sensory neurons that provide a relay to the brain. The position of the head in space is similarly monitored by hair cells in the fluid-filled semicircular canals of the inner ear.

For both birds and primates, the major sense organ is the eye, which provides much of the needed information about the environment. The outer, transparent layer of the eye is the cornea, behind which lies the lens. The image is focused on the light-sensitive retina by the cornea and by changes in the shape of the lens produced by the ciliary muscles. Light passes through the eyeball to the retina, which contains densely packed photoreceptor cells, the rods and the cones. Rods, which are more light-sensitive than cones, are responsible for night vision. Cones provide greater resolution than rods and are responsible for color vision. Birds and primates have areas of the retina specialized for acute vision; such an area, where cones are concentrated, is known as a fovea. Photoreception depends on the stimulation of light-sensitive pigments, consisting of retinal and a protein known as an opsin, in the rods and the cones. This stimulation causes changes in the membrane permeability of the photoreceptors, altering their release of neurotransmitters to the bipolar cells, and thereby altering the bipolar cells' release of neurotransmitters to the ganglion cells. Changes in the pattern in which action potentials are fired by the ganglion cells, whose axons form the optic nerve, convey the information to the brain. Considerable processing of the information occurs in the retina before it is transmitted to the brain.

Skeletal muscles are the effectors of the somatic nervous system. Each muscle is made up of muscle fibers—long, multinucleate cells bound together by connective tissues. Every fiber is surrounded by an outer membrane, the sarcolemma. Each muscle cell contains 1,000 to 2,000 small strands, the myofibrils, running parallel to the length of the cell. Each myofibril is surrounded by specialized endoplasmic reticulum, the sarcoplasmic reticulum, and is traversed by transverse tubules, which are formed by an invagination of the sarcolemma. Myofibrils are made up of units called sarcomeres, which consist of alternating thin and thick filaments. Contraction occurs when the filaments slide past one another. The thin filaments are composed of actin, troponin, and tropomyosin; and the thick filaments, of myosin. The globular heads of the myosin molecules serve as binding sites linking the thick and thin filaments during contraction and as enzymes for the splitting of ATP, which provides energy for muscle contraction.

Skeletal muscle fibers contract in response to stimulation by a motor neuron, whose axon terminals release acetylcholine, causing depolarization of the sarcolemma. The resulting action potential travels along and into the muscle fiber by way of the membrane of the transverse tubules. This causes the sarcoplasmic reticulum to release stored calcium ions (Ca^{2+}) into the fiber cytoplasm. The sudden increase in the intracellular Ca^{2+} concentration permits the interaction between actin and myosin that brings about contraction of the sarcomeres.

QUESTIONS

1. Distinguish among the following terms: interoceptor/proprioceptor/exteroceptor; cochlea/organ of Corti; taste/smell; rods/cones; bipolar cell/ganglion cell; muscle fiber/muscle cell/myofibrils; thick filament/thin filament; troponin/tropomyosin; sarcomere/sarcolemma/sarcoplasmic reticulum; synapse/neuromuscular junction.

2. What are the five classes of stimuli to which our sense organs respond?

3. Why does food seem flavorless when you have a cold?

4. Consider the process that occurs when another person speaks a word and you then hear the word. Describe the steps involved from the time the air leaves the lungs of the speaker until you are aware of the word spoken.

5. Why is hearing considered a form of mechanoreception?

6. What advantage do mammals gain by having such a long basilar membrane in the organ of Corti?

7. Label the drawing at the right.

8. Why don't we see objects upside down?

9. Draw a diagram of the sarcomere relaxed and contracted. What happens to the size of the I and A bands and the H zone as a sarcomere contracts? Explain how this provides the essential clue to the contractile mechanism.

10. What is the role of the T system in muscle contraction? What is the significance of the fact that the solution of ions and molecules in the extracellular environment flows through its tubules?

11. Explain the events whereby an action potential in a motor neuron causes a muscle to contract.

12. In the 1960s, the advertisement for a lecture by a whimsical biologist began as follows: "Available *now*. **Linear Motor.** Rugged and dependable: design optimized by worldwide field testing over an extended period." The copy ended with "Good to eat." What was the subject of the lecture?

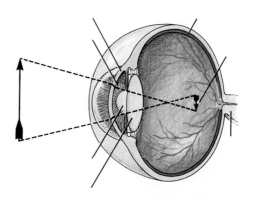

CHAPTER 43

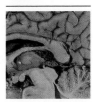

Integration and Control IV: The Vertebrate Brain

43-1 *In this section of tissue from the gray matter of a human brain, you can see the cell bodies of a number of neurons, embedded in a dense meshwork of axons. All of the information processing that occurs in the brain depends on interconnected networks of neurons, of which there are estimated to be some 100 billion in the human brain.*

In the three preceding chapters, we have considered a number of the structures, molecules, and cellular processes by which relevant information is obtained and transmitted through the animal body, leading to adjustments in the internal environment and action in the external environment. As animals have become more complex in the course of their evolutionary history, the task of integration and control has similarly become increasingly complex. That it is accomplished so successfully in all vertebrates can be attributed to an accompanying evolutionary trend, the increased centralization of integration and control in one dominant processing center, the brain.

This trend has produced a structure that, in humans, weighs about 1,400 grams (3 pounds), has the consistency of semisoft cheese, and is, by far, the most complex and highly organized structure on this planet. Its functions are essential not only for the integration and control of the multitude of physiological activities that occur throughout the body but also for those processes that we identify as "mind"—consciousness, perception and understanding of information from the external environment, thought, memory, and the variety of emotions that characterize human experience. Although our knowledge of the structure and function of the vertebrate brain—and particularly of the human brain—is increasing rapidly, it is such a complex system that many think we shall never fully understand it.

The soft substance of the brain, upon which all of its functions depend, is well protected by the bones of the skull. Like the spinal cord, it consists of white matter—the fiber tracts, made white by their lipid-rich myelin sheaths—and gray matter. The gray matter contains not only the cell bodies of as many as 100 billion neurons but also numerous supporting cells of the neuroglia. In some areas of the human brain, neurons and neuroglial cells are so densely packed that a single cubic centimeter of gray matter contains some 6 million cell bodies, with each neuron making synaptic connections to as many as 80,000 others.

THE STRUCTURAL ORGANIZATION OF THE BRAIN: AN EVOLUTIONARY PERSPECTIVE

The vertebrate brain had its evolutionary beginning as a series of three bulges at the anterior end of the hollow, dorsal neural tube. In human embryonic development, as we shall see in Chapter 45, this history is repeated as a groove at the surface of the early embryo closes, giving rise to the tubular structure from which the central nervous system—the brain and spinal cord—will develop. The cavities, known as ventricles, within the three anterior bulges persist in the mature brain and are filled with the same cerebrospinal fluid that fills the interior of the spinal cord, which develops from the posterior portion of the neural tube. In lower vertebrates, the three anterior bulges retain their linear arrangement,

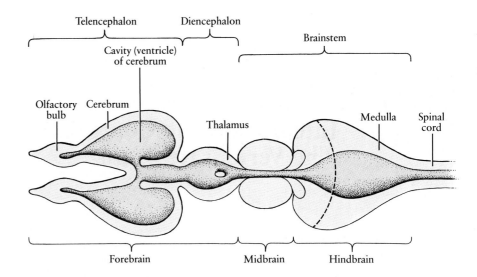

43–2 *A dorsal view showing the linear organization of the vertebrate brain. The brain has been cut horizontally to show the ventricles (cavities within the brain), which are continuous with the interior of the spinal cord. Like the central canal of the spinal cord, the ventricles are filled with fluid. The location of the cerebellum, which is a dorsal projection from the hindbrain, is indicated by the dashed line. The pons, another major structure of the hindbrain, is located ventrally and is not visible from this perspective.*

forming the hindbrain, the midbrain, and the forebrain (Figure 43–2). In birds and mammals, they become folded over one another in the course of development, but they can still be identified as distinct regions. Thus the terms hindbrain, midbrain, and forebrain are used to describe the principal regions of all vertebrate brains, including the human (Table 43–1).

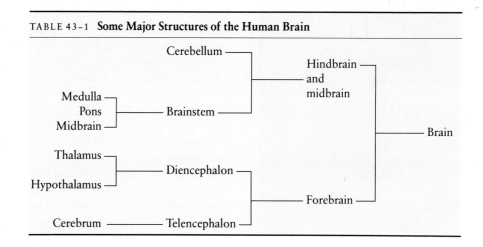

TABLE 43-1 **Some Major Structures of the Human Brain**

Hindbrain and Midbrain

The hindbrain and midbrain in birds and mammals can be seen as a knobby extension of the spinal cord, the **brainstem,** and a convoluted structure known as the **cerebellum.** The brainstem is the old brain, evolutionarily speaking, and is surprisingly similar from fish to *Homo sapiens* (Figure 43–3). Like the spinal cord, it contains nuclei (clusters of neuron cell bodies) involved with reflexes. Centers in the medulla, the posterior portion of the brainstem, control heartbeat and respiration, among other functions, which is why a blow to the base of the skull is so dangerous. The brainstem also contains sensory and motor neurons that serve the skin, muscles, and other structures of the head, as well as all the nerve fibers that pass between the spinal cord and the higher brain centers. Many of these fiber tracts cross over in the brainstem, so that the right side of the brain receives messages from and sends signals to the left side of the body, and vice versa. This crossing over is one of the major organizational features of all vertebrate nervous systems, but its evolutionary origin and significance remain unknown.

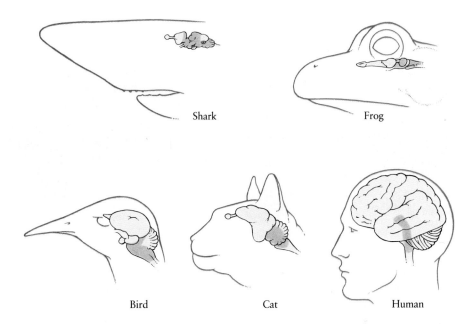

43–3 *The brains of five vertebrates. The brainstems (indicated in dark orange) include the medulla, pons, and midbrain. They are approximately the same in the different vertebrate groups. However, the cerebrum has become progressively larger in the course of evolution, and its two segments have folded upward, forming the two cerebral hemispheres. The cerebral cortex, the outer surface of the cerebral hemispheres, reaches its greatest development in the primates, particularly* Homo sapiens. *The olfactory bulb, readily visible in most vertebrate brains as a stalked knob at the anterior tip, is hidden in the human brain by the much more developed cerebrum.*

The cerebellum, a dorsal outgrowth of the primitive hindbrain, is concerned with the execution and fine-tuning of complex patterns of muscular movement. It is much larger in homeotherms than in the more slow-moving fish and reptiles and reaches its greatest relative size in birds, in which it is associated with the exquisite coordination necessary for flight. A ventral enlargement of the hindbrain, the pons ("bridge") contains fibers that provide communication between the left and right portions of the cerebellum, as well as ascending and descending fiber tracts. Auditory information is also relayed through the pons.

In the lower vertebrates, a major part of the midbrain is made up of the optic lobes, which receive fibers of the optic nerves. In mammals, the analysis of visual information has become a function of the forebrain, and the midbrain serves primarily as a relay-and-reflex center.

Forebrain

As you can see in Figure 43–2, the primitive forebrain is divided into two major parts; these are called the **diencephalon** and the **telencephalon.** The diencephalon, which contains the thalamus and the hypothalamus, is a major coordinating center of the brain. The thalamus, two egg-shaped masses of gray matter tucked within the cerebrum, constitutes the main relay center between the brainstem and the higher brain centers. Its nuclei process and sort sensory information. The hypothalamus, lying just below the thalamus, contains nuclei responsible for coordinating the activities associated with sex, hunger, thirst, pleasure, pain, and anger. As we have seen in earlier chapters, it contains the mammalian thermostat (page 784) and is the source of the hormones ADH and oxytocin, which are stored in and released from the posterior lobe of the pituitary (page 828). Most important, it is the major center for integration of the nervous and endocrine systems, acting through its release of peptide hormones that regulate the secretion of tropic hormones from the anterior pituitary (page 826).

The telencephalon ("end brain") is the most anterior portion of the brain and the structure that has changed the most in the course of vertebrate evolution. In the most primitive vertebrates, the fishes, it is concerned almost entirely with olfactory information and is known as the rhinencephalon, or "smell brain." In

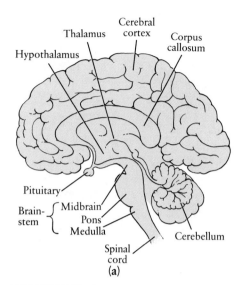
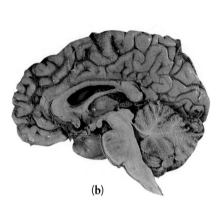
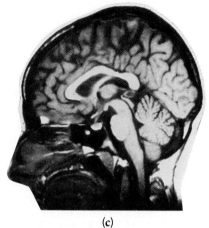

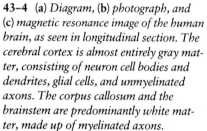

43-4 (a) *Diagram,* (b) *photograph, and* (c) *magnetic resonance image of the human brain, as seen in longitudinal section. The cerebral cortex is almost entirely gray matter, consisting of neuron cell bodies and dendrites, glial cells, and unmyelinated axons. The corpus callosum and the brainstem are predominantly white matter, made up of myelinated axons.*

The image in (c) *is that of an intact, living brain. Magnetic resonance imaging (MRI) is one of several new diagnostic tools available to neurologists. Like computerized axial tomography (CAT scans), MRI provides detailed images of soft tissues without requiring the injection of dyes or radioactive substances. The image results from a computerized summation of radio frequency signals emitted by the nuclei of hydrogen, sodium, and phosphorus atoms in the tissues when they are exposed to an intense magnetic field.*

reptiles, and especially birds, the most prominent structure of the telencephalon is the corpus striatum, which is involved in the control of complicated stereotyped behavior. In mammals, the **cerebrum,** the central portion of the telencephalon, is folded up into the two cerebral hemispheres and is greatly increased in size in relation to other parts of the brain. This increase reaches its greatest extent (so far) in the human brain, in which the many folds and convolutions of the surface of the cerebrum, the **cerebral cortex,** greatly increase its surface area. In humans, the cerebrum occupies 80 percent of the total brain volume. The wrinkling and folding of the cortex allows its enormous area of 2,500 square centimeters to fit within the confines of the skull. The cerebral hemispheres are connected to each other by a tightly packed, relatively large mass of fibers called the **corpus callosum.**

BRAIN CIRCUITS

In the brain, as elsewhere in the body, there is a division of labor, with different parts of the brain performing different, specific functions. However, integration and control of the multitude of processes occurring in an animal's body depend on coordination of all of the activities occurring in the different parts of the brain. Information is exchanged between different regions of the brain by way of diffuse tracts of bundled axons. Thus, a local network of neurons in one region of the brain can have its activity modified by—and can modify the activity of— networks of neurons located elsewhere in the brain. Two examples of such integration involve the reticular activating system and the limbic system.

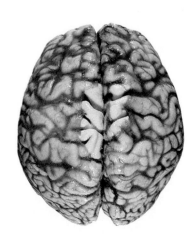

43-5 *A brain, viewed from above. The many convolutions of the cerebral cortex are clearly visible. By these, you can immediately distinguish this brain as human. The brain and spinal cord are enclosed in three layers of membranous tissue, known as the meninges. The outermost layer, the dura mater, which has begun to peel away from the surface of the brain, is visible in this photograph.*

The Reticular Activating System

The reticular activating system is made up of the reticular formation, a core of tissue running through the brainstem (Figure 43-6), and neurons in the thalamus that function as an extension of this system. It is of particular interest because it is involved with arousal and with that hard-to-define state we know as consciousness. All the sensory systems have fibers that feed into this system, which apparently filters incoming stimuli and discriminates the important from the unimportant. Stimulation of the reticular activating system, either artificially or by incoming sensory impulses, results in increased electrical activity in other areas of the brain.

The existence of such a filtering system is well verified by ordinary experience. A person may sleep through the familiar blare of a subway train or a loud radio or TV program but wake instantly at the cry of the baby or the stealthy turn of a doorknob. Similarly, we may be unaware of the contents of a dimly overheard conversation until something important—our own name, for instance—is mentioned, and then our degree of attention increases.

The Limbic System

The limbic system is a network of mostly subcortical ("below the cortex") neurons that form a loop around the upper part of the diencephalon, linking the hypothalamus to the cerebral cortex and to other structures as well. It is thought to be the circuit by which drives and emotions, such as hunger, thirst, and desire for pleasure, are translated into complex actions, such as seeking food, drinking water, or courting a mate. As we shall see later in this chapter, it is also a principal circuit in the consolidation of memory. The structures of the limbic system are phylogenetically primitive, corresponding to the telencephalon (rhinencephalon) of reptiles.

THE CEREBRAL CORTEX

The cerebral cortex is a thin layer of gray matter about 1.5 to 4 millimeters thick, covering the surface of the cerebral hemispheres. It is the most recent development in the evolution of the vertebrate brain. Fish and amphibians have no cerebral cortex, and reptiles and birds have only a rudimentary indication of a cortex. More primitive mammals, such as rats, have a relatively smooth cortex. Among the primates, however, the cortex becomes increasingly complex. Of the approximately 100 billion nerve cells in the human brain, about 10 billion are estimated to be in the cerebral cortex.

In *Homo sapiens* and other primates, each of the cerebral hemispheres is divided into lobes by two deep fissures, or grooves, in the surface. The principal fissures are the central sulcus, which runs down the side of each hemisphere, and the lateral sulcus. There are four lobes—frontal, parietal, temporal, and occipital—on each hemisphere (Figure 43-7).

Although much of the functioning of the cerebral cortex remains poorly understood, it is the most thoroughly studied region of the human brain—and, inherently, the most fascinating. We shall therefore consider it in some detail, devoting particular attention to some of the more provocative findings about its divisions of labor.

Motor and Sensory Cortices

Certain areas of the cerebral cortex have been mapped in terms of the functions they perform (Figure 43-8). Some of the information is based upon observation of human patients in whom particular areas of the cortex have been destroyed by

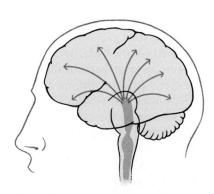

43-6 *The reticular formation is a diffuse network of neurons in the brainstem. Here, incoming stimuli are monitored and analyzed, leading to modulation of the activity of other areas of the brain. The reticular activating system includes the reticular formation and its thalamic extension. It is concerned with general alertness and the direction of attention.*

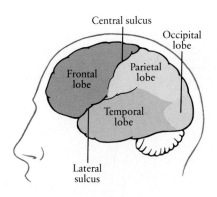

43-7 *The principal sulci (fissures) and lobes of the human cerebral cortex.*

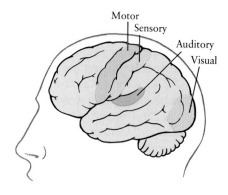

43-8 *The human cerebral cortex, showing the location of the motor and sensory areas, on either side of the central sulcus, and the auditory and visual zones. The motor and sensory cortices span the brain like earphones. Functionally, the left and right motor and sensory cortices are mirror images, with the left cortex receiving signals from and sending signals to the right side of the body, and vice versa.*

disease or accident, and some of it is derived from surgical procedures performed on experimental animals. Other studies involve stimulating particular areas of the cortex and observing what takes place in various parts of the body, or stimulating various sensory receptors and noting electrical discharges in parts of the cortex. Such investigations have been carried out in experimental animals as well as in humans undergoing brain surgery. It should be noted that touching the brain produces no sensation of pain because it has no sensory receptors.

These studies have shown that the area just anterior to the central sulcus, in the frontal lobe, contains neurons concerned with integration of activities performed by the skeletal muscles. Each point on the motor cortex, as it is called, is involved in the movement of a different part of the body. The relative amount of cortex allocated to a particular group of muscles varies from animal to animal. For instance, in humans, the areas controlling the hand and fingers are very large (Figure 43-9), whereas the area controlling a cat's paw is somewhat smaller and that for a horse's hoof is very small indeed.

Immediately posterior to the central sulcus, in the parietal lobe, is the sensory cortex. It is involved with the reception of tactile (touch) stimuli, as well as stimuli related to taste, temperature, and pain. Its greatest representation is for those parts of the body that are most richly endowed with sensory receptors: fingertips, tongue, lips, face, genitalia.

In the temporal lobe, partially buried within the lateral sulcus, is the auditory cortex (see Figure 43-8). This region of the cortex is the processing center for the

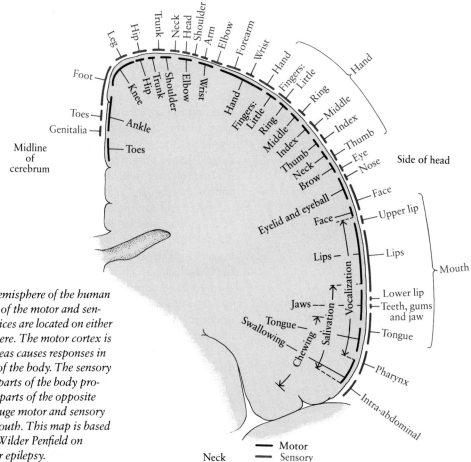

43-9 *A combined cross section of one hemisphere of the human cerebrum, indicating the functional areas of the motor and sensory cortices. The motor and sensory cortices are located on either side of the central sulcus in each hemisphere. The motor cortex is indicated in black; stimulation of these areas causes responses in corresponding parts of the opposite side of the body. The sensory cortex is in color; stimulation of various parts of the body produces electrical activity in corresponding parts of the opposite side of this cortex. Notice the relatively huge motor and sensory areas associated with the hand and the mouth. This map is based largely on studies done by neurosurgeon Wilder Penfield on patients undergoing surgical treatment for epilepsy.*

signals relayed from the sensory neurons of the ear. As you will recall from the last chapter, different hair cells of the ear, and thus different sensory neurons, are stimulated by different frequencies (pitches) of sound. The signals from the different sensory neurons are relayed to different regions of the auditory cortex, which can be mapped according to the frequencies eliciting the strongest response.

The visual cortex occupies the occipital lobe. By using a tiny point of light to stimulate very small regions of the retina, one after the other, investigators have been able to show that each region of the retina is represented by several corresponding but larger regions of the visual cortex. The fovea (page 866), which represents about 1 percent of the area of the human retina, projects to nearly 50 percent of the visual cortex. This enormous over-representation, combined with the very substantial portions of the motor and sensory cortices devoted to the hands, provide additional evidence of the importance of eye-hand coordination in primate evolution.

The Perception of Form

How do we perceive form? It was long thought that we perceive form simply because of the way the visual image is arranged on the visual cortex. However, numerous studies conducted over the past 30 years have revealed that a quite different mechanism is involved. Each of the regions of the visual cortex to which different regions of the retina project contains a variety of cells, different groups of which respond to different types of visual stimuli.

This feature of visual processing was first revealed in experiments performed by David H. Hubel and Torsten N. Wiesel at Harvard University. They used microelectrodes to record the responses of individual cells in the visual cortex of the cat to different visual stimuli and obtained some very surprising results (Figure 43–10). One class of cortical neurons, for example, responded only to a horizontal bar; as the bar was tipped away from the horizontal, discharges from the cells slowed and ceased. Another class responded best to a vertical bar, ceasing to fire

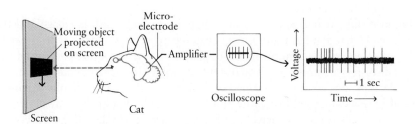

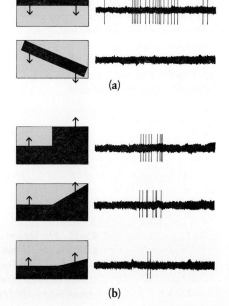

43–10 *The experimental method of Hubel and Wiesel. A cat with a small electrode recording from a single cell in its visual cortex observes a screen on which a moving shape is projected. (a) and (b) represent tracings from two different neurons; the arrows indicate the direction of movement of the shape on the screen. The first neuron (a) responds when the cat looks at a horizontal bar moving downward; if the moving bar is tilted rather than horizontal, the action potentials cease. The second neuron (b) responds to the upward movement of a right angle. Note the reduction in response as the angle increases.*

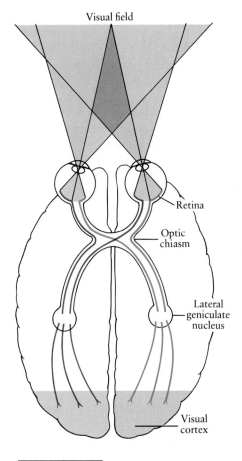

43-11 *The neural pathways leading from the retina to the visual cortex, as viewed from below. Axons of the ganglion cells travel from the retina to the lateral geniculate nuclei by way of the optic nerves. In the lateral geniculate nuclei, the ganglion cell axons synapse with other neurons that carry the signals to neurons of the visual cortex. The optic chiasm is a structure formed by the crossing over of axons conveying information from half of the visual field of each eye. Because of this crossing over, the right and left areas of the visual cortex each "see" only the opposite half of the visual field.*

as the bar was turned toward the horizontal. A third class fired only when the bar was moved from left to right; another only in response to a right-to-left movement. Another class of cells turned out to be a right-angle detector; such cells responded only to an angle moving across the visual field and were most excited when the angle was a right angle.

As we saw in the last chapter, considerable processing of visual information occurs in the retina before its transmission to the ganglion cells, which carry the information into the brain. In a region of the thalamus known as the lateral geniculate nucleus, the ganglion cells synapse with other neurons that relay the signals to the visual cortex (Figure 43-11). The lateral geniculate nucleus, however, is not merely a relay station; it is the site of further processing of the visual information. Its neurons apparently sort the signals representing different aspects of the visual image and then send them on to specific neurons of the visual cortex. What we see is not a direct image but rather a mental picture computed by the brain from a vast amount of coded information concerning the individual features that collectively form shapes, spatial relationships, and patterns of movement, light and dark, and color.

Left Brain/Right Brain

For more than 100 years it has been known that injury to the left side of the brain often results in impairment or loss of speech (aphasia), whereas a corresponding injury to the right side of the brain usually does not. Two areas in the left cerebral hemisphere concerned with speech have been mapped, primarily through work with patients who have had a stroke (an occlusion of the blood supply to a particular portion of the brain). These are known as Broca's area and Wernicke's area (Figure 43-12), each named for the nineteenth-century neurologist who first identified it. About 90 percent of all right-handed people and 65 percent of all left-handed people have these speech areas in the left cerebral cortex.

Broca's area is located just anterior to the region of the motor cortex that controls movements of the muscles of the lips, tongue, jaw, and vocal cords. Damage to this area results in slow and labored speech—if speech is possible at all—but does not affect comprehension. Wernicke's area is adjacent to and partially surrounds the auditory cortex. Localized damage in Wernicke's area results in speech that is fluent but often meaningless, and comprehension of both spoken and written words is impaired.

Because the left hemisphere in most people controls both language and the more capable hand, it has traditionally been regarded as dominant, with the right hemisphere described as mute, minor, or passive. However, severe perceptual and spatial disorders are now known to result from injury to the right hemisphere. For example, a person who has sustained such an injury may have difficulty orienting himself in space, may get lost easily and have difficulty finding his way in new or unfamiliar areas, and may have trouble recognizing familiar faces or voices.

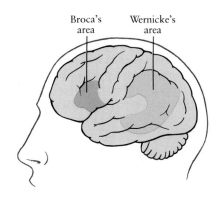

43-12 *The human cerebral cortex, showing the areas associated with language. Damage to Broca's area—the more anterior area —affects speech. Damage to Wernicke's area affects comprehension of language. These two areas of the left cerebral cortex are connected by a fiber tract.*

Musical talent also appears to reside in the right hemisphere: damage to the right hemisphere may result in loss of musical ability, leaving speech unimpaired. Alexander Luria, the eminent Russian neuropsychologist, described a composer whose best work was done after he became aphasic following a massive left hemisphere stroke.

The acquisition of different functions by the two cerebral hemispheres is seen as another way of increasing the functional capacity of the brain without increasing the size of the skull (which is calculated to be as large as possible in relation to the size of the birth canal). There is some evidence that this lateralization of function—that is, the differentiation of function between the two cerebral hemispheres—is part of the developmental process. It is well known, for instance, that in a young child areas of the right hemisphere can take over following damage to the left hemisphere, and completely normal speech may develop. The ability of the right hemisphere to assume this left-hemisphere function is closely correlated with the age at which the injury occurs.

It appears that songbirds also have some lateralization of function in their brains. Studies of songbirds indicate that their musical ability resides in the left hemisphere. Fernando Nottebohm at Rockefeller University has shown, for instance, that canaries with lesions in the left hemisphere sing a grossly distorted song. Canaries with right-hemisphere lesions show only minor changes in song production. Also, as in children with left-hemisphere damage, the right hemisphere of canaries can, with time, take over the functions of the damaged left hemisphere.

The song nuclei—the regions of the canary brain controlling song—can be readily identified anatomically. Nottebohm has found that the size of these regions varies from bird to bird and from season to season. Males with small song centers generally have small repertoires, and the song centers of females, which sing very little, are one-fourth the size of those of males. The size of the nuclei increases in the early spring and then decreases in the summer and fall.

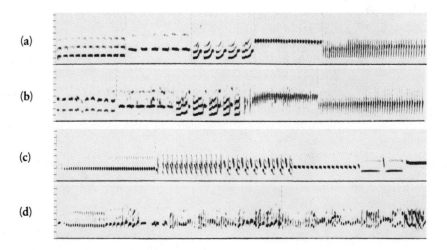

43–13 *Sound spectrographs of the song of* (a) *a normal male canary and* (b) *the same canary after damage to the vocal center in the right hemisphere. As you can see, there is little change in the bird's song. By contrast, sound spectrographs of the song of another male canary* (c) *before and* (d) *after damage to the vocal center in the left hemisphere reveal a dramatic change in the bird's song.*

Split Brain

As we mentioned previously, the two cerebral hemispheres are connected by the corpus callosum. In some cases of epilepsy, severing the corpus callosum lessens the severity of the epileptic attacks. In the 1960s, Roger Sperry and his coworkers at the California Institute of Technology launched a series of studies on such patients, who came to be known as split-brain patients. The name arose because, as Sperry and his associates showed, once the corpus callosum was severed, the

Electrical Activity of the Brain

Many of the neurons of the brain fire action potentials spontaneously and continuously. This electrical activity can be monitored by measuring the difference in electric potential between an electrode placed on a specific area of the scalp and a "neutral" electrode placed elsewhere on the body or between pairs of electrodes on the head. The voltages that can be detected at the scalp are considerably less than the change in potential that occurs in a single action potential, which is about 110 millivolts. From the scalp of a normal adult, about 300 microvolts—only 0.3 millivolt—is the maximum that can be recorded, and thus extremely sensitive recording equipment is required.

An electroencephalogram (EEG) is a record of the electrical activity of the brain. EEG recordings represent the combined activity of large numbers of neurons. Electroencephalography is a useful tool for the diagnosis and monitoring of epilepsy (which is an electrical storm in the brain), cerebral tumors, and brain damage, because characteristic EEG patterns can be correlated with certain levels and types of brain activity. Electroencephalography is also used to detect brain death, which is indicated by the total cessation of electrical activity.

One of the characteristic EEG patterns is the alpha wave, a slow, fairly irregular wave with a frequency of about 8 to 12 cycles per second, recorded at the back of the head. Alpha waves are usually produced during periods of relaxation. In most persons they are conspicuous only when the eyes are shut. About two-thirds of the general population have alpha waves that are disrupted by attention. Of the remaining third, about one-half have almost no alpha waves at all and about one-half have persistent alpha waves that are not easily disrupted by attention. Another EEG pattern is the beta wave. Beta waves are lower amplitude (lower voltage) than alpha waves, but their frequency is greater, from 18 to 32 cycles per second. Beta waves occur in bursts and are associated with mental activity and excitement.

Sleep is characterized by repeated cycles in which the sleeping person passes through distinct stages during which the EEG waves get larger and slower. In the course of eight hours of sleep, a young adult typically passes through five such cycles. During each cycle there is usually a period of rapid, low-amplitude waves similar to those seen in alert persons. This stage, called paradoxical sleep, is associated with rapid eye movements (REMs), in which the eyes make rapid, coordinated movements.

REM sleep differs from the other sleep stages. For example, although there is electroencephalographic evidence of alertness, the muscles are more relaxed than in ordinary deep sleep. Further, although the sleeper is harder to awaken from REM than from deep sleep, once awakened, he or she is more alert. Fluctuations in heart rate, blood pressure, and respiration are also seen during REM sleep. Many male subjects experience erections of the penis during REM sleep, and women have similar erections of the clitoris with secretions of vaginal fluid. A person awakened during a period of REM sleep nearly always reports that he or she has been dreaming.

The significance of REM sleep is not known. At one time, REM sleep was thought to be the period when all dreaming occurred. On the basis of more recent work, it has been suggested that dreaming can

two hemispheres of the brain in these patients were functionally separated. In general, it was found that split-brain patients are able to carry out their normal activities, but under controlled experimental conditions, they behave as if they have two separate brains. If such patients are asked to identify by touch objects that they cannot see, for instance, they can name those they can feel with the right hand but not those they can feel only with the left hand. This is apparently because information from the right hand goes to the left cerebral hemisphere, where the speech centers of the brain are located, and information from the left hand goes to the right brain, which is generally mute. If, for instance, a patient's left hand is given a plastic object shaped like the number 2, which the patient cannot see, he or she is unable to identify the object verbally but can readily tell the experimenter what it is by extending two fingers.

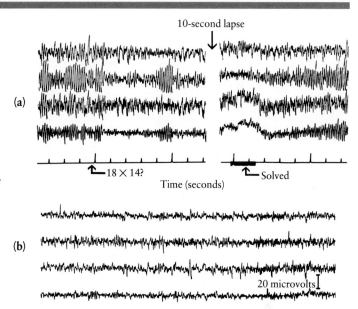

(a) *EEG of a student with conspicuous alpha rhythm. Electrodes were placed at four positions on the head. When the student was asked to multiply 18 × 14, the alpha rhythm was suppressed and then resumed after the problem was solved (10 seconds of the recording are omitted).* (b) *EEG of a student with a complete absence of alpha rhythms. According to a recent study, people who have almost no alpha rhythms under any conditions think almost exclusively by visual imagery, whereas people with persistent alpha rhythm tend to be abstract thinkers.*

occur at any time during sleep but that conditions for recalling dreams are most favorable when subjects are awakened during REM sleep. It has also been proposed that REM sleep serves as an information-processing period during which data from the previous waking cycle are sorted, processed, and stored. Supporting the hypothesis that this stage of sleep is essential to the memory process are studies showing that laboratory rats forget tasks they have learned if they are deprived of REM sleep. Similarly, evidence indicates that the student who stays up all night cramming for an exam will not have as good a recollection of the material studied as the student who studies and then sleeps.

The EEG measures the electrical activity of the cerebral cortex as recorded at the surface of the scalp. Within the last decade, technological advances have made possible the detection and localization of electrical activity deep within the human brain. The principle involved is well known to students of physics: any flow of electricity is accompanied by a magnetic field. The instrument used is called the SQUID (for Superconducting Quantum Interference Device). Brain tissue is essentially transparent to magnetic fields, so that by aiming two or more magnetic detectors at a source within the brain, it is possible to record and plot the source of activity from subcortical structures. This technique is presently used principally by clinical neurologists to locate the small areas within the brains of epileptics where their seizures originate. In addition, it appears to be an important new tool for correlating discrete areas of electrical activity in the brain with particular mental processes.

In split-brain patients, if a picture is briefly flashed before them while their eyes are held in a fixed position, the right hemisphere sees only the left side of the picture and the left hemisphere sees only the right side (see Figure 43–11). If a word is flashed to the left hemisphere, the patient is able to identify and write the word correctly. If the word is flashed to the right hemisphere, the patient is unable to speak or write the word. However, such a person is quite able to select an object corresponding to the word with his left hand from a group of objects hidden from the left hemisphere by a screen bisecting the visual field.

Sperry's findings created a flurry of interest in the psychological consequences of these "two realms of consciousness," but many authorities now believe that the popular speculations about them have far outstripped the hard evidence.

Intrinsic Processing Areas

The unmapped areas of the cerebral cortex were once known as the "association," or "silent" cortex. These terms were introduced in the course of the early mapping studies, when these areas did not respond in any specific way to stimulation and were thought to function as a sort of giant switchboard, interconnecting the motor and sensory cortices. More recent studies have shown, however, that the organization of the brain is more vertical than horizontal. For example, severing the motor from the sensory cortex by deep vertical cuts appears to have little, if any, effect on an animal's behavior. Anatomical studies support the interpretation that communication between the sensory and motor areas of the cortex takes place primarily via lower brain centers, particularly the thalamus.

The regions of the cortex traditionally considered to form the association cortex—all of the areas not mapped in Figures 43-8 and 43-12—are now more accurately described as intrinsic processing areas. This means that although they receive and process information from neurons in other areas of the brain, they do not receive directly relayed sensory information of the type, for example, that travels from the retina to the visual cortex by way of the lateral geniculate nuclei. Similarly, they transmit information to neurons in other areas of the brain but not directly to neurons leading out of the brain. Certain intrinsic processing areas do, however, receive information from the visual, auditory, or sensory cortices, and others transmit information to the motor cortex.

Mapping of the intrinsic processing areas is a considerably more complex undertaking than the mapping of the sensory and motor cortices, and this work is still in its earliest stages. However, using a variety of techniques, including the actual physical tracing of neurons and their connections, neurobiologists are beginning to gain some insight into the activities of these vast areas of the cortex. For example, it is now known that the posterior region of the parietal lobe and the lower portion of the temporal lobe receive signals transmitted by neurons of the adjacent visual cortex and are involved in the further processing of visual information (Figure 43-14). As you will recall (page 887), the visual cortex contains sets of very precisely tuned neurons, each receiving information about one specific feature of an object in the visual field. Current evidence indicates that the processing of all of these bits of information into a mental image of the whole object occurs in the lower portion of the temporal lobe. However, processing of the bits of information that locate the object in space—and identify its spatial relationships to other objects—occurs in the posterior region of the parietal lobe. A number of studies suggest that similar divisions of labor may also be involved in the processing of other types of sensory information.

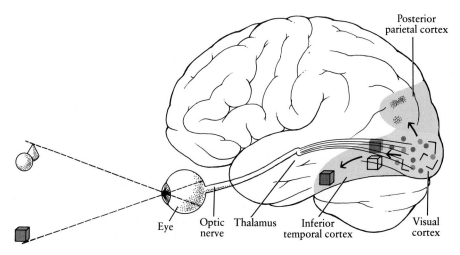

43-14 *The processing of visual information involves not only the visual cortex but also adjacent intrinsic processing areas in the parietal and temporal lobes. Information about the components of a visual image, received in the visual cortex, is transmitted to the temporal lobe, where complex analysis occurs that leads to a perception of a whole image. Simultaneously, other coded information is transmitted to the parietal lobe, where spatial relationships of the object with other objects in the visual field are analyzed.*

The proportion of the cerebral cortex devoted to intrinsic processing is much higher in primates than in other mammals and is very large in humans. This suggests that, in addition to their role in the processing of sensory information, these areas have something to do with what is special about the human mind. Also, about half of the total area involved is in the frontal lobes, the part of the brain that has developed most rapidly during the recent evolution of *Homo sapiens*. The frontal lobes are responsible for our high forehead, as compared with the beetle brow of our most immediate ancestors. Public appraisal of the function of the frontal lobes is reflected in the terms "high brow" and "low brow." The present scientific consensus is that the intrinsic processing areas are concerned with the integration of sensory information with emotion and its retention in memory, with the organization of ideas, which is, of course, a key component of learning, and with long-range planning and "intention." Although it will undoubtedly be a long time, if ever, before we understand the activities in these regions of the brain—which, of course, underlie our very capacity to understand—a variety of studies are beginning to shed light on the kinds of processes that may be occurring there.

LEARNING AND MEMORY

For neurobiologists, perhaps the greatest challenge is to understand the mechanisms of learning and memory. If we define learning as a change in behavior based on experience, and if the functions of the brain—the mind—are to be explained in terms of the structures of which it is composed, then learning—and the memory on which it depends—must involve changes in these structures. But what are these changes, and where and how do they take place?

Almost 60 years ago, the late Karl Lashley set out to locate the physical change, the trace of memory, which he called the engram. He taught rats and other animals to solve particular problems and then performed operations to see whether he could remove the portion of the cortex containing that particular engram. But he never found the engram. As long as he left enough brain tissue to enable the animal to respond to the test procedures at all, he left memory as well. And the amount of memory that remained was generally proportional to the amount of remaining brain tissue. Lashley concluded that memory is "nowhere and everywhere present." With the much more sophisticated techniques now available for brain research, contemporary neurobiologists are just beginning to unlock its hiding places.

There are two distinct types of memory, short-term and long-term. A simple example of short-term memory is looking up an unfamiliar number in a phone book; you usually remember it just long enough to dial it. (In fact, the capacity of short-term memory is about seven items, the number of digits necessary for a local call.) If you call the number enough times, it is transferred to long-term memory. This laying down of long-term memory is analogous to the establishment of a footpath. The more frequently the path is traveled, the better established it becomes. The analogy is strengthened by the familiar experience of consciously retrieving a name, for instance, by seeking out related information that puts one "on the right track." As we shall see, it is now thought that the establishment of such a pathway involves alterations in the synapses by which neurons communicate with one another.

The concept of two kinds of memory is supported by experience with patients with memory deficits; it is possible to lose one kind of memory and not the other. A blow to the head can result, for instance, in the soap-opera kind of amnesia for prior events while not interfering with short-term memory or the establishment of new long-term memories. (Typically, "lost" memories return, indicating that what has been lost is not the memory itself but rather the capacity to retrieve it.)

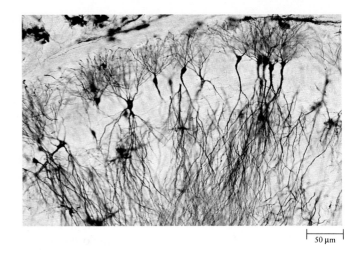

43-15 *Neurons of the hippocampus, one of the principal structures involved in the consolidation of memory.*

Conversely, injury to the hippocampus, which is part of the limbic system, does not affect already established long-term memories but does interfere with the transfer of short-term memories to long-term memory. A patient with bilateral destruction of the hippocampus can remember where he lived as a child but not where he lives now, for example. He can carry on an apparently normal conversation, but if the person he was talking to leaves the room and returns a minute or two later, he will not recognize him.

Anatomical Pathways of Memory

Studies in experimental animals and with persons who have experienced memory loss as a result of disease, injury, or as a consequence of brain surgery required to treat disease or injury, have revealed that the hippocampus is only one of several structures involved in memory. Figure 43-16 shows the principal regions in the human brain presently thought to be involved in the consolidation and storage of memory. These regions include the hippocampus ("seahorse," given this name by early neuroanatomists because of its shape) and amygdala ("almond"), both located on the inner surface of the temporal lobe; the thalamus and another structure of the diencephalon, the mammillary body; the basal forebrain, one of the ancient parts of the telencephalon; and a portion of the frontal lobe known as the prefrontal cortex.

According to current hypotheses, information is transmitted along independent pathways from the various sensory cortical areas to the hippocampus and amygdala, from which independent pathways carry the information to the thalamus and the mammillary body. Neurons of the thalamus and mammillary body, in turn, conduct the information to the basal forebrain and to the prefrontal cortex. Parallel circuits transmit processed information in the opposite direction, in what is thought to be a positive feedback process. The basal forebrain, which degenerates in Alzheimer's disease, is a principal source of the neurotransmitter acetylcholine in the brain. The acetylcholine that its neurons release in the feedback circuit is apparently vital for the processes that occur in other parts of the circuit, particularly the amygdala and the hippocampus.

Experimental studies indicate that, in addition to its role as a relay station in these circuits, the amygdala is also the region in which information from different senses is linked—so that, for example, when you think of the seashore, you can remember not only the visual image of sand and waves but also the sound of the lapping waves, the smell and taste of salt spray, and the feel of sand between your toes. And, apparently because of connections between the amygdala and the hypothalamus, the memories have emotional content—the details of which depend on your particular experiences at the seashore. Other fibers of the amygdala appear to communicate back to the primary sensory cortical areas. It is hypothesized that the "storage" of sensory memories in the primary sensory areas is an important element in the subsequent recognition of similar sensory input.

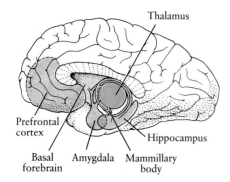

43-16 *Structures of the human brain involved in the consolidation and storage of memory, as shown in longitudinal section. Damage to any of these structures results in memory loss, the details of which vary according to the structure affected. For example, the memory loss associated with strokes typically involves damage to the prefrontal cortex, the thalamus, or the posterior portion of the hippocampus. Korsakoff's syndrome, an amnesia that develops in some chronic alcoholics, involves degeneration in the thalamus and mammillary body. The memory loss associated with Alzheimer's disease involves neurons in the basal forebrain. Inflammation or a temporary interruption of the oxygen supply to the brain can cause damage to the amygdala and the anterior portion of the hippocampus that also results in memory loss.*

Current evidence indicates that the traversing of all of these pathways, including the feedback pathways, is required for the consolidation of long-term memory. Although specific sensory memories appear to be stored in the sensory cortices, more complex memories may be stored elsewhere. The neurosurgeon Wilder Penfield, who was largely responsible for mapping the sensory and motor areas of the cortex, found that in a small percentage of his patients stimulation of certain areas of the cortex—especially in the temporal lobe—evoked particular experiences so vividly that patients reported that they felt they were actually reliving past events. Although there is currently some dispute as to whether Penfield actually evoked memories (as opposed to fantasies or dreams), it has been hypothesized that he may have made contact with memory pathways in the temporal lobe.

The pathways we have traced are those involved in "recognition" memory, which may be a distinct process from the memory that is involved in automatic motor responses to particular stimuli. This second type of memory, which has been termed "habit" or "procedural" memory, underlies such activities as the regurgitation of food by an adult bird in response to the open craw of its offspring or the shifting of your weight as you round a curve on a bicycle. It is thought to depend on a distinct set of pathways, involving the corpus striatum (page 884), an ancient part of the telencephalon.

Synaptic Modification

Although work with human patients and experimental animals is slowly elucidating the pathways through which information travels in the establishment of memory, it leaves unanswered the question of the changes at the cellular and molecular levels that form the "stuff" of memory. Possible answers to this question are beginning to emerge from studies performed by many groups of investigators in which invertebrate animals—or even isolated neurons or groups of neurons—are used as the model organisms. Invertebrates are useful because, first of all, their neurons are large and unmyelinated. Moreover, their nervous systems have far fewer neurons, numbering in the thousands rather than in the billions. Thus, invertebrate behavior circuits are relatively easy to trace—easy, that is, in contrast to the bewildering complexity of the circuits in the vertebrate nervous system and of the behaviors associated with their functioning.

A notable example of such work is that of Eric Kandel and his associates at Columbia University on the gill withdrawal reflex of the sea hare *(Aplysia)* (Figure 43–17). When this mollusk is touched gently on the underside, it quickly withdraws its siphon and delicate gills in a protective reaction. The stimulus, it has been found, causes 24 sensory neurons in the area to fire. These, in turn, activate interneurons and six motor neurons that control the movement of the gill and cause it to contract.

43–17 *The sea hare* Aplysia, *a shell-less mollusk that is shedding new light on the process of learning. The neurons of Aplysia, like those of the squid and other invertebrates, are quite large and their axons are unmyelinated. Moreover, its nervous system has far fewer neurons than a vertebrate nervous system. Individual neurons can be identified, their pattern of organization mapped, and microelectrodes inserted into them. Thus, investigators can trace the pathways followed by nerve impulses in response to particular stimuli and monitor the modifications in transmission associated with learning.*

Alzheimer's Disease

Most of us take for granted the reliability of memory, day in and day out, as we engage in the many activities—both mental and physical—that make up our lives. Alzheimer's disease, which is estimated to afflict some 1.5 to 2 million older persons in this country alone, makes clear what a precious gift memory is and how devastating its loss can be.

Alzheimer's disease was first described in 1907 by the German neurologist Alois Alzheimer, who recognized it as a distinct pathology afflicting a very small number of people in their forties and fifties. Characterized by progressive memory loss, the disease ultimately leads to severe dementia, including the inability to think, speak, or perform even the most basic tasks of personal care. Death generally follows within 3 to 10 years. By the 1970s, it had become clear that many cases of "senility" in elderly persons, previously assumed to be an inevitable consequence of age, were actually the same disease first identified by Alzheimer.

The biological changes characteristic of Alzheimer's disease are revealed most strikingly in autopsy studies of brain tissue. Three abnormalities are found consistently: accumulations of tangled and twisted protein filaments within neuron cell bodies; structures known as neuritic plaques, which are clusters of degenerated axon terminals associated with a protein known as amyloid; and accumulations of this same protein adjacent to and within the walls of blood vessels. Although these abnormalities are found in various regions of the cerebral cortex, they are most apparent in the structures associated with memory—the hippocampus and the amygdala. In addition, there is a loss of neurons whose cell bodies are located in nuclei of the basal forebrain, which, as we have noted, are a major source of acetylcholine. The axons of these neurons extend not only to the hippocampus and amygdala but also into many areas of the cerebral cortex. The death of these neurons thus

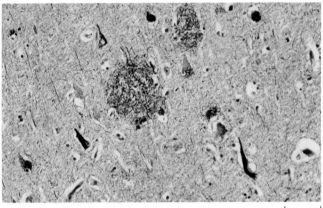

A section of tissue from the hippocampus of an Alzheimer's disease victim. The large round structure in the center (as well as the oval structure at the top) is a neuritic plaque, an aggregation of degenerating axons and amyloid protein. The dark, kite-shaped objects are the cell bodies of neurons containing abnormal tangles of protein filaments. These filaments are insoluble in water and impervious to chemical or enzymatic breakdown; they persist long after the cell containing them has died.

If *Aplysia* is touched repeatedly, it becomes habituated to the stimulus and ceases to withdraw. Habituation, which is regarded as a very simple form of learning, is associated with a gradual decrease in the amount of neurotransmitter released by the repeatedly stimulated sensory neurons. This decrease is reflected, in turn, by a decline in the response of the motor neurons controlling the gill. However, if the stimulus is stronger—for example, a jab rather than a gentle stroking—*Aplysia* becomes sensitized to it; the motor response becomes more rapid and emphatic. At the synapse, the effect of sensitization is opposite to that of habituation; that is, there is a gradual increase in the amount of neurotransmitter released by the sensory neurons.

reduces the supply of acetylcholine in the regions of the brain where their axons terminate.

Whether the abnormalities observed with Alzheimer's are the cause of the disease or, instead, its consequence is not yet known. It is also unclear whether Alzheimer's is one disease, with one underlying cause, or whether it is a family of diseases, with several different causes that lead to the same set of pathological changes. There is strong evidence that genetic factors are involved in early-onset Alzheimer's disease and less conclusive evidence that they may also be involved in a significant proportion of the cases that occur much later in life. As a result, considerable research has been devoted to the identification of the gene coding for the amyloid protein associated with the disease. This protein, which consists of 42 amino acids, is synthesized as part of a much larger molecule that exhibits many of the characteristics of a membrane glycoprotein. Knowledge of the amino acid sequence has made possible the synthesis of gene probes, and these probes, along with the analysis of restriction enzyme fragments, have located this gene on chromosome 21. As you will recall (page 385), Down's syndrome is caused by an extra copy of chromosome 21, and perhaps significantly, many persons with Down's syndrome ultimately develop Alzheimer's disease.

Now that the gene coding for the amyloid protein has been identified, studies are underway to explore its function and regulation and to determine whether the protein is a normal or abnormal gene product. The role of external factors—toxic agents, infectious agents, and immunological responses to infection—in the disease process is also being explored. One of the most intriguing suggestions that other factors may be involved is provided by an outbreak of neurological disorders that occurred on the island of Guam in the early 1950s, causing the death of 20 percent of the native population over the age of 25. The disorders included amyotrophic lateral sclerosis (Lou Gehrig's disease), Parkinson's disease, and an Alzheimer-like dementia. Animal studies suggest that this outbreak was triggered by the ingestion of large amounts of an unusual amino acid present in high concentrations in a cycad (page 505) found on Guam. During the Japanese occupation of the island during World War II, food shortages were severe, and the native population subsisted principally on foods prepared from the cycad seeds. After the war ended, the cycad ceased to be a major food source, and since the 1950s, the incidence of neurological disorders has dropped dramatically.

The brain, as we have seen, is an enormously complex structure, in which numerous intricate, interdependent processes occur. To even begin to understand its normal functioning is extraordinarily difficult. To determine the cause or causes of a disease such as Alzheimer's is a staggering task. An article in the journal *Science* in April of 1986 estimated that it could take as long as 5 to 10 years to locate genetic markers for the hereditary form of Alzheimer's disease. And yet, a series of articles published in the same journal less than a year later reported not only the location of such genetic markers but also the cloning of the gene coding for the amyloid protein. It is therefore possible that a great deal more will be known by the time you read this.

Numerous studies with *Aplysia*, with another mollusk known as *Hermissenda*, and with preparations of tissue from the hippocampus of various mammals all support the hypothesis that alterations in the strength of synaptic transmission are critical in memory and learning. These alterations are thought to depend on changes in both the presynaptic and postsynaptic cells. One important element may be the opening or blocking of ion channels that influence the release of neurotransmitter by the presynaptic cell and the degree of depolarization or hyperpolarization of the postsynaptic cell in its resting state. Various models of the possible mechanisms by which these alterations occur and are made to endure are currently being intensively investigated. Among the factors that may be involved are second messengers, including both calcium ion and the ubiquitous cyclic AMP, changes in the pattern of protein synthesis, possibly affecting the numbers and kinds of membrane receptors, and shifts in the location of membrane receptors.

Although the problems of memory and learning are still as bewildering (and as fascinating) as those of human heredity were 40 years ago, neurobiologists appear to be on the threshold of new levels of understanding. Some scientists—Kandel is one of them—believe that the answers will come through a simple model, the equivalent of *Drosophila* and the bacteriophage T4. Others contend that the enormous complexities of the vertebrate brain will never be understood in terms of simple invertebrate models and single cells but rather that the secrets lie in the vast communication network itself. Stay tuned.

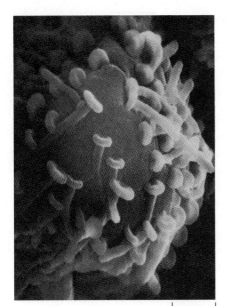

43-18 *Synaptic knobs of the sea hare* Aplysia, *as shown by the scanning electron microscope. These knobs, which are axon terminals of a number of different presynaptic neurons, are all converging on the cell body of a single postsynaptic neuron. Research with* Aplysia *has demonstrated that changes in the strength of synaptic transmissions play a key role in learning.*

SUMMARY

One of the principal trends in vertebrate evolution has been the increasing centralization of control in one dominant processing center, the brain, which reaches its greatest complexity in *Homo sapiens*. The vertebrate brain develops from three bulges at the anterior end of the hollow, dorsal neural tube, giving rise to the hindbrain, midbrain, and forebrain. The hindbrain consists of the medulla, pons, and cerebellum, the brain structure associated with coordination of fine-tuned movements. The medulla, pons, and midbrain make up the brainstem, which controls vital functions such as heartbeat and respiration and serves as the relay between the spinal cord and the rest of the brain.

The posterior part of the forebrain, the diencephalon, comprises the thalamus, the main relay center between the brainstem and higher brain centers, and the hypothalamus, which contains nuclei associated with basic drives and emotions and is the center for the integration of the nervous and endocrine systems. The anterior part of the forebrain, the telencephalon, consists, in humans, of the two cerebral hemispheres connected by the corpus callosum and covered by the much convoluted cerebral cortex.

Coordination of the activities occurring in different parts of the brain depends on the exchange of information between local networks of neurons. Two major brain circuits involved in such exchanges are the reticular activating system, concerned with arousal and attention, and the limbic system, concerned with translating drives and emotions into actions.

In both experimental animals and humans, it has been possible to correlate particular areas of the cerebral cortex with particular functions. These areas include the motor cortex, sensory cortex, and parts of the cortex concerned with vision, hearing, and speech. In the motor and sensory cortices, the two cerebral hemispheres are mirror images of one another, with the right hemisphere controlling and receiving information from the left side of the body, and vice versa. However, the speech centers are found only in one hemisphere, nearly always the left one, and other faculties, such as spatial orientation and musical ability, appear to be associated with the right hemisphere. Usually the functions of the two hemispheres are integrated, but studies of patients whose corpus callosum has been severed indicate that the two hemispheres can function independently and confirm that they differ in their capacities.

Most of the human cortex has no direct sensory or motor function and consists of areas that receive signals from and transmit signals to neurons in other areas of the brain. Some of these areas participate in the further processing of information transmitted from the primary sensory, auditory, and visual cortices.

Others function in the integration of sensory information with memory and emotion, in the organization of ideas, and in long-term planning. About half of the total area devoted to such intrinsic processing is located in the frontal lobes, the part of the brain that developed most rapidly in human evolution.

Memory and learning are thought to involve both the processing of information through specific anatomical circuits and modifications of synaptic activity. Two types of memory have been identified: short-term memory and long-term memory. The consolidation of long-term "recognition" memory appears to depend on a circuit that includes the hippocampus, amygdala, diencephalon, basal forebrain, and prefrontal cortex. Feedback from this circuit to the primary sensory cortices is thought to play a role in subsequent recognition of similar sensory input. "Habit" or "procedural" memory appears to depend on a different circuit that includes the corpus striatum. Studies with invertebrate model systems indicate that, at the cellular level, memory and learning involve changes at synapses, including alterations in neurotransmitter release by presynaptic cells and the subsequent response by postsynaptic cells.

QUESTIONS

1. Distinguish among the following: brainstem/cerebellum/cerebrum; diencephalon/telencephalon; thalamus/hypothalamus; reticular activating system/limbic system; motor cortex/sensory cortex; Broca's area/Wernicke's area; left brain/right brain; hippocampus/amygdala.

2. Diagram the major structures of the vertebrate brain.

3. Relate the relative sizes of various portions of the brains of the animals in Figure 43–3 with the habitats in which they live and the ways in which they live, obtain food, and escape predators.

4. Diagram the major structures of the human brain.

5. Sketch the human cerebral cortex and indicate on it the areas that have been mapped.

6. What functions might be affected by damage (from stroke, accident, or disease) to the cerebellum? To the reticular formation? To the dorsal portion of the cerebral cortex anterior to the central sulcus?

7. Monkeys that have suffered damage to the amygdala are unable to remember whether an object, however familiar, is edible or inedible. Each time they encounter an object—for example, a banana—they not only look at it, but feel it, smell it, and taste it, before deciding whether to eat it or not. What is a probable explanation of this behavior?

8. Consider the axon of a presynaptic cell that, as a result of repeated stimulation, releases smaller quantities of excitatory neurotransmitter in response to arriving action potentials. What effect would you expect this to have on the initiation of action potentials in the postsynaptic cell?

9. Consider a postsynaptic cell in which membrane ion channels are blocked, with the result that the resting potential of the cell is less negative than it was when the ion channels were open. Will such a cell require a larger or smaller input of excitatory neurotransmitter in order to fire?

CHAPTER **44**

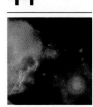

The Continuity of Life I: Reproduction

In the preceding chapters of this section, we have been primarily concerned with the ways in which the individual vertebrate body maintains itself. In this chapter and the next, we shall consider the process by which new individuals are produced, maintaining the continuity of the species, and how these individuals develop to become reasonable facsimiles of their parents.

As we saw in Chapters 25 through 28, there is enormous diversity in reproductive and life cycle patterns in the animal kingdom. Most vertebrates—and all mammals—reproduce sexually. As you will recall (page 249), sexual reproduction involves two events: meiosis and fertilization. In vertebrates, which are almost always diploid, meiosis produces gametes, the only haploid forms in the life cycle. The gametes are specialized for motility (sperm) or for production and storage of nutrients (eggs). They are produced in the gonads of individuals of the two separate sexes, male and female. In many invertebrates (insects, in particular), the generations are nonoverlapping, as is the case in annual plants. In vertebrates, however, parents not only survive after their young are produced but often are essential to the survival of the young. This evolutionary trend toward increasing parental care becomes pronounced among birds and reaches its fullest expression among certain of the mammals, ourselves included.

In most species of fish and in amphibians, as in many invertebrates, fertilization is external. Among organisms that lay amniote eggs (reptiles, birds, and monotreme mammals), fertilization is internal. The outer protective shell, produced as

44-1 *Lioness and cub.*

the egg moves through the female reproductive tract, is laid down after the egg cell is fertilized, enclosing the embryo and its membranes. Fertilization is also internal among marsupial and placental mammals, in which the embryo develops within the mother and is nourished by her.

In the following pages, we are going to describe sexual reproduction in mammals, using *Homo sapiens* as our representative organism. We shall first trace the development of the male gametes, then the development of the female gametes, and, finally, we shall describe the special structures and activities that provide for fertilization and the subsequent implantation of the developing embryo.

THE MALE REPRODUCTIVE SYSTEM

Sperm cells—the male gametes—are produced in the **testes** (singular, testis). The testes develop in the abdominal cavity of the male embryo and, in the human male, descend into an external sac, the scrotum. This descent usually occurs before birth. The function of the scrotum, apparently, is to keep the testes in an environment cooler than the abdominal cavity. A temperature 3°C lower than that of the body is necessary for the sperm to develop. Sperm are not produced in an undescended testis, and even temporary immersion of the testes in warm water—as in a hot bath—has been known to produce temporary sterility. (This effect is not reliable enough, however, to recommend its use as a birth control measure.) When the temperature outside the scrotal sac is warm, the sac is thin and hangs loose in multiple folds. When the outside temperature is cold, the muscles under the skin of the scrotum contract, drawing the testes close to the body. In this way, a fairly constant testicular temperature is maintained.

Spermatogenesis

Each testis (Figure 44–2) is subdivided into about 250 compartments (lobules), and each of these is packed with tightly coiled seminiferous ("seed-bearing") tubules. These are the sperm-producing regions of the testes. Between the tubules are the interstitial cells, the sources of the male sex hormone testosterone. Each seminiferous tubule is about 80 centimeters long, and the two testes together contain a total of about 500 meters of tubules. The sperm are produced continuously within the tubules.

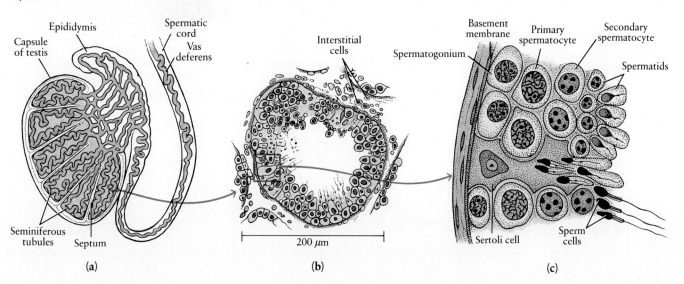

44–2 *The testis* (a) *consists mostly of tightly packed coils of seminiferous tubules, containing* (b) *sperm cells in various stages of development. The entire developmental sequence takes eight to nine weeks. The interstitial cells, which are found in connective tissues between the tubules, are the source of the male hormone testosterone.* (c) *As shown in an idealized cross section of a portion of a seminiferous tubule, spermatogonia develop into cells known as primary spermatocytes. In the first meiotic division, these divide into two equal-sized cells, the secondary spermatocytes. In the second meiotic division, four equal-sized spermatids are formed. These differentiate into functional sperm cells. The Sertoli cells support and nourish the developing sperm. The sperm cells leave the testis through the epididymis and the vas deferens.*

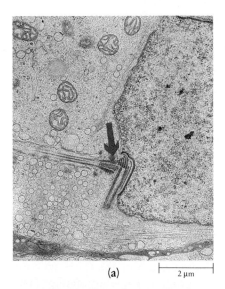

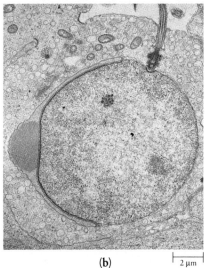

44–3 *Mammalian spermatids in the process of differentiation.* **(a)** *The nucleus almost fills the right half of the electron micrograph. The two centrioles have moved to a position just behind the nucleus (see arrow), and the flagellum has begun to form from one of them. To the left of the nucleus are several mitochondria.* **(b)** *The nucleus occupies the center of this micrograph. At the upper right, the flagellum is forming and, at the left, you can see the beginning of the acrosome. Note the complex connecting piece that fits into a notch at the posterior end of the nucleus and links the flagellum to the nucleus.*

The tubules contain two types of cells: spermatogenic (sperm-producing) cells and Sertoli cells. The spermatogenic cells pass through several stages of differentiation. Once production of sperm begins at puberty in the human male, it goes on continuously; thus, in a single seminiferous tubule, it is possible to find cells in all the different stages of spermatogenesis. It is unusual, however, to find all of the stages in a single cross section, because spermatogenesis characteristically occurs in waves that travel down the tubules.

Cells in the first stage of spermatogenesis, the spermatogonia, line the basement membrane of each seminiferous tubule. Spermatogonia are diploid and have, in the human, 44 autosomes and 2 sex chromosomes, an *X* and a *Y*. Spermatogonia divide continuously. Some of the cells produced by these mitotic divisions remain undifferentiated, whereas others, in the course of their successive mitotic divisions, move away from the basement membrane and begin to differentiate, giving rise to primary spermatocytes. Primary spermatocytes undergo the first meiotic division* to produce two secondary spermatocytes, each of which contains 22 autosomes and either an *X* chromosome or a *Y* chromosome; each of the 23 chromosomes consists of two chromatids. The secondary spermatocytes undergo the second meiotic division to produce spermatids, each of which contains the haploid number of single chromosomes. Spermatids develop without further division into sperm cells, or spermatozoa. It takes eight to nine weeks for a spermatogonium to differentiate into four sperm cells. During this time the developing cells receive nutrients from adjacent Sertoli cells.

Differentiation of Spermatids

A spermatid is a small spherical or polygonal ("many-sided") cell that develops into a sperm cell. The sequence of changes by which the quite unremarkable spermatid becomes the highly specialized, very extraordinary sperm cell is an excellent example of cell differentiation. Differentiation is a process that, as we shall see in the next chapter, is an essential component of embryonic development.

The first visible sign of differentiation of a spermatid is the appearance, within the Golgi complex, of vesicles containing small, dark granules. These vesicles enlarge and coalesce into a single vesicle, the acrosome. The acrosome contains enzymes that will help the sperm to penetrate the protective layer surrounding the egg. The position of the acrosomal vesicle determines the polarity of the sperm; that is, it establishes where the anterior end, or "head," is going to be.

During the early stages of acrosome formation, the cell's pair of centrioles moves to the cell membrane at the end of the cell opposite the acrosome. One of these centrioles appears to initiate the assembly of tubulin dimers (page 78) into microtubules—the beginning of the sperm flagellum. The centrioles then move back toward the nucleus, carrying the cell membrane inward with them (Figure 44–3). One centriole eventually forms part of a connecting piece within the neck of the sperm, linking the flagellum to the nucleus. The other centriole, the one that gave rise to the flagellum, disintegrates as the connecting piece develops.

As the flagellum, or tail, grows, it becomes apparent that its axial filament has the 9 + 2 structure characteristic of eukaryotic cilia and flagella. Mitochondria aggregate about its basal end, forming a continuous spiral, providing a ready energy source (ATP) for the flagellar movement. The rest of the axial filament, almost to its tip, is surrounded by nine additional protein fibers tightly coiled in a helix that forms a fibrous sheath. These fibers, which are somewhat thicker than the microtubules within the flagellum, presumably play some role in sperm motility.

---

* For a review of meiosis, see pages 253 to 257. The stages of spermatogenesis are diagrammed in Figure 12–12 on page 258.

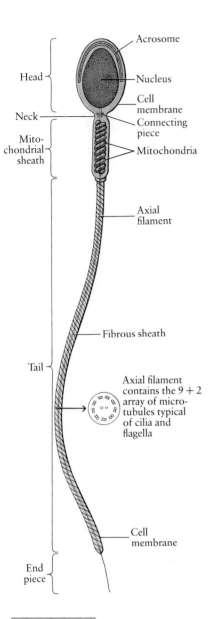

44-4 *Diagram of a human sperm cell. The mature cell consists primarily of the nucleus, carrying the "payload" of tightly condensed DNA and associated protein, the very powerful tail (flagellum), and mitochondria, which provide the power for sperm motility. The acrosome is a specialized lysosome (see page 119) containing enzymes that help the sperm penetrate the protective layers of an unfertilized secondary oocyte.*

During this period, the nucleus condenses, apparently by eliminating water. Once the tail has formed, the cell rapidly elongates. Longitudinal bundles of microtubules can be seen in the cell at this time, and they may play a role in changing the shape of the cell. As the cell lengthens, the bulk of the cytoplasm, together with the Golgi complex, is sloughed away.

In its final form, the sperm cell consists of the acrosome, the tightly condensed nucleus, the connecting piece in the neck, the mitochondrial sheath, and the long, powerful flagellum itself, all bounded by the cell membrane (Figure 44-4). In the fully differentiated sperm, all other functions have been subordinated to the task of providing motility for delivery of the "payload," the DNA and its associated protein, condensed and coiled in the sperm head. A young adult human male may produce several hundred million sperm per day; a ram may produce several billion.

Pathway of the Sperm

The pathway of the sperm can be traced in Figure 44-5. From the testis, the sperm are carried to the **epididymis,** which consists of a coiled tube 7 meters long, overlying the testis. It is surrounded by a thin, circular layer of smooth muscle fibers. The sperm are nonmotile when they enter the epididymis and gain motility only after some 18 hours there. (Maximum motility, however, occurs only after they have entered the female reproductive tract.)

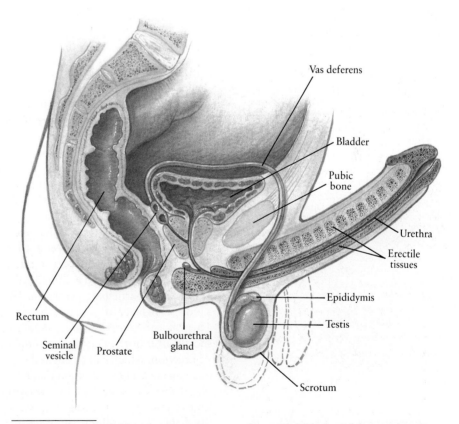

44-5 *Diagram of the human male reproductive tract, showing the penis and scrotum before (dashed lines) and during erection. Sperm cells formed in the seminiferous tubules of the testis enter the epididymis. From there they move to the vas deferens, where most of them are stored. The vas deferens merges with a duct from the seminal vesicle and then, within the prostate gland, joins the urethra. The sperm cells are mixed with fluids, mostly from the seminal vesicles and prostate gland. The resulting mixture, the semen, is released through the urethra of the penis. The urethra is also the passageway for urine, which is stored in the bladder.*

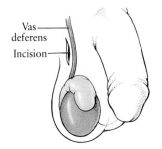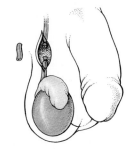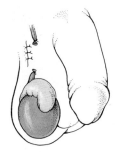

44-6 *In a vasectomy, the vas deferens on each side is severed, and the cut ends are folded back and tied off, preventing the release of sperm from the testis. The sperm cells are reabsorbed by the body, and the ejaculate is normal except for the absence of sperm.*

This is a relatively safe and almost painless procedure that does not require a general anesthetic or hospitalization. Because the nerves and blood vessels between the testes and the rest of the body are left intact, the procedure does not affect hormone levels, sexual potency, or performance. Its chief drawback is that it is generally not reversible.

From the epididymis, the sperm pass to the **vas deferens** (plural, vasa deferentia), where most of them are stored. A vas deferens, an extension of the tightly coiled tubules of the epididymis, leads from each testis into the abdominal cavity. The vasa deferentia are covered with a heavy, three-layered coat of smooth muscle whose contractions propel the sperm along. Each vas deferens and its accompanying nerves, arteries, veins, and connective-tissue wrapping constitute a spermatic cord. The spermatic cords, one from each testis, are found along the same path the testes took during their descent in the embryo. Severing of the vasa deferentia (vasectomy) offers a relatively safe and almost painless means of birth control (Figure 44-6).

Within the posterior wall of the abdominal cavity, the vasa deferentia loop around the bladder, where they merge with the ducts of the **seminal vesicles**. The vas deferens from each testis then enters the **prostate gland** and merges with the urethra, which extends the length of the **penis**. The urethra serves both for the excretion of urine and the ejaculation of sperm.

Erection of the Penis and Orgasm in the Male

The function of the penis is to deposit sperm cells within the reproductive tract of the female. The penis, in various forms, has evolved independently in a number of species of insects and in other invertebrates. It is found among some reptiles and birds; all flightless birds have penes and so do all ducks, flightless or not. In most reptiles and birds, however, one opening, the cloaca, serves as the passage for eggs or sperm and also for the elimination of wastes. These animals mate by juxtaposition of their cloacae. Only among mammals is the penis found in all species.

The human penis is formed of three cylindrical masses of spongy erectile tissue, each of which contains a large number of small spaces, each about the size of a pinhead. Two of these masses are in the dorsal portion of the penis, and the third lies beneath them, surrounding the urethra (Figure 44-7). This third mass is enlarged at the distal (far) end to form the glans penis, which is a smooth protective cap over the spongy tissues. At the basal end, it is enlarged to form the bulb of the penis, which is embedded below the pelvic cavity and is surrounded by muscles that participate in orgasm. The exterior portion of the penis is covered by a loose, thin layer of skin, which at the end forms an encircling fold over the glans. This fold, the foreskin, is sometimes surgically removed (circumcision), usually shortly after birth. The urethra terminates in a slitlike opening in the glans.

Erection, which can be elicited by a variety of stimuli, occurs as a consequence of an increased flow of blood that fills the spongy, erectile tissues of the penis. The blood flow is controlled by parasympathetic nerve fibers to the blood vessels leading into the erectile tissues. As the tissues become distended, they compress the veins and so inhibit the flow of blood out of the tissues. With continued stimulation, the penis and the underlying bulb become hard and enlarged.

Erection is accompanied by the discharge into the urethra of a small amount of fluid from the bulbourethral glands, pea-shaped organs at the base of the penis. This fluid serves as a lubricant, facilitating the movement of spermatozoa along the male urethra and aiding penetration of the penis into the female. Continued stimulation of mechanoreceptors in the penis and scrotum (such as may be produced by repeated thrusting of the penis in the vagina) sends an escalating

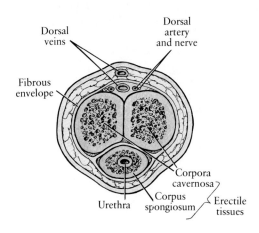

44-7 *A cross section of a human penis. The penis is formed of three cylindrical masses of spongy erectile tissue that contain a large number of small spaces, each about the size of a pinhead. Erection of the penis is caused by dilation of the blood vessels carrying blood to the spongy tissues.*

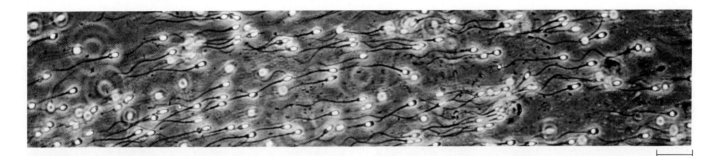

$\vdash\!\!-\!\!\dashv$ 25 μm

44-8 *Human sperm. About 300 to 400 million sperm cells are present in the ejaculate of an average, healthy adult male.*

series of nerve impulses through reflex arcs in the lower spinal cord to motor neurons innervating different muscles of the reproductive system. Among the first muscles to contract are those in the scrotum, raising the testes close to the body, and those encircling each epididymis and vas deferens, moving the spermatozoa toward and into the urethra. As this occurs, the seminal vesicles secrete a fructose-rich fluid that nourishes the sperm cells. This fluid contains a high concentration of prostaglandins (page 833), which stimulate contractions in the musculature of the uterus and oviducts and so may assist the sperm in reaching the egg. The prostate gland adds a thin, milky, alkaline fluid that helps neutralize the normally acidic pH of the female reproductive tract. Finally, contractions are initiated in the muscles surrounding the bulb; these contractions propel the sperm and accompanying fluid out through the urethra (ejaculation) and produce some of the sensations associated with orgasm.

The sperm, along with the secretions from the seminal vesicles, the prostate gland, and the bulbourethral glands, constitute semen. The volume of semen measures from 3 to 4 milliliters per ejaculation. Even though sperm constitute less than 10 percent of the semen, about 300 to 400 million sperm cells are present in each ejaculate of a normal adult male. The mortality of sperm in the female reproductive tract is staggering—of the original 300 to 400 million cells, only a few hundred survive to reach the oviducts, where fertilization occurs. Although only one sperm cell needs to make contact with an oocyte for fertilization to occur, the odds are such that fertility requires the release of enormous numbers at one time. Males that produce fewer than 20 million sperm per milliliter of fluid are generally sterile.

The Role of Hormones

In addition to producing the sperm cells, the testes are also the major source of male hormones, known collectively as **androgens.** The principal androgen, testosterone, is necessary for the formation of sperm cells. It is a steroid, produced primarily by the interstitial cells of the testes (see Figure 44-2b). Other androgens are produced in the adrenal cortex (see page 832).

Androgens are first produced in early embryonic development, causing the male fetus to develop as a male rather than a female. After birth, androgen production continues at a very low level until the boy is about 10 years old. Then there is a surge in testosterone, resulting in the onset of sperm production (which marks the beginning of puberty) accompanied by enlargement of the penis and testes and also of the prostate and other accessory organs. In the healthy human male, a high level of testosterone production continues into the fourth decade of life, when it begins gradually to decline.

As can be readily observed, testosterone also has effects on other parts of the body not directly involved in the production and deposition of sperm. In the human male, these effects include growth of the larynx and an accompanying deepening of the voice, an increase in skeletal size, and a characteristic distribu-

44–9 *Secondary sex characteristics among vertebrates. The male and female animals in each of these photos are actively courting.*

tion of body hair. Androgens stimulate the biosynthesis of proteins and so of muscle tissue. They stimulate the apocrine sweat glands, whose secretions attract bacteria and so produce the body odors associated with sweat after puberty. And they may cause the sebaceous glands of the skin to become overactive, resulting in acne. Such characteristics, associated with sex hormones but not directly involved in reproduction, are known as **secondary sex characteristics.**

In other animals, testosterone is responsible for the lion's mane, the powerful musculature and fiery disposition of the stallion, the cock's comb and spurs, and the bright plumage of many adult male birds (Figure 44–9). It is also responsible for a variety of behavior patterns such as the scent marking of dogs, the courting behavior of sage grouse, and various forms of aggression toward other males found in a great many vertebrate species.

Since almost the beginnings of agriculture, domestic animals have been castrated in order to make them fatter, less tough, and more manageable. Eunuchs (human castrates) were traditionally, and for obvious reasons, used as harem guards, and as recently as the eighteenth century, selected boys were castrated before puberty in order to retain the purity of their soprano voices for church and opera choirs. As a consequence of these early practices, the effects of testosterone —or, more precisely, of its absence—were the first hormonal influences to be recognized and studied.

Regulation of Hormone Production

The production of testosterone is regulated by a negative feedback system involving, among other components, a gonadotropic hormone called luteinizing hormone (LH). LH is produced by the pituitary gland, under the influence of the hypothalamus. It is carried by the blood to the interstitial tissues of the testes, where it stimulates the output of testosterone. As the blood level of testosterone increases, the release of LH from the pituitary is slowed (Figure 44–10).

The testes are also under the influence of a second pituitary hormone, follicle-stimulating hormone (FSH). It acts on the Sertoli cells of the testes and, through them, on the developing sperm. Among the factors involved in the regulation of FSH production is a protein hormone, known as inhibin, that is secreted by the Sertoli cells. This hormone, as you might expect, inhibits FSH production. Its discovery is of more than academic interest; it would appear a likely candidate for the long-sought male contraceptive.

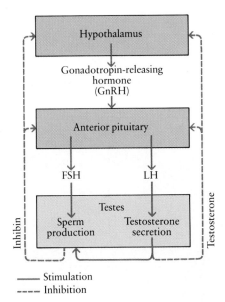

44-10 *Hormonal regulation of testicular function depends on negative feedback. The hypothalamus stimulates the anterior lobe of the pituitary to produce LH (luteinizing hormone), which, in turn, stimulates production and release of testosterone from the interstitial cells of the testes. The production of gonadotropin-releasing hormone (GnRH) by the hypothalamus is inhibited by the resulting increased concentration of testosterone and may also be inhibited by the increased concentration of LH. In addition, testosterone is thought to act directly on the pituitary to suppress the release of LH. As a consequence of these combined inhibitory effects, the secretion of LH by the pituitary is decreased.*

Similarly, in another negative feedback loop, FSH (follicle-stimulating hormone) acts on the Sertoli cells, which produce, in turn, the hormone inhibin. This hormone specifically inhibits FSH production. FSH itself may also act on the hypothalamus to reduce the production of GnRH.

The combined action of testosterone and FSH is required to initiate spermatogenesis.

In the human male, the rates of testosterone release are fairly constant. In many animals, however, male hormone production is triggered by changes in temperature, daylight, or other environmental cues and changes seasonally. Testosterone production may also be affected by social circumstances. Studies on bulls, for example, have shown that after a bull sees a cow, his blood level of LH rises as much as seventeenfold; within about half an hour, the blood level of testosterone reaches its peak. In many animal societies, including wolves and cape hunting dogs (the wild dogs of the African plains), socially inferior males may never become sexually mature, presumably because of depressed testosterone production. Human testosterone production also may vary according to the emotional climate. In a study made among Army GIs during the Vietnam War, testosterone levels in recruits in basic training and among combat troops were markedly lower than the normal levels of testosterone found in men in behind-the-lines assignments.

Testosterone production is also dramatically affected by the synthetic compounds known as anabolic steroids. These drugs, which are chemical variants of testosterone, were originally developed in Germany in the 1930s in an attempt to produce the muscle-building effects of the natural hormone without its masculinizing effects. Because of their chemical similarity to testosterone, anabolic steroids function as inhibitors in the negative feedback system regulating testosterone production. In adult males, their use can reduce testosterone levels by as much as 85 percent, causing shrinkage of the testes and growth of the breasts. Long-term use of these drugs also greatly increases the risk of kidney and liver damage, liver cancer, and heart disease. In adolescents, anabolic steroids can lead to premature baldness and failure to attain full height.

THE FEMALE REPRODUCTIVE SYSTEM

The female reproductive system is shown in Figure 44–11. The gamete-producing organs are the **ovaries,** each a solid mass of cells about 3 centimeters long. They are suspended in the abdominal cavity by ligaments (bands of connective tissue) and mesenteries. The oocytes, from which the eggs develop, are in the outer layer of the ovary.

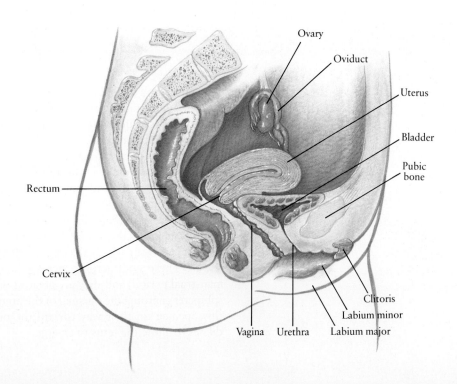

44-11 *The female reproductive organs. Notice that the uterus lies at right angles to the vagina. This is one of the consequences of the bipedalism and upright posture of Homo sapiens and one of the reasons that childbirth is more difficult for the human female than for other mammals.*

Sex and the Brain

As we saw in the last chapter, male songbirds have a song center that is located in the left cerebral hemisphere. Enlargement of the song center, which is associated with the establishment of territories and the initiation of courtship behavior, is produced by an increase in circulating testosterone. If a male songbird is castrated, the song center does not enlarge and the bird never sings. If the castrated bird is injected with testosterone, the song center enlarges and he begins to sing. Female birds normally do not sing, and both hemispheres are the same size. However, if adult female songbirds are given testosterone, their song centers enlarge and they begin to sing, although their musical repertoires never become as large or varied as those of the males. Thus, in songbirds, there are differences between the brains of male and female animals, and these differences can be traced, in part, to the influence of testosterone on the brain itself.

Rats don't sing, but they do have certain patterns of clearly defined behavior that differ between the sexes. The most obvious—and highly necessary—differences are in mating behavior. When a female rat in estrus (page 915) is exposed to a sexually competent male, she crouches and arches her back, exposing her genitals, in a posture called lordosis. The male mounts her, grabs her flanks, and makes pelvic thrusts that result in ejaculation. Lordosis is eliminated by removal of the ovaries and is restored by administration of estrogens. In fact, lordosis in a female rat depends on the presence of estrogens in cells of the hypothalamus.

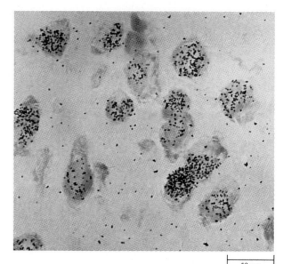

This autoradiograph of a section of brain tissue from an adult female rat shows the localization of tritium-labeled estradiol in the nuclei of cells of the hypothalamus. The black dots represent the presence of estradiol bound to an estrogen receptor. In rats and many other species, the hypothalamus has been shown to play a key role in the hormonal regulation of sexual behaviors.

Other important structures include the oviducts (sometimes called the Fallopian tubes or the uterine tubes), the uterus, the vagina, and the vulva. The **uterus** is a hollow, muscular, pear-shaped organ slightly smaller in size than a clenched fist (about 7.5 centimeters long and 5 centimeters wide) in the nonpregnant female. It lies almost horizontally in the abdominal cavity and is on top of the bladder. The uterus is lined by the endometrium, which has two principal layers, one of which is shed at menstruation and another from which the shed layer is regenerated. The smooth muscles in the walls of the uterus move in continuous waves. This motion possibly increases the motility of both the sperm on its journey to the oviduct and the oocyte, from which the egg develops, as it passes from the oviduct to the uterus. These contractions increase when the endometrium is shed during a menstrual period, and they are greatest when a woman is in labor. The muscular sphincter guarding the opening of the uterus is the **cervix**. The sperm pass through this opening on their way toward the oocyte. At the time of birth, the cervix dilates to allow the fetus to emerge.

In adult male rats, the normal repertoire of mating behavior ceases if the rat is castrated; the behavior is restored by testosterone injections. Studies using radioactive testosterone have shown that this hormone can also be localized in neurons in certain areas of the brain, particularly in a group of neurons in the hypothalamus. (This same area of the hypothalamus is the source of gonadotropin-releasing hormone.) Injection of testosterone to this area produces an increase in electrical activity of these neurons and an increase in sexual activity. Electrical stimulation of this area also produces an increase in sexual activity. However, administration of estrogens to these same castrated male rats does not produce lordosis or any other sex-specific female behavior. In short, there is a difference between the male and female brain in its response to hormones.

This differentiation between the male and female brain occurs during a critical period in early development. Rats have a 21-day gestation period. About eight days prior to birth, the embryonic male testes begin to secrete testosterone, and this secretion continues until the tenth day after birth. If male rats are castrated on the day of birth (thus depriving them of testosterone for about half the normal period), they never develop male sexual behavior. If they are given testosterone as adults, they do not respond to it. However, if these same adult rats are given estrogens, they show the lordosis behavior in the presence of normal males. By contrast, if male rats are castrated on the tenth day after birth or thereafter, they will not respond to the administration of estrogens. Thus, it is concluded that in rats sex hormones affect the brain in two stages: (1) an early period, during which differentiation of brain cells occurs, and (2) a sexually mature period, in which the differentiated brain cells respond to the hormones by evoking particular patterns of behavior.

Scientists are interested in what other types of behavior might be influenced by sex differences in the cellular organization of the brain. In rhesus monkeys, for instance, the critical burst of testosterone production occurs entirely in utero, around the middle of the 168-day gestation period. Female rhesus monkeys that are masculinized by high doses of testosterone given to their mothers during the critical period show an increase in rough-and-tumble play, an increase in aggressive behavior, and a decrease in maternal imitative behavior. The critical period in humans—the period of the surge in testosterone production—occurs about the second to third month in embryonic development. The question of what aspects of human behavior—if any—may be influenced by sex differences in the brain is a fascinating one to scientists and nonscientists alike. It is, however, generally agreed that even if such differences exist, they are largely overcome by the powerful forces of environment and culture.

The **vagina** is a muscular tube about 7.5 centimeters long that leads from the cervix of the uterus to the outside of the body. It is the receptive organ for the penis and is also the birth canal. Its exterior opening is between the urethra, the tube leading from the bladder, and the anus. The lining of the vagina is rich in glycogen, which bacteria normally present in the vagina convert to lactic acid. As a consequence, the vaginal tract is mildly acidic, with a pH between 4 and 5.

The external genital organs of the female are collectively known as the **vulva**. The **clitoris**, most of which is embedded in the surrounding tissue, is about 2 centimeters long and is homologous with the penis of the male (in the early embryo, the structures are identical). Like the penis, it is composed chiefly of erectile tissue. The clitoris has two bulbs (analogous to the penile bulb of the male) that lie on either side of the opening of the vagina. The **labia** (singular, labium) are folds of skin. The labia majora are fleshy and, in the adult, covered with pubic hair. They enclose and protect the underlying, more delicate structures. (Embryonically, they are homologous with the scrotum in the male.) The labia minora are thin and membranous.

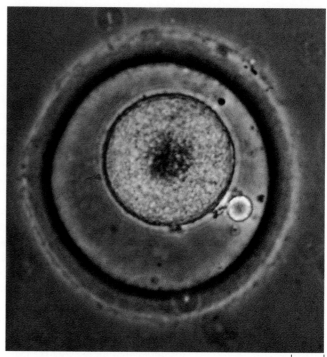

44-12 *An unfertilized human secondary oocyte. The dark strands within the nucleus are chromosomes. A polar body is at the lower right.*

Oogenesis

In human females, the primary oocytes begin to form about the third month of fetal development. By the time of birth, the two ovaries contain some 2 million primary oocytes, which have reached prophase of the first meiotic division. These primary oocytes remain in prophase until the female matures sexually. Then, under the influence of hormones, the first meiotic division of a primary oocyte resumes, resulting in a secondary oocyte and a polar body. The first meiotic division is completed at about the time of **ovulation** (the release of the oocyte from the ovary). Of the original 2 million primary oocytes, about 300 to 400 reach maturity, usually one at a time, about every 28 days from puberty to menopause, which typically occurs at about age 50. Given this timetable, a moment's reflection will reveal that more than 50 years may elapse between the beginning and the end of the first meiotic division in a particular oocyte.

Maturation of the oocyte involves both meiosis and a great increase in size. This size increase reflects the accumulation of stored food reserves and metabolic machinery, such as messenger RNA and enzymes, required for the early stages of development. As we saw in Figure 12–13 on page 258, oocytes do not divide into equal-sized cells, as spermatocytes do. Instead, one very large cell (100 micrometers in diameter in humans) is produced. The other nuclei are, in effect, discarded.

When a primary oocyte is ready to complete meiosis, the nuclear envelope fragments, and the chromosomes move to the surface of the cell. As the nucleus divides, the cytoplasm of the oocyte bulges out. One set of chromosomes moves into the bulge, which then pinches off into a small cell, the first polar body. The rest of the cellular material forms the large secondary oocyte. The first meiotic division is completed a few hours before ovulation. The second meiotic division does not take place until after fertilization. This division produces the ovum and another small polar body. As a consequence of these unequal cell divisions, most of the accumulated food reserves of the oocyte are passed on to a single ovum. The first polar body may also divide, although there is no functional reason for it to do so. All the polar bodies eventually die.

44-13 *Oocytes develop near the surface of the ovary within follicles. After a secondary oocyte is discharged from a follicle (ovulation), the remaining cells of the ruptured follicle give rise to the corpus luteum, which secretes estrogens and progesterone. If the ovum is not fertilized, the corpus luteum is reabsorbed within two weeks. If the ovum is fertilized, the corpus luteum persists for about three months, continuing its production of estrogens and progesterone. These hormones, which are later produced in large quantities by the placenta, maintain the uterus during pregnancy.*

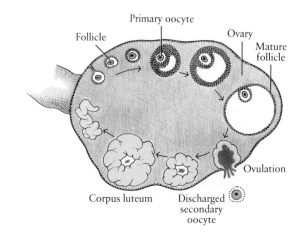

Oocytes develop near the surface of the ovary. An oocyte and the specialized cells surrounding it are known as an **ovarian follicle** (Figure 44–13). The cells of the follicle supply nutrients to the growing oocyte and also secrete estrogens, the hormones that support the continued growth of the follicle and initiate the buildup of the endometrium. During the final stages of its growth, the follicle moves to the surface and produces a thin, blisterlike elevation that eventually bursts, releasing the oocyte (Figure 44–14). Usually a number of follicles begin to enlarge simultaneously, but only one becomes mature enough to release its oocyte, and the others regress.

44-14 *Ovulation.* (a) *A secondary oocyte, visible only as a tiny speck, begins to emerge from an ovarian follicle.* (b, c) *With explosive force, the oocyte bursts out of the follicle, surrounded by a halo of material known as the corona radiata.* (d) *Completely free of the follicle and much of the material it carried out with it, the oocyte begins its journey to an oviduct.*

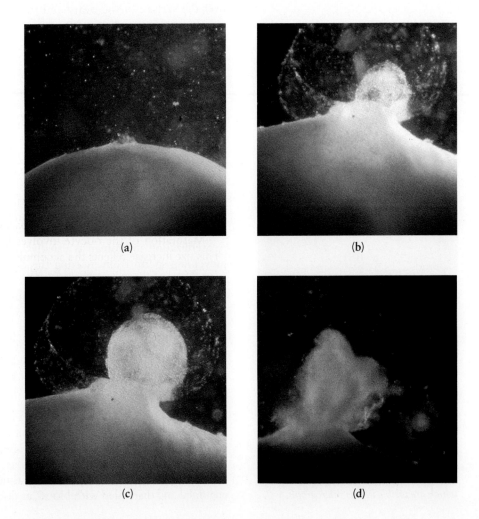

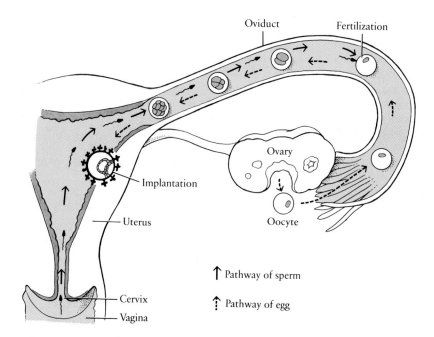

44-15 *Fertilization of the egg by sperm. About once a month in the nonpregnant female of reproductive age, an oocyte is ejected from an ovary and is swept into the adjacent oviduct. Fertilization, when it occurs, normally takes place within an oviduct, after which the young embryo passes down the oviduct and becomes implanted in the lining of the uterus. Muscular movements of the oviduct, plus the beating of the cilia that line it, propel the egg cell down the oviduct toward the uterus. If the oocyte is not fertilized within about 36 hours, it dies, usually by the time it reaches the uterus. A sperm cell has a life expectancy of about 48 hours within the female reproductive tract.*

Pathway of the Oocyte

When the oocyte is released from the follicle at ovulation, it is swept into the adjacent oviduct by the movement of the funnel-shaped opening of the oviduct over the surface of the ovary and by the beating of cilia that line the fingerlike projections surrounding this opening (Figure 44–15). This mechanism is so effective that women who have only one ovary and only one oviduct—and these on opposite sides of the body—have become pregnant.

The oocyte then moves slowly down the oviduct, propelled by peristaltic waves produced by the smooth muscles of the walls. The journey from the ovary to the uterus takes about three days. An unfertilized oocyte lives only about 72 hours after it is ejected from the follicle, but it is apparently capable of being fertilized for less than half of that time. So, fertilization, if it is to occur, must occur in an oviduct. If the egg cell is fertilized, the young embryo becomes implanted in the endometrium three to four days after it reaches the uterus, six or seven days after the egg cell was fertilized. If the oocyte is not fertilized, it dies, and the endometrial lining of the uterus is shed at menstruation. Fertilized eggs implanted in the endometrium are sometimes lost in abnormal menstrual flow, but it is difficult to estimate the number of such very short-lived pregnancies.

Sterilization in women is usually carried out by severing the oviducts, thus preventing sperm from meeting the oocyte. This procedure is known as tubal ligation. Passage of the oocyte and its fertilization are also prevented when the oviducts are blocked from natural causes, such as the formation of scar tissue following infection. Until recently, it was virtually impossible for women with such blockages to bear children. In the procedure known as in vitro fertilization, however, the oocyte is removed surgically from the ovary just prior to ovulation, fertilized with the husband's sperm in a laboratory dish, and inserted in the uterus at the time it would have arrived had it been fertilized naturally in the oviduct. In about 15 to 20 percent of the cases in which in vitro fertilization is attempted, the fertilized egg survives the procedure and subsequent development proceeds normally.

Orgasm in the Female

Under the influence of a variety of stimuli, the clitoris and its bulbs become engorged and distended with blood, as does the penis of the male. This process is

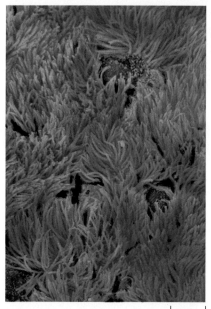

44-16 *Scanning electron micrograph of the inner lining of a mammalian oviduct. The cells of the lining of the oviduct have numerous microvilli, and the surface of the lining is carpeted with cilia. Interspersed among the ciliated cells are mucus-secreting goblet cells, a few of which are visible in this micrograph.*

somewhat slower in women than in men, largely because the valves trapping the blood in the female sexual structures tend to leak. The distension of the tissues is accompanied by the secretion into the vagina of a fluid that both lubricates its walls and neutralizes its acidic, and therefore spermicidal, environment.

Orgasm in the female, as in the male, is marked by rhythmic muscular contractions, followed by expulsion into the veins of the blood trapped in the engorged tissues. Homologous muscles produce orgasm in the two sexes, but in the female there is no ejaculation of fluid through the urethra or the vagina. At orgasm, the cervix drops down into the upper portion of the vagina, where the semen tends to form a pool. The female orgasm also may produce contractions in the oviducts that propel the sperm upward. It has been calculated that it would take a sperm cell at least two hours to make its way up the oviducts under its own power, but sperm have been found in the oviducts as soon as five minutes after intercourse. Orgasm in the female, however, is not necessary for conception.

Hormonal Regulation in Females

The Menstrual Cycle

The production of oocytes in all vertebrate females is cyclic. It involves both the interplay of hormones and changes in the follicle cells and the lining of the uterus. This recurring pattern of varying hormone levels and tissue changes is known in humans as the **menstrual cycle**. It is timed and controlled by the hypothalamus. The hormones that participate in the extremely complex feedback system regulating the menstrual cycle include the estrogens and progesterone (the female sex hormones), the pituitary gonadotropins FSH and LH, and gonadotropin-releasing hormone (GnRH) from the hypothalamus. In low concentrations, estrogens act through negative feedback to inhibit the production of FSH and GnRH (and so of LH). In high concentrations, estrogens act through positive feedback to increase the sensitivity of the pituitary to GnRH and may also stimulate the secretion of GnRH; the result is an increase in the synthesis of LH and FSH by the pituitary. In high concentrations, progesterone, in the presence of estrogens, inhibits the secretion of GnRH and thus the production of LH and FSH.

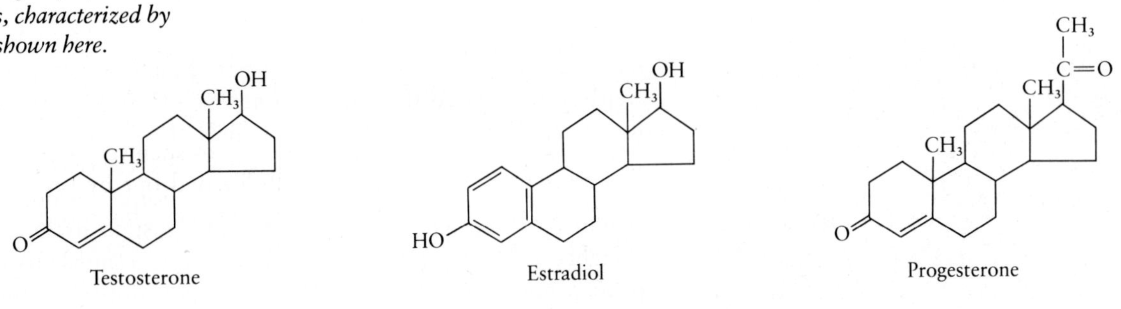

44-17 *The sex hormones testosterone, estradiol, and progesterone. Although there are a number of estrogens, estradiol is the most important. Note that the chemical structures of these three hormones differ only very slightly, in contrast to the great differences in their physiological effects—another example of the extreme specificity of biochemical actions. All of these belong to the group of molecules known as steroids, characterized by the four-ring structure shown here.*

Testosterone   Estradiol   Progesterone

At the beginning of the cycle, during menstruation, hormone levels are low (Figure 44-18). After a few days, an oocyte and its follicle begin to mature under the influence of the gonadotropins FSH and LH. As the follicle enlarges, it secretes increased amounts of estrogens, which stimulate the regrowth of the endometrium in preparation for implantation of a fertilized egg cell. The rapid rise in estrogen levels near the midpoint of the cycle triggers a sharp increase in the release of LH by the pituitary. The spurt of high LH stimulates the follicle to release the oocyte, which begins its passage to the uterus. Under the continued stimulus of LH, the cells of the emptied follicle grow larger and fill the cavity, producing the **corpus luteum** ("yellow body"). As the cells of the corpus luteum

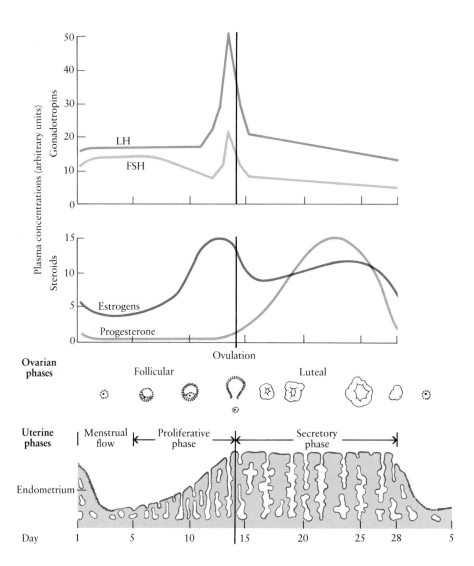

44–18 *Diagram of the events that take place during the menstrual cycle, which involves changes in hormone levels, in structures at the surface of the ovary, and in the uterine lining. The cycle begins with the first day of menstrual flow, which is caused by the shedding of the endometrium, the lining of the uterine wall. The increase of FSH and LH at the beginning of the cycle promotes the growth of the ovarian follicle and its secretion of estrogens. Under the influence of estrogens, the endometrium regrows. The sudden rise in estrogens just before midcycle triggers a sharp increase in the release of LH from the pituitary, which stimulates the release of the oocyte (ovulation). (It is not known what role, if any, is played by the simultaneous increase in FSH.) Following ovulation, LH and FSH levels drop. The follicle is converted to the corpus luteum, which secretes estrogens and also progesterone. Progesterone further stimulates the endometrium, preparing it for implantation of the young embryo. If pregnancy does not occur, the corpus luteum degenerates, the production of progesterone and estrogens falls, the endometrium begins to slough off, FSH and LH concentrations increase once more, and the cycle begins anew.*

increase in size, they begin to synthesize significant amounts of progesterone as well as estrogens. As the progesterone levels increase, estrogens and progesterone together inhibit the production of GnRH and so of the gonadotropic hormones LH and FSH from the pituitary. If pregnancy does not occur, the corpus luteum is reabsorbed and the production of ovarian hormones drops. Without hormonal support, the endometrium can no longer sustain itself, and a portion of it is sloughed off in the menstrual fluid. Then, in response to the now low level of ovarian hormones, the level of pituitary gonadotropic hormones begins to rise again, followed by development of a new follicle and a rise in estrogens as the next monthly cycle begins.

The cycle usually lasts about 28 days, but individual variation is common. Even in women with cycles of average length, ovulation does not always occur at the same time in the cycle (which is the reason the "rhythm method" is an unreliable means of birth control). Although the menstrual cycle does not require an environmental cue, as do reproductive cycles in many other vertebrates, it is clearly under the influence of external factors to some extent. For example, some women find that emotional upset delays a menstrual period or eliminates it completely.

The onset of menstruation marks the beginning of puberty in the human female. The average age of puberty is 12.3 years, but the normal range is very wide. The increased production of female sex hormones preceding puberty induces the development of secondary sex characteristics, such as enlargement of the hips and breasts.

Estrus

Females of almost all mammalian species except *Homo sapiens* will mate only during **estrus,** the fertile period during which oocytes are released. Estrus may occur only once a year (as in wolves or deer), about once a month (as in cows and horses), or every few days (as in rats and mice).

The periods of estrus may last from only a few hours to three or four weeks. In animals, such as dogs, that produce eggs over a period of days, the eggs may be fertilized at different times and by different males, which explains in part why a mixed-breed litter can contain such an astonishing variety of siblings. In some mammals, such as cats, rabbits, and minks, although the egg is mature and the female receptive during estrus, ovulation occurs only under the stimulus of copulation. This is obviously a very efficient system, ensuring maximum economy in the utilization of gametes. There is suggestive evidence that in some women, also, ovulation may be triggered by sexual intercourse.

The human female appears to be one of the few female animals receptive to mating during infertile periods. Some anthropologists speculate that this receptivity coevolved with the establishment of strong pair-bond relationships between human or prehuman males and females. A consequence of this pair-bond relationship is a society based on a family unit, in contrast to most other primates, in which the social and breeding unit is a troop or band. The establishment of such family units is seen as the original basis for the traditional division of labor between the sexes, with the female concentrating on childbearing and the home and the male on hunting, protecting the family unit, and defending territory. If the anthropologists are right, this behavioral adaptation on the part of the human female has profoundly influenced the shape of human civilization.

CONTRACEPTIVE TECHNIQUES

Statistics show that 25 percent of women who conceive become pregnant within one month of trying to do so, 63 percent within six months, 75 percent within nine months, and 90 percent within 18 months. Using no contraceptive methods, 80 percent of women of childbearing age having regular sexual intercourse become pregnant within a year. However, a variety of contraceptive techniques is available for couples who wish to prevent or defer pregnancy. In Table 44–1, they are rated in order of effectiveness, in terms of the average number of pregnancies per year among women of childbearing age using them. In most cases, two figures are given for effectiveness. The first, lower, figure is an "ideal" figure, obtained when the method is used consistently and correctly. The second figure is an average figure, reflecting actual experience.

For many years the most widely used contraceptive techniques were barrier methods, such as the diaphragm and the condom. In the 1960s and 1970s, many couples abandoned barrier methods, and "the pill," a combination of synthetic estrogens and progesterone, came into wide use. When taken daily, it keeps the level of these hormones in the blood high enough to shut off production of the pituitary hormones FSH and LH. Without FSH the ovarian follicles do not ripen, and in the absence of LH no ovulation occurs, so pregnancy is not possible.

TABLE 44-1 **Methods of Birth Control Currently Available**

METHOD	MODE OF ACTION	EFFECTIVENESS (PREGNANCIES PER 100 WOMEN PER YEAR)	ACTION NEEDED AT TIME OF INTERCOURSE	REQUIRES INSTRUCTION IN USE	POSSIBLE UNDESIRABLE EFFECTS
Vasectomy	Prevents release of sperm	0	None	No	Usually produces irreversible sterility
Tubal ligation	Prevents passage of oocyte to uterus	0	None	No	Usually produces irreversible sterility
"The pill" (estrogens and progesterone)	Inhibits secretion of FSH and LH, thereby preventing follicle maturation and ovulation	0–10	None	Yes, timing	Early—some water retention, breast tenderness, nausea; late—increased risk of cardiovascular disease
"Minipill" (progesterone alone)	Causes changes in cervix and uterus that prevent conception	3–10	None	Yes, timing	Break-through bleeding, ectopic (tubal) pregnancy
"Morning-after pill" (50 × normal dose of estrogens)	Arrests pregnancy, probably by preventing implantation	?	None	Yes; prescribed only in emergencies, for example, for rape victims	Breast swelling, water retention, abdominal pain, nausea
Diaphragm with spermicidal jelly	Prevents sperm from entering uterus, jelly kills sperm	2–20	Yes, insertion before intercourse	Yes, must be inserted correctly each time; leave in at least 6 hours after	None usually, may cause irritation
Condom (worn by male)	Prevents sperm from entering vagina	3–36	Yes, male must put on after erection	Not usually	Some loss of sensation in male
Intrauterine device (wound with copper or containing progesterone that is slowly released)	Prevents fertilization or implantation	4–5	None; may be left in place for as long as 4 years	No, but must be inserted by a physician and periodically checked	Menstrual discomfort, displacement or loss of device, uterine infection, ectopic pregnancy
Cervical cap with spermicidal jelly	Prevents sperm from entering uterus, jelly kills sperm	4–?	Yes, insertion before intercourse	Yes, must be inserted correctly each time; more difficult to insert than diaphragm	None usually, may cause irritation
Contraceptive sponge (contains spermicide)	Prevents sperm from entering uterus; kills sperm	13–20	Yes, insertion before intercourse	Yes, must be inserted correctly each time; leave in at least 6 hours after	None usually, may cause irritation
Vaginal foam, jelly alone	Spermicidal, mechanical barrier to sperm	20–30	Yes, requires application before intercourse	Yes, must use within 30 minutes of intercourse; leave in at least 6 hours after	None usually, may cause irritation
Rhythm	Abstinence during probable time of ovulation	14–47	None	Yes, must know when to abstain, based on daily temperatures and/or cyclic changes in vaginal mucus	Requires abstinence during part of cycle
Douche	Washes out sperm that are still in the vagina	?–85	Yes, immediately after	No	May propel sperm through cervix into uterus
Withdrawal	Removes penis from vagina before ejaculation	?–93	Yes, withdrawal	No	Frustration in some

Recently, however, the diaphragm and the condom have once more become popular. In combination with spermicidal jellies, these two methods, and particularly the condom, provide a barrier not only against sperm but also against many infectious agents. As Table 44–1 reveals, however, their protection is far from absolute.

SUMMARY

In vertebrates, reproduction is characteristically sexual and involves two parents. The male and female gametes, sperm and eggs, are formed by meiosis in the gonads (the testes and the ovaries). Sperm cells are produced in the seminiferous tubules of the testes. The spermatogonia become primary spermatocytes; then, after the first meiotic division, secondary spermatocytes; and, following the second meiotic division, spermatids, which then differentiate into sperm cells. These sperm cells enter the epididymis, a tightly coiled tube overlying the testis, where they are partially mobilized. The epididymis is continuous with the vas deferens, which leads along the posterior wall of the abdominal cavity, around the bladder, and into the prostate gland. Just before entering the prostate, the two vasa deferentia merge with ducts of the seminal vesicles and then, within the prostate, with the urethra, which leads out through the penis.

The penis is composed largely of spongy erectile tissue that can become engorged with blood, enlarging and hardening. At the time of ejaculation, sperm are propelled along the vasa deferentia by contractions of a surrounding coat of smooth muscle. Secretions from the seminal vesicles, the prostate, and the bulbourethral glands are added to the sperm as they move toward the urethra. The resulting mixture, the semen, is expelled from the urethra by muscular contractions involving, among other structures, the base of the penis. These muscular contractions also contribute to the sensations of orgasm.

Production of sperm and the development of male secondary sex characteristics are under the control of hormones, including gonadotropin-releasing hormone (GnRH), the gonadotropins LH (luteinizing hormone) and FSH (follicle-stimulating hormone), and testosterone (the principal androgen). LH acts on the interstitial cells, located between the seminiferous tubules, to stimulate the production of testosterone. FSH and testosterone stimulate the production of sperm. Production of LH and FSH is regulated by a negative feedback system involving both the hypothalamus and the pituitary.

The female gamete-producing organs are the ovaries. The primary oocytes develop within nests of cells called follicles. The first meiotic division begins in the female fetus and is completed at ovulation. The second meiotic division is completed at fertilization. On the average, one secondary oocyte is released every 28 days and travels down the oviduct to the uterus. If it is fertilized (which usually takes place in an oviduct), the embryo becomes implanted in the lining of the uterus (the endometrium). If it is not fertilized, it degenerates and the endometrial lining is shed at menstruation.

The production of oocytes and the preparation of the endometrium for implantation of the embryo are cyclic. The reproductive cycle, which is known in humans as the menstrual cycle, is controlled by hormones, including gonadotropin-releasing hormone (GnRH), the gonadotropic hormones FSH and LH, and estrogens and progesterone (the female sex hormones). Before ovulation, FSH and LH from the pituitary stimulate the ripening of the follicle and the secretion of estrogens. After ovulation, the corpus luteum, which forms from the emptied follicle, produces both estrogens and progesterone. Progesterone and estrogens both stimulate the growth of the endometrium.

The part of the female reproductive cycle during which mating occurs in most mammals is known as estrus. Although ovulation is cyclic in human females, they, unlike most other animals, are continuously receptive to sexual intercourse, a behavior pattern that may have coevolved with strengthened pair bonding.

QUESTIONS

1. Distinguish among the following: spermatocytes/spermatids/sperm cells; oocyte/ovum; ovarian follicle/corpus luteum.

2. Consider the kinds of organisms that generally have external fertilization and those that generally have internal fertilization. What differences in their life styles require these differing mechanisms?

3. How is the shelled egg of a hen fertilized?

4. What would constitute the ejaculate of a man who has undergone a vasectomy? Would a vasectomy affect the structures associated with orgasm? Why or why not?

5. Describe the effects of luteinizing hormone (LH) and follicle-stimulating hormone (FSH) in the human male and female.

6. Explain the changes in the plasma concentrations of the gonadotropins and sex hormones during the menstrual cycle as shown on the graphs below.

7. During which days in the menstrual cycle is a woman most likely to become pregnant? (Include data on the longevity of eggs and sperm in making this calculation.) Why is the use of the calendar method of birth control much less effective than other methods?

8. What would be an appropriate method of contraception for a couple who have infrequent intercourse (once a month)? For a couple who have frequent intercourse (three times a week), with plans to have children eventually? For a couple who have frequent intercourse but do not wish to have any children? Under what circumstances, if any, would you elect to have a vasectomy or a tubal ligation?

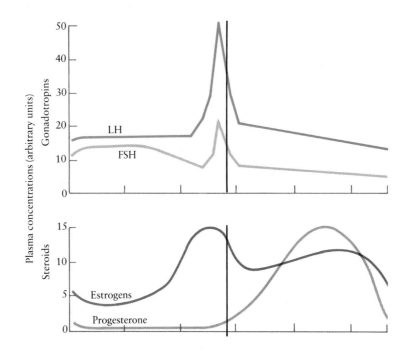

CHAPTER 45

The Continuity of Life II: Development

Each of our lives began very modestly with the fusion of a tiny, motile sperm cell and a larger, but structurally simpler egg. If the newly fertilized egg of an animal—any animal—is examined under the electron microscope, there is no hint whatsoever of the awesome potential with which this single cell is endowed. How do the complex structures of the embryo and, subsequently, of the adult animal develop from this one, apparently simple cell? This question is one of the most fundamental in all of biology, and although it has commanded the attention of scientists for more than 100 years, a complete and comprehensive answer remains elusive.

During its development, the fertilized egg is transformed into a complete organism, closely resembling its parents and consisting, depending on the species, of hundreds to billions of cells. This process involves growth, differentiation, and morphogenesis—that is, increase in size, specialization of cells, tissues, and organs, and shaping of the adult body form. The course of development in many species has been closely observed and meticulously documented, illuminating the sequence of developmental events. Only in recent years, however, have the tools become available to begin deciphering the underlying cellular and molecular events that give rise to and control the observed processes.

Development, like other biological processes, is most clearly explained through concrete examples. Thus, in this chapter we shall first consider in some detail the course of embryonic development in three representative organisms—the sea urchin, the amphibian, and the chick—each of which has made major contributions to scientific understanding. With the general principles and overall pathways of development established, we shall then return to the fertilized human egg, which we left in the last chapter as it began its journey down the oviduct to the uterus, and consider the marvelous series of events that precede and lead to the birth of each new human infant.

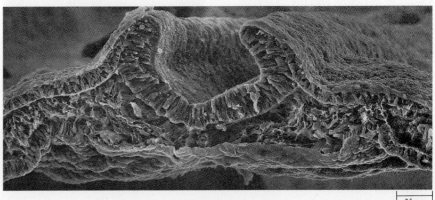

45–1 *One of the distinguishing features of phylum Chordata, to which we and all other vertebrates belong, is a dorsal, hollow nerve cord. It develops from a groove, known as the neural groove, on the dorsal surface of the early embryo. As you can see in this scanning electron micrograph of the neural groove of a chick embryo, the tissue on either side of the groove curves upward, forming two arches that ultimately meet to form the hollow neural tube. As we shall see later in this chapter, the curvature is caused by changes in the shape of the individual cells.*

25 µm

DEVELOPMENT OF THE SEA URCHIN

Let us begin by following the course of development in the sea urchin, an echinoderm. As we saw in Chapter 26 (page 544), the earliest stages of embryonic development in echinoderms are, in important respects, similar to those observed in vertebrates and other chordates, indicating a closer evolutionary relationship than one would expect on the basis of the adult forms. Of all the invertebrates, echinoderms are the most useful as model organisms for the study of early vertebrate development. Sea urchins, in particular, have long been a favorite of embryologists because (1) their eggs, which are produced in great numbers, are fertilized and develop externally, making it possible to study their development under relatively simple laboratory conditions; (2) both egg and developing embryo are almost transparent, so that it is possible to observe many of the early events without disrupting them; and (3) the process is rapid. In about 48 hours, the zygote develops into a free-swimming larval form, known as the **pluteus**. Moreover, sea urchins are abundant in pleasant places such as Woods Hole, Massachusetts, where biologists like to spend the summer.

Fertilization and Activation of the Egg

Development begins with fertilization of the egg by the sperm (Figure 45–2). Like the mammalian sperm cell (page 903), the sperm cell of the sea urchin consists of an acrosome at its anterior tip, a highly condensed, tightly packed nucleus, a small amount of cytoplasm, and a long flagellum, all surrounded by a cell membrane. The egg cell, which is much larger than the sperm cell (see Figure 5–10b, page 109), is surrounded by an outer membrane, known as the **vitelline envelope**, attached to the cell membrane. Embedded in the vitelline envelope and displayed on its surface are species-specific protein receptors that participate in the binding of the sperm to the egg. Surrounding the vitelline envelope are additional layers of jelly.

When the sperm meets the egg, enzymes released from the acrosome dissolve the jelly layers at the point of contact, creating a path through which the sperm cell can move. Other molecules from the acrosome are the three-dimensional complement of the protein receptors displayed on the vitelline envelope. These molecules coat the outer surface of the sperm head; when they bind to their receptors, the sperm is able to penetrate the vitelline envelope. Ultimately, the cell membranes of the sperm and the egg make contact, they fuse, and the sperm nucleus enters the cytoplasm of the egg.

Fertilization—the fusion of sperm and egg—has at least four consequences. First, changes take place on the surface of the fertilized egg to prevent the entry of additional sperm. Within a second of sperm contact, the cell membrane of the

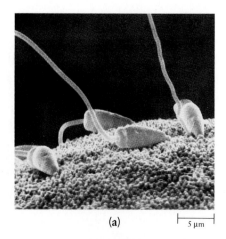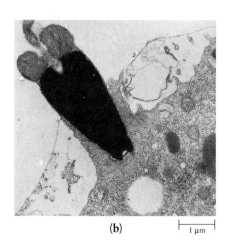

45-2 *Egg and sperm cells from a sea urchin:* **(a)** *sperm cells on the egg surface,* and **(b)** *a sperm penetrating the egg. The dark, torpedo-shaped body is the sperm nucleus.*

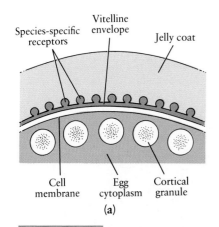
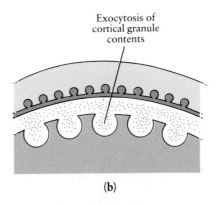
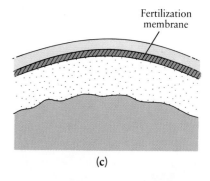

45–3 *Formation of the fertilization membrane.* (a) *The surface layers of an unfertilized egg consist of a jelly coat, the vitelline envelope, which bears species-specific protein receptors, and the cell membrane. Within the cytoplasm, near the cell membrane, are membrane-bound vesicles known as cortical granules. Contact of a sperm cell with the receptors on the vitelline envelope triggers an increase in the Ca^{2+} concentration in the cytoplasm. This causes the cortical granules to fuse with the cell membrane* (b), *releasing their contents, including enzymes and polysaccharides, into the space between the cell membrane and the vitelline envelope. The enzymes break down the protein receptors on the vitelline envelope and the substances that hold it to the cell membrane. The polysaccharides raise the osmotic potential of the space beneath the vitelline envelope, causing water to flow inward, thus lifting the vitelline envelope off the surface of the egg* (c). *The vitelline envelope "hardens" to form the fertilization membrane.*

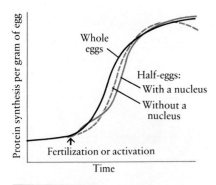

45–4 *Rates of protein synthesis during the first few hours after activation in whole eggs, half-eggs with a nucleus, and half-eggs without a nucleus. The initial pattern of protein synthesis is the same whether the activated egg contains a nucleus or not, indicating that the mRNA for early protein synthesis is present in the egg before fertilization.*

egg, which normally maintains an internal electric potential of about −60 millivolts, is depolarized. Sodium ions (Na^+) flow into the cytoplasm, temporarily reversing the polarity of the cell. This change in electric potential sweeps, wavelike, around the surface of the egg and makes the cell membrane suddenly unresponsive to the advances of other sperm. Simultaneously, calcium ions (Ca^{2+}) sequestered within the endoplasmic reticulum are released into the cytoplasm. The increase in the concentration of free Ca^{2+} ions triggers a series of reactions that cause the vitelline envelope to lift off the surface of the egg (Figure 45–3). The protective outer membrane that forms from the vitelline envelope is known as the **fertilization membrane.**

Second, the egg is activated metabolically, as evidenced by a dramatic increase in protein synthesis and a rise in oxygen consumption.

Third, the genetic material of the male is introduced into the female gamete—the haploid sperm nucleus fuses with the haploid egg nucleus to produce a diploid zygote nucleus. The genotype of the new individual is thus established.

Fourth, the egg begins to divide by mitosis, with the centrioles of the sperm participating in the organization of the mitotic spindle. The developmental chain of events is thus set in motion.

In many species, although activation of the egg and mitosis follow fertilization, they can also proceed without it. In the sea urchin egg, for instance, even exposure to a hypertonic solution can start the developmental chain of events. Frogs' eggs can be activated and will divide after being pricked with a glass needle or given a mild electric shock. In fact, in some species it has been possible to demonstrate that activation of the egg does not require any nucleus at all, male or female. Sea urchin eggs can be divided in two, half with a nucleus, half without. Each half can then be activated artificially. Protein production in the half without a nucleus is as great initially, on a gram-for-gram basis, as in the half with a nucleus or in the whole egg (Figure 45–4), although, of course, the enucleated half does not long survive. The egg, during the course of its maturation within the ovary, transcribes the genetic information needed for the early rounds of biosynthetic activity following fertilization, but the transcribed mRNA is inactive—stored in the cytoplasm of the egg—until fertilization initiates translation.

From Zygote to Pluteus

The initial stages of development in the sea urchin are shown in Figure 45–5, on the next page. The zygote divides about once an hour for 10 hours, in the process known as **cleavage.** The embryo at this stage is called a **morula.** As the cells continue to divide, Na^+ ions are pumped from them into the extracellular spaces, followed by an osmotic flow of water. This creates a fluid-filled cavity, known as the **blastocoel,** in the center of the embryo. When the blastocoel is fully formed, the embryo—now a fluid-filled sphere of a single layer of cells—is called the **blastula,** and its cells are known as blastomeres. The sea urchin blastula is about the same size as the egg cell from which it developed. Cleavage has not changed the total volume but has greatly altered the surface-to-volume ratio (page 103) and also the ratio of volume of the nucleus to that of the cytoplasm of the individual cells.

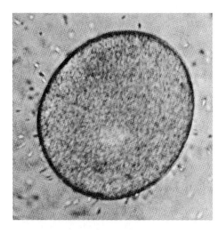

(a) *Numerous spermatozoa can be seen surrounding an unfertilized egg.*

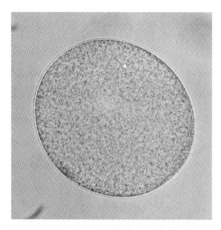

(b) *Fertilized egg; the fertilization membrane has just begun to form. The light area slightly above the center is the diploid nucleus.*

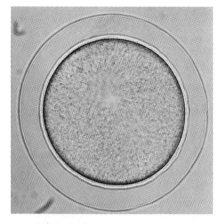

(c) *The fertilization membrane is fully formed. The egg has begun to divide; if you look closely, you can see that there are two nuclei.*

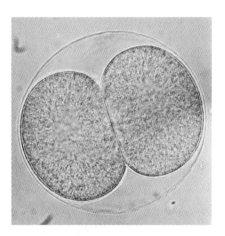

(d) *The first division.*

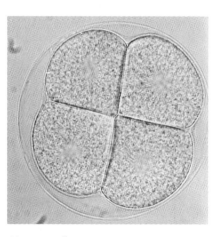

(e) *Four-cell stage.*

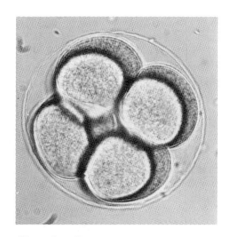

(f) *Eight-cell stage.*

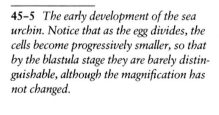

45–5 *The early development of the sea urchin. Notice that as the egg divides, the cells become progressively smaller, so that by the blastula stage they are barely distinguishable, although the magnification has not changed.*

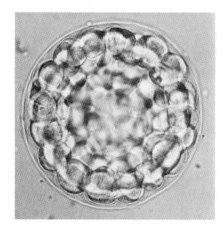

(g) *The blastocoel forms.*

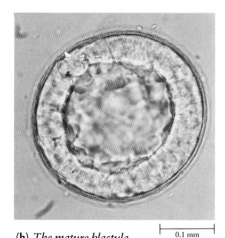

(h) *The mature blastula.*

0.1 mm

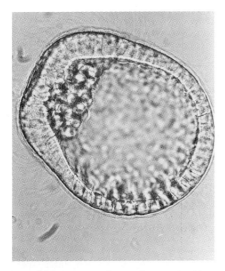

(a) *The beginning of gastrulation; the blastopore has begun to form at the upper left, and cells near the blastopore have begun to migrate across the blastocoel.*

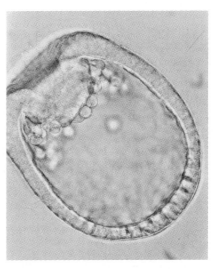

(b) *The outer cell layer begins to fold inward at the blastopore, forming the archenteron.*

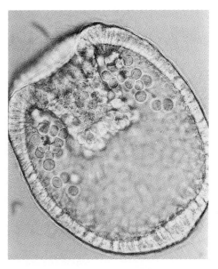

(c) *The outer layer of cells continues to move across the blastocoel.*

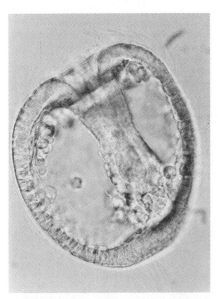

(d) *The mature gastrula.*

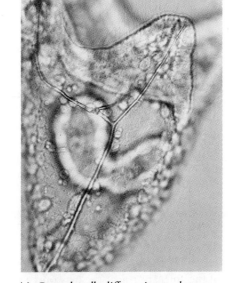

(e) *Gastrula cells differentiate and organize to form the pluteus larva.*

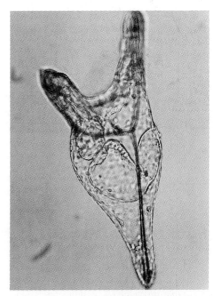

(f) *Within 48 hours after fertilization, the egg has developed into a free-swimming multicellular organism, the pluteus.*

45-6 *Gastrulation in the sea urchin.*

The formation of the blastula is followed by a process known as **gastrulation** (from *gaster*, the Greek word for "stomach"), which gives rise to the primitive gut. Gastrulation in the sea urchin (Figure 45-6) begins with the formation of the **blastopore**, an opening into the blastula. Cells near the blastopore break loose and, with the aid of contractile pseudopodia with sticky tips, move over the interior surface of the blastula toward the opposite pole. These cells are known as the primary mesenchyme. Next, the entire cell layer closest to the blastopore turns inward, moving through the blastocoel to the opposite pole, forming a new cavity, the **archenteron**. The archenteron will ultimately develop into the digestive tract,

The Cytoplasmic Determination of Germ Cells

Early in the development of virtually all animals, the germ cells—that is, the cells that will ultimately differentiate into spermatogonia or oogonia—are set aside. Their fate is determined well before differentiation of any other cell types has begun. This early segregation of the germ cells is thought to reduce the number of cell divisions that they must undergo prior to spermatogenesis or oogenesis and, therefore, to reduce the likelihood of errors in DNA replication. At a later stage in development, the germ cells migrate from the site at which they were set aside to the site at which the gonads will form.

In insects that undergo a complete metamorphosis between a nonreproductive larval stage and a reproductive adult stage (see page 576), a total reorganization of the cells and tissues occurs. Consequently, early determination of the future germ cells and protection of their genetic integrity are of particular importance. The mechanism by which this is accomplished has been studied most extensively in *Drosophila*. As we saw in Section 3, more is known about the genetics of *Drosophila* than of any other animal. In laboratories throughout the world, populations of *Drosophila* are maintained for which the genetic makeup is known in extraordinary detail. Such populations make possible a variety of experiments producing subsequent generations in which the precise ancestry of the individual flies—or even portions of the flies—can be identified.

During cleavage in the fertilized egg of *Drosophila* and other insects, mitosis is not initially accompanied by cytokinesis. Numerous rounds of mitosis occur until the number of nuclei required for the cells of the blastula are produced. The nuclei then migrate to the periphery of the cell, followed by a wave of cytokinesis that produces the individual cells of the blastula. The blastula, like the egg from which it formed, is elongated, with a clear anterior-posterior axis. It consists of a single layer of cells—except at its posterior pole, where a group of cells visibly distinct from the other cells of the blastula form a second, outer layer. These cells, known as the pole cells, are the future germ cells. As the experiment diagrammed here shows, their fate is determined by a region of cytoplasm, known as the polar plasm, located at the posterior pole of the egg. In this experiment, polar plasm from an early *Drosophila* embryo of genotype A is injected into the anterior end of another embryo of a different genotype, B. When cytokinesis occurs in the genotype B embryo, pole cells form at both its anterior and posterior poles. Further development in this embryo is thwarted, but the pole cells at its anterior can be removed and injected into the posterior region of still another embryo, representing a third genotype, C. The fly that develops from this embryo is of genotype C. Breeding experiments, however, reveal that it produces gametes of both genotype C and genotype B. The gametes of genotype B are the pro-

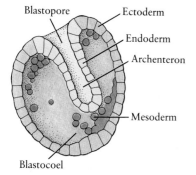

45–7 *The sea urchin gastrula. Gastrulation produces a three-layered embryo. The archenteron becomes the digestive tract, and the blastopore becomes the anus. Ultimately, the blastocoel is almost entirely obliterated. In this and subsequent illustrations, ectoderm is blue, endoderm is yellow, and mesoderm is red.*

and the blastopore will become the anus. As you will recall (page 545), formation of the anus at or near the blastopore is the defining characteristic of the deuterostomes, which include the echinoderms and the chordates.

As a result of the movements that take place at gastrulation, three embryonic tissue layers have been formed: an outside layer, the ectoderm; a middle layer, the mesoderm, which formed from the primary mesenchyme cells; and an inner layer, the endoderm (Figure 45–7). Also, the anterior-posterior axis of the embryo has become obvious.

Once gastrulation is completed, evidence of cell differentiation can be observed. Cells in the mesoderm begin to secrete calcium-containing granules that develop into tiny three-armed spicules. These become the supporting skeleton for the pluteus. At the point at which the endoderm touches the opposite surface of the blastocoel, ectodermal cells curve inward to form the mouth of the larva. The archenteron subdivides into stomach, intestine, and anus, and a ring of long cilia forms near the mouth region. The single fertilized egg cell has become a number of specialized, differentiated cells, performing specific functions, such as digestion; producing new organelles, such as cilia; and secreting new products, such as the calcium-containing skeleton.

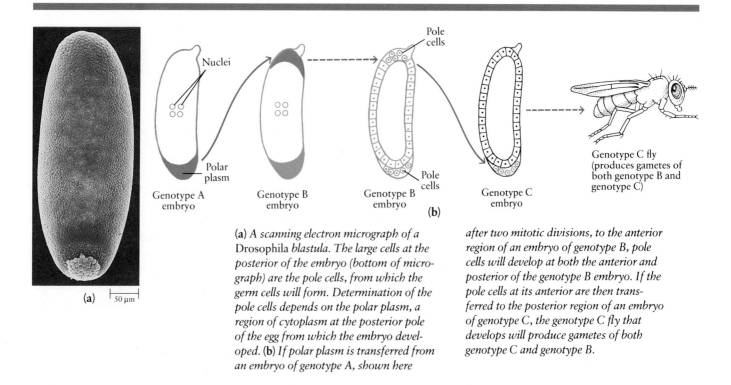

(a) A scanning electron micrograph of a Drosophila blastula. The large cells at the posterior of the embryo (bottom of micrograph) are the pole cells, from which the germ cells will form. Determination of the pole cells depends on the polar plasm, a region of cytoplasm at the posterior pole of the egg from which the embryo developed. (b) If polar plasm is transferred from an embryo of genotype A, shown here after two mitotic divisions, to the anterior region of an embryo of genotype B, pole cells will develop at both the anterior and posterior of the genotype B embryo. If the pole cells at its anterior are then transferred to the posterior region of an embryo of genotype C, the genotype C fly that develops will produce gametes of both genotype C and genotype B.

geny of the pole cells that developed at the anterior end of the genotype B embryo in response to the injection of polar plasm from the genotype A embryo.

The substance or substances in the polar plasm of the *Drosophila* egg that determine the formation of pole cells and, subsequently, of the germ cells, are localized within small granules. These granules, which are present in the cytoplasm before fertilization, are rich in RNA of maternal origin—coded by the genes of the female fly that laid the egg. Recent studies in amphibians suggest that a similar mechanism, involving RNA and other cytoplasmic determinants laid down in the unfertilized egg, may also be involved in the differentiation and setting apart of the germ cells of vertebrates.

The Influence of the Cytoplasm

The sea urchin egg contains a relatively small amount of stored food (yolk), which tends to be concentrated in the lower half of the egg. This lower half is called the vegetal half; the upper half is called the animal half. At cleavage, the first two cell divisions run longitudinally from the pole of the animal half to the pole of the vegetal half, perpendicular to one another—the way you would quarter an apple. If these four cells are separated, each can develop into a normal pluteus larva. During the third cleavage division, however, an important event takes place—the cytoplasm of the animal half and that of the vegetal half are separated as the four cells split across the equator, producing an eight-cell embryo (see Figure 45–5f). If the eight-cell embryo is separated into two halves longitudinally, normal development will occur, but if the embryo is divided along the third cleavage plane, the two halves develop abnormally. The top (animal) half has large tufts of cilia but no gut. The bottom (vegetal) half is almost all gut, with no mouth, shortened arms, no cilia, and only a few, if any, spicules (see Figure 45–8 on the next page).

The explanation for these observations lies in chemical factors, known as cytoplasmic determinants, that are distributed in two gradients, an animal gra-

45–8 *When an eight-cell sea urchin embryo is separated into two halves by moving a glass needle between the cells, the future development of the resulting halves depends on the plane of division. (a) If the division is longitudinal, producing embryos containing cells from both the animal and vegetal poles, subsequent development is normal. (b) If, however, the division is at the equator of the embryo, the four cells from the animal pole produce a ciliated embryo, permanently arrested at the blastula stage. The four cells from the vegetal pole give rise to an embryo that looks relatively normal but lacks critical structures, such as a mouth, cilia, and spicules.*

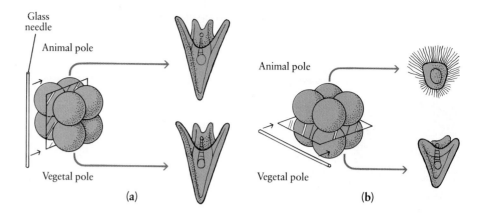

dient and a vegetal gradient, within the cytoplasm of the egg. Current evidence suggests that the cytoplasmic determinants are most likely messenger RNA or protein molecules, localized within the cytoplasm by the cytoskeleton. During cleavage, these molecules are "captured" within the cytoplasm of individual blastomeres. The particular cytoplasmic determinants contained within a cell then affect its subsequent pattern of development. As microsurgical experiments on sea urchin and other embryos make clear, it is not just the nucleus that controls the differentiation of cells. The cytoplasm surrounding the nucleus also has profound effects.

Other experiments have demonstrated that interactions among the individual blastomeres are also important in development. If you put the early blastula of a sea urchin in calcium-free sea water and agitate the water, the cells will separate. The cells still look the same, and neither their genetic material nor their cytoplasm has been altered in any crucial way. But they simply stop dividing and very shortly thereafter they deteriorate and die. Returning individual cells to normal sea water does not restore their developmental potential. Suppose, however, the cells are returned to normal sea water and, by gently stirring, are brought into contact with one another. Under these conditions the cells reaggregate, arrange themselves in their former pattern, and once more begin to divide.

DEVELOPMENT OF THE AMPHIBIAN

The eggs of frogs and many other amphibians are laid in shallow water and, like sea urchin eggs, are fertilized externally; hence, they too can be readily observed. The amphibian egg, however, contains a much larger amount of yolk. The animal half and the vegetal half of the egg differ markedly in appearance. For example, in frogs of the genus *Rana*, the yolk is massed in the lower hemisphere of the unfertilized egg, and the upper two-thirds of the egg is covered by a heavily pigmented layer of cytoplasm (Figure 45–9a). Immediately following fertilization, there is a massive reorganization of the cytoplasm. When the sperm penetrates the

45–9 *Formation of the gray crescent in a frog's egg. (a) Before fertilization, the upper two-thirds of the egg is "capped" by a heavily pigmented layer of cytoplasm. The bulk of the yolk is massed in the lower hemisphere. The nucleus of the egg is near the pole of the upper hemisphere. (b) The egg has been fertilized; the sperm nucleus has entered at the right and is moving toward the center of the egg. The trail that marks its path is caused by the disruption of pigment granules. The whole pigment cap has rotated toward the point of sperm penetration. The region on the other side of the egg, from which the pigmented layer has moved away, becomes the gray crescent. Accompanying these movements of the pigmented layer are rearrangements of the cytoplasm in more interior regions of the egg.*

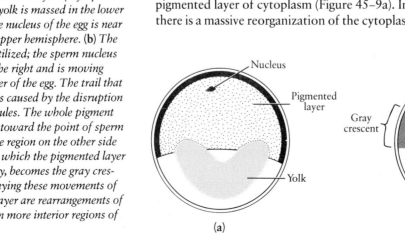

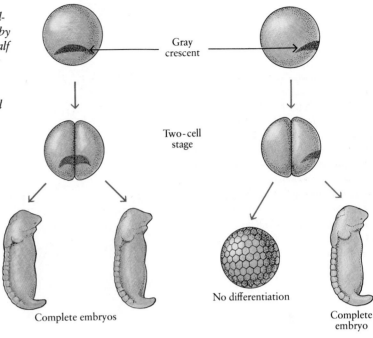

45–10 *The importance of the gray crescent cytoplasm in development was demonstrated by separating the two cells formed by the first cleavage of the egg. When the egg on the left divided, half of the gray crescent cytoplasm passed into each of the two new cells. When these cells were separated from each other, each formed a complete embryo. The first cleavage of the egg on the right resulted in all the gray crescent cytoplasm going to one cell and none to the other. When these cells were separated, the one without the crescent did not develop.*

egg, the pigment cap rotates toward the point of sperm penetration, and a gray crescent appears on the side of the egg opposite the point of sperm entry (Figure 45–9b). Formation of the gray crescent is complete within 20 minutes.

In a series of brilliant experiments by Hans Spemann in the early 1900s, it was shown that the cytoplasm associated with the gray crescent is of critical importance in the later development of the embryo. For instance, if cells of an amphibian embryo are divided at the two-cell stage, each blastomere may develop into a normal embryo. Whether or not both cells develop normally depends on whether or not each has received some of the cytoplasm containing the gray crescent (Figure 45–10). These experiments provide further evidence that differences in the cytoplasm, and particularly in localized factors, play a major role in determining the course of early development.

Cleavage and Blastula Formation

Cleavage in the amphibian egg differs from that in the sea urchin egg chiefly because of differences in the yolk content. When yolk is absent or present only in small amounts, as in the sea urchin, cleavage is uniform, producing cells of similar size. When larger amounts of yolk are present, as in frogs, the egg divides unevenly (Figure 45–11), forming larger cells in the vegetal hemisphere. As with the sea urchin egg, cleavage results in the formation of a blastula, but the amphibian blastocoel is small and usually off-center. The blastopore appears as a crescent-shaped slit; it always forms at the boundary between the gray crescent and the vegetal hemisphere.

Gastrulation and Neural Tube Formation

Gastrulation in the amphibian differs from that in the sea urchin in detail but not in principle. In the sea urchin, one wall of the blastula moves inward, forming a tubular extension toward the other wall. In the yolky frog's egg, the process of inward movement is characterized by an extensive series of coordinated cellular movements—migration of single cells and sheets of cells—and the folding of cell layers.

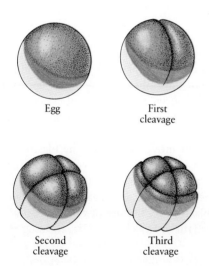

45–11 *Eggs, such as those of the frog, that contain a large amount of yolk concentrated in one hemisphere cleave unequally. The first longitudinal cleavage usually bisects the gray crescent. The second cleavage, also longitudinal, is at right angles to the first. Thus the egg is split through the poles to produce four cells shaped like the segments of an orange. The third cleavage separates the lower, yolkier (vegetal) part of the embryo from the upper, less yolky (animal) part. As you can see, the four cells in the animal hemisphere are much smaller than the four in the vegetal hemisphere.*

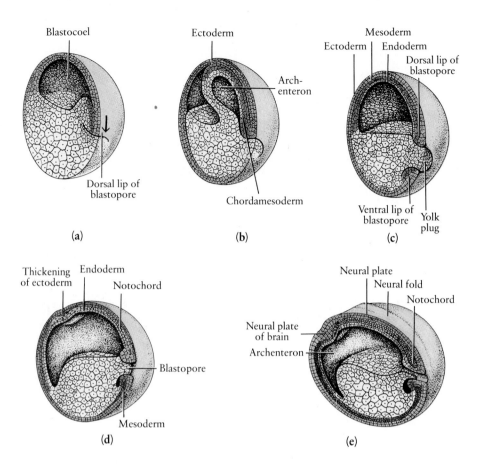

45–12 *Development of the frog gastrula. (a) The blastopore forms in the blastula, and cells from the outer surface begin moving across its dorsal lip to the interior. (b) As the cellular migrations progress, the blastocoel is obliterated and the archenteron created. (c) Three embryonic tissue layers are established: ectoderm, endoderm, and mesoderm. (d) The mesoderm directly over the roof of the archenteron, known as the chordamesoderm, differentiates to become the notochord (shown in green). The ectoderm on the dorsal surface overlying the notochord thickens and flattens, forming the neural ectoderm (shown in purple), from which the neural plate and neural folds begin to form (e).*

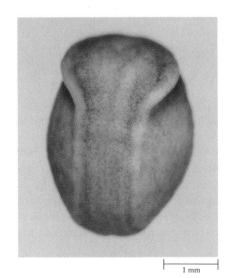

45–13 *The neural plate of a newt embryo. Newts are amphibians that, like other salamanders, have tails throughout their lives. The wide portion of the neural plate, at the anterior of the embryo, will give rise to the brain; the narrow portion will give rise to the spinal cord.*

The formation of the blastopore marks the initiation of gastrulation (Figure 45–12). Cells at the **dorsal lip** of the blastopore change shape and sink below the surface as they move to the interior. These cells are replaced by cells that move from their original positions on the surface of the embryo to the dorsal lip. As a result, the entire sheet of cells that forms the surface of the amphibian embryo appears to stream toward and over the lip of the blastopore as if it were being hauled over a pulley.

Once inside the embryo, the migrating cells move away from the blastopore and deeper into the interior. The direction in which they move is the future anterior-posterior axis of the animal. The advancing cells form the walls of an increasingly spacious archenteron, and the blastocoel disappears. The archenteron remains open to the outside at the blastopore, the site of the future anus. Yolk-laden cells are visible within the boundaries of the blastopore, forming the yolk plug.

In the course of gastrulation, the primary embryonic tissues—endoderm, mesoderm, and ectoderm—become arranged in a three-layered pattern. The floor of the archenteron is composed of yolk-laden endodermal cells. Its roof consists of endodermal cells that have been pushed and pulled into the interior by a sheet of mesodermal cells that lies above it along the axis of the embryo. This sheet of mesoderm includes cells destined to form the notochord (page 591) and is called the **chordamesoderm.** At the sides of the archenteron, other mesodermal cells have slipped between the ectoderm and endoderm, forming the **lateral plate mesoderm.** On the dorsal surface of the embryo, lying above the chordamesoderm, is a sheet of ectoderm that will ultimately give rise to the brain and spinal cord; this tissue is known as **neural ectoderm.** The ectoderm covering the rest of the gastrula, known as **epidermal ectoderm,** will give rise to the epidermis of the skin.

By the end of gastrulation, the first visible signs of differentiation have begun to appear. The chordamesoderm has formed the notochord, and the neural ectoderm has begun to thicken, forming the neural plate (Figure 45–13). The ridges of

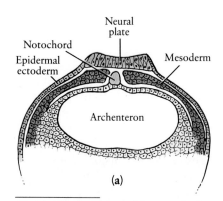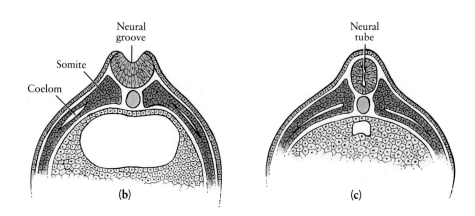

45–14 *The formation of the neural tube from the neural plate in the frog.* **(a, b)** *The thickened elevations of neural ectoderm on the right and left sides of the neural plate curve inward, forming the neural groove. The ridges bordering the neural groove then meet and fuse.* **(c)** *Finally, the resulting neural tube pinches off from the epidermal ectoderm.*

the neural plate curve upward and inward, forming the neural groove. Ultimately these ridges meet and fuse to form the **neural tube**, which pinches off from the rest of the ectoderm (Figure 45-14). At about the same time, the two longitudinal strips of chordamesoderm on either side of the differentiated notochord split into segments, forming the blocks of tissue called **somites**. In the lateral plate mesoderm, the coelom (page 545) forms between two layers of tissue. Thus the principal features of the vertebrate have been established. The endoderm, which contains the yolk, continues to take up much of the body mass until the stored nutrients are finally absorbed by the larva, the tadpole.

The movements of cells and tissues in gastrulation are completely regular and predictable, embryo after embryo. By applying harmless dyes to the surface of the late blastula, embryologists have developed fate maps (Figure 45-15a) that identify the groups of cells that give rise to the tissue layers of the mature gastrula. During subsequent development, these embryonic tissue layers differentiate to form the specialized tissues and organs of the adult animal (Figure 45-15b).

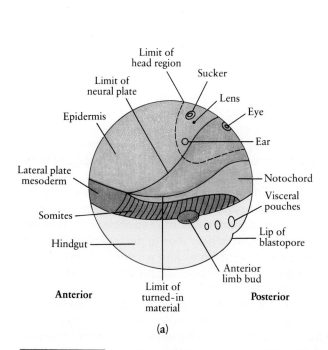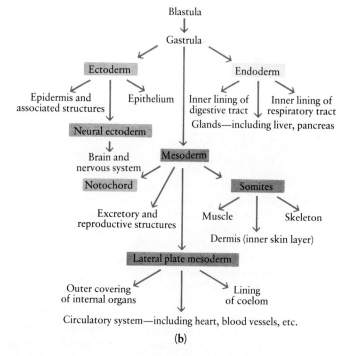

45–15 **(a)** *A fate map of the amphibian, showing the groups of cells on the surface of the blastula that, in the course of gastrulation and subsequent development, give rise to the tissues of the mature embryo (the tadpole). A comparison of the location of particular groups of cells on the blastula with their ultimate location within the tadpole makes clear the extent of the cellular reorganization that occurs during amphibian development.* **(b)** *The tissue layers that form as a result of gastrulation later give rise to the specialized cells and tissues of the adult animal. This pattern of differentiation is characteristic of all vertebrates, ourselves included.*

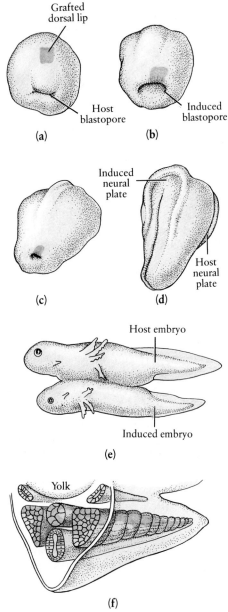

45-16 *Experiment showing the activity of the organizer.* (a) *The dorsal lip of the blastopore from one amphibian embryo is grafted onto another embryo.* (b) *The grafted dorsal lip induces the formation of a second blastopore, through which* (c) *the grafted tissue moves into the interior of the embryo at gastrulation. Note that in* (b) *and* (c) *the embryo is viewed from a different angle than in* (a), *with the result that the host blastopore is not visible. As development proceeds, two neural plates are formed* (d), *and a double embryo is produced* (e). *In* (f) *you see the structure of the secondary embryo under the yolk of the host embryo. The dark brown cells are derived from the graft, and the light tan cells are host material that has been induced to undergo these differentiations.*

The Role of Tissue Interactions

Differentiation, as we saw in Chapter 18, is the result of the selective activation and inactivation of specific genes in the nucleus of a cell. Although molecular geneticists are gradually identifying the mechanisms that, at the level of the DNA molecule, control such activation and inactivation, the linkage of their discoveries to the observed events of development remains one of the major challenges of embryology. However, a variety of experiments, such as the transplantation of the nuclei of tadpole intestinal cells into enucleated eggs (see page 359), have demonstrated that, for some cell types, differentiation does not become irreversible until fairly late in the developmental process. Although the cells have become differentiated, they retain their full developmental potential. For many cell types, however, the developmental potential is gradually limited as gastrulation proceeds and sheets of cells become properly positioned in the embryo. This process, in which the fate of a cell becomes fixed, depends on a progressive series of interactions between different tissue types. In the amphibian, the first tissue to develop a unique character is the dorsal lip of the blastopore. It is the key to the events that follow.

The Organizer

The tissue that lies at the dorsal lip of the blastopore of the amphibian is one of the most thoroughly investigated of all embryonic tissues. Spemann, in a continuation of the studies described earlier (see Figure 45-10), discovered that he could divide the developing blastula in two and sometimes obtain two normal embryos. Once formation of the blastula was complete and gastrulation had begun, however, he found that he could obtain only one normal embryo from the operation. The embryo that was normal always developed from the half that contained the dorsal lip of the blastopore. (This was actually an extension of his previous experiments on the gray crescent; as you will recall, the dorsal lip forms in the blastula at the position once occupied in the egg by the gray crescent.)

Later, in 1924, Hilde Mangold, a student of Spemann, excised the dorsal lip from one salamander embryo and grafted it into the future belly region of a salamander embryo of a different species. When the embryo that had received the graft developed, it was found that a second embryo had formed within its tissues, a sort of Siamese twin to the first. Because the two salamanders differed in color, it was possible to distinguish the tissues of the host embryo from those of the transplant. The transplanted dorsal lip had given rise to chordamesoderm and a notochord, as it would have had it not been removed in the first place. This in itself was noteworthy because previous experiments had shown that other portions of an embryo of similar age, if transplanted, would develop strictly according to the site to which they were transplanted. Even more remarkable, however, was the fact that the rest of the second embryo was made up of tissues of the host. As countless repetitions of this experiment have shown, the transplanted dorsal lip has the effect of organizing tissues of the host into a second embryo (Figure 45-16). This is how the dorsal lip of the blastopore (the prospective chordamesoderm) came to be called the **organizer.**

Despite years of study, we still do not know the nature of the organizer or how it exerts its effect. However, it is known that in chordate embryos of every species examined, the chordamesoderm serves this same function of primary organizer, inducing differentiation of the overlying ectoderm and setting in motion the other events of development.

Induction and Inducers

The action of Spemann's organizer is just one of many tissue interactions that take place in development. This general process is known as **embryonic induction.**

Embryonic induction occurs when two different types of tissue come in contact with one another in the course of development and one tissue induces the other to differentiate.

Numerous experiments have shown that induction usually takes place, even when direct contact is prevented between the two tissues, if chemical exchanges are allowed, as by placing a porous filter between them. If chemical exchanges are prevented, as by a piece of metal foil, induction does not occur. Although the chemical substance or substances exchanged have not yet been identified, they are apparently of a very general nature. Experiments have shown that the capacity of the organizer to induce neural plate formation is not species-specific. One can, for example, provoke the formation of a secondary neural tube in a chick embryo by means of a graft of tissue from a rabbit embryo. Moreover, a number of chemicals, some of which never occur in embryos, will induce neural plate formation in embryonic ectoderm. It is not, then, that the inducing substance endows the ectoderm with the capacity to form nerve tissue; rather, it merely evokes a potential that is already present in the responding tissue and initiates a new pattern of gene activity.

DEVELOPMENT OF THE CHICK

The hen's egg (Figure 45–17), familiar to us all, is different in many respects from the eggs previously examined. First, and most obvious, it is surrounded by a shell, which permits it to develop on land. Second, related to its terrestrial development, it is surrounded by a system of membranes. (Such eggs, as noted on page 596, are called amniote eggs.) Third, it contains a large amount of yolk. This large quantity of stored food makes possible a longer period of development before emergence of the immature animal. As a result, the chick, although still a juvenile at the time of hatching, is much farther along in development than the pluteus larva or the tadpole.

The yolk of the hen's egg, like that of the eggs of other birds and also of reptiles, is so large and dense that cleavage does not involve most of the egg mass. The only part of the fertilized egg that cleaves is a thin layer of cytoplasm that sits like a cap on top of the yolk and contains the nucleus. Cleavage of this thin layer produces a lozenge-shaped blastula known as a **blastodisc** (Figure 45–18). Birds, reptiles, and monotreme mammals develop from a blastodisc.

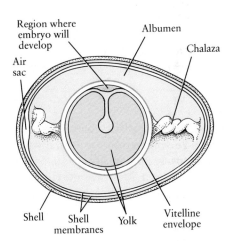

45–17 *Anatomy of a chicken egg. At the time of its release from the ovary in ovulation, the egg consists of the vitelline envelope and its contents. Fertilization, if it occurs, takes place in the upper end of the oviduct. As the egg moves through the oviduct, layers of albumen (egg white) synthesized by and released from cells of the oviduct are added to it. The first of these, the chalaza, is the ropelike structure often observed when an egg is opened. The chalaza suspends the egg cell and, much like a twisted rubber band, rotates it, ensuring that the embryo, developing on one side of the fertilized egg, is always upward, close to the warmth of the mother hen. Two shell membranes and the shell itself are added before the egg is laid.*

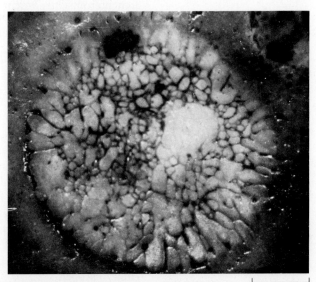

45–18 *The blastodisc of a chick embryo.*

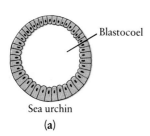
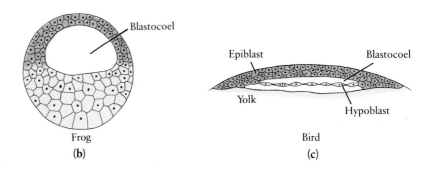

45–19 *Blastulas of* (a) *a sea urchin,* (b) *a frog, and* (c) *a bird. In the sea urchin and the frog, the blastula is a hollow sphere of cells. In the bird, it takes the form of a blastodisc—a flattened ball of cells—that sits on the surface of the yolk. The blastocoel is located between the epiblast and the hypoblast.*

If you break open a fertilized chick egg when it is first laid, you can see the blastodisc as a white mass about 2 millimeters in diameter on top of the yolk. Microscopic examination shows that this mass is made up of many cells (close to 100,000) and that the cells are in two layers: an upper layer, known as the **epiblast,** and a lower layer, the **hypoblast.** The space between them is the blastocoel (Figure 45–19).

If the egg is left intact and development continues, within a short period of time a visible line, known as the **primitive streak,** appears on the surface of the blastodisc. The primitive streak is, in effect, an elongated blastopore. At the beginning of gastrulation (Figure 45–20a), cells of the epiblast that will give rise to the mesoderm and endoderm of the embryo sink inward at the primitive streak and begin to spread out between the epiblast and the hypoblast. (Although some of the cells of the hypoblast contribute to the endoderm of the embryo proper, most of the hypoblast cells are subsequently involved in the formation of membranes surrounding the developing embryo.) As gastrulation continues (Figure 45–20b), the developing mesoderm spreads outward, forming a continuous middle layer. The notochord begins to form from a strip of chordamesoderm cells running from anterior to posterior in the middle of the mesodermal layer, and the coelom forms in the lateral plate mesoderm. Neural ectoderm begins to differentiate above the chordamesoderm. As the cells giving rise to the mesoderm move from the epiblast to the interior, they pull the future epidermal ectoderm and neural ectoderm from the sides of the blastodisc, forming a complete ectodermal covering over the surface of the embryo. As the process proceeds (Figure 45–20c), the embryo becomes tubular and lifts off the surface. At this stage, the similarity of the chick embryo to the frog embryo becomes apparent; both consist of concentric circles of ectoderm, mesoderm, and endoderm surrounding the archenteron. The notochord has differentiated, the somites have formed, and the sides of the neural plate are elevated. Gradually, the ridges of the neural plate come together, fusing to form the hollow, dorsal neural tube (Figure 45–20d). The basic body plan of the vertebrate-to-be is established.

Extraembryonic Membranes of the Chick

Eggs develop in water. In the amniote egg, this water is contained within a set of membranes formed from the tissues of the embryo, the **extraembryonic membranes.** These membranes begin as extensions of the blastodisc, and each is formed from a combination of two of the primary tissue types. The endoderm of the extraembryonic membranes is derived from cells of the hypoblast, and the mesoderm and ectoderm from cells originally located in the epiblast.

As Figure 45–21a (page 934) shows, the yolk gradually becomes surrounded by a membrane of mesoderm and endoderm, known as the **yolk sac.** The function of this membrane is nutritive; its endodermal cells digest the yolk, and blood vessels formed in its mesodermal component carry food molecules from the yolk into the embryo proper.

45–20 *Gastrulation and formation of the neural tube in the chick.* **(a)** *At the beginning of gastrulation, cells that will give rise to the mesoderm and endoderm of the embryo begin their migration through the primitive streak.* **(b)** *As gastrulation continues, the notochord begins to develop, and the lateral plate mesoderm splits into two layers to form the coelom. Neural ectoderm and epidermal ectoderm begin to differentiate.* **(c)** *The embryo becomes tubular and lifts off the surface of the yolk. The notochord has differentiated, the somites have formed, and the sides of the neural plate have begun to elevate.* **(d)** *The ridges of the neural plate come together, fusing to form the hollow, dorsal neural tube. As we shall see shortly, the portions of the three primary embryonic tissues located below the tubular embryo also undergo a series of foldings, producing a set of membranes that enclose and protect the embryo as it continues its development.*

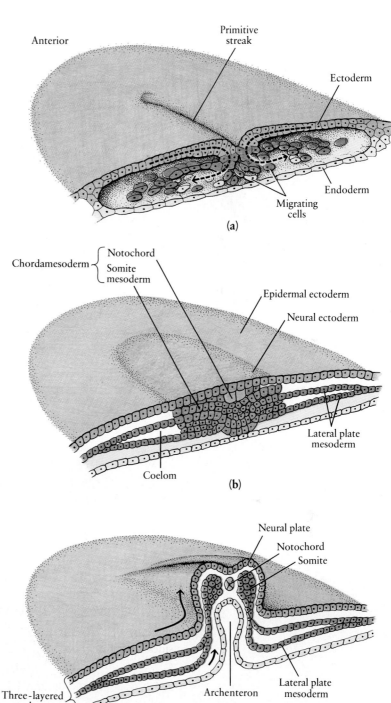

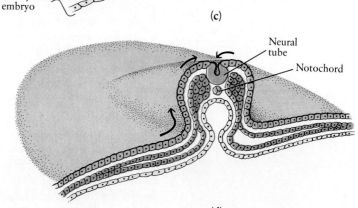

The four extraembryonic membranes:
Yolk sac (endoderm and mesoderm)
Allantois (endoderm and mesoderm)
Amnion (mesoderm and ectoderm)
Chorion (mesoderm and ectoderm)

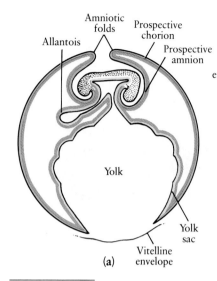
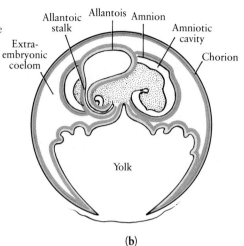
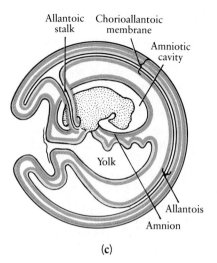

45-21 *Development of the extraembryonic membranes of the chick. As the embryo becomes tubular and lifts from the yolk (see Figure 45-20c), a series of membranes begin to form from the tissues at its base.* **(a)** *One membrane, the yolk sac, grows around and almost completely encloses the yolk. A second, the allantois, arises as an outgrowth of the rear of the gut. The third and fourth are elevated over the embryo by a folding process during which the membrane is doubled. When the folds fuse, two separate membranes are formed* **(b)**. *The inner one is the amnion, and the outer one is the chorion. The chorion eventually fuses with the allantois to form the chorioallantoic membrane* **(c)**, *which, in the later stages of development, encloses embryo, yolk, and all the other structures. Note that each membrane is composed of two primary embryonic tissues.*

When the developing embryo lifts from the surface of the blastodisc (see Figure 45-20c), the ectoderm and mesoderm located at its base begin to fold upward, around and over the tubular embryo. These folds of tissue, the amniotic folds, eventually fuse with each other over the embryo, enclosing it in a fluid-filled cavity, the **amniotic cavity** (Figure 45-21b). The saline fluid in the cavity serves to buffer the embryo against mechanical and thermal shocks. Fusion of the amniotic folds creates two membranes, each composed of a layer of ectoderm and mesoderm, separated by the extraembryonic coelom. The inner membrane is the **amnion,** and the outer membrane is the **chorion.**

As development proceeds and the embryo becomes increasingly demarcated from the yolk mass (although always attached to it by the yolk sac stalk), a ventral and posterior pouch is formed from the primitive hindgut. This pouch is the **allantois,** and, as you can see in Figure 45-21b, its wall is composed of endoderm and mesoderm. At first, the function of the allantois is excretion. Since bird and reptile eggs are closed off from their environment by the shell, there is no way to dispose of the products of nitrogen metabolism, most of which are toxic. Evolution has provided a dual answer to this problem. First, the nitrogenous wastes are converted to uric acid, which, as we saw in Chapter 37, is insoluble and of low toxicity. The uric acid is then voided from the embryo into the allantois, which essentially serves as an embryonic garbage bag. When a chick hatches, the accumulated uric acid can be found adhering to the abandoned shell.

As development proceeds further, the allantois expands in volume and pushes out into the extraembryonic coelom, gradually enveloping the embryo proper as it does so. Eventually, as Figure 45-21c shows, the allantois effectively obliterates the extraembryonic coelom, and its walls fuse with the chorion. Since the allantois is plentifully supplied with blood vessels, the **chorioallantoic membrane,** formed by the fusion of the allantoic wall with the chorion, acts as an efficient respiratory membrane for the embryo during its later development.

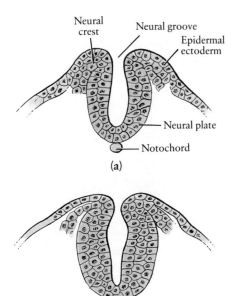
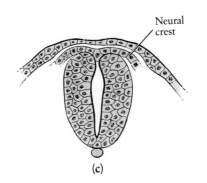
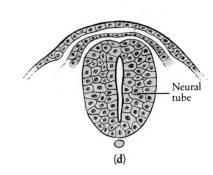

45-22 *During the formation of the neural tube, some of the neural ectoderm cells at the crests of the neural folds are pinched out as the folds come together. These cells form a variety of structures in the peripheral nervous system. Eventually, some of them will migrate down toward the notochord and aggregate into ganglia. There they develop into sensory neurons, sending fibers up to connect with the dorsal part of the spinal cord and out into the surrounding tissues.*

Organogenesis: The Formation of Organ Systems

While the extraembryonic membranes are forming, the development of the embryo itself continues. These later stages of development, following cleavage and gastrulation, are generally termed **organogenesis,** the formation of the organ systems. Organogenesis begins with the inductive interaction between ectoderm and underlying chordamesoderm. Each of the three primary tissues formed during gastrulation then proceeds to undergo growth, differentiation, and morphogenesis. This process is essentially the same in all vertebrates.

Differentiation of the Ectoderm

The neural tube stretches out along the dorsal surface, growing and becoming longer and thinner as the embryo increases in size. The cells in the neural tube at first appear to be all alike, but some of them, the future motor neurons, begin to extend long processes that grow out beyond the neural tube and invade the peripheral organs and tissues.

When the neural tube is first forming, some of the ectodermal cells at the crests of the neural folds are left behind in the surrounding tissue, as shown in Figure 45-1 (page 919). Some of these cells from the neural crest, the future sensory neurons, migrate into positions around the neural tube (Figure 45-22) and form connections between the dorsal part of the tube and the surrounding tissues. Others become Schwann cells, which surround and insulate the nerve fibers of the peripheral nervous system, some become the pigment cells (melanocytes) found at the base of the epidermis, and some form the adrenal medulla. Meanwhile, mesodermal cells have begun to migrate toward the neural tube and underlying notochord and to aggregate around them. Here they differentiate into cartilage and give rise to the hollow vertebral column. If a portion of neural tube is transplanted to another region of the mesoderm, mesodermal cells will similarly aggregate around it as a result of tissue interaction.

As we noted in Chapter 43, the brain begins as three bulges in the foremost part of the neural tube (Figure 45-23). Almost immediately, two saclike protrusions, the optic vesicles, appear on the sides of the anterior region of the forming brain. These spherical vesicles enlarge until they come into contact with the epidermal ectoderm, while still remaining connected to the neural tube by the narrowing optic stalk. Within this stalk, the optic nerve will develop.

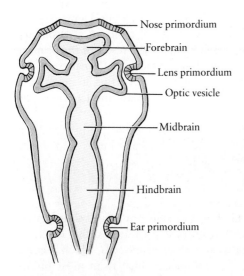

45-23 *The development of the brain and associated sense organs. At the anterior end of the neural tube, local swellings produce three distinct bulges—the forebrain, the midbrain, and the hindbrain. The forebrain then bulges laterally, and two saclike optic vesicles appear. At the same time, the epidermal ectoderm folds inward to meet the optic vesicles. The earliest stages—primordia—of ears and nostrils also appear first as infoldings of the epidermal ectoderm.*

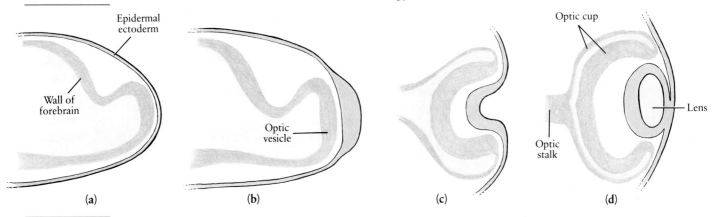

45-24 *Eyes develop as a result of interactions between the epidermal ectoderm and the optic vesicles. When the optic vesicle reaches the epidermal ectoderm, the epidermal ectoderm begins to thicken and differentiate. The optic vesicle flattens out and invaginates, becoming the double-walled optic cup. The invaginated wall will become the retina; the outer, thinner wall develops into the pigment layer. The rim of the optic cup will later form the edge of the pupil. The thickened layer of epidermal ectoderm pinches off to become the transparent lens, and the overlying epidermal ectoderm, which also becomes transparent, forms the cornea. The connection with the brain remains as the optic stalk, within which is the optic nerve. Note that the retina is thus a differentiated extension of the brain.*

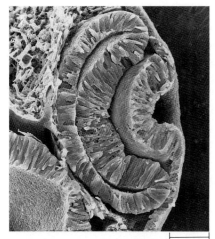

45-25 *The developing eye of a chick embryo, as revealed by the scanning electron microscope. As the cells of the thick, invaginated layer of the optic cup differentiate into rods and cones and bipolar, horizontal, amacrine, and ganglion cells, the multiple connections that characterize the retina (see Figure 42–13, page 869) will become established.*

When the optic vesicle comes into contact with the inner surface of the epidermal ectoderm, the external surface of the vesicle flattens out and pushes inward. The vesicle thus becomes a double-walled cup. The rim of the cup becomes the edge of the pupil. The opening of the cup is large at first, but the rim bends inward and converges, so that the opening of the pupil becomes smaller. At the point where the optic vesicle touches the epidermal ectoderm, the epidermal ectoderm begins to differentiate into a lens, becoming transparent (Figure 45–24).

The differentiation of the lens of the eye is an example of a **secondary induction**. The inducing tissue (the optic vesicle) is itself the result of the primary induction of dorsal ectoderm by dorsal mesoderm. In fact, the formation of the complete eye involves at least six separate inductions, occurring in an orderly sequence, each one linked to the one before, with primary induction by the organizer as the starting point.

Other infoldings and differentiations of the epidermal ectoderm in the region of the head, associated with the presence of the embryonic brain, lead to the elaboration of the organs of olfaction and hearing.

Differentiation of the Mesoderm

The series of somites that came to lie on each side of the notochord shortly after gastrulation (Figure 45–27) now differentiate into three kinds of cells: (1) sclerotome cells, which later form skeletal elements; (2) dermatome cells, which become part of the developing skin; and (3) myotome cells, which form most of the musculature. In aquatic vertebrates, this segmental pattern, with its simple double series of back muscles, is retained in adulthood; it is well suited to the side-to-side motion of the body in swimming and crawling. In terrestrial vertebrates, however, this initial segmental arrangement is almost obliterated by modifications and additions of the muscles associated with locomotion on land.

Lateral plate mesoderm adjacent to the somites forms the kidneys, the gonads, and the ducts of the excretory and reproductive systems. Earlier in development, the lateral plate mesoderm in the ventral portion of the embryo split into two sheets, creating the coelom (see Figure 45–20b). One of these sheets forms the lining of the thoracic and abdominal cavities, and the other becomes the outer layer of the internal organs.

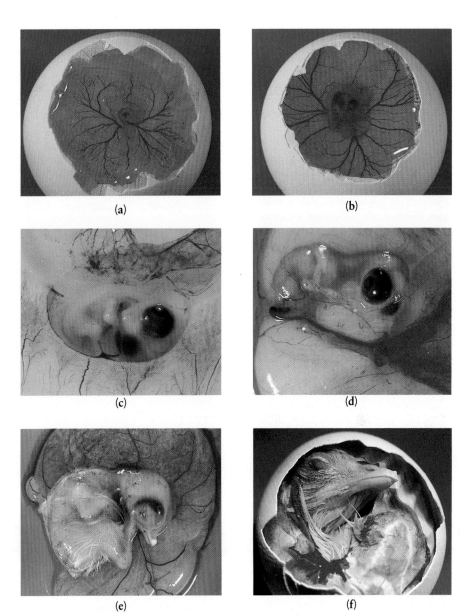

45–26 *Major stages of organogenesis and subsequent development in the chick embryo. (a) At 4 days, large blood vessels have grown out from the embryo into the yolk sac. Note the heart, above the junction of the blood vessels with the embryo. (b) At 5 days, the yolk sac has grown further around the yolk. Its endodermal cells secrete lipase (page 722), an enzyme that digests fat in the yolk. Nutrient molecules are carried to the embryo via the mesodermal blood vessels. (c) By 8 days, the large, pigmented eyes are clearly visible. (d) At 11 days, the limbs are developing. The brain is visible through the transparent skull. (e) By 14 days, feathers are beginning to appear. (f) By 21 days, just prior to hatching, a white knob is fully formed on the end of the beak; this is the egg tooth, which is used in pecking open the shell.*

45–27 *Somites of a chick embryo, flanking the neural tube. The formation of somites from the chordamesoderm on either side of the notochord proceeds from the anterior of the embryo (top of micrograph). Three pairs of somites are fully formed, and a fourth pair is beginning to separate from the posterior chordamesoderm. The wormlike extensions from the upper edge of the neural tube are groups of neural crest cells (Figure 45–22) beginning their downward migration.*

Differentiation of the Endoderm

The endoderm differentiates into tissues of the respiratory and digestive tracts and a number of related organs. Early in development, endodermal pouches develop at the anterior end of the archenteron. They push laterally until they meet the epidermal ectoderm, which folds inward to meet them. This produces a series of grooves on the surface of the embryo. In aquatic vertebrates, endoderm and ectoderm fuse and a perforation forms, around which gill filaments develop. In terrestrial vertebrates, including humans, portions of these endodermal pouches develop into the auditory tubes (see Figure 42–6, page 864), tonsils, parathyroid glands, and the thymus gland. Posterior to this region, the lungs develop as similar outpocketings, branching into two sacs. More posterior still, outpocketings from the primitive gut begin to differentiate into liver, gallbladder, and pancreas.

No organ system is derived from only one type of tissue. For example, the lining of the intestine is of endodermal origin; these lining cells secrete the digestive juices and absorb the digested materials. However, as we saw in Chapter 34, the functional structure of the intestine also includes muscles, connective tissue, blood vessels, nerves, and an outer wrapping. These are made up of tissues derived from mesoderm and ectoderm.

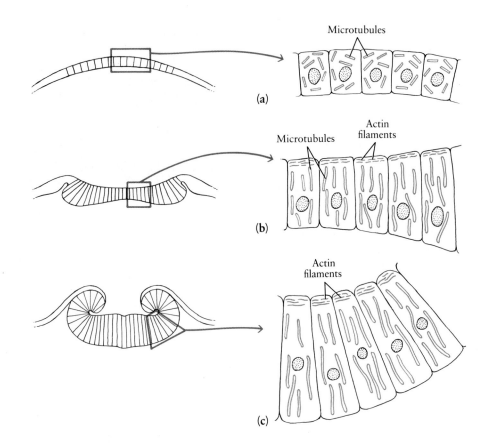

45–28 *Formation of the vertebrate neural tube depends on changes in the shape of individual cells brought about by movements of elements of the cytoskeleton.* (a) *Newly differentiated cells of neural ectoderm are cuboidal, and the microtubules within the cells appear to be randomly arranged.* (b) *As formation of the neural plate proceeds, the microtubules become arranged parallel to the dorsal-ventral axis of the embryo. As the microtubules lengthen through the addition of tubulin dimers (page 78), the cells become columnar. Just below the upper surface of the membrane of each cell is a band of actin filaments, running parallel to the left-right axis of the embryo. When this band contracts* (c), *the cells become wedge-shaped, forcing the entire sheet of cells to curve upward. A similar process is involved in the development of other curved or tubular structures that consist of sheets of cells.*

Morphogenesis: The Shaping of Body Form

Each part of the body has a characteristic shape and structure. For example, the spinal cord is basically a hollow tube, the liver is composed of distinctive lobules, the salivary glands consist of groups of secretory cells clustered around ducts, and the lungs consist of microscopic air spaces lined by extremely thin sheets of cells. Only a few cellular processes, repeated over and over in various permutations and combinations, appear to be responsible for shaping the various structures of the body. These processes include (1) increases or decreases in the rates of cell growth and division, (2) changes in the adhesion of cells to neighboring cells, (3) the deposition of extracellular materials, and (4) changes in cell shape brought about by extension or contraction. For example, formation of the neural tube from the flattened neural plate is accomplished by changes in the shape of the individual cells. As shown in Figure 45–28, cells in the neural plate first become elongated, a change that is associated with the alignment of microtubules parallel to the dorsal-ventral axis of the embryo. Following elongation, a layer of actin filaments, located just below the cell membrane in the dorsal portion of each cell, contracts. This causes the cells to become wedge-shaped, which, in turn, causes the edges of the neural plate to begin curving upward.

Such cellular processes give rise not only to the internal structures that are similar from one vertebrate to another but also to the distinctive structures that characterize different groups of vertebrates. Among the most striking features of terrestrial vertebrates are four limbs. As we saw in Figure 20–7 (page 414), the limbs of reptiles, birds, and mammals all have the same basic structure, and yet there are obvious—and significant—differences between, for example, the forelimbs of a bird (wings) and the forelimbs of a human (arms). Similar differences are apparent between the forelimbs of a bird and its hindlimbs (legs). The developmental process that gives rise to these differences is known as **pattern formation**. It has been studied principally through experimental manipulations of the developing chick wing.

The first visible sign of wing development is the appearance of a wing bud, which consists of a central mass of mesodermal cells covered by a thin layer of

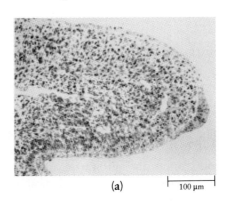 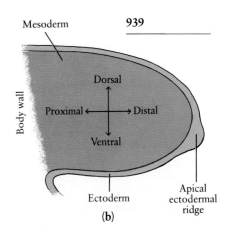

45–29 (a) *The wing bud of a chick embryo, viewed from the posterior. This micrograph was taken three and a half days after the wing bud began its development. Note the thickened region, the apical ectodermal ridge, at its tip.* (b) *The tissues of the wing bud consist of an outer layer of ectoderm, from which the skin and feathers will form, and an inner layer of mesoderm, from which all of the bones, cartilage, tendons, and muscles of the wing will develop.*

ectoderm. As the wing bud begins to grow outward, a ridge of specialized tissue, the apical ectodermal ridge, forms at its tip (Figure 45–29). If this ridge is removed, elongation will not occur. Immediately beneath the apical ectodermal ridge is a region of actively dividing mesodermal cells, known as the progress zone, which moves outward with the apical ridge as the wing bud elongates. The progress zone produces the cells from which the structures of the completed wing are formed. As the wing bud elongates and the progress zone moves outward, newly divided cells are left behind and begin to differentiate into the characteristic structures of the wing from its base at the body wall (the proximal end) to its tip (the distal end).

Numerous experiments have been performed in which the apical ectodermal ridge, the mesodermal cells of the progress zone, or both have been grafted between wing buds at different stages of development. These experiments have shown that the length of time that cells spend near the apical ectodermal ridge before they are left behind is a critical factor in determining the structures into which they differentiate. The cells that are left behind first spend less time under the influence of substances secreted from the apical ectodermal ridge, and they become the proximal structures of the wing. Those that are left behind later spend more time in the region of the apical ectodermal ridge, and they become the distal structures of the wing.

A wing is characterized not only by a specific series of structures from its proximal end to its distal end but also by a specific series of structures along the axis from anterior to posterior (Figure 45–30). A variety of grafting experiments have shown that differentiation along this axis is controlled by chemical substances produced by cells at the posterior margin of the wing. For example, if a piece of tissue is transplanted from the posterior margin of one wing bud to the anterior margin of another, the wing that develops is a kind of Siamese twin in which the bones are duplicated in a mirror image of the normal wing structure (see Figure 45–31 on the next page).

45–30 *A dorsal view of the wing of a chick embryo, 10 days after the wing bud first began its development. In the course of the evolution of the bird wing, two of the five digits of the basic vertebrate "hand" have been lost—the digits corresponding to the thumb and little finger. Thus, the digits of the wing are numbered, from anterior (top of micrograph) to posterior (bottom of micrograph), as 2, 3, and 4.*

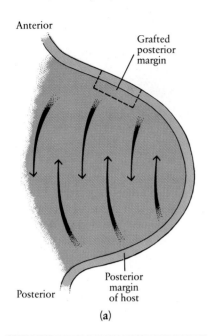

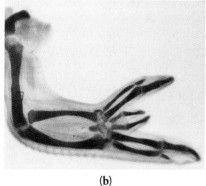

45–31 *Development of the correct structures on the anterior-to-posterior axis of the chick wing depends on a concentration gradient of a chemical substance, now known to be retinoic acid, released from cells at the posterior margin of the wing bud.* **(a)** *If cells from the posterior margin of one wing bud are grafted onto the anterior margin of another wing bud, and the posterior margin of the host wing bud is left intact, a dual gradient is established.* **(b)** *The result is a wing with six digits, arranged, from anterior to posterior, in the sequence 4, 3, 2, 2, 3, 4.*

The substance released from the posterior margin of the wing bud has recently been identified as retinoic acid, a member of the same chemical family that includes the carotenoids, vitamin A, and the retinal of the vertebrate eye. Retinoic acid diffuses from the posterior margin toward the anterior margin of the wing bud, creating a concentration gradient from high (posterior margin) to low (anterior margin). The receptors for retinoic acid have now been located on the surface of cells of the developing wing bud. Structurally the retinoic acid receptors are very similar to the receptors for steroid hormones and thyroid hormone, and it appears that their mechanism of action is also similar (see page 836). The particular sequence of cellular events that is initiated by the binding of retinoic acid to its receptors is thought to be influenced by the number of receptors activated, which is, of course, related to the location of the cell within the concentration gradient.

The identification of retinoic acid as a signalling molecule in wing development and the isolation and characterization of its receptor suggest that the integration and control of development are, in principle, similar to the integration and control of other physiological processes. A sense of high adventure now prevails as developmental biologists seek to identify other substances involved in the regulation of development, the mechanisms by which these substances turn genes on and off, the genes themselves, and the proteins for which they code.

DEVELOPMENT OF THE HUMAN EMBRYO

The basic patterns of development are remarkably similar throughout the animal kingdom, and particularly among vertebrates. Thus, experimental studies of the development of sea urchins, frogs, chicks, and mammals such as mice have provided a solid foundation for our understanding of human development—an area of biology in which experimental manipulation is, throughout the world, strictly prohibited.

Human oocytes, as we saw in the preceding chapter, mature in the ovary and are released at approximately 28-day intervals. Fertilization usually takes place in an oviduct. As in other species, fertilization results in (1) changes in the outer surface of the egg that prevent entry of other sperm cells, (2) metabolic activation of the egg, (3) introduction of the genetic material of the father, and (4) cleavage. After fertilization, the egg continues its passage down the oviduct, where the first cell divisions take place (Figure 45–32). At about 36 hours after fertilization, the fertilized egg divides to form two cells; at 60 hours, the two cells divide to form four cells. At three days, the four cells divide to form eight. In this early stage, all of the cells are of equal size, as they are in the sea urchin. During this period, the embryo is entirely on its own, following its own genetic program and supplying its own metabolic requirements. All it needs is a suitable fluid environment containing progesterone.

By about five days after fertilization, the blastula consists of some 120 cells, and a fluid-filled blastocoel has begun to form within it. The structure of the mammalian blastula, which is known as a **blastocyst,** is rather different from that of the blastulas we have examined previously. A cross section of the blastocyst looks somewhat like a ring with the stone on the inside. The inner cell mass (the stone) is a ball of cells at one pole of the blastula that will develop into the embryo proper. The ring is called the **trophoblast** (from the Greek word *trophe*, "to nourish"). It is composed of a double layer of cells and completely encloses the developing embryo. The trophoblast is the precursor of the chorion. (In the chick, as you will recall, the chorion develops from mesoderm and ectoderm formed at the base of the embryo during gastrulation. It is as if the mammalian embryo has skipped a step, arriving prepackaged.)

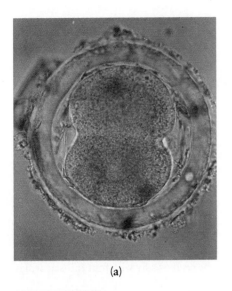 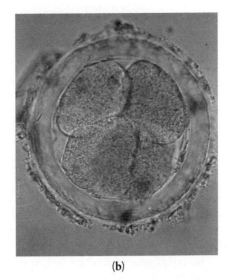 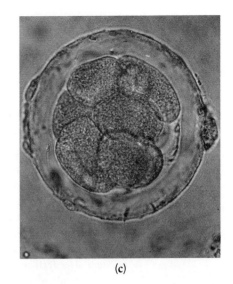

(a)   (b)   (c)

45–32 *A human embryo at (a) the two-cell stage, (b) the four-cell stage, and (c) the eight-cell stage. Although the number of cells has doubled with each cleavage, the volume of the embryo has not increased. Thus, the cells are still easily contained within the fertilization membrane. Sometimes the cells separate at the two-cell stage, resulting in identical twins.*

About three days after the embryo reaches the uterus (about six days after fertilization), the trophoblast makes contact with the tissues of the uterus and releases a protein hormone, chorionic gonadotropin. This hormone protects the pregnancy by stimulating the corpus luteum to continue its production of estrogens and progesterone, thus preventing menstruation (see page 913). Most pregnancy tests involve the detection of chorionic gonadotropin in blood or urine; with the development of monoclonal antibodies (page 802) against human chorionic gonadotropin, these tests have become much more sensitive and reliable.

The trophoblast cells, multiplying rapidly by now, chemically induce changes in the endometrium and invade it. As the embryo penetrates the endometrial tissues—the process known as **implantation** (Figure 45–33)—it becomes surrounded by ruptured blood vessels and the nutrient-filled blood escaping from them. The trophoblast thickens and develops fingerlike projections that invade the uterine lining to develop into the ectodermal component of the chorion.

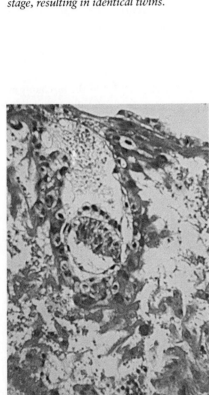 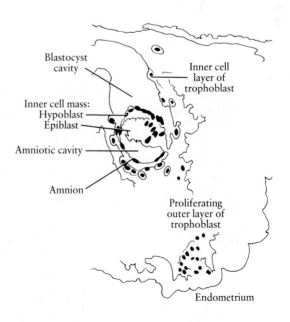

45–33 *Implantation of a human embryo. The tiny embryo invades the lining of the uterus within a week after fertilization. Subsequently, the placenta begins to form. This organ is the site at which oxygen, nutrients, and other materials are exchanged between the mother and embryo; it is also the source of hormones that help to maintain pregnancy. Implantation usually occurs three to four days after the young embryo reaches the uterus. Once implantation has been achieved, the embryo is no longer independent but is parasitic on the mother.*

Genetic Control of Development: The Homeobox

Underlying all of the developmental processes we have examined are, of course, genes—sequences of information encoded by DNA molecules. The cytoplasmic determinants that have such a profound influence on early development are the direct products of maternal genes, which are transcribed during the maturation of the oocyte and translated into protein after fertilization. These cytoplasmic factors, as well as all of the other substances involved in the induction of tissue differentiation and pattern formation, cause changes in the expression of specific genes in the nuclei of the embryonic cells they affect. The first steps in identifying these genes and in determining their functions in development are being made possible by the ubiquitous fruit fly, *Drosophila*.

Drosophila, like all insects, exhibits a striking segmentation, both in the larval stages and in the adult. Moreover, the adult is characterized by a series of complex appendages, including mouthparts, antennae, wings, and legs, precisely located on particular segments. Two distinct sets of genes controlling pattern formation in *Drosophila* have now been identified. One set, known as segmentation genes, controls the number and sequence of segments in the larva and, subsequently, in the three regions (head, thorax, and abdomen) of the adult fly. Mutations in these genes can result in larvae and flies in which certain segments are missing, or, alternatively, are duplicated. Another set, known as homeotic genes, controls the identity of segments; that is, they control the way in which a given segment develops and the appendages that it bears. Mutations of homeotic genes can result in bizarre disruptions of the normal developmental pattern. For example, mutations of the genes of the antennapedia complex, which controls the development of antennae on the head and of legs on the three segments of the thorax, can result in the development of legs on the head—where the antennae should have developed. Another group of homeotic genes, the bithorax complex, is involved in the control of the development of both legs and wings on the thoracic segments. In a normal fruit fly, the first thoracic segment bears a pair of legs, the second segment bears a pair of legs and a pair of wings, and the third segment bears a pair of legs and a pair of greatly reduced wings, known as halteres. Although the halteres play no role in the propulsion required for flight, they are critical in enabling the fly to maintain its balance while airborne. Mutations in two of the loci of the bithorax complex produce flies in which the third thoracic segment develops as a second thoracic segment; the result is a fly with four full-size wings and no halteres.

At present, we know very little about the functions of the proteins coded by the homeotic genes. Some of these proteins undoubtedly regulate the expression of other genes that have more specific roles in development. Others may be involved in the regulation of basic cellular activities, such as cell growth and division. For example, the gene known as *engrailed*, which is involved in the division of each segment into anterior and posterior compartments, appears to act by limiting the growth of populations of cells. In *engrailed* mutants, cells of the posterior part of each segment grow into the anterior part, which normally they would not penetrate.

Despite our lack of knowledge of the mechanisms by which homeotic genes exert their effects, many of the genes themselves have now been located on the *Drosophila* chromosomes and have been cloned and sequenced. In 1983, it was discovered that more than a dozen of the *Drosophila* homeotic genes contain a common DNA sequence of 180 nucleotides. This sequence is known as the homeobox. The nucleotide sequence of the homeobox has now been identified in the DNA of a variety of animals, ranging from worms to humans, all of which exhibit a segmental pattern at some stage of their development. The polypeptide strand dictated by the nucleotide sequence of the homeobox contains many basic amino acids, suggesting that it may function as a regulatory molecule that binds to DNA (which, as you will recall, is acidic), altering the course of gene expression. These discoveries suggest that even though the adult forms of different animals vary greatly, certain "master" genes play a critical role in the development of body organization and pattern throughout the animal kingdom.

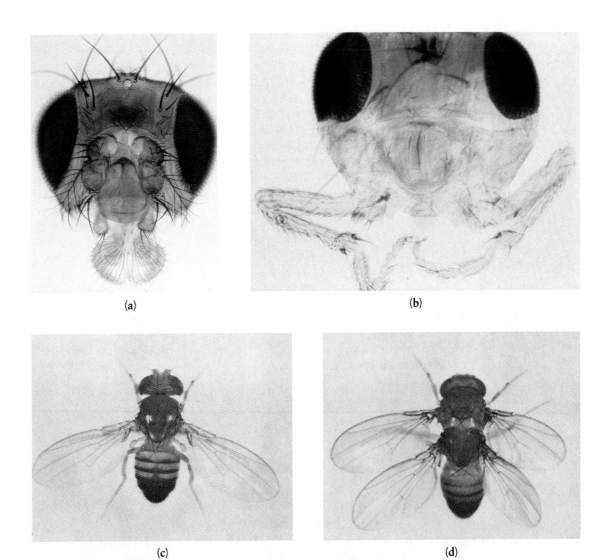

The effects of mutations in two groups of homeotic genes of Drosophila. *(a) Head of a normal fly, and (b) head of a mutant fly in which the antennae are replaced by legs. (c) A normal fly with one pair of wings and one pair of halteres, and (d) a mutant fly with two pairs of wings and no halteres.*

Extraembryonic Membranes

As the embryo becomes implanted, the extraembryonic membranes begin to develop. These have interesting similarities to and differences from the avian-reptilian membranes discussed previously. In the first place, the yolk sac, which develops first, has no yolk. This lack of yolk is a secondary evolutionary development. The monotremes produce eggs with yolks. The marsupials also produce eggs with yolks, but the yolks are discarded at the first cleavage. The placental mammals produce eggs with no yolks; instead there is a prominent cavity where the yolk "used to be." However, cleavage and the cellular migrations of gastrulation proceed as if the yolk were still there.

In humans and other mammals, the allantois buds off from the hindgut and eventually forms the primitive urinary bladder. Wastes are not segregated as uric acid and stored, as they are in the chick, but are transported as urea and ammonia to the maternal bloodstream. In the course of development, the endodermal component of the allantois becomes greatly reduced, but the allantoic mesoderm spreads out; it is the source of the major blood vessels on the embryonic side of the placenta. The allantoic stalk eventually becomes the umbilical cord, the embryo's principal connection with the uterine tissues. The blood vessels of the allantois transport oxygen and nutrients that have diffused in from the mother's circulatory system and carry off carbon dioxide and other wastes.

The third extraembryonic membrane is the amnion. The space between the amnion and the embryo is the amniotic cavity, and, as in the chick, it is filled with a saline fluid. The fourth membrane is the chorion, which is a combination of ectoderm from the trophoblast and mesoderm grown out from the embryo itself. By about the fourteenth day, chorionic villi begin to form: the formation of the villi represents the beginning of the mature placenta.

In amniocentesis (page 387), one of the two prenatal genetic testing procedures in current use, fluid samples are taken from the amniotic cavity. These samples contain cells that have been sloughed off by the embryo. The major drawback of amniocentesis is that it cannot be performed until the sixteenth week of pregnancy. A newer procedure, chorionic villus biopsy, can be performed as early as the eighth or ninth week of pregnancy. In this procedure, a small sample of tissue is removed from the chorion, which is, of course, genetically identical to the embryo itself.

The Placenta

By the end of the third week after conception, the placenta covers 20 percent of the uterus. It is a disc-shaped mass of spongy tissue through which all exchanges between mother and embryo take place. The placenta is formed as a result of the interactions of a maternal tissue, the endometrium, with the extraembryonic chorion, and has a rich blood supply from both. However, the embryonic and maternal circulatory systems are not directly connected (Figure 45–34), so maternal and embryonic blood cells do not mix. Molecules, including food and oxygen, diffuse from the maternal bloodstream through the placental tissue and into the blood vessels that carry them into the embryo. Similarly, carbon dioxide and other waste products from the embryo are picked up from the placenta by the maternal bloodstream and carried away for disposal through the mother's lungs and kidneys. The mother's blood volume begins to increase in response to these extra demands. Her appetite increases and the absorption of certain nutrients, such as calcium and iron, is increased.

From this point in development until birth, the embryo remains securely attached to the placenta by the umbilical cord. This permits the embryo to float freely in its sac of amniotic fluid.

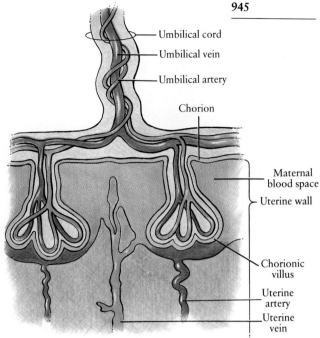

45–34 *From the placenta, numerous fingerlike chorionic villi project into the maternal blood space in the wall of the uterus. This space is kept charged with blood from branches of the uterine artery. Across the thin barrier separating fetal from maternal blood, exchange of materials takes place: soluble food substances, oxygen, water, and salts pass into the umbilical vein from the mother's blood; carbon dioxide and nitrogenous waste, brought to the placenta in the umbilical arteries, pass into the mother's blood. The placenta is thus the excretory organ of the embryo as well as its respiratory surface and its source of nourishment.*

The chorionic component of the placenta is, from the perspective of the mother's immune system, a graft of foreign tissue. As such, it has the potential to evoke a cell-mediated immune response (page 801), and this sometimes happens, leading to miscarriage. In most pregnancies, however, there is no immune response to the embryonic tissues. A selective immune suppression occurs that allows the embryo to develop unharmed while full protection of both mother and embryo from infectious agents is maintained. The mechanism by which this is accomplished is not yet understood.

As the placenta matures, it begins producing large amounts of estrogens and progesterone (Figure 45–35). By the end of the third month of pregnancy, the placenta has completely replaced the corpus luteum as a source of estrogens and progesterone. Chorionic gonadotropin is no longer produced by the chorion, and the corpus luteum degenerates. It is hypothesized that many of the miscarriages that occur in the third month result from the degeneration of the corpus luteum before the placenta is producing adequate levels of hormones to maintain the pregnancy.

The First Trimester

During the second week of its life, the embryo grows to 1.5 millimeters in length, and its major body axis begins to develop. (In this and subsequent measurements, the embryo is measured from crown to rump.) As it elongates, a primitive streak forms, very similar in appearance to the primitive streak of the chick. Cell migrations through the primitive streak establish the three-layered embryo.

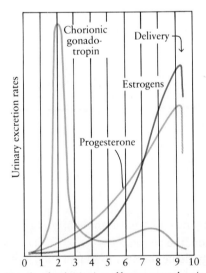

45–35 *Levels of estrogens, progesterone, and chorionic gonadotropin excreted in the urine during pregnancy. Urinary excretion rates reflect the concentrations of these hormones in the blood.*

During the third week, the embryo grows to 2.3 millimeters long, and most of its major organ systems begin to form. The neural groove is the beginning of the central nervous system, which is the first organ system to develop. By 22 days, the very rudimentary heart, still only a tube, begins to flutter and then to pulsate. From this time on, the heart will not stop its 100,000 beats per day until the death of the individual. Soon after, the eyes begin to form. Also, by this time, about 100 cells have been set aside in the yolk sac as germ cells, from which the egg or sperm cells of the individual will eventually develop. These cells begin to crawl, amoeba-like, toward the site at which the gonads will develop.

By the end of the first month, the embryo is 5 millimeters in length and has increased its mass 7,000 times. The neural groove has closed, and the embryo is now C-shaped. At this stage, it can be clearly seen that the tissues lateral to the

 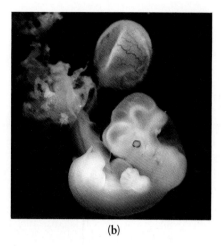 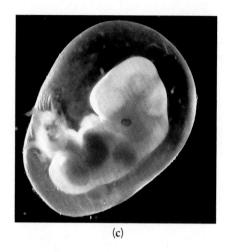

45–36 *Human embryos, early in their development.* **(a)** *In this four-week-old embryo, the neural groove is still open. The protuberances near the top are the limb buds of the future arms; the limb buds of the legs, which develop more slowly, are visible at the bottom. This embryo is 5 millimeters long.* **(b)** *By five weeks, the embryo is 10 millimeters in length. Its developing brain, an eye, hands, and a long tail can be seen. The yolk sac is at the top.* **(c)** *At six weeks, the embryo has grown to 15 millimeters in length. As it floats securely in the amniotic cavity, its heart beats rapidly. The brain continues to grow, and the eyes are more developed. The dark red object in the abdominal region is the liver. Skin folds have formed that will give rise to the outer ears.*

notochord are arranged in paired somites. The embryo typically has 40 pairs of somites, from which muscles, bones, and connective tissues will develop. The heart, even as it beats, develops from a simple set of paired, contracting tubes into a four-chambered vessel.

By 38 days, the germ cells reach their destination, the developing gonads. Although the rudimentary gonads have begun to form by this time, male and female embryos are still anatomically identical. Whether the infant develops as a male or as a female appears to be determined by a single gene situated on the Y chromosome. This gene, which was first identified late in 1987, codes for a DNA-binding regulatory protein. It is thought to be the master switch of sexual development, turning on other genes that lead to the differentiation of the primitive gonads into testes and of the germ cells into spermatogonia. In the absence of this gene, the gonads develop as ovaries and the germ cells become oogonia. More recently, a gene has been located on the X chromosome that, although not identical to the Y chromosome gene, is very similar in its nucleotide sequence. It is not yet known whether this is an inactive pseudogene (see page 374) or an active gene coding for a protein that acts either in conjunction with or in opposition to the product of the gene on the Y chromosome. In addition to the action of a master gene or genes, interactions between the germ cells and the primitive gonads are necessary for organogenesis to proceed. As the testes form, they begin their secretion of androgens, and under the influence of androgens, the external genitalia and other structures become masculinized. If a female embryo is exposed to androgens, it will become similarly masculinized, although, gonadally speaking, it will be female.

During the second month, the embryo increases in mass about 500 times. By the end of this period, it weighs about 1 gram, slightly less than the weight of an aspirin tablet, and is about 3 centimeters long. Despite its small size, it is almost human-looking, and from this time on it is generally referred to as a **fetus**. Its head is still relatively large, because of the early and rapid development of the brain, but the relative size of the head will continue to be reduced throughout gestation (and throughout childhood as well).

Arms, legs, elbows, knees, fingers, and toes are all forming during this time (Figure 45–37). As another reminder of our ancestry, there is a temporary tail. The tail reaches its greatest length in the second month and then, as its growth rate slows, becomes surrounded by other tissues and disappears. Gallbladder and pancreas are present at this stage, and there is clear differentiation of the divisions of the digestive tract. The liver now constitutes about 10 percent of the body of the fetus and is its main blood-forming organ. By the end of the second month, the major steps in organ development are more or less complete. In fact, the rest of development is mostly concerned with growth and the maturation of physiological processes.

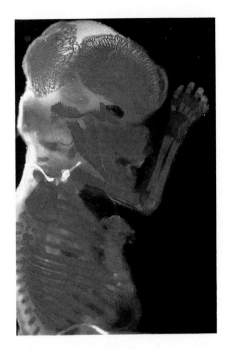

45–37 By the seventh week of embryonic life, the skeleton begins to turn from cartilage to bone, as revealed by differential staining. A gradual layering of membranes, one on top of another, will give rise to the skull, which protects the developing brain.

The first two months are the most sensitive period in human development as far as the possible influence of external factors is concerned. For example, when the arms and legs are mere rudiments (fourth and fifth weeks), a number of substances can upset the normal course of events and result in limb abnormalities. The experience with the presumably safe tranquilizer thalidomide in the early 1960s is a tragic and familiar example. Because of the widespread publicity about the "thalidomide babies," now adults, there was an increase in research into teratogens (substances that cause fetal deformities). A large number of drugs have been found to cause birth defects, and pregnant women and their doctors have become far more cautious about the use of any medication during these critical first months. Similarly, exposure to x-rays at doses that would not affect an adult or even an older fetus may produce permanent abnormalities. Heavy alcohol consumption and smoking also affect normal organogenesis; frequently, alcoholic mothers have miscarriages during the first trimester.

Infections may also affect the development of the embryo. Rubella (German measles) is a very mild disease in children and adults. Yet when contracted by the mother during the fourth through the twelfth weeks of pregnancy, it can have damaging effects on the formation of the heart, the lens of the eye, the inner ear, and the brain, depending on exactly when the infection occurs in relation to embryonic development.

During the third month, the fetus begins to move its arms and kick its legs, and the mother may become aware of its movements. Reflexes, such as the startle reflex and (by the end of the third month) sucking, first appear at this time. Its face becomes expressive; the fetus can squint, frown, or look surprised. Its respiratory organs are fairly well formed by this time but, of course, are not yet functional. The external sex organs begin to develop.

By the end of the third month, the fetus is about 9 centimeters long from the top of its head to its buttocks and weighs about 15 grams (0.5 ounce). It can suck

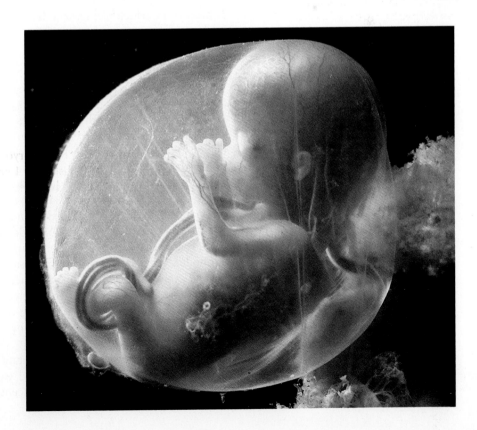

45–38 Although the head of a 12-week-old fetus is disproportionately large, its features are clearly human. Its fingers and toes are fully developed, and outer ears and eyelids have formed. The lids are fused and will remain closed for the next three months. The umbilical cord, connecting the fetus to the placenta, contains a vein and two arteries.

and swallow, and occasionally does swallow some of the fluid that surrounds it in the amniotic sac. The finger, palm, and toe prints are now so well developed that they can be clearly distinguished by ordinary fingerprinting methods. The kidneys and other structures of the urinary system develop rapidly, although waste products are still disposed of through the placenta. By the end of this period—the first trimester of development—all the major organ systems have been laid down.

The Second Trimester

During the fourth month, movements of the fetus become obvious to the mother. Its bony skeleton is forming and can be seen with x-rays. The body is becoming covered with a protective cheesy coating. The four-month-old fetus is about 14 centimeters long and weighs about 115 grams (4 ounces).

By the end of the fifth month, the placenta covers about 50 percent of the uterus. The fetus has grown to almost 20 centimeters and now weighs 250 grams. It has acquired hair on its head, and its body is covered with a fuzzy, soft hair called the lanugo, from the Latin word for "down." Its heart, which beats between 120 and 160 times per minute, can be heard with a stethoscope. The five-month-old fetus is already discarding some of its cells and replacing them with new ones, a process that will continue throughout its lifetime. Nevertheless, a five-month-old fetus cannot yet survive outside of the uterus. The youngest fetus on record to survive was about 23 weeks old and required continuous assistance in breathing, taking food, and maintaining its body temperature.

During the sixth month, the fetus has a sitting height of 30 to 36 centimeters and weighs about 680 grams. By the end of the sixth month, it could survive outside the mother's body, although probably only with respiratory assistance in an incubator. Its skin is red and wrinkled, and although teeth are only rarely visible at birth, they are already forming dentine. The cheesy body covering, which helps protect the fetus against abrasions, is now abundant. Reflexes are more vigorous. In the intestines is a pasty green mass of dead cells and bile, known as meconium, which will remain there until after birth.

The Final Trimester

During the final trimester, the fetus increases greatly in size and weight. In fact, it normally doubles in size just during the last two months. During this period, many nerve tracts are forming, and the number of brain cells is increasing at a very rapid rate. By the seventh month, brain waves can be recorded, through the abdomen of the mother, from the cerebral cortex of the fetus. Numerous studies have demonstrated that the protein intake of the mother is of critical importance during this period if the child is to have full development of its brain.

As the fetal period progresses, the physiology of the fetus becomes increasingly like the physiology of the adult, and so agents that affect the mother also threaten the maturing fetus. A depressingly familiar example is found in the infants born with cocaine, heroin, or methadone addiction.

During the last month of pregnancy, the baby usually begins to acquire antibodies from its mother, a process that continues after birth through the mother's milk (see page 727). Antibodies, which are far too large to diffuse across the placenta, are conveyed by highly selective active transport. The immunity they confer is only temporary. Within one to two months after birth, the maternal antibodies will be gradually replaced by antibodies manufactured by the baby's own immune system.

During this last month, the growth rate of the baby begins to slow down. (If it continued at the same rate, the child would weigh 90 kilograms—about 200 pounds—by its first birthday.) The placenta begins to regress and becomes tough and fibrous.

45-39 *A fetus at 18 weeks of age. When the thumb comes close to the mouth, the head turns, and the lips and tongue begin sucking motions—an important reflex for survival after birth.*

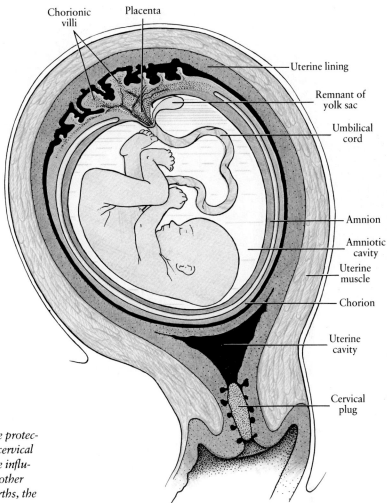

45-40 *A human fetus, shortly before birth, showing the protective membranes and uterine tissues surrounding it. The cervical plug is composed largely of mucus. It develops under the influence of progesterone and serves to exclude bacteria and other infectious agents from the uterus. In 95 percent of all births, the fetus is in this head-down position.*

Weight at birth is the major factor in infant mortality. Infants weighing less than 2,000 grams (4 pounds, 6 ounces) at birth are at high risk of death or severe brain damage. Infants weighing less than 2,500 grams (5 pounds, 8 ounces) are considered to be low weight, and they are 40 times as likely to die within a month of birth as heavier infants. Two-thirds of all babies who die are low weight. From the 1950s until the early 1980s, infant mortality in the United States declined steadily, but it has now stabilized at 10.6 deaths for each 1,000 live births, placing the United States seventeenth in infant mortality worldwide. In other developed countries, however, infant mortality has continued to decline, reaching a level of 6.0 and 6.6 deaths per 1,000 births in Finland and Japan, respectively. The principal cause of low birth weight is inadequate maternal nutrition and prenatal care, which costs, on the average, $500 to $800 per pregnancy—less than the cost per day of hospitalization of a low-weight infant in an intensive care nursery. In the United States at the present time, a black infant is more than twice as likely to be born underweight as a white infant and twice as likely to die within four weeks of birth. This disturbing statistic reflects a lack of prenatal care and counseling that could, for the population group at highest risk—young, unmarried mothers —lead to heavier and healthier babies at birth.

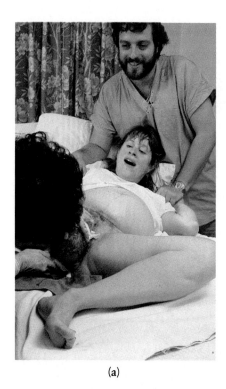 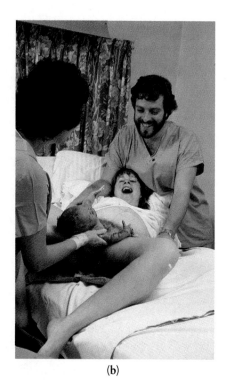 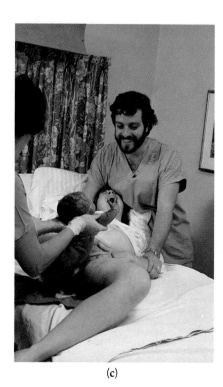

(a) (b) (c)

45–41 *The birth of a baby. When the umbilical cord—the baby's lifeline throughout its development—is severed, the infant will begin its separate existence.*

Birth

The date of birth is calculated as about 266 days after conception or 280 days after the beginning of the last menstrual period. Babies are rarely born on the scheduled day, but some 75 percent are born within two weeks of that day.

Labor is divided into three stages: dilation, expulsion, and placental stages. Dilation, which usually lasts from 2 to 16 hours (it is longer with the first baby than with subsequent births), begins with the onset of contractions of the uterus. It ends with the full dilation, or opening, of the cervix. At the beginning, uterine contractions occur at intervals of about 15 to 20 minutes and are relatively mild. By the end of the dilation stage, contractions are stronger and occur about every 1 to 2 minutes. At this point, the opening of the cervix is about 10 centimeters. Rupture of the amniotic sac, with the expulsion of fluids, usually occurs during this stage.

The second, or expulsion, stage lasts 2 to 60 minutes. It begins with the full dilation of the cervix and the appearance of the head in the cervix, called crowning. Contractions at this stage last from 50 to 90 seconds and are 1 or 2 minutes apart.

The third, or placental, stage begins immediately after the baby is born. It involves contractions of the uterus and the expelling of fluid, blood, and finally the placenta with the umbilical cord attached (also called the afterbirth). The placenta now weighs about 500 grams, about one-sixth of the weight of the infant. Minor uterine contractions continue; they help to stop the flow of blood and to return the uterus to its prepregnancy size and condition.

The baby emerges from the warm, protective enclosure in which it has been nourished and permitted to grow for nine months. The umbilical cord—until that moment, its lifeline—is severed. The baby cries as it takes its first breath, starts to breathe regularly, and so begins its independent existence.

45–42 *Within five to six minutes after vaginal delivery, the human infant is alert and its pupils are dilated, even in the presence of bright lights. This alertness, which may play a role in the bonding of infant and mother in the first hour of life, is the result of a surge of adrenaline and noradrenaline similar to that in "fight-or-flight" reactions (page 847). Birth puts great stress on an infant, as it is pushed and shoved by the uterine contractions, and temporary shortages of oxygen may occur as the blood vessels of the umbilical cord are pinched closed. Reflexes, similar to the diving reflex of marine mammals (page 744), alter the patterns of blood distribution, protecting vital organs, particularly the brain, heart, and lungs, which must begin functioning immediately after birth.*

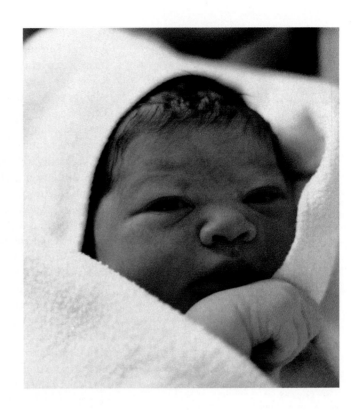

EPILOGUE

When did this particular human life begin? When the sperm encountered the egg? When the embryo became a fetus, visibly human? At the first heartbeat? The first brain wave? When the infant became viable as an independent entity? In the past, these matters were discussed by philosophers and theologians who were concerned with the question of when the soul enters the body. These issues have been revived in the ethical and legal controversies concerning abortion. In the evolutionary sense, however, none of these events marks the beginning of life. Life began more than 3 billion years ago and has been passed on since that time from organism to organism, generation after generation, to the present, and stretches on into the future, farther than the mind's eye can see. Each new organism is thus a temporary participant in the continuum of life. So is each sperm, each egg, indeed, in a sense, each living cell. Every individual, however, is a unique blend of heredity and experience, never to be duplicated and therefore irreplaceable; but from the perspective of the biological continuum, a human life lasts no longer than the blink of an eye.

SUMMARY

In the process of development, a fertilized egg becomes a complete organism, consisting of hundreds to billions of cells and closely resembling its parent organisms. This process involves growth, differentiation, and morphogenesis.

Development in most species of animals begins with fertilization—the fusion of egg and sperm. Fertilization results in (1) changes in the protective membranes of the egg that prevent fertilization by another sperm cell, (2) metabolic activation of the egg cell, (3) introduction of the genetic material of the male parent into the egg, and (4) cell division.

Development takes place in three stages: cleavage, gastrulation, and organogenesis. During cleavage, the original egg cell divides, with very slight, if any, change in overall volume. In eggs containing little or no yolk, such as those of the sea urchin, cleavage results in cells of approximately equal size. In yolkier eggs, such as those of the frog, cleavage is unequal, with fewer and larger cells in the yolky (vegetal) hemisphere. In eggs with a great deal of yolk, such as the hen's egg, cleavage is limited to a small, nonyolky disc at the top of the egg. In all cases, when cleavage is complete the embryo consists of a cluster of cells, the blastula, with a central cavity, the blastocoel.

Gastrulation involves the movement of cells into new relative positions and results in the establishment of three tissue layers: endoderm, mesoderm, and ectoderm. At the beginning of gastrulation, an opening forms, the blastopore. Cells from the outer surface of the embryo migrate through the blastopore into the blastocoel and form a new cavity, the archenteron, which will be the primitive gut. In birds and mammals, the homologue of the blastopore is the primitive streak. By the time the movements of the cells in gastrulation are complete, the mesoderm has differentiated into chordamesoderm and lateral plate mesoderm. The ectoderm on the dorsal surface overlying the chordamesoderm differentiates into neural ectoderm, flattening into the neural plate, which then curves upward and folds to form the neural tube. The chordamesoderm gives rise to the notochord and to the somites, which become embryonic muscle tissue. The lateral plate mesoderm splits into two layers, creating the fluid-filled coelom.

Each of the three primary tissue layers established during gastrulation gives rise to particular tissues and organs. The epidermis and the nervous system arise from ectoderm. The notochord, muscles, bones and cartilage, heart and blood vessels, excretory organs and gonads, the inner lining of the skin, and the outer linings of the digestive and respiratory tracts arise from mesoderm. The lungs, salivary glands, pancreas, liver, and inner lining of the digestive tract develop from endoderm.

The small area of tissue on the dorsal lip of the blastopore prior to gastrulation —the future chordamesoderm—is known as the organizer because, as shown by experiments on amphibians, it induces the cells overlying it to form the neural tube. If it is implanted in another embryo, it will induce the formation of a second neural tube in that embryo. Other organ systems also develop by this same process of embryonic induction, in which one tissue induces changes in the growth or migration of an adjacent tissue with which it comes in contact. A variety of factors are thought to be involved in the initial differentiation of the cells of the dorsal lip and in subsequent interactions. These include localized cytoplasmic determinants, particularly mRNA and proteins of maternal origin, substances that diffuse from specific cells and influence other cells, and cell adhesion molecules.

In amniote eggs, such as those of reptiles, birds, and mammals, the embryo, as it develops, forms four extraembryonic membranes: yolk sac, allantois, chorion, and amnion. In reptiles and birds, these membranes perform essential roles in supplying the developing embryo with food molecules and oxygen, in removing nitrogenous waste products, and in protecting the embryo from abrasion. In mammals, the yolk sac is the site in which germ cells are sequestered prior to their migration to the developing gonads, the allantois develops into the umbilical cord, the chorion forms the structures of the fetal side of the placenta, and the amnion, as in birds and reptiles, encloses the embryo in a fluid-filled cavity.

The mammalian embryo at the blastula stage is known as a blastocyst, and it consists of an inner cell mass and an outer layer of cells, the trophoblast. During the earliest stages of development, which take place in an oviduct, there is a large increase in the number of cells but little or no increase in the total size of the embryo. When the embryo descends into the uterus, at about the sixth day of development in humans, the trophoblast develops rapidly and invades the maternal tissues. The trophoblast becomes the chorion and interacts with tissues of the endometrium to form the placenta, through which the embryo receives its food and oxygen and excretes carbon dioxide and other waste products. The embryo is attached to the placenta by the umbilical cord.

When the human embryo is about 2 weeks old, a primitive streak forms, followed by the development of a neural plate and neural groove, which folds to form the neural tube. Although the embryo is still very small (about 2.5 millimeters long), most of the major organs have begun to form in these very early weeks, which is why damage caused to the embryo by viral infection, x-rays, or drugs during this period can be widespread. By the end of the second month, the embryo, now called a fetus, is almost human-looking, although it only weighs about 1 gram. By the end of the third month, all of the organ systems have been laid down. During the second trimester, development of the organ systems continues, and during the final trimester, there is a great increase in size and weight. Birth occurs, on the average, 266 days after fertilization.

QUESTIONS

1. Distinguish among the following: blastula/gastrula; blastopore/primitive streak; blastocoel/archenteron; ectoderm/mesoderm/endoderm; chordamesoderm/lateral plate mesoderm; fertilization/implantation; extraembryonic membranes/placenta; embryo/fetus; dilation/expulsion/placental stage.

2. Describe, in general terms, the end results of each of the following events: fertilization, cleavage, gastrulation, organogenesis.

3. Describe the similarities and differences in the blastulas of a sea urchin, an amphibian, a bird, and a mammal.

4. Cytoplasmic factors that influence the subsequent fate of cells may be localized or distributed in gradients. Give at least one example of each type, and describe its particular effects on the developing embryo.

5. The differentiation of functional sperm cells from spermatids, discussed in the preceding chapter, is under the control of genes on the Y chromosome. Sperm cells, however, are haploid and carry either an X chromosome or a Y chromosome. Explain, on the basis of principles discussed in this chapter, how it is possible for a spermatid containing an X chromosome (and thus no Y chromosome) to differentiate.

6. Follow the course of a single cell from its place of origin in the fertilized egg of a frog to its position in the eye cup of an early frog embryo. List, for each stage, all of the influences on this cell that might affect its history.

7. Follow the course of a single cell from its place of origin in the fertilized egg of a chick to its position in the third digit of the developing wing. List, for each stage, all of the influences on this cell that might affect its history.

8. In mammals, the fetus is basically feminine and fetal androgens are necessary to produce male characteristics. In birds, however, the fetus is basically masculine and fetal estrogens are necessary to produce female characteristics. Why would the arrangement in birds be unworkable in mammals?

9. If you look closely at Figure 45–34, you will notice that the umbilical arteries carry deoxygenated blood and the umbilical vein carries oxygenated blood. Why is this the case?

SUGGESTIONS FOR FURTHER READING

Books

BLOOM, FLOYD E., LAURA HOFSTADTER, and ARLYNE LAZERSON: *Brain, Mind, and Behavior*, W. H. Freeman and Company, New York, 1984.

A readable companion to the television series, beautifully illustrated.

BRECHER, EDWARD M., et al.: *Licit and Illicit Drugs: The Consumers Union Report on Narcotics, Stimulants, Depressants, Inhalants, Hallucinogens, and Marijuana—Including Caffeine, Nicotine, and Alcohol*, Little, Brown and Company, Boston, 1973.*

A well-balanced, straightforward, and nonsensational review.

BROWDER, L. W.: *Developmental Biology*, 2d ed., Saunders College/Holt, Rinehart and Winston, Philadelphia, 1984.

A general, comprehensive coverage of basic embryology and modern developmental biology.

---

* Available in paperback.

ECKERT, ROGER, and DAVID RANDALL: *Animal Physiology: Mechanisms and Adaptations*, 3d ed., W. H. Freeman and Company, New York, 1988.

> The emphasis is on basic principles. Well-written and handsomely illustrated.

GANONG, W. T.: *Review of Medical Physiology*, 12th ed., Lange Medical Publications, Los Angeles, 1987.*

> A concise, accurate, up-to-date summary of human physiology. An excellent reference book.

GORDON, MALCOLM S., et al.: *Animal Physiology: Principles and Adaptations*, 4th ed., The Macmillan Company, New York, 1982.

> A good animal physiology textbook, intended for an advanced course. The authors emphasize function as it relates to the survival of organisms in their natural environments.

KARP, GERALD, and N. J. BERRILL: *Development*, 2d ed., McGraw-Hill Book Company, New York, 1981.

> A general textbook of developmental biology, with clear explanations and thorough coverage.

KESSEL, RICHARD G., and RANDY H. KARDON: *Tissues and Organs: A Text-Atlas of Scanning Electron Microscopy*, W. H. Freeman and Company, San Francisco, 1979.*

> A collection of more than 700 outstanding scanning electron micrographs of vertebrate tissues and organs, beautifully reproduced. The accompanying text summarizes our knowledge of each organ system and its component parts.

LUCIANO, DOROTHY S., ARTHUR J. VANDER, and JAMES H. SHERMAN: *Human Anatomy and Physiology*, 2d ed., McGraw-Hill Book Company, New York, 1983.

> An introductory anatomy and physiology textbook. The text is clearly written, and the detailed illustrations—particularly of the skeletal, muscular, nervous, and circulatory systems—are outstanding.

NEWSHOLME, ERIC, and TONY LEECH: *The Runner: Energy and Endurance*, Fitness Books, Roosevelt, N.J., 1984.*

> A lively introduction to the energetics of the working human body. Equally useful as a primer for runners who want to know more about their own physiology and as an introduction to the biochemistry of carbohydrate and fat metabolism.

NILSSON, LENNART, AXEL INGLEMAN-SUNDBERG, and CLAES WIRSÉN: *A Child Is Born: The Drama of Life before Birth*, Dell Publishing, Inc., New York, 1986.*

> This is an account of the history of life before birth. The book describes in detail the development of the unborn child from the moment of fertilization and also the changes in the mother during pregnancy. There are magnificent color photographs of the developing fetus.

NILSSON, LENNART: *The Body Victorious*, Delacorte Press, New York, 1987.

> An account of the human immune system—its development, its normal functioning, and its disorders—filled with more of the stunning color photographs and micrographs for which Nilsson is noted. Moreover, if you should be in any doubt as to the physiological consequences of cigarette smoking, you owe it to yourself to examine the relevant photographs in this book.

ROMER, ALFRED: *The Vertebrate Story*, 4th ed., The University of Chicago Press, Chicago, 1971.*

> The history of vertebrate evolution, written by an expert but as readable as a novel.

ROMER, ALFRED, and THOMAS S. PARSONS: *The Vertebrate Body*, 6th ed., Saunders College/Holt, Rinehart and Winston, Philadelphia, 1985.

> A thorough account of comparative vertebrate anatomy. The discussion of embryonic development is outstanding.

SCHMIDT-NIELSEN, KNUT: *Animal Physiology: Adaptation and Environment*, 3d ed., Cambridge University Press, New York, 1983.

> Schmidt-Nielsen is concerned with underlying principles of animal physiology—the problems animals have to solve in order to survive. The emphasis is on comparative physiology, and the lucid exposition is illuminated by many interesting examples.

SCHMIDT-NIELSEN, KNUT: *Desert Animals*, Oxford University Press, New York, 1964.

> Although considered the definitive work on the physiological problems relating to heat and water, this readable book also contains numerous anecdotes—such as that about Dr. Blagden—and many fascinating personal observations.

SCIENTIFIC AMERICAN: *The Brain*, W. H. Freeman and Company, New York, 1979.*

> A reprint of the September 1979 issue of Scientific American. Its 11 chapters by different authors are devoted to current knowledge of neurons and their organization and functions within the brain. The many drawings and micrographs are excellent.

SCIENTIFIC AMERICAN: *Progress in Neuroscience*, W. H. Freeman and Company, New York, 1985.*

> A collection of articles from Scientific American, originally published between 1979 and 1984, covering such topics as the role of calcium ion at the synapse, the development of the nervous system, hearing, and the genetic control of behavior.

SCIENTIFIC AMERICAN: *What Science Knows about AIDS*, W. H. Freeman and Company, New York, 1988.*

> A reprint of the October 1988 issue of Scientific American. Its 10 chapters by different authors provide a thorough review of our current knowledge of human immunodeficiency virus (HIV), the causative agent of AIDS; of the disease process and epidemiology of AIDS; and of the prospects and problems in the development of effective treatments and vaccines. Highly recommended.

SHEPHERD, GORDON M.: *Neurobiology*, 2d ed., Oxford University Press, New York, 1988.

> A general textbook that provides a broad coverage of contemporary neurobiology.

SMITH, HOMER W.: *From Fish to Philosopher*, Doubleday & Company, Inc., Garden City, N.Y., 1959.*

---

* Available in paperback.

Smith was an eminent specialist in the physiology of the kidney. Writing for the general public, he explains the role of this remarkable organ in the story of how, in the course of evolution, organisms have increasingly freed themselves from their environments.

THOMPSON, R. F.: *Introduction to Physiological Psychology,* 2d ed., Harper & Row, Publishers, Inc., New York, 1988.

Intended for the undergraduate student, this text presents an up-to-date survey of the biological foundations of psychology.

VANDER, ARTHUR J., JAMES H. SHERMAN, and DOROTHY S. LUCIANO: *Human Physiology: The Mechanisms of Body Function,* 4th ed., McGraw-Hill Book Company, New York, 1985.

Most highly recommended. The text is a model of clarity, and the diagrams, many of which we have borrowed, are splendid.

WALBOT, V., and N. HOLDER: *Developmental Biology,* Random House, Inc., New York, 1987.

A modern molecular and genetic approach to developmental biology. Highly recommended.

WEST, JOHN B.: *Everest—The Testing Place,* McGraw-Hill Book Company, New York, 1985.

An account, written for the general reader, of the adventures and findings of the American Medical Expedition to Everest (see page 746). Professor West, one of the leaders of the expedition, is a lucid and delightful writer on both human physiology and mountaineering.

Articles

ADA, GORDON L., and NOSSAL, GUSTAV: "The Clonal Selection Theory," *Scientific American,* August 1987, pages 62–69.

AOKI, CHIYE, and PHILIP SIEKEVITZ: "Plasticity in Brain Development," *Scientific American,* December 1988, pages 56–64.

AXELROD, JULIUS, and TERRY D. REISINE: "Stress Hormones: Their Interaction and Regulation," *Science,* vol. 224, pages 452–459, 1984.

BARNES, DEBORAH M.: "Steroids May Influence Changes in Mood," *Science,* vol. 232, pages 1344–1345, 1986.

BEACONSFIELD, PETER, GEORGE BIRDWOOD, and REBECCA BEACONSFIELD: "The Placenta," *Scientific American,* August 1980, pages 95–102.

BERRIDGE, MICHAEL J.: "The Molecular Basis of Communication within the Cell," *Scientific American,* October 1985, pages 142–152.

BRAMBLE, DENNIS M., and DAVID R. CARRIER: "Running and Breathing in Mammals," *Science,* vol. 219, pages 251–256, 1983.

BROWN, MICHAEL S., and JOSEPH L. GOLDSTEIN: "How LDL Receptors Influence Cholesterol and Atherosclerosis," *Scientific American,* November 1984, pages 58–66.

BUISSERET, PAUL D.: "Allergy," *Scientific American,* August 1982, pages 86–95.

CANTIN, MARC, and JACQUES GENEST: "The Heart as an Endocrine Gland," *Scientific American,* February 1986, pages 76–81.

CAPLAN, ARNOLD I.: "Cartilage," *Scientific American,* October 1984, pages 84–94.

CARMICHAEL, STEPHEN W., and HANS WINKLER: "The Adrenal Chromaffin Cell," *Scientific American,* August 1985, pages 40–49.

CERAMI, ANTHONY, HELEN VLASSARA, and MICHAEL BROWNLEE: "Glucose and Aging," *Scientific American,* May 1987, pages 90–96.

COHEN, IRUN R.: "The Self, the World and Autoimmunity," *Scientific American,* April 1988, pages 52–60.

COLLIER, R. JOHN, and DONALD A. KAPLAN: "Immunotoxins," *Scientific American,* July 1984, pages 56–64.

COTMAN, CARL W., and MANUEL NIETO-SAMPEDRO: "Cell Biology of Synaptic Plasticity," *Science,* vol. 225, pages 1287–1294, 1984.

CRAWSHAW, LARRY I., BRENDA P. MOFFITT, DANIEL E. LEMONS, and JOHN A. DOWNEY: "The Evolutionary Development of Vertebrate Thermoregulation," *American Scientist,* vol. 69, pages 543–550, 1981.

DOOLITTLE, RUSSELL F.: "Fibrinogen and Fibrin," *Scientific American,* December 1981, pages 126–135.

DUNANT, YVES, and MAURICE ISRAËL: "The Release of Acetylcholine," *Scientific American,* April 1985, pages 58–66.

EASTMAN, JOSEPH T., and ARTHUR L. DeVRIES: "Antarctic Fishes," *Scientific American,* November 1986, pages 106–114.

EDELMAN, GERALD M.: "Cell-Adhesion Molecules: A Molecular Basis for Animal Form," *Scientific American,* April 1984, pages 118–129.

EDELSON, RICHARD L., and JOSEPH M. FINK: "The Immunologic Function of Skin," *Scientific American,* June 1985, pages 46–53.

EISENBERG, EVAN, and TERRELL L. HILL: "Muscle Contraction and Free Energy Transduction in Biological Systems," *Science,* vol. 227, pages 999–1006, 1985.

FEDER, MARTIN E., and WARREN W. BURGGREN: "Skin Breathing in Vertebrates," *Scientific American,* November 1985, pages 126–142.

FINE, ALAN: "Transplantation in the Central Nervous System," *Scientific American,* August 1986, pages 52–58B.

FRENCH, ALAN R.: "The Patterns of Mammalian Hibernation," *American Scientist,* vol. 76, pages 568–575, 1988.

GALLO, ROBERT C.: "The AIDS Virus," *Scientific American,* January 1987, pages 46-56.

GARCIA-BELLIDO, ANTONIO, PETER A. LAWRENCE, and GINES MORATA: "Compartments in Animal Development," *Scientific American,* July 1979, pages 102-110.

GEHRING, WALTER J.: "The Molecular Basis of Development," *Scientific American,* October 1985, pages 152B-162.

GLICKSTEIN, MITCHELL: "The Discovery of the Visual Cortex," *Scientific American,* September 1988, pages 118-127.

GOLDE, DAVID W., and JUDITH C. GASSON: "Hormones that Stimulate the Growth of Blood Cells," *Scientific American,* July 1988, pages 62-70.

GOLDSTEIN, GARY W., and A. LORRIS BETZ: "The Blood-Brain Barrier," *Scientific American,* September 1986, pages 74-83.

GOTTLIEB, DAVID I.: "GABAergic Neurons," *Scientific American,* February 1988, pages 82-89.

HALL, BRIAN K.: "The Embryonic Development of Bone," *American Scientist,* vol. 76, pages 174-181, 1988.

HELLER, H. CRAIG, LARRY I. CRAWSHAW, and HAROLD T. HAMMEL: "The Thermostat of Vertebrate Animals," *Scientific American,* August 1978, pages 102-113.

HUDSPETH, A. J.: "The Hair Cells of the Inner Ear," *Scientific American,* January 1983, pages 54-64.

HYNES, RICHARD O.: "Fibronectins," *Scientific American,* June 1986, pages 42-51.

JACOBS, BARRY L.: "How Hallucinogenic Drugs Work," *American Scientist,* vol. 75, pages 386-392, 1987.

KOLATA, GINA: "Genes and Biological Clocks," *Science,* vol. 230, pages 1151-1152, 1985.

KORETZ, JANE F., and GEORGE H. HANDELMAN: "How the Human Eye Focuses," *Scientific American,* July 1988, pages 92-99.

LAGERCRANTZ, HUGO, and THEODORE A. SLOTKIN: "The 'Stress' of Being Born," *Scientific American,* April 1986, pages 100-107.

LAURENCE, JEFFREY: "The Immune System in AIDS," *Scientific American,* December 1985, pages 84-93.

MARRACK, PHILIPPA, and JOHN KAPPLER: "The T Cell and Its Receptor," *Scientific American,* February 1986, pages 36-45.

MARX, JEAN L.: "How the Brain Controls Birdsong," *Science,* vol. 217, pages 1125-1126, 1982.

MARX, JEAN L.: "The Immune System 'Belongs in the Body,' " *Science,* vol. 227, pages 1190-1192, 1985.

MASLAND, RICHARD H.: "The Functional Architecture of the Retina," *Scientific American,* December 1986, pages 102-111.

MILLER, C. ARDEN: "Infant Mortality in the U.S.," *Scientific American,* July 1985, pages 31-37.

MILSTEIN, CESAR: "Monoclonal Antibodies," *Scientific American,* October 1980, pages 66-74.

MISHKIN, MORTIMER, and TIM APPENZELLER: "The Anatomy of Memory," *Scientific American,* June 1987, pages 80-89.

MOOG, FLORENCE: "The Lining of the Small Intestine," *Scientific American,* November 1981, pages 154-176.

MORELL, PIERRE, and WILLIAM T. NORTON: "Myelin," *Scientific American,* May 1980, pages 89-116.

MORRISON, ADRIAN R.: "A Window on the Sleeping Brain," *Scientific American,* April 1983, pages 94-102.

NADEL, ETHAN R.: "Physiological Adaptations to Aerobic Training," *American Scientist,* vol. 73, pages 334-343, 1985.

ORCI, LELIO, JEAN-DOMINIQUE VASSALLI, and ALAIN PERRELET: "The Insulin Factory," *Scientific American,* September 1988, pages 85-94.

PERUTZ, M. F.: "Hemoglobin Structure and Respiratory Transport," *Scientific American,* December 1978, pages 92-125.

POGGIO, TOMASO, and CHRISTOF KOCH: "Synapses That Compute Motion," *Scientific American,* May 1987, pages 46-52.

REICHARDT, LOUIS F.: "Immunological Approaches to the Nervous System," *Science,* vol. 225, pages 1294-1299, 1984.

ROBERTS, LESLIE: "Zeroing in on the Sex Switch," *Science,* vol. 239, pages 21-23, 1988.

ROBINSON, THOMAS F., STEPHEN M. FACTOR, and EDMUND H. SONNENBLICK: "The Heart as a Suction Pump," *Scientific American,* June 1986, pages 84-91.

RODGER, JOHN C., and BELINDA L. DRAKE: "The Enigma of the Fetal Graft," *American Scientist,* vol. 75, pages 51-57, 1987.

ROSE, NOEL R.: "Autoimmune Diseases," *Scientific American,* February 1981, pages 80-103.

SCHMIDT-NIELSEN, KNUT: "Countercurrent Systems in Animals," *Scientific American,* May 1981, pages 118-128.

SCHNAPF, JULIE L., and DENIS A. BAYLOR: "How Photoreceptor Cells Respond to Light," *Scientific American,* April 1987, pages 40-47.

SNYDER, SOLOMON H.: "The Molecular Basis of Communication between Cells," *Scientific American,* October 1985, pages 132-141.

STALLONES, RUEL A.: "The Rise and Fall of Ischemic Heart Disease," *Scientific American,* November 1980, pages 53-59.

STRICKER, EDWARD M., and JOSEPH G. VERBALIS: "Hormones and Behavior: The Biology of Thirst and Sodium Appetite," *American Scientist,* vol. 76, pages 261-267, 1988.

STRYER, LUBERT: "The Molecules of Visual Excitation," *Scientific American,* July 1987, pages 42-50.

TAMARKIN, LAWRENCE, CURTIS J. BAIRD, and O. F. X. ALMEIDA: "Melatonin: A Coordinating Signal for Mammalian Reproduction?" *Science,* vol. 227, pages 714-720, 1985.

TONEGAWA, SUSUMU: "The Molecules of the Immune System," *Scientific American,* October 1985, pages 122-131.

UNANUE, EMIL R., and PAUL M. ALLEN: "The Basis for the Immunoregulatory Role of Macrophages and Other Accessory Cells," *Science,* vol. 236, pages 551–557, 1987.

WASSARMAN, PAUL M.: "The Biology and Chemistry of Fertilization," *Science,* vol. 235, pages 553–560, 1987.

WASSARMAN, PAUL M.: "Fertilization in Mammals," *Scientific American,* December 1988, pages 78–84.

WINGFIELD, JOHN C., et al.: "Testosterone and Aggression in Birds," *American Scientist,* vol. 75, pages 602–608, 1987.

WOLPERT, LEWIS: "Pattern Formation in Biological Development," *Scientific American,* October 1978, pages 154–164.

WURTMAN, RICHARD J.: "Alzheimer's Disease," *Scientific American,* January 1985, pages 62–74.

YOUNG, JOHN D., and ZANVIL A. COHN: "How Killer Cells Kill," *Scientific American,* January 1988, pages 38–44.

ZAPOL, WARREN M.: "Diving Adaptations of the Weddell Seal," *Scientific American,* June 1987, pages 100–105.

ZUCKER, MARJORIE B.: "The Functioning of Blood Platelets," *Scientific American,* June 1980, pages 86–103.

APPENDIX A

Metric Table

	QUANTITY	NUMERICAL VALUE	ENGLISH EQUIVALENT	CONVERTING ENGLISH TO METRIC
Length	kilometer (km)	1,000 (10^3) meters	1 km = 0.62 mile	1 mile = 1.609 km
	meter (m)	100 centimeters	1 m = 1.09 yards	1 yard = 0.914 m
			= 3.28 feet	1 foot = 0.305 m
	centimeter (cm)	0.01 (10^{-2}) meter	1 cm = 0.394 inch	1 foot = 30.5 cm
				1 inch = 2.54 cm
	millimeter (mm)	0.001 (10^{-3}) meter	1 mm = 0.039 inch	1 inch = 25.4 mm
	micrometer (μm)	0.000001 (10^{-6}) meter		
	nanometer (nm)	0.000000001 (10^{-9}) meter		
	angstrom (Å)	0.0000000001 (10^{-10}) meter		
Area	square kilometer (km^2)	100 hectares	1 km^2 = 0.3861 square mile	1 square mile = 2.590 km^2
	hectare (ha)	10,000 square meters	1 ha = 2.471 acres	1 acre = 0.4047 ha
	square meter (m^2)	10,000 square centimeters	1 m^2 = 1.1960 square yards	1 square yard = 0.8361 m^2
			= 10.764 square feet	1 square foot = 0.0929 m^2
	square centimeter (cm^2)	100 square millimeters	1 cm^2 = 0.155 square inch	1 square inch = 6.4516 cm^2
Mass	metric ton (t)	1,000 kilograms = 1,000,000 grams	1 t = 1.103 tons	1 ton = 0.907 t
	kilogram (kg)	1,000 grams	1 kg = 2.205 pounds	1 pound = 0.4536 kg
	gram (g)	1,000 milligrams	1 g = 0.0353 ounce	1 ounce = 28.35 g
	milligram (mg)	0.001 gram		
	microgram (μg)	0.000001 gram		
Time	second (sec)	1,000 milliseconds		
	millisecond (msec)	0.001 second		
	microsecond (μsec)	0.000001 second		
Volume (solids)	1 cubic meter (m^3)	1,000,000 cubic centimeters	1 m^3 = 1.3080 cubic yards	1 cubic yard = 0.7646 m^3
			= 35.315 cubic feet	1 cubic foot = 0.0283 m^3
	1 cubic centimeter (cm^3)	1,000 cubic millimeters	1 cm^3 = 0.0610 cubic inch	1 cubic inch = 16.387 cm^3
Volume (liquids)	kiloliter (kl)	1,000 liters	1 kl = 264.17 gallons	1 gal = 3.785 l
	liter (l)	1,000 milliliters	1 l = 1.06 quarts	1 qt = 0.94 l
				1 pt = 0.47 l
	milliliter (ml)	0.001 liter	1 ml = 0.034 fluid ounce	1 fluid ounce = 29.57 ml
	microliter (μl)	0.000001 liter		

APPENDIX B

Temperature Conversion Scale

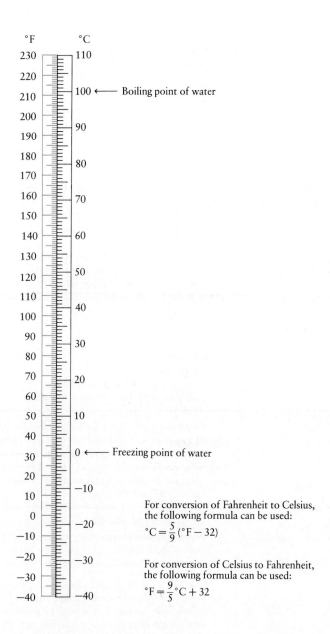

For conversion of Fahrenheit to Celsius, the following formula can be used:

$$°C = \frac{5}{9}(°F - 32)$$

For conversion of Celsius to Fahrenheit, the following formula can be used:

$$°F = \frac{9}{5}°C + 32$$

APPENDIX C

Classification of Organisms

There are several ways to classify organisms. The one presented here follows the overall scheme described at the end of Chapter 20. Organisms are divided into five major groups, or kingdoms: Monera, Protista, Fungi, Plantae, and Animalia.

The chief taxonomic categories are kingdom, division or phylum, class, order, family, genus, species. (The taxonomic categories of division and phylum are equivalent. The term "division" is generally used in the classification of prokaryotes, algae, fungi, and plants, whereas "phylum" is used in the classification of protozoa and animals.) The following classification includes all of the generally accepted divisions and phyla. Certain classes and orders, particularly those mentioned in this book, are also included, but the listing is far from complete. The number of species given for each group is an estimated number of living (that is, contemporary) species described and named to date.

KINGDOM MONERA

Monerans (prokaryotes) are cells that lack a nuclear envelope, chloroplasts and other plastids, mitochondria, and 9 + 2 flagella. Prokaryotes are unicellular but sometimes occur as filaments or other superficially multicellular bodies. Their predominant mode of nutrition is heterotrophic, by absorption, but some groups are autotrophic, either photosynthetic or chemosynthetic. Reproduction is primarily asexual, by binary fission or budding, but genetic exchanges occur in some as a result of conjugation, transformation, transduction, and exchanges of plasmids. Motile forms move by means of bacterial flagella or by gliding.

The classification of prokaryotes is not hierarchical, and the schemes that are most widely used do not reflect evolutionary relationships. Kingdom Monera contains representatives of two distinct lineages, the Archaebacteria and the Eubacteria. Archaebacteria include the methanogens, thermoacidophiles, and extreme halophiles. Among the principal lineages of Eubacteria are the green bacteria (sulfur and nonsulfur), purple bacteria (sulfur and nonsulfur) and related forms, spirochetes, cyanobacteria, and gram-positive bacteria. About 2,700 distinct species of prokaryotes are recognized.

KINGDOM PROTISTA

Eukaryotic organisms, including unicellular and multicellular photosynthetic autotrophs (algae), multinucleate or multicellular heterotrophs (slime molds and water molds), and unicellular or simple colonial heterotrophs (protozoa). Their modes of nutrition include photosynthesis, absorption, and ingestion. Reproduction is both sexual (in some forms) and asexual. They move by 9 + 2 flagella (or cilia) or pseudopodia or are nonmotile. About 60,000 living species and another 60,000 extinct species known only from their fossils.

DIVISION

DIVISION EUGLENOPHYTA: euglenoids. Unicellular photosynthetic (or sometimes secondarily heterotrophic) organisms with chlorophylls *a* and *b*. They store food as paramylon, an unusual carbohydrate. Euglenoids usually have a single apical flagellum and a contractile vacuole. Sexual reproduction is unknown. Euglenoids occur mostly in fresh water. There are some 1,000 species.

DIVISION CHRYSOPHYTA: diatoms, golden-brown algae, and yellow-green algae. Unicellular photosynthetic organisms with chlorophylls *a* and *c* and the accessory pigment fucoxanthin. Food stored as the carbohydrate laminarin or as large oil droplets. Cell walls consisting mainly of cellulose, sometimes heavily impregnated with siliceous materials. There may be as many as 13,000 living species.

CLASS

Class Bacillariophyceae: diatoms. Chrysophyta with double siliceous shells, the two halves of which fit together like a pillbox. They are sometimes motile by the secretion of mucilage fibrils along a specialized groove, the raphe. There are nearly 10,000 living species and at least 15,000 extinct species.

Class Chrysophyceae: golden-brown algae. A diverse group of organisms, including flagellated, amoeboid, and nonmotile forms, some naked and others with a cell wall that may be ornamented with siliceous scales. About 3,000 species.

DIVISION

DIVISION DINOFLAGELLATA: "spinning" flagellates. Unicellular photosynthetic organisms with chlorophylls *a* and *c*. Food is stored as starch. Cell walls contain cellulose. Most of the organisms in this division have two lateral flagella, one of which beats in a groove that encircles the organism. They probably have no form of sexual reproduction, and their mitosis is unique. Nearly 2,000 living species.

DIVISION CHLOROPHYTA: green algae. Unicellular, colonial, coenocytic, or multicellular, characterized by chlorophylls *a* and *b* and various carotenoids. The carbohydrate food reserve is starch. The cell walls consist of polysaccharides, including cellulose in some forms. Motile cells have two lateral or apical flagella. True multicellular genera do not exhibit complex patterns of differentiation. Multicellularity has arisen at least three times, and quite possibly more often. There are about 9,000 known species and possibly many more.

CLASS

Class Chlorophyceae: Unicellular, colonial, or multicellular green algae, found predominantly in fresh water. Cell division involves a system of microtubules parallel to the plane of cell division, and the nuclear envelope persists during mitosis. Sexual reproduction involves the formation of a dormant zygote that subsequently undergoes meiosis, producing the haploid cells in which the organisms spend most of the life cycle.

Class Charophyceae: Unicellular or multicellular green algae, found predominantly in fresh water. Cell division involves a system of microtubules perpendicular to the plane of cell division; the nuclear envelope breaks down during the course of mitosis. Sexual reproduction involves the formation of a dormant zygote that subsequently undergoes meiosis, producing the haploid cells in which the organisms spend most of the life cycle. Certain members of this class resemble plants more closely than do any other organisms.

Class Ulvophyceae: Coenocytic or multicellular green algae, found predominantly in salt water. Cell division involves a system of microtubules perpendicular to the plane of cell division, but the nuclear envelope persists during mitosis. Sexual reproduction often involves alternation of generations, with meiosis in the diploid sporophyte producing haploid spores that germinate to produce the haploid gametophyte.

DIVISION

DIVISION PHAEOPHYTA: brown algae. Multicellular marine organisms characterized by the presence of chlorophylls *a* and *c* and the pigment fucoxanthin. Their food reserve is the carbohydrate laminarin. Motile cells are biflagellate, with one forward flagellum and one trailing flagellum. A considerable amount of tissue differentiation is found in some of the kelps, with specialized conducting cells in some genera that transport photosynthate to dimly lighted regions of the alga. There is, however, no differentiation into leaves, roots, and stem, as in plants. About 1,500 species.

DIVISION RHODOPHYTA: red algae. Primarily marine organisms characterized by the presence of chlorophyll *a* and red pigments known as phycobilins. Their carbohydrate reserve is a special type of starch (floridean). No motile cells are present at any stage in the complex life cycle. The algal body is built up of closely packed filaments in a gelatinous matrix and is not differentiated into leaves, roots, and stem. It lacks specialized conducting cells. There are some 4,000 species.

DIVISION MYXOMYCOTA: plasmodial slime molds. Heterotrophic amoeboid organisms that form a coenocytic plasmodium that creeps along as a mass and eventually differentiates into sporangia, each of which is coenocytic and eventually gives rise to many spores. Predominant mode of nutrition is by ingestion. More than 550 species.

DIVISION ACRASIOMYCOTA: cellular slime molds. Heterotrophic organisms in which there are separate amoebas that eventually swarm together to form a mass but retain their identity within this mass, which eventually differentiates into a compound sporangium. Principal mode of nutrition is by ingestion. Seven genera and about 65 species.

DIVISION CHYTRIDIOMYCOTA: chytrids (water molds). Coenocytic aquatic heterotrophs with a vegetative body, or thallus, usually differentiated into rhizoids and a sporangium. Nutrition by absorption. Cell walls contain chitin. Asexual reproduction, with sexual reproduction in some forms. Both spores and gametes are flagellated. About 900 species.

DIVISION OOMYCOTA: oomycetes (water molds). Coenocytic filamentous heterotrophs, primarily aquatic. Nutrition by absorption. Their cell walls are composed of glucose polymers, including cellulose. Both asexual and sexual reproduction, with flagellated asexual spores and nonmotile gametes. The coenocytic filaments (hyphae) are diploid, and the haploid gametes are produced by meiosis. About 800 species.

PHYLUM

PHYLUM MASTIGOPHORA: mastigophorans (flagellates). Unicellular heterotrophs with flagella, most of which are symbiotic forms such as *Trichonympha* and *Trypanosoma*, the cause of sleeping sickness. Some free-living species. Reproduction usually asexual by binary fission. About 1,500 species.

PHYLUM SARCODINA: sarcodines. Unicellular heterotrophs with pseudopodia, such as amoebas. No stiffening pellicle; some produce shells. Reproduction may be asexual or sexual. About 11,500 living species and some 33,000 fossil species.

PHYLUM CILIOPHORA: ciliates. Unicellular heterotrophs with cilia, including *Paramecium* and *Stentor*. Reproduction is asexual, but genetic exchange through the phenomenon of conjugation is also common. About 8,000 species.

PHYLUM OPALINIDA: opalinids. Parasitic protists found mainly in the digestive tracts of frogs and toads. Covered uniformly with cilia or flagella. Reproduction asexual by fission or sexual, with flagellated gametes. About 400 species.

PHYLUM SPOROZOA: sporozoa. Parasitic protists; usually without locomotor organelles during a major part of their complex life cycle. Includes *Plasmodium*, several species of which cause malaria. About 5,000 species.

KINGDOM FUNGI

Eukaryotic filamentous or, rarely, unicellular organisms. The filamentous forms consist basically of a continuous mycelium; this mycelium becomes septate (partitioned off) in certain groups and at certain stages of the life cycle. Chitin is present in the cell walls of all fungi. Fungi are saprobic or parasitic heterotrophs, with nutrition by absorption. Reproductive cycles often include both sexual and asexual phases. Most fungi are haploid, with the zygote the only diploid stage in the life cycle. No flagella or cilia at any stage of the life cycle. Some 100,000 species of fungi have been named.

DIVISION

DIVISION ZYGOMYCOTA: terrestrial fungi, such as black bread mold, with the hyphae septate only during the formation of reproductive bodies. The division includes about 600 described species, of which about 30 occur as components of the endomycorrhizae that are found in about 80 percent of all vascular plants.

DIVISION ASCOMYCOTA: terrestrial and aquatic fungi, including *Neurospora*, powdery mildews, morels, and truffles. The hyphae are septate but the septa perforated; complete septa cut off the reproductive bodies, such as spores or gametangia. Sexual reproduction involves the formation of a characteristic cell, the ascus, in which meiosis takes place and within which spores are formed. The hyphae in many ascomycetes are packed together into complex "fruiting bodies." Yeasts are unicellular ascomycetes that reproduce asexually by budding. About 30,000 species, in addition to 25,000 species that occur in lichens.

DIVISION BASIDIOMYCOTA: terrestrial fungi, including the mushrooms and toadstools, with the hyphae septate but the septa perforated; complete septa cut off reproductive bodies. Sexual reproduction involves formation of basidia, in which meiosis takes place and on which the spores are borne. There are some 25,000 species.

DIVISION DEUTEROMYCOTA: Fungi Imperfecti. Mainly fungi in which the sexual cycle has not been observed. The deuteromycetes are classified by their asexual spore-bearing structures. There are some 25,000 species, including *Penicillium*, the original source of penicillin, fungi that cause athlete's foot and other skin diseases, and many of the molds that give cheeses, such as Roquefort and Camembert, their special flavor.

KINGDOM PLANTAE

Multicellular photosynthetic eukaryotes, primarily adapted to life on land. The photosynthetic pigment is chlorophyll *a*, with chlorophyll *b* and a number of carotenoids serving as accessory pigments. The cell walls contain cellulose. There is considerable differentiation of tissues and organs. Reproduction is primarily sexual with alternating gametophytic and sporophytic phases; the gametophytic phase has been progressively reduced in the course of evolution. The egg- and sperm-producing structures are multicellular and are surrounded by a sterile (nonreproductive) jacket layer; the zygote develops into an embryo, or young sporophyte, while encased in the archegonium (seedless plants) or embryo sac (seed plants). The living members of the plant kingdom include the bryophytes and nine divisions of vascular plants—plants with complex differentiation of the sporophyte into leaves, roots, and stems and with well-developed strands of conducting tissue for the transport of water and organic materials.

DIVISION

DIVISION BRYOPHYTA: liverworts, hornworts, and mosses. Multicellular plants with photosynthetic pigments and food reserves similar to those of the green algae. They have multicellular gametangia with a sterile jacket one-cell-layer thick. The sperm are biflagellate and motile. Gametophytes and sporophytes both exhibit complex multicellular patterns of development, but conducting tissues are usually completely absent and are not well differentiated when present; true roots, leaves, and stems are absent. Most of the photosynthesis is carried out by the gametophyte, upon which the sporophyte is nutritionally dependent, at least initially. There are some 16,000 species.

CLASS

Class Hepaticae: liverworts. The gametophytes are either thallose (not differentiated into roots, leaves, and stem) or "leafy," and the sporophytes are relatively simple in construction. About 6,000 species.

Class Anthocerotae: hornworts. The gametophytes are thallose. The sporophyte grows from a basal meristem for as long as conditions are favorable. Stomata are present on the sporophyte. About 100 species.

Class Musci: mosses. The gametophytes are "leafy." The sporophytes have complex patterns of spore discharge. Stomata are present on the sporophyte. About 9,500 species.

DIVISION

DIVISION PSILOPHYTA: whisk ferns. Homosporous vascular plants with or without microphylls. The sporophytes are extremely simple, and there is no differentiation between root and shoot. The sperm are motile. Two genera, with several species.

DIVISION LYCOPHYTA: club mosses. Homosporous and heterosporous vascular plants with microphylls; extremely diverse in appearance. All lycophytes have motile sperm. There are five genera, with about 1,000 species.

DIVISION SPHENOPHYTA: horsetails. Homosporous vascular plants with jointed stems marked by conspicuous nodes and elevated siliceous ribs. Sporangia are borne in a cone at the apex of the stem. Leaves are scalelike. Sperm are motile. Although now thought to have evolved from a megaphyll, the leaves of the horsetails are structurally indistinguishable from microphylls. There is one genus, *Equisetum*, with 15 living species.

DIVISION PTEROPHYTA: ferns. They are mostly homosporous, although some are heterosporous. All possess megaphylls. The gametophyte is more or less free-living and usually photosynthetic. Multicellular gametangia and free-swimming sperm are present. About 12,000 species.

DIVISION CONIFEROPHYTA: conifers. Seed plants with active cambial growth and simple, needle-like leaves; the ovules are not enclosed and the sperm are not flagellated. There are some 50 genera, with about 550 species; the most familiar group of gymnosperms.

DIVISION CYCADOPHYTA: cycads. Seed plants with sluggish cambial growth and pinnately compound, palmlike or fernlike leaves. The ovules are not enclosed. The sperm are flagellated and motile but are carried to the ovule in a pollen tube. Cycads are gymnosperms. There are 10 genera, with about 100 species.

DIVISION GINKGOPHYTA: ginkgo. Seed plants with active cambial growth and fan-shaped leaves with open dichotomous venation. The ovules are not enclosed and are fleshy at maturity. Sperm are carried to the ovule in a pollen tube but are flagellated and motile. They are gymnosperms. There is one species only.

DIVISION GNETOPHYTA: gnetophytes. Seed plants with many angiospermlike features. They are the only gymnosperms in which vessels are present in the xylem. Motile sperm are absent. There are three very distinctive genera, with about 70 species.

DIVISION ANTHOPHYTA: flowering plants (angiosperms). Seed plants in which the ovules are enclosed in a carpel (in all but a very few genera), and the seeds at maturity are borne within fruits. They are extremely diverse vegetatively but are characterized by the flower, which is basically insect-pollinated. Other modes of pollination, such as wind pollination, have been derived in a number of different lines. The gametophytes are much reduced, with the female gametophyte often consisting of only seven cells at maturity. Double fertilization involving the two nonmotile sperm nuclei of the mature male gametophyte gives rise to the zygote and to the primary endosperm nucleus; the former becomes the embryo and the latter a special nutritive tissue, the endosperm. About 235,000 species.

	CLASS	*Class Monocotyledones:* monocots. Flower parts are usually in threes, leaf venation is usually parallel, vascular bundles in the young stem are scattered, true secondary growth is not present, and there is one cotyledon. About 65,000 species.
		Class Dicotyledones: dicots. Flower parts are usually in fours or fives, leaf venation is usually netlike, the vascular bundles in the young stem are in a ring, there is true secondary growth with vascular cambium commonly present, and there are two cotyledons. About 170,000 species.
KINGDOM ANIMALIA		Eukaryotic multicellular organisms. Their principal mode of nutrition is by ingestion. Many animals are motile, and they generally lack the rigid cell walls characteristic of plants. Considerable cellular migration and reorganization of tissues often occur during the course of embryonic development. Their reproduction is primarily sexual, with male and female diploid organisms producing haploid gametes that fuse to form the zygote. More than 1.5 million living species have been described, and the actual number may be more than 50 million.
PHYLUM		PHYLUM PORIFERA: sponges. Simple multicellular animals, largely marine, with stiff skeletons, and bodies perforated by many pores that admit water containing food particles. All have choanocytes, "collar cells." About 5,000 species.
		PHYLUM MESOZOA: extremely simple wormlike animals, all parasites of marine invertebrates. The body consists of 20 to 30 cells, organized in two layers. About 50 species.
		PHYLUM CNIDARIA: polyps and jellyfishes. Radially symmetrical animals with a gastrovascular cavity and two-layered bodies of a jellylike consistency. Reproduction is asexual or sexual. They are the only organisms with cnidocytes, special stinging cells. All are aquatic and most are marine. About 9,000 species.
	CLASS	*Class Hydrozoa: Hydra, Obelia,* and other *Hydra*-like animals. They are often colonial and frequently have a regular alternation of asexual and sexual forms. The polyp form is dominant.
		Class Scyphozoa ("cup animals"): marine jellyfishes, including *Aurelia.* The medusa form is dominant. They have true muscle cells.
		Class Anthozoa ("flower animals"): sea anemones, colonial corals, and related forms. They have no medusa stage.
PHYLUM	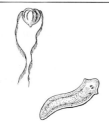	PHYLUM CTENOPHORA: comb jellies and sea walnuts. They are free-swimming, often almost spherical animals, with a gastrovascular cavity. They are translucent, gelatinous, delicately colored, and often bioluminescent. They possess eight bands of cilia, for locomotion. About 90 species.
		PHYLUM PLATYHELMINTHES: flatworms. Bilaterally symmetrical with three embryonic tissue layers. The digestive cavity has only one opening. They have no coelom or pseudocoelom and no circulatory system. They have complex hermaphroditic reproductive systems and excrete by means of special flame cells. About 13,000 species.
	CLASS	*Class Turbellaria:* planarians and other nonparasitic flatworms. They are ciliated, carnivorous, and have ocelli ("eyespots").
		Class Trematoda: flukes. They are parasitic flatworms with digestive cavities.
		Class Cestoda: tapeworms. They are parasitic flatworms with no digestive cavities; they absorb nourishment through body surfaces.
PHYLUM		PHYLUM GNATHOSTOMULIDA: tiny acoelomate marine worms, characterized by a unique pair of hard jaws. The digestive cavity has only one opening. About 80 species.
		PHYLUM RHYNCHOCOELA: proboscis, nemertine, or ribbon worms. These acoelomate worms are nonparasitic, usually marine, and have a tubelike gut with mouth and anus, a retractile proboscis armed with a hook for capturing prey, and simple circulatory and reproductive systems. About 650 species.

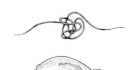

PHYLUM NEMATODA: roundworms. The phylum includes minute free-living forms, such as vinegar eels, and plant and animal parasites, such as hookworms. They have a pseudocoelom and are characterized by elongated, cylindrical, bilaterally symmetrical bodies. About 12,000 species have been described and named, and there may be as many as 400,000 to 500,000 species.

PHYLUM NEMATOMORPHA: horsehair worms. They are extremely slender, brown or black pseudocoelomate worms up to 1 meter long. Adults are free-living, but the larvae are parasitic in arthropods. About 230 species.

PHYLUM ACANTHOCEPHALA: spiny-headed worms. They are parasitic pseudocoelomate worms with no digestive tract and a head armed with many recurved spines. About 500 species.

PHYLUM KINORHYNCHA: tiny pseudocoelomate worms that burrow in muddy ocean shores. They are short-bodied, covered with spines, and have a spiny retractile proboscis. About 100 species.

PHYLUM GASTROTRICHA: microscopic, wormlike pseudocoelomate animals that move by longitudinal bands of cilia. About 400 species.

PHYLUM LORICIFERA: microscopic pseudocoelomate animals of the ocean bottom, characterized by a plate-covered body, numerous spines projecting from the head, and a retractile tube for a mouth. The first of the three known species was discovered in 1982; there may be many more.

PHYLUM ROTIFERA: microscopic, wormlike or spherical pseudocoelomate animals, "wheel animalcules." They have a complete digestive tract, flame cells, and a circle of cilia on the head, the beating of which suggests a wheel; males are minute and either degenerate or unknown in many species. About 1,500 to 2,000 species.

PHYLUM ENTOPROCTA: microscopic, stalked, sessile animals that superficially resemble hydrozoans but are much more complex, having three embryonic tissue layers, a pseudocoelom, and a complete digestive tract. They were long misclassified with the coelomate bryozoans (ectoprocts), which they also resemble. About 75 species.

PHYLUM MOLLUSCA: unsegmented coelomate animals, with a head, a mantle, and a muscular foot, variously modified. They are mostly aquatic; soft-bodied, often with one or more hard shells, and a heart with three chambers. All mollusks, except bivalves, have a radula (rasplike organ used for scraping or marine drilling). At least 47,000 living species, and perhaps many more; an additional 35,000 species are known only from their fossils.

CLASS

Class Aplacophora: solenogasters. Wormlike marine animals, with no clearly defined shell, mantle, or foot. The presence of a radula identifies them as mollusks. About 250 species.

Class Polyplacophora: chitons. The mollusks with the closest resemblance to the hypothetical primitive form, they have an elongated body covered with a mantle in which are embedded eight dorsal shell plates. About 600 species.

Class Monoplacophora: *Neopilina*. Mostly deep-sea mollusks with a large, single dorsal shell and multiple pairs of gills, nephridia, and retractor muscles. Two genera with eight species.

Class Scaphopoda: tooth or tusk shells. They are marine mollusks with a conical tubular shell. About 350 species.

Class Bivalvia: two-shelled mollusks, including clams, oysters, mussels, scallops. They usually have a hatchet-shaped foot and no distinct head. Generally sessile. At least 7,500 species.

Class Gastropoda: asymmetrical mollusks, including snails, whelks, slugs. They usually have a spiral shell and a head with one or two pairs of tentacles. At least 37,500 living species and about 15,000 fossil species.

Class Cephalopoda: octopuses, squids, *Nautilus*. They are characterized by a "head-foot" with eight or ten arms or many tentacles, mouth with two horny jaws, and well-developed eyes and nervous system. The shell is external *(Nautilus)*, internal (squid), or absent (octopus). All except *Nautilus* have ink glands. About 600 species.

PHYLUM

PHYLUM ANNELIDA: segmented worms. They usually have a well-developed coelom, a one-way digestive tract, head, circulatory system, nephridia, and well-defined nervous system. About 9,000 species.

CLASS

Class Oligochaeta: soil, freshwater, and marine annelids, including *Lumbricus* and other earthworms. They have scanty bristles and usually a poorly differentiated head. About 3,000 species.

Class Polychaeta: mainly marine worms, such as *Nereis*. They have a distinct head with tentacles, antennae, and specialized mouthparts. Parapodia are often brightly colored. About 6,000 species.

Class Hirudinea: leeches. They have a posterior sucker and usually an anterior sucker surrounding the mouth. They are freshwater, marine, and terrestrial; either free-living or parasitic. About 300 species.

PHYLUM

PHYLUM SIPUNCULA: peanut worms. Unsegmented marine worms with a stout body, a long, retractile proboscis, and a trochophore larva that resembles those of the polychaete annelids. About 300 species.

PHYLUM ECHIURA: spoon worms. Marine worms with a nonretractile proboscis that contracts to form a structure resembling a spoon. Their embryonic development and trochophore larvae resemble those of the polychaete annelids. About 100 species.

PHYLUM PRIAPULIDA: predatory burrowing marine worms, characterized by a retractile, spine-bearing proboscis. They resemble the pseudocoelomate kinorhynchs, but may have a true coelom. Only 15 species.

PHYLUM POGONOPHORA: beard worms. Unsegmented except at the posterior end, these slender marine worms live in long tubes in deep-sea sediments. Although they have no mouth or digestive tract, the anterior region of the body bears a crown of tentacles. About 100 species.

PHYLUM PENTASTOMIDA: wormlike parasites of vertebrate respiratory systems, sometimes with two pairs of short appendages at the anterior end of the body. They lack circulatory, respiratory, and excretory systems, but the nervous system resembles those of annelids and arthropods. About 70 species.

PHYLUM TARDIGRADA: water bears. Tiny segmented animals, with a thin cuticle and four pairs of stubby legs. They are common in fresh water and in the moisture film on mosses; when water is unavailable, they enter a state of suspended animation. About 350 species.

PHYLUM ONYCHOPHORA: *Peripatus*. Caterpillarlike animals with many short, unjointed pairs of legs. Their relatively soft bodies, segmentally arranged nephridia, muscular body walls, and ciliated reproductive tracts resemble those of the annelids; their antennae and eyes resemble those of both the polychaete annelids and the arthropods; their jaws, protective cuticle, brain, and circulatory and respiratory systems resemble those of arthropods. Most give birth to live young, and in some species the embryo is nourished through a placenta. About 70 species.

PHYLUM ARTHROPODA: The largest phylum in the animal kingdom, arthropods are segmented animals with paired, jointed appendages, a hard jointed exoskeleton, a complete digestive tract, reduced coelom, no nephridia, a dorsal brain, and a ventral nerve cord with paired ganglia in each segment. More than 1 million species have been classified to date.

CLASS

Class Merostomata: horseshoe crabs. Aquatic arthropods with chelicerae (pincers or fangs), pedipalps, compound eyes, four pairs of walking legs, and book gills. Only four species.

Class Pycnogonida: sea spiders. Aquatic chelicerates with slender bodies and four or, rarely, five pairs of legs, which are often very long. About 500 species.

Class Arachnida: spiders, mites, ticks, scorpions. Most members are terrestrial, air-breathing; four pairs of walking legs; chelicerae may be pincers or fangs; pedipalps are usually sensory. About 57,000 species.

Class Crustacea: lobsters, crayfish, crabs, shrimps. Crustaceans are mostly aquatic, with compound eyes, two pairs of antennae, one pair of mandibles, and typically two pairs of maxillae. The thoracic segments have appendages, and the abdominal segments are with or without appendages. About 25,000 species.

Class Chilopoda: centipedes. They have a head and 15 to 177 trunk segments, each with one pair of jointed appendages. About 3,000 species.

Class Diplopoda: millipedes. They have a head and a trunk with 20 to 200 body rings, each with two pairs of appendages. About 7,500 species.

Class Pauropoda: tiny, soft-bodied arthropods that resemble millipedes but have only 11 or 12 segments and 9 or 10 pairs of legs. They have branched antennae. About 300 species.

Class Symphyla: garden centipedes and their relatives. Soft-bodied arthropods with one pair of antennae, 12 pairs of jointed legs with claws, and a pair of unjointed posterior appendages. About 130 species.

Class Insecta: insects, including bees, ants, beetles, butterflies, fleas, lice, flies, etc. Most insects are terrestrial and breathe by means of tracheae. The body has three distinct parts (head, thorax, and abdomen); the head bears compound eyes and one pair of antennae; the thorax bears three pairs of legs and usually two pairs of wings. About 1 million species.

PHYLUM BRACHIOPCDA: lamp shells. Marine animals with two hard shells (one dorsal and one ventral), they superficially resemble clams. Fixed by a stalk or one shell in adult life, they feed by means of a lophophore. About 250 living species; 30,000 extinct species.

PHYLUM PHORONIDA: sedentary, elongated worms that secrete and live in a leathery tube. They have a U-shaped digestive tract and a ring of ciliated tentacles (the lophophore) with which they feed. Marine. Only 18 species.

PHYLUM BRYOZOA (ECTOPROCTA): "moss" animals. These microscopic aquatic organisms are characterized by the lophophore, a crown of hollow, ciliated tentacles, with which they feed and by a U-shaped digestive tract. They usually form fixed and branching colonies. Coelomates, they superficially resemble the pseudocoelomate entoprocts. Some species retain larvae in a special brood pouch. About 4,000 species.

PHYLUM ECHINODERMATA: starfish and sea urchins. Echinoderms are radially symmetrical in the adult stage, with well-developed coelomic cavities, an endoskeleton of calcareous ossicles and spines, and a unique water vascular system. They have tube feet. All are marine. About 6,000 living species and 20,000 species known only from their fossils.

Class Crinoidea: sea lilies and feather stars. Sessile animals, they often have a jointed stalk for attachment, and they have 10 arms bearing many slender lateral branches. Most species are fossils; only 20 living species.

Class Stelleroidea: starfish and brittle stars. They have 5 to 50 arms, an oral surface directed downward, and rows of tube feet on each arm. Brittle stars have greatly elongated, highly flexible slender arms and rapid horizontal locomotion.

Class Echinoidea: sea urchins and sand dollars. Skeletal plates form a rigid external covering that bears many movable spines.

Class Concentricycloidea: sea daisies. The microscopic members of this newly established class have five plates on the dorsal surface, two rings of tube feet on the ventral surface, and no digestive tract.

Class Holothuroidea: sea cucumbers. They have a sausage-shaped or wormlike elongated body.

PHYLUM CHAETOGNATHA: arrow worms. Free-swimming, planktonic marine worms, they have a coelom, a complete digestive tract, and a mouth with strong sickle-shaped hooks on each side. About 60 species.

PHYLUM HEMICHORDATA: acorn worms. The body is divided into three regions—proboscis, collar, and trunk. The coelomic cavities provide a hydrostatic skeleton similar to the water vascular system of echinoderms, and their larvae resemble starfish larvae. They have both ventral and dorsal nerve cords, and the anterior portion of the dorsal cord is hollow in some species. They also have a pharynx with gill slits. About 80 species.

PHYLUM CHORDATA: animals having at some stage a notochord, pharyngeal gill slits (or pouches), a hollow nerve cord on the dorsal side, and a tail. About 43,000 species.

SUBPHYLUM CEPHALOCHORDATA: lancelets. This small subphylum contains only *Branchiostoma* and related forms. They are somewhat fishlike marine animals with a permanent notochord the whole length of the body, a nerve cord, a pharynx with gill slits, and no cartilage or bone. About 28 species.

SUBPHYLUM UROCHORDATA: tunicates. Adults are saclike, usually sessile, often forming branching colonies. They feed by ciliary currents, have gill slits, a reduced nervous system, and no notochord. Larvae are active, with well-developed nervous system and notochord. They are marine. About 1,300 species.

SUBPHYLUM VERTEBRATA: vertebrates. The most important subphylum of Chordata. In the vertebrates the notochord is an embryonic structure; it is typically replaced in the course of development by cartilage or bone, forming the segmented vertebral column, or backbone. A cranium, or skull, surrounds a well-developed brain. Vertebrates usually have a tail. About 41,700 species.

CLASS

Class Agnatha: lampreys and hagfish. These are eel-like aquatic vertebrates without limbs, with a jawless sucking mouth, and no bones, scales, or fins. About 60 species.

Class Chondrichthyes: sharks, rays, skates, and other cartilaginous fishes. They have complicated copulatory organs, scales, and no air bladders. They are almost exclusively marine. About 625 species.

Class Osteichthyes: bony fishes, including nearly all modern freshwater fishes, such as sturgeon, trout, perch, anglerfish, lungfish, and some almost extinct groups. They usually have an air bladder or (rarely) a lung. More than 19,000 species.

Class Amphibia: salamanders, frogs, and toads. They usually respire by gills in the larval stage and by lungs in the adult stage. They have incomplete double circulation and a usually naked skin. The limbs are legs. They were the first vertebrates to inhabit the land and were ancestors of the reptiles. Their eggs are unprotected by a shell and embryonic membranes. About 2,500 species.

Class Reptilia: turtles, lizards, snakes, crocodiles; includes many extinct species such as the dinosaurs. Reptiles breathe by lungs and have incomplete double circulation. Their skin is usually covered with scales. The four limbs are legs (absent in snakes and some lizards). They are ectotherms. Most live and reproduce on land, although some are aquatic. The embryo is enclosed in an egg shell and has protective membranes. About 6,000 species.

Class Aves: birds. Birds are homeothermic animals with complete double circulation and a skin covered with feathers. The forelimbs are wings. The embryo is enclosed in an egg shell with protective membranes. About 9,000 living species; an estimated 14,000 species are extinct.

Class Mammalia: mammals. Mammals are homeothermic animals with complete double circulation. Their skin is usually covered with hair. The young are nourished with milk secreted by the mother. They have four limbs, usually legs (forelimbs, sometimes arms, wings, or fins), a diaphragm used in respiration, a lower jaw made up of a single pair of bones, three bones in each middle ear connecting eardrum and inner ear, and almost always seven vertebrae in the neck. About 4,500 species.

SUBCLASS

Subclass Prototheria: monotremes. These are the oviparous (egg-laying) mammals with imperfect temperature regulation. There are only three living species: the duckbilled platypus and spiny anteaters of Australia and New Guinea.

Subclass Metatheria: marsupials, including kangaroos, opossums, and others. Marsupials are viviparous mammals, usually with a yolk-sac placenta; the young are born in a very immature state and are carried in an external pouch of the mother for some time after birth. They are found chiefly in Australia and South America. About 260 species.

Subclass Eutheria: mammals with a well-developed chorioallantoic placenta. This subclass comprises the great majority of living mammals. There are 16 orders of Eutheria:

Order

Insectivora: shrews, moles, hedgehogs, etc.
Chiroptera: bats. Aerial mammals with the forelimbs wings.
Dermoptera: "flying" lemurs.
Edentata: toothless mammals—anteaters, sloths, armadillos, etc.
Lagomorpha: rabbits and hares.
Carnivora: carnivorous animals—cats, dogs, bears, weasels, seals, etc.
Tubulidentata: aardvarks.
Pholidota: pangolins.
Rodentia: rodents—rats, mice, squirrels, etc.
Artiodactyla: even-toed ungulates (hoofed mammals)—cattle, deer, camels, hippopotamuses, etc.
Perissodactyla: odd-toed ungulates—horses, zebras, rhinoceroses, etc.
Proboscidea: elephants.
Hyracoidea: hyraxes.
Sirenia: manatees, dugongs, and sea cows. Large aquatic mammals with the forelimbs finlike, the hind limbs absent.
Cetacea: whales, dolphins, and porpoises. Aquatic mammals with the forelimbs fins, the hind limbs absent.
Primates: prosimians, monkeys, apes, and humans.

Glossary

This list does not include units of measure or names of taxonomic groups, which can be found in Appendixes A and C, or terms that are used only once in the text and defined there.

abdomen: In vertebrates, the portion of the trunk containing visceral organs other than heart and lungs; in arthropods, the posterior portion of the body, made up of similar segments and containing the reproductive organs and part of the digestive tract.

abscisic acid (ABA) [L. *ab*, away, off + *scissio*, dividing]: A plant hormone with a variety of inhibitory effects; brings about dormancy in buds, maintains dormancy in many types of seeds, and effects stomatal closing; also known as the "stress hormone."

abscission [L. *ab*, away, off + *scissio*, dividing]: In plants, the dropping of leaves, flowers, fruits, or stems at the end of a growing season, as the result of formation of a two-layered zone of specialized cells (the abscission zone) and the action of a hormone (ethylene).

absorption [L. *absorbere*, to swallow down]: The movement of water and dissolved substances into a cell, tissue, or organism.

absorption spectrum: The characteristic pattern of the wavelengths (colors) of light that a particular pigment absorbs.

acetylcholine (asset-ill-**coal**-een): One of the principal chemicals (neurotransmitters) responsible for the transmission of nerve impulses across synapses.

acid [L. *acidus*, sour]: A substance that causes an increase in the number of hydrogen ions (H^+) in a solution and a decrease in the number of hydroxide ions (OH^-); having a pH of less than 7; the opposite of a base.

actin [Gk. *aktis*, a ray]: A protein, composed of globular subunits, that forms filaments that are among the principal components of the cytoskeleton. Also one of the two major proteins of muscle (the other is myosin); the principal constituent of the thin filaments.

action potential: A transient change in electric potential across a membrane; in nerve cells, results in conduction of a nerve impulse; in muscle cells, results in contraction.

action spectrum: The characteristic pattern of the wavelengths (colors) of light that elicit a particular reaction or response.

activation energy: The energy that must be possessed by atoms or molecules in order to react.

active site: The region of an enzyme surface that binds the substrate during the reaction catalyzed by the enzyme.

active transport: The energy-requiring transport of a solute across a cell membrane (or a membrane of an organelle) from a region of lower concentration to a region of higher concentration (that is, against a concentration gradient).

adaptation [L. *adaptare*, to fit]: (1) The evolution of features that make a group of organisms better suited to live and reproduce in their environment. (2) A peculiarity of structure, physiology, or behavior that aids the organism in its environment.

adaptive radiation: The evolution from a primitive and unspecialized ancestor of a number of divergent forms, each specialized to fill a distinct ecological niche; associated with the opening up of a new biological frontier.

adenosine diphosphate (ADP): A nucleotide consisting of adenine, ribose, and two phosphate groups; formed by the removal of one phosphate from an ATP molecule.

adenosine monophosphate (AMP): A nucleotide consisting of adenine, ribose, and one phosphate group; can be formed by the removal of two phosphates from an ATP molecule; in its cyclic form, functions as a "second messenger" for a number of vertebrate hormones and neurotransmitters.

adenosine triphosphate (ATP): The nucleotide that provides the energy currency for cell metabolism; composed of adenine, ribose, and three phosphate groups. On hydrolysis, ATP loses one phosphate group and one hydrogen ion to become adenosine diphosphate (ADP), releasing energy in the process. ATP is formed from ADP and inorganic phosphate in an enzymatic reaction that traps energy released by catabolism or energy captured in photosynthesis.

ADH: Abbreviation of antidiuretic hormone.

adhesion [L. *adhaerere*, to stick to]: The holding together of molecules of different substances.

ADP: Abbreviation of adenosine diphosphate.

adrenal gland [L. *ad*, near + *renes*, kidney]: A vertebrate endocrine gland. The cortex (outer surface) is the source of cortisol, aldosterone, and other steroid hormones; the medulla (inner core) secretes adrenaline and noradrenaline.

adrenaline: A hormone, produced by the medulla of the adrenal gland, that increases the concentration of sugar in the blood, raises blood pressure and heartbeat rate, and increases muscular power and resistance to fatigue; also a neurotransmitter across synaptic junctions. Also called epinephrine.

adventitious [L. *adventicius*, not properly belonging to]: Referring to a structure arising from an unusual place, such as roots growing from stems or leaves.

aerobic [Gk. *aēr*, air + *bios*, life]: Any biological process that can occur in the presence of molecular oxygen (O_2).

afferent [L. *ad*, near + *ferre*, to carry]: Bringing inward to a central part, applied to nerves and blood vessels.

agar: A gelatinous material prepared from certain red algae that is used to solidify nutrient media for growing microorganisms.

aldosterone [Gk. *aldainō*, to nourish + *stereō*, solid]: A hormone produced by the adrenal cortex that affects the concentration of ions in the blood; it stimulates the reabsorption of sodium and the excretion of potassium by the kidney.

alga, *pl.* **algae** (**al**-gah, **al**-jee): A unicellular or simple multicellular eukaryotic photosynthetic organism lacking multicellular sex organs.

alkaline: Pertaining to substances that increase the number of hydroxide ions (OH^-) in a solution; having a pH greater than 7; basic; opposite of acidic.

allantois [Gk. *allant*, sausage]: One of the four extraembryonic membranes that form during the development of reptiles, birds, and mammals.

allele frequency: The proportion of a particular allele in a population.

alleles (al-eels) [Gk. *allelon*, of one another]: Two or more different forms of a gene. Alleles occupy the same position (locus) on homologous chromosomes and are separated from each other at meiosis.

allopatric speciation [Gk. *allos*, other + *patra*, fatherland, country]: Speciation that occurs as the result of the geographic separation of a population of organisms.

allosteric interaction [Gk. *allos*, other + *stereō*, solid, shape]: An interaction involving an enzyme that has two binding sites, the active site and a site into which another molecule, an allosteric effector, fits; the binding of the effector changes the shape of the enzyme and activates or inactivates it. Allosteric interactions also play a role in transport processes involving integral membrane proteins.

alternation of generations: A sexual life cycle in which a haploid *(n)* phase alternates with a diploid *(2n)* phase. The gametophyte *(n)* produces gametes *(n)* by mitosis. The fusion of gametes yields zygotes *(2n)*. Each zygote develops into a sporophyte *(2n)* that forms haploid spores *(n)* by meiosis. Each haploid spore forms a new gametophyte, completing the cycle.

altruism: Self-sacrifice for the benefit of others; any form of behavior that increases the fitness of the recipient while reducing the fitness of the altruistic individual.

alveolus, *pl.* **alveoli** [L. dim. of *alveus*, cavity, hollow]: One of the many small air sacs within the lungs in which the bronchioles terminate. The thin walls of the alveoli contain numerous capillaries and are the site of gas exchange between the air in the alveoli and the blood in the capillaries.

amino acids (am-ee-no) [Gk. *Ammon*, referring to the Egyptian sun god, near whose temple ammonium salts were first prepared from camel dung]: Organic molecules containing nitrogen in the form of $-NH_2$ and a carboxyl group, $-COOH$, bonded to the same carbon atom; the "building blocks" of protein molecules.

ammonification: The process by which decomposers break down proteins and amino acids, releasing the excess nitrogen in the form of ammonia (NH_3) or ammonium ion (NH_4^+).

amnion (am-neon) [Gk. dim. of *amnos*, lamb]: One of the four extraembryonic membranes that form during the development of reptiles, birds, and mammals; it encloses a fluid-filled space, the amniotic cavity, that surrounds the developing embryo.

amniote egg: An egg that is isolated and protected from the environment during the period of its development by a series of extraembryonic membranes and, often, a more or less impervious shell; the amniote eggs of birds and many reptiles are completely self-sufficient, requiring only oxygen from the outside.

amoeboid [Gk. *amoibē*, change]: Moving or feeding by means of pseudopodia (temporary cytoplasmic protrusions from the cell body).

AMP: Abbreviation of adenosine monophosphate.

anabolism [Gk. *ana*, up + *-bolism* (as in metabolism)]: Within a cell or organism, the sum of all biosynthetic reactions (that is, chemical reactions in which larger molecules are formed from smaller ones).

anaerobe [Gk. *an*, without + *aēr*, air + *bios*, life]: Cell that can live without free oxygen; obligate anaerobes cannot live in the presence of oxygen; facultative anaerobes can live with or without oxygen.

anaerobic [Gk. *an*, without + *aēr*, air + *bios*, life]: Applied to a process that can occur without oxygen, such as fermentation; also applied to organisms that can live without free oxygen.

analogous [Gk. *analogos*, proportionate]: Applied to structures similar in function but different in evolutionary origin, such as the wing of a bird and the wing of an insect.

anaphase (anna-phase) [Gk. *ana*, up + *phasis*, form]: In mitosis and meiosis II, the stage in which the chromatids of each chromosome separate and move to opposite poles; in meiosis I, the stage in which homologous chromosomes separate and move to opposite poles.

androgens [Gk. *andros*, man + *genos*, origin, descent]: Male sex hormones; any chemical with actions similar to those of testosterone.

angiosperms (an-jee-o-sperms) [Gk. *angeion*, vessel + *sperma*, seed]: The flowering plants. Literally, a seed borne in a vessel; thus, any plant whose seeds are borne within a matured ovary (fruit).

anisogamy [Gk. *aniso*, unequal + *gamos*, marriage]: Sexual reproduction in which one gamete is larger than the other; both gametes are motile.

annual plant [L. *annus*, year]: A plant that completes its life cycle (from seed germination to seed production) and dies within a single growing season.

antennae: Long, paired sensory appendages on the head of many arthropods.

anterior [L. *ante*, before, toward, in front of]: The front end of an organism.

anther [Gk. *anthos*, flower]: In flowering plants, the pollen-bearing portion of a stamen.

antheridium, *pl.* **antheridia:** In bryophytes and some vascular plants, the multicellular sperm-producing organ.

anthropoid [Gk. *anthropos*, man, human]: A higher primate; includes monkeys, apes, and humans.

antibiotic [Gk. *anti*, against + *bios*, life]: An organic compound, inhibitory or toxic to other species, that is formed and secreted by an organism.

antibody [Gk. *anti*, against]: A globular protein, synthesized by a B lymphocyte, that is complementary to a foreign substance (antigen) with which it combines specifically.

anticodon: In a tRNA molecule, the three-nucleotide sequence that base pairs with the mRNA codon for the amino acid carried by that particular tRNA; the anticodon is complementary to the mRNA codon.

antidiuretic hormone (ADH) [Gk. *anti*, against + *diurgos*, thoroughly wet + *hormaein*, to excite]: A peptide hormone synthesized in the hypothalamus that inhibits urine excretion by inducing the reabsorption of water from the nephrons of the kidneys; also called vasopressin.

antigen [Gk. *anti*, against + *genos*, origin, descent]: A foreign substance, usually a protein or polysaccharide, that, when bound to a complementary antibody displayed on the surface of a B lymphocyte or to a complementary T-cell receptor, stimulates an immune response.

aorta (a-ore-ta) [Gk. *aeirein*, to lift, heave]: The major artery in blood-circulating systems; the aorta sends blood to the other body tissues.

apical dominance [L. *apex*, top]: In plants, the hormone-mediated influence of a terminal bud in suppressing the growth of axillary buds.

apical meristem [L. *apex*, top + Gk. *meristos*, divided]: In vascular plants, the growing point at the tip of the root or stem.

arboreal [L. *arbor*, tree]: Tree-dwelling.

archegonium, *pl.* **archegonia** [Gk. *archegonos*, first of a race]: In bryophytes and some vascular plants, the multicellular egg-producing organ.

archenteron [Gk. *arch*, first, or main + **enteron**, gut]: The main cavity within the early embryo (gastrula) of many animals; lined with endoderm, it opens to the outside by means of the blastopore and ultimately becomes the digestive tract.

artery: A vessel carrying blood from the heart to the tissues; arteries are usually thick-walled, elastic, and muscular. A small artery is known as an arteriole.

artificial selection: The breeding of selected organisms for the purpose of producing descendants with desired characteristics.

ascus, *pl.* **asci** (as-kus, as-i) [Gk. *askos*, wineskin, bladder]: In the fungi of division Ascomycota, a specialized cell within which two haploid nuclei fuse to produce a zygote that immediately divides by meiosis; at maturity, an ascus contains ascospores.

asexual reproduction: Any reproductive process, such as budding or the division of a cell or body into two or more approximately equal parts, that does not involve the union of gametes.

atmospheric pressure [Gk. *atmos*, vapor + *sphaira*, globe]: The weight of the earth's atmosphere over a unit area of the earth's surface.

atom [Gk. *atomos*, indivisible]: The smallest particle into which a chemical element can be divided and still retain the properties characteristic of the

Glossary

element; consists of a central core, the nucleus, containing protons and neutrons, and electrons that move around the nucleus.

atomic number: The number of protons in the nucleus of an atom; equal to the number of electrons in the neutral atom.

atomic weight: The average weight of all the isotopes of an element relative to the weight of an atom of the most common isotope of carbon (^{12}C), which is by convention assigned the integral value of 12; approximately equal to the number of protons plus neutrons in the nucleus of an atom.

ATP: Abbreviation of adenosine triphosphate, the principal energy-carrying compound of the cell.

ATP synthetase: The enzyme complex in the inner membrane of the mitochondrion and the thylakoid membrane of the chloroplast through which protons flow down the gradient established in the first stage of chemiosmotic coupling; the site of formation of ATP from ADP and inorganic phosphate during oxidative phosphorylation and photophosphorylation.

atrioventricular node [L. *atrium*, yard, court, hall + *ventriculus*, the stomach + *nodus*, knot]: A group of slow-conducting fibers in the atrium of the vertebrate heart that are stimulated by impulses originating in the sinoatrial node (the pacemaker) and that conduct impulses to the bundle of His, a group of fibers that stimulate contraction of the ventricles.

atrium, *pl.* **atria** (a-tree-um) [L., yard, court, hall]: A thin-walled chamber of the heart that receives blood and passes it on to a thick, muscular ventricle.

autonomic [Gk. *autos*, self + *nomos*, usage, law]: Self-controlling, independent of outside influences.

autonomic nervous system [Gk. *autos*, self + *nomos*, usage, law]: In the peripheral nervous system of vertebrates, the neurons and ganglia that are not ordinarily under voluntary control; innervates the heart, glands, visceral organs, and smooth muscle; subdivided into the sympathetic and parasympathetic divisions.

autosome [Gk. *autos*, self + *soma*, body]: Any chromosome other than the sex chromosomes. Humans have 22 pairs of autosomes and one pair of sex chromosomes.

autotroph [Gk. *autos*, self + *trophos*, feeder]: An organism that is able to synthesize all needed organic molecules from simple inorganic substances (e.g., H_2O, CO_2, NH_3) and some energy source (e.g., sunlight); in contrast to heterotroph. Plants, algae, and some groups of prokaryotes are autotrophs.

auxin [Gk. *auxein*, to increase + *in*, of, or belonging to]: One of a group of plant hormones with a variety of growth-regulating effects, including promotion of cell elongation.

auxotroph [L. *auxillium*, help + Gk. *trophos*, feeder]: A mutant with a defect in the enzymatic pathway for the synthesis of a particular molecule, which must therefore be supplied for normal growth.

axil [Gk. *axilla*, armpit]: The upper angle between a twig or leaf and the stem from which it grows.

axillary [Gk. *axilla*, armpit]: In botany, term applied to buds or branches occurring in the axil of a leaf.

axis: An imaginary line passing through a body or organ around which parts are symmetrically aligned.

axon [Gk. *axon*, axle]: A long process of a neuron, or nerve cell, that is capable of rapidly conducting nerve impulses over great distances.

B lymphocyte: A type of white blood cell capable of becoming an antibody-secreting plasma cell; a B cell.

bacteriophage [L. *bacterium* + Gk. *phagein*, to eat]: A virus that parasitizes a bacterial cell.

bark: In plants, all tissues outside the vascular cambium in a woody stem.

basal body [Gk. *basis*, foundation]: A cytoplasmic organelle of animals and some protists, from which cilia or flagella arise; identical in structure to the centriole, which is involved in mitosis and meiosis in animals and some protists.

base: A substance that causes an increase in the number of hydroxide ions (OH^-) in a solution and a decrease in the number of hydrogen ions (H^+); having a pH of more than 7; the opposite of an acid. *See* Alkaline.

base-pairing principle: In the formation of nucleic acids, the requirement that adenine must always pair with thymine (or uracil) and guanine with cytosine.

basidium, *pl.* **basidia** (ba-sid-ium) [L., a little pedestal]: A specialized reproductive cell of the fungi of division Basidiomycota, often club-shaped, in which nuclear fusion and meiosis occur; homologous with the ascus.

behavior: All of the acts an organism performs, as in, for example, seeking a suitable habitat, obtaining food, avoiding predators, and seeking a mate and reproducing.

biennial [L. *biennium*, a space of two years; *bi*, twice + *annus*, year]: Occurring once in two years; a plant that requires two years to complete its reproductive cycle; vegetative growth occurs in the first year, sexual reproduction and death in the second.

bilateral symmetry [L. *bi*, twice, two + *lateris*, side; Gk. *summetros*, symmetry]: A body form in which the right and left halves of an organism are approximate mirror images of each other.

bile: A yellow secretion of the vertebrate liver, temporarily stored in the gallbladder and composed of organic salts that emulsify fats in the small intestine.

binary fission [L. *binarius*, consisting of two things or parts + *fissus*, split]: Asexual reproduction by division of the cell or body into two equal, or nearly equal, parts.

binomial system [L. *bi*, twice, two + Gk. *nomos*, usage, law]: A system of naming organisms in which the name consists of two parts, with the first designating genus and the second, species; originated by Linnaeus.

biogeochemical cycle [Gk. *bios*, life + *geō*, earth + *chēmeia*, alchemy; *kyklos*, circle, wheel]: The cyclic path of an inorganic substance, such as carbon or nitrogen, through an ecosystem. Its geological components are the atmosphere, the crust of the earth, and the oceans, lakes, and rivers; its biological components are producers, consumers, and detritivores, including decomposers.

biological clock [Gk. *bios*, life + *logos*, discourse]: Proposed internal factor(s) in organisms that governs functions that occur rhythmically in the absence of external stimuli.

biomass [Gk. *bios*, life]: Total weight of all organisms (or some group of organisms) living in a particular habitat or place.

biome: One of the major types of distinctive plant formations; for example, the grassland biome, the tropical rain forest biome, etc.

biosphere [Gk. *bios*, life + *sphaira*, globe]: The zones of air, land, and water at the surface of the earth occupied by living things.

biosynthesis [Gk. *bios*, life + *synthesis*, a putting together]: Formation by living organisms of organic compounds from elements or simple compounds.

blade: (1) The broad, expanded part of a leaf. (2) The broad, expanded photosynthetic part of the thallus of a multicellular alga or a simple plant.

blastocoel [Gk. *blastos*, sprout + *koilos*, a hollow]: The fluid-filled cavity in the interior of a blastula.

blastocyst [Gk. *blastos*, sprout + *kystis*, sac]: The blastula stage of a developing mammal; consists of an inner cell mass that will give rise to the embryo proper and a double layer of cells, the trophoblast, that is the precursor of the chorion.

blastodisc [Gk. *blastos*, sprout + *discos*, a round plate]: Disklike area on the surface of a large, yolky egg that undergoes cleavage and gives rise to the embryo.

blastomere [Gk. *blastos*, sprout + *meris*, part of, portion]: One of many cells produced by cleavage of the fertilized egg.

blastopore [Gk. *blastos*, sprout + *poros*, a way, means, path]: In the gastrula stage of an embryo, the opening that connects the archenteron with the outside; represents the future mouth in some animals (protostomes), the future anus in others (deuterostomes).

blastula [Gk. *blastos*, sprout]: An animal embryo after cleavage and before gastrulation; usually consists of a fluid-filled sphere, the walls of which are composed of a single layer of cells.

bond strength: The strength with which a chemical bond holds two atoms together; conventionally measured in terms of the amount of energy, in kilocalories per mole, required to break the bond.

botany [Gk. *botanikos*, of herbs]: The study of plants.

Bowman's capsule: In the vertebrate kidney, the bulbous unit of the nephron, which surrounds the glomerulus. In filtration, the initial process in urine formation, blood plasma is forced from the glomerular capillaries into Bowman's capsule.

brainstem: The most posterior portion of the vertebrate brain; includes medulla, pons, and midbrain.

bronchus, *pl.* **bronchi** (bronk-us, bronk-eye) [Gk. *bronchos*, windpipe]: One of a pair of respiratory tubes branching into either lung at the lower end of the trachea; it subdivides into progressively finer passageways, the bronchioles, culminating in the alveoli.

bud: (1) In plants, an embryonic shoot, including rudimentary leaves, often protected by special bud scales. (2) In animals, an asexually produced outgrowth that develops into a new individual.

buffer: A combination of H^+-donor and H^+-acceptor forms of a weak acid or a weak base; a buffer prevents appreciable changes of pH in solutions to which small amounts of acids or bases are added.

bulb: A modified bud with thickened leaves adapted for underground food storage.

bulk flow: The overall movement of a fluid induced by gravity, pressure, or an interplay of both.

bundle of His: In the vertebrate heart, a group of muscle fibers that carry impulses from the atrioventricular node to the walls of the ventricles; the only electrical bridge between the atria and the ventricles.

C_3 pathway: *See* Calvin cycle.

C_4 pathway: The set of reactions by which some plants initially fix carbon in the four-carbon compound oxaloacetic acid; the carbon dioxide is later released in the interior of the leaf and enters the Calvin cycle. Also known as the Hatch-Slack pathway.

callus [L. *callos*, hard skin]: In plants, undifferentiated tissue; a term used in tissue culture, grafting, and wound healing.

calorie [L. *calor*, heat]: The amount of energy in the form of heat required to raise the temperature of 1 gram of water 1°C; in making metabolic measurements the kilocalorie (Calorie) is generally used. A Calorie is the amount of heat required to raise the temperature of 1 kilogram of water 1°C.

Calvin cycle: The set of reactions in which carbon dioxide is reduced to carbohydrate during the second stage of photosynthesis.

calyx [Gk. *kalyx*, a husk, cup]: Collectively, the sepals of a flower.

CAM photosynthesis: *See* Crassulacean acid metabolism.

capillaries [L. *capillaris*, relating to hair]: Smallest thin-walled blood vessels through which exchanges between blood and the tissues occur; connect arteries with veins.

capillary action: The movement of water or any liquid along a surface; results from the combined effect of cohesion and adhesion.

capsid: The protein coat surrounding the nucleic acid core of a virus.

capsule (kap-sul) [L. *capsula*, a little chest]: (1) A slimy layer around the cells of certain bacteria. (2) The sporangium of a bryophyte.

carbohydrate [L. *carbo*, charcoal + *hydro*, water]: An organic compound consisting of a chain or ring of carbon atoms to which hydrogen and oxygen are attached in a ratio of approximately 2:1; carbohydrates include sugars, starch, glycogen, cellulose, etc.

carbon cycle: Worldwide circulation and reutilization of carbon atoms, chiefly due to metabolic processes of living organisms. Inorganic carbon, in the form of carbon dioxide, is incorporated into organic compounds by photosynthetic organisms; when the organic compounds are broken down in respiration, carbon dioxide is released. Large quantities of carbon are "stored" in the seas and the atmosphere, as well as in fossil fuel deposits.

carbon fixation: The second stage of photosynthesis; energy stored in ATP and NADPH by the energy-capturing reactions of the first stage is used to reduce carbon from carbon dioxide to simple sugars.

cardiovascular system [Gk. *kardio*, heart + L. *vasculum*, a small vessel]: In animals, the heart and blood vessels.

carnivore [L. *caro, carnis*, flesh + *voro*, to devour]: Predator that obtains its nutrients and energy by eating meat.

carotenoids [L. *carota*, carrot]: A class of pigments that includes the carotenes (yellows, oranges, and reds) and the xanthophylls (yellow); accessory pigments in photosynthesis.

carpel [Gk. *karpos*, fruit]: A leaflike floral structure enclosing the ovule or ovules of angiosperms, typically divided into ovary, style, and stigma; a flower may have one or more carpels, either single or fused. A single carpel or a group of fused carpels is also known as a pistil.

carrying capacity: In ecology, the average number of individuals of a particular population that the environment can support under a particular set of conditions.

cartilage [L. *cartilago*, gristle]: A connective tissue in skeletons of vertebrates; forms much of the skeleton of adult lower vertebrates and immature higher vertebrates.

Casparian strip (after Robert Caspary, German botanist): In the roots of plants, a thickened, waxy strip that extends around and seals the walls of endodermal cells, restricting the diffusion of solutes across the endodermis into the vascular tissues of the root.

catabolism [Gk. *katabole*, throwing down]: Within a cell or organism, the sum of all chemical reactions in which large molecules are broken down into smaller parts.

catalyst [Gk. *katalysis*, dissolution]: A substance that lowers the activation energy of a chemical reaction by forming a temporary association with the reacting molecules; as a result, the rate of the reaction is accelerated. Enzymes are catalysts.

category [Gk. *katēgoria*, category]: In a hierarchical classification system, the level at which a particular group is ranked.

cell [L. *cella*, a chamber]: The structural unit of organisms, surrounded by a membrane and composed of cytoplasm and, in eukaryotes, one or more nuclei. In most plants, fungi, and bacteria, there is a cell wall outside the membrane.

cell cycle: A regular, timed sequence of the events of cell growth and division through which dividing cells pass.

cell membrane: The outer membrane of the cell; also called the plasma membrane.

cell plate: In the dividing cells of most plants (and in some algae), a flattened structure that forms at the equator of the mitotic spindle in early telophase; gives rise to the middle lamella.

cell theory: All living things are composed of cells; cells arise only from other cells. No exception has been found to these two principles since they were first proposed well over a century ago.

cellulose [L. *cellula*, a little cell]: The chief constituent of the cell wall in all plants and some protists; an insoluble complex carbohydrate formed of microfibrils of glucose molecules.

cell wall: A plastic or rigid structure, produced by the cell and located outside the cell membrane in most plants, algae, fungi, and prokaryotes; in plant cells, it consists mostly of cellulose.

central nervous system: In vertebrates, the brain and spinal cord; in invertebrates it usually consists of one or more cords of nervous tissue plus their associated ganglia.

centriole (sen-tree-ole) [Gk. *kentron*, center]: A cytoplasmic organelle identical in structure to a basal body; flagellated cells and all animal cells, including those without flagella, have centrioles at the spindle poles during division.

centromere (sen-tro-mere) [Gk. *kentron*, center + *meros*, a part]: Region of constriction of chromosome that holds sister chromatids together.

cerebellum [L. dim. of *cerebrum*, brain]: A subdivision of the vertebrate brain that lies above the brainstem and behind and below the cerebrum; functions in coordinating muscular activities and maintaining equilibrium.

cerebral cortex [L. *cerebrum*, brain]: A thin layer of neurons and glial cells forming the upper surface of the cerebrum, well developed only in mammals; the seat of conscious sensations and voluntary muscular activity.

cerebrum [L., brain]: The portion of the vertebrate brain occupying the upper part of the skull, consisting of two cerebral hemispheres united by the corpus callosum; coordinates most activities.

character displacement: A phenomenon in which species that live together in the same environment tend to diverge in those characteristics that overlap; exemplified by Darwin's finches.

chelicera, *pl.* **chelicerae** [Gk. *cheilos*, the edge, lips + *cheir*, arm]: First pair of appendages in horseshoe crabs, sea spiders, and arachnids; usually take the form of pincers or fangs.

chemical reaction: An interaction among atoms, ions, or molecules that results in the formation of new combinations of atoms, ions, or molecules; the making or breaking of chemical bonds.

chemiosmotic coupling: The mechanism by which ADP is phosphorylated to ATP in mitochondria and chloroplasts. The energy released as electrons pass down an electron transport chain is used to establish a proton gradient across an inner membrane of the organelle; when protons subsequently flow down this electrochemical gradient, the potential energy released is captured in the terminal phosphate bonds of ATP.

chemoreceptor: A sensory cell or organ that responds to the presence of a specific chemical stimulus; includes smell and taste receptors.

chemosynthetic: Applied to autotrophic bacteria that use the energy released by specific inorganic reactions to power their life processes, including the synthesis of organic molecules.

chemotactic [Gk. *cheimō*, storm + *taxis*, arrangement, order]: Of an organism, capable of responding to a chemical stimulus by moving toward or away from it.

chiasma, *pl.* **chiasmata** (kye-az-ma) [Gk., a cross]: Connection between paired homologous chromosomes at meiosis; the site at which crossing over and genetic recombination occur.

chitin (kye-tin) [Gk. *chitōn*, a tunic, undergarment]: A tough, resistant, nitrogen-containing polysaccharide present in the exoskeleton of arthropods, the epidermal cuticle or other surface structures of many other invertebrates, and in the cell walls of fungi.

chlorophyll [Gk. *chloros*, green + *phyllon*, leaf]: A class of green pigments that are the receptors of light energy in photosynthesis.

chloroplast [Gk. *chloros*, green + *plastos*, formed]: A membrane-bound, chlorophyll-containing organelle in eukaryotes (algae and plants) that is the site of photosynthesis.

chorion (core-ee-on) [Gk., skin, leather]: The outermost extraembryonic membrane of developing reptiles, birds, and mammals; in placental mammals it contributes to the structure of the placenta.

chromatid (crow-ma-tid) [Gk. *chrōma*, color]: Either of the two strands of a replicated chromosome, which are joined at the centromere.

chromatin [Gk. *chrōma*, color]: The deeply staining complex of DNA and histone proteins of which eukaryotic chromosomes are composed.

chromosome [Gk. *chrōma*, color + *soma*, body]: The structure that carries the genes. Eukaryotic chromosomes are visualized as threads or rods of chromatin, which appear in a contracted form during mitosis and meiosis, and are otherwise enclosed in a nucleus. Prokaryotic chromosomes consist of a closed circle of DNA with which a variety of proteins are associated. Viral chromosomes are linear or circular molecules of DNA or RNA.

chromosome map: A diagram of the linear order of the genes on a chromosome.

cilium, *pl.* **cilia** (silly-um) [L., eyelash]: A short, thin structure embedded in the surface of some eukaryotic cells, usually in large numbers and arranged in rows; has a highly characteristic internal structure of two inner microtubules surrounded by nine pairs of outer microtubules; involved in locomotion and the movement of substances across the cell surface.

circadian rhythms [L. *circa*, about + *dies*, day]: Regular rhythms of growth or activity that occur on an approximately 24-hour cycle.

cladogenesis [Gk. *clados*, branch + *genesis*, origin]: The splitting of an evolutionary lineage into two or more separate lineages; one of the principal patterns of evolutionary change; also known as splitting evolution.

class: A taxonomic grouping of related, similar orders; category above order and below phylum.

cleavage: The successive cell divisions of a fertilized egg of an animal to form a multicellular blastula.

cline [Gk. *klinein*, to lean]: A graded series of changes in some characteristic within a species, correlated with some gradual change in temperature, humidity, or other environmental factor over the geographic range of the species.

cloaca [L., sewer]: The exit chamber from the digestive system in reptiles and birds; also may serve as the exit for the reproductive and urinary systems.

clone [Gk. *klon*, twig]: A line of cells, all of which have arisen from the same single cell by repeated cell divisions; a population of individuals derived by asexual reproduction from a single ancestor.

cnidocyte (ni-do-site) [Gk. *knide*, nettle + *kytos*, vessel]: A stinging cell containing a nematocyst; characteristic of cnidarians.

coadaptive gene complex: A group of genes that collectively produce coordinated phenotypic characteristics; when linked together on one chromosome, known as a supergene.

cochlea [Gk. *kochlias*, snail]: Part of the inner ear of mammals; concerned with hearing.

codominance: In genetics, the phenomenon in which the effects of both alleles at a particular locus are apparent in the phenotype of the heterozygote.

codon (code-on): Basic unit ("letter") of the genetic code; three adjacent nucleotides in a molecule of DNA or mRNA that form the code for a specific amino acid or for polypeptide chain termination.

coelom (see-loam) [Gk. *koilos*, a hollow]: A body cavity formed between layers of mesoderm and in which the digestive tract and other internal organs are suspended.

coenocytic (see-no-sit-ik) [Gk. *koinos*, shared in common + *kytos*, a vessel]: An organism or part of an organism consisting of many nuclei within a common cytoplasm.

coenzyme [L. *co*, together + Gk. *en*, in + *zyme*, leaven]: A nonprotein organic molecule that plays an accessory role in enzyme-catalyzed processes, often by acting as a donor or acceptor of a substance involved in the reaction. NAD^+, FAD, and coenzyme A are common coenzymes.

coevolution [L. *co*, together + *e-*, out + *volvere*, to roll]: The simultaneous evolution of adaptations in two or more populations that interact so closely that each is a strong selective force on the other.

cofactor: A nonprotein component that plays an accessory role in enzyme-catalyzed processes; some cofactors are ions, and others are coenzymes.

cohesion [L. *cohaerere*, to stick together]: The attraction or holding together of molecules of the same substance.

cohesion-tension theory: A theory accounting for the upward movement of water in plants. According to this theory, transpiration of a water molecule results in a negative (below 1 atmosphere) pressure in the leaf cells, inducing the entrance from the vascular tissue of another water molecule, which, because of the cohesive property of water, pulls with it a chain of water molecules extending up from the cells of the root tip.

coleoptile (coal-ee-op-tile) [Gk. *koleon*, sheath + *ptilon*, feather]: The sheath enclosing the apical meristem and leaf primordia of a germinating monocot.

collagen [Gk. *kolla*, glue]: A fibrous protein in bones, tendons, and other connective tissues.

collenchyma [Gk. *kolla*, glue]: In plants, a type of supporting cell with an irregularly thickened primary cell wall; alive at maturity.

colony: A group of organisms of the same species living together in close association.

commensalism [L. *com*, together + *mensa*, table]: See Symbiosis.

community: All of the populations of organisms inhabiting a common environment and interacting with one another.

companion cell: In angiosperms, a specialized parenchyma cell associated with a sieve-tube member and arising from the same mother cell as the sieve-tube member.

competition: Interaction between members of the same population or of two or more populations using the same resource, often present in limited supply; interference competition involves fighting or other direct interactions, whereas exploitative competition involves the removal or preemption of a resource.

competitive exclusion: The hypothesis that two species with identical ecological requirements cannot coexist stably in the same locality and the species that is more efficient in utilizing the available resources will exclude the other; also known as Gause's principle, after the Russian biologist G. F. Gause.

complementary DNA (cDNA): DNA molecules synthesized by reverse transcriptase from an RNA template.

compound [L. *componere*, to put together]: A chemical substance composed of two or more kinds of atoms in definite ratios.

compound eye: In arthropods, a complex eye composed of many separate elements, each with light-sensitive cells and a lens that can form an image.

condensation [L. *co*, together + *densare*, to make dense]: A type of chemical reaction in which two molecules join to form one larger molecule, simultaneously splitting out a molecule of water. The biosynthetic reactions in which monomers (e.g., monosaccharides, amino acids) are joined to form polymers (e.g., polysaccharides, polypeptides) are condensation reactions.

conditioning: A form of learning in which one stimulus (the conditional stimulus) comes to be associated with and elicit the same response as another stimulus (the unconditional stimulus).

cone: (1) In plants, the reproductive structure of a conifer. (2) In vertebrates, a type of photoreceptor cell in the retina, concerned with the perception of color and with the most acute discrimination of detail.

conjugation [L. *conjugatio*, a joining, connection]: The sexual process in some unicellular organisms by which genetic material is transferred from one cell to another by cell-to-cell contact.

connective tissues: Supporting or packing tissues that lie between groups of nerves, glands, and muscle cells, and beneath epithelial cells, in which the cells are irregularly distributed through a relatively large amount of extracellular material; include bone, cartilage, blood, and lymph.

consumer, in ecological systems: A heterotroph that derives its energy from living or freshly killed organisms or parts thereof. Primary consumers are herbivores; higher-level consumers are carnivores.

continental drift: The gradual movement of the earth's continents that has occurred over hundreds of millions of years.

continuous variation: A gradation of small differences in a particular trait, such as height, within a population; occurs in traits that are controlled by a number of genes.

convergent evolution [L. *convergere*, to turn together; *evolutio*, to unfold]: The independent development of similarities between unrelated groups, such as porpoises and sharks, resulting from adaptation to similar environments.

cork [L. *cortex*, bark]: A secondary tissue that is a major constituent of bark in woody and some herbaceous plants; made up of flattened cells, dead at maturity; restricts gas and water exchange and protects the vascular tissues from injury.

cork cambium [L. *cortex*, bark + *cambium*, exchange]: The lateral meristem that produces cork.

corolla (ko-role-a) [L. dim. of *corona*, wreath, crown]: Petals, collectively; usually the conspicuously colored flower parts.

corpus callosum [L., callous body]: In the vertebrate brain, a tightly packed mass of myelinated nerve fibers connecting the two cerebral hemispheres.

corpus luteum [L., yellowish body]: An ovarian structure that secretes estrogens and progesterone, which maintain the uterus during pregnancy. It develops from the remaining cells of the ruptured follicle following ovulation.

cortex [L., bark]: (1) The outer, as opposed to the inner, part of an organ, as in the adrenal gland. (2) In a stem or root, the primary tissue bounded externally by the epidermis and internally by the central cylinder of vascular tissue.

cosmid: In recombinant DNA technology, a vector constructed of a DNA segment flanked by cohesive regions (COS regions) of the bacteriophage lambda.

cotyledon (cottle-ee-don) [Gk. *kotyledon*, a cup-shaped hollow]: A leaflike structure of the embryo of a seed plant; contains stored food used during germination.

countercurrent exchange: An anatomical device for manipulating gradients so as to maximize uptake (or minimize loss) of O_2, heat, etc.

coupled reactions: In cells, the linking of endergonic (energy-requiring) reactions to exergonic (energy-releasing) reactions that provide enough energy to drive the endergonic reactions forward.

covalent bond [L. *con*, together + *valere*, to be strong]: A chemical bond formed as a result of the sharing of one or more pairs of electrons.

Crassulacean acid metabolism: A process by which some species of plants in hot, dry climates take in carbon dioxide during the night, fixing it in organic acids; the carbon dioxide is released during the day and used immediately in the Calvin cycle.

cristae: The "shelves" formed by the intricate folding of the inner membrane of the mitochondrion.

cross-fertilization: Fusion of gametes formed by different individuals; as opposed to self-fertilization.

crossing over: During meiosis, the exchange of genetic material between paired chromatids of homologous chromosomes.

cuticle (ku-tik-l) [L. *cuticula*, dim. of *cutis*, the skin]: (1) In plants, a layer of waxy substance (cutin) on the outer surface of epidermal cell walls. (2) In animals, the noncellular, outermost layer of many invertebrates.

cyclic AMP: A form of adenosine monophosphate (AMP) in which the atoms of the phosphate group form a ring; functions in chemical communication in slime molds, in positive regulation of operons, and as a "second messenger" for a number of vertebrate hormones and neurotransmitters.

cytochromes [Gk. *kytos*, vessel + *chrōma*, color]: Heme-containing proteins that participate in electron transport chains; involved in cellular respiration and photosynthesis.

cytokinesis [Gk. *kytos*, vessel + *kinesis*, motion]: Division of the cytoplasm of a cell following nuclear division.

cytokinin [Gk. *kytos*, vessel + *kinesis*, motion]: One of a group of chemically related plant hormones that promote cell division, among other effects.

cytoplasm (sight-o-plazm) [Gk. *kytos*, vessel + *plasma*, anything molded]: The living matter within a cell, excluding the genetic material.

cytoskeleton: A network of filamentous protein structures within the cytoplasm that maintains the shape of the cell, anchors its organelles, and is involved in cell motility; includes microtubules, actin filaments, and intermediate filaments.

deciduous [L. *decidere*, to fall off]: Refers to plants that shed their leaves at a certain season.

decomposers: Specialized detritivores, usually bacteria or fungi, that consume such substances as cellulose and nitrogenous waste products. Their metabolic processes release inorganic nutrients, which are then available for reuse by plants and other organisms.

denaturation: The loss of the native configuration of a macromolecule resulting, for instance, from heat treatment, extreme pH changes, chemical treatment, or other denaturing agents. It is usually accompanied by a loss of biological activity.

dendrite [Gk. *dendron*, tree]: A process of a neuron, typically branched, that receives stimuli from other cells.

denitrification: The process by which certain bacteria living in poorly aerated soils break down nitrates, using the oxygen for their own respiration and releasing nitrogen back into the atmosphere.

density-dependent factors: Factors affecting the birth rate or mortality rate of a population, the effects of which vary with the density (number of individuals per unit area or volume) of the population; include resources for which members of the same or different populations compete, predation, and disease.

density-independent factors: Factors affecting the birth rate or mortality rate of a population, the effects of which are independent of the density of the population; often involve weather-related events.

deoxyribonucleic acid (DNA) (dee-ox-y-rye-bo-new-**clay**-ick): The carrier of genetic information in cells, composed of two complementary chains of nucleotides wound in a double helix; capable of self-replication as well as coding for RNA synthesis.

dermis [Gk. *derma*, skin]: The inner layer of the skin, beneath the epidermis.

desmosome [Gk. *desmos*, bond + *soma*, body]: A type of cell-cell junction that provides mechanical strength in animal tissues; consists of a plaque of dense fibrous material between adjacent cells, with clusters of filaments looping in and out from the cytoplasm of the two cells.

detritivores [L. *detritus*, worn down, worn away + *voro*, to devour]: Organisms that live on dead and discarded organic matter; include large scavengers, smaller animals such as earthworms and some insects, as well as decomposers (fungi and bacteria).

deuterostome [Gk. *deuteros*, second + *stoma*, mouth]: An animal in which the anus forms at or near the blastopore in the developing embryo and the mouth forms secondarily elsewhere; echinoderms and chordates are deuterostomes. Deuterostomes are also characterized by radial cleavage during the earliest stages of development and by enterocoelous formation of the coelom.

development: The progressive production of the phenotypic characteristics of a multicellular organism, beginning with the fertilization of an egg.

diaphragm [Gk. *diaphrassein*, to barricade]: In mammals, a sheetlike tissue (tendon and muscle) forming the partition between the abdominal and thoracic cavities; functions in breathing.

dicotyledon (**dye**-cottle-ee-don) [Gk. *di*, double, two + *kotyledon*, a cup-shaped hollow]: A member of the class of flowering plants having two seed leaves, or cotyledons, among other distinguishing features; often abbreviated as dicot.

diencephalon [Gk. *di*, two + *enkephalos*, brain]: One of the two principal subdivisions of the vertebrate forebrain; the posterior portion of the forebrain, it contains the thalamus and the hypothalamus.

differentiation: The developmental process by which a relatively unspecialized cell or tissue undergoes a progressive (usually irreversible) change to a more specialized cell or tissue.

diffusion [L. *diffundere*, to pour out]: The net movement of suspended or dissolved particles down a concentration gradient as a result of the random spontaneous movements of individual particles; the process tends to distribute the particles uniformly throughout a medium.

digestion [L. *digestio*, separating out, dividing]: The breakdown of complex, usually insoluble foods into molecules that can be absorbed into the body and used by the cells.

dikaryon (dye-**care**-ee-on) [Gk. *di*, two + *karyon*, kernel]: A cell or organism with paired but not fused nuclei derived from different parents; found among the fungi of divisions Ascomycota and Basidiomycota.

dioecious (dye-**ee**-shus) [Gk. *di*, two + *oikos*, house]: In angiosperms, having the male (staminate) and female (carpellate) flowers on different individuals of the same species.

diploid [Gk. *di*, double, two + *ploion*, vessel]: The condition in which each autosome is represented twice ($2n$); in contrast to haploid (n).

disaccharide [Gk. *di*, two + *sakcharon*, sugar]: A carbohydrate molecule composed of two monosaccharide monomers; examples are sucrose, maltose, and lactose.

diurnal [L. *diurnus*, of the day]: Applied to organisms that are active during the daylight hours.

division: A taxonomic grouping of related, similar classes; a high-level category below kingdom and above class. Division is generally used in the classification of prokaryotes, algae, fungi, and plants, whereas an equivalent category, phylum, is used in the classification of protozoa and animals.

DNA: Abbreviation of deoxyribonucleic acid.

dominant allele: An allele whose phenotypic effect is the same in both the heterozygous and homozygous conditions.

dormancy [L. *dormire*, to sleep]: A period during which growth ceases and metabolic activity is greatly reduced; dormancy is broken when certain requirements, for example, of temperature, moisture, or day length, are met.

dorsal [L. *dorsum*, the back]: Pertaining to or situated near the back; opposite of ventral.

dorsal lip: The tissue on the dorsal edge of the blastopore of the vertebrate embryo; the prospective chordamesoderm, it functions as an organizer, inducing undifferentiated cells to follow a specific course of development.

double fertilization: A phenomenon unique to the angiosperms, in which the egg and one sperm nucleus fuse (resulting in a $2n$ fertilized egg, the zygote) and simultaneously the second sperm nucleus fuses with the two polar nuclei (resulting in a $3n$ endosperm nucleus).

duodenum (duo-**dee**-num) [L. *duodeni*, twelve each—from its length, about 12 fingers' breadth]: The upper portion of the small intestine in vertebrates, where food is digested into molecules that can be absorbed by intestinal cells.

ecological niche: A description of the roles and associations of a particular species in the community of which it is a part; the way in which an organism interacts with all of the biotic and abiotic factors in its environment.

ecological pyramid: A graphic representation of the quantitative relationships of numbers of organisms, biomass, or energy flow between the trophic levels of an ecosystem. Because large amounts of energy and biomass are dissipated at every trophic level, these diagrams nearly always take the form of pyramids.

ecological succession: The gradual process by which the species composition of a community changes.

ecology [Gk. *oikos*, home + *logos*, a discourse]: The study of the interactions of organisms with their physical environment and with each other and of the results of such interactions.

ecosystem [Gk. *oikos*, home + *systema*, that which is put together]: The organisms in a community plus the associated abiotic factors with which they interact.

ecotype [Gk. *oikos*, home + L. *typus*, image]: A locally adapted variant of a species, differing genetically from other ecotypes of the same species.

ectoderm [Gk. *ecto*, outside + *derma*, skin]: One of the three embryonic tissue layers of animals; it gives rise to the outer covering of the body, the sensory receptors, and the nervous system.

ectotherm [Gk. *ecto*, outside + *therme*, heat]: An organism, such as a reptile, that maintains its body temperature by taking in heat from the environment or giving it off to the environment. *See also* Poikilotherm.

effector [L. *ex*, out of + *facere*, to make]: Cell, tissue, or organ (such as muscle or gland) capable of producing a response to a stimulus.

efferent [L. *ex*, out of + *ferre*, to bear]: Carrying away from a center, applied to nerves and blood vessels.

egg: A female gamete, which usually contains abundant cytoplasm and yolk; nonmotile and often larger than a male gamete.

electric potential: The difference in the amount of electric charge between a region of positive charge and a region of negative charge. The establishment of electric potentials across cell and organelle membranes makes possible a number of phenomena, including the chemiosmotic synthesis of ATP, the conduction of nerve impulses, and muscle contraction.

electron: A subatomic particle with a negative electric charge equal in magnitude to the positive charge of the proton but with a much smaller mass; normally found within orbitals surrounding the atom's positively charged nucleus.

electron acceptor: Substance that accepts or receives electrons in an oxidation-reduction reaction, becoming reduced in the process.

electron carrier: A specialized molecule, such as a cytochrome, that can lose and gain electrons reversibly, alternately becoming oxidized and reduced.

electron donor: Substance that donates or gives up electrons in an oxidation-reduction reaction, becoming oxidized in the process.

electron transport: The movement of electrons down a series of electron-carrier molecules that hold electrons at slightly different energy levels; as electrons move down the chain, the energy released is used to form ATP from ADP and phosphate. Electron transport plays an essential role in the final stage of cellular respiration and in the energy-capturing reactions of photosynthesis.

element: A substance composed only of atoms of the same atomic number and that cannot be decomposed by ordinary chemical means.

embryo [Gk. *en*, in + *bryein*, to swell]: The early developmental stage of an organism produced from a fertilized egg; a young organism before it emerges from the seed, egg, or body of its mother. In humans, refers to the first two months of intrauterine life. *See* Fetus.

embryo sac: The female gametophyte of a flowering plant, contained within an ovule; typically consists of seven cells with a total of eight haploid nuclei.

endergonic [Gk. *endon*, within + *ergon*, work]: Energy-requiring, as in a chemical reaction; applied to an "uphill" process.

endocrine gland [Gk. *endon*, within + *krinein*, to separate]: Ductless gland whose secretions (hormones) are released into the extracellular spaces, from which they diffuse into the circulatory system; in vertebrates, includes pituitary, sex glands, adrenal, thyroid, and others.

endocytosis [Gk. *endon*, within + *kytos*, vessel]: A cellular process in which material to be taken into the cell induces the membrane to form a vacuole enclosing the material; the vacuole is released into the cytoplasm. Includes phagocytosis (endocytosis of solid particles), pinocytosis (endocytosis of liquids), and receptor-mediated endocytosis.

endoderm [Gk. *endon*, within + *derma*, skin]: One of the three embryonic tissue layers of animals; it gives rise to the epithelium that lines certain internal structures, such as most of the digestive tract and its outgrowths, most of the respiratory tract, and the urinary bladder, liver, pancreas, and some endocrine glands.

endodermis [Gk. *endon*, within + *derma*, skin]: In plants, a layer of specialized cells, one cell thick, that lies between the cortex and the vascular tissues in young roots. The Casparian strip of the endodermis prevents diffusion of solutes across the root.

endometrium [Gk. *endon*, within + *metrios*, of the womb]: The glandular lining of the uterus in mammals; thickens in response to secretion of estrogens and progesterone; one of its two principal layers is sloughed off in menstruation.

endoplasmic reticulum [Gk. *endon*, within + *plasma*, from cytoplasm; L. *reticulum*, network]: An extensive system of membranes present in most eukaryotic cells, dividing the cytoplasm into compartments and channels; often coated with ribosomes.

endorphin: One of a group of small peptides with morphine-like properties; produced by the vertebrate brain.

endosperm [Gk. *endon*, within + *sperma*, seed]: In plants, a 3*n* tissue containing stored food that develops from the union of a sperm nucleus and the two nuclei of the central cell of the female gametophyte; found only in angiosperms.

endothelium [Gk. *endon*, + *thele*, nipple]: A type of epithelial tissue that forms the walls of the capillaries and the inner lining of arteries and veins.

endotherm [Gk. *endon*, within + *therme*, heat]: An organism, such as a bird or a mammal, that maintains its body temperature internally through metabolic processes. *See also* Homeotherm.

enterocoelous [Gk. *enteron*, gut + *koilos*, a hollow]: Formation of the coelom during embryonic development as cavities within mesoderm originating from outpocketings of the primitive gut; characteristic of deuterostomes.

entropy [Gk. *en*, in + *trope*, turning]: A measure of the randomness or disorder of a system.

enzyme [Gk. *en*, in + *zyme*, leaven]: A globular protein molecule that accelerates a specific chemical reaction.

epidermis [Gk. *epi*, on or over + *derma*, skin]: In plants and animals, the outermost layers of cells.

epinephrine: *See* Adrenaline.

episome: A plasmid that has become incorporated into a bacterial chromosome.

epistasis [Gk., a stopping]: Interaction between two nonallelic genes in which one of them interferes with or modifies the phenotypic expression of the other.

epithelial tissue [Gk. *epi*, on or over + *thele*, nipple]: In animals, a type of tissue that covers a body or structure or lines a cavity; epithelial cells form one or more regular layers with little intercellular material.

equilibrium [L. *aequus*, equal + *libra*, balance]: The state of a system in which no further net change is occurring; result of counterbalancing forward and backward processes.

erythrocyte (eh-**rith**-ro-site) [Gk. *erythros*, red + *kytos*, vessel]: Red blood cell, the carrier of hemoglobin.

estrogens [Gk. *oistros*, frenzy + *genos*, origin, descent]: Female sex hormones, which are the predominant secretions of the ovarian follicle during the preovulatory phase of the menstrual cycle; also produced by the corpus luteum and the placenta.

estrus [Gk. *oistros*, frenzy]: The mating period in female mammals, characterized by ovulation and intensified sexual activity.

ethology [Gk. *ethos*, habit, custom + *logos*, discourse]: The comparative study of patterns of animal behavior, with emphasis on their adaptive significance and evolutionary origin.

ethylene: A simple hydrocarbon ($H_2C = CH_2$) that functions as a plant hormone; plays a role in fruit ripening and leaf abscission.

eukaryote (you-**car**-ry-oat) [Gk. *eu*, good + *karyon*, nut, kernel]: A cell having a membrane-bound nucleus, membrane-bound organelles, and chromosomes in which DNA is combined with histone proteins; an organism composed of such cells.

eusocial [Gk. *eu*, good + L. *socius*, companion]: Applied to animal societies, such as those of certain insects, in which sterile individuals work on behalf of reproductive individuals.

evolution [L. *e-*, out + *volvere*, to roll]: Changes in the gene pool from one generation to the next as a consequence of processes such as mutation, natural selection, nonrandom mating, and genetic drift.

exergonic [Gk. *ex*, out of + *ergon*, work]: Energy-yielding, as in a chemical reaction; applied to a "downhill" process.

exocrine glands [Gk. *ex*, out of + *krinein*, to separate]: Glands, such as digestive glands and sweat glands, that secrete their products into ducts.

exocytosis [Gk. *ex*, out of + *kytos*, vessel]: A cellular process in which particulate matter or dissolved substances are enclosed in a vacuole and transported to the cell surface; there, the membrane of the vacuole fuses with the cell membrane, expelling the vacuole's contents to the outside.

exon: A segment of DNA that is transcribed into RNA and expressed; dictates the amino acid sequence of part of a polypeptide.

exoskeleton: The outer supporting covering of the body; common in arthropods.

exponential growth: In populations, the increasingly accelerated rate of growth due to the increasing number of individuals being added to the reproductive base. Exponential growth is very seldom approached or sustained in natural populations.

expressivity: In genetics, the degree to which a particular genotype is expressed in the phenotype of individuals with that genotype.

extinct [L. *exstinctus*, to be extinguished]: No longer existing.

extraembryonic membranes: In reptiles, birds, and mammals, membranes formed from embryonic tissues that lie outside the embryo proper, protecting it and aiding metabolism; include amnion, chorion, allantois, and yolk sac.

F_1 (first filial generation): The offspring resulting from the crossing of plants or animals of a parental generation.

F_2 (second filial generation): resulting from crossing members of the F_1 generation among themselves.

facilitated diffusion: The transport of substances across a cell or organelle membrane from a region of higher concentration to a region of lower concentration by protein molecules embedded in the membrane; driven by the concentration gradient.

Fallopian tube: *See* Oviduct.

family: A taxonomic grouping of related, similar genera; the category below order and above genus.

fatty acid: A molecule consisting of a —COOH group and a long hydrocarbon chain; fatty acids are components of fats, oils, phospholipids, glycolipids, and waxes.

feedback systems: Control mechanisms whereby an increase or decrease in the level of a particular factor inhibits or stimulates the production, utilization, or release of that factor; important in the regulation of enzyme and hormone levels, ion concentrations, temperature, and many other factors.

fermentation: The breakdown of organic compounds in the absence of oxygen; yields less energy than aerobic processes.

fertilization: The fusion of two haploid gamete nuclei to form a diploid zygote nucleus.

fetus [L., pregnant]: An unborn or unhatched vertebrate that has passed through the earliest developmental stages; a developing human from about the second month of gestation until birth.

fibril [L. *fibra*, fiber]: Any minute, threadlike structure within a cell.

fibrous protein: Insoluble structural protein in which the polypeptide chain is coiled along one dimension. Fibrous proteins constitute the main structural elements of many animal tissues.

filament [L. *filare*, to spin]: (1) A chain of cells. (2) In plants, the stalk of a stamen.

filtration: The first stage of kidney function; blood plasma is forced, under pressure, out of the glomerular capillaries into Bowman's capsule, through which it enters the renal tubule.

fission: *See* Binary fission.

fitness: The genetic contribution of an individual to succeeding generations relative to the contributions of other individuals in the population.

fixed action pattern: A behavior that appears substantially complete the first time the organism encounters the relevant stimulus; tends to be highly stereotyped, rigid, and predictable.

flagellum, *pl.* **flagella** (fla-**jell**-um) [L. *flagellum*, whip]: A long, threadlike organelle found in eukaryotes and used in locomotion and feeding; has an internal structure of nine pairs of microtubules encircling two central microtubules.

flower: The reproductive structure of angiosperms; a complete flower includes sepals, petals, stamens (male structures), and carpels (female structures).

food chain: A sequence of organisms related to one another as prey and predator.

food web: A set of interactions among organisms, including producers, consumers (herbivores and carnivores), and detritivores, through which energy and materials move within a community or ecosystem.

fossil [L. *fossilis*, dug up]: The remains of an organism, or direct evidence of its presence (such as tracks). May be an unaltered hard part (tooth or bone), a mold in a rock, petrification (wood or bone), unaltered or partially altered soft parts (a frozen mammoth).

founder effect: Type of genetic drift that occurs as the result of the founding of a population by a small number of individuals.

fovea [L., pit]: A small area in the center of the retina in which cones are concentrated; the area of sharpest vision.

free energy change: The total energy change that results from a chemical reaction or other process (such as evaporation); takes into account changes in both heat and entropy.

frequency-dependent selection: A type of natural selection that decreases the frequency of more common phenotypes in a population and increases the frequency of less common phenotypes.

fruit [L. *fructus*, fruit]: In angiosperms, a matured, ripened ovary or group of ovaries and associated structures; contains the seeds.

function [L. *fungor*, to busy oneself]: Characteristic role or action of a structure or process in the normal metabolism or behavior of an organism.

gametangium, *pl.* **gametangia** [Gk. *gamein*, to marry + L. *tangere*, to touch]: A unicellular or multicellular structure in which gametes are produced.

gamete (**gam**-meet) [Gk., wife]: A haploid reproductive cell whose nucleus fuses with that of another gamete of an opposite mating type or sex (fertilization); the resulting cell (zygote) may develop into a new diploid individual or, in some protists and fungi, may undergo meiosis to form haploid somatic cells.

gametophyte: In organisms that have alternation of haploid and diploid generations (all plants and some green algae), the haploid *(n)* gamete-producing generation.

ganglion, *pl.* **ganglia** (**gang**-lee-on) [Gk. *ganglion*, a swelling]: Aggregation of nerve cell bodies; in vertebrates, refers to an aggregation of nerve cell bodies located outside the central nervous system.

gap junction: A junction between adjacent animal cells that allows the passage of materials between the cells.

gastric [Gk. *gaster*, stomach]: Pertaining to the stomach.

gastrovascular cavity [Gk. *gaster*, stomach + L. *vasculum*, a small vessel]: A digestive cavity with only one opening, characteristic of the phyla Cnidaria (jellyfish, hydra, corals, etc.) and Ctenophora (comb jellies, sea walnuts); water circulating through the cavity supplies dissolved oxygen and carries away carbon dioxide and other waste products.

gastrula [Gk. *gaster*, stomach]: An animal embryo in the process of gastrulation; the stage of development during which the blastula, with its single layer of cells, turns into a three-layered embryo, made up of ectoderm, mesoderm, and endoderm, often enclosing an archenteron.

Gause's principle: *See* Competitive exclusion.

gene [Gk. *genos*, birth, race; L. *genus*, birth, race, origin]: A unit of heredity in the chromosome; a sequence of nucleotides in a DNA molecule that performs a specific function, such as coding for an RNA molecule or a polypeptide.

gene flow: The movement of alleles into or out of a population.

gene pool: All the alleles of all the genes of all the individuals in a population.

genetic code: The system of nucleotide triplets in DNA and RNA that carries genetic information; referred to as a code because it determines the amino acid sequence in the enzymes and other protein molecules synthesized by the organism.

genetic drift: Evolution (change in allele frequencies) owing to chance processes.

genetic isolation: The absence of genetic exchange between populations or species as a result of geographic separation or of premating or postmating mechanisms (behavioral, anatomical, or physiological) that prevent reproduction.

genome: The complete set of chromosomes, with their associated genes.

genomic DNA (gDNA): DNA fragments produced by the action of restriction enzymes on the DNA of a cell or organism.

genotype (jean-o-type): The genetic constitution of an individual cell or organism with reference to a single trait or a set of traits; the sum total of all the genes present in an individual.

genus, *pl.* **genera** (jean-us) [L. *genus*, race, origin]: A taxonomic grouping of closely related species.

geologic eras: See Table 24–1, pages 496–497.

germ cells [L. *germinare*, to bud]: Gametes or the cells that give rise directly to gametes

germination [L. *germinare*, to bud]: In plants, the resumption of growth or the development from seed or spore.

germ layer: A layer of distinctive cells in an embryo; an embryonic tissue layer. The majority of multicellular animals have three germ layers: ectoderm, mesoderm, and endoderm.

gibberellins (jibb-e-**rell**-ins) [Fr. *gibberella*, genus of fungi]: A group of chemically related plant growth hormones, whose most characteristic effect is stem elongation in dwarf plants and bolting.

gill: The respiratory organ of aquatic animals, usually a thin-walled projection from some part of the external body surface or, in vertebrates, from some part of the digestive tract.

gland [L. *glans, glandis*, acorn]: A structure composed of modified epithelial cells specialized to produce one or more secretions that are discharged to the outside of the gland.

globular protein [L. dim. of *globus*, a ball]: A polypeptide chain folded into a roughly spherical shape.

glomerulus (glom-**mare**-u-lus) [L. *glomus*, ball]: In the vertebrate kidney, a cluster of capillaries enclosed by Bowman's capsule; blood plasma minus large molecules filters through the walls of the glomerular capillaries into the renal tubule.

glucagon [Gk. *glukus*, sweet + *agō*, to lead toward]: Hormone produced in the pancreas that acts to raise the concentration of blood sugar.

glucose [Gk. *glukus*, sweet]: A six-carbon sugar ($C_6H_{12}O_6$); the most common monosaccharide in animals.

glycogen [Gk. *glukus*, sweet + *genos*, race or descent]: A complex carbohydrate (polysaccharide); one of the main stored food substances of most animals and fungi; it is converted into glucose by hydrolysis.

glycolipids [Gk. *glukus*, sweet + *lipos*, fat]: Organic molecules similar in structure to fats, but in which a short carbohydrate chain rather than a fatty acid is attached to the third carbon of the glycerol molecule; as a result, the molecule has a hydrophilic "head" and a hydrophobic "tail." Glycolipids are important constituents of cell and organelle membranes.

glycolysis (gly-**coll**-y-sis) [Gk. *glukus*, sweet + *lysis*, loosening]: The process by which a glucose molecule is changed anaerobically to two molecules of pyruvic acid with the liberation of a small amount of useful energy; catalyzed by cytoplasmic enzymes.

Golgi complex (**goal**-jee): An organelle present in many eukaryotic cells; consists of flat, membrane-bound sacs, tubules, and vesicles. It functions as a processing, packaging, and distribution center for substances that the cell manufactures.

gonad [Gk. *gone*, seed]: Gamete-producing organ of multicellular animals; ovary or testis.

granulocyte [L., *grānum*, grain or seed + Gk. *kytos*, vessel]: A type of phagocytic white blood cell involved in the inflammatory response; characterized by numerous lysosomes that give the cell a granular appearance under the light microscope. Granulocytes are classified on the basis of their staining properties as neutrophils, eosinophils, or basophils.

granum, *pl.* **grana** [L., grain or seed]: In chloroplasts, stacked membrane-bound disks (thylakoids) that contain chlorophylls and carotenoids and are the sites of the light-trapping reactions of photosynthesis.

gravitropism [L. *gravis*, heavy + Gk. *trope*, turning]: The direction of growth or movement in which the force of gravity is the determining factor; also called geotropism.

gross productivity: A measure of the rate at which energy is assimilated by the organisms in a trophic level, a community, or an ecosystem.

ground tissues: In leaves and young roots and stems, all tissues other than the epidermis and the vascular tissues.

guard cells: Specialized epidermal cells surrounding a pore, or stoma, in a leaf or green stem; changes in turgor of a pair of guard cells cause opening and closing of the pore.

gymnosperm [Gk. *gymnos*, naked + *sperma*, seed]: A seed plant in which the seeds are not enclosed in an ovary; the conifers are the most familiar group.

habitat [L. *habitare*, to live in]: The place in which individuals of a particular species can usually be found.

habituation [L. *habitus*, condition]: A response to a repeated stimulus in which the stimulus comes to be ignored and a previous behavior pattern is restored; one of the simplest forms of learning.

half-life: The average time required for the disappearance or decay of one-half of any amount of a given substance.

haploid [Gk. *haploos*, single + *ploion*, vessel]: Having only one set of chromosomes (n), in contrast to diploid $(2n)$; characteristic of eukaryotic gametes, of gametophytes in plants, and of some protists and fungi.

Hardy-Weinberg equilibrium: The steady-state relationship between relative frequencies of two or more alleles in an idealized population; both the allele frequencies and the genotype frequencies will remain constant from generation to generation in a population breeding at random in the absence of evolutionary forces.

haustorium, *pl.* **haustoria** [L. *haustus*, from *haurire*, to drink, draw]: A projection from a parasitic oomycete, fungus, or plant that functions as a penetrating and absorbing organ.

heat of vaporization: The amount of heat required to change a given amount of a liquid into a gas; 540 calories are required to change 1 gram of liquid water into vapor.

heme [Gk. *haima*, blood]: The iron-containing group of heme proteins such as hemoglobin and the cytochromes.

hemocoel [Gk. *haima*, blood + *koilos*, a hollow]: A blood-filled space within the tissues; characteristic of animals with an incomplete circulatory system, such as mollusks and arthropods.

hemoglobin [Gk. *haima*, blood + L. *globus*, a ball]: The iron-containing protein in vertebrate blood that carries oxygen.

hemophilia [Gk. *haima*, blood + *philios*, friendly]: A group of hereditary diseases characterized by failure of the blood to clot and consequent excessive bleeding from even minor wounds.

hepatic [Gk. *hēpatikos*, liver]: Pertaining to the liver.

herbaceous (her-**bay**-shus) [L. *herba*, grass]: In plants, nonwoody.

herbivore [L. *herba*, grass + *vorare*, to devour]: A consumer that eats plants or other photosynthetic organisms to obtain its food and energy.

heredity [L. *herres, heredis*, heir]: The transmission of characteristics from parent to offspring.

hermaphrodite [Gk. *Hermes* and *Aphrodite*]: An organism possessing both male and female reproductive organs; hermaphrodites may or may not be self-fertilizing.

heterosis [Gk. *heteros*, other, different]: Hybrid vigor; the overall superiority of the hybrid over either parent.

heterotroph [Gk. *heteros*, other, different + *trophos*, feeder]: An organism that must feed on organic materials formed by other organisms in order to obtain energy and small building-block molecules; in contrast to autotroph. Animals, fungi, and many unicellular organisms are heterotrophs.

heterozygote [Gk. *heteros*, other + *zugōtos*, a pair]: A diploid organism that carries two different alleles at one or more genetic loci.

heterozygote superiority: The greater fitness of an organism heterozygous at a given genetic locus as compared with either homozygote.

hibernation [L. *hiberna*, winter]: A period of dormancy and inactivity, varying in length, depending on the species, and occurring in dry or cold seasons. During hibernation, metabolic processes are greatly slowed and, even in mammals, body temperature may drop to just above freezing.

histones: A group of five relatively small, basic polypeptide molecules found bound to the DNA of eukaryotic cells.

homeostasis (home-e-o-**stay**-sis) [Gk. *homos*, same or similar + *stasis*, standing]: Maintenance of a relatively stable internal physiological environment or internal equilibrium in an organism.

homeotherm [Gk. *homos*, same or similar + *therme*, heat]: An organism, such as a bird or mammal, capable of maintaining a stable body temperature independent of the environment.

hominid [L. *homo*, man]: Humans and closely related primates; includes modern and fossil forms, such as the australopithecines, but not the apes.

hominoid [L. *homo*, man]: Hominids and the apes.

homologues [Gk. *homologia*, agreement]: Chromosomes that carry corresponding genes and associate in pairs in the first stage of meiosis; each member of the pair is derived from a different parent.

homology [Gk. *homologia*, agreement]: Similarity in structure and/or position, assumed to result from a common ancestry, regardless of function, such as the wing of a bird and the foreleg of a mammal.

homozygote [Gk. *homos*, same or similar + *zugōtos*, a pair]: A diploid organism that carries identical alleles at one or more genetic loci.

hormone [Gk. *hormaein*, to excite]: An organic molecule secreted, usually in minute amounts, in one part of an organism that regulates the function of another tissue or organ.

host: (1) An organism on or in which a parasite lives. (2) A recipient of grafted tissue.

hybrid [L. *hybrida*, the offspring of a tame sow and a wild boar]: (1) Offspring of two parents that differ in one or more inheritable characteristics. (2) Offspring of two different varieties or of two different species.

hydrocarbon [L. *hydro*, water + *carbo*, charcoal]: An organic compound consisting of only carbon and hydrogen.

hydrogen bond: A weak molecular bond linking a hydrogen atom that is covalently bonded to another atom (usually oxygen, nitrogen, or fluorine) to another oxygen, nitrogen, or fluorine atom of the same or another molecule.

hydrolysis [L. *hydro*, water + Gk. *lysis*, loosening]: Splitting of one molecule into two by addition of H^+ and OH^- ions from water.

hydrophilic [L. *hydro*, water + Gk. *philios*, friendly]: Having an affinity for water; applied to polar molecules or polar regions of large molecules.

hydrophobic [L. *hydro*, water + Gk. *phobos*, fearing]: Having no affinity for water; applied to nonpolar molecules or nonpolar regions of molecules.

hypertonic [Gk. *hyper*, above + *tonos*, tension]: Of two solutions of different concentration, the solution that contains the higher concentration of solute particles; water moves across a selectively permeable membrane into a hypertonic solution.

hypha [Gk. *hyphe*, web]: A single tubular filament of a fungus or an oomycete; the hyphae together make up the mycelium, the matlike "body" of a fungus.

hypothalamus [Gk. *hypo*, under + *thalamos*, inner room]: The region of the vertebrate brain just below the thalamus; responsible for the integration of many basic behavioral patterns that involve correlation of neural and endocrine functions.

hypothesis [Gk. *hypo*, under + *tithenai*, to put]: A temporary working explanation or supposition based on accumulated facts and suggesting some general principle or relation of cause and effect; a postulated solution to a scientific problem that must be tested and if not validated, discarded.

hypotonic [Gk. *hypo*, under + *tonos*, tension]: Of two solutions of different concentration, the solution that contains the lower concentration of solute particles; water moves across a selectively permeable membrane from a hypotonic solution.

imbibition [L. *imbibere*, to drink in]: The capillary movement of water into germinating seeds and into substances such as wood and gelatin, which swell as a result.

immune response: A highly specific defensive reaction of the body to invasion by a foreign substance or organism; consists of a primary response in which the invader is recognized as foreign, or "not-self," and eliminated and a secondary response to subsequent attacks by the same invader. Mediated by two types of lymphocytes: B lymphocytes, which mature in the bone marrow and are responsible for antibody production, and T lymphocytes, which mature in the thymus and are responsible for cell-mediated immunity.

immunoglobulins: Complex, highly specific globular proteins synthesized by B lymphocytes; include both circulating antibodies and antibodies displayed on the surface of B lymphocytes prior to activation.

imprinting: A rapid and extremely narrow form of learning, common in birds and important in species recognition, that occurs during a very short critical period in the early life of an animal; depends on exposure to particular characteristics of the parent or parents.

inbreeding: The mating of individuals that are closely related genetically.

inclusive fitness: The relative number of an individual's alleles that are passed on from generation to generation, either as a result of his or her own reproductive success, or that of related individuals.

incomplete dominance: In genetics, the phenomenon in which the effects of both alleles at a particular locus are apparent in the phenotype of the heterozygote.

independent assortment: *See* Mendel's second law.

induction [L. *inducere*, to induce]: (1) In genetics, the phenomenon in which the presence of a substrate initiates transcription and translation of the genes coding for the enzymes required for its metabolism. (2) In embryonic development, the process in which one tissue or body part causes the differentiation of another tissue or body part.

inflammatory response: A nonspecific defensive reaction of the body to invasion by a foreign substance or organism; involves phagocytosis by white blood cells and is often accompanied by accumulation of pus and an increase in the local temperature.

innate releasing mechanism: In ethology, an area within an animal's brain that is hypothesized to respond to a specific stimulus, setting in motion, or "releasing," the sequence of movements that constitute a fixed action pattern.

insertion sequences: Relatively short sequences of DNA that can produce copies of themselves that become incorporated at other sites in the same chromosome or in other chromosomes; also known as simple transposons.

insulin: A peptide hormone, produced by the vertebrate pancreas, that acts to lower the concentration of glucose in the blood.

interferon: A protein made by virus-infected cells that inhibits viral multiplication.

intermediate filaments: Fibrous protein filaments that form part of the cytoskeleton; found in greatest density in cells subject to mechanical stress.

interneuron: Neuron that transmits signals from one neuron to another within a local region of the central nervous system; may receive signals from and transmit signals to many different neurons.

interphase: The portion of the cell cycle that occurs before mitosis or meiosis can take place; includes the G_1, S, and G_2 phases.

intron: A segment of DNA that is transcribed into RNA but is removed enzymatically from the RNA molecule before the mRNA enters the cytoplasm and is translated; also known as an intervening sequence.

invagination [L. *in*, in + *vagina*, sheath]: The local infolding of a layer of tissue, especially in animal embryos, so as to form a depression or pocket opening to the outside.

inversion: A chromosomal aberration in which a double break occurs and a segment is turned 180° before it is reincorporated into the chromosome.

ion (eye-on): Any atom or small molecule containing an unequal number of electrons and protons and therefore carrying a net positive or net negative charge.

ionic bond: A chemical bond formed as a result of the mutual attraction of ions of opposite charge.

isogamy [Gk. *isos*, equal + *gamos*, marriage]: Sexual reproduction in which both gametes are motile and are structurally alike.

isolating mechanisms: Mechanisms that prevent genetic exchange between individuals of different populations or species; they prevent mating or successful reproduction even when mating occurs; may be behavioral, anatomical, or physiological.

isotonic [Gk. *isos*, equal + *tonos*, tension]: Having the same concentration of solutes as another solution. If two isotonic solutions are separated by a selectively permeable membrane, there will be no net flow of water across the membrane.

isotope [Gk. *isos*, equal + *topos*, place]: Atom of an element that differs from other atoms of the same element in the number of neutrons in the atomic nucleus; isotopes thus differ in atomic weight. Some isotopes are unstable and emit radiation.

karyotype [Gk. *kara*, the head + *typos*, stamp or print]: The general appearance of the chromosomes of an organism with regard to number, size, and shape.

keratin [Gk. *karas*, horn]: One of a group of tough, fibrous proteins formed by certain epidermal tissues and especially abundant in skin, claws, hair, feathers, and hooves.

kidney: In vertebrates, the organ that regulates the balance of water and solutes in the blood and the excretion of nitrogenous wastes in the form of urine.

kinetic energy [Gk. *kinetikos*, putting in motion]: Energy of motion.

kinetochore [Gk. *kinetikos*, putting in motion + *choros*, chorus]: Disk-shaped protein structure within the centromere to which spindle fibers are attached during mitosis or meiosis.

kingdom: A taxonomic grouping of related, similar phyla or divisions; the highest-level category in biological classification.

kin selection: The differential reproduction of lineages of related individuals—that is, different groups of related individuals of a species reproduce at different rates; leads to an increase in the frequency of alleles shared by members of the groups with the greatest reproductive success.

Krebs cycle: Stage of cellular respiration in which acetyl groups are broken down into carbon dioxide; molecules reduced in the process can be used in ATP formation.

lagging strand: In DNA replication, the 3' to 5' strand of the DNA double helix, synthesized as a series of Okazaki fragments in the 5' to 3' direction; these segments are subsequently linked to one another in condensation reactions catalyzed by the enzyme DNA ligase.

lamella (lah-mell-ah) [L. dim. of *lamina*, plate or leaf]: Layer, thin sheet.

larva [L., ghost]: An immature animal that is anatomically very different from the adult; examples are caterpillars and tadpoles.

lateral meristem [L. *latus, lateris*, side + Gk. *meristos*, divided]: In vascular plants, one of the two rings of tissue (vascular cambium and cork cambium) that produce new cells for secondary growth.

leaching: The dissolving of minerals and other elements in soil or rocks by the downward movement of water.

leading strand: In DNA replication, the 5' to 3' strand of the DNA double helix, which is synthesized continuously.

learning: The process that leads to modification in individual behavior as the result of experience.

leucoplast [Gk. *leukos*, white + *plastes*, molder]: In plant cells, a colorless organelle that serves as a starch repository; usually found in cells not exposed to light, such as those in roots and the internal tissues of stems.

leukocyte [Gk. *leukos*, white + *kytos*, vessel]: White blood cell; principal types include granulocytes, monocytes and macrophages, and lymphocytes.

lichen: Organism composed of a fungus and a green alga or a cyanobacterium that are symbiotically associated.

life cycle: The entire span of existence of any organism from time of zygote formation (or asexual reproduction) until it itself reproduces.

limbic system [L. *limbus*, border]: Neuron network forming a loop around the inside of the brain and connecting the hypothalamus to the cerebral cortex; thought to be circuit by which drives and emotions are translated into complex actions and to play a role in the consolidation of memory.

linkage: The tendency for certain alleles to be inherited together because they are located on the same chromosome.

linkage group: A pair of homologous chromosomes.

lipid [Gk. *lipos*, fat]: One of a large variety of organic substances that are insoluble in polar solvents, such as water, but that dissolve readily in nonpolar organic solvents; includes fats, oils, waxes, steroids, phospholipids, glycolipids, and carotenes.

locus, *pl.* **loci** [L., place]: In genetics, the position of a gene in a chromosome. For any given locus, there may be a number of possible alleles.

logistic growth: A pattern of population growth in which growth is rapid when the population is small, gradually slows as the population approaches the carrying capacity of its environment, and then oscillates as the population stabilizes at or near its maximum size; it is one of the simplest growth patterns observed for populations in nature.

loop of Henle (after F. G. J. Henle, German pathologist): A hairpin-shaped portion of the renal tubule of mammals in which a hypertonic urine is formed by processes of diffusion and active transport.

lumen [L., light]: The cavity of a tubular structure, such as endoplasmic reticulum or a blood vessel.

lymph [L. *lympha*, water]: Colorless fluid derived from blood by filtration through capillary walls in the tissues; carried in special lymph ducts.

lymphatic system: The system through which lymph circulates; consists of lymph capillaries, which begin blindly in the tissues, and a network of progressively larger vessels that empty into the vena cava; also includes the lymph nodes, spleen, thymus, and tonsils.

lymph node [L. *lympha*, water + *nodus*, knot]: A mass of spongy tissues, separated into compartments; located throughout the lymphatic system, lymph nodes remove dead cells, debris, and foreign particles from the circulation; also are sites at which foreign antigens are displayed to immunologically active cells.

lymphocyte [L. *lympha*, water + Gk. *kytos*, vessel]: A type of white blood cell involved in the immune response; B lymphocytes differentiate into antibody-producing plasma cells, whereas cytotoxic T lymphocytes lyse diseased eukaryotic cells; other T lymphocytes interact with both cytotoxic T lymphocytes and with B lymphocytes.

lysis [Gk., a loosening]: Disintegration of a cell by rupture of its cell membrane.

lysogenic bacteria (lye-so-jenn-ick) [Gk. *lysis*, a loosening + *genos*, race or descent]: Bacteria carrying a bacteriophage integrated into the bacterial chromosome. The virus may subsequently set up an active cycle of infection, causing lysis of the bacterial cells.

lysosome [Gk. *lysis*, loosening + *soma*, body]: A membrane-bound organelle in which hydrolytic enzymes are segregated.

macromolecule [Gk. *makros*, large + L. dim. of *moles*, mass]: An extremely large molecule; refers specifically to proteins, nucleic acids, polysaccharides, and complexes of these.

macrophage [Gk. *makros*, large + *phagein*, to eat]: A type of phagocytic white blood cell important in both the inflammatory and immune responses.

major histocompatibility complex (MHC): In mammals, a group of at least 20 different genes, each with multiple alleles, coding for the protein components of the antigens that are displayed on nucleated cells and that serve to identify "self."

mandibles [L. *mandibula*, jaw]: In crustaceans, insects, and myriapods, the appendages immediately posterior to the antennae; used to seize, hold, bite, or chew food.

mantle: In mollusks, the outermost layer of the body wall or a soft extension of it; usually secretes a shell.

marine [L. *marini(us)*, from *mare*, the sea]: Living in salt water.

marsupial [Gk. *marsypos*, pouch, little bag]: A mammal in which the female has a ventral pouch or folds surrounding the nipples; the premature young leave the uterus and crawl into the pouch, where each one attaches itself by the mouth to a nipple until development is completed.

matrix: The dense solution in the interior of the mitochondrion, surrounding the cristae; contains enzymes, phosphates, coenzymes, and other molecules involved in cellular respiration.

mechanoreceptor: A sensory cell or organ that receives mechanical stimuli such as those involved in touch, pressure, hearing, and balance.

medulla (med-**dull**-a) [L., the innermost part]: (1) The inner, as opposed to the outer, part of an organ, as in the adrenal gland. (2) The most posterior region of the vertebrate brain; connects with the spinal cord.

medusa: The free-swimming, bell- or umbrella-shaped stage in the life cycle of many cnidarians; a jellyfish.

megaspore [Gk. *megas*, great, large + *spora*, a sowing]: In plants, a haploid (n) spore that develops into a female gametophyte.

meiosis (my-o-sis) [Gk. *meioun*, to make smaller]: The two successive nuclear divisions in which a single diploid (2n) cell forms four haploid (n) nuclei, and segregation, crossing over, and reassortment of the alleles occur; gametes or spores may be produced as a result of meiosis.

Mendel's first law: The factors for a pair of alternative characters are separate, and only one may be carried in a particular gamete (genetic segregation). In modern form: Alleles segregate in meiosis.

Mendel's second law: The inheritance of a pair of factors for one trait is independent of the simultaneous inheritance of factors for other traits, such factors "assorting independently" as though there were no other factors present (later modified by the discovery of linkage). Modern form: The alleles of unlinked genes assort independently.

menstrual cycle [L. *mensis*, month]: In humans and certain other primates, the cyclic, hormone-regulated changes in the condition of the uterine lining; marked by the periodic discharge of blood and disintegrated uterine lining through the vagina. Mammals with a menstrual cycle lack a well-defined period of estrus.

meristem [Gk. *merizein*, to divide]: The undifferentiated plant tissue, including a mass of rapidly dividing cells, from which new tissues arise.

mesenteries [Gk. *mesos*, middle + *enteron*, gut]: Double layers of mesoderm that suspend the digestive tract and other internal organs within the coelom.

mesoderm [Gk. *mesos*, middle + *derma*, skin]: In animals, the middle layer of the three embryonic tissue layers. In vertebrates, includes the chorda-mesoderm, which gives rise to the notochord and skeletal muscle, and the lateral plate mesoderm, which gives rise to the circulatory system, most of the excretory and reproductive systems, the lining of the coelom, and the outer covering of the internal organs.

mesophyll [Gk. *mesos*, middle + *phyllon*, leaf]: The internal tissue of a leaf, sandwiched between two layers of epidermal cells; consists of palisade parenchyma and spongy parenchyma cells.

messenger RNA (mRNA): A class of RNA molecules, each of which is complementary to one strand of DNA and which serves to carry the genetic information from the chromosome to the ribosomes, where it is translated into protein.

metabolism [Gk. *metabole*, change]: The sum of all chemical reactions occurring within a cell or organism.

metamere [Gk. *meta*, middle + *meros*, part]: One of a linear series of similar body segments.

metamorphosis [Gk. *metamorphoun*, to transform]: Abrupt transition from larval to adult form, such as the transition from tadpole to adult frog.

metaphase [Gk. *meta*, middle + *phasis*, form]: The stage of mitosis or meiosis during which the chromosomes lie in the equatorial plane of the spindle.

microbe [Gk. *mikros*, small + *bios*, life]: A microscopic organism.

micronutrient [Gk. *mikros*, small + L. *nutrire*, to nourish]: An inorganic nutrient required in only minute amounts for plant growth, such as iron, chlorine, copper, manganese, zinc, molybdenum, and boron.

microspore [Gk. *mikros*, small + *spora*, a sowing]: In plants, a haploid (n) spore that develops into a male gametophyte; in seed plants, it becomes a pollen grain.

microtubule [Gk. *mikros*, small + L. dim. of *tubus*, tube]: An extremely small hollow tube composed of two types of globular protein subunits. Among their many functions, microtubules make up the internal structure of cilia and flagella.

middle lamella: In plants, distinct layer between adjacent cell walls, rich in pectins and other polysaccharides; derived from the cell plate.

mimicry [Gk. *mimos*, mime]: The superficial resemblance in form, color, or behavior of certain organisms (mimics) to other more powerful or more protected ones (models), resulting in protection, concealment, or some other advantage for the mimic.

mineral: A naturally occurring element or inorganic compound.

mitochondrion, pl. **mitochondria** [Gk. *mitos*, thread + *chondros*, cartilage or grain]: An organelle, bound by a double membrane, in which the reactions of the Krebs cycle, terminal electron transport, and oxidative phosphorylation take place, resulting in the formation of CO_2, H_2O, and ATP from acetyl CoA and ADP. Mitochondria are the organelles in which most of the ATP of the eukaryotic cell is produced.

mitosis [Gk. *mitos*, thread]: Nuclear division characterized by chromosome replication and formation of two identical daughter nuclei.

mole [L. *moles*, mass]: The amount of an element equivalent to its atomic weight expressed in grams, or the amount of a substance equivalent to its molecular weight expressed in grams.

molecular weight: The sum of the atomic weights of the constituent atoms in a molecule.

molecule [L. dim. of *moles*, mass]: A particle consisting of two or more atoms held together by chemical bonds; the smallest unit of a compound that displays the properties of the compound.

molting: Shedding of all or part of an organism's outer covering; in arthropods, periodic shedding of the exoskeleton to permit an increase in size.

monocotyledon [Gk. *monos*, single + *kotyledon*, a cup-shaped hollow]: A member of the class of flowering plants having one seed leaf, or cotyledon, among other distinguishing features; often abbreviated as monocot.

monocyte [Gk. *monos*, single + *kytos*, vessel]: A type of circulating white blood cell that, in the presence of infectious organisms or other foreign invaders, becomes transformed into a macrophage.

monoecious (mo-nee-shus) [Gk. *monos*, single + *oikos*, house]: In angiosperms, having the male and female structures (the stamens and the carpels, respectively) on the same individual but on different flowers.

monomer [Gk. *monos*, single + *meros*, part]: A simple, relatively small molecule that can be linked to others to form a polymer.

monosaccharide [Gk. *monos*, single + *sakcharon*, sugar]: A simple sugar, such as glucose, fructose, ribose.

monotreme [Gk. *monos*, single + *trēma*, hole]: A nonplacental mammal, such as the duckbilled platypus, in which the female lays shelled eggs and nurses the young.

morphogenesis [Gk. *morphe*, form + *genesis*, origin]: The development of size, form, and other structural features of organisms.

morphological [Gk. *morphe,* form + *logos,* discourse]: Pertaining to form and structure, at any level of organization.

motor neuron: Neuron that conducts nerve impulses from the central nervous system to an effector, which is typically a muscle or a gland.

muscle fiber: Muscle cell; a long, cylindrical, multinucleated cell containing numerous myofibrils, which is capable of contraction when stimulated.

mutagen [L. *mutare,* to change + *genus,* source or origin]: A chemical or physical agent that increases the mutation rate.

mutant [L. *mutare,* to change]: An organism carrying a gene that has undergone a mutation.

mutation [L. *mutare,* to change]: The change of a gene from one allelic form to another; an inheritable change in the DNA sequence of a chromosome.

mutualism [L. *mutuus,* lent, borrowed]: *See* Symbiosis.

mycelium [Gk. *mykes,* fungus]: The mass of hyphae forming the body of a fungus.

mycorrhizae [Gk. *mykes,* fungus + *rhiza,* root]: Symbiotic associations between particular species of fungi and the roots of vascular plants.

myelin sheath [Gk. *myelinos,* full of marrow]: A lipid-rich layer surrounding the long axons of neurons in the vertebrate nervous system; in the peripheral nervous system, made up of the membranes of Schwann cells.

myofibril [Gk. *mys,* muscle + L. *fibra,* fiber]: Contractile element of a muscle fiber, made up of thick and thin filaments arranged in sarcomeres.

myoglobin [Gk. *mys,* muscle + L. *globus,* a ball]: An oxygen-binding, heme-containing globular protein found in muscles.

myosin [Gk. *mys,* muscle]: One of the principal proteins in muscle; makes up the thick filaments.

NAD: Abbreviation of nicotinamide adenine dinucleotide, a coenzyme that functions as an electron acceptor.

natural selection: A process of interaction between organisms and their environment that results in a differential rate of reproduction of different phenotypes in the population; can result in changes in the relative frequencies of alleles and genotypes in the population—that is, in evolution.

nectar [Gk. *nektar,* the drink of the gods]: A sugary fluid that attracts insects to plants.

negative feedback: A control mechanism whereby an increase in some substance inhibits the process leading to the increase; also known as feedback inhibition.

nematocyst [Gk. *nema, nematos,* thread + *kyst,* bladder]: A threadlike stinger, containing a poisonous or paralyzing substance, found in the cnidocyte of cnidarians.

nephridium, *pl.* **nephridia** [Gk. *nephros,* kidney]: A tubular excretory structure found in many invertebrates.

nephron [Gk. *nephros,* kidney]: The functional unit of the kidney in reptiles, birds, and mammals; a human kidney contains about 1 million nephrons.

nerve: A group or bundle of nerve fibers with accompanying connective tissue, located in the peripheral nervous system. A bundle of nerve fibers within the central nervous system is known as a tract.

nerve fiber: A filamentous process extending from the cell body of a neuron and conducting the nerve impulse; an axon.

nerve impulse: A rapid, transient, self-propagating change in electric potential across the membrane of an axon.

nervous system: All the nerve cells of an animal; the receptor-conductor-effector system; in humans, the nervous system consists of the central nervous system (brain and spinal cord) and the peripheral nervous system.

net productivity: In a trophic level, a community, or an ecosystem, the amount of energy (in calories) stored in chemical compounds or the increase in biomass (in grams or metric tons) in a particular period of time; it is the difference between gross productivity and the energy used by the organisms in respiration.

neural groove: Dorsal, longitudinal groove that forms in a vertebrate embryo; bordered by two neural folds; preceded by the neural-plate stage and followed by the neural-tube stage.

neural plate: Thickened strip of ectoderm in early vertebrate embryos that forms along the dorsal side of the body and gives rise to the central nervous system.

neural tube: Primitive, hollow, dorsal nervous system of the early vertebrate embryo; formed by fusion of neural folds around the neural groove.

neuromodulator: A chemical agent that is released by a neuron and diffuses through a local region of the central nervous system, acting on neurons within that region; generally has the effect of modulating the response to neurotransmitters.

neuron [Gk., nerve]: Nerve cell, including cell body, dendrites, and axon.

neurosecretory cell: A neuron that releases one or more hormones into the circulatory system.

neurotransmitter: A chemical agent that is released by a neuron at a synapse, diffuses across the synaptic cleft, and acts upon a postsynaptic neuron or muscle or gland cell and alters its electrical state or activity.

neutron (new-tron): An uncharged particle with a mass slightly greater than that of a proton. Found in the atomic nucleus of all elements except hydrogen, in which the nucleus consists of a single proton.

niche: *See* Ecological niche.

nitrification: The oxidation of ammonia or ammonium to nitrites and nitrates, as by nitrifying bacteria.

nitrogen cycle: Worldwide circulation and reutilization of nitrogen atoms, chiefly due to metabolic processes of living organisms; plants take up inorganic nitrogen and convert it into organic compounds (chiefly proteins), which are assimilated into the bodies of one or more animals; bacterial and fungal action on nitrogenous waste products and dead organisms return nitrogen atoms to the inorganic state.

nitrogen fixation: Incorporation of atmospheric nitrogen into inorganic nitrogen compounds available to plants, a process that can be carried out only by some soil bacteria, many free-living and symbiotic cyanobacteria, and certain symbiotic bacteria in association with legumes.

nitrogenous base: A nitrogen-containing molecule having basic properties (tendency to acquire an H^+ ion); a purine or pyrimidine.

nocturnal [L. *nocturnus,* of night]: Applied to organisms that are active during the hours of darkness.

node [L. *nodus,* knot]: In plants, a joint of a stem; the place where branches and leaves are joined to the stem.

nondisjunction [L. *non,* not + *disjungere,* to separate]: The failure of chromatids to separate during meiosis, resulting in one or more extra chromosomes in some gametes and correspondingly fewer in others.

noradrenaline: A hormone, produced by the medulla of the adrenal gland, that increases the concentration of sugar in the blood, raises blood pressure and heartbeat rate, and increases muscular power and resistance to fatigue; also one of the principal neurotransmitters; also called norepinephrine.

norepinephrine: *See* noradrenaline.

notochord [Gk. *noto,* back + L. *chorda,* cord]: A dorsal rodlike structure that runs the length of the body and serves as the internal skeleton in the embryos of all chordates; in most adult chordates the notochord is replaced by a vertebral column that forms around (but not from) the notochord.

nuclear envelope [L. *nucleus,* a kernel]: The double membrane surrounding the nucleus within a eukaryotic cell.

nucleic acid: A macromolecule consisting of nucleotides; the principal types are deoxyribonucleic acid (DNA) and ribonucleic acid (RNA).

nucleoid: In prokaryotic cells, the region of the cell in which the chromosome is localized.

nucleolus (new-**klee**-o-lus) [L., a small kernel]: A small, dense region visible in the nucleus of nondividing eukaryotic cells; consists of rRNA molecules, ribosomal proteins, and loops of chromatin from which the rRNA molecules are transcribed.

nucleosome [L. *nucleus*, a kernel + Gk. *soma*, body]: A complex of DNA and histone proteins that forms the fundamental packaging unit of eukaryotic DNA; its structure resembles a bead on a string.

nucleotide [L. *nucleus*, a kernel]: A molecule composed of a phosphate group, a five-carbon sugar (either ribose or deoxyribose), and a purine or pyrimidine base; nucleotides are the building blocks of nucleic acids.

nucleus [L., a kernel]: (1) The central core of an atom, containing protons and neutrons, around which electrons move. (2) The membrane-bound structure characteristic of eukaryotic cells that contains the genetic information in the form of DNA organized into chromosomes. (3) A group of nerve cell bodies within the central nervous system.

ocellus, *pl.* **ocelli** [L. dim. of *oculus*, eye]: A simple light receptor common among invertebrates.

Okazaki fragments (after R. Okazaki, Japanese geneticist): In DNA replication, the discontinuous segments in which the 3′ to 5′ strand (the lagging strand) of the DNA double helix is synthesized; typically 1,000 to 2,000 nucleotides long in prokaryotes, and 100 to 200 nucleotides long in eukaryotes.

olfactory [L. *olfacere*, to smell]: Pertaining to smell.

oligosaccharins [Gk. *oligo*, few + *sakcharon*, sugar]: Short carbohydrate chains released from plant cell walls in response to a variety of stimuli, including injury; hypothesized to play a role in regulation of plant growth and development.

ommatidium, *pl.* **ommatidia** [Gk. *ommos*, eye]: The single visual unit in the compound eye of arthropods; contains light-sensitive cells and a lens able to form an image.

omnivore [L. *omnis*, all + *vorare*, to devour]: An organism that "eats everything"; for example, an animal that eats both plants and meat.

oncogene [Gk. *onkos*, tumor + *genos*, birth, race]: One of a group of eukaryotic genes that closely resemble normal genes of the cells in which they are found and that are thought to play a role in the development of cancer; their gene products appear to be regulatory proteins, involved in the control of either cell growth or cell division.

oocyte (o-uh-sight) [Gk. *oion*, egg + *kytos*, vessel]: A cell that gives rise by meiosis to an ovum.

oogamy (oh-og-amy) [Gk. *oion*, egg + *gamos*, marriage]: Sexual reproduction in which one of the gametes, usually the larger, is not motile.

operator: A segment of DNA that interacts with a repressor protein to regulate the transcription of the structural genes of an operon.

operon [L. *opus, operis*, work]: In the bacterial chromosome, a segment of DNA consisting of a promoter, an operator, and a group of adjacent structural genes; the structural genes, which code for products related to a particular biochemical pathway, are transcribed onto a single mRNA molecule, and their transcription is regulated by a single repressor protein.

opportunistic species: Species characterized by high reproduction rates, rapid development, early reproduction, small body size, and uncertain adult survival.

orbital [L. *orbis*, circle, disk]: In the current model of atomic structure, the volume of space surrounding the atomic nucleus in which an electron will be found 90 percent of the time.

order: A taxonomic grouping of related, similar families; the category below class and above family.

organ [Gk. *organon*, tool]: A body part composed of several tissues grouped together in a structural and functional unit.

organelle [Gk. *organon*, instrument, tool]: A formed body in the cytoplasm of a cell.

organic [Gk. *organon*, instrument, tool]: Pertaining to (1) organisms or living things generally, or (2) compounds formed by living organisms, or (3) the chemistry of compounds containing carbon.

organism [Gk. *organon*, instrument, tool]: Any living creature, either unicellular or multicellular.

organizer [Gk. *organon*, instrument, tool]: In vertebrates, the part of an embryo capable of inducing undifferentiated cells to follow a specific course of development; in particular, the dorsal lip of the blastopore.

osmosis [Gk. *osmos*, impulse, thrust]: The diffusion of water across a selectively permeable membrane (a membrane that permits the free passage of water but prevents or retards the passage of a solute). In the absence of other factors that affect the water potential, the net movement of water is from the side containing a lower concentration of solute to the side containing a higher concentration.

osmotic potential [Gk. *osmos*, impulse, thrust]: The tendency of water to move across a selectively permeable membrane into a solution; it is determined by measuring the pressure required to stop the osmotic movement of water into the solution; the higher the solute concentration, the greater the osmotic potential of the solution.

ovary [L. *ovum*, egg]: (1) In animals, the egg-producing organ. (2) In flowering plants, the enlarged basal portion of a carpel or a fused carpel, containing the ovule or ovules; the ovary matures to become the fruit.

oviduct [L. *ovum*, egg + *ductus*, duct]: The tube serving to transport the eggs to the outside or to the uterus; also called uterine tube or Fallopian tube (in humans).

ovulation: In animals, release of an egg or eggs from the ovary.

ovule [L. dim. of *ovum*, egg]: In seed plants, a structure composed of a protective outer coat, a tissue specialized for food storage, and a female gametophyte with an egg cell; becomes a seed after fertilization.

ovum, *pl.* **ova** [L., egg]: The egg cell; female gamete.

oxidation: Gain of oxygen, loss of hydrogen, or loss of an electron by an atom, ion, or molecule. Oxidation and reduction take place simultaneously, with the electron lost by one reactant being transferred to another reactant.

oxidative phosphorylation: The process by which the energy released as electrons pass down the mitochondrial electron transport chain in the final stage of cellular respiration is used to phosphorylate (add a phosphate group to) ADP molecules, thereby yielding ATP molecules.

pacemaker: *See* Sinoatrial node.

paleontology [Gk. *palaios*, old + *onta*, things that exist + *logos*, discourse]: The study of the life of past geologic times, principally by means of fossils.

palisade cells [L. *palus*, stake + *cella*, a chamber]: In plant leaves, the columnar, chloroplast-containing parenchyma cells of the mesophyll.

pancreas (pang-kree-us) [Gk. *pan*, all + *kreas*, meat, flesh]: In vertebrates, a small, complex gland located between the stomach and the duodenum, which produces digestive enzymes and the hormones insulin and glucagon.

parasite [Gk. *para*, beside, akin to + *sitos*, food]: An organism that lives on or in an organism of a different species and derives nutrients from it.

parasitism: *See* Symbiosis.

parasympathetic division [Gk. *para*, beside, akin to]: A subdivision of the autonomic nervous system of vertebrates, with centers located in the brain and in the most anterior and most posterior parts of the spinal cord; stimulates digestion; generally inhibits other functions and restores the body to normal following emergencies.

parenchyma (pah-renk-ee-ma) [Gk. *para*, beside, akin to + *en*, in + *chein*, to pour]: A plant tissue composed of living, thin-walled, randomly arranged cells with large vacuoles; usually photosynthetic or storage tissue.

parthenogenesis [Gk. *parthenon*, virgin + *genesis*, birth]: The development of an organism from an unfertilized egg.

pellicle [L. dim. of *pellis*, skin]: A flexible series of protein strips inside the cell membrane of many protists.

penetrance: In genetics, the proportion of individuals with a particular genotype that show the phenotype ascribed to that genotype.

peptide bond [Gk. *pepto*, to soften, digest]: The type of bond formed when two amino acids are joined end to end; the acidic group (—COOH) of one amino acid is linked covalently to the basic group (—NH_2) of the next, and a molecule of water (H_2O) is removed.

perennial [L. *per,* through + *annus,* year]: A plant that persists in whole or in part from year to year and usually produces reproductive structures in more than one year.

pericycle [Gk. *peri,* around + *kyklos,* circle]: One or more layers of cells completely surrounding the vascular tissues of the root; branch roots arise from the pericycle.

peripheral nervous system [Gk. *peripherein,* to carry around]: All of the neurons and axons outside the central nervous system, including both motor neurons and sensory neurons; consists of the somatic nervous system and the autonomic nervous system.

peristalsis [Gk. *peristellein,* to wrap around]: Successive waves of muscular contraction in the walls of a tubular structure, such as the digestive tract or an oviduct; moves the contents, such as food or an egg cell, through the tube.

peritoneum [Gk. *peritonos,* stretched over]: A membrane that lines the body cavity and forms the external covering of the visceral organs.

permeable [L. *permeare,* to pass through]: Penetrable by molecules, ions, or atoms; usually applied to membranes that let given solutes pass through.

peroxisome: A membrane-bound organelle in which enzymes catalyzing peroxide-forming and peroxide-destroying reactions are segregated; in plant cells, the site of photorespiration.

petiole (pet-ee-ole) [Fr., from L. *petiolus,* dim. of *pes, pedis,* a foot]: The stalk of a leaf, connecting the blade of the leaf with the branch or stem.

pH: A symbol denoting the concentration of hydrogen ions in a solution; pH values range from 0 to 14; the lower the value, the more acidic a solution, that is, the more hydrogen ions it contains; pH 7 is neutral, less than 7 is acidic, more than 7 is alkaline.

phagocytosis [Gk. *phagein,* to eat + *kytos,* vessel]: Cell "eating." See Endocytosis.

phenotype [Gk. *phainein,* to show + *typos,* stamp, print]: Observable characteristics of an organism, resulting from interactions between the genotype and the environment.

pheromone (fair-o-moan) [Gk. *phero,* to bear, carry]: Substance secreted by an animal that influences the behavior or development of other animals of the same species, such as the sex attractants of moths, the queen substance of honey bees.

phloem (flow-em) [Gk. *phloos,* bark]: Vascular tissue of higher plants; conducts sugars and other organic molecules from the leaves to other parts of the plant; in angiosperms, composed of sieve-tube members, companion cells, other parenchyma cells, and fibers.

phospholipids: Organic molecules similar in structure to fats, but in which a phosphate group rather than a fatty acid is attached to the third carbon of the glycerol molecule; as a result, the molecule has a hydrophilic "head" and a hydrophobic "tail." Phospholipids form the basic structure of cell and organelle membranes.

phosphorylation: Addition of a phosphate group or groups to a molecule.

photon [Gk. *photos,* light]: The elementary particle of light and other electromagnetic radiations.

photoperiodism [Gk. *photos,* light]: The response to relative day and night length, a mechanism by which organisms measure seasonal change.

photophosphorylation [Gk. *photos,* light + *phosphoros,* bringing light]: The process by which the energy released as electrons pass down the electron transport chain between photosystems II and I during photosynthesis is used to phosphorylate ADP to ATP.

photoreceptor [Gk. *photos,* light]: A cell or organ capable of detecting light.

photorespiration [Gk. *photos,* light + L. *respirare,* to breathe]: The oxidation of carbohydrates in the presence of light and oxygen; occurs when the carbon dioxide concentration in the leaf is low in relation to the oxygen concentration.

photosynthesis [Gk. *photos,* light + *syn,* together + *tithenai,* to place]: The conversion of light energy to chemical energy; the synthesis of organic compounds from carbon dioxide and water in the presence of chlorophyll, using light energy.

phototropism [Gk. *photos,* light + *trope,* turning]: Movement in which the direction of the light is the determining factor, such as the growth of a plant toward a light source; a curving response to light.

phyletic change [Gk. *phylon,* race, tribe]: The changes taking place in a single lineage of organisms over a long period of time; one of the principal patterns of evolutionary change.

phylogeny [Gk. *phylon,* race, tribe]: Evolutionary history of a taxonomic group. Phylogenies are often depicted as "evolutionary trees."

phylum, *pl.* phyla [Gk. *phylon,* race, tribe]: A taxonomic grouping of related, similar classes; a high-level category below kingdom and above class. Phylum is generally used in the classification of protozoa and animals, whereas an equivalent category, division, is used in the classification of prokaryotes, algae, fungi, and plants.

physiology [Gk. *physis,* nature + *logos,* a discourse]: The study of function in cells, organs, or entire organisms; the processes of life.

phytochrome [Gk. *phyton,* plant + *chrōma,* color]: A plant pigment that is a photoreceptor for red or far-red light and is involved with a number of developmental processes, such as flowering, dormancy, leaf formation, and seed germination.

phytoplankton [Gk. *phyton,* plant + *planktos,* wandering]: Aquatic, free-floating, microscopic, photosynthetic organisms.

pigment [L. *pigmentum,* paint]: A colored substance that absorbs light over a narrow band of wavelengths.

pinocytosis [Gk. *pinein,* to drink + *kytos,* vessel]: Cell "drinking." See Endocytosis.

pituitary [L. *pituita,* phlegm]: Endocrine gland in vertebrates; the anterior lobe is the source of tropic hormones, growth hormone, and prolactin and is regulated by secretions of the hypothalamus; the posterior lobe stores and releases oxytocin and ADH produced by the hypothalamus.

placenta [Gk. *plax,* a flat object]: A tissue formed as the result of interactions between the inner lining of the mammalian uterus and the extraembryonic chorion; serves as the connection through which exchanges of nutrients and wastes occur between the blood of the mother and that of the embryo.

plankton [Gk. *planktos,* wandering]: Small (mostly microscopic) aquatic and marine organisms found in the upper levels of the water, where light is abundant; includes both photosynthetic (phytoplankton) and heterotrophic (zooplankton) forms.

planula [L. dim. of *planus,* a wanderer]: The ciliated, free-swimming type of larva formed by many cnidarians.

plasma [Gk., form or mold]: The clear, colorless fluid component of vertebrate blood, containing dissolved ions, molecules, and plasma proteins; blood minus the blood cells.

plasma cell: An antibody-producing cell resulting from the differentiation and proliferation of a B lymphocyte that has interacted with an antigen complementary to the antibodies displayed on its surface; a mature plasma cell can produce from 3,000 to 30,000 antibody molecules per second.

plasma membrane: The membrane surrounding the cytoplasm of a cell; the cell membrane.

plasmid: In prokaryotes, an extrachromosomal, independently replicating, small, circular DNA molecule.

plasmodesma, *pl.* plasmodesmata [Gk. *plassein,* to mold + *desmos,* band, bond]: In plants, a minute, cytoplasmic thread that extends through pores in cell walls and connects the cytoplasm of adjacent cells.

plastid [Gk. *plastos,* formed or molded]: A cytoplasmic, often pigmented, organelle in plant cells; includes leucoplasts, chromoplasts, and chloroplasts.

platelet (plate-let) [Gk. *platus,* flat]: In mammals, a round or biconcave disk suspended in the blood and involved in the formation of blood clots.

pleiotropy (plee-o-trope-ee) [Gk. *pleios,* more + *trope,* a turning]: The capacity of a gene to affect a number of different phenotypic characteristics.

poikilotherm [Gk. *poikilos,* changeable + *therme,* heat]: An organism with a body temperature that varies with that of the environment.

polar [L. *polus*, end of axis]: Having parts or areas with opposed or contrasting properties, such as positive and negative charges, head and tail.

polar body: Minute, nonfunctioning cell produced during those meiotic divisions that lead to egg cells; contains a nucleus but very little cytoplasm.

polar covalent bond: A covalent bond in which the electrons are shared unequally between the two atoms; the resulting polar molecule has regions of slightly negative and slightly positive charge.

pollen [L., fine dust]: In seed plants, spores consisting of an immature male gametophyte and a protective outer covering.

pollination [L. *pollen*, fine dust]: The transfer of pollen from the anther to a receptive surface of a flower.

polygenic inheritance [Gk. *polus*, many + *genos*, race, descent]: The determination of a given characteristic, such as weight or height, by the interaction of many genes.

polymer [Gk. *polus*, many + *meris*, part or portion]: A large molecule composed of many similar or identical molecular subunits.

polymorphism [Gk. *polus*, many + *morphe*, form]: The presence in a single population of two or more phenotypically distinct forms of a trait.

polyp [Gk. *polus*, many + *pous*, foot]: The sessile stage in the life cycle of cnidarians.

polypeptide [Gk. *polus*, many + *pepto*, to soften, digest]: A molecule consisting of a long chain of amino acids linked together by peptide bonds.

polyploid [Gk. *polus*, many + *ploion*, vessel]: Cell with more than two complete sets of chromosomes per nucleus.

polyribosome: Two or more ribosomes together with a molecule of mRNA that they are simultaneously translating; a polysome.

polysaccharide [Gk. *polus*, many + *sakcharon*, sugar]: A carbohydrate polymer composed of monosaccharide monomers in long chains; includes starch, cellulose.

polysome: *See* Polyribosome.

population: Any group of individuals of one species that occupy a given area at the same time; in genetic terms, an interbreeding group of organisms.

population bottleneck: Type of genetic drift that occurs as the result of a population being drastically reduced in numbers by an event having little to do with the usual forces of natural selection.

portal system [L. *porta*, gate]: In the circulatory system, a circuit in which blood flows through two distinct capillary beds, connected by either veins or arteries, before entering the veins that return it to the heart.

posterior: Of or pertaining to the rear, or tail, end.

potential energy: Energy in a potentially usable form that is not, for the moment, being used; often called "energy of position."

predator [L. *praedari*, to prey upon; from *prehendere*, to grasp, seize]: An organism that eats other living organisms.

pressure-flow hypothesis: A hypothesis accounting for sap flow through the phloem system. According to this hypothesis, the solution containing nutrient sugars moves through the sieve tubes by bulk flow, moving into and out of the sieve tubes by active transport and diffusion.

prey [L. *prehendere*, to grasp, seize]: An organism eaten by another organism.

primary growth: In plants, growth originating in the apical meristem of the shoots and roots, as contrasted with secondary growth; results in an increase in length.

primary structure of a protein: The amino acid sequence of a protein.

primate: A member of the order of mammals that includes anthropoids and prosimians.

primitive [L. *primus*, first]: Not specialized; at an early stage of evolution or development.

primitive streak [L. *primus*, first]: The thickened, dorsal, longitudinal strip of ectoderm and mesoderm in early avian, reptilian, and mammalian embryos; equivalent to the blastopore in other forms.

procambium [L. *pro*, before + *cambium*, exchange]: In plants, a primary meristematic tissue; gives rise to vascular tissues of the primary plant body and to the vascular cambium.

producer, in ecological systems: An autotrophic organism, usually a photosynthesizer, that contributes to the net primary productivity of a community.

progesterone [L. *progerere*, to carry forth or out + *steiras*, barren]: In mammals, a steroid hormone produced by the corpus luteum that prepares the uterus for implantation of the embryo; also produced by the placenta during pregnancy.

prokaryote [L. *pro*, before + Gk. *karyon*, nut, kernel]: A cell lacking a membrane-bound nucleus and membrane-bound organelles; a bacterium or a cyanobacterium.

promoter: Specific segment of DNA to which RNA polymerase attaches to initiate transcription of mRNA from an operon.

prophage: A bacterial virus (bacteriophage) integrated into a host chromosome.

prophase [Gk. *pro*, before + *phasis*, form]: An early stage in nuclear division, characterized by the condensing of the chromosomes and their movement toward the equator of the spindle. Homologous chromosomes pair up during meiotic prophase.

proprioceptor [L. *proprius*, one's own]: Receptor that senses movements, position of the body, or muscle strength.

prosimian [L. *pro*, before + *simia*, ape]: A lower primate; includes lemurs, lorises, tarsiers, and bush babies, as well as many fossil forms.

prostaglandins [Gk. *prostas*, a porch or vestibule + L. *glans*, acorn]: A group of fatty acids that function as chemical messengers; synthesized in most, possibly all, cells of the body; thought to play key roles in fertilization and in triggering the onset of both menstruation and labor.

prostate gland [Gk. *prostas*, a porch or vestibule + L. *glans*, acorn]: A mass of muscle and glandular tissue surrounding the base of the urethra in male mammals; the vasa deferentia merge with ducts from the seminal vesicles, enter the prostate gland, and there merge with the urethra. The prostate gland secretes an alkaline fluid that has a stimulating effect on the sperm as they are released.

protein [Gk. *proteios*, primary]: A complex organic compound composed of one or more polypeptide chains, each made up of many (about 100 or more) amino acids linked together by peptide bonds.

proton: A subatomic particle with a single positive charge equal in magnitude to the charge of an electron and with a mass slightly less than that of a neutron; a component of every atomic nucleus.

protoplasm [Gk. *protos*, first + *plasma*, anything molded]: Living matter.

protostome [Gk. *protos*, first + *stoma*, mouth]: An animal in which the mouth forms at or near the blastopore in the developing embryo; mollusks, annelids, and arthropods are protostomes. Protostomes are also characterized by spiral cleavage during the earliest stages of development and by schizocoelous formation of the coelom.

provirus: A virus of a eukaryote that has become integrated into a host chromosome.

pseudocoelom [Gk. *pseudes*, false + *koilos*, a hollow]: A body cavity consisting of a fluid-filled space between the endoderm and the mesoderm; characteristic of the nematodes.

pseudopodium [Gk. *pseudes*, false + *pous*, pod-, foot]: A temporary cytoplasmic protrusion from an amoeboid cell, which functions in locomotion or in feeding by phagocytosis.

pulmonary [L. *pulmonis*, lung]: Pertaining to the lungs.

pulmonary artery [L. *pulmonis*, lung]: In birds and mammals, an artery that carries deoxygenated blood from the right ventricle of the heart to the lungs, where it is oxygenated.

pulmonary vein [L. *pulmonis*, lung]: In birds and mammals, a vein that carries oxygenated blood from the lungs to the left atrium of the heart, from which blood is pumped into the left ventricle and from there to the body tissues.

punctuated equilibrium: A model of the mechanism of evolutionary change that proposes that long periods of no change ("stasis") are punctuated by periods of rapid speciation, with natural selection acting on species as well as on individuals.

Punnett square: The checkerboard diagram used for analysis of allele segregation.

pupa [L., girl, doll]: A developmental stage of some insects, in which the organism is nonfeeding, immotile, and sometimes encapsulated or in a cocoon; the pupal stage occurs between the larval and adult phases.

purine [Gk. *purinos*, fiery, sparkling]: A nitrogenous base, such as adenine or guanine, with a characteristic two-ring structure; one of the components of nucleic acids.

pyramid, ecological: *See* Ecological pyramid.

pyramid of energy: A diagram of the energy flow between the trophic levels of an ecosystem; plants or other autotrophs (at the base of the pyramid) represent the greatest amount of energy, herbivores next, then primary carnivores, secondary carnivores, etc.

pyrimidine: A nitrogenous base, such as cytosine, thymine, or uracil, with a characteristic single-ring structure; one of the components of nucleic acids.

quaternary structure of a protein: The overall structure of a globular protein molecule that consists of two or more polypeptide chains.

queen: In social insects (ants, termites, and some species of bees and wasps), the fertile, or fully developed, female whose function is to lay eggs.

radial symmetry [L. *radius*, a spoke of a wheel + Gk. *summetros*, symmetry]: The regular arrangement of parts around a central axis such that any plane passing through the central axis divides the organism into halves that are approximate mirror images; seen in cnidarians, ctenophorans, and adult echinoderms.

radiation [L. *radius*, a spoke of a wheel, hence, a ray]: Energy emitted in the form of waves or particles.

radioactive isotope: An isotope with an unstable nucleus that stabilizes itself by emitting radiation.

receptor: A protein or glycoprotein molecule with a specific three-dimensional structure, to which a substance (for example, a hormone, a neurotransmitter, or an antigen) with a complementary structure can bind; typically displayed on the surface of a membrane. Binding of a complementary molecule to a receptor may trigger a transport process or a change in processes occurring within the cell.

recessive allele [L. *recedere*, to recede]: An allele whose phenotypic effect is masked in the heterozygote by that of another, dominant allele.

reciprocal altruism: Performance of an altruistic act with the expectation that the favor will be returned.

recognition sequence: A specific sequence of nucleotides at which a restriction enzyme cleaves a DNA molecule.

recombinant DNA: DNA formed either naturally or in the laboratory by the joining of segments of DNA from different sources.

recombination: The formation of new gene combinations; in eukaryotes, may be accomplished by new associations of chromosomes produced during sexual reproduction or crossing over; in prokaryotes, may be accomplished through transformation, conjugation, or transduction.

reduction [L. *reducere*, to lead back]: Loss of oxygen, gain of hydrogen, or gain of an electron by an atom, ion, or molecule; oxidation and reduction take place simultaneously, with the electron lost by one reactant being transferred to another.

reflex [L. *reflectere*, to bend back]: Unit of action of the nervous system involving a sensory neuron, often one or more interneurons, and one or more motor neurons.

relay neuron: Neuron that transmits signals between different regions of the central nervous system.

releaser: In ethology, a stimulus that functions as a communication signal between members of the same species and that sets in motion, or "releases," the sequence of movements that constitute a fixed action pattern.

renal [L. *renes*, kidneys]: Pertaining to the kidney.

replication fork: In DNA synthesis, the Y-shaped structure formed at the point where the two strands of the original molecule are being separated and the complementary strands are being synthesized.

repressor [L. *reprimere*, to press back, keep back]: In genetics, a protein that binds to the operator, preventing RNA polymerase from attaching to the promoter and transcribing the structural genes of the operon; coded by a segment of DNA known as the regulator.

resolving power [L. *resolvere*, to loosen, unbind]: The ability of a lens to distinguish two lines as separate.

respiration [L. *respirare*, to breathe]: (1) In aerobic organisms, the intake of oxygen and the liberation of carbon dioxide. (2) In cells, the oxygen-requiring stage in the breakdown and release of energy from fuel molecules.

resting potential: The difference in electric potential (about 70 millivolts) across the membrane of an axon at rest.

restriction enzymes: Enzymes that cleave the DNA double helix at specific nucleotide sequences.

reticular activating system [L. *reticulum*, a network]: A brain circuit involved with alertness and direction of attention to selected events; includes the reticular formation, a core of tissue that runs centrally through the brainstem, and neurons in the thalamus.

reticulum [L., network]: A fine network (e.g., endoplasmic reticulum).

retina [L. dim. of *rete*, net]: The light-sensitive layer of the vertebrate eye; contains several layers of neurons and photoreceptor cells (rods and cones); receives the image formed by the lens and transmits it to the brain via the optic nerve.

retrovirus [L., turning back]: An RNA virus that codes for an enzyme, reverse transcriptase, that transcribes the RNA into DNA.

reverse transcriptase: An enzyme that transcribes RNA into DNA; found only in association with retroviruses.

rhizoid [Gk. *rhiza*, root]: Rootlike anchoring structure in fungi and nonvascular plants.

rhizome [Gk. *rhizoma*, mass of roots]: In vascular plants, a horizontal stem growing along or below the surface of the soil; may be enlarged for storage or may function in vegetative reproduction.

ribonucleic acid (RNA) (rye-bo-new-clay-ick): A class of nucleic acids characterized by the presence of the sugar ribose and the pyrimidine uracil; includes mRNA, tRNA, and rRNA. RNA is the genetic material of many viruses.

ribosomal RNA (rRNA): A class of RNA molecules found, along with characteristic proteins, in ribosomes; transcribed from DNA of the chromatin loops that form the nucleolus.

ribosome: A small organelle composed of protein and ribonucleic acid; the site of translation in protein synthesis; in eukaryotic cells, often bound to the endoplasmic reticulum. Many ribosomes attached to a single strand of mRNA are called a polyribosome, or polysome.

RNA: Abbreviation of ribonucleic acid.

rod: Photoreceptor cell found in the vertebrate retina; sensitive to very dim light, responsible for "night vision."

root: The descending axis of a plant, normally below ground and serving both to anchor the plant and to take up and conduct water and minerals.

root hair: An extremely fine cytoplasmic extension of an epidermal cell of a young root; root hairs greatly increase the surface area for the uptake of water and minerals.

saprobe [Gk. *sapros*, rotten, putrid + *bios*, life]: An organism that feeds on nonliving organic matter.

sarcolemma [Gk. *sarx*, the flesh + *lemma*, husk]: The specialized cell membrane surrounding a muscle cell (muscle fiber); capable of propagating action potentials.

sarcomere [Gk. *sarx*, the flesh + *meris*, part of, portion]: Functional and structural unit of contraction in striated muscle.

sarcoplasmic reticulum [Gk. *sarx*, the flesh + *plasma*, from cytoplasm + L. *reticulum*, network]: The specialized endoplasmic reticulum that encases each myofibril of a muscle cell.

schizocoelous [Gk. *schizo*, to split + *koilos*, a hollow]: Formation of the coelom during embryonic development by a splitting of the mesoderm; characteristic of protostomes.

sclerenchyma [Gk. *skleros*, hard]: In plants, a type of supporting cell with thick, often lignified, secondary walls; may be alive or dead at maturity; includes fibers and sclereids.

secondary sex characteristics: Characteristics of animals that distinguish between the two sexes but that do not produce or convey gametes; includes facial hair of the human male and enlarged hips and breasts of the female.

secondary structure of a protein: The simple structure (often a helix, a sheet, or a cable) resulting from the spontaneous folding of a polypeptide chain as it is formed; maintained by hydrogen bonds and other weak forces.

secretion [L. *secernere*, to sever, separate]: (1) Product of any cell, gland, or tissue that is released through the cell membrane and that performs its function outside the cell that produced it. (2) The stage of kidney function in which, through active transport processes, molecules remaining in the blood plasma are selectively removed from the peritubular capillaries and pumped into the filtrate in the renal tubule.

seed: A complex structure formed by the maturation of the ovule of seed plants following fertilization; upon germination, a seed develops into a new sporophyte; generally consists of seed coat, embryo, and a food reserve.

segregation: *See* Mendel's first law.

selectively permeable [L. *seligere*, to gather apart + *permeare*, to go through]: Applied to membranes that permit passage of water and some solutes but block passage of most solutes; semipermeable.

self-fertilization: The union of egg and sperm produced by a single hermaphroditic organism.

self-pollination: The transfer of pollen from anther to stigma in the same flower or to another flower of the same plant, leading to self-fertilization.

semen [L., seed]: Product of the male reproductive system; includes sperm and the sperm-carrying fluids.

seminal vesicles [L. *semen*, seed + *vesicula*, a little bladder]: In male mammals, small vesicles, the ducts of which merge with the vasa deferentia as they enter the prostate gland; they produce an alkaline, fructose-containing fluid that suspends and nourishes the sperm cells.

sensory neuron: A neuron that conducts impulses from a sensory receptor to the central nervous system or central ganglion.

sensory receptor: A cell, tissue, or organ that detects internal or external stimuli.

septum [L., fence]: A partition, or cross wall, that divides a structure, such as a fungal hypha, into compartments.

sessile [L. *sedere*, to sit]: Attached; not free to move about.

sex chromosomes: Chromosomes that are different in the two sexes and that are involved in sex determination.

sex-linked trait: An inherited trait, such as color discrimination, determined by a gene located on a sex chromosome and that therefore shows a different pattern of inheritance in males and females.

sexual reproduction: Reproduction involving meiosis and fertilization.

sexual selection: A type of natural selection that acts on characteristics of direct consequence in obtaining a mate and successfully reproducing; thought to be the chief cause of sexual dimorphism, the striking phenotypic differences between the males and females of many species.

shoot: The aboveground portions, such as the stem and leaves, of a vascular plant.

sieve cell: A long, slender cell of the phloem of gymnosperms; involved in transport of sugars synthesized in the leaves to other parts of the plant.

sieve tube: A series of sugar-conducting cells (sieve-tube members) found in the phloem of angiosperms.

sinoatrial node [L. *sinus*, fold, hollow + *atrium*, yard, court, hall + *nodus*, knot]: Area of the vertebrate heart that initiates the heartbeat; located where the superior vena cava enters the right atrium; the pacemaker.

smooth muscle: Nonstriated muscle; lines the walls of internal organs and arteries and is under involuntary control.

social dominance: A hierarchical pattern of social organization involving domination of some members of a group by other members in a relatively orderly and long-lasting pattern.

society [L. *socius*, companion]: An organization of individuals of the same species in which there are divisions of resources, divisions of labor, and mutual dependence; a society is held together by stimuli exchanged among members of the group.

sociobiology: The study of the biological basis of social behavior.

solution: A homogeneous mixture of the molecules of two or more substances; the substance present in the greatest amount (usually a liquid) is called the solvent, and the substances present in lesser amounts are called solutes.

somatic cells [Gk. *soma*, body]: The differentiated cells composing body tissues of multicellular plants and animals; all body cells except those giving rise to gametes.

somatic nervous system [Gk. *soma*, body]: In vertebrates, the motor and sensory neurons of the peripheral nervous system that control skeletal muscle; the "voluntary" system, as contrasted with the "involuntary," or autonomic, nervous system.

somite: One of the blocks, or segments, of tissue into which the chorda-mesoderm is divided during differentiation of the vertebrate embryo.

specialized: (1) Of cells, having particular functions in a multicellular organism. (2) Of organisms, having special adaptations to a particular habitat or mode of life.

speciation: The process by which new species are formed.

species, *pl.* **species** [L., kind, sort]: A group of organisms that actually (or potentially) interbreed in nature and are reproductively isolated from all other such groups; a taxonomic grouping of anatomically similar individuals (the category below genus).

species-specific: Characteristic of (and limited to) a particular species.

specific: Unique; for example, the proteins in a given organism, the enzyme catalyzing a given reaction, or the antibody to a given antigen.

specific heat: The amount of heat (in calories) required to raise the temperature of 1 gram of a substance 1°C. The specific heat of water is 1 calorie per gram.

sperm [Gk. *sperma*, seed]: A mature male sex cell, or gamete, usually motile and smaller than the female gamete.

spermatid [Gk. *sperma*, seed]: Each of four haploid (*n*) cells resulting from the meiotic divisions of a spermatocyte; each spermatid becomes differentiated into a sperm cell.

spermatocytes [Gk. *sperma*, seed + *kytos*, vessel]: The diploid (2*n*) cells formed by the enlargement of the spermatogonia; they give rise by meiotic division to the spermatids.

spermatogonia [Gk. *sperma*, seed + *gonos*, a child, the young]: The unspecialized diploid (2*n*) cells on the walls of the testes that, by meiotic division, become spermatocytes, then spermatids, then sperm cells.

sphincter [Gk. *sphinktēr*, a band]: A circular muscle surrounding the opening of a tubular structure or the juncture of different regions of a tubular structure (e.g., the pyloric sphincter, at the juncture of the stomach and the small intestine); contraction of the sphincter closes the passageway, and relaxation opens it.

spinal cord: Part of the vertebrate central nervous system; consists of a thick, dorsal, longitudinal bundle of nerve fibers extending posteriorly from the brain.

spindle: In dividing cells, the structure formed of microtubules that extends from pole to pole; the spindle fibers appear to maneuver the chromosomes into position during metaphase and to pull the newly separated chromosomes toward the poles during anaphase.

spiracle [L. *spirare*, to breathe]: One of the external openings of the respiratory system in terrestrial arthropods.

splitting evolution: *See* Cladogenesis.

sporangiophore (spo-ran-ji-o-for) [Gk. *spora*, seed + *phore*, from *phorein*, to bear]: A specialized hypha or a branch bearing one or more sporangia.

sporangium, *pl.* **sporangia** [Gk. *spora*, seed]: A unicellular or multicellular structure in which spores are produced.

spore [Gk. *spora*, seed]: An asexual reproductive or resting cell capable of developing into a new organism without fusion with another cell; in contrast to a gamete.

sporophyll [Gk. *spora*, seed + *phyllon*, the leaves]: Spore-bearing leaf. The carpels and stamens of flowers are modified sporophylls.

sporophyte [Gk. *spora*, seed + *phytos*, growing]: In organisms that have alternation of haploid and diploid generations (all plants and some green algae), the diploid ($2n$) spore-producing generation.

stamen [L., a thread]: The male structure of a flower, which produces microspores or pollen; usually consists of a stalk, the filament, bearing a pollen-producing anther at its tip.

starch [M.E. *sterchen*, to stiffen]: A class of complex, insoluble carbohydrates, the chief food-storage substances of plants; composed of 1,000 or more glucose units and readily broken down enzymatically into these units.

statocyst [Gk. *statos*, standing + *kystis*, sac]: An organ of balance, consisting of a vesicle containing granules of sand (statoliths) or some other material that stimulates sensory cells when the organism moves.

stem: The aboveground part of the axis of vascular plants, as well as anatomically similar portions below ground (such as rhizomes).

stem cells: The common, self-regenerating cells in the marrow of long bones that give rise, by differentiation and division, to red blood cells and all of the different types of white blood cells.

stereoscopic vision [Gk. *stereōs*, solid + *optikos*, pertaining to the eye]: Ability to perceive a single, three-dimensional image from the simultaneous but separate images delivered to the brain by each eye.

steroid: One of a group of lipids having four linked carbon rings and, often, a hydrocarbon tail; cholesterol, sex hormones, and the hormones of the adrenal cortex are steroids.

stigma [Gk. *stigme*, a prick mark, puncture]: In plants, the region of a carpel serving as a receptive surface for pollen grains, which germinate on it.

stimulus [L., goad, incentive]: Any internal or external change or signal that influences the activity of an organism or of part of an organism.

stoma, *pl.* **stomata** [Gk., mouth]: A minute opening in the epidermis of leaves and stems, bordered by guard cells, through which gases pass.

strategy [Gk. *strategein*, to maneuver]: A group of related traits, evolved under the influence of natural selection, that solve particular problems encountered by living organisms; often includes anatomical, physiological, and behavioral characteristics.

striated muscle [L., from *striare*, to groove]: Skeletal voluntary muscle and cardiac muscle. The name derives from the striped appearance, which reflects the arrangement of contractile elements.

stroma [Gk., a bed, from *stronnymi*, to spread out]: A dense solution that fills the interior of the chloroplast and surrounds the thylakoids.

structural gene: Any gene that codes for a protein; in distinction to regulatory genes.

style [L. *stilus*, stake, stalk]: In angiosperms, the stalk of a carpel, down which the pollen tube grows.

substrate [L. *substratus*, strewn under]: (1) The foundation to which an organism is attached. (2) A substance on which an enzyme acts.

succession: *See* Ecological succession.

sucrose: Cane sugar; a common disaccharide found in many plants; a molecule of glucose linked to a molecule of fructose.

sugar: Any monosaccharide or disaccharide.

supergene: *See* Coadaptive gene complex.

surface tension: A tautness of the surface of a liquid, caused by the cohesion of the molecules of liquid. Water has an extremely high surface tension.

symbiosis [Gk. *syn*, together with + *bioonai*, to live]: An intimate and protracted association between two or more organisms of different species. Includes mutualism, in which the association is beneficial to both; commensalism, in which one benefits and the other is neither harmed nor benefited; and parasitism, in which one benefits and the other is harmed.

sympathetic division: A subdivision of the autonomic nervous system, with centers in the midportion of the spinal cord; slows digestion; generally excites other functions.

sympatric speciation [Gk. *syn*, together with + *patra*, fatherland, country]: Speciation that occurs without geographic isolation of a population of organisms; usually occurs as the result of hybridization accompanied by polyploidy; may occur in some cases as a result of disruptive selection.

synapse [Gk. *synapsis*, a union]: A specialized junction between two neurons where the activity in one influences the activity in another; may be chemical or electrical, excitatory or inhibitory.

syngamy (sin-gamy) [Gk. *syn*, with + *gamos*, a marriage]: The union of gametes in sexual reproduction; fertilization.

synthesis [Gk. *syntheke*, a putting together]: The formation of a more complex substance from simpler ones.

synthetic theory: The currently prevailing theory of the mechanism of evolutionary change; combines the Darwinian two-step model of variation and selection with the principles of Mendelian genetics.

systematics [Gk. *systema*, that which is put together]: Scientific study of the kinds and diversity of organisms and of the relationships among them.

T lymphocyte: A type of white blood cell arising from precursors in the thymus gland and, upon maturation, involved in cell-mediated immunity and interactions with B lymphocytes; a T cell.

tagmosis [Gk. *tagma*, arrangement, order + *-osis*, process]: The formation of groups of segments (metameres) into body regions (tagmata) with functional differences.

taxon, *pl.* **taxa** [Gk. *taxis*, arrange, put in order]: A particular group, ranked at a particular categorical level, in a hierarchical classification scheme; for example, *Drosophila* is a taxon at the categorical level of genus.

taxonomy [Gk. *taxis*, arrange, put in order + *nomos*, law]: The study of the classification of organisms; the ordering of organisms into a hierarchy that reflects their essential similarities and differences.

telencephalon [Gk. *tēl*, far off + *enkephalos*, brain]: One of the two principal subdivisions of the vertebrate forebrain; the anterior portion of the forebrain, it contains the cerebrum and the olfactory bulbs.

telophase [Gk. *telos*, end + *phasis*, form]: The last stage in mitosis and meiosis, during which the chromosomes become reorganized into two new nuclei.

temperate bacteriophage: A bacterial virus that may become incorporated into the host-cell chromosome.

template: A pattern or mold guiding the formation of a negative or complement.

tentacles [L. *tentare*, to touch]: Long, flexible protrusions located about the mouth of many invertebrates; usually prehensile or tactile.

territory: An area or space occupied and defended by an individual or a group; trespassers are attacked (and usually defeated); may be the site of breeding, nesting, food gathering, or any combination thereof.

tertiary structure of a protein: A complex structure, usually globular, resulting from further folding of the secondary structure of a protein;

forms spontaneously due to attractions and repulsions among amino acids with different charges on their R groups.

testcross: A mating between a phenotypically dominant individual and a homozygous recessive "tester" to determine the genetic constitution of the dominant phenotype, that is, whether it is homozygous or heterozygous for the relevant gene.

testis, *pl.* **testes** [L., witness]: The sperm-producing organ; also the source of the male sex hormone testosterone.

testosterone [Gk. *testis,* testicle + *steiras,* barren]: A steroid hormone secreted by the testes in higher vertebrates and stimulating the development and maintenance of male sex characteristics and the production of sperm; the principal androgen.

tetrad [Gk. *tetras,* four]: In genetics, a pair of homologous chromosomes that have replicated and come together in prophase I of meiosis; consists of four chromatids.

thalamus [Gk. *thalamos,* chamber]: A part of the vertebrate forebrain just posterior to and tucked below the cerebrum; the main relay center between the brainstem and the higher brain centers.

thallus [Gk. *thallos,* a young twig]: A simple plant or algal body without true roots, leaves, or stems.

theory [Gk. *theorein,* to look at]: A generalization based on many observations and experiments; a verified hypothesis.

thermodynamics [Gk. *therme,* heat + *dynamis,* power]: The study of transformations of energy. The first law of thermodynamics states that, in all processes, the total energy of a system plus its surroundings remains constant. The second law states that all natural processes tend to proceed in such a direction that the disorder or randomness of the system increases.

thorax [Gk., breastplate]: (1) In vertebrates, that portion of the trunk containing the heart and lungs. (2) In crustaceans and insects, the fused, leg-bearing segments between head and abdomen.

thylakoid [Gk. *thylakos,* a small bag]: A flattened sac, or vesicle, that forms part of the internal membrane structure of the chloroplast; the site of the light-trapping reactions of photosynthesis and of photophosphorylation; stacks of thylakoids collectively form the grana.

thyroid [Gk. *thyra,* a door]: An endocrine gland of vertebrates, located in the neck; source of an iodine-containing hormone (thyroxine) that increases the metabolic rate and affects growth.

tight junction: A junction between adjacent animal cells that prevents materials from leaking through the tissue; for example, intestinal epithelial cells are surrounded by tight junctions.

tissue [L. *texere,* to weave]: A group of similar cells organized into a structural and functional unit.

tonoplast [Gk. *tonos,* stretching, tension + *plastos,* formed, molded]: In plant cells, the membrane surrounding the vacuole.

trachea, *pl.* **tracheae** (trake-ee-a) [Gk. *tracheia,* rough]: An air-conducting tube. (1) In insects and some other terrestrial arthropods, a system of chitin-lined air ducts. (2) In terrestrial vertebrates, the windpipe.

tracheid (tray-key-idd) [Gk. *tracheia,* rough]: In vascular plants, an elongated, thick-walled conducting and supporting cell of xylem, characterized by tapering ends and pitted walls without true perforations.

tract: A group or bundle of nerve fibers with accompanying connective tissue, located within the central nervous system.

transcription [L. *trans,* across + *scribere,* to write]: The enzymatic process by which the genetic information contained in one strand of DNA is used to specify a complementary sequence of bases in an RNA molecule.

transduction [L. *trans,* across + *ducere,* to lead]: (1) The transfer of genetic material (DNA) from one cell to another by a virus. (2) The conversion of one form of energy into another form of energy; for example, the conversion of the energy of a chemical stimulus into the energy of an action potential.

transfer RNA (tRNA) [L. *trans,* across + *ferre,* to bear or carry]: A class of small RNAs (about 80 nucleotides each) with two functional sites; one recognizes a specific activated amino acid; the other carries the nucleotide triplet (anticodon) for that amino acid. Each type of tRNA accepts a specific activated amino acid and transfers it to a growing polypeptide chain as specified by the nucleotide sequence of the mRNA being translated.

transformation [L. *trans,* across + *formare,* to shape]: A genetic change produced by the incorporation into a cell of DNA from the external medium.

translation [L. *trans,* across + *latus,* that which is carried]: The process by which the genetic information present in a strand of mRNA directs the sequence of amino acids during protein synthesis.

translocation [L. *trans,* across + *locare,* to put or place]: (1) In plants, the transport of the products of photosynthesis from a leaf to another part of the plant. (2) In genetics, the breaking off of a piece of chromosome with its reattachment to a nonhomologous chromosome.

transpiration [L. *trans,* across + *spirare,* to breathe]: In plants, the loss of water vapor from the stomata.

transposon [L. *transponere,* to change the position of]: A DNA sequence carrying one or more genes that is capable of moving from one location in the chromosomes to another. Simple transposons, also known as insertion sequences, carry only the genes essential for transposition; complex transposons carry genes that code for additional proteins.

tritium: A radioactive isotope (^{3}H) of hydrogen with a half-life of 12.5 years.

trophic level [Gk. *trophos,* feeder]: The position of a species in the food web or chain, that is, its feeding level; a step in the movement of biomass or energy through an ecosystem.

trophoblast [Gk. *trophos,* feeder + *blastos,* sprout]: In the early mammalian embryo (the blastocyst), a double layer of cells that surrounds the inner cell mass and subsequently gives rise to the chorion.

tropic [Gk. *trope,* a turning]: Pertaining to behavior or action brought about by specific stimuli, for example, phototropic ("light-oriented") motion, gonadotropic ("stimulating the gonads") hormone.

tuber [L. *tuber,* bump, swelling]: A much-enlarged, short, fleshy underground stem, such as that of the potato.

turgor [L. *turgere,* to swell]: The pressure exerted on the inside of a plant cell wall by the fluid contents of the cell; the interior of the cell is hypertonic in relation to the fluids surrounding it and so gains water by osmosis.

urea [Gk. *ouron,* urine]: An organic compound formed in the vertebrate liver; principal form of disposal of nitrogenous wastes by mammals.

ureter [Gk. from *ourein,* to urinate]: The tube carrying urine from the kidney to the cloaca (in reptiles and birds) or to the bladder (in amphibians and mammals).

urethra [Gk. from *ourein,* to urinate]: The tube carrying urine from the bladder to the exterior of mammals.

uric acid [Gk. *ouron,* urine]: An insoluble nitrogenous waste product that is the principal excretory product in birds, reptiles, and insects.

urine [Gk. *ouron,* urine]: The liquid waste filtered from the blood by the kidney and stored in the bladder pending elimination through the urethra.

uterine tube: *See* Oviduct.

uterus [L., womb]: The muscular, expanded portion of the female reproductive tract modified for the storage of eggs or for housing and nourishing the developing embryo.

vacuole [L. *vacuus,* empty]: A membrane-bound, fluid-filled sac within the cytoplasm of a cell.

vagus nerve [L. *vagus,* wandering]: A nerve arising from the medulla of the vertebrate brain that innnervates the heart and visceral organs; carries parasympathetic fibers.

vaporization [L. *vapor,* steam]: The change from a liquid to a gas; evaporation.

vascular [L. *vasculum*, a small vessel]: Containing or concerning vessels that conduct fluid.

vascular bundle: In plants, a group of longitudinal supporting and conducting tissues (xylem and phloem).

vascular cambium [L. *vasculum*, a small vessel + *cambium*, exchange]: In plants, a cylindrical sheath of meristematic cells that divide mitotically, producing secondary phloem to one side and secondary xylem to the other, but always with a cambial cell remaining.

vas deferens, *pl.* **vasa deferentia** (vass **deff**-er-ens) [L. *vas*, a vessel + *deferre*, to carry down]: In mammals, the tube carrying sperm from a testis to the urethra.

vector [L., carrier]: In recombinant DNA, a small, self-replicating DNA molecule, or a portion thereof, into which a DNA segment can be spliced and introduced into a cell; generally a plasmid, a bacteriophage, or a cosmid.

vein [L. *vena*, a blood vessel]: (1) In plants, a vascular bundle forming part of the framework of the conducting and supporting tissue of a leaf. (2) In animals, a blood vessel carrying blood from the tissues to the heart. A small vein is known as a venule.

vena cava (vee-na **cah**-va) [L., blood vessel + hollow]: A large vein that brings blood from the tissues to the right atrium of the four-chambered mammalian heart. The superior vena cava collects blood from the forelimbs, head, and anterior or upper trunk; the inferior vena cava collects blood from the posterior body region.

ventral [L. *venter*, belly]: Pertaining to the undersurface of an animal that holds its body in a horizontal position; to the front surface of an animal that holds its body erect.

ventricle [L. *ventriculus*, the stomach]: A muscular chamber of the heart that receives blood from an atrium and pumps blood out of the heart, either to the lungs or to the body tissues.

vertebral column [L. *vertebra*, joint]: The backbone; in nearly all vertebrates, it forms the supporting axis of the body and protects the spinal cord.

vesicle [L. *vesicula*, a little bladder]: A small, intracellular membrane-bound sac.

vessel [L. *vas*, a vessel]: A tubelike element of the xylem of angiosperms; composed of dead cells (vessel members) arranged end to end. Its function is to conduct water and minerals from the soil.

viable [L. *vita*, life]: Able to live.

villus, *pl.* **villi** [L., a tuft of hair]: In vertebrates, one of the minute, fingerlike projections lining the small intestine that serve to increase the absorptive surface area of the intestine.

virus [L., slimy, liquid, poison]: A submicroscopic, noncellular particle composed of a nucleic acid core and a protein coat; parasitic; reproduces only within a host cell.

viscera [L., internal organs]: The collective term for the internal organs of an animal.

vitamin [L. *vita*, life]: Any of a number of unrelated organic substances that cannot be synthesized by a particular organism and are essential in minute quantities for normal growth and function.

water cycle: Worldwide circulation of water molecules, powered by the sun. Water evaporates from oceans, lakes, rivers, and, in smaller amounts, soil surfaces and bodies of organisms; water returns to the earth in the form of rain and snow. Of the water falling on land, some flows into rivers that pour water back into the oceans and some percolates down through the soil until it reaches a zone where all pores and cracks in the rock are filled with water (groundwater); the deep groundwater eventually reaches the oceans, completing the cycle.

water potential: The potential energy of water molecules; regardless of the reason (e.g., gravity, pressure, concentration of solute particles) for the water potential, water moves from a region where water potential is greater to a region where water potential is lower.

wild type: In genetics, the phenotype that is characteristic of the vast majority of individuals of a species in a natural environment.

worker: A member of the nonreproductive laboring caste of social insects.

xanthophyll [Gk. *xanthos*, yellow + *phyllon*, leaf]: In algae and plants, one of a group of yellow pigments; a member of the carotenoid group.

xylem [Gk. *xylon*, wood]: A complex vascular tissue through which most of the water and minerals are conducted from the roots to other parts of the plant; consists of tracheids or vessel members, parenchyma cells, and fibers; constitutes the wood of trees and shrubs.

yolk: The stored food in egg cells that nourishes the embryo.

yolk sac: In developing reptiles and birds, the extraembryonic membrane that surrounds and encloses the yolk; performs a nutritive function. In mammals, the extraembryonic membrane in which the germ cells are set aside very early in development.

zoology [Gk. *zoe*, life + *logos*, a discourse]: The study of animals.

zooplankton [Gk. *zoe*, life + *plankton*, wanderer]: A collective term for the nonphotosynthetic organisms present in plankton.

zygote (zi-goat) [Gk. *zygon*, yolk, pair]: The diploid ($2n$) cell resulting from the fusion of male and female gametes (fertilization); a zygote may either develop into a diploid individual by mitotic divisions or may undergo meiosis to form haploid (n) individuals that divide mitotically to form a population of cells.

Illustration Acknowledgments

Page ix © Rita Summers/Colorado Nature Photographic Studio; **Page vii** *(top)* © Stephen Dalton/NHPA; *(middle, left to right)* © Brian Parker/Tom Stack & Associates; © Kim Taylor/Bruce Coleman Ltd.; © Larry West; **Page ix** © E. R. Degginger/Earth Scenes; © James L. Castner; © Larry West; **Page x** *(left to right)* Lennart Nilsson, THE BODY VICTORIOUS. New York: Delacorte Press. Boehringer Ingelheim International GmbH; © E. R. Degginger/Animals Animals; Antone G. Jacobson

I-1 © Francisco Erize/Bruce Coleman; **Page 1** © Raymond A. Mendez/Animals Animals; **I-2** The Royal College of Surgeons of England; **I-3** © Chip & Rosa Maria Peterson; **I-4** (a), (b) © J. Fennell/Bruce Coleman; (c) © W. H. Hodge/Peter Arnold; **I-5** American Museum of Natural History; **I-6** © Breck P. Kent/Animals Animals; **I-7** (a) Christopher Ralling; (b) Medical Illustration Unit, The Royal College of Surgeons of England; **I-9** © Frans Lanting/Bruce Coleman Ltd.; **I-10** (a) The Granger Collection; (b) Ann Ronan Picture Library; **Page 8** The Royal College of Surgeons of England; **I-11** (a) Field Museum, Photo Researchers; (b) © S. Robinson/NHPA; **I-12** (a) Rare Books Division, New York Public Library; (b) John Mais; **I-13** The Bettmann Archive; **I-14** © Laura Riley/Bruce Coleman; **I-15** (a) © Eric V. Grávé; (b) © J. Robert Waaland/Biological Photo Service; **I-16** © Larry West; **I-17** © Bruce Coleman; **I-18** © E. S. Ross; **I-19** Terry Erwin & Linda Sims, Smithsonian Institution; **I-20** (a) © Clem Haagner/Bruce Coleman; (b) © Raymond A. Mendez/Animals Animals

Page 20 National Optical Astronomy Observatories; **Page 21 & 1-1** © Sea Studio, Inc./Peter Arnold; **Page 23** © Jen & Des Bartlett/Bruce Coleman; **1-2** © John Cancalosi; **Page 26** (a) © John D. Cunningham/Visuals Unlimited; (b) © M. Walker/NHPA; (c) © Mary M. Thacher/Photo Researchers; **Page 27** (d) © Mitch Reardon/Photo Researchers; (e) © Jeff Foott; (f) © Dwayne M. Reed; (g) © Stephen Dalton/Photo Researchers; **1-5** © Bruce Coleman; **1-7** (b) © Charles M. Falco/Science Source, Photo Researchers; **1-8** © Jen & Des Bartlett/Bruce Coleman; **1-13** © Runk & Schoenberger/Grant Heilman Photography; **1-14** H. Berg & G. Forté; **1-15** (a) © Frieder Sauer/Bruce Coleman; (b) George I. Schwartz; (c) © Biology Media/Photo Researchers; (d) © Manfred Kage/Peter Arnold

Page 40 © Michael Medford/Wheeler Pictures; **2-1** © E. R. Degginger/Bruce Coleman; **2-4** Fritz Polking; **2-5** © Runk & Shoenberger/Grant Heilman Photography; **2-7** (b) © Brian Milne/Earth Scenes; **Page 51** (b) © John D. Cunningham/Visuals Unlimited; (c) © Michael Medford/Wheeler Pictures; **2-11** Jeanne M. Riddle

Page 55 © Herbert B. Parsons/Photo NATS; **3-1** © E. R. Degginger/Earth Scenes; **3-3** After DuPraw, E. J. (1968). *Cell and molecular biology*. New York: Academic Press, Inc.; **Page 58** John M. Sieburth; **3-7** © Charles & Elizabeth Schwartz/Animals Animals; **3-9** (a), (b), (c) After Lehninger, A. L. (1975). *Biochemistry*, 2d ed. New York: Worth Publishers, Inc.; (d) L. M. Biedler; (e) J. C. Warren; **3-10** (c) R. D. Preston; **3-11** (b) © Herbert B. Parsons/Photo NATS; **3-13** © Caroline Kroeger/Animals Animals; **3-16** B. E. Juniper; **Page 71** (a), (b) © Sloop-Ober/Visuals Unlimited; **3-20** Sequence information from Lehninger, A. L. (1982). *Principles of biochemistry* (p. 135). New York: Worth Publishers, Inc.; **3-21** (b) After Wilson, E. O., et al. (1977). *Life, cells, organisms, populations*. Sunderland, MA: Sinauer Associates, Inc.; **3-22** (b) Computer graphics modeling and photography by Arthur J. Olson, Ph.D., Research Institute of Scripps Clinic, La Jolla, CA 92037, © 1988; **3-23** (a) After Alberts, B., Bray, D., Lewis, J., Raff, M., Roberts, K. & Watson, J. D. (1983). *Molecular biology of the cell*. New York: Garland Publishing Company; (b) Daniel Friend; **3-24** (a) © Anthony Bannister/NHPA; (b) © Robert L. Dunne/Bruce Coleman; **3-25** (a) After Karp, G. (1979). *Cell biology*. New York: McGraw-Hill Book Company; (b) © Manfred Kage/Peter Arnold; **3-27** Adapted from Dickerson, R. E. & Geis, I. (1969). *The structure and action of proteins*. Menlo Park, CA: W. A. Benjamin, Inc. Copyright 1969 by Dickerson & Geis; **3-28** (a), (b) Margaret Clark; **Page 83** After Lehninger, A. L. (1982). *op. cit.*

Page 84 © David M. Phillips; **4-1** Brent McCown; **4-2** Big Bear Solar Observatory; **Page 86** Pasteur Institute & The Rockefeller University Press; **4-4** © S. Johannson & Frank Lane/Bruce Coleman; **4-5** Sidney W. Fox; **4-6** © S. M. Awramik/Biological Photo Service; **4-8** (b) A. Ryter; **4-9** (b) Lang, N. J. (1965). *Journal of Phycology, 1,* 127–134; **4-10** (b) George Palade; **4-12** (b) Michael A. Walsh; **4-13** (b) Keith Porter; **4-14** After Alberts, et al. (1983). *op. cit.*; **4-15** (a), (b), (c) David M. Phillips; **4-17** After Alberts, et al. (1983). *op. cit.*; **4-18** *Ibid*; **4-19** (a)–(d) Keith Roberts & James Barnett

Page 102 Osborn, M. (October 1985). The molecules of life. *Scientific American*; **5-2** (a) © Eric V. Grávé/Photo Researchers; (b) © M. Schliwa/Visuals Unlimited; **5-3** © J. Robert Waaland/Biological Photo Service; **5-4** J. D. Robertson; **5-6** Adapted in part from *Scientific American* (February 1984), p. 81, and in part from Darnell, J., Lodisch, H. & Baltimore, D. (1986). *Molecular cell biology*. New York: W. H. Freeman and Company; **5-7** (a, *photo*) Myron C. Ledbetter; (b) After Albershamm, P. (April 1975). *Scientific American*; **5-8** (a) Daniel Friend; (b) Nigel Unwin; (c) Barbara J. Stevens & Hewson Swift; **5-9** Ursula Goodenough; **5-10** (a) © Doug Wechsler; (b) Mia Tegner & David Epel; **5-13** Osborn, M. (October 1985). The molecules of life. *Scientific American*; **5-14** (a), (b), (c) *Ibid*; **Page 114** Adapted from Darnell, et al. (1986). *op. cit.*; **5-17** Peter Webster; **5-18** (a), (b) Don Fawcett; **5-19** Don Fawcett/Photo Researchers; **5-20** (b) Flickinger, C. J. (1975). *Journal of Cell Biology, 49,* 221.; **5-22** Birgit Satir; **5-23** (a) Don Fawcett; (b) G. Decker; **5-24** (b) Keith Porter; **5-25** (a) Roland R. Dute; (b) Myron C. Ledbetter; **5-26** David Stetler; **5-28** © Manfred Kage/Peter Arnold; **5-29** Don Fawcett; **5-30** Gregory Antipa; **5-31** (b) Peter Satir; **5-32** © David M. Phillips/Visuals Unlimited

Page 127 © M. I. Walker/NHPA; **6-2** Keith R. Porter; **6-3** © Frederick J. Dodd/Peter Arnold; **6-5** (a) G. M. Hughes; (b) After Schmidt-Nielson, K. (1979). *Animal physiology*, 2d ed. Cambridge: Cambridge University Press; **6-6** (a) © M. I. Walker/NHPA; (b), (c) © Thomas Eisner; **6-7** After Lehninger, A. L. (1975). *Biochemistry*, 2d ed. New York: Worth Publishers, Inc.; **6-9** (b) Daniel Branton; **6-10** Adapted from Raven, P. E., Eichhorn, S. E. & Evert, R. F. (1986). *Biology of plants*, 4th ed. (Fig. 4-9). New York: Worth Publishers, Inc.; **6-11** Adapted from Alberts, et al.

(1983). *Molecular biology of the cell* (p. 289). New York: Garland Publishing Company; **6-12** Keith R. Porter/Photo Researchers; **6-14** Adapted from Raven, *et. al.* (1986). *op. cit.;* **6-15** Adapted from *Scientific American* (May 1984), p. 54; **6-16** (a)-(d) Gregory Antipa; **6-17** (a)-(d) Perry, M. M. & Gilbert, A. B. (1979). *Journal of Cell Science, 39,* 257–272; **6-18** (a) Ray F. Evert; (b) Peter K. Hepler; **Page 141** (a), (b), (c) K. T. Raper; (d)-(g) © David Scharf/Peter Arnold; **6-19** (a) N. Bernard Gilula; (b) Adapted in part from Darnell et al. (1986). *Molecular cell biology* (Fig. 14-63). New York: W. H. Freeman and Company; in part from *Scientific American* (May 1978), p. 150; and in part from *Scientific American* (October 1985), p. 106.

Page 144 Carolina Biological Supply Company; **7-1** John Mais; **7-3** L. P. Wisniewski & K. Hirschhorn; **7-5** © David M. Phillips/Visuals Unlimited; **7-6** Carolina Biological Supply Company; **7-7** After Alberts, *et al.* (1983). *Molecular biology of the cell* (Fig. 11-11, p. 619). New York: Garland Publishing Company; **7-8** (a) Adapted from Alberts, *et al., op. cit.* (Fig. 11-47, p. 652); (b) M. J. Schibler; **7-9** (a) Andrew S. Bajer; (b), (c) Adapted from Alberts, *et al., op. cit.* (Fig. 11-48, p. 652); **7-10, 7-11,** and **7-12** Andrew S. Bajer; **7-13** (a)-(d) Carolina Biological Supply Company; **7-14** (a), (b) Beams, H. W. & Kessel, R. G. (1976). *American Scientist, 64,* 279; **7-15** James Cronshaw

Page 158 © Grant Heilman Photography; **Page 159 & 8-1** © Bruce Coleman; **Page 161** © Larry West; **8-2** © Zig Leszczynski/Animals Animals; **8-3** ©Alain Eurard/Photo Researchers; **Page 163** Lotte Jacobi; **8-5** (a)-(d) Miami Seaquarium; **8-7** Jeremy Pickett-Heaps; **8-8** After Lehninger, A. L. (1975). *Biochemistry,* 2d ed., New York: Worth Publishers, Inc.; **8-10** (a) Ibid; (b) William Goddard III; **8-16** Adapted from Lehninger, A. L. (1982). *Principles of biochemistry* (pp. 212 & 219). New York: Worth Publishers, Inc.; **8-18** V. Lennard; **8-19** After Watson, J. D., *et al.* (1970). *Molecular biology of the gene,* 2d ed., Menlo Park, CA: The Benjamin/Cummings Publishing Company; **8-22** After Lehninger (1975). *op. cit.;* **Page 179** (a) © Robert Pearcy/Animals Animals; (b) © Bill Curtsinger/Rapho, Photo Researchers; **8-24** (a) © Larry West; (b) © M. P. Price/Bruce Coleman; **Page 185** © R. D. Estes

Page 186 © Grant Heilman Photography; **9-1** (a) © Bob Evans/Peter Arnold; (b) Gary Robinson; (c) Bray, R. (1978). *Science, 200,* 333–334, © 1978 by AAAS; **9-5** (b) © Grant Heilman Photography; **9-7** Adapted from Darnell, *et al.* (1986). *Molecular cell biology.* New York: W. H. Freeman and Company; **Page 193** R. H. Kirschner; **9-8** (b) Lester J. Reed from Boyer, P. D., ed. (1970). *The enzymes, Vol. 1.* New York: Academic Press, Inc., **9-9** Adapted from Vander, A. J., Sherman, J. H. & Luciano, D. (1969). *Human physiology.* New York: McGraw-Hill Book Company; **9-10** After Lehninger, A. L. (1975). *Biochemistry,* 2d ed. New York: Worth Publishers, Inc.; **9-13** After Takano, T., Kallai, O. B., Swanson, R. and Dickerson, R. E. (1973). *Journal of biological chemistry, 248,* 5244; **9-14** Adapted from Lehninger, A. L. (1982). *Principles of biochemistry.* New York: Worth Publishers, Inc.; **9-15** After Alberts, B., *et al.* (1983). *Molecular biology of the cell.* New York: Garland Publishing Company; **9-16** (a) After Lehninger, A. L. (1982). *op. cit.;* (b) John N. Telford; **9-18** After Lehninger, A. L. (1975). *op. cit.;* **Page 202** (a), (b) Lieber, C. S. (March 1976). The metabolism of alcohol. *Scientific American;* **9-21** After Lehninger, A. L. (1975). *op. cit.*

Page 206 © Paul W. Johnson & J. McN. Sieburth/Biological Photo Service; **10-1** © J. Metzner/Peter Arnold; **10-4** (a), (b) Pearse, V. & Buchsbaum, R. (1987). *Living invertebrates.* Pacific Grove, CA: The Boxwood Press; photo by Karl J. Marschall; **10-8** Micrograph by Oxford Scientific Films/Bruce Coleman; **10-10** (a) A. D. Greenwood; (b) L. K. Shumway; **10-13** After Lehninger, A. L. (1975). *Biochemistry,* 2d ed. New York: Worth Publishers, Inc.; **Page 217** © Paul W. Johnson & J. McN. Sieburth/Biological Photo Service; **Page 221** Stoeckenius, W. (June 1976). The purple membrane of salt-loving bacteria. *Scientific American;* **10-17** Pallard, S. G. & Kozlowski, T. T. (1980). *New Phytologist, 85,* 363-368; **10-21** Ray F. Evert

Page 232 Edward Hicks, *Noah's Ark* (detail), The Philadelphia Museum of Art: Bequest of Lisa Norris Elkins; **Page 233 & 11-1** Computer Graphics Laboratory, University of California, San Francisco. © Regents of the University of California; **Page 235** Courtesy LKB Productions, Sweden; **11-2** (a), (b), (c) The Granger Collection; **11-3** The Bettmann Archive; **11-4** The Granger Collection; **11-6** Adapted from von Frisch, K. (1964). *Biology.* translated by Jane Oppenheimer. New York: Harper & Row Publishers, Inc.; **11-12** Bill Ratcliffe; **11-13** (a) Dr. V. Orel, The Moravian Museum; (b) Courtesy LKB Productions, Sweden

Page 249 Ripon Microslides; **12-1** (a) © John Bova/Photo Researchers; (b) Arnold Sparrow, Brookhaven National Laboratory; **12-9** (a), (b) William Marks; **12-10** (a), (b) Mary E. Clutter; **12-12** After DuPraw, E. J. (1968). *Cell and molecular biology.* New York: Academic Press, Inc.; **Page 259** After Moore, J. L. (1963). *Heredity and development.* New York: Oxford University Press; **12-14** B. John; **Page 262** Ripon Microslides

Page 263 Ralph G. Somes, University of Connecticut; **13-1** Columbiana Collection, Rare Book and Manuscript Library, Columbia University; **13-4** © Jeremy Burgess/Science Photo Library, Photo Researchers; **Page 267** © George F. Godfrey/Animals Animals; **13-7** © John Chiasson/Gamma-Liaison; **13-9** (a) © Richard Kolar/Animals Animals; (b) © Grant Heilman Photography; (c), (d) © Jane Burton/Bruce Coleman; **13-10** (a)-(d) Ralph G. Somes, University of Connecticut **13-11** After Ayala, F. J. & Kiger, J. A., (1980). *Modern genetics.* Menlo Park, CA: The Benjamin/Cummings Publishing Company; **13-13** (a) After E. D. Merrell, 1964; (b) *Journal of Heredity, 15* (1914); **13-14** Photograph by F. B. Hutt (1930). *Journal of Genetics, 22,* 126; **13-15** B. John; **13-20** Beth Myers, courtesy of William Marks

Page 281 Nelson Max & Richard Dickerson; **14-1** Kleinschmidt, A. K., Land, D., Jacherts, D. & Zahn, R. K. (1962). *Biochemica Biophysica Acta, 61,* 857–864; **14-2** (a), (b) © Bruce Iverson; **14-3** After Koob, D. D. & Bogs, W. E. (1972). *The nature of life.* Reading, MA: Addison-Wesley Publishing Co.; **14-5** From Cairns, J., Stent, G. S., & Watson, J. D., eds. (1966). *Phage and the origins of molecular biology.* Cold Spring Harbor, NY: Cold Spring Harbor Laboratory of Quantitative Biology; **14-6** M. Wurtz, Biozentrum, University of Basel/Science Photo Library, Photo Researchers; **14-8** ©Lee D. Simon/Science Photo Library, Photo Researchers; **14-9** (a) From Watson, J. D. (1968). *The double helix.* New York: Atheneum Publishers; (b) Vittorio Luzatti; **14-10** (b) Nelson Max & Richard Dickerson; **Page 291** Watson (1968). *op. cit.;* **14-13** After Lehninger, A. L. (1975). *Biochemistry,* 2d ed. New York: Worth Publishers, Inc.; **14-14** Ibid; **14-16** (a-d) Bernhard Hirt; **14-17** Blumenthal, A. B., Kreigstein, H. J. & Hogness, D. S. (1973). *Cold Spring Harbor Symposium on Quantitative Biology, 38,* 205; **Page 300** C. M. Plork, Museum of Comparative Zoology, Harvard University

Page 301 © K. G. Murti/Visuals Unlimited; **15-1** © Gary R. Robinson/Visuals Unlimited; **15-6** (b) © K. G. Murti/Visuals Unlimited; **15-14** Hans Ris; **15-15** (b) Miller, O. L., Hamkalo, B. A. & Thomas, C. A. (1970). *Science, 69,* 392–395. © 1970 by AAAS

Page 319 Paris Match; **16-1** Jack Griffith; **16-5** Adapted from Darnell, J., Lodish, H. & Baltimore, G. (1986). *Molecular cell biology* (p. 278). New York: Scientific American Books; **16-6** Paris Match; **16-11** (a) Palchaudhuri, S., Bell, E. & Salton, M. R. J. (1975). *Infection and Immunity, 11,* 1141; (b), (c), (d) T. Kakefuda; **16-12** Judith Carnahan & Charles Brinton, Jr.; **16-13** After Ayala, F. J. & Kiger, J. A., (1980). *Modern genetics.* Menlo Park, CA: Benjamin/Cummings Publishing Co., Inc.; **16-14** Ibid; **16-15** Cohen, S. N. (December 27, 1969). *Nature;* **16-16** (a)-(d) B. Menge, J. V. B. Brock, H. Wunderli, K. Lickfield, M. Wurtz & E. Kellenberger; **16-20** Based on Alberts, *et al.* (1983). *Molecular biology of the cell.* New York: Garland Publishing Company; **16-22** Watson, *et al.,* (1987). *Molecular biology of the gene* (p. 335). Menlo Park, CA: Benjamin/Cummings Publishing Co., Inc.; **16-23** Stanley Falkow

Page 340 Eli Lilly and Co.; **17-1** John C. Fiddes & Howard M. Goodman; **17-2** Adapted from Alberts, *et al.* (1983). *Molecular biology of the cell.* New York: Garland Publishing Company; **17-3** Adapted from Lehninger, A. L. (1982). *Principles of biochemistry* (p. 863, Fig. 28-4). New York: Worth Publishers, Inc.; **17-5** Adapted from Watson, J. D., Tooze, J. & Durtz, D. T. (1983). *Recombinant DNA* (p. 64, Fig. 5-5). New York: W. H. Freeman and Company; **17-6** (a) Stanley N. Cohen; **17-7** (a), (b), (c) Huntington Potter & David Dressler, *Life Magazine* © 1980 Time Inc.; **17-8** Adapted from Fristrom, J. W. & Spieth, P. T. (1980). *Principles of genetics* (p. 346, Fig. 12-20). New York & Concord: Chiron Press; **17-9** Adapted from Alberts, *et al.* (1983). *Molecular biology of the cell* (Fig.

Illustration Acknowledgments

4-56). New York: Garland Publishing Company; **17-10** (a) Jack Griffith; (b), (c) Adapted from Darnell, *et al.* (1986). *Molecular cell biology* (p. 247, Fig. 17-22). New York: W. H. Freeman and Company; (d) Daniel Nathans; **17-11** After Lehninger, A. L. (1982). *op. cit.;* **17-12** After *City of Hope Quarterly, 7,* 2. (Winter, 1978); **17-13** Eli Lilly and Co.; **17-14** Eugene W. Nester; **17-15** Keith Wood, University of California, San Diego

Page 355 Robert Noonan; **18-1** E. J. DuPraw; **18-2** Rich, A., *et al.* (1981). *Science, 211,* 171–176. © 1981 by AAAS; **18-3** (a) Victoria Foe; **18-4** (a) Barbara Hamkalo; (b), (c) Adapted from Alberts, *et al.* (1983). *Molecular biology of the cell* (Fig. 8-5). New York: Garland Publishing Company; **18-5** Victoria Foe; **18-6** After Alberts, *et al.* (1983). *op. cit.;* **18-8** James German; **18-9** George T. Rudkin; **18-10** (*photo*) James German; (*art*) After Chambon, P. (May 1981). Split genes. *Scientific American,* 60–71; **18-11** After Lewin, R. (1981). *Science, 212,* 28–32; **18-12** H. C. MacGregor; **18-13** (a), (b) Ullrich Scheer, W. W. Franke & M. F. Trendelenburg; **Page 366** Don Fawcett; **18-14** After Rahbar, S. (Winter 1982). Abnormalities of human hemoglobin. *City of Hope Quarterly, 11,* 2.; **18-15** After Goodenough, U. (1983). *Genetics,* 3rd ed. New York: Saunders College/Holt, Rinehart & Winston; **Page 370** Grabowski, P. J. & Cech, T. R. (1981). *Cell, 23,* 467–476; **18-18** After Lehninger, A. L. (1982). *Principles of biochemistry.* New York: Worth Publishers, Inc.; **18-19** *Ibid;* **18-20** UPI/Bettmann Newsphotos; **18-21** Adapted from Darnell, *et al.* (1986). *Molecular cell biology.* New York: W. H. Freeman and Company; **Page 375** Dr. Mary Clutter, Cold Spring Harbor Laboratory Research Archives; **18-22** (a), (b) R. D. Goldman; **18-23** (a) after Jon W. Gordon & Frank Ruddle; (b) Jon W. Gordon & Frank Ruddle; **18-24** R. L. Brinster; **18-25** Robert Noonan; **Page 381** R. Portman & M. L. Birnstiel

Page 382 © Tony Mendoza/The Picture Cube; **19-1** Gernsheim Collection, Humanities Research Center, University of Texas at Austin; **19-3** After Yunis, J. J. (1976). *Science, 191,* 1268–1270. © 1976 by AAAS; **19-4** (a) © Tony Mendoza/The Picture Cube; **19-7** (a), (b) Jorge J. Yunis; **19-8** (b) Courtesy Nell Ubbelohde & William Stryk; **19-9** After Stryer, L. (1988). *Biochemistry,* 3rd ed. (p. 512, Fig. 21-21). New York: W. H. Freeman and Company; **19-10** (a) Courtesy Jan Chalker; (b) John S. O'Brien; **19-11** Margaret Clark; **19-12** Scala/Art Resource; **19-13** Richmond Products, Boca Raton; **19-15** After Lerner, I. M. (1968). *Heredity, evolution, and society.* New York: W. H. Freeman and Company; **19-17** (a), (b) Steve Uzzell III; **Page 396** Lifecodes Corporation

Page 404 Donald R. Perry; **Page 405 & 20-1** © Michael Fogden/Animals Animals; **Page 407** © Brian Parker/Tom Stack & Associates; **20-2** (a) © Larry West; (b) © L. Campbell/NHPA; (c) John Shaw/NHPA; **20-3** (a) © John H. Gerard/DPI; (b) © Hans Reinhard/Bruce Coleman; **20-4** The Bettmann Archive; **20-6** (a) © Leonard Lee Rue III/Photo Researchers; (b) © Brian Parker/Tom Stack & Associates; (c) © Kevin Schafer/Tom Stack & Associates; (d) © Dale & Marian Zimmerman/Animals Animals; (e) © Tom McHugh/Photo Researchers; **20-8** R. M. Kristensen; **20-9 & 20-10** After Dobzhansky, T. (1977). *Evolution.* New York: W. H. Freeman and Company; **20-11** (a) After Lehninger, A. L. (1982). *Principles of biochemistry.* New York: Worth Publishers, Inc.; (b) Data provided by Lai-Su L. Yeh, Protein Identification Resource, National Biomedical Research Foundation, Georgetown University Medical Center; **20-12** Adapted from Sibley & Ahlquist (February 1986). *Scientific American, 85;* **Page 422** (a) © Ralph A. Reinhold/Animals Animals; (b) © Zig Leszczynski/Animals Animals; (c) Adapted from O'Brien, S. J., *et. al.* (September 12, 1985). *Nature, 317,* 141; **20-13** Adapted from Sibley & Ahlquist (February 1986). *Scientific American, 92;* **20-14** (a) © Manfred Kage/Peter Arnold; (b) © Eric V. Gravé; (c) © Robert P. Carr/Bruce Coleman; (d) © Marion Patterson/Black Star; (e) © Dwight R. Kuhn

Page 428 Hans Reichenbach; **21-1** © Paul Chesley/Photographers Aspen; **21-2** (a), (b), (c) David Greenwood; **21-3** Adapted from Fox, G. E. *et. al.* (1980). *Science, 209,* 459; **21-4** (*photo*) Lee D. Simon; **21-5** Victor Lorian & B. Atkinson; **21-6** (a) R. G. E. Murray; (b) I. D. J. Burdett & R. G. E. Murray; **21-7** John Swanson; **21-8** Hoeniger, J. F. M. (1965). *Journal of General Microbiology, 40,* 29; **21-9** (a) J. Adler; (b) After DePamphilis, M. L. & Adler, J. (1971). *Journal of Bacteriology, 105,* 395; **Page 435** D. L. Balkwill & D. Maratea; **21-10** (a) Eric V. Gravé; (b) R. S. Wolfe; **21-11** J. F. M. Hoeniger; **21-12** (a) Center for Disease Control; (b), (c) Listgarten, M. A. & Socransky, S. S. (1964). *Journal of Bacteriology, 10,* 127–138; **21-13** Virus Laboratory, Parke, Davis & Co.; **21-14** Hoeniger, J. F. M. & Headly, C. L. (1968). *Journal of Bacteriology, 96,* 1835–1947; **21-15** Hans Reichenbach; **21-16** Stolp, H. & Starr, M. P. (1963). *Antoine von Leeuwenhoek, 29,* 217–248; **21-17** (a) Jerry D. Davis; (b) Germaine Cohen-Bazire; **21-18** (photo) M. Jost; **21-19** © Robert & Linda Mitchell; **Page 443** (a) Whately, J. & Lewin, R. A. (1977). *New Phytologist, 79,* 303–313; (b) F. R. Turner, Indiana University; **21-20** Jack Griffith; **21-21** (a), (b) Almeida, J. D. & Howatson, A. F. (1963). *Journal of Cell Biology, 16,* 616; (c) R. C. Williams & H. W. Fisher; (e) Lee D. Simon; **21-22** Jurgen Kartenbeck; **21-23** T. Koller & S. M. Sogo, Swiss Federal Institute of Technology, Zurich; **21-24** Stanley Prusiner; **21-25** Ny Carlsberg Glyptotek; **21-26** (a), (b) Computer graphics modeling and photography by Arthur J. Olson, Ph.D., Research Institute of Scripps Clinic, La Jolla, CA 92037 © 1988.

Page 452 © Kim Taylor/Bruce Coleman Ltd.; **22-1** © Eric V. Gravé; **22-2** (a) William J. Larsen; (b) D. A. Stetler & W. M. Laetsch; **22-3** (a), (b) R. W. Greene; **22-6** © D. P. Wilson/Eric & David Hosking; **22-9** (a) © Eric V. Gravé; (b) H. N. Guttmann; **22-10** (a) © Eric V. Gravé; (b) © D. P. Wilson/Eric & David Hosking; **22-11** (a) Kent Cambridge Scientific Instruments; (b) G. A. Fryxell; **22-12** (a), (b) © D. P. Wilson/Eric & David Hosking; **22-13** (a)–(d) Ronald Hoham,Colgate University; **22-14** (a) © M. I. Walker; (b) George I. Schwartz; (c) © Kim Taylor/Bruce Coleman Ltd.; **22-15** (a) © Grant Heilman Photography; (b) © Oxford Scientific Films, Animals Animals; (c) © D. P. Wilson/Eric & David Hosking; **22-17** (*photo*) Trond Braten, EM Laboratory for Biosciences, Olso; **22-19** (a), (b) © D. P. Wilson/Eric & David Hosking; **22-20** (a) © Larry West; (b) © John Shaw/NHPA; **22-21** John W. Taylor; **22-22** Alma W. Barksdale; **22-23** After Niederhauser, J. S. & Cobb, W. (May 1959). *Scientific American;* **22-24** (a) © Eric V. Gravé; (b) © Robert L. Owen; **22-25** (a), (b) © Eric V. Gravé/Photo Researchers; **22-26** © Eric V. Gravé; **Page 472** D. Kubai & H. Ris; **22-28** G. A. Horridge & S. L. Tamm; **22-30** (a)–(d) © Eric V. Gravé

Page 479 © John Shaw/NHPA; **23-1** © E. S. Ross; **23-3** © C. W. Perkins/Earth Scenes; **23-4** John H. Troughton; **23-5** H. C. Hoch & D. P. Maxwell; **23-6** (c) Carolina Biological Supply Company; **Page 484** John Hodgin from Buller, A. H. R. *Researches on Fungi, Vol. 6.* New York: Longman, Inc.; **23-7** (a) © Alvin E. Staffan; (b) © John Shaw; **23-9** (a) C. Bracker; (b) Drawing by B. O. Dodge, from E. L. Tatum; **23-11** (a) © Agenzia Fotografica, Luisa Ricciarini, Milan; (b) © E. S. Ross; (c) Donald Simons; **23-12** G. L. Barron; **Page 489** (a) N. Allin & G. L. Barron, University of Guelph; (b) David Pramer; **23-13** (a) © E. S. Ross; (b) © J. N. A. Lott, McMaster Univ./Biological Photo Service; (c) © John Shaw/NHPA; **23-14** (a), (b) Vernon Ahmadjian; **23-15** (a) R. J. Molina, J. M. Trappe & G. S. Strickler; (b) B. Zak, U. S. Forest Service

Page 493 © Larry West; **24-1** © James L. Castner; **24-2** (a) L. E. Graham; (b) K. J. Nicklas; **24-3** L. E. Graham; **24-4** R. E. Magill, Botanical Research Institute, Pretoria; **24-5** (a), (b) © D. S. Neuberger; **24-6** (a) © E. S. Ross; (b), (c) © Robert A. Ross; **24-7** © D. S. Neuberger; **24-9** (a) © G. I. Bernard/Earth Scenes; (b), (c) © Ray F. Evert; **24-10 & 24-11** After Raven, P. H., Evert, R. F. & Eichhorn, S.E. (1987). *Biology of plants,* 4th ed. New York: Worth Publishers, Inc.; **24-12** (a., (b) © E. S. Ross; (c) © R. Carr; **24-13** (a) © Marilyn Wood/Photo NATS; (b) © B. Miller/Biological Photo Service; **24-14** © Liz & Tony Bomford/Survival Anglia Ltd., **Page 504** Kristine Rasmussen & Stuart Naquin; **24-15** (a) D. A. Steingraeber; (b) © G. R. Roberts; (c) © William H. Harlow/Photo Researchers; (d) © Michael Fogden/Earth Scenes; **24-17** (a) © Heather Angel; (b) © Photo Researchers; **24-19** © Grant Heilman Photography; **Page 510** © Stephen J. Krasemann/Photo Researchers; **24-24** © Larry West; **24-26** (a) © Robert & Linda Mitchell; (b), (c) © E. S. Ross; (d) W. A. Calder; (e) Donna Howell; **24-27** © Sandy Gregg/Imagery; (b) © G. I. Bernard/Oxford Scientific Films, Animals Animals; **24-28** © Patti Murray/Earth Scenes

Page 518 © H. Wes Pratt/Biological Photo Service; **25-1** © C. E. Mills/Biological Photo Service; **25-2 & 25-3** Nat Fain; **25-4** (a) Adolf Seilacher, University of Tübingen; (b) M. F. Glaessner; **25-5** (a), (b), (e) S. Conway Morris; (c), (d) H. B. Whittington; **25-7** © Marty Snyderman; **25-9** © Manfred Kage/Peter Arnold; **25-10** Heather Angel; **25-11** Robert B. Short; **25-16** After Buchsbaum, R. & Milne, L. J. (1962). *The lower animals: Living invertebrates of the world.* Garden City, NY: Doubleday & Co., Inc.; **25-17** (a) © D. P. Wilson/Eric & David Hosking; (b) © Runk

& Schoenberger/Grant Heilman Photography; **25-18** © Shaw Photos/Bruce Coleman; **25-20** © H. Wes Pratt/Biological Photo Service; **Page 530** (a) © Allan Power/Bruce Coleman; (b) © Ian Took/Biofotos; **25-21** © Sea Studios Inc./Peter Arnold; **25-24** (a) Maria Wimmer; **25-25** © Scott Johnson/Animals Animals; **25-27** After Roberts, M.B.V. (1971). *Biology: A functional approach.* New York: The Ronald Press Company; **25-29** (a) © J. H. Robinson/Photo Researchers; (b) © Eric V. Gravé/Photo Researchers; (c) © Walker England/Photo Researchers; **Page 537** (a) Ming Wong; (b) © Frank L. Lambrecht/Visuals Unlimited; **25-30** (a) W. Sterrer/Bermuda Biological Station; (b) © Kathie Atkinson/Oxford Scientific Films, Animals Animals; **25-31** © Kim Taylor/Bruce Coleman; **25-32** (c) © Larry Jensen/Visuals Unlimited; **25-33** (a) R. P. Higgins; (b) © M. I. Walker/Photo Researchers; (c) © Kim Taylor/Bruce Coleman; **25-34** © D. Wrobel/Biological Photo Service

Page 543 Doug Wechsler; **26-4** (a) After Roberts, M.B.V. (1971). *Biology: A functional approach.* New York: The Ronald Press Company; (b) A. Solem, Field Museum of Natural History; **26-5** (b), (d) Specimens courtesy of Edward I. Coher; (c) After a photograph supplied by Russel H. Jensen, Delaware Museum of Natural History; **26-6** © James Carmichael Jr./NHPA; **26-7** (a), (b) © Doug Wechsler; **26-9** Carl Roessler/Animals Animals; **26-10** © G. I. Bernard/Oxford Scientific Films, Animals Animals; **26-11** (c) © D. P. Wilson/Eric & David Hosking; **26-15** © Oxford Scientific Films, Animals Animals; **26-16** © Sea Studios, Inc./Peter Arnold; **26-17** © Russ Kinne/Photo Researchers; **26-18** (a) © Visuals Unlimited; (b) © Ralph Buchsbaum; (c) © Bradford Calloway; **26-19** © D. Foster, WHOI/Visuals Unlimited; **26-20** J. Teague Self; **26-21** Robert O. Schuster; **26-22** © Raymond A. Mendez/Animals Animals; **26-23** Russell Zimmer; **26-24** Specimen supplied by Edwad I. Coher; **26-25** © Dr. Steven K. Webster/Biological Photo Service; **26-26** Judith Winston, American Museum of Natural History

Page 565 © Robert & Linda Mitchell; **27-1** © Stephen Dalton/Oxford Scientific Films, Animals Animals; **27-2** © Ken Lucas/Biological Photo Service; **27-3** (a) © Doug Wechsler; (b) © Jack Dermid; **27-4** After Wilson, E. O. et. al. (1977). *Life: Cells, organisms, populations.* Sunderland, MA: Sinauer Associates, Inc.; **27-7** © Heather Angel; **27-9** (a) © Zig Leszczynski/Animals Animals; (b) © Karl Switak/NHPA; (c) © Michael Fogden/Animals Animals; (d) © Peter J. Bryant/Biological Photo Service; (e) © Stephen Dalton/NHPA; **27-11** (a) © A. Kerstitch/Taurus; (b) T. E. Adams; (c) © E. S. Ross; **27-13** (a) © Heather Angel; (b) © Robert & Linda Mitchell; (c) © Kjell B. Sandved/Bruce Coleman; **27-15** (photo) © Jane Burton/Bruce Coleman Ltd.; **27-16** (a) © E. S. Ross; (b) © John R. Macgregor/Peter Arnold; (c), (d) © E. S. Ross; **27-17** (a)–(h) © Peter J. Bryant/Biological Photo Service; **27-18** After Gordon, M. S. (1972). *Animal physiology: Principles and adaptations,* 2nd ed. New York: The Macmillan Company; **27-19** John Mais; **Page 581** James E. Lloyd; **27-20** After Roberts, M.B.V. (1971). *Biology: A functional approach.* New York: The Ronald Press Company; **27-21** © Don & Esther Phillips/Tom Stack & Associates; **27-23** Ross Hutchins; **27-25** © E. S. Ross; **27-26** © Alvin E. Staffan

Page 587 © Frans Lanting; **28-1** © Fred Bavendam/Peter Arnold; **28-2** © Jeff Rotman; **28-4** © Heather Angel; **28-5** (a) © Heather Angel; (b) © Fred Bavendam/Peter Arnold; **28-6** (a), (b) F. W. E. Rowe, Australian Museum; **28-7** © Steve Earley/Animals Animals; **28-8** (a), (b) © D. P. Wilson/Eric & David Hosking; **28-9** © Heather Angel; **28-10** (b) © Heather Angel; **28-11** (c) © R. L. Sefton/Bruce Coleman; **28-12** James Hanken; **28-13** © Breck P. Kent/Animals Animals; **28-14** © Zig Leszczynski/Animals Animals; **28-15** (a)–(d) Redrawn from Lewin, R. (1982). *The thread of life.* Washington, D.C.: Smithsonian Books; (e) © Oxford Scientific Films, Animals Animals; **28-16** © American Museum of Natural History; **28-17** © Tom McHugh/Photo Researchers; **28-18** © Kim Taylor/Bruce Coleman; **28-20** (a) © Nielsen/Imagery; (b) © Gordon Langsbury/Bruce Coleman; **28-21** (a) © Zig Leszczynski/Animals Animals; (b) Lee R. Crist, Member of Golf Collectors' Society; **28-22** (a) American Museum of Natural History; (b) © Barbara Laing/Picture Group; **28-23** © Anthony Mercieca/Photo Researchers; **28-24** (a) The New York Zoological Society; (b) © Jack Dermid; **28-26** (a) © Michael Fogden/Bruce Coleman; (b) © Kenneth W. Fink/Bruce Coleman; (c) © Harry Engels/Photo Researchers; (d) © Jen & Des Bartlett/Bruce Coleman Ltd.; (e) © Frans Lanting; (f) © Jeff Rotman; (g) © C. Allan Morgan/Peter Arnold

Page 610 © Heather Angel; **Page 611 & 29-1** © Bruce D. Thomas; **Page 613** © E. R. Degginger/Earth Scenes; **29-3** (a)–(d) Dennis Kunkel/Phototake; **29-4** (a) © David L.Mulcahy; (b) Myron C. Ledbetter; **29-5** (a) © Louise K. Broman/Photo Researchers; (b) © James L. Castner; (c) © E. R. Degginger/Earth Scenes; (d) © John C. Coulter/Visuals Unlimited; (e) © Doug Wechsler; (f) © E. S. Ross; (g) © Virginia P. Weinland/Photo Researchers; **29-6** (c) © Ray F. Evert; **29-7** (a)–(d) © Ray F. Evert; **Page 622** © Ray F. Evert; **29-12** (a) © W. H. Hodge/Peter Arnold; (b) © Larry West; (c) © Renee Purse/Photo Researchers

Page 625 © James L. Castner; **30-4** © Manfred Kage/Peter Arnold; **30-5** © Ray F. Evert; **30-6** (a) © G. J. James/Biological Photo Service; (b) © James L. Castner; (c) © John D. Cunningham/Visuals Unlimited; **30-8** © Ray F. Evert; **30-9** © Robert Mitchell/Earth Scenes; **30-10** After Ray, P. (1971). *The living plant,* 2d ed. New York: Holt, Rinehart & Winston, Inc.; **30-11, 30-12 & 30-13** © Ray F. Evert; **30-14** After Ray (1971), *op. cit.*; **30-16** (a) © Jack Dermid; (b) © Robert Mitchell; **30-17** (a) © Biophoto Associates/Photo Researchers; (b), (c) © Ray F. Evert; (d) © Randy Moore/Visuals Unlimited; **30-18** (a), (b), (c) © Ray F. Evert; **30-19** (d) © J. N. A. Lott, McMaster University/Biological Photo Service; **30-20 & 30-21** © Ray F. Evert; **30-24** © G. I. Bernard/Oxford Scientific Films, Earth Scenes; **30-25** Frank C. Vasek; **30-27** From Ketchum, R. (1970). *The secret life of the forest,* American Heritage Press; **Page 646** (a) © Bruce Coleman; (b) C. W. Ferguson, Laboratory of Tree-Ring Research, University of Arizona

Page 650 © Breck P. Kent/Earth Scenes; **31-2** © James L. Castner; **31-3** John Shaw, National Audubon Society Collection/Photo Researchers; **31-4** After Richardson, M. (1968). *Translocation in plants.* New York: St. Martin's Press (After Stoat & Hoagland, 1939); **31-5** John L. Troughton; **31-6** Adapted from Northington, D. K. & Goodin, J. R. (1984). *The botanical world,* Mosby; **31-7** After Penny, M. G. & Bowling, D. J. F. (1974) *Planta,* 119, 17–25; **31-8** Jerry D. Davis; **31-9** © Roger Archibald/Earth Scenes; **31-10** (a), © R. K. Burnard/BPS; (b) © C. W. May/BPS; **Page 658** (a) © Grant Heilman Photography; **Page 659** (b) Troughton, J. H. & Donaldson, L. (1972). *Probing plant structures.* New York: McGraw-Hill; **31-11** (a), (b) © Science Photo Library/Photo Researchers; (c) George A. Schaefers; **Page 661** (b), (c) Walter Eschrich & Eberhard Fritz; **31-15** © Robert P. Carr/Bruce Coleman; **31-16** S. A. Wilde; **31-17** © Runk & Schoenerger/Grant Heilman Photography; **Page 666** (a) © Jane Burton/Bruce Coleman; (b) © Breck P. Kent/Earth Scenes; **Page 667** (c), (d) © Runk & Schoenerger/Grant Heilman Photography; **31-18** (a) R. R. Herbert, R. I. Holsten & R. W. F. Hardy, E. I. duPont de Nemours & Co.; (b) R. R. Herbert, J. G. Griswell & R. W. F. Hardy, Central Research & Development Department, E. I. duPont de Nemours & Co.; (c) Mary Alice Webb; **31-19** © Karlene V. Schwartz

Page 671 © Larry West; **32-1** © Jack Dermid; **32-5** © Runk & Schoenberger/Grant Heilman Photography; **32-6** (a) © J. N. A. Lott, McMaster Univ/BPS; (b), (c) Nitsch, J. P. (1950). *American Journal of Botany,* 37(3); **32-9** H. R. Chen; **Page 678** Philip Harrington; **32-10** (a) © Ray F. Evert; (b) © Grant Heilman Photography; **32-12** © Larry West; **32-14** S. W. Wittwer & Michigan Agricultural Experiment Station; **32-15** After Wilson, E.O. et. al. (1977). *Life: Cells, organisms, populations.* Sunderland, MA: Sinauer Associates, Inc.; **32-16** D. E. Varner; **32-17** Kiem Trân Thanh Vân, CNRS, Laboratoire de Phytotron; **32-18 & 32-19** Randy Moore, **32-20**(a), (b) After Naylor, A. W. (May 1952). The control of flowering. *Scientific American;* **32-21** © John R. MacGregor/Peter Arnold; **32-24** © Breck P. Kent/EarthScenes; **32-25** (a), (b) © Jack Dermid; **32-27** © E. S. Ross; **32-28** (a) © Jeff Foott/Bruce Coleman; (b) © E. S. Ross; **32-29** (a), (b) © James L. Castner; **Page 696** After Wilson (1977), *op. cit.*

Page 698 Richard Clarkson/Time Magazine; **Page 699 & 33-1** © Biophoto Associates, Photo Researchers; **Page 701** © Lennart Nilsson, BEHOLD MAN. Boston: Little, Brown & Company; **33-4** © Veronika Burmeister/Visuals Unlimited; **33-5** (a), (b) L. A. Staehelin & B. E. Hull, University of Colorado, Boulder; **33-6** Keith R. Porter; (c) © John D. Cunningham/Visuals Unlimited; **33-8** (a) Hugh E. Huxley; (b) Don Fawcett; **33-11** After Eckert, R. & Randall, D. (1978). *Animal physiology.* New York: W. H. Freeman & Company; **33-13** © Heather Angel; **33-16** © Lennart Nilsson, BEHOLD MAN. Boston: Little, Brown & Company

Illustration Acknowledgments

Page 715 © Manfred Kage/Peter Arnold; **34-4** After Howells, W. W. (1967). *Mankind in the making*, rev. ed. Garden City, NY: Doubleday & Company; **34-5** (a) I. Kaufman Arenberg; **34-8** (a) Jeanne M. Riddle; **34-9** (b) © Manfred Kage/Peter Arnold; **34-10** (b) © Keith R. Porter/Photo Researchers; **Page 724** © Frank Sitemon/The Picture Cube; **34-12** © Don Fawcett/Photo Researchers; **34-13** © Paulette Brunner/Tom Stack & Associates; **34-14** © David Hiser/Photographers Aspen; **34-15** Keith R. Porter

Page 732 © Manfred Kage/Peter Arnold; **35-1** © Frans Lanting; **35-3** © Thomas Eisner; **35-4** © Stephen Dalton/Animals Animals; **35-7** (b) R. R. Duncker; **Page 739** (a), (b), (c) Myron Melamed; **35-10** (a) © Manfred Kage/Peter Arnold; (b) E. R. Weibel; **35-11** After Beck, W. S. (1971). *Human design*. New York: Harcourt Brace Jovanovich; **35-13** William Goddard III, California Institute of Technology; **35-15** After Schmidt-Nielsen, K. (1972). *How animals work*. New York: Cambridge University Press; **Page 744** Anker Odum, from *Scenic wonders of Canada*, © 1976 by Reader's Digest Association (Canada) Ltd.; **Page 746** John B. West

Page 749, 36-1, 36-2, 36-3, & 36-4 © Lennart Nilsson, THE BODY VICTORIOUS. New York: Delacorte Press. Boehringer Ingelheim International GmbH; **36-6** Kessel, R. G. & Kardon, R. H. (1979). *Tissues and organs: A text-atlas of scanning electron microscopy*. New York: W. H. Freeman & Co.; **36-7** Don Fawcett; **36-11** Robert Rosenblum; **36-12** After Vander, A. J., Sherman, J. H. & Luciano, D. S. (1969). *Human physiology*. New York: McGraw-Hill Book Company; **36-14** After Roberts, M.B.V. (1971). *Biology: A functional approach*. New York: The Ronald Press Company

37-1 © Bob & Clara Calhoun/Bruce Coleman Ltd.; **37-4 & 37-5** Adapted from Schmidt-Nielsen, K. (1983). *Animal physiology: Adaptation and environment*, 3rd ed. Cambridge University Press; **37-6** From Lewis Carroll, *Alice's adventure in wonderland*, illustrated by John Tenniel; **37-11** From Kessel, R. G. & Karden, R. H. (1979). *Tissues and organs: A text-atlas of scanning electron microscopy*. New York: W. H. Freeman & Company; **37-14** Don Fawcett, from Bloom, W. & Fawcett, D. W. (1968). *Histology*, 9th ed. Philadelphia: W. B. Saunders Company

Page 777 © E. S. Ross; **38-1** © Doug Allen/Oxford Scientific Films, Animals Animals; **38-3** After Gordon, M. S. (1972). *Animal physiology: Principles and adaptations*, 2d ed. New York: The Macmillan Company; **38-4** © Joe McDonald/Visuals Unlimited; **38-5** After Gordon (1972), *op. cit.*; **38-6** (a) © John Gerlach/Visuals Unlimited; **38-7** After Schmidt-Nielsen, K. (1972). *How animals work*. New York: Cambridge University Press; **Page 783** © Jen & Des Bartlett/Bruce Coleman; **38-8** After Vander, A. J., Sherman, J. H. & Luciano, D. S. (1969). *Human physiology*. New York: McGraw-Hill Book Company; **38-9** (a) © Helen Williams/Photo Researchers; (b) © James Simons/Bruce Coleman; (c) © Leonard Lee Rue III/Bruce Coleman; **38-10** (a) © Richard Kolar/Animals Animals; (b) © Lee Lyon/Bruce Coleman Ltd; (c) © Anthony Bannister/Animals Animals; **38-12** (a) © Hans Reinhard/Bruce Coleman; (b) © Jeff Foott; (c) © Jane Burton/Bruce Coleman; **38-14** © E. S. Ross

Page 791 © Boehringer Ingelheim International GmgH, photo Lennart Nilsson; **39-1** © Lennart Nilsson, BEHOLD MAN. Boston: Little, Brown & Company; **39-2** © John D. Cunningham/Visuals Unlimited; **39-3** © Boehringer Ingelheim International GmbH, photo Lennart Nilsson; **39-6** (a), (b) Joseph Feldman; **Page 797** © Bernard Pierre Wolff/Photo Researchers; **39-8** (a) After Niels Kai Jerne (June 1973). The immune system. *Scientific American*; (b) Computer graphics modeling and photography by Arthur J. Olson, Ph.D., Research Institute of Scripps Clinic, La Jolla, CA 92037 © 1987; **39-9** (a), (b), (c) A. G. Amit; **39-10** After Lerner, R. A. & Dixon, F. J. (June 1973), *Scientific American*; **39-11** © Andrejs Liepins/Science Photo Library, Photo Researchers; **Page 803** Adapted from *Scientific American* (October 1980), 68; **39-12 & 39-13** Adapted from *Scientific American* (February 1986), 43; **39-14** Don Wiley & Jack Strominger, Harvard University; **Page 807** (a) Adapted from a chart based on information supplied by Mary-Claire King in Diamond, J. (18 June 1987), *Nature*, 552; (b) New York Times Picture Service; **39-15** Adapted from *Scientific American* (October 1985), 124 & (December 1985), 89; **39-16** R. Cotran, Harvard Medical School; **39-17** Adapted from *Scientific American* (October 1985), 124 & (December 1985), 89; **39-18** (a), (b) (c) © Boehringer Ingelheim International GmbH, photo Lennart Nilsson; **39-21** © Lennart Nilsson, THE BODY VICTORIOUS. New York: Delacorte Press; **39-22** Adapted from *Scientific American* (January 1987), 48 & cover: **39-23** Hans Gelderblom; **39-24** (a), (b) G. Janossy & M. Bofill, Royal Free Hospital School of Medicine; **39-25** © Nigel Dennis/NHPA

Page 821 © E. R. Degginger/Animals Animals; **40-1** Adapted from Guttman & Hopkins; **40-2** © Manfred Kage/Peter Arnold; **40-4** (a), (b), (c) © E. R. Degginger/Animals Animals; **Page 830** (a) Philip Osdoby; (b) Alan Boyde; **40-12** After Wilson, E. O. et. al. (1977). *Life: Cells, organisms, populations*. Sunderland, MA: Sinauer Associates; **40-13 & 40-14** Adapted from Vander, A. J., Sherman, J. H., & Luciano, D.S. (1985). *Human physiology*, 4th ed. New York: McGraw-Hill Book Company

41-3 (b) Cedric S. Raine; **41-5** Peter S. Amenta, Hahneman University, Philadelphia; **41-9** American Museum of Natural History; **41-10** After Eckert, R. & Randall, D. (1978). *Animal physiology*. New York: W.H. Freeman & Company; **41-12** (b) David Hubel; **41-13, 41-14 & 41-15** After Kuffler, S. W. & Nichols, J. G. (1976) *From neuron to brain*. Sunderland, MA: Sinauer Associates; **41-16** After Novak, C. R. (1975). *The human nervous system: Principles of neurobiology*. New York: McGraw-Hill Book Company; **41-17** Adapted from Darnell, J., Lodish, H., & Baltimore, D. (1986). *Molecular cell biology*. New York: W. H. Freeman; **41-18** (a) Cedric S. Raine; **41-20** Adapted from Williams & Warwick (1975). *Functional neuroanatomy of man*. Philadelphia: W. B. Saunders

Page 858 © Andy Purcell/Bruce Coleman Ltd.; **42-1** © W. Garst/Tom Stack & Associates; **42-2** (a), (b) © Heather Angel; **42-3** After Beck. W. S. (1971). *Human design*. New York: Harcourt Brace Jovanovich; **42-4** © Bruce D. Thomas; **Page 863** © Gunter Ziesler/Bruce Coleman Ltd.; **42-8** (a), (b) Robert S. Preston, courtesy of J. E. Hawkins, Kresge Hearing Research Institute, The University of Michigan Medical School; **42-11** (a) © Zip Leszczynski/Animals Animals; (b) © Andy Purcell/Bruce Coleman Ltd.; (c) © Mary Clay/Tom Stack & Associates; (d) © Tom McHugh/Photo Researchers; **42-12** E. R. Lewis; **Page 870** (a), (b) © Kim Taylor/Bruce Coleman Ltd.; **42-15** © Stephen Krasemann/NHPA, **42-16** (b) from Hogan, M. J., Alvarado, J. A., & Weddell, J. E. (1971). *Histology of the human eye: An atlas and textbook*. W. B. Saunders Company; **42-19 & 42-20** Hugh E. Huxley; **42-23** (b), (c) After Porter, K. R. & Bonneville, M. (1968). *Fine structure of cells and tissues*. Philadelphia: Lea & Febiger; **42-24** (a), (b) John Heuser; **42-25** © Lennart Nilsson, BEHOLD MAN. Boston: Little, Brown & Company

Page 881 © Martin M. Rotker/Taurus Photos; **43-1** © Manfred Kage/Peter Arnold; **43-2** After Davis, P.W. & Solomon, E. P. (1974). *The world of biology*. New York: McGraw-Hill Book Company; **43-4** (b) © Martin M. Rotker/Taurus Photos; (c) © G. E. Medical Systems; **43-5** © Martin M. Rotker/Taurus Photos; **43-13** Fernando Nottebohm; **Page 891** Klemm, W. R. (1972). *Science, the brain, and our future*. New York: Pegasus; **43-14** Adapted from *Scientific American* (June 1987), 83; **43-15** Christine Gall, University of California, Irvine; **43-16** Adapted from *Scientific American* (June 1987), 82; **43-17** © Steve Earley/Animals Animals; **Page 896** Daniel P. Perl; **43-18** E. R. Lewis

Page 900 © C. Edelmann, La Villette/Photo Researchers; **44-1** © Günter Ziesler/Bruce Coleman Ltd.; **44-3** (a), (b) David M. Phillips; **44-4** After Roberts, M. B. V. (1971). *Biology: A functional approach*. New York: The Ronald Press Company; **44-8** © Lennart Nilsson, BEHOLD MAN. Boston: Little, Brown & Company; **44-9** (a) © Norman Tomalin/Bruce Coleman; (b) © Leonard Lee Rue III/Bruce Coleman; (c) © David Hughes/Bruce Coleman; (d) © Kim Taylor/Bruce Coleman Ltd.; **44-10** After Vander, A. J., Sherman, J. H. & Luciano, D. S. (1980). *Human physiology*. New York: McGraw-Hill Book Company; **Page 908** Joan I. Morrell & Donald Pfaff, The Rockefeller University; **44-12** Rugh, R. & Shettles, L. B. (1971). *From conception to birth: The drama of life's beginnings*. New York: Harper & Row; **44-14** © C. Edelmann, La Villette/Photo Researchers; **44-16** © D. M. Phillips/Visuals Unlimited; **44-18 & Page 918** After Vander et al. (1980), *op cit.*

Page 919 Antone G. Jacobson; 45-1 Kathryn W. Tosney; 45-2 (a) William Byrd; (b) Everett Anderson; 45-3 Adapted from Alberts, et al. (1983). *Molecular biology of the cell*. New York: Garland Publishing Company; 45-4 After Baker, J. J. W. & Allen, G. E. (1977). *The study of biology*, 3rd ed. Reading, MA: Addison-Wesley Publishing Company, Inc.: 45-5 & 45-6 Tryggve Gustafson; 45-7 After Wilson, E. O. et al. (1978). *Life on earth*, 2d ed. Sunderland, MA: Sinauer Associates; Page 925 (a) Turner, F. R. & Mahowald, A. P. (1976). Scanning electron microscopy of *Drosophila* embryogenesis: I. The structure of the egg envelopes and the formation of the cellular blastoderm. *Developmental Biology, 50*, 95-108; (b) Adapted from Alberts, et al. (1983), op. cit.; 45-11 & 45-12 After Huettner, 1949; 45-13 Antone G. Jacobson; 45-14 After Huettner, 1949; 45-16 After Waddington, 1966; 45-18 M. P. Olsen; 45-19 & 45-20 After Wilson, et al. (1978), op. cit.; 45-21 After Torrey, 1962; 45-23 After Sussman, M. (1964). *Animal growth and development*, 2d ed. Englewood Cliffs, NJ: Prentice-Hall, Inc.; 45-24 After Bodemer, C. W. (1968). *Modern embryology*. New York: Holt, Rinehart & Winston, Inc.; 45-25 S. R. Hilfer; 45-26 (a)-(f) © Runk & Schoenberger/Grant Heilman Photography; 45-27 Kathryn W. Tosney; 45-28 Adapted from Karp, G. & Berrill, N. J. (1981). *Development*, 2d ed. New York: McGraw-Hill; 45-29 (a) Martin Raff; (b) Adapted from Alberts, et al. (1983), op. cit.; 45-30 & 45-31 Lewis Wolpert; 45-32 (a), (b), (c) © Petit Format, Nestle/Science Source, Photo Researchers; 45-33 Eugene M. Long; Page 943 (a), (b) Walter J. Gehring, University of Basel; (c), (d) E. B. Lewis; 45-36 (a), (b), (c) © Lennart Nilsson, BEHOLD MAN. Boston: Little, Brown & Company; 45-37 © London Scientific Photos; 45-38 & 45-39 © Lennart Nilsson, BEHOLD MAN. Boston: Little, Brown & Company; 45-41 (a), (b), (c) © Jeffrey Reed/Medichrome; 45-42 © International Stock Photo

Page 958 © Bryn Campbell/Biofotos; Page 959 & 46-1 © E. S. Ross; Page 961 © Bruce D. Thomas; 46-2 © Zig Leszczynski/Animals Animals; 46-3 (a) © Michael W. F. Tweedie/Bruce Coleman; (b) © Michael W. F. Tweedie/Photo Researchers; 46-4 © R. E. Pelham/Bruce Coleman; 46-6 American Museum of Natural History; 46-7 (a) © Jane Burton/Bruce Coleman Ltd.; (b) © Leonard Lee Rue III/Bruce Coleman; (c) © G. R. Roberts; 46-8 (a) © Kenneth W. Fink/Photo Researchers; (b) Steven C. Kaufman/Peter Arnold; (c) © A. B. Joyce/Photo Researchers; 46-9 © Jeff Foott; 46-10 Painting by Charles R. Knight, American Museum of Natural History; 46-11 (a)-(e) from Rugh, R. *Experimental embryology: Techniques and procedures*, 7th ed. Minneapolis: Burgess Publishing Co.; 46-12 Redrawn from Luria, Gould & Singer (1981). *A view of life*. Benjamin/Cummings; Page 971 © Calvin Larsen/Photo Researchers; 46-13 © Bruce D. Thomas; 46-14 Office of Public Information, Columbia University

Page 974 © Jen & Des Bartlett/Bruce Coleman Ltd.; 47-1 Hulton Picture Company/The Bettmann Archive; Page 975 © Stephen Dalton/Photo Researchers; 47-2 © George Holton/Photo Researchers; 47-4 (a), (b) John Mais; 47-5 After Mather & Harrison (1949). *Heredity, 3*, 977; 47-6 R. C. Lewontin; 47-7 Courtesy POSSUM; 47-10 Edmund B. Gerard; 47-11 Victor McKusick; 47-12 © Anthro-Photo; 47-13 © Jim Brandenburg; 47-15 © Daniel J. Harper; 47-16 © Jen & Des Bartlett/Bruce Coleman Ltd.; Page 987 from Lewis Carroll. *Through the looking glass*; 47-17 (b) © John Colwell/Grant Heilman Photography

Page 991 © Frans Lanting; 48-1 After Harlow, J.R. (1976), *Scientific American*; 48-2 (a) © Heather Angel; (b) © Kim Taylor/Bruce Coleman Ltd.; 48-3 (a), (b) © Heather Angel; 48-4 (a) © Kenneth W. Fink/Photo Researchers; (b) © Kent & Donna Dannen/Photo Researchers; 48-5 Karu Kurita © 1982 DISCOVER Publications; 48-6 After Charles W. Brown, Santa Rosa Junior College; 48-7 (a) © Mart R. Gross; 48-8 (a) © E. R. Degginger/Animals Animals; (b) Adapted from Futuyama (1979). *Evolutionary biology* (After Clarke 1962, based on data of Popham 1942); 48-9 (a) © A. J. Deane/Bruce Coleman Ltd.; (b) © Jonathan Scott/Seaphot Ltd.: Planet Earth Pictures; (c) © Avon & Tilford/Ardea Photographics; (d) © Frans Lanting; Page 1001 © M. Andersson/VIREO; 48-10 (a) © Alan D. Cruikshank/Photo Researchers; (b) Alan Root/Bruce Coleman; 48-11 From Gould, S. J. (1977). *Ever since Darwin*. New York: W. W. Norton & Company; 48-12 After Clausen, J. & Hiesey, W. M. (1950). *Publication 615*, Carnegie Institute of Washington; 48-13 (a) © Lincoln Brower; (b) © Frans Lanting; (c), (d) © Dwight R. Kuhn; 48-14 (a)-(d) © Lincoln Brower; 48-15 (a)-(d) © E. S. Ross; (e) © Hans Pfletschinger/Peter Arnold; 48-16 (a) © Chip Clark; (b) © Robert & Linda Mitchell; 48-17 (a) © C. Haagner/Bruce Coleman; (b) M. P. Kahl/Bruce Coleman; 48-18 (a), (b) © E. S. Ross; 48-19 © E. R. Degginger/Bruce Coleman

Page 1010 © Jen & Des Bartlett/Bruce Coleman; 49-1 (a), (b) © Joan Baron; 49-2 After Wilson, E. O. & Bossert, W. H. (1971). *A primer of population biology*. Sunderland, MA: Sinauer Associates; Page 1013 John B. Shelton; 49-4 After Raven, P. H. et al. (1986). *Biology of plants*. New York: Worth Publishers, Inc.; 49-6 © Jen & Des Bartlett/Bruce Coleman; 49-8 After Wallace, V. & Srb, A. (1964). *Adaptation*. Englewood Cliffs, NJ: Prentice-Hall, Inc.; Page 1017 United States Department of Agriculture; 49-9 (a), (b), (e), (f) © Tui De Roy; (c) © Tui De Roy/Bruce Coleman; (d) D. Cavagnaro/Peter Arnold; 49-10 & 49-11 After Lack, D. (1961). *Darwin's finches*. New York: Harper & Row, Inc.; 49-12 Peter R. Grant, *Ecology and evolution of Darwin's finches* © 1986 Princeton University Press. Plate 34 (Photo by P. T. Boag) reprinted with permission of Princeton University Press; 49-13 After Raven, P. H. et al (1986), op. cit.; 49-14 After Dobzhansky, T., Ayala, F. J., Stebbins, G. L. & Valentine, J.W. (1977). *Evolution*. New York: W. H. Freeman & Co. (after Simpson & Beck 1965); 49-15 © Bill Thompson/Woodfin Camp & Associates; 49-16 & 49-17 Redrawn from Simpson, G. G. (1951). *Horses*. New York: Oxford University Press; 49-18 After Stanley, S. M. (1979). *Macroevolution: Pattern and process*. New York: W. H. Freeman & Co.

Page 1031 Gallery of Prehistoric Art, Astoria, NY; 50-1 © A. Christiansen/Frank Lane Picture Agency; 50-4 © E. S. Ross; (b) © H. Albrecht/Bruce Coleman; 50-5 *National Geographic* (November 1985); 50-6 (a), (b) © Tom McHugh/Photo Researchers; 50-7 After Howells, W. W. (1967). *Mankind in the making*, rev. ed. Garden City, NY: Doubleday & Co., Inc.; 50-8 © Townsend P. Dickinson/Comstock; 50-9 Ralph Morse, LIFE Magazine © 1965 Time Inc. 50-10 © Mickey Gibson/Tom Stack & Associates; 50-11 © Peter Veit/DRK Photo; 50-12 (a), (b), (c) © Peter Davey/Bruce Coleman; (d) Teleki/Baldwin; (e), (f) © Richard Wrangham/Anthro-Photo; (g) © Linda Koebner/Bruce Coleman; 50-13 © David L. Brill; Page 1039 © John Reader; 50-14 © David L. Brill, National Museum of Tanzania, Dar es Salaam; 50-15 (a), (b) The Cleveland Museum of Natural History; 50-16 *National Geographic* (November 1985); 50-17 © John Reader; 50-18 © Michael Holford; 50-20 © A. Walker; 50-21 © Michael Holford; 50-23 (a) Ralph S. Solecki; (b) © David L. Brill; 50-24 © Michael Holford; Pages 1048-1049 Gallery of Prehistoric Art, Astoria, NY; 50-26 © James D. Wilson/Woodfin Camp & Associates

Page 1052 © Jim Brandenburg; 51-1 A. Christiansen/Frank Lane Picture Agency; 51-2 © Stephen Dalton/Photo Researchers; 51-3 P. Kirk Visscher; 51-4 Redrawn from Lorenz & Tinbergen, 1937; 51-6 © George D. Lepp/Comstock; 51-7 Nina Leen, LIFE Magazine © 1965 Time Inc.; 51-8 After Mazudazo Konish; 51-9 (a), (b) © C. R. Carpenter; 51-10 (a) © E. S. Ross; (b) © Alan Blank/Bruce Coleman; 51-11 (a) © E. S. Ross; (b) © Treat Davidson/NAS, Photo Researchers; 51-12 (a) © E. S. Ross; (b) © Colin G. Butler/Bruce Coleman; 51-13 (a) Christopher Springman © National Geographic Society; (b) © Raymond A. Mendez/Animals Animals; 51-14 (a) © Jim Brandenburg; (b) T. W. Ransom; 51-15 (a) © Leonard Lee Rue III/NAS, Photo Researchers; (b) © Robert Dunne/Photo Researchers; (c) © Sullivan & Rogers/Bruce Coleman; (d) © Zig Leszczynski/Animals Animals; 51-16 © Chesher/Photo Researchers; Page 1067 Shirley Baty; 51-17 After Kreb, J. R. & Davies, N. B. (1981). *An introduction to behavioral ecology*. Sunderland, MA: Sinauer Associates; 51-18 (a), (c) © Patricia D. Moehlman; (b) After Krebs & Davies (1981), op. cit.; 51-19 © Charlie Heidecker/Visuals Unlimited; 51-20 © Joseph Popp/Anthro-Photo; 51-21 © Sarah Blaffer Hrdy/Anthro-Photo; 51-22 © Steven C. Kaufman/Peter Arnold; Page 1073 (a), (b) © Hans & Judy Beste/Animals Animals; 51-23 © David Bygott/Anthro-Photo; Page 1076 Andy Blaustein

Page 1084 © M. Philip Kahl, Jr./Photo Researchers; Page 1085 & 52-1 © Günter Ziesler/Bruce Coleman Ltd.; Page 1087 © M. A. Chappell/Animals Animals; 52-3 © John M. Burnley/Photo Researchers; Page 1091 Adapted from May & Anderson (1987). *Nature, 326*, 140; 52-7 After *Nature, 298* (August 26, 1982); 52-8 Adapted from McNaughton & Wolfe (1979). *General ecology*, 2d ed. Holt, Rinehart, & Winston; 52-9 (a) © Joseph Van Wormer/Bruce Coleman Ltd.; (b) © Mark N. Boulton/Bruce Coleman Ltd.; (c) © Jeff Foott; 52-10 (a), (b) McNaughton & Wolfe

(1979), *op. cit.* (After Lack 1954); **52-11** After Kendeigh, S. C. (1961). *Animal ecology*. Englewood Cliffs, NJ: Prentice-Hall Inc. (After Shelford, 1911); **52-12** © Jane Burton/Bruce Coleman Ltd.; **52-13** (a) © Russ Kinne/Photo Researchers; (b) © C.A. Henley; (c) © Jane Burton/Bruce Coleman; (d) © Cosmos Blank/Photo Researchers; (e) © Eric Hosking/Bruce Coleman; (f) © E. S. Ross; **52-14** From Sarukhan & Harper, 1973; **52-15** © Jack Dermid; **52-16** © M. A. Chappell/Animals Animals

Page 1106 © James L. Castner; **53-1** © William E. Townsend, Jr./Photo Researchers; **53-2** Adapted from Jared Diamond (May-June 1978). *American Scientist*, 326; **53-5** After MacArthur, 1958; **53-6** © Joan Baron; **53-7** After Lack, D. (1961). *Darwin's finches*. New York: Harper & Row; **53-8** Adapted from Ricklefs, R. E. (1979). *Ecology*, 2d ed. New York: Chiron Press Inc.; **53-9** Joseph Connell; **53-11** © Maslowski/Photo Researchers; **53-12** (a), (b), (c) © Thomas Eisner; **53-13** (a) © Fred Whitehead/Animals Animals; (c), (d) After Fenton, M. B. & Fullard, J. H. (1981). Moth hearing and the feeding strategies of bats. *American Scientist*, 69, 266-275; **53-14** (a) © Michael Fogden/Bruce Coleman; (b), (c) © James L. Castner; (d) © Peter Ward/Bruce Coleman; (e) © E. S. Ross; **53-15** (a), (b) Australian Department of Lands; **53-17** (a), (b) Joe Lubchenco; **53-18** Australian News and Information Service; **53-19** (a) © R. Mariscal/Bruce Coleman; (b) © Douglas Faulkner; (c) © E. S. Ross; (d) © P. Ward/Bruce Coleman; **53-20** (a)-(d) © Daniel Janzen; **53-21** After Ricklefs (1979), *op. cit.*; **53-22** Adapted from Connell, J. H. (March 24, 1978). *Science*, 199, 1303; **53-23** (a) Roger Del Moral; (b) Shirley Baty; **Page 1127** (a), (b), (c) © Townsend Dickenson/Comstock; **53-24** (a), (b) Wayne Sousa; **53-25** (a), (b) Teresa Turner

Page 1131 © McDougal Tiger Tops/Ardea Photographics; **54-1** NASA; **Page 1135** NASA; **54-6** © Jack Dermid; **Page 1141** Robert R. Hessler; **54-9** (a), (b) © McDougal Tiger Tops/Ardea Photographics; **54-10** (a) © G. D. Plage/Bruce Coleman Ltd.; (b) © M. A. Chappell/Animals Animals; (c) © F. W. Lane/Bruce Coleman; **Page 1144** Irven de Vore/Anthro-Photo; **54-16** G. E. Likens; **54-17** (b) Michigan Department of Natural Resources; **54-18** © J. Donosa/Sygma

Page 1154 & 55-1 © Wolfgang Kaehler; **55-2** (a) © Michael P. Gadomski/Earth Scenes; (b) © Nielsen/Imagery; (c) © Alan G. Nelson/Animals Animals; **55-3** (a), (b) © D. P. Wilson/Eric & David Hosking; (c) © Tom Stack/Tom Stack & Associates; (d) © Claudia Mills/Biological Photo Service; **55-5** (a) © Jack Dermid; (b) © Frans Lanting; (c) © Stephen J. Krasemann/Photo Researchers; (d) © Breck P. Kent/Animals Animals; (e) © Jeff Foott; **55-6** (a) © Jack Dermid; (b) © Rip Griffith/Photo Researchers; **55-7 & 55-8** After Arms, K. & Camp, P.S. (1979). *Biology*, 2d ed. New York: Sanders College Publishing; **55-9** After A. W. Küchler; **55-10** (a) © Rod Planck/Tom Stack & Associates; (b) © Robert P. Carr/Bruce Coleman; (c) © Patti Murray/Earth Scenes; (d) © Breck P. Kent/Animals Animals; (e) © Wolfgang Kaehler; (f) © Leonard Lee Rue III/Animals Animals; **55-11** (a), (b) © Les Blacklock; (c) © Johnny Johnson/Animals Animals; (d) © Erwin & Peggy Bauer/Bruce Coleman; (e) © Leonard Lee Rue III/Animals Animals; (f) © Ralph A. Reinhold/Animals Animals; **55-12** (a) © Bill Ruth/Bruce Coleman; (b) © Bob & Clara Calhoun/Bruce Coleman; (c) © Kennan Ward; (d) © Bob & Clara Calhoun/Bruce Coleman; (e) © E. R. Degginger/Animals Animals; **55-13** (a) Pat Caulfield; (b) © Paul Rogers/Imagery; (c) © Franz J. Camenzind/Seaphot Ltd.: Planet Earth Pictures; (d) © Joe McDonald/Animals Animals; **55-14** © Peter Ward/Bruce Coleman; **55-15** (a) Jack Wilburn/Earth Scenes; (b) © R.J. Erwin/Photo Researchers; (c) © Ardea Photographics; **55-16** (a) © Bruce Coleman; (b) © Max Thompson/NAS, Photo Researchers; (c) © Ric Ergenbright; (d) © Greg Vaughn/Tom Stack & Associates; (e) J. Reveal; **55-17** (a) ©Jeff Foott/Bruce Coleman; (b), (d) © E. R. Degginger/Bruce Coleman; (c) © Stephen J. Krasemann/Peter Arnold; **55-18** (a) © E. S. Ross; (b) C. W. Rettenmeyer; (c) © Kevin Schafer/Tom Stack & Associates; **55-19** (a) © James H. Carmichael/Bruce Coleman; (b) © L. & D. Klein/Photo Researchers; (c) © Leonard Lee Rue III/Bruce Coleman; (d) © E. S. Ross; (e) © Wolfgang Kaehler; **55-20** (a), (b) © R. O. Bierregaard

Index

Page numbers in boldface type indicate major text discussions.

ABA (see Abscisic acid)
Abalones, 550
A band, 874
Abdomen, 566, 568, 570, 571, 574, 584, 585
Abdominal cavity, 701, 713, 718
  development of, in chick embryo, 936
  and gonads, 902, 904, 917
Abdominal wall, 716
Abortion, 951
  elective, 397–398
  spontaneous, 385
Abscisic acid (ABA), 669, 673, **680**
  and gibberellins, 681
  and stomatal movement, 654
Abscission, **679, 694–695**
Absolute temperature (ΔT), **165**, 183
Absolute zero, 778
Absorption, 769, 770
  in digestion, 715, 720, **723**
  of light, in homeostasis, 781
Absorption spectrum
  of chlorophyll, 212
  of phytochromes, 681
  of pigments, 210–211
Acacias, 1110, 1121
Accessory pigments, 440, 459
Accessory proteins, of cytoskeleton, 114
Accommodation, 867, 868
Acetabularia, 109–110
Acetaldehyde, 191
Acetic acid, and catabolism of ethanol, 202
Acetylcholine, 137, 853, 854, 856, 876, 877, 878
  in Alzheimer's disease, 897
  as neurotransmitter, 857
  and parasympathetic nervous system, 846, **847**
  receptor for, 137
  and regulation of transport, 179–180
Acetyl CoA, 194, 729
  in catabolism, 203
  in glycolysis, 194, 200, 201
  in Krebs cycle, 203
  in prokaryotic replication, 325
Acetyl groups
  and cell respiration, 204
  in Krebs cycle, 194, 200
Achenes, 620–621
Acid(s), 47–52, 54
  enzyme regulation of, 176
Acid growth, 674–675, 693
Acid rain, 50–51
Acoelomates, 520, 532, 533, **538**, 539, 542, 544
Acorn worms, 587, **591**, 604
Acquired immune deficiency syndrome (see AIDS)
Acrasin, 141, 838
Acrasiomycetes, 455, 466, 467, 469, 477
Acromegaly, 825
Acrosome, 902, 903, 920
ACTH, 74, 824, 825, 831, 833, 838
  and endorphins, 854
Actin, **703**, 874, 875, 876
  in cardiac muscle, 708
  and cellular mobility, 121
  in cytokinesis, 153
  and cytoplasmic streaming, 121
  filaments, 110, 114, 121, 124, 130, 133, 146, 938
  gene family, 368
  in muscle cells, 518
  in muscle contraction, 706
  and red blood cell, 114
  in vertebrate muscle, 713
Actinomycetes, 430, 449, **668**

Actinosphaerium, 103, 470
Action potentials, **848–850**, 851, 856, 857, 879
  in brain, 890
  molecular basis of, **848–850**
  in motor neurons, 877
  in muscle contraction, 873
  in sensory perception, 859, 864
  at synapses, **852**
Action spectrum, 213, 681
Active site, **170**, 171, 176–180, 183–184, 799
Active transport, 135, **136**, 143, 655, 663, 669–670, 674–675, 721, 723, 754, 771, 772, 774
  across the placenta, 948
  energetics of, 178–182, 184
  in loop of Henle, 773
  and mitochondrion, 136
  in plants, 669–670
  of potassium (K^+) ion in plants, 653
  in roots, 633
Acuity, visual, 580
Adaptation, 228, 413, 427, 536, 567, 575, 579, 586, 609, **629–630**, 642, **1001–1005**, 1007, 1009, 1110, 1113, 1129
  and adaptive radiation, 1022–1023
  to altitude, 733
  in angiosperms, 620–624
  in animals, 519, 520, 710–711
  and competition, 1110, 1113
  and directional selection, 995, **997**
  and "diving reflex," 733, **744**
  in *Equus*, **1025–1027**
  evolutionary, 245, 493, **970–972**, 973
  in fruits, 620–621
  imperfections in, **970–972**
  in marsupials vs. mammals, **1023**
  and natural selection, 7, **1001–1005**
  in Pacific Northwest forests, 1166
  in plants, 493, 495
  and predation, 1116–1117
  in prokaryotes, 437
  in reptiles, **1024**
  for rocky seashores, 1157
  to seasonal change, 621–624
  in seedlings, 687
  and sexual selection, **998–1001**, 1009
  to temperature extremes, 777–778, **787–790**
  and territoriality, 1064
Adaptive radiation, **1022–1023**, 1030, 1031
  of mammals, 1050
Adaptive value of behaviors, 1053–1054, 1077
Adenine, 80–81, 305, 314, 341, 429
  in ATP, 180, 184
  and cytokinins, 676
  in DNA, 172, 283–284, 288–292, 296, 298
  in NAD, 171–172
  and poly-A tail, 369
  in RNA, 172
Adenosine diphosphate (see ADP)
Adenosine monophosphate (see AMP)
Adenosine triphosphate (see ATP)
Adenovirus, 444–445
Adenylate cyclase, 837, 861
ADH (antidiuretic hormone), 712, 775, 776, 824, 827, 828, 839, 883
  and hypotonic urine, 773–774
  and regulation of kidney function, 774
Adhesion, 42, 54, 105, 938
Addiction, 855, 948
Addison's disease, 775
A-DNA, 356, 379
ADP, 137
  from ATP in energy reactions, 181–183
  and electron transport, 195–200
  F_1 unit, 199

in glycolysis, 186–190, 200–205
in Krebs cycle, 195–196
in muscle contraction, 875
in nitrogen fixation, 668
phosphorylation of, 220–221
in the stages of photosynthesis, 216, 218–221, 229
Adrenal cortex, 70, 775–776, 824, **829–832**
  and glucocorticoids, 829, **830–831**
  hormone production, regulation of, 827
  hormones of, 829
  and mineralocorticoids, 829, **830–831**
  negative feedback control in, 827
  and pituitary gland, 823
  sex hormones and, 832, **905**
  and stress, 833
Adrenal gland, 119, 823
  and heartbeat, 756–757
  hormone receptors in, 756
Adrenaline, 181, 712, 764, 823, 824, 832, 833, 837–838, 839
  and ADH, 775
  and blood pressure, 759
  in infants, 951
  and regulation of heartbeat, 757
  and temperature regulation, 786
Adrenal medulla, 712, 757, 822, 824, **832**, 833, 847
  development of, 935
  and endorphins, 855, 856
  and enkephalins, 854, 855
Adrenocorticotropic hormone (see ACTH)
Adventitious roots, 635, 636, 642–643, 694–695
  and auxin, 674
Aerobes
  evolution of, 206
  prokaryotes, 431
Aerobic respiration, 190, 193–194
Afferent arteriole, of kidney, 770–771
Africa
  and human evolution, 1047, 1050, 1051
  savannas of, 1170
African green monkey, 817
Afterbirth, 950
Agassiz, Louis, 4, 519
Age
  and Alzheimer's disease, 896
  and hibernation, 786
  of parents, and Down's syndrome, 386
Age structure, 1087, **1092–1093**, 1104
  of United States, 1092–1093
  and world population explosion, **1097**
Agglutination, 798, 809, 810–811
Aggregate fruits, 620
Aggression, 826
  in dominance hierarchies, 1062
  and male sex hormones, 906
  in sticklebacks, 1054–1055
  and territoriality, 1063–1065
Agnaths, 593 (see also Hagfish; Lampreys)
Agriculture, 622
  and destruction of rain forests, 1176–1177
  ecological problems of, 659, **1148–1149**
  and human population growth, 1096–1097
  and plant variation, 643
Agrobacterium tumefaciens, 352–354
Ahlquist, Jon E., 420, 423
AIDS, 752, **813–819**
  and blood supply, 392
  infection, **815–816**
  prospects for treatment, **817–818**
  and recombinant DNA, 342
  reverse transcription in, 445
  transmission of, **817**, 1087, **1090–1091**
  virus, 121, **814–816**
AIDS-related complex (ARC), 816
Air, 733, 736, 779

I-1

Air pollution
  and acid rain, 50-51
  and ginkgo, 505
  and lichens, 488
Air pressure
  in alveoli, 741
  and diffusion, **733**
  measurement of, 128
Air roots (see Pneumatophores)
Air sacs, 598, 737
Alanine, 73
Alarm calls, 1055, 1059
Albedo, 781
Albersheim, Peter, 682
Alberts, Bruce M., 370
Albinism, **389**
  and phenylketonuria, 388
Albino, coats, in rabbits, 268-269
Albumen, 596, 750
  in egg, 931
  selection for, 995
Alcohol(s), 56 (*see also* Cholesterol, Ethanol, Steroids)
  and cirrhosis, 202
  fermentation, 190-191
Alcoholism, and Korsakoff's syndrome, 894
Aldehyde group, 56-57
  in chlorophyll, 211
Aldolase, 188
Aldoses, 60
Aldosterone, **775**, 776, 824, 829, 832
Alertness, 885, 890, 951
Aleurone cells, 681
Aleurone layer, 622
Alga(e), 13, 120, 203, 411, **455-465**, 474, 476-477, 542 (*see also* Cyanobacteria; Lichens; Seaweeds)
  accessory pigments of, 456, 459
  alternation of generations in, 463-464, 465
  brown, 455, 465
  cell division in, 457, 459, 460, 461, 462, 465
  characteristics of, 456, 457
  coralline, 465
  cytoplasmic streaming in, 121
  filamentous, 212
  freshwater, 463, 465
  golden-brown, 456, 457
  green, 120, 148, 461-464, 530
  at intertidal zones, 1124-1125, 1128-1129
  marine, 455, 456-457, 463, 465, 1156
  origins of, 497
  photosynthetic, 206, 211, 213, 222
  red, 122, 455, 456, 457, 465
  reproduction in, 459, 460, 462-464, 465
  volvocine line, 461, 462
Alkaptonuria, 301, 388
Allantois, **934**, 952
  in human embryo, 944
Allele(s), 240, 243-245, 247, 260, 266-267, 274-277, 279
  for behavioral traits, 1052-1053, 1076
  for blood groups, 810-811, 997
  and directional selection, 997
  distribution of, 243
  diversity of, 247
  dominant, 240, 242, 247, 266, 268-271
  exchange of, 275-276, 279
  and founder effect, 983
  frequency (*see* Allele frequency)
  and gene flow, 982
  and genetic disease, 400
  and genetic drift, 982-983
  for hemophilia, 400
  interactions among, **268**, 279
  and kin selection, 1066-1070
  for lactose intolerance, 727
  for MHC antigens, 805, 806
  multiple, **268**, 279, 980
  mutation and, 978, 981, 987
  mutations in, 391, 395, 400
  and natural selection, 1008
  neutral, 978
  and nonrandom mating, 984
  and phenotype matching, 1077
  for phenylketonuria, 388-389
  and population bottlenecks, 983
  recessive, 240-243, 247, 266-268, 279
  recombination of, 244-245, **274-275**, 276-277, 279

  segregation of, 274
  sex-linked, 266, 268
  and sexual reproduction, 986-987
  for sickle-cell anemia, 388-389
  wild-type, 266
Allele frequency, **974**, 989, 990
  of blood groups, **996**
  and diploidy, 983, 986-987
  and gene pools, **974**
  and genetic drift, 982-983
  and Hardy-Weinberg equilibrium, **979-981**
  in human populations, 989
  and outbreeding, 984-985
  and sexual reproduction, 984
Allele interaction(s), **268**, 279
  codominance, **268**, 279
  dominant-recessive, 279
  incomplete dominance, **268**, 279
  multiple alleles, **268**, 279
Allergens, 835
Allergic reactions, 830, 836
Allergies, 812-813, 819
  evolution of, 813
  and IgE antibody, 799
  and immune system, 792
Allolactose, 323-325
Allopatric speciation, **1010-1011**, 1029
  in punctuated equilibria, 1028
  vs. sympatric, 1010, 1014-1015
Allopolyploidy, **1012**
Allosteric binding sites, 176-177, 179, 184
Allosteric effectors, 325
  in enzyme activation and inactivation, 176-177, 184
  in glycolysis, 188
  in transcription, 325
Allosteric inhibitors, 177
Allosteric interactions, 164, **176-177**
  and glycolysis, 188
  and membrane transport proteins, **178-180**, 181
Alpha-amylase, 682
Alpha chain, 804
  for MHC antigens, 805
  for T-cell receptors, 805
Alpha-fructose, 63
Alpha globin, 359, 367
Alpha-glucose, 60-61, **63**, 65
Alpha helix
  of cell membrane proteins, 106
  of protein structure, 74-75
Alpha-tubulin, 78, 110
Alpha waves, 891
Alpine forests, 1166, 1167, 1177
Alternation of generations, 252, 463-464, 467, 477, 494, 495, 507
  in angiosperms, 512, 614
  in brown algae, 465
  in bryophytes, 499
  in Charophyceae, 516
  evolution of, 494, 495
  in green algae, 463-464
  in plants, 252
  and polyp/medusa system, 526
  in red algae, 465
Alternation of reproduction, 1103
Alternative life-history characteristics, **1099-1105**
Alternative splicing, 380
  and biosynthesis of polypeptides, 380
  introns, 369, 370
Altitude, 733, **746**
  adaptations to, 733, **746**
  and competition, 1107
  ecotypes and, 1003
  and temperature, 1161, 1177
Altruism, 1055, **1059**, 1067, 1078 (*see also* Reciprocal altruism)
Alu sequence, 365
Alvarez, Walter, 1024
Alveoli, 738, **740**, 747
  capillaries of, 758
Alzheimer, Alois, 896
Alzheimer's disease, 385, 397-398, **896-897**
  Down's syndrome and, 897
  memory loss in, 894, **896-897**
  and prions, 447
Amacrine cells, 869, 936
Amanita, 487
Amines, 853, 854

Amino acid(s), 55, 57, 82, 305, 307, 309, 311-315, 317, 418-419, 766, 1145 (*see also specific acids*)
  acidic, 73
  AMP complex, 312
  in antibodies, 798-799
  and auxotrophs, 173
  basic, 73
  charge of, 304
  derivatives of, 853-854
  in DNA, 281
  and enzyme regulation, 176
  in enzyme structure, 170-171
  essential, 73, 76, 730
  and hormone secretion, 824
  linkage to tRNA, 312
  and nitrogen, 76
  nonpolar, 73
  nutritional requirements for, 74, 728, 730
  origin of, 87
  polar, 73
  and prokaryote classification, 429
  and proteins, 70, **72**
  reabsorption of, through kidney, **770**, 776
  R groups of, 171
  sequences of, 74, 307, 316
  sequencing of, 316, **418-420**, 427, 429
  and symbiosis, 725
  synthesis of, 303
  transcription for, 320
Amino acid-AMP-enzyme complex, 312
Aminoacyl site(s), 312-314
Aminoacyl-tRNA synthetase, 311, 315, 317
Amino group, 48
Aminopeptidase, 722
Ammonia, 429, 439, 944, 1147, 1153
  in deamination, 766
  in digestion of cellulose, 724
  in forest ecosystems, 1150
  molecular structure, 44
  in nitrogen fixation, 429, 439, 1147
  nitrogen in, 76
  vs. water, 76
Ammonification, **1147-1148**, 1153
Ammonium cyanate, 11
Ammonium ion (NH_4^+), 1147, 1153
  in ammonification, 1147, 1153
  and chemosynthetic bacteria, 429
  in forest ecosystems, 1150
  in nitrogen fixation, 439
  plant requirements for, 664
  in symbioses, 668-669
Amnesias, 893, 894
  in Alzheimer's disease, **896-897**
Amniocentesis, 387-388, 394, 400, **944**
Amnion, 576, **932**, 952
  in human embryo, 944
Amniote egg, 596, 604, 900-901, 905, 908, 912, 917, 952
  in chickens, 931, 932
  and reptilian evolution, 1022
Amniotic cavity, **934**
  in human embryo, 944, 946
Amniotic fluid, 387, 400, 944
Amniotic sac, 950
Amoeba(s), 122, 259, 454, 469, 470, 477
  behavior patterns, 474
  and cellular slime molds, 467, 837, 838
  chemotaxis in, 474
  classification, 13
  pathogenic, 470
  phagocytosis and, 138
  and slime molds, 141
Amoeba proteus, 37, 121
Amoebocytes, 523, 524, 531, 542, 588
Amoeboid algae, 460
Amoeboid organisms, 455
Amoore, John, 861
AMP, 181, 312
  cyclic (*see* Cyclic AMP)
Amphetamines, 853
Amphibians, 587, **595**, 597, 604
  brain of, 885
  cleavage and blastula formation in, **927**
  embryonic development of, 919, **926-931**
  evolution of, 505
  gastrulation in, 927-929
  gills of, 737
  heart of, 754, 763

I-2

induction in, 930-931
lungs of, 736
melanocyte-stimulating hormone in, 826
and melatonin, 833
neural tube formation in, 927-929
organizer in, **930**
ovum of, 926
startle behavior in, 1054
tissue interactions in embryo, **930-931**
vomeronasal organ, 861
Ampicillin, 336
Amplification cascade, 751, 837-838
Amplitude, visual, in microscopy, 99
Ampulla, 588
Amygdala, 899
 and memory, **894**, 896
Amylase, 718, 730
Amyloid protein, 896-897
Amylopectin, 64-65
Amyloplasts, 683-684
Amylose, 64-65
Amyotropic lateral sclerosis (Lou Gehrig's disease), 897
Anabaena azollae, 92
Anabolic pathways, **203-204**
Anabolic steroids, 907
Anabolism, **168**, 183
 in cell, 192
 and organic molecules, 203
 pathways, **203-204**
Anaerobes, 90, 429, 439, 442
 and digestion in herbivores, 724
 evolution of, 206
Anaerobic respiration, 190, **191**, 192-194 (*see also* Fermentation)
 acetaldehyde and, 190
 ATP and, 190-191
 glycogen in, 190
 lactic acid and, 190-191
 and NAD+, 190-191
 and NADH, 190-191
 pyruvic acid in, 190-191
Analgesics, 855
Analogous structures, 414
Analogy, 414, 416, 427
Anaphase,
 of meiosis, 254, 256
 of mitosis, 150, **151**, 155
Anaphylactic shock, 813
Anatomy, 414, 418
 and cellular transport, 130
Ancestor(s), 413, 427
Anderson, Roy, 1090-1091
Andersson, Malte, 1000-1001
Androgens, **905-906**, 917
 in embryo, 946
 in interstitial cells, 901
Anemia, 304
 and cyanocobalamin, 724
 hemolytic, 811-812
 sickle cell (*see* Sickle cell anemia)
Anger, and stomach motility, 720
Angina pectoris, 760
Angiosperms, 411, 425, 497, 498, 50l, **509-515**, 516, 517, 578, **613-624**
 adaptations in, **621-624**
 classification of, 511
 dicots, 511
 evolution of, 509, 510, 515, 621
 fertilization in, 511
 flowers, **511-515**
 fruits, 515, 619-621
 growth in, **621-624**
 vs. gymnosperms, 638-639
 and insect pollination, 512-513
 in lakes and ponds, 1155
 life cycle of, 512
 monocots, 511
 numbers of, 511
 pollination, **512-515**
 and predation, 515
 reproductive strategies, 509
 seeds and fruits in, **619-621**
 sexual reproduction in, **613-619**
 toxic compounds in, 515
 vascular system of, 509
Ångstrom, 95
Animal cell, 110, **112**, 117, 125, 257

cell cycle, 146
cytokinesis in, 153
cytoskeleton, **112**
division of, 155
mitosis in, **150-152**
nuclear envelope of, 112
nucleolus of, 108, **112**, 124
nucleus of, **112**
Animalia, 423-427, **518-697**
 cell type, 125
 diversification in, 716
 kingdom, 14
Animals, **14**, 411, 424, **519-697**, 1139-1142, 1143, 1150, 1153
 behavior (*see* Animal behavior)
 bilateral symmetry in, 520, **531-605**
 classification of, 454, 520
 colonial, 562
 distribution of, and biogeography, 966
 diversity in, 519
 earliest, 519
 evolutionary relationships among, 521
 evolution of, 92, 454 516, (*see also* Evolution)
 growth in, vs. plants, 630
 multicellularity, 94
 number of species, 518
 origin and classification of, **519-522**
 radially symmetrical, 520, **524-531**
 stages of socialization, 1059-1061
 tissues, 142-143
 water requirements in, 650, 765
Animal behavior, **1052-1079**
 associative learning in, **1053-1056**
 in birds, **1063-1065**
 causation and, 1053-1054
 conflicts of interest in, **1071-1074**
 dominance hierarchies, **1062-1063**
 evolution of, **1052-1079**
 fixed action patterns in, **1054-1055**
 genetic basis of, **1052-1053**
 and human behavior, 1075
 imitative learning in, **1056-1057**
 imprinting in, **1056-1057**
 insect societies, **1059-1061**
 kin selection, **1065-1070**
 learning in, **1055-1058**
 reciprocal altruism in, **1074-1075**
 and selfish gene, **1070-1074**
 stages of socialization, 1059-1061
 territoriality, **1063-1070**
 vertebrate societies, **1062-1065**
Animal gradient, 925-926
Aniridia, and Wilms' tumor, 387, 399-400
Anisogamy, 463
Ankyrin, 114
Annelids, 520, 532, 533, 544, 545, **554-558**, 560, 561, 564, 565, 566, 567, 568
 vs. arthropods, 554
 evolutionary relationships among, 554
Annual rings, **644-645**, 646
Annuals (plants), **621**, **623**, 624, 1100
 in deciduous forest, 1164
 in deserts, 1175
 in grasslands, 1170
Antagonistic muscle groups, 706-707
Antarctic, ozone layer in, 1134
Antelopes, 1110
Antennae, 557, 558, 561, 565, 566, 571, 572, 574, 582, 584, 585, 1060
Antennae pigments, 440
Antennapedia complex, 942
Antheridia, 495, 499
Anthers, 238-239, 249, 511, 614, 615, 616, 624
Anthozoans, 526, **529-530**
 asexual reproduction in, 529
 colonial, 530
 vs. hydrozoans, 529
 polyp forms in, 529
 sexual reproduction in, 529
Anthropoids, **1033**, **1034**, 1051
Antibiotics, 352, 432, 449, 488, 795 (*see also* Neomycin; Penicillin; Streptomycin; Sulfanilamide; Tetracycline)
 fungi as sources of, 480, 484
 resistance to, 964
Antibody(ies), 70, 72, 76, 111, 342, 352, 804
 action of, **797-798**
 for AIDS virus, 815

B lymphocytes and, 818-819 (*see also* Immunoglobulins)
formation of, **795-801**
genetic code for, **371-372**, 380-381, 784, 801, 809
in mother's milk, 727, 948
production of, during infection, 809
and Rh disease, 812
structure of, **798-799**
and T-cell receptors, 809
vs. viral mutations, 444-445
Antibody-coding gene, **371-372**, 380-381
Anticoagulants, 558, 752
Anticodons, 311-312, 315
 tRNA, 312-313
Antidepressants, 853
Antidiuretic hormone (*see* ADH)
Antifreeze protein, 44
Antigens, **371**, **795-798**, 799, 800, **804**, **818-819**, 813
 and allergies, 813
 binding of, 798, 800, 802, 804
 on cancer cells, 808
 in cell-mediated immunity, 801, 802
 and clonal selection, 800
 and cytotoxic T cell, 806-807
 and major histocompatibility complex, 805-809
 on red blood cells, 810, 811-812
 and Rh factor, 811-812
 and T lymphocytes, 804, 819
Antihistamines, 813
Anti-inflammatory drugs, 831
Antimicrobial drugs, 449
Antiparallel strands, of DNA, 289, 290, 296
Antipodal cells, 617
Antiport, 137
Antiviral drugs, 450-451
Ants, 573, 576
 and acacias, mutualism, 1121-1122
 and aphids, 1121-1122
 social behavior in, **1059-1060**
Anus, 538, 544, 545, 715, 725, 909, 924
Anvil (incus), 864
Aorta(s), **753**
 blood pressure in, 759-760
 and human heart, 755, 786
 pressure receptors in, 774
 and vascular circuits, 758-759
Apertures, of cell membrane, 130, 133
Apes, 496, 1034, **1035-1036**, 1051
Aphasia, 888-889
Aphids, 660, 1121
Apical dominance, 641, 675
 and auxin, 675, **694**
 and ethylene, 679, 694, 695
 reversal of, and cytokinin, 677
Apical meristems, 647, 674-675, 694
 and flowering, 688
 and growth, 630
 and root growth, 634-635
 of roots, 634, 648
 and shoot growth, 618, **619**, **640-641**, 642, 648
Aplysia (*see* Sea hare)
Apocrine glands, 863, 906 (*see also* Sweat glands)
Appendages, 558, 564, 565, 566, 570, 571, 573, 574, 579, 585
Appendix, 724-725
Aquatic environments, **1154-1160**
Arachidonic acid, 833, 835
Arachnids, 566, 569, **570-571** (*see also* Mites; Scorpions; Spiders)
Arboreal life, and primate evolution, 1031-1033, 1034-1035, 1036, 1051
ARC, 816
Archaebacteria, 430, 432, 450, 453
Archaeopteryx, 598, 599, 967
Archaic *Homo sapiens*, 1045, **1046**, 1051
Archegonia, 251, 495, 499, 506
Archenteron, 952
 of amphibian, 927-929
 of chick embryo, 932, 937
 of sea urchin, 923, 924
Arctic (*see also* Tundra)
 ozone layer in, 1134
 radioactive pollution in, 1151
Ardrey, Robert, 1044
Arginine, 73, 173, 303-304, 310
 and auxotrophs, 173
Argon, 30
Aristotle, 410, 413, 578

I-3

Aristotle *(Continued)*
 early ideas on taxonomy, 410, 413
 on heredity, 236
 and origins of biology, 1, 5
 and spontaneous generation, 86
Arm, evolution of, 1032–1033
Armadillos, **967–968**
Arms, 601
 in human embryo, 946
Arousal, 885, 898
Arrow worms, 587, **590**, 604
Arteriole(s), **753**, 758, 760
 and blood pressure, **759–762**
 and capillary beds, 759
 and cardiovascular regulation, 761, 763
 and heart disease, 761
 in nephron, 770
 and temperature regulation. 786
Artery(ies), **753**, 761, 763
 and blood pressure, **759–762**, 764
 carotid, 859
 coronary, 71
 heat conservation in, 758
 hepatic, 759
 and portal systems, 759
 pulmonary, 755, 758
 structure of, **753**
Arthritis, rheumatoid, 118
Arthropods, 66, 497, 520, 540, 544, 560, 561, 564, **565–586**
 adaptations in, 1006
 vs. annelids, 554, 566, 567
 behavior, **579–585**
 characteristics of, **565–568**
 classifications of, **568–578**
 evolution of, 564
 exoskeleton of, 480
 vs. mammals, 585
 vs. mollusks, 567–568
 nervous system of, 842
 number of species, 565
 and pentastomids, 560
 premating isolation mechanisms of, 1016
 respiration in, 741
 senses, **579–585**
 structure, **567–568**
 success of, **578–585**, 586
 terrestrial, 568
Artificial fertilization, 236
Artificial selection, 989, 991, 1016
 vs. natural, 976–977, 990
Artiodactyls, 602, 603
Ascocarps, 303, 485, 486
Ascomycetes, 481, **484–486**, 491
 and antibiotics, 484
 cell walls, 482
 dikaryons, 481
 diseases caused by, 481
 and ectomycorrhizae, 491
 and Fungi Imperfecti, 482
 in lichens, 484, 489
 life cycle of, 485
 Neurospora, 481, 484
 parasitism in, 486
 reproduction in, 485–486
 spores, 481
Ascorbic acid, 729
Ascospores, 485, 486
Ascus, 302–303, 485, 486, 491 (*see also* Spores, case)
Asexual reproduction, 648, 984 (*see also* Parthenogenesis; Vegetative Reproduction)
 in algae, 459, 460, 462
 in black bread mold, 483
 in bryophytes, 500
 in fungi, 481, 485, 486, 491
 in hybrid plant populations, 1014
 in *Plasmodium*, 473
 in protozoa, 468, 469, 470
 vs. sexual, 986, 987, **1103–1104**, 1105
 in water molds, 467, 468
 in yeasts, 486
Aspartic acid, 73
 in C_4 photosynthesis, 224
Aspirin, 835, 836
Assimilation, **1146–1147**
Association cortex, 892
Associative learning, **1055–1056**, 1077

Assortment
 independent, 270, 274, 276, 279
 random, in meiosis, 253, 201
Aster, 149, 150, 152, 154, 254
Asteroid hypothesis of mass extinctions, 1024–1025
Asthma, 836
Astronomy, and history of science, 9
Atherosclerosis, 70–71, 558, **760–761**
Athlete's foot, 488
Atmosphere, 1152
 and biogeochemical cycles, 1145
 boundaries of, 1132–1133, 1134–1135
 climate, **1136–1137**
 ecosystems and, **1132–1135**
 layers of, **1132–1133, 1134**
 nitrogen in, 1146–1147
 oxygen in, 452
 ozone layer, **1134–1135**
 primitive, 87, 100
 temperature of, and latitude, 1161
 weather, **1136–1137**
 wind, **1136–1137**
Atmosphere (unit of measurement), 128, 733
Atmospheric boundaries, 1132–1133
Atmospheric pressure, **733**
 and human respiration, 741
Atom(s), 23, 38
 and charge, 23, 34
 and electrons, **28**
 and living systems, 35–37
 and molecules, **23–38**
 structure of, **28–34**
Atomic number, **24**, 38
Atomic structure, **28–34**
 electrons in, 28
 of elements, 24
 models of, 28
 of organisms, 36–37
Atomic weight, 24, 38, 163
ATP, 136, 137, 180–184, 203–205, 307, 312, 653, 655
 and active transport in plants, 653, 655
 adenine in, 172, 180, 184
 and ADP in energy reactions, 183
 in aerobic respiration, 193
 in anaerobic respiration, 190–191
 in bioluminescence, 182, 352–353
 biosynthesis of, 199
 in carnivorous plant movements, 693
 in cellular respiration, 167
 in chloroplasts, 220, 229
 conservation of, 188
 in cyclic AMP formation, 837, 861
 in electron transport, 195–200
 F_1 unit, 199
 in glycolysis, 186–190
 hydrolysis of, 181–182
 in Krebs cycle, 194–196
 in muscle contraction, 875
 and nitrogen fixation, 668
 phosphate bonds of, 186, 189, 190, 204–205
 phosphate groups of, 137, 180–183, 189
 in phosphorus cycle, 1146
 in photosynthesis, 216, 218–223, 229
 in plant coiling, 691
 potential energy in, 181–182
 regulation of synthesis, 200
 and ribose, 180, 184
 and sperm motility, 902
 structure of, 180–181
ATPase(s), 181, 199, 878
ATP synthetase, 205
 and cell respiration, 192
 complex, 192
 and electron transport, 198–199
 and F_1 and F_O factors, 198
 and phosphorylation of ADP, 220, 221, 229
 vesicles and, 199
Atriopore, 591, 592
Atrioventricular node, **756–758**, 763
Atrium(a), 547, 591, 592, 754–758, 760, 763–764, 775–776
Attention, **885**, 890, 898
Auditory cortex, 886, 892, 898
Auditory tubes, 864, 937
Australia
 and adaptations, **1023**
 biogeography of, 966

 marsupials in, 1010, **1012**, 1013
Australopithecines, **1038–1042**, 1043, 1044, 1051
 evolution of, 1047
 vs. *Homo*, 1042, 1045
 at Laetoli, **1039**
Autoimmune diseases, 812, 819, 830
Autonomic nervous system, **712**, 719, 842, 843, **845–847**, 856, 873
 and blood flow, **757**
 and cardiac muscle, 878
 functions of, 847
 and homeostasis, 847
 and intestinal tract, 723
 neurotransmitters in, 846, 847
 regulation of blood glucose, 726
 regulation of heartbeat, 757
 vs. somatic nervous system, 845, 846
Autopolyploidy, **1012–1013**
Autoradiography, 908, 914
Autosomal dominants, in genetic disease, 391
Autosomal recessives, in genetic disease, 388–389
Autosome(s), 264–265, 278
 abnormalities, 385–386
 and albinism, 389
 and color-blindness, 392
 and Down's syndrome, 385
 and genetic disease, 390
 and Huntington's disease, 391
 number of, 383, 399
 and spermatogenesis, 902
Autotomization, 572, 589
Autotrophs, 89, 90, 94, 101, 436, 450, 452, 518, 1131, 1152
 chemosynthetic, 206, 436, **439**
 and ecosystem energy flow, 1131, 1152
 and organic molecules, 203
 photosynthetic, **440–443, 455–465**, 476–477
Auxins, **673–677, 694–695**
 and cell elongation, 680
 and circadian rhythms, 689
 and cytokinins, **676–677**
 and ethylene, 679
 and flowering, 688
 and gibberellins, 681, 688
 and gravitropism, 682–683
 and leaf abscission, 679
 and plant growth, **673–677**
 and touch response in plants, 691
Auxotrophs, 173
Avery, O. T., 281, 284, 299
Aves, 416
Avogadro's number, 48–49
Avoidance mechanisms, 1114–1116
Axelrod, Daniel, 509
Axial filament, **437**, 450
 of sperm cell, 902
Axillary buds, 641–642, 648, 694
 and apical dominance, **675**, 677
 and ethylene, 694
Axils, 641, 648, 528
Axis, of earth, and distribution of life, 1136, 1160–1177
Axolotls, 595, 735
Axons, 121, **708**, 709, 713–714, 842, 844, 856, 87 881
 in *Aplysia*, 895
 of autonomic nervous system, 846, 847, 856
 in brain circuits, 884
 in cerebral cortex, 884
 giant, 553
 in gray matter, 881
 membrane of, 849–852, 856
 in nerve impulse transmission, **848–851**
 in neuromuscular junction, 876, 877
 of parasympathetic division vs. sympathetic, 847
 terminals, 852, 853
 in visual cortex, 888
Axon hillock, 856, 877
Axopod, 103
Azospirillum, 668

Baboons, 1063
"Baby boom," 1093
Bacilli, 429, **436**, 437–438
Backflow, prevention of
 in cardiovascular circulation, 755–756, 762
 in lymph node, 794

Background extinction rate, 1024
Bacteria, 91, 407–408, 1142, 1144, 1153
    actinomycetes, 430
    aerobic, evolution of, 452
    archaebacteria, 430, 450
    bacilli, 429, 436
    capsules in, 433
    cell division in, 145
    chemosynthetic, 89
    chromosome of, **319–320**
    classifications of, **13**
    cocci, 429, 436
    conjugation, 352
    cyanobacteria, 430, **440–443**, 450
    as detritivores, 1146
    drug resistance in, **964–965**
    earliest, 89, 90
    endospore-forming, 430
    evolution of, **764–765**
    genes in, 314
    genetic recombination of, 336
    genetic transformation of, 282, 329, 331, 336, 376
    gliding, 436, 438, 450
    gram-negative, 450
    gram-positive, 430, 450
    green, 430, 440, 450
    halobacteria, 221, 430
    infection of, 305
    lactic acid bacteria, 430
    locomotion, 131
    magnetic, 435
    methanogens, 430
    mycoplasmas, 430
    myxobacteria, 438
    in nitrogen cycle, 1144, 1146
    nitrogen-fixing, 670
    and penicillin, 178
    photosynthetic, 217, 219, 430
    and plant nutrition, 662, 670
    protein synthesis in, 315
    pseudomonads, 430
    purple, 430, 440, 450
    purple sulfur, 217
    replication of, **319–320**
    Rhizobia, 430
    sensory responses, 131
    spirilla, 429, 436
    spirochetes, 430, 450
    and sulfanilamide, 177
    symbiotic, and nitrogen fixation, 665–670
    thermophilic, 428, 430
Bacteriochlorophylls, **440**
Bacteriophage(s), 304–305, 331–332, 373, 445
    DNA of, 308
    in genetic research, 281, **284**, 285, 295
    infection of bacteria, 305
    lambda phage, **334–335**, 337–338, 341, **344–345**, 346, 353, 373
    nontemperate, 333–334
    φ X174, 330, 349, **350**
    prophage, **331–335**, 338
    in recombinant DNA, 341, **344–345**, 346, 353
    temperate, **331–333**, 338
    T-even bacteriophages, 330
Bacteriorhodopsin, 221
Balance, 863, 864, 879
Balanced polymorphism, **992–993**
Banana slug, 985
Banding patterns on chromosomes, 277–278, 383, 399, 422
Bands (social groups), 1034, 1037
Barbs, protist, 122
Bark, **644–645**
Barnacles, 571, 572, 579, 585, 1112–1113
Barometer, 733
Barr body, 267, 359
Basal bodies, 123, 125
    and centrioles, 146, 149
    in *Chlamydomonas*, 146
    of prokaryotic flagellum, 434
Basal forebrain, and memory, 894, 896, 899
Basal metabolism, 752
Base(s), **47–52**, 54
    enzyme regulation of, 176
Basement membrane, 702, 740
Base pairing
    and mRNA transcription, 306

    rules of, 310–317
Basidia, 486, 487, 491
Basidiocarp, 486–487
Basidiomycetes, 481, **486–488**, 491
    cell walls, 482
    club, 491
    corn smut, 487
    cytoplasm of, 482
    earthstars, 488
    and ectomycorrhizae, 491
    fairy rings, 486
    gill fungi, 487
    hyphae, 482, 486, 487
    jelly fungi, 488
    life cycle, 487
    mushrooms, 481, **486–487**
    mycelium of, 486
    parasitism in, 488
    pathogenic, 481
    puffballs, 488
    reproduction in, 486
    rusts, 481, 488
    shelf fungus, 487
    stinkhorns, 488
    toadstools, 481
    white rot fungus, 488
Basidiospores, 486, 487
Basilar membrane, 864, 865
Basophils, 792, 812–813 (*see also* Immune system)
Bates, Henry Walter, 965, 972, 1004
Batesian mimicry, **1004–1005**, 1115
Bateson, William, 263, 268–270, 974
Bats, 136, 115–1116
B-complex vitamins, 194, 724, **729**
B-DNA, 356, 379
Beadle, George, 301–304, 316
"Beads-on-a-string" model of linking DNA, 357
Beagle, voyage of, **1–3**, 3–6, 8, 966
Beak size, in Darwin's finches, 1019–1020, 1021, 1110–1111
Bear(s), 27, 422, 1008
Beardworms, 520, 560, 564
Bedrock, 1154
Bees, 573, 576, 779, 1004
    eusocial vs. subsocial, 1061
    social behavior in, 1059–1060
Beetles, 14, 567, 573, 574, 965
Behavior
    and animal evolution, **1052–1079**
    avoidance, in *Paramecium*, 475–476
    of bowerbirds, 1073
    causation and, 1053–1054
    circuits, in invertebrates, 895–898
    conflicts of interest, 1071–1074
    and DNA, 1052–1053
    dominance hierarchies, **1062–1065**
    fixed action patterns, **1054–1055**
    genetic basis of, **1052–1053**
    and growth, 630
    in honey bees, 1060–1061
    human (*see* Human behavior)
    and human evolution, 1050
    in insect societies, 1059–1061
    kin selection, **1065–1070**
    learning, 893–898, **1055–1058**
    outbreeding, 985
    patterns of, 1052
    and predation, 1109, 1116–1117
    programmed, **585**, 586
    in protists, 474–476, 477, 993
    reciprocal altruism, **1074–1075**
    rituals, 1016
    and selfish gene, **1020–1074**
    and sex hormones, 906, 908–909
    social, 1058–1059
    stages of socialization, **1059–1060**
    stereotyped, **893–898**
    and temperature regulation, 777, 781, 789–790
    territoriality, **1063–1065**
    and vertebrate societies, 1062–1065
Bell, Graham, 986–987
Belt desmosomes, 703
Belt, Thomas, 1122
Beltian bodies, 1122
Belts (air currents), 1137
Beltsville group, 685–686
"Bends," 733, 744

Benthic life, 1156
Beri-beri, 729
"Best man" hypothesis, 986–987
Beta-carotene, 211, 461, 625, 871
Beta chains
    for Class II MHC antigen, 805
    for T-cell receptor, 804
Beta-endorphins, 854, 855
Beta-fructose, 63
Beta-galactosidase, 321–323, 325, 351
    and lactose, 321–322
    in prokaryotic metabolism, 321–322
    rate of synthesis, 322
Beta globin, 359, 367–368
    and sickle cell anemia, 390, 394, 397
Beta-glucose, 60–61, **63**, 65
Beta pleated sheet, 74–75
Beta pseudogene, 368
Beta tubulin, 78, 110
Beta waves, 890
B horizon, 663
Bicarbonate (HCO_3^-) ion, 743, 748, 766
Biceps, 708
Bicuspids, 716–717
Bidirectional replication of DNA, 295, 320
    in *E. coli*, 326
    in eukaryotes, 358
    vs. rolling-circle replication, 327
Biennials, **621–622**, **623**, 624, 681, 694
"Big bang" reproducers, 1099, 1100
"Big bang" theory, 23, 307
Bilaterality, 587
Bilayer, cell membrane, 105
Bile, 716, 722–723, 730
Bilirubin, 726
Binary fission, **437**, 450, 453, 468–469
Binding
    and allosteric interactions, 176
    of enzyme and substrate, 170
Binding site(s), 176–177
    allosteric, 176–177, 179
    antigen-antibody, 802, 818
    for CAP-cAMP complex, 325
    of F_1 unit for ATP and ADP, 199
    for messenger RNA, 311
    of regulatory proteins, 360
Binomial system, for naming organisms, 35, 409
Biochemical pathways, and evolution, 969
Biofeedback, 845
Biogenic amines, 853
Biogeochemical cycles, 1153
    agricultural ecosystems and hunger, 1148–1149
    and Chernobyl reactor, 1151–1152
    concentration of elements in, **1150–1152**
    DDT in, 1150–1151
    in ecosystems, **1145–1152**
    nitrogen cycle, **1146–1149**
    phosphorus cycle, **1146**
    recycling in forest ecologies, 1149–1150
Biogeography, **966**, **967**, **973**, **1123–1124**
Biological clocks, 689–691, 695, 833, **834**
Biological determinism, 1075
Biological roles, of male and female, 235
Biology
    and evolution, **961**
    history of, **1–14**
    of populations, **960–973**
    universality of principles of, 263
Bioluminescence, 460, 531, 580
    ATP and, 352–353
    in mushrooms, 182
    in tobacco plant, 352–353
Biomass, 1139–1141
    and energy transfer, 1143
    and productivity, 1139–1141
    pyramids for, 1145
Biomes, **1161–1163**, 1177
Biosonar, **1115–1116**
Biosphere, 12, **37–38**, 85, 161, 1087, **1154–1177**
    and acid rain, **50–51**
    biomes, **1161–1163**
    coniferous forests, **1166–1168**
    and conservation biology, 1114, **1126–1127**
    deserts, **1172–1173**
    destruction of tropical forests, 1150, 1176–1177
    and human population growth, 1092, 1093
    and increased CO_2, **226–227**

I-5

Biosphere *(Continued)*
  lakes, **1155**
  land, **1160–1177**
  Mediterranean scrub, **1171**
  oceans, **1155–1157**
  origin of, 101
  and ozone layer, **1134–1135**
  Pacific Northwest forests, **1166–1168**
  and plants, 515
  ponds, 1155
  rivers and streams, **1154–1155**
  savannas, **1170**
  seashore, **1157–1160**
  taiga, **1166**
  temperate forests, **1164–1166**
  temperate grasslands, **1169–1170**
  temperature and greenhouse effect, 1177
  tropical forests, **1174–1177**
  tropical grasslands, **1170**
  tropical rain forests, **1174–1176**
  tundra, **1168–1169**
Biosynthesis, 183–184
  of DNA nucleotides, 298
  of glucose, 337
  of hemoglobin, 359
  of histones, 356
  of lactose, 321–322
  precursor compounds, 203
  of proteins, 322, 325, 337
  of transport compounds, 203, 205
  of tryptophan, 321–323, 325
Biotechnology, **350–352**
Biotic environment, 1131, 1152
Biotin, 303, 729
Bipedalism, 597, 907, 1039, 1044
Bipolar cells, 869, 870, 871, 936
Birds, 496, 587, **598–600**, 604, **753**
  adaptive radiation in, 1022
  care of offspring in, 900, **1033**, **1100**
  cerebral cortex of, 885
  circulatory system of, 763
  cloaca of, 904
  clutch size, **995**, **1099**
  courtship rituals, **872**, **1073**
  display behavior in, 1063–1064
  egg of, 900–901
    and DDT, 1151
  flight in, and brain, 883
  "following response" in, 1056
  gaping behavior in, 1051–1052, 1054
  homeothermy, **781–783**, 786
  hypertonic urine in, 766, 775
  learning in, 1056–1057
  lungs of, 737
  migration in, 1168
  olfaction in, 1033
  penes, 904
  plumage, and courtship, 872, **1000–1001**
  as pollinators, 514
  predation in, 992
  vs. reptiles, 598
  scavenging in, 1142
  sexual selection in, **999–1001**
  song learning, 1020, **1056–1057**
  songs, and territoriality, 1078
  territoriality in, **1063–1065**
  tropical rain forests, 1175
  vision in, 870
Birth
  and "diving reflex," 744
  human, **950–951**, 953
Birth control, 1097 *(see also* Contraceptive techniques)
Birth defects, 386, 947
Birth rate, 1088
  of human population, 1092–1093, **1096–1097**, 1098
Birth weight, and survival curves, **949**
Biston betularia (see Peppered moth)
Bivalves 545, 546, **549–550**, 563 *(see also* Mussels; Oysters; Scallops)
  vs. brachiopods, 562
  reproduction in, 550
Blackman, F. F., 215
Black bread mold, 481, 482, 483
Bladder, 770–771, 903, 908, 909, 917
Blades
  in algae, 465
  evolution of, 1047
  in grass plants, 642, 628
Blagden, Charles, 777, 785, 788
Blastocoel, 545, 952
  in amphibian, 927, 928
  in chicken, 932
  in human embryo, 940
  in sea urchin, **921**, **922**, 923, 924
Blastocyst, **940**, 952
Blastodisc, 931, 932, 934
Blastomeres, 921, 926, 927
Blastopore, 544, 545, 562, 563, 587, 952
  of amphibian, 928, 930
  dorsal lip of, **930–931**
  vs. primitive streak, 932
  of sea urchin, **923**, 924
Blastula, 526, 544, 545, 952
  in amphibian, 927–930, 932
  in *Drosophila*, 924–925
  in human embryo, 940
  in sea urchin, 921, **922**, 923, 932
Blaustein, Andrew, 1076
Blending inheritance, **237–238**, 247
Blind spot, 869
Blood, 52, 547, 558, 737, 746
  and AIDS virus, 819
  and animal biology, 700–701
  calcium in, 824
  cell, 104 *(see also* Red blood cells; White blood cells)
  circulation of, **749–759**
  clotting, 216, 392–393, 751–752, 760, 792, 838 *(see also* Hemophilia)
  composition of, 749, **750–752**, 765
  flow, **759–760**
    and water potential, 129
  gas exchange in, 737, 741–743
  groups *(see* Blood groups)
  and immune response, 792
  inflammatory response, 791–792
  vs. lymphatic system, 749, 750–752, 763
  osmotic potential of, 762, 764
  pressure *(see* Blood pressure)
  and respiration, 740
  temperature regulation in, 782, 785
  tonicity of, 132
  transfusions, 810–812
  types *(see* Blood groups)
  vessels *(see* Blood vessels)
  volume *(see* Blood volume)
Blood groups, genetics of, 268, **810–812**
  determination of, 810–811, 819
  evolution of, **996**
Bloodletting, 558
Blood pressure, 712, **759–762**, 764
  and atherosclerosis, 761
  and cardiac peptide, 752, **759**, **762**
  and circadian rhythms, 834
  diastolic, 759
  hormonal regulation of, 759
  hypertension, 761
  and kidney function, 775–776
  measurement of, 754
  reflex, 762
  regulation of, 756, **759**, **762**, 828, 832
  serum, 752
  systolic, 759
Bloodstream *(see also* Circulation)
  in digestion, 730
  and infection, 819
  and lymphatic system, 749, 750–752, 763
  and respiration, 738, 747
Blood vessels, 130, 547, 556, 744, 705, **753–754**, 763–764 *(see also specific vessels)*
  development of, 952
  diseases of, 760
  in homeostasis, 786–788, 790
  hormone receptors in, 756
  in lymph node, 794
  parietal, 556
  and prostaglandins, 835
  and regulation of blood pressure, **759–762**
  structure of, **753**
Blood volume
  feedback control of, 712
  and heart disease, 761
  and kidney function, 756, 775–776
  of mother and fetus, 944
  and water balance, 824
"Blooms" of algae, 440, 1095
Blubber, 779
Bluebirds, 1114
Blue-green algae *(see* Cyanobacteria)
Blue light
  and phototropism in plants, 673
  and stomatal opening, 654
Blushing, 760
B-lymphocyte(s), **795–801**, 802, 803, 818–819
  and AIDS virus, 815
  clonal selection of, **799–800**
  differentiation of, **795**, 800
  and IgD antibodies, 799
  immune system, **793**
  life history, 795–797
  and T lymphocytes, 804, 808, 809
B-lymphocyte precursors, 818
Body, 94, 746, 859
  animal, 710–715
  and climate, 785
  compartments of, **768–769**
  of earthworms, 734, 737
  fluids, regulation of, **775–776**
  human, composition of, 94
  of neuron, 708
  orientation of, 859
  of plants, **625–648**
  shaping of *(see* Morphogenesis)
  size, 779, **995**
    and brain capacity, 1045
    natural selection for, 1020–1021
  temperature of, **778**, **784**
    and circadian rhythms, **834**
    and endocrine system, 828
  of vertebrates, **702–709**
  water regulation in, 775–776
Body lice, 437
Body odors, 906
Body plan, 520, 542
  of annelids, 554
  of arthropods, **565–568**
  of chick embryo, 932
  of cnidarians, 525
  of earthworm, 555
  of echinoderms, 604
  of flowering plants, **625–626**
  of horseshoe crab, 569
  of mollusks, 546, 554, 563
  of pseudocoelomates, 540–541
  of sea urchins and sand dollars, 589
  of sponges, 523
  of starfish, 588
  two-layered, 542
Bohr, Niels, 28
Bollworms, 1017
Bolting and flowering, 681, 694–695
Bond(s), chemical, **30**, 38, 57
  amino acid–transfer RNA, 312
  and bioactive molecules, 35
  and charge, 31, 38
  covalent, **30–34**, 38, 181
  disulfide bridge, 75
  double, 33–34, 38
  and electrons, **31–33**
  hydrogen, 33, 41, 288–290, 294, 296, 308, 311, 312, 334, 420
  ionic, 30–31
  peptide, 74–75, 175, 313
  phosphate, 186, 189, 190, 204, 205, 298–299
  polar-covalent, 32
  single, 32, 34
  strength of, 57
  triple, 33–34, 38
Bonding
  infant-mother, 951, 1037
    evolution of, 1033
  temporary, 1036–1037
Bone(s)
  cells, 704 *(see also* Osteoblasts; Osteocytes)
  chips, 707
  compact, 705
  density, regulation of, **830–831**
  in embryonic development, 946–947, **952**
  formation of, and mineral intake, 728
  growth, 824, 825
  hormonal regulation in, 829
  marrow *(see* Bone marrow)

muscles, 706 (see also Skeletal muscle)
number of, 705
periosteum, 705
spongy, 705
tendons in, 706
and vitamin D, 729
Bone marrow, 830
  cells, 147
  formation of white blood cells in, 751, 792, 801, 818
  and immune system, 792, 793, 801, 818
  megakaryocytes in, 751
  and T lymphocytes, 801
Bonner, James, 685–686
Bony fish, 767
Booby, blue-footed, 27
Boron, 657
"Bottle-baby disease," 727
Botulism, 878
Bowerbirds, 1063, **1073**, 1078
Bowman's capsule, **770**, 771–773, 776
Brachiation, 1032, 1035
Brachiopods (see Lamp shells)
Brachiostoma, 735
Brachydactylism, 979
Brain, 568, 578, 593, 601, 604, 843, 844
  in apes, 1035
  in archaic *Homo sapiens*, 1046
  in arthropods, 842
  in australopithecines, 1039, 1040
  blood supply to, 759
  and cardiac regulation, 762
  of cephalopods, 553
  cerebral cortex, **885–893**
  circuits (see Brain circuits)
  cochlear nucleus of, 864
  and control of respiration, 745, 747
  damage, 890
  development of
    in amphibian, 928
    in chick embryo, 935, 936
    in human embryo, 946–948
  effect of AIDS on, 816
  electrical activity of, **890–891**
  and endocrine system, **826**
  evolution of, 531, 533
    of brain capacity, 1039, 1042, 1045, 1046, 1051
    in vertebrates, **881–884**
  and glucocorticoids, 830
  and glucose metabolism, 833
  of *Homo erectus*, 1044
  of *Homo habilis*, 1042
  of *Homo neanderthalensis*, 1046
  hormone receptors in, 756
  innate releasing mechanisms in, 1054, 1078
  integration and control in, **881–899**
  and learning, **893–898**, 1054–1055, 1078
  and memory, **893–898**
  and nervous system, 708–709
  of octopus, 552
  of onychophorans, 561
  organization of, 892–893, 909
  and regulation of heartbeat, 764
  sexual differentiation in, **908–909**
  structure of, **881–884**
  in vertebrates, 701, 712, 713, 842
  waves (see Alpha waves; Beta waves; Brain waves)
Brain case, 601
Brain circuits, **884–885**, 898
  limbic system, **885**
  and memory, 899
  reticular activating system, **885**
Brain hormone, 578
Brainstem, 844, **882–883**, 885, 898 (see also Cerebellum; Medulla; Pons)
  and opiates, 854
  and respiration, 745, 748
Brain waves, **890–891**, 948
Branches, 630, 640–641, 648
Branching patterns, 415–416 (see also Cladistics)
Branching points, 421 (see also Cladistics)
Branching sequences, 427
Branchiostoma, 520, **591–592**, 604
Branch roots, 634, 635, 648, 695
Breast
  and anabolic steroids, 907
  cancer of, 739

and female sex characteristics, 915
Breastbone, 599 (see also Keel)
Breast-feeding, 1096
Breathing, 747, 950 (see also Respiration)
Breathing tubes, 566 [(see also Trachea(e)]
Breeding
  of dogs, 976
  of *Drosophila*, 976–977
  and natural selection, 991
Breeding experiments, 1053 (see also Artificial selection)
Breeding sites, 1016
Breeding territories, 1063
"Breeding true," in genetics, 238–239, 242, 266
Brightness, visual, in microscopy, 99
Brinster, Ralph, 378
Bristle(s), 555 [see also Seta(e)]
  in *Drosophila*, **976–977**
  in honey bee, 1060
Bristlecone pine, 646
British soldier lichen, 490
Brittle stars, 587, **589**, 604
Broca's area, 888
Bronchi, 738, 739, 740, 747
Bronchioles, 813, 835
  in human respiratory system, 738, 740, 747
Bronowski, Jacob, 16
Brood chamber, 563
"Brood pouches," 550
Brower, Jane, 1004
Brown algae, 455, **465**, 477, 1157
Brown bear, 1008
Browsing, 1110, 1121, 1165, 1166
  in evolution of the horse, 1025–1028
  vs. grazing, and dentition, 968
Bruchid weevils, 1114
Brush border, 774
Bryophytes, 496, **498–500**, 516
  and air pollution, 498
  Anthocerotae, 498
  asexual reproduction in, 500
  classification of, 497
  evolution of, 516
  Hepaticae, 498
  life cycle, 499–500
  Musci, 498
  photosynthesis in, 498
  reproduction, 499–500, 516
  vs. vascular plants, 498, 499
Bryozoans, 520, 541, 562, 563, 564
Buccal cavity, 471, 475
Buccal glands, 717–718
Büchner, Eduard, 11
Büchner, Hans, 11
Budding, in asexual reproduction, 259, 437, 445–446, 450, 486, 528, 529
Buds, 695
  and abscisic acid, 680
  axillary, 641–642, 648
  of flowers, 614
  and meristematic tissue, 648
  and shoots, 640, 648
  terminal, 641
Budworms, 1017
Buffers, **52**, 54
Buffon, Georges-Louis Leclerc de, 2
Bulb, penile, 830, 905, 917
Bulbourethral glands, 904, 905, 917
Bulk flow, of fluids, 128, **129**, 142, 747
  vs. diffusion in respiration, 734
  and respiration in vertebrate lung, 736
  and translocation, 662
  and water transport in plants, 670
Bulk transport, 133
Bull's-horn acacia, 1122
Bundle of His, **756–757**
Bundle-sheath cells, 224–226
Burgess Shale, 520, 521
Burials, 1046, 1051
Burnet, Macfarlane, 800
Burns, and infection, 810
Burrowing, 555, 557
Burrows, 567, 571, 1064, 1116
Bush, Guy, 1014–1015
Bush babies, 1034, 1051
Bushes, 1164
Butler, Samuel, 1070
Butterflies, 573, 577

Cabbage, 575
Cacti, 414, 1173
  adaptation in, 629
  vs. euphorbs, 1007
Cactus moth, 1117–1118
Caffeine, 515, 853
Calcarea, 523
Calciferol, 729
Calcitonin, 369, 824, 829, 830, 839
Calcitonin-gene-related protein, 369
Calcium, 728, 944
Calcium carbonate, 523, 529
  in algae, 465
  in algal cell walls, 457
  in sarcodine shells, 470
Calcium ion (Ca^{2+}), **31**, 695, 834
  absorption, in intestine, 831
  and auxin-cytokinin interactions, 677
  and biogeochemical cycles, 1145
  in blood, 829, 830
  in bone, 829, **830–831**
  and fertilization of sea urchin ovum, 921
  and gravitropism, 682–683
  hormonal regulation of, 824, 829
  in human diet, 831
  in learning, 898
  in muscle contraction, 873, 876, 877, 880
  in phototropism, 695
  in plant cell growth, **675**
  plant requirements for, 657
  in soil, 658
Calcium phosphate, 716
California condor, 423, 1104
California poppy, 425
California vole, 1114
Calling, 583
Callose, **638**
Callus, 353, 677, 678
Calmodulin, 675
Calories, 43, 68, 1139, 1143
  vs. bulk in nutrition, 726
  and ecosystem productivity, 1139, 1143
  energy cost of, 1139, 1143
  nutritional, 43, 68, 76
  in Western diets, 730
Calorimeter, 59
Calvin, Melvin, 222
Calvin cycle, **222–223**, 224–226, 228–230
  lack of, in halobacteria, 430, 440
  net energy gain from, 223
  and the three-carbon pathway, 222–223
Calyptra, 500
Calyx, **614**
Cambial cells, **648**
Cambium (see Cork Cambium; Procambium; Vascular cambium)
Cambrian period, 497, 519, 520, 548
Camel, 789, 790
Camembert, 488
Camera eye, 866
Camouflage, 826, 959, 972
  in corixid bugs, 998
  as frequency-dependent selection, 998
  in moths, 993
  in polar bears, 961, 1108
  and predation, 1116–1117, 1121
cAMP (see cyclic AMP)
CAM photosynthesis, 1173
CAM plants, 655 (see also Crassulacean acid metabolism)
Canary, 889
Cancer
  of breast, 739
  carcinogens, 376
  cell division in, 147–148
  and chromosome deletion, 387
  of colon, and diet, 730
  genes for, 375–377
  genetic origins of, 381
  genetic regulation and, 375–377
  and growth-promoting proteins, 373–374, 376
  and immune response, 808–809
  and interferons, 793
  and lymph nodes, 794
  of lung, 739
  and monoclonal antibodies, 802
  mutagens, 376
  mutations and, 376, 381

I-7

Cancer *(Continued)*
 vs. normal cells, 808
 oncogenes, 376–377, 381
 promoter sequences, 377, 378, 379
 and proteins, 375–377
 and radioisotopes, 25
 recombinant DNA and, 342, 348, 376–379
 and regulatory proteins, 360, 373
 research, 378–379, 537, 798–789
 and retroviruses, 376–377, 445
 and reverse transcription, 376
 Rous sarcoma virus, 376
 and sickle cell anemia, 78
 of stomach, and blood groups, 996
 viral vectors for, 375–377, 381
 and Wilms' tumor, 386–387
Cancer research
 phagocytosis and, 378
 promoter sequences, 377, 378, 379
 recombinant DNA in, 376–379
 SV40 in, 377
 thallium in, 25
 transposons in, 377
 viruses in, 377
Candelabra model of human evolution, 1047
Candelas, 215
Canines (teeth), 601, **716–717**
Cann, Rebecca, 1049–1050
Cannabis sativa, 575
Canopy, forest, 407, 1164, 1174
CAP, 325, 337
 in CAP-cAMP complex, 325, 337
 in repression, 325
Capillaries, 114, 733, 738, **753**, 761, 818
 and agglutination, 811
 alveolar, 740, 742
 and the bends, 733, 738
 blood flow at, 763–764
 blood pressure, 759
 diffusion at, **753–754**
 exchange of substances at, 749
 gas exchange at, **753–754**
 glomerular, 771
 and immune response, 792
 and lymph system, 762–763, 795
 peritubular, 776
 and portal systems, **759**
 vs. red blood cells, **750**
 and sickle cell anemia, 78
 of skin, 710
 of small intestine, 721
 structure of, 753
 and temperature regulation, 786
 of tissue, and respiration, 742
 in vertebrates, 701
 and water gain, 764
Capillary action, 42, 54, 650
Capillary beds, **759**
Capsid, viral, 330, 338, 342, 346, 373, 443–445, 450–451
Capsule, in pathogenic microbes, 433
Capuchin monkeys, 1035
Carapace, 569, 571, 596
Carbohydrates, 55, **59–70**, 1138
 in algal food storage, 457, 465
 catabolism of, **186–205**
 in diet, and disease, 730
 in food chain, 1138
 human requirements for, 726–728
 metabolism of, 824
 processing, in liver, 725
 and the products of photosynthesis, 228
 transport of, in algae, 465
Carbon, 747
 amount of, in air, 733
 atomic structure of, 28
 in biogeochemical cycles, 1145
 catabolism of, **186–205**
 covalent bonding of, 32, 58
 double bonding of, 33
 elimination of, 737
 fixation of, 216, **222–228**, 229, 230
 four-carbon pathway in photosynthesis, **223–228**
 in glycolysis, 187–188
 importance for living systems, 32, 35, 58
 and origin of life, 87, 89
 photosynthesis, 576
 and respiration, 747

 role in organic molecules, **55–82**
 vs. silicon, 58
 three-carbon pathway in photosynthesis, **222–223**
 transport of in blood, **743**
Carbon cycle, **226**, 227
Carbon dioxide, 33, 504, 628, 1177
 in anaerobic respiration, 439
 in blood, 743, 749
  regulation of, 745
 in cellular respiration, 191
 in cellular waste products, 766
 concentration
  in atmosphere, 226
  gradient, and the four-carbon pathway, 225
 and diffusion, 130
 double bonding in, 33
 in glycolysis, 186
 and greenhouse effect, 1177
 and hemoglobin affinity, 743
 nonpolarity of, 33
 partial pressure of, 746
 in photosynthesis, 206, 209, 223–224, 226, 230, 504
 in plants, 650, 669
 in plasma, 748
 and rate limitation in photosynthesis, 215
 receptors for, 745
 reduction of, in photosynthesis, 219, 222, 229
 and respiration, 732–734, 740, 743, 745, 747–748
 symmetry of, 33
Carbon dioxide "blanket," and the greenhouse effect, 226
Carbon-fixing reactions
 in chemosynthetic ecosystems, 1140
 in photosynthesis, 216, 219, **222–228**, 229
 in prokaryotes, 450
Carbonic acid, 743
Carbonic anhydrase, **743**
Carboniferous period, 497, 503, 504, 597
Carbon skeleton, and glyceraldehyde phosphate, 228
Carbonic acid, 169
 and pH buffers, 52
Carbonyl group(s), and peptide bonds, 74
Carboxyl group(s), 48, 56–57, 313
Carboxypeptidase, 171, 722
Carcinogens, and mutagens, 376
Cardiac function, 769
Cardiac glycosides, 575, 1003–1004
Cardiac muscle, 823, 842, 851, 856, 873
 and autonomic nervous system, 706, 708, 712, 842, 851, 856, 878
 circulation to, 758
 contraction of, 875, 877–878
 stimulation of, 756
Cardiac output, **756**
Cardiac peptide, 756, 759, 775–776
Cardiovascular disease, 760
 and anabolic steroids, 902
 and contraceptive drugs, 916
Cardiovascular regulating center, **762**, 764
Cardiovascular system, **752–758**, 763–764
 circuits of, **758–759**
 closed, 753, 763
 open, 753
 pulmonary, **758–759**
 systemic, **758–759**
Care of offspring, 110–1101
 in birds, 1063
 natural selection for, 1072
 in primates, 1033–1034
Caribou, 1151, 1168, 1169
Carnivora, 412
Carnivores, 526, 529, 533, 550, 552, 561, 568, 573, 580, 595, 597, 599, 600, 602, 1031, 1153, 1170
 dentition of, 717
 and energy flow, 1141–1142
 in food chain, 1143
 hominids as, 1044
 plants *(see* Carnivorous plants)
 as secondary consumers, 1087
Carnivorous plants, **666–667**
 movement in, 692–693
Carotenoids, 211, 212, 213, 228, 428, 729, 871
 in algae, 456, 457, 465
 in cyanobacteria, 428, 440–442
 in photosynthetic bacteria, 440–442
 in thermophilic bacteria, 428
Carotid arteries

 blood pressure regulation in, 762
 interoceptors in, 774, 859
 and regulation of respiration, 745
Carotid body, 747
Carpels, 511, 513, 613–615, 617, 619–621, 624
Carpellate flowers, 614
Carrier(s), of genetic disorders
 of Down's syndrome, 385–386
 of hemophilia, 382, 392–393
Carrier-assisted transport, **134–135**, 142
 and organelles, 135
 polarity and, 134
 and structure of proteins, 135, 142
Carrier proteins, 135, 142
 permeases, 178–179
Carrying capacity, **1089–1091**, 1104–1105, 1116–1119
Cartilage, 553, 704
 in chick embryo, 935, 952
 in human embryo, 947
 in knees, 707
 in trachea, 740
Cartilaginous fish, 767
Cartmill, Matt, 1044
Casparian strip, **632**, 633, 648
Caste systems, 1062
Castration, 906, 908–909
Cats, 267, 273, **412–413**
Catabolic pathways, 203, **204**
 in cell, 192
 and Krebs cycle, 203
Catabolism, **168**, 182–183
 acetyl CoA and, 203
 of carbohydrates, **186–205**
 of ethanol, 191, 202
 of glycogen, 68
 pathways, 203-**204**
 of phenylalanine, 388
 of polysaccharides, 203
Catabolite activator protein *(see* CAP)
Catalysis
 in glycolysis, 187
 in synthesis of ATP, 199
Catalyst(s), **169**, 170
 in biosynthesis of ATP, 199
 feedback inhibition, 177
 regulation of catalysis, 175–178
Catalytic intron, 370
Catarrhine anthropoids, 1034
Catastrophism, 3–4
Catbird, 423
Catecholamines, 824, 832, 836
Caterpillars, 576, 577
Cattails, 511
Causation, in animal behavior, **1053–1054**, 1077
Cave art, 1047, **1048–1049**
Cave dwellers, 1045, 1046, 1047, 1048–1049, 1051
Cavy *(see* Patagonian hare)
cDNA, **342**, 343, 346–347, 353
Cech, T. C., 370
Cecropia moth, 585
Cecum, 724
Celiac ganglion, 846
Cell(s), 533 *(see also specific cells and tissues)*
 anabolism in, 192
 catabolism in, 192
 compartments, 127, 129, 130, 192–193
 composition, 106
 continuity between, 10
 cycle, 146–148, 154–155, 356
 daughter, 144, 150–155, 328
 differentiation of, 359, 902, 928–929
 diffusion in, 142
 diploid, 250
 discovery of, 10
 functional integrity of, 142
 glycolysis in, **186–205**
 growth, 147–148
 haploid, 250–251, 261
 inhibition of, 147–148
 junctions, **140–142**
 metabolism, 103, 110, 125, 127, 128, 283
 motility, 110, **111**, 114, 121–122
 movement in, 121–122, **127–143**
 observation of, 99
 organization, 36, 102–126
 origin of, 88–89, 90, 149, 150
 and origin of viruses, 444

I-8

pH and, 127
properties of, 36–38, 88, 104
shape, **103, 112–113**, 114, 149–150, 938
  of bacteria, **436–437**
  and morphogenesis, 938
size, 101–103, 124, 479
skeleton, 704
specialization of, 36, 479
structure, 91–92, 103, 104, 110–121, 142
surface/volume ratio of, 102, 103, 124, 142, 479
theory of, 90, 100
transformation, 376
versatility of, 192
water and, 96, 128, 142
Cell body, of neuron, 708, 713, 842
Cell-cell communication, 140, 142, 143
  hormones in, 140–141
  in immune response, 808
  lipids in, 67
  and memory and learning, 898
  in plant tissues, 141–143
Cell-cell junctions, **140–142**
Cell cycle, 146–148, 154–155, 294, 306, 356
  G_1 (gap) phase, **146–148**, 150
  G_2 (gap) phase, **146–148**, 150
  regulation of, 146, 148, 155
  R point, 148
  in single-celled organisms, 147
  S (synthesis) phase, **146**, 155, 306
Cell division, 103, 106, 109–110, 111, 112–113, 121, 123, 155, 297
  in algae, 457, 459, 460, 461, 462–464, 465
  in animal cell, 154, 155
  in bacterium, 145
  in cancer, 147–148
  and cell membrane, 145, 152, 154
  and cell wall, 154, 149
  cellulose in, 154
  and chloroplasts, 146
  in ciliates, 471
  and circadian rhythms, 689
  in colonial organisms, 461–462
  and cytokinins, **679**, 694–695
  cytoplasm in, 155
  cytoskeleton in, 149–150
  in *Drosophila*, 146
  endoplasmic reticulum in, 146
  energy in, 112–113, 119, **127–143**
  energy transformations in, 161, 183–184
  environment and, 147
  enzymes in, 146, 153
  in eukaryotes, 144, **145**, 155
  after fertilization, 920–921, 931, 950
  Golgi complex in, 146
  histones in, 148
  in human egg, 940
  lysosomes in, 146
  in meiosis, 251–255
  metabolism of **167–180**
  microtubules in, 146, 148–150
  mitochondrion in, 146
  and morphogenesis, 938
  nuclear envelope in, 150, 151, 152
  nucleolus in, 150, 152
  nucleus in, 145, 150–155
  in oocyte, 910
  organelles in, 144–146, 155
  pH and, 147
  planes of cell division, 456
  in plants, 494
  plastids in, 146
  polysaccharides in, 153–154
  prokaryotic, **145**, 154, 428, 431, 437
  protein biosynthesis and, 78, 146
  in protists, 144
  in protozoa, 470
  rate of, 147
  receptors, 148
  ribosome, 146
  in sea urchin zygote, 921–922
  simple, 144
  in single-celled organisms, 144
  size of cell during, 146
  in spermatogonia, 902
  S (synthesis) phase of cell cycle and, **146**, 155
  temperature and, 146–147
  tubulin dimers in, 149, 152
  vacuoles and, 146

Cell elongation, 694
  and auxins, 680
  and cytokinins, 676
  and gibberellins, 680
  in plants, 671, **674–675**
Cell-mediated endocytosis, lipoproteins and, 139
Cell-mediated immune response, **801–808**, 809, 819, 945
Cell membrane(s), 102, 103, **104–106**, 111, **112–113**, 114, 118, 124, 127–129, 135–140, 673, 695
  apertures in, 130, 135
  attachment site, 335
  and cell division, 145, 152, 154
  and cell transport, **127–143**
  and cell variety, 125
  and circadian rhythms, 691
  cytoplasmic face of, 105, 124
  and endocytosis, 138–139, 143
  and energy reactions, 168
  ethylene production and, 679
  evolution of, 370, 969
  and exocytosis, 143
  exterior face of, 105, 124
  and hormones, **837–838**
  interior surface of, 135, 138–139
  internal, 213
  lipid barrier in, 134–136
  phospholipids in, 69, 104–105
  and photosynthesis, 206, 213, 228
  and phytochromes, 687
  and potassium (K^+) ion transport in plants, 655
  prokaryotic, **431–433**, 441, 450
  of protozoans, 471
  receptor sites in, 131
  regulation of transport, **127–143**, 180
  of root cells, and absorption, 632–633
  of sarcodines, 470
  of sea urchin ovum, 921
  selective permeability of, 127–128, 132–133, 142
  of sperm cell, 902, 903
  and sporulation, 437–438
  of stomata, 653
  structure and composition of, 104, 105–106
  and tight junctions, 703
  "totipotent," 359
  transport across, **127–142**, 143
  transport proteins in, 184
  vacuoles and, 139
  vesicles and, 138–140
Cell plate, 153–154
Cell theory, 90, 100, 112
Cellular metabolism, 784, 789
Cellular movement, 121–122, **127–143**
Cellular respiration, 161, 167–168, **186–205**, 711
  and acetyl groups, 204
  aerobic, 190
  and altitude, 746
  anaerobic, 190, **191**, 192–194
  and ATP synthetase, 192, 205
  carbon dioxide and, 191
  coenzymes in, 191–192
  and cytoplasm, 193
  efficiency of, 201
  energy levels in, 186
  enzymes in, 187–188
  in eukaryotes, 191, 204
  evolution of, 493
  free-energy change in, 187
  and mitochondrion, 167, 191–193, 197–201
  phosphates in, 191–192
  selective permeability and, 191, 201
  and thyroxine, 828
  water in, 191
Cellular slime molds, 455, 466, 467
  cyclic AMP in, 837, 838
Cellular transport, **127–143**, 130, 154
  active, 135, **136**, 143
  blood vessels and, 130
  bulk transport, 133
  carrier-assisted transport, **134–135**, 142
  cell-cell junctions, **140–142**
  chloride ions (Cl^-) in, 134
  cotransport, 136
  countercurrent exchange, 124, 130
  diffusion, 128, **129–135**, 136, 142
  endocytosis, **138–140**, 142–143
  energy and, 135–136
  extracellular, 124, 154

of hydrophilic molecules, 134, 136–137
of hydrophobic molecules, 134, 136–137
at mitochondrion, 136, 205
phagocytosis, **138–140**, 143
pinocytosis, **138–140**, 143
and potassium (K^+) ions, 136
potential energy in, 135–136, 180
proteins in, 106, 124, 135–137
rate of, 178–179
receptor-mediated, 137, **138–140**, 143
regulation of, 142–143, 180
of sodium, 134, 136
sodium-potassium pump, 136–137, 143
temperature and, 175–176, 178
and thermodynamics, 180
and vacuoles, **138–140**
vesicle-mediated, **138–140**
Cellulase, 352, 679
Cellulose, 59, **65**, 66, 228, 352, 426, 592
  in algae, 456, 457, 465
  in cell wall, **106–107**, 113, 124–125, 154, 456, 457, 465
  in *Coleochaete*, 494
  digestion of, 438, 469, 550
    in herbivores, 724
  and lignins, 622, 637
  microfibrils, 113, 653
  and plant cell structure, 78
  structure of, 66
  tubes (*see* Infection threads)
  of water molds, 469
Cell wall, 90, 92, 103, 106–107, **112**, 118, 124–125, 140, 168, 178, 256, 426, 578
  algae, 456–462, 465
  bacterial, and penicillin, 178
  calcium (Ca^{2+}) ion and, 677
  in cell division, 149, 154
  cellulose, **106–107**, 113, 124–125, 154
  of cyanobacteria, 441–442
  growth and, 674–675
  of *Heliobacterium chlorum*, 442
  of oomycetes, 467
  pressure, 134
  primary, 106–107
    in collenchyma cells, 636–637
    in tracheids, 639
  in prokaryotes, 107, **432–433**, 450
  regulation of transport in, 633
  of roots, 632
  secondary, 106–107
    in sclereids, 637
    in tracheids, 639
    in vessel members, 639
  silicon in, 459–460
  structure, 106
  synthesis, 118
  and turgor, **134**, 142
Cenozoic era, 496, 513
Centimeter, 95
Centipedes, 573, 586
"Central dogma," of DNA biology, **306**, 342
Central nervous system, 70l, **708–709**, 713, 842, **843–844**, 856
  and AIDS, 816
  development of, in human embryo, 945
  dorsal, evolution of, 701
  and endorphins, 854
  in human beings, **792**
  receptors for opiates, 854
  and sensory perception, 859
  and water intoxication, 769
Central sulcus, 885–886
Centrifugation
  differential, 193
  preparative, 192
  purified fractions, 192
  supernatant, 192–193
  ultracentrifugation, 192, 292–293, 311
  zonal, 193
Centrioles, **123**, 125–126, 146, 149, 150–152, 154, 155, 921
  of algae, 457, 465
  and basal bodies, 146, 149
  in *Chlamydomonas*, 146
  in daughter cell, 146
  in flowering plants, 146, 150
  in fungi, 146, 150
  in meiosis, 254–255

Centrioles (Continued)
  in nematodes, 146, 150
  replication of, 152
  in spermatid differentiation, 902
  in spindle formation, 146
Centromere(s), 148-149, 150, 155, 275
  in anaphase of mitosis, 151
  and DNA replication, 359-360
  in karyotypes, 384
  in meiosis, 255, 261
  and simple-sequence DNA, 369, 380
Centromere region, in meiosis, 253-255
Century plant, 1172
Cepaea, 992-993
Cephalization, 532, 542, 842
Cephalochordates (see Lancelets)
Cephalopods, 545-547, **551-553**, 563 (see also Cuttlefish; Octopuses; Squids)
  courtship and mating in, 553
  evolution of, 551
  intelligence of, 551
  locomotion in, 546
Cephalothorax, 566, 568, 570, 571, 572, 585
Cereals, 1148
Cerebellum, **882-883**, 898
Cerebral cortex, 601, **884**, **885-893**, 898 (see also Association cortex; Auditory cortex; Motor cortex; Sensory cortex; Visual cortex)
  Alzheimer's in, **896-897**
  electric activity in, **890-891**
  evolution of, 883-884, 885
  of fetus, 948
  intrinsic processing centers in, **892-893**
  and language, **888-891**
  left brain, **888-891**
  limbic system and, 885
  lobes, **885**
  motor and sensory, **885-888**
  prefrontal, **894**
  right brain, **889-891**
  sulci, 885-886
Cerebral hemispheres, 898 (see also Cerebral cortex; Corpus callosum; Left brain; Right brain)
  dominance in, 888-889
  evolution of, 883, 884, **885**, 886
  and language, 888-891
  "split brain," 889-891
Cerebral hemorrhage, 760
Cerebrospinal fluid, 881-882
Cerebrum, 883, **884**, 886, 898, (see also Cerebral hemispheres)
Cervical cap, 916
Cervical dilation, 950
Cervical plug, 949
Cervix, **908**, 909, 913, 950
Cesium chloride (CsCl), 292-293
Cesium-137, 1152
Cestoda (see Tapeworms)
C_4 photosynthesis, **223-228**, 352, 678, 1173
  vs. C_3 photosynthesis, **226-227**
  evolution of, 226
  grasses and plants with, **223-228**, 230
  temperature range for, 227
C_4 plants, **223-228**, 229-230, 650
  transpiration in, vs. CAM plants, 655
CGRP, 369
Chaetognatha, 520 (see also Arrow worms)
Chalaza, 931
Chance
  and evolution, 973, 979, 982
  in Hardy-Weinberg equilibrium, 979, 990
  in meiotic recombination, 253
  and mutations, 982
Change, evolutionary, 981
  founder effect, **983**
  gene flow, **982**
  mutations, **981-982**
  nonrandom mating, 984
  population bottlenecks, **983**
Change in entropy (ΔS), **165**, 183
Change in heat content (ΔH), 183
  in energy reactions, **164-165**
Chaos chaos, 474
Chaparral, 1171
Character displacement, **1110**, 1111, 1129
Characteristics, inherited, 235, 306
  alternative combinations of, 235
  blending, 268

dominant, 239-243, 245
hereditary, 235, 238
in human beings, 271
novel combinations of, 242
recessive, 239-243, 245
skipping of generations, 237
Chargaff, Erwin, 286, 288, 289
Charge
  and amino acids, 304
  atomic, 23-24
  and chemical bonds, 31, 38
  and functional groups, 56
  and hydrophilic molecules, 46
  and molecules, 62
Charophyceae, 461, 462, 493, 516
Chase, Martha, 285, 286
Cheating, 1072-1074, 1075, 1079
Cheeses, 429-430, 481, 488
Cheetah, 412
Chelicera(e), 566, 568, 569, 570, 585 (see also Fangs; Pincers)
Chelicerates, 566, **568-571**, 585, 589
Chemical bonds
  and ATP, 181
  covalent bonds, 181
  and energy transformation, 165-169, 180-181
  as potential energy, 163-169
Chemical defenses, in plants, 1003-1004
Chemical energy, 161, 165-166
Chemical equations, **34-35**, 38, 37
Chemical evolution, 87-88, 142
  of ATP, 182
  of DNA, 362-370
  of enzymatic regulation, 176
  of NAD, 182
  of organic molecules, 142
Chemical properties
  and atomic bonding, 34
  and electrons, 24, 38
  of functional groups, 56-57
Chemical reactions, **34-35**, 38, 166, 167
  and cell theory, 90, 112
  combination, 34, 38
  condensation (see Condensation reaction)
  direction of, 34, 181
  dissociation, 34, 38
  and electrons, 24, 30
  exchange, 34-35, 38
  hydrolysis, 64
  in photosynthesis, **206-230**
  temperature and, 64
Chemical signals, 140
Chemiosmotic coupling, 197-200, 205
  cotransport systems in, 200
  at mitochondrial matrix, 197-199
  and photosynthesis, 218-220, 229
  proton gradient in, 197-198
  and pyruvic acid, 200
  selective permeability and, 197
Chemiosmotic power
  in flagella, 200, 432
  in living systems, 200
  and photosynthesis, 200, 203
  and stomatal movements, 653
Chemistry
  laws of, and biology, 35, 38, 10-11
Chemoreceptors, 535, 551, 552, 585, 588, 859, **860-862**, 879 (see also Smell; Taste)
  in carotid arteries, 745
  in mammals, 863
Chemosynthesis, 89-90, 94, 101
  in prokaryotes, 436, **439**, 450
Chemosynthetic ecosystems, **1140-1141**
Chemotaxis
  in amoebas, 474
  in *Paramecium*, 475, 476
Chernobyl reactor and ecology, **1151-1152**
Chestnut blight, 481, 484
Chiasma, 275
  in meiosis, 253-254
Chick (embryo)
  development of, 919, **931-940**
  differentiation of ectoderm in, **935-936**
  differentiation of endoderm in, **937**
  differentiation of mesoderm in, **936**
  extraembryonic membranes of, **932-934**
  vs. frog embryo, **832**
  morphogenesis in, **938-940**

organogenesis in, **935-937**
Chickens
  comb shape, 269
  learning in, 1055-1056
  pecking order in, 1062-1063
Chief cells, 720
Childbearing, 915
Childbirth, 828, 907, 1006
Chilopoda, 573, 586
Chimpanzees, 420, 1035, **1036-1037**, 1051
  vs. australopithecines, **1041**
  reciprocal altruism in, 1075
Chironomus, 108
Chitin, 66, 106, 125, 426, 480, 546, 557, 566, 570, 574, 734
  in arthropod exoskeletons, 480
  in fungi, 491
  in water mold cell walls, 469
Chitons, 546, 548
Chlamydomonas, 92-94, 104, 108, 121, 461, 462
  basal bodies in, 146
  centrioles in, 146
  life cycle of, 250-251, 259
  photosynthesis in, 213
Chloride ion (Cl^-), **31**
  in loop of Henle, 772-773
  nutritional requirements for, 728
  plant requirement for, 657
  and potassium (K^+) ion transport in plants, 653
  and touch response in plants, 692
  in transport, 134
Chlorine
  and biogeochemical cycles, 1145
  in ozone layer, 1134
Chlorofluorocarbons, **1134-1135**
Chlorophyceae, 461-462
Chlorophyll, 29, 31, 90, 92, 113, 120
  absorption spectrum of, 212
  in algae, 456-459, 461, 465
  chlorophyll a, 211-213, 216, 218, 220, 228, 457-459, 461, 465, 625
  chlorophyll b, 211-212, 228, 457-459, 461, 625
  chlorophyll c, 457, 459, 465
  in cyanobacteria, 431, 441-442
  and leaf senescence, 679
  magnesium in, 656, 670
  oxidation of, 216, 218
  in photosynthesis, 206, **211-213**, 216, 221, 228
  in photosynthetic bacteria, 441-442
Chlorophyll a, 211-213, 216, 218, 220, 228, 457-459, 461, 465, 625
  vs. bacteriochlorophylls, 440, 442
Chlorophyll b, 211-213, 228, 457-459, 461, 625
Chlorophyta, 454, 455, 456, 457, **461-464**, 476 (see also Green algae)
Chloroplasts, 93, 108, **113**, 114, 120, 125, 167, 424, 426, 454, 476, 515, 625-626, 627
  and active transport in plants, 653
  in algae, 456, 458, 460, 463
  binary fission in, 453
  of bundle-sheath cell, 224
  and cell division, 146
  development of, 120
  DNA of, 362
  in euglenoids, 458
  evolution of, 146, 362, 440, 450, 453, 456
  exchange of materials in, 127
  in ferns, 503
  inner membrane of, 214, 228
  intermembrane space of, 214
  internal membranes of, 211, 228 (see also Thylakoids)
  lack of, in prokaryotes, 440-441
  of mesophyll cell, 224
  outer membrane of, 214, 228
  in photosynthesis, 212-213, 216, 219-220, 222, 224, 228-229
  replication of, 453
  structure of, **214**
  vesicle of, 120
Chlorpromazine, 853
CHNOPS, **35**, 55, 96
Choanocytes, 522, 523, 524, 542
Choanoflagellates, 522
Choking, 719
Cholecystokinin, 723, 823, 855
Cholesterol, **70-71**, 82, 431, 450, 723, 728, 750
  and atherosclerosis, 730, 761

in cell membrane, 104–106, 124
and diet, 730
in gallstones, 722, 723
and heart attack, 730
metabolism of in liver, 726
receptor-mediated endocytosis of, 138
regulation of, 71
Chondrichthyans (*see* Fishes, cartilaginous)
Chordamesoderm
  in amphibians, 928–929, 952
  in chick embryo, 932, 935, 937
  organizer and, 930
Chordates, 520, 544, 587, **591–604** (*see also* Lancelets; Tunicates; Vertebrates)
  embryonic development of, 919–920
Chorioallantoic membrane, 934
Chorion, 388, 944–945, 952, 953
  in chick embryo, 934
  in human embryo, 940, 944, 945
Chorionic gonadotropin, 941, 945
Chorionic villus biopsy, 944
C horizon, 663
Choroid, 866
Chorusing, 1064
Chromatids, 155, 275–276, 355
  and DNA, 357, 358
  in Down's syndrome, 385
  in karyotypes, 384
  in meiosis, 253–256, 260–261
  in mitosis, 148–150, 151–152
  "sister," 155, 253–254, 256, 261
  in spermatogenesis, 902
Chromatin, 93, 108, **112**, 124, 127, 263–264, 267, 278, 355–357
  condensation of, 253, 357
  euchromatin, 359, 360, 380
  and gene regulation, 380
  heterochromatin, 359, 360, 380
  and histones, 356
  in meiosis, 253
  in mitosis, 146, 149, 155
  in nucleolus, 366
  staining, 359
Chromatophores, 552
Chromoplasts, 120, 125, 626
Chromosomal insertion, 331–333, 338
Chromosome, 125, 145, 148–154, 247, 275–276, 278, 316, 337–338
  abnormalities in, **277–278, 385–387**, 399
  and asynchronous replication, 326
  bacterial, **319–320**
  banding patterns on, 277–278, 383, 399, 422
  in cell nucleus, 108
  and cellular nucleus, 356
  complement, 145, 148
  condensation (*see* Condensation)
  crossing over in (*see* Crossing over)
  daughter, 256
  deletion in, 277, 279, 376, 386–387, 394, 399
  discovery of, 109
  duplication in, 277, 279
  of *E. coli*, **319–320**
  episomes in, **326**, 335, 337
  eukaryotic, 281, 294, 355–358, 379–381, 453
  extra, 399
  folding of, 358
  fragmentation of, 277
  gene(s) and, **260–261, 274–277**
  and gene families, 367
  and gene replication, 359, 360
  giant, 277–279, 359–360
  homologous, 274–275, 279 (*see also* Homologues)
    in meiosis, 250, 253–256, 260–261
  homologous recombination in, 336–337
  in hybrids, 1012
  independent assortment and, 260
  insertions in, 331–333, 338
  inversion in, 277, 279
  of lambda phage, 334–335
  location of genes on, **263–266**, 274–275, 277–278
  looped domains in, **358**
  mapping, **275–277**, 302, **328–329**, 338
  in meiosis, 249–250, 253, 255–256, 259–261, 264, 355, 358
  in mitosis, 355, 358
  movement of, 120, 123, 152
  number, 249
    in human beings, 145, 264, 382, 399

in oogenesis, 909–910
and plasmids, 326
and polyploidy, 1012
prokaryotic, 145, 358, 379–381, 431, 437–438, 472
and prophages, **331–334**, 335, 338
"puffs," and DNA synthesis, 360
reassortment of, in meiosis, 261
recombination of, in meiosis, 250, 253
replication of, 154–155, **254, 281–300, 319–320**, 358, 359, 360
sex chromosomes, 264–266, 278
and simple-sequence DNA, 364
and spermatogenesis, 902
staining of, 383
structure of, 91, 124, 146–148, 355–358, 364
Sutton's hypothesis, 264
and synchronous replication, 326
theta, 295
translocation in, 277, 279, 376, 385–386, 400
  and Down's syndrome, 385–386, 400
and transposons, 335–336
viral, 330–334, 444
X chromosome, 264, 266, 279, 359, 872
Y chromosome, 264–266
Chromosome complement, 145, 148
Chromosome mapping, 276–279
Chrysophytes, 455–457, **459–460**, 476
Chthalamus stellatus barnacles, 1112–1113
Chylomicrons, 721, 723
Chymotrypsin, 171, 175, 722
Chymotrypsinogen, 175, 722
Chytrids, 455, 469, 477, 482
Cicada, 66
Cigarettes, 71, 739, 761
Cilia, **78**, 94, 110, **121–123**, 125, 149, 151, 426, 468, 471, 473, 475, 541, 546, 866, 912, 924
  ATPases in, 181
  and cigarette smoke, 738–740
  and defense against infections, 791
  and embryonic development, 924–925
  fused, 531, 542
  and sense of smell, 861
Ciliary bodies, 866, 867
Ciliary muscles, 867
Ciliates, 102, 122, 455, 468, 469, **470–471**, 477, 533
  evolution of, 470
  genetic transfer in, 471
  locomotion in, 470–471
  macronuclei, 471–472
  micronuclei, 471–472
  trichocysts, 471
Ciliophora, 455, 469, 470, 477
Circadian rhythms, **689–691**, 695, 824, 834, 839
Circular DNA, 319–320
Circulation, 588, **749–764**
  of blood, 750–752
  blood pressure, **759–762**
  blood vessels, **753–754**
  cardiovascular system, **752–759**
  closed, 763
  of earthworm, 556
  the heart, **754–758**
  lymphatic system, **759–763**, 764
  in mammals, **761**
  systemic, 754
  vascular circuitry, **758–759**
Circulatory system, 547, 568, 585, **749–762**, 763–764
  closed, 547, 554, 556, 564
  double, 763
  of earthworm, 556
  of onychophorans, 561
  open, 547, 563
  pulmonary, 763
  and respiration, 734
  of ribbon worms, 538, 542
  systemic, 763
Circumcision, 904
Circumnutation, 691–692
Cirrhosis, 202
Cirri, 470–471
Cisternae, 116, 117
Citric acid
  in Krebs cycle, 194
Civilization
  and forest ecosystems, 1165
  and microevolution, 962–963
  and sexual receptivity, 915
Cladistics, **415–416**, 418, 427

methodology of, 416
Cladogenesis, **1029–1030**
  in Darwin's finches, 1021
  in evolution of the horse, 1027
  in fossil record, 1021–1022
  in marsupials, **1023**
  and punctuated equilibrium, 1028
Cladogram, **417**
Cladophora, 545, 546, 549, 563
Class, 410, 427
Classical conditioning, **1055**, 1074
Classical genetics, 199, 281
Classification (*see also* Linnaeus; Taxonomy)
  of animals, 454
  biological, 407, 411
  of fungi, 454, **481–488**
  hierarchical, **410–412**
  need for, **407**
  of organisms, **407–427**
  of prokaryotes, **450**, 454
    Gram staining, 432
  of protists, **454–455**
  of plants, 454
  schemes for, 427
Clathrin, 138–139
Clausen, Jens, 1002–1003
Claviceps purpura, 486
Clavicles, 763
Claws, 565, 572, 573, 599
Cleavage, 587, 952
  in amphibian, 927
  in chick embryo, 931, 935
  and cytoplasmic determinants, 926
  in human embryo, 940–941, 944
  radial, 544, 562, 563, 564, 587
  in sea urchin embryo, 921, 925
  spiral, 544, 563, 587
  unequal, 927, 952
Clements, F. E., 1125
Climate
  and body shape, 785
  and distribution of life, 1161–1163
  and evolution of bipedalism, 1044
  and evolution of the horse, 1027
  and growth rate of plants, 645–646, 671
  and ocean currents, 1156–1157, **1158**
  patterns, and ecosystems, 1131, **1136–1137**
  and water transport in plants, 654, 655
Climax community, **1125**
Clines, 1002–1003, 1009
Clitellum, 557
Clitoris, 760, **909**, 912–913
Cloaca, 766, 904
Clonal selection theory of antibody formation, **799–801**, 818–819
Clones, **343–345**, 346–348, 351, 353, 819
  of cytotoxic T cells, 808
  in plant growth, 642–643
Cloning, 340, 342, **343–345**, 346, 351, 353, 394, 802–803
Closed circulatory system, 753, 763
Closed systems, 166
Clostridium botulinum, 438, 442
Clotting of blood, **751–752**, 763
Clouds, and solar energy, 1133
Club mosses, 498, 502, 503, 507
Clumped dispersion, 1093–1094
Clustering, in honey bees, 1061
Clutch size, in birds, 995, 1099
Cnidarians, 520, **525–530**, 533, 534, 539, 542, 632 (*see also* Anthozoans; Hydrozoans; Scyphozoans)
  classification of, 526
  haploid and diploid forms of, 526
  larvae of [*see* Planula(e)]
  life cycle of, 526
Cnidocytes, 525, 526, 528, 542
Coadaptations, **1099**, 1104
Coadaptive gene complexes, **993**, 1008
Coagulation of blood, 751–752, 729
Coal, 57, 503, **504**
Coated pits, 138
Coated vesicles, 139
Cobalt (Co^{2+}) ion
  and biogeochemical cycles, 1145
  plant requirements for, 657
Cocaine, 515, 948
Cocci, 429, 436
Cochlea, 863, 864, 865

I-11

Cocklebur, 684–686
Cocoons, 557, 576
Coding sequences, 320
Codium magnum, 461, 462
Codominance, genetic, **268**, 279
 in blood groups, 810
Codon(s), 307, 312
 initiator, 312–313, 350–351
 messenger RNA, 309–315, 317
 in prokaryotic replication, 320–321
 termination, 309, 313–314
Coelom, 520, 522, 532, 544–545, 547, 554–555, 559, 562–564, 567, 587, 588, 701
 development of
  in amphibian embryo, 929, 952
  in chick embryo, 932-**934**, 936
 in vertebrates, 713
Coelomates, 520, 532, 587 (*see also* Deuterostomes; Protostomes)
Coelomic cavities, 591, 604
Coelomic fluid, 544
Coenocytic organisms, 462, 463, 480, 482
 fungi, 480
 organization of, 462, 463
 slime molds, 466
 water molds, 467, 469
Coenzyme(s), 171–173, 183–184
 acetyl CoA, 194, 220–223, 325
 coenzyme A, 194
 coenzyme Q, 197, 200–201
 in prokaryotes, 430
 recycling of, 172–173
 in respiration, 191–192
 vitamins, 728
Coevolution, **1003–1005**, 1007, 1009
Coexistence, 1111, 1113 (*see also* Parasitism; Resource partitioning)
 and ecological niches, **1109–1113**, 1129
Cofactors, **170**, 171, 175, 183–184
 ions as, 171
Cohen, Stanley, 344
Cohesion, 42, 54, 652, 669
Cohesion-tension theory, **651–652**, 669
Cohort, 1093
Coho salmon, 997–998
Coiling, in plants, 647, 691–693
Colchicine, 384, 1012
Cold
 adaptations to, **787–788**, 790
 and ADH, 775
"Cold-bloodedness," 780 (*see also* Ectothermy; Endothermy; Homeothermy; Poikilothermy)
Cole, Lamont, 1143
Coleochaete, 493, 494, 516, 625,
Coleoptiles, 619, 631, 682–683
Coleus, 640–641, 675
Collagen, 77, 115, 304, 704, 713, 753, 830
 in bone, 704
 in connective tissue, 713
 gene for, 362
 and vitamin C, 729
Collar, 591
 bone, 763 (*see also* Clavicles)
 viral, 445
Collateral ligaments, 707
Collecting duct, **770**, 772–773, 775–776, 832
Collenchyma cell, **647**, 648
 vs. sclerenchyma, **636, 637**
Colon, cancer of, and diet, 730
Colonial organisms, 461, 462, **527–529**, 530, 541 (*see also* Colonies)
 in algae, 463
 cell division in, 461–462
 reproduction in, 462–463
Colonies
 algae, 456, 461, 462
 bacterial, 436, 442
 eusocial vs. subsocial, 1060–1061
 volvocine, 461
Colonization, 1017
 and destruction of habitats, 1126, 1130
 of the earth by human beings, 1047
 and ecological succession, 1124–1125, 1128–1129
 and extinction, 1123–1124
Colony-stimulating factors, 830–831
Color, in flower evolution, 513, 514, 517
Color blindness, 267, 391, 400
 gene for, 391–392

 and sex-linked trait, 391–392
Color vision, 872
 evolution of in primates, 1033–1034, 1051
Columbine, 514
Columnar cells, 702
Comb (of beehive), 1060–1061
Comb (of rooster), 906
Combat
 in cephalopods, 553
 in mollusks, 547
 and territoriality, 1066–1067
Combination, chemical, **34**, 38
 in organic molecules, 58
Comb jellies, 524, 531, 542
Combustion, 208
Commensalism, 447, 451, 1120, 1130
Common cold, 444, 448
Common giraffe, 1011
Communal nesting sites, 1074
Communal territories, 1065
Communication
 between areas of the brain, 892–893
 among cells, 67, 140–142, 143, 822, 838, 841
 chemical, 822
 and memory, 893
 by pheromones, **585**, 586
 among plants, **693–694**, 695, 696
 in plant tissues, 141–143
 as releasers, 1054, 1078
 and social structures, 1058, 1078
 by sound, **583–584**, 586
 and tongue, evolution of, 717
Communities, **1106–1130**, 1161, 1177
 competition in, **1106–1114**
 composition of, and environment, 1123–1129, 1130, 1177
 and convergent evolution, 1161, 1177
 dynamism of, 1129–1130
 ecological succession, **1124–1129**
 interactions within, **1106–1130**
 intermediate disturbance hypothesis, **1124–1125**
 island biogeography model, **1123–1124**
 predation in, **1114–1119**
 productivity of, 1138
 stability in, 1123–1129, 1130
 structure of, 1106–1107, 1113, 1130
 symbiosis in, **1119–1122**
Compact bone, 705
Companion cells, **638–639**, 647, 648, 662–663
Compartmentalization, 555
 of biochemical functions in cell, 192–193
Compartments
 of the body, **768–769**, 775
 intracellular, 127, 129, 130, 192–193
Competition, **1106–1114**, 1164
 competitive exclusion, 1108–1109, 1129
 current debate, 1106–1108
 ecological niches, **1106**
 and evolution, 12, **1106–1114**
 within families, 1071–1074
 exploitative, 1106
 experimental approaches to, 1112–1113
 and "fitness," 1106
 interference, 1106
 interspecific, 1106
 intraspecific, 1106
 as a limit on population, 1088, 1090, 1098
 for light in the forest, 1164
 among males for females, 580
 for mates, and the selfish gene theory, 1072–1074, 1078
 and niche overlap, 1111
 and origin of life, 90
 past competition and resource partitioning, **1110–1111**
 physiological limits and, 1113
 prevalence of, 1113
 and prokaryote evolution, 439
 resource partitioning, **1106, 1107, 1108, 1109–1111**
 and sexual selection, **998–1001**
 in tidal zones, 1158
 for water in savannas, 1170
 "winner takes all," 1113–1114
Competitive inhibition, **177–178**, 184
Complement, 797–798, 1012
Complementary base-pairing, 286–288, **289**
 in recombinant DNA, 353

Complementary DNA, **342**, 343, 346–347, 353
Complementary strand, 358
 in DNA replication, 289, 292, 295–296, 297, 299
 in rolling-circle replication, 338
Complete flowers, 614
Complex transposons, 336, 338
Complexity
 and natural selection, 1005–1006
 organ and tissue levels of, 533
Composite flowers, 616
Compound eyes, 1060
Compound leaves, 626, 630
Compounds, chemical, 34, 38
 vs. elements, 34
Comprehension, and Wernicke's area, 888
Computer, in phenetics, 420
Concealment, 1116–1117
Concentration
 of elements, in ecosystems, 1150–1152
 enzyme and substrate, 174–175
 gradient, 129–131, 134–136, 142–143, 180, 849–850
  and prokaryotic chemotaxis, 434
  in plants, 652, 654
  and potassium (K^+) ion transport, 655
  of retinoic acid in embryo, 940
  of sucrose, in plants, 662
  of water in leaf, 652
 of ions in blood, 745–747, 748
 vs. partial pressure of gases, 733
 and plant biology, 653–654, 656
 products and reactants, 174
 regulation of, in body solutions, 765–766, 775
 relative, and cells, 127, 129
 of solutions, 128, 132, 314
 of substrate, and enzyme reactions, 174–175
 and thermodynamics, 180
 of urine, 773
 water, 128, 132, 133
Conciliatory behavior, 1063
Condensation
 of chromosome, 146, **148–151**, 155, 355, 357, 358, **359–360**, 380
 in meiosis, 255
Condensation reactions, **64**
 in DNA replication, 297–299
 in fats, 67–68
 and lambda phage replication, 334
 in polymers, 64
 in proteins, 72–73
 in recombinant DNA, 343
Conditional response, **1055**
Conditional stimulus, **1055**
Condom, 915, **916**, 197
Conduction, **778–779**
 in cardiac pacemaker, 757
 and heat loss, 789
 of nerve impulse, 857
Cones (visual), 866, 868–872, 936, 1033
Cones (of trees), 505–507
Conflicts of interest, genetic, **1071–1074**
Conformation
 of enzyme, 175–179, 183
 of molecules in energy reactions, 181
 of proteins, 777–778
Conidia, 481, 485, 488
Conidiophores, 488
Congenital defects, and chromosome deletions, 386–387
Coniferous forests, 663, **1164–1166**, 1177
Conifers, 497–498, 504–505, 509, 516–517, 1166–1168, 1177
 evolution of, 510
 leaves, vs. deciduous, 508, 623
 photosynthesis in, 629
Connecting piece, of sperm cell, 902–903
Connective tissue, 140, 702, **704–708**, 710, 713, 753, 813
 blood as, 753
 development in human embryo, 946
Connell, Joseph H., 1111, 1112–1113
Connexons, 142
Conjugation, **326–330**, 331, 337, 426, 435, 471
 in bacteria, 352
 in *E. coli*, 327
 F⁻ (female) cell in, 327–328, 338
 F⁺ (male) cell in, 327–328, 338
 and genetic recombination, 336–337

I-12

and plasmid, 326–330
in protozoa, 468–469
and sexual recombination, 327
vs. transduction, 334
Consciousness, and reticular activating system, 885
Conservation biology, 1114, **1126–1127**
Conservation
of energy, 786–787
in evolution, 1006
in hypotonic urine, **772–774**
of water, 776, 785, 790
Conservative replication, 293
Consolidation of memory, 885, **893–895**
Constancy, 514–515
Constant (C) genes, 372, 854
Constant (C) region, 371–372
of AIDS virus, 815
of antibodies, 798–799, 818
of T-cell receptors, 819
Constriction groove, 153
Consumers, 1138, **1139–1141**, 1142, 1145, 1153 (*see also* Detritivores)
Contact inhibition, of cell growth, 147–148
Contamination, 86, 996
Continental drift, **1012–1013**
Continental shelves, 1157
Continents, and biomes, **1160**
Continuity of living systems, 10, 90
Contraceptive techniques, 901, 904, 906, 914, **915–917**
Contractile assemblies, 122
Contractile proteins, 70, 72, 527, 706, 875
Contractile vacuoles, 133, 458, 471
Contraction, muscular, 537, 950
Control (*see also* Homeostasis; Integration; Regulation)
areas, in ecology, 1112–1113
in the brain, **881–899**
experimental, 15, 131
of respiration, 745
Control protein, for cell cycle, 148
Convection, 789, **779**
Convergent evolution, **1007**, 1009, 1161, 1171
Cooperation, 1059, 1078, 1079
in dominance hierarchies, 1063
in eusocial species, 1060
and reciprocal altruism, 1074–1075, 1079
Coordination, and cerebellum, 898
Copepods, 572
Copernicus, Nicolas, 7, 9
Copper (Cu^{2+}) ion in plants, 657
Copulation
and dominance hierarchies, 1062
in earthworms, 557
in fireflies, 580
and life-history patterns, 1099–1100
and ovulation, 915
and "selfish gene" theory, 1072–1074
Copulatory sac, 535
CoQ (*see* Coenzyme Q)
Coralline algae, 465
Coral reefs, 465, 529, **530, 1124**
construction of, 526
productivity of, 1157
Corals, 524, 529, 530, 542
Core, viral, 445
Corepressor, 323–324, 337
Corey, Robert, 74
Cork, 645
Cork cambium, 644–645, 647
Cork cells, 647, 648
Corliss, John B., 1140
Corn, 93, 140, 224, 226, 622
diseases, 487
fertilization in, 617
"high-lysine," 76
Corn Belt, 1170
Cornea, 579, 866–868
development of, 936
Corn smut, 487
Corolla, 614, 616
Corona radiata, 911
Coronary arteries, 71, 728, 758
Corpora allata, 578
Corpora cardiaca, 578
Corpus callosum, **884**, 889, 898
Corpus luteum, 824, 911, 913–914, 917, 941, 945
Corpus striatum, 884, 894, 895

Cortex, cerebral (*see* Cerebral cortex)
Cortex (of ciliates), 471
Cortex (of kidney), 770
Cortex (of plant tissues), 647, 648
root, **632**, 634, 635, 648, 666–667
stem, 640, 644–645, 648
Cortex (of spore coat), 438
Cortical area (of lymph nodes), 794
Cortical follicles, 816
Cortical granules, 921
Corticosteroids, 829, 831, 834
Cortisol, 73, 824, 825, 829, 830–831, 833, 839
Corynebacterium diphtheriae, 429, 448
Cosmids, **344–345**, 353
COS region(s)
of lambda phage, **334**
in recombinant DNA, 345, 353
Cotransport, 137, 773
in chemiosmotic coupling, 200
in plants, 662
Cottonwood, 222
Cotyledons, 508, 618, 619, 628, 632
Cougar, 412
Countercurrent exchange, 130, 774
and diffusion, 129, 136
in gills, 735
in homeostasis, **787–788**, 790
in loop of Henle, 774
in multicellular animals, 124
Countercurrent multiplier, 774
Coupled reactions, **181–182**, 184
Courtship, 553, 826, 986, 1015, 1020, 1052 (*see also* Mating; Sexual Selection)
in arachnids, 571
in birds, 872, **1073**
in fireflies, 580
and male sex hormones, 906
rituals, and sexual selection, 999–1000
and the selfish gene theory, 1072
and territoriality, 1063
Covalent bonds, **30–34**, 38, 181, 664
double, 33–34
and hydrophilic molecules, 46
and nitrogen fixation, 664
nonpolar, 32–33
and organic molecules, 57, 58
polar, 32
single, 32, 34
strength of, 57–58
triple, 33–34
of water, 41, 54
Covering tissues, 532
Cowpox, 796
Crabgrass, 227
Crabs, 552, 571, 572, 585
"Cramming," 891
Cranium, 593, 604 (*see also* Brain; Brain case; Skull)
"Crash" (population), 1089, 1099, 1104
Crassulacean acid, **655**
Crassulacean acid metabolism (CAM) plants, 655
Crawling, 936
Crayfish, 571
Creationism, 1, 2, 6, **965–966** (*see also* Special Creation)
Creosote bush, 642–643, 1094
Cretaceous period, 496, 509, 1031
and asteroid hypothesis, 1024–1025
and formation of continents, 1012
and iridium anomaly, 1024–1025
and mass extinctions, 1024–1025
Creutzfeldt-Jakob disease, 447
Crick, Francis, 287, **288–289**, 291, 292–293, 298–300, 306–307, 309, 317, 356, 358, 379, 383, 399
Crickets, 583
Crinoidea (*see* Feather stars)
Cristae, 119, 458
in respiration, 191–193
Critical period (in embryo development) **1056–1057**, 1078
Crocodile, 414, 596–597, 969
Cro-Magnons, 1047, 1051
Crop, 555, 574
Crop plants, improving, 1149
Crossbreeding, 238, 976–977 (*see also* Artifical selection)
Crosses, 239–242, 247, 266–268, 270 (*see also* Testcross)
of *Neurospora*, 303

Cross-fertilization, 240
Crossing over, 275–276, 279
homologous recombination in, 336–337
in meiosis, 253–255, 259, 261
in sexual reproduction, 984
in vertebrate nervous system, 882
Crossings, artificial, 237
Crossover, 276
percentage of (*see* Recombination, frequency of)
points, 253
studies, 276
Cross-pollination, 238
Cross walls (*see* Partitions; Septa)
Crow, 423
Crowding, and population explosion, 1097
Crown gall disease, 352–354
Crowning, 950
Crown (of tooth), 716
Crown (of trees), 1164, 1174
Cruciate ligaments, 707
Crust, geologic, 85, 1140
Crustaceans, 26, 554, 565, 566, 567, **571–572**, 585, 589
Crustose lichens, 490
Ctenophores, 530, 524–525, **531**, 532, 542 (*see also* Comb jellies; Sea walnuts)
C_3 grasses, 227, 228
C_3 plants, **223–228**
C_3 photosynthesis, **223–228**, 655
grasses and plants with, **223–228**, 230
temperature range for, 227
and transpiration, 650
vs. C_4 photosynthesis, **226–227**
Cuboidal cells, 702
Cultivation, 1139, 1148
Culture
and birth rate, 1096–1097
evolution of, 1045
and human behavior, 1075
vs. selfish gene, 1075
Culture plate, 346
Curare, 878
Currents, and life forms, 1155, **1156–1158**
Cusps, of teeth, 716
Cuticle (exoskeletal), 566–568, 572, 574, 582 (*see also* Cutin; Exoskeleton)
Cuticle (of plants), 214, 493, 495, 509, 627, 629, 648, 654
in coniferous leaves, 1166
lack of, in pond weeds, 1155
in stems, 636
Cutin, 495, **627**
Cuttlefish, 545, 551, 553
Cuvier, Georges, 3–5
Cyanobacteria, 91–92, 101, 430, 431, **436**, 440–443, 450
in evolution, 452, 456
lichens, 488–490, 492
photosynthesis in, 213, 217, 452
phycobilins, 456
pigmentations of, 428
symbiotic, 453
Cyanocobalamin, 729
Cyanogen bromide, 351
Cycads, 496–498, 505, 897
Cyclic AMP, 141, 325, 337, 839
and acrasin, 838
in CAP-cAMP complex, 325, 337
in *E. coli*, 838
in hormone action, 837–838
and learning, 898
in repression, 325
and slime molds, 837, 838
and smell, 861
Cyclic electron flow, **216**, 229
Cyclic GMP, 854, 871
Cyclosporin, 488, 810, 819
Cyclostomes (*see* Agnaths)
Cygnets, 1066–1067
Cylinders (in vascular plants), 501
Cypresses, 505
Cysteine, 73, 75, 285, 310
and lead poisoning, 177
Cystic fibrosis, 397
Cysts (*Trichinella*), 539–540
Cytochrome(s), **196–200**
in electron transport, 196–197, 200
and nutrition, 728

I-13

Cytochrome(s) *(Continued)*
   and photosynthetic phosphorylation, 219
Cytochrome *c*, **196–200**, 418–420
Cytokinesis, 145, 146, **153–154**, 155, 315, 359, 251
   and actin filaments, 146
   in animal cells, **153**
   in *Coleochaete*, 494
   constriction groove in, 153
   in *Drosophila*, 924
   in green algae, 462
   in meiosis, 255–256, 258–259
   in mitosis, 146, **153–154**
   in plant cell, **153–154**
   and polyploidy, 1012
   in protozoa, 468
Cytokinins, 673, **676–679**, **694–695**
   and apical dominance, 675
   and auxins, **676–677**, 681
   effects of on cell division, **694–695**
   and leaf senescence, 677
Cytology, 109, 247
   and genetics, **260–261**, 263–264
Cytoplasm, 94, 102, 104, 108, 109, **110–115**, 121, 124, 144–146, 150-151, 153–154, 461, 464
   in animal cells, **112**
   in cell division, 155
   in cell respiration, 193
   cytoplasmic bridge, 327
   cytoplasmic streaming, 121, 130, 142
   determinants of embryonic development, **924–925**, 952
   and desmosomes, 703
   and energy transformations, 168
   and gene transcription, 380
   and glycolysis, 187, 189, 200, 204
   grey crescent, 926–927
   introns and exons in, 380
   of leaf cells, 627
   in meiosis, 256, 258
   and messenger RNA transcription, 368–369
   and photosynthesis, 213–214
   in plant cells, **113**
   polar plasm, **924–925**
   in prokaryotic cell, 431
   and respiration of cell, 193
   and ribosome assembly, 366
   RNA in, 305, 308
   and vesicle-mediated endocytosis, 138–140
   in volvocine line, 462
Cytoplasmic bridge, 327
Cytoplasmic channels, 472
Cytoplasmic determinants, 952
   of germ cells, **924–925**
   of embryonic development, 925–926, 927, 942–943
Cytoplasmic streaming, 121
Cytosine, 80–81, 296, 299
   in DNA, 283–284, 288–290, 292
   and prokaryotic taxonomy, 429
   in recombinant DNA, 341
   and gene regulation, **360–361**
Cytoskeleton, 110–111, 115, 121, 124, 154
   accessory proteins of, 114
   in animal cell, **112**
   and cell division, 149–150
   and cytoplasmic streaming, 130
   in energy transformations, 168
   and morphogenesis in vertebrate embryo, 938
   in plant cell, **113**
Cytotoxic T cells, 801, 819
   and AIDS virus, 804
   and cancer, 809
   function of, 806–807
   and T8 glycoprotein, 805
   and transplant rejections, 810
Cytotoxins, 807

Dandelion, 1103
Daphnia, 1103 (*see also* Water fleas)
Dark-field microscope, 99, 100, 101
Darkness, measurement of, in plants, 685–686
"Dark" reactions, of photosynthesis (*see* Light-independent reactions)
Darnell, James, 366
Dart, Raymond, 1038, 1044
Darwin, Charles, **1–10**, 14–15, 27, 221, 237–238, 245–246, 282, 306, 409, 416, 555, 847, 1010, 1017, 1020–1021, 1028–1029, 1088 (*see also* Evolution; Natural Selection; *Origin of Species*)
   and carnivorous plants, 666
   and evolutionary theory, **961–962**, 963, 965–972, 968, 998, 1065–1066, 1113–1114
   and photoperiodism, **689**
   and phototropism, 672–673
   and population genetics, 974–976, 989
   and sexual selection, 991, 1000–1001
   and special creation, 965–967
Darwin, Erasmus, 2
Darwin, Francis, 672–673
Darwin's finches
   character displacement in, 1110–1111
   natural selection in, 1017–1019, **1020–1021**
Dating
   of geological eras, **970**
   radioisotope, 25
dATP, 298
Daughter cells, 144, 150–155, 315, 328, 459, 460, 462
Daughter chromosome, 256
Daughter nucleus, 252
Day length, 685
Daylight, annual cycles of, 1136
   and regulation of male sex hormones, 907
Day-neutral plants, 684, 695
DDT, in biogeochemical cycles, 1150–1151
Deamination, 203, 766
Death rate, 1102, 1104–1105, 1088
   and human population explosion, 1097
Death Valley, 172
Decay, 439, 1167
Deciduous forests, **623**, 624, **663**, **1164**, 1171
   decay in, vs. coniferous forests, 1167
   layers of growth in, 1164
Decomposers, 1131, **1142–1143**, 1153
   in forest ecosystems, 1150
   in lakes and ponds, 1155
   and phosphorus cycle, 1146
Decomposition, 1131, **1142–1143**
   and cycling of materials, 1131, 1153
   in forest ecosystems, 1150
   fungi in, 480, 492
   of soil, 429
     and methanogens, 439
     prokaryotes in, 439
   and soil nitrogen, 1131, 1146
   in taiga, 1155
   in temperate forests, 1166
   in tropical rain forests, 1176
   in water, 1155
Decompression sickness, 744
Deep-sea diving, adaptations to, 733
Deep-sea oases, **1140–1141**
Defenses, 696
   against microorganisms (*see* Immune system)
   in plants, 682, 693–694
   of territory in vertebrates, **1064–1066**
Deficiencies, genetic, 400
Deforestation, 1150 (*see also* Tropical rain forests)
Deformity, congenital
   and inherited traits, 273
de Graaf, Régnier, 237
Degrees of relatedness, and kin selection, 1067–1069
Dehiscent fruits, 620–621
Delbrück, Max, 284, 309
Dehydration, 832
   and kidney function, 774
   microscopic specimens, 98
   and thirst, 769
Deletion, chromosomal, 277, 279, 376, 386–387, 399
   and aniridia, 399
   in Duchenne muscular dystrophy, 394
   Wilms' tumor, 399
Dellaporta, Stephen, 365
Delta globin, 367
de Mairan, Jean-Jacques, 689
Dementia, 816, 896–897
Denaturation, 396
   of DNA, 345, 347, 354, 363–364
   of enzymes, 175, 177–178, 777–778
Dendrites, **708**, 713, 842, 852, 855, 856, 877
   in cerebral cortex, 884
Dendrochronology, 646
Denitrification, 1148
Density (physical), 44–45, 54
Density (population), 1087, **1093–1094**, 1114 (*see also* Population dynamics)
   fluctuations in, 1095–1099
Density-dependent (population) factors, **1098**, 1105
Density-dependent inhibition, 147–148
Density-independent (population) factors, **1098**, 1104
Denticles, 593
Dentine, 716, 948
Dentition (*see also* Browsing; Grazing; Teeth)
   of australopithecines, 1039–1040
   of early human beings, 1042, 1045, 1046, 1051
   of horse, evolution of, 968, 1025–1027
   in mammals, **716–717**
Deoxyadenosine triphosphate, 298
Deoxyguanosine triphosphate, 298
Deoxyribonuclease, 722
Deoxyribonucleic acid (*see* DNA)
Deoxyribose, 282–283, 290, 299, 305
Depolarization, 897, 921
Depression, 831, 853
Depth of respiration, 743–747, 748
Dermal system, in angiosperms, **625**, 648
Dermatitis, 729
Dermatome cells, 936
Dermis, 710 (*see also* Epidermis; Skin)
de Saussure, Nicholas Theodore, 206
Descartes, René, 10
Deserts, **1172–1173**, 1177
   latitude, 1137
   spread of, and human population explosion, 1177
Desert scrub, 1139
Desmosomes, **703**, 721
Desmotubules, 140
Determination of germ cells, **924–925**
Detorsion, 550
Detritivores, **1142–1143**, 1145, 1153, 1159
Detritus, 1142
Deuterium, 25
Deuteromycetes, 488, 491
Deuterostomes, 520, 544, 545, 563, 564, **587–604**, 924
   embryonic development of, 545
   primitive, 562
   vs. protostomes, 544
Development, 544, 563, **919–953**
   in amphibians, **926–931**
   and cell differentiation, 902
   in chick embryo, **931–940**
   in coelomates, 545
   and cytoplasm, **924–925**
   echinoderms vs. chordates, 926
   of egg cells (*see* Oogenesis)
   genetic regulation of, 940, **941–942**, 951
   homeobox and, **942–943**
   of human embryo, **940–951**
   in invertebrates, 920–926
   larval, 550
   and lateralization of function, 889
   and limits on natural selection, 1009
   and living systems, 27
   of plants, **625–648**
   in sea urchin, 919, **920–926**
   and sex differences, 909
   of sperm cells (*see* Spermatogenesis)
Developmental potential, in embryonic cells, 930
Devonian period, 3, 497, 594
de Vries, Hugo, 244, 245, 263, 315, 974
Dexterity, 1032, 1037
DGTP, 298
Diabetes mellitus, 728, 766, 812, 832, 837,
Diagnosis
   of genetic disease, 340, 382, 387–390, **394–398** (*see also* Genetic disease)
   karyotyping in, 383
   prenatal, 387–388
Diagnostic marker(s), 395
   "anonymous" markers, 395
   in cystic fibrosis, 399
   in Huntington's disease, 397
   in recombinant DNA, 399–400
Diamond, Jared, 1106
Diaphragm (muscular wall), 718, 745, 748
   and human respiration, **741**
   in vertebrates, 701, 713
Diaphragm (contraceptive technique), 915–917
Diarrhea, 469, 725, 727
Diastolic blood pressure, 759
Diatomaceous earth, 460
Diatoms, 455–457, **459–460**, 476, 540, 1156
Dicots, 498, 613–615, 618, 619, 623

apical dominance in, 675
vs. monocots, 511, 628, 631–633, 635
woody, and growth, 648
Dicytostelium discoideum (*see* Slime mold)
Didinium, 122, 139
Diencephalon, **883**, 885, 898 (*see also* Hypothalamus; Thalamus)
and memory, 894, 899
Diet
health hazards in, 730
of insects, and evolution, 578, 586
and water balance, 767
Differential centrifugation, 193
Differential-interference microscope, 99, 100, 101
Differentiation, 359, 751, 919, 930
in amphibian embryo, 924
in angiosperms, 619
in chick embryo, 935–937, 938
of ecotypes, in divergent evolution, 1008
genetic basis of, 930
of germ cells, 925
of gonads, in human beings, 946
in plant cells, 625, 634, 635, 640, 647, 671, 676–678, 681-682
in sea urchin zygote, 924
sexual, origins of, 463
in spermatogenesis, **902-903**
of T lymphocytes, 804
Diffraction, in microscopy, 99
Diffusion, 128, **129-135**, 142, 533, 534, 747
vs. air pressure, **733**
vs. bulk flow, in respiration, 737
at capillaries, 753
carbon dioxide in, 130
in cells, 142
facilitated, 135–136, 142–143
of gases, 568–569
and gas exchange in animals, 547, 556
and gas exchange in human respiration, 740
and gas exchange in plants, 669
in gills, 735
of hormones, 821, 823
oxygen in, 130
in photosynthesis, 214
and placenta, 945
and polarity of water, 130
rate, 129, 142
and respiration
and animals, 736–737
in earthworms, 734, 737
in roots, 633
simple, 135–136, 142
and thermodynamics, 180
and urine formation, 772
water, 128, 130
and water transport in plants, 650
Digestion, 546, 570, 588, **715-730**, 846, 847 (*see also* Digestive system; Digestive tract)
of cellulose, 550
in earthworms, 555
evolution of, 715
large intestine, 725
oral cavity, 716–718
and parasympathetic nervous system, 712
pharynx and esophagus, 718–720
and regulation of blood flow, 759
small intestine, 725
stomach, 720
Digestive cavity, 533, 534
Digestive enzymes, 115–116, 171, 703, 713, 716, **722**, 730
activation of, 175
and intestinal microvilli, **721**
in pancreas, 725
Digestive glands, 547, 572, 588, 823
Digestive organs, 546
Digestive structures, 532
Digestive system, 574–575, 710, **715-730** (*see also* Digestive tract)
Digestive tract, 547, 558, 572, 574, 952 (*see also* Digestion; Digestive system; Gut)
of annelids, 564
development of
in chick embryo, 937
in human embryo, 946
in sea urchin embryo, 923–924
of earthworm, 555
human, **715**

layers of, 715–716
major accessory glands, **725-726**
mouth-to-anus, 538, 542
mucosa, 715–716
muscularis externa, 715–716
one-way, 539, 542
and respiratory system, 718
serosa, 715–716
submucosa, 715–716
and temperature regulation, 782
in vertebrates, **715-725**
and water gain, 769
Digital dexterity, 1032, 1037
as adaptation, 1002
Digits, 601
in primate evolution, 1031, 1032–1033, 1037, 1051
vs. vertebrate "hand," 939
vs. wing, in embryo, 939, 940
Dihydroxyacetone, 60
Dihydroxyacetone phosphate, 188–189
Dikaryons, 481, 485, 486, 491
Dikaryosis, 426
Dik-dik, 1110
Dilation (of pupil), 951
Dilation (stage of labor), 950
Dimer(s), 76 (*see also* Tubulin)
Dinitrogen pentoxide, 165
Dinoflagellates, 455–457, 460, 472, 476, 530
Dinosaurs, 597, 598, 600
and breakup of Pangaea, 1012, 1022
extinction of, 496, 510, **1024**
first appearance, 496
and mammalian evolution, 1031
Dinucleotides, 171
Dioecious flowers, 614
Dipeptidase, 722
Dipeptides, 313
Diphosphoglycerate, 189
Diphtheria, 429, 448
Diploblasty, 525
Diplococci, 436
Diploid number, of chromosomes, **249-250**, 252, 256, 276, 383, 399, 460, 463, 579, 921, **986-987**, 990
and allele frequency, 983, 986–987
and gene pool, 976, 983
and genetic variability, promotion of, **986-987**, 990
and kin selection, 1068
vs. polyploidy, 1012
in queen bee, 1060–1061
in spermatogonia, 902
in sporophyte life cycle, 250
Diploid phase, in alternation of generations, 250, 252
Diplopoda, 566 (*see also* Millipedes)
Diptera, 573 (*see also* Flies; Gnats; Mosquitos)
Directional selection, 994–995, **997**, **1000-1001**, 1009
and male ornamentation, 1000–1001
and phyletic change, 1021
and sexual selection, 1000
Direct repeats, 335, 338, 374
Disaccharides, 59, **61**, 64
Discrimination (intraspecific), 1056–1057, 1078
Disease(s), 428, 433, 438–439 (*see also specific disease*; Genetic diseases; Immune system)
autoimmune, 830
of blood vessels, **760-761**
cell destruction in, 448
and competition, 1114
ergotism, 486
and extinction of early *Homo sapiens*, 1050
fungal, 481
genetic, 340
heart, **760-761**
and host defenses, 448, 451
human, 488
microbial, **448-450**
and parasitism, 469, 1120
in plants, 469, 487, 488, 492
prevention and control of, **448-450**
protozoans and, 469
response to, **791-819**
and toxins, 448
Disks, 593
Disparlure, 585
Dispersal, 579, 586
abilities, 1124, 1126
mechanisms, in angiosperms, 620–621

of species, 576
Dispersion patterns, of populations, 1087, 1099, **1093-1094**, 1103-1104
Dispersive replication, 293
Display behavior, in males, 1063–1064
in bowerbirds, 1073
and selfish gene theory, 1072
Disruptive selection, 994–995, **996-997**, 1008–1009, **1014-1015**
Dissociation, chemical, 34, 38
Distal tubule (of kidney), 832, **770**, 772–773
formation of urine in, **772**
Distribution
of alleles, 243
of blood types, **996**
of energy and biomass, 1136, 1152
of genetic characteristics, 994–1001
height, 272
of life on land, 1160–1161
of organisms, and biogeography, 966
of phenotypes, and geography, **1002-1003**
random, 129
Distribution patterns (of populations), 1112
Disturbances (*see also* Intermediate disturbance hypothesis)
and forest composition, 1164
frequency of, and species diversity, 1124
and population [*see* Density-independent (population) factors]
Disulfide bridges, 75, 371
Diurnal movements, of plants, 689
Diurnal primates, 1034
Divergent evolution, **1008**, 1009
Diversity, **407-427**
of animals, **518-519**, 520
of arthropod behavior, 586
of birds, 600
and coral reefs, 530
and evolution, 12
gene for, 372
of insects, 578
of living systems, **405-427**
of mammals, 600
predation and, 1119, 1127
of prokaryotes, 428, **436-437**, 579
of reptiles, 604
of species, 973 (*see also* Evolution; Natural Selection; Species)
Diversity genes, for T-cell receptor, 804
Diving reflex
in diving animals, 733, **744**
and infant stress at birth, 951
Division (biological classification), 410, 413, 427 (*see also* Phylum)
Division, cell (*see* Cell division)
Division, as reproduction, 529 (*see also* Meiosis; Mitosis)
Division of labor, 558, 562, 576
in brain, 889–891, **892-893**
in eusocial vs. presocial species, 1060
between the sexes, 915
in societies, 1058
in vascular plants, 625
DNA, 11, 80–81, 301, 305, 307, 314–317
adenine in, 172, 282–284, 288–292, 296, 298–299
A-DNA, 356, 379
amino acids in, 281
amount of, 363, 380
antiparallel strands of, 289–290, 296
bacteriophage, 308
B-DNA, 356, 379
"beads-on-a-string" model, 357
and behavior, 1052–1053
bidirectional replication (*see* Bidirectional replication)
cDNA, **342**, 343, 346–347, 353
in cell division, 145
and cell variety, 125
of chloroplast, **362**
of ciliates, 471
circular, 319–320, 337
cleaving, **340-341**, 346, 348, 353
cloning, 353, **343-345**, 346, 351
as code, 301, 305
complementary strand, 289, 292, 295, 296, 297, 299, 327–328, 338, 358
in conjugation, 327–328
in corn, 93

I-15

DNA (Continued)
  in cyanobacteria, 442
  and diversity, 298
  donor strand, in conjugation, 327–328
  double helix, **281–300**, 305, 308, 320, 356
    and heredity, 235
  duplication of, and genetic variation, **988–989**
  in *E. coli*, 319–320
  eukaryotic, 145, 452
  and evolution, 9, 11
  evolution of, 362, 370, 423
  function, 80, 287–289
  gDNA, **340–341**, 343, 346, 353, 354
  genetic "instructions" of, 301
  and heredity, 235
  histones and, 357, 358, 379
  homologous recombination of, 336–338
  homologous segments of, 420
  and human evolution, 1048–1050
  hybridization, 308, 363–364, 365, 372, 429
  inactive strand of, 306
  information carrier, 298
  intermediate-repeat, **365**, 380
  lagging strands, 320
  leading strand, 320
  length of, 355–356
  linked, 330–331
  linking, 357
  "meaningless," 355, 363, 366, 380
  and meiosis, 359
  of mitochondria, 316, **362**, 423, 452–453
  nature of, 282
  in nucleus of cell, 423
  origin of, 87–88, 370
  phosphate bonds in, 298–299
  phosphate groups in, 283–294, 288, 290, 298, 299
  and plant cell division, 676
  prokaryotic, 319–320, 337, 450, 472
  prokaryotic vs. eukaryotic, 125, 145
  and prokaryotic operon, 323
  proteins associated with, 355, 379
  protein structure and, 304
  purines, 283–285, 287, 289, 290, 299
  pyrimidines, 283–285, 287–290, 299
  as radioactive tracer, 420–421
  recombinant, 319, **340–352**, 376–379 (*see* Recombinant DNA)
  repetition of, 355, 360, **363–365**, 380, 420
  replication of (*see* DNA replication)
  ribose, 282–283, 290, 299
  vs. RNA, 296, 299, 305
  and RNA transcription, 306
  rolling-circle replication of, **327–328**, 338
  "selfish," 335
  sequence of bases in, 305
  and sex determination in human beings, 946
  simple-sequence, **364**, 380
  single-copy, **366**, 380, 420
  "single-strand switch," 337
  size, 288
  in sperm cell, 903
  stability of, 287
  and steroid hormones, 837
  structure, 287–289, 305, 357
  sugar-phosphate units of, 289
  supercoiling, 294, 297, 299
  template strand, 306–307, 314–315, 317, 358, 368
  30-nanometer fiber, 357
  thymine in, 283–284, 288–290, 292, 296, 299
  transcription, **301–317**, 359
  transcription units, 366
  viral, 333, 443–447
  viruses, 338, 350, 373
  Z-DNA, 356
DNA "fingerprinting," **396**
DNA ligase, 296, **297**, 299, 345, 358
  and lambda phage, 334
DNA polymerase(s), 294–299, 307, 358
  in diagnosis, 396
  and prokaryotic cell, 320
DNA "puffs," 360
DNA repair enzymes, 298
DNA replication, 144, 145, 146, 148, 155, **281–300**, 320
  accuracy of, 298
  bidirectional, 295
  complementary base-pairing, 286–288, **289**
  complementary strand, 289, 292, 295–297, 299

condensation reactions in, 297, 299
conservative, 293
direction of, 290, 296–297
dispersive, 293
enzymes in, 294, 296–299, 358
in eukaryotes, 295, 297
initiator proteins, 294
lagging strand, **296**, 297, 299, 320
leading strand, **296**, 297, 299
mechanics of, **292–298**
nucleotides, 281, 283–284, 289–290, 292–299
Okazaki fragments, **296–297**, 299
origin of replication, **294**, 295–296, 299
in prokaryotes, 295, 297
"proofreading" in, **297–298**
rate of, 294
replication bubble, 294–295, 297
replication forks, 294–297, 299
and RNA, 296, 299
semiconservative, **292**, 293, 294, 299, 309
single-stranded binding proteins, 197, 294, 299
and S phase of the cell cycle, 294
supercoiling, 294, 297, 299
template strand, 292–294, 296–297, 299
in viruses, 284
Watson-Crick model, **288–290**, 293, 298
DNA-RNA hybrids, 345
DNA viruses, 338, 350, 373
Dobzhansky, Theodosius, 973
Dodder, 94, 691–692
Dogfish, 593
Dogs
  and bears, 415, 1008
  hearing in, 865
  latent variability in, 976
Dolphins, 603
Domains
  of antibodies, 798
  of proteins, 363
Domestication, 1121, 1146
Dominance, in cerebral hemispheres, 888–889
Dominance hierarchies, 1036, **1062–1063**
  and breeding population, 1062–1063
  in chickens, 1062
  evolution of, 1044
  and mating behavior, 1072
  ritualized behavior in, 1062
  and selfish gene theory, 1072, 1074
  and waiting behavior, **1074**
  in wolves, 1063
Dominant alleles
  in brachydactylism, 979
  and Hardy-Weinberg equilibrium, 979
  in Huntington's disease, 391
Dominant traits, 266, 268–271, 276
Donis-Keller, Helen, 399
Donor (F^+) cell, in conjugation, 327–328, 338
Dopamine, 853
Dormancy, 560, 624, **639**, 671, 694–695 (*see also* Hibernation)
  and abscisic acid, **680**
  in angiosperms, **621–624**
  in bacteria, 429
  in buds, 648
  clines and, 1002
  and cold, 778
  evolution of, in plants, 623–624
  in seeds, **623–624**
    and phytochromes, 687
  and shoot growth, 641
  in spores, 437–438
Dorsal lip of blastopore ("organizer"), 928, 930–931, 952
Dorsal nerve chord, 919
Dorsal neural tube, 898
  and evolution of the human brain, 881
  and spinal cord, 881
Dorsal root ganglia, 843, 844
Double circulatory system, 763
"Double fertilization," 617, 619, 629
Double helix, **281–300**, 308, 356
  and heredity, 235
  in prokaryotic transcription, 320
Down's syndrome, **385–386**, 399
Dreaming, 890–891
Drinking, 789
Drip tips, 629
Drones, 1060–1061

Drop line, 571
Drosophila, 123, 295, 302, 375, 408–409, 834, 993
  artifical selection in, **976–977**
  bristle number in, **976–977**, 993
  cell division in, 146
  chromosome mapping in, 277
  determination of germ cells in, 924
  disruptive selection in, **996**
  "fitness" for genetic research, 263
  and gene-linkage studies, 395
  genetic regulation of development in, 942–943
  in genetic research, 263–266, 268, 274, 277–279
  gene transfer in, 379
  giant chromosome of, 360
  heterozygosity in, 978
  wild-type, 265
Drought
  and grasslands, 1169
  response to, in plants, 654
Drought-deciduous plants, 1173
Drowning, 744
Drug resistance, 449 (*see also* R plasmid)
  as directional selection, 997
  and *E. coli*, 327
  enzymes in, 329
  evolution of, **964–965**, 973
  and plasmids, 326, 329, 336
  in *Shigella*, 329
  transfer of, 329, 337
  transposons and, 336
  viruses and, 329
Drug resistance plasmid (*see* R plasmid)
Drugs
  and birth defects, 947, 953
  and heart disease, 761
  and neuromuscular junction, 878
  pain-killing, 854
Drupes, 620
Duchenne muscular dystrophy, **393–394**, 400
  diagnosis of, **397**
Duck, classification of, 423
Duodenal ulcers, 996
Duodenum, **720**, 722, 730
Duplicate DNA, 988–989
Duplication, in chromosome, 277, 279
DuPraw, E. J., 16, 246, 355–356
Dura mater, 884
Dust, in atmosphere, 1133
Dutch elm disease, 481, 484
Dwarfism, 351, **391**, 398, 681, 829, 983
  as autosomal dominant, 391
Dynamic equilibrium, 129
Dynamism, of communities and populations, 1129–1130
Dysentery, amoebic, 470
Dysmenorrhea, 835
Dystrophin, 394·397

Ear
  balance mechanisms in, 859
  and biosonar in bats, 1115–1116
  canal, 864
  development in chick embryo, 935
  development in human embryo, 946–947
  human, 859, **863–865**, 879
  of insects, 567
  in moths, and predation, 1115–1116
  sensory neurons of, 887
Eardrum, 584, 864 (*see also* Tympanum)
Earth, **84–85**, 1152
  age of, 2–3, 5, 9, 85, 101, **970–971**
  ecosystems of, **1131–1153**
  "fitness" of, for life, 89
  formation of, 84
  and origin of life, 89
  position of, and environment, 1177
Earthquakes, 1013
Earthworms, 554, **555–557**, 564
  circulatory system of, **556**, 752
  as detritivores, 1142
  ecological importance of, **555**
  and forest floor, 1166
  gas transport in, 734, 737
  nervous system of, 842
Eccles, John, 851
Ecdysone, 360, 578
Echinoderms, 520, 544, **587–590**, 591, 604, 826 (*see also* Brittle stars; Concentricycloidea; Feather

stars; Sand dollars; Sea cucumbers; Sea lilies; Sea urchins)
Echinoidea, 587
Echolocation, 1115–1116
E. coli (*see Escherichia coli*)
Ecological modeling, **1090–1091**
Ecological niches, **1109–1113**, 1138
    fundamental niche, **1113**
    realized niche, **1113**
Ecological problems
    acid rain, **50–51**
    conservation biology, 1114, **1126–1127**
    decreased minerals in soil, 1149–1150
    decrease in ozone layer, **1134–1135**
    destruction of tropical rain forests, 1150, 1176–1177
    human population growth and, 1092, 1093
    increased CO_2, **226–227**
    of modern agriculture, 659, **1148–1149**
Ecological succession, **1125–1129**, 1130
    facilitation hypothesis, **1125**, **1128**
    inhibition hypothesis, 1125, 1128
    tolerance hypothesis, 1125
Ecology, **429**, **1085–1105** (*see also* Biosphere; Ecosystems; Environment)
    and competition, 1106
    coral reefs and, 529
    earthworms in, 555
    and evolution, 961
    experimental approaches to, 1106–1107
    fungi and, 492
    importance of plants in, 613
    microorganisms in, 447–450
    nitrogen-fixing bacteria in, 441
    and nitrogen-fixing cyanobacteria, 441
    and prokaryotes, 439
    and sexual selection, 999
    and stabilizing selection, 995
    symbiosis in, **447–448**
Economics, and mating systems, 999
*Eco*RI restriction enzyme, 341, 344, 346, 348
Ecosystems, **1131–1153**
    agriculture and, **1148–1149**
    biogeochemical cycles in, **1145–1152**
    chemosynthetic, **1140–1141**
    Chernobyl reactor and, **1151–1152**
    and climate, 1131, **1136–1137**
    communities in, 1130
    concentration of elements and, **1150–1152**
    consumers in, **1139–1141**
    detritivores, **1142–1143**
    ecotypes, **1002–1003**, 1010
    efficiency of energy transfer in, **1143**
    energy cost of food gathering, **1144**
    energy flow through, 161, 1085, 1087, 1090, **1131–1132**, **1137–1145**
    energy transfer and ecosystem structure, **1143**
    forest, recycling in, **1149–1150**
    Green Revolution, **1148–1150**
    and hunger, **1148–1149**
    influence of the atmosphere, **1132–1135**
    nitrogen cycle, **1146–1149**
    ozone layer, threat to, **1134–1135**
    phosphorus cycle, **1146**
    and populations, 1087
    producers in, **1138–1139**
    productivity of, 1138
    prokaryotes, 430, 439
    and solar energy, 1131, **1132–1137**
    trophic levels, **1137–1143**
    weather and, **1136–1137**
    wind and, **1136–1137**
Ecotypes, **1002–1003**, 1009
    adaptation and, **1002–1003**, 1009
    and brown bear, 1008
    differentiation in, 1008
    and speciation, 1003
Ectoderm, 525, 532, 542, 545, 952
    in aquatic vertebrates, 937
    in chick embryo, 932, 934, 937
    differentiation of
        in amphibian embryo, 928
        in chick embryo, **935–936**
        in sea urchin, **924**
    in human embryo, 944
    induction in, 931
    neural, 928
Ectomycorrhizae, 491–492

Ectoprocts (*see* Bryozoans)
Ectothermy, 597, 702, 780, 789–790
Edelman, Gerald, 798
Edema, 762, 769
Edentata, 602
Ediacaran fossils, 520
Editing, of mRNA, 368
EEG, 890–891
Eels, 594
Effector cells, 841
Effector hormones, in plants, **694**, 695
Effectors, 323, 709, 845–846, 856
    allosteric, 325
    and enzyme energetics, 176–177, 179, 184
Efferent arteriole (of nephron), 776
Efficiency, 162, 320, 1141, 1153
    of cells, 174, 183
    in conservation of nutrients in forest, 1150
    of energy transfer in trophic levels, 1143
    of prokaryotic transcription, 321–322
Egg(s)
    case, 571
    retrieval of, in greylag geese, 1054
    temperature regulation of, in birds, 783
Egg cells, 103, 109, 138, 139, 153, 237, 240, 251, 252, 258, 519, 576, 586, 600, 951–952 (*see also* Oocyte; Ovum)
    activation of, in sea urchin, 920–921
    amniote egg, 596, 604, 900–901, 905, 908, 912, 917, 931, 932, 952, 1022
    in angiosperms, 617, 624
    in gymnosperms, 506
Egg depositors, 565
Egg mass, 553
Egg-producing organs, 588
Egg shell, 577
Egg tooth, 937
Einstein, Albert, 16, 163
Ejaculation, 258–259, 904–905, 917
EKG (*see* Electrocardiogram)
Elastic fibers, 704
Elastin, 77
Eldredge, Niles, 1028
Electrical activity, in brain, **890–891**
Electrical energy, 162, 186, 207
Electrical impulses, 136, 140, 142
    and gap junctions, 142
    in nerve conduction, 714, 847–848
    in plants, 692–693, 696
    and sodium-potassium pump, 136
Electric potential, 656, 848–849, 920 (*see also* Resting potential)
Electric ray, 186
Electrocardiogram (EKG), 757
Electrochemical gradient
    and active transport, in plants, 653
    and electron transport, 198–199, 205
    in photosynthesis, 218, 220, 229
    and potassium (K^+) ion in plants, 654–655
Electroencephalogram (EEG), 890–891
Electromagnetic radiation, 25
    speed of, 207
    wavelengths of, 207, 209
Electromagnetic spectrum, 207, 209, 210
Electron(s), **24**, 38
    arrangement of, 29–30
    in atomic structure, 28
    and chemical bonds, 31–33
    chemical importance of, 24, 38
    flow of (*see* Electron flow, cyclic)
    movements of, 28–35
    "shielding," 32
Electron acceptors, 171–173
    in electron transport, 195–197, 203
    and enzymatic reactions, 171–173
    in glycolysis, 187
    in Krebs cycle, 200, 204
    primary, in photosynthesis, 216, 218–219, 228–229
    in prokaryotes, 430
Electron carriers, 184
    in electron transport, 195, 197–201, 204
    in photosynthesis, 216, 219
Electron donors
    in green bacteria, 440, 450
    in photosynthetic nonsulfur bacteria, 440, 450
    sulfur as, 440
Electron flow, cyclic, **219**
Electron microscope, **14**, **95–99**, 305, 423, 454

Electron "shuttle," 200–201
Electron transport, 187, **193–205**, 418
    ADP in, 195–200
    ATP in, 195–201
    ATP synthetase, 205
    coenzyme Q in, 197
    cytochromes in, 196–197, 200
    electrochemical gradient, 198–199, 205
    electron acceptors, 195–197, 203
    electron carriers, 195–201, 204
    energy in, 201, 204–205
    energy levels and, 186
    $FADH_2$ in, 195–196, 200–201
    FMN in, 197
    integral membrane proteins, 200
    at mitochondrion, 191, 196–199
    and NADH, 196–198
    oxidative phosphorylation in, **196–197**, 198–200
    pH in, 198
    phosphates in, 197, 199, 200
    in photosynthesis, 206, 218–220, 229
    in prokaryotes, 431, 450
    and proton pump, 197–199, 205
    and vitamin E, 729
Electrophoresis, 307, **340**, 344, 347–349, 353–354, 396, 422, 978
Electroreceptors, 859
Elementary particles, 23
Elements, **23**
    CHNOPS, **35**
    vs. compounds, 34
    concentration of, in biomass, 1150–1152
    electrons in, 30
    isotopic forms, 25
    polar-covalent bonding in, 32
    structure, 24
Eleodes longicollis, 1114–1115
Elephant, 9, 603, 865
Elevation, and temperature, 1161
Elimination (digestive), 737
Elimination (extinction), 1020, 1108, 1114, 1116–1117, 1120 (*see also* Competitive Exclusion; Extinction; Island biogeographical hypothesis)
El Niño, **1020**, 1158
Elongation
    in plant cells, 634, 640
    stage of protein biosynthesis, 312–313
Emboli, **760–761**
Embryo, 103, 236, 258, 385, 544, 576, 596, 600, 625, 634, 905, 912
    and amniote egg, 900–901
    of angiosperm, 614, 617–618, **619**, 622, 624
    anterior-posterior axis of, 924, 928, 940
    cytoplasmic determinants and, 925
    development, 415, 427 (*see also* Development)
    in amphibian, 919, **926–931**
    of brain, 881
    in chick, 919, **931–940**
    in human being, 919, **940–951**
    in sea urchin, 919, **920–926**
    in diagnosis of genetic disease, 385
    and gene regulation, 367, 380
    genetic regulation of, 942–943
    of gymnosperms, 507
    immune system of, 812
    implantation of, 917
    layers of, 924
    mammal, homologies in, 969
    of plants, 625, 647–648, 506, 629–631, 634, 647–648
    and auxin production, 675
    and gibberellin, 681
    hormones in, 695
    RNA content in, 305
    sex organs in, 909
Embryonic induction, **930–931**, 952
Embryo sac, 614, 617
Emigration, 1088, 1092, 1104
    and gene flow, 982
    and Hardy-Weinberg equilibrium, 979, 989
    as limitation of population, 1099
    and population growth rate, 1088
Emotions, 854
    and hypothalamus, **883**
    and limbic system, 898
    and intrinsic processing centers, 893–899
    and memory, 894
    and menstrual cycle, 914

Emus, 967
Enamel, 716
Encapsulation, 282, 283
Endergonic reactions, 164, 174, 181, 183-184, 298
Endocrine glands, **828-839** (*see also* Endocrine system; Hormones; *specific glands*)
  adrenal, 823
    cortex, 824, **829-832**, 833, 839
    medulla, 824, **832**, 833, 839
  composition of, 822
  corpus luteum, 824
  ductless, 823
  vs. exocrine glands, 823
  gonads, 839
  hormones in, 823-824
  hypothalamus, 823, 824, **826-828**, 839
  ovary, 824
  pancreas, 823, 824, **832-833**, 839
  parathyroid, 824, **829**, 839
  pineal, 824, **833**, 839
  pituitary, 823, 824, **825-826**, 833
  regulation of, 824
  reproductive organs, 823
  testis, 824
  thyroid, 823, 824, **828-829**, 839
Endocrine system, **711-712**, 714, **821-840**, 899 (*see also* Endocrine glands; Hormones)
  cellular communication in, 808
  feedback control in, 712-713
  and hypothalamus, 883
  integration and control in, 856, 883, 898
  and nervous system, **711-713**, 821-822, 847
  and pancreas, 725
  and stomach, 720
  and temperature regulation, **784-786**
Endocytosis, 138-140, 142-143, 342, 837, 853
  coated pits in, 138
  cell-mediated, 139
  cell membrane and, 138-139, 143
  phagocytosis, **138-140**
  pinocytosis, **138-140**
  receptor-mediated, 138-140, 143, 839
  and vacuoles, 143
  vesicle-mediated, 138-140
Endoderm, 525, 532, 542, 545, 937, 952
  in amphibians, 928-929
  in chick embryo, 932, **933**, 937, 952
  in sea urchin, 924
Endodermis, 655
  and mineral uptake in plants, 655
  of roots, **632**, 633, 634, 648
Endoflagella, 437
Endometrium, **908**, 911, 912, 917, 941, 953
  and development of placenta, 944
  and implantation of ovary, 941
  in menstrual cycle, 913-914
Endomycorrhizae, 491-492
Endoperoxides, 835
Endoplasmic reticulum, 114, **115-117**, 118, 202, 458, 853
  in animal cell, **112**
  in cell division, 146
  and cell variety, 125
  in energy reactions, in cell, 168
  function, 127
  in liver cells, 726
  of muscle cells, 706
  in pancreatic cells, 725
  in plant cell, **113**, 140
  and plasmodesmata, 140
  rough, 115, 116-117, 118, 124
  smooth, 115, 116-117, 124
  transitional, 117, 118
Endorphins, **854-855**
Endoskeleton, 701, 713 (*see also* Skeleton, internal)
  advantages of, 593
  and muscle, 706-708
Endosperm, 617-618, 622, 624
  and gibberellins, 681
  and monocots, 631
Endospore-forming bacteria, 430, 431, 438
Endospores, 438
Endosymbioses, 452-453
Endothelium, 754
  of arteries, **753**
  of blood vessels, 792, 795
  of capillaries, **754**
  and glucose transport, 754

  and immune response, 792, 795
Endothermic reactions, 165
Endothermy, 597-599, **780**
  vs. ectothermy, 702
  vs. homeothermy, 782
  in mammals, 701-702
Energetics, **159-184** (*see also* Energy; Energy Cost)
  of active transport, 178-182, 184
  in biology, 12
  of cell, 112-113, 119, **127-143**, 154
  coupled reactions, **181-182**
  of DNA replication, 298-299
  energy reactions, **164-165**, 183
Energy, **159-184**, 1131-1132, 1153 (*see also* Energetics; Energy cost)
  of activation, 168, 169, 180-181, 183
  and ATP, in photosynthesis, 228-229
  in biology, 12
  in cell, 112-113, 119, **127-143**
  and cellular transport, 135-136
  chemical energy, 161, 165-166
  of chemosynthetic ecosystems, **1140-1141**
  and circulation, **749-764**
  in condensation and hydrolysis, 64
  conservation of, 165
  conversion of, 27, 36, **159-184**
  cost of food gathering, 1144
  and digestion in vertebrates, **715-730**
  of DNA replication, 298-299
  and ecosystems, 161, 1085, 1087, **1137-1145**
  efficiency of transfer in, **1143**
  electrical, 162, 186
  and electron transport, 201, 204-205
  endergonic reactions, 164, 174, 181, 183-184
  endothermic reactions, 165
  and entropy, 164-166
  expenditure, vs. oxygen consumption, 732
  flow of, 167
  free energy change, **165**, 183
  and kidney, 771, 774
  kinetic, 28, 43, 128, 162, 168-169
  of light, 206, 218, 228
  and living systems, 161, 166, 183, 187
  mechanical, 128, 161, 163, 165-166, 186
  of metabolism, **710-711**, **749-764**
  motion as, 28, 43, 161
  and NADH, 203
  of NADPH, and photosynthesis, 229
  in organic molecules, 57-59
  origin of, 165
  and origin of life, 87, 100
  and osmosis, 133
  in photosynthesis, 206, 228-229
  in plant cells, 24, 168
  of position, 28
  potential, 135-136, 180
    in photosynthesis, 218-220, 229
  radiant, 161, 165, 167
  reactions (*see* Endergonic reactions; Exergonic reactions)
  and respiration, **732-748**
  and sexual selection, 998-999
  solar, and biosphere, 515, **1122-1137**
  storage, vs. exercise, 742
  storage of, in organisms, 1139
  of sun, 84-85, 161, 167
  and temperature regulation, 777-790
  transfer, and ecosystem structure, **1145**
  transport, in living systems, 61
  in trophic levels, **1137-1143**
Energy-capturing reactions, **216-221**, 222, 228, 229
Energy cost
  and dominance hierarchies, 1063
  of food-gathering, 1141
  of homeothermy, 786
  of leaf-shedding, 623, 1164
  of respiration, 736
  of temperature regulation, 782, 786
Energy levels, atomic, **28-33**, 38, 162, 183-185
  and covalent bonds, 32-33
  and electron transport, 195-196
  and bioactive elements, 35
  and energy transformations, 181
  and glycolysis, 186
  and oxidation/reduction, 162, 165
  and photosynthesis, 210, 213, 216, 218, 228
  and respiration, 186
Energy of activation, 168-169, 180, 181, 183

Energy transformations, **159-184**, 186-187, 206
  enzymes in, 59
  in glycolysis, 189-190, **200-202**
  in Krebs cycle, 196
  from light to chemical energy, in photosynthesis, 211, 213, 222, 229
Engrailed gene, 942
Englemann, T. W., 212
Engram, 893
Enkephalins, 854, 855
Enterocoelous formation, 545, 562, 564, 587
Enterokinase, 722
Entomorpha, 1119
Entoprocts, 520, 541, 543
  vs. bryozoans, 541
  vs. hydrozoans, 541
  reproduction in, 541
Entrainment, **690**
Entropy, 165-166, 183, 710-711
  and biology, 12
  change in, **165**, 183
  and energy reactions, 165
  laws of thermodynamics, **162-166**, 183
  potential, 162-163
  and temperature, 183
Envelope, nuclear, 366
Environment, 407, 413, 710, 991, **992-993**, 994, **1008-1009**, 1088-1089 (*see also* Biosphere; Ecosystems)
  abiotic, 1131
  and adaptation, **970-972**, **1002-1005**, 1007, **1009**
  balanced polymorphism and, 982-983
  and behavior, 1053, 1077
  and biological organization, 37
  biotic, 1131
  and bud growth, 641
  carrying capacity of, 1089
  and communitites, 1106, 1129
  destruction of
    and acid rain, **50-51**
    and conservation biology, 1114, **1126-1127**
    and human population growth, 1092, 1093
    and increased CO_2, **226-227**
    and ozone layer, **1134-1135**
    in tropical rain forests, 1150, 1176-1177
  disturbances in (*see* Density-independent [population] factors)
  and ecological niches, 1109
  effect of, on cell division, 147
  and energy conversion, 27
  and evolution, 416, **962**
  "fitness" of, 211
  genes and, 268, **271**, 279
  and genetic recombination, 462
  internal (*see* Homeostasis)
  and limits on populations, 1095
  and menstrual cycle, 914
  and natural selection, 7, 991, 994, 1008
  and phenotype, 994
  photoperiodism, **684-689**
  and plant growth, 645-646, 671, **684-689**
  pollution, **50-51**
  and prokaryotes, 429
  punctuated equilibria, 1028
  and sexual reproduction, 986-987
  and speciation, 965, **1012-1021**
  and stabilizing selection, 995
  and stomatal movements, 654
  and water balance, 767-768, 775
Environmental pollution, **50-51**
Enzymatic pathways, 173-174, 192
Enzymes, **11**, 106, 110, 112, 115, 125-126, 183-184, 301, 320-322, 418, 540
  activation of, 175
  active site of, **170**, 171, 176-180, 183-184
  antiviral, 793
  biosynthesis of, 301
  in blood plasma, 763
  cascades, 837
  in catabolism of phenylalanine, 388
  in cell, 88, 124
  in cell division, 146, 153
  and cell membrane, 178
  change or loss of function in, 304
  coenzymes, 171-173, 183-184
  conformation of, 175-179, 183
  deficiencies in, 400
  denaturation of, 175, 177-178

I-18

digestive, 115–116, 171, 175, 574, 715, 730, 823
DNA repair, 298
in DNA replication, 294, 296–299, 358
and DNA sequencing, 340, 354
and drug resistance, 329
*Eco*RI, 341, 344, 346, 348
and energy transformations, 59
in ethanol catabolism, 202
in fermentation, 190–191
"ferments," 190
genetic regulation of, 302–303
in glycolysis, 187–188
*Hin*dIII, 341, 348
and hormone metabolism, 823
*Hpa*I, 341
hydrolytic, **118**, 125
inducible, 321–322
in lambda phage replication, 334–335
in lysosomes, 173
lytic, 118, 122
in mRNA transcription, 368
and metabolic pathways, 167–168
in methylation, 341
mitochondrial, 453
necessity for, in cells, 168, **169–180**
"one-gene–one-enzyme" hypothesis, **301–304**, 316
and oocyte penetration, 902
peptide bonds and, 175
pH and, 175–176
in phagocytosis, 138–139
in photosynthesis, 215, 219, 222, 229
primary structure of, 302
production of, 301
in prokaryotic transcription, 320–322
as proteins, 70, 72, 76
rate of reactions, 777
regulation of, in cell (*see* Regulation of Enzymes)
regulation of acids and bases, 176
repressible, 322
and repression, 327
and respiration, 191–193
restriction, **340–341**, 344–350, 365, 372
in RNA synthesis, 358
in rolling-circle replication, 328
segregation in cell, 173
specificity of, 170, 302
structure and function, 170
and temperature, 184
and transport proteins, 135
and viruses, 373–374
Eocene epoch, 496, 1012, 1034, 1051
Eohippus, 409, **968** (*see* Hyracotherium)
Eosinophils, 792
Epiblast, 932
Epidemics, 813
Epidemiology, 817
Epidermal cells, 214, 529, 666 (*see also* Dermal system; Dermis; Skin)
in angiosperms, 647
and cork, 644, 648
differentiation of, 634
of leaf, 627, 641, 648
of roots, **632**, 633–635
secondary growth and, 645
of stems, 636–637
as touch sensors, 691
and transpiration, 652–653
Epidermal ectoderm, 928–929, 935–937
Epidermal hairs, 629, 641
Epidermal pores, 495
Epidermis, 214, 508–509, 525, 527, 528, 531, 542, 563, 566, 640, 702, **703**, 710
development of, 952
of skin, 703, 710
Epididymis, **901–903**, 904–905, 917
Epiglottis, 718
Epilepsy, 889–891
Epilimnion, 46
Epiphytes, 1175
Episomes, **326**, 335, 337
Epithelial cells, 94, 111, 523, 542, 702
of digestive tract, 823
of glands, 822–823
and immune response, 813
of intestine, 721
of stomach, 720
Epitheliomuscular cells, 527, 528
Epithelium, 140, 532, **702–703**, 713, 735 (*see also* Epithelial cells)
in alveoli, 740
in bone, 705
vs. connective tissue, 704
in desmosomes and tight junctions, **703**
of digestive tract, 715
in eye, 868
in gills, 735
of glands, 839
and immune response, 791
of intestine, 555, 723
and respiration, 737, 738, 740
in sebaceous glands, 710
in skin, 710
Epithet, of binomial, 408–409, 426
Epsilon globin, 367
Equator
of cell, 151–155
and climate, 1136–1137
Equatorial plane
of cell, 155
in meiosis, 254, 256
Equilibrium (physical)
in atmosphere, 1133–1134
chemical, **174**, 175
dynamic, 129
ionic, 47
and potential energy, 174
of water potential, 134
Equilibrium (populations), 1095 (*see also* Hardy-Weinberg equilibrium)
in communities, 1123
island biogeography hypothesis, **1123–1124, 1126–1127**, 1130
life-history patterns, 1099
Equilibrium hypothesis of island biogeography, **1123–1124, 1126–1127**, 1130
Equisetum, 502
Equus, evolution of, **1025–1027**
Eras, geologic, **970**
Erectile tissue, 917
in clitoris, 909
in penis, **904**, 917
Erection, 760, 903, **904–905, 912–913**
Erector muscles, 710, 786
Ergot, 481, 486
Erithacus rubecula (*see* Robin, English)
Erosion, 665
Erwin, Terry, 15
Erythrocytes (*see* Red blood cells)
Escherich, Theodor, 409
Escherichia coli, 35, 115, 305, 307, 308, 310, 311, 312, 316, 337-338, 409, 429–430, 440, 669
bidirectional replication in, 320
biosynthesis of proteins in, 314
cell of, **431–432**
chromosome of, **319–320**, 337
conjugation in, 327
DNA of, 319–320
and drug resistance, 327
efficiency of, 320
flagellum of, 434
in genetic research, 284–286, 292, 295
infection cycle in, 330
lac operon of, 838
lactose metabolism in, 321–322, 324–326
lambda phage in, 334–335
molecular biology of, **319–320**, 337
mutants of, 322
mutation rate in, 437
name of, 409
phage infection of, 308
properties of, **35–36**
proteins in, 72
in recombinant DNA, 340–341, 343, 350–353
rolling-circle replication in, 327
shape, 436
symbiotic, 448, 725
viruses and, 330
Esophagus, 547, 551, 555, 568, 715, **718–720**, 730, 740
Essential amino acids, 76, 728
Establishment phase of population growth, 1088, 1089
Estradiol, 908, **913**, 914
Estrogen(s), 824, 836, 916–917, 941, 945
in human pregnancy, 941, 945
in oocyte development, 911
in ovulation, **913**, 914
in "the pill," 915–917
and sex differentiation in brain, 908–909
Estrus, 908, **915**, 917, 1036, 1071
Estuaries, 1134, 1159–1160
Ethanol, 204
abuse, 202
catabolism, 191, 202, 203
and cirrhosis, 190–191
fermentation of, 190–191, 202
and NAD^+, 202
Ethical dilemmas, in genetic counseling, 384, 388, 397–398
Ethology, 1052–1079
Ethyl alcohol (*see* Ethanol)
Ethylene, 673, **679**, 694–695
and auxins, 694–695
and inhibition of apical dominance, **675**
and leaf abscission, **679**
and touch response in plants, 691
Etiolation, 687
Eubacteria, 430, 440, 450, 453
Eucalyptus, 70
Euchromatin, 359, 360, 380
Eugenics, 987
Euglena, 455, 458–459, 474
Euglenoids, 458, 459, 476
Euglenophyta, 455, 456, 457, **458–459**, 476
Eukaryotes, **13**, 423, 426, 452
compartmentalization of, 479
diversification of, 497
genetic variability in, 986–987
habituation in, 474
photosynthetic, 424
vs. prokaryotes, 431, 434, 440
unicellular, 424
Eukaryotic cell, **90–92**, 93, 101–102, 104, 106, 108, 115–118, 122–125, 167, 219, 228, 249, 305, 309, 311, 316, 355–356
bidirectional synthesis in, 358
cell division in, 144, **145**, 155
chromosome of, 281, 294, 355–358, 379–381, 453
DNA and histone proteins in, 452
DNA biosynthesis in, 145, 146, 148, 358, **359–362**, 380
evolution of, 375, 452–454, 476, 479
evolutionary advantages of, 454
fungi as, 482
genes of, 340
genetic regulation in, **359–362**
genetic transcription, 355, 358–360
genome, 347
genome of, **363–368**, 375
homologous recombination in, 336
life cycle of, 259
Mendelian genetics in, 361
microscopy and, 96
molecular genetics of, **355–381**
mRNA transcription in, **368–370**
nuclear envelope in, 452, 453
nucleus of, 452
organelles in, 452, 453
vs. prokaryotic cell, 355, 452
protists, 452, 476
rate of replication, 380
and recombinant DNA, 342
regulation of exchange in, 127
replication of chromosome, 358
respiration in, 191, 204
RNA transcription in, 358, 359–361
single-celled, 144
size of, 95–96
and T lymphocytes, 801
transposons in, 371, **374–375**, 381
unicellular, 259, 482
viruses, 373–374
Eumetazoa, 520
Euphorbs, 1007
Eusocial species, 1060, 1069, 1078
Eustachian tube, 864
Evaporation, 43–44, 161, 736, 768
heat loss through, 789–790
and temperature regulation, **785**, 789
and water movement in plants, 652, 669
Eve, 237
Evening primrose, 244
Evidence
for evolution, **961–973**

I-19

Evidence *(Continued)*
  in science, 15
Evolution, 238, 247, 407–408, 416
  and adaptation, 245
  of aerobic bacteria, 452
  agents of, 981–984
  of allergies, 813
  of alternation of generations, 494–495
  of amniote egg, 596
  of angiosperms, 510
  of animals, 518, 520, 544, 545, 710–711
    arboreal life and, 1031–1033
  of arthropods, 567, 578
  of ATP, 182
  of behavior, **1052–1079**
  biogeography and, **966**, 967
  and biological interaction, 37
  of biomes, 1161
  of birds, 599
  of blood groups, **997**
  of the brain, 531, **881–884**, 885, 887, 893, 898–899, 1045
  and breakup of Pangaea, **1012–1013**
  of cell-mediated immune response, 809
  of cell membrane, 370
  of cell types, 92
  of cellular energetics, 161, 167
  of C₄ photosynthesis, 226
  change and, 246
  chemical, 87–88, 142
  of chloroplasts, 146, 442, 453
  of cilia and flagella, 123
  and cladogenesis, 1021–1022
  of coelom, 544
  coevolution, 1007, 1009
  of color vision, 872
  and competition, 12
  constraints on, **1006**, 1009
  and continental drift, **1012–1013**
  convergent, 1008, 1009
  and cyanobacteria, 442, 456
  in Darwin's finches, 1019
  of digestive tract, 532, 715
  divergent, 1008, 1009
  of DNA, 362, 370
  and DNA mutations, 287
  of echinoderms, 587
  and ecology, **961–973**
  and energy conservation, 59
  and environment, 416
  of eukaryotes, 375, 452–453, 1120–1121
  evidence for, **961–973**
  of eye-hand coordination, 1002
  of fishes, 587, 594
  of flight, **598–600**
  of flowers, **512–515**, 517
  of four-carbon pathway of photosynthesis, 228
  of fruits, 515
  of fungi, 482
  of gene pools, 989
  genetic basis for, **974–990**
  and genetic recombination, 326, 337
  and genetic variation, **975–989**
  and geology, 2–7
  of glycolysis, 191
  and Hardy-Weinberg equilibrium, 979–981
  of hemoglobin family, 367
  history, 407, 412–413, 415, 424, 427 *(see also Phylogeny)*
  of hominids, **1031–1051**
  of *Homo sapiens*, 362, **1031–1051**
  of hormones, 838
  of horse, **1025–1027**, 1028
  of hypertonic urine in mammals, 772
  of immune response, 791
  of insects, 578
  and introns, 362
  of invertebrates, 538
  of jaws, 594
  and kidney function, 765
  of mammals, 600–601, 1010, 1023
  of marsupials, 600, 1010, 1023, 1031
  and mass extinctions, 1025
  mechanism of *(see Evolution, theory of; Natural Selection)*
  and Mendelian principles, 245
  of metabolic rates in animals, **754–755**, 763
  of mitochondria, 146, 362, 452–453
  of monotremes, 600, 1031
  of multicellular organisms, 454
  of NAD, 182
  and natural selection, **991–1009** *(see also Natural selection)*
  of nervous system, 535
  novelties in, 417
  of oomycetes, 482
  "opportunism" of, 767
  of organic molecules, 142
  and origin of life, 88
  of oxygen debt, 190
  of pancreas, 725
  and photosynthesis, 206, 218
  and phyletic change, 1021
  of placental mammals, 600, 1010, 1023, 1031
  of plants, 501–502, **510**, 613, 615, 621, 625
  in populations, 974, 989
  and predation, 1114–1119, 1129
  of primates, **1031–1051**
  "progress" in, **1005–1006**
  of prokaryotes, 428, 430, 452, 454
  properties of life, 26–27
  of proteins, 326, 363, 370
  of protists, 452–455
  and punctuated equilibria, 1028
  rate of, 1028
  and recombinant DNA, 354
  relationships, 418, 420
  of reptiles, **597–598**
  of respiratory systems, **734–737**, 747
  separation in, 421
  of sexual reproduction, 1102
  of social behavior, 974, 989, 1059
  and surface-to-volume ratio, 520
  and survival, 425
  theory of *(see Evolution, theory of)*
  of transposons, 375
  of vertebrates, 594
  of viruses, 444
  of vision, 870
Evolution, theory of *(see also Darwin, Charles; Natural Selection; Origin of Species)*
  and altruism, 1059
  Darwinian, **5–10**, 245, 306
  DNA, RNA, and confirmation of, **306**
  evidence for, **961–973**
  and fossils, 3–4, **966–968**
  geology and, 2–7
  history of, **1–10**, 27
  and Mendelian principles, 245
  and paleontology, 1028
  phenotype in, 993
  pre-Darwinian, **1–5**
  and punctuated equilibrium model, **1028–1029**
  recent, 972–973
  and sterile caste problem, 1058–1059, 1065
  synthetic theory, **972–973**
Evolutionary systematics, **412–414**, 427
Exchange, chemical, 34–35, 38
Excision
  of exons, 369
  of introns, 369, 370
  random, and recombinant DNA, 348
Excretion, 546, 575, 712, **765–776**, 934
  in human embryo, 944–945, 953
  kidney, 766
  and urea, 766
Excretory organs, 546, 567, 952 *(see also Excretory system)*
Excretory system, 542, 547, 554, 574–575, 936
Exercise, 190, 732, 742, 745, 759
Exergonic reaction, 164–167, 169, 174, 181, 183–184, 298
  in sodium-potassium transport, 136
Exocrine glands, 823, 839
Exocytosis, **138–140**, 142, 143, 852
  cell membrane and, 143
  and vacuoles, 143
Exons, **361–362**, 380
  excision of, 369
  and functional regions, 363
  and single-copy DNA, 366
  splicing of, 369–370
Exoskeleton, 66, **566–567**, 568, 578, 585–586, 734, 735 *(see also Skeleton, external)*
Exothermic reactions, 165
Exothermy, 778

Expectation, in reciprocal altruism, 1074
Experience
  and heredity, 951
  in learning, 893, 1055, 1078
  vs. selfish gene in human beings, 1075
Experimental design, 15, 86, 87, 109, 131, 208, 212, 309, 552, 581, 660–661, 678, 802, 887, 930, 963, 964–965, 1000–1001, 1002– 1003, 1004–1005, 1053–1057
  artifical selection, 976–977
  for competition in the wild, 1112–1113
  conclusions, 15
  controls in, 15, 131
  Darwin, 672
  DNA sequencing, 978
  in ecology, 1087, 1106–1107, 1119, 1129
  "fitness" of material, 238
  hypothesis in, 85, 131
  and intermediate disturbances hypothesis, 1124–1125
  intervention vs. observation, 1106–1107, 1119, 1129
  in island biogeography model, 1123–1124
  probability, 245
  Went, 672–673
Expiration, 738, 741, 768
Exploitative competition, 1106
Exponential growth, 1088–1090, 1104–1105, 1148
  of AIDS, 1091
Expressivity, **271**, 279
Expulsion (in labor), 950
External fertilization, **920**, 926
Exteroceptors, 859–860
Extinctions, 598, 1030, 1087, 1104, 1113, 1130
  and agriculture, 1148
  of birds, 496
  and California condor, 1104
  of dinosaurs, 496, 510,
  in fossil record, 1024–1025
  of hominids, 1043
  and human evolution, 1047
  and human responsibility, 1176–1177
  vs. immigration, 1123–1124 *(see also Island biogeography model)*
  and isolated populations, 1011
  of mammals, 496
  mass extinctions, 1176–1177
  of plants, 497
  rate of, 1024
  theories of, 44
  and whooping crane, 1104
Extracellular fluid, 765, 769
Extracellular matrix, 704, 713
Extraembryonic chorion, 944
Extraembryonic coelom, 934
Extraembryonic membranes
  of chick embryo, **932–934**, 952
  of human embryo, **944–945**, 952
Extremities (bodily), adaptations in, 1002
Eye(s), 879
  of annelids, 554
  arthropod, 566
  camera, 581
  compound, **579–582**, 586
  development of
    in chick embryo, **936–937**
    in human embryo, 945–946
  evolution of, 866
  of frog, 870
  frontally directed, 1033
  human
    vs. bee, 859, 1060
    vs. frog, 870
    and visible light, 209
  laterally directed, 1033
  mammalian, 601
  of mollusks, 549
  of octopus, 552–553
  retina, **868–872**
  simple, 582 *(see also Ocelli)*
  vertebrate, 535, 580, **866–873**
Eye-hand coordination, 870, 877, 1002, 1032
Eyelashes, 789
Eyelids, 947
Eyespots, 588

Face, development of, 947
Facilitated diffusion, 135–136, 142–143

Facilitation hypothesis, **1125**, 1128, 1130
Factor VIII, in hemophilia, 392, 397, 752
FAD, 204, 729
   in Krebs cycle, 195, 200
$FADH_2$, 204
   and electron transport, 195–197
   in Krebs cycle, 195–196, 200–201
Fainting, 759–760
Fairy rings, 486
Fall overturn, 46
Family, 410, 412, 427, 915
Family tree, 423
Famine, 1149
Fangs, 566, 568, 585 (*see also* Chelicerae; Pincers)
Far-red light, and phytochrome, 686–687, 695
Farsightedness, 867
Fascicle, 505
Fast-twitch fibers, 878
Fat(s), **67–70**, **228**, **578** (*see also* Fatty acids, Lipids)
   absorption of, in lymphatic system, 763
   and calories, 68
   chemical bonding in, 34
   condensation reactions in, 67–68
   as connective tissue, 710
   in diet, 730
   digestion of, 720, 722–733, 730
   as energy storage, 67
   and fatty acids, 67
   and glucagon, 832
   and glucocorticoids, 830
   and heart disease, 761
   human requirements for, 726, 730
   insolubility of, 47
   as insulation, 68–69, 779, 786, 790
   as insulation, 68–69
   molecular structure of, 67
   saturated, 67–68
   and seasonal adaptation, **787–789**, 790
   subcutaneous, 710
   transport in bloodstream, 750
   unsaturated, 67–68
Fat cells, 723, 729
Fate maps, 929
Fatty acids, 55, 67–69, 202
   absorption of, 723
   breakdown of, 825
   in chemoreception, 863
   digestion of, 721
   essential, 730
   and fats, 67
   and growth hormone, 833
   in muscle fibers, 878
   properties, 82
   prostaglandins as, 833
   saturated, 67
   unsaturated, 67
Fear, 720, 760, 847
Feathers, 598, 599, 702, 779, 786–787, 790, 937, 938 (*see also* Insulation)
Feather stars, 587, 589, 604
Feces, 547, 724–725, 768, 775
Feedback inhibition, **176**, 177, 184, 712–713
   and aldosterone, 775
   in blood pressure regulation, 762
   of hormone production, 823, 825, 827, 831, 839
   in liver regulation of glucose, 730
   and recognition memory, 899
   in vertebrates, 714
Feeding zones, 1109, 1110
Felidae, 412
Felis, 412
Female, 235–236, 243, 258
   biological role of, 235
   cell (*see* F^- cell, Conjugation)
   and color blindness, 391–392
   genetic disorders in, 400
Female choice, and sexual selection, 984, **998–1001**
Female cooperatives, in hominid evolution, 1044
Female hormones
   and atherosclerosis, 761
   regulation of, **913–915**
   reproductive system, **907–915**, 917
Fermentation, 11, 190–191, 204, 206
Femur, 705, 707, 1041
Fenton, M. Brock, 1115–1116
Fermentation, 11, 190–191, 204, 206
   in herbivore digestion, 724
   in urkaryotes, 453

and yeasts, 486
"Ferments," 190
Ferns, 256, 496–498, **502–503**, 504, 526, 1164
   fossils of, 502
   leaves of, 503
   life cycle of, **250–252**
   reproduction in, **502**, **503**
   sporophyte of, 250
Fertility, of soil, 663
Fertility factor (*see* F plasmid)
Fertilization, 236, 238, 426, 900, 901, **912**
   and allele frequency, 984, 989
   in angiosperms, 511, 615, **617**
   artificial, 236
   in chickens, 931
   and contraception, 915–917
   external, 519, 523, 538, 543, 550, 557, 558, 559, 588, 595, 900, 920, 926
   and gamete fusion, 252
   and genetic variability, 984
   in human beings, 940–941
   internal, 519, 523, 535, 542, 543, 550, 551, 553, 577, 595, 900
   and life cycle, 250
   and meiosis in oocyte, 917
   and nitrogen-fixing symbionts, 665
   in plants, 499, 502
   and sea urchins, 920–921, **922**, 951
   in sexual reproduction, 249–251, 261
   of self, 590
Fertilization membrane, **921–922**, 941
Fertilization tubes, in water molds, 467
Fertilizers (synthetic), 669, 1149
Fetus
   addictions in, 948
   amniocentesis of, 387–388, 400
   deformities of, 947
   heart of, 947–949
   and hemoglobin, 367, 743
   human, **946**
   male sex hormones in, 905
   prenatal detection of disease, 387
   spontaneous abortion of, 385
Feulgen, Robert, 282–283
Feulgen stain, 282–283
Fever, 784, 792–793, 795
F factor (*see* F plasmid)
F factor sequence, 327–328
Fiber (dietary), 730
Fiber cells (in plants), **637**, 647
Fibers
   in connective tissue, 704
   elastic, 704
   muscle, **706–708** (*see also* Muscle fibers)
   nerve (*see* Axons)
   reticular, 704
Fibrils, 77, 426
   contractile, 426
   of spirochetes, 437
Fibrin, **751–752**, 761
Fibrinogen, 750, **751**
Fibroblast, 100
   mouse, 146
Fibrosis, 394
Fibrous proteins, 77, 82, 111
Fibrous root systems, 635
Fibrous sheath, 902
Fiddle heads, 503
Fidelity
   and selfish gene, 1072
   as survival strategy, 1074
Field layer, 1164
"Fight or flight" reactions, 846–847, 857
   and infant alertness, 951
Filaments (*see* Actin filaments; Contractile filaments; Intermediate filaments; Myosin filaments)
   and bacterial shapes, 436
   in desmosomes, **703**
   in flowers, 511
   fungal, 480 (*see also* Hyphae)
   in prokaryote flagellum, 434
   of sarcomeres, 874
   of stamen, 614, 624
Filariasis, 540
Filter-feeding, 549
Filtration, in kidney, 771, 776
Fingers, 416
   development of, in human embryo, 946–948

evolution of, in primates, 1032
and motor cortex, 886
Fire
   control of, and human evolution, 1045, 1051
   in grasslands, 1169–1170
Fire blight, 439
Fireflies, **580–581**
Firs, 505
"First family," 1040 (*see also* Hadar fossils)
First filial generation (*see* F_1 generation)
Fischer, Emil, 170
Fish(es), 587, **593–595**, 604 (*see also specific fishes*)
   bony, 593–595, 604
   brain of, 883
   cartilaginous, 593, 594, 595 (*see also* Dogfish; Sharks; Skates; Rays)
   in coral reefs, 530
   earliest, 497
   freshwater, 594
   heart, 763
   kidney in, 767
   lunged, 594–595
   respiration in, 734–737
   rhinencephalon, 883
   saltwater, 594
   startle behavior in, 1053–1054
   temperature regulation in, 780–781
   and transition to land, **593–595**
Fishing industries, and ocean currents, 1157
Fish oils, 728
Fission, 535
   binary, in protozoa, 568–569
"Fitness," 974–975, **1059**, 1066 (*see also* Inclusive fitness; Reproductive success)
   and behavioral evolution, 1076
   and competition, 1106, 1129
   of *Drosophila* for genetic research, 263
   of environment for photosynthesis, 211
   and frequency-dependent selection, **997**, 1009
   of light, **209–211**
   of material, 238, 284
   and natural selection, 991–992
   and sexual selection, 1001
   and social behavior, 1059, 1078
   of water, 54
Fitz Roy, Captain, 1, 2, 5
Fixation
   of carbon, 216, 223–225, 228, 230
   of microscopic specimens, 96, 98, 101
   of nitrogen, 352
Fixatives, 96, 98
Fixed action patterns, **1054–1055**, 1077
Flagella, 92, 110, **122–123**, 125, 131, 149, 153, 251, 426
   algal, 456–457
   apical, 457
   ATPases in, 181
   vs. axial filaments, 437
   chemiosmotic power and, 200
   of dinoflagellates, 460
   of euglenoids, 458–459
   lateral, 457
   of opalinids, 473
   prokaryotic, **433–436**, 450
   protozoan, 468–470
   of slime molds, 469
   of sperm cell, 902–903
   of sponge, 522
   of water molds, 467, 469
Flagellated cells, 522
Flagellates, 455, 477
Flagellin, 433–434
Flakes, 1042, 1047
Flame cells, 537, 542
Flamingo, 423
Flatworms, 122, 520, 526, 532, **533–537**, 542 (*see also* Tapeworms; Trematodes; Turbellarians)
   free-living, 534
   life cycle of, 537
   marine, 534
   parasitic, 542
Flavin(s)
   and active transport in plants, 655
   in phototropism, 655, 673
Flavin adenine dinucleotide (*see* FAD)
Flavin mononucleotide (*see* FMN)
Flavoproteins, 195
Flemming, Walter, 109

I-21

Flicker-fusion rates, 580
Flies, 573 (see also Drosophila)
Flight
  evolution of, in birds, **598-600**
  and homeothermy, 782
  insects and, 573, 576, 579, 586
  reptiles and, 597
Flightless birds, adaptations in, 1006
Floral parts, 513, **613-614** (see also Carpels; Petals; Sepals; Stamens)
  evolution of, 613-614
  fusion of, 513
  senescence, and ethylene, 679
Florets, 514, 616
Florida scrub jay, 1070, 1074
Floridean starch, 457
Flour, 622
Flower clusters, 616, 670
Flowering, 694-695
  and bolting, and gibberellins, 681
  clines and, 1002
  hormones for, **688**
  and photoperiodism, 684-686, 695
  and phytochrome, 695
Flowering plants, 504 (see also Angiosperms; Flowers)
  body plan, 626
  lack of centriole, 146
  polyploidy in, 1014
  reproductive isolation in, 1016
Flowers, 503, **511-515**, 611, **613-624**, 626
  adaptations in, 515, 971
  and apical meristem, 640-641, 626
  and auxin, 675
  and circadian rhythms, 689-690
  coevolution with pollinators, 1003
  color
    evolution of, 517
    and pollinators, 513
  constancy and, 514-515
  defenses against predation, 517
  evolution of, **512-515**, 517
  and evolution of insects, 513
  fertilization in, 511
  floral parts, 513 (see also Floral parts)
  imperfect, 511
  odors, 515
  ovaries in, 513
  perfect, 511
  pollination, **512-515**
  reproduction in, 512
  reproductive structures, 517
  shape of, and reproductive isolation, 517, 1016
  structure of, 511
  symmetry in, 513
Fluctuations, in population, 1095
Fluid-mosaic model of cell membrane, 106, 124
Fluids
  bulk flow of, 128, **129**, 142
  movement of, and gravity, 142
  net flow of, 133
Flukes (see Trematodes)
Fluorescence, 214
fMet, 312-313
FMN, 195
  in electron transport, 197
  in Krebs cycle, 195, 200-201
F_O factor
  in ATP synthetase, 198
  in electron transport at mitochondrion, 198-199
Folic acid, 177, 729
Follicles
  lymphatic, 794-795
  ovarian, 825-826
Follicle-stimulating hormone (see FSH)
Following response, 1056
F_1 factor
  in ATP synthetase, 198
  in electron transport at mitochondrion, 198-199
F_1 (first filial) generation, **239-244**, 260, 265
  and principle of segregation, 239
Food (see also Diet; Nutrition)
  competition for, 1107-1108
  and ecological niche, 1109
  human, 488
  multicellular organisms and, 714
  and prey communities, 1118-1119, 1129

production of, ecological problems, 659, **1148-1149**
processing of, 533
storage of
  in algae, 476, 477
  in plants, 632, 636-640
Food chains, 459, 1134, **1137-1138, 1150-1152**, 1158
Food gathering, **1144**
Food sharing, 1074-1075
Food supply, as population limitation, 1088, 1089
Food vacuole, 471
Food webs, 1138, 1140-1141, 1143
"Foolish seedling" disease, 680-681
Foraging, 1114-1119, 1129, 1143
Foramen magnum, 1039
Foraminiferans, 470
Forebrain, 898 (see also Cerebral cortex; Cerebral hemispheres; Corpus callosum; Diencephalon; Hypothalamus; Telencephalon; Thalamus)
  development of, in chick embryo, 935
  evolution of, **883-884**
  and memory, 899
Foregut, 572
Forehead, 601, 893
Forelimb, vertebrate, 413, 969
Forensic medicine, 398
Forests
  alpine, 1166
  coniferous, **1166-1168**
  deciduous, 1177
  destruction of, 226, 1126-1127
    and mass extinctions, 1176-1177
  detritivores vs. herbivores in, 1143
  herbivores in, 1143
  layers of growth in, 1164
  mixed west-coast, 1166
  monsoon, 1174
  nitrogen recycling in, **1148-1150**
  Pacific northwest, **1166-1168**
  subtropical mixed, 1164, 1166
  taiga, **1166**
  temperate, **1164-1166**
  tropical, **1174-1177**
  tropical mixed, **1174**
  tropical rain forests, **1174-1176**
Form, perception of, **887-888**
Formulas, 727
Formyl, 313
Forsythia, 120
Fossey, Dian, 1036
Fossil(s), 85, 89, 90, 91-92, 100, 415, 521, 567, 587, 597, 992, 1008
  adaptive radiation in, 1022-1023
  of angiosperms, 509
  of apes, 1035
  Archaeopteryx, 967
  of australopithecines, 1038-1042
  and bipedalism, 1044
  of birds, 598, 599
  carbon and, 57
  cladogenesis and, 1021-1022
  and continental drift, 1013, 1029
  dating of, 25, **970-971**
  earliest, 497
  of Eohippus, 968
  and evolution, 973
  and evolutionary theory, 3-5, 966-968
  extinctions in, 1024-1025
  of ferns, 502
  and geological record, **970-971**
  of green algae, 494
  at Hadar, 1040-1041
  of hominids, 1038-1042
  of *Homo sapiens*, 1046-1047
  of horses, **1025-1027**
  at Laetoli, **1039**, 1041-1042
  of monkeys, 1034
  phyletic change in, **1021**
  of plants, 491, 496, 502-503, 613
  and punctuated equilibria, **1028-1029**
  speciation in, **1021-1025**
Fossil fuels, 504
  and fertilizers, 669
  and "greenhouse effect," 226, 1144, 1177
Fossil record, 412, 421-422, 427, 510
Founder effect, **983**, 990, **1011**
  and cladogenesis, **1022**

Four-carbon pathway, **223-228** (see C_4 plants)
  evolution of, 228
Four-chambered heart, evolution of, 763
Fovea, 866, 868, 870
  evolution of, in primates, 1033, 1051
  in visual cortex, **887**
Fox, Sidney W., 88-89
F (sex factor) plasmid, 326-329, 337
Fragmentation (reproduction), 437, 450, 529, 535, 538
Frame-shifts, **316**
Franklin, Rosalind, 288-289, 291
Free energy change, **165**, 167, 169, 183
  in cellular respiration, 187
  and chemical equilibrium, 174
  and energy reactions, 181
  in glycolysis, 187, 204-205
  and metabolism, 165, 167, 169
Free-moving limbs, evolution of, 1032-1033, 1051
Free water, external fertilization in, 499, 502
Freeze-drying, 661
Freeze-fracture technique, 135, 221, 441, 771
Freeze-thaw process, in tundra, 1168
Freezing, 44
Freshwater environments, 1154-1155, 1177 (see also Rivers; Streams)
Freshwater organisms, 132, 440, 476-477 (see also Rivers; Streams)
  adaptations in, 775
  diatoms, 457
  euglenoids, 457
  golden-brown algae, 457
  green algae, 457, 461
  protozoa, 470
  water balance in, 775
  water molds, 467
Frequency
  of alleles (see Allele frequency)
  of genotype (see Genotype frequency)
Frequency (sound), 583-584, 887
  in bat sonar, 1115-1116
Frequency-dependent selection, **997-998**, 1009
Freud, Sigmund, 985
Friction, and homeostasis, 790
Frizzle trait, 273
Frogs, 420, 595
  development in, 919, **926-931**
  embryo, 919, **926-931**
  reproduction in, 595
  respiration in, 736
Frontal lobes, 885-886, 899
  evolution of, 892
  intrinsic processing centers and, 885-886
Frontally directed eyes, 1033
Fructose, **59-61**, 62-63
  alpha-fructose, 63
  beta-fructose, 63
  in biosynthesis of sucrose, 170
  in glycolysis, 188
  in hydrolysis of sucrose, 170
  and photosynthesis, 228
  in seminal fluid, 905
  structure, 62-63
Fructose diphosphate, 223
Fructose 1,6-diphosphate, 188
Fructose 6-phosphate, 188
Fruit fly (see Drosophila)
Fruiting bodies, 438
Fruits, 511, 515, 626, 694
  abscission of, and auxins, 695
  aggregate, 620
  of angiosperms, 509, 613-614, **619-621**, 624
  and cell types in angiosperms, 647
  citrus, 134
  dehiscent, 620-621
  development of, 675, 694
  evolution of, 515
  growth, and cytokinins, 694
  indehiscent, 620-621
  multiple, 620
  production of auxins in, 675
  ripening, and ethylene, 679, 694-695
  sclereids in, 637
  simple, 620
FSH, 824-826, **906-907**, 917
  in menstrual cycle, **913-914**
  and "the pill," 915
F_2 (second filial) generation, 239-245, 265

dominant-recessive ratio in, 245
and principle of segregation, 239
Fuchsia, 246
Fuchsin, 282
Fucoxanthin, 456-457, 459, 465
Fucus, 465
Fuel supplies, 1148
Fugitive species, 1089
Fullard, James H., 1115-1116
Functional groups, 56-57
  aldehyde, 56
  carboxyl, 56
  and charge, 56
  hydroxyl, 56
Functional protein groups, 320, 337
Functional region, 138, 363
Functional segment (*see* Domain)
Fundamental niches, 1113, 1129
Fungi, **91**, 92, 101, 106, 122, 124, 125, 250, 411, 424-425, **479-492**, 1142, 1153, 1166
  and antibiotics, 449
  ascomycetes, **484-486**, 491
  basidiomycetes, **486-488**, 491
  cell walls of, 479, 480
  characteristics of, **480-481**
  chitin vs. cellulose, 480
  classification of, 454, **481-488**
  as coenocytic organisms, 480, 482
  commercial uses of, 481
  and decomposition, 480, 492, 1142, 1153
  as detritivores, 1142, 1153
  deuteromycetes, **488**, 491
  dikaryons, 481, 491
  diseases caused by, 481
  dispersal mechanisms in, 484
  earliest, 497
  ecological importance, 492
  and ectomycorrhizae, 491
  and endomycorrhizae, 491
  evolution of, 482
  and forest floor, 1166
  gametangia, 481
  gametes, 481
  haustoria, 480, 490
  as heterotrophs, 480, 491
  hyphae, 481, 486, 491
  kingdom, 13, 424-427
  lack of centriole, 146, 150
  and lichens, **488-490**, 491, 492
  meiosis in, 462, 491
  mushrooms, **486-487**
  mycelia, 480-481
  and mycorrhizae, 488, **490-491**, 492
  Neurospora as, 303
  and nitrogen cycle, 1146
  parasitism in, 480, 488, 492
  and photosynthesis, 488
  and plant diseases, 492, 678
  and plant nutrition, 662
  as predators, **489**
  reproduction in, 481
  as saprobes, 480, 488
  sporangia, 481, 491
  spores, 462, 480-481, 484, 491
  symbiosis with plants, **488-491**, 664-665, 670
  syngamy in, 462
  in taiga, 1166
  zygomycetes, 491
Fungi Imperfecti, 481, 488, 489, 491
  Penicillium, 481
  sexual cycles, lack of, 481
Fungus infections, 448
Fur, 702, 779, 787, 790 (*see also* Insulation)
Fur seal, 1007
Fusion (*see also* Fertilization)
  atomic, 163
  of experimental cells, 803
  of gametes, in meiosis, 252

GABA (gamma-aminobutyric acid), 853, 856
Galactose, 61, 320
  in glycolysis, 190
  in prokaryotic transcription, 325
Galapagos finches, 6, 966, 1022 (*see also* Darwin's finches)
Galapagos Islands, 6, 8, 27, 966, 1017-1021
Galapagos Rift, 1140
Galapagos tortoise, 6

Galilei, Galileo, 7, 9
Gallbladder, 715, **722-723**, 937, 946
Gall midge, 1103
Gallstones, 722
Galvani, Luigi, 847
Gametangia, 467, 481, 483, 485, 486,
Gamete(s), **237**, 249-250, 252-253, 260-261, 315, 426, 591 (*see also* Chromosomes; Fertilization; Genes)
  of algae, 460, 462-464, 465
  of angiosperms, 615
  of anthozoans, 529
  of ascomycetes, 486
  of brown algae, 465
  of cnidarians, 526
  defective, in genetic disease, 385
  diploidy in, 1013, 1015
  DNA of, 286
  and early theories of heredity, 237
  evolution of, 463
  female, 463, 907, 917
  of ferns, 250-251
  formation of, 250, 252
  in fungi, 481
  fusion of, 463-464 (*see also* Syngamy)
  genetic investment in, 1071
  genetic recombinations in, 259
  and germ cells, 924-925
  haploidy in, 250, 252, 519
  and heredity, 250, 252, 253
  human, 252-253, 258-259
  in hybrids, 1015
  male, 463, 900-902, 917
    vs. female, 1071
  of mollusks, 546
  motility of, 463
  mutation in, and gene flow, 982
  and parthenogenesis, 1103
  paternal, 278
  of *Plasmodium*, 473, 474
  in protozoa, 468
  and segregation of alleles, 240-241
  and sexual selection, 999
  of sponges, 524
  vs. spores, 463
Gametophytes, 250-252, 256
  of algae, 463, 464, 465
  in angiosperms, 614-615, 617, 624
  evolution of, 501
  in flowers, 511
  in gymnosperms, 507
  of pines, 506-508
  shrinkage of, in evolution, 501
Gamma globin, 367, 799
Gamma rays, 207
Gamow, George, 307, 309
Ganglia, 533, 535, 551, 557, 564, 568, 585, 842-843
  adrenal medulla as, 712
  of autonomic nervous system, 846-847, 856
  celiac, 846
  cerebral, 535, 549, 551
  development of, in chick embryo, 935
  dorsal, 568
  pedal, 549, 551
  pleural, 551
  subesophageal, 551
  supraesophageal, 551
  in sympathetic nervous system, 846, 857
  visceral, 549, 551
Ganglion cells, 868, 869, 870, 888, 936
Gap junctions, 142-143, 756, 851-852
  and sodium-potassium pump, 142
Gap phases of cell cycle (*see* G_1 phase; G_2 phase)
Garden of Eden deep-sea oasis, **1140-1141**
Garden pea (*Pisum sativum*), 238, 239
Garrod, Sir Archibald, 301
Gases
  in atmosphere, 1133
  diffusion of, in leaves, 669-670
  in plants, 628, 648
  in polychaetes, 558
Gas exchange
  in animals, 547
  at capillaries, 754
  evolution of, 734-737, 747
  in human respiratory system, 738, 740, **741-745**
  in respiration, **733**, 734-736, **737**, **741-745**
Gas transport, 734, **741-745**

Gastric glands, 720
Gastric juices, 720, 722, 730
Gastric pits, 720
Gastrin, 720, **723**, 823, 855
Gastrodermis, 525, 527, 528, 531, 542
Gastropods, 545, 546, **550-551**, 563 (*see also* Abalones; Periwinkles; Slugs; Snails; Whales)
  predation in, 545
  reproduction in, 551
Gastrotrichs, 520, 540, 541, 543
Gastrovascular cavity, 525, 528, 529, 531, 542
Gastrula, 923, 924, 928
Gastrulation
  in amphibian, **927-929**, 952
  in chick embryo, **932-933**, **935**, **936**
  and differentiation in embryonic tissues, 930
  in human embryo, 944
Gas vesicles, 440
Gated ion channels, 850
Gause, G. F., 1108, 1112, 1129
Gazelles, 1110
gDNA, **340-341**, 343, 346, 353, 354
GDP, 861, 872
Geiger counter, 25
Gemmae, 500
Gemmules, 523
Gene(s), 250, 251, 273, 275-276, 279, 418
  of AIDS virus, 818
  alternative splicing of, 380
  for Alzheimer's disease, 897
  amplification, 396
  for antibodies, **371-372**, 380-381, 819
  bacterial, 314
  for behavior, 1053
  beta pseudogene, 368
  for blood groups, 810
  and cancer, 375-377, 381
  changes in order of, 277
  and chromosome, 260-261, **274-277**
  for collagen, 362
  and color blindness, 391-392, 400
  and color vision, 872
  combinations of, 243
  constant (C), 372
  development of concept of, **268-273**
  diversity (D), 372
  for drug resistance, 965
  for dystrophin, 394
  and early theories about heredity, 238
  environment and, 268, **271**, 279
  epistasis, **270-271**, 279
  and errors in metabolism, 301
  eukaryotic, 340
  expressivity, **271**, 279
  families, 365, **367-368**, 380
  for hirudin, 558
  for histones, 365, 366, 367, 380
  independent assortment of, 242-243
  for insecticide resistance, 964
  interactions (*see* Gene Interactions)
  joining gene, 372
  "jumping" gene, 336, 375
  library, 395
  linear array of, 276-277, 279, 328
  linkage, **274**, 279
    in sickle cell anemia, 395
  location of, **263-268**, 274-275, 277-278
  locus of, 275, 279
  manipulation of, **340-354**
  Mendelian genetics, 240, 244-245, 247, 361
  movable genetic elements, **371-375**, 380
  mutations in, 268, 367-368, 388, 981-982
  oncogenes, 376-377, 381
  "one-gene–one-enzyme" hypothesis, **301-304**, 316
  operator, **323**, 324-325, 337
  overlapping, 249, 250
  penetrance, **271**, 279
  vs. phenotypes, **993**, 1008
  phenotypic effects of, 268, 273
  pleiotropy, **273**, 276
  polygenes, 271
  in populations, **974-990**
  prokaryotic, 328
  promoter, 320, 323, 325, 335, 337, 377-379
  for proteins, **301-304**, 355, 363, 366, 367, 368
  pseudogenes, 367, 368, 374
  reassortment of, 238
  recombination of (*see* Genetic recombination)

I-23

Gene(s) (Continued)
  regulation of, in eukaryotes, 355, **359–362**, 380
  regulation of, in prokaryotes, 337
  regulation of enzymes, 302–303
  regulator, 322–324, 337, 378–379
  resistance gene, 326, 329, 336, **964–965**
  "selfish," **1070–1074**
  separation on chromosome, 276–277
  and sex determination in human beings, 946
  sex-linkage, **264–266, 268**
  spacer sequences in, 365–366
  structural, 320, 323, 337, 362, 368, 379–380
  for T-cell receptor, 804
  transfer of (see Gene Transfer)
  and transposons, **335–337,** 379
  variable, 372
  vectors, 331–334, 338
Genealogy, 415–416, 427
Gene amplification, 396
Gene duplication, 989
Gene families, 365, **367–368,** 380
  mutations in, 367, 368, 380
Gene flow, **982,** 990
  and cladogenesis, 1022
  in Darwin's finches, 1010
  and evolution, 990
  and geographic isolation, 1010–1011
  and human evolution, 1047
  vs. natural selection, 990
Gene interactions, **263–279**
  codominance, **268,** 279
  epistasis, **270–271,** 279
  incomplete dominance, **268,** 279
  novel phenotypes, **269–270,** 279, 993
  polygenic inheritance, **271–272,** 279
Gene loci, in heterozygote superiority, 988
Gene modification, 989
Gene pool, **974,** 1052
  and allele frequency, 979–981, 990
  altruism and, 1066–1067
  and behavior, 1052
  competition for inclusion, 1072–1074, 1078
  diploidy and, 986–987
  and directional selection, **997**
  and evolution, 991
  founder effect, 983
  gene flow, 982–983
  genetic drift, 982
  Hardy-Weinberg equilibrium and, **979–981**
  heterozygote superiority and, 987–988
  and kin selection, 1066
  latent variability in, 976–977
  laws of probability and, 979, 982–983
  multiple alleles, **980**
  mutations in, 978, 981–982
  and natural selection, limits on, 1006, 1009
  and nonrandom mating, 984, 990
  and outbreeding, 981–985
  and population bottlenecks, 983, 990
  and populations, **974,** 989
  speciation, 1010
  steady state in, 979–981
  variability in (see Genetic variability)
Generations, alternation of (see Alternation of generations)
Gene regulation, 355, **359–362,** 380, 1092
  and life-history patterns, **1099**
  methylation in, **361–362,** 380
  and mortality patterns, 1105
  negative regulation, 322, 325, 337
  positive regulation, 322, 325, 337
  precision of, 326
  and replication, 358, 380
  RNA polymerase and, 368
  and viroids, 446–447
Gene replicas, 1070–1071
Genetic code, **307–309**
  breaking of, **307–309**
  for insulin, 359
  for proteins, 355, 363, 366, 367, 368
  RNA and, 380
  translation of, **301–317**
  universality of, **316,** 355
Genetic counseling, 394
Genetic defects, 382, **385–398**
  correction of, 379, 400
  diagnosis and detection, **394–398**
  expression of, 400

Genetic disease(s), 340, **385–398**
  albinism, 389
  Alzheimer's disease, 897
  cystic fibrosis, 397, 399
  Down's syndrome, **385–388,** 399
  dwarfism, **391**
  enzyme deficiencies, 400
  hemophilia, 382, 392–393, 753
  Huntington's disease, **391**
  muscular dystrophy, 393–394, 397, 400
  phenylketonuria, 388–389
  sickle-cell anemia, 388, 390, 394–395, 397, 400
  Tay-Sachs disease, 389
Genetic drift, **982–983,** 990
  founder effect, 983
  and human evolution, 1050
  population bottlenecks, 983
Genetic engineering, **340–351** (see also Recombinant DNA)
Genetic exchange
  in ciliates, 471
  in prokaryotes, 450
  in protozoa, 469, 471
Genetic isolation
  Darwin's finches, 1017–1021
  postmating isolating mechanisms, 1016
  premating isolating mechanisms, 1016
  and sexual selection, 1015–1016
  and speciation, **1015–1016,** 1029
Genetic material, 144, 155 (see also DNA)
Genetic notation, 240
Genetic recombination
  and allele frequency, 979–981, 989–990
  chromosomal insertion, 331–333, 338
  conjugation, **326–330,** 331, 337
  "errors" in, 380
  and evolution, 326, 337
  frequency of, 276–277
  homologous, 336–338
  infection and, 336
  introns, 363
  and lambda phage, 337
  and living systems, 36
  and nuclear pores, 380
  offspring of, 276
  and plasmids, 337
  in prokaryotes, 337, **437,** 450
  and prophages, 331–334, 338
  reversible integration, 337
  in rolling-circle replication, 327–337, 338
  in sexual reproduction, 984
  "single-strand switch," 337
  strategies for, **336–337**
  transduction, **332–334,** 336–338
  transformation, 282, 329, 331, 336, 376
  transposons and, **335–337,** 379
  viruses and, 336
Genetic research
  bacteriophages in, 281, **284,** 285, 295
  E. coli in, 284–286, 292
  pneumococci, 299
  radioactive labeling, 285–286, 292–293, 299
  viruses in, 281, 284–286
Genetics, 233, **235–248,** 263, 281
  and basis for evolution, **974–990**
  classical, 281, 299
  cytology and, **260–261**
  and detection of diseases, 394–399
  early, 237, 247
  "golden age" of, 263
  and heart disease, 761
  human, **266, 340–354,** 382–400
  and laws of probability, 243, **244–245**
  and medical science, 382
  Mendelian, **238–248,** 249, 361
  microscopy and, 237
  modern, 238
  molecular, 281, 299, 308, **319–338**
    "central dogma" of, 306
  non-Mendelian, 362
  origins, **235–248**
Genetic variability, 471
  and diploidy, 986–987, 990
  and evolutionary "progress," 1006
  and frequency-dependent selection, 998
  in heterosis, 988–990
  and heterozygote superiority, 987–988
  laws of probability and, 979

maintenance of, **979–981, 991–1009**
  mutations and, **981–982**
  natural selection and, 987–988, **991–1109**
  in plant populations, 998
  preservation of, **984–988**
  in prokaryotes, 450
  quantification of, **978**
  sexual reproduction and, 984, 990
Gene transfer, 331–333, 338, **352–353,** 354, **377–378,** 379–381
  chromosomal insertions, 331–333, 338
  conjugation, **326–330,** 331, 337
  and homologous recombination, 336–338
  and nuclear pores, 380
  plasmids and, **326–329**
  in rolling-circle replication, 327–328, 338
  "single-strand switch," 337
  transduction, **332–334,** 336–338
  transformation, 282, 329, 331, 336, 376
  and transposons, **335–337,** 379
Genitalia, and reproductive isolation, 1016, **1017**
Genome
  and antibody formation, 801
  eukaryotic, 347, 420
  human, 396–397
    mapping, **398–399,** 400
  "meaningless" DNA in, 380
  sequencing, **398–399,** 400
  structure, 380
  single-copy, 423
Genomic DNA (see gDNA)
Genotype, 240–241, 243, 247, 271, 306, 342, 991, **993–994,** 1008
  vs. ecotypes, 1003
  equilibrium, 979
  and founder effect, 983
  and gene flow, 982
  and genetic mutations, 981–982
  and Hardy-Weinberg equilibrium, 989–990
  and kin selection, 1068–1070
  and multiple alleles, **980**
  and mutations, 978, **981–982**
  and natural selection, 1008
  vs. phenotype, **993,** 994
  reorganization of, 975
  and selfish gene, 1070
  survival of, 974–975
  variety of, and sexual reproduction, 986
  and waiting behavior, 1074
Genotypic ratio, 241
Genus, 408–410, 412–413, 424, 426–427
Geographic barriers, and speciation, 1010, 1011, **1012–1013**
Geographic distribution, 1009 (see also Clines)
  of blood groups, **996**
  and genetic variation, 1009
  and phenotypic variation, **1002–1003**
  and speciation, 1003
Geographic isolation, 1010, 1012–1013, 1029, 1034
  and Darwin's finches, **1019–1020**
  and evolution of monkeys, 1034
  and speciation, 1010, **1012–1014**
Geology
  and biogeochemical cycles, 1145
  and evolutionary theory, 2–7, 1028
  and fossil record, **966–968, 970–971,** 1011, **1012–1013,** 1028 (see also Fossils)
Geothermal energy, and chemosynthetic ecosystems, 1140–1141
Gerenuks, 1110
Germ cells, 924–925, 952
  segregation of, in human beings, 945–946
German measles (see Rubella)
Germinal center, 816
Germination
  and angiosperms, 623
  and environmental pollution, 50
  in epiphytes, 1175
  and gibberellins, 680–682
  and phytochromes, 695
  of seeds, 630–631, 647–648
  of spore coat, 438
  of spores, 462
  vs. vegetative reproduction, 642–643
  and water, 42
Germ layers, 532, 533
Ghost fish, 44
Giant chromosome, 277–279

and genetic studies, 359–360
supergenes in, 993
Giantism, 825
Giardia, 469
Gibberellic acid (GA$_2$), **680–681**
Gibberellins, 673, **680–682**, 694–695
and flowering, 688
Gibbons, 1035, 1036
Gibbs, Josiah Willard, 165
Gilbert, Walter, 348–349
Gill fungi, 480, 487
Gills, 130, 546–550, 553, 563, 565, 572, 588, 594, 747
arches, 594
book gills, 566, 569, 585
and circulatory system, 754
cover, 735
development in aquatic vertebrates, 937
evolution of, 734–735
vs. lungs, 737, 747
and poikilothermy, 780
pouches, 593, 969
slits, 591–592, 604, 735
Gill withdrawal reflex, 895–896
Ginkgo, 497–498, 505
Giraffes, 236, 1110
Gizzard, 555, 716
Glaciations, 510
and evolution of the bear, 1008
and *Homo neanderthalensis*, 1046, 1051
and mixed west-coast forests, 1166
and tundra, 1168
Gland cells, 527
in bony fish, 767
Glands, **703**, 713, 821 (*see also specific glands*; Digestive system; Endocrine glands; Endocrine system; Hormones)
and autonomic nervous system, 712, 842, 845, 856
endocrine, 711 (*see also* Endocrine glands; Endocrine system)
exocrine, 823
pancreas, 725
sweat, 786
thyroid, 784, 786
Glans penis, 904
Glass sponges (*see* Hexactinellida)
Glial cells, 708, 714, 842, 843, 851 (*see also* Neuroglia; Schwann cells)
in cerebral cortex, 884 (*see also* Neuroglia)
Gliding bacteria, 438, 436, 450, 456
and *Heliobacterium chlorum*, 443
Globular proteins, 75–76, 78, 82
Globular tertiary structure
of cell membrane, 106
Globulins, 750
Glomerulus, **770**, 776
Glottis, 736
Glucagon, 64, 823–824, 832, 833, 838–839
and regulation of blood glucose, 726
synthesis of, in pancreas, 725
Glucocorticoids, 824, 829, 830–**831**, 839
and negative feedback control, 831
side effects of, 831
Glucose, 117, 321, 326, 337, 730, 782, 786, 847
absorption from digestive tract, 833
alpha-glucose, 60–61, **63**, 65
beta-glucose, 60–61, **63**, 65
cell entry, and rate limitation, 136
and digestion, 730
and energy storage, 180
excretion of, 766
and glucocorticoids, 830
and glycogen, 64, 837–838
hormonal control of, 824–825, 832
and hormone secretion, 824
in hydrolysis and formation of sucrose, 170, 182
in laminarin, 457
in liver cells, 136
liver regulation of, 725–726
and muscle contraction, 878
oxidation and combustion of, 61, 164–167, **186–205**, 768, 790, 824, 833 (*see also* Glycolysis)
pancreas and regulation of, 725
in photosynthesis, 223, 228–229
positive regulation of, 337
production, 167, 837–838
reabsorption through kidney, 770

regulation of, in blood, **726**, 832
structure of, 62–63
transport, 136–137, 754
and water balance, 768
Glucose 1-phosphate, 837
Glucose 6-phosphate, 188, 190, 203, 832
Glutamic acid, 73
Glutamine, 73, 310, 314, 799
Glutaraldehyde, 98
Glyceraldehyde, 60
Glyceraldehyde phosphate, 188–189
as energy source for living systems, 228
and photosythesis, 223, 228–229
as product of Calvin cycle, 223
Glycerol, 82, 203, 721–723
as "antifreeze," 44
and molecular structure of fat, 67–68
and phospholipids, 69
Glycine, 73, 77
Glycogen, **64–65**, 117, 136–137, 190, 824, 832, 833, 909
and anaerobic respiration, 190
catabolism of, 69
and digestion, 730
and energy storage, 67
and glucose, 64
in liver, 726
and muscle contraction, 878
and photosynthesis, 228
and regulation of blood glucose, 726
as storage of food reserves, 518
Glycolic acid, 225
Glycolipids, 67, 69, 82, 105, 117–118, 124
in cellular membranes, 69
Glycolysis, 117, 161, 167, 168, **186–205**
acetyl CoA in, 194, 200, 201
ADP and ATP in, 186–190, 200–205
aldolase in, 188
allosteric interactions in, 188
anaerobic, 878
carbon and, 187–188
carbon dioxide and, 186
coenzyme A in, 194
cytoplasm in, 193
in diving mammals, **744**
efficiency of, 187, 201
energy levels and, 186
enzymes in, 187–188
evolution of, 191
and exercise, 190
free energy in, 187, 204–205
fructose in, 188
ions in, 187–189
and NAD$^+$, 187–190, 200–201
and NADH, 188–190, 200–201
net energy harvest of, 189–190, **200–202**
oxidation of, 164, **186–205**
oxidative phosphorylation in, 189
photosynthesis, 223, 228
pyruvic acid in, 187, 190, 200, 204
reduction of, 186
regulation of, 188
steps of, **188**
in urkaryotes, 453
and water, 191
Glycoprotein antigens, 805
Glycoproteins, 117–118, 815,
on AIDS virus, 824
and cell-mediated immunity, 801, 819
in epithelial tissues, 702
GM$_2$ ganglioside, 389
Gnathostomulids, 520, **538**, 540, 542
Gnats, 573
Gnetophytes, 498, 505
Gnetum, 26
GnRH, 826, 909, 917
and menstrual cycle, **913–914**
Gnu, 409
Goblet cells, 620, 716, **721**, 739, 912
Goiter, 829
"Golden age" of genetics, 263
Golden-brown algae, 456–457, **459–460**, 476
Golgi complex, 108, 112–113, 116, **117–118**, 125, 153, 902–903
and cell division, 146
and vesicle-mediated transport, 138–139
Gonadotropin-releasing hormone (*see* GnRH)
Gonadotropins, 825, 906, 913, 917

Gonads, 900, 917
embryonic development of, 936, 945–946, 952
germ cells, and development of, 924–925
hormones, regulation of, 827
hormones in, 825
negative feedback control in, 827
Gondwana, 1012–1013, 1034
G$_1$ (gap) phase, of cell cycle, **146**-148, 150
Gonium, 461, 462
Gonopores, 557, 572
Goose, 423, 1056
Goose pimples, 786
Gordian worms, 540
Gordon, John W., 378
Gorillas, 1032, 1035–1036
Gould, Stephen Jay, 1006, 1028, 1050, 1103,
Gout, 118
Gradients
carbon dioxide, in photosynthesis, 225
cesium chloride density, 292–293
concentration, 129–131, 134–136, 142–143, 180, 547
electrochemical, 198–199, 205
in photosynthesis, 218, 220, 229
light, 1158
potential energy, in photosynthesis, 220
proton, in chemiosmotic coupling, 197–198
in photosynthesis, 218, 221, 229
sucrose density gradient, 193
temperature, and homeostasis, 780–781
voltage, 198
water potential, 132–133
Grafts, 643
experimental, 939, 940
and mutualism between plants, 1121
of tissue, 810
Grains, 511, **622** (*see also* Grasses)
aleurone layer in, 681
and gibberellins, 681, 694–695
Gram, Hans Christian, 432
Gram-negative bacteria, 432–434, 450
Gram-positive bacteria, 432–434, 442, 450
Gram staining, 432
Grana, 120
in photosynthesis, 213–214, 224
Grant, Peter R., 1020
Granulocytes, **792**, 818, 831
Grapes, and fermentation, 190–191
Grasses, 511, **622**, 642–643 (*see also* Grains)
adaptation in, 642–643
in field layer, 1164
and gibberellins, 681, 694–695
on sandy beaches, 1159
in savannas, 1170
structure of, 642
symbiosis in, 668
in temperate grasslands, 1169–1170
Grasshopper, 275, 575, 583
meiosis in, 254, 260
Grasshopper mouse, 1114–1115
Grasslands, 496, 663, **1177** (*see also* Tundra)
ecosystem pyramids for, **1145**
savanna, **1170**
temperate, **1169–1170**
tropical, **1170**
Gravitropism, **682–684**, 694–695
Gravity, 529, 578
detection of, in plants, 684
and fluid movement, 142
and osmosis, 133
and plant growth, 671, 674, **682–684**
Gray crescnt, 926–927, 930
Gray matter, 843, 881, **884**, 885
Grazing
vs. browsing, 968
and dentition, 717, 968
and destruction of tropical rain forests, 1176
and evolution of horses, 1026–1027, 1028
and resource partitioning, 1110
in savannas, 1170
and spread of the Sahara, 1172
Great Barrier Reef, 530
Great Basin, 1172
Great Plains, 1170
Green algae, 455, 461–462, 476, 499 (*see also* Cyanobacteria)
alternation of generations in, 463–464, 507
Charophyceae, 494

I-25

Green algae (Continued)
  chlorophyll in, 461
  chloroplasts, 463
  coenocytic organization of, 462–463
  Coleochaete, 493–494
  colonies of, 461–462, 463
  cytoplasmic fusion in, 464
  in evolution, 458, 461, 462
  in evolution of plants, 493–495
  freshwater, 463
  gametes of, 462–464
  gametophytes of, 463, 464
  heteromorphism, 464
  isomorphism, 464
  and lichens, 488–490, 492
  life cycle of, **462–465**
  multicellular forms, 461, 462
  multinucleate, 462, 463
  reproductive cycles of, 476–477
  sporophytes, 463–464
  symbiosis in, 458, 461
  unicellular forms, 461
Green bacteria, 430, 431, **440**, 442, 450
Greenhouse effect, 226, **1177**, 1178
Green nonsulfur bacteria, 430, 440
Green Revolution, **1148–1149**
Green snow, 461
Green stems, 636, 640, 648
Green sulfur bacteria, 430, 440
Greylag geese, 1054
Griffin, Donald, 1115
Griffith, Frederick, 281
Grooming, 1069
Gross, Mart R., 997
Gross productivity, of ecosystems, 1138–1139
Ground finches, 1019–1020, 1111
Ground hemlock, 107
Groundhog, 409
Ground layer, 1164, 1168
Ground meristems, 618, 625–626, 634–635, 647
Ground substance, in bone, 704
Ground tissue system, in plants, **626, 648**
  in leaves, 627, 648
  in pith of roots, 633
  in stems, 636–637
Groundwater, 53, 1153
Group selection theory, 1065, 1078
Growing seasons
  in deciduous forests, 1164
  in deserts, 1173
  and greenhouse effect, 1177
Growth, 84
  and anabolic steroids, 907
  cellular, 147–148 (*see also* Cells)
  in human embryo, **940–951**
  of insect larva, **576**, 578
  and male hormones, 905
  as property of living systems, 27
  in vertebrate embryo, 919, **926–953**
Growth (of plants), 519, **625–648**, 694–696 (*see also* Abscisic acid; Auxins; Cytokinins; Ethylene; Gibberellins; Growth-inhibiting hormones; Growth-promoting hormones)
  in angiosperms, 621–624
  vs. animal growth, 630
  and hormones, **671–684**
  primary, 630, 645, 648
  rate of, 645, **671**
  regulation of, **671–696**
  of roots, 631, **634–636**, 674
  secondary, 644–648
  of shoots, **640–643**
  vegetative reproduction in, 642–643
Growth (of populations)
  and asexual reproduction, 1102–1103
  and carrying capacity, **1089**
  human population explosion, **1096–1097**
  logistic, 1089–1091
  and parthenogenesis, **1102**
  population cycles, **1098–1099**
  of populations, patterns of, 1088
Growth hormone, 824–825, 827, 833, 836, 839
Growth-inhibiting hormones, in plants, 680 (*see also* Abscisic acid; Gibberellins)
Growth patterns, of populations, **1088–1091**, 1104
Growth-promoting hormones, in plants, 673–679 (*see also* Auxins; Cytokinins; Ethylene)

Growth-promoting proteins and cancer, 373–374, 376
Growth rate
  vs. competitive fitness, 1108
  in plants, 645, 671
  in populations, 1088–1089, 1092–1093
Growth rings (*see* Annual rings)
Growth zones, 634, 640
Grubs, 576
GTP, 312, 861, 871–872
  in energy reactions, 180
G_2 (gap) phase, of cell cycle, 146, 147–148, 150, 154–155
Guanine, 80–81, 180, 429, 568
  in DNA, 283–284, 288–290, 292, 296, 298
Guano, 766
Guanosine diphosphate (*see* GDP)
Guanosine triphoshate (*see* GTP)
Guard cells, at stomata, 627–628, 653–654, 669
Guillemin, Roger, 827
Gulf Stream, 1157
Gulls, 1114
Gurdon, J. B., 359
Gusella, James F., 395, 397
Gut, 554, 563, 715 (*see also* Digestive tract)
  differentiation of, in chick embryo, 937, 952 (*see* Gastrulation)
  mouth-to-anus, 567
  tubular, 567
Guthrie, Woody, 391
Guttation, 651
Gymnosperms, 496–497, 501, 504, **505–509**, 516
  conifers, 505–506
  cycads, 505
  diversification of, 505
  and evolution of angiosperms, 509
  Ginkgo, 505
  gnetophytes, 505
  heterospory, 507
  integument in, 507
  leaves in, 508
  megaspores, formation of, 508
  ovule, 507
  pollination in, 512
  seeds, **507–508**
  sieve cells in, 638–639
  tracheids in, 639
Gypsy moth, 585, 108

"Habit" memory, 895, 899
Habitats (*see also* Aquatic environments; Biosphere; Clines; Ecology; Ecosystems; Ecotypes; Forest ecosystems; Oceans)
  and adaptation, **1002–1003**
  and balanced polymorphism, 992
  and biosphere, **1154–1177**
  of bluebirds, 1114
  and competition, 1107
  and Darwin's finches, **1023**
  destruction of, 1126–1127
  high altitude, 1161
  high latitude, 1161
  and insects, 586
  and marsupial adaptation, **1019**
  selection of, 526
Habituation, 474
  in *Aplysia*, 896
  and learning, 1055, 1077
Hadar fossils, 1042, 1050–1051
Hagfish, 593, 767
Hair(s), 702
  axillary, 863
  bodily, 601
  development of, in human embryo, 948
  epidermal (in leaves), 629, 641
  follicle, 859–860
  and male hormones, 905–906
  as mechanoreceptors, 860
  of polar bear, 961
  pubic, 863
  tactile, 556, 566, 584
  and temperature regulation, 786
Hair cells, 529, 863–865, 887
Haldane, J.B.S., 574, 974
Half-lives, radioactive, **971**
Hallucinogens, 487
Halobacteria, 221, 430
Halophytes, **658–659**

Halteres, 942–943
Hamilton, W. D., 1059, 1065–67, 1069
Hammer (maleus), 864
Hämmerling, Joachim, 109–110
Hamner, Karl C., 685–686
Hand
  evolution of, 1032–1033, 1051
  development of, in human embryo, 939, 946
  and motor cortex, 886
  vs. wings, 939
Hand ax, 1045, 1051
Hand-eye coordination, 870, 887, 1002, 1032
Handling ability, evolution of, 1032 (*see also* Dexterity; Digits; Hand-eye coordination)
Haplodiploidy, 1068
Haploid number of chromosomes, 249–257, 260–262, 462, 463 (*see also* Alternation of generations; Egg cells; Gametophyte; Meiosis; Sperm cells)
  in angiosperms, 617
  in basidiomycetes, 486
  in cells, 250–251, 261
  in drone bees, 1060
  in gametophytes, 251 (*see* Gametophytes)
  nucleus, 253, 255, 256, 258
  phase, of life cycle, 250, 252
  in plant cells, 256
  reproductive cells, 252
  in spermatids, 902
  spores, 251
Haplotypes, 806–807
Hapsburg lip, 235
Hardy, G. H., **979–981**
Hardy-Weinberg equation, **979–981**, 989–990
  application of, 981
Hardy-Weinberg equilibrium, **979–981**, 982, 989
  derivation of, 979–980
  and gene flow, 980
  and genetic drift, 982–983
  and multiple alleles, 980
  and nonrandom mating, **984**
  significance of, 981
Hares, 602
Harper, John L., 650
Harvest mice, 161
Hatch-Slack pathway (*see* Four-carbon pathway)
Haustoria, 468, 480, 490, 692
Hawaiian Islands, cladogenesis in, 1022
Hawk, 423
Hay fever, 812–813
HDL, 71, 726, 761
Head, 566, 574, 585
  development of, in fetus, 947
  orientation of, in primates, 1033
  of viruses, 445
Head-foot, 546, 563
Hearing, 879
  and cerebral cortex, **886–887**
  development of, in embryo, 936
  discrimination of sound, 865
  in mammals, **863–865**
  in moths, 1115–1116
  and pons, 883
  and predation in bats, 1115–1116
  receptors for
    in human beings, 859–864
    in other mammals, 863
  and song learning in birds, 1057
Heart, 547, 556, **749–764**, 937, 952 (*see also* Blood; Blood vessels; Circulation; Circulatory system)
  of arthropods, 568
  accessory, 547
  beat (*see* Heartbeat)
  and cardiac peptide, 775–776
  chambers of, 547, 763
  and circulatory system (*see* Circulatory system)
  development of, in human embryo, 945–946, 948
  diseases of (*see* Heart attack; Heart disease)
  as endocrine organ, 775
  evolution of, **754–758**, 763
  and glucocorticoids, 830
  human, **755–757**
  "lymph" heart, 763
  pressure receptors in, 774
  rate, 764
    and circadian rhythms, 834
    and diving, 744
    and REM sleep, 890

I-26

and regulation of blood pressure, 762
secretion of hormones in, 756
Heart attack, 760–761
vs. choking, 719
Heartbeat, 752–753, **755–756**, 847, 945
and brainstem, 882, 898
and hibernation, 786–787
rate of, 764
and circadian rhythms, 834
and diving, 744
and REM sleep, 890
regulation of, 756–758, 763–764, 832
Heart disease, 70–71, **760–761**
and atherosclerosis, 70–71
and cigarettes, 71
detection of, 25
Heart murmur, 756
Heartwood, 645
Heat, **43–44**, 54, 100–101, 161–162, 214
adaptations to, **788–789**, 790, 1002–1003
animal, 224
balance, **778–780**, 786
and body size, 779
capacity, 43
change in (ΔH), **164–165**, 183
comparative, 43
and conduction, 789
conservation of, 1002–1003
as energy, 161–162
exchange, between organism and environment, 778–779
of fusion, 44, 54
gain, 789
in atmosphere, 1133
and greenhouse effect, 1177
loss, 789
in atmosphere, 1133
in organic reactions, 59
ozone layer and, 1134
sources of, 778
specific, 43, 54
transfer, 778–779
of vaporization, 43–44, 54
Heat of fusion, 44, 54
Heat of vaporization, 43–44, 54
Heavy chains, and antibody formation, **798**, 801, 818
Heavy nitrogen, 293
Height distribution, 272
Heimlich maneuver, **719**
Helicase, 294–295, 297, 299
Heliobacterium chlorum, **442–443**, 456
Heliozoans, 470
Helium, 29–30, 733
Helper T cells, **801**, 819
and AIDS virus, 815–816
and cytotoxic T cells and B lymphocytes, 808
and macrophages and B lymphocytes, 809
and T4 glycoprotein, 804
Helping behavior
and kin selection, **1069–1070**
Heme, 78–79, 82
Heme group
of cytochrome *c*, 196
of hemoglobin, 363, 367
in respiration, 742
Hemichordates, 520, 587, 591, 604 (*see also* Acorn worms)
Hemispheres, cerebral (*see* Cerebral hemispheres)
Hemizygosity, 266
Hemlocks, 505
Hemocoel, 547, 568
Hemocyanin, vs. hemoglobin, 742
Hemoglobin, **70–72**, **78–80**, 82, 104, 114, 115, 304–314, 747–749, 763
affinity for oxygen, **742**
alpha globin, 359, 367
beta globin, 359, 362, 363, 367–368
beta pseudogene, 368
biosynthesis of, 359
and the cytochromes, 196
delta globin, 367
epsilon globin, 367
evolution of, 367
family, **367**
gamma globin, 367
heme group, 363, 367
hemoglobin S, 80
metabolism of, in liver, 726

and myoglobin, 367
and nutrition, 728
and oxygen affinity, **742**
and oxygen transfer, **720**
polypeptides and, 367
rabbit beta globin, 378–379
in red blood cells, 168, **750**
in respiration, **741–744**
and sickle cell anemia, 80, 316, 340, 388, 390, 397,400
structure of, **304**, **742**
Hemoglobin family, 367
Hemolytic anemia, 811–812
Hemophilia, 382, **392–393**, 400, **752**
and AIDS, 814, 817–818
diagnosis of, 397
hemophilia A, 392
Hemophilus, 341
Hens, pecking order, **1062**
Hepatic portal system, 759
Hepatic portal vein, and regulation of glucose, 726
Hepatitis, 752
Hepatitis B virus, 796
Herbaceous plants, **621–622**
in deciduous forest, 1164
in tundra, 1168
Herbivores, 549, 550, 573, 1110, 1139–1142, 1152
and energy flow, 1139–1142, 1152
in forest communities, 1143
in grasslands, 1145, 1170
plant defenses against, 1121–1122
predation and, 1116–1117
as primary consumers, 1085
and resource partitioning, 1110
in savanna, 1170
terrestrial, **724–725**
Herbs, 511, 621
Hereditary characteristics (*see* Characteristics, inherited; Traits)
Hereditary disease [*see also* Genetic disease(s)]
Hereditary information, 144, 155 [*see also* DNA; Gene(s); Inheritance]
Heredity, 100, 109, 110, 124, 235, 238, 264
in alkaptonuria, 301
altered patterns of, 277
and blood groups, 811
and cell nucleus, 109–110
and cell theory, 90
and chemical laws, 11
early ideas, **236**
and evolutionary theory, 972–973
vs. experience, and learning, 951
and genetic variability, 976–977
and heart disease, 71, 761
molecular basis of, **281–300**
and natural selection, 7
nature of, 235
proteins and, 74, 281, 285, 299
and Rh factor, 811
and social organization, 235
transmission of, 81, 287
Hermaphroditism, 524, 531, 535, 536, 540, 542, 550, 551, 557, 562, 563, 590
advantages of, 524, 985
sequential, 541, 543, 551, 571
simultaneous, 543, 551, 571
Heroin, 855
fetal addiction to, 948
Herpes simplex virus, 448
Hershey, A. D., 284–286
Hershey-Chase experiments, 285–286
Hertwig, Oscar, 109, 263
Heterochromatin, 359, 360, 380–381
Heterogamy, 264
Heteromorphic generations, 495
Heteromorphism, 464, 465
Heterosis, 988, 990
Heterospory, 501, 507
Heterotrophs, 89, 90, 94, 101, 118, 228, 476, 477, 578, 1131, 1138, 1152 (*see also* Carnivores; Consumers; Herbivores; Omnivores)
and asteroid hypothesis, 1024
characteristics of, 469
classification of, **454–455**
dinoflagellates, 460
and ecosystem energy flow, 1131, 1152
energy sources of, 193
in food chain, 1138, 1152

fungi as, 480, 491
Hfr cell, 327–329, 338
mass extinctions of, 1024
multicellular, **466–468**, 542
multinucleate, **466–468**
and organic molecules, 203
origin of, 459
and origin of animals, 454
parasitism in, 452
phagocytosis in, 138–139
prokaryotic, 435, **438–439**, 450
protists, 452
protozoans, **468–476**
unicellular, **468–476**
Heterotrophy, 426, 518
by absorption, 426
by ingestion, 426
Heterozygosity, 240–241, 243, 245, 247, 268, 304, 978
in diploidy, 986–987
in *Drosophila*, 978
and genetic defects, 400
and genetic variability, 987–988
and Hardy-Weinberg equilibrium, 979–980
heterosis as, 988
heterozygote superiority, **987–988**, 990
in human populations, 978
for MHC antigens, 806
and phenylketonuria, 388–389
and polymorphism, 984
and sickle cell anemia, 388–390, 988
Tay-Sachs disease, 389
Heterozygote superiority, **987–988**, 990
and genetic variability, 987–988
heterosis as, 988
in sickle-cell anemia, 988
Hexactinellida, 523
Hexokinase, 188
Hexoses, 60
Hibernation, **786–787**, 790
Hiesey, William, 1002
High-density lipoproteins (*see* HDL)
"High-energy" bonds, 181
Higher primates (*see* Anthropoids)
High frequency of recombination cell (*see* Hfr cell)
High selection lines, 977
High-voltage electron microscope, 110
Himalayan coat color, 268
Hindbrain
development of, in chick embryo, 935
evolution of, **882–883**, 898
Hindgut, 572, 574, 585
*Hind*III restriction enzyme, 341, 348
in diagnosis of Huntington's disease, 395
Hipparion, 25
Hippocampus, 899
and Alzheimer's disease, 896
and learning, 894, 897
and memory, 894, 897
Hippocrates, 236
Hippopotamuses, 603,
Hirudin, 558
Histamine, 792, 813, 818, 823
Histidine, 73, 310
Histoautoradiography, 661
Histones, 91, 146, 356–357
biosynthesis of, 356
and cell division, 146, 148
and chromatin, 356
and DNA, 357, 358, 379, 452
and energy organelles, 362
genetic code for, 365, 366, 367, 380
lack of, in prokaryotic DNA, 450
in viral DNA, 444
Hives, 813
HIV virus, **814–816**, 819, 1090–1091 (*see also* AIDS)
vs. SIV virus, 817
Holdfasts, 465, 1157
Holophyly, 416, 418
Homeobox, **942–943**
Homeostasis, 84, 714
and autonomic nervous system, 847 (*see also* Nervous System)
and evolution, 711
excretion, **765–776**
and feedback control, 712–713 (*see also* Feedback inhibition; Negative feedback; Positive feedback)

I-27

Homeostasis (Continued)
  immune response, **791-819** (*see also* Immune system)
  in multicellular organisms, 709, 714
  as property of living systems, 26
  temperature regulation, **777-790**
  in vertebrates, **711**
  water balance, **765-776**
Homeothermy
  and clines, **780-782**, **786**, 789-790, 1002
  coevolution in, 1007
  and hindbrain, 883
Homeotic genes, 942
Home range, 1063
Hominids, 496, 1038-1047, 1051
  and apes, **1034-1036**
  archaic *Homo sapiens*, **1046**
  australopithecines, **1038-1042**
  care of young, evolution of, 1033
  cave art, **1048-1049**
  controversies, **1043-1044**
  earliest, 1038
  emergence of, **1038-1051**
  evolution of, **1031-1051**
  Hadar fossils, 1042, 1050-1051
  Homo erectus, **1044-1045**
  Homo habilis, **1042**
  Homo sapiens, 1047 (*see also Homo sapiens*)
  Homo sapiens neanderthalensis, **1046-1047**
  Homo sapiens sapiens, **1047**
  Laetoli fossils, **1039**
  and monkeys, **1034-1035**
  origin of modern humans, **1047-1050**
  primate evolution and, 1031-1038
  and prosimians, 1034
  social behavior in primates, 1036-1038
  upright posture, **1033**
  visual acuity in, 1033
Hominoids, **1035-1038**, 1044, 1051
Homo erectus, **1044-1045**
  evolution of in Africa, 1047
  extinction of, 1050
Homo (genus), 93, 410, 1038, 1042-1043 (*see also* Archaic *Homo sapiens*; Hominids; Hominoids; *Homo erectus*, *Homo habilis*, *Homo sapiens*; *Homo sapiens neanderthalensis*; *Homo sapiens sapiens*)
Homogamy, 264
Homogenization of tissue, 192-193
Homogentisate, 388
Homo habilis, **1042**, 1043, 1045, 1051
Homologous recombination, **336-338**
  in crossing over, 336-337
  in eukaryotes, 336
  in meiosis, 336-337
  "single-strand switch," 337
Homologous structures, 414
Homologues (of chromosomes), 274-275, 279, 414
  in Down's syndrome, 385
  and evolution, **969**, 973
  in grasshopper, 275
  homologous recombination in, 337-338
  in karyotyping, 383, 399
  lack of, in hybrids, 1014
  in meiosis, 250, 252-256, 259-261
  nondisjunction in, 385, 399
Homology, 414-416, 427
  and phylogeny, **413-414**
Homo sapiens, 316, 411, 420, 587, **701**, **713**, **1043**, 1051 (*see also* Human beings)
  archaic, **1046**
  brain of, evolution, **881-884**, 898
  cardiovascular system, 758-759
  cave art, **1048-1049**
  cerebrum, 884
  characteristics of, **701-702**
  circulatory system of, **749-762**, **763-764**
  dentition of, **716-717**
  digestion in, **715-730**
  early theories of, 4
  embryonic development of, 919, **940-951**
  evolution of, 362, **1031-1051**
  excretion in, **765-776**
  functions of, 710-714
  heart of, **754-758**
  heterozygosity, **978**
  homeostasis in, 711 (*see* Homeostasis)

vs. *Homo erectus*, 1045, 1051
and *Homo sapiens neanderthalensis*, **1046-1047**
and *Homo sapiens sapiens*, **1047**
as hunters, 1117-1119
immune system, **793-819**
intrinsic processing centers in, 893
kidneys of, **769-776**
levels of organization in, 709-710
lymphatic system of, 763
mating systems of, 999
mother's milk, 727
musculature of, 706-708
nervous system of, 708-709, **711-712**, **841-857** (*see* Nervous system *and specific nervous systems*)
nutritional requirements of, 726
origin of modern humans, **1047-1050**
population growth of, 1096-1097
rate of mutations in, 982
reproduction in, **900-918**
  male, **901-907**
  female, **907-915**
respiratory system of, **738-740**
skeleton of, **704-705**
spiteful behavior in, 1059
symbiosis in, **447-450**
uniform age of mortality in, 1092
water balance in, **765-776**
Homo sapiens neanderthalensis, **1046-1047**, 1050-1051
Homo sapiens sapiens, 1042, 1051
Homospory, 501, 503
Homozygosity, **240**, 241-243, 247, 260, 274, 276, 304, 990
  in diploidy, 986-987
  in dwarfism, 983
  in genetic disease, 389-390
  and Hardy-Weinberg equation, 979-980
  and heterozygote superiority, **987-988**, 990
  in polydactyly, 983
  and population bottlenecks, 983
  in sickle-cell anemia, 390
  and social structure, 983
  in Tay-Sachs disease, 983
Homunculus, 237
Honey, 1060-1061
Honey bees, 1078
  annual cycle in, **1061**
  behavioral alleles in, 1053
  kin selection in, 1068
  as pollinators, 514
  queen, **1060-1061**
  social behavior in, **1060-1061**
Honeydew, 1121
Hoof-and-mouth disease, 796
Hoofs, evolution of, in horses, 1025-1027
Hooke, Robert, 10, 94
Hooker, Joseph, 8
Hooknose male salmon, 997, 998
Hooks, 536
  flagellar, 434
Hookworms, 539, 540
"Hopeful monster" theory, 982
Horizontal cells, 869, 871, 936
Hormone-receptor complex, 836-837, 839
Hormones, 70-71, 74, 106, 115, 304, 418, 703, 711-713, **831-840** (*see also* Endocrine glands; Endocrine system)
  and Addison's disease, 775
  and ADH, 775
  of adrenal cortex, **829-832**
  adrenaline, 837-838
  of adrenal medulla, **832**
  amino acid derivatives, 823
  and amplification cascade, 751, 837-838
  in blood plasma, 763
  and blood pressure regulation, 759
  at capillaries, 754, 764
  catecholamines, 836
  in cell-cell communication, 140-141
  chemical composition of, 824
  and cholesterol, 71
  and circadian rhythms, 834
  degradation of, 823
  in digestion, 730
  and embryo implantation, 941
  and enzyme cascades, 837
  in estrus, 915

feedback control of, 712-713 (*see also* Feedback inhibition)
female (*see* Sex hormones, female)
gastrointestinal, 722, **723**
in heart, 756
and heartbeat regulation, 764
of hypothalamus, **826-828**, 883
and insect metamorphosis, 578, 586
intracellular receptors for, **836**
and kidney function, **774-775**
male (*see* Sex hormones, male)
mammalian, 838
mechanism of action, **836-838**, 839
membrane receptors for, 836, **837-838**
in menstrual cycle, 913-915
and molting, 578
mRNA and, **836-837**
negative feedback control of, 823, 825, 827, 831, 839
vs. neurotransmitters, 822
of pancreas, **823-833**
of parathyroid glands, 829
peptides, 823, 836
of pineal gland, **833**
of pituitary gland, **825-826**
and placenta, **941**
in plants [*see* Hormones (plant)]
and population limitations, 1099
positive feedback control of, 831
in pregnancy, 911, 913-915
production of, 823
prostaglandins, 833-836
receptors for, 821, 836-838
regulation of, in liver, 726
in regulation of blood glucose (*see* Glucose; Glycolysis)
regulation of production, 823, 834
in reproductive system
  female, **913-915**
  male, **905-907**, 908-909, 917
sex, 70
and sex differences in the brain, 908-909
sex hormones (*see* Sex hormones)
steroids, 117-134, 823, 829-832, 836-837
and temperature regulation, 784-786, 790
of thyroid gland, **828-829**
thyroid hormone, 836
transport of, 821-823
and vesicle-mediated transport, 140
Hormones (plant), **671-684**, **694-695** (*see also* Abscisic acid; Auxins; Cytokinins; Ethylene)
  in agriculture, 643
  and growth, **671-673**, **673-684**
  interactions, 673, 675
  and phototropism, 671-673
  and plant response to stimuli, 694
  and water movement in plants, 654
Hornworts, 496, 498-500
Horse, 409
  evolution of, **968**, **1025-1027**
Horsehair worms, 520, 540, 542
Horseshoe crabs, 566, 569, 585
Horsetails, 497-498, 502, 504
Host cell, 328, 331-335, 338, 444, 451
Host defenses, and microbial infection, 448, 451 (*see also* Immune system)
Host-parasite relations (*see* Symbiosis)
House sparrow, 1002-1003
Houseworking bee, 1060
Howard, Eliot, 1063
Howler monkeys, 1035
*Hpa*I restriction enzyme, 341, 394-395
Hrdy, Sarah Blaffer, 1071
Hubby, J. L., 978
Hubel, David H., 887
Huddling, 1062
Human behavior (*see also* Behavior)
  vs. animal behavior, 1075
  biology of, **1075-1076**
  and circadian rhythms, 834
  and evolution, 1050
  vs. hominid behavior, 1036-1038
Human beings, 601, 604 (*see also* Hominids; *Homo sapiens*)
  vs. animals, **601**, **604**
  cell nucleus of, 383
  dominant and recessive alleles in, 268
  eye of, 209

gametes, 252–253, 258–259
genetics of, 266, **340–354**
genome, 380, 396–399
heredity and social organization, 235
inherited characteristics, 271
karyotype of, **383–388**
life cycle of, 252
number of chromosomes in, 264
origin of, **1047–1050**
and symbiosis, 537, 539
Human population
  explosion of, **1096–1097**
    and agricultural revolution, **1148–1149**
    and destruction of rainforests, 1176–1177
    and spread of Sahara, 1172
  history of, 1096
  theories of, 7 (*see also* Biosphere; Ecosystems; Population dynamics; Populations)
Human society, and reciprocal altruism, 1074, 1079
Humboldt current, 1157
Humidity, and transpiration, 652
Hummingbird, 61, 67
Humulin, 354
Humus, 663–664, 670
Hunger
  and limbic system, 885
  and population explosion, 1097, **1148–1149**
  sensory receptors for, 859
Hunter-gatherers, 1096, 1144
Hunting, 915
Huntington's disease, 391, 398, 853
  as autosomal dominant, 391
  diagnosis, 395, 397
Hutchinson, G. E., 961, 1099
Hutton, James, 2–3, 5
Huxley, Thomas H., 968, 1027
Hybridization, 308, 380
  and cancer research, 378
  denaturation, 363–364
  DNA-DNA, **420–423**, 427
  in genealogy studies, 420–421
  and monoclonal antibodies, 802, 803
  nucleic acid, 340, **345**, 346–347, 353–354
  and repetitious DNA, 363–364, 365, 372
  RNA and DNA, 308
Hybrid plant cultivation, 678
Hybrids
  allopolyploidy in, 1012
  RNA-DNA, 308
  between subspecies, 1011, 1014
Hybrid vigor (*see* Heterosis)
Hydra, 259, 527
  digestion in, 527
  mortality rate of, 1092
  nerve impulse in, 841
  nervous system of, 527, 841–842
Hydrocarbons, **55**, 56, 61
Hydrochloric acid (HCl), 47, 720
Hydrogen
  in anaerobic respiration, 439
  in atmosphere, 733
  atomic structure, 23–24
  biological importance of, 35
  bomb, 163
  and cohesion-tension theory, 651
  covalent bonds in, 32
  ion, 48, 49 (*see* Hydrogen ions)
  isotopes of, 35
  and origin of life, 87
Hydrogen bonds, 210, 294, 296
  and cohesion-tension theory, 651
  and COS regions of lambda phage, 334
  in DNA, 288–290, 299
  and evaporation, 44
  in insoluble compounds, 47
  and protein structure, 74–75
  in water, 33, **41–43**, 53
Hydrogen ions (H$^+$), 48, 49, 204, 748
  in blood, regulation of, 745–746
  in electron transport, 197
  in enzymatic reactions, 172
  in glycolysis, 187–189
  in Krebs cycle, 194
  peroxide, 119
  in photosynthesis, 218, 220, 229
  in plant growth, 674–675
  in plant movements, 693
  in respiration, 743, 745

  and stomatal movements, 653–654
Hydrogen sulfide, 217, 436, 439, 1140
Hydrolysis, 64–65, 722
  of ATP, 181–182, 875
  in enzyme regulation, 175
  of sucrose, 170
Hydrolytic enzymes, in plants, **694–695**
  and gibberellins, 681–682
Hydronium ion (H$_3$O$^+$), 47, 54
Hydrophilic molecules, **46**, 54
  and charge, 46
  covalent bonds in, 46
  and transport, 134, 136–137
Hydrophobic molecules, **47**, 54, 134, 136–137
Hydrostatic pressure, 128, 133–134, 604
  in arteries, 764
  of blood, vs. osmotic potential, 762
  and lymphatic system, 762
  and water gain, 769
Hydroxide ion (OH$^-$), 47, 54, 56
Hydroxyl group, 56–57, 305
  of nucleotides, 80
Hydrozoans, 526, **527–528**, 529, 541 (*see also Hydra*)
Hygienic honey bees, 1053
Hymenopterans, 573, 576, **1059–1061**, 1059 (*see also* Ants; Bees; Wasps)
Hyperactivity, 831
Hyperparathyroidism, 829–830
Hyperpolarization
  of postsynaptic cells, 856
  of synaptic cells, and learning, 897
  in vision, 872
Hypertension, **761**
  and dietary salt, 730
Hypertonicity, 132–134, 142
  of bacterial cells, 432
  of blood, 725
  of urine in mammals, 776
Hyperventilation, 746–747
Hyphae, 467–468, 481, 483, 485–487
  cell walls of, 480
  in ectomycorrhize, 491
  in endomycorrhizae, 491
  of fungi, 480–481, 489, 491
  haustoria (*see* Haustoria)
  of *Penicillium*, 488
  rhizoids, 482
  septate, 482
  sporangiophores, 481–482
Hyphal sheath, 491
Hypoblast, 932
Hypocotyl, 631
Hypolimnion, 46
Hypothalamic releasing hormones, 855
Hypothalamus, 712, 823–824, **826–828**, 831–832, 853, **883**, 898
  and ADH, 774, 776, 827–828
  and childbirth, 828
  control of pituitary by, 825, **826–827**
  and female sex hormones, 913
  and limbic system, 885
  and male sex hormones, **906**, 907–909, 917
  negative feedback control in, 827
  portal system of, **827**
  and prolactin production, 825
  and sexual behavior, regulation of, 914
  and temperature regulation, 713, 784, 790
Hypotheses, in science, 14–15, 17, 85, 412 (*see also specific hypotheses*)
  of branching sequences, in cladistics, 417 (*see also* Cladogram)
  in ecology, 1087, 1104
  and experimental design, 131, 1001
  testing of, 417, 1106–1107
Hypothyroidism, 829
Hypotonicity, 132–134, 142
Hyracotherium, 409, **1025–1027** (*see also* Eohippus; Horse)
Hyrax, 409

IAA (indoleacetic acid), 673–674, **676–677** (*see also* Auxins)
I band, 874
Ibis, 423
Ice, 44, 1155
Ice Ages, and evolution of plants, **510**
Identical repeats, 335

Igneous rock, 1160
Ikatura, Keiichi, **350–351**
Imbibition, 42
Imitative learning, **1057–1058**, 1078
Immigration, 989, 1123–1124, 1130 (*see also* Island biogeography model)
  and community composition, 1123–1124, 1130
  and destruction of habitat, 1126–1127
  and extinction, 1123–1124
  and gene flow, 982
  and genetic variability, 989
  and Hardy-Weinberg equation, 979
  and population growth rate, 1088–1092
  and speciation, in Darwin's finches, 1017
Immune disorders, 812–818, 819 (*see also* AIDS; Allergies; Autoimmune diseases)
Immune response, **791–818**, 830–831, 835–836, 855 (*see also* Immune system)
  and blood, 749, **750–751**
  cancer and, **808–809**
  and cellular communication, 808
  evolution of, **819**
  immune system, **793–819**
  inflammatory response, **791–793**
  and lymphatic system, 763
  nonspecific defenses, **791–793**
  primary, 793
  and prostaglandins, 839
  secondary, 793
Immune system, 711, **793–819**
  AIDS, 813–819
  allergies, 812–818
  antibody coding in, 372
  autoimmune diseases, 812–818
  B lymphocytes and antibodies, **795–801**
  and cancer, **809**
  cell-mediated immunity, **801–808**, 819
  and depletion of ozone layer, 1134
  development of, in human infant, 948
  disorders of, **812–818**
  fever, 784
  human, 793
  immunoglobulins, 70, 72, 76
  memory cell response, 819
  mucosal immunity, 799
  T lymphocytes, **801–808**
Immunity, **791–819**
  cell-mediated, 801–808, 819
  in fetus, 948
  memory-cell response, 819
  and mother's milk, 727
  mucosal, 819
Immunization, 449, 797 (*see also* Inoculation; Vaccination)
Immunofluorescence microscopy, 111
Immunoglobulins, 70, 72, 76, 750, **795–801**, 812, 818 (*see also* Antibodies)
Immunology, 797, 818
Immunosuppression, 831
Immunosuppressive drugs, 810, 813, 819, 834
Imperfect fungus, 488
Impetigo, 448
Implantation, of embryo, 901, 912, 916, 941
  and progesterone, 914
Imprinting, **1056–1057**, 1078
  in white-crowned sparrow, 1056–1057
Inbreeding, 977, 982
  anatomical arrangements against, 984–985
  and heterosis, 988
  and kin selection, 1077
  and mutation, 982
  taboos against, 985
  and sterility, 977, 985
Incest taboos, 985
Incisors, 601–602, **716–717**
Inclusive fitness, **974, 1069**, 1075, **1076–1077**, 1078
  and altruism, 1079
  and kin recognition, 1075, **1076–1077**
  and selfish gene, **1070–1075**
Incomplete dominance, **268**, 279
Incomplete penetrance, 271, 279
Indehiscent fruits, 620–621 (*see also* Achenes; Nuts)
Independent assortment, **241–243**, 245, 247, 260–261, 270, 274, 276, 279
  of behavioral genes, 1053
  in meiosis, 260–261
  principle of, **241–243**, 247
  and sexual reproduction, 984

Independent effectors, 527
India, 1092, 1097
Indian pipe, 94
Indoleacetic acid (*see* IAA)
Induced-fit hypothesis, 170–171
Inducer protein, 323–325, 337
Inducers, embryonic, **930–931**, 952
Inducible enzymes, 321–322
Inducible operons, 323–325, 337
Induction, 337
Induction (embryonic), **930–931**, 952
  in chick embryo, 935–936
  primary, 930–931, 935–936
  secondary, 936
Industrialization, **962–965**, 973
Industrial melanism, **962–963**
  as directional selection, 997
Industrial revolution, and population growth, 1096–1097
Infant diarrhea, 449
Infanticide, 1071
Infant mortality rates, 949
Infection, 751
  by AIDS virus, **815–816**
  and antibodies, 819 (*see* Immune response; Immune system; Immunoglobulins)
  and birth defects, 947, 953
  and burns, 808
  and contraceptive techniques, 917
  and cytotoxic T cells, 808
  epidemics, 813
  and inflammatory response, 818 (*see also* Inflammatory response)
  and overcrowding, 437
  smallpox, 796–797
Infection cycle, 285, 286, 330–333, 338, 451
  in *E. coli*, 330
  and genetic recombination, 336
  lysogenic cells, **331–332**, 333
  lysogeny, 322, 330–331, **332**, 333
  lytic cycle, 330–335, 338
  microbial, 448
  phage, 344
  and prokaryotic nucleoid, 330
  retroviral, **342**
  susceptibility, 385
  and vacuoles, 330
  and viruses, 322, 330–335, 338, 432, 445–446, 451
Infection rate, of AIDS virus, **1090–1091**
Infection threads, 666–668
Infectious diseases, 1087, **1090–1091**
  epidemics, 813
  as limitation to population growth, 1084
  in plants, 678
Inflammatory response, **791–793**, 808, 818, 830–831, 835–836, 855
  and allergies, 813
  complement in, 798
  vs. immune response, 808
  and microbial infection, 448
Influorescence, 616, 670
Influenza, 444–445, 813
Infrared light, 211, 872
Ingenhousz, Jan, 208, 217
Ingram, Vernon, 304
Ingrowths, 734
Inheritance, 238, 264, 413 (*see also* Genetics; Heredity)
  blending, **237–238**, 247
  early theories, 4–5
  and evolutionary theory, **962**, 973
  mechanism of, 247
  Mendelian, 361
  and natural selection, 7, 238
  non-Mendelian, 362
  polygenic, **271–272**, 279
  unit concept of, 247
  of variations, 247
Inhibin, 906–907
Inhibition
  of animal activity, 568
  of cell growth, 147–148
  in cerebral ganglia, 557
  competitive inhibition, **177**, 178, 184
  of enzyme activity, **177**
  irreversible inhibition, **177–178**, 184
  noncompetitive inhibition, **177–178**, 184
  reversible, 176–178

  in seed coat, 623
  of self-fertilization, 990
Inhibition hypothesis, **1125, 1128**, 1130
Inhibitor proteins, 177
Initiation, stage of mRNA translation, 312–313
Initiation codon, for DNA replication, 350–351
Initiation complex, 312, 314
Initiator proteins, for DNA replication, 294
Injury, response to (*see also* Immune response; Immune system; Inflammatory response)
  blood clotting, 753
  in cerebral hemispheres, 889
  edema in, 762
  in knee, 707
  and memory loss, 894
  in plants, 638–639, 647, 696
Innate releasing mechanisms, 1054, 1077
Inner cell membrane, of prokaryotes, 432
Inner ear, 865
Inoculation, 796
Inorganic molecules, 11 (*see also* Elements; Minerals)
  movement of in ecosystems, 1153 (*see* Biogeochemical cycles)
  oxidation of, in chemosynthetic autotrophs, 439
Input, of minerals, 1149
Insecticides, resistance to, **964**
  as directional selection, 997
  and malaria, 474
Insectivores, vs. early primates, 1031
Insects, 14, 554, 565, 566, 572, **573–578**, 586
  vs. animals, 14
  characteristics of, **574**
  chitin, 480
  determination of germ cells in, 924
  evolution of, 513
  and evolution of flowers, **513, 614**
  excretion in, 766
  exoskeleton of, 480
  life cycle of, **575–578**, 586
  microevolution in, **964**
  pollination by, 512–514
  reproduction in, 579
  resistance to insecticides in, **964**
  respiration in, 734
  social behavior in, **1059–1061**, 1078
Insertion sequences, 335, 374 (*see also* Transposons)
Insertion site, 329
  of prophage, 334
Insolubility, 47
Inspiration, 738, 741
Instars, 576, 577, 580
Insulation, 598, 599, **779, 786–787**, 790
  in camels, 789
  fats as, 68–69
Insulin, 72, 75–76, 138, 304, 349, 351, 822–824, 832–833, 837–839, 855 (*see also* Humulin)
  genetic code for, 359
  and regulation of blood glucose, 726
  secretion of, in pancreas, 725
  structure, 75
Integral membrane proteins, 104–106, 114, 124, 135, 178–179
  and cellular transport, 135, 142
  and electron transport, 200
  and membrane transport proteins, 178–179
  permeases, 135, 178–179
  in red blood cells, 114
Integration, 711–714
  of memory with sensory information, 898–899
  of muscular activities, 886
  of neuroendocrine system, at hypothalamus, 883
  in vertebrate brain, **881–899**
Integuments
  of angiosperm ovule, 619
  of gymnosperms, 506–507
Intelligence
  in chimpanzees, 1037
  evolution of, 1044
  in *Homo sapiens*, 702
Intelligence quotient (IQ), 994
Intentions, and intrinsic processing areas, 893
Interbreeding, 1014–1015, 1029
  and populations, 974, 989
  vs. reproductive isolation, 1010–1030
Intercellular spaces, 654
Intercostal muscles, in respiration, 741, 745, 748
Intercourse, 900–903, **904**, 905–918
  and contraception, **915–917**

  and ovulation, 915
Interference competition, 1106
Interferons, **793**, 818, 823, 855
  of cytotoxic T cells, 807
  and monoclonal antibodies, 802
Intergenerational conflict, 1071
Interglacials, 510
Interior surface
  of cell membrane, 135, 138–139
Interleukins, 821, 823, 835–836, 855
Intermediate disturbance hypothesis, **1124–1125**, 1130
Intermediate filaments, 110, 111, **112**, 124
  and nucleolus, 111
Intermediate-repeat DNA, **365**, 380
Internal fertilization, 900
Internal membrane
  and cellular transport, 142, 197–198
Internal transport, 547
Interneurons, 708, **709**, 843–844
Internodes, 626, 641, 648
Interoceptors, 859
Interphase, **146**, 149, 155
  of meiosis, 253, 255, 261
  and DNA transcription, 359
Intersexual selection, 998–999
Interspecific competition, **1106–1114**
  in barnacles, 1113
  and ecological niches, 1111
  and intermediate disturbance hypothesis, 1124
Interspecific interactions, **1106–1122**
Interstitial cells, 824, **901**, 905, 906–907, 917
Interstitial fluid, 750, 754, 763, 775, 793–794
  as a body compartment, **768–769**
  and loop of Henle, 773–774
  and lymphatic system, 763
  and respiration, 737
  and substance exchange at capillaries, 763
Intertidal zones, 1113, 1119, 1158
  competition in, 1124–1125, 1127
Intervening sequence (*see* Intron)
Intestines
  absorption of calcium from, 831
  development
    in sea urchin embryo, 924
    in vertebrate embryo, 937
  in digestion, 730
  of earthworm, 555
  human, 544
  and immune system, 795
  motility, and nervous system, 723
Intoxication, 202
Intracellular fluids, 775
  as a body compartment, 768–769
Intraspecific competition, 1106 (*see also* Dominance hierarchies; Inclusive fitness; Sexual selection)
Intraspecific discrimination, **1056**, 1078 (*see also* Kin selection)
Intrauterine devices (IUDs), **916**
Intravenous drug use, and AIDS, 817
Intrinsic processing centers, **892–893**, 898–899
Introns, **361–363**, 380, 420
  and alternative splicing, 369, 370
  and cellular nucleus, 380
  excision of, 369, 370
  and meosis, 363
  in protein coding, 366
  and recombination, 363
  and viroids, 446–447
  and viruses, 362
Inversion, of chromosomal segments, 277, 279
  and protection of supergenes, 993
Invertebrates, 414, **519–697** (*see also specific invertebrates*)
  asexual reproduction in, 1103
  in coral reefs, 530
  embryonic development in, **920–926**
  evolution of, 538
  lower, **522–543**
  nervous systems of, 841–842
Inverted repeats, 335–336, 338, 374
In vitro fertilization, 912
Involuntary controls of respiration, 747
Involuntary muscles, 706 (*see also* Muscles)
Involuntary nervous system, 712 (*see* Autonomic nervous system)
Iodine, 728, 828, 829
Ions, 31

I-30

and action potentials, 848–850, 851–852
in blood plasma, 763
calcium (Ca^{2+}), **31**
chloride (Cl^-), **31**, 134
as cofactors, 171
equilibrium in, 47
in glycolysis, 187–189
hydrogen (H^+), 48, 49, 119, 172, 187–189, 194, 197, 204
hydronium (H_3O^+), 47, 54
hydroxide (OH^-), 47, 54, 56
and learning, 897–898
magnesium (Mg^{2+}), **31**, 171
and nerve impulse transmission, 848–852
potassium (K^+), **31**, 136
and proton pump, 197–199, 205
reabsorption of, in kidney, 770
regulation of, 775
sodium (Na^+), **31**
sodium-potassium pump, 849–852
Ionic bonds, **30–31**, 38
biological importance of, 31
in water, 41
Ionization, of water, 47–52, 54
IQ, heredity vs. environment in, **994**
Iridium anomaly, **1024–1025**
Iris (of eye), 866
Iron
in biogeochemical cycles, 1145
in cytochromes, 196
in hemoglobin, and respiration, 742
nutritional requirements for, 728
during human pregnancy, 944
plant requirements for, 657
in soil, 664
Irreversible inhibition, **177–178**, 184
penicillin and, 178
Irrigation, 659
Ischemia, 760–761
Island biogeography model, **1123–1124**, 1130
and conservation biology, 1126–1127
"Islands" (ecological)
and ecological research, **1123–1124**
size of, and species diversity, 1123
and speciation, 1010–1011
Islet cells, 823, 832, 839
Islets of Langerhans, 725
Isocitric acids, 655
Isogametes, 464
Isogamy, 463
Isoleucine, 73, 75, 76
Isomerase, 189
Isomorphism, 464–465
Isopentenyl adenine, 676 (see also Cytokinins)
Isotonicity, 132–134, 142, 767
Isotopes (see also Radioactive labeling, Radioactive probes), **24**, 38
and cancer research, 25
dating by, 25
and elements, 25
heavy isotopes, 292, 293, 299
^{18}O, and van Niel's hypothesis, 217
Isozymes, 978
IUDs, **916**

"Jack" males in salmon, 997–998
Jacob, François, 322–323
Jaffe, M. J., 691
Janzen, David, 1121
Java man, 1044
Jaws (see also Mandibles), 501, 538, 541, 561, 565, 585, 590, 594, 597
Jeffreys, Alec, 396
Jellyfish, 524, 526, 528, 529, 542
Jelly fungi, 488
Jenner, Edward, 796
Jet lag, 834
Jet propulsion, in cephalopods, 546, 553
Jews, and Tay-Sachs disease, 389
Johanson, Donald, 1038, 1040, **1041**
Joining genes, 372
Joints, **706–707**
Joshua trees, 1172
"Jumping" gene, 336, 375
Junctions
cell-cell, **140–142**
Junipers, 505
Juniper savanna, 1172

Jurassic period, 496, 598, 1031
Juvenile dependency, 1033
Juvenile hormone, 578
Juvenile-onset diabetes, 812

Kandel, Eric, 895, 898
Kangaroo rat, 600, 765, 768
Kaposi's sarcoma, 814, 816
Karyotype, **383–388**, 399
and amniocentesis, 387
of Down's syndrome, 385
human, **383–388**
preparation of, **384**
Katydids, 583, 959
Keck, David, 1002–1003
Keels, 598, 599
Kelps, 455, 465, 477
Kelvin, Lord, 970
Kelvin scale, 778
Kentucky bluegrass, 1012, 1014
Kepler, Johannes, 7
Keratin, 74, 77, 82, 111, 710, 791
Kernels, 622
Ketones, 57
Ketoses, 60
Kettle holes, 1168
Kettlewell, H. B. D., 962
Khorana, H.G., 310
Kidneys, 749, **769–776**
and ADH, 828
blood supply to, 759
in calcium regulation, 829
cells, 774
cushioning of, 68
development of
in chick embryo, 936
in human embryo, 948
and evolution, 765
excretion in, 763, 775
function of, **771–774**
hormones and, 774–775, 824, 829
hormone receptors in, 756
and osmosis, 134–135
regulation of, 775
substances regulated by, 766
and temperature regulation, 785
and transition to land, 765, 767–768
transplant rejection in, 810
and water balance, 775
Kiem Tran Thanh Van, 682
Killed-virus vaccines, 449
"Killer cells," 807, 811
Kimeu, Kamoya, 1044
Kinase, 182
Kinetic energy, **28**, 43, 128, 162
and atmospheric temperature, 1133
and intracellular energy transport, 168–169
Kinetin, 676 (see also Cytokinins)
and IAA, 676, 677
Kinetochore, 148, **149**, 150–151
in meiosis, 254, 256
Kinetochore fibers, 149, 150, 151, 152
King Clone, 643
Kingdom(s), **12–14**, 94, 410, **423–426**, 427
animal, 410, 423, 425
mineral, 410, 423
monophyletic, 424
plant, 410, 423
Kingfishers, 42
King penguin, 1007
Kinorhynchs, 520, 540, 541, 543, 559
Kin recognition, **1065–1070**, 1078
and reciprocal altruism, 1075, 1078
in tadpoles, **1076–1077**
tests for, **1069–1070**
Knee, 707, 946
Kneecap (see Patella)
Köhler, Georges, 802
Korsakoff's syndrome, 894
Koshland, Daniel E., Jr., 131
Krebs cycle, 187, 191, 193, **194–196**, 200–204, 222–223
acetyl CoA in, 196
acetyl groups in, 204
ADP and ATP in, 194–196
and catabolism, 203
citric acid in, 194
electron acceptors in, 195–197, 203

FAD in, 195, 200
$FADH_2$ in, 195–196, 200–201
flavoproteins in, 195
FMN in, 195, 200–201
and NAD^+, 194–195
and NADH, 194–195
net energy harvest of, 196
pyruvic acid in, 194–196
riboflavin in, 195
thyamine in, 729
Krypton, 733
K-selected life-history patterns, 1099
Küchler, A. W., 1162
Kunkel, Louis, 394

Labia, 909
Labium, 575
Labor, stages of, 908, 950
lac operon, 323, 325–326, 337
mutations in, 323
protein coding in, 326
and recombinant DNA, 351
Lactase, 722–723, **727**
Lactic acid, 204, 744, 878
and anaerobic respiration, 190–191
and fermentation, 204
in vagina, 909
Lactic acid bacteria, 428, 430
Lactose, 59, 61, 722, 727
and beta-galactosidase, 321–322
prokaryotic metabolism of, 321–326, 337
Ladybirds, 574
Laetoli, fossils at, **1039**, 1041–1042
Lagging strands of DNA
and prokaryotic cell, 320
in replication, **296**, 297, 299
Lagomorphs, 415, 602 (see also Hares, Rabbits)
Lakes
net primary productivity of, 1139
organisms in. 1155
seasonal cycle of, 45
Lamarck, Jean Baptiste, 4–5, 7, 306
Lambda integrase, 335
Lambda phage, **334–335**, 338, 373
chromosome of, 334–335
COS region of, **334**
and DNA ligase, 334
enzymes in, 334–335
in host cell, 334–335
recognition site, 335
and recombination, 337
replication of, 334–335
"sticky ends" of, 334–335
and temperate bacteriophages, **331–333**, 334, 338
Lamellae, 106, 107, 154, 871,
of gills, 130
middle, 106, 154
Laminarin, 457, 465
Lampreys, 593
Lamp shells, 520, 562, 564
Lancelets, **591–592**, 604
Land (see also Biosphere; Ecology; Ecosystems)
biomes, **1161–1163**
creation of, by coral, 530
coniferous forests, **1166–1168**
desert, **1172–1173**
destruction of rain forests, **1174–1176**
life on, **1160–1177**
Mediterranean scrub, 1171
Pacific northwest forests, **1166–1168**
savanna, 1170
taiga, **1166**
temperate forests, **1164–1166**
temperate grasslands, **1169–1170**
tropical forests, **1174–1177**
tropical grasslands, 1170
tundra, **1168–1169**
Landsteiner, Karl, 810–811
Land-use planning, 1127
Language, 604, **888–891**, 890
Lanugo, 948
Large intestine, 715, 725, 730
Larva(e), 519, 533, 571, 575–580, 586–587, 590–592, 595, 604
of bryozoans, 563
of ctenophores, 531
of frog (see Tadpole)
of honey bee, 1060

Larvae(s) *(Continued)*
  of pentastomids, 560
  of sea urchin *(see* Pluteus)
  of spiny-headed worms, 540
  of tapeworms, 536
  trochophore, **554**, 558, 559, 564
Laryngitis, 738
Larynx, 738, 740
  and male sex hormones, 905
Lashley, Karl, 893
Latent variability, 976–977
Lateral geniculate nuclei, **888**
Lateralization of function, 889
Laterally directed eyes, evolution of, 1033
Lateral meristems, 644
Lateral plate mesoderm, 928–929, 932–933, 936, 952
Lateral roots *(see* Branch roots)
Lateral sulcus, 886
Laterites, 1176
Latitude, 1137, 1161, 1172
Laurasia, 1012–1013
Lava, 85, 87
Lavoisier, Antoine, 209
Laws
  of chemistry, 35, 3810–3811
  of physics, 35, 38
  of probability
    in genetics, **244–245**
    and genetic variability, 882–883, 979
    in human evolution, 1050
  of thermodynamics, **162–166**, 183, 710–711
  in science, 14–16
Layers
  of atmosphere, **1132–1135**
  of rock *(see* Strata)
LDL, 71, 138, 723, 726, 761
Leaching of nutrients, 663
  from forest ecosystems, 1150
Lead, 177
"Leader" protein, 116
"Leader" sequence, 320–321
Leading strand of DNA, **296**, 297, 299
  and prokaryotic replication, 320
Leaf drop, ethylene and, 679 *(see also* Leaf-shedding)
Leaf hairs, and transpiration, 652
Leaf litter, 663, 1164, 1166, 1176
Leaf rolling, 654
Leaf scar, and abscission, 679
Leaf-shedding, 1164 *(see also* Deciduous forest ecosystems)
Leakey, Louis, 1038, 1040–1042
Leakey, Mary, 1035, 1038–1041
Leakey, Richard, 1038, 1040–1042, 1044
Learning, 552, **893–899**, **1055–1058**, 1077–1078
  altitudes, 746
  in *Aplysia*, 895–898
  associative, **1055–1056**
  and birdsongs, **1056–1057**
  vs. fixed action patterns, 1055
  vs. habituation, 896
  imitative, 1057–1058
  imprinting, 1056–1057
  and intrinsic processing areas, 893
  and kin recognition, 1077–1078
  and memory, **893–898**
  and organization of ideas, 893
  periods of, in primates, 1033
  and phenotype matching, 1077
  synaptic modifications in, **895–898**
Leaves, 626, **627–630**, 647
  abscission of, **679**, 680, 694–695
  adaptations in, **629–630**
  and apical meristem, 640
  auxin production of, 675
  axils of, 641, 648
  buttresses, 641
  and cell types, in angiosperms, 647
  compound, 626
  conifer, 505, 508, 623, 1166
  deciduous, 1164, 624
  deciduous vs. conifer, 623
  development, in embryo, **640–641**
  diffusion of gases at, 669
  dormancy in, 621
    and abscission, 680
  evaporation in, 651–652
  evolution of, 627
  of ferns, 503

  vs. floral parts, 613
  ground tissue of, 648
  primordia, 640–641, 648
  rainforest, 1174
  and roots, 626
  senescence, 677, **679**, 694
  and shoots, 640
  simple, 626
  and stems, 626
  structure, **627–630**
  succulent, 629–630
  temperature of, 650
  and tendrils, 642
  and translocation, 662
Lectins, 666
Lederberg, Esther, 964–965
Lederberg, Joshua, 964–965
Leeches, 557, **558**, 564
Leeuwenhoek, Anton van, 104, 237, 448, 469
Left brain, **888–891**, 898 *(see also* Cerebral hemispheres; Right brain; Split brain)
  and handedness, 888
  and speech, 888–889, 898
  and split brain, **889–891**
Left-handedness, 888
Legs, 574, 582, 596, 601
  development of, 946
  feeding, 571
  jointed, 565
  swimming, 571
  walking, 570–572
Legumes, 620
  nitrogen transport in, 665–666, **668**, 670
  predation and, 1114
  "sleep movements" in, 690
Lemurs, 1034, 1051
Lens, 866, 867
  development of, 936
Leopards, 412
  clouded, 412
  snow, 412
Lepidopterans, 573 *(see also* Butterflies; Moths)
Lettvin, Jerome, 870
Leucine, 73, 76, 309
Leucine-enkephalin, 855
Leucoplasts, **120**, 125, 626
Leukemia
  and interferons, 793
  and strontium-90 in environment, 1151
Leukocytes *(see* White blood cells)
Leukotrienes, 835–836
Levels of organization, biological, 35–38
  ecological properties of, 1087–1088
Levene, P. A., 283–284, 287–288
Lewontin, R. C., 978
LH, 824–826, **906–907**, 917
  in menstrual cycle, **913**, 914
  and "the pill," 915
Lianas, 1174–1175
Lichens, **488–490**, 492, 1119
  and air pollution, 488
  and ascomycetes, 484, 489
  British soldier, 490
  classification, 489
  crustose, 490
  and cyanobacteria, 488, 489
  foliose, 490
  fruticose, 490
  and fungi, 488
  and green algae, 488–490
  and peppered moth, 962–963
  and photosynthesis, 492
Licking, 785
Lieber, Charles S., 202
Life *(see also specific life forms;* Biosphere; Ecology; Ecosystems; Land; Living systems; Oceans; Populations)
  benthic, 1156
  biomes, **1161–1163**
  carbon and, 32
  cellular basis of, 969
  chemiosmotic power in, 200
  continuity of
    and development, **919–953**
    and reproduction, **900–918**
  cycle, 250–252, 256, 261, 427
  distribution of on land, 1160–1161
  diversity of, **405–427**

  and evolutionary theory, 969 *(see also* Evolution, theory of)
  in forests, **1164–1176**
  in lakes, 1155
  on land, **1160–1177**
  in oceans, **1155–1157**
  organization of, 142
  origins of, **13**, **23**, **85–88**, 89, 92, 951
  pelagic, 1156
  in ponds, 1155
  properties of, 26–27, **84–101**
  in rivers and streams, **1154–1155**
  at seashores, **1157–1160**
  and solar energy, **1132–1137**
  and spontaneous generation, 86
  and temperature, 777–790
  and water, 40
Life cycles, 250–252, 256, 261, 427, 519
  of angiosperms, 621–624
  of animals, 252
  of basidiomycetes, 487
  of bryophytes, 499–500
  of *Chlamydomonas*, 462–463
  of cnidarians, 526, 542
  and dispersion patterns, 1094
  fertilization and, 250
  of frogs, 595
  of green algae, **462–465**
  of honey bees, 1060
  of insects, **575–578**, 586
  insects vs. invertebrates, **575–578**
  meiosis and **250–252**
  of pentastomids, 560
  of *Plasmodium*, **473**
Life expectancy
  and blood groups, 996
  and reproduction, 1092, 1100
Life-history patterns, **1099–1104**, 1105
Life span, 1151
  of arthropods, 585
  and capacity for learning, 1055
  men vs. women, 761
  vs. reproductive span, in age structure, 1092
Ligaments
  connective tissue in, 704
  in knees, 706–707
  of ovaries, 907
Light, **206–230**
  absorption of, 213
  bioeffects of, 210
  brightness of, 208
  colors of, 207
  "corpuscles" of, 207
  critical wavelength of, 207–208
  as energy, 161–162
  and energy flow in ecosystems, 1137
  "fitness" of, for photosynthesis, **209–211**
  and heat loss, 779
  and homeotherms, 781
  intensity, and photosynthesis, 215
  and melatonin production, 833
  in microscopy, 96–100
  nature of, **207–211**
  in the oceans, 1156
  particle model of, 208–209
  photons, 208–209
  photosynthesis, effect on, **215–216**
  and photoperiodism, 695
  and phototropism, **671–673**
  and pineal gland, 824, 833
  and plant coiling, 671
  and plant growth, 645–646, **671–673**, 688
  and primary productivity of ecosystems, 1137
  red, 207
  and reproductive isolation, 1016
  speed of, 207
  and stomatal movements, 669
  ultraviolet, 274, 331, **332**
  violet, 207
  visible spectrum, 207, 210
  and vision, 866–869, 872
  and water translocation in plants, 654
  wavelength of, 96, 208, 211
  wave model of, 207–209
  white, 207
Light chains, of antibodies, **798**, 801, 818
Light/dark cycle, 688, 833
Light/dark reactions, and stomatal movements, 654

Light-dependent reactions of photosynthesis, 215–216
Light-independent reactions of photosynthesis, 215–216
Light microscopy, 95–101, 214, 423
"Light" reactions of photosynthesis (see Light-independent reactions)
Light-sensitive cells, 833
Light-trapping reactions, **218–219**
Lignin, 107, 637, 639
Lilies, 151, 511, 514
Limb buds, 946
Limbic system, 884, **885**, 898
  and consolidation of memory, 894
  and drives, emotions, and actions, 898
Limbs, 571
  development of in embryo, 937–938, 946
  evolution of, 1006
  free-moving vs. single-plane, 1051, 1052–1053
Lime, 663
Limestone, 529, 1171 (see also Calcium carbonate)
Limiting factor curves, 1095
Limiting factors, 1087, **1095–1098**, 1104
  density-dependent, **1098**
  density-independent, **1098**
Limnetic zone, 1155
Limulus, 596
Linear array
  of genes on chromosome, 276–277, 279
  of prokaryotic chromosome, 328
Lining tissues, 532
Linkage of genes, **274**, 279
Linked DNA, 333
"Linker" DNA, 357
Linnaeus, Carolus, 2, **408**, **410**, **412–413**, 423, 975
Lions, 412, 602
  familial conflicts of interest in, 1071
  waiting behavior in, 1074
Lipases, 722, 730, 937
Lipid(s), 55, **67–70**, 82, 228, 457
  barrier, 134, 136
  and cell division, 146
  membrane, 117
  metabolism of, 118, 824
  phosphate groups in, 69
  synthesis, 117–118
Lipid barrier, of cell membrane, 134–136
Lipid bilayer, of cell membrane
  and cell transport, 130, 134
  and enzyme activity, 178
  in prokaryotic cell membrane, 450
Lipoproteins
  and cell-mediated endocytosis, 139
  HDL, 71
  LDL, 71, 138
  in plasma, 726
Lithosphere, 1145
Littoral zone, 1155
Liver, 715–716, 730
  adrenaline and, 837–838
  alcoholic hepatitis, 202
  blood supply to, 759
  cell, 117, 142, 147
  and cholesterol metabolism, 70–71, 723, 726
  cirrhosis, 202
  and degradation of hormones, 823
  development of, in embryo, 937, 946, 952
  in digestion, 722, **725–726**
  and ethanol metabolism, 202
  glucose transport in, 136
  glycogen storage in, 64
  and hemoglobin metabolism in, 726
  and hormone regulation, 726
  of rat, 202
  and regulation of blood glucose, 725
Liverworts, 496, 498–500, 1164
Living standards, and population explosion, 1149
Living systems
  adaptation in, 27, 228
  atomic structure of, **28–34**
  development in, 27
  energy use, 27, 161, 183, 187
  flow of energy through, 1137
  growth in, 27
  heredity and, 235
  vs. nonliving, 26–27, 35–36, 37, 127
  organization of, 84, 127, 161, 183
  origin of, 85–88

properties, 26–27, 35–37, 84–101
and second law of thermodynamics, 166
"stratgeies" of, 59
structure of, 127
and water, 40–41
Lizards, 596–597
Lloyd, James, 581
Loam, 664
Lobes, of cerebral cortex, 885
Lobsters, **571–572**, 585
Lobules, of tesis, 901
Lockjaw, 429
Locomotion, 518, 531, 533
  bacterial, 131
  in brittle stars, 589
  in cephalopods, 546
  in earthworms, 555, 708
  in echinoderms, 604
  and embryonic development, 936
  in octopuses, 552
  in planarians, 533
  in protozoa, 468–469, 475, 477
  and receptors, 131
  in reptiles, 597
  in starfish, 588
Locus, of genes, 275, 279
Locusts, 853
Logarithm, 48
Logistic model of growth, **1089–1091**, 1104
Long-day plants, 684, 686, 695
Long-range planning, in human brain, 893, 899
Long-spurred violet, 408
Long-tailed windowbirds, 1000–1001
Long terminal repeat (see LTR)
Long-term memory, **893–895**, 899
Looped domain, of chromosome, 358
Loop of Henle, 770, **772–774**, 776
Lophophorates, 520, **562–563**, 564 (see also Bryozoans; Lamp shells; Phoronid worms)
  deuterostome characteristics in, 562, 564
Lordosis, in rats, 908–909
Lorenz, Konrad, 1052, 1054–1055, 1056
Loricifera, 415, 520, 540, 543
Lorises, 1034, 1051
Lovejoy, Owen, 1044
Low density lipoprotein (see LDL)
Lower back pain, 1006
Low-pressure areas, 1137
Low selection lines, 977
L ring, of prokaryote flagellum, 434
LSD (lysergic acid diethylamide), 486, 853
LTR, 373
Lubchenco, Jane, 1119
Lubricating fluid, 706
Luciferase, **352–353**, 354
Luciferin, 352–353
Lucretius, 2
"Lucy," 1000, **1001**
Lumen
  of blood vessels, 753–754
  of endoplasmic reticulum, 116
"Lumpers," taxonomists, 412
Luna, 425
Lungfish, 594–595, 736
Lungs, 544, 747
  book, 566, 568, 570, 585, 594, 604
  bulk flow in, 736
  cancer of, **739**
  and circulatory system, 755–756, **758–759**, 761
  development of, in embryo, 937, 952
  evolution of, 747
  vs. gills, **734–737**
  in human respiratory system, 738–740
  size of, in mammals, 744
  ventilation in, 736
Lupine, 425
Lupus erythematosus, 812
Luria, Alexander, 889
Luria, Salvador, 284
Luteinizing hormone (see LH)
Luther, Martin, 7
Lwoff, André, 322–333, **332**
Lycophyta, 497–498, 501–502
Lycopodium, 502
Lycopod trees, 504
Lyell, Charles, 5, 8
Lymph, 132, 704, 749, 775, 792–795
Lymphatic ducts, **762–763**

Lymphatic system, **762–764**
  and digestion of fat, 721, 723
  human, 763
Lymph capillaries, 763
"Lymph heart," 763
Lymph nodes, 763–764, **793**, 794–795, 816
Lymphocyte(s), 763, 792, 794, 798, 801, 831 (see also B lymphocyte; T lymphocyte)
  and antibody gene, 371–372, 381
  and cDNA, 342
Lymphokines, 807, 823, 831
Lymph vessels, 723, 730, 763, 769, 793–795
Lynx, 412, 1118–1119
Lyon, Mary, 267
Lyon hypothesis, **267**
Lysergic acid diethylamide (see LSD)
Lysine, 73, 76, 178, 622, 728
Lysogenic cells, **331–332**, 333
Lysogeny, 322, 330–331, **332**, 333
  and temperance, 331
Lysosomes, 112, **118**, 119, 125, 138, 751, 792
  in cell division, 146
  and enzymatic pathways, 173
  isolation of, 192
  in Tay-Sachs disease, 389
Lysozyme, 432, 799
Lytic cycle, 330–335, 338
  and ultraviolet light, 331, **332**
  and x-ray, 331
Lytic enzymes, 79

Macaques, learning in, 1057–1058
MacArthur, Robert, 1099, 1102, 1109, 1123
Macroevolution, **965–972**, 973, 1010, 1025, 1029–1030
  adaptations and, **970–972**
  and biogeography, **966**
  in fossil record, **966–968**
  and homology, **969**
  number of species, **965**
  rate of, 973
Macromolecules, 117–118
  and biological organization, 36
  and fungi, 13
  importance for living systems, 36
Macronuclei
  of ciliates, 469, 471, 472
  of *Paramecium*, 471
Macronutrients, **656–657**
Macrophages, 121, 138, 791, **792**, 793–794, 830–831, 836
  and AIDS virus, 815–816
  in cell-mediated immunity, 801
  and cytotoxic T cells, 807–808
  and helper T cells and B lymphocytes, 808–809
  in inflammatory response, 818–819
  inflammatory vs. immune response, 808
  in transplant rejection, 810
Maggots, 86, 576
Magnesium
  in biogeochemical cycles, 1145
  in chlorophyll molecule, 656, 670
Magnesium (Mg^{2+}) ion, **31**, 171, 657
  regulation of, 766
  in soil, 662
Magnetic field, 891 (see also MRI; SQUID)
Magnetic resonance imaging (see MRI)
Magnetism, and bacterial orientation, 435
Magnetite, 435
Magnetoreceptors, 859
Major histocompatibility complex (MHC) antigens, **805**, 806–810, 819
  genetic code for antigens, 805
Malaria, 469, 473, 474, 797
  research on, vs. cancer research, 537
  and sickle cell anemia, 988
Male, 235, 236, 243
  biological role of, 235
  and color blindness, **391–392**
  competition in, 998–1001
  gametes, 900–901
  genetic disease in, 386, 391–394, 400
  height distribution, 272
  hormes in, **905–907**
  infanticide in, 1071
  ornamentation in, **1000–1001**
  puberty, 902

I-33

Male (Continued)
  reproductive system of, **901–907**
  sex chromosome abnormalities in, 386, 391–394
Male dominance, 1036
  evolution of, 1044
Malic acid, 224, 655
Mallee fowl, 783
Malnutrition, 728, 1148 (see Hunger; Populations)
Malpighian tubules, 568, 572, 574, 575, 585
Maltase, 722, **723**
Malthus, Thomas, 7–8, 991
Maltose, 59, 64, 722
Mammalia, 412, 416
Mammals, 587, **600–604**
  adaptive radiation of, and evolution, 1022, 1050–1051
  care of young in, 900
  circulatory system, 763
  digestion in, 715–730
  embryo of, 952–953
  endothermy in, 701–702
  extinctions of, 496
  grazing, 496
  heart, 754–758
  homeothermy in, 781–782
  hypertonic urine in, 772
  metabolism of, 763
  reproduction in, **900–918**
  temperature regulation in, 789–790
Mammary glands, 601
  and hormones, 828
  milk-secreting cells, 828
Mammillary body(ies), 894
Mammoth, woolly
  and elephant, 9
Mandibles, 571, 572, 575, 585 (see also Jaws)
Mandibulates
  acquatic (see Crustaceans)
  terrestrial, 566, **572–578**, 585 (see also Insects; Myriapods)
Manganese (Mn^{2+}) ion, 657
Mangold, Hilda, 930
Mangrove forests, 1160
Mannose, 190
Mantle, 546–550, 553, 563
  of lymph node, 794
Mantle cavity, 546–551, 553, 563
Mapping
  chromosome, **275–277**, **328–329**, 338, 383
  human genome, **398–399**, 401
Map unit, 276–277
Margins, of lakes and ponds, 1155
Marine environments, **1150–1160** (see also Oceans, Seashore)
Marine microorganisms, 465, 476, 477 (see also Cephalopods; Fishes; Mollusks; Sharks; Whales and specific marine animals)
  algae, 463
  brown algae, 455, 457
  golden-brown algae, 457, 459–460
  green algae, 455, 457, 461
  phytoplankton, 456, 460
  plankton, 456, 459
  red algae, 455, 457
  yellow-green algae
Marler, Peter, 1056
Marmosets, 1035
Marrow, 705
Mars, 235
Marsh, O. C., 409
Marsh, Othniel G., 968, 1027
Marshall Islands, 530
Marshes
  inhabitants of, 1155
  net primary productivity of, 1139
Marsupials, 600–601
  amniote egg in, 901, 944
  biogeography and, 966
  evolution of, 1010, **1023**, 1031
  vs. placental mammals, **1023**
Masked puddle frog, 407
Mass, conservation of, 163
Mass extinctions, 1024–1025, 1030
  and human responsibility, 1176–1177
Mast cells, 836
  and allergies, 812–813
  and basophils, 792
Master genes

and animal development, 942
and sex determination in human beings, 946
Mastigophorans, 455, 468–470, 477
Mate preference, 998–1000, **1001** (see also Sexual selection)
  competition for, 998–1001
  and Darwin's finches, 1016, 1020
  in females, and the selfish gene, 998, 1072
  and imprinting, 1056
  as premating isolating mechanism, 1016, 1020
  and territoriality, 1063–1064
Materialism, 17–18
Maternal blood space, 945
Maternal care
  evolution of, and arboreal life, 1033
  in sponges, 524
Maternal genes, and homeobox, 942–943
Maternal nutrition, and birth weight, 949
Mathematics
  and analysis of experimental results, 238
Mating, 553, 1063 (see also Behavior; Courtship; Nonrandom Mating; Random Mating; Sexual Reproduction; Sexual Selection)
  in bowerbirds, **1073**
  competition for, 998
  and estrus, in mammals, 915, 917
  on fireflies, 580
  as isolating mechanisms, 1016
  length of, selection for, 995
  in lobsters, 572
  in rats, 908–909
  and selfish gene, 1072–1074
  and speciation, 1013–1017
  in sticklebacks, 1054–1055
  in water molds, 467
Mating calls, 1016
Mating substances, 585
Matrix, extracellular, 713
Matrix, of mitochondrion
  and chemiosmotic coupling, 197–199
  and electron transport, 205
  and respiration, 191–193
Matthaei, Heinrich, 307
Maturation, rapid vs. slow, and population growth, **1099–1105**
Mauthner cells, 1054
Maxam, Allan, 348–349
Maxillae, 575
Maxwell, James Clerk, 207, 209
May, Robert M., 1090–1091, 1144
Mayr, Ernst, 5, 9, 407, 408, 973, 1022, 1028
McClintock, Barbara, 375, 863
Measles, 796
Measurement, in science, 17
Mechanical energy, 128, 161, 163, 165–166, 186
Mechanists, 10
Mechanoreception, 863–865, 879 (see also Balance; Hearing)
Mechanoreceptors, 556, 859, 860, 904
Medical research, 537
Medical science, and genetics, 382
Mediterranean scrub, 1171, 1177
Medium, minimal, 303
Medulla (of adrenal gland) (see Adrenal medulla)
Medulla (of brain), 764, 882–883, 898
  and regulation of heart, **762**, 764
Medulla (of kidney), **770**
Medullary cords, 794
Medusa(e), 525–526, 542
  vs. polyps, 525
  sexual reproduction in, 526
Megakaryotes, 751
Megaphylls, 501, 508–509
Megasporangia, 507
Megaspores, 507–508, 511
Meiosis, **249–261**, 264, 278, 279, 286, 306, 426
  in algae, 460, 462–465
  and allele frequency, 989
  cell division in, 251–255
  in *Coleochaete*, 494
  and condensation of chromosome, 355, 358
  and diploid number of chromosomes, 249–250, 264
  division of chromosome in, 264
  DNA transcription in, 359
  in fungi, 462, 491
  gamete fusion in, 252
  and genetic recombination, 261

and genetic variation, 984, 989
in gymnosperms, 507
homologous recombination in, 336–337
in human beings, **258**
independent assortment in, 260–261
and introns, 363
and life cyle, **250–252**
meiosis I, **253–255**, 258, 261
meiosis II, 253, **255–256**, 258, 261
vs. mitosis, **252–253**, 257
Neurospora crassa, 302
nondisjunction in, 385, 399
and nucleolus, 366
in oogenesis, 910
phases of, **253–258**
  anaphase I, 253, 254
  anaphase II, 256
  interphase, 253, 255, 261
  metaphase I, 253, 254, 259
  metaphase II, 256
  prophase I, 253, 254, 275
  prophase II, 255
  telophase I, 253, 255
  telophase II, 256
in plants, 462
and polyploidy, 1012–1013
in protists, 462
in protozoa, 468–471
random assortment in, 253, 261
"reductive division" in, 250
sex chromosome segregation during, 278
and sexual reproduction, **249–262**, 900
as source of variation, 261, 984, 989
in spermatogenesis, **901**, 902
in vertebrate gametes, 917
Meissner's corpuscles, 860
Melanin, 388
  in albinism, 389
  in phenylketonuria, 388
Melanocytes (see Pigment cells)
Melatonin, 824, 833, 839
Membrane, 88, 90–93, 102
  of cell (see Cell membrane)
  of chloroplast, 211, 214, 218
  of endoplasmic reticulum, **115–117**
  internal, and cell transport, 127, 142
  of mitochondrion, 119, 197–201
  nuclear, 108
  of organelles, 135
  photosynthetic, **213–214**
  of plasmodesmata, 140
  of plastid, 120
  selective permeability of, 124
  synthesis of, 117
  thylakoid, **113**
  transport, 124, 178–180, 181
  of vacuole, 115
  and vesicle-mediated endocytosis, 139
Membrane, cell, 102, 103, **104–106**, 111, **112–113**, 114, 118
  in cell division, 153–155
  and cell variety, 125
  cholesterol in, 70
  cytoplasmic face of, 105, 124
  exterior face of, 105, 124
  lipids in, 69
  proteins in, 106, 178–181
  structure and composition of, 104, 105–106
  transport across, **127–143**
Membrane glycoproteins, 804, 897
Membrane ion channel, 857
Membranelles, 452, 470–471
Membrane proteins, 70, 72, 106
Membrane receptors, 862
  and viral research, 451
Membrane transport proteins, 178–181
  and allosteric interactions, 178–181
  and electron transport, 198
  integral membrane proteins, 104–106, 114, 124, 135, 142, 178–179
Membranous fragments, of nuclear envelope, 151
Memory, 898, 899 (see also "Habit" memory; Long-term memory; Recognition memory; Short-term memory)
  in Alzheimer's disease, **896–897**
  anatomical pathways for, 893, **894–895**
  consolidation of, 885
  and emotions, 894

and intrinsic processing centers, 899
and learning, 893-898
long-term, **893-898**
  cellular basis for, 895-898
and REM sleep, 891
and sensory information, 893, 894, 898-899
short-term, **893-895**
synaptic modifications and, **895-898**
Memory cells, **795-796**, 804, 807-809 (*see also* Immune system)
  in allergies, 812
  and antibody activity, 819
  differentiation of, 808, 809
  in immune response, 819
"Memory" in plants, 691
Memory loss, 893-894, 896-897 (*see also* Amnesias)
Mendel, Gregor, **238-242**, **244-247**, 260-261, 263-265, 271, 274, 278-279, 282, 284, 301, 979, 981, 984
  contribution to genetics, **238-247**
  experimental method of, **238**
  hypotheses of, 238, 240-241
  influence of, 246
  laws of, **239-243**
    and modifications, 268-279, 362
  and modern evolutionary theory, 972-973
  principle of independent assortment, **241-243**
  principle of segregation, **239-241**, 247
  principles of, 268
  use of quantitation, 238-239, 244-245
Mendelian genetics, 268, 973, 979, 981, 989
  of eukaryote, 361
  and population genetics, 974
  and synthetic theory of evolution, 972-973
Meninges, 884
Meningitis, 432-433
Menisci, 707
Menopause, 761, 831, 910
Menstrual cycle, 913-915, 917
  environmental influences on, 914
  and estrus, 915
  length of, 914
  and prostaglandins, 835
  synchronization of, 863
Menstrual fluid, 914
Menstruation
  and contraception, 908, 912, 916, 817
  prevention of, during pregnancy, 941
Mental retardation
  in Down's syndrome, 385-386
  in Duchenne muscular dystrophy, 393-394
  in phenylketonuria, 388-389
Menthol, 734
Mercury, 42
  in barometers, 732
  in measurement of blood pressure, 759
Meristematic cells, 647
Meristematic tissues, 678
  apical, 630, 647-648
  buds, 648
  ground, 647-648, 625
  primary, 647-648, 625
  vascular cambium, 647, 648
Merkel cells, 860
Merozoites, 473-474
Mescaline, 515
Meselson, Matthew, 258, 292-293
Meselson-Stahl experiments, 258
Mesenteries, 532, 544, 563
  in human digestive tract, 716, **718**, **907**
Mesoderm, 532, 542, 544-545, 563
  embryonic, 939, 944, 952
    in amphibians, 928-929
    in chick embryo, 932-937, 938
    differentiation of, 936, 937
    in sea urchin, **924**
    in vertebrates, 532, 701
Mesoglea, 525, 528, 531, 542
Mesophyll cells, 224-226, 627, 647-648
Mesosphere, 1132, **1133**
Mesozoans, 520, **524**, 542
Mesozoic era, 496, 510, 598
  and mammalian evolution, 1031
Messenger, artificial, 310
Messenger hypothesis, and RNA, 308
Messenger molecules, 142
Messenger RNA, **306-307**, **309-317**
  and antibody biosynthesis, 372

artificial, 310
binding of, 311, 320, 368-369
biosynthesis of (*see* transcription)
and cytoplasmic determinants in embryo, **926**, 952
and DNA, **306-307**
editing of, 368
and enzyme production in grains, 682
and exons, 380
and hormone actions, 836-837
and introns, 361, 380
and leaf senescence, 677
nucleolus and, 366
and oligosaccharides, 682
in oocyte, 910
and plant cell growth, 675
in recombinant DNA, 342, 345-347, 351, 353
retrovirus and, 374
and reverse transcriptase, 353
and ribosome, 368-369
in sea urchin ovum, 921
sequential reading of, 310
transcription, **320-323**, 325, **368-370**, 372, 380, 836, 839
translation of, 368
viral, 353
viral DNA and RNA as, 445
and viral infection, 793
and viroids, 446
Metabolic rate
  in endothermy, 702
  evolution of, 755, 763
  in homeothermy, 782
  and temperature regulation, 782
Metabolic pathways, **167-180**, 183, 228, 540 (*see also* Anabolism; Catabolism; Metabolism)
  and predation, 1114
  and prokaryotic classification, 429
Metabolism, **167-169**, 183
  at altitude, 746
  basal, 732
  of cell, 102, 104, 110, 127, 136, **167-180**, 283
  and circulation, **749-764**
  cost of, vs. net productivity, 1138, 1153
  and digestion, **715-730**
  and energy, 203, 710-711
  of glucose, 875
  and homeostasis, **777-790**
  hormonal regulation of, 824, 829
  inborn errors of, 301
  of lactose, in prokaryotes, 321-326, 337
  of minerals, 829
  and oxygen absorption, 743
  pathways for, **167-180**, 183
  rate of
    in endothermy, 702
    and evolution, 755, 763
    in homeothermy, 782
    and temperature regulation, 782
  and respiration, **732-748**
Metals
  photoelectric effects of, 207
Metameres, 554
Metamorphosis, **576**, 579-580, 586-597, 592, 925
  and hormones, 578, 586
  incomplete, 595
Metaphase
  of mitosis, 150, **151**, 152, 155
Meter, 95
Methadone addiction, 948
Methane, **32-33**, 87, 430, 439
Methanogens, 90, 430, 432, 439
Methionine, 73, 76, 285, 312-313, 351
Methionine-enkephalin, 855
Method, scientific, 15-18
Methyl groups, 57
Methylating enzyme, 341
Methylation
  enzymes in, 341
  and gene regulation, 360-361, 380
  of nucleotides, **341**, 353
Methylcytosine, 360
Methylguanine, 368-369
Methyl-guanine cap, 368, 369, 380
MHC antigens (*see* Major histocompatibility complex)
  tests for, 811, 819
  and tissue grafts, 810
Micrasterias, 206

Microbiology, 428
Microevolution, 962-965, 973
  in drug resistance in bacteria, **964-965**
  in insecticide resistance, **964**
  vs. macroevolution, 1010, 1029
  in the peppered moth, **962-965**
  and speciation, 1010, 1029
Microfibrils, 66, 106-107, **113**, 653
Microfilaments (*see* Actin filaments, Cellulose microfibrils)
Microfossils, 85, 89, 90, 91-92, 100, 452
Microhabitat, 1110
Micronuclei, 469, 471, 472
Micronutrients, **656-657**
Microorganisms (*see also specific microorganism*)
  and disease, 439, **448**
  and human ecology, **447-450**
Microphylls, 501
Microscopy, 86, **94-99**, 101, 104, 109, 110, 212, 237, 247
  dark field, 99, 100, 101
  electron, **14**, 95-99, 305, 423
  freeze-fracture tecnnique, 135
  high-voltage electron, 110
  immunofluorescence, 111
  light, 95-101
  phase-contrast, 99, 100, 101
  scanning electron, 95-98, 101
  transmission electron, 95-98, 101, 110
  and video, 99, 101
Microsporangia, 507, 511
Microspores, 506-507
Microsurgery, 109
Microtubule organizing center, 149
Microtubules, 78, 103, 110, 111, **112-113**, 121-126, 184, 795
  alpha-tubulin, 78
  in animal cells, 78, 112
  "arms" of, 152
  ATPases in, 181
  beta-tubulin, 78
  and cell division, 146, 148-150
  and cellular motility, 121-122
  composition of, 78
  and enzymes, 146, 153
  and morphogenesis of vertebrate embryo, 938
  in olfactory cells, 862
  organization of, 149
  in plant cells, 113
  in sarcodines, 470
  in sperm cells, 902-903
  and spindle, 149, 153, 254, 256, 472
Microvilli, **720-721**, 782, 862
  in oviduct, 912
  of T lymphocytes, 801
Midbrain, 898, 935
  evolution of, **882-883**
Middle ear, 864
Middle lamella, 106, 154, 639
Midgut, 574
Miescher, Friedrich, 282
Migration
  of birds, 1168
  of cells, 927-929, 952
    in embryonic development, 935-937, 945-946
  of early human beings, 1047
  and homeothermy, 782
  of marsupials, 1013
Milk, 703, 713, **727** (*see also* Mother's milk)
  in human diet, 831
  "letting down," 828
  production of, 824, 825, 836
Milkweed, 515
  coevolution in, **1003-1005**, 1009
Miller, Stanley, 87, 90
Millipedes, 554, 566-567, 573, 586, 1006
Milstein, Cesar, 802
Mimicry, **1003-1005**, 1009, 1116
  in ant-acacia symbiosis, 1116
  Batesian, **1004-1005**
  in cleaner fish, 1121
  in moths, 1116, 1122
  Müllerian, **1004-1005**
  and natural selection, 972
Mind, **881-899** (*see also* Brain; Emotions; Learning; Memory)
Mineral budgets, **1149-1150**
Mineralocorticoids, 829, **832**

I-35

Minerals (*see also* Biogeochemical cycles; Elements)
  availability of, 1139, 1150
  and carnivorous plants, 666
  concentration of in biomass, 1050-1152
  content of, in soil, 663-664
  distribution of, 1160
  human requirements for, 730
  metabolism of, 829
  movement of, in plants, **650-657**, 669-670
  nutritional requirements for, 728
    in plants vs. animals, **656-657**, 1139, 1150
  uptake of, in plants, **655-656**
"Mini-islands," and speciation, 1015
Mining, and disruptive selection, 996
"Minipill," 916
Miocene apes, 1035
Miocene epoch, 496
"Miracle drugs," 964 (*see also* Antibiotics; Resistance)
Mirsky, Alfred, 286
Miscarriage, 945-946
Mitchell, Peter, 197
Mites, 566, 585
Mitochondria, 92, 110, 116, **119**, 120, 426, 476
  active transport at, 136
  in animal cells, **112**
  binary fission in, 453
  and cell division, 146
  and cellular respiration, 167, 191-193, 204-205
  and cell variety, 125
  and chemiosmotic coupling, 197-199
  in *Chlamydomonas*, 108
  in corn, 93
  DNA of, 316, **362**, 396, 452-453, 1048-1050
  electron "shuttle" in, 200-201
  electron transport at, 191, 196-199, 200-201, 204-205
  enzyme synthesis in, 453
  and ethanol, 202
  in *Euglena*, 458
  evolution of, 146, 362, 452-453
  F_1 and F_O in, 198-199
  in glycolysis, 201
  inner compartment of, 191
  inner membranes of, 220
  intermembrane space, 197-198
  in intestine, 721
  isolation of, 192
  in kidney cells, 774
  as level of organization, 36
  matrix, 214, 220
  membrane of
    and electron transport, 191, 197-201, 204
    and photosynthesis, 220
  in motor neurons, 877
  and photosynthesis, 206, 214, 218-219, 229
  in plant cells, **113**
  replication of, 453
  ribosomes of, 452
  selective permeability of, 191, 201
  and sodium-potassium pump, 136
  and sperm motility, 902
  and temperature regulation, 786
  transport processes, 204-205
Mitochondrial clock, 1048-1050
Mitochondrial DNA, 1048-1050
"Mitochondrial Eve," 1050
Mitochondrial sheath, of sperm cell, 902-903
Mitosis, **145**, 146-147, **148-153**, 154-155, 249, 251-253, 256, 259- 261, 286-297, 302, 306, 315
  in algae, 462, 465
  anaphase, 150, **151**, 155
  in animal cells, **150-152**
  in ascomycetes, 486
  and condensation of chromosome, 355, 358
  in *Drosophila*, 924
  evolution of, 472
  and heredity, 247
  interphase, **146**, 149, 155
  vs. meiosis, **252-253**, 257, 984
  membranous fragments, of nuclear envelope, 151
  metaphase, 150, **151**, 152, 155, 254, 260, 384
  in mosses, 500
  nondisjunction in, 385
  and nucleolus, 366
  nucleus in, 150, 152, 153
  in ovum, at fertilization, 921
  phases, **150-152**
  and polyploidy, 1012
  in prokaryotes, 472
  prophase, **150-151**, 152, 155, 384
  in protists, 472
  in protozoa, 468-470
  in spermatogonia, 902
  spindle fibers in, 472, 921
  telophase, 150, **152**, 153, 155
  zygotic, 359
Mixed west-coast forests, 1166-1167, 1177
M line, 874
Mobility, cellular
  and actin filaments, 121
Mockingbird, 423
Models
  of atom, **28-29**
  ecological, **1090-91**, 1106-1107, 1112-1113, 1119, 1123
  for fixed action patterns, 1054-1055
  vs. mimics, 1004-1005
  of molecular structure, 41, 56, **62-63**
Moehlman, Patricia, 1069
Mojave desert, 1172
Molars, 601, 602, 716-717
Molds, 13
Mole (animal), 27
Mole (measurement), **48-49**
Molecular biology, 80, 533
  of *E. coli*, 319-320, 337
  and evolutionary theory, **969**, 972-973
  of human evolution, 1048-1050
  of prokaryotes, **319-320**, 337, 357-358
  of viruses, **319-338**
Molecular clock, 419-420, 423
  vs. mitochondrial clock, 1048-1050
Molecular genetics, 281, 299
  and genotype, 342
  of prokaryotes, **319-338**
  of viruses, **319-338**
Molecular taxonomy, 430
Molecular weight, 48
Molecule(s), 30, 38, **55-82**
  atomic structure of, **23-38**
  ball-and-stick model of, 41, 56, 62
  bioactive, 35
  charge of, 62
  conformation of, in energy reactions, 181
  double-stranded, 420, 421
  hybrid, 421
  as level of organization, 36
  models for, 41, 56, **62-63**
  motion of, and temperature, 129
  mutual repulsion of, and cellular energetics, 180-181
  organic molecules, 87, 101, 142, 168, 176, 203
  polarity of, 33, 44
  shape and structure, **55-82**
  single-stranded, 420
  structural formulae for, 62-63
  weight of, 48
Mollusks, 122, 520-521, 532, 544, **545-554**, 563, 564, 567, 568
  behavior of, 545, 564
  circulatory system of, 752-753
  classifications of, 545, 548
  evolution of, 553
  hemocyanin in, 742
  nudibranch, 526
  supply systems in, **547**
Molting, 66
  of arthropods, **567**
  hormonal regulation of, 578
  of insects, 576, 578
  of lobsters, 572
Molting hormone (*see* Ecdysone)
Molybdenum, 657
Monarch butterfly, coevolution in, **1003-1005**, 1009
Monarch caterpillars, and milkweed, 1003-1004
Monera, 13, 94, 125, 423, 425-427, **428-451** (*see also* Prokaryotes)
Monkeys, **1034-1035**
  social behavior in, **1036-1038**
Monoclonal antibodies, 800, 802-803, 941
Monocots, 498
  vs. dicots, 511, 628, 631, 633, 635
  floral parts of, 613-615, 619
Monocytes, **792**, 793, 830, 831
Monod, Jacques, 322-333
Monoecious flowers, 614, 616
Monogamy, 999-1000, 1069
  in hominoids, 1036
  and sexual selection, 999-1000
  in silver-backed jackals, 1069
  in turtle doves, 1074
Monomers, 59
Mononucleate cells, 708
Monophyletic ideal, **412-413**
Monophyletic taxon, 412
Monoplacophorans, 546, 548
Monosaccharides, **59-61**, 64, 203
Monosynaptic reflex arc, 844
Monotremes, 600-601
  amniote egg in, 900-901, 944
  blastodisc in, 931
  evolution of, 600, 1031
Monsoon forest, 1174-1177
Montagu, Mary Wortley, 796-797
Morality, and science, 18
Morchella esculenta, 485
Morels, 481, 485
Morgan, Thomas Hunt, 263-266, 268, 274-275, 278, 302, 375, 382, 974
"Morning-after pill," 916
Morphine, 854, 855
Morphogenesis
  in angiosperms, 619
  and cell growth, 938, 951
  in chick embryo, **938-940**
  of primary tissues in embryo, 919, 935
Mortality, infant, 949
Mortality patterns, 1087, **1092, 1104**, 1134 (*see also* Death rate)
  and depletion of ozone layer, 1134
  and human population explosion, 1097
  and limitations on population growth, 1098
Mortality rate, from heart disease, 761
Mosaic flies, 834
Mosquitoes, 409, 573, 584
"Moss animals" (*see* Bryozoans)
Mosses, 115, 496, 498-500, 1164
  life cycle, 499-500
Mother's milk, 948
Moths, 573, 584
  and predation by bats, 1115-1116
Motility, 426
  of animals, 518
  of bryozoan larvae, 563
  of echinoderms, 587
  vs. growth, 630
  of *Hydra*, 527
  of medusa, 525
  of mollusks, 545
  in plants, 671
  in prokaryotes, 434, 436, 450
  of sperm cells, 902-903, 908
Motor cortex, **885-888**, 898
  and intrinsic processing centers, 892
Motion
  as energy, 161
  and living systems, 36
Motor cells, 692
Motor fibers, 844
Motor neurons, 708, **709**, 712, 843-846, 856, 876-878, 882
  development of, in embryo, 935
  and habituation, 896
  and learning, 896
  in respiration, 745, 748
  of somatic nervous system, 845
Motor organs, 546
Motor response
  human, **858-880**
  integration of, with sensory perception, 879
  and sensory response, 131
Motor unit, 877, **878-879**
Mountain belts, 1012
Mountain lilac, 1014
Mountain ranges, and rainfall, 1160
Mount St. Helen's, 1025, 1098, 1125
Mouth, 538, 544, 545 (*see also* Lips; Teeth; Tongue)
  development of, 924
  in vertebrate digestion, 715-717, 730
Mouthparts, 571, 574, 578, 1060
Movable genetic elements, 338, **371-375**
  and genetic recombination (*see* Genetic recombination)
  lambda phage, **334-335**, 338

plasmids, 326–329, **331**, 337–338
and reverse transcriptase, 373–375
transposons, **335–337**
viruses as, 374, 381
Movement
of cell (*see* Cell mobility)
of molecules, 129
of solutes, 128
and uncertainty principle, 129
of water, 128, 132, 142
MRI, 884
M ring, 434
mRNA (*see* Messenger RNA)
Mucilage ducts, 640
Mucosa, **715–716**, 720, 722
Mucosal immunity, 799
Mucous membranes, in immune response, 702, 791
Mucous sheath, 557
Mucus, 115, 533, 702–703, 713
in allergies, 813
and cilia, 94
and contraception techniques, 916
and defense against infection, 791
in feces, 725
and gliding bacteria, 436, 438, 450
in intestine, 722
in oviduct, 912
in respiratory system, 738–740
in saliva, 717
in stomach, 720
vesicles, 118
Muddy seashore, 1157, 1177
Mud flats, 1159–1160
Mud puppy, 595
Mules, 1012, 1015–1016
Müller, F., 1004
Muller, H. J., 274
Müllerian mimicry, **1004–1005**
vs. fruiting bodies, 438
and plant evolution, 493
problems of, 710–711
Multicellular organisms, **92–93**, 101, 455, 476, 518, 523, 701, 714
algae, 461–462, 465
countercurrent exchange in, 124, 130
earliest, 497
evolution of, 454
green algae, 462–463
vs. multinucleate organisms, 479
vs. unicellular organisms, 821
Multicellularity, **36**, 424, 426
Multimeric proteins, 76
Multinucleate cells, 483
Multinucleate organisms, 462, 463, 479
Multiple alleles, 980
Multiple epidermis, 628
Multiple fruits, 620
Multiple myeloma, 798, 800, 802–803
Multiple sclerosis, 812
Mumps, 796
Muscle(s), 518, 532, 539, 555, 572, 591, 598, 702, **706–708**
adductor, 549
arteriolar, 759
cardiac (*see* Cardiac muscle)
cells, 122, 190–191, 518, 531, 723
contraction (*see* Muscle contraction)
development of in embryo, 936, 946, 952
erector, 710
fatigue, 190–191
fibers, 190–191
glycogen storage in, 64
involuntary, 706
longitudinal, 539, 542, 555
and male hormones, 906
in male reproductive system, 905
movement of, and brain, 883, 886
and movement of lymph, 763
and oxygen debt, 190–191
segmented, 593
skeletal, 707–708, 858, **873–876**, 880
smooth, **706–708**, 713, 715
striated, **706–708**, 713
and temperature regulation, 786, 790
voluntary, 706
Muscle contraction, **873–879**, 880
action potential in, 873, 877
Ca^{2+} in, 873, 876–877, 880

mechanics of, **874–876**
and neuromuscular junction, **876–878**
regulation of, 876
speed of, 878
timing of, 879
Muscle fibers, 878
Muscle tissue, 140, 702, **706–708**, 713
cardiac, 706
erector, 710, 723
involuntary, 706
and male hormones, 906
skeletal, **707–708**
smooth, **706–708**, 715
striated, **706–708**, 713
in vertebrates, 713
Muscular dystrophy, 393–394, 397
Muscularis externa, **715–716**
Mushrooms, 13, 481, **486–487**
Amanita bisporigera, 487
Agaricus campestris, 487
bioluminescent, 182
cap, 487
common field mushroom, 487
gill fungi, 487
gills in, 487
hallucinogenic, 487
poisonous, 487
skirt, 487
Musical ability, and brain, 898
Mussels, 122, 460, 549
Mustard family, 515
Mutagens, 274
and carcinogens, 376
Mutants, **244**
of *Drosophila*, 265, 274
of *E. coli*, 322
in genetic research, 328
of *Neurospora*, 303, 316
Mutation(s), **244–246**, 247, 273, 279, 303–304, 315–317, 338, 395, 419
AIDS virus and, 815, 817, 818
and antibodies, 801, 819
of auxotrophs, 173
and cancer, 376, 381
in color blindness, 400
in *Drosophila*, 265, 274
duplication of DNA in, 988–989
and enzyme deficiencies, 400
and evolutionary change, 246, 981–982, 990
and evolutionary theory, 245, **962**
of gene, 268
and gene families, 367, 368, 380
and genetic disease, 391, 400
and genetic variation, 247, 978, **981–982**
and Hardy-Weinberg equilibrium, 979, 989
in hemophilia, 400
in homeotic genes, 942–943
and human evolution, 1050
in *lac* operon, 323
in mitochondrial DNA, 1050
in muscular dystrophy, 400
in peppered moth, 962–963
of pneumococcus, 282
point mutation, 316–317
in prokaryotes, 437, 450
rate of, 420, 982
redefinition of, **314–316**, 317
and RNA viruses, 373
in rolling-circle replication, 328
and sickle cell anemia, 395
as a source of genetic variation, 247, 978, **981–982**
and transposon, 335–336, 338, 374
in viruses, 445
"Mutation theory" of cancer
Mute swans, **1066–1067**
Mutual dependence, and social behavior, 1058
Mutualism, 447, 451, 1120–1122, 1130
Mutual repulsion, 168–169
and cellular energetics, 180–181
Muzzle-nuzzle gesture, 1063
Myasthenia gravis, 812
Mycelium, 480, 481, 483, 485, 486–487
Mycobacterium tuberculosis, 429, 448
Mycoplasmas, 430–432
Mycorrhizae, **490–492**
ecological importance of, 490–491
and fungi, 488
hyphal sheath of, 491

importance of in evolution, 491–492
mutualism, with roots, 1121
and plant nutrition, 492, 670
and plant survival, 490–491
and root hairs, 632
symbiosis with, 490–491, **664–665**
and water uptake, 491
Myelin, 70
Myelin sheath, 842–843, 851, 857, 877, 881, 884
Myofibrils, 873–875
Myoglobin, 367, **745**, 780, 878
in diving mammals, 744
function of, in respiration, **745**
structure of, 745
Myonemes, 452, 471, 527
Myosin, 713, 874–876, 878
in cardiac muscle, 708
in cell motility, 121–122
filaments, 125
in muscle cells, 518
and muscle contraction, 706
in vertebrate muscle, 713
Myotome cells, 936
Myotomes, 592
Myriapods, **572–573**
Myxobacteria, 438
Myxoma virus, 1120
Myxomycota, 455, 466–467, 469, 47

N-acetyl-hexosaminidase, 389
NAD^+, 171–172, 183
and niacin, 729
and ribose, 172–173
structure and function, 171–172
NAD^+, 172–173, 204
in aerobic respiration, 193–194
in anaerobic respiration, 190–191
in ethanol metabolism, 202
in glycolysis, 187–190, 200–201
in Krebs cycle, 194–195
oxidation, 203
in photosynthesis, 216
NADH, 172–173, 216
in aerobic respiration, 193–194
in anaerobic respiration, 190–191
in electron transport, 196–198
in energy metabolism, 203
in ethanol catabolism, 202
in glycolysis, 188–190, 200–201
in Krebs cycle, 194–195
$NADP^+$, in photosynthesis, 216, 218, 219, 229
NADPH, in photosynthesis, 216, 218–219, 222–223, 229
Nails, 1033, 1051
Naked mole rat, 1062, 1069, 1078
Naming, binomial system for, 35, 408, **410**, 426
Nanometer, 57, 95
Nanoplankton, 459
Naphthalenacetic acid, 673–674 (*see also* Auxins)
Naphthoquinone, 729
Nasal cavities, 738
Natural selection, 7, 238, 287, 973 (*see also* Evolution)
adaptation in, 7, **970–972, 1001–1005** (*see also* Adaptation)
and allele distribution, 981
vs. artificial, 976–977, 991
and balanced polymorphism, **992–993**
and behavior, 976, 1052
in blood groups, **997**
and character displacement, 1110–1111, 1129
and chemical evolution, 87
Darwinian theories of **1–10**
in Darwin's finches, 1020–1021
and diploidy, 986–987
directional, **997**
disruptive selection, **996–997**
and drug resistance in bacteria, **964–965**
of early cells, 90
and ecotypes, 1002–1003
and energy conservation, 59
and environment, 7
in evolution of the horse, 1025–1027
for families, 1065–1067
frequency-dependent, **997–998**
and gene flow, 982
in heterosis, 998, 990
heterozygote superiority, 987–988, 990

I-37

Natural selection (Continued)
  history of, **1-10**
  in hominids, 1043
  and inherited variations, 306
  insecticide resistance in, **964**
  for intelligence in human beings, 1045
  and kin selection, 1065-1074
  of life-history patterns, 1099-1100
  limitations on, 1006
  maintenance of variability in, **991-993**
  and modes of reproduction, 1105
  against olfaction in primates, 1033, 1051
  in peppered moth, **962-963**
  and phenotype, 977, **993-994**
  in population genetics, 991
  and population growth, 1105
  predation and, 1114
  premating isolating mechanisms, 1016
  in protists, 476
  sexual selection, **998-1001**
  and sickle cell anemia, 390
  and social behavior, 1059
  and species, **7, 9, 12** (see also Speciation)
  stabilizing selection, **995**, 1008
  types of, **994-1001**
Nautilus, chambered, 551
Neanderthal man (see Homo neanderthalensis)
Nearsightedness, 867
Nectar, 513, 514, 689, 1061
Nectaries, 513, 1122
Negative feedback, 712, 917
  in blood pressure regulation, 762-764
  and menstrual cycle, 913
  and regulation of male hormones, 906-907
Negative regulation, of genetic transcription, 322, 325, 337
Negative tropisms, 671
Neisseria gonorrhoeae, 326
Neisseria meningitides, 433
Nematocyst, 525, 526, 528-529
Nematodes, 122, 489, 520, **532-540**, 542
  and disease, 543
  ecological importance of, 543
  lack of centriole, 146, 150
  reproduction in, 539
Nemiroff, Martin, 744
Neo-Darwinian synthesis (see Evolution, theory of)
Neofelis, 412 (see also Leopard, clouded)
Neomycin, 329
Neon, 30, 733
Nephridia, 547-548, 550, 554-556, 561, 563-564
Nephrons, **770-771**, 828, 832
  and cardiac peptide, 775-776
Nerve cells, 136-137, 140, 147, 178-180, 542, 713 (see also Neurons)
  energetics of, 178-180
  and sodium-potassium pump, 136
  replication of, 147
Nerve cords, 533, 535, 542, 549, 551, 555-557, 585, 588
  dorsal, 591, 592, 604
  hollow, 591, 592, 604
  ventral, 564
Nerve endings, 835, 859-860 (see also Neuromuscular junction; Synapses)
Nerve fibers, 708 (see Axons)
  of brainstem, 844
  myelinated, 851, 857
  of spinal cord, 843, 843
Nerve gases, 178
Nerve impulse, **708-709**, 821, **847-851**, 852, 856, **858-873**
  action potential, **848-850**, 851, 859
  conduction of, 857
  vs. electric current, 848
  vs. electrical signal in plants, 692-693
  propagation of, 851, 856
  transmission of, 822
Nerve net, 527 (see also Nervous system)
Nerves, 518, 557, 709, 842, 844, 846 (see also Nervous system; Neurons)
Nerve tissues, 140, 532, 705, **708-709**
Nerve tracts, 949
Nervous system, 397, 425-426, **841-857**, 898 (see also Brain; Central Nervous System; Neurons; Sensory Receptors)
  and adrenal medulla, 832
  and AIDS, 819

of annelids, 564
of arthropods, 568, 579, 585, 842
autonomic, 842, 845, 873 (see also Autonomic Nervous System)
cellular communication in, 820
central, 842, **843-844** (see also Central Nervous System)
in cnidarians, 525, 542
crossing over in, 882
development of, in embryo, 952
and digestion, 730
of earthworm, 842, **556-557**
and endocrine system, **711-712**, 821, 822, 847 (see also Endocrine System; Hormones)
evolution of, 535, **841-842**
feedback control in, 712-713, 714
of gastropods, 551
of hemichordates, 591, 604
and hormone secretion, 824
of *Hydra*, 527, 841-842
integration and control in, **711-712**, 856-857
integration of, at hypothalamus, 893
integration of information in, 855-856
of mollusks, 549, 564
of octopuses, 552, 564
organization of, **842-847**, 856
parasympathetic (see Parasympathetic nervous system)
peripheral, **844-847**, 856 (see also Peripheral Nervous System)
of planarians, 533, **535**, 842
and regulation of heartbeat, **756**, 764
somatic, 842, 845, 873 (see also Somatic nervous system)
specialization in, 841
and stomach, 720
sympathetic (see Sympathetic nervous system)
temperature regulation in, **784-786**, 790
in vertebrates, **842-857**
Nested sets, and cladistics, 416-417
Nesting, 1063, 1072-1073, 1074
Nestlings, 1052
Net flow, of fluids, 133
Net movement, 129
  of water molecules, 142
Net primary production, 1138, 1157
Net productivity, 1138-1139, 1143, 1148
Neural crest cells, 937
Neural ectoderm, 928-929, 932-**933**, 935, 937, 952
Neural folds, 928
Neural groove, 919, 929, 945-946, 953
Neural plate, **928-929**, 930-931, 932, 938, 952-953
Neural tube, 927-929, 932-**933**, 935, 937, 938, 952
Neuroendocrine system, **821-822**
Neuroglia, 708, 842, 843, 881, 884
Neuromodulators, 854-857
Neuromotor mechanisms, 518
Neuromuscular junction, **876-878**
Neurons, 140, 535, 702, 713, 821-822, 841-842, 855-857, 881 (see also Axons; Dendrites; Nerve cells; Nervous system)
  in brain circuits, 884
  in cerebral cortex, 889
  and ejaculation, 905
  in grey matter, 884
  interneurons, 708-709
  motor (see Motor neurons)
  nuclei, in brainstem, 882
  preganglionic vs. postganglionic, 845
  relay, 708-709, 843
  respiratory, 748
  sensory (see Sensory neurons)
  in sexual behavior, 905
  subcortical, 885
  in synapse, 708-709
  types of, **708-709**
  vertebrate, **708**
  in visual cortex, 807-808, 892
Neurosecretory cells, 578, 588, 822, 823, 826, 827, 839
Neurospora, 481, 484, 486
Neurospora crassa, 173, 301-303, 482
  life cycle of, 250
Neurotransmitters, 822, 832, 835, 836, 841, 846-847, 850, 852, **853-856**, 857
  in brain, 899
  and drugs, 853
  vs. hormones, 822

learning and, 899
memory and, 896
mode of action, **853-855**
at neuromuscular junction, 877
photoreceptor cells, 868
Neutral alleles, 978
Neutralists, 978
Neutron(s), **23-24**, 38, 163
Neutrophils, **792-793**
Newton, Isaac, 207
New World monkeys, 1034-1035, 1051
N-formylmethionine, 312-313
Niacin, 172, 729
Niche, ecological (see Ecological niches)
Niche overlap, 1111, 1129
Nicotiana tabacum, 353
Nicotinamide, 172
Nicotinamide adenine dinucleotide (see NAD)
Nicotine, 172, 515
Nicotinic acid, 172 (see also Niacin)
Night blindness, 729
Nirenberg, Marshall, 307
Nitella, 127
Nitrate (NO_3^-) ions, 657, 664, 1148, 1150, 1153
Nitrates, 76
Nitrification, **1146-1147**, 1153
Nitrifying bacteria, **1148-1150**
Nitrite (NO_2^-) ions, 439, 1147
Nitrites, 76
Nitrogen, 76, 87, 516 (see also Nitrification; Nitrogen cycle; Nitrogen fixation)
  and amino acids, 76
  amount of in air, 733
  fixation, 352, 429, 430, 441 (see also Nitrogen fixation)
  in forest ecosystems, 1145, 1146-1150
  heavy, 293
  loss of, in plants, 664
  in metabolic waste, 766
  plant requirements for, 664
  in respiratory gases, 743
Nitrogenase, **667**, 669
Nitrogen cycle, 1145, **1146-1149** (see also Biogeochemical cycles)
  nitrogen fixation in, **1147**
  prokaryotes in, 439
  recycling, in forest ecosystems, **1149-1150**
Nitrogen fixation, 678, 1148-1150, 1153
  in bacteria (see Nitrogen-fixing bacteria)
  and nitrogen cycle, **1147**
  and photosynthesis, 429
  and plant nutrition, 670
  in plant symbioses, 664, **665**, 666 (see also Mycorrhizae)
  in prokaryotes, 429
Nitrogen-fixing bacteria, 429-430, 1134
  Cyanobacteria, 441
  and depletion of ozone layer, 1148-1149
  Heliobacterium chlorum, 442
  and nitrogen cycle, 1148-1149
  Rhizobia, 430
  roots, 1121
Nitrogen-fixing nodules, in roots, 665, 668
Nitrogenous bases, 80, 82, 118
  of DNA, 283-284, 287-288, 299
  of NAD, 172
Noah's ark model of human evolution, 1047-1048
"Noble" gases, 30
Nocturnalism, 600, 768, 789, 1031, 1034, 1173
Nodes
  of myelin sheath, 851
  in plants, 626, 642
  in shoots, 641, 648, **649**
Nomad societies, 1096
Nomenclature, 410
  binomial system of, 408, 410, 426
Noncompetitive inhibition, **177**, 178, 184
Nondisjunction, 385, 399
  and Down's syndrome, 385
  and polyploidy, 1012-1013
Non-histone proteins, 431
Nonliving systems, vs. living, 127
Non-Mendelian inheritance, 362
Nonmotility, 426
Nonrandom mating, 984, 990
Nonspecific immune response, **791-793** (see Inflammatory response; Interferons)
  anatomic barriers, 791

inflammatory response, 791–793
interferons, 793
Nonspecific mineral requirements, 656
Nontemperate bacteriophages, 333–334
Noradrenaline, 746, 759, 824, 832–833, 839, 853, 857, 951
  in infant alertness, 951
  as neurotransmitter, 857
  in sympathetic nervous system, 846–847
Northern coniferous forests (see Taiga)
Northern elephant seal, 983
Northern hemisphere, 1136, 1156–1157
North Pole, 1136
Nose, 738
Nostoc, 489
Nostrils, 736, 935
Notochord, 591–593, 604, 928–930, 932–936, 946, 952
Nottebohm, Fernando, 889
"Nuclear clock," 1048–1049
Nuclear envelope, 91, 108, 112, 116, 124, 426
  in animal cell, 112
  in cell division, 150–152
  in cell variety, 125
  in ciliates, 472
  double membrane system, 127
  in eukaryotes, 452–453
  and exchange of substances, 127
  and intermediate filaments, 111
  and meiosis, 254–255, 258
  in mitosis, 150–152
  and nucleus, 366
  in protozoa, 470
Nuclear family
  conflicts of interest in, 1070
  evolution of, 1044
Nuclear membrane, 108
Nuclear pores
  and gene transfer, 380
  and ribosome assembly, 366
Nuclear testing, **1150–1151**
Nuclear winter, 1025
Nuclei (in nervous system), 842 (see also Nervous system; Neurons)
Nucleic acid base-pairing
  and recombinant DNA, 245
Nucleic acids, 55, 80, 420, 427
Nucleoid, 90, 102, 337
  and infection cycle, 336
  in prokaryotes, 431
  and prokaryotic replication, 319–320
Nucleolus, 155, 472
  in animal cell, 106, **112**, **124**
  in cell division, 150, 152
  function, 366
  during meiosis, 254, 368
  during mitosis, 366
  and nucleus, 366
  structure, 366
Nucleosomes, **357**, 379, 444
Nucleotide(s), 55, 80, 82, 87, 305, 307, 309–312, 314, 316–317, 420
  addition of, 315, 317
  and behavior, 1052–1053
  biosynthesis of, for DNA, 298
  deletion of, 315, 317
  disruption in sequence of, 315, 317
  DNA and RNA, 308, 317
  in DNA replication, 281, 283–284, 289–290, 292–299
  and gene regulation, 380
  sequence of, 307, 316
  sequencing of, **340–341**, **420**, 427, 429, 817, 942, 978
  substitution of, 316
  "synonym" sequences, 309
Nucleus (atomic), **23–24**, 38, 163
  in atomic structure, 28
  and oxidation-reduction reactions, 166
  polar-covalent bonding and, 32
Nucleus (cellular), **90–92**, 93, 101, 102, **108–110**, 124, 127, 146, 155, 267, 383
  in animal cell, **112**
  in cell division, 155
  chromosomes in, 356
  in coenocytic organisms, 462
    daughter, 145, 153–155
    in meiosis, 252–253

diploid, 249, 253, 483
division of, 253, 462 (see also Meiosis; Mitosis)
envelope (see Nuclear envelope)
of eukaryotes, 452
and exchange of materials, 127
function, 109
of fungi, 485, 487
haploid, 253, 255–256, 258, 260
and heredity, 109–110
and introns, 380
membrane (see Nuclear membrane)
and messenger RNA transcription, 369
in mitosis, 150, 152–153
of muscle cells, 706
of neutrophils, 792
vs. nucleolus, 366
of oocyte, 910
of ovum, 922, 926
in plant cell, **113**
pores of, 366, 380
post-meiotic, 258
and rRNA genes, 365-**366**
of sperm cells, 902–903
Numerical phenetics, 415, **416**, 427
"Nurse" cells, 138
Nursing, 702
  evolution of in primates, 1033
  genetic conflicts of interest in, 1071
  and monogamy, 1072
Nutrient canals, 705
Nutrient cycles, 1176
Nutrients, in forest ecosystems, 1147–1150
Nutrition
  in arachnids, 570
  in chemosynthetic autotrophs, **439** (see also Autotrophs)
  in heterotrophs, 438–439 (see also Heterotrophs)
  human, 70–71, 76, 730
  of human embryos, 932, 937, 941, 944–945
  in mammals, 601
  modes of, 424, 426, 518, 555
  of mothers, and birth weight, 949
  in photosynthetic autotrophs, 440–443 (see also Autotrophs)
  in plants, **650–670**
  and population explosion, 1097
  in prokaryotes, **438–442**
  in reptiles, 597
Nutritional requirements, 726–729
Nutritive cell, 527–528
Nuts, 511, 621, 637
Nymphs, 576, 586

Oats, 226
Obelia, 527
Obesity, 728–729
Objectivity, in science, 17
Obligate anaerobes, 429, 439, 442
Obligate parasitism, 330
Observation
  vs. experimentation in ecology, 1087, 1106–1107, 1112
  in science, 14–17
Occipital lobe, 885
Ocean, 87
Ocean currents, 1156–1157, **1158**, 1160
Oceans, **1155–1157**, 1177 (see also Marine environments; Marine organisms)
  currents of, 1156–1157, **1158**, 1160
  and evolution, 1012
  net primary productivity of, 1139
  and sunlight, 1133
Ocelli, 528, 533, 535, 542, 582 (see also Eyes, simple)
  evolution of, 529
Ochoa, Severo, 307
Octopuses, 545, 551, **552**, 553, 563–564
Odors, discrimination of, 861 (see also Smell)
Offspring
  conflicts of interest with, 1071
  and "fitness," 974–975
  and Hardy-Weinberg equation, 989
  live vs. eggs, 702
  numbers of, in primates, 1033
  size and number, and population growth, 1099, 1105
  small vs. large, **1102**
  variability in, 986
Offspring, care of, 561, 575, 589, 900, 917, 999, 1052

alternative patterns of, **1099–1104**
  in birds, 598
  in bivalves, 550
  in California condor, 1104
  among cephalopods, 553
  evolution of in primates, 1033, 1034
  in insects, 576
  live vs. eggs, 702
  in mammals, 600, 702
  in sponges, 524
O'Hara, Richard, 1076
Oils, **67–70**
  as algal food reserves, 457, 465
  atomic bonding in, 34
Okazaki, Reiiji, **296–297**
Okazaki fragments, **296–297**, 299
  and eukaryotic replication, 358
  in prokaryotic replication, 320
Olduvai Gorge, 1040, 1042
Old World monkeys, 1034–1035, 1051
Oleic acid, 67–68
Olfaction (see also Odors; Smell)
  development of, in embryo, 936
  selection against, in primates, 1033, 1052
Olfactory bulb, 883
Olfactory cells, 862
Olfactory epithelium, 861, 862
Oligocene epoch, 496, 1026, 1034
Oligochaetes, 554, 555, 564 (see also Earthworms)
Oligonucleotides, synthetic, 343, 346, 353
Oligosaccharides, 673, **682**, 683, 688, 694–695
Ommatidia, 579, 580
  response time of, 580
Omnivores, 550, 601, 1107, 1166
Onchocerca, 540
"Oncogene hypothesis" of cancer, 376
Oncogenes, 376–377, 381
One-celled organisms (see also Prokaryotes, Protists, Single-celled organisms)
  algae, 460, 461
  cell cycle in, 147
  cell division in, 144
  euglenoids, 458
  in evolution, 461
  green algae, 461
  heterotrophs, **468–476**
  protozoa, **468–476**
  yeasts as, 486
"One-gene–one-enzyme" hypothesis, **301–304**, 316
One-way digestive tract, 538, 715
Onion root, 146, 150, 151
Onychophorans, 520, 560, 561, 564
Oocysts, 473
Oocytes, 294, **907**, 908, **909**, 910–914, 940
  and contraceptive techniques, 916
  in estrus, 915
  and hormones, 913–914
  human, **907**, 908, **909**, 910–914
  pathway of, **912–913**
  primary, 258, 917
  production of, **910–911**, 924
  and prostaglandins, 835
  secondary, 258, 917
Oogamy, 463, 467, 469, 477, 494, 516
Oogenesis, **910–911**, 924
Oogonia, 258, 925, 946
Oomycetes, 455, 467–469, 477
  vs. fungi, 482
  and potato blight, 468, 678, 682, 1149
Opalina, 473
Opalinids, 455, 468, 469, 473, 477
Oparin, A. I., 87
Open circulatory system, vs. closed, 752–753
Operant conditioning, **1055–1056**, 1077
Operator gene, **323**, 324–325, 337
Operculum, 551, 569
Operons, **322–326**, 337, 368, 380
  inducible, 323–325, 337
  lac, 323, 325, 326, 337
  operator gene, **323**, 324–325
  prokaryotic, **322–326**
  repressible, 323–325, 337
  and RNA polymerase, 323–325
  trp operon, 323, 325, 337
Opiates, **854–855**
  addiction to, 854–855
Opisthonecta, 144
Opines, 352

Opium, 515, 854
Opportunistic life-history patterns, 1099, 1105
Opportunistic species, 1089, 1103, 1105, 1149
Opposing thumb, 1032
Opossum, 117, 600
Opsin, 871, 872
Optic chiasm, 888
Optic cup, 936
Optic lobes, 552, 883
Optic nerves, 868, 869, 883, 935–936
Optic stalk, 935–936
Optic vesicles, 935, 936
Optimum tolerance range, 1095
Oral cavity, 715, **716–718**
Oral groove
  of *Didinium*, 139
  of *Paramecium*, 471, 475
Orangutans, 1032, 1035–1036, 1044
Orbital, atomic, **29**
Orbital model
  of atomic structure, **28–29**
  of molecular structure, 62–63
Orchids, 490, 511, 515, 678, 971
Order (biological level of classification), 410, 427
Order, and science, 14
Ordovician period, 497, 523, 567
Organelles, 91–92, 101–102, 104, 106, 110, 114–120, 124
  in animal cells, **112**
  and biochemical pathways, 192–193
  carrier-assisted transport in, 135
  in cell division, 144–146, 155
  DNA and, 362
  energy flow in, 168
  enzymic pathways in, 173
  in eukaryotes, 452, 453
  membranes of, 127, 135
  vs. organs, 140
  and osmosis, 132–133
  in plant cells, **113**
  in prokaryotes, 431, 450
  of water molds, 467
Organic molecules, 11, **55–82**, 87–89, 101, 1131, 1153
  and anabolism, 203
  and autotrophs, 203
  biogeochemical cycles of, **1139–1141**
  carbon in, **55–82**
  combination in, 58
  covalent bonds in, 57–58
  cycles of, in ecosystems, **1139–1141**
  decomposition of, 176
  and energy transformation, 168
  evolution of, 142
  and heterotrophs, 203
  and photosynthesis, 228, 1137
  production of, in biosphere, 1138–1140
  and prokaryote nutrition, 438
  recycling of, in ecosystems, **1146–1149**
  translocation of, in plants, **657–662**, 670
Organisms, 407
  atomic structure and, 35
  classification of, **407–427**
  coenocytic, 462, 463
  development of embryo of, **919–953**
  diversity, and evolution, 961
  energy and metabolism in, 710–711
  functions of, **710–714**, 715
  homeostasis in, 711 (*see also* Homeostasis)
  integration and control in, 711–715
  organization of, **36–37**
  populations of, **1087–1105**
  relationships of, 423
  respiration in, **732–748**
  single-celled, 407
  varieties of, 426
Organization
  biological levels of, 35–38
  of cells, **102–106**, 424
  coenocytic, 462–463
  levels of, in vertebrates, **709–710**
  of living systems, 26, 38, 84, 127
  of proteins, **74–80**
Organizer, **930–931**, 952
Organ of Corti, 865, 866, 879
Organogenesis, 946
  in chick embryo, **935–937**
  differentiation of tissues in, 935–937

in human embryo, **945–948**, 952, 953
  sex determination in, 946
Organ systems, 544 (*see also specific system*)
  development of (*see* Organogenesis)
  digestive (*see* Digestive system)
  in vertebrates, 713
Organ tissue, 140
Orgasm, 847
  in female, **912–913**, 917
  in male, **904–905**, 917
Orientation, and fixed action patterns, 1054
Origin of replication, **294**, 295–296, 299, 320
Origin of Species, 7–8, 15, 85, 282, 961–962, 967, 969, 972, 976, 1028
Origins of life, 497
  and energy, 87, 100
  and volcanic eruptions, 85, 87
  and water, 87–89
Ornamentation of males, **998, 1000–1001, 1073**
Osculum, 522, 524
Osmium tetroxide, 98
Osmoregulation, **656–657**, 658–659
Osmosis, 124, 128, **132–134**, 142
  and cell lysis, 432
  central pore and, 133
  energy and, 133
  in gills, 767
  and gravity, 133
  in kidney, 134, 135
  and organelles, 132–133
  osmotic potential, 132, **133–134**
  osmotic pressure, 133
  in plants, **669–670**
  and pressure-flow hypothesis, **662–663**
  and roots, 633
  and stomatal movement, 653
  and touch response in plants, 692
  in tubules, 132
  and urine formation, 772–773
  and water movement in plants, 650, 669
  and water potential, 132–133
Osmotic potential
  of blood, 750, 762, 764
  on interstitial fluid, 750
  in ovum at fertilization, 921
  in plants, 656, 670
  of plasma, 750
  and water exchange, 775
Osmotic receptors, 774
Ossicles, 590, 705
Osteichthyans (*see* Fishes, bony)
Osteoblasts, 704, **830–831**
Osteocytes, 704
Osteoporosis, 830–831
Ostrich, 31, 967
Ostrum, John, 599
Outbreeding, 1074
  and genetic variability, 984–985, 990
  in gram-negative prokaryotes, 432
  and self-sterility alleles, 990
  waiting behavior and, 1074
Outer membrane, of nuclear envelope, 112
Outgrowths, 734
Outpocketings, of gut, 545, 563
Output, of minerals, 1150
Ovalbumin, 361–362
Oval window membrane, 864, 865
Ovarian follicles, 237, 911, 912, 913–915, 917
Ovary(ies), 138, 238, 258, 598, 824, **907**, 909–912, 917
  of angiosperms, 614, 616, 620–622, 624
  of chicken, 931
  development of, in human embryo, 940, 946
  in flowers, 511, 513
  hormone production, control of, 827
  hormone regulation of, 825
  human, **907**, 909–912, 914, 940, 946
  and menstrual cycle, 914
  plant, and auxin, 675, 695
  and sex hormones, 70
Overturn, 46, 1155
Oviducts, 535, 596, 905, 911–913, 917, 940, 952
Ovulation, 258, 824, **910**, 911, 912, 917, 918
  and estrogens, 915
  and intercourse, 915
  and "the pill," 914
Ovule, 238, 506, 507, 508
  of angiosperms, 614–617, 619, 620, 624

in flowers, 511
  of pines, 506
Ovum, 237, 247, 258, 264, 910 (*see also* Egg cell(s))
  activation of, **920–922**
  of amphibian, 926–927
  binding with sperm cell, **920**
  changes in at fertilization, 921, 940, 951
  of chicken, 931
  formation of, 258
  human, activation of, 940
  mitosis in, 921
  nucleus of, 250
  and reproductive isolation, 1016
Oxaloacetic acid, 194, 195
  in four-carbon pathway, 223–225, 229
Oxidation, 164, 166, 183, 217
  of ammonia and ammonium ion, 1147
  in chemosynthetic autotrophs, 439
  of chlorophyll, 216–218
  of food molecules, 61
  of glucose, 61, 164, **186–205**, 790
  of hydrogen sulfide, 436
  of inorganic compounds, 439
  of nutrients, and water balance, 768, 775
  of polysaccharides, 64
  and respiration, 733–734, 747
  of vitamin A, 212
Oxidation-reduction reactions, 163, **165–167**, 186 (*see also* Glycolysis, Respiration)
  and atomic nucleus, 166
  in energy metabolism, 203
  and NAD$^+$, 203
  of pyruvic acid, 193
Oxidative phosphorylation, 199, 203, 205, 878
  in electron transport, **196–197**, 198–200
  in glycolysis, 189
  in photosynthesis, 219–221, 229 (*see also* Photophosphorylation)
  regulation of, 200
Oxygen, 87, 204, 209, 235
  at altitude, 746
  in atmosphere, 452, 733
  biological importance of, 33, 35
  in blood, **750**, 763–768
  in cell respiration, 453
  in chemiosmotic coupling, 196–198
  and chemosynthetic prokaryotes, 440, 450
  consumption of, and physical work, 209
  debt, 190–191, 835, 878
  and diffusion, 130
  double bonds in, 33
  and electron transport, 196–198
  exchange, in leaves, 628
  free, origin of, 206
  in glycolysis, 201
  and hemoglobin, **742–743**, 749, **750**, 763,
  and myoglobin, **745**
  and ozone layer, 1133–1134
  and photosynthesis, 167
  and photosynthetic anaerobes, 440
  polar properties of, 33
  in red blood cells, **750**, 763
  in respiration, 732–738, 740–745
  sickle cell anemia, 390
  in van Niel's hypothesis, 217
Oxygen debt, 190–191
Oxytocin, 824, 827, 828, 839, 883
Oysters, 545, 549, 563
  mortality rate in, 1092
Ozone
  in atmosphere, 733
  layer, 1132, 1133, **1134–1135**

PABA, 177
Pacemaker, cardiac, **756–757**, 763, 878 (*see also* Sinoatrial node)
Pacific northwest forests, **1166–1168**
Pacinian corpuscles, 860
Paddles, 565
Pain
  control of, 854, 855
  receptors for, 859, 860
  and sensory cortex, 886
Paine, R. T., 1119
Pair-bond relationships, 915, 918, 1072
Pais, Abraham, 16
Paleoanthropology, **1038–1051**
Paleobiology, and fossil record, 1021–1025

Paleocene epoch, 496, 1034
Paleolithic art, 1042, **1048–1049**
Paleontology, 3–4, 15, 420, **1021–1025**, 1028, 1030
Paleozoic era, 497, 505, 510
Palisade cells, 214, 627, 648
Palmiter, Richard, 378
Palolo worms, 210
Palpi, 575, 585
Pampas, 1169
Pancreas, 116, 127, **715–716**, 730, 824, **832–833**
  and biosynthesis of chymotrypsin, 175
  in digestion, **725**
  as endocrine gland, 823
  as exocrine gland, 823
  and glucagon, 64
  hormones from, 855
  and regulation of blood glucose, 726
Pancreatic amylase, 722, **725**
Pandas, 422
Pandorina, 37, 461, 462
Pangaea, 1011, **1012–1013**
Panthera, 412
Panting, 785–786
Pantothenic acid, 194, 729
para-Aminobenzoic acid (*see* PABA)
Paradoxical sleep, 890
para-Hydroxyphenylpyruvate, 332
Parallax, 867
Paramecium, **13**, 102, 110, 122, 132–133, 139, 144, 471
  avoidance behavior in, **475–476**
  competition between, 1108
  locomotion in, 473
  macro- and micronuclei, 471
  temperature sensitivity in, 475, 476
Paramylon, 457–458, 476
Paraphyletic taxon, 418
Parapodia, 454, 558
Parasitism, 792
  and adaptation, 1006
  in dodder, 691–692
  in *E. coli*, 448
  in fungi, 480, 482, 486, 488, 492 (*see also*
    Mycorrhizae)
  in gastropods, 550
  in horsehair worms, 543
  in leeches, 558
  in mycoplasmas, 432
  in nematodes, 539–540, 542
  vs. predation, **1120**
  in prokaryotes, 432–439, 442, 451
  in protozoa, 468, 469, 470
  in tapeworms, 533, 536
  in viruses, 330
Parasympathetic nervous system, 764, 842, 845–847,
    856, 857
  cardiovascular regulation in, 762
  and erection of penis, 904
  and regulation of digestion, 723
  regulation of heartbeat, **757**
  vs. sympathetic nervous system, 846, 847, 857
Parathyroid glands, 824, **829**, 937
Parathyroid hormone, 824, 829–831, 839
Parazoa, 520, 522
Parenchyma cells, 626, 627, 632, 636, 645, 648
  secondary growth and, 645
  and water movement, 652
Parent cell, in meiosis, 252
Parenting, 917
Parietal cells, 718
Parietal lobe, 885–886, 892
Parkinson's disease, 853, 897
Parotid glands, 718
Parthenogenesis, 541, 543, 557, 560, **1103**
Partial pressure, 733, 740, 742, 743, 745, 747
Partitioning, 529, 555, 562
  of resources (*see* Resource partitioning)
Partitions (*see also* Septae), 554
Passive diffusion
  and absorption in small intestine, 723
  in loop of Henle, 772–773
Pasteur, Louis, 11, **86–87**, 236, 797
Patagonian hare, 966
Patella, 707
Paternity, tests for, 811
Pathogenic microbes, 439, 448–451
  capsules in, 433
  toxins in, 448, 451

Pathways
  for oocyte, **903–905**
  for sperm, **903–905**
Pattern formation, in embryo, **938**
Patterns A–D, of Huntington's disease, 395, 397
Patterns of water circulation, 1156–1157
Pauling, Linus, 74, 288–289, 291, 304, 340
Pauropods, 573, 586
Pavlov, Ivan, 1055, 1078
P_{CO_2}, 740, 743, 746, 747
PCR, in diagnosis, 396
Peacock, 77
Peanut worms, 520, 559, 564
Peat, 504
Pecking
  and learning, 1055–1056
  order, 1062–1063, 1078
Pectin(s), 66, 154
  in cell walls, 457
  in collenchyma cells, 637
Pedicel, 562
Pedigree(s)
  for color blindness, 392
  for hemophilia, 393
  human, 382
  for Huntington's disease, 395
Pedipalps, 568–570, 585
Peking man, 1040–1041
Pelagic life, **1156–1157**
Pelecypoda (*see* Bivalves)
Pelican, 423
Pellagra, 729
Pellicle, 457
Pelvis, hominid vs. chimpanzee, 1041
Penetrance, **271**
  incomplete, 271, 279
Penfield, Wilder, 886, 895
Penicillin, 178, 449, 488
  resistance to, evolution of, 964–965
Penicillium, 481, 488
Penis, 535, 760, 903, **904**, 905, 909, 912
  vs. clitoris, 909, 912
  and contraception, 916
  erection and orgasm, **904–905**
Pentastomids, 520, 560, 564
Pentose(s), 60, 190
PEP, 223–224, 229
PEP carboxylase, 224–225, 230
Peppered moth, **962–965**, 973, 993, 1028
Pepsin, 176, 720, 722
Pepsinogen, 720, 722
Peptide bonds, **74**
  and carbonyl groups, 74
  and enzyme activity, 175
  and protein structure, 74
Peptide chain, 313
Peptide hormones, 883
Peptides, 826, 836, 839
  in digestion, 722
  hormones as, 823, 824
Peptidoglycans, 434, 450
  in prokaryotic cell walls, 107, 432
  in spore coat, 438
Per capita rate of increase, 1088, 1089, 1104
Perception, **887–889**, 892
Perch, 594
Perennials, 621, **622**, 624, 641, 1168, 1173
Perfect flower, 614
Per gene, 834
Pericycle, 647, 648
  and cell types in angiosperms, 647
  in roots, **633**, 634
  and vascular cambium, 644
Periods, geologic, 970
Periosteum, 705
Peripheral membrane proteins, 105–106, 124
Peripheral nervous system, 712, 842, **844–847**, 856,
    935
  development of, in chick embryo, 935
  in digestive tract, 716, 718
  neurotransmitters in, 847
  in stomach, 720
  in yoga, 845
Peripheral proteins, 114
Perissodactyls, 602
Peristalsis, 912
Peritoneum, 716, **718**
Peritubular capillaries, 770–773, 776

Periwinkles, 550
Permafrost, 1168
Permeability
  of cell membrane, 127–128, 132–133, 142
  selective, 112, 124, 127–128, 142, 191, 197, 201
Permeases, 135, 178–179 (*see also* Carrier proteins,
    Transport proteins)
Permian period, 497, 503–504, 505, 509, 597, 1024
Peroxisomes, **118–119**, 125, 494
  in animal cell, 112
  of photosynthetic cells, 225
Perspiration, 703, 713, 784, 786
Pesticides, 1148
Petals, 238, 511, 513, 613–614, 619, 624
Petioles, 628, 679, 692
Petroleum, 504
Peyer's patches, 795
P_{fr}, **686–687** (*see* Phytochromes)
PGA, in three-carbon pathway, 222–224
pH, **48–49**, 52, 54, 184
  in biological systems, 52, 54
  of blood, and altitude, 746
  and cell growth in plants, 675
  and cell metabolism, 127
  and digestion, 715
  effect on cell division, 147
  and electron transport, 198
  and enzyme regulation, 175–176
  of female reproductive tract, 905, 909, 913
  of gastric juice, 720
  of intestinal enzymes, 722
  and oxygen debt, 190
  and rate of membrane transport, 178
  of soil, 664, 670
  of stomach, 791
Phaeophyta, 455–457, 465, 477
Phage φX174, 349, **350**
Phagocytic cells, 797
Phagocytosis, 118, 121, **138–140**, 143, 281, 798, 818
  and cancer research, 378
  enzymes in, 138–139
  in heterotrophs, 138–139
  and inflammatory response, 792
  and lymphokines, 807
  and macrophages, 808
  and pathogenic bacteria, 433
  in protists, 138–139
  in white blood cells, 750–751
Pharynx, 534, 539, 541, 555, 591, 592, 604, 864
  in digestions, 715, 717, **718–720**
  in respiration, 736, 738
Phase-contrast microscopy, 99–101
Phases of meiosis, **253–257**
Phases of mitosis, **150–152**
Phellem (*see* Cork cell)
Phenetics, 415, 418
Phenobarbitol, 117
Phenotype, **240–244**, 247, 266, 269, 271, 306, 315
  in balanced polymorphism, 992–993
  and behavior, 993
  and "central dogma" of molecular biology, 342
  and chromosome abnormalities, 386–387
  dominant, 241, 247
  effects of gene on, 268, 273
  and genetic interaction, 993–994
  vs. genotype, 993–994
  in learning, 1077
  and Mendelian genetics, **240–244**, 247
  and mutations, 982
  and natural selection, 977, **993–994**, **995–1006**
  and nonrandom mating, 984
  novel, **269**, 279
  phenotype matching, 1077
  phenotypic ratio, 279
  recessive, 244, 247, 279
  and sexual reproduction, 994
  and speciation, 1010–1011
Phenotypic ratio, 239, 242–243, 247, 279
  independent assortment and, 243, 247
Phenylalanine, 73, 76, 309, 388–389 (*see also*
    Phenylketonuria)
Phenylketonuria, 388–389, 400
  and albinism, 388
  and alkaptonuria, 388
  incidence, 400
Pheromones, 585, 863
  in bees (*see* Queen substance)
  in naked mole rats, 1078

I-41

Pheromones (Continued)
  as premating isolation mechanisms, 1016–1017
Phloem, 501, 625, 637, 644
  and cell types in angiosperms, 647–648
  origin of, 634, 647
  and pressure-flow hypothesis, 662
  in roots, **633**, 634
  secondary, 675, 681
  in stems, 640
  in translocation, 660, 670
  in transport of sugars, 650
  in vascular cambium, 645
Phoromids, 520, 562, 564
Phosphatase, 722
Phosphate(s), 312
  in blood, 829
  in cellular respiration, 191–192
  in electron transport, 197, 199, 200
  hormonal control of metabolism in, 829
Phosphate bonds
  in ATP, 186, 189, 190, 204–205
  in DNA, 298–299
Phosphate groups, 57, 82, 184
  in ATP, 137, 180–183, 189
  in DNA, 283, 284, 288, 290, 298, 299
  enzyme reactions, 171–172
  in glycolysis, 189
  and lipid structure, 69
Phosphoenolpyruvate (see PEP)
Phosphofructokinase, 188
Phosphoglucoisomerase, 188
Phosphoglycerate (see PGA)
Phospholipids, **67, 69**, 82, 835
  in cell membrane, 69, 104–105
  and glycerol, 69
Phosphorus, 35, 1145
  cycle, **1146**
  and mycorrhizae, 491
  nutritional requirements for, 728
  plant requirements for, 657
Phosphorylase *a*, 837, 838
Phosphorylation, 182, 189, 861, 878
  of ADP to ATP, in photosynthesis, 220–221
  oxidative, 196–200, 203, 205, 219–221, 229
  of pyruvic acid to PEP, 225
Photoelectricity, 207
Photon(s), 29, 208–209, 216, 219, 869
Photonegativity, 535
Photoperiodism, **684–689**, 695, 833, 863 (see also Biological clock; Circadian Rhythms; Phytochromes)
  and flowering, 688
  and phytochrome, 686
  and "sleep movements," 689
Photoperiods, 685
Photophobia, 729
Photophosphorylation, 218, **219–221**, 229, 653
Photoreception, 588, 866–873, 879 (see also Vision)
Photoreceptor cells, 529, 458, 459, 484, 859 (see also Cones; Rods)
Photoreceptor organ, 529
Photorespiration, 225, 230, 494
Photosensitivity, 556
Photosynthesis, 87–89, 92–93, 101, 114, 125, 161–162, 165, 167, 200, 203, **206–231**, 426, **440–443**, 454, 529
  action spectrum of, 212
  and asteroid hypothesis, 1024
  and bioenergetics, 161–162
  in bryophytes, 498
  Calvin cycle in, **222–223**, 224–226, 228–230
  and CAM plants, 655
  and carbon dioxide fixation, 504
  and carbon formation, 516
  C_4 photosynthesis, 224–228, 352
  and chemiosmotic power, 200, 203
  and circadian rhythms, 689
  components of, **209**
  in coral reefs, 530
  C_3 photosynthesis, **224–228**
  in cyanobacteria, 452
  efficiency of, 230
  and energy flow in the biosphere, 1137
  and environment, 206
  equation for, 206, 209, 228
  evolution of, 208, 218, 495
  four-carbon pathway, **223–228**
  and fungi, 488

and gas exchange, 209
in leaf cells, 648
in leaves, **627–630**, 647
and lichens, 492
light, effect of, **215–216**
light-dependent reactions in, 215–216
light-independent reactions in, 215–216
in Monera, 94
and organic compounds, 56
origin of, 206
oxidative phosphorylation in, 219–221, 229
and oxygen, 167
and ozone layer, 1133
photophosphorylation in, 218, **219–221**, 229
pigments in, 211–213
in plant cell, 113
and plant evolution, 625
products of, **228**
in prokaryotes, 429–431, 436, **440–443**, 450
in protists, 452
rate of, 212, 215
  in C_4 plants, 226
  in C_3 plants, 226
stages of, **215–216**, 229
temperature, effect of, **215–216**
three-carbon pathway, 222–223
in trophic levels, 1085
Photosynthesis, light, and life, **206–231**
Photosynthetic autotrophs, **455–465**
  characteristics of, **456–457**
  chrysophytes, **459–460**
  division of labor, 456
  pigmentation in, **456–457**
  storage of food, 456, 457
Photosynthetic cells, 211, 225
Photosynthetic membranes, **213–214**, 441
Photosynthetic organisms, 1125–1127
  and ecological succession, 1125–1127
  and ecosystem energy flow, 1131, 1137
  in food chains, 1138
  in oceans, 1155–1156
  in rivers and streams, 1155
Photosynthetic pigments, 431, **440–442**
Photosystems, 216
  Photosystem I, 216, 218–219, 229
  Photosystem II, 216, 218–220, 228–229
Phototaxis, 474
Phototropisms, 484, **671–673**, 694–695
Phycobilins, **440–441**, 442, 456, 457, 465
Phycocyanin, 440
Phycoerythrin, 440
Phyla, 410, 413, 427, 520 (see also Division)
Phyletic change, 1029–1030
  and adaptive radiation, 1002
  in australopithecines, 1040, 1043
  vs. cladogenesis, 1022
  in evolution of the horse, 1027
  in fossil record, 1022
  in hominids, 1043
  vs. punctuated equilibria, 1028
Phylogenetic relationships, 415 (see also Classification; Taxonomy)
Phylogenetic tree, 415, 417, 544
Phylogeny, 412–413, 416, 418, 427
  conventional, 418, 419
  homology and, **413–414**
Physical work
  and oxygen, 209
Physics, laws of, and biology, 10–12
Phytoalexins, 682
Phytochromes, **686–687**, 690–691
Phytophthora infestans, 467–468
Phytoplankton, 226, 456, 476, 1134, 1145, 1155
Pigmentation
  natural selection for, 728
  and temperature regulation, 781
Pigment cap, 926–927
Pigment cells, 529, 579, 710, 935
Pigment layer of eye, 936
Pigments, 120, **211–213**, 216, 221, 228, 552
  absorption spectrum and, 210, 211
  and albinism, 389
  algal, 456–457
  antennae, 228
  in bacteria, 428, 431, 440
  beta-carotene, 211
  and color blindness, 392
  in photoperiodism, 695

in photosynthesis, 456–459 (see also Photosynthetic pigments)
of photosynthetic protists, 476–477
in phototropisms, 673
P_{700}, in energy-capturing reactions, 216, 218–219, 229
P_{680}, in energy-capturing reactions, 216, 218, 228
in respiration, 741–745, 747 (see also Hemoglobin; Myoglobin)
riboflavin and, 195
visual, and vitamin A, 729
Pilbeam, David, 1035, 1044
Pili, prokaryotic, 327, **432–435**, 450
Pilin, 435
"The pill," **916–917**
Pill bug, 409, 571, 572
Pilobolus, 484
Piltdown forgery, 1044
Pincers, 565, 566, 568, 572, 585 (see also Chelicerae)
Pineal gland, 824, **833**, 839
Pine barrens, 1164
Pines, 491, 505, **506–508**
Pinocytic vessels, 754
Pinocytosis, **138–140**
Pinworms, 539
Pisaster, 1106, **1119**
Pistil, 511
Pitch (of sound), 583
Pitcher plant, 666–667
Pith, **633**, 640, 647–648
Pith rays, 644, 647
Pituitary dwarfism, 825
Pituitary gland, 759, 823–824, **825–826**, 831, 883
  and ADH, 774, 776
  anterior lobe, 825–827
  control of, by hypothalamus, **826–827**
  effect of stress on, 833
  and endorphins, 854, 855
  hormones of, 825–826, 827, 855
  intermediate lobe, 826
  menstrual cycle, 913
  negative feedback control in, 827
  posterior lobe, 826, 828
  and regulation of TSH, 712–713
  and sex hormones in male, **906**, 907, 917
PKU (see Phenylketonuria)
Placenta, 387, **561, 600**, 911, **944–945**
  active transfer across, 948
  expulsion of, 950
  human, **944–945**, 952, 953
Placental mammals, 600–601, 1023, 1031
Placental stage (of labor), 950
Placoderms, 595
Plains, 1169
Planarians, 533, 534
  freshwater, 533, 534
  nervous system of, 842
  reproduction in, 535
Planetary model of atom, 28
Plankton, 456, 531, 541, 572, 590, 592, 604
Plantae, 14, 125, 423–427
Plant cell, 103, 106, 107, 110, **113**, 114, 116–117, 140, **153–155**
  cell cycle in, 146
  cell plate, 153–154, 213
  cellulose, 78, **106–107**, 113, 124–125, 154
  cell wall (see Cell wall)
  chloroplasts (see Chloroplasts)
  cytokinesis in, 153–154
  cytoskeleton of, **113**
  desmotubules, 140
  energy use in, 168
  growth of, 107
  mitosis in, **154**
  mobility of, 121
  nucleus of, **113**
  peroxisomes in, 118–119
  plasmodesmata, 107, **113**, 140–143
  plastids, 120, 125, 146
  turgor in, **134**, 142
  vacuoles in, 134
  vesicles in, 153, 154–155
  wall pressure, 134
Plants, 84–85, 203, 209, 407, 411, 424, **493–517**, 542
  adaptations of, 495, 516
  alternation of generations in, 463, 494, 495, 516
  ancestors, 493–495
  angiosperms, 498, 505, **509–515**

I-42

asexual reproduction in, 500, 1103
biology of, **611–696**
biosphere, 515, 516
body of, 624, **625–648**
bryophytes vs. vascular, 496
and carbon formation, 516
in Carboniferous period, 504
cell division in, 494
cells and tissues of **625–627**
cell walls, 480
classification of, 454, **497–515**
coevolution with insects, **578–579**
development of, **625–648**
dicots, 498
diseases in, 487, 488
distribution of, and biogeography, 966
earliest, 454, 455, 461, 496, 497, 613
ecological importance of, 613
in energy cycle, 1143
evolution of, 92, **493–497**, 501–502, 503–506, 509–510, 516
and evolution of animals, 516
ferns, 498, **502–503**
fertilization in, 499, 502
fibers, 637
flowering, 407, 424, 496–497, 498, **613–624** (see also Angiosperms)
food storage in, 493
fossils of, 496
growth of, 630, **671–696**
gymnosperms, 497–498, **505–509**
hormones, effects of, **694**
and inorganic substances, 516
meiosis in, 462
microfibrils in, 66
monocots, 498
multicellularity of, 94
and mycorrhizae, 492
in nitrogen cycle, 1147–1149
nutrition of, 208, 492, **662–669** (see also Photosynthesis)
and organic compounds, 516
origins of, 454, 455, 461, 496
parasites of, 486, 492
parthenogenesis in, 1103
and photoperiodism (see Photoperiodism)
pollination and, 517
polyploidy in, 1012–1014
as primary producers, 1138, 1139, 1140, 1152
regulation of growth, **671–696**
reproduction in, 501–502, 516, 517
seeds, **502, 503–515**
spores, 462
starch in, 14
symbioses in, **664–670**
sympatric speciation in, 1014
transport processes, 496, **650–670**
vascular, 496, **501–515**, 578–579 (see Vascular plants)
Planula(e), 526, 529, 533, 542
Plasma, 704, **750**, 763, 765, **768–769**, 772
and blood types, 810–811
as body compartment, 775
and Bowman's capsule, 778
and formation of urine, 773
gas exchange in, 741, 742
regulation of composition, 770–771
Plasma cells, **795**, 798, **800–801**, 804, 808, 809, 819
and active T lymphocytes, 804, 808, 809
and antibody formation, 795, **800–801**
and multiple myeloma, 798, 802
Plasma membrane (see Cell membrane)
Plasma proteins, 725, 750
in blood clotting, 750, **751**
and osmotic potential of blood, 762–763
Plasmids, 331, 337–338, 431, 437
asynchronous replication of, 326
and chromosome, 326
and conjugation, **326–330**
and drug resistance, 336–337, 965
episomes, **326**, 335, 337
F plasmid, 326–329, 337
pSC101, 344
in recombinant DNA, **344**, 350–354
and recombination, 337
R plasmid, 326, 329, 336
R6 plasmid, 329
sex factor, 326–329

synchronous replication of, 326
Ti plasmid, **352–353**, 354
vs. transposons, 336
as vectors in research, 378
Plasmodesma(ta), 107, **113**, 140–143, 224
endoplasmic reticulum and, 140
membrane of, 140
of roots, 633
Plasmodial slime molds, 455, 466–467
Plasmodium, 466, 467
Plasmodium, **473–474**
Plasmolysis, 134
Plastids, **120**, 125, 146, 493, 626, 632, 683–684
Platelets, 750, **751–752**, 763, 835, 839,
Plate tectonics, **1012–1013**, 1160
Platyhelminthes (see Flatworms)
Platypus, 600
Platyrrhine anthropoids, 1034–1035
Pleiotropy, **273**, 279, 1035
Pleistocene epoch, 496, 510, 1096, 1166
Pleura, 740
Pleurisy, 740
Pliocene epoch, 496, **1013**
Plumage, 872
in bowerbirds, 1073
and sex characteristics, 906
and sexual selection, 999, **1000–1001**
Pluteus, 920, 921–924, 925
Pneumatophores, 636
Pneumococcus, 281–283
in genetic research, 299
mutations of, 282
Pneumonia, 282, 429, 436, 816
Pods, 238, 620
Pogonophora (see Beard worms)
Poikilothermy, **780–781**, 789–790
Poison, 538
Poison glands, 570, 590
Poison ivy, 516
Polar bear, 961, **1008**
Polar bodies, 258, **910**
Polar-covalent bonds, 32–33, 38
Polar fibers
formation of, 150
spindle and, 149–151, 153
Polar front, 1137
Polarity
and action potential, 856
in bacteria, 435
and carrier-assisted transport, 134
and diffusion, 130
and nerve impulse, 848–851, 852, 854, 856
of ovum at fertilization, 921
of sperm, 902
of water, 33, 41, 44, 46, 54
Polar nucleus, 617, 624
Polar plasm, 924–925
Pole cells, 924–925
Poles
of cell, 150–152, 155
in meiosis, 255, 256
of spindle, 149
Poles (magnetic) and world weather patterns, 1136–1137
Poliomyelitis, 449, 796, 813–819
Pollen, 237–239, 507, 508, 511, 614, 812, 971
and allergies, 812
and outbreeding in plants, 984–985
storage of, 1060–1061
transport of, by bees, 1060
Pollen grains, 506, 507, 613–614, **615**
Pollen tubes, 508, 511, 614, 615, 617, 624
Pollination, 238, 517
and adaptation of flowers, 515
of angiosperms, **512–515**
and insects, **512–514**
insects vs. wind, 513
by wind, 507
Pollinators, 512–515, 517
circadian rhythms of, 689–690
coevolution with flowers, 1003
Pollution, 50–51
and biogeochemical cycles, **1150–1152**
as disruptive selection, 996
and evolution, **962–965**
and immune response, 791
and ozone layer, 1134–1135
and population growth, 1095

Poly-AC, 310
Poly-AG, 310
Polyandry, 999–1000
Poly-A tail, 369, 374, 380
Polychaetes, 554, **558**, 561, 564
Polydactyly, 271, 983
Polygene(s), 271
Polygenic inheritance, **271–272**, 279, 1053, 1077
Polygyny, 989–1000
Polymer(s), **59**, 64, 88
Polymerase chain reaction (see PCR)
Polymorphisms, in populations, 984, 992–993, **996**, 998
"Polymorphs," 792
Polyps, 525–530, 572
asexual reproduction in, 526
vs. medusa, 525
Polypeptide(s), **74**, 75, 82, 313, 316
and alternative gene splicing, 380
and cell membrane structure, 106
chains, 82, 371–372
in hemoglobin family, 367
hormones as, 824
radioactive, 309
transcription for, 369
Polypeptide chains, 82, 304–305, 309–310, 312–314, 316, 371–372
in enzyme structure, 170
heavy, 371–372
light, 371–372
Polyphyletic taxon, 413
Polyploidy, **1012–1014**, 1015, 1029
in cells, 249
in number of chromosomes, 252
Polyribosomes, 314, 795
Polysaccharide(s), 59, 64–65, 426
catabolism of, 203
in cell division, 153–154
in cell walls, of algae, 456, 457
encapsulation, of microbes, 282–283, 433
noncellulose, 426
in prokaryotic cell wall, 432
storage of, 64
structural, 65
Polysomes, 314, 795
Polysynaptic reflex arc, 844
Poly-T tail, 369
Poly-U, 307
Polyunsaturated fats, 728
Pomes, 620
Ponderosa pine, 505
Ponds, **1155**
Pons, 882–883, 898
Population bottlenecks, **983**, 990
Population cycles, **1098–1099**, 1104, 1118–1119
Population density, **1085**, 1087, **1093–1094**
Population dynamics, **1087–1105** (see also Populations)
age structure, **1092–1093**
and carrying capacity, 1089–1091
density, **1093–1094**
dispersion, **1093–1094**
mortality patterns, **1092**
patterns of population growth, 1088–1091
properties of populations, **1087–1094**
Population explosion, 449, **1096–1097**
Population genetics, **974–990** (see also Communities; Population dynamics; Populations
of human evolution, 1050
and kin selection, 1066–1067
and phyletic change, 1021
Populations, 408, **974** (see also Population dynamics)
bottlenecks, **983**
care of offspring and, 1099–1105
clumped dispersions in, 1093
and communities, **1106–1130**
"crashes," 1089
creation of, 1010–1011
density-dependent factors, **1098**
density-independent factors, **1098**
dynamics, **1087–1105**
dynamism of, 1129–1130
and energy transfer, 1145
exponential growth of, **1088**
extinction of, 1011
gene flow in, 982
genetic drift, 982–983
genetic variability in (see Genetic variability)

I-43

Populations (Continued)
  heterozygy, 978
  human population explosion, **1096–1097**
  isolation of, **1011**
  latent variability in, **976–977**
  life-history patterns, **1099–1104**
  limitations on, **1095–1099**
  logistic growth of, **1089–1091**
  and natural selection, **991–1009**
  per capita rate of increase of, 1088
  population cycles, **1098–1099**
  predation in, **1116–1119**
  properties of, **1087–1094**
  random dispersion of, 1093
  regular dispersion, 1093
  regulation of **1095–1099**
  and speciation, 1011
Pore
  of anther, 614
  central, and osmosis, 133
  genital, 536
  nuclear, 104, 106, 108
  of pollen grain, 615
  of taste buds, 717
Poriphera (see Sponges)
Porphyrin ring, 78–79, 211
  in cytochromes, 196
  in hemoglobin, and respiration, 742
Portal systems, **759**, 826, 827
  in nephron, 770
Portuguese man-of-war, 528
Position receptors, 859
Positive feedback, 712
  in memory pathways, 894–895
  in menstrual cycle, 913
Positive regulation, of genetic transcription, 322, 325, 337
Positive tropism, 671
Postganglionic fibers, 846, 857
Postganglionic neurons, 845, 846, 856
Postmating isolating mechanisms, 1015, 1016, **1017**
Postpartum taboos, 1097
Postsynaptic cells, 852, 854, 856
  hyperpolarization of, 756
  and learning, 897–898
  polarization of, 855
Posture, upright, 601 (see also Bipedalism)
Potassium, 31
  and aldosterone, 775–776
  in biogeochemical cycles, 1145
  and cell transport, 136
  hormonal regulation of, 832
  and hormone secretion, 824
  ion (K^+), **31**, 766
    and action potential, 849
    and osmosis in plants, 656–657, 669
    plant requirements for, 657–658
    in soil, 664
    and stomatal movements, 653, 654
    and touch response in plants, 692
Potassium-argon dating, 1040
Potato, 625–626
Potato blight, 468, 678, 682, 1149
Potato spindle-tuber disease, 446–447
Potential
  osmotic, 132, **133–134**, 670, 775
  water, 128, 129, 134, 670
Potential energy, **28**, 128–129, 162–163, 205
  in anaerobic respiration, 191
  and ATP, 181–182
  and cellular transport, 135–136, 180
  and chemical bonds, 163–169
  and electron transport, 195, 197, 198, 205
  in energy transformations, 186–187
  and entropy, 164–166
  and equilibrium, 174
  in glycolysis, 187, 190
  gradient, in photosynthesis, 220
Potentilla glandulosa, **1003**
P_{O_2}, 733, 740, 742 (see also Oxygen, partial pressure of)
Pouch, marsupial, 600
Powdery mildews, 481
P-protein, **638–639**
Prairie, 1169–1170
Praying mantis, 582, 583
Precambrian era, 497
Precipitation, 1139, 1177

Precursor compounds, 203, 205
Predation, 518, 531, 538, 545, 551, 552, 558, 570, 572, 580–582, 588, 590, 599, 858, **1114–1119**
  and altruism, 1055
  in amoebas, 474
  and balanced polymorphism, 992–993
  and behavior, 1054
  in chimpanzees, 1036–1037
  and competition, 1106, 1107
  defenses against, in plants, 515, 517, 692–694
  and dentition, 717
  and evolution of respiration, 735
  by fungi, 489
  and inhibition hypothesis, 1128
  as a limitation on population, 1098, 1099
  and natural selection, 992–993, 998, 1000
  by plants, 692–693
Predator-prey interactions, **1114–1119**, 1129
Predators
  and adaptation, 1006
  and carrying capacity, 1116–1119
  and evolution, 1114–1119
  and frequency-dependent selection, 998
  scavengers as, 1142
  in food chain, 1138, 1142
Prediction, 1104
  of infection rates, **1090–1091**
  in Mendelian genetics, 240
  of population growth, 1092
  in science, 15, 18
  of weather cycles, 1158
Prefrontal cortex, 894, 899
Preganglionic fibers, 846, 857
Preganglionic neurons, 845
Pregnancy, 911, 912, 941
  birth, **950**
  contraception and, 916
  placenta, **944–945**
  tests for, 802
Prehensile limbs, 1034–1035
Premating isolating mechanisms, 1015, 1016–1017
Premolars, 601, 602, 716, 717
Prenatal testing, 388
Preprophase band, 154
Presenting, 1063
Presocial species, 1060
Pressure
  air, 584
  atmospheric, 733
  cell wall, 134
  glomerular, 770–771
  hydrostatic, 128, 133–134 (see Hydrostatic pressure)
  in lungs, 747
  osmotic, 132–132
  water, 584
  and water movement, 128, 142
Pressure-flow hypothesis, **662–663**, 670
Pressure receptors, and blood volume, 774
Presynaptic cells, 852, 897–898
Prey, 1036–1037, 1114–1119, 1129 (see also Predation; Predator-Prey Interactions; Predators)
  in food chain, 1137–1138, 1143
  and frequency-dependent selection, 998
Priapulids, 520, 559, 564
Prickly-pear cactus, 1117–1118
Priestley, Joseph, 208–209
Primary cell wall, 106–107, 636–637, 639
Primary consumers, 1085
  in forests, 1164
  in oceans, 1156
  at seashore, 1157
Primary electron acceptor, in photosynthesis, 216, 218–219, 228–229
Primary growth, 647–648 (see also Branches; Buds; Bulbs; Leaves; Shoots)
  in angiosperms, 619
  of roots, **634**
  vs. secondary growth, 645
  of shoot system, **640–643**
Primary induction, 930–931, 935–936
Primary meristems, 618, 634, 635, 647
Primary mesenchyme, 923, 924
Primary oocytes, 910
Primary phloem, 635, 640 (see also Phloem)
Primary producers, **1138–1139**, 1140–1141, **1145**, 1153
Primary productivity, 1139, 1177

Primary roots, 632
Primary spermatocytes, **901**, 902, 917
Primary structure, of proteins, 74, 79, 302
Primary vascular system, 640
Primary xylem, 634, 635, 640 (see also Xylem)
Primates
  blood groups in, 996
  cerebral cortex of, 885
  evolution of, **1031–1038**
  hand and arm of, **1032–1033**
  intrinsic processing centers in, 883
  learning in, 1055
  monkeys, **1034–1035**
  phlyogenetic tree, 1032
  social behavior in, **1036–1038**
  upright posture in, **1033**
  visual acuity in, **1033**
Primers, 296–297, 396
Primitive streak, 932, **933**, 945, 952
Primordia, 640–641, 648, 935
P ring, 434
Prions, **446–447**, 451
Prism, 207
Probability
  and Hardy-Weinberg equilibrium, 979, 982
  laws of, **244–245**
  product rule of, 244–245
  same size and, 245
  sum rule of, 244–245
Probe, radioactive, 364
Proboscis, 538, 540, 542, 559, 591
Procambium, 618, 625, 634
  and cell types in angiosperms, 647
  in roots, 634
  and secondary growth, 644
Prochloron, **442**, 443, 456, 458
Proconsul africanus, 1035
Prodigal life-history patterns, 1099
Producers, **1138–1139**, 1153 (see also Primary productivity)
Profundal zone, 1155
Progesterone, 824, 836, 949
  and development of embryo, 940–941
  in female reproductive system, 917
  and menstrual cycle, **913**, 914
  in "the pill," 915–916
  in pregnancy, 945
Proglottid, 536
Programmed behavior, 689–690
Progress zone, 939
Prokaryotes, **13**, **90–92**, 93, 101, 102, 104, 106, 124, 305, 309, 311, 313, 316, 411, 420, 423, 426, **428–451**
  anaerobic, 429
  autotrophs, 515
  cell division in, **145**, 154
  cell of, 115, 319–320, **431–436**
  chemosynthetic autotrophs, **439**
  chromosome of, 145, 319–320
  cilia in, 122
  classification of, **429–430**, 454
  cyanobacteria, **440–443**
  diversity of, **436–437**
  DNA of, 145, 319–320, 450, 472
  DNA polymerase in, 320
  earliest, 497
  and ecology, 439, **447–450**
  electron transport in, 450
  vs. eukaryotes, 355, 379, 431, 434, 452
  in evolution, 452, 454
  gene, linearity of, 328
  genetic regulation in, 320–323, 337
  genetic transcription in, 320–322, 355, 358–360, 380
  genetic variability in, 428, 450
  heterotrophs, **438–439**
  metabolism of lactose, 321–326, 337
  molecular biology of, 357–358
  molecular genetics of, **319–338**
  nucleoid in, 337
  nutrition in, **438–442**
  Okazaki fragments in, 320
  operon of, **322–326**, 337
  photosynthetic autotrophs, **439**
  photosynthesis in, 456
  in recombinant DNA, 343
  regulation of exchange in, 127
  replication, 325, 337, 358

I-44

resting forms in, **437–438**
ribosome, 320, 337
RNA polymerase in, 320, 337
size of, 95–96
sporulation in, 437–438, 462
symbioses in, **447–448**, 453, 456, 458
transcription in, 320–322, 325, 337
Prolactin, 824, 825, 836, 839
Proline, 73, 303
Promoter gene, 310, 320, 323, 325, 337, 377–379
    and cancer research, 377, 378, 379
    in transposons, 335, 374
Prophages, **331–334**, 338, 448
    and chromosomes, **331–334**
    insertion sites, 329, 334
Prophase, **150–151**, 152, 155
    prophase I of meiosis, 275
Proprioceptors, **582**, 583, 586, 859
Prop roots, 636
Prosimians, **1034**, 1051
Prostacyclins, 835
Prostaglandins, 70, 831, **833–836**, 839, 905
    formation of, 833
    and inflammatory resposne, 835, 836
    and immune response, 835, 836
    and menstrual cycle, 835
    and uterus, 835
Prostate gland, 903, **904**, 905, 917
Protective layer, and abscission, 679
Protein(s), 55, 305, 311, 313, 316–317, 427 (*see also specific types*)
    accessory, 114
    alpha helix, 74–75
    and alternative gene splicing, 380
    amino acids in, 70, **72**
    antifreeze, 44
    assembly, 305, 308
    association with DNA, 355, 358, 379
    beta pleated sheet, 74–75
    biosynthesis of, **301–317** (*see also* Messenger RNA)
    and cancer, 375–377
    CAP, 325, 337
    CAP-cAMP complex, 325, 337
    carrier, 135, 142
    in cell division, 78, 146
    in cell membrane, 70, 72, 106, 124
    in cellular transport, 135, 142
    change or loss of function in, 304
    in cilia and flagella, 122–123
    collagen, 77, 115, 362
    composition of, 306, 317
    condensation reactions in, proteins, 72–73
    contractile, 70, 72
    control, for cell cycle, 148
    corepressor, 323, 324, 337
    in cytochromes, 196
    digestion of, 720, 722, 730
    domains, 363
    in *E. coli*, 72
    effectors, 176–177, 179, 184, 323, 325
    enzymes as, 70, 72, 76
    evolution of, 326, 370
    fibrous, 77, 82, 111
    filaments, 121
    flavoproteins, 195
    functional groups, 320, 337
    functional regions of, 138, 363
    functions of, 305
    genetic code for, **301–304**, 320–321, 355, 359, 363, 366, 367, 368
    globular, 75–76, 78, 82
    and glucagon, 832
    and glucocorticoids, 830
    growth-promoting, 373–374, 376
    and heredity, 281, 285, 299
    homologous, 420, 422
    hydrogen bonds in, 74–75
    inducer, 323–325, 337
    inhibitor, 177
    initiator, for DNA replication, 294
    insulin, 138
    integral membrane, 104–106, 114, 124, 135, 142, 178–179, 200
    introns in coding, 366
    membrane transport, 178–181
    metabolism of, 824
    multimeric, 76
    in muscle cells, 122

nutritional requirements for, 730
origin of, 89
in pancreas, 725
peptide bonds in, 74
peripheral membrane, 106, 124
permeases, 135, 178–179
positive regulation of, 322, 325, 337
primary structure of, 74, 79
protective, 70, 72
quaternary structure of, 76, 79, 170–171
receptor, 138
in red blood cells, 114
regulatory, **70, 72**, 322, 325, 358, **360–361**, 373, 375–377, 380
repressor, **323**, 324–325, 337
riboflavin, 195
and ribose, 172
ribosomal, 115, 366
RNA and biosynthesis, 306
secondary structure of, 74–75, 79
selectivity of, 135, 136
single-stranded binding proteins, 294, 297, 299
storage, 70, 72
storage of, 67, 120
structural, 70, 72, 304
structure, **74–80**, 135, 142, **304–305**, 419
subunits, 146
synthesis of, 115, 116, 117, 172, 306, **309–314**, 315–317
    and learning, 898
tertiary structure of, 75–76, 79
translation for, **301–317**, 368–369
transport, 70, 72, 82, 106, 124, 135–137, 181, 184, 325
viral, 342
in virus coat, 305, 308
Protein filaments, 121, 896
Proteinoid microsphere, 88–89
Protein subunits, 146
Proteus mirabilis, 432
Prothoracic gland, 578
Prothrombin, **751–752**
Protists, 13, 94, 122, 424–427, **452–477**, 515
    autotrophs, 452
    behavior in, **474–476**, 477
    cell division in, 144
    cell type, 125
    cilia and flagella in, 122
    ciliates, 452
    classification of, **454–455**
    and digestion, in herbivores, 724
    and evolution, 519
    evolution of, **452–455**
    heteromorphs, 452
    life cycle of, 250
    natural selection and, 476
    parasitism in, 452
    phagocytosis in, 138–139
    photosynthetic, **454–455**
    reproduction in, 476, 477
    spores in, 462
    symbioses in, 724
Protoavis, 598, 599
Protocorm, 490
Protoderm, 618, 625, 634, 635, 641, 647
Protogenes, 989
Proton(s), **23–24**, 38, 163, 218, 220, 229
    gradient, 197
    and polar-covalent bonding, 32
    pump, 197–199, 205
Protonema(ta), 499, 500,
Proton gradient, 197–200, 218, 221, 229
Proton pump
    and cell elongation in plants, 674–675
    and electron transport, 197–199, 205
    in photosynthesis, 220–221, 229
Protoplasm, 466, 467
Protostomes, 520, **544–564**
    vs. deuterostomes, 544
    embryonic development in, 545
Protozoans, 411, 423, 455, **468–476**, 477
    behavior patterns in, **474–476**
    cell division in, 470
    characteristics of, 469
    ciliates, **470–471**
    classification, 468, **469**
    conjugation in, 468
    evolution of, 477

parasitism in, 468, 469, 470
reproduction in, 468, 469, 470
symbioses in, 469
Provirus, AIDS as, 815–817
Proviruses, 373–374, 381
Proximal tubule, 770–772
    active transport at, 774
    and formation of urine, 772–773
Proximate causation, **1053–1054**, 1077
Prudent life-history patterns, 1099
pSC101 plasmid, 344
Pseudocoelomates, 520, 532, 538, **539–541**, 542–543, 544, 559
Pseudogenes, 367, 368, 374, 946
Pseudomonads, 430
Pseudomonas aeruginosa, 425
Pseudopodia, 37, 468, 469, 470, 527, 750–751, 792, 923
Pseudotrebouxia, 489
P_{700}, in energy-capturing reactions, 216, 218–219, 229
Psilocybin, 487
Psilophyta, 497, 498, 501, 502
P (peptide) site, 312–313
P_{680}, in energy-capturing reactions, 216, 218, 228
PSTV, 446–447
Pterobranchs, 591
Pterophyta, 497, 498, 502
Puberty, 833, 902, 905, 906, 910, 915
Pubic hair, 909
Puffballs, 488
"Puffs," chromosome, 360
Pulmonary circulation, 755–756, **758–759**, 763
Pulp cavity, 716
Pulse, 753
Punctuated equilibria, **1028–1029**, 1030
Punctuationalists, 1030, 1043
Punnett, Reginald, 283, 268, 269
Punnett square, 240, 243–245, 260, 263–266, 980
Pupa(e), 576, 577, 578, 580, 586, 1053, 1060
Pupil, 866, 936, 951
Purified fractions, 192
Purines, 80–81, 118–119
    in DNA, 283–285, 287, 289, 290, 299
Purple bacteria, 430, 431, **440**, 450
Purple nonsulfur bacteria, 430, **440**, 453
Purple sulfur bacteria, 430, **440**
Pus, 757
Pyloric sphincter, 720, 722
Pyramid, of energy flow, **1143**
Pyrenoid, 458
Pyridoxine, 729
Pyrimidines, 80–81, 283–285, 287–290, 299, 305
Pyruvic acid
    in aerobic respiration, 193–194
    in anaerobic respiration, 190–191
    in glycolysis, 187, 190, 200, 204
    in Krebs cycle, 194–196
    oxidation of, 193
    in photosynthesis, 224–225

Quantum, 29
Quaternary period, 496
Quaternary structure, of proteins, 76, 79
    and enzyme activity, 170–171
Queen bee, **1060–1061**, 1068, 1078
Queen conch, 166
Queen rat, 1062
Queen substance, 1061
Quinine, 515

Rabbit, 97, 268–269
    vs. myxoma virus, 1120
Rabbit beta globin, 378–379
Raccoon, 422
Race
    and heart disease, 761
    origin of, 1047
    vs. species and subspecies, 1010, 1014
    and sympatric speciation, 1014–1015
Radiant energy, 161, 165, 167, 1133
Radiation, 274
    absorption of, in atmosphere, 1133
    electromagnetic, 25
    as heat loss, 789
Radicle, **634**
Radioactive fallout, **1150–1152**, 1153
Radioactive isotopes (*see* Isotopes)

I-45

Radioactive labeling, 285
  in genetic research, 285–286, 292–293, 299
  in recombinant DNA, 348–349
Radioactive probes
  in diagnosis of genetic disease, 394, **397–399**, 400
  in genetic diagnosis, 396
  in recombinant DNA, **346–347**, 354
  tritium, 364
Radioactivity, 25, 85, 87, 100, 308, 982, 1151–1153
Radioisotope(s), decay of, 417 (*see also* Isotopes)
  in geological dating, **770–771**, 971
  in plant research, **660–661**
Radiolabelling, 908–909
Radiometric clocks, 971
Radio waves, 207
Radon, 733
Radula, **546–547**, 548, 563
Rainfall, 1165, 1172, 1174
  DDT in, 1151
  and leaching of nutrients, 1150
  and mineral budget of ecosystems, 1150
  and plant growth, 646
  and population dispersion, 1094
  world patterns of, 1136–1137, 1152
Rain forest, tropical, 405, 407, 629, **1174–1176**
Rain shadow, **1137**, 1170, 1174
Random assortment, of alleles in meiosis, 253, 261
Random dispersion, 1093–1094
Random mating, 979–980, 984, 989
Random motion, 129, 164–165, 183, 733
"Random tick" hypothesis, 419–420
Range of tolerance, 1095
Rapid eye movements (REMs), 890
Rat, 202
Rate
  of biochemical reactions, 169, 174–175, 176–181
  of biosynthesis of lactose, 321–322
  of cell division, 147
  of cell transport, 178–179
  of countercurrent exchange, 142
  of diffusion, 129, 136
  of DNA replication, 294
  of molecular motion, 129
  and uncertainty principle, 129
Rate limitation
  on entry of glucose into cell, 136
  of photosynthesis, 215–216
Raub, David M., 1024
Rays (animal), 593
Rays (in plant tissue), 644, 647
Reabsorption, in kidney, **771**, 774, 776
Reaction cascade, in blood clotting, **751–752**, 763
Reading frames, 310, 316
  in recombinant DNA, 350
  shifts in, during protein biosynthesis, **316**
Realized niche, **1113**, 1129
Receptor-mediated endocytosis, 138–140, 143
  and AIDS infection, 815
  and cell membrane, 135
Receptor protein, 71–72, 76
  for cholesterol, 71
  functional regions of, 138
Receptors, 106, 124, 184
  for abscisic acid, 654
  acetylcholine, 137
  and blood pressure regulation, 762
  for cardiac peptides, 756, 775
  clustering of, 138
  and endocytosis, 138–140, 143
  and enzyme inhibition, 179–181
  for estrogen, 908, 914
  functional regions of, 138
  for interferons, 793
  and learning, 898
  and locomotion, 131
  membrane, and cell division, 148
  for MHC antigens, 808–809
  for neurotransmitters, 857
  for retinoic acid, 940
  sensory, 708–709 (*see also* Sensory receptors)
  T-cell, **804**, 819
  temperature, 784, 790
  and transport, 137
Receptor sites
  in cell membrane, 131
  and endocytosis, 138–140
Recessive allele
  in albinism, 389

  in hemophilia, 382, 400
  and mutations, 400
  and phenylketonuria, 388–389
  and sex-linked characteristics, 382, 400
  and sickle-cell anemia, 33, **390**
  Tay-Sachs, 389
Recessive traits, 265–268, 274, 279
Recipient (F⁻) cell, in conjugation, 327–328, 338
Recipients, of social actions, 1059, 1078
Reciprocal altruism, **1074–1075**, 1079
Recognition, 894, 1075, 1076–1077, 1079
"Recognition" memory, 895, 899
Recognition sequence, 341, 344, 345, 353, 374
  and RFLPs, 400
Recognition site, 344–345, 347, 374
  of lambda phage, 335
  in sickle cell anemia, 395
Recolonization, 1124–1125, 1128–1129, 1130
  and ecological succession, 1124–1125, 1128–1129
  and landscape structure, 1126–1127
Recombinant DNA, 319, **340–354**, 376–379, 382, 400
  and AIDS, 342, 817
  in Alzheimer's research, 897
  and cancer research, 342, 348, 376–379
  cleaving DNA, **340–341**, 346, 348, 353
  cloning, **343–345**, 346, 351, 353
  complementary base-pairing in, 353
  condensation reactions in, 343
  COS regions, 345, 353
  cytosine, 341
  denaturation, 345, 347, 354, 363–364
  and diagnosis of genetic diseases, 400
  E. coli in, 340–341, 343, 350–353
  economic importance of, 352
  electrophoresis, **340**, 344, 347–349, 353–354
  eukaryotic cells in, 342
  evolution and, 354
  and Factor VIII, 752
  future of, 399
  in hemophilia treatment, 392
  hybridization, **340**, **345**, 346–347
  and interferon synthesis, 793
  isolation, 346, 348
  lac operon in, 351
  lambda phage in, 341, **344–345**
  messenger RNA in, 342, 345–347, 351
  nucleic acid base-pairing in, 245
  plasmids in, **344**, 350–354
  prokaryotes in, 343
  radioactive labeling in, 348–349
  radioactive probes in, **346–347**, 354
  random excision, 348
  reading frames in, 350
  restriction fragments in, 346–349
  retroviruses in, **342**, 353
  RNA in, **340–353**
  sequencing DNA, 340, 346, **348–349**, 353–354
  somatostatin, synthesis of, **350–351**
  somatotropin, synthesis of, 351–352, 354
  "sticky ends" in, 340–342, 344–345, 353
  Ti plasmid in, **352–353**, 354
  and vaccines, 796–797
  vectors in, **343–345**, 346–347, 352–354
  viruses in, 340–346, 352–353, 451
Recombination, genetic, 331, 337–338, 519, 526, 563
  of alleles, **274–275**, 276–277, 279
  chromosomal insertions, 331–333, 338
  conjugation, **326–330**, 331, 337
  "errors," 380
  and evolution, 326, 337
  frequency of, 276, 277
  homologous recombination, 336–338
  infection cycle and, 336
  in rolling-circle replication of DNA, **327–328**, 338
  introns, 363
  and lambda phage, 337
  and plasmids, 337
  in prokaryotes, 437
  and prophages, **331–334**, 338
  reverse integration, 337
  and sexual reproduction, 994 (*see also* Sexual reproduction)
  "single-strand switch," **337**
  in sporulation, 463–464
  strategies for, **337–338**
  transduction, **332–334**, 336–338
  transformation, 282, 329, 331, 336, 376

  and transposons, **335–337**, 379
  viruses and, 336
Rectum, 715
Recuperation (of populations), 1103–1104
Red algae, 455–457, **465**, 477
Red blood cells, 114, 135, 742, 749, **750**, 751–753, 763
  and actin filaments, 114
  antigens on, 810–811, 819
  carbonic anhydrase in, 743
  and hemoglobin, 168, 747
  number, 147
  origin of, 792
  replication of, 147
  in respiration, 747
Red fibers, 878
Redi, Francesco, 86
Red light, 686–687, 695
Redox (*see* Oxidation-reduction reactions)
"Red Queen" hypothesis, 987
Red rat snake, 161
Red snow, 461
Red tides, 460
Reduction, **165–167**, 183
  of carbon dioxide, in photosynthesis, 219, 222, 229
  of glucose, 186
"Reduction division," in meiosis, 250
Reductionists, 11
Redwoods, 505
Reef fish, 186
Reefs, 1139
Reflex arcs, **709**, 844, 905
Reflexes, 762, 882, 947, 948, 951
Regeneration, 538, 572, 588
Regular dispersion, 1093–1094
Regulation, genetic, 355, **359–362**, 380
  and cancer, 375–377
  of embryonic development, **942–943**
  in eukaryotes, **359–362**
  methylation in, **360–361**, 380
  negative regulation, 322, 325, 337
  positive regulation, 322, 325, 337
  precision of, 326
  in prokaryotes, 320–323, 337
  and replication, 358, 380
  and sex determination in embryo, 946
Regulation of catalysis, 175–178
Regulation of cell cycle, 146, 148, 155
Regulation of cell transport, 142–143
  acetylcholine and, 179–180
Regulation of enzymes, 175–178, 184
  allosteric interactions, **176–177**, 178–181
  competitive inhibition, **177**, 178, 184
  irreversible inhibition, **177–178**, 184
  noncompetitive inhibition, **177–178**, 184
  pH and, 175–176
  receptors and, 179–181
  reversible inhibition, 176–178
Regulation of exchange
  in cell, 127
  in prokaryotes, 127
Regulation of glycolysis, 188
Regulatory gene, 351–353
  in eukaryotes, 378–379
  in prokaryotes, 322–324, 337
Regulatory proteins, 70, 72
  in eukaryotes, 358, **360–361**, 373, 375–377, 380
  in prokaryotes, 322, 325
Regurgitation, 1052, 1063
Reindeer, 1151–1152, 1169
Rejection, 810, 819
Relations, and reproductive success, **1065–1070**
Relative concentration, and cells, 127–129
Relativity theory, 163
Relay loops, and feedback control, 713
Relay neurons, 708, **709**, 843, 844
Release factor, 313
Releasers (behavioral), 1054, 1077
Releasing hormones, 839
REM sleep, 890
Renal tubule, **770**, 771, **772–774**, 776
Rennin, 352
Reorganization of cytoplasm, in embryo, 926–927
Repeats
  direct, 335, 338, 374
  long terminal, 373
  inverted, 335–336, 338, 374
Repetition, of DNA, 355, 360, **363–365**, 380

Replica plating, 964–965
Replication, **281–300**, 358, 380
 of AIDS virus, 815
 asynchronous, 326
 of bacterial chromosome, **328–329**
 bidirectional, 320
 of centriole, 152
 of chloroplasts, 453
 of chromosome, 155, **319–320**
  in meiosis, 253
 of DNA, 144–148, 155, **281–300**, 320
 "eye," 326
 forks, 294–297, 299, 320, 358
 and gene regulation, 358, 380
 of lambda phage, 334–335
 of linked DNA, 333
 of mitochondria, 453
 nerve cells, 147
 Okazaki framents in, 320, 358
 origin, 320, 358
 of plasmids, 326
 prokaryotic, 320–322, 325, 358
 rate of, in eukaryotes, 358
 of red blood cells, 147
 rolling-circle, **327–328**, 338
 semiconservative, 309, 358
 synchronous, 326
 theta, 320
 of viral DNA and RNA, 451
 of viroids, 446
 of viruses, 330–331, 445–446
Replication bubble, 294–295, 297, 326
Replication forks, 294–297, 299
 in eukaryotes, 358
 in prokaryotes, 320
Repressible enzyme, 322
Repressible operon, 323–325, 337
Reproduction, 586
 asexual, 259, 519, 523, 526, 528, 535, 537, 538,
  542, 562, 563, 588, 648, **1102–1103** (see also
  Asexual reproduction)
 in bryophytes, 499, 500
 in cell, 88
 in colonial organisms, 461–462
 early vs. late, **1100**
 in *E. coli*, 327
 enzymes and, 329
 in female, **907–915**
 in flower, 238
 in fungi, 481
 in honey bees, 1061
 and life-history patterns, **1099–1104**
 and living systems, 11, 26, 84
 in male, **901–907**
 patterns of, 519, 541
 in plants, 671
 in plasmids, 326, 329, 336
 and population growth, **1099–1104**
 in prokaryotes, **437–438**, 450
 in protists, 476, 477
 in protozoa, 469–470
 and selfish gene, 1071
 sexual, 519, 523, 526, 528, 531, 535, 538, 539, 542,
  563 (see also Sexual Reproduction)
 in *Shigella*, 329
 stability in, and evolution, 962
 transposons and, 336
 vegetative, **642–643**
 in viruses, 329
Reproductive isolation, 1010, **1013, 1016,** 1020
Reproductive organs, 546, 567 (see also Gonads)
Reproductive potential, 1088, 1104
Reproductive span, 1092, 1096, 1100
Reproductive system (see also Reproduction)
 development of, in embryo, 936
 human, **901–916**
Reptiles, 416, 418, 587, **596–598,** 600, 604, 775, 884,
 885
 cladogenesis in, 1022
 cloaca of, 904
 eggs of, 900
 evolution of, 505
 and evolution of mammals, 1031, 1050
 lungs of, 736
 temperature regulation in, 781–782
 vomeronasal organ, 861
Resistance, 964–965, 973
Resistance gene, 352

Resolution
 of human eye, 94
 of microscopes, 95–97, 99
 power of, 101
Resorption lacunae, 831
Resource partitioning, **1109–1111,** 1113
Resources, 1106–1108
 competition for, 1111
 and dispersion of populations, 1094
 division of, in societies, 1058
 and dominance hierarchies, 1062
 and facilitation hypothesis, 1125
 and limitations on populations, 1088, 1090, 1095,
  1105
 and mating behavior, 1072, 1078
Respiration, cellular (see Cellular respiration)
Respiration, in organisms, 209, 226, **732–748,** 882,
 890 (see also Respiratory system)
 in arthropods, 568, 734
 and circadian rhythms, 834
 control of, 745–747
 diffusion and air pressure in, **733**
 in earthworms, 556
 and energy loss, 1140, 1145
 evolution of, **734–737**
 gas exchange in, **741–745**
 hemoglobin in, 741–743
 in human beings, **738–740**
 in lampreys, 593
 mechanics of, 741
 principles of, in large mammals, 737
 rate of, 748
 regulation of, in brainstem, 898
 water loss in, 775
Respiratory neurons, 745, 748
Respiratory pigments, **741–745** (see also Hemoglobin;
 Myoglobin)
Respiratory system, 574–575, **732–748** (see also
 Respiration)
 development of, in embryo, 937
 vs. digestive system, 718
 evolution of, **734–737**
 gas exchange in, **741–745**
 gills, 734–737
 human, 738–740
 in large mammals, **737**
 of onychophorans, 561
 tracheal, 568, 574, 734, 737
Response to stimulus (see also Behavior; Nervous
 System; Sensory Neurons)
 in bacteria, 131
 integration and control of, 711–712
 in plants, 671, 694–695
 and vertebrate nervous system, 708–709
Resting forms in prokaryotes, 437–438
Resting potential, 848, 850, 856
Restricted transduction (see Specialized transduction)
Restriction enzymes, **340–341,** 344–350, 353–354,
 365, 372, 1050
 in Alzheimer's disease research, 899
 in "DNA fingerprinting," 396
 in genetic diagnosis, 395–397
Restriction-fragment-length-polymorphisms,
 394–397
Restriction fragments, 346–349
Reticular activating system, 884, **885,** 898
Reticular fibers, 704
Reticulum cell, 794, 816
Retina, 866, 867, **868–872,** 869
 development in chick embryo, **936**
 evolution of, in primates, 1033
 information processing for, **869–879**
 nerve impulse in, 212
 structure, 868
 and visual cortex, **887–888**
Retinal, 212, 221, 871, 872
Retinoic acid, 940
Retinular cells, 579
Retractile filaments, 522
Retroviruses, 373–377, 381, 445
 and AIDS, 814, 819
 and cancer, 376–377
 and messenger RNA, 374
 in recombinant DNA, **342,** 353
Reverse transcriptase, **340–342,** 353, 373–375
 and movable genetic elements, 381
Reverse transcription
 in AIDS, 445, 814–815, 817

 and cancer, 376, 445
 in transposons, 374
Reversible inhibition, 176–178
RFLP(s), **394–397**
R groups, of amino acids, 171
Rhabdom, 579
Rh disease, 811–812
Rhea darwinii, 409
Rheumatic fever, 448
Rheumatoid arthritis, 812, 836
Rh factor, 811–812, 819 (see also Blood groups)
Rhinencephalon, 883, 885
Rhinoceros, 77, 1110
Rhinovirus, 445, 450
Rhizobia, 430, **664–668,** 670
Rhizoids, 251, 467, 482, 483, 498
Rhizomes, 259, 502, 642–643, 648, 1014
Rhizopus, 481, 482
Rhodophyta, 456, 457, **465,** 477
Rhodopsin, 871, 872
Rhynia major, 501
Rhythm, 583
Rhythm method, 914, 916
Ribbon worms, 520, 532, **538,** 542
Rib cage, 741
Riboflavin, 195, 729
Ribonucleic acid (see RNA)
Ribonucleotides, 306–307
 assembly of, 368
Ribose, 55, 59–60, 80, 296, 305
 and ATP, 180, 184
 and NAD, 172–173
 and protein biosynthesis, 172
Ribosomal RNA, **310–311,** 316–317, 420
 genetic code for, 362, 365, 367, 380
 and nucleolus, 366
 and plant cell growth, 675
 and prokaryotic classification, 430
 transcription, 368
Ribosomal subunit, 108
 in animal cell, **112**
Ribosome, 91, 108, **115,** 117, **118,** 124, 258, 305,
 313–314, 317, 1126
 in animal cell, **112**
 assembly, and nuclear pores, 366
 in cell division, 146
 and cell variety, 125
 construction of, 108
 of *E. coli,* 310–311
 and evolution, 969
 in messenger RNA binding, 320, 368, 369
 of mitochondria, 452
 in plant cell, **113**
 of prokaryotes, 431
 ribonucleotide assembly, 368
 ribosomal proteins, 366
 RNA, 366
 subunits, 108, **112,** 311–313, 366
Ribulose, 60
Ribulose biphosphate (see RuBP)
Rice, 76, 226, 441, 622
Rickets, 729
Rickettsiae, 437
Rift Valley, 1039
Right-handedness, 888
Right hemisphere, **888–891,** 898
Rigor mortis, 875
Ring structure, 60
Ring-tailed lemurs, 1034
Ringworm, 481, 488
Ripening
 of fruit, and ethylene, 679, 694 (see also Ethylene;
  Fruits; Hormones)
 of ovarian follicle, 917
Ritualized behavior, 1062, 1063, 1065, 1072 (see also
 Behavior; Courtship; Dominance Hierarchies;
 Mating)
River blindness, 540
Rivers, as ecosystem, 1177
RNA, 80–82, 305, 307, 310–311, 317 (see also spe-
 cific types)
 and adenine, 172
 and assembly of proteins, 305–306, 308
 catalyst, 370
 in cell nucleus, 108
 complementarity to DNA. 306
 direction of transcription of, 306, 310, 313
 vs. DNA, 305–306

I-47

RNA (Continued)
  in DNA replication, 296, 299
  evolution of, 370
  function, 80
  and genetic code, 380
  and germ cell determination in embryo, 925, 952
  as messenger, **306-307**, 308
  mitochondrial, 316
  mRNA (see Messenger RNA)
  origin of, 87
  and origin of life, 88, 370
  in ovum, 921
  and phosphorus cycle, 1146
  in protein biosynthesis (see Messenger RNA)
  radioactive, 308
  in recombinant DNA, **340-354**
  retroviruses, 373-377
  role of, **305-307**
  rRNA (see Ribosomal RNA)
  structure of, 305
  transcription of, 306-307, 309, 310, 313, 317
  transcripts, 380
  tRNA (see Transfer RNA)
  varieties of, 306
  viral, as mRNA, 443-445
  viroids, 445
  viruses, 338, 373
RNA catalyst, 370
RNA polymerase, 306, 314
  and catalysis of RNA transcription, 307, 317
  in eukaryotes, 380
  in gene transcription, 368
  and operon, 323-325
  in prokaryotes, 320, 337
  and viral RNA, 373
RNA precursor, 365
RNA primase, 296-297, 299
RNA viruses, 338, 350
  and mutations, 373
Robin, 409
Rock, 85
  age of, and evolution, 967, **970-971**
  and soil characteristics, 663, 670
Rocky seashores, 1157, 1158
Rod, of flagellum, 434
Rodents, 602, 717, 1031
Rods (vision cells) 866, 868-872, 936
  development of in chick embryo, 936
  evolution of in primates, 1033
Roeder, Kenneth, 1116
Roles, ecological, 424, 1106
Rolling-circle replication of DNA, **327-328**, 338
Root, of tooth, 716
Root cap, 634, 635, 647-648
  gravitropism in, 683-684
Root cells, 665
Root core cells, 683-684
Root hairs, **632**, 634, 650
  and symbiotic nitrogen fixation, **666-667**
  and transpiration, 652
Root pressure, 651
Roots, 120, 674, 694-695
  adventitious, and auxin, 694-695
  branch roots, 634, 635, 648
  and cell types in angiosperms, 647-648
  and dormancy, 621
  and fungi (see also Fungi; Mycorrhizae; Nitrogen fixation; Symbioses)
  and gravitropism, 683-684
  growth of, 683, 684, 695
  and mycorrhizae, 1121
  and nitrogen cycle, 1147, 1150
  and nitrogen-fixing prokaryotes, 430, 1121
  primary growth, **634**
  prop roots, 636
  secondary growth of, 645
  and soil, 631
  vs. stems and leaves, 626
  and symbiotic nitrogen fixation, **666-667**
  and synthesis of cytokinins, 677
  taproots, **635**
  water uptake in, 655
Root tip, 634, 635, 640
Roquefort cheese, 488
Rosettes, 621, 681
Rotation of earth, 1156
  and distribution of life, 1160
  and ocean currents, 1156

and weather cycles, 1136-1137
Rothenbuhler, Walter, 1053
Rotifers, 520, 540, 541, 543
Rough endoplasmic reticulum, 124
Round window membrane, 864, 865
Roundworms, 122, 489, 539, 540 (see also Nematodes)
Rous sarcoma virus, 376
R (drug resistance) plasmid, 326, 329, 336
R point of cell cycle, 148
rRNA (see Ribosomal RNA)
r-selected life-history patterns, 1099
Rubella (German measles), 947
Rubin, Gerald M., 379
RuBP, 222-226, 229-230
RuBP carboxylase, 222, 224-226
Ruddle, Frank H., 378
Ruminants, 439, 724
Runners, 641, 648, 1003, 1102
Running water, as ecosystem, **1154-1155**, 1177
Rusts, 481, 488
Rye, 486, 622

Sabin vaccine, 449, 796
Sagebrush, 1172
Saguaro cactus, 1172
Sahara desert, 1172
St. Anthony's fire, 486
Salamanders, 40, 109, 364, 595
Saliva, 703, 713, 785
  composition of, **717-718**
  in digestion, 717-718
  in herbivores, 724
  in immune response, 791
Salivary amylase, 718, 722
Salivary glands, 715, 717-718, 722, 952
Salk vaccine, 449, 796
Salmon, 594
Salmonella
  motility of, 131
Salt (see also Sodium; Sodium Chloride)
  dietary, 730
  excretion of, in skin, 769
  and kidney disease, 769
Salt glands, 659
Salt marshes, 1159-1160
Saltmarsh song sparrows, 1092
Sami (Lapps), 1152
Sand, 664
Sand dollars, 587, **589-599**, 604
Sanders, Andrew, 643
Sanger, Frederick, 349
Sap, 129
Saprobes, 467
  fungi as, 480, 482, 488
  prokaryotic, 438-439
Sapwood, 645
Sarcodines, 455, 460, 468, 469, **470**, 477
Sarcolemma, 873, 876-878
Sarcomeres, 122, 874, 875
Sarcoplasmic reticulum, 873, 876
Saturated fat(s), 82, 761
Saturated fatty acids, 67, 71
Savanna, 1169, **1170**, 1177
Scala naturae, 1, 4-5
Scale insects, 964
Scales (animal), 594, 596, 598, 702
Scales (plant), 506, 629
Scallops, 549
Scanning electron microscope, 95-96, 97, 98, 101
Scaphopoda (see Tusk shells)
Scarlet cup, 485
Scarlet fever, 448
Scavengers, 550, 1042, 1142
Scent marking, 906, 1064
Schally, Andrew, 827
Schistosoma, 537
Schistosomiasis, **537**, 797
Schizocoelus formation, 544, 545, 587
Schleiden, Matthias, 10
Schoener, Thomas, 1113
Schwann, Theodor, 10
Schwann cells, 708, 842, 843, 877, 935 (see also Neuroglia; Neurons)
Science, nature of, **14-18**
Scientific method, 15-16, 18
Sclera, 866

Sclereids, **637**, 647-648
Sclerenchyma cells, **636-637**, 648
Sclerospongiae, 523
Sclerotome cells, 936
Scorpions, 566, 570, 585
Scotch pine, 507
Scrapie, 447, 451
Scrotum, 901, 903, 904, 909
Scurvy, 729
Scutellum, 619, 631
Scyphozoans, 526, 528-529
Sea, 1156, 1177
Sea anemones, 524, 526, 529, 542
Sea cucumbers, 587, 590, 604
Sea hare (*Aplysia*), **895-898**
Sea lettuce, 463-464
Sea levels, 1177
Sea lilies, 587, 589, 604
Seals, 69, 733, **744**
Sea mat, 563
Seashore, **1157-1160**, 1177
Seasons, 45, 1112
  adaptations to, in plants, 621-624, 629, 790
  and birdsongs, 889
  and bolting and flowering, 681
  circadian rhythms in, 690
  and dispersion of population, 1094
  and earth's rotation, 1136, 1152
  and latitude, 1161, 1177
  and male hormone regulation, 907
  and photoperiodism, **684-688**
  and plant growth, 646
  and reproductive isolation, 1016
  and woody plants, 675
Sea spiders, 566, 569, 585
Sea squirts, 591
Sea urchins, 37, 109, 263, 587, **589-590**, 604, 919, **920-926**
Sea walnuts, 524, 531, 542
Seaweeds, 456, 462, 465, 477
Sebaceous glands, 710
Secondary cell wall, 106, 107, 637, 639
Secondary consumers, 1087, 1140-1141, 1153, 1164
Secondary growth, in plants, **644-648**, 675
Secondary induction, 936
Secondary infections, 1090
Secondary oocyte, 910, 911
Secondary phloem, 644, 647-648 (see also Phloem)
  and auxins, 675
  and gibberellins, 681
Secondary plant substances, 515
Secondary sex characteristics, **905-906**, 915, 917
Secondary spermatocytes, **901**, 902, 917
Secondary structure, of proteins, **74-75**, 79
Secondary substances, and coevolution, 1003-1005
Secondary tissues, **644-648**
Secondary xylem, 644, 646, 647, 648 (see also Xylem)
  and auxins, 675, 681
  and gibberellins, 681
Second filial generation (see F_2 generation)
"Second messengers," 837, 839, 854, 857, 871, 897-898
  and learning, 897-898
  and neuromodulators, 855, 857
"Second wind," 191
Secretin, 722, **723**, 823
Sectioning, 98
Sedimentation, 311
Seed coats, 619, 622, **623-624**, 630, 632, 634, 647, 681
Seedlings, 506, 695
  adaptation in, 687
  in angiosperms, 512
  development of, and phytochromes, 687
  dicot vs. monocot, 631
  differentation in, 634
  and gibberellins, 681-682
Seed plants, 503, 504 (see also Angiosperms; Gymnosperms)
Seeds, 695
  and alternation of generations, 507
  of angiosperms, 498, 613, **619-621**, 624
  and auxin, 675, 695
  and beak size in Darwin's finches, 1021
  coat of (see Seed coats)
  dispersal, and life-history patterns, 1102
  dormancy in, 621, **623-624**, 680, 695
  evolution of, 502

germination of, 630–631, 647–648 (*see also* Germination)
and gibberellins, 681–682
of *Ginkgo*, 505
of gymnosperms, 497–498, 506, **507–508**
and phytochromes, **687**, 695
of pine, 508
vs. runners in reproduction, 1102
and sclereids, 637
size, selection for, **1102**
vs. spores, 507
toxic, 515
transportation of, 515
and vegetative reproduction, 642–643
vs. zygotes, 507
Segmental pattern of development, 942
Segmentation, 562, 565, 566, 585
Segmentation genes, 942
Segments, 571, 572
Segregation, principle of, **239–241**, 245, 247, 261, 274
consequences of, **240–241**
and meiosis, 261
Seismosaurus, 597
Selection (*see also* Natural selection)
experiments, 989 (*see* Artificial selection)
of habitat, 586
of T lymphocytes, 804–805
Selectionism, 978
Selection processes, **994–1001**
directional, 944, **997**, 1009
disruptive, 994, 996–997, 1008–1009
stabilizing, **995**, 1008
Selective blocking, 343
Selective breeding, 236
Selective immune suppression, 945
Selective permeability, 127–128, 132–133, 142
of cell membrane, 112, 124, 127–128, 132–133, 142
and cellular respiration, 191, 201
and chemiosmotic coupling, 197
at loop of Henle, 772, 773
of mitochondrial membrane, 191, 201
Selectivity of proteins, 135, 136
"Self" antigens, 819
Selfish behavior, 1059, 1078
Selfish DNA, 335 (*see also* Selfish gene)
Selfish gene, **1070–1074**, 1078
and human behavior, **1075–1076**
Self-pollination, 238–240, 242, 984
Self recognition, in T lymphocytes, **804–805**
Self-replication, 235
Self-splicing introns, and viroids, 446–447
Self-sterility alleles, 985, 990, 998
Semen, 236, 819, 903, **905**, 913, 917
prostaglandins in, 833, 835
DNA fingerprinting, 396
Semibalanus balanoides, 1142–1143
Semicircular canals, 854, 859, 863
Semiconservative replication, **292**, 293–294, 299, 309, 358
Semidesert, 1172
Semidesert scrub, 1139
Semidormancy, 1170
Seminal fluid, 237, 905
Seminal vesicles, 833, 903, **904**, 905, 917
Seminiferous tubules, **901–902**, 903, 917
Senescence, **694**
and abscission, 679
of floral parts, and ethylene, 679
of leaf, and cytokinins, 677
of leaf, and ethylese, **694**
Senility, 385
vs. Alzheimer's disease, 896
Sensilla, **582**, 583, 584 (*see also* Touch receptors)
Sensitive plant, **692**, 693, 695
Sensitization, 896
Sensors for touch, 691
Sensory cortex, **885–888**, 898
and intrinsic processing centers, 892
and memory pathways, 894–895
and perception of form, 887–889
Sensory fibers, 844
Sensory information, and memory, **892–893**, 899
Sensory nerve endings, 702, 710
Sensory neurons, 844, 849, 858, 859, 882, 886–887
development of, in chick embryo, 935
of ear, 887

in vertebrates, **708–709**, 712
Sensory organs, 528, 529, 542, 546, 568, 579, 585, 586, 588
Sensory perception, 579
human, **858–880**
and motor response, integration of, 879
Sensory receptors, 527, 528, 531, 542, 553, 554, 556, 564, 582, 649, 708–709, 844, 886
chemoreception, **859**, 860–862
in cnidocytes, 525
in complex animals, 518
in ear, 864, 865
electroreceptors, 859
exteroreceptors, 859–860
hearing, 864, 865
human, **858–873**, **879**
hunger, 859
in *Hydra*, 527
interoceptors, 859
magnetoreceptors, 859
mechanoreceptors, **859**, 863–865
and nausea, 859
pain receptors, 859
photoreceptors, 859, 866, 867
in planarians, 535
proprioceptors, 859
sensitivity of, 527
smell, 862
stretch, 859
taste bud, 717
temperature receptors, 859
thirst, 859
Sensory response
in bacteria, 131
vs. gravitropism, 683
and living systems, 36
and motor response, 131
Sepals, 511, 613, **614**, 624
Sepkowski, J. John, Jr., 1024
Septa, 481, 483, 485, 486, 491, 554 (*see also* Cross walls; Partitions)
Septate junctions, 703
Sequencing DNA, 340, 346, **348–349**, 353–354
Sequencing studies, 380
Sequential reactions, 167–180
Sequoia, giant, 14
Serine, 73, 131, 310
Serosa, **715–716**, 718
Serotonin, 853
Serpent stars (*see* Brittle stars)
Sertoli cells, **901–902**, 906–907
Serum, 752, 811
Sessile animals, 1156
barnacles, 571
bivalves, 549
echinoderms, 587, 589
mollusks, 545
polyps, 525
sponges, 522, 523
Seta(e), 548, 558, 560, 582 (*see also* Bristles)
7-Methylguanine (*see* Methylguanine)
Sex, determination of, **264**, 946
Sex characteristics, 824, **905–906**, 917
Sex chromosomes, 264–266, 279, 399
abnormalities in, **386–387**
extra, 399
number, 382, 399
segregation of, during meiosis, 278
Sex drives, and hypothalamus, 883
Sexes
separation of, 538–541, 543, 550, 551, 553, 558, 559, 560, 568, 571, 576, 588, 591
Sex factor plasmid (*see* F plasmid)
Sex hormones, 70, 712
female, 824, 832, **913–915**
and growth of vocal cords, 738
in interstitial cells, **901**
male, 824, 832, **906–907**
negative feedback in male, 917
Sex-linked characteristics, 400
color blindness, 391–392
hemophilia, 382, 392–393
muscular dystrophy, 393–394
Sex-linked genes, **264–266**, 268
Sexual behavior
competition and, **1071–1074** (*see also* Competition; Sexual Selection)
hormonal regulation of, 908

and hypothalamus, 883, 914
Sexual differences, in brain, **908–909**
Sexual differentiation, origins of, 463
Sexual dimorphism, **999–1000**, 1036
Sexually transmitted diseases, 814, 817
Sexual practices, and spread of AIDS, 819, 1090–1091
Sexual receptivity, 915
Sexual recombination, 327, 1049
Sexual reproduction, 238, **249–262**, 259, 261, 483, 485–487, **613–619**, 984, 990
in algae, 459, 460, 462, 463, 464, 465,
and alternation of generations, 252
vs. asexual, 984, 986, 987, 1102, 1105
conflicts of interest and, 1071–1074
consequences of, 259
energy cost of, 986, 998–999
extensiveness of, 259
in ferns, 502
in fungi, 481, 485–486, 491
and genetic variability, 259, 984, 990
and human evolution, 1049
and hybridization, 1015
inefficiency of, 259
meiosis and, **249–262**
in plants, **613–624**
and prostaglandins, **835**
in protozoa, 468–470
seasonal, 502
and sexual dimorphism, 1000
success of, 984, 986, 987
in vertebrates, **900–918**
in water molds, 467
in yeasts, 486
Sexual selection, **998–1001**, 1009, 1015
in bowerbirds, **1073**
and care of offspring, 999
and conflicts of interest, **1071–1074**
and directional selection, 1000
and dominance hierarchies, 1063
in frigate birds, 1015
and imprinting, 1056
and mating systems, 998–1001
and selfish gene, 1072–1074
Shadowing, 98
Shanidar cave, 1046
Sharks, 593, 767
Sheaths, 224
of bacteria, 436
of cyanobacteria, 440
in grass plants, 642
of virus, 445
Shelf fungus, 487
Shell membranes, 931
Shell patterns, 992–993
Shells, 545, 546, 548, 550, 551, 553, 562, 563
diatom, 459
internal, 551
of sarcodines, 470
secretion of by mantle, 563
of snail, 992
Shells, of eggs, 900–901, 931, 995
Shigella, drug-resistance in, 329
Shock, 760, 813
Shoot apex, 674
Shoots (*see also* Shoot system)
adaptations in, 642
and buds, 648
and cell types in angiosperms, 647–648
and gibberellins, 694–695
growth, and cytokinins, 694
primary growth of, **640–643**
vs, roots, 640
specialization in, 640–643, 648
Shoot system, 626, 630, 648
Shoot tip, 640–641
Short-day plants, 684–686, 695
Short-grass prairie, 1169
Short-term memory, **893–895**, 899
Shrews, 600
Shrimp, 571, 585
Shrubland, 1139
Shrub layer, 1164
Shrubs, 511
Sibley, Charles G., 420, 423
Siblings
bonds between, 1033, 1037
among chimpanzees, 1037

I-49

Siblings (Continued)
  and inclusive fitness, 1069–1070
  and kin selection, **1067–1069**, 1070
Sickle cell anemia, **78–80**, 82, 304, 314, 316, 388, 390, **394–395**, 400
  capillaries in, 78
  diagnosis of, 397
  hemoglobin and, 80, 340
  heterozygote superiority in, 988
  mutations and, 395
Sickle cell beta globin, 397
Sierra Nevada, 1137, 1172
Sieve cells, 501, 508, 509, 638
Sieve plates, **638**, 639
Sieve-tube members, 501, **638–639**, 647, 648, 662
Sieve-tube membrane, 662
Sieve tubes, **638**, 647–648, 660–663, 670
Sieve-tube sap, 638, 660
Sight, in primates, 1033 (see also Vision)
Signal sequence, 116
Sign stimuli, 1054
Silica, 523
Silicon
  in algae, 457, 459–460
  vs. carbon, 58
  in diatom shells, 459–460
  in horsetails, 502
Silk, 77, 570, 577
Silt, 664
Silurian period, 497
Silver-backed jackals, 1069
Simberloff, Daniel, 1123
Simian immunodeficiency virus (SIV), 817
Simian virus 40, 348, 373, 376, 377, 444
Simple diffusion, 135–136, 142
Simple epithelium, 702
Simple fruit, 620
Simple leaves, 626
Simple-sequence DNA, **364**, 380
Simple transposons, 335, 338
  as "selfish" DNA, 335
Simpson, George Gaylord, 973, 1022, 1027
Sims, Linda, 15
Singing, group, 583 (see also Songbirds; Song learning; Song nuclei)
Single bonds, atomic, 32, 34
Single-celled organisms (see also One-celled organisms; Prokaryotes; Protists)
  cell cycle in, 147
  cell division in, 144
Single-copy DNA, **366**, 380, 420
Single-plane motion in limbs, 1025
Single-species hypothesis of human evolution, 1043
Single-stranded binding proteins, 294, 297, 299
"Single-strand switch," 337
Sinoatrial node, **756**, 763
Siphon, 546, 553
Sirenin, 467
Sitka willow, 694
SIV, 817
6-Benzylamino purine (BAP), 676
Six-carbon sugars, 228
Skates, 593
Skeletal muscle, **707–708**, **712**, 858
  and blood flow, 762
  cerebral cortex, 886
  contraction of, **874–875**
  and exercise, 759
  and myoglobin, 745
  and somatic nervous system, 842, 845, 856
  structure of, **873–875**
Skeletal supports, 595, 604
Skeleton, 578
  bony, 594
  calcium, 590, 604
  cartilaginous, 593
  coral, 529, 530
  development of, in embryo, 924, 936, 946–948
  exterior (see Exoskeleton)
  growth, and male hormones, 905
  human, **704–708**
  hydrostatic, 539, 542, 544, 555, 563
  interior, 588, 604 (see also Endoskeleton)
  Lucy, 1041
  muscle of (see Skeletal muscle)
  Neanderthal, 1046
Skin, 709, **710**
  in amphibians, 754, 826

and camouflage, 826
color, 728
connective fibers in, 704, 710
desmosomes and tight junctions in, 703
development of, in embryo, 936
epidermis of, **703**, 710
human, 859–860
in immune response, 791, 819
and polysynaptic reflex arc, 844
in reptiles, 826
respiration through, 735, 736
sensitivity of, 860
and temperature regulation, 784, **786**, 787
and vitamin D, 728
water loss from, 775
Skin cancer, 1134
Skirts, 487
Skull plates, 593
Skulls, 705, 881, 889, 1042
  of apes, 1036
  of australopithecines, 1042–1045
  bones of, 601
  development of, in embryo, 937, 947
  of monkeys, 1034
  vertebrate, 701, 713
Skunk cabbage, 182
Sleep, **787–788**, 890–891 (see also REM sleep)
Sleeping sickness, 469
Sleep movements, in plants, 689–691
Sliding filament model of muscle contraction, 874
Slime, 436, 438, 450 (see also P-protein)
Slime molds, 121, 140, 466–467, 469, 472, 474, 476, 477 (see also Cellular slime molds; Plasmodial slime molds)
Sloth, three-toed, 602
Slow-conducting fibers, 757
Slow-twitch fibers, 878
Slugs, 545, 550, 551
Small intestine, 715, **720–725** (see also Duodenum)
Smallpox, 796–797
Smell, 861–863
  and aldehyde groups, 56
  evolution of, 860
  in human beings, 601
  and nocturnal animals, 860
  odors, discrimination of, 861
  olfactory epithelium, 861
  receptors for, 859, 861, 862
  sense of, in fish, 861
Smith, William, 3, 967
Smoking
  and birth defects, 947
  and heart disease, 761
  and lung cancer, 739
Smooth endoplasmic reticulum, 124
Smooth muscle, 706, **708**, 713, 716, 873
  in atherosclerosis, 764
  and autonomic nervous system, 712, 842, 845, 856
  and blood pressure regulation, 759, 764
  in blood vessels, 753
  in bronchioles, 740
  and cardiovascular regulation, 762
  contraction of, 875–877
  in digestive tract, 716
  in esophagus, 718
  and parasympathetic nervous system, 847
  postsynaptic, 877
  and prostaglandins, 835, 839
  in small intestine, 721
  in vas deferens, 917
Smuts, 481, 488
Snail fever (see Schistosomiasis)
Snails, 545, 546, 550, 551, 563
  as host of *Schistosoma*, 537
  land-dwelling, 550
  respiration in, 736–737
  shell color in, **992–993**
  terrestrial, 572
Snakes, 596, 597, 969
Snapdragons, 268, 515
Snow, 1136, 1149–1150, 1154
Snow algae, 461
Snow geese, 984
Snowshoe hare, 1118–1119
Social behavior, **1058–1059**, 1078 (see also Behavior; Communication; Courtship; Dominance; Mating; Ritualized behavior; Sexual behavior; Societies)

in bowerbirds, 1071–1073
donors and recipients on, 1059
in hominids, **1036–1038**
human, **1075–1076**
in insects, **1059–1061**
kin selection, **1065–1070**
and population dispersion, 1094
reciprocal altruism, **1074–1075**
and selfish gene, **1070–1074**
stages of, 1059–1060
territoriality, **1063–1065**
types of, **1059**
waiting, **1074**
Social Darwinism, 975
Social dominance, 907
  and breeding population, 1062–1063
  in chickens, 1062
  dominance hierarchies, 1036, **1062–1063**
  evolution of, 1044
  and mating behavior, 1072
  ritualized behavior in, 1062
  and selfish gene theory, 1072, 1074
  and waiting behavior, **1074**
  in wolves, 1063
Social evolution, 1059–1060
Societies, 1078
  behavior in, **1058–1059**
  communication in, 1058
  dominance hierarchies (see Dominance hierarchies)
  and homozygosity, 983
  and human behavior, **1075–1076**
  imitative learning in, 1057–1058
  in insects, **1059–1061**
  kin selection in, 1065–1070
  and mating, 999 (see Mating)
  and outbreeding, 985
  and selfish gene, 1070–1074
  stages of socialization in, **1059–1060**
  territoriality (see Territoriality)
  vertebrate, **1062–1065**
  waiting behavior in, **1074**
Sodium bicarbonate, 718, 722
Sodium chloride, 775
Sodium hydroxide, 47
Sodium ion (Na^+), **31**, 766
  and action potentials, 650, 849, 851
  and aldosterone, 775–776
  in biogeochemical cycles, 1145
  excretion, and circadian rhythms, 834
  in fertilization of ovum, 921
  hormonal regulation of, 832
  in loop of Henle, 772–773
  nutritional requirements for, **656**, 657, 658, 728
  transport of, 134, 136
  in urine formation, 772
Sodium-potassium pump, **136**, 137, 143
  of axon membrane, 850, 851
  electrical impulses at, 136
  in halophytes, 658
  and mitochondrion, 136
Soil, 663–664 (see also Decomposition; Detritivores; Fungi; Nitrogen cycle; Saprobes)
  decomposition of, 429
  layers of, 663–664
  loss of nitrates from, 1148
  nitrogen in, **1147–1150**, 1153
  and root growth, 631
  of temperate forest ecosystems, 1166
  of tropical rainforests, 1176
Solar constant, 1132
Solar energy, 1131, 1136, 1152 (see also Light; Photosynthesis; Sun)
  in ecosystems, **1132–1137**
  vs. fossil fuels, **1144**
  and water cycle, 1133
Solar plexus, 846
Solar system, 7, 84–85
Soldier termites, 582
Solenogasters, 548
Solutes, **44**, 46
  concentration and, 128, 132, 314
  movement of, 128
  and osmotic receptors, 774
  random distribution of, 129
  regulation of, in body, **765–776**
  transport molecules and, 137
  in urine formation, 771–774, 775
  (see also Diffusion; Osmosis)

Solutions, **44**, 128
  concentration of, 128, 132
  (*see also* Concentration; Diffusion; Liquids; Osmosis; Water)
Solvent(s), **44**
  water as, 44, 54
Somatic cell, 102, 145, 267, 286, 299, 315
  chromosome number of, 249
Somatic nervous system, **712**, 842, **845**, 856, 873
  vs. autonomic nervous system, 845, 846
Somatostatin, **350–351**, 823, 824, 827, 832, 833, 838, 839
  and glucose regulation in blood, 726
  synthesis of, in pancreas, 725
Somatotropin, 824, 825, 827, 839
  synthesis of, 351–352, 354, 378–379
Somites, 929, 932, 936, **937**, 946, 952
Songbirds, 889, 908
Song learning, 1056–1057, 1078
Song nuclei, 889
Sonogram, 387
Sonoran desert, 1172
Sorghum, 226
Sori, 503
Sound, 584
  discrimination of, 865 (*see also* Auditory Cortex; Hearing)
  frequencies of, 865
  perception of, 864
  processing of, in brain (*see* Auditory Cortex; Hearing)
Sound receptors, 584
Sound spectrography, 886, 1056–1057
Source-to-sink translocation, **662–663**, 670
Sousa, Wayne, 1124–1125, 1127
Southern, E. M., 347
Southern blotting, **347**
Southern Hemisphere, 1136, 1156–1157
Sowbug, 209 (*see also* Pillbugs)
Soybean, 120
Space-filling model of molecular structure, 41, 62
Spacer sequences, 365–366
Spallanzani, Lazzaro, 1115
Spatial organization, in brain, 887–889, 890–891, 892, 898 (*see also* Parietal lobe; Right brain)
Special creation, 1, 2, 6, 9, **965–966**, 967, 970
Specialization, 586
  and adaptation, 1006
  of cells, 101
  among decomposers, 1145
  in embryonic development, 929
  in plants, **626**, 640–643, 647–648, 676–678
  prokaryotes, 439
  of tissues, 532
  of vertebrate cells, 702–709
Specialized transduction, **333–334**, 338
Speciation, 408, 996, 1029 (*see also* Species)
  and adaptation, **970–972**
  adaptive radiation and, **1022–1023**
  allopatric, **1010–1012**
  and biogeography, **966**, 967
  cladogenesis, **1021–1022**
  and competition, **1106–1114**
  and continental drift, **1012–1014**
  Darwin's finches, **1017–1021**
  disruptive selection and, **1014–1015**
  vs. divergent evolution, **996**, **1007**, 1009
  and ecotypes, 1003
  and evolutionary theory, 962, 965
  and extinction, **1024–1025**
  in the fossil record, 966–968, 1021–1025
  genetic isolation and, **1015–1016**
  and geologic isolation, 1010, **1012–1014**
  on "islands," 1010–1011, 1015 (*see* Island biogeography model)
  macroevolution and, **965–972**
  microevolution and, **962–965**
  modes of, **1010–1015**
  vs. mutation, 1028–1029
  origin of, **1010–1030**
  phyletic change, **1021**
  polyploidy, **1012–1014**
  and predation, **1114–1119**
  and punctuated equilibria, **1028–1029**
  and reproductive isolation, **1012–1014**
  sympatric, **1012–1015**
Species, 2–10, **407–410**, 412, 424, 426, **965**, 973 (*see also* Natural Selection; Speciation)

adaptation between, **1003–1005**
animal, 408
clusters of, 1014
in communities, **1106–1130**
and competition, **1106–1114**
continuity of, 900
in cultivated areas, 1148
defined, **407–410**
divergence in, 1111
elimination of, 1108
eusocial, 1060
and evolution, **1–10**
formation of (*see* Speciation)
and fragmentation of habitat, 1126–1127
fugitive, **1089**
and geologic isolation, 1010, **1012–1014**
and island biogeography hyothesis, 1123–1124, 1126–1127
macroevolution in, **965–972**
microevolution in, **962–965**
modes of speication, **1010–1015**
naming, **408–410**
and natural selection, 7, 9, 12
origin of, **1010–1030**
and predation, 1106, **1119**
presocial and subsocial, 1060
vs. race and subspecies, 1014
and reproductive isolation, 1012–1014
speciation (*see* Speciation)
Specific heat, 43, 54
Specimen, microscopic, 95–99, 101
  freeze-fracture technique, 135
  preparation of, 96–99
  viewing, 95–96
Spectrin, **114**
Spectrophotometry, 210, 687
Spectrum, visible, 872
Speech, 738 (*see also* Broca's area; Communication; Wernicke's area)
  and cerebral cortex, 888
  and left brain, 898, 890
  speech centers, 888, 890, 898
  and split brain, 898
Spemann, Hans, 927, 930
Sperm cells, 237–238, 241, 247, 251, 252, 254, 258, 264, 519, 524, 535, 900, **901**
  of angiosperms, 615
  biflagellate, 499
  and contraceptive techniques, 916
  and DNA fingerprinting, 396
  in fertilization, **911–912**
  early views of, 237
  in flowers, 511
  formation of, 258
  genetic variation and, 259
  guinea pig, 108
  and hormones, 905, 906
  human, 122, **901–907**, 916
  life expectancy, **912**
  multiflagellate, 503
  and non-Mendelian inheritance, 362
  nonmotile, 506
  nucleus of, 250
  number of, in human ejaculate, 258
  origin of, 122
  and ovum, 920, 926
  pathway for, **903–905**
  sea urchin, 107
  structure of, **902–903**
Spermathecas, 557
Spermatic cords, 904
Spermatids, 258, **901–903**, 917
Spermatocytes, 294, **901**, 917
  primary, 258
  secondary, 258
Spermatogenesis, 824, **901–903**
  differentiation of spermatids in, **902–903**
  and germ cells, 924
  hormonal regulation of, 907
Spermatogenic cells, 902
Spermatogonia, 258, **901**, 902, 917, **924–925**, 946
Spermatophores, 553
Spermatozoa (*see* Sperm cells)
Spermicidal jelly, 916–917
Spermists, 237
Sperm nucleus, 617, 624
Sperry, Roger, 889, 891
Sphagnum, 498

S phase, of cell cycle, 155
  and biosynthesis of histiones, 356
  in cell division, **146**
  and DNA replication, 294
Sphenophyta, 497, 498, 501, 502,
Sphincters, 716, 720, 847
Spicules, 523, 548, 924
Spider monkeys, 1035
Spiders, 566, 568, 585
Spinal column, 701, 713 (*see* Vertebral column)
Spinal cord, 708–709, 712, 881
  and brainstem, 881, 882, 898
  and control of respiration, 745, 748
  development of, in embryo, 881, 928, 935
  human, 873
  meninges, 884
  and opiate receptors, 854
  and respiration, 745, 748
  in vertebrates, 842, 843, 844
  white matter of, 881
Spindle, cellular, 123, 148, **149**, 150–155
  ATPases in, 181
  fibers, 149–155, 254–256
  formation of, 125, 149
  in meiosis, 256
  microtubules, 153, 254, 256
  in karyotyping, 384
  polar fibers of, 149–151, 153
  pole of, 149
Spindle fibers, 472
  in meiosis, 256
  in mitosis, 472
  in slime molds, 472
Spines, 563, 588, 629–630
Spinnerets, 570
Spiny-headed worms, 520, 540, 542
Spiracles, 568, 574, 575, 734, 964
Spirilla, 429, 436
Spirochetes, 430, **437**, 450
Spirogyra, 462, 463
Spiteful behavior, 1059, 1078
Spleen
  formation of red blood cells in, 751
  and immune response, 793–794, 795, 803
Splicing of exons, 369–370
Split brain, **889–891**
"Splitters," taxonomists, 412
Sponges, 520, **522–524**, 542
Spongin, 523
Spongy bone, 705
Spongy parenchyma, 627, 648
Spontaneous generation, **86–87**, 236
Spontaneous mutation, 982
Spoon worms, 520, 559, 564
Sporangia, 251, 466
  in black bread mold, 483
  of ferns, 503
  fungal, 481, 482, 491
  of *Pilobolus*, 484
  of water molds, 467, 468
Sporangiophores, 468, 481–483, 484
Spore coat, 437–438
Spores, 250–252, 256, 462, 477
  in algae, 463
  asexual, 481, 485
  of basidiomycetes, 486
  of black bread mold, 483
  case, 302 (*see also* Ascus)
  conidia, 485
  diploidy in, 463
  in ferns, 503
  formation of, 256, 429, 437–438, 450
  fungal, 481, 491
  genetic recombination in, 463–464
  germination of, 462, 463
  in plants, 462
  vs. seeds, 507
  sexual, 481
  of slime molds, 466, 469
Sporophylls, 503, 507, 511
Sporophyte, 250–252, 507
  of algae, 463, 464, 465
  of angiosperms, 617, 624, 630
  diploid phase, 252
  of ferns, 250, 503
  of gymnosperms, 507
Sporozoans, 455, 468, 469, 477
Sporozoites, 473, 474

I-51

Sporulation, **437–438**, 450
Spot desmosomes, 703
Spradling, Allan C., 379
Spring overturn, 46
Springs, 1154
Spruces, 51, 505
Squamous cells, 702
SQUID, 891
Squids, 545, 546, 551
    giant axons of, 848
Squirrel, 1055
S ring, 433
Stability
    atomic, 25, 38
    of bioactive molecules, 35
    chemical, and evolution, 88
    in organic molecules, 58
    of proteins, 75
    reproductive, and evolution, 962
Stabilizing selection, 994, **995**, 1008
"Staggered cuts," 335
Stahl, Franklin W., 292–293, 358
Staining
    of chromosomes, 383, 384, 399
    of microsocpic specimens, 96, 98, 99, 101, 109
Stamens, 511, 613–614, 616, 619, 624
Staminate flowers, 614
Stamping grounds, 1064
Standard of living, 1097, 1177
Standing water, 1155, 1177 (see also Lakes; Ponds)
Stanley, Steven M., 1028
Staphylococci, 436
Staphylococcus aureus, **432**
Starch, 59, **64–66**, 730
    in algae, 457, 461
    breakdown by saliva, 718
    in chloroplasts, 108
    Floridean, 457
    in glycolysis, 190
    granules, 82, 120
    metabolism of, in vertebrates, 730
    and pancreatic amylase, 722, **725**
    in photosynthesis, 224
Starfish, 587, **588**, 589, 604
Starlings, 423, 1114
Start codon, 320–321
Startle behavior, 1053–1054
Startle reflex, 947
Starvation, 769
Stasis, evolutionary, 1028
Statistics, in science, 15
Statocysts, **528–529**, 542, 549, 683
Statoliths, 529, 683
Stearic acid, 67–68
Stebbins, Ledyard, 973
Stein, Jeffrey, 1141
Stelleroidea, 587, **588–589** (see also Brittle stars; Starfish)
Stem cells, 831
    differentiation on, 792, 818
    and red blood cells, 792
    and T lymphocytes, 810
    and white blood cells, 818
Stem cuttings, 643
Stems, 625–626, 627, **636–640**, 647–648
    arrangement of, **640**
    and cell types in angiosperms, 647–648
    dormancy in, 621
    in ferns, 502, 503
    green stems, 648
    growth of, 630, **640–643**
    hyperelongation of, 681
    and roots, 640
    vs. roots and leaves, 626
    secondary growth in, 644
    and shoot system, 640
    structure of, **636–640**
    vascular tissues of, **638–639**, 648
Stent, Gunther, 291
Stentor, 452
Steppes, 1169–1170
Stereoscopic vision, 1033–1034, 1051
Stereotyped behavior, 475, 884
Stereotyped responses, 475
Sterile castes, 1058–1059, 1065, 1069, 1078
Sterility, 901, 904–905, 912, 916
    and chromosomal abnormalities, 386, 399
    and hybrids, 1014, 1016

and inbreeding, 977, 985
and tubal ligation, 916
and vasectomy, 916
Sterilization, 904, 912, 916, 987
Steroids, 829, 836, 837, 839
    in allergies, 813
    and cholesterol, 723
    hormones, 70, 74, 82, 117, 134, 467, 824
        female, **913**
        male, 905
    lack of in prokaryotes, 431
    vs. retinoic acid, 940
Steward, F. C., 359
Stickleback, 1054–1055
"Sticky ends," 338
    of lambda chromosome, 334–335
    in recombinant DNA, 340–342, 344–345, 353
Stigma (eyespot), 458
Stigma (of flower), 238–239, 511, 613–615, 617, 624
Sting, 1060
Stinkhorns, 488
Stipe, 465
Stirrup (stapes), 864, 965
Stoeckenius, Walther, 221
Stolons, 483
Stoma(ta), 493, 544, 627, **628**, 629, 648, **652–655**, 1164
    and abscisic acid, 694–695
    of angiosperms, 509
    of conifer leaves, 508–509
    crassulacean metabolism in, 655
    crypt of, 629
    factors influences, **654–655**
    gas diffusion, regulation in, 669
    movements, mechanism of, **653**
    in stems, 636
    transpiration, role in, 650
    water movement and, **652–655**
Stomach, 547, 588, 715–719, **720**, 924 (see also Digestion; Digestive system)
    acid, 722
    enzymes in, 715, 718, **720**, 722, 733
    hormones in, 723, 730
    immune response in, 791
Stomach acid, 722
Stomatal crypt, 629
Stop codon, 320–321
Storage
    of fats, 67–70
    of oils, 67–70
    of proteins, 67
    of sugars, 64, 67
Storage proteins, 70, 72
Stork, 423
Strata, geologic, 3, 967, **970–971**
Strategies
    for genetic recombination, 331, **337–338**
    for living systems, 59
Stratified epithelium, 702
Stratosphere, 1132, **1133**
Strawberries, 642
Streams
    life in, **1154–1155**, 1177
    net primary productivity of, 1139
Streptococci, 436
Streptococcus pneumoniae, 282, 429
Streptomycin
    resistance to, 329
Stress, 833, 855
    and ADH, 775
    of birth, 951
    and glucocorticoids, 830
    and heart disease, 761
    and punctuated equilibria, 1028
    release of hormones during, 839
    and susceptibility to illness, 830
    and sympathetic nervous system, 712
Stress fibers, 111
Stress hormone (see Abscisic acid)
Stretch receptors, 745, 762
Striated muscle, **706–708**, 713, 718
Stroke, 760–761
    and aphasia, 888
    and memory loss, 894
Stroke volume, **756**
Stroma, 213–214, 216, 220, 222, 229
Strontium-90, **1150–1151**
Structural formula(e)

atomic, 33
molecular, 62–63
Structural genes, 337, 362, 368, 379–380
    eukaryotic, 362, 368, 379–380
    origin of, 989
    prokaryotic, 320, 323
    transcription of, 323, 325, 368
Structural proteins, 70, 72, 304
Structure of proteins, 304–305
    and carrier-assisted transport, 135, 142
    primary, 74, 79
    quaternary 76, 79, 170–171
    secondary, 74–75, 79
    tertiary, 75–76, 79
Sturtevant, A.H., 275–277
Styles, 511, **614–617**, 624
Stylets, 539, 561, 660
Subatomic particles, 23
Subclavian veins, 763
Subcortical neurons, 885
Subcutaneous fat, 68, 710
Subjectivity
    in science, 17
    of taxonomic methods, 415, 418
Submission, social, 1062–1063 (see Dominance hierarchies)
Submucosa, **715–716**, 720
Subsocial species, 1060
Subsoil, 663 (see also Permafrost)
Subspecies, 1010, 1011
Substrate, **170**, 171, 174–178, 183–184, 522
    concentration of, 174–175
Subtidal zones, 1158
Subtropical mixed forest, 1164
Succulent plants, 629–630
Suckers, 536, 552, 553, 558, 588
Sucking reflex, 947–948
Sucking tubes, 565
Suckling, and population growth, 1054, 1096, 1100
Sucrase, 170, 722, 723
Sucrose, 59, 61, 64, 228, 660, 722
    density gradient, 193, 662
    hydrolysis of, 170
    in sap, 129
    in sugar cane, 182
    translocation of, in plants, 638, **657–662**, 670
Sugar(s), **59–66** (see also Disaccharides; Monosaccharides; Polysaccharides; Sucrose)
    and calories, 68
    six-carbon, 228
    translocation of, in plants, **657–662**, 670
    transport of, in plants, 660
    and yeasts, 486
Sugar cane, 182, 226
Sugar-phosphate units, of DNA, 289
Sulci, 885
Sulfa drugs, 449
Sulfanilamide, 177
Sulfate, 438
    and chemosynthetic organisms, 438
    reduction of, in ecosystems, 1140
Sulfhydryl groups
    and lead poisoning, 177
Sulfur, 35, 516
    biogeochemical cycles of, 1145
    and chemosynthetic organisms, 436, 440, 450
    and green bacteria, 440
    plant requirements for, 657
    and purple bacteria, 440
Summer stagnation, 46
Sun, 159, 183, 1152 (see also Light; Photosynthesis; Solar energy)
    energy of, 84–85, 161–162, 167
    formation of, 84–85
    and heat gain in ecosystems, 778–779, 789
    and origin of life, 90, 100
    and photosynthesis, 209, 226 (see Photosynthesis)
    and productivity of ecosystems, 1139
    and temperature regulation in reptiles, 781
    and transpiration, 652
    and water cycle, 53
Sundew, 666, 692–693, 695
Supercoiling, of DNA, 294, 297, 299 (see also Topoisomerases)
Supergenes, 993
Superior vena cava, 763
Superkingdoms, 430
Supernatant, 192–193

Suppressor T cells, 801, 819 (see also Cytotoxic T cells)
　and AIDS virus, 816
　and B lymphocytes and cytotoxic T cells, 808
　and T8 glycoprotein, 804
Supratidal zone, 1158
Surface, cellular, 142
　of red blood cells, 750
　and vesicle-mediated transport, 138-140
　and volume, 102-104, 142
Surface area, 544, 547
　and gills, 734-735, 747
　and oxygen diffusion in reptiles, 747
　of red blood cells, 750
　and respiration, 734
　of small intestine, 720
Surface tension, 42, 54
Surface-to-volume ratio, 520, 789, 878
　of cells, 102, 103, 124, 878
　and conservation of energy, 787
　and diffusion, 142
　in earthworm respiration, 737
　and heat transfer, 779
　in gills, 734
　of larger organisms, 711
"Survival of the fittest," **974-975**
Survivorship curves, 1092, 1102
Suspensors, 618, 619
Suspensory ligaments, 866, 867
Sutherland, Earl W., 838
Sutton, Walter S., 260-261, 263-264
Sutton's hypothesis, 260-261, 264, 266, 274, 277
SV40, 348, 373, 376, 377, 444
Swallowing, **718-720**, 736, 738, 740, 948
Swamps, 1139
Sweat and water balance, 769
Sweat glands, 710, 863
Sweet pea, 270
Swim bladder, 594 (see also Air bladder)
Swimmerets, 572
"Swimmer's ear," 425
Swimming, 434, 591, 936, 1058
Swimming bells, 528
Swiss starlings, **995**
Symbiosis, **447-448**, 451, 452-453, **664-670**, **1119-1122**, 1130
　ants and acacias, 1121-1122
　in chemosynthetic ecosystems, **1140-1141**
　commensalism, 447-448
　in coral reefs, 530, 1120
　and cyanobacteria, 453
　in digestion, 723-724, 730
　in evolution, 452-453, 476
　and fungi, **488-491**
　green algae in, 461
　and immune response, 791
　in lichens, 492
　mutualism, 447-448, 1120-1122
　and mycorrhizae, **490-491**, 492
　parasitism, 447-448, 1120
　in plants, **664-670**
　in protozoa, 464
Symmetry
　bilateral, 531, 533, 535, 542, 550, 587, 590, 604
　in flowers, 513
　radial, 525, 531, 542, 587, 604
Sympathetic nervous sytem, 712, **757**, 842, 845, 846, 847, 856, 857
　and bloodflow regulation, 759
　and cardiovascular regulation, 762
　and digestive system, regulation of, 723
　and endocrine system, 832
　vs. parasympathetic nervous system, 846-847, 857
Sympatric speciation, **1012-1015**, 1029 (see also Disruptive selection; Polyploidy)
Symphylans, 573, 586
Symport, 137
Synapses, 708, 841, **851-856**, 857
　chemical, 851-852, 853
　electrical, 851, 852
　and learning, 893, 899
　and memory, 893, 894-895, **895-898**
　modifications at, 893
　at neuromuscular junction, 876
　in peripheral nervous system, 844, 845, 846
　transmission at, 898
Synaptic cleft, 852, 855, 857, 877
Synaptic knobs, 852, 898

Synaptic vesicles, 852, 853, 877
Synchronous replication, of plasmids, 326
Syndrome, 385
　Down's syndrome, 385
Synergid, 617
Syngamy, 426, 460
　in fungi, 462
　in protists, 462
　in protozoa, 468, 469
Synthesis phase (see S phase)
Synthetic theory of evolution, **972-973**, 974
　and natural selection, 1008
　and punctuated equilibrium hypothesis, 1029-1030
Syphilis, 437
Systematics
　evolutionary, **412-414**, 427
　traditional, 416
Systemic circulation, 754, 756, **758-759**, 761, 763
Systolic blood pressure, **759**
Szent-Györgyi, Albert, 218

Tadpole, 929, **1076-1077**
Tagma(ta), 568
Tagmosis, 558, 564, 565, 572
Taiga, **1139, 1166**, 1177
Tail(s), 559, 592, 595, 598, 604
　in human embryo, 946
　as prehensile limb, 1034-1035
　of sperm cell, 902-903
　of virus, 445
Tall-grass prairie, 1169
"Tangled bank" hypothesis, 986-987
Tannins, 692
Tapeworms (see also Beef tapeworm), **536**, 542
Taproot system, 635
Tardigrades, 520, 560, 561, 564
Target cell, 140
Target sequence (site), 335-336, 338
Target tissues, 712-713, 822
Tarsiers, 1031-1032, 1034
Taste, **601**, 860
　in human beings, 601
　receptors for, 859, 860
　and sensory cortex, 886
Taste buds, **717**, 860
Taste cells, 557
Taste organs, 585 (see also Chemoreceptors)
Tasting, 574
Tatum, Edward L., 301-304, 316
Taung child, 1038, 1044
Taxa, 410, 412-416, 424, 482
　as historical unit, 413
　holophyletic, 427
　monophyletic, 413, 416, 427
　paraphyletic, 418
　placement of, 423
　polyphyletic, 413
Taxonomic studies, 420
Taxonomy, 308, 410, 412, 427
　alternative methodologies of, **415-418**
　of birds, 423
　cladistics, **415-416**, 427
　evolutionary, 415
　hierarchical, 427
　human, 1043-1051
　methods, **414-418**
　molecular, **418-423**, 427
　phenetics, 415-418
　of prokaryotes, 450
　traditional, 414-415
Tay-Sachs disease, **389**, 400
　incidence, 389
　and population bottleneck, 983
　recessive allele for, 400
T-cell receptors, **804**, 819
T cells (see T lymphocytes)
Tears, and immune response, 791
Technology
　and cost of food production, 1144
　and science, 14-15
Tectonic plates, 1140
Teeth, 594, 601, 602, 716, 948 (see also Canines; Dentition; Incisors; Molars; Premolars)
　radular, 547
T8 glycoprotein, 804, 808 (see also Cytotoxic T cells; Glycoproteins; Immune system; T lymphocytes)
Telencephalon, 883-884, 885, 895, 898 (see also

Cerebral cortex; Cerebral hemispheres; Corpus callosum)
Telophase, 150, **152**, 153, 155
Temperance, **331-333**, 338
　and lysogeny, 331
Temperate bacteriophage, **331-333**, 334, 338
Temperate forests, 1139, **1164-1166**, 1177
Temperate grasslands, 1139, **1169-1171**, 1177
Temperate mixed forests, 1164, **1177**
Temperature, 43, 1152
　absolute, and energy reactions, **165**, 183
　and active transport in plants, 669
　annual changes in, 1136
　in atmospheric layers, 1132-1133
　and behavior, 789
　and biological systems, 43
　and blood flow, 759
　body, 916 (see also Body temperature)
　and cell division, 146-147
　and chemical reactions, 43
　and climate, 1136-1137
　in deserts, 1172-1173
　and entropy, 183
　and enzyme reactions, 175-176, 184
　feedback control of, 713
　and "frizzle" trait, 273
　and gene expression, 271
　and germination, 623
　gradients, and homeostasis, 780-782
　and greenhouse effect, 1177
　and "hardening," 44
　in the hemispheres, 1136
　and inflammatory response, 792
　and male hormone regulation, 907
　and molecular motion, 129
　and *Paramecium*, 475, 476
　and photosynthesis, **215-216**
　and plant growth, 645-646, 671
　range, for life, **777-778**, 789
　and rate of membrane transport, 178
　regulation of, **777-790**
　and reproductive isolation, 1016
　and scrotal sac, 901
　and sensory cortex, 886
　and stomatal movements, 654
　and tidal zonation, 1158, 1160
　and transpiration, 652
Temperature-dependent reactions, in photosynthesis, 215
Temperature-independent reactions, in photosynthesis, 215
Temperature receptors, 859
Template strand, 358
　in conjugation, 328
　in DNA replication, 292, 294-297, 299
　and messenger RNA synthesis, 368
Temporal escape strategies, 1116
Temporal lobe
　and auditory cortex, 892
　and instrinsic processing centers, 885-886, 892
　and memory, 894, 895
　and visual cortex, 892
Tendons, 704, 706
Tendrils, 630, 642, 691-693, 695
Ten percent "rule of thumb," for energy efficiency, 1143, 1151, 1153
Tension, 652, 669
Tentacles, 525, 526, 528, 529, 531, 548, 557, 558, 560, 562, 564, 695
Teratogens, **947**
Terminal, of neuron, 708
Terminal bud, 641
Terminal electron transport, 191, 204
Terminal regions, 111
Termination, of protein biosynthesis, 312-313
　signals for, 314, 317
Terminators, 310
Termites, 469, 576, 582, 1059-1060, 1069
Territoriality, **1063-1065, 1066-1067**, 1078
　in chimpanzees, 1037
　and regular dispersion, 1094
　and sexual selection, 1000-1004
Territorial boundaries, 1064
Territorial defenses, **1064-1065, 1066-1067**
Territories, **1063-1065, 1066-1067**
　and helpers, 1070
　and the "selfish gene," 1074
Tertiary period, 496, 1012-1013, 1025

Tertiary structure, 184
   and enzyme activity, 170–171, 175
Testcross, 241, 243, 247, 265–266, 274, 276
Testes, 535, 824, 825, **901**, 903, 904, 905, 906, **907**, **917**
   development of in human embryo, 946
   hormone production, control of, 927
   and sex hormones, 70, 258
Testicles, 901
Testosterone, 70, 117, 824, **905**, 906–908, **913**, 917
   and interstitial cells, **901**
   in mammals, 906
   and sex differences in brain, 907–908
Tests (shells), 469, 470
Tetracycline, 344
   resistance, 329
Tetrad, 253–254
Tetrahydrocannabinol, 515
Tetrahymena furgasoni, 118, 370
Tetramers, 76
Tetranucleotide theory, 283–284
Tetraploidy, 1013
T-even bacteriophages, 285–286, 330, 341, 444–445
T4 bacteriophage, 330
T4 glycoprotein, 804, **815**, 819 (*see also* Helper T cells; Immune response; T lymphocytes)
Thalamus, **883, 885,** 888, **892–893,** 898
Thalidomide, 947
Thallium, 25
Thallus, 463, 465, 467
Theca, 460
Thecodonts, 597
Theophrastus, 236
Theory, in science, 9, 14–17
Thermoacidophiles, 430
Thermocline, 46
Thermodynamics, **162–166**
   and cell transport, 180
   and diffusion, 180
   laws of, 12, **162–166**, 183, 710–711, 776
   and living systems, 166
Thermonuclear reaction, 84–85
   in sun, 161–163
Thermophilic bacteria, 428
Thermosphere, 1132, **1133**
Thermostat, 784–786, 790, 883
Theropods, 599
"Theta" chromosome, 295
Theta replication, 320
Thirst, 769, 785, 788–789, 859
30-nanometer fiber, of DNA, 357
32-cell stage, of embryo, 931
Thomas, Lewis, 16
Thoracic cavity, 701, 713, 718, 740–741, 747, 762, 936
Thoracic duct, 763
Thorax, 566, 571, 574, 584, 585
Thorns, 629, 1121
Three-carbon pathway, 222–223 (*see also* Calvin cycle)
Three-chambered heart, 754–755, 763
Threonine, 73, 76, 310
Thrombi, **760–761**
Thrombin, **751–752**
Thromboplastin, **835**
Thrush (bird), 423, 1107
Thrush (disease), 481, 488
Thumb, 601
   evolution of, 1032
Thylakoid membrane, 113
Thylakoids, 120, 442
   inner compartments of, 213
   membrane, **113**
   in photosynthesis, **213–214,** 216, 218, 220–222, 224, 228–229
Thylakoid space, 213–214, 220, 229
Thymidine kinase, 378–379
Thymine, 80–81, 305–306, 310, 314, 351, 429
   in DNA, 283–284, 288–290, 292, 296, 299
Thymus, 819
   development in vertebrate embryo, 937
   and immune system, 793
   and MHC proteins, 804
   and T lymphocytes, 805, 819
Thyroid gland, 713, 784, 786, 828, 824, **828–829**
   hormone production, control of, 827
   and negative feedback control, 827
   and pituitary, 825

Thyroid hormone (*see* Thyroxine)
Thyroid-releasing hormone (*see* TSH)
Thyroid-stimulating hormone, 713, 784 (*see* TSH)
Thyrotropin, 825, 826
Thyrotropin-releasing hormone (*see* TRH)
Thyroxine (thyroid hormone), 713, 728, 784, 786, 824–825, 828, 836, 839, 940
Tibia, 707
Ticks, 570, 585
Tidal, 1158
Tideland communities, **1157–1160**
Tide line, 1159
Tiger, 412
Tight junctions, **703,** 721
Time, biological, 93
"Time's arrow," 164
Tinbergen, Niko, 872, 1052, 1054–1055
Ti plasmid, **352–353,** 354
Tips of chromosomes, 364
Tissue(s), 140
   animal, 142–143
   connective, 140
   embryonic, 525, 532, 542, **930–931** (*see also* Ectoderm; Endoderm; Mesoderm)
   epithelial, 140
   exchanges with blood, 749
   homogenization, 192–193
   human (*see* Bone; Connective Tissue; Epithelial Tissue; Muscle Tissue; Nerve Tissue)
   and human heart, 756
   interactions during development, **930–931**
   layers, 520, 532, 929
   muscle, 140
   nerve, 140
   organ, 140
   plant, 141–143 (*see also* Ground tissue; Phloem; Vascular tissue; Xylem)
   and respiration, 737
   systems, in angiosperms, **625–627,** 648
      transplants, 810–812
   turnover, 830
   in vertebrates, **702–710,** 713
Tissue factor, 751
Tissue systems, in angiosperms, **625–627,** 648
   dermal, 648
   differentiation of, 636
   ground, 648
   in roots, 634, 635
   vascular, 648
Titmice, 1057
T lymphocyte precursors, 804
T lymphocytes, **801, 804–805,** 818, 836
   functions of, 806–808
   and immune system, **793, 801–808,** 819
   and immunosuppressive drugs, 810
   life history of, 801, 804–805
   and monoclonal antibodies, 802
   selection of, 804–805
   and thymus gland, 804
Toads, 595, 1054
Toadstools, 481
Tobacco, 172, 307, 352–353, 678
Tobacco mosaic virus (TMV), 291, 305, 307, 444
Tocopherol, 729
Toes, 601, 946, 947, 948
Togetherness, 778–779
Tolerance hypothesis, 1125
Tomatoes, 678
Tone deafness, 583
Tonegawa, Susumu, 372–373
Tongs, 565
Tongue, 717, 718
Tonicity
   of blood, 132
   hypertonicity, 132–134, 142
   hypotonicity, 132–134, 142
   isotonicity, 132–134, 142
Tonsils, 793–795, 937
Tonoplast, 115
Tool use, 1002, 1037, 1042, 1044, 1073, 1159
   in early *Homo sapiens*, 1046, 1047
   in *Homo erectus*, 1044–1045
Tooth shells, (*see* Tusk shells)
Top soil, 663, 1166
Topoisomerases, **294,** 297, 299
Torsion, 550, 551
Touch, 588, 886
   response to, in plants, **691–693**

Touch receptors, 551, **582,** 586, 859, 860, 1060 (*see also* Mechanoreceptors; Sensilla)
Toxic waste, 488
Toxins, 72, 451
   in cdinarians, 526
   in dinoflagellates, 460
   in mushrooms, 487
   of octopus, 552, 553
"Tracers," 25
Trachea(e), 736
   of arthropods, 734, 737
   and evolution of lungs, 736
   human
      and cigarette smoke, 739
      in respiration, 738, 740, 747
   in mammals, 718–719
Tracheids, 107, 501, 509, **639,** 647–648
   in confier leaf, 508–509
   in secondary growth, 647
   water conduction in, 652, 660, 669
Tracheophytes, 497
Tracts, 842, 843
Trailer sequence, 320–321
Traits
   and clines, **1002**
   dominant, 260, 266, 268–271, 276
   hereditary, 249
   and life-history patterns, **1099,** 1105
   polygenic, 271–272, 279
   recessive, 260, 265–268, 274, 279
   sex-linked, 266–267, 278
   types of selection for, **995**
Transacetylase, 325
Transcription
   for amino acids, 320
   of DNA, 359
   efficiency of, 321–322
   in eukaryotes, 355, 358–360, 362, 368
   of Messenger RNA, **301–317, 320–323,** 325, **368–370,** 372, 380
   and polypeptides, 369
   in prokaryotes, 320–322, 325, 337, 355, 358–360, 380
   and promoter sequences, 368
   and regulatory proteins, 360–361
   reverse, 306, 340–342, 353, 373–376
   of RNA (*see* RNA)
   RNA polymerase, 368
   of structural genes, 320, 323, 325, 368
   units, 366
   of viral DNA, 330
Transcription units, 366
Transduction, **332–334,** 336–338, 859
   vs. conjugation, 337
   general, **332–333,** 338
   and genetic recombination, 336–337
   olfactor, 861
   in prokaryotes, 437
   specialized, **333–334,** 338
Transfer, genetic, 331, 337–338
   and chromosomal insertions, 331–333, 338
   conjugation, **326–330,** 331, 337
   homologous recombination, 336–338
   and nuclear pores, 380
   in rolling-circle replication, **327–328,** 338
   "single-strand switch," 337
   transduction, **332–334,** 336–338
   transformation, 282, 329, 331, 336, 376
   and transposons, **337–338,** 379
Transfer RNA, 310, **311, 312–317,** 420
   aminoacyl-tRNA synthetase, 311, 315, 317
   anticodons for, 312–313
   genetic code for, 362, 380
   initiator, 312
   linkage to amino acids, 312
   in prokaryote classification, 430
   shape of, 311, 317
   transcription for, 368
   varieties of, 311
Transformation, 282, 329, 331, 336, 376
   in bacteria, 282, 329, 331, 336
   and homologous recombination, 336
   in prokaryotes, 437
Transformations of energy (*see* Energy transformations)
Transforming factor, 281–284, 371
Transfusions, 810–812, 814, 817–818, 819
"Transgenic" mice, 378

Transition region, 640
Translation, 306, 309–310, **312–314**, 317
   of messenger RNA, 324, 368
Translocation (genetic), 277, 279, 376, 385, 399, 670
   and Down's syndrome, 385–386
Translocation (in plants), 661
   phloem in, 660
   in plants, **657–662**
   pressure-flow hypothesis, 662
   rate of, 660
Transmission electron microscope, 95–96, 97, 98, 101, 110
Transpiration, **650**, 669, 1164 (*see also* Transpiration stream)
   and drought, 654
   influences on, **625–655**
   and water uptake, 651
Transpiration stream, **650–652**, 653, 655, 662, 670
Transplantation, 488, **810–812** (*see also* Transfusions)
   of organs, 488
   of tissue, 810–812, 819
Transport, across cell membranes, **127–143**
   active, 135, **136**, 143
   bulk transport, 133
   carrier-assisted, **134–135**, 142
   cellular (*see* Cellular transport)
   chloride ions (Cl-) in, 134
   cotransport, 13
   countercurrent exchange, 124, 130
   diffusion, 128, **129–135**, 136, 142
   endocytosis, **138–140**, 142–143
   energetics of, 178–182, 184
   energy and, 135–136
   extracellular, 124, 154
   glucose, 136–137
   of hydrophilic molecules, 134, 136–137
   of hydrophobic molecules, 134, 136–137
   kinetic energy and, 168–169
   and mitochondrion, 136
   molecules, 137
   phagocytosis, **138–140**, 143
   pinocytosis, **138–140**, 143
   polarity and diffusion, 130
   potassium (K+) ions in, 136
   potential energy and, 135–136, 180
   proteins in, 106, 124, 135–137, 184
   rate of, 178–179
   receptor-mediated, 137, **138–140**, 143
   regulation of, 142–143, 178–180
   of sodium, 134, 136
   sodium-potassium pump, 136–137, 143
   temperature and, 175–176, 178
   thermodynamics and, 180
   and vacuoles, **138–140**
   vesicle-mediated, **130–140**
Transport (in plants), **636–640, 650–670**
   rate of, 662
   regulation of, **632–633**
Transport membrane, 124
Transport molecules, 137
Transport proteins, 70, 72, 82, 106, 124, **135–137**, 184, 325
   antiport, 137
   cotransport, 137
   and hypertonic urine, 772–774
   in plants, 662
   sodium-potassium pump, **136**, 137, 143
   symport, 137
   uniport, 137
Transposase, 335–336, 338, 374
Transposition, genetic, 801, 819
Transposons, **335–337**, 338
   and chromosome, 335–336
   complex, 336–338
   direct repeats, 374
   and drug resistance, 337
   in eukaryotes, **374–375**, 371, 381
   and genetic recombination, 337, 379
   and gene transfer, 379
   identical repeats, 335
   insertion sequences, 335, 374
   inverted repeats, 335–336, 338, 374
   and mutations, 335–336, 338, 374
   vs. plasmids, 336
   promoter gene in, 335, 374
   recognition sequence in, 374
   replica, 374
   and reverse transcription, 374
   simple, 335, 338
   as vectors in cancer research, 379
Transverse tubules, 873
Transverse wall, 438
Trapping, 599
Tree fern, 503
Tree finches, **1019–1020**
Tree layer, 1164
Trees, 1165–1166, 1170–1172, 1174–1175 (*see also* Forest ecosystems)
Trees, hardwood, 511
Trehalose, 61
Trematodes, **536–537**, 542
TRH, 784, 826
Trial and error learning (*see* Operant conditioning)
Triassic period, 496, 597
Tributaries, 1155
Trichinella, 539–540
Trichocysts, 471
Trichonympha, 469
Tricladida, 534
Trigger hairs, 693
Triglycerides, **67–68** (*see also* Fats, Oils)
Triiodothyronine, 828
Trillium erectum (*see* Wake-robin)
Trilobites, 3
Trimer(s), 76
Trimesters, **945–949**
Trioses, 60
Triphosphates, 181–184, 307, 358
   in DNA replication, 298, 299
Triple fusion, 617, 624
Triplet code hypothesis, 350
Triploblasty, 532, 533
Triploidy, 617, 624
Tritium, 25, 364
Trivers, Robert, 1074
tRNA (*see* Transfer RNA)
Troops, 1036
Trophic levels, of food web, **1137–1143**, **1145**, 1153, 1085, 1097
Trophoblasts, **940**, 944, 952–953
Tropical forests, 1174–1177
Tropical grasslands, 1139, **1170**, 1177
Tropical mixed forest, 1174, 1177
Tropical rain forests, 405, 407, **1124, 1139, 1174–1177**
   destruction of, 1126–1127
   and intermediate disturbance hypothesis, 1124
   and island biogeography model, 1126–1127
   retention of nutrients in, 1150
Tropical seasonal forest, 1139
Tropic hormones, 825, 827, 839, 883
Tropics, 226
Tropisms, **671–672**
Tropomyosin, 876
Troponin, 876
Troposphere, **1132–1133**
Trout, 130, 594
trp operon, 323, 325, 337
Truffles, 481, 485, 491
Trunk, 591
Trunks (in plants), 625, **645**
Trypanosoma, 469
Trypsin, 175, 722
Trypsinogen, 722
Tryptophan, 73, 76, 674, 728 (*see also trp* operon)
   biosynthesis of, 321–323, 325
TSH, 713, 784, 824, 825, 828
T system, 873, 874, 876
T3 proteins, 804
Tubal ligation, 912, 916
Tube cell, 615
Tube nucleus, 615, 616
Tuber, 120
Tuberculosis, 429, 448
Tubers, 445, 626, 636, 641, 648
Tube worms, **1140–1141**
Tubules, 122, 534, 574, 575, 747
   in arthropod respiration, 734
   in Golgi Complex, 122
   and osmosis, 132
Tubulin dimers
   in cell division, 149, 152
   in development of neural plate, 938
   in sperm cells, 902
Tufts, flagellar, 434
Tumbling, prokaryotic movement, 434
Tumors, 890
Tundra, **1139, 1168–1169**, 1177
Tunic, 592
Tunicates, **591–592**, 604
Turbellarians, **533–535**, 542
Turgor, 696
   and cell elongation in plants, 674–675
   in cell wall, **134**, 142
   and regulation of gas diffusion, 669
   of stems, 636
   and touch response in plants, 692–693
   and water stress, 654, 669
Turner, Teresa, 1128
"Turnover tissues," 630
Turtle doves, 1074
Turtles, 596, 597, 768
Tusks, 717
Tusk shells, 548
Twining, 691–693
Two-cell stage of embryo, 941
Two-chambered heart, **754**, 763
2,4-D (auxin), 674 (*see also* Auxins)
Tympanic air sacs, 584
Tympanic membrane, 864, 865
Tympanic organs, 584, 586
Tympanum, 584 (*see also* Eardrum)
Type A blood, 268
Type AB blood, 268
Type B blood, 268
Type O blood, 268
Typhus, 437
Tyrosine, 73, 388, 832
   and albinism, 389
   in phenylketonuria, 388

Ulna, 1032
Ultimate causation, **1053–1054**, 1077
Ultracentrifuge, 192, 292–293, 311
Ultrasound
   in detection of genetic defects, 387
Ultraviolet light, 872, 1060
   in atmosphere, 1133–1135
   and lytic cycle, 331, **332**
   and mutations, 982
   and ozone layer, 1134–1135
   and vitamin D, 728
Ulva, 462, 463, 464, 1128
Umbilical arteries, 945
Umbilical cord, 944, 947, 950, 952, 953
Uncertainty principle, 29, 129
Unconditional response, **1055**
Unconditional stimulus, 1055
Understanding, 893
"Undertaker" bees, 1053
Unequal cleavage, 927, 952
Ungulates, **1110**, 1031
Unhygienic bees, 1053
Unicellular organisms, 455, 480, 821, 838
   vs. multicellular, 821
   protists, **452–477**
Uniformitarianism, 2–3, 5
Uniport, 137
Unit characters, of phenetics, 416
United States, age structure, 1092–1093
Unity, of living systems, 123
"Universal donor" and "recipient" blood groups, 811
Universe
   "big bang" theory, 23, 307
   as closed energy system, 166
   origin of, 23
Unsaturated fats, 67–68, 82
Unsaturated fatty acids, 67–68, 82
Upright posture, 907, **1033**, 1038 (*see also* Bipedalism)
   in australopithecines, 1038–1039, 1040
   consequences of, 1006
   evolution of, **1033**, 1035, 1046
Upwelling, 1157
Uracil, 80–81, 296, 305–307, 309–310, 314, 351
   as radioactive tracer, 308
Urea, 11, 749, 775–776
   and breakdown of cellulose, 724
   in embryonic excretion, 944
   excretion of, **766**
   and formation of urine, **772**, 773
Ureter, 770–771
Urethra, 770–771,
   of penis, 903–905, 917

Uric acid, 568, 575, **766**, 775, 934
Urinary flow, and ADH, 775
Urinary system, 711, **771**
Urine, 134, 775–776, 903, 904
  hypertonic, **772–774**, 776
  in kidney, 770
  water loss through, 767–768, **770–772**
Urkaryotes, 453
Urochordates (*see* Tunicates)
Uterine contractions, 950
Uterine tubes (*see* Oviducts)
Uterus, 561, 600, 824, 828, 835, 907–909, 941, 945, 950, 952
  and contraceptive techniques, 916
  contractions of, 824, 828, 950
  and female sex hormones, **913**
  fertilization, 912
  implantation of embryo in, 941, 952
  maternal blood space, 945
  and pregnancy, 911
  effects of prostaglandins on, 835, 905
  wall of, in menstrual cycle, 913–914 (*see also* Endometrium)

Vaccination, 352, 796–797, 819
Vaccines, 352, 474
  against AIDS, 817–818, 819
  against poliomyelitis, 449, 813
  against smallpox, 796–797
  against viruses, 445
Vacuoles, 93, **113**, **115**, 120, 121, 124, 125
  in cell division, 146
  in cell elongation, 674
  and cell membrane, 139
  and cellular transport, **138–140**
  central, 214
  contractile, 133
  in endocytosis, 143
  in energy transformations, 168
  in exocytosis, 133
  in infection cycle, 330
  in palisade cells, 648
  in *Paramecium*, 767
  phagocytosis in, 118, 751
  in *Pilobolus*, 484
  in plant cells, 134
  in water molds, 467
Vagina, **907**, 908, **909**, 916
  and bipedalism, 907
  and contraception, 916
  and intercourse, 904
  and orgasm, 913
  pH of, 905, 908, 913
Vaginal foam or jelly, 916
Vaginal lubricant, 913
Vagus nerve, 757
Valine, 73, 76, 310, 314
Valonia, 462, 463
Value judgments, in science, 17–18
Valves, 755
  in aorta, 755
  in heart, 755–756
  in lymph nodes, 794
Van Helmont, Jan Baptista, 208, 236
Van Leeuwenhoek, Anton, 104, 237, 448, 469
Van Niel, C. B., 217
Van Niel's hypothesis, 217
Vaporization, 43
Variable genes, 372, 804
Variable (V) regions, 371–372, 815, 818
  of antibodies, 798, 801, 818
  and genetic code for antibodies, 819
  of T-cell receptor, 819
Variation
  and adaptation, **962**, 972, 973
  in behavior, 1052 (*see* Behavior)
  and evolution, 26
  among families, 1065–1066
  genetic, 245, 271, 279
    and meiosis, 261
  hereditary, 246
    persistence of, 246
  in natural populations, 246
  and natural selection, 7
  in populations (*see* Genetic variation)
Variolation, 796
Vascular bundles, 214, 224, 501
  in dicots vs. monocots, 628, 640

in plants, 648
Vascular cambium, 644, 675, 681
  and cell types in angiospores, 647
  and secondary growth, 644–646, 648
Vascular circuitry, **758–759**
Vascular cylinder
  of roots, **632**, **633**, 634
  of stems, 640
Vascularization, 810
Vascular plants, 496, **501–515**, 625
  angiosperms, 505, **509–515**
  vs. bryophytes, 498, 499
  classification of, 497, 516
  conducting systems in, 501
  diversification of, 516
  earliest, 497
  evolution of, 497, 501–502, 516
  ferns, **502–503**
  fossils, 491
  gymnosperms, **505–509**
  and mycorrhizae, 490–491
  reproductive cycle in, 501–502
  response to touch in, 695
  Rhynia major, 501
  seed-bearing, **503–515**
  seedless, 502–503, 516
Vascular tissue system, 501, **625–627**, 648 (*see also* Phloem; Xylem)
  of roots, **633**
  of shoots, 640–641
Vas deferens, 535, **901–902**, 903, **904**, 905, 917
Vasectomy, **904**, 916
Vasoconstriction, 759, 760, 764, 770–771, 786
Vasodilation, 759, 760, 761, 764, 786
Vasopressin, 824, 828 (*see also* ADH)
Vectors
  plasmids as, 378
  in recombinant DNA, **343–345**, 346–347, 352–354
  transposons as, 379
  viruses as, 331–334, 338, 377
Vegetables, 511
Vegetal gradient, 925–926
Vegetal hemisphere, of ovum, 924, 926–927, 952
Vegetal poles, 926
Vegetarianism, 76, 728
Vegetation, and energy flow, 1137
Vegetative cell, 438
Vegetative parts, 621, 624 (*see also* Leaves; Roots; Stems)
Vegetative reproduction, **642–643**, 1103
Veins, **627**, **629**, **753**, 762, 763
  and blood pressure, 760, 761, 763
  of conifer leaf, 508–509
  in dicots vs. monocots, 628
  and heat conservation, 788
  in leaves, **628**
  and lymph vessels, 763
  in plants, 648
  and portal systems, 759, 827
  pulmonary, **753**, 755, **758**
Venae cavae, **753**, **755**, 756, **758**, 759, 762
Venn diagram, 410
Ventilation lungs, 736
Ventricles, 547
  of brain, 881–882
  of heart, **754**, 755–757, 758, 764
Venules, **753**, 758, 761, 763
Venus, 235
Venus flytrap, **667**, 675, 692–693, 695
Vertebral column, 592, 604, 701, 713, (*see also* Backbone; Spinal column)
  embryonic, 935
  evolution, 1006
Vertebrates, 3, 407, 532, 540, 587, 591, **592–604**
  brain of, **881–899**
  cephalization in, 842
  characteristics, **701–714**
  circulatory system, **749–764**
  color vision in, 872
  evolutionary homologies in, 969
  filter feeding in, 594
  hormone transport in, 821, 822, 839
  muscle fibers in, 878
  nervous system of, **841–857**
  reproduction in, **900–918**
  social behavior in, **1062–1065**, 1078 (*see also* Behavior)
  territoriality in, **1063–1065**, 1066–1067

Vesicles, **115**, 116–118, 124, 127, **138–140**, 142, 213
  in acrosome, 902
  in animal cell, 112
  and ATP synthetase, 199
  and cell membrane, 139
  in cell plate, 154–155
  of choloroplast, 120
  coated, 139
  and cytokinesis, 153
  in endocytosis, 138–140
  in energy transformations, 168
  secretion, 118
  transport, 115, 116, 117, **138–140**
Vesicle-mediated endocytosis, **138–140**
  and membranes, 139
Vesicle-mediated transport, 115–117, **138–140**
  and Golgi complex, 138–139
  and hormones, 140–141
Vessel members, 501, **638**, 647–648, 669
Vessels, in xylem, 639, 652
Viability of fetus, 948
Viceroy butterfly, 1004
Victoria, Queen, 382, 393
  as carrier of hemophilia gene, 382, 393
Video imaging, and microscopy, 99, 101
Villi, **720–721**, 723, 730, 782
Vines, 1175
Viola, 408
Viral antigens, 808
"Viral theory" of cancer, 376
Virchow, Rudolf, 10
Viroids, **446–447**, 451
Virulence, 282, 433
Viruses, 305, 308, **330–335**, 338, 428, **443–450**, 451, 806–808, 814–816
  AIDS, **792**, 818
  assembly of, in host, 445–446
  bacteriophage, 281–285, 295, 304, 305, 331–332, 334, 373
  and cancer, 375–377, 381
  capsid of, 330, 338, 373, 443–445, 450–451
  and cell-mediated immunity, 801
  chromosome of, 330, 331–334
  and cytotoxic T cells, 806–808
  in disease, 448–450
  DNA of, 333
  in DNA research, 281, 284–286
  DNA viruses, 338, 350
  and drug resistance, 329
  eukaryotic, 373–374
  and genetic recombination, 336
  and hepatitis B, 796
  HIV virus, **814**
  infection by, 322, 330–335, 338, **445–446**
  and interferons, 818
  and introns, 362
  lambda phage, **334–335**, 338
  and memory-cell response, 819
  molecular genetics of, **330–338**
  as movable genetic elements, 374, 381
  nontemperate bacteriophages, 333–334
  as obligate parasite, 330
  φX174 phage, 330
  and prions, **446–447**
  prophage, **331**, 333–335, 338
  protein coat of, 305
  proviruses, 373–374, 381
  in recombinant DNA, 340–346, 352–353
  replication of, 284, 330–331
  and resistance genes, 329
  RNA, 373, 350
  RNA viruses, 306, 338
  Rous sarcoma virus, 376
  specificity, 445
  SV40, 348, 373, 376–377
  temperate bacteriophage, **331–333**, 334, 338
  T-even bacteriophages, 330, 341
  T4 bacteriophage, 330
  tobacco mosaic, 291, 305, 307
  and vaccines, 445, 451, 796
  as vectors, 331–334
  as vectors in cancer research, 377
  and viroids, **446–447**
Visceral mass, 546, 550, 551, 563
Visceral peritoneum (*see* Serosa)
Vision, 210, **579–582**, **866–873**, 879, 898
  acuity in, **1033**, 1051
  in animals, 867

in birds, 870
blind spot in, 869
and cerebral cortex, **887-888**
cells for (*see* Cones; Rods)
color, 868, 872
evolution of, 870
focus, 867
in frogs, 870
human, 601, 872
information processing, **869-870**
and intrinsic processing centers, **892**
non-color, 868
receptor cells for, 859, 866
resolution in, 868
retina, **868-872**
stereoscopic, 552, 867, 1033, 1051
visual pigments, **871-872**
Visual acuity, **1033**, 1051
Visual cortex, **887-888**, 898
and instrinsic processing center, **892**
perception of form in, **887-888**
Visual field, 580, 867
Visual pigments, **871-872**
Visual processing, **887-888**, 892
Visual purple (*see* Rhodopsin)
Visual signals, 1016
Vitalism, 10-11
"Vital spirit" theory, 10-11
Vitamin(s), **728-729**, 730
A, oxidation of, 212, 726, 728, 729, 871
B-complex, 194, 729
and coenzymes, 171-172, 184
and digestion, 730
and ethanol, 202
human requirements, 726
reabsorption of, 771
riboflavin, 195
symbiotic synthesis of, 724-725
Vitamin A, 212, 726, 728, **729**, 871
Vitamin B_1, **729**
Vitamin B_2, 729 (*see* Riboflavin)
Vitamin C, 729
Vitamin D, **728**, 729, 824, 829, 831
Vitamin deficiencies, 728
Vitamin D_3, 726, 729
Vitamin E, 726, 729
Vitamin K, 439, 724, 725, 729
Vitelline envelope, 920-921, 930
Vocal cords, 738
Voice, 738, 905
Volcanic islands, 1012
Volcanoes, 1140
Voles, 1098
Voltage-gated ion channels, 851
Voltage gradient, 198
Volume
of cell, and surface area, 102
of lungs, 736
Voluntary muscles, 706
Voluntary nervous system, 712 (*see* Somatic nervous system)
Volvocine line of algae, 461-462
Volvox, 461-462
Vomeronasal organ, 861
Vomit, 720
Von Frisch, Karl, 1052, 1055
Vorticella, 425
Voyage of the Beagle, **966**
Vulva, **908-909**
Vultures, 423

Waiting behavior, 1074, 1079
Wake-robin, 249
Wald, George, 210
Walker, Alan, 1040
Walking-stick insect, 972
Wall, cellular (*see* Cell wall)
Wallaby, 966
Wallace, Alfred Russel, 8, 15, 961, 962, 965, 972, 1004
Wall pressure, 134
Warbler finch, **1019-1020**
"Warm-bloodedness," 780 (*see also* Ectothermy; Endothermy; Homeothermy; Poikilothermy)
Warning behavior, 1055, 1059, 1075-1076, 1077
Warning characteristics, 1004
Wart hogs, 17
Wasps, 573, 576, 1059-1060

Water, **40-54**, 518, 525 (*see also* Fluids)
abundance, 40-55
and adaptation, 624
vs. ammonia, 76
balance, 656
biogeochemical cycles of, 1145
in blood, 763
in cell, 96, 128, 142
in cell metabolism, 128
cohesion-tension theory, 651-652
compartments, **768-769**
concentration, 128, 132-133
conduction in, **779**
conservation of, in body, 775
consumption of, 788-789
content, in organisms, 43
covalent bonds in, 33, 41, 54
density of, 42, 46, 54
and diffusion, 128, 130
in digestive organs, 720, 725, 730
distribution of, 53
"fitness" for life, 54
freezing point, 43
fresh, 549, 550, 1154
gain, 767-769, 775
in glycolysis, 182
in homeostasis, 790
hydrogen bonds in, 33, **41-43**, 54
ionic bonds in, 41
ionic equilibrium in, 47
ionization of, 47-52, 54
loss, in photosynthesis, 230
in plants, 628
melting point, 44
movement of, 128, 132, 142
net movement, 142
and origin of life, 87-89, 1154
and photosynthesis, 217, 219-220, 229
and plants, **650-654**
growth, 671
reproduction, 499
in roots, 632, **633**
storage, 629, 636
transport, 669-670
water balance, 656
in plasma, 763
polarity, 33, 41, 44, 46, 54, 130
reabsorption of, 776
in respiratory gas, **743**
saltwater, 549, 550
as solvent, 44, 54
specific heat, 43
structure, 41, 44
surface tension, 42
and temperature regulation, 785-786
transpiration, 628, 651-652
uptake of, in roots, **633**, 650-651
and urea, **776**
in urine, 773
volume of, in animal body, 775
water cycle, 53-54
Water balance, **766-769**
Water bears (*see* Tardigrades)
Water buttercup, 271
Water cycle, 53-54, 145, 1133
Waterfleas, 571
Water intoxication, 769
Water loss, 775, **767-769**, 772, 775, 1164
Water molds, 455, 467-469, 476, 477
Water potential, 128, 129
and bloodflow, 129
and cellular respiration
equilibrium in, 129
gradient, 132-133
in plants, 652, 663, 670
and osmosis, 132-133
Water stress, 225, 669
Water table, 53
Water vapor, 648, 650, 669, 1133, 1153
Water vascular system, **588**, 589, 604
Watson, James, 3-7, 287, **288-291**, 292-293, 298-300, 317, 356, 358, 375, 379, 383
Watson-Crick model, **288-290**, 293, 298, 301
Waxes, 67, 70, 82, 1060
Weapons, and hominid evolution, 1044
Weather
and ecosystems, 1131, **1136-1137**
as limit on population, 1098, 1099

and troposphere, 1133
warming and cooling cycles, 1158
Weaver, Rufus B., 844
Webs, 570-571
Wegener, Alfred, 1012
Weinberg, G., 979-981
Wells, Martin. 552
Went, Frits W., 672-673
Wernicke's area, 888
Western yellow pine, 505
Weston, W. H., 489
Wet lands, destruction of, 1160
Wexler, Nancy, 395
Whales, 496, 603, 733, 741, **744**, 1007
Wheat, 159-161, 168, 226, **622**, 1014, 1148-1149
cell, 149
genetics of, 272
polyploidy in, 1014
Wheat germ, 622
Whelks, 550, 962
Whipworm, 539
Whisk ferns, 498, 502
White, Ray, 399
White, Tim, 1042
White blood cells, 118, 456-457, 791-792, 798, 818, 892 (*see also* B lymphocytes; Leukocytes)
and circadian rhythms, 834
and hormone production, 823
in immune response, 433
White-crowned sparrow, 1056
White ferns, 878
Whitefish, 152
White matter, 843 (*see also* Cerebral cortex; Nerve cord)
in human brain, 881, 884
White rot fungus, 488
Whooping crane, 1104
Whorls, 614, 641
Wildebeest, 409
Wild pansy, 408
"Wild-type" allele, 266
Wiesel, Torsten N, 887
Wilkinson, Gerald, 1074
Willows, 491
Wilms' tumor, 399
and chromosomal deletion, 386-387, 399-400
Wilson, E. O., 1099, 1102, 1123
Wilting, 692
Wind
and ocean currents, 1156, **1158**
patterns, 1131, **1136-1137**, 1152, 1160
as pollinator, 507
Winding, 695
Windpipe (*see* Tracheae)
Wine, 190-191, 486
Wing buds, 938, **939**
Wings, 566, 573, 574
aerodynamics of, 599
development of in embryo, 938, 943
Winter-hardy plants, 44
Wintering, 1061
Winter stratification, 46
Wishbone, 599
Withdrawal, 916
Wöhler, Friedrich, 11
Wolpoff, Milford, 1050
Wolves, 1063, 1116-1117
Woodchuck, 409
Wood fibers, 509
Woodland, 1139
Woodland warblers, 1109
Wood lice, 409 (*see* Pillbugs)
Woodpecker, 1002
Woodpecker finch, 6, 1002
Woody perennials, 622
Woody plants, 644-648, 675
Work
and thermodynamics, 162-164
Worker bees, 1053, 1058, 1060-1061, 1068, 1069, 1078
World Health Organization (WHO), 720, 727, 797
World population, **1096-1097**
Worms, 533-540, 542-543
segmented, **544-558**, 564 (*see also* Annelids)
Wynne-Edwards, V. C., 1065

Xanthophylls, 440, 456
X chromosome, 264, 266-267, 279, 359, 383, 400

X chromosome *(Continued)*
  and color blindness, 391–392
  and hemophilia, 392
  mutations in, 400
  sex-linked characteristics in, 391–394
Xenon, 733
X-linked characteristics
  hemophilia, 392–393
  muscular dystrophy, 393–394
XO (male) chromosome pair, 264
X-ray diffraction, 288–289, 356
X-rays, 274, 661, 947, 953
  and lytic cycle, 331
  and mutations, 982
XX (female) chromosome pair, 264, 278, 399
XY (male) chromosome pair, 264, 278, 286
Xylem, 501, 625, **638-639**, 644, 647–648
  and auxins, 681
  and cell types in angiosperms, 647–648
  conduction of water in, 669
  evolution of, 509
  and gibberellins, 681
  of green stems, 640
  mineral transport in, 655
  origin of, 634
  of roots, **633**

  secondary, 675
  in secondary growth, 645
  transport in, 650, 660
Xylem pressure, 651–652, 655

Y chromosome, 264, 266–267, 383, 391–392, 399, 946
Yam, 410
Yeasts, 11, 13, 190–191, 480, 481, 486
  and fermentation, 486
Yellow-green algae, **459–460**
Yellow-jacket, 1005
Yoga, and peripheral nervous system, 845
Yolk, 596, 925, 926–927, 929, 930–934, 937, 944, 952
  selection for, 995
Yolk plug, 928
Yolk sac, **932, 934**, 937, 944–946, 952

Z-DNA, 356
Zeatin, 676 (*see also* Cytokinins)
Zebras, 1110
Zinc (Zn^{2+}) ion, 657
Zinjanthropus (*see Australopithecus boisei*)
Z line, 874, 875

Zonal centrifugation, 193
Zonation, 1158
Zone of elongation, 634
Zone of intolerance, 1095
Zones of physiological stress, 1095
Zone of saturation, 53
Zoophytes, 522
Zooplankton, 1145, 1166
Zygomycetes, 481, **482–484**, 491
Zygosporangia, 483
Zygospores, 462, 468, 481, 482, 491
Zygote, 264, 526, 528, 529, 544
  of algae, 460, 462–464
  of angiosperms, 617, 619
  in black bread mold, 483
  of *Chlamydomonas*, 462, 463
  of fungi, 482, 491
  and meiosis, 250, 261
  and mitosis, 359
  of opalinids, 473
  of *Plasmodium*, 473
  of protozoa, 468, 470
  of sea urchins, 920, **921–924**, 922
  vs. seeds, 507
  of slime molds, 467
  of yeasts, 486